Formulas/Equations

Distance Formula If $P_1 = (x_1, y_1)$ and $P_2 = (x_2, y_2)$, the distance from P_1 to P_2 is

$$d(P_1, P_2) = \sqrt{(x_2 - x_1)^2 + (y_2 - y_1)^2}$$

Equation of a Circle The equation of a circle of radius r with center at (h, k) is

$$(x - h)^2 + (y - k)^2 = r^2$$

Slope of a Line The slope m of the line containing the points $P_1 = (x_1, y_1)$ and $P_2 = (x_2, y_2)$ is

$$m = \frac{y_2 - y_1}{x_2 - x_1} \qquad \text{if } x_1 \neq x_2$$

$$m \text{ is undefined} \qquad \text{if } x_1 = x_2$$

Point-Slope Equation of a Line The equation of a line with slope m containing the point (x_1, y_1) is

$$y - y_1 = m(x - x_1)$$

Slope-Intercept Equation of a Line The equation of a line with slope m and y-intercept b is

$$y = mx + b$$

Quadratic Formula The solutions of the equation $ax^2 + bx + c = 0$, $a \neq 0$, are

$$x = \frac{-b \pm \sqrt{b^2 - 4ac}}{2a}$$

If $b^2 - 4ac > 0$, there are two real unequal solutions.
If $b^2 - 4ac = 0$, there is a repeated real solution.
If $b^2 - 4ac < 0$, there are two complex solutions.

Functions

Constant Function $f(x) = b$

Linear Function $f(x) = mx + b$, m is slope, b is y-intercept

Quadratic Function $f(x) = ax^2 + bx + c$

Polynomial Function $f(x) = a_n x^n + a_{n-1} x^{n-1} + \cdots + a_1 x + a_0$

Rational Function $R(x) = \dfrac{p(x)}{q(x)} = \dfrac{a_n x^n + a_{n-1} x^{n-1} + \cdots + a_1 x + a_0}{b_m x^m + b_{m-1} x^{m-1} + \cdots + b_1 x + b_0}$

Exponential Function $f(x) = a^x$, $a > 0$, $a \neq 1$

Logarithmic Function $f(x) = \log_a x$, $a > 0$, $a \neq 1$

Precalculus

Fourth Edition

Michael Sullivan
Chicago State University

Prentice Hall, Upper Saddle River, NJ 07458

Library of Congress Cataloging-in-Publication Data

Sullivan, Michael, 1942–
   Precalculus / Michael Sullivan. — 4th ed.
     p. cm.
   Includes index.
   ISBN 0-13-228594-0
   1. Algebra. 2. Trigonometry. I. Title.
QA154.2.S846 1996
512′.13—dc20   95-52939
                    CIP

Editor-in Chief: Jerome Grant
Acquisitions Editor: Sally Denlow
Editor-in Chief of Development: Ray Mullaney
Development Editor: Charles Fenn
Director of Production and Manufacturing: David W. Riccardi
Production Editor: Robert C. Walters
Marketing Manager: Jolene Howard
Copy Editor: William Thomas
Interior Designer: Rosemarie Votta
Cover Designer: Rosemarie Votta
Cover Art: Riverwatch Marketing
Creative Director: Paula Maylahn
Art Director: Amy Rosen
Assistant to Art Director: Rod Hernandez
Manufacturing Manager: Alan Fischer
Photo Researcher: Rona Tuccillo
Photo Editor: Lorinda Morris-Nantz
Supplements Editor: Audra J. Walsh
Production Assistant: Joanne Wendelken

 © 1996, 1993, 1990, 1987 by Prentice-Hall, Inc.
Simon & Schuster/A Viacom Company
Upper Saddle River, NJ 07458

Printed in the United States of America

10 9 8 7 6 5 4 3 2 1

ISBN 0-13-228594-0

Prentice-Hall International (UK) Limited, London
Prentice-Hall of Australia Pty. Limited, Sydney
Prentice-Hall Canada Inc. Toronto
Prentice-Hall Hispanoamericana, S.A., Mexico
Prentice-Hall of India Private Limited, New Delhi
Prentice-Hall of Japan, Inc., Tokyo
Simon & Schuster Asia Pte. Ltd., Singapore
Editora Prentice-Hall do Brasil, Ltda., Rio de Janeiro

For my Parents ... Thanks

Contents

As a professor at an urban public university for over 30 years, I am aware of the varied needs of students—students who range from having little mathematical background and a fear of mathematics courses to those who have had a strong education and are extremely motivated. For some of your students, this will be their last course in mathematics, while others may decide to further their mathematical education. I have written this text for both groups. As the author of precalculus, engineering calculus, finite math and business calculus texts and as a teacher, I understand what students must know if they are to be successful in upper level courses, but as a father of four, I also understand the realities of college life.

In the Fourth Edition

The elements of the previous editions that have proved so successful remain in this edition. Nevertheless, many changes, some obvious, others subtle, have been made. Virtually every change is the result of thoughtful comments and suggestions from colleagues and students who have used previous editions. As a result of this input, for which I am sincerely grateful, this edition will be an improved teaching device for professors and a better learning tool for students.

Revised/Reorganized Material *Exponential and Logarithmic Functions*—the first section on exponential functions is followed by sections on logarithmic functions and their properties so that the connection between the two functions is clear. A new section on exponential and logarithmic equations carefully distinguishes between exact solutions obtained by properties and the approximate solutions obtained through the use of a graphing device.

Trigonometric Functions—this chapter has been enhanced by new examples and exercises involving problems usually seen in calculus.

Graphs of the Trigonometric Functions—this chapter has been reorganized and expanded. The graphs of the sine and cosine functions are treated first, followed by sinusoidal graphs, and a new section devoted to applications involving damped vibration and harmonic motion. The remainder of the chapter deals with the graphs of the other trigonometric functions, followed by a discussion of the inverse trigonometric functions.

Systems of Inequalities—this section has been expanded to include both linear and nonlinear inequalities. Linear Programming now appears after Systems of Inequalities providing an application of linear inequalities.

Induction, Sequences, Sets, Counting and Probability—New to this edition are three sections on Sets, Counting and Probability. This material is combined in one chapter with sequences, induction, and the binomial theorem. Separate sections for the arithmetic and geometric sequences form part of the reorganization of the chapter.

Please refer to the Overview located in the Student's Preface.

Functional Use of Color The new design uses color effectively and functionally to help the student identify definitions, theorems, formulas and procedures. Included in the student's edition is an overview of how these colorized elements can be helpful when studying. Many new illustrations have been included to provide a dynamic realism to selected examples and exercises. All the graphing utility and line art has been computer generated for consistency and accuracy. For the purpose of clarity, all line art utilizes two colors. With over 1200 pieces of art throughout the text, my aim is to help the student visualize mathematics and to show how visualization can be used to solve mathematical problems.

Examples In addition to the solid traditional examples found in earlier editions, I have added many new examples (now totaling over 600) to the text. I have increased the use of application and technology examples to help provide motivation and realism in the text. Examples are worked out in appropriate detail, starting with simple, reasonable problems and working gradually up to more challenging ones. Many examples involve applications that will be seen in calculus or in other disciplines.

Applications Every opportunity has been taken to present understandable, realistic applications consistent with the abilities of the student, drawing from such sources as tax rate tables, the Guinness Book of World Records, and newspaper articles. For added interest, some of the applied exercises have been adapted from textbooks the students may be using in other courses (such as economics, chemistry, physics, etc.)

Technology This edition continues to encourage the use of state of the art technology when it is required or helpful. Whether a computer program or a graphing calculator, these products can provide students with new insights to the visualization of mathematics. A new ***Appendix on Graphing Utilities*** illustrates some of the features of using a graphing utility. Many new examples and exercises that require a graphing utility have been added. ![icon] will be used when the use of a graphing utility is required or suggested. Careful thought has been given to these examples and exercises: most ask for conclusions that cannot be obtained through usual algebra processes. This serves to reinforce the power of technology in solving problems. If you or your school would like to require a graphing utility and fully utilize this power, I have also written *Precalculus: Enhanced with Graphing Utilities.* To receive an examination copy contact your Prentice Hall representative.

Communication and Critical Thinking As in previous editions, I illustrate the role writing, verbalizing, research and critical thinking has in mathematics. In this edition, I have created specially designed exercises to encourage students to actively participate in these activities. These exercises are designated by a 🖊. The recommendations of organizations such as the National Council of Teachers of Mathematics, American Association of Two Year Colleges and the Mathematics Association of America support these kinds of exercises. I hope you find them as positive as I have.

Collaborative Projects In each chapter, a full page has been devoted to collaborative learning. These multi-tasked projects, written by Hester Lewellen, one of the co-authors of the University of Chicago High School Mathematics Project, will help your students get together and problem solve in a very real manner. Some of the projects have suggestions for utilizing a graphing utility; all require critical thinking and communication.

Exercises The exercises added to this edition are mostly of three types: visual, where students are asked to draw conclusions about a graph; technological, where a graphing utility is required 📱; and open-ended, where critical thinking, writing, research or collaborative effort is required 🖊. For this edition, about one third of the traditional exercises were also modified. The text now contains over 5500 exercises, of which over 900 are applied problems. Exercise sets begin with problems designed to build confidence, continue with problems which relate to worked out examples in the text and conclude with problems that are more challenging. Many of the problems, especially those at the beginning are visual in character—such as showing a graph and asking for conclusions.

Answers are given in the back of the book for all the odd-numbered problems. Fully worked out solutions for the odd-numbered problems are found in the **Student's Solutions Manual** while the even solutions are found in the **Instructor's Solutions Manual.**

Instructor's Supplementary Aids

Instructor's Solutions Manual Contains complete step-by-step worked out solutions to all the even numbered exercises in the textbook. Also included are strategies for using the collaborative learning projects found in each chapter. Transparency masters which duplicate important illustrations in the text can be found as well.

Video Tutors A new videotape series which has been created to accompany Precalculus, Fourth Edition by Michael Sullivan. These videos contain a half hour long review of the most important topics for each chapter. Entertaining and educational, these videos provide an alternative process which can add to your students' success in this course. Also included is a graphing calculator tutorial and a review of basic mathematical concepts. Written by a mathematics teacher and a tutor, these videos concentrate on those topics that "get them every time."

Written Test Item File Features six tests per chapter plus four forms of a final examination, prepared and ready to be photocopied. Of these tests, 3 are multiple choice and 3 are free response.

PH Custom Test (computerized testing generator - Mac and IBM) This versatile testing system allows the instructor to easily create up to 99 versions of a customized test. Users may add their own test items and edit existing items in WYSIWYG format. Each objective in the text has at least one multiple choice and free response algorithm. Free upon adoption. Allows for on-line testing and grading.

Student's Supplementary Aids

Student's Solutions Manual Contains complete step-by-step worked out solutions to all the odd numbered exercises in the textbook. This is terrific for getting instant feedback on whether your students are proceeding correctly while checking their work. ISBN: 013-370164-6

Visual Precalculus A software package for IBM compatible computers which consists of two parts. Part One contains routines to graph and evaluate functions, graph conic sections, investigate series, carry out synthetic division, and illuminate important concepts with animation. Part Two contains routines to solve triangles, graph systems of linear equations and inequalities, evaluate matrix expressions, apply Gaussian elimination to reduce or invert matrices and graphically solve linear programming problems. Those routines will provide additional insights into the material covered within the text. ISBN: 013-456450-2

X (Plore) A powerful (yet inexpensive) fully programmable symbolic and numeric mathematical processor for IBM and Macintosh computers. This program will allow your students to evaluate expressions, graph curves, solve equations and use matrices. This software package may also be used for calculus or differential equations. If your students do not own a graphing calculator, this is a good option. ISBN: 013-014225-X

New York Times Supplement A free newspaper from Prentice Hall and the New York Times which includes interesting and current articles on mathematics in the world around us. Great for getting students to talk and write about mathematics! This supplement is created new each year.

For any of the above supplements, please contact your Prentice Hall representative for your own complimentary copy.

Acknowledgments

Textbooks are written by an author, but evolve from an idea into final form through the efforts of many people. Special thanks to Don Dellen, who first suggested and then saw to completion this book and the other books in my Precalculus series.

With the completion of the fourth edition of this text, there are many people I would like to thank for their input, encouragement, patience and support. The following are contributors to the first three editions and the current revision. They have my deepest thanks and appreciation. I apologize for any omissions . . .

James Africh, Brother Rice High School
Steve Agronsky, Cal Poly State University
Joby Milo Anthony, University of Central Florida
James E. Arnold, University of Wisconsin, Milwaukee
Agnes Azzolino, Middlesex County College
Wilson P. Banks, Illinois State University
Dale R. Bedgood, East Texas State University

William H. Beyer, University of Akron
Richelle Blair, Lakeland Community College
Trudy Bratten, Grossmont College
William J. Cable, University of Wisconsin—Stevens Point
Lois Calamia, Brookdale Community College
Roger Carlsen, Moraine Valley Community College
Denise Corbett, East Carolina University
Theodore C. Coskey, South Seattle Community College
John Davenport, East Texas State University
Duane E. Deal, Ball State University
Vivian Dennis, Eastfield College
Karen R. Dougan, University of Florida
Louise Dyson, Clark College
Paul D. East, Lexington Community College
Don Edmondson, University of Texas, Austin
Christopher Ennis, University of Minnesota
Garret J. Etgen, University of Houston
W. A. Ferguson, University of Illinois, Urbana/Champaign
Iris B. Fetts, Clemson University
Mason Flake, student at Edison Community College
Merle Friel, Humboldt State University
Richard A. Fritz, Moraine Valley Community College
Dewey Furness, Ricke College
Wayne Gibson, Rancho Santiago College
Joan Goliday, Santa Fe Community College
Frederic Gooding, Goucher College
Ken Gurganus, University of North Carolina
James E. Hall, University of Wisconsin, Madison
Judy Hall, West Virginia University
Edward R. Hancock, DeVry Institute of Technology
Brother Herron, Brother Rice High School
Kim Hughes, California State College, San Bernardino
Ron Jamison, Brigham Young University
Richard A. Jensen, Manatee Community College
Sandra G. Johnson, St. Cloud State University
Moana H. Karsteter, Tallahassee Community College
Arthur Kaufman, College of Staten Island
Thomas Kearns, North Kentucky University
Keith Kuchar, Manatee Community College
Tor Kwembe, Chicago State University
Linda J. Kyle, Tarrant County Jr. College
H. E. Lacey, Texas A & M University
Christopher Lattin, Oakton Community College
Adele LeGere, Oakton Community College
Stanley Lukawecki, Clemson University
Virginia McCarthy, Iowa State University
James McCollow, DeVry Institute of Technology
Laurence Maher, North Texas State University
James Maxwell, Oklahoma State University, Stillwater
Carolyn Meitler, Concordia University
Eldon Miller, University of Mississippi
James Miller, West Virginia University
Michael Miller, Iowa State University
Jane Murphy, Middlesex Community College
James Nymann, University of Texas, El Paso
Sharon O'Donnell, Chicago State University
Seth F. Oppenheimer, Mississippi State University
E. James Peake, Iowa State University
Thomas Radin, San Joaquin Delta College
Ken A. Rager, Metropolitan State College
Elsi Reinhardt, Truckee Meadows Community College
Jane Ringwald, Iowa State University
Stephen Rodi, Austin Community College
Howard L. Rolf, Baylor University
Edward Rozema, University of Tennessee at Chattanooga
Dennis C. Runde, Manatee Community College

John Sanders, Chicago State University
Susan Sandmeyer, Jamestown Community College
A.K. Shamma, University of West Florida
Martin Sherry, Lower Columbia College
Timothy Sipka, Alma College
John Spellman, Southwest Texas State University
Becky Stamper, Western Kentucky University
Neil Stephens, Hinsdale South High School
Tommy Thompson, Brookhaven College
Richard J. Tondra, Iowa State University
Marvel Townsend, University of Florida
Jim Trudnowski, Carroll College
Richard G. Vinson, University of Southern Alabama
Darlene Whitkenack, Northern Illinois University
Chris Wilson, West Virginia University
Carlton Woods, Auburn University
George Zazi, Chicago State University

   Recognition and thanks are due particularly to the following individuals for their valuable assistance in the preparation of this edition: Jerome Grant for his support and commitment; Sally Denlow, for her genuine interest and insightful direction; Bob Walters for his organizational skill as production supervisor; Jolene Howard for her innovative marketing efforts; Ray Mullaney for his specific editorial comments; Charles Fenn for his helpful suggestions in preparing this revision; the entire Prentice-Hall sales staff for their confidence; and to Katy Murphy and Michael Sullivan III for checking my answers to all the exercises. Special thanks to Michael Sullivan III for producing all the graphing utility art.

                                                                Michael Sullivan

Preface

As you begin your study of precalculus, you might feel overwhelmed by the number of theorems, definitions, procedures and equations that confront you. You may even wonder whether you can learn all this material in a single course. For many of you, this may be your last mathematics course, while for others, just the first in a series of many. Don't worry—either way, this text was written with you in mind. While I write math texts in upper level courses, I also am the father of four college students who came home with frustration and questions—*I know what you are going through!*

This text was designed to help you—the student—master the terminology and basic concepts of precalculus. These aims have helped to shape every aspect of the book. Many learning aids are built into the format of the text to make your study of this material easier and more rewarding. This book is meant to be a "machine for learning," one that can help you to focus your efforts and get the most from the time and energy you invest.

Here are some hints I give my students at the beginning of the course:

1. Take advantage of the feature PREPARING FOR THIS CHAPTER. At the beginning of each chapter, I have prepared a list of topics to review. Be sure to take the time to do this. It will help you proceed quicker and more confidently through the chapter.
2. Read the material in the book before the lecture. Knowing what to expect and what is in the book, you can take fewer notes and spend more time listening and understanding the lecture.
3. After each lecture, rewrite your notes as you re-read the book, jotting down any additional facts that seem helpful. Be sure to do the Now Work Problem x as you proceed through a section. After completing a section, be sure to do the assigned problems. Answers to the Odd ones are in the back of the book.
4. If you are confused about something, visit your instructor during office hours immediately, before you fall behind. Bring your attempted solutions to problems with you to show your instructor where you are having trouble.
5. To prepare for an exam, review your notes. Then proceed through the Chapter Review. It contains a capsule summary of all the important material of the chapter. If you are uncertain of any concept, go back into the chapter and study it further. Be sure to do the Review Exercises for practice.

Remember the two "golden rules" of precalculus:

1. DON'T GET BEHIND! The course moves too fast, and it's hard to catch up.
2. WORK LOTS OF PROBLEMS. Everyone needs to practice, and problems show where you need more work. If you can't solve the homework problems without help, you won't be able to do them on exams.

I encourage you to examine the following overview for some hints on how to use this text.

Best Wishes!

Michael Sullivan

OVERVIEW

Beginning each chapter, after Chapter One, there is a list of concepts from previous chapters you should understand before moving on to the new chapter. By reviewing this list and making sure you feel comfortable with the material covered, you will begin to notice that learning algebra is a building process. This review makes sure your foundation is laid!

Each chapter opens with a photo and application which illustrates algebra in use around us. The photo is accompanied by a mathematical problem which you will learn how to solve during the course of the chapter. This allows you to understand the kind of issues that will be covered as well as how they might be used in the 'real world'.

Also on the first page of the chapter, you will find an outline of the topics to be covered within. This is a good way to organize your notes for studying.

Most chapters open with a brief historical discussion of "where this material came from". It is helpful to understand how others created and used these ideas to solve their everyday problems.

New terms appear in boldface type where they are defined.

Major definitions appear in large type enclosed within a color screen. These are important vocabulary items for you to know.

Reproduced chapter page

Chapter 2

FUNCTIONS AND THEIR GRAPHS

PREPARING FOR THIS CHAPTER

Before getting started on this chapter, review the following concepts:
Domain of a variable (p. 3)
Graphs of certain equations (Example 5, p. 58; Example 6, p. 60; Example 7, p. 61; Example 8, p. 62)
Tests for symmetry of an equation (p. 60)
Procedure for finding intercepts of an equation (p. 59)
Steps for setting up applied problems (pp. 24–25)

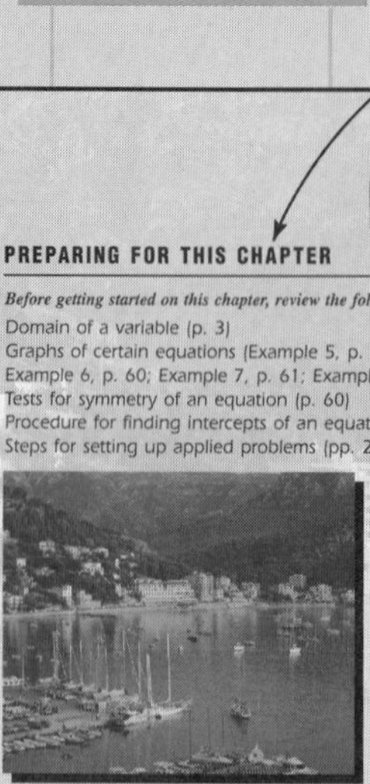

Preview Getting from an Island to Town

An island is 2 miles from the nearest point P on a straight shoreline. A town is 12 miles down the shore from P.
(a) If a person can row a boat at an average speed of 5 miles per hour and the same person can walk 2 miles per hour, express the time T it takes to go from the island to town as a function of the distance x from P to where the person lands the boat.
(b) How long will it take to travel from the island to town if the person lands the boat 4 miles from P?
(c) How long will it take if the person lands the boat 8 miles from P? [Example 9 in Section 2.1]
(d) Is there a place to land the boat so that the travel time is least? Do you think this place is closer to town or closer to P? Discuss the possibilities. Give reasons. [Problems 63 and 64 in Exercise 2.1]

Perhaps the most central idea in mathematics is the notion of a *function*. This important chapter deals with what a function is, how to graph functions, how to perform operations on functions, and how functions are used in applications.

The word *function* apparently was introduced by René Descartes in 1637. For him, a function simply meant any positive integral power of a variable *x*. Gottfried Wilhelm von Leibniz (1646–1716), who always emphasized the geometric side of mathematics, used the word function to denote any quantity associated with a curve, such as the coordinates of a point on the curve. Leonhard Euler (1707–1783) employed the word to mean any equation or formula involving variables and constants. His idea of a function is similar to the one most often used today in courses that precede calculus. Later, the use of functions in investigating heat flow equations led to a very broad definition, due to Lejeune Dirichlet (1805–1859), which describes a function as a rule or correspondence between two sets. It is his definition that we use here.

2.1 Functions

In many applications, a correspondence often exists between two sets of numbers. For example, the revenue R resulting from the sale of x items selling for $10 each is $R = 10x$ dollars. If we know how many items have been sold, then we can calculate the revenue by using the rule $R = 10x$. This rule is an example of a *function*.

As another example, if an object is dropped from a height of 64 feet above the ground, the distance s (in feet) of the object from the ground after t seconds is given (approximately) by the formula $s = 64 - 16t^2$. When $t = 0$ seconds, the object is $s = 64$ feet above the ground. After 1 second, the object is $s = 64 - 16(1)^2 = 48$ feet above the ground. After 2 seconds, the object strikes the ground. The formula $s = 64 - 16t^2$ provides a way of finding the distance s when the time t ($0 \le t \le 2$) is prescribed. There is a correspondence between each time t in the interval $0 \le t \le 2$ and the distance s. We say that the distance s is a *function* of the time t because:

1. There is a correspondence between the set of times and the set of distances.
2. There is exactly one distance s obtained for a prescribed time t in the interval $0 \le t \le 2$.

Let's now look at the definition of a function.

Definition of Function

Let X and Y be two nonempty sets of real numbers.* A **function** from X into Y is a rule or a correspondence that associates with each element of X a unique element of Y. The set X is called the **domain** of the function. For each element x in X, the corresponding element y in Y is called the **value** of the function at x, or the **image** of x. The set of all images of the elements of the domain is called the **range** of the function.

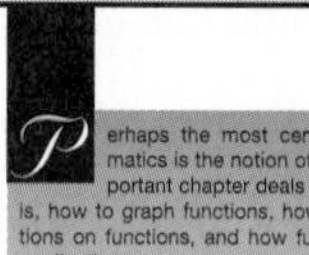

Refer to Figure 1. Since there may be some elements in Y that are not the image of some x in X, it follows that the range of a function may be a subset of Y.

The rule (or correspondence) referred to in the definition of a function is most often given as an equation in two variables, usually denoted x and y.

FIGURE 1

*The two sets X and Y can also be sets of complex numbers, and then we have defined a complex function. In the broad definition (due to Lejeune Dirichlet), X and Y can be any sets.

Historical Features place the mathematics you are learning in a historical context. By learning how others have used similar concepts you will understand how it may be used in your own life. Sometimes there are extra historical exercises for you to solve on your own!

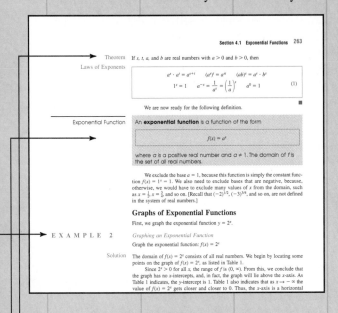

Theorems are noted with the word "Theorem" in blue. These are especially important to study. When the proof of a theorem is given, the word "Proof" also appears.

All important formulas are enclosed by a box and shown in color. This is done to alert you to important concepts.

Examples are easy to locate and are titled to tell you what they are about. Although the solution is entirely worked out, you should go through it using pencil and paper as you go. If you get stuck on a homework problem, chances are you can find an example that models the problem. Many solutions contain a Check. It is always a good practice to Check solutions when you can.

The Now Work Problem xx feature asks you to do a particular problem before you go on in the section. This is to ensure that you have mastered the material just presented before going on. By following the practice of doing these Now Work Problems, you will gain confidence and save time.

Important Procedures and STEPS are noted in the left column and are separated from the body of the text by two horizontal color rules. You will need to know these procedures and steps to do your homework problems and prepare for exams.

Hints or warnings are offered where appropriate. Sometimes there are short cuts or pitfalls that students should know about—I've included them.

The "graphing calculator" icon can be seen throughout the text and in the end of section and review exercises. These indicate that they require the use of a graphing calculator or computer software program.

Many students possess a graphing calculator or have access to computer software that does graphing. To help you understand what such devices are capable of, I have placed Appendix B at the end of the book, called Graphing Utilities, that explains some of the capabilities of a grapher. Many examples and Exercises in the book require a grapher. Careful thought has been given to these examples and exercises: most ask for conclusions that cannot be obtained through usual algebra processes. This should help you to see the power of technology in solving problems.

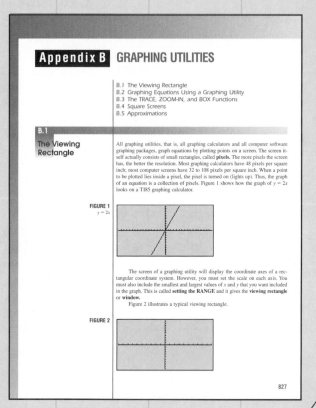

Mission Possible: In the "real world" colleagues often collaborate to solve more difficult problems—or problems that may have more than one answer. Every chapter has a "Mission Possible" for you and your classmates to collaborate on. All of these projects will require you to verbally communicate or write up your answers. Good communication skills are very important to becoming successful—no matter what your future holds.

The Chapter Review is for your use in checking your understanding of the chapter materials. "Things to Know" is the best place to start. Check your understanding of the concepts listed there. Then demonstrate to yourself that you know "How to" solve the items contained within that section. "Fill in the Blanks" will determine your comfort with vocabulary. "True/False" is a stickler for knowing definitions! If you are uncertain of any concept, go back into the chapter and study it further. Be sure to do the Review Exercises for practice. These reviews are for your success in this course—Make good use of them.

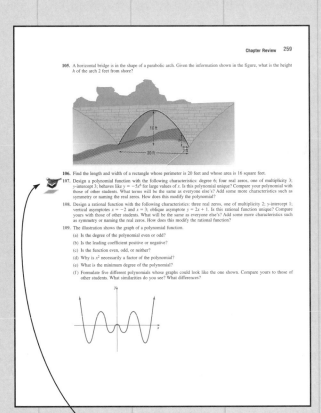

The "pencil and book" icon is used to indicate open-ended questions for discussion, writing, group or research projects.

---

170 Chapter 2 Functions and Their Graphs

37. Water is poured into a container in the shape of a right circular cone with radius 4 feet and height 16 feet (see the figure). Express the volume V of water in the cone as a function of the height h of the water. [*Hint:* The volume V of a cone of radius r and height h is $V = \frac{1}{3}\pi r^2 h$.]

38. *Federal Income Tax* Two 1994 Tax Rate Schedules are given in the accompanying table. If x equals the amount on Form 1040, line 37, and y equals the tax due, construct a function f for each schedule.

1994 TAX RATE SCHEDULES

SCHEDULE X—USE IF YOUR FILING STATUS IS SINGLE				SCHEDULE Y-1—USE IF YOUR FILING STATUS IS MARRIED FILING JOINTLY OR QUALIFYING WIDOW(ER)			
If the amount on Form 1040, line 37, is: Over—	But not over—	Enter on Form 1040, line 38	of the amount over—	If the amount on Form 1040, line 37, is: Over—	But not over—	Enter on Form 1040, line 38	of the amount over—
$0	$22,750	..... 15%	$0	$0	$38,000	..... 15%	$0
22,750	55,100	$3,412.50 + 28%	22,750	38,000	91,850	$5,700.00 + 28%	38,000
55,100	115,000	12,470.50 + 31%	55,100	91,850	140,000	20,778.00 + 31%	91,850
115,000	250,000	31,039.50 + 36%	115,000	140,000	250,000	35,704.50 + 36%	140,000
250,000	.....	79,639.50 + 39.6%	250,000	250,000	.....	75,304.50 + 39.6%	250,000

Chapter Review

THINGS TO KNOW

Function

A rule or correspondence between two sets of real numbers so that each number x in the first set, the domain, has corresponding to it exactly one number y in the second set. The range is the set of y values of the function for the x values in the domain.
x is the independent variable; y is the dependent variable.
A function f may be defined implicitly by an equation involving x and y or explicitly by writing $y = f(x)$.
A function can also be characterized as a set of ordered pairs (x, y) or $(x, f(x))$ in which no two pairs have the same first element.

Function notation

$y = f(x)$
f is a symbol for the function or rule that defines the function.
x is the argument, or independent variable.
y is the dependent variable.
$f(x)$ is the value of the function at x, or the image of x.

Domain

If unspecified, the domain of a function f is the largest set of real numbers for which the rule defines a real number.

Vertical-line test

A set of points in the plane is the graph of a function if and only if every vertical line intersects the graph in at most one point.

Even function f

$f(-x) = f(x)$ for every x in the domain ($-x$ must also be in the domain).

---

478 Chapter 7 Analytic Trigonometry

Product-to-sum formulas
$$\sin\alpha\sin\beta = \frac{1}{2}[\cos(\alpha-\beta) - \cos(\alpha+\beta)]$$
$$\cos\alpha\cos\beta = \frac{1}{2}[\cos(\alpha-\beta) + \cos(\alpha+\beta)]$$
$$\sin\alpha\cos\beta = \frac{1}{2}[\sin(\alpha+\beta) + \sin(\alpha-\beta)]$$

Sum-to-product formulas
$$\sin\alpha + \sin\beta = 2\sin\frac{\alpha+\beta}{2}\cos\frac{\alpha-\beta}{2}$$
$$\sin\alpha - \sin\beta = 2\sin\frac{\alpha-\beta}{2}\cos\frac{\alpha+\beta}{2}$$
$$\cos\alpha + \cos\beta = 2\cos\frac{\alpha+\beta}{2}\cos\frac{\alpha-\beta}{2}$$
$$\cos\alpha - \cos\beta = -2\sin\frac{\alpha+\beta}{2}\sin\frac{\alpha-\beta}{2}$$

HOW TO

Establish identities
Solve a trigonometric equation

FILL-IN-THE-BLANK ITEMS

1. Suppose that f and g are two functions with the same domain. If $f(x) = g(x)$ for every x in the domain, the equation is called a(n) _____. Otherwise, it is called a(n) _____ equation.
2. $\cos(\alpha + \beta) = \cos\alpha\cos\beta$ _____ $\sin\alpha\sin\beta$
3. $\sin(\alpha + \beta) = \sin\alpha\cos\beta$ _____ $\cos\alpha\sin\beta$
4. $\cos 2\theta = \cos^2\theta -$ _____ $=$ _____ $- 1 = 1 -$ _____
5. $\sin^2\frac{\alpha}{2} =$ _____$\frac{}{2}$

TRUE/FALSE ITEMS

T F 1. $\sin(-\theta) + \sin\theta = 0$ for all θ.
T F 2. $\sin(\alpha + \beta) = \sin\alpha + \sin\beta + 2\sin\alpha\sin\beta$.
T F 3. $\cos 2\theta$ has three equivalent forms: $\cos^2\theta - \sin^2\theta$, $1 - 2\sin^2\theta$, and $2\cos^2\theta - 1$.
T F 4. $\cos\frac{\alpha}{2} = \pm\frac{\sqrt{1+\cos\alpha}}{2}$, where the $+$ or $-$ sign depends on the angle $\alpha/2$.
T F 5. Most trigonometric equations have unique solutions.
T F 6. The equation $\tan\theta = \pi/2$ has no solution.

REVIEW EXERCISES

In Problems 1–32, establish each identity.
1. $\tan\theta\cot\theta - \sin^2\theta = \cos^2\theta$
2. $\sin\theta\csc\theta - \sin^2\theta = \cos^2\theta$
3. $\cos^2\theta(1 + \tan^2\theta) = 1$
4. $(1 - \cos^2\theta)(1 + \cot^2\theta) = 1$

Chapter Review 259

105. A horizontal bridge is in the shape of a parabolic arch. Given the information shown in the figure, what is the height h of the arch 2 feet from shore?

106. Find the length and width of a rectangle whose perimeter is 20 feet and whose area is 16 square feet.

107. Design a polynomial function with the following characteristics: degree 6; four real zeros, one of multiplicity 3; y-intercept 3; behaves like $y = -5x^6$ for large values of x. Is this polynomial unique? Compare your polynomial with those of other students. What terms will be the same as everyone else's? Add some more characteristics such as symmetry or naming the real zeros. How does this modify the polynomial?

108. Design a rational function with the following characteristics: three real zeros, one of multiplicity 2; y-intercept 1; vertical asymptotes $x = -2$ and $x = 3$; oblique asymptote $y = 2x + 1$. Is this rational function unique? Compare yours with those of other students. What will be the same as everyone else's? Add some more characteristics such as symmetry or naming the real zeros. How does this modify the rational function?

109. The illustration shows the graph of a polynomial function.
   (a) Is the degree of the polynomial even or odd?
   (b) Is the leading coefficient positive or negative?
   (c) Is the function even, odd, or neither?
   (d) Why is x^2 necessarily a factor of the polynomial?
   (e) What is the minimum degree of the polynomial?
   (f) Formulate five different polynomials whose graphs could look like the one shown. Compare yours to those of other students. What similarities do you see? What differences?

Student's Supplementary Aids

You may find yourself seeking out extra help with this course. Many students have found the following items to be useful in becoming successful in precalculus. Your college bookstore should have these items available, but if not, they can order them for you.

Student's Solutions Manual Contains complete step-by-step worked out solutions to all the odd numbered exercises in the textbook. This is terrific for getting instant feedback on whether you are proceeding correctly while solving problems. ISBN: 013-456435-9

Visual Precalculus A software package for IBM compatible computers which consists of two parts. Part One contains routines to graph and evaluate functions, graph conic sections, investigate series, carry out synthetic division, and illuminate important concepts with animation. Part Two contains routines to solve triangles, graph systems of linear equations and inequalities, evaluate matrix expressions, apply Gaussian elimination to reduce or invert matrices and graphically solve linear programming problems. These routines will provide additional insights into the material covered within the text. ISBN: 013-456450-2

X (Plore) A powerful (yet inexpensive) fully programmable symbolic and numeric mathematical processor for IBM and Macintosh computers. This program will allow you to evaluate expressions, graph curves, solve equations and manipulate matrices. This software package may also be used for calculus or differential equations. If your students do not own a graphing calculator—this is a good option. ISBN: 013-014225-X

New York Times Supplement A free newspaper from Prentice Hall and the New York Times which includes interesting and current articles on mathematics in the world around us. Great for getting students to talk and write about mathematics! This supplement is created new each year.

Photo Credits

Chapter 1	The Sears Tower	Four by Five/Superstock
Chapter 2	Port of Soller	Pedro Coll/The Stock Market
Chapter 3	Golden Gate Bridge	Deborah Davis/PhotoEdit
Chapter 4	Sobriety Testing	Bachmann/Photrl
Chapter 5	Gibbs Hill Lighthouse	Marvullo/The Stock Market
Chapter 6	Cadillac Mountain, Acadia National Park	Larry Ulrich/Tony Stone Images
Chapter 7	Shotput	Focus on Sports
Chapter 8	Leaning Tower of Pisa	Sarah Stone/Tony Stone Images
Chapter 9	Satellite Station	Four by Five/Superstock
Chapter 10	Track Run	David Madison Photography
Chapter 11	Stained Glass Window	Wolfgang Koohler
Chapter 12	Corporate Jet	Joe Towers/The Stock Market

Chapter 1

Preview How Far Can You See?

The tallest inhabited building in North America is the Sears Tower in Chicago.* If the observation tower is 1454 feet above ground level, how far can a person standing in the observation tower see?

[Problem 101 in Exercise 1.1]

*Guinness Book of World Records.

The investigation of equations and their solutions has played a central role in algebra for many hundreds of years. In fact, the Babylonians in 200 BC had a well-developed algebra that included a solution for quadratic equations. Of late, the study of inequalities (Section 1.4) has been equally important. The idea of using a system of rectangular coordinates also dates back to about 200 BC, when such a system was used for surveying and city planning. Sporadic use of rectangular coordinates continued until the 1600's. By that time, algebra had developed sufficiently so that René Descartes (1596–1650) and Pierre de Fermat (1601–1665) were able to take the crucial step of using rectangular coordinates to translate geometry problems into algebra problems, and vice versa. This step was supremely important. It allowed both geometers and algebraists to gain critical new insights into their subjects and made possible the development of calculus.

1.1

Review Topics from Algebra and Geometry

Sets

When we want to treat a collection of similar but distinct objects as a whole, we use the idea of a **set.** For example, the set of *digits* consists of the collection of numbers 0, 1, 2, 3, 4, 5, 6, 7, 8, and 9. If we use the symbol D to denote the set of digits, then we can write

$$D = \{0, 1, 2, 3, 4, 5, 6, 7, 8, 9\}$$

In this notation, the braces $\{\ \}$ are used to enclose the objects, or **elements,** in the set. This method of denoting a set is called the **roster method.** A second way to denote a set is to use **set-builder notation,** where the set D of digits is written as

$$D = \{\quad x \quad | \quad x \text{ is a digit}\}$$

Read as "D is the set of all x such that x is a digit."

E X A M P L E   1   *Using Set-builder Notation and the Roster Method*

(a) $E = \{x | x \text{ is an even digit}\} = \{0, 2, 4, 6, 8\}$

(b) $O = \{x | x \text{ is an odd digit}\} = \{1, 3, 5, 7, 9\}$ ∎

In listing the elements of a set, we do not list an element more than once because the elements of a set are distinct. Also, the order in which the elements are listed is not relevant. Thus, for example, $\{2, 3\}$ and $\{3, 2\}$ both represent the same set.

If every element of a set A is also an element of a set B, then we say that A is a **subset** of B. If two sets A and B have the same elements, then we say that A **equals** B. For example, $\{1, 2, 3\}$ is a subset of $\{1, 2, 3, 4, 5\}$; and $\{1, 2, 3\}$ equals $\{2, 3, 1\}$.

Real Numbers

Real numbers are represented by symbols such as

$$25, \quad 0, \quad -3, \quad \tfrac{1}{2}, \quad -\tfrac{5}{4}, \quad 0.125, \quad \sqrt{2}, \quad \pi, \quad \sqrt[3]{-2}, \quad 0.666\ldots$$

The set of **counting numbers,** or **natural numbers,** is the set $\{1, 2, 3, 4, \ldots\}$. (The three dots, called an **ellipsis,** indicate that the pattern continues indefinitely.) The set of **integers** is the set $\{\ldots, -3, -2, -1, 0, 1, 2, 3, \ldots\}$. A **rational number** is a number that can be expressed as a quotient a/b of two integers, where

FIGURE 1

$$\pi = \frac{C}{d}$$

the integer b cannot be 0. Examples of rational numbers are $\frac{3}{4}$, $\frac{5}{2}$, $\frac{0}{4}$, and $-\frac{2}{3}$. Since $a/1 = a$ for any integer a, every integer is also a rational number. Real numbers that are not rational are called **irrational.** Examples of irrational numbers are $\sqrt{2}$ and π (the Greek letter pi), which equals the constant ratio of the circumference to the diameter of a circle. See Figure 1.

Real numbers can be represented as **decimals.** Rational real numbers have decimal representations that either **terminate** or are nonterminating with **repeating** blocks of digits. For example, $\frac{3}{4} = 0.75$, which terminates; and $\frac{2}{3} = 0.666\ldots$, in which the digit 6 repeats indefinitely. Irrational real numbers have decimal representations that neither repeat nor terminate. For example, $\sqrt{2} = 1.414213\ldots$ and $\pi = 3.14159\ldots$. In practice, irrational numbers are generally represented by approximations. We use the symbol $\approx$ (read as "approximately equal to") to write $\sqrt{2} \approx 1.1414$ and $\pi \approx 3.1416$.

Often, letters are used to represent numbers. If the letter used is to represent *any* number from a given set of numbers, it is referred to as a **variable.** A **constant** is either a fixed number, such as 5, $\sqrt{2}$, and so on, or a letter that represents a fixed (possibly unspecified) number. In general, we will follow the practice of using letters near the beginning of the alphabet, such as a, b, and c, for constants and using those near the end, such as x, y, and z, as variables.

In working with expressions or formulas involving variables, the variables may only be allowed to take on values from a certain set of numbers, called the **domain of the variable.** For example, in the expression $1/x$, the variable x cannot take on the value 0, since division by 0 is not allowed.

It can be shown that there is a one-to-one correspondence between real numbers and points on a line. That is, every real number corresponds to a point on the line and, conversely, each point on the line has a unique real number associated with it. We establish this correspondence of real numbers with points on a line in the following manner.

We start with a line that is, for convenience, drawn horizontally. Pick a point on the line and label it O, for **origin.** Then pick another point some fixed distance to the right of O and label it U, for **unit.** The fixed distance, which may be 1 inch, 1 centimeter, 1 light-year, or any unit distance, determines the **scale.** We associate the real number 0 with the origin O and the number 1 with the point U. Refer to Figure 2. The point to the right of U that is twice as far from O as U is associated with the number 2. The point to the right of U that is three times as far from O as U is associated with the number 3. The point midway between O and U is assigned the number 0.5, or $\frac{1}{2}$. Corresponding points to the left of the origin O are assigned the numbers $-\frac{1}{2}$, -1, -2, -3, and so on. The real number x associated with a point P is called the **coordinate** of P, and the line whose points have been assigned coordinates is called the **real number line.** Notice in Figure 2 that we placed an arrowhead on the right end of the line to indicate the direction in which the assigned numbers increase. Figure 2 also shows the points associated with the irrational numbers $\sqrt{2}$ and π.

The real number line divides the real numbers into three classes: the **negative real numbers** are the coordinates of points to the left of the origin O; the real number **zero** is the coordinate of the origin O; the **positive real numbers** are the coordinates of points to the right of the origin O.

Let a and b be two real numbers. If the difference $a - b$ is positive, then we say that a is **greater than** b and write $a > b$. Alternatively, if $a - b$ is positive, we can also say that b is **less than** a and write $b < a$. Thus, $a > b$ and $b < a$ are equivalent statements.

FIGURE 2

Real number line

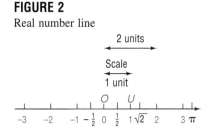

On the real number line, if $a > b$, the point with coordinate a is to the right of the point with coordinate b. For example, $0 > -1$, $\pi > 3$, and $\sqrt{2} < 2$. Furthermore,

> $a > 0$ is equivalent to a is positive
>
> $a < 0$ is equivalent to a is negative

If the difference $a - b$ of two real numbers is positive or 0, that is, if $a > b$ or $a = b$, then we say that a is **greater than or equal to** b and write $a \geq b$. Alternatively, if $a \geq b$, we can also say that b is **less than or equal to** a and write $b \leq a$.

Statements of the form $a < b$ or $b > a$ are called **strict inequalities;** statements of the form $a \leq b$ or $b \geq a$ are called **nonstrict inequalities.** The symbols $>$, $<$, $\geq$, and $\leq$ are called **inequality signs.**

If x is a real number and $x \geq 0$, then x is either positive or 0. As a result, we describe the inequality $x \geq 0$ by saying that x is nonnegative.

Inequalities are useful in representing certain subsets of real numbers. In so doing, though, other variations of the inequality notation may be used.

E X A M P L E   2   *Graphing Inequalities*

(a) In the inequality $x > 4$, x is any number greater than 4. In Figure 3, we use a left parenthesis to indicate that the number 4 is not part of the graph.

FIGURE 3
$x > 4$

(b) In the inequality $4 < x \leq 6$, x is any number between 4 and 6, including 6 but excluding 4. In Figure 4, we use a right bracket to indicate that 6 is part of the graph.

FIGURE 4
$x > 4$ and $x \leq 6$

■

■   Now work Problem 15 (in the exercise set at the end of this section).

Let a and b represent two real numbers with $a < b$: A **closed interval,** denoted by **[a, b]**, consists of all real numbers x for which $a \leq x \leq b$. An **open interval,** denoted by **(a, b)**, consists of all real numbers x for which $a < x < b$. The **half-open,** or **half-closed, intervals** are **(a, b]**, consisting of all real numbers x for which $a < x \leq b$, and **[a, b)**, consisting of all real numbers x for which $a \leq x < b$. In each of these definitions, a is called the **left end point** and b the **right end point** of the interval. Figure 5 illustrates each type of interval.

FIGURE 5

(a)   Closed interval   (b)   Open interval   (c)   Half-open (half-closed) intervals

The symbol ∞ (read as "infinity") is not a real number, but a notational device used to indicate unboundedness in the positive direction. The symbol −∞ (read as "minus infinity") also is not a real number, but a notational device used to indicate unboundedness in the negative direction. Using the symbols ∞ and −∞, we can define five other kinds of intervals:

$[a, \infty)$ consists of all real numbers x for which $a \le x < \infty$ $(x \ge a)$

(a, ∞) consists of all real numbers x for which $a < x < \infty$ $(x > a)$

$(-\infty, a]$ consists of all real numbers x for which $-\infty < x \le a$ $(x \le a)$

$(-\infty, a)$ consists of all real numbers x for which $-\infty < x < a$ $(x < a)$

$(-\infty, \infty)$ consists of all real numbers x for which $-\infty < x < \infty$ (all real numbers)

Figure 6 illustrates these types of intervals.

FIGURE 6

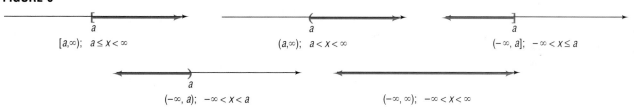

$[a, \infty);\ \ a \le x < \infty$ $(a, \infty);\ \ a < x < \infty$ $(-\infty, a];\ \ -\infty < x \le a$

$(-\infty, a);\ \ -\infty < x < a$ $(-\infty, \infty);\ \ -\infty < x < \infty$

■ Now work Problems 21 and 25.

The *absolute value* of a number a is the distance from the point whose coordinate is a to the origin. For example, the point whose coordinate is −4 is 4 units from the origin. The point whose coordinate is 3 is 3 units from the origin. See Figure 7. Thus, the absolute value of −4 is 4, and the absolute value of 3 is 3.

FIGURE 7

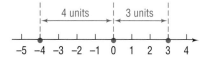

A more formal definition of absolute value is given next.

Absolute Value

> The **absolute value** of a real number a, denoted by the symbol $|a|$, is defined by the rules
>
> $$|a| = a \quad \text{if } a \ge 0 \quad \text{and} \quad |a| = -a \quad \text{if } a < 0$$

For example, since −4 < 0, then the second rule must be used to get $|-4| = -(-4) = 4$.

E X A M P L E   3   *Computing Absolute Value*

(a) $|8| = 8$ (b) $|0| = 0$ (c) $|-15| = 15$ ■

■   Now work Problem 29.

Look again at Figure 7. The distance from the point whose coordinate is -4 to the point whose coordinate is 3 is 7 units. This distance is the difference $3 - (-4)$, obtained by subtracting the smaller coordinate from the larger. However, since $|3 - (-4)| = |7| = 7$ and $|-4 - 3| = |-7| = 7$, we can use the absolute value to calculate the distance between two points without being concerned about which coordinate is smaller.

Distance between P and Q

If P and Q are two points on a real number line with coordinates a and b, respectively, the **distance between P and Q,** denoted by $d(P, Q)$, is

$$d(P, Q) = |b - a|$$

Since $|b - a| = |a - b|$, it follows that $d(P, Q) = d(Q, P)$.

EXAMPLE   4

Finding Distance on a Number Line

Let P, Q, and R be points on the real number line with coordinates $-5, 7$, and -3, respectively. Find the distance:

(a) Between P and Q (b) Between Q and R

Solution

(a) $d(P, Q) = |7 - (-5)| = |12| = 12$ (See Figure 8.)

(b) $d(Q, R) = |-3 - 7| = |-10| = 10$

FIGURE 8

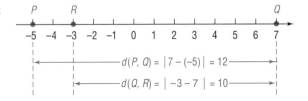

■

Exponents

Integer exponents provide a shorthand device for representing repeated multiplications of a real number.

If a is a real number and n is a positive integer, then the symbol a^n represents the product of n factors of a. That is,

$$a^n = \underbrace{a \cdot a \cdot \ldots \cdot a}_{n \text{ factors}}$$

where it is understood that $a^1 = a$. Thus, $a^2 = a \cdot a$, $a^3 = a \cdot a \cdot a$, and so on. In the expression a^n, a is called the **base** and n is called the **exponent,** or **power.** We read a^n as "a raised to the power n" or as "a to the nth power." We usually read a^2 as "a squared" and a^3 as "a cubed."

Care must be taken when parentheses are used in conjunction with exponents. For example, $-2^4 = -(2 \cdot 2 \cdot 2 \cdot 2) = -16$, whereas $(-2)^4 = (-2) \cdot (-2) \cdot (-2) \cdot (-2) = 16$. Notice the difference: The exponent applies only to the number or parenthetical expression immediately preceding it.

If $a \neq 0$, we define

$$a^0 = 1 \qquad \text{if } a \neq 0$$

If $a \neq 0$ and if n is a positive integer, then we define

$$a^{-n} = \frac{1}{a^n} \qquad \text{if } a \neq 0$$

With these definitions, the symbol a^n is defined for any integer n.

The following properties, called the **laws of exponents**, can be proved using the preceding definitions. In the list, a and b are real numbers, and m and n are integers.

Laws of Exponents

$$a^m a^n = a^{m+n} \qquad (a^m)^n = a^{mn} \qquad (ab)^n = a^n b^n$$

$$\frac{a^m}{a^n} = a^{m-n} = \frac{1}{a^{n-m}}, \qquad \text{if } a \neq 0 \qquad \left(\frac{a}{b}\right)^n = \frac{a^n}{b^n}, \qquad \text{if } b \neq 0$$

E X A M P L E   5   *Using the Laws of Exponents*

Write each expression so that all exponents are positive.

(a) $\dfrac{x^5 y^{-2}}{x^3 y}, \quad x \neq 0, y \neq 0$ (b) $\dfrac{xy}{x^{-1} - y^{-1}}, \quad x \neq 0, y \neq 0$

Solution   (a) $\dfrac{x^5 y^{-2}}{x^3 y} = \dfrac{x^5}{x^3} \cdot \dfrac{y^{-2}}{y} = x^{5-3} \cdot y^{-2-1} = x^2 y^{-3} = x^2 \cdot \dfrac{1}{y^3} = \dfrac{x^2}{y^3}$

(b) $\dfrac{xy}{x^{-1} - y^{-1}} = \dfrac{xy}{\dfrac{1}{x} - \dfrac{1}{y}} = \dfrac{xy}{\dfrac{y-x}{xy}} = \dfrac{(xy)(xy)}{y-x} = \dfrac{x^2 y^2}{y-x}$ ■

■ Now work Problem 57.

The **principal nth root of a number** a, symbolized by $\sqrt[n]{a}$, is defined as follows:

$$\sqrt[n]{a} = b \quad \text{means} \quad a = b^n \qquad \text{where } a \geq 0 \text{ and } b \geq 0 \text{ if } n \text{ is even and } a, b \text{ are any real numbers if } n \text{ is odd}$$

Notice that if a is negative and n is even, then $\sqrt[n]{a}$ is not defined. When it is defined, the principal nth root of a number is unique.

The symbol $\sqrt[n]{a}$ for the principal nth root of a is sometimes called a **radical;** the integer n is called the **index,** and a is called the **radicand.** If the index of a radical is 2, we call $\sqrt[2]{a}$ the **square root** of a and omit the index 2 by simply writing $\sqrt{a}$. If the index is 3, we call $\sqrt[3]{a}$ the **cube root** of a.

E X A M P L E   6   *Simplifying Principal nth Roots*

(a) $\sqrt[3]{8} = 2$ because $8 = 2^3$ (b) $\sqrt{64} = 8$ because $64 = 8^2$

(c) $\sqrt[3]{-64} = -4$ because $-64 = (-4)^3$ (d) $\sqrt[4]{\frac{1}{16}} = \frac{1}{2}$ because $\frac{1}{16} = (\frac{1}{2})^4$

(e) $\sqrt{0} = 0$ because $0 = 0^2$ ∎

These are examples of **perfect roots.** Thus, 8 and -64 are perfect cubes, since $8 = 2^3$ and $-64 = (-4)^3$; 64 and 0 are perfect squares, since $64 = 8^2$ and $0 = 0^2$; and $\frac{1}{2}$ is a perfect 4th root of $\frac{1}{16}$, since $\frac{1}{16} = (\frac{1}{2})^4$.

In general, if $n \geq 2$ is a positive integer and a is a real number,

$$\sqrt[n]{a^n} = a \qquad \text{if } n \text{ is odd} \qquad \text{(1a)}$$
$$\sqrt[n]{a^n} = |a| \qquad \text{if } n \text{ is even} \qquad \text{(1b)}$$

Notice the need for the absolute value in equation (1b). If n is even, then a^n is positive whether $a > 0$ or $a < 0$. But if n is even, the principal nth root must be nonnegative. Hence, the reason for using the absolute value—it gives a nonnegative result.

E X A M P L E   7   *Simplifying Radicals*

(a) $\sqrt{8} = \sqrt{4 \cdot 2} = \sqrt{4} \cdot \sqrt{2} = 2\sqrt{2}$
(b) $\sqrt[3]{-16} = \sqrt[3]{-8 \cdot 2} = \sqrt[3]{-8} \cdot \sqrt[3]{2} = -2\sqrt[3]{2}$
(c) $\sqrt{x^2} = |x|$ ∎

∎ Now work Problem 49.

Radicals are used to define **rational exponents.** If a is a real number and $n \geq 2$ is an integer, then

$$a^{1/n} = \sqrt[n]{a}$$

provided $\sqrt[n]{a}$ exists.

If a is a real number and m and n are integers containing no common factors with $n \geq 2$, then

$$a^{m/n} = \sqrt[n]{a^m} = (\sqrt[n]{a})^m \qquad \text{(2)}$$

provided $\sqrt[n]{a}$ exists.

In simplifying $a^{m/n}$, either $\sqrt[n]{a^m}$ or $(\sqrt[n]{a})^m$ may be used. Generally, taking the root first, as in $(\sqrt[n]{a})^m$, is preferred.

E X A M P L E   8

Using Equation (2)

(a) $8^{2/3} = (\sqrt[3]{8})^2 = 2^2 = 4$ (b) $16^{3/2} = (\sqrt{16})^3 = 4^3 = 64$

(c) $(-8x^5)^{1/3} = \sqrt[3]{-8x^3 \cdot x^2} = \sqrt[3]{(-2x)^3 \cdot x^2} = \sqrt[3]{(-2x)^3}\sqrt[3]{x^2}$

$$= -2x\sqrt[3]{x^2} \qquad \blacksquare$$

A more detailed discussion of radicals and rational exponents is given in Appendix A, Section A.2.

■ Now work Problem 47.

Polynomials

Monomial

A **monomial** in one variable is the product of a constant times a variable raised to a nonnegative integer power. Thus, a monomial is of the form

$$ax^k$$

where a is a constant, x is a variable, and $k \geq 0$ is an integer.

Two monomials ax^k and bx^k, when added or subtracted, can be combined into a single monomial by using the distributive property. For example,

$$2x^2 + 5x^2 = (2 + 5)x^2 = 7x^2 \quad \text{and} \quad 8x^3 - 5x^3 = (8 - 5)x^3 = 3x^3$$

Polynomial

A **polynomial** in one variable is an algebraic expression of the form

$$a_nx^n + a_{n-1}x^{n-1} + \cdots + a_1x + a_0$$

where $a_n, a_{n-1}, \ldots, a_1, a_0$ are constants,* called the **coefficients** of the polynomial, $n \geq 0$ is an integer, and x is a variable. If $a_n \neq 0$, it is called the **leading coefficient,** and n is called the **degree** of the polynomial.

The monomials that make up a polynomial are called its **terms.** If all the coefficients are 0, the polynomial is called the **zero polynomial,** which has no degree.

---

*The notation a_n is read as "*a* sub *n*." The number *n* is called a **subscript** and should not be confused with an exponent. We use subscripts to distinguish one constant from another when a large or undetermined number of constants is required.

Polynomials are usually written in **standard form,** beginning with the nonzero term of highest degree and continuing with terms in descending order according to degree. Examples of polynomials are

POLYNOMIAL	COEFFICIENTS	DEGREE
$3x^2 - 5 = 3x^2 + 0 \cdot x + (-5)$	$3, 0, -5$	2
$8 - 2x + x^2 = 1 \cdot x^2 - 2x + 8$	$1, -2, 8$	2
$5x + \sqrt{2} = 5x^1 + \sqrt{2}$	$5, \sqrt{2}$	1
$3 = 3 \cdot 1 = 3 \cdot x^0$	3	0
0	0	No degree

Although we have been using x to represent the variable, letters such as y or z are also commonly used. Thus,

$3x^4 - x^2 + 2$ is a polynomial (in x) of degree 4.

$9y^3 - 2y^2 + y - 3$ is a polynomial (in y) of degree 3.

$z^5 + \pi$ is a polynomial (in z) of degree 5.

A more detailed discussion of polynomials is given in Appendix A, Section A.1.

Pythagorean Theorem

FIGURE 9

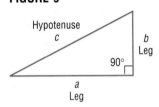

The *Pythagorean Theorem* is a statement about *right triangles*. A **right triangle** is one that contains a **right angle,** that is, an angle of 90°. The side of the triangle opposite the 90° angle is called the **hypotenuse;** the remaining two sides are called **legs.** In Figure 9 we have used c to represent the length of the hypotenuse and a and b to represent the lengths of the legs. Notice the use of the symbol $\ulcorner$ to show the 90° angle. We now state the Pythagorean Theorem.

Pythagorean Theorem

In a right triangle, the square of the length of the hypotenuse is equal to the sum of the squares of the lengths of the legs. That is, in the right triangle shown in Figure 9,

$$c^2 = a^2 + b^2 \tag{3}$$

∎

E X A M P L E   9   *Finding the Hypotenuse of a Right Triangle*

In a right triangle, one leg is of length 4 and the other is of length 3. What is the length of the hypotenuse?

Solution   Since the triangle is a right triangle, we use the Pythagorean Theorem with $a = 4$ and $b = 3$ to find the length c of the hypotenuse. Thus, from equation (3), we have

$$c^2 = a^2 + b^2$$
$$c^2 = 4^2 + 3^2 = 16 + 9 = 25$$
$$c = 5$$

∎

∎ Now work Problem 69.

The converse of the Pythagorean Theorem is also true.

Converse of the Pythagorean
Theorem

In a triangle, if the square of the length of one side equals the sum of the squares of the lengths of the other two sides, then the triangle is a right triangle. The 90° angle is opposite the longest side. ∎

E X A M P L E   1 0

Verifying That a Triangle Is a Right Triangle

Show that a triangle whose sides are of lengths 5, 12, and 13 is a right triangle. Identify the hypotenuse.

FIGURE 10

Solution

We square the lengths of the sides:

$$25, \quad 144, \quad 169$$

Notice that the sum of the first two squares (25 and 144) equals the third square (169). Hence, the triangle is a right triangle. The longest side, 13, is the hypotenuse. See Figure 10. ∎

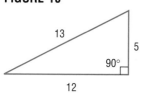

∎ Now work Problem 79.

Geometry Formulas

Certain formulas from geometry are useful in solving algebra problems. We list some of these formulas next.

For a rectangle of length l and width w,

$$\text{Area} = lw \qquad \text{Perimeter} = 2l + 2w$$

For a triangle with base b and altitude h,

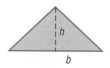

$$\text{Area} = \frac{1}{2}bh$$

For a circle of radius r (diameter $d = 2r$),

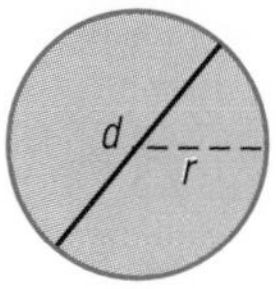

$$\text{Area} = \pi r^2 \qquad \text{Circumference} = 2\pi r = \pi d$$

For a rectangular box of length l, width w, and height h,

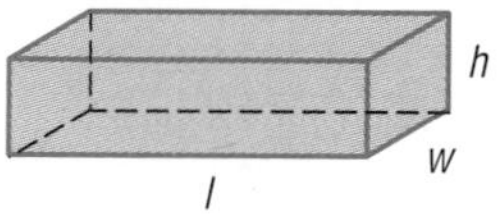

$$\text{Volume} = lwh$$

Calculators

Calculators are finite machines. As a result, they are incapable of displaying decimals that contain a large number of digits. For example, some calculators are capable of displaying only eight digits. When a number requires more than eight digits, the calculator either truncates or rounds. To see how your calculator handles decimals, divide 2 by 3. How many digits do you see? Is the last digit a 6 or a 7? If it is a 6, your calculator truncates; if it is a 7, your calculator rounds.

There are different kinds of calculators. An **arithmetic** calculator can only add, subtract, multiply, and divide numbers; therefore, this type is not adequate for this course. **Scientific** calculators have all the capabilities of arithmetic calculators and also contain **function keys** labeled ln, log, sin, cos, tan, x^y, inv, and so on. As you proceed through this text, you will discover how to use many of the function keys. **Graphing** calculators have all the capabilities of scientific calculators and contain a screen on which graphs can be displayed.

For those who have access to a graphing calculator, we have included comments, examples, and exercises marked with a ▨, indicating that a graphing calculator is required. We have also included an appendix that explains some of the capabilities of a graphing calculator. The ▨ comments, examples, and exercises may be omitted without loss of continuity, if so desired.

Arithmetic operations on your calculator are generally shown as keys labeled as follows:

| + | Addition | − | Subtraction | × | Multiplication | ÷ | Division |

Your calculator also has the following key:

| AC | or | C | To clear the memory and the display window of previous entries.

The next example illustrates a use of the keys for parentheses,

(and)

E X A M P L E   1 1   *Using a Calculator*

Evaluate: $\dfrac{22 + 8}{8 + 2}$

Solution   Remember, we treat this expression as if parentheses enclose the numerator and the denominator.

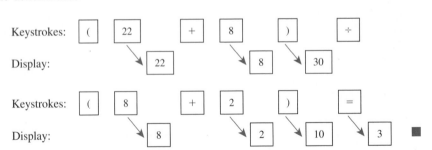

Be careful! If you work the problem in Example 11 without using parentheses, you obtain $22 + 8/8 + 2 = 22 + 1 + 2 = 25$.

■   Now work Problem 93.

1.1

Exercise 1.1

In Problems 1–10, replace the question mark by $<$, $>$, or $=$, whichever is correct.

1. $\frac{1}{2}$? 0   **2.** 5 ? 6   **3.** -1 ? -2 **4.** -3 ? $-\frac{5}{2}$ **5.** π ? 3.14

6. $\sqrt{2}$? 1.41   **7.** $\frac{1}{2}$? 0.5   **8.** $\frac{1}{3}$? 0.33   **9.** $\frac{2}{3}$? 0.67   **10.** $\frac{1}{4}$? 0.25

11. On the real number line, label the points with coordinates 0, 1, -1, $\frac{5}{2}$, -2.5, $\frac{3}{4}$, and 0.25.

12. Repeat Problem 11 for the coordinates 0, -2, 2, -1.5, $\frac{3}{2}$, $\frac{1}{3}$, and $\frac{2}{3}$.

In Problems 13–20, write each statement as an inequality.

13. x is positive   **14.** z is negative

15. x is less than 2   **16.** y is greater than -5

17. x is less than or equal to 1   **18.** x is greater than or equal to 2

19. x is less than 5 and x is greater than 2

20. y is less than or equal to 2 and y is greater than 0

In Problems 21–24, write each inequality using interval notation, and illustrate each inequality using the real number line.

21. $0 \le x \le 4$ **22.** $-1 < x < 5$ **23.** $4 \le x < 6$ **24.** $-2 < x \le 0$

In Problems 25–28, write each interval as an inequality involving x, and illustrate each inequality using the real number line.

25. $[2, 5]$ **26.** $(1, 2)$ **27.** $[4, \infty)$ **28.** $(-\infty, 2]$

In Problems 29–32, find the value of each expression if $x = 2$ and $y = -3$.

29. $|x + y|$ **30.** $|x - y|$ **31.** $|x| + |y|$ **32.** $|x| - |y|$

In Problems 33–52, simplify each expression.

33. 3^0 **34.** 3^2 **35.** 4^{-2} **36.** $(-3)^2$

37. $(\frac{2}{3})^2$ **38.** $(\frac{-4}{5})^3$ **39.** $3^{-6} \cdot 3^4$ **40.** $4^{-2} \cdot 4^3$

41. $(\frac{2}{3})^{-2}$ **42.** $(\frac{3}{2})^{-3}$ **43.** $\dfrac{2^3 \cdot 3^2}{2 \cdot 3^{-2}}$ **44.** $\dfrac{3^{-2} \cdot 5^3}{3 \cdot 5}$

45. $9^{3/2}$ **46.** $16^{3/4}$ **47.** $(-8)^{4/3}$ **48.** $(-27)^{2/3}$

49. $\sqrt{32}$ **50.** $\sqrt[3]{24}$ **51.** $\sqrt[3]{-\frac{8}{27}}$ **52.** $\sqrt{\frac{4}{9}}$

In Problems 53–68, simplify each expression so that all exponents are positive. Whenever an exponent is negative or 0, we assume that the base does not equal 0.

53. $x^0 y^2$ **54.** $x^{-1} y$ **55.** $x^{-2} y$ **56.** $x^4 y^0$

57. $\dfrac{x^{-2} y^3}{xy^4}$ **58.** $\dfrac{x^{-2} y}{xy^2}$ **59.** $\left(\dfrac{4x}{5y}\right)^{-2}$ **60.** $(xy)^{-2}$

61. $\dfrac{x^{-1} y^{-2} z}{x^2 y z^3}$ **62.** $\dfrac{3x^{-2} y z^2}{x^4 y^{-3} z}$ **63.** $\dfrac{(-2)^3 x^4 (yz)^2}{3^2 xy^3 z^4}$ **64.** $\dfrac{4x^{-2}(yz)^{-1}}{(-5)^2 x^4 y^2 z^{-2}}$

65. $\dfrac{x^{-2}}{x^{-2} + y^{-2}}$ **66.** $\dfrac{x^{-1} + y^{-1}}{x^{-1} - y^{-1}}$ **67.** $\left(\dfrac{3x^{-1}}{4y^{-1}}\right)^{-2}$ **68.** $\left(\dfrac{5x^{-2}}{6y^{-2}}\right)^{-3}$

In Problems 69–78, a and b are the lengths of the legs of a right triangle and c is the length of the hypotenuse. Find the missing length.

69. $a = 5, b = 12, c = ?$ **70.** $a = 6, b = 8, c = ?$ **71.** $a = 10, b = 24, c = ?$

72. $a = 4, b = 3, c = ?$ **73.** $a = 7, b = 24, c = ?$ **74.** $a = 14, b = 48, c = ?$

75. $a = 3, c = 5, b = ?$ **76.** $b = 6, c = 10, a = ?$ **77.** $b = 7, c = 25, a = ?$

78. $a = 10, c = 13, b = ?$

In Problems 79–84, the lengths of the sides of a triangle are given. Determine which are right triangles. For those that are right triangles, identify the hypotenuse.

79. 3, 4, 5   **80.** 6, 8, 10   **81.** 4, 5, 6

82. 2, 2, 3   **83.** 7, 24, 25   **84.** 10, 24, 26

In Problems 85–96, use a calculator to approximate each expression. Round off your answer to two decimal places.

85. $(8.51)^2$ **86.** $(9.62)^2$ **87.** $4.1 + (3.2)(8.3)$ **88.** $(8.1)(4.2) + 6.1$

89. $(8.6)^2 + (6.1)^2$ **90.** $(3.1)^2 + (9.6)^2$ **91.** $8.6 + \dfrac{10.2}{4.2}$ **92.** $9.1 - \dfrac{8.2}{10.2}$

93. $\dfrac{2.3 - 9.25}{8.91 + 5.4}$ **94.** $\dfrac{4.73 - 2.4}{81 + 6.39}$ **95.** $\dfrac{\pi + 8}{10.2 + 8.6}$ **96.** $\dfrac{21.3 - \pi}{6.1 + 8.8}$

97. *Geometry* Find the diagonal of a rectangle whose length is 8 inches and whose width is 5 inches.

98. *Geometry* Find the length of a rectangle of width 3 inches if its diagonal is 20 inches long.

99. *Finding the Length of a Guy Wire* A radio transmission tower is 100 feet high. How long does a guy wire need to be if it is to connect a point halfway up the tower to a point 30 feet from the base?

100. Answer Problem 99 if the guy wire is attached to the top of the tower.

101. *How Far Can You See?* The tallest inhabited building in North America is the Sears Tower in Chicago.* If the observation tower is 1454 feet above ground level, use the figure to determine how far a person standing in the observation tower can see (with the aid of a telescope). Use 3960 miles for the radius of Earth. [*Note:*   1 mile = 5280 feet]

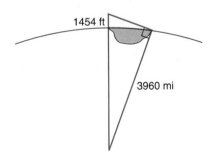

In Problems 102–104, use the fact that the radius of Earth is 3960 miles.

102. *How Far Can You See?* The coming tower of the USS *Silversides,* a World War II submarine now permanently stationed in Muskegon, Michigan, is approximately 20 feet above sea level. How far can one see from the conning tower?

103. *How Far Can You See?* A person who is 6 feet tall is standing on the beach in Fort Lauderdale, Florida and looks out onto the Atlantic Ocean. Suddenly, a ship appears on the horizon. How far is the ship from shore?

*Source: Guinness Book of World Records.

104. *How Far Can You See?* The deck of a destroyer is 100 feet above sea level. How far can a person see from the deck? How far can a person see from the bridge, which is 150 feet above sea level?

105. If $a \leq b$ and $c > 0$, show that $ac \leq bc$. [*Hint:*   Since $a \leq b$, it follows that $a - b \leq 0$. Now multiply each side by c.]

106. If $a \leq b$ and $c < 0$, show that $ac \geq bc$.

107. If $a < b$, show that $a < (a + b)/2 < b$. The number $(a + b)/2$ is called the **arithmetic mean** of a and b.

108. Refer to Problem 107. Show that the arithmetic mean of a and b is equidistant from a and b.

109. Are there any real numbers that are both rational and irrational? Are there any real numbers that are neither? Explain your reasoning.

110. Explain why the sum of a rational number and an irrational number must be irrational.

111. What rational number does the repeating decimal 0.9999 . . . equal?

112. Is there a positive real number "closest" to 0?

113. I'm thinking of a number! It lies between 1 and 10; its square is rational and lies between 1 and 10. The number is larger than π. Correct to two decimal places, name the number. Now think of your own number, describe it, and challenge a fellow student to name it.

114. Write a brief paragraph that illustrates the similarities and differences between "less than" ($<$) and "less than or equal" ($\leq$).

115. *The Gibb's Hill Lighthouse, Southampton, Bermuda,* in operation since 1846, stands 117 feet high on a hill 245 feet high, so its beam of light is 362 feet above sea level. A brochure states that the light itself can be seen on the horizon about 26 miles distant. Verify the correctness of this information. The brochure further states that ships 40 miles away can see the light and planes flying at 10,000 feet can see it 120 miles away. Verify the accuracy of these statements. What assumption did the brochure make about the height of the ship?

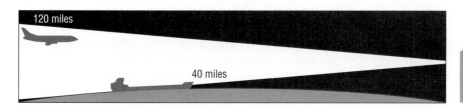

120 miles

40 miles

116. You have 1000 feet of flexible pool siding and wish to construct a swimming pool. Experiment with rectangular-shaped pools with perimeters of 1000 feet. How do their areas vary? What is the shape of the rectangle with the largest area? Now compute the area enclosed by a circular pool with a perimeter (circumference) of 1000 feet. What would be your choice of shape for the pool? If rectangular, what is your preference for dimensions? Justify your choice. If your only consideration is to have a pool that encloses the most area, what shape should you use?

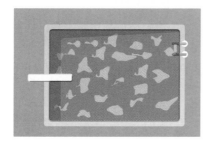

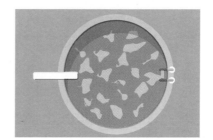

1.2

Equations

An **equation in one variable** is a statement in which two expressions, at least one containing the variable, are equal. The expressions are called the **sides** of the equation. Since an equation is a statement, it may or may not be true, depending on the value of the variable. Unless otherwise restricted, the admissible values of the variable are those in the domain of the variable. Those admissible values of the variable, if any, that result in a true statement are called **solutions,** or **roots,** of the equation. To **solve an equation** means to find all the solutions of the equation.

For example, the following are all equations in one variable, x:

$$x + 5 = 9 \qquad x^2 + 5x = 2x - 2 \qquad \frac{x^2 - 4}{x + 1} = 0 \qquad x^2 + 9 = 5$$

The first of these statements, $x + 5 = 9$, is true when $x = 4$ and false for any other choice of x. Thus, 4 is a solution of the equation $x + 5 = 9$. We also say that 4 **satisfies** the equation $x + 5 = 9$, because, when x is replaced by 4, a true statement results.

Sometimes an equation will have more than one solution. For example, the equation

$$\frac{x^2 - 4}{x + 1} = 0$$

has either $x = -2$ or $x = 2$ as a solution.

Sometimes we will write the solutions of an equation in set notation. This set is called the **solution set** of the equation. For example, the solution set of the equation $x^2 - 9 = 0$ is $\{-3, 3\}$.

Unless indicated otherwise, we will limit ourselves to real solutions. Some equations have no real solution. For example, $x^2 + 9 = 5$ has no real solution, because there is no real number whose square when added to 9 equals 5.

An equation that is satisfied for every choice of the variable for which both sides are defined is called an **identity.** For example, the equation

$$3x + 5 = x + 3 + 2x + 2$$

is an identity, because this statement is true for any real number x.

Two or more equations that have precisely the same solutions are called **equivalent equations.** For example, all the following equations are equivalent, because each has only the solution $x = 5$:

$$2x + 3 = 13$$
$$2x = 10$$
$$x = 5$$

These three equations illustrate one method for solving many types of equations: Replace the original equation by an equivalent equation, and continue until an equation with an obvious solution, such as $x = 5$, is reached. The question, though, is: "How do I obtain an equivalent equation?" In general, there are five ways to do so.

Procedures That Result
in Equivalent Equations

1. Interchange the two sides of the equation:

$$\text{Replace} \qquad 3 = x \quad \text{by} \quad x = 3$$

2. Simplify the sides of the equation by combining like terms, eliminating parentheses, and so on:

$$\text{Replace} \quad (x + 2) + 6 = 2x + (x + 1)$$
$$\text{by} \quad x + 8 = 3x + 1$$

3. Add or subtract the same expression on both sides of the equation:

$$\text{Replace} \quad 3x - 5 = 4$$
$$\text{by} \quad (3x - 5) + 5 = 4 + 5$$

4. Multiply or divide both sides of the equation by the same nonzero expression:

$$\text{Replace} \quad \frac{3x}{x - 1} = \frac{6}{x - 1} \quad x \neq 1$$
$$\text{by} \quad \frac{3x}{x - 1} \cdot (x - 1) = \frac{6}{x - 1} \cdot (x - 1)$$

5. If one side of the equation is 0 and the other side can be factored, then we may use the product law* and set each factor equal to 0:

$$\text{Replace} \quad x(x - 3) = 0$$
$$\text{by} \quad x = 0 \quad \text{or} \quad x - 3 = 0$$

Whenever it is possible to solve an equation in your head, do so. For example:

The solution of $2x = 8$ is $x = 4$.

The solution of $3x - 15 = 0$ is $x = 5$.

Often, though, some rearrangement is necessary.

E X A M P L E   1   *Solving an Equation*

Solve the equation: $(x + 1)(2x) = (x + 1)(2)$

Solution   We begin by collecting all terms on the left side:

$$(x + 1)(2x) = (x + 1)(2)$$
$$(x + 1)(2x) - (x + 1)(2) = 0$$
$$(x + 1)(2x - 2) = 0 \quad \text{Factor.}$$
$$x + 1 = 0 \quad \text{or} \quad 2x - 2 = 0 \quad \text{Apply the product law.}$$
$$x = -1 \qquad\qquad 2x = 2$$
$$x = 1$$

The solution set is $\{-1, 1\}$.   ■

E X A M P L E   2   *Solving an Equation*

Solve the equation: $\dfrac{3x}{x - 1} + 2 = \dfrac{3}{x - 1}$

---

*The product law states that if $ab = 0$ then $a = 0$ or $b = 0$ or both equal 0.

Solution First, we note that the domain of the variable is $\{x|x \neq 1\}$. Since the two quotients in the equation have the same denominator, $x - 1$, we can simplify by multiplying both sides by $x - 1$. The resulting equation is equivalent to the original equation, since we are multiplying by $x - 1$, which is not 0 (remember, $x \neq 1$).

$$\frac{3x}{x - 1} + 2 = \frac{3}{x - 1}$$

$$\left(\frac{3x}{x - 1} + 2\right) \cdot (x - 1) = \frac{3}{x - 1} \cdot (x - 1) \qquad \text{Multiply both sides by } x - 1; \\ \text{cancel on the right.}$$

$$\frac{3x}{x - 1} \cdot (x - 1) + 2 \cdot (x - 1) = 3 \qquad \text{Use the distributive property on} \\ \text{the left side; cancel on the left.}$$

$$3x + (2x - 2) = 3 \qquad \text{Simplify.}$$

$$5x - 2 = 3$$

$$5x = 5 \qquad \text{Add 2 to each side.}$$

$$x = 1 \qquad \text{Divide both sides by 5.}$$

The solution appears to be 1. But recall that $x = 1$ is not in the domain of the variable. Thus, the equation has no solution. ■

■ Now work Problem 19.

Steps for Solving Equations

STEP 1: List any restrictions on the domain of the variable.
STEP 2: Simplify the equation by replacing the original equation by a succession of equivalent equations following the procedures listed earlier.
STEP 3: If the result of Step 2 is a product of factors equal to 0, use the product law and set each factor equal to 0 (procedure 5).
STEP 4: Check your solution(s).

E X A M P L E   3

Solving an Equation

Solve the equation: $x^3 = 25x$

Solution We first rearrange the equation to get 0 on the right side:

$$x^3 = 25x$$

$$x^3 - 25x = 0$$

We notice that x is a factor of each term on the left:

$$x(x^2 - 25) = 0$$

$$x(x + 5)(x - 5) = 0 \quad \text{Difference of two squares}$$

$$x = 0 \quad \text{or} \quad x + 5 = 0 \quad \text{or} \quad x - 5 = 0 \quad \text{Set each factor equal to 0.}$$

$$x = -5 \quad \text{or} \quad x = 5 \quad \text{Solve.}$$

The solution set is $\{-5, 0, 5\}$. ■

■ Now work Problem 21.

Because there are two points whose distance from the origin is 5 units, -5 and 5, the equation $|x| = 5$ will have the solution set $\{-5, 5\}$.

E X A M P L E   4   *Solving an Equation Involving Absolute Value*

Solve the equation: $|x + 4| = 13$

Solution   There are two possibilities:

$$x + 4 = 13 \quad \text{or} \quad x + 4 = -13$$
$$x = 9 \qquad\qquad x = -17$$

Thus, the solution set is $\{-17, 9\}$.   ■

■   Now work Problem 39.

Quadratic Equations

It is assumed that you are familiar with solving **quadratic equations** that is, equations equivalent to one written in the **standard form** $ax^2 + bx + c = 0$, where a, b, and c are real numbers and $a \neq 0$.

When a quadratic equation is written in standard form, $ax^2 + bx + c = 0$, it may be possible to factor the expression on the left side as the product of two first-degree polynomials. See Appendix A, Section A.1, for a discussion of factoring.

E X A M P L E   5   *Solving a Quadratic Equation by Factoring*

Solve the equation: $x^2 = 12 - x$

Solution   We put the equation in standard form by adding $x - 12$ to each side:

$$x^2 = 12 - x$$
$$x^2 + x - 12 = 0$$

The left side may now be factored as

$$(x + 4)(x - 3) = 0 \quad \text{The factors 4 and } -3 \text{ of 12 have the sum 1.}$$

so that

$$x + 4 = 0 \quad \text{or} \quad x - 3 = 0$$
$$x = -4 \qquad\qquad x = 3$$

The solution set is $\{-4, 3\}$.   ■

When the left side factors into two linear equations with the same solution, the quadratic equation is said to have a **repeated solution.** We also call this solution a **root of multiplicity 2,** or a **double root.**

E X A M P L E   6   *Solving a Quadratic Equation by Factoring*

Solve the equation: $x^2 - 6x + 9 = 0$

Solution   This equation is already in standard form, and the left side can be factored:

$$x^2 - 6x + 9 = 0$$
$$(x - 3)(x - 3) = 0$$

so that

$$x = 3 \quad \text{or} \quad x = 3$$

The equation has only the repeated solution 3.   ■

■ Now work Problems 61 and 67.

Quadratic equations also can be solved by using the *quadratic formula.* See Appendix A, Section A.3, for the derivation.

Theorem If $b^2 - 4ac \geq 0$, the real solution(s) of the quadratic equation

$$ax^2 + bx + c = 0 \qquad a \neq 0 \tag{1}$$

is (are) given by the **quadratic formula:**

Quadratic Formula

$$x = \frac{-b \pm \sqrt{b^2 - 4ac}}{2a} \tag{2}$$

■

The quantity $b^2 - 4ac$ is called the **discriminant** of the quadratic equation, because its value tells us whether the equation has real solutions. In fact, it also tells us how many solutions to expect.

Discriminant of a Quadratic Equation

For a quadratic equation $ax^2 + bx + c = 0$,
1. If $b^2 - 4ac > 0$, there are two unequal real solutions.
2. If $b^2 - 4ac = 0$, there is a repeated real solution, a root of multiplicity two.
3. If $b^2 - 4ac < 0$, there is no real solution.

We shall consider quadratic equations whose discriminant is negative in Section 1.5.

EXAMPLE   7

Solving a Quadratic Equation Using the Quadratic Formula

Use the quadratic formula to find the real solutions, if any, of the equation: $3x^2 - 5x + 1 = 0$

Solution The equation is in standard form, so we compare it to $ax^2 + bx + c = 0$ to find *a, b,* and *c:*

$$3x^2 - 5x + 1 = 0$$
$$ax^2 + bx + c = 0$$

With $a = 3$, $b = -5$, and $c = 1$, we evaluate the discriminant $b^2 - 4ac$:

$$b^2 - 4ac = (-5)^2 - 4(3)(1) = 25 - 12 = 13$$

Since $b^2 - 4ac > 0$, there are two real solutions, which can be found using the quadratic formula:

$$x = \frac{-b \pm \sqrt{b^2 - 4ac}}{2a} = \frac{5 \pm \sqrt{13}}{6}$$

The solution set $\{(5 - \sqrt{13})/6, (5 + \sqrt{13})/6\}$.   ■

■   Now work Problem 73.

E X A M P L E   8   *Solving a Quadratic Equation Using the Quadratic Formula*

Use the quadratic formula to find the real solutions, if any, of the equation: $3x^2 + 2 = 4x$

Solution   The equation, as given, is not in standard form.

$$3x^2 + 2 = 4x$$

$$3x^2 - 4x + 2 = 0 \quad \text{Put in standard form.}$$

$$ax^2 + bx + c = 0 \quad \text{Compare to standard form.}$$

With $a = 3$, $b = -4$, and $c = 2$, we find

$$b^2 - 4ac = 16 - 24 = -8$$

Since $b^2 - 4ac < 0$, the equation has no real solution.   ■

1.2

Exercise 1.2

In Problems 1–72, solve each equation.

1. $6 - x = 2x + 9$

2. $3 - 2x = 2 - x$

3. $2(3 + 2x) = 3(x - 4)$

4. $3(2 - x) = 2x - 1$

5. $8x - (2x + 1) = 3x - 10$

6. $5 - (2x - 1) = 10$

7. $\frac{1}{2}x - 4 = \frac{3}{4}x$

8. $1 - \frac{1}{2}x = 5$

9. $0.9t = 0.4 + 0.1t$

10. $0.9t = 1 + t$

11. $\frac{2}{y} + \frac{4}{y} = 3$

12. $\frac{4}{y} - 5 = \frac{5}{2y}$

13. $(x + 7)(x - 1) = (x + 1)^2$

14. $(x + 2)(x - 3) = (x - 3)^2$

15. $x(2x - 3) = (2x + 1)(x - 4)$

16. $x(1 + 2x) = (2x - 1)(x - 2)$

17. $z(z^2 + 1) = 3 + z^3$

18. $w(4 - w^2) = 8 - w^3$

19. $\frac{x}{x - 3} + 3 = \frac{3}{x - 3}$

20. $\frac{3x}{x + 2} = \frac{-6}{x + 2} - 2$

21. $x^2 = 9x$

22. $x^3 = x^2$

23. $t^3 - 9t^2 = 0$

24. $4z^3 - 8z^2 = 0$

25. $\frac{2x}{x^2 - 4} = \frac{4}{x^2 - 4} - \frac{1}{x + 2}$

26. $\frac{x}{x^2 - 9} + \frac{1}{x + 3} = \frac{3}{x^2 - 9}$

27. $\frac{x}{x + 2} = \frac{1}{2}$

28. $\frac{3x}{x - 1} = 2$

29. $\frac{3}{2x - 3} = \frac{2}{x + 5}$

30. $\frac{-2}{x + 4} = \frac{-3}{x + 1}$

31. $(x + 2)(3x) = (x + 2)(6)$

32. $(x - 5)(2x) = (x - 5)(4)$

33. $\frac{6t + 7}{4t - 1} = \frac{3t + 8}{2t - 4}$

34. $\frac{8w + 5}{10w - 7} = \frac{4w - 3}{5w + 7}$

35. $\frac{2}{x - 2} = \frac{3}{x + 5} + \frac{10}{(x + 5)(x - 2)}$

36. $\frac{1}{2x + 3} + \frac{1}{x - 1} = \frac{1}{(2x + 3)(x - 1)}$

37. $|2x| = 6$

38. $|3x| = 12$

39. $|2x + 3| = 5$

40. $|3x - 1| = 2$

41. $|1 - 4t| = 5$

42. $|1 - 2z| = 3$

43. $|-2x| = 8$ **44.** $|-x| = 1$ **45.** $|-2|x = 4$

46. $|3|x = 9$ **47.** $\frac{2}{3}|x| = 8$ **48.** $\frac{3}{4}|x| = 9$

49. $\left|\dfrac{x}{3} + \dfrac{2}{5}\right| = 2$ **50.** $\left|\dfrac{x}{2} - \dfrac{1}{3}\right| = 1$ **51.** $|x - 2| = -\frac{1}{2}$

52. $|2 - x| = -1$ **53.** $|x^2 - 4| = 0$ **54.** $|x^2 - 9| = 0$

55. $|x^2 - 2x| = 3$ **56.** $|x^2 + x| = 12$ **57.** $|x^2 + x - 1| = 1$

58. $|x^2 + 3x - 2| = 2$ **59.** $x^2 = 4x$ **60.** $x^2 = -8x$

61. $z^2 + 4z - 12 = 0$ **62.** $v^2 + 7v + 12 = 0$ **63.** $2x^2 - 5x - 3 = 0$

64. $3x^2 + 5x + 2 = 0$ **65.** $x(x - 7) + 12 = 0$ **66.** $x(x + 1) = 12$

67. $4x^2 + 9 = 12x$ **68.** $25x^2 + 16 = 40x$ **69.** $6x - 5 = \dfrac{6}{x}$

70. $x + \dfrac{12}{x} = 7$ **71.** $\dfrac{4(x - 2)}{x - 3} + \dfrac{3}{x} = \dfrac{-3}{x(x - 3)}$ **72.** $\dfrac{5}{x + 4} = 4 + \dfrac{3}{x - 2}$

In Problems 73–82, find the real solutions, if any, of each equation. Use the quadratic formula.

73. $x^2 - 4x + 2 = 0$ **74.** $x^2 + 4x + 2 = 0$ **75.** $x^2 - 5x - 1 = 0$

76. $x^2 + 5x + 3 = 0$ **77.** $2x^2 - 5x + 3 = 0$ **78.** $2x^2 + 5x + 3 = 0$

79. $4y^2 - y + 2 = 0$ **80.** $4t^2 + t + 1 = 0$ **81.** $4x^2 = 1 - 2x$

82. $2x^2 = 1 - 2x$

In Problems 83–86, find the real solutions, if any, of each equation. Use the quadratic formula and express any solutions rounded off to two decimal places.

83. $x^2 - 4x + 2 = 0$ **84.** $x^2 + 4x + 2 = 0$

85. $x^2 + \sqrt{3}x - 3 = 0$ **86.** $x^2 + \sqrt{2}x - 2 = 0$

In Problems 87–92, use the discriminant to determine whether each quadratic equation has two unequal real solutions, a repeated real solution, or no real solution, without solving the equation.

87. $x^2 - 5x + 7 = 0$ **88.** $x^2 + 5x + 7 = 0$ **89.** $9x^2 - 30x + 25 = 0$

90. $25x^2 - 20x + 4 = 0$ **91.** $3x^2 + 5x - 2 = 0$ **92.** $2x^2 - 3x - 4 = 0$

93. *Physics* A ball is thrown vertically upward from the top of a building 96 feet tall with an initial velocity of 80 feet per second. The distance s (in feet) of the ball from the ground after t seconds is $s = 96 + 80t - 16t^2$.
(a) After how many seconds does the ball strike the ground?
(b) After how many seconds will the ball pass the top of the building on its way down?

94. *Physics* An object is propelled vertically upward with an initial velocity of 20 meters per second. The distance s (in meters) of the object from the ground after t seconds is $s = -4.9t^2 + 20t$.
(a) When will the object be 15 meters above the ground?
(b) When will it strike the ground?
(c) Will the object reach a height of 100 meters?
(d) What is the maximum height?

95. Show that the sum of the roots of a quadratic equation is $-b/a$.

96. Show that the product of the roots of a quadratic equation is c/a.

97. Find k such that the equation $kx^2 + x + k = 0$ has a repeated real solution.

98. Find k such that the equation $x^2 - kx + 4 = 0$ has a repeated real solution.

99. Show that the real solutions of the equation $ax^2 + bx + c = 0$ are the negatives of the real solutions of the equation $ax^2 - bx + c = 0$. Assume that $b^2 - 4ac \geq 0$.

100. Show that the real solutions of the equation $ax^2 + bx + c = 0$ are the reciprocals of the real solutions of the equation $cx^2 + bx + a = 0$. Assume that $b^2 - 4ac \geq 0$.

101. *Summing Consecutive Integers* The sum of the consecutive integers 1, 2, 3, . . . , n is given by the formula $\frac{1}{2}n(n+1)$. How many consecutive integers, starting with 1, must be added to get a sum of 666?

102. *Geometry* If a polygon of n sides has $\frac{1}{2}n(n-3)$ diagonals, how many sides will a polygon with 65 diagonals have? Is there a polygon with 80 diagonals?

103. Explain what is wrong in the following steps:

$$x = 2 \tag{1}$$
$$3x - 2x = 2 \tag{2}$$
$$3x = 2x + 2 \tag{3}$$
$$x^2 + 3x = x^2 + 2x + 2 \tag{4}$$
$$x^2 + 3x - 10 = x^2 + 2x - 8 \tag{5}$$
$$(x - 2)(x + 5) = (x - 2)(x + 4) \tag{6}$$
$$x + 5 = x + 4 \tag{7}$$
$$1 = 0 \tag{8}$$

104. Which of the following pairs of equations are equivalent? Explain.

(a) $x^2 = 9$; $x = 3$ (b) $x = \sqrt{9}$; $x = 3$ (c) $(x - 1)(x - 2) = (x - 1)^2$; $x - 2 = x - 1$

105. The equation

$$\frac{5}{x + 3} + 3 = \frac{8 + x}{x + 3}$$

has no solution, yet when we go through the process of solving it we obtain $x = -3$. Write a brief paragraph to explain what causes this to happen.

106. Make up an equation that has no solution and give it to a fellow student to solve. Ask the fellow student to write a critique of your equation.

107. Describe three ways in which you might solve a quadratic equation. State your preferred method; explain why you chose it.

108. Explain the benefits of evaluating the discriminant of a quadratic equation before attempting to solve it.

109. Make up three quadratic equations: one having two distinct solutions, one having no real solution, and one having exactly one real solution.

110. The word *quadratic* seems to imply for (*quad*), yet a quadratic equation is an equation that involves a polynomial of degree 2. Investigate the origin of the term *quadratic* as it is used in the expression *quadratic equation*. Write a brief essay on your findings.

1.3

Setting Up Equations: Applications

The previous section provides the tools for solving equations. But, unfortunately, applied problems do not come in the form, "Solve the equation. . . ." Instead, they are narratives that supply information—hopefully, enough to answer the question that inevitably arises. Thus, to solve applied problems we must be able to translate the verbal description into the language of mathematics. We do this by using symbols (usually letters of the alphabet) to represent unknown quantities and then finding relationships (such as equations) that involve these symbols. The process of doing this is called **mathematical modeling.**

Any solution to the mathematical problem must be checked against the mathematical problem, the verbal description, and the real problem.

Let's look at a few examples that will help you to translate certain words into mathematical symbols.

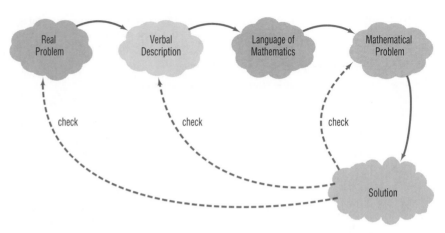

E X A M P L E   1

Translating Verbal Descriptions into Mathematical Expressions

(a) The area of a rectangle is the product of its length times its width.

   Translation: If A is used to represent the area, l the length, and w the width, then $A = lw$.

(b) For uniform motion, the velocity of an object equals the distance traveled divided by the time required.

   Translation: If v is the velocity, s the distance, and t the time, then $v = s/t$.

(c) A total of $5000 is invested, some in stocks and some in bonds. If the amount invested in stocks is x, express the amount invested in bonds in terms of x.

   Translation: If y is the amount invested in bonds, then $x + y = 5000$. Thus, if x is the amount invested in stocks, then the amount invested in bonds is $y = 5000 - x$.

(d) Let x denote a number.

   The number 5 times as large as x is $5x$.

   The number 3 less than x is $x - 3$.

   The number that exceeds x by 4 is $x + 4$.

   The number that, when added to x, gives 5 is $5 - x$.   ■

■   Now work Problem 1.

   Mathematical equations that represent real situations should be consistent in terms of the units used. In Example 1(a), if l is measured in feet, then w also must be expressed in feet, and A will be expressed in square feet. In Example 1(b), if v is measured in miles per hour, then the distance s must be expressed in miles and the time t must be expressed in hours. It is a good practice to check units to be sure that they are consistent and make sense.

   Although each situation has its own unique features, we can provide an outline of the steps to follow in setting up applied problems.

Steps for Setting Up
Applied Problems

STEP 1:   Read the problem carefully, perhaps two or three times. Pay particular attention to the question being asked in order to identify what you are looking for. If you can, determine realistic possibilities for the answer.

STEP 2:   Assign a letter (variable) to represent what you are looking for, and, if necessary, express any remaining unknown quantities in terms of this variable.

STEP 3:   Make a list of all the known facts, and write down any relationships among them, especially any that involve the variable. These may take the form of an equation (or, later, an inequality) involving the variable. If possible, draw an appropriately labeled diagram to assist you. Sometimes, a table or chart helps.

STEP 4:   Solve the equation for the variable, and then answer the question asked in the problem.

STEP 5:   Check the answer with the facts in the problem. If it agrees, congratulations! If it does not agree, try again.

Let's look at an example.

E X A M P L E   2

Determining an Hourly Wage

Colleen grossed $435 one week by working 52 hours. Her employer pays time-and-a-half for all hours worked in excess of 40 hours. With this information, can you determine Colleen's regular hourly wage?

Solution

STEP 1:   We are looking for an hourly wage. Our answer will be in dollars per hour.

STEP 2:   Let x represent the regular hourly wage; x is measured in dollars per hour.

STEP 3:   We set up a table:

	HOURS WORKED	HOURLY WAGE	SALARY
Regular	40	x	$40x$
Overtime	12	$1.5x$	$12(1.5x) = 18x$

The sum of regular salary plus overtime salary will equal $435. Thus, from the table, $40x + 18x = 435$.

STEP 4:

$$40x + 18x = 435$$
$$58x = 435$$
$$x = 7.50$$

Thus, Colleen's regular hourly wage is $7.50 per hour.

STEP 5:   Forty hours yields a salary of $40(7.50) = 300, and 12 hours of overtime yields a salary of $12(1.5)(7.50) = 135, for a total of $435. ■

■   Now work Problem 15.

Interest

The next example involves **interest.** Interest is money paid for the use of money. The total amount borrowed (whether by an individual from a bank in the form of a loan or by a bank from an individual in the form of a savings account) is called the **principal.** The **rate of interest,** expressed as a percent, is the amount charged for the use of the principal for a given period of time, usually on a yearly (that is, per annum) basis.

If a principal of P dollars is borrowed for a period of t years at a per annum interest rate r, expressed as a decimal, the interest I charged is

Simple Interest Formula

$$I = Prt \tag{1}$$

Interest charged according to formula (1) is called **simple interest.**

E X A M P L E   3

Financial Planning

An investor with $70,000 decides to place part of her money in corporate bonds paying 12% per year and the rest in a Certificate of Deposit paying 8% per year. If she wishes to obtain an overall return of 9% per year, how much should she place in each investment?

Solution

STEP 1: The question is asking for two dollar amounts: the principal to invest in the corporate bonds and the principal to invest in the Certificate of Deposit.

STEP 2: We let x represent the amount (in dollars) to be invested in the bonds. Then $70,000 - x$ is the amount that will be invested in the certificate. (Do you see why?)

STEP 3: We set up a table:

	PRINCIPAL $	RATE	TIME yr	INTEREST $
Bonds	x	12% = 0.12	1	$0.12x$
Certificate	$70,000 - x$	8% = 0.08	1	$0.08(70,000 - x)$
Total	70,000	9% = 0.09	1	$0.09(70,000) = 6300$

Since the total interest from the investments is to equal $0.09(70,000) = 6300$, we must have the equation

$$0.12x + 0.08(70,000 - x) = 6300$$

(Note that the units are consistent: the unit is dollars on each side.)

STEP 4:
$$0.12x + 5600 - 0.08x = 6300$$
$$0.04x = 700$$
$$x = 17,500$$

Thus, the investor should place $17,500 in the bonds and $70,000 - $17,500 = $52,500 in the certificate.

STEP 5: The interest on the bonds after 1 year is 0.12($17,500) = $2100; the interest on the certificate after 1 year is 0.08($52,500) = $4200. The total annual interest is $6300, the required amount. ∎

■ Now work Problem 23.

Uniform Motion

The next two examples deal with moving objects.

If an object moves at an average velocity v, the distance s covered in time t is given by the formula

$$s = vt \qquad (2)$$

That is, Distance = Velocity · Time. Objects that are moving in accordance with formula (2) are said to be in **uniform motion.**

E X A M P L E   4

Physics: Uniform Motion

A friend of yours, who is a long-distance runner, runs at an average velocity of 8 miles per hour. Two hours after your friend leaves your house, you leave in your car and follow the same route as your friend. If your average velocity is 40 miles per hour, how long will it be before you reach your friend? How far will each of you be from your house?

Solution   Refer to Figure 11. We use t to represent the time (in hours) that it takes the car to catch up with the runner. When this occurs, the total time elapsed for the runner is $t + 2$ hours.

FIGURE 11

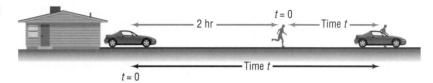

Set up the following table:

	VELOCITY mi/hr	TIME hr	DISTANCE mi
Runner	8	$t + 2$	$8(t + 2)$
Car	40	t	$40t$

Since the distance traveled is the same, we are led to the following equation:

$$8(t + 2) = 40t$$
$$8t + 16 = 40t$$
$$32t = 16$$
$$t = \frac{1}{2} \text{ hour}$$

It will take you $\frac{1}{2}$ hour to reach your friend. Each of you will have gone 20 miles.

Check: In 2.5 hours, the runner travels a distance of $(2.5)(8) = 20$ miles. In $\frac{1}{2}$ hour, the car travels a distance of $\left(\frac{1}{2}\right)(40) = 20$ miles.   ■

■   Now work Problem 51.

E X A M P L E   5   *Physics: Uniform Motion*

A motorboat heads upstream a distance of 24 miles on a river whose current is running at 3 miles per hour. The trip up and back takes 6 hours. Assuming that the motorboat maintained a constant speed relative to the water, what was its speed?

Solution   See Figure 12. We use v to represent the constant speed of the motorboat relative to the water. Then the true speed going upstream is $v - 3$ miles per hour, and the true speed going downstream is $v + 3$ miles per hour. Since Distance = Velocity × Time, then Time = Distance/Velocity. We set up a table.

FIGURE 12

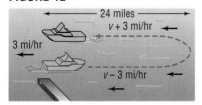

	VELOCITY mi/hr	DISTANCE mi	TIME = DISTANCE/VELOCITY hr
Upstream	$v - 3$	24	$\dfrac{24}{v - 3}$
Downstream	$v + 3$	24	$\dfrac{24}{v + 3}$

Since the total time up and back is 6 hours, we have

$$\frac{24}{v-3} + \frac{24}{v+3} = 6$$

$$\frac{24(v+3) + 24(v-3)}{(v-3)(v+3)} = 6$$

$$\frac{48v}{v^2-9} = 6$$

$$48v = 6(v^2-9)$$

$$6v^2 - 48v - 54 = 0$$

$$v^2 - 8v - 9 = 0$$

$$(v-9)(v+1) = 0$$

$$v = 9 \quad \text{or} \quad v = -1$$

We discard the solution $v = -1$ mile per hour, so the speed of the motorboat relative to the water is 9 miles per hour.　■

Other Applied Problems

The next two examples illustrate problems that you will probably see again in a slightly different form if you study calculus.

E X A M P L E　6　　*Preview of a Calculus Problem*

From each corner of a square piece of sheet metal, remove a square of side 9 centimeters. Turn up the edges to form an open box. If the box is to hold 144 cubic centimeters, what should be the dimensions of the piece of sheet metal?

Solution　We use Figure 13 as a guide. We have labeled by x the length of a side of the square piece of sheet metal. The box will be of height 9 centimeters and its square base will have $x - 18$ as the length of a side. The volume (Length × Width × Height) of the box is therefore

$$9(x-18)(x-18) = 9(x-18)^2$$

FIGURE 13

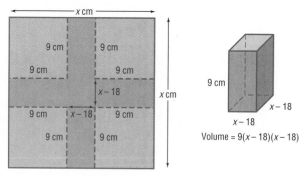

Since the volume of the box is to be 144 cubic centimeters, we have

$$9(x-18)^2 = 144$$

$$(x-18)^2 = 16$$

$$x - 18 = \pm 4$$

$$x = 18 \pm 4$$

$$x = 22 \quad \text{or} \quad x = 14$$

We discard the solution $x = 14$ (do you see why?) and conclude that the sheet metal should be 22 centimeters by 22 centimeters.

Check: If we begin with a piece of sheet metal 22 centimeters by 22 centimeters, cut out a 9 centimeter square from each corner, and fold up the edges, we get a box whose dimensions are 9 by 4 by 4, with volume $9 \times 4 \times 4 = 144$ cubic centimeters, as required. ∎

■ Now work Problem 31.

E X A M P L E   7   *Preview of a Calculus Problem*

A piece of wire 8 feet in length is to be cut into two pieces. Each piece will then be bent into a square. Where should the cut in the wire be made if the sum of the areas of these squares is to be 2 square feet?

Solution   We use Figure 14 as a guide. We have labeled by x the length of one of the pieces of wire after it has been cut. The remaining piece will be of length $8 - x$. If each length is bent into a square, then one of the squares has a side of length $x/4$ and the other a side of length $(8 - x)/4$. Since the sum of the areas of these two squares is 2, we have the equation

FIGURE 14

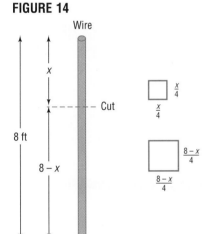

$$\left(\frac{x}{4}\right)^2 + \left(\frac{8 - x}{4}\right)^2 = 2$$

$$\frac{x^2}{16} + \frac{64 - 16x + x^2}{16} = 2$$

$$2x^2 - 16x + 64 = 32$$

$$2x^2 - 16x + 32 = 0 \quad \text{Put in standard form.}$$

$$x^2 - 8x + 16 = 0 \quad \text{Divide by 2.}$$

$$(x - 4)^2 = 0 \quad \text{Factor}$$

$$x = 4$$

Since $x = 4$, $8 - x = 4$, and the original piece of wire should be cut into two pieces, each of length 4 feet.

Check: If the length of each piece of wire is 4 feet, then each piece can be formed into a square whose side is 1 foot. The area of each square is then 1 square foot, so the sum of the areas is 2 square feet, as required. ∎

1.3

Exercise 1.3

In Problems 1–10, translate each sentence into a mathematical equation. Be sure to identify the meaning of all symbols.

1. *Geometry* The area of a circle is the product of the number π times the square of the radius.
2. *Geometry* The circumference of a circle is the product of the number π times twice the radius.
3. *Geometry* The area of a square is the square of the length of a side.
4. *Geometry* The perimeter of a square is four times the length of a side.
5. *Physics* Force equals the product of mass times acceleration.

6. *Physics* Pressure is force per unit area.

7. *Physics* Work equals force times distance.

8. *Physics* Kinetic energy is one-half the product of the mass times the square of the velocity.

9. *Business* The total variable cost of manufacturing x dishwashers is $150 per dishwasher times the number of dishwashers manufactured.

10. *Business* The total revenue derived from selling x dishwashers is $250 per dishwasher times the number of dishwashers sold.

11. *Finance* A total of $20,000 is to be invested, some in bonds and some in Certificates of Deposit (CD's). If the amount invested in bonds is to exceed that in CD's by $2000, how much will be invested in each type of instrument?

12. *Finance* A total of $10,000 is to be divided up between Katy and Mike, with Mike to receive $2000 less than Katy. How much will each receive?

13. *Finance* An inheritance of $900,000 is to be divided among Katy, Mike, and Dan in the following manner: Mike is to receive $\frac{3}{4}$ of what Katy gets, while Dan gets $\frac{1}{2}$ of what Katy gets. How much does each receive?

14. *Sharing the Cost of a Pizza* Mike and Colleen agree to share the cost of an $18 pizza based on how much each ate. If Colleen ate $\frac{2}{3}$ the amount Mike ate, how much should each pay?

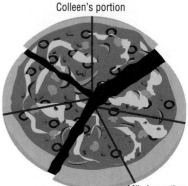

Colleen's portion

Mike's portion

15. *Computing Hourly Wages* A worker who is paid time-and-a-half for hours worked in excess of 40 hours had gross weekly wages of $442 for 48 hours worked. What is the regular hourly rate?

16. *Computing Hourly Wages* Colleen is paid time-and-a-half for hours worked in excess of 40 hours and double-time for hours worked on Sunday. If Colleen had gross weekly wages of $342 for working 50 hours, 4 of which were on Sunday, what is her regular hourly rate?

17. *Football* In an NFL football game, one team scored a total of 41 points, including one safety (2 points) and two field goals (3 points each). After scoring a touchdown (6 points), a team is given the chance to score 1 extra point. The team missed 2 extra points after scoring touchdowns. How many touchdowns did they get?

18. *Basketball* In a basketball game, one team scored a total of 70 points and made three times as many field goals (2 points each) as free throws (1 point each). How many field goals did they have?

19. *Geometry* The perimeter of a rectangle is 60 feet. Find its length and width if the length is 8 feet longer than the width.

20. *Geometry* The perimeter of a rectangle is 42 meters. Find its length and width if the length is twice the width.

21. *Enclosing a Garden* A gardener has 46 feet of fencing to be used to enclose a rectangular garden that has a border 2 feet wide surrounding it (see the figure).
    (a) If the length of the garden is to be twice its width, what will be the dimensions of the garden?
    (b) What is the area of the garden?
    (c) If the length and width of the garden were to be the same, what would be the dimensions of the garden?
    (d) What would be the area of the square garden?

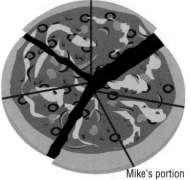

22. *Chemistry: Sugar Molecules* A sugar molecule has twice as many atoms of hydrogen as it does oxygen and one more atom of carbon than oxygen. If a sugar molecule has a total of 45 atoms, how many are oxygen? How many are hydrogen?

23. *Financial Planning* A recent retiree requires $6000 per year in extra income. She has $50,000 to invest and can invest in B-rated bonds paying 15% per year or in a Certificate of Deposit (CD) paying 7% per year. How much money should be invested in each to realize exactly $6000 in interest per year?

24. *Financial Planning* After 2 years, the retiree referred to in Problem 23 finds she now will require $7000 per year. Assuming that the remaining information is the same, how should the money be reinvested?

25. *Banking* A bank loaned out $12,000, part of it at the rate of 8% per year and the rest at the rate of 18% per year. If the interest received totaled $1000, how much was loaned at 8%?

26. *Banking* A loan officer at a bank has $1,000,000 to lend and is required to obtain an average return of 18% per year. If she can lend at the rate of 19% or the rate of 16%, how much can she lend at the 16% rate and still meet her requirement?

27. *Dimensions of a Window* The area of the opening of a rectangular window is to be 143 square feet. If the length is to be 2 feet more than the width, what are the dimensions?

28. *Dimensions of a Window* The area of a rectangular opening is to be 306 square centimeters. If the length exceeds the width by 1 centimeter, what are the dimensions?

29. *Geometry* Find the dimensions of a rectangle whose perimeter is 26 meters and whose area is 40 square meters.

30. *Watering a Field* An adjustable water sprinkler that sprays water in a circular pattern is placed at the center of a square field whose area is 1250 square feet (see the figure). What is the shortest radius setting that can be used if the field is to be completely enclosed within the circle?

31. *Constructing a Box* An open box is to be constructed from a square piece of sheet metal by removing a square of side 1 foot from each corner and turning up the edges. If the box is to hold 4 cubic feet, what should be the dimensions of the sheet metal?

32. *Constructing a Box* Rework Problem 31 if the piece of sheet metal is a rectangle whose length is twice its width.

33. *Physics* A ball is thrown vertically upward from the top of a building 96 feet tall with an initial velocity of 80 feet per second. The distance s (in feet) of the ball from the ground after t seconds is $s = 96 + 80t - 16t^2$.
(a) After how many seconds does the ball strike the ground?
(b) After how many seconds will the ball pass the top of the building on its way down?

34. *Constructing a Coffee Can* A 39 ounce can of Hills Bros.® coffee requires 188.5 square inches of aluminum. If its height is 7 inches, what is its radius? (The surface area A of a right circular cylinder is $A = 2\pi r^2 + 2\pi rh$, where r is the radius and h is the height.)

35. *Business: Discount Pricing* A builder of tract homes reduced the price of a model by 15%. If the new price is $125,000, what was its original price? How much can be saved by purchasing the model?

36. *Business: Discount Pricing* A car dealer, at a year-end clearance, reduces the list price of last year's models by 15%. If a certain four-door model has a discounted price of $8000, what was its list price? How much can be saved by purchasing last year's model?

37. *Business: Marking Up the Price of Books* A college book store marks up the price it pays the publisher for a book by 25%. If the selling price of a book is $56.00, how much did the book store pay for the book?

38. *Personal Finance: Cost of a Car* The suggested list price of a new car is $12,000. The dealer's cost is 85% of list. How much will you pay if the dealer is willing to accept $100 over cost for the car?

39. *Working Together on a Job* Mike can deliver his newspapers in 30 minutes. It takes Danny 20 minutes to do the same route. How long would it take them to deliver the newspapers if they work together?

40. *Working Together on a Job* A painter by himself can paint four rooms in 10 hours. If he hires a helper, they can do the same job together in 6 hours. If he lets the helper work alone, how long will it take for the helper to paint four rooms?

41. *Computing Grades* Going into the final exam, which will count as two tests, Colleen has test scores of 80, 83, 71, 61, and 95. What score does Colleen need on the final in order to have an average score of 80?

42. *Computing Grades* Going into the final exam, which will count as two-thirds of the final grade, Dan has test scores of 86, 80, 84, and 90. What score does Dan need on the final in order to earn a B, which requires an average score of 80? What does he need to earn an A, which requires an average of 90?

43. *Football* A tight end can run the 100 yard dash in 12 seconds. A defensive back can do it in 10 seconds. The tight end catches a pass at his own 20 yard line with the defensive back at the 15 yard line. (See the figure.) If no other players are nearby, at what yard line will the defensive back catch up to the tight end?

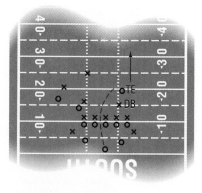

44. *Computing Business Expenses* Debbie, an outside saleswoman, uses her car for both business and pleasure. Last year, she traveled 30,000 miles, using 900 gallons of gasoline. Her car gets 40 miles per gallon on the highway and 25 in the city. She can deduct all highway travel, but no city travel, on her taxes. How many miles should Debbie be allowed as a business expense?

45. *Constructing a Border around a Garden* A landscaper, who just completed a rectangular flower garden measuring 6 feet by 10 feet, orders 1 cubic yard of premixed cement, all of which is to be used to create a border of uniform width around the garden. If the border is to have a depth of 3 inches, how wide will the border be? (1 cubic yard = 27 cubic feet)

46. *Physics* An object is propelled vertically upward with an initial velocity of 20 meters per second. The distance s (in meters) of the object from the ground after t seconds is $s = -4.9t^2 + 20t$.
    (a) When will the object be 15 meters above the ground?
    (b) When will it strike the ground?
    (c) Will the object reach a height of 100 meters?
    (d) What is the maximum height?

47. *Reducing the Size of a Candy Bar* A jumbo chocolate bar with a rectangular shape measures 12 centimeters in length, 7 centimeters in width, and 3 centimeters in thickness. Due to escalating costs of cocoa, management decides to reduce the volume of the bar by 10%. To accomplish this reduction, management decides the new bar should have the same 3 centimeter thickness, but the length and width each should be reduced an equal number of centimeters. What should be the dimensions of the new candy bar?

48. *Reducing the Size of a Candy Bar* Rework Problem 47 if the reduction is to be 20%.

49. *Constructing a Border around a Pool* A pool in the shape of a circle measures 10 feet across. One cubic yard of concrete is to be used to create a circular border of uniform width around the pool. If the border is to have a depth of 3 inches, how wide will the border be? (1 cubic yard = 27 cubic feet)

50. *Constructing a Border around a Pool* Rework Problem 49 if the depth of the border is 4 inches.

51. *Physics: Uniform Motion* A motorboat can maintain a constant speed of 16 miles per hour relative to the water. The boat makes a trip upstream to a certain point in 20 minutes; the return trip takes 15 minutes. What is the speed of the current? (See the figure.)

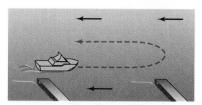

52. *Purity of Gold* The purity of gold is measured in karats, with pure gold being 24 karats. Other purities of gold are expressed as proportional parts of pure gold. Thus, 18 karat gold is $\frac{18}{24}$, or 75% pure gold; 12 karat gold is $\frac{12}{24}$, or 50% pure gold; and so on. How much 12 karat gold should be mixed with pure gold to obtain 60 grams of 16 karat gold?

53. *Running a Race* Mike can run the mile in 6 minutes, and Dan can run the mile in 9 minutes. If Mike gives Dan a head start of 1 minute, how far from the start will Mike pass Dan? (See the figure.) How long does it take?

54. *Physics: Uniform Motion* A motorboat heads upstream on a river that has a current of 3 miles per hour. The trip upstream takes 5 hours, while the return trip takes 2.5 hours. What is the speed of the motorboat? (Assume that the motorboat maintains a constant speed relative to the water.)

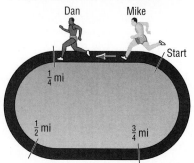

55. *Rescue at Sea* A ship that is in danger of sinking radios the Coast Guard for assistance. When the rescue craft leaves the Coast Guard station, the ship is 60 miles away and heading directly toward the station. If the average speed of the ship is 10 miles per hour and the average speed of the rescue craft is 20 miles per hour, how long will it take for the rescue craft to reach the ship? (See the figure.)

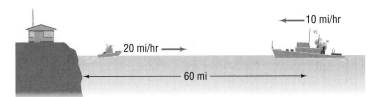

56. *Physics: Uniform Motion* Two cars enter the Florida Turnpike at Commercial Boulevard at 8:00 AM, each heading for Wildwood. One car's average speed is 10 miles per hour more than the other's. The faster car arrives at Wildwood at 11:00 AM, $\frac{1}{2}$ hour before the other car. What is the average speed of each car? How far did each travel?

57. *Emptying Oil Tankers* An oil tanker can be emptied by the main pump in 4 hours. An auxiliary pump can empty the tanker in 9 hours. If the main pump is started at 9 AM, when should the auxiliary pump be started so that the tanker is emptied by noon?

58. *Cement Mix* A 20 pound bag of Economy brand cement mix contains 25% cement and 75% sand. How much pure cement must be added to produce a cement mix that is 40% cement?

59. *Emptying a Tub* A bathroom tub will fill in 15 minutes with both faucets open and the stopper in place. With both faucets closed and the stopper removed, the tub will empty in 20 minutes. How long will it take for the tub to fill if both faucets are open and the stopper is removed?

60. *Range of an Airplane* An air rescue plane averages 300 miles per hour in still air. It carries enough fuel for 5 hours of flying time. If, upon takeoff, it encounters a wind of 30 miles per hour, how far can it fly and return safely? (Assume that the wind remains constant.)

61. *Home Equity Loans* Suppose that you obtain a home equity loan of $100,000 that requires only a monthly interest payment at 10% per annum, with the principal due after 5 years. You decide to invest part of the loan in a 5 year CD that pays 9% compounded and is paid monthly and part in a B+ rated bond due in 5 years that pays 12% compounded and is paid monthly. What is the most you can invest in the CD to ensure that the monthly home equity loan payment is made?

62. *Comparing Olympic Heroes* In the 1984 Olympics, Carl Lewis of the United States won the gold medal in the 100 meter race with a time of 9.99 seconds. In the 1896 Olympics, Thomas Burke, also of the United States, won the gold medal in the 100 meter race in 12.0 seconds. If they ran in the same race, repeating their respective times, by how many meters would Lewis beat Burke?

 63. *Computing Average Speed* In going from Chicago to Atlanta, a car averages 45 miles per hour, and in going from Atlanta to Miami, it averages 55 miles per hour. If Atlanta is halfway between Chicago and Miami, what is the average speed from Chicago to Miami? Discuss an intuitive solution. Write a paragraph defending your intuitive solution. Then solve the problem algebraically. Is your intuitive solution the same as the algebraic one? If not, find the flaw.

64. *Speed of a Plane* On a recent flight from Phoenix to Kansas City, a distance of 919 nautical miles, the plane arrived 20 minutes early. On leaving the aircraft, I asked the captain, "What was our tail wind?" He replied, "I don't know, but our ground speed was 550 knots." How can you determine if enough information is provided to find the tail wind? If possible, find the tail wind. (1 knot = 1 nautical mile per hour)

65. *Critical Thinking* You are the manager of a clothing store and have just purchased 100 dress shirts for $20.00 each. After 1 month of selling the shirts at the regular price, you plan to have a sale giving 40% off the original selling price. However, you still want to make a profit of $4 on each shirt at the sale price. What should you price the shirts at initially to ensure this? If, instead of 40% off at the sale, you give 50% off, by how much is your profit reduced?

66. *Critical Thinking* Make up a word problem that requires solving a linear equation as part of its solution. Exchange problems with a friend. Write a critique of your friend's problem.

67. *Critical Thinking* Without solving, explain what is wrong with the following mixture problem: How many liters of 25% ethanol should be added to 20 liters of 48% ethanol to obtain a solution of 58% ethanol? Now go through an algebraic solution. What happens?

1.4

Inequalities

An **inequality in one variable** is a statement involving two expressions, at least one containing the variable, separated by one of the inequality symbols $<$, $\leq$, $>$, or $\geq$. To **solve an inequality** means to find all values of the variable for which the statement is true. These values are called **solutions** of the inequality.

For example, the following are all inequalities involving one variable, x:

$$x + 5 < 8 \qquad 2x - 3 \geq 4 \qquad x^2 - 1 \leq 3 \qquad \frac{x + 1}{x - 2} > 0$$

In working with inequalities, we will need to know certain properties that they obey.

We begin with the **trichotomy property,** which states that either two numbers are equal or one of them is less than the other.

For any pair of real numbers a and b,

Trichotomy Property

$$a < b \quad \text{or} \quad a = b \quad \text{or} \quad b < a$$

If $b = 0$, the trichotomy property states that, for any real number a,

$$a < 0 \quad \text{or} \quad a = 0 \quad \text{or} \quad a > 0$$

That is, any real number is negative or 0 or positive, a fact we have already noted.

For any real number a, we have

Nonnegative Property

$$a^2 \geq 0$$

In the following properties, a, b, and c are real numbers.

Transitive Property of Inequalities

If $a < b$ and $b < c$, then $a < c$.
If $a > b$ and $b > c$, then $a > c$.

Addition Property of Inequalities

If $a < b$, then $a + c < b + c$.
If $a > b$, then $a + c > b + c$.

The **addition property** states that the sense, or direction, of an inequality remains unchanged if the same number is added to each side.

Multiplication Properties of Inequalities

If $a < b$ and if $c > 0$, than $ac < bc$.
If $a < b$ and if $c < 0$, then $ac > bc$.

If $a > b$ and if $c > 0$, then $ac > bc$.
If $a > b$ and if $c < 0$, then $ac < bc$.

The **multiplication properties** state that the sense, or direction, of an inequality *remains the same* if each side is multiplied by a *positive* real number, while the direction is *reversed* if each side is multiplied by a *negative* real number.

The **reciprocal property** states that the reciprocal of a positive real number is positive and the reciprocal of a negative real number is negative.

Reciprocal Property for Inequalities	If $a > 0$, then $\dfrac{1}{a} > 0$. If $a < 0$, then $\dfrac{1}{a} < 0$.

Solving Inequalities

Two inequalities having exactly the same solution set are called **equivalent inequalities.**

As with equations, one method for solving an inequality is to replace it by a series of equivalent inequalities, until an inequality with an obvious solution, such as $x < 3$, is obtained. We obtain equivalent inequalities by applying some of the same operations as those used to find equivalent equations. The addition property and the multiplication properties form the basis for the following procedure.

Procedures That Leave the Inequality Symbol Unchanged

1. Simplify both sides of the inequality by combining like terms and eliminating parentheses:

$$\text{Replace} \quad (x + 2) + 6 > 2x + (x + 1)$$
$$\text{by} \qquad x + 8 > 3x + 1$$

2. Add or subtract the same expression on both sides of the inequality:

$$\text{Replace} \qquad 3x - 5 < 4$$
$$\text{by} \qquad (3x - 5) + 5 < 4 + 5$$

3. Multiply or divide both sides of the inequality by the same *positive* expression:

$$\text{Replace} \quad 4x > 16 \quad \text{by} \quad \frac{4x}{4} > \frac{16}{4}$$

Procedures That Reverse the Sense or Direction of the Inequality Symbol

1. Interchange the two sides of the inequality:

$$\text{Replace} \quad 3 < x \quad \text{by} \quad x > 3$$

2. Multiply or divide both sides of the inequality by the same *negative* expression:

$$\text{Replace} \quad -2x > 6 \quad \text{by} \quad \frac{-2x}{-2} < \frac{6}{-2}$$

E X A M P L E   1 *Solving an Inequality*

Solve the inequality $4x + 7 \geq 2x - 3$, and graph the solution set.

Solution

$$\begin{aligned}
4x + 7 &\geq 2x - 3 \\
4x + 7 - 7 &\geq 2x - 3 - 7 && \text{Subtract 7 from both sides.} \\
4x &\geq 2x - 10 && \text{Simplify.} \\
4x - 2x &\geq 2x - 10 - 2x && \text{Subtract } 2x \text{ from both sides.} \\
2x &\geq -10 && \text{Simplify.} \\
\frac{2x}{2} &\geq \frac{-10}{2} && \text{Divide both sides by 2. (The sense of the inequality is unchanged.)} \\
x &\geq -5 && \text{Simplify.}
\end{aligned}$$

FIGURE 15
$-5 \leq x < \infty$ or $[-5, \infty)$

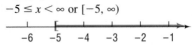

The solution set is $\{x| -5 \leq x < \infty\}$ or, using interval notation, all numbers in the interval $[-5, \infty)$.

See Figure 15 for the graph. ■

■ Now work Problem 1.

E X A M P L E 2 *Solving a Combined Inequality*

Solve the inequality $-5 < 3x - 2 < 1$, and graph the solution set.

Solution The inequality

$$-5 < 3x - 2 < 1$$

is equivalent to the two inequalities

$$-5 < 3x - 2 \quad \text{and} \quad 3x - 2 < 1$$

We will solve each of these inequalities separately. For the first inequality,

$$\begin{aligned}
-5 &< 3x - 2 \\
-5 + 2 &< 3x - 2 + 2 && \text{Add 2 to both sides.} \\
-3 &< 3x && \text{Simplify.} \\
\frac{-3}{3} &< \frac{3x}{3} && \text{Divide both sides by 3.} \\
-1 &< x && \text{Simplify.}
\end{aligned}$$

The second inequality is solved as follows:

$$\begin{aligned}
3x - 2 &< 1 \\
3x - 2 + 2 &< 1 + 2 && \text{Add 2 to both sides.} \\
3x &< 3 && \text{Simplify.} \\
\frac{3x}{3} &< \frac{3}{3} && \text{Divide both sides by 3.} \\
x &< 1 && \text{Simplify.}
\end{aligned}$$

The solution set of the original pair of inequalities consists of all x for which

$$-1 < x \quad \text{and} \quad x < 1$$

FIGURE 16
$-1 < x < 1$ or $(-1, 1)$

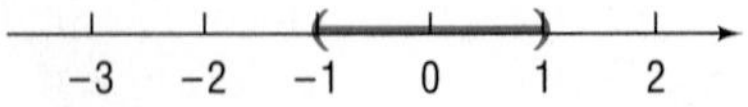

This may be written more compactly as $\{x|-1 < x < 1\}$. In interval notation, the solution is $(-1, 1)$. See Figure 16 for the graph. ■

We observe in this process that the two inequalities we solved required exactly the same steps. A shortcut to solving the original inequality is to deal with the two inequalities at the same time, as follows:

$$-5 < \quad 3x - 2 \quad < 1$$
$$-5 + 2 < 3x - 2 + 2 < 1 + 2 \quad \text{Add 2 to each part.}$$
$$-3 < \quad 3x \quad < 3 \quad \text{Simplify.}$$
$$\frac{-3}{3} < \quad \frac{3x}{3} \quad < \frac{3}{3} \quad \text{Divide each part by 3.}$$
$$-1 < \quad x \quad < 1 \quad \text{Simplify.}$$

■ Now work Problem 11.

To solve inequalities that contain polynomials of degree 2 and higher as well as some that contain rational expressions, we rearrange them so that the polynomial or rational expression is on the left side and 0 is on the right side. An example will show you why.

E X A M P L E   3

Solving a Quadratic Inequality

Solve the inequality $x^2 + x - 12 > 0$, and graph the solution set.

Solution   We factor the left side, obtaining

$$x^2 + x - 12 > 0$$
$$(x + 4)(x - 3) > 0$$

The product of two real numbers is positive either when both factors are positive or when both factors are negative.

BOTH NEGATIVE	**OR**	**BOTH POSITIVE**
$x + 4 < 0$ and $x - 3 < 0$		$x + 4 > 0$ and $x - 3 > 0$
$x < -4$ and $x < 3$		$x > -4$ and $x > 3$
The numbers x that are less than -4 and at the same time less than 3 are simply		The numbers x that are greater than -4 and at the same time greater than 3 are simply
$x < -4$	or	$x > 3$

FIGURE 17
$-\infty < x < -4$ or $3 < x < \infty$;
$(-\infty, -4)$ or $(3, \infty)$

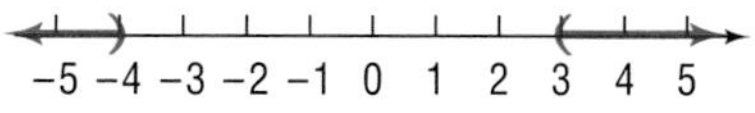

The solution set is $\{x \mid -\infty < x < -4 \text{ or } 3 < x < \infty\}$. In interval notation, we write the solution as $(-\infty, -4)$ or $(3, \infty)$. See Figure 17. ■

We can also obtain the solution to the inequality of Example 3 by another method. The left-hand side of the inequality is factored so that it becomes $(x + 4)(x - 3) > 0$, as before. We then use the real number line to construct a graph that uses the solutions to the equation

$$x^2 + x - 12 = (x + 4)(x - 3) = 0$$

which are $x = -4$ and $x = 3$. These numbers separate the real number line into three intervals: $-\infty < x < -4$, $-4 < x < 3$, and $3 < x < \infty$. See Figure 18(a).

FIGURE 18

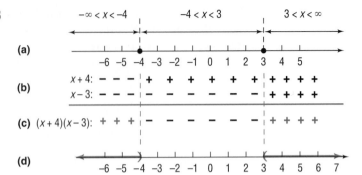

Now, if $x < -4$, then $x + 4 < 0$. We indicate this fact about the expression $x + 4$ by placing minus signs $(- - -)$ to the left of -4. See Figure 18(b). If $x > -4$, then $x + 4 > 0$. We indicate this fact about $x + 4$ by placing plus signs $(+ + +)$ to the right of -4.

Similarly, if $x < 3$, then $x - 3 < 0$. We indicate this fact about $x - 3$ by placing minus signs to the left of 3. If $x > 3$, then $x - 3 > 0$. We indicate this fact about $x - 3$ by placing plus signs to the right of 3. See Figure 18(b).

Next we prepare Figure 18(c) as follows: Since we know that the expressions $x + 4$ and $x - 3$ are both negative for $x < -4$, it follows that their product is positive for $x < -4$. Since we know that $x - 3$ is negative and $x + 4$ is positive for $-4 < x < 3$, it follows that their product is negative for $-4 < x < 3$. Finally, since both expressions are positive for $x > 3$, their product is positive for $x > 3$. We place plus and minus signs as shown in Figure 18(c) to indicate these facts.

Finally, in Figure 18(d) we show on the number line where the expression $(x + 4)(x - 3)$ is positive. The solution to the original inequality $(x + 4)(x - 3) = x^2 + x - 12 > 0$ is $\{x \mid -\infty < x < -4 \text{ or } 3 < x < \infty\}$, as before.

The preceding discussion demonstrates that the sign of each factor of an expression is the same on each interval that the real number line was divided into. Consequently, an alternative, and simpler, approach to obtaining Figure 18(c) would be to select a **test number** in each interval and use it to evaluate each factor to see if it is positive or negative. You may choose any number in the interval as a test number.

In Figure 19(a), the test numbers -5, 1, 4 that we selected are in blue. For $x + 4$, we insert $- - -$ under $-\infty < x < -4$ because, for the test number -5, the value of $x + 4$ is $-5 + 4 = -1$, a negative number. Continuing across, we in-

FIGURE 19

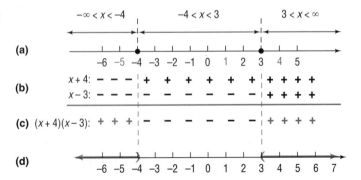

sert + + + under $-4 < x < 3$ because, for the test number 1, the value of $x + 4$ is $1 + 4 = 5$, a positive number. This process is continued for each factor and for each interval to obtain Figure 19(b). The rest of Figure 19 is obtained as before.

We shall employ the method of using a test number to solve inequalities. Here is another example showing all the details.

E X A M P L E   4   *Solving a Quadratic Inequality*

Solve the inequality $x^2 \leq 4x + 12$, and graph the solution set.

Solution   First, we rearrange the inequality so that 0 is on the right side:

$$x^2 \leq 4x + 12$$
$$x^2 - 4x - 12 \leq 0$$
$$(x + 2)(x - 6) \leq 0 \qquad \text{Factor.}$$

Next, we set the left side equal to 0 and solve the resulting equation:

$$(x + 2)(x - 6) = 0$$

The solutions of the equation are -2 and 6, and they separate the real number line into three intervals:

$$-\infty < x < -2 \qquad -2 < x < 6 \qquad 6 < x < \infty$$

See Figure 20.

FIGURE 20

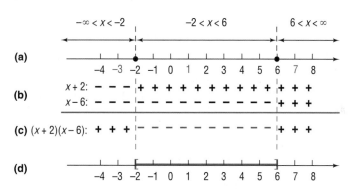

Now, for the test number -3, we find that $x + 2 = -3 + 2 = -1$, a negative number, so we place minus signs under the interval $-\infty < x < -2$. For the test number 1, we find $x + 2 = 1 + 2 = 3$, a positive number, so we place plus signs under the interval $-2 < x < 6$. Continuing in this fashion, we obtain Figure 20(b).

Next we enter the signs of the product $(x + 2)(x - 6)$ in Figure 20(c). Since we want to know where the product $(x + 2)(x - 6)$ is negative, we conclude that the solutions are numbers x for which $-2 < x < 6$. However, because the original inequality is nonstrict, numbers x that satisfy the equation $x^2 = 4x + 12$ are also solutions of the inequality $x^2 \leq 4x + 12$. Thus, we include -2 and 6, and the solution set of the given inequality is $\{x \mid -2 \leq x \leq 6\}$, that is, all x in $[-2, 6]$. See Figure 20(d). ∎

■ Now work Problems 15 and 21.

We have been solving inequalities by rearranging the inequality so that 0 is on the right side, setting the left side equal to 0, and solving the resulting equation. The solutions are then used to separate the real number line into intervals. But what if the resulting equation has no real solution? In this case, we rely on the following result.

Theorem   If a polynomial equation has no real solutions, the polynomial is either always positive or always negative.   ■

For example, the equation

$$x^2 + 5x + 8 = 0$$

has no real solutions. (Do you see why? Its discriminant, $b^2 - 4ac = 25 - 32 = -7$, is negative.) The value of $x^2 + 5x + 8$ is therefore always positive or always negative. To see which is true, we test its value at some number (0 is the easiest). Because $0^2 + 5(0) + 8 = 8$ is positive, we conclude that $x^2 + 5x + 8 > 0$ for all x.

E X A M P L E   5

Solving a Polynomial Inequality

Solve the inequality $x^4 \leq x$, and graph the solution set.

Solution   We rewrite the inequality so that 0 is on the right side:

$$x^4 \leq x$$
$$x^4 - x \leq 0$$

Then we proceed to factor the left side:

$$x^4 - x \leq 0$$
$$x(x^3 - 1) \leq 0$$
$$x(x - 1)(x^2 + x + 1) \leq 0$$

The solutions of the equation

$$x(x - 1)(x^2 + x + 1) = 0$$

are just 0 and 1, since the equation $x^2 + x + 1 = 0$ has no real solutions. We note that the expression $x^2 + x + 1$ is always positive. (Do you see why?) Next, we use 0 and 1 to separate the real number line into three intervals:

$$-\infty < x < 0 \qquad 0 < x < 1 \qquad 1 < x < \infty$$

Then we construct Figure 21 showing the signs of x, $x - 1$, and $x^2 + x + 1$. Note that, since $x^2 + x + 1 > 0$ for all x, we enter plus signs next to it.

FIGURE 21

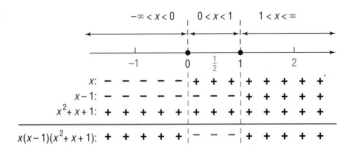

FIGURE 22

$0 \le x \le 1$ or $[0, 1]$

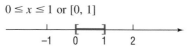

Since we want to know where $x^4 \le x$ or, equivalently, where $x(x-1)(x^2+x+1) \le 0$, we conclude from Figure 21 that the solution set is $\{x|0 \le x \le 1\}$, that is, all x in $[0, 1]$. See Figure 22.   ∎

■ Now work Problem 33.

E X A M P L E   6

Solving a Rational Inequality

Solve the inequality $\dfrac{4x+5}{x+2} \ge 3$, and graph the solution set.

Solution   We first note that the domain of the variable consists of all real numbers except -2. We rearrange terms so that 0 is on the right side:

$$\frac{4x+5}{x+2} \ge 3$$

$$\frac{4x+5}{x+2} - 3 \ge 0$$

$$\frac{4x+5-3(x+2)}{x+2} \ge 0 \quad \text{Rewrite using } x+2 \text{ as the denominator.}$$

$$\frac{x-1}{x+2} \ge 0 \quad \text{Simplify.}$$

FIGURE 23

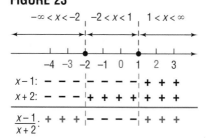

The sign of a rational expression depends on the sign of its numerator and the sign of its denominator. Thus, for a rational expression, we separate the real number line into intervals using the numbers obtained by setting the numerator and the denominator equal to 0. For this example, they are -2 and 1. See Figure 23.

The bottom line in Figure 23 reveals the numbers x for which $(x-1)/(x+2)$ is positive. However, we want to know where the expression $(x-1)/(x+2)$ is positive or 0. Since $(x-1)/(x+2) = 0$ only if $x = 1$, we conclude that the solution set is $\{x|-\infty < x < -2 \text{ or } 1 \le x < \infty\}$, that is, all x in $(-\infty, -2)$ or $[1, \infty)$. See Figure 24.   ∎

FIGURE 24

$-\infty < x < -2$ or $1 \le x < \infty$

$(-\infty, -2)$ or $[1, \infty)$

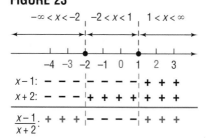

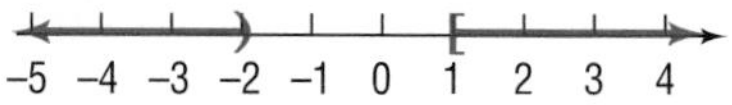

In Example 6, you may wonder why we did not first multiply both sides of the inequality by $x + 2$ to clear the denominator. The reason is that we do not know whether $x + 2$ is positive or negative and, as a result, we do not know whether to reverse the sense of the inequality symbol after multiplying by $x + 2$. However, there is nothing to prevent us from multiplying both sides by $(x + 2)^2$, which is always positive, since $x \ne -2$. (Do you see why?) Then

$$\frac{4x+5}{x+2} \ge 3 \qquad x \ne -2$$

$$\frac{4x+5}{x+2}(x+2)^2 \ge 3(x+2)^2$$

$$(4x+5)(x+2) \ge 3(x^2+4x+4)$$

$$4x^2+13x+10 \ge 3x^2+12x+12$$

$$x^2+x-2 \ge 0$$

$$(x+2)(x-1) \ge 0 \qquad x \ne -2$$

This last expression leads to the same solution set obtained in Example 6.

■ Now work Problem 41.

Let's look at an inequality involving absolute value.

EXAMPLE 7   *Solving an Inequality Involving Absolute Value*

Solve the inequality $|x| < 4$, and graph the solution set.

Solution

FIGURE 25 $-4 < x < 4$ or $(-4, 4)$

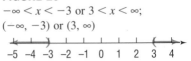

We are looking for all points whose coordinate x is a distance less than 4 units from the origin. See Figure 25 for an illustration. Because any x between -4 and 4 satisfies the condition $|x| < 4$, the solution set consists of all numbers x for which $-4 < x < 4$, that is, all x in $(-4, 4)$. ■

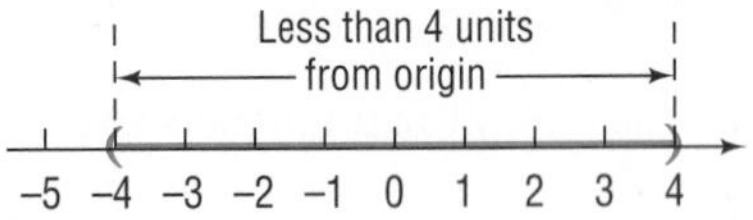

EXAMPLE 8   *Solving an Inequality Involving Absolute Value*

Solve the inequality $|x| > 3$, and graph the solution set.

Solution

FIGURE 26
$-\infty < x < -3$ or $3 < x < \infty$;
$(-\infty, -3)$ or $(3, \infty)$

We are looking for all points whose coordinate x is a distance greater than 3 units from the origin. Figure 26 illustrates the situation. We conclude that any x less than -3 or greater than 3 satisfies the condition $|x| > 3$. Consequently, the solution set consists of all numbers x for which $-\infty < x < -3$ or $3 < x < \infty$, that is, all x in $(-\infty, -3)$ or $(3, \infty)$. ■

We are led to the following results:

Theorem   If a is any positive number, then

$|u| < a$ is equivalent to	$-a < u < a$	(1)
$|u| \leq a$ is equivalent to	$-a \leq u \leq a$	(2)
$|u| > a$ is equivalent to	$u < -a$ or $u > a$	(3)
$|u| \geq a$ is equivalent to	$u \leq -a$ or $u \geq a$	(4)

■

EXAMPLE 9   *Solving an Inequality Involving Absolute Value*

Solve the inequality $|2x + 4| \leq 3$, and graph the solution set.

Solution

$$|2x + 4| \leq 3$$ This follows the form of equation (2); the expression $u = 2x + 4$ is inside the absolute value bars.

$$-3 \leq 2x + 4 \leq 3$$ Apply equation (2).

$$-3 - 4 \leq 2x + 4 - 4 \leq 3 - 4$$ Subtract 4 from each part.

$$-7 \leq 2x \leq -1$$ Simplify.

$$\frac{-7}{2} \leq \frac{2x}{2} \leq -\frac{1}{2}$$ Divide each part by 2.

$$-\frac{7}{2} \leq x \leq -\frac{1}{2}$$ Simplify.

FIGURE 27
$-\frac{7}{2} \leq x \leq -\frac{1}{2}$ or $\left[-\frac{7}{2}, -\frac{1}{2}\right]$

The solution set is $\{x|\ -\frac{7}{2} \leq x \leq -\frac{1}{2}\}$, that is, all x in $[-\frac{7}{2}, -\frac{1}{2}]$. See Figure 27. ■

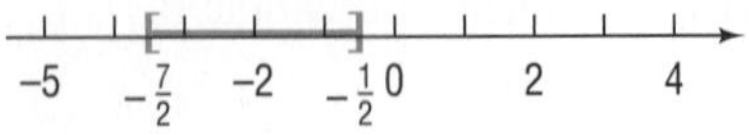

■ Now work Problem 49.

E X A M P L E   1 0

Solving an Inequality Involving Absolute Value

Solution

Solve the inequality $|2x - 5| > 3$, and graph the solution set.

$|2x - 5| > 3$ This follows the form of equation (3); the expression
$u = 2x - 5$ is inside the absolute value bars.

$2x - 5 < -3$	or	$2x - 5 > 3$	Apply equation (3).
$2x - 5 + 5 < -3 + 5$	or	$2x - 5 + 5 > 3 + 5$	Add 5 to each part.
$2x < 2$	or	$2x > 8$	Simplify.
$\dfrac{2x}{2} < \dfrac{2}{2}$	or	$\dfrac{2x}{2} > \dfrac{8}{2}$	Divide each part by 2.

FIGURE 28
$-\infty < x < 1$ or $4 < x < \infty$;
$(-\infty, 1)$ or $(4, \infty)$

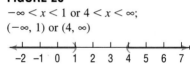

$$x < 1 \qquad \text{or} \qquad x > 4 \qquad \text{Simplify.}$$

The solution set is $\{x \mid -\infty < x < 1 \text{ or } 4 < x < \infty\}$, that is, all x in $(-\infty, 1)$ or $(4, \infty)$. See Figure 28.   ■

Warning: A common error to be avoided is to attempt to write the solution $x < 1$ or $x > 4$ as $1 > x > 4$, which is incorrect, since there are no numbers x for which $x < 1$ *and* $x > 4$. Another common error is to "mix" the symbols and write $1 < x > 4$, which, of course,

1.4

Exercise 1.4

In Problems 1–56, solve each inequality. Graph the solution set.

1. $3x - 1 \geq 3 + x$ 2. $2x - 2 \geq 3 + x$ 3. $-2(x + 3) < 8$

4. $-3(1 - x) < 12$ 5. $4 - 3(1 - x) \leq 3$ 6. $8 - 4(2 - x) \leq -2x$

7. $\frac{1}{2}(x - 4) > x + 8$ 8. $3x + 4 > \frac{1}{2}(x - 2)$ 9. $\frac{x}{2} \geq 1 - \frac{x}{4}$

10. $\frac{x}{3} \geq 2 + \frac{x}{6}$ 11. $0 \leq 2x - 6 \leq 4$ 12. $4 \leq 2x + 2 \leq 10$

13. $-5 \leq 4 - 3x \leq 2$ 14. $-3 \leq 2 - 2x \leq 9$ 15. $(x - 5)(x + 2) < 0$

16. $(x - 1)(x + 2) < 0$ 17. $-x^2 + 9 > 0$ 18. $-x^2 + 1 > 0$

19. $x^2 + x > 12$ 20. $x^2 + 7x < -12$ 21. $x(x - 7) > 8$

22. $x(x + 1) > 2$ 23. $4x^2 + 9 < 6x$ 24. $25x^2 + 16 < 40x$

25. $(x - 1)(x^2 + x + 1) > 0$ 26. $(x + 2)(x^2 - x + 1) > 0$ 27. $(x - 1)(x - 2)(x - 3) < 0$

28. $(x + 1)(x + 2)(x + 3) < 0$ 29. $-x^3 + 2x^2 + 3x < 0$ 30. $-x^3 - 2x^2 + 8x < 0$

31. $x^4 > x^2$ 32. $x^3 < 4x$ 33. $x^3 > x^2$

34. $x^3 < 3x^2$ 35. $\dfrac{x + 1}{1 - x} < 0$ 36. $\dfrac{3 - x}{x + 1} < 0$

37. $\dfrac{(x - 1)(x + 1)}{x} < 0$ 38. $\dfrac{(x - 3)(x + 2)}{x - 1} < 0$ 39. $\dfrac{(x - 2)^2}{x^2 - 1} \geq 0$

40. $\dfrac{x+5}{x^2-4} \geq 0$

41. $\dfrac{x+4}{x-2} \leq 1$

42. $\dfrac{x+2}{x-4} \geq 1$

43. $\dfrac{2x+5}{x+1} > \dfrac{x+1}{x-1}$

44. $\dfrac{1}{x+2} > \dfrac{3}{x+1}$

45. $|2x| < 8$

46. $|3x| < 12$

47. $|3x| > 12$

48. $|2x| > 6$

49. $|x-2| < 1$

50. $|x+4| < 2$

51. $|3t-2| \leq 4$

52. $|2u+5| \leq 7$

53. $|x-3| \geq 2$

54. $|x+3| \geq 2$

55. $|1-4x| < 5$

56. $|1-2x| < 3$

57. Express the fact that x differs from 2 by less than $\frac{1}{2}$ as an inequality involving an absolute value. Solve for x.

58. Express the fact that x differs from -1 by less than 1 as an inequality involving an absolute value. Solve for x.

59. Express the fact that x differs from -3 by more than 2 as an inequality involving an absolute value. Solve for x.

60. Express the fact that x differs from 2 by more than 3 as an inequality involving an absolute value. Solve for x.

61. *Body Temperature* "Normal" human body temperature is 98.6°F. If a temperature x that differs from normal by at least 1.5° is considered unhealthy, write the condition for an unhealthy temperature x as an inequality involving an absolute value, and solve for x.

62. *Household Voltage* In the United States, normal household voltage is 115 volts. However, it is not uncommon for actual voltage to differ from normal voltage by at most 5 volts. Express this situation as an inequality involving an absolute value. Use x as the actual voltage and solve for x.

63. *Electricity Rates* Commonwealth Edison Company's summer charge for electricity is 10.819¢ per kilowatt-hour.[*] In addition, each monthly bill contains a customer charge of $9.06. If last summer's bills ranged from a low of $82.14 to a high of $279.63, over what range did usage vary (in kilowatt-hours)?

64. *Water Bills* The Village of Oak Lawn charges homeowners $17.76 per quarter year plus $1.34 per 1000 gallons for water usage in excess of 12,000 gallons.[†] In 1991, one homeowner's quarterly bill ranged from a high of $49.92 to a low of $28.48. Over what range did water usage vary?

65. *Markup of a New Car* The markup over dealer's cost of a new car ranges from 12% to 18%. If the sticker price is $8800, over what range will the dealer's cost vary?

66. *IQ Tests* A standard intelligence test has an average score of 100. According to statistical theory, of the people who take the test, the 2.5% with the highest scores will have scores of more than 1.95σ above the average, where σ (sigma, a number called the *standard deviation*) depends on the nature of the test. If $\sigma = 12$ for this test and there is (in principle) no upper limit to the score possible on the test, write the interval of possible test scores of the people in the top 2.5%.

67. In your Economics 101 class, you have scores of 68, 82, 87, and 89 on the first four of five tests. To get a grade of B, the average of the five test scores must be greater than or equal to 80 and less than 90. Solve an inequality to find the range of the score you need on the last test to get a B.

68. Repeat Problem 67 if the fifth test counts double.

69. *Physics* A ball is thrown vertically upward with an initial velocity of 80 feet per second. The distance s (in feet) of the ball from the ground after t seconds is $s = 80t - 16t^2$. For what time interval is the ball more than 96 feet above the ground? (See the figure.)

70. *Physics* Rework Problem 69 to find when the ball is less than 64 feet above the ground.

71. *Business* The monthly revenue achieved by selling x wristwatches is figured to be $x(40 - 0.2x)$ dollars. The wholesale cost of each watch is $28. How many watches must be sold each month to achieve a profit (revenue − cost) of at least $100?

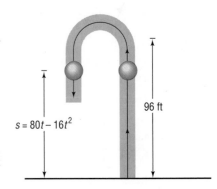

$s = 80t - 16t^2$

96 ft

[*]*Source:* Commonwealth Edison Co., Chicago, Illinois, 1991.
[†]*Source:* Village of Oak Lawn, Illinois, 1991.

72. *Business* The monthly revenue achieved by selling x boxes of candy is figured to be $x(5 - 0.05x)$ dollars. The wholesale cost of each box of candy is $1.50. How many boxes must be sold each month to achieve a profit of at least $60?

73. If $a > 0$, show that the solution set of the inequality

$$x^2 < a$$

consists of all numbers x for which

$$-\sqrt{a} < x < \sqrt{a}$$

74. If $a > 0$, show that the solution set of the inequality

$$x^2 > a$$

consists of all numbers x for which

$$x > \sqrt{a} \quad \text{or} \quad x < -\sqrt{a}$$

In Problems 75–82, use the results found in Problems 72 and 74 to solve each inequality.

75. $x^2 < 1$ 76. $x^2 < 4$ 77. $x^2 \geq 9$ 78. $x^2 \geq 1$

79. $x^2 \leq 16$ 80. $x^2 \leq 9$ 81. $x^2 > 4$ 82. $x^2 > 16$

83. Find k such that the equation $x^2 + kx + 1 = 0$ has no real solution.

84. Find k such that the equation $kx^2 + 2x + 1 = 0$ has two distinct real solutions.

85. Make up an inequality that has no solution. Make up one that has exactly one solution.

86. The inequality $x^2 + 1 < -5$ has no solution. Explain why.

1.5

Complex Numbers

One property of a real number is that its square is nonnegative. For example, there is no real number x for which

$$x^2 = -1$$

To remedy this situation, we introduce a number called the **imaginary unit,** which we denote by i and whose square is -1. Thus,

$$i^2 = -1$$

This should not surprise you. If our universe were to consist only of integers, there would be no number x for which $2x = 1$. This unfortunate circumstance was remedied by introducing numbers such as $\frac{1}{2}$ and $\frac{2}{3}$, the *rational numbers.* If our universe were to consist only of rational numbers, there would be no number x whose square equals 2. That is, there would be no number x for which $x^2 = 2$. To remedy this, we introduced numbers such as $\sqrt{2}$ and $\sqrt[3]{5}$, the *irrational numbers.* The *real numbers,* you will recall, consist of the rational numbers and the irrational numbers. Now, if our universe were to consist only of real numbers, then there would be no number x whose square is -1. To remedy this, we introduce a number i, whose square is -1.

In the progression outlined, each time we encountered a situation that was unsuitable, we introduced a new number system to remedy this situation. And each new number system contained the earlier number system as a subset. The number system that results from introducing the number i is called the **complex number system.**

Complex Numbers

Complex numbers are numbers of the form $a + bi$, where a and b are real numbers. The real number a is called the **real part** of the number $a + bi$; the real number b is called the **imaginary part** of $a + bi$.

For example, the complex number $-5 + 6i$ has the real part -5 and the imaginary part 6.

When a complex number is written in the form $a + bi$, where a and b are real numbers, we say it is in **standard form.** However, if the imaginary part of a complex number is negative, such as in the complex number $3 + (-2)i$, we agree to write it instead in the form $3 - 2i$.

Also, the complex number $a + 0i$ is usually written merely as a. This serves to remind us that the real numbers are a subset of the complex numbers. The complex number $0 + bi$ is usually written as bi. Sometimes the complex number bi is called a **pure imaginary number.**

Equality, addition, subtraction, and multiplication of complex numbers are defined so as to preserve the familiar rules of algebra for real numbers. Thus, two complex numbers are equal if and only if their real parts are equal and their imaginary parts are equal. That is,

Equality of Complex
Numbers

$$a + bi = c + di \quad \text{if and only if} \quad a = c \text{ and } b = d \qquad (1)$$

Two complex numbers are added by forming the complex number whose real part is the sum of the real parts and whose imaginary part is the sum of the imaginary parts. That is,

Sum of Complex Numbers

$$(a + bi) + (c + di) = (a + c) + (b + d)i \qquad (2)$$

To subtract two complex numbers, we follow the rule

Difference of Complex
Numbers

$$(a + bi) - (c + di) = (a - c) + (b - d)i \qquad (3)$$

E X A M P L E   1   *Adding and Subtracting Complex Numbers*

(a) $(3 + 5i) + (-2 + 3i) = [3 + (-2)] + (5 + 3)i = 1 + 8i$

(b) $(6 + 4i) - (3 + 6i) = (6 - 3) + (4 - 6)i = 3 + (-2)i = 3 - 2i$ ■

■ Now work Problem 5.

Products of complex numbers are calculated as illustrated in Example 2.

E X A M P L E   2   *Multiplying Complex Numbers*

$$(5 + 3i) \cdot (2 + 7i) = 5 \cdot (2 + 7i) + 3i(2 + 7i) = 10 + 35i + 6i + 21i^2$$

↑   ↑

Distributive property   Distributive property

$$= 10 + 41i + 21(-1)$$

↑

$i^2 = -1$

$$= -11 + 41i \qquad ■$$

Based on the procedure of Example 2, we define the **product** of two complex numbers by the formula

Product of Complex
Numbers

$$(a + bi) \cdot (c + di) = (ac - bd) + (ad + bc)i \qquad (4)$$

Do not bother to memorize formula (4). Instead, whenever it is necessary to multiply two complex numbers, follow the usual rules for multiplying two binomials, as in Example 2, remembering that $i^2 = -1$. For example,

$$(2i)(2i) = 4i^2 = -4$$

$$(2 + i)(1 - i) = 2 - 2i + i - i^2 = 3 - i$$

■ Now work Problem 11.

Algebraic properties for addition and multiplication, such as the commutative, associative, and distributive properties, hold for complex numbers. Of these, the property that every nonzero complex number has a multiplicative inverse, or reciprocal, requires a closer look.

Conjugates

Conjugate

If $z = a + bi$ is a complex number, then its **conjugate,** denoted by **z̄,** is defined as

$$\bar{z} = \overline{a + bi} = a - bi$$

For example, $\overline{2 + 3i} = 2 - 3i$ and $\overline{-6 - 2i} = -6 + 2i$.

E X A M P L E   3   *Multiplying a Complex Number by Its Conjugate*

Find the product of the complex number $z = 3 + 4i$ and its conjugate $\bar{z}$.

Solution   Since $\bar{z} = 3 - 4i$, we have

$$z\bar{z} = (3 + 4i)(3 - 4i) = 9 + 12i - 12i - 16i^2 = 9 + 16 = 25 \qquad ■$$

The result obtained in Example 3 has an important generalization:

Theorem The product of a complex number and its conjugate is a nonnegative real number. Thus, if $z = a + bi$, then

$$z\bar{z} = a^2 + b^2 \qquad (5)$$

Proof: If $z = a + bi$, then

$$z\bar{z} = (a + bi)(a - bi) = a^2 - (bi)^2 = a^2 - b^2i^2 = a^2 + b^2 \qquad \blacksquare$$

To express the reciprocal of a nonzero complex number z in standard form, multiply the numerator and denominator by its conjugate $\bar{z}$. Thus, if $z = a + bi$ is a nonzero complex number, then

$$\frac{1}{a + bi} = \frac{1}{z} = \frac{1}{z} \cdot \frac{\bar{z}}{\bar{z}} = \frac{\bar{z}}{z\bar{z}} = \frac{a - bi}{(a + bi)(a - bi)}$$

$$= \frac{a - bi}{a^2 + b^2}$$

$$\uparrow$$

Use (5).

$$= \frac{a}{a^2 + b^2} - \frac{b}{a^2 + b^2}i$$

E X A M P L E 4 *Writing the Reciprocal of a Complex Number in Standard Form*

Write $\dfrac{1}{3 + 4i}$ in standard form $a + bi$; that is, find the reciprocal of $3 + 4i$.

Solution The idea is to multiply the numerator and denominator by the conjugate of $3 + 4i$, that is, the complex number $3 - 4i$. The result is

$$\frac{1}{3 + 4i} = \frac{1}{3 + 4i} \cdot \frac{3 - 4i}{3 - 4i} = \frac{3 - 4i}{9 + 16} = \frac{3}{25} - \frac{4}{25}i \qquad \blacksquare$$

To express the quotient of two complex numbers in standard form, we multiply the numerator and denominator of the quotient by the conjugate of the denominator.

E X A M P L E 5 *Writing the Quotient of Complex Numbers in Standard Form*

Write each of the following in standard form:

(a) $\dfrac{1 + 4i}{5 - 12i}$ (b) $\dfrac{2 - 3i}{4 - 3i}$

Solution (a) $\dfrac{1 + 4i}{5 - 12i} = \dfrac{1 + 4i}{5 - 12i} \cdot \dfrac{5 + 12i}{5 + 12i} = \dfrac{5 + 20i + 12i + 48i^2}{25 + 144}$

$$= \frac{-43 + 32i}{169} = \frac{-43}{169} + \frac{32}{169}i$$

(b) $\dfrac{2 - 3i}{4 - 3i} = \dfrac{2 - 3i}{4 - 3i} \cdot \dfrac{4 + 3i}{4 + 3i} = \dfrac{8 - 12i + 6i - 9i^2}{16 + 9} = \dfrac{17 - 6i}{25} = \dfrac{17}{25} - \dfrac{6}{25}i$

$$\blacksquare$$

■ Now work Problem 19.

EXAMPLE 6   *Writing Other Expressions in Standard Form*

If $z = 2 - 3i$ and $w = 5 + 2i$, write each of the following expressions in standard form:

(a) $\dfrac{z}{w}$ (b) $\overline{z + w}$ (c) $z + \overline{z}$

Solution   (a) $\dfrac{z}{w} = \dfrac{z \cdot \overline{w}}{w \cdot \overline{w}} = \dfrac{(2 - 3i)(5 - 2i)}{(5 + 2i)(5 - 2i)} = \dfrac{10 - 15i - 4i + 6i^2}{25 + 4}$

$\quad\quad = \dfrac{4 - 19i}{29} = \dfrac{4}{29} - \dfrac{19}{29}i$

(b) $\overline{z + w} = \overline{(2 - 3i) + (5 + 2i)} = \overline{7 - i} = 7 + i$

(c) $z + \overline{z} = (2 - 3i) + (2 + 3i) = 4$ ■

The conjugate of a complex number has certain general properties that we shall find useful later.

For a real number $a = a + 0i$, the conjugate is $\overline{a} = \overline{a + 0i} = a - 0i = a$. That is,

Theorem   The conjugate of a real number is the real number itself.   ■

Other properties of the conjugate that are direct consequences of the definition are given next. In each statement, z and w represent complex numbers.

Theorem   The conjugate of the conjugate of a complex number is the complex number itself:

$$(\overline{\overline{z}}) = z \qquad (6)$$

The conjugate of the sum of two complex numbers equals the sum of their conjugates:

$$\overline{z + w} = \overline{z} + \overline{w} \qquad (7)$$

The conjugate of the product of two complex numbers equals the product of their conjugates:

$$\overline{z \cdot w} = \overline{z} \cdot \overline{w} \qquad (8)$$

We leave the proofs of equations (6), (7), and (8) as exercises.

Powers of i

The **powers of i** follow a pattern that is useful to know:

$$i^1 = i \qquad\qquad i^5 = i^4 \cdot i = 1 \cdot i = i$$
$$i^2 = -1 \qquad\qquad i^6 = i^4 \cdot i^2 = -1$$
$$i^3 = i^2 \cdot i = -i \qquad\qquad i^7 = i^4 \cdot i^3 = -i$$
$$i^4 = i^2 \cdot i^2 = (-1)(-1) = 1 \qquad i^8 = i^4 \cdot i^4 = 1$$

And so on. Thus, the powers of i repeat with every fourth power.

E X A M P L E   7   *Evaluating Powers of i*

(a) $i^{27} = i^{24} \cdot i^3 = (i^4)^6 \cdot i^3 = 1^6 \cdot i^3 = -i$

(b) $i^{101} = i^{100} \cdot i^1 = (i^4)^{25} \cdot i = 1^{25} \cdot i = i$ ∎

E X A M P L E   8   *Writing the Power of a Complex Number in Standard Form*

Write $(2 + i)^3$ in standard form.

Solution   We use the special product formula for $(x + a)^3$:

$$(x + a)^3 = x^3 + 3ax^2 + 3a^2x + a^3$$

Thus,

$$(2 + i)^3 = 2^3 + 3 \cdot i \cdot 2^2 + 3 \cdot i^2 \cdot 2 + i^3$$
$$= 8 + 12i + 6(-1) + (-i)$$
$$= 2 + 11i$$ ∎

∎ Now work Problem 33.

Quadratic Equations with a Negative Discriminant

Quadratic equations with a negative discriminant have no real number solution. However, if we extend our number system to allow complex numbers, quadratic equations will always have a solution. Since the solution to a quadratic equation involves the square root of the discriminant, we begin with a discussion of square roots of negative numbers.

Principal Square Root of $-N$

If N is a positive real number, we define the **principal square root of $-N$,** denoted by $\sqrt{-N}$, as

$$\sqrt{-N} = \sqrt{N}\,i$$

where i is the imaginary unit and $i^2 = -1$.

E X A M P L E   9

Evaluating the Square Root of Negative Numbers

(a) $\sqrt{-1} = \sqrt{1}i = i$ (b) $\sqrt{-4} = \sqrt{4}i = 2i$

(c) $\sqrt{-8} = \sqrt{8}i = 2\sqrt{2}i$ ∎

E X A M P L E   1 0

Solving Equations

Solve each equation in the complex number system.

(a) $x^2 = 4$ (b) $x^2 = -9$

Solution   (a)
$$x^2 = 4$$
$$x = \pm\sqrt{4} = \pm 2$$

The equation has two solutions, -2 and 2.

(b)
$$x^2 = -9$$
$$x = \pm\sqrt{-9} = \pm\sqrt{9}i = \pm 3i$$

The equation has two solutions, $-3i$ and $3i$.   ∎

∎   Now work Problem 45.

Warning: When working with square roots of negative numbers, do not set the square root of a product equal to the product of the square roots (which can be done with positive numbers). To see why, look at this calculation: We know that $\sqrt{100} = 10$. However, it is also true that $100 = (-25)(-4)$, so

$$10 = \sqrt{100} = \sqrt{(-25)(-4)} \underset{\uparrow}{=} \sqrt{-25}\sqrt{-4}$$

Here is the error.

$$= (\sqrt{25}i)(\sqrt{4}i) = (5i)(2i) = 10i^2 = -10$$

Because we have defined the square root of a negative number, we now can restate the quadratic formula without restriction.

Theorem   In the complex number system, the solutions of the quadratic equation $ax^2 + bx + c = 0$, where a, b, and c are real numbers and $a \neq 0$, are given by the formula

Quadratic Formula
$$x = \frac{-b \pm \sqrt{b^2 - 4ac}}{2a} \tag{9}$$

E X A M P L E   1 1

Solving Quadratic Equations in the Complex Number System

Solve the equation $x^2 - 4x + 8 = 0$ in the complex number system.

Solution   Here $a = 1$, $b = -4$, $c = 8$, and $b^2 - 4ac = 16 - 4(8) = -16$. Using equation (9), we find

$$x = \frac{4 \pm \sqrt{-16}}{2} = \frac{4 \pm \sqrt{16}i}{2} = \frac{4 \pm 4i}{2} = 2 \pm 2i$$

The equation has the solution set $\{2 - 2i, 2 + 2i\}$.

Check:

$$2 + 2i; \quad (2 + 2i)^2 - 4(2 + 2i) + 8 \; = 4 + 8i + 4i^2 - 8 - 8i + 8$$
$$= 4 - 4 = 0$$
$$2 - 2i; \quad (2 - 2i)^2 - 4(2 - 2i) + 8 \; = 4 - 8i + 4i^2 - 8 + 8i + 8$$
$$= 4 - 4 = 0$$ ■

■ Now work Problem 51.

The discriminant, $b^2 - 4ac$, of a quadratic equation still serves as a way to determine the character of the solutions.

Discriminant of a Quadratic Equation

In the complex number system, consider a quadratic equation $ax^2 + bx + c = 0$ with real coefficients.
1. If $b^2 - 4ac > 0$, the equation has two unequal real solutions.
2. If $b^2 - 4ac = 0$, the equation has a repeated real solution—a double root.
3. If $b^2 - 4ac < 0$, the equation has two complex solutions that are not real. The solutions are conjugates of each other.

1.5

Exercise 1.5

In Problems 1–38, write each expression in the standard form a + bi.

1. $(2 - 3i) + (6 + 8i)$ **2.** $(4 + 5i) + (-8 + 2i)$ **3.** $(-3 + 2i) - (4 - 4i)$ **4.** $(3 - 4i) - (-3 - 4i)$

5. $(2 - 5i) - (8 + 6i)$ **6.** $(-8 + 4i) - (2 - 2i)$ **7.** $3(2 - 6i)$ **8.** $-4(2 + 8i)$

9. $2i(2 - 3i)$ **10.** $3i(-3 + 4i)$ **11.** $(3 - 4i)(2 + i)$ **12.** $(5 + 3i)(2 - i)$

13. $(-6 + i)(-6 - i)$ **14.** $(-3 + i)(3 + i)$ **15.** $\dfrac{10}{3 - 4i}$ **16.** $\dfrac{13}{5 - 12i}$

17. $\dfrac{2 + i}{i}$ **18.** $\dfrac{2 - i}{-2i}$ **19.** $\dfrac{6 - i}{1 + i}$ **20.** $\dfrac{2 + 3i}{1 - i}$

21. $\left(\dfrac{1}{2} + \dfrac{\sqrt{3}}{2}i\right)^2$ **22.** $\left(\dfrac{\sqrt{3}}{2} - \dfrac{1}{2}i\right)^2$ **23.** $(1 + i)^2$ **24.** $(1 - i)^2$

25. i^{23} **26.** i^{14} **27.** i^{-15} **28.** i^{-23}

29. $i^6 - 5$ **30.** $4 + i^3$ **31.** $6i^3 - 4i^5$ **32.** $4i^3 - 2i^2 + 1$

33. $(1 + i)^3$ **34.** $(3i)^4 + 1$ **35.** $i^7(1 + i^2)$ **36.** $2i^4(1 + i^2)$

37. $i^6 + i^4 + i^2 + 1$ **38.** $i^7 + i^5 + i^3 + i$

In Problems 39–44, perform the indicated operations and express your answer in the form a + bi.

39. $\sqrt{-4}$ **40.** $\sqrt{-9}$ **41.** $\sqrt{-25}$

42. $\sqrt{-64}$ **43.** $\sqrt{(3 + 4i)(4i - 3)}$ **44.** $\sqrt{(4 + 3i)(3i - 4)}$

In Problems 45–64, solve each equation in the complex number system.

45. $x^2 + 4 = 0$ **46.** $x^2 - 4 = 0$ **47.** $x^2 - 16 = 0$ **48.** $x^2 + 25 = 0$

49. $x^2 - 6x + 13 = 0$ **50.** $x^2 + 4x + 8 = 0$ **51.** $x^2 - 6x + 10 = 0$ **52.** $x^2 - 2x + 5 = 0$

53. $8x^2 - 4x + 1 = 0$ **54.** $10x^2 + 6x + 1 = 0$ **55.** $5x^2 + 2x + 1 = 0$ **56.** $13x^2 + 6x + 1 = 0$

57. $x^2 + x + 1 = 0$ **58.** $x^2 - x + 1 = 0$ **59.** $x^3 - 8 = 0$ **60.** $x^3 + 27 = 0$

61. $x^4 - 16 = 0$ **62.** $x^4 - 1 = 0$ **63.** $x^4 + 13x^2 + 36 = 0$ **64.** $x^4 + 3x^2 - 4 = 0$

In Problems 65–70, without solving, determine the character of the solutions of each equation in the complex number system.

65. $3x^2 - 3x + 4 = 0$ **66.** $2x^2 - 4x + 1 = 0$ **67.** $2x^2 + 3x - 4 = 0$

68. $x^2 + 2x + 6 = 0$ **69.** $9x^2 - 12x + 4 = 0$ **70.** $4x^2 + 12x + 9 = 0$

71. $2 + 3i$ is a solution of a quadratic equation with real coefficients. Find the other solution.

72. $4 - i$ is a solution of a quadratic equation with real coefficients. Find the other solution.

In Problems 73–76, $z = 3 - 4i$ and $w = 8 + 3i$. Write each expression in the standard form $a + bi$.

73. $z + \bar{z}$ **74.** $w - \bar{w}$ **75.** $z\bar{z}$ **76.** $\overline{z - w}$

77. Use $z = a + bi$ to show that $z + \bar{z} = 2a$ and that $z - \bar{z} = 2bi$.

78. Use $z = a + bi$ to show that $(\bar{\bar{z}}) = z$.

79. Use $z = a + bi$ and $w = c + di$ to show that $\overline{z + w} = \bar{z} + \bar{w}$.

80. Use $z = a + bi$ and $w = c + di$ to show that $\overline{z \cdot w} = \bar{z} \cdot \bar{w}$.

81. Explain to a friend how you would add two complex numbers and how you would multiply two complex numbers. Explain any differences in the two explanations.

82. Write a brief paragraph that compares the method used to rationalize the denominator of a rational expression and the method used to write a complex number in standard form.

1.6

Rectangular Coordinates and Graphs

FIGURE 29

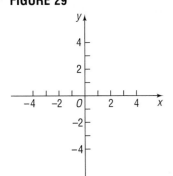

We locate a point on the real number line by assigning it a single real number, called the *coordinate of the point*. For work in a two-dimensional plane, we locate points by using two numbers.

We begin with two real number lines located in the same plane: one horizontal and the other vertical. We call the horizontal line the **x-axis,** the vertical line the **y-axis,** and the point of intersection the **origin O.** We assign coordinates to every point on these number lines, as described earlier (Section 1.1) and shown in Figure 29, using a convenient scale. In mathematics we usually use the same scale on each axes; in applications, a different scale is often used on each axis. The origin O has a value of 0 on both the x-axis and the y-axis. We follow the usual convention that points on the x-axis to the right of O are associated with positive real numbers, and those to the left of O are associated with negative real numbers. Those on the y-axis above O are associated with positive real numbers, and those below O are associated with negative real numbers. In Figure 29, the x-axis and y-axis are labeled as x and y, respectively, and we have used an arrow at the end of each axis to denote the positive direction.

The coordinate system described here is called a **rectangular,** or **Cartesian,*** **coordinate system.** The plane formed by the x-axis and y-axis is sometimes called the **xy-plane,** and the x-axis and y-axis are referred to as the **coordinate axes.**

Any point P in the xy-plane can then be located by using an **ordered pair** (x, y) of real numbers. Let x denote the signed distance of P from the y-axis (*signed* in the sense that if P is to the right of the y-axis then $x > 0$, and if P is to the left of the y-axis, then $x < 0$); and let y denote the signed distance of P from the x-axis. The ordered pair (x, y), also called the **coordinates** of P, then gives us enough information to locate the point P in the plane.

For example, to locate the point whose coordinates are $(-3, 1)$, go 3 units along the x-axis to the left of O and then go straight up 1 unit. We **plot** this point

*Named after René Descartes (1596–1650), a French mathematician, philosopher, and theologian.

FIGURE 30

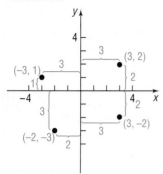

by placing a dot at this location. See Figure 30, in which the points with coordinates $(-3, 1)$, $(-2, -3)$, $(3, -2)$, and $(3, 2)$ are plotted.

The origin has coordinates $(0, 0)$. Any point on the x-axis has coordinates of the form $(x, 0)$, and any point on the y-axis has coordinates of the form $(0, y)$.

If (x, y) are the coordinates of a point P, then x is called the **x-coordinate,** or **abscissa,** of P and y is the **y-coordinate,** or **ordinate,** of P. We identify the point P by its coordinates (x, y) by writing $P = (x, y)$. Usually, we will simply say "the point (x, y)," rather than "the point whose coordinates are (x, y)."

The coordinate axes divide the xy-plane into four sections, called **quadrants,** as shown in Figure 31. In quadrant I, both the x-coordinate and the y-coordinate of all points are positive; in quadrant II, x is negative and y is positive; in quadrant III, both x and y are negative; and in quadrant IV, x is positive and y is negative. Points on the coordinate axes belong to no quadrant.

FIGURE 31

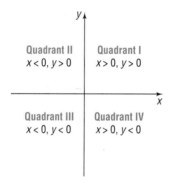

Comment: On a graphing calculator, you can set the scale on each axis. Once this has been done, you obtain the **viewing rectangle.** See Figure 32 for a typical viewing rectangle. You should now read Section B.1, The Viewing Rectangle, in Appendix B.

FIGURE 32

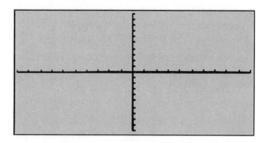

If the same units of measurement, such as inches, centimeters, and so on, are used for both the x-axis and the y-axis, then all distances in the xy-plane can be measured using this unit of measurement. The *distance formula* provides a straightforward method for computing the distance between two points in the xy-plane.

Theorem   The distance between two points $P_1 = (x_1, y_1)$ and $P_2 = (x_2, y_2)$, denoted by $d(P_1, P_2)$, is

Distance Formula

$$d(P_1, P_2) = \sqrt{(x_2 - x_1)^2 + (y_2 - y_1)^2} \tag{1}$$

EXAMPLE 1

Finding the Distance Between Two Points

Find the distance d between the points $(-4, 5)$ and $(3, 2)$.

Solution   Using the distance formula (1), the solution is obtained as follows:

$$d = \sqrt{[3 - (-4)]^2 + (2 - 5)^2} = \sqrt{7^2 + (-3)^2}$$
$$= \sqrt{49 + 9} = \sqrt{58} \approx 7.62 \qquad \blacksquare$$

Proof of the Distance Formula   Let (x_1, y_1) denote the coordinates of point P_1, and let (x_2, y_2) denote the coordinates of point P_2. Assume that the line joining P_1 and P_2 is neither horizontal nor vertical. Refer to Figure 33(a). The coordinates of P_3 are (x_2, y_1). The horizontal distance from P_1 to P_3 is the absolute value of the difference of the x-coordinates, or $|x_2 - x_1|$. The vertical distance from P_3 to P_2 is the absolute value of the difference of the y-coordinates, or $|y_2 - y_1|$. See Figure 33(b). The distance $d(P_1, P_2)$ that we seek is the length of the hypotenuse of the right triangle, so, by the Pythagorean Theorem, it follows that

$$[d(P_1, P_2)]^2 = |x_2 - x_1|^2 + |y_2 - y_1|^2$$
$$= (x_2 - x_1)^2 + (y_2 - y_1)^2$$
$$d(P_1, P_2) = \sqrt{(x_2 - x_1)^2 + (y_2 - y_1)^2}$$

FIGURE 33

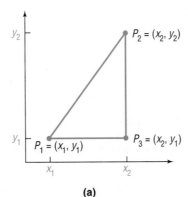

(a)

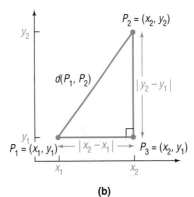

(b)

Now, if the line joining P_1 and P_2 is horizontal, then the y-coordinate of P_1 equals the y-coordinate of P_2; that is, $y_1 = y_2$. Refer to Figure 34(a). In this case, the distance formula (1) still works, because, for $y_1 = y_2$, it reduces to

$$d(P_1, P_2) = \sqrt{(x_2 - x_1)^2 + 0^2} = \sqrt{(x_2 - x_1)^2} = |x_2 - x_1|$$

A similar argument holds if the line joining P_1 and P_2 is vertical. See Figure 34(b). Thus, the distance formula is valid in all cases.

FIGURE 34

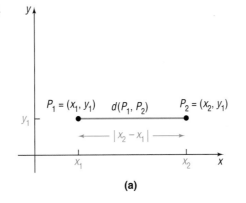

(a)

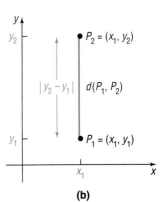

(b)

The distance between two points $P_1 = (x_1, y_1)$ and $P_2 = (x_2, y_2)$ is never a negative number. Furthermore, the distance between two points is 0 only when the points are identical—that is, when $x_1 = x_2$ and $y_1 = y_2$. Also, because $(x_2 - x_1)^2 = (x_1 - x_2)^2$ and $(y_2 - y_1)^2 = (y_1 - y_2)^2$, it makes no difference whether the distance is computed from P_1 to P_2 or from P_2 to P_1—that is, $d(P_1, P_2) = d(P_2, P_1)$.

Rectangular coordinates enable us to translate geometry problems into algebra problems, and vice versa. The next example shows how algebra (the distance formula) can be used to solve some geometry problems.

E X A M P L E   2

Using Algebra to Solve Geometry Problems

Consider the three points $A = (-2, 1)$, $B = (2, 3)$ and $C = (3, 1)$.

(a) Plot each point and form the triangle ABC.

(b) Find the length of each side of the triangle.

(c) Verify that the triangle is a right triangle.

(d) Find the area of the triangle.

FIGURE 35

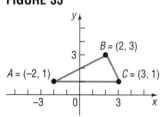

Solution
(a) Points A, B, C, and triangle ABC are plotted in Figure 35.

(b) $d(A, B) = \sqrt{[2 - (-2)]^2 + (3 - 1)^2} = \sqrt{16 + 4} = \sqrt{20} = 2\sqrt{5}$

$d(B, C) = \sqrt{(3 - 2)^2 + (1 - 3)^2} = \sqrt{1 + 4} = \sqrt{5}$

$d(A, C) = \sqrt{[3 - (-2)]^2 + (1 - 1)^2} = \sqrt{5^2 + 0^2} = 5$

(c) To show that the triangle is a right triangle, we need to show that the sum of the squares of the lengths of the two smaller sides equals the square of the length of the longest side. (Why is this sufficient?) Looking at Figure 35, it seems reasonable to conjecture that the right angle is at vertex B. Thus, we shall check to see whether

$$[d(A, B)]^2 + [d(B, C)]^2 = [d(A, C)]^2$$

We find

$$[d(A, B)]^2 + [d(B, C)]^2 = (2\sqrt{5})^2 + (\sqrt{5})^2$$
$$= 20 + 5 = 25 = [d(A, C)]^2$$

so it follows from the converse of the Pythagorean Theorem that triangle ABC is a right triangle.

(d) Because the right angle is at B, the sides AB and BC form the base and altitude of the triangle. Its area is therefore

FIGURE 36

$$\text{Area} = \tfrac{1}{2}(\text{Base})(\text{Altitude}) = \tfrac{1}{2}(2\sqrt{5})(\sqrt{5}) = 5 \text{ square units} \qquad \blacksquare$$

 Now work Problem 19.

We now derive a formula for the coordinates of the **midpoint of a line segment.** Let $P_1 = (x_1, y_1)$ and $P_2 = (x_2, y_2)$ be the end points of a line segment, and let $M = (x, y)$ be the point on the line segment that is the same distance from P_1 as it is from P_2. See Figure 36. The triangles P_1AM and MBP_2 are congruent.*

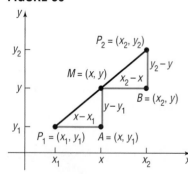

*The following statement is a postulate from geometry. Two triangles are congruent if their sides are the same length (SSS), or if two sides and the included angle are the same (SAS), or if two angles and the included side are the same (ASA).

[Do you see why? Angle AP_1M = Angle BMP_2,* Angle P_1MA = Angle MP_2B, and $d(P_1, M) = d(M, P_2)$ is given. Thus, we have Angle–Side–Angle.] Hence, corresponding sides are equal in length. That is,

$$x - x_1 = x_2 - x \quad \text{and} \quad y - y_1 = y_2 - y$$
$$2x = x_1 + x_2 \qquad\qquad 2y = y_1 + y_2$$
$$x = \frac{x_1 + x_2}{2} \qquad\qquad y = \frac{y_1 + y_2}{2}$$

Theorem   The midpoint (x, y) of the line segment from $P_1 = (x_1, y_1)$ to $P_2 = (x_2, y_2)$ is given by

Midpoint Formula

$$(x, y) = \left(\frac{x_1 + x_2}{2}, \frac{y_1 + y_2}{2}\right) \tag{2}$$

Thus, to find the midpoint of a line segment, we average the x-coordinates and the y-coordinates of the end points.

E X A M P L E   3   *Finding the Midpoint of a Line Segment*

Find the midpoint of the line segment from $P_1 = (-5, 3)$ to $P_2 = (3, 1)$. Plot the points P_1 and P_2 and their midpoint. Check your answer.

Solution   We apply the midpoint formula (2) using $x_1 = -5$, $x_2 = 3$, $y_1 = 3$, and $y_2 = 1$. See Figure 37. Then the coordinates (x, y) of the midpoint M are

$$x = \frac{x_1 + x_2}{2} = \frac{-5 + 3}{2} = -1 \quad \text{and} \quad y = \frac{y_1 + y_2}{2} = \frac{3 + 1}{2} = 2$$

FIGURE 37

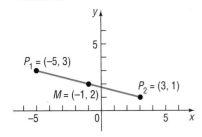

That is, $M = (-1, 2)$.

Check: Because M is the midpoint, we check the answer by verifying that $d(P_1, M) = d(M, P_2)$:

$$d(P_1, M) = \sqrt{[-1 - (-5)]^2 + (2 - 3)^2} = \sqrt{16 + 1} = \sqrt{17}$$
$$d(M, P_2) = \sqrt{[3 - (-1)]^2 + (1 - 2)^2} = \sqrt{16 + 1} = \sqrt{17} \qquad \blacksquare$$

■ Now work Problem 5.

Graphs of Equations

The **graph of an equation** in two variables x and y consists of the set of points in the xy-plane whose coordinates (x, y) satisfy the equation.

Let's look at some examples.

---

*Another postulate from geometry states that the transversal $\overline{P_1P_2}$ forms equal corresponding angles with the parallel lines $\overline{P_1A}$ and $\overline{MB}$.

E X A M P L E   4   *Graphing an Equation*

Graph the equation $y = 2x + 5$

Solution   We want to find all points (x, y) that satisfy the equation. To locate some of these points (and thus get an idea of the pattern of the graph), we assign some numbers to x and find corresponding values for y:

IF	THEN	POINT ON GRAPH
$x = 0$	$y = 2(0) + 5 = 5$	$(0, 5)$
$x = 1$	$y = 2(1) + 5 = 7$	$(1, 7)$
$x = -5$	$y = 2(-5) + 5 = -5$	$(-5, -5)$
$x = 10$	$y = 2(10) + 5 = 25$	$(10, 25)$

By plotting these points and then connecting them, we obtain the graph of the equation (a straight line), as shown in Figure 38. ■

FIGURE 38
$y = 2x + 5$

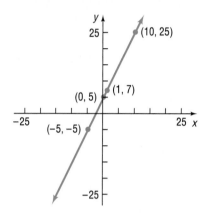

E X A M P L E   5   *Graphing an Equation*

Graph the equation: $y = x^2$

Solution   Table 1 provides several points on the graph. In Figure 39, we plot these points and connect them with a smooth curve to obtain the graph (a *parabola*).

TABLE 1

x	$y = x^2$	(x, y)
-4	16	$(-4, 16)$
-3	9	$(-3, 9)$
-2	4	$(-2, 4)$
-1	1	$(-1, 1)$
0	0	$(0, 0)$
1	1	$(1, 1)$
2	4	$(2, 4)$
3	9	$(3, 9)$
4	16	$(4, 16)$

FIGURE 39
$y = x^2$

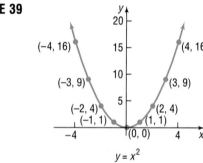

The graphs of the equations shown in Figures 38 and 39 do not show all points. For example, in Figure 38, the point $(20, 45)$ is a part of the graph of $y = 2x + 5$, but it is not shown. Since the graph of $y = 2x + 5$ could be extended out as far as we please, we use arrows to indicate that the pattern shown continues.

Thus, it is important when illustrating a graph to present enough of the graph so that any viewer of the illustration will "see" the rest of it as an obvious continuation of what is actually there. This is referred to as a **complete graph.**

So, one way to obtain a complete graph of an equation is to plot a sufficient number of points on the graph until a pattern becomes evident. Then these points are connected with a smooth curve following the suggested pattern. But how many points are sufficient? Sometimes knowledge about the equation tells us. For example, we will learn in the next section that, if an equation is of the form $y = mx + b$, then its graph is a straight line. In this case, two points would suffice to obtain the graph.

One purpose of this book is to investigate the properties of equations in order to decide whether a graph is complete. In this section, we shall graph equations by plotting a sufficient number of points on the graph until a pattern becomes evident; then we connect these points with a smooth curve following the suggested pattern. Shortly, we shall investigate various techniques that will enable us to graph an equation without plotting so many points.

FIGURE 40

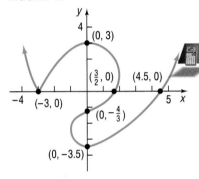

 Comment: Another way to obtain the graph of an equation is to use a graphing utility. Read Section B.2, Graphing Equations Using a Graphing Utility, in Appendix B.

Two techniques that reduce the number of points required to graph an equation involve finding *intercepts* and checking for *symmetry.*

The points, if any, at which a graph intersects the coordinate axes are called the **intercepts.** The x-coordinate of a point at which the graph crosses or touches the x-axis is an **x-intercept,** and the y-coordinate of a point at which the graph crosses or touches the y-axis is a **y-intercept.** For example, the graph in Figure 40 has three x-intercepts, -3, $\frac{3}{2}$, and 4.5, and three y-intercepts, -3.5, $-\frac{4}{3}$, and 3.

Procedure for Finding Intercepts

1. To find the x-intercept(s), if any, of the graph of an equation, let $y = 0$ in the equation and solve for x.
2. To find the y-intercept(s), if any, of the graph of an equation, let $x = 0$ in the equation and solve for y.

Comment: For many equations, finding intercepts may not be so easy. In such cases, a graphing utility can be used. Read Section B.3, The TRACE, ZOOM-IN, and BOX Functions in Appendix B to find out how a graphing utility locates intercepts.

Another useful tool for graphing equations involves *symmetry,* particularly symmetry with respect to the x-axis, the y-axis, and the origin.

Symmetry with Respect to the x-Axis

A graph is said to be **symmetric with respect to the x-axis** if, for every point (x, y) on the graph, the point $(x, -y)$ is also on the graph.

Symmetry with Respect to the y-Axis

A graph is said to be **symmetric with respect to the y-axis** if, for every point (x, y) on the graph, the point $(-x, y)$ is also on the graph.

Symmetry with Respect to the Origin

A graph is said to be **symmetric with respect to the origin** if, for every point (x, y) on the graph, the point $(-x, -y)$ is also on the graph.

Figure 41 illustrates the definition. Notice that, when a graph is symmetric with respect to the x-axis, the part of the graph above the x-axis is a reflection of the part below it, and vice versa. When a graph is symmetric with respect to the y-axis, the part of the graph to the right of the y-axis is a reflection of the part to the left of it, and vice versa. Notice that symmetry with respect to the origin may be viewed in two ways:

1. As a reflection about the y-axis, followed by a reflection about the x-axis.
2. As a projection along a line through the origin so that the distances from the origin are equal.

FIGURE 41

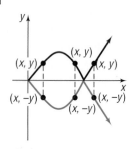

Symmetric with respect to
the x-axis

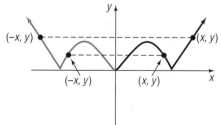

Symmetric with respect to the y-axis

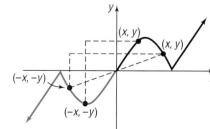

Symmetric with respect to
the origin

■   Now work Problem 29.

When the graph of an equation is symmetric, the number of points that you need to plot in order to see the pattern is reduced. For example, if the graph of an equation is symmetric with respect to the y-axis, then, once points to the right of the y-axis are plotted, an equal number of points on the graph can be obtained by reflecting them about the y-axis. Thus, before we graph an equation, we first want to determine whether it has any symmetry. The following tests are used for this purpose.

Tests for Symmetry | To test the graph of an equation for symmetry with respect to the:

x-axis   Replace y by $-y$ in the equation. If an equivalent equation results, the graph of the equation is symmetric with respect to the x-axis.

y-axis   Replace x by $-x$ in the equation. If an equivalent equation results, the graph of the equation is symmetric with respect to the y-axis.

Origin   Replace x by $-x$ and y by $-y$ in the equation. If an equivalent equation results, the graph of the equation is symmetric with respect to the origin.

E X A M P L E   6

Graphing an Equation by Finding Intercepts and Checking for Symmetry

Graph the equation $y = x^3$. Find any intercepts and check for symmetry first.

Solution | First, we seek the intercepts. When $x = 0$, then $y = 0$; and when $y = 0$, then $x = 0$. Thus, the origin $(0, 0)$ is the only intercept. Now we test for symmetry:

x-Axis   Replace y by $-y$. Since the result, $-y = x^3$, is not equivalent to $y = x^3$, the graph is not symmetric with respect to the x-axis.

y-Axis: Replace x by $-x$. Since the result, $y = -x^3$, is not equivalent to $y = x^3$, the graph is not symmetric with respect to the y-axis.

Origin: Replace x by $-x$ and y by $-y$. Since the result, $-y = -x^3$, is equivalent to $y = x^3$, the graph is symmetric with respect to the origin.

Because of the symmetry, we need to locate points on the graph only for $x \geq 0$, such as (0, 0), (1, 1), and (2, 8). Figure 42 shows the graph.

FIGURE 42
Symmetry with respect to the origin

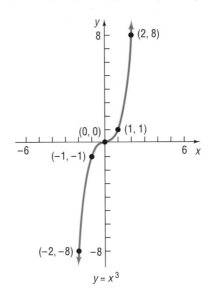

$y = x^3$

■   Now work Problem 35.

E X A M P L E   7   *Graphing an Equation*

Graph the equation $x = y^2$. Find any intercepts and check for symmetry first.

Solution   The lone intercept is (0, 0). The graph is symmetric with respect to the x-axis. (Do you see why? Replace y by $-y$.) Figure 43 shows the graph.   ■

FIGURE 43
Symmetry with respect to the x-axis

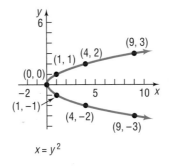

$x = y^2$

FIGURE 44
$y = \sqrt{x}$

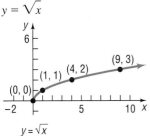

$y = \sqrt{x}$

If we restrict y so that $y \geq 0$, the equation $x = y^2$, $y \geq 0$, may be written equivalently as $y = \sqrt{x}$. The portion of the graph of $x = y^2$ in quadrant I is therefore the graph of $y = \sqrt{x}$. See Figure 44.

Comment: To see the graph of the equation $x = y^2$ on a graphing calculator, you will need to graph two equations, $y = \sqrt{x}$ and $y = -\sqrt{x}$. We discuss why in the next chapter. See Figure 45.

FIGURE 45

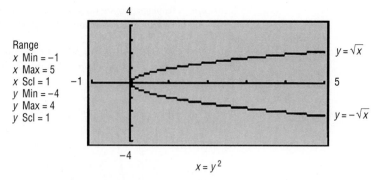

Range
x Min = −1
x Max = 5
x Scl = 1
y Min = −4
y Max = 4
y Scl = 1

$x = y^2$

E X A M P L E   8   *Graphing an Equation*

Graph the equation: $y = \dfrac{1}{x}$

Find any intercepts and check for symmetry first.

Solution   We check for intercepts first. If we let $x = 0$, we obtain a 0 denominator, which is not allowed. Hence, there is no y-intercept. If we let $y = 0$, we get the equation $1/x = 0$, which has no solution. Hence, there is no x-intercept. Thus, the graph of $y = 1/x$ does not cross the coordinate axes.

Next we check for symmetry:

x-Axis: Replacing y by $-y$ yields $-y = 1/x$, which is not equivalent to $y = 1/x$.

y-Axis: Replacing x by $-x$ yields $y = -1/x$, which is not equivalent to $y = 1/x$.

Origin: Replacing x by $-x$ and y by $-y$ yields $-y = -1/x$, which is equivalent to $y = 1/x$.

The graph is symmetric with respect to the origin.

Finally, we set up Table 2, listing several points on the graph. Because of the symmetry with respect to the origin, we use only positive values of x. From Table 2, we infer that, if x is a large and positive number, then $y = 1/x$ is a positive number close to 0. We also infer that, if x is a positive number close to 0, then $y = 1/x$ is a large and positive number. Armed with this information, we can graph the equation. Figure 46 illustrates some of these points and the graph of $y = 1/x$. Observe how the absence of intercepts and the existence of symmetry with respect to the origin were utilized.

TABLE 2

x	$y = 1/x$	(x, y)
$\frac{1}{10}$	10	$(\frac{1}{10}, 10)$
$\frac{1}{3}$	3	$(\frac{1}{3}, 3)$
$\frac{1}{2}$	2	$(\frac{1}{2}, 2)$
1	1	$(1, 1)$
2	$\frac{1}{2}$	$(2, \frac{1}{2})$
3	$\frac{1}{3}$	$(3, \frac{1}{3})$
10	$\frac{1}{10}$	$(10, \frac{1}{10})$

FIGURE 46

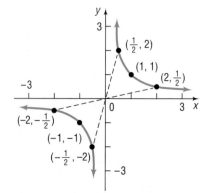

$y = \dfrac{1}{x}$

Comment: Figure 47 shows the graph of $y = 1/x$ using a graphing utility. We infer from the graph that there are no intercepts. We may also infer from the graph that there is possible symmetry with respect to the origin. The TRACE function provides further evidence of this symmetry. As the graph is TRACEd, the points $(-x, -y)$ and (x, y) are obtained. For example, the pair of points $(-0.95238, -1.05)$ and $(0.95238, 1.05)$ both lie on the graph.

FIGURE 47

$y = \dfrac{1}{x}$

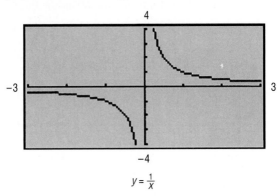

$y = \frac{1}{x}$

■ Now work Problem 77.

Circles

One advantage of a coordinate system is that it enables us to translate a geometric statement into an algebraic statement, and vice versa. Consider, for example, the following geometric statement that defines a circle.

Circle

A **circle** is a set of points in the xy-plane that are a fixed distance r from a fixed point (h, k). The fixed distance r is called the **radius,** and the fixed point (h, k) is called the **center** of the circle.

FIGURE 48

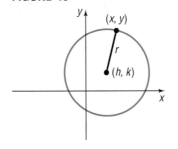

Figure 48 shows the graph of a circle. Is there an equation having this graph? If so, what is the equation? To find the equation, we let (x, y) represent the coordinates of any point on a circle with radius r and center (h, k). Then the distance between the points (x, y) and (h, k) must always equal r. That is, by the distance formula,

$$\sqrt{(x - h)^2 + (y - k)^2} = r$$

or, equivalently,

$$(x - h)^2 + (y - k)^2 = r^2$$

The **standard form of an equation of a circle** with radius r and center (h, k) is

Standard Form
of an Equation of a Circle

$$(x - h)^2 + (y - k)^2 = r^2 \qquad (3)$$

Conversely, by reversing the steps, we conclude: The graph of any equation of the form of equation (3) is that of a circle with radius r and center (h, k).

E X A M P L E   9   *Graphing a Circle*

Graph the equation: $(x + 3)^2 + (y - 2)^2 = 16$

Solution   By comparing the given equation to the standard form of the equation of a circle, we conclude that the graph of the given equation is a circle. Moreover, the comparison yields information about the circle:

FIGURE 49

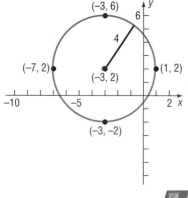

$$(x + 3)^2 + (y - 2)^2 = 16$$
$$[x - (-3)]^2 + (y - 2)^2 = 4^2$$
$$(x - h)^2 + (y - k)^2 = r^2$$

We see that $h = -3$, $k = 2$, and $r = 4$. Hence, the circle has center $(-3, 2)$ and a radius of 4 units. To graph this circle, we first plot the center $(-3, 2)$. Since the radius is 4, we can locate four points on the circle by going out 4 units to the left, to the right, up, and down from the center. These four points can then be used as guides to obtain the graph. See Figure 49.   ■

■   Now work Problem 61.

Comment: Now read Section B.4, Square Screens, in Appendix B.

E X A M P L E   1 0   *Using a Graphing Utility to Graph a Circle*

Graph the equation: $x^2 + y^2 = 4$

Solution   This is the equation of a circle with center at the origin and radius 2. To graph this equation, we must first solve for y.

$$x^2 + y^2 = 4$$
$$y^2 = 4 - x^2$$
$$y = \pm \sqrt{4 - x^2}$$

There are two equations to graph: first, we graph $y = \sqrt{4 - x^2}$ and then $y = -\sqrt{4 - x^2}$ on the same square screen. (Your circle will appear oval if you do not use a square screen.) See Figure 50.   ■

FIGURE 50
$x^2 + y^2 = 4$

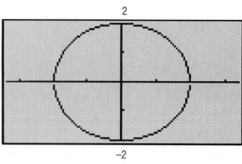

$x^2 + y^2 = 4$

E X A M P L E   1 1   *Writing the Standard Form of the Equation of a Circle*

Write the standard form of the equation of the circle with radius 3 and center $(1, -2)$.

Solution   Using the form of equation (3) and substituting the values $r = 3$, $h = 1$, and $k = -2$, we have

$$(x - h)^2 + (y - k)^2 = r^2$$
$$(x - 1)^2 + (y + 2)^2 = 9$$ ∎

The standard form of an equation of a circle of radius r with center at the origin $(0, 0)$ is

$$x^2 + y^2 = r^2$$

If the radius $r = 1$, the circle whose center is at the origin is called the **unit circle** and has the equation

$$x^2 + y^2 = 1$$

FIGURE 51
Unit circle $x^2 + y^2 = 1$

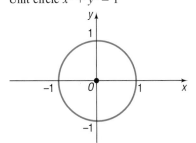

See Figure 51.
If we eliminate the parentheses from the standard form of the equation of the circle obtained in Example 11, we get

$$(x - 1)^2 + (y + 2)^2 = 9$$
$$x^2 - 2x + 1 + y^2 + 4y + 4 = 9$$

which we find, upon simplifying, is equivalent to

$$x^2 + y^2 - 2x + 4y - 4 = 0$$

By completing the squares on both the x- and y-terms, it can be shown that any equation of the form

$$x^2 + y^2 + ax + by + c = 0$$

has a graph that is a circle, or a point, or has no graph at all. For example, the graph of the equation $x^2 + y^2 = 0$ is the single point $(0, 0)$. The equation $x^2 + y^2 + 5 = 0$, or $x^2 + y^2 = -5$, has no graph, because sums of squares of real numbers are never negative. When its graph is a circle, the equation

General Form
of the Equation of a Circle

$$x^2 + y^2 + ax + by + c = 0$$

is referred to as the **general form of the equation of a circle.**

■   Now work Problem 45.

The next example shows how to transform an equation in the general form to an equivalent equation in standard form. As we said earlier, the idea is to use the method of completing the square on both the x- and y-terms.

E X A M P L E   1 2   *Graphing a Circle Whose Equation Is in General Form*
Graph the equation: $x^2 + y^2 + 4x - 6y + 12 = 0$

FIGURE 52
$x^2 + y^2 + 4x - 6y + 12 = 0$

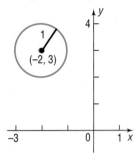

Solution   We rearrange the equation as follows:

$$(x^2 + 4x) + (y^2 - 6y) = -12$$

Next, we complete the square of each expression in parentheses. Remember that any number added on the left also must be added on the right:

$$(x^2 + 4x + 4) + (y^2 - 6y + 9) = \quad -12 + 4 + 9$$
$$(x + 2)^2 + (y - 3)^2 = \quad 1$$

We recognize this equation as the standard form of the equation of a circle with radius 1 and center $(-2, 3)$. Figure 52 illustrates the graph.   ■

■ Now work Problem 63.

1.6

Exercise 1.6

In Problems 1 and 2, plot each point in the xy-plane. Tell in which quadrant or on what coordinate axis each point lies.

1. (a) $A = (-3, 2)$ (b) $B = (6, 0)$ (c) $C = (-2, -2)$
   (d) $D = (6, 5)$ (e) $E = (0, -3)$ (f) $F = (6, -3)$

2. (a) $A = (1, 4)$ (b) $B = (-3, -4)$ (c) $C = (-3, 4)$
   (d) $D = (4, 1)$ (e) $E = (0, 1)$ (f) $F = (-3, 0)$

3. Plot the points $(2, 0)$, $(2, -3)$, $(2, 4)$, $(2, 1)$, and $(2, -1)$. Describe the set of all points of the form $(2, y)$, where y is a real number.

4. Plot the points $(0, 3)$, $(1, 3)$, $(-2, 3)$, $(5, 3)$, and $(-4, 3)$. Describe the set of all points of the form $(x, 3)$, where x is a real number.

In Problems 5–18, find the distance $d(P_1, P_2)$ between the points P_1 and P_2. Also find the midpoint of the line segment joining P_1 and P_2.

5.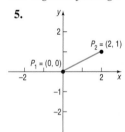

6.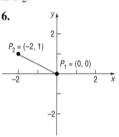

7.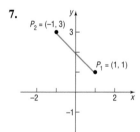

8.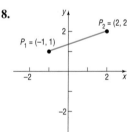

9. $P_1 = (3, -4)$; $P_2 = (3, 1)$
11. $P_1 = (-3, 2)$; $P_2 = (6, 0)$
13. $P_1 = (4, -3)$; $P_2 = (6, 1)$
15. $P_1 = (-0.2, 0.3)$; $P_2 = (2.3, 1.1)$
17. $P_1 = (a, b)$; $P_2 = (0, 0)$

10. $P_1 = (-1, 0)$; $P_2 = (2, 1)$
12. $P_1 = (2, -3)$; $P_2 = (4, 2)$
14. $P_1 = (-4, -3)$; $P_2 = (2, 2)$
16. $P_1 = (1.2, 2.3)$; $P_2 = (-0.3, 1.1)$
18. $P_1 = (a, a)$; $P_2 = (0, 0)$

In Problems 19–22, plot each point and form the triangle ABC. Verify that the triangle is a right triangle. Find its area.

19. $A = (-2, 5)$; $B = (1, 3)$; $C = (-1, 0)$
21. $A = (-5, 3)$; $B = (6, 0)$; $C = (5, 5)$

20. $A = (-2, 5)$; $B = (12, 3)$; $C = (10, -11)$
22. $A = (-6, 3)$; $B = (3, -5)$; $C = (-1, 5)$

23. Find all points having an *x*-coordinate of 2 whose distance from the point $(-2, -1)$ is 5.

24. Find all points having a *y*-coordinate of -3 whose distance from the point $(1, 2)$ is 13.

25. Find all points on the *x*-axis that are 5 units from the point $(2, -3)$.

26. Find all points on the *y*-axis that are 5 units from the point $(-4, 4)$.

In Problems 27–34, plot each point. Then plot the point that is symmetric to it with respect to:
(a) The x-axis   (b) The y-axis   (c) The origin

27. $(3, 4)$ **28.** $(5, 3)$ **29.** $(-2, 1)$ **30.** $(4, -2)$

31. $(1, 1)$ **32.** $(-1, -1)$ **33.** $(-3, -4)$ **34.** $(4, 0)$

In Problems 35–44, the graph of an equation is given.
(a) List the intercepts of the graph.
(b) Based on the graph, tell whether the graph is symmetric with respect to the x-axis, y-axis, and/or origin.

35.
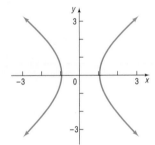

36.
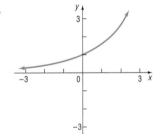

37.
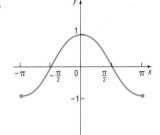

38.
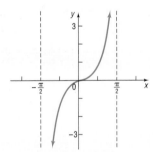

39.
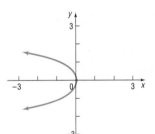

40.
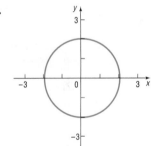

41.
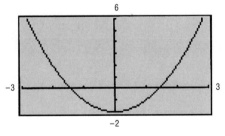

42.
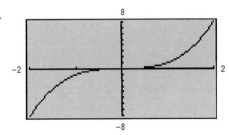

43.
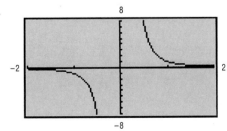

44.
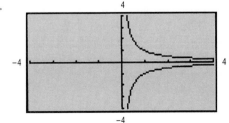

In Problems 45–50, write the standard form of the equation and the general form of the equation of each circle of radius r and center (h, k).

45. $r = 2$; $(h, k) = (0, 2)$ **46.** $r = 3$; $(h, k) = (1, 0)$ **47.** $r = 5$; $(h, k) = (4, -3)$

48. $r = 4$; $(h, k) = (2, -3)$ **49.** $r = 2$; $(h, k) = (0, 0)$ **50.** $r = 3$; $(h, k) = (0, 0)$

In Problems 51–54, find the center and radius of each circle. Write the standard form of the equation.

51. 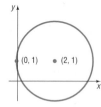   **52.** 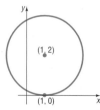   **53.** 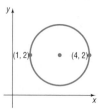   **54.**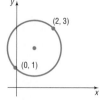

In Problems 55–58, match each graph with the correct equation.

(a) $(x - 3)^2 + (y + 3)^2 = 9$ (c) $(x - 1)^2 + (y + 2)^2 = 4$

(b) $(x + 1)^2 + (y - 2)^2 = 4$ (d) $(x + 3)^2 + (y - 3)^2 = 9$

55. 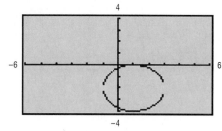   **56.**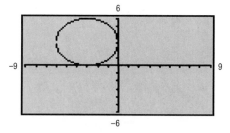

57. 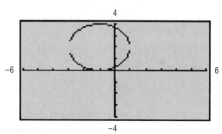   **58.**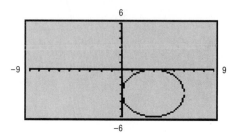

In Problems 59–68, find the center (h, k) and radius r of each circle. Graph the circle.

59. $x^2 + y^2 = 4$ **60.** $x^2 + (y - 1)^2 = 1$ **61.** $(x - 3)^2 + y^2 = 4$

62. $(x + 1)^2 + (y - 1)^2 = 2$ **63.** $x^2 + y^2 + 4x - 4y - 1 = 0$ **64.** $x^2 + y^2 - 6x + 2y + 9 = 0$

65. $x^2 + y^2 - x + 2y + 1 = 0$ **66.** $x^2 + y^2 + x + y - \frac{1}{2} = 0$ **67.** $2x^2 + 2y^2 - 12x + 8y - 24 = 0$

68. $2x^2 + 2y^2 + 8x + 7 = 0$

In Problems 69–72, use the accompanying graph.

69. Extend the graph to make it symmetric with respect to the x-axis.

70. Extend the graph to make it symmetric with respect to the y-axis.

71. Extend the graph to make it symmetric with respect to the origin.

72. Extend the graph to make it symmetric with respect to the x-axis, y-axis, and origin.

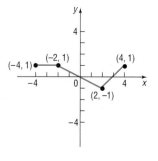

In Problems 73–86, list the intercepts and test for symmetry.

73. $x^2 = y$ **74.** $y^2 = x$ **75.** $y = 3x$ **76.** $y = -5x$

77. $x^2 + y - 9 = 0$ **78.** $y^2 - x - 4 = 0$ **79.** $4x^2 + 9y^2 = 36$ **80.** $x^2 + 4y^2 = 4$

81. $y = x^3 - 27$ **82.** $y = x^4 - 1$ **83.** $y = x^2 - 3x - 4$ **84.** $y = x^2 + 4$

85. $y = \dfrac{x}{x^2 + 9}$ **86.** $y = \dfrac{x^2 - 4}{x}$

87. *Baseball* A major league baseball "diamond" is actually a square, 90 feet on a side (see the figure). Overlay a rectangular coordinate system on a major league baseball diamond so that the origin is at home plate, the positive x-axis lies in the direction from home plate to first base, and the positive y-axis lies in the direction from home plate to third base.
(a) What are the coordinates of first base, second base, and third base? Use feet as the unit of measurement.
(b) If the right fielder is located at (310, 15), how far is it from there to second base?
(c) If the center fielder is located at (300, 300), how far is it from there to third base?

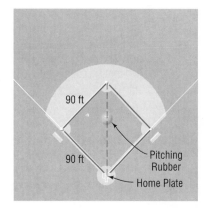

88. *Little League Baseball* The layout of a Little League playing field is a square, 60 feet on a side.* Overlay a rectangular coordinate system on a Little League baseball diamond so that the origin is at home plate, the positive x-axis lies in the direction from home plate to first base, and the positive y-axis lies in the direction from home plate to third base.
(a) What are the coordinates of first base, second base, and third base? Use feet as the unit of measurement.
(b) If the right fielder is located at (180, 20), how far is it from there to second base?
(c) If the center fielder is located at (220, 220), how far is it from there to third base?

89. An automobile and a truck leave an intersection at the same time. The automobile heads east at an average speed of 40 miles per hour, while the truck heads south at an average speed of 30 miles per hour. Find an expression for their distance apart d (in miles) at the end of t hours.

90. A hot air balloon, headed due east at an average speed of 15 miles per hour and at a constant altitude of 100 feet, passes over an intersection (see the figure). Find an expression for its distance d (measured in feet) from the intersection t seconds later.

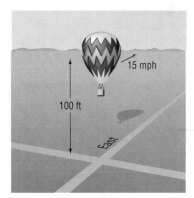

91. *Weather Satellites* Earth is represented on a map of a portion of the solar system so that its surface is the circle with equation $x^2 + y^2 + 2x + 4y - 4091 = 0$. A satellite circles 0.6 unit above Earth with the center of its circular orbit at the center of Earth. Find the equation for the orbit of the satellite on this map.

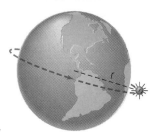

Source: Little League Baseball, Official Regulations and Playing Rules, 1991.

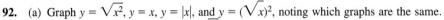

In Problem 92, you may use a graphing calculator, but it is not required.

92. (a) Graph $y = \sqrt{x^2}$, $y = x$, $y = |x|$, and $y = (\sqrt{x})^2$, noting which graphs are the same.
(b) Explain why the graphs of $y = \sqrt{x^2}$ and $y = |x|$ are the same.
(c) Explain why the graphs of $y = x$ and $y = (\sqrt{x})^2$ are not the same.
(d) Explain why the graphs of $y = \sqrt{x^2}$ and $y = x$ are not the same.

93. Make up an equation with the intercepts $(2, 0)$, $(4, 0)$, and $(0, 1)$. Compare your equation with a friend's equation. Comment on any similarities.

94. An equation is being tested for symmetry with respect to the x-axis, the y-axis, and the origin. Explain why, if two of these symmetries are present, the remaining one must also be present.

95. Draw a graph that contains the points $(-2, -1)$, $(0, 1)$, $(1, 3)$, and $(3, 5)$. Compare your graph with those of other students. Are most of the graphs almost straight lines? How many are "curved"? Discuss the various ways these points might be connected.

1.7

The Straight Line

In this section we study a certain type of equation that contains two variables, called a *linear equation,* and its graph, a *straight line.*

Slope of a Line

FIGURE 53

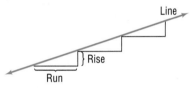

Consider the staircase illustrated in Figure 53. Each step contains exactly the same horizontal **run** and the same vertical **rise.** The ratio of the rise to the run, called the *slope,* is a numerical measure of the steepness of the staircase. For example, if the run is increased and the rise remains the same, the staircase becomes less steep. If the run is kept the same, but the rise is increased, the staircase becomes more steep. This important characteristic of a line is best defined using rectangular coordinates.

Slope of a Line

Let $P = (x_1, y_1)$ and $Q = (x_2, y_2)$ be two distinct points with $x_1 \neq x_2$. The **slope m** of the nonvertical line L containing P and Q is defined by the formula

$$m = \frac{y_2 - y_1}{x_2 - x_1} \qquad x_1 \neq x_2 \qquad (1)$$

If $x_1 = x_2$, L is a **vertical line** and the slope m of L is **undefined** (since this results in division by 0).

Figure 54(a) provides an illustration of the slope of a nonvertical line; Figure 54(b) illustrates a vertical line.

FIGURE 54

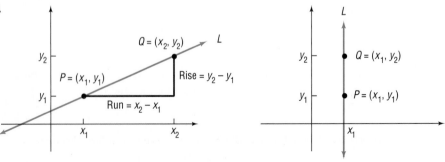

As Figure 54(a) illustrates, the slope m of a nonvertical line may be viewed as

$$m = \frac{y_2 - y_1}{x_2 - x_1} = \frac{\text{Rise}}{\text{Run}}$$

We can also express the slope m of a nonvertical line as

$$m = \frac{y_2 - y_1}{x_2 - x_1} = \frac{\text{Change in } y}{\text{Change in } x} = \frac{\Delta y}{\Delta x}$$

That is, the slope m of a nonvertical line L is the ratio of the change in the y-coordinates from P to Q, $\Delta y = y_2 - y_1$, to the change in the x-coordinates from P to Q, $\Delta x = x_2 - x_1$.

Two comments about computing the slope of a nonvertical line may prove helpful:

1. Any two distinct points on the line can be used to compute the slope of the line. (See Figure 55 for justification.)

FIGURE 55

Triangles ABC and PQR are similar (equal angles). Hence, ratios of corresponding sides are proportional. Thus:

Slope using P and $Q = \dfrac{y_2 - y_1}{x_2 - x_1}$

$= $ Slope using A and $B = \dfrac{d(B, C)}{d(A, C)}$

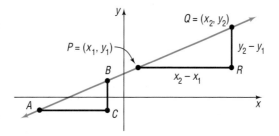

2. The slope of a line may be computed from $P = (x_1, y_1)$ to $Q = (x_2, y_2)$ or from Q to P, because

$$\frac{y_2 - y_1}{x_2 - x_1} = \frac{y_1 - y_2}{x_1 - x_2}$$

■ Now work Problem 1.

To get a better idea of the meaning of the slope m of a line L, consider the following example.

E X A M P L E   1   *Finding the Slopes of Various Lines Containing the Same Point (2, 3)*

Compute the slopes of the lines L_1, L_2, L_3, and L_4 containing the following pairs of points. Graph all four lines on the same set of coordinate axes.

$$
\begin{array}{lll}
L_1: & P = (2, 3) & Q_1 = (-1, -2) \\
L_2: & P = (2, 3) & Q_2 = (3, -1) \\
L_3: & P = (2, 3) & Q_3 = (5, 3) \\
L_4: & P = (2, 3) & Q_4 = (2, 5)
\end{array}
$$

Solution   Let m_1, m_2, m_3, and m_4 denote the slopes of the lines L_1, L_2, L_3, and L_4, respectively. Then

$$m_1 = \frac{-2 - 3}{-1 - 2} = \frac{-5}{-3} = \frac{5}{3} \quad \text{\small A rise of 5 divided by a run of 3}$$

$$m_2 = \frac{-1 - 3}{3 - 2} = \frac{-4}{1} = -4$$

$$m_3 = \frac{3 - 3}{5 - 2} = \frac{0}{3} = 0$$

m_4 is undefined

The graphs of these lines are given in Figure 56.

FIGURE 56

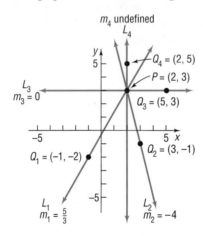

Figure 56 illustrates the following facts:

1. When the slope of a line is positive, the line slants upward from left to right (L_1).
2. When the slope of a line is negative, the line slants downward from left to right (L_2).
3. When the slope is 0, the line is horizontal (L_3).
4. When the slope is undefined, the line is vertical (L_4).

FIGURE 57

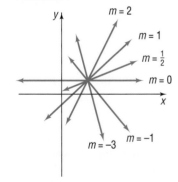

Figure 57 illustrates some additional facts about the slope of a line. Note that the closer the line is to the vertical position, the greater the magnitude of the slope.

Seeing the Concept On the same square screen, graph the following equations:

$y = 0$ Slope of line is 0.

$y = \frac{1}{2}x$ Slope of line is $\frac{1}{2}$.

$y = x$ Slope of line is 1.

$y = 2x$ Slope of line is 2.

$y = 6x$ Slope of line is 6.

On the same screen, now graph the following equations:

$y = 0$ Slope of line is 0.

$y = -\frac{1}{2}x$ Slope of line is $-\frac{1}{2}$.

$y = -x$ Slope of line is -1.

$y = -2x$ Slope of line is -2.

$y = -6x$ Slope of line is -6.

The next example illustrates how the slope of a line can be used to graph the line.

E X A M P L E   2

Graphing a Line Given a Point and a Slope

Draw a graph of the line that passes through the point (3, 2) and has a slope of:

(a) $\frac{3}{4}$ (b) $-\frac{4}{5}$

Solution   (a) Slope = Rise/Run. The fact that the slope is $\frac{3}{4}$ means that for every horizontal movement (run) of 4 units to the right there will be a vertical movement (rise) of 3 units. If we start at the given point (3, 2) and move 4 units to the right and 3 units up, we reach the point (7, 5). By drawing the line through this point and the point (3, 2), we have the graph. See Figure 58.

FIGURE 58
Slope = $\frac{3}{4}$

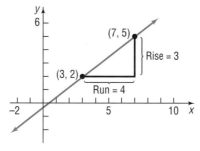

slope = $\frac{3}{4}$

(b) The fact that the slope is

$$-\frac{4}{5} = \frac{-4}{5} = \frac{\text{Rise}}{\text{Run}}$$

means that for every horizontal movement of 5 units to the right there will be a corresponding vertical movement of −4 units (a downward movement). If we start at the given point (3, 2) and move 5 units to the right and then 4 units down, we arrive at the point (8, −2). By drawing the line through these points, we have the graph. See Figure 59.

FIGURE 59
Slope = $-\frac{4}{5}$

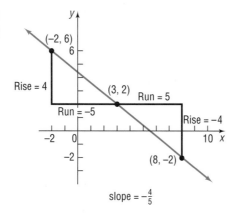

slope = $-\frac{4}{5}$

Alternatively, we can set

$$-\frac{4}{5} = \frac{4}{-5} = \frac{\text{Rise}}{\text{Run}}$$

so that for every horizontal movement of -5 units (a movement to the left) there will be a corresponding vertical movement of 4 units (upward). This approach brings us to the point $(-2, 6)$, which is also on the graph shown in Figure 59. ■

■   Now work Problem 17.

Equations of Lines

Now that we have discussed the slope of a line, we are ready to derive equations of lines. As we shall see, there are several forms of the equation of a line. Let's start with an example.

E X A M P L E   3   *Graphing a Line*

Graph the equation: $x = 3$

Solution   We are looking for all points (x, y) in the plane for which $x = 3$. Thus, no matter what y-coordinate is used, the corresponding x-coordinate always equals 3. Consequently, the graph of the equation $x = 3$ is a vertical line with x-intercept 3 and undefined slope. See Figure 60.

FIGURE 60
$x = 3$

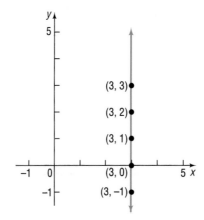

As suggested by Example 3, we have the following result:

Theorem   A vertical line is given by an equation of the form

Equation of a Vertical Line

$$x = a$$

where a is the x-intercept. ■

Comment: To graph an equation using a graphing utility, we need to express the equation in the form $y =$ expression in x. But $x = 3$ cannot be put in this form. To overcome this, most graphing utilities have special ways for drawing vertical lines. LINE, PLOT, and VERT are among the more common ones. Consult your manual to determine the correct methodology for your graphing utility.

Now let L be a nonvertical line with slope m and containing the point (x_1, y_1). See Figure 61. For any other point (x, y) on L, we have

FIGURE 61

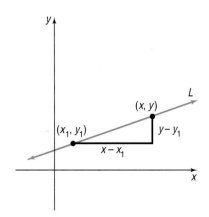

$$m = \frac{y - y_1}{x - x_1} \quad \text{or} \quad y - y_1 = m(x - x_1)$$

Theorem   An equation of a nonvertical line of slope m that passes through the point (x_1, y_1) is

Point–Slope Form of an
Equation of a Line

$$y - y_1 = m(x - x_1) \tag{2}$$

■

E X A M P L E   4   *Using the Point–Slope Form of a Line*

An equation of the line with slope 4 and passing through the point $(1, 2)$ can be found by using the point–slope form with $m = 4$, $x_1 = 1$, and $y_1 = 2$:

$$y - y_1 = m(x - x_1)$$
$$y - 2 = 4(x - 1)$$
$$y = 4x - 2$$

See Figure 62.   ■

FIGURE 62
$y = 4x - 2$

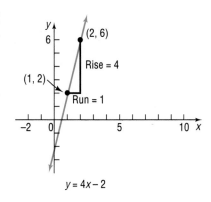

$y = 4x - 2$

E X A M P L E   5   *Finding the Equation of a Horizontal Line*

Find an equation of the horizontal line passing through the point $(3, 2)$.

Solution   The slope of a horizontal line is 0. To get an equation, we use the point–slope form with $m = 0$, $x_1 = 3$, and $y_1 = 2$:

$$y - y_1 = m(x - x_1)$$
$$y - 2 = 0 \cdot (x - 3)$$
$$y - 2 = 0$$
$$y = 2$$

See Figure 63 for the graph.

FIGURE 63
$y = 2$

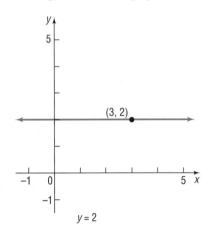

As suggested by Example 5, we have the following result:

Theorem   A horizontal line is given by an equation of the form

Equation of a Horizontal Line

$$y = b$$

where b is the y-intercept.

E X A M P L E   6

Finding an Equation of a Line Given Two Points

Find an equation of the line L passing through the points $(2, 3)$ and $(-4, 5)$. Graph the line L.

FIGURE 64
$y - 3 = -\frac{1}{3}(x - 2)$

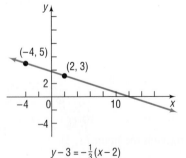

$y - 3 = -\frac{1}{3}(x - 2)$

Solution   Since two points are given, we first compute the slope of the line:

$$m = \frac{5 - 3}{-4 - 2} = \frac{2}{-6} = \frac{-1}{3}$$

We use the point $(2, 3)$ and the fact that the slope $m = -\frac{1}{3}$ to get the point–slope form of the equation of the line:

$$y - 3 = -\frac{1}{3}(x - 2)$$

See Figure 64 for the graph.

In the solution in Example 6, we could have used the other point $(-4, 5)$, instead of the point $(2, 3)$. The equation that results, although it looks different, is equivalent to the equation we obtained in the example. (Try it for yourself.)

Another form of the equation of the line in Example 6 can be obtained by multiplying both sides of the point–slope equation by 3 and collecting terms:

$$y - 3 = -\frac{1}{3}(x - 2)$$

$$3(y - 3) = 3(-\frac{1}{3})(x - 2) \quad \text{Multiply by 3.}$$

$$3y - 9 = -1(x - 2)$$

$$3y - 9 = -x + 2$$

$$x + 3y - 11 = 0$$

General Form of an Equation of a Line

The equation of a line L is in **general form** when it is written as

$$Ax + By + C = 0 \tag{3}$$

where A, B, and C are three real numbers and A and B are not both 0.

■ Now work Problems 31 and 35.

Every line has an equation that is equivalent to an equation written in general form. For example, a vertical line whose equation is

$$x = a$$

can be written in the general form

$$1 \cdot x + 0 \cdot y - a = 0 \quad A = 1, B = 0, C = -a$$

A horizontal line whose equation is

$$y = b$$

can be written in the general form

$$0 \cdot x + 1 \cdot y - b = 0 \quad A = 0, B = 1, C = -b$$

Lines that are neither vertical nor horizontal have general equations of the form

$$Ax + By + C = 0 \quad A \neq 0 \text{ and } B \neq 0$$

Because the equation of every line can be written in general form, any equation equivalent to (3) is called a **linear equation.**

Another useful equation of a line is obtained when the slope m and y-intercept b are known. In this event, we know both the slope m of the line and a point $(0, b)$ on the line; thus, we may use the point–slope form, equation (2), to obtain the following equation:

$$y - b = m(x - 0) \quad \text{or} \quad y = mx + b$$

Theorem   An equation of a line L with slope m and y-intercept b is

---

Slope–Intercept Form
of an Equation of a Line

$$y = mx + b \qquad (4)$$

■

Exploration: On the same screen, graph the following lines (see Figure 65):

$$y = 2$$
$$y = x + 2$$
$$y = -x + 2$$
$$y = 3x + 2$$
$$y = -3x + 2$$

What do you conclude about the lines $y = mx + 2$?

FIGURE 65
$y = mx + 2$

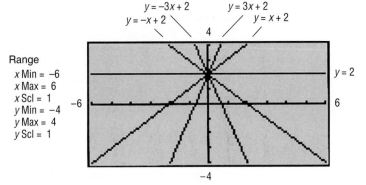

Range
x Min = −6
x Max = 6
x Scl = 1
y Min = −4
y Max = 4
y Scl = 1

On the same screen, now graph the following lines (see Figure 66):

$$y = 2x$$
$$y = 2x + 1$$
$$y = 2x - 1$$
$$y = 2x + 4$$
$$y = 2x - 4$$

What do you conclude about the lines $y = 2x + b$?

FIGURE 66
$y = 2x + b$

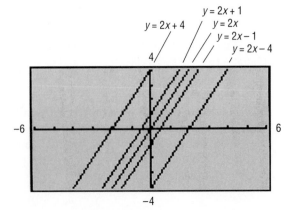

ISSION POSSIBLE

Chapter 1

PREDICTING THE FUTURE OF OLYMPIC TRACK EVENTS

Some people think that women athletes are beginning to "catch up" to men in the Olympic Track Events. In fact, researchers at UCLA published data in 1992 that seemed to indicate that within 65 years the top men and women runners would be able to compete on an equal basis. Other researchers disagreed. Here are some data on the 200 meter run with winning times for the Olympics in the year given.

	MEN'S TIME IN SECONDS	WOMEN'S TIME IN SECONDS
1948	21.1	24.4
1952	20.7	23.7
1956	20.6	23.4
1960	20.5	24.0
1964	20.3	23.0
1968	19.83	22.5
1972	20.00	22.40
1976	20.23	22.37
1980	20.19	22.03
1984	19.80	21.81
1988	19.75	21.34
1992	19.73	21.72

(You may notice that the introduction of better timing devices meant more accurate measures starting in 1968 for the men and 1972 for the women.)

1. To begin your investigation of the UCLA conjecture, make a graph of these data. To get the kind of accuracy you need, you should use graph paper. You will need to use only the first quadrant. On your x-axis place the Olympic years from 1948 to 2048, counting by 4's. On your y-axis place the numbers from 15 to 25, which represent the seconds; if you are using graph paper, allow about four squares per one second. Use X's to represent the men's times on the graph and O's to represent the women's times. These should form what we call a *scatter plot* not a neat line or curve.
2. Using a ruler or straightedge, draw a line that represents roughly the slope and direction indicated by the men's scores. Do the same for the women's scores. Write a sentence or two explaining why you think your line is a good representation of the scatter plot.
3. Next find the equation for each line, using your graph to estimate the slope and y-intercept for each. Try to be as accurate as possible, remembering your units and using the points where the line crosses intersections of the grid. The slope will be in seconds per year.
4. Do your two lines appear to cross? In what year do they cross? If you solve the two equations algebraically, do you get the same answer?
5. (Optional) Some graphing calculators enable you to do this problem in a statistics mode. They will plot the individual point that you type in and find a linear regression that represents the data. If you have a graphing calculator, find out whether their equations match yours.
6. Make a group decision about whether you think that women's times will catch up to men's times in the future. Write two or three sentences explaining why you believe they will or will not.

When the equation of a line is written in slope–intercept form, it is easy to find the slope m and y-intercept b of the line. For example, suppose that the equation of a line is

$$y = -2x + 3$$

Compare it to $y = mx + b$:

$$\begin{array}{c} y = -2x + 3 \\ \quad\;\; \uparrow \quad\; \uparrow \\ y = \;\; mx + b \end{array}$$

The slope of this line is -2 and its y-intercept is 3.

Let's look at another example.

EXAMPLE 7 *Finding the Slope and y-Intercept of a Line*

Find the slope m and y-intercept b of the line $2x + 4y - 8 = 0$. Graph the line.

Solution To obtain the slope and y-intercept, we transform the equation into its slope–intercept form. Thus, we need to solve for y:

$$\begin{aligned} 2x + 4y - 8 &= 0 \\ 4y &= -2x + 8 \\ y &= -\tfrac{1}{2}x + 2 \end{aligned}$$

The coefficient of x, $-\frac{1}{2}$, is the slope, and the y-intercept is 2. We can graph the line in two ways:

1. Use the fact that the y-intercept is 2 and the slope is $-\frac{1}{2}$. Then, starting at the point $(0, 2)$, go to the right 2 units and then down 1 unit to the point $(2, 1)$. See Figure 67.

Or

2. Locate the intercepts. Because the y-intercept is 2, we know one intercept is $(0, 2)$. To obtain the x-intercept, let $y = 0$ and solve for x. When $y = 0$, we have

$$\begin{aligned} 2x + 4 \cdot 0 - 8 &= 0 \\ 2x - 8 &= 0 \\ x &= 4 \end{aligned}$$

Thus, the intercepts are $(4, 0)$ and $(0, 2)$. See Figure 68.

FIGURE 67
$2x + 4y - 8 = 0$

FIGURE 68
$2x + 4y - 8 = 0$

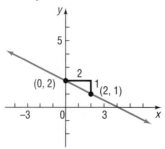

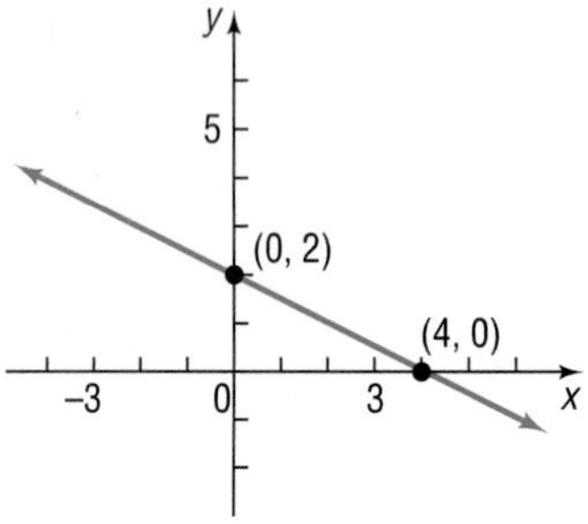

■ Now work Problem 63.

Parallel and Perpendicular Lines

When two lines (in the plane) have no points in common, they are said to be **parallel.** Look at Figure 69. There we have drawn two lines and right triangles by drawing sides parallel to the coordinate axes. These lines are parallel if and only if the right triangles are similar. (Do you see why? Two angles are equal.) But the triangles are similar if and only if the ratios of corresponding sides are equal.

FIGURE 69

The lines are parallel if and only if their slopes are equal.

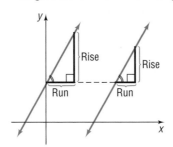

This suggests the following result:

Theorem   Two distinct nonvertical lines are parallel if and only if their slopes are equal. ■

The use of the words "if and only if" in the preceding theorem means that actually two statements are being made, one the converse of the other.

If two distinct nonvertical lines are parallel, then their slopes are equal.

If two distinct nonvertical lines have equal slopes, then they are parallel.

E X A M P L E   8   *Showing That Two Lines Are Parallel*

Show that the lines given by the following equations are parallel:

$$L: \quad 2x + 3y - 6 = 0 \qquad M: \quad 4x + 6y = 0$$

FIGURE 70
Parallel lines

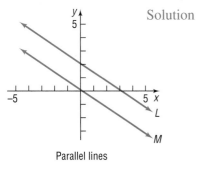

Parallel lines

Solution   To determine whether these lines have equal slopes, we write each equation in slope–intercept form:

$L: \quad 2x + 3y - 6 = 0$	$M: \quad 4x + 6y = 0$
$3y = -2x + 6$	$6y = -4x$
$y = -\frac{2}{3}x + 2$	$y = -\frac{2}{3}x$
Slope $= -\frac{2}{3}$	Slope $= -\frac{2}{3}$

Because these lines have the same slope, $-\frac{2}{3}$, but different y-intercepts, the lines are parallel. See Figure 70. ■

E X A M P L E   9   *Finding a Line That Is Parallel to a Given Line*

Find an equation for the line that contains the point $(2, -3)$ and is parallel to the line $2x + y - 6 = 0$.

Solution   The slope of the line we seek equals the slope of the line $2x + y - 6 = 0$, since the two lines are to be parallel. Thus, we begin by writing the equation of the line $2x + y - 6 = 0$ in slope–intercept form:

$$2x + y - 6 = 0$$
$$y = -2x + 6$$

The slope is -2. Since the line we seek contains the point $(2, -3)$, we use the point–slope form to obtain

$$y + 3 = -2(x - 2)$$

$$2x + y - 1 = 0 \qquad \text{General form}$$

$$y = -2x + 1 \qquad \text{Slope–intercept form}$$

This line is parallel to the line $2x + y - 6 = 0$ and contains the point $(2, -3)$. See Figure 71.

FIGURE 71

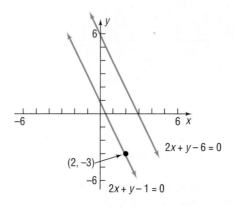

◼   Now work Problem 43.

When two lines intersect at a right angle (90°), they are said to be **perpendicular.** See Figure 72.

FIGURE 72
Perpendicular lines

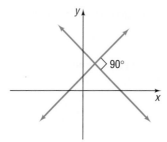

The following result gives a condition, in terms of their slopes, for two lines to be perpendicular:

Theorem   Two nonvertical lines are perpendicular if and only if the product of their slopes is -1.   ◼

Here, we shall prove the "only if" part of the statement:

If two nonvertical lines are perpendicular, then the product of their slopes is -1.

In Problem 100, you are asked to prove the "if" part of the theorem; that is,

If two nonvertical lines have slopes whose product is -1, then the lines are perpendicular.

FIGURE 73

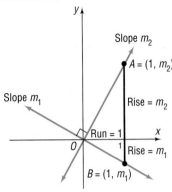

Proof   Let m_1 and m_2 denote the slopes of the two lines. There is no loss in generality (that is, neither the angle nor the slopes are affected) if we situate the lines so that they meet at the origin. See Figure 73. The point $A = (1, m_2)$ is on the line having slope m_2, and the point $B = (1, m_1)$ is on the line having slope m_1. (Do you see why this must be true?)

Suppose that the lines are perpendicular. Then triangle OAB is a right triangle. As a result of the Pythagorean Theorem, it follows that

$$[d(O, A)]^2 + [d(O, B)]^2 = [d(A, B)]^2 \tag{5}$$

By the distance formula, we can write each of these distances as

$$[d(O, A)]^2 = (1 - 0)^2 + (m_2 - 0)^2 = 1 + m_2^2$$
$$[d(O, B)]^2 = (1 - 0)^2 + (m_1 - 0)^2 = 1 + m_1^2$$
$$[d(A, B)]^2 = (1 - 1)^2 + (m_2 - m_1)^2 = m_2^2 - 2m_1m_2 + m_1^2$$

Using these facts in equation (5), we get

$$(1 + m_2^2) + (1 + m_1^2) = m_2^2 - 2m_1m_2 + m_1^2$$

which, upon simplification, can be written as

$$m_1m_2 = -1$$

Thus, if the lines are perpendicular, the product of their slopes is -1.   ■

You may find it easier to remember the condition for two nonvertical lines to be perpendicular by observing that the equality $m_1m_2 = -1$ means that m_1 and m_2 are negative reciprocals of each other; that is, either $m_1 = -1/m_2$ or $m_2 = -1/m_1$.

E X A M P L E   1 0   *Finding the Slope of a Line Perpendicular to Another Line*

If a line has slope $\frac{3}{2}$, any line having slope $-\frac{2}{3}$ is perpendicular to it.   ■

E X A M P L E   1 1   *Finding the Equation of a Line Perpendicular to Another Line*

Find an equation of the line passing through the point $(1, -2)$ and perpendicular to the line $x + 3y - 6 = 0$. Graph the two lines.

Solution   We first write the equation of the given line in slope–intercept form to find its slope:

$$x + 3y - 6 = 0$$
$$3y = -x + 6$$
$$y = -\tfrac{1}{3}x + 2$$

The given line has slope $-\frac{1}{3}$. Any line perpendicular to this line will have slope 3. Because we require the point $(1, -2)$ to be on this line with slope 3, we use the point–slope form of the equation of a line:

$$y - (-2) = 3(x - 1)$$
$$y + 2 = 3(x - 1)$$

This equation is equivalent to the forms

$$3x - y - 5 = 0 \qquad \text{General form}$$
$$y = 3x - 5 \qquad \text{Slope–intercept form}$$

FIGURE 74

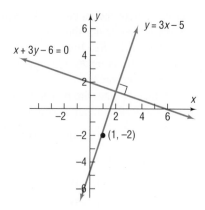

See Figure 74.

■   Now work Problem 51.

Warning: Be sure to use a square screen when you graph perpendicular lines. Otherwise, the angle between the two lines will appear distorted.

1.7

Exercise 1.7

In Problems 1–4, find the slope of the line.

1.

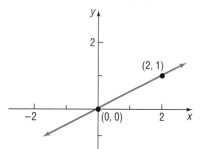

2.

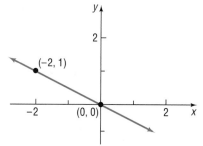

3.

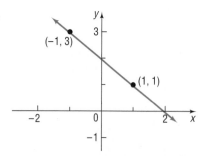

4.

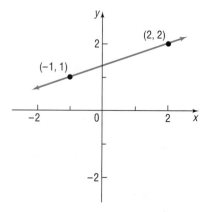

In Problems 5–14, plot each pair of points and determine the slope of the line containing them. Graph the line.

5. (2, 3); (4, 0)   **6.** (4, 2); (3, 4)   **7.** (−2, 3); (2, 1)   **8.** (−1, 1); (2, 3)

9. (−3, −1); (2, −1)   **10.** (4, 2); (−5, 2)   **11.** (−1, 2); (−1, −2)   **12.** (2, 0); (2, 2)

13. ($\sqrt{2}$, 3); (1, $\sqrt{3}$)   **14.** (−2$\sqrt{2}$, 0); (4, $\sqrt{5}$)

In Problems 15–22, graph the line passing through the point P and having slope m.

15. $P = (1, 2)$; $m = 3$ **16.** $P = (2, 1)$; $m = 4$ **17.** $P = (2, 4)$; $m = \frac{-3}{4}$

18. $P = (1, 3)$; $m = \frac{-2}{5}$ **19.** $P = (-1, 3)$; $m = 0$ **20.** $P = (2, -4)$; $m = 0$

21. $P = (0, 3)$; slope undefined   **22.** $P = (-2, 0)$; slope undefined

In Problems 23–30, find an equation of each line. Express your answer using either the general form or the slope–intercept form of the equation of a line, whichever you prefer.

23.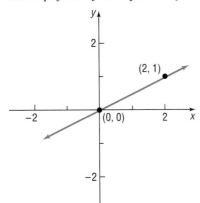

24.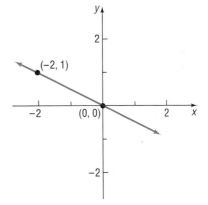

25.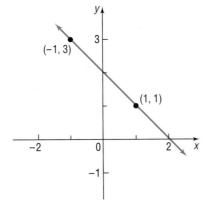

26.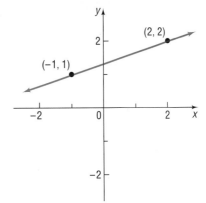

27.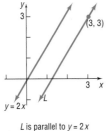

L is parallel to $y = 2x$

28.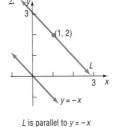

L is parallel to $y = -x$

29.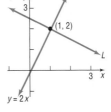

L is perpendicular to $y = 2x$

30.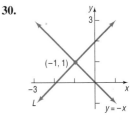

L is perpendicular to $y = -x$

In Problems 31–54, find an equation for the line with the given properties. Express your answer using either the general form or the slope–intercept form of the equation of a line, whichever you prefer.

31. Slope = 3; passing through $(-2, 3)$ **32.** Slope = 2; passing through $(4, -3)$

33. Slope = $-\frac{2}{3}$; passing through $(1, -1)$ **34.** Slope = $\frac{1}{2}$; passing through $(3, 1)$

35. Passing through $(1, 3)$ and $(-1, 2)$ **36.** Passing through $(-3, 4)$ and $(2, 5)$

37. Slope = -3; y-intercept = 3   **38.** Slope = -2; y-intercept = -2

39. x-intercept = 2; y-intercept = -1 **40.** x-intercept = -4; y-intercept = 4

41. Slope undefined; passing through $(2, 4)$ **42.** Slope undefined; passing through $(3, 8)$

43. Parallel to the line $y = 2x$; passing through $(-1, 2)$

44. Parallel to the line $y = -3x$; passing through $(-1, 2)$

45. Parallel to the line $2x - y + 2 = 0$; passing through $(0, 0)$

46. Parallel to the line $x - 2y + 5 = 0$; passing through $(0, 0)$

47. Parallel to the line $x = 5$; passing through $(4, 2)$

48. Parallel to the line $y = 5$; passing through $(4, 2)$

49. Perpendicular to the line $y = \frac{1}{2}x + 4$; passing through $(1, -2)$

50. Perpendicular to the line $y = 2x - 3$; passing through $(1, -2)$

51. Perpendicular to the line $2x + y - 2 = 0$; passing through $(-3, 0)$

52. Perpendicular to the line $x - 2y + 5 = 0$; passing through $(0, 4)$

53. Perpendicular to the line $x = 8$; passing through $(3, 4)$

54. Perpendicular to the line $y = 8$; passing through $(3, 4)$

In Problems 55–74, find the slope and y-intercept of each line. Graph the line.

55. $y = 2x + 3$ **56.** $y = -3x + 4$ **57.** $\frac{1}{2}y = x - 1$ **58.** $\frac{1}{3}x + y = 2$ **59.** $y = \frac{1}{2}x + 2$

60. $y = 2x + \frac{1}{2}$ **61.** $x + 2y = 4$ **62.** $-x + 3y = 6$ **63.** $2x - 3y = 6$ **64.** $3x + 2y = 6$

65. $x + y = 1$ **66.** $x - y = 2$ **67.** $x = -4$ **68.** $y = -1$ **69.** $y = 5$

70. $x = 2$ **71.** $y - x = 0$ **72.** $x + y = 0$ **73.** $2y - 3x = 0$ **74.** $3x + 2y = 0$

75. Find an equation of the x-axis.

76. Find an equation of the y-axis.

In Problems 77–80, match each graph with the correct equation:
(a) $y = x$ (b) $y = 2x$ (c) $y = x/2$ (d) $y = 4x$

77.

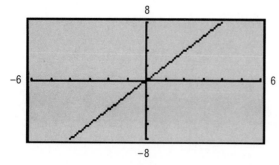

78.

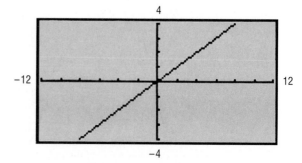

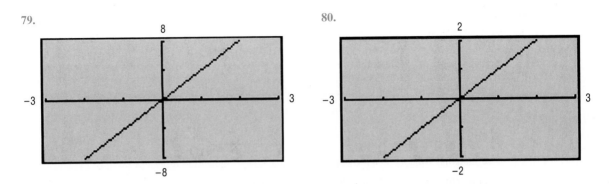

79.

80.

In Problems 81–84, write an equation of each line. Express your answer using either the general form or the slope–intercept form of the equation of a line, whichever you prefer.

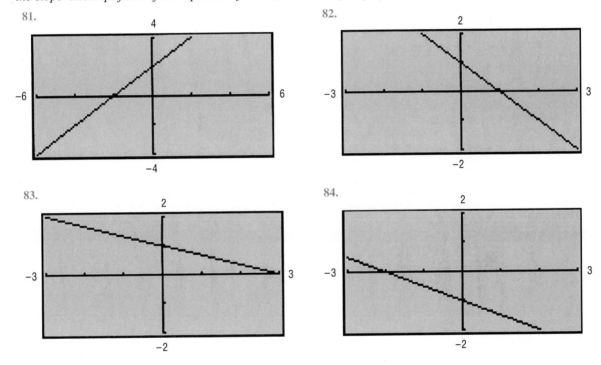

81.

82.

83.

84.

85. *Measuring Temperature* The relationship between Celsius (°C) and Fahrenheit (°F) degrees of measuring temperature is linear. Find an equation relating °C and °F if 0°C corresponds to 32°F and 100°C corresponds to 212°F. Use the equation to find the Celsius measure of 70°F.

86. *Measuring Temperature* The Kelvin (K) scale for measuring temperature is obtained by adding 273 to the Celsius temperature.
    (a) Write an equation relating K and °C
    (b) Write an equation relating K and °F (see Problem 85).

87. *Business: Computing Profit* Each Sunday, a newspaper agency sells x copies of certain newspaper for $1.00 per copy. The cost to the agency of each newspaper is $0.50. The agency pays a fixed cost for storage, delivery, and so on, of $100 per Sunday.
    (a) Write an equation that relates the profit P, in dollars, to the number x of copies sold. Graph this equation.
    (b) What is the profit to the agency if 1000 copies are sold?
    (c) What is the profit to the agency if 5000 copies are sold?

88. *Business: Computing Profit* Repeat Problem 87 if the cost to the agency is $0.45 per copy and the fixed cost is $125 per Sunday.

89. *Cost of Electricity* In 1991, Commonwealth Edison Company supplied electricity in the summer months to residential customers for a monthly customer charge of $9.06 plus 10.819¢ per kilowatt-hour supplied in the month.* Write an equation that relates the monthly charge C, in dollars, to the number x of kilowatt-hours in the month. Graph this equation. What is the monthly charge for using 300 kilowatt-hours? For using 900 kilowatt-hours?

90. Show that an equation for a line with nonzero x- and y-intercepts can be written as

$$\frac{x}{a} + \frac{y}{b} = 1$$

where a is the x-intercept and b is the y-intercept. This is called the **intercept form** of the equation of a line.

91. The **tangent line** to a circle may be defined as the line that intersects the circle in a single point, called the **point of tangency** (see the figure). If the equation of the circle is $x^2 + y^2 = r^2$ and the equation of the tangent line is $y = mx + b$, show that:
(a) $r^2(1 + m^2) = b^2$ [*Hint:* The quadratic equation $x^2 + (mx + b)^2 = r^2$ has exactly one solution.]
(b) The point of tangency is $(-r^2m/b, r^2/b)$.
(c) The tangent line is perpendicular to the line containing the center of the circle and the point of tangency.

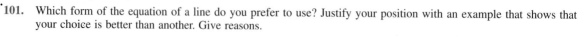

92. The Greek method for finding the equation of the tangent line to a circle used the fact that at any point on a circle the line containing the radius and the tangent line are perpendicular (see Problem 91). Use this method to find an equation of the tangent line to the circle $x^2 + y^2 = 9$ at the point $(1, 2\sqrt{2})$.

93. Use the Greek method described in Problem 92 to find an equation of the tangent line to the circle $x^2 + y^2 - 4x + 6y + 4 = 0$ at the point $(3, 2\sqrt{2} - 3)$.

94. Refer to Problem 91. The line $x - 2y + 4 = 0$ is tangent to a circle at $(0, 2)$. The line $y = 2x - 7$ is tangent to the same circle at $(3, -1)$. Find the center of the circle.

95. Find an equation of the line containing the centers of the two circles

$$x^2 + y^2 - 4x + 6y + 4 = 0 \quad \text{and} \quad x^2 + y^2 + 6x + 4y + 9 = 0$$

96. Show that the line containing the points (a, b) and (b, a) is perpendicular to the line $y = x$. Also show that the midpoint of (a, b) and (b, a) lies on the line $y = x$.

97. The equation $2x - y + C = 0$ defines a **family of lines,** one line for each value of C. On one set of coordinate axes, graph the members of the family when $C = -4$, $C = 0$, and $C = 2$. Can you draw a conclusion from the graph about each member of the family?

98. Rework Problem 97 for the family of lines $Cx + y + 4 = 0$.

99. If a circle of radius 2 is made to roll along the x-axis, what is an equation for the path of the center of the circle?

100. Prove that if two nonvertical lines have slopes whose product is negative, then the lines are perpendicular. [*Hint:* Refer to Figure 73.]

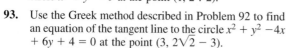 **101.** Which form of the equation of a line do you prefer to use? Justify your position with an example that shows that your choice is better than another. Give reasons.

102. Can every line be written in slope–intercept form? Explain.

103. Does every line have two distinct intercepts? Explain. Are there lines that have no intercepts? Explain.

104. What can you say about two lines that have equal slopes and equal y-intercepts?

Source: Commonwealth Edison Co., Chicago, Illinois, 1991.

105. What can you say about two lines with the same *x*-intercept and the same *y*-intercept? Assume that the *x*-intercept is not 0.

106. If two lines have the same slope, but different *x*-intercepts, can they have the same *y*-intercept?

107. If two lines have the same *y*-intercept, but different slopes, can they have the same *x*-intercept? What is the only way this can happen?

108. The accepted symbol used to denote the slope of a line is the letter *m*. Investigate the origin of this symbolism. Begin by consulting a French dictionary and looking up the French word *monter*. Write a brief essay on your findings.

109. The term *grade* is used to describe the inclination of a road. How does this term relate to the notion of slope of a line? Is a 4% grade very steep? Investigate the grades of some mountainous roads and determine their slopes. Write a brief essay on your findings.

110. *Carpentry* Carpenters use the term *pitch* to describe the steepness of staircases and roofs. How does pitch relate to slope? Investigate typical pitches used for stairs and for roofs. Write a brief essay on your findings.

Chapter Review

THINGS TO KNOW

Absolute value	$\lvert a \rvert = a$ if $a \geq 0$, $\quad \lvert a \rvert = -a$ if $a < 0$.
Principal *n*th root of *a*	$\sqrt[n]{a} = b$ means $a = b^n$, where $a \geq 0$ and $b \geq 0$ if *n* is even and *a, b,* are any real numbers if *n* is odd.
Polynomial	Algebraic expression of the form $a_n x^n + a_{n-1}x^{n-1} + \cdots + a_1 x + a_0$; if $a_n \neq 0$, then *n* is the degree of the polynomial
Pythagorean Theorem	In a right triangle, the square of the length of the hypotenuse is equal to the sum of the squares of the lengths of the legs.

FORMULAS

Quadratic equation and quadratic formula	If $ax^2 + bx + c = 0$, $a \neq 0$, then $x = \dfrac{-b \pm \sqrt{b^2 - 4ac}}{2a}$.

Discriminant	If $b^2 - 4ac > 0$, there are two distinct real solutions. If $b^2 - 4ac = 0$, there is one repeated real solution. If $b^2 - 4ac < 0$, there are two distinct complex solutions that are not real; the solutions are conjugates of each other.
Absolute value	If $|u| = a$, $a > 0$, then $u = -a$ or $u = a$. If $|u| \leq a$, $a > 0$, then $-a \leq u \leq a$. If $|u| \geq a$, $a > 0$, then $u \leq -a$ or $u \geq a$.
Distance formula	$d = \sqrt{(x_2 - x_1)^2 + (y_2 - y_1)^2}$
Midpoint formula	$(x, y) = \left(\dfrac{x_1 + x_2}{2}, \dfrac{y_1 + y_2}{2} \right)$
Slope	$m = \dfrac{y_2 - y_1}{x_2 - x_1}$, if $x_1 \neq x_2$; undefined if $x_1 = x_2$
Parallel lines	Equal slopes ($m_1 = m_2$)
Perpendicular lines	Product of slopes is -1 ($m_1 \cdot m_2 = -1$)

EQUATIONS

Vertical line	$x = a$
Horizontal line	$y = b$
Point–slope form of an equation of a line	$y - y_1 = m(x - x_1)$; m is the slope of the line, (x_1, y_1) is a point on the line
General form of an equation of a line	$Ax + By + C = 0$, A, B not both 0
Slope–intercept form of an equation of a line	$y = mx + b$; m is the slope of the line, b is the y-intercept
Standard form of an equation of a circle	$(x - h)^2 + (y - k)^2 = r^2$; r is the radius of the circle, (h, k) is the center of the circle
General form of an equation of a circle	$x^2 + y^2 + ax + by + c = 0$

HOW TO

Solve equations	Find the center and radius of a circle, given the equation
Solve inequalities	Obtain the equation of a circle
Solve applied problems	Graph circles
Use the distance formula	Find the slope and intercepts of a line, given the equation
Graph equations by plotting points	Graph lines
Find the intercepts of a graph	Obtain the equation of a line
Test an equation for symmetry	

FILL-IN-THE BLANK ITEMS

1. Two equations (or inequalities) that have precisely the same solution set are called _____.

2. An equation that is satisfied for every choice of the variable for which both sides are meaningful is called a(n) _____.

3. The quantity $b^2 - 4ac$ is called the _____ of a quadratic equation. If it is _____, the equation has no real solution.

4. If $a < 0$, then $|a| =$ _____.

5. $\sqrt{x^2} =$ _____.

6. If each side of an inequality is multiplied by a(n) _____ number, then the direction of the inequality is reversed.

7. In the complex number $5 + 2i$, the number 5 is called the _____ part; the number 2 is called the _____ part; and the number i is called the _____ _____.

8. If, for every point (x, y) on a graph, the point $(-x, y)$ is also on the graph, then the graph is symmetric with respect to the _____.

9. The set of points in the xy-plane that are a fixed distance from a fixed point is called a(n) _____. The fixed distance is called the _____; the fixed point is called the _____.

10. The slope of a vertical line is _____; the slope of a horizontal line is _____.

11. Two nonvertical lines have slopes m_1 and m_2, respectively. The lines are parallel if _____; the lines are perpendicular if _____ _____.

TRUE/FALSE ITEMS

T   F   1. Rational numbers are also real numbers.

T   F   2. Irrational numbers have decimals that neither repeat nor terminate.

T   F   3. The conjugate of $2 + \sqrt{5}i$ is $-2 + \sqrt{5}i$.

T   F   4. The circumference of a circle of diameter d is πd.

T   F   5. If the discriminant of a quadratic equation is positive, then the equation has two complex solutions that are conjugates of each other.

T   F   6. The expression $x^2 + x + 1$ is positive for any real number x.

7. If $a < b$ and $c < 0$, which of the following statements are true?

T   F   (a) $a \pm c < b \pm c$

T   F   (b) $a \cdot c < b \cdot c$

T   F   (c) $a/c > b/c$

T   F   8. Vertical lines have undefined slope.

T   F   9. The slope of the line $2y = 3x + 5$ is 3.

T   F   10. Perpendicular lines have slopes that are reciprocals of one another.

T   F   11. The radius of the circle $x^2 + y^2 = 9$ is 3.

REVIEW EXERCISES

In Problems 1–24, find all the real solutions, if any, of each equation. (Where they appear, a, b, m, and n are constants.)

1. $2 - \dfrac{x}{3} = 5$

2. $\dfrac{x}{4} - 2 = 4$

3. $-2(5 - 3x) + 8 = 4 + 5x$

4. $(6 - 3x) - 2(1 + x) = 6x$

5. $\dfrac{3x}{4} - \dfrac{x}{3} = \dfrac{1}{12}$

6. $\dfrac{4 - 2x}{3} + \dfrac{1}{6} = 2x$

7. $\dfrac{x}{x - 1} = \dfrac{5}{4}$

8. $\dfrac{4x - 5}{3 - 7x} = 4$

9. $x(1 - x) = 6$

10. $x(1 + x) = 2$

11. $\dfrac{1}{2}\left(x - \dfrac{1}{3}\right) = \dfrac{3}{4} - \dfrac{x}{6}$

12. $\dfrac{1 - 3x}{4} = \dfrac{x + 6}{3} + \dfrac{1}{2}$

13. $(x - 1)(2x + 3) = 3$

14. $x(2 - x) = 3(x - 4)$

15. $2x + 3 = 4x^2$

16. $1 + 6x = 4x^2$

17. $\sqrt[3]{x^2 - 1} = 2$

18. $\sqrt{1 + x^3} = 3$

19. $x(x + 1) + 2 = 0$

20. $3x^2 - x + 1 = 0$

21. $|2x - 3| = 5$

22. $|3x + 1| = 5$

23. $10a^2x^2 - 2abx - 36b^2 = 0$

24. $\dfrac{1}{x - m} + \dfrac{1}{x - n} = \dfrac{2}{x}$

In Problems 25–44, solve each inequality.

25. $\dfrac{2x - 3}{5} + 1 \leq \dfrac{x}{2}$

26. $\dfrac{5 - x}{3} \leq 6x - 1$

27. $-9 \leq \dfrac{2x + 3}{-4} \leq 7$

28. $-4 < \dfrac{2x - 2}{3} < 6$

29. $6 > \dfrac{3 - 3x}{12} > 2$

30. $6 > \dfrac{5 - 3x}{2} \geq -3$

31. $2x^2 + 5x - 12 < 0$

32. $3x^2 - 2x - 1 \geq 0$

33. $\dfrac{6}{x + 2} \geq 1$

34. $\dfrac{-2}{1 - 3x} < -1$

35. $\dfrac{2x - 3}{1 - x} < 2$

36. $\dfrac{3 - 2x}{2x + 5} \geq 2$

37. $\dfrac{(x - 2)(x - 1)}{x - 3} > 0$

38. $\dfrac{x + 1}{x(x - 5)} \leq 0$

39. $\dfrac{x^2 - 8x + 12}{x^2 - 16} > 0$

40. $\dfrac{x(x^2 + x - 2)}{x^2 + 9x + 20} \leq 0$

41. $|3x + 4| < \frac{1}{2}$

42. $|1 - 2x| < \frac{1}{3}$

43. $|2x - 5| \geq 7$

44. $|3x + 1| \geq 2$

In Problems 45–52, solve each equation in the complex number system.

45. $x^2 + x + 1 = 0$

46. $x^2 - x + 1 = 0$

47. $2x^2 + x - 2 = 0$

48. $3x^2 - 2x - 1 = 0$

49. $x^2 + 3 = x$

50. $2x^2 + 1 = 2x$

51. $x(1 - x) = 6$

52. $x(1 + x) = 2$

In Problems 53–58, write each expression so that all exponents are positive.

53. $\dfrac{x^{-2}}{y^{-2}}$

54. $\left(\dfrac{x^{-1}}{y^{-3}}\right)^2$

55. $\dfrac{(x^2y)^{-4}}{(xy)^{-3}}$

56. $\dfrac{\left(\dfrac{x}{y}\right)^2}{\left(\dfrac{y}{x}\right)^{-1}}$

57. $(25x^{-4/3}y^{-2/3})^{3/2}$

58. $(16x^{-2/3}y^{4/3})^{-3/2}$

In Problems 59–68, use the complex number system and write each expression in the standard form a + bi.

59. $(6 - 3i) - (2 + 4i)$

60. $(8 + 3i) + (-6 - 2i)$

61. $4(3 - i) + 3(-5 + 2i)$

62. $2(1 + i) - 3(2 - 3i)$

63. $\dfrac{3}{3 + i}$

64. $\dfrac{4}{2 - i}$

65. i^{68}

66. i^{21}

67. $(2 + 3i)^3$

68. $(3 - 2i)^3$

In Problems 69–78, find a general equation of the line having the given characteristics.

69. Slope $= -2$;   passing through $(2, -1)$

70. Slope $= 0$;   passing through $(-3, 4)$

71. Slope undefined;   passing through $(-3, 4)$

72. x-intercept $= 2$;   passing through $(4, -5)$

73. y-intercept $= -2$;   passing through $(5, -3)$

74. Passing through $(3, -4)$ and $(2, 1)$

75. Parallel to the line $2x - 3y + 4 = 0$;   passing through $(-5, 3)$

76. Parallel to the line $x + y - 2 = 0$;   passing through $(1, -3)$

77. Perpendicular to the line $x + y - 2 = 0$;   passing through $(1, -3)$

78. Perpendicular to the line $3x - y + 4 = 0$;   passing through $(-2, 2)$

In Problems 79–84, graph each line and label the x- and y-intercepts.

79. $4x - 5y + 20 = 0$

80. $3x + 4y - 12 = 0$

81. $\frac{1}{2}x - \frac{1}{3}y + \frac{1}{6} = 0$

82. $-\frac{3}{4}x + \frac{1}{2}y = 0$

83. $\sqrt{2}x + \sqrt{3}y = \sqrt{6}$

84. $\frac{x}{3} + \frac{y}{4} = 1$

In Problems 85–88, find the center and radius of each circle.

85. $x^2 + y^2 - 2x + 4y - 4 = 0$

86. $x^2 + y^2 + 4x - 4y - 1 = 0$

87. $3x^2 + 3y^2 - 6x + 12y = 0$

88. $2x^2 + 2y^2 - 4x = 0$

In Problems 89–96, list the x- and y-intercepts and test for symmetry.

89. $2x = 3y^2$

90. $y = 5x$

91. $4x^2 + y^2 = 1$

92. $x^2 - 9y^2 = 9$

93. $y = x^4 + 2x^2 + 1$

94. $y = x^3 - x$

95. $x^2 + x + y^2 + 2y = 0$

96. $x^2 + 4x + y^2 - 2y = 0$

97. *How Far Can a Pilot See?* On a recent flight to San Francisco, the pilot announced that we were 139 miles from the city, flying at an altitude of 35,000 feet. The pilot claimed that he could see the Golden Gate bridge and beyond. Was he telling the truth? How far could he see?

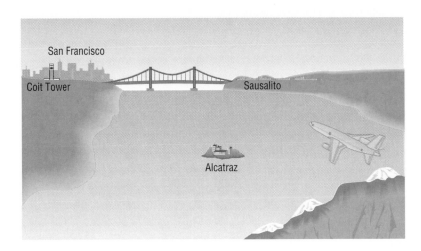

98. *Physics: Uniform Motion* A man is walking at an average speed of 4 miles per hour alongside a railroad track. A freight train, going in the same direction at an average speed of 30 miles per hour, requires 5 seconds to pass the man. How long is the freight train? Give your answer in feet.

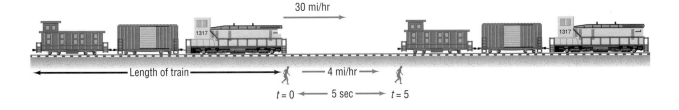

99. *Extent of Search and Rescue* A search plane has a cruising speed of 250 miles per hour and carries enough fuel for at most 5 hours of flying. If there is a wind that averages 30 miles per hour and the direction of search is with the wind one way and against it the other, how far can the search plane travel?

100. *Extent of Search and Rescue* If the search plane described in Problem 99 is able to add a supplementary fuel tank that allows for an additional 2 hours of flying, how much farther can the plane extend its search?

101. *Evening Up a Race* In a 100 meter race, Mike crosses the finish line 5 meters ahead of Dan. To even things up, Mike suggests to Dan that they race again, this time with Mike lining up 5 meters behind the start.

(a) Assuming that Mike and Dan run at the same pace as before, does the second race end in a tie?

(b) If not, who wins?

(c) By how many meters does he win?

(d) How far back should Mike start so that the race ends in a tie?

After running the race a second time, Dan, to even things up, suggests to Mike that he (Dan) line up 5 meters in front of the start.

(e) Assuming again that they run at the same pace as in the first race, does the third race result in a tie?

(f) If not, who wins?

(g) By how many meters?

(h) How far up should Dan start so that the race ends in a tie?

102. Explain the differences among the following three problems. Are there any similarities in their solution?

(a) Write the expression as a single quotient: $\dfrac{x}{x-2} + \dfrac{x}{x^2-4}$

(b) Solve: $\dfrac{x}{x-2} + \dfrac{x}{x^2-4} = 0$

(c) Solve: $\dfrac{x}{x-2} + \dfrac{x}{x^2-4} < 0$

103. Make up four problems that you might be asked to do given the two points $(-3, 4)$ and $(6, 1)$. Each problem should involve a different concept. Be sure that your directions are clearly stated.

104. Describe each of the following graphs. Give justification.

(a) $x = 0$ (b) $y = 0$ (c) $x + y = 0$ (d) $xy = 0$ (e) $x^2 + y^2 = 0$

PREPARING FOR THIS CHAPTER

Before getting started on this chapter, review the following concepts:

Domain of a variable (p. 3)
Graphs of certain equations (Example 5, p. 58;
Example 6, p. 60; Example 7, p. 61; Example 8, p. 62)
Tests for symmetry of an equation (p. 60)
Procedure for finding intercepts of an equation (p. 59)
Steps for setting up applied problems (pp. 24–25)

FUNCTIONS AND THEIR GRAPHS

Preview   Getting from an Island to Town

An island is 2 miles from the nearest point P on a straight shoreline. A town is 12 miles down the shore from P.

(a) *If a person can row a boat at an average speed of 5 miles per hour and the same person can walk 2 miles per hour, express the time T it takes to go from the island to town as a function of the distance x from P to where the person lands the boat.*

(b) *How long will it take to travel from the island to town if the person lands the boat 4 miles from P?*

(c) *How long will it take if the person lands the boat 8 miles from P? [Example 9 in Section 2.1]*

(d) *Is there a place to land the boat so that the travel time is least? Do you think this place is closer to town or closer to P? Discuss the possibilities. Give reasons. [Problems 63 and 64 in Exercise 2.1]* ∎

$\mathcal{P}$erhaps the most central idea in mathematics is the notion of a *function*. This important chapter deals with what a function is, how to graph functions, how to perform operations on functions, and how functions are used in applications.

The word *function* apparently was introduced by René Descartes in 1637. For him, a function simply meant any positive integral power of a variable x. Gottfried Wilhelm von Leibniz (1646–1716), who always emphasized the geometric side of mathematics, used the word function to denote any quantity associated with a curve, such as the coordinates of a point on the curve. Leonhard Euler (1707–1783) employed the word to mean any equation or formula involving variables and constants. His idea of a function is similar to the one most often used today in courses that precede calculus. Later, the use of functions in investigating heat flow equations led to a very broad definition, due to Lejeune Dirichlet (1805–1859), which describes a function as a rule or correspondence between two sets. It is his definition that we use here.

2.1

Functions

In many applications, a correspondence often exists between two sets of numbers. For example, the revenue R resulting from the sale of x items selling for \$10 each is $R = 10x$ dollars. If we know how many items have been sold, then we can calculate the revenue by using the rule $R = 10x$. This rule is an example of a *function*.

As another example, if an object is dropped from a height of 64 feet above the ground, the distance s (in feet) of the object from the ground after t seconds is given (approximately) by the formula $s = 64 - 16t^2$. When $t = 0$ seconds, the object is $s = 64$ feet above the ground. After 1 second, the object is $s = 64 - 16(1)^2 = 48$ feet above the ground. After 2 seconds, the object strikes the ground. The formula $s = 64 - 16t^2$ provides a way of finding the distance s when the time t ($0 \le t \le 2$) is prescribed. There is a correspondence between each time t in the interval $0 \le t \le 2$ and the distance s. We say that the distance s is a *function* of the time t because:

1. There is a correspondence between the set of times and the set of distances.
2. There is exactly one distance s obtained for a prescribed time t in the interval $0 \le t \le 2$.

Let's now look at the definition of a function.

Definition of Function

Function

> Let X and Y be two nonempty sets of real numbers.* A **function** from X into Y is a rule or a correspondence that associates with each element of X a unique element of Y. The set X is called the **domain** of the function. For each element x in X, the corresponding element y in Y is called the **value** of the function at x, or the **image** of x. The set of all images of the elements of the domain is called the **range** of the function.

FIGURE 1

Refer to Figure 1. Since there may be some elements in Y that are not the image of some x in X, it follows that the range of a function may be a subset of Y.

The rule (or correspondence) referred to in the definition of a function is most often given as an equation in two variables, usually denoted x and y.

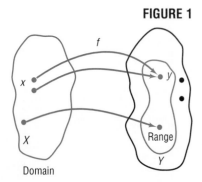

Domain

*The two sets X and Y can also be sets of complex numbers, and then we have defined a complex function. In the broad definition (due to Lejeune Dirichlet), X and Y can be any two sets.

Comment: When we select a viewing rectangle to graph a function, the values of Xmin, Xmax give the domain we wish to view, while Ymin, Ymax give the range we wish to view. These settings usually do not represent the actual domain and range of the function.

E X A M P L E   1

Example of a Function

Consider the function defined by the equation

$$y = 2x - 5 \qquad 1 \le x \le 6$$

The domain $1 \le x \le 6$ specifies that the number x is restricted to the real numbers from 1 to 6, inclusive. The rule $y = 2x - 5$ specifies that the number x is to be multipled by 2 and then 5 is to be subtracted from the result to get y. For example, the value of the function at $x = \frac{3}{2}$ (that is, the image of $x = \frac{3}{2}$) is $y = 2 \cdot \frac{3}{2} - 5 = -2$. ∎

Functions are often denoted by letters such as f, F, g, G, and so on. If f is a function, then for each number x in its domain the corresponding image in the range is designated by the symbol $f(x)$, read as "f of x" or as "f at x." We refer to $f(x)$ as the **value of f at the number x.** Thus, $f(x)$ is the number that results when x is given and the rule for f is applied; $f(x)$ does *not* mean "f times x." For example, the function given in Example 1 may be written as $f(x) = 2x - 5$, $1 \le x \le 6$.

Figure 2 illustrates some other functions. Note that for each of the functions illustrated, to each x in the domain, there is one value in the range.

FIGURE 2

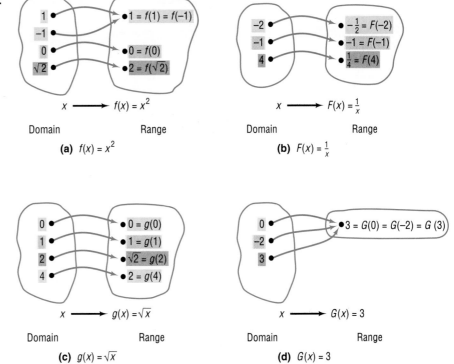

(a) $f(x) = x^2$

(b) $F(x) = \frac{1}{x}$

(c) $g(x) = \sqrt{x}$

(d) $G(x) = 3$

E X A M P L E   2

Finding Values of a Function

For the function

$$f(x) = x^2 + 3x - 4 \qquad -5 \le x \le 5$$

find the value of f at:

(a) $x = 0$ (b) $x = 1$ (c) $x = -4$ (d) $x = 5$

Solution   (a) The value of f at $x = 0$ is found by replacing x by 0 in the stated rule. Thus,

$$f(0) = 0^2 + 3(0) - 4 = -4$$

(b) $f(1) = 1^2 + 3(1) - 4 = 1 + 3 - 4 = 0$

(c) $f(-4) = (-4)^2 + 3(-4) - 4 = 16 - 12 - 4 = 0$

(d) $f(5) = 5^2 + 3(5) - 4 = 25 + 15 - 4 = 36$ ∎

∎   Now work Problem 3.

In general, when the rule that defines a function f is given by an equation in x and y, we say that the function f is given **implicitly.** If it is possible to solve the equation for y in terms of x, then we write $y = f(x)$ and say that the function is given **explicitly.** In fact, we usually write "the function $y = f(x)$" when we mean "the function f defined by the equation $y = f(x)$." Although this usage is not entirely correct, it is rather common and should not cause any confusion. For example:

IMPLICIT FORM	EXPLICIT FORM
$3x + y = 5$	$y = f(x) = -3x + 5$
$x^2 - y = 6$	$y = f(x) = x^2 - 6$
$xy = 4$	$y = f(x) = 4/x$

Not all equations in x and y define a function $y = f(x)$. If an equation is solved for y and two or more values of y can be obtained for a given x, then the equation does not define a function $y = f(x)$. For example, consider the equation $x^2 + y^2 = 1$, which defines a circle. If we solve for y, we obtain $y = \pm\sqrt{1 - x^2}$ so that two values of y will result for numbers x between -1 and 1. Thus, $x^2 + y^2 = 1$ does not define a function.

 Comment: The explicit form of a function is the form required by a graphing calculator. Now do you see why it is necessary to graph a circle in two "pieces"?

We list below a summary of some important facts to remember about a function f.

Summary of Important Facts
about Functions

1. $f(x)$ is the image of x, or the value of f at x, when the rule f is applied to an x in the domain.
2. To each x in the domain of f, there is one and only one image $f(x)$ in the range.
3. f is the symbol we use to denote the function. It is symbolic of the domain and the rule we use to get from an x in the domain to $f(x)$ in the range.

Calculators

Most calculators have special keys that enable you to find the value of many functions. On your calculator, you should be able to find the square function, $f(x) = x^2$; the square root function, $f(x) = \sqrt{x}$; the reciprocal function, $f(x) = 1/x$; and many others that will be discussed later in this book (such as $\ln x$ and $\log x$). When you enter x and then press one of these function keys, you get the value of that function at x. Try it with the functions listed in Example 3.

E X A M P L E   3   *Finding Values of a Function on a Calculator*

(a) $f(x) = x^2$; $f(1.234) = 1.522756$

(b) $F(x) = 1/x$; $F(1.234) = 0.8103727$

(c) $g(x) = \sqrt{x}$; $g(1.234) = 1.1108555$ ■

Domain of a Function

Often, the domain of a function f is not specified; instead, only a rule or equation defining the function is given. In such cases, we agree that the domain of f is the largest set of real numbers for which the rule makes sense or, more precisely, for which the value $f(x)$ is a real number. Thus, the domain of f is the same as the domain of the variable x in the expression $f(x)$.

E X A M P L E   4   *Finding the Domain of a Function*

Find the domain of each of the following functions:

(a) $f(x) = \dfrac{3x}{x^2 - 4}$ (b) $g(x) = \sqrt{4 - 3x}$

Solution   (a) The rule f tells us to divide $3x$ by $x^2 - 4$. Since division by 0 is not allowed, the denominator $x^2 - 4$ can never be 0. Thus, x can never equal 2 or -2. The domain of the function f is $\{x | x \neq -2, x \neq 2\}$.

(b) The rule g tells us to take the square root of $4 - 3x$. But only nonnegative numbers have real square roots. Hence, we require that

$$4 - 3x \geq 0$$
$$-3x \geq -4$$
$$x \leq \tfrac{4}{3}$$

The domain of g is $\{x | -\infty < x \leq \tfrac{4}{3}\}$ or the interval $(-\infty, \tfrac{4}{3}]$.   ■

■   Now work Problem 41.

If x is in the domain of a function f, we shall say that **f is defined at x,** or **$f(x)$ exists. If x is not in the domain of f, we say that f is not defined at x,** or $f(x)$ **does not exist.** For example, if $f(x) = x/(x^2 - 1)$, then $f(0)$ exists, but $f(1)$ and $f(-1)$ do not exist. (Do you see why?)

We have not said much about finding the range of a function. The reason is that when a function is defined by an equation, it is often difficult to find the range. Therefore, we shall usually be content to find just the domain of a function when only the rule for the function is given. We shall express the domain of a function using interval notation, set notation, or words, whichever is most convenient.

When we use functions in applications, the domain may be restricted by physical or geometric considerations. For example, the domain of the function f defined by $f(x) = x^2$ is the set of all real numbers. However, if f is used as the rule for obtaining the area of a square when the length x of a side is known, then we must restrict the domain of f to the positive real numbers, since the length of a side can never be 0 or negative.

Independent Variable; Dependent Variable

Consider a function $y = f(x)$. The variable x is called the **independent variable,** because it can be assigned any of the permissible numbers from the domain. The variable y is called the **dependent variable,** because its value depends on x.

Any symbol can be used to represent the independent and dependent variables. For example, if f is the *cube function,* then f can be defined by $f(x) = x^3$ or $f(t) = t^3$ or $f(z) = z^3$. All three rules are identical: each tells us to cube the independent variable. In practice, the symbols used for the independent and dependent variables are based on common usage.

E X A M P L E   5   *Construction Cost*

The cost per square foot to build a house is \$110. Express the cost C as a function of x, the number of square feet. What is the cost to build a 2000 square foot house?

Solution   The cost C of building a house containing x square feet is $110x$ dollars. A function expressing this relationship is

$$C(x) = 110x$$

where x is the independent variable and C is the dependent variable. In this setting, the domain is $\{x|x > 0\}$ since a house cannot have 0 or negative square feet. The cost to build a 2000 square foot house is

$$C(2000) = 110(2000) = \$220{,}000 \qquad \blacksquare$$

It is worth observing that in the solution to Example 5 we used the symbol C in two ways; it is used to name the function, and it is used to symbolize the dependent variable. This double use is common in applications and should not cause any difficulty.

E X A M P L E   6   *Area of a Circle*

Express the area of a circle as a function of its radius.

Solution   We know that the formula for the area A of a circle of radius r is $A = \pi r^2$. If we use r to represent the independent variable and A to represent the dependent variable, the function expressing this relationship is

$$A(r) = \pi r^2$$

In this setting, the domain is $\{r|r > 0\}$. (Do you see why?) $\qquad \blacksquare$

■   Now work Problem 61.

The Graph of a Function

In applications, a graph often demonstrates more clearly the relationship between two variables than, say, an equation or table would. For example, Figure 3 shows the price per share (vertical axis) of McDonald's Corp. stock at the end of each week from Nov. 4, 1994 to Jan. 27, 1995 (horizontal axis). We can see from the graph that the price of the stock was falling over the few days preceding Nov. 25 and was rising over the days from Jan. 20 through Jan. 27. The graph also shows that the lowest price during this period occurred on Dec. 9 while the highest occurred on Jan. 27. Equations and tables, on the other hand, usually require some calculations and interpretation before this kind of information can be "seen."

FIGURE 3

Weekly closing prices of McDonald's Corp. stock

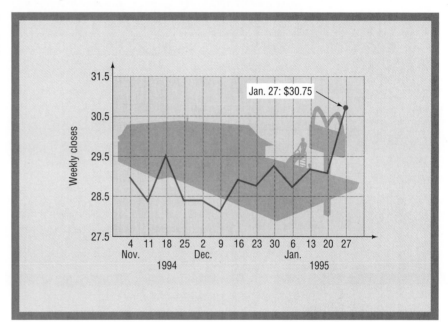

Look again at Figure 3. The graph shows that, for each time on the horizontal axis, there is only one price on the vertical axis. Thus, the graph represents a function, although the exact rule for getting from time to price is not given.

When the rule that defines a function f is given by an equation in x and y, the **graph of** f is the graph of the equation, that is, the set of points (x, y) in the xy-plane that satisfies the equation.

Not every collection of points in the xy-plane represents the graph of a function. Remember, for a function f, each number x in the domain of f has one and only one image $f(x)$. Thus, the graph of a function f cannot contain two points with the same x-coordinate and different y-coordinates. Therefore, the graph of a function must satisfy the following **vertical-line test:**

Theorem
Vertical-Line Test

A set of points in the xy-plane is the graph of a function if and only if a vertical line intersects the graph in at most one point. ∎

It follows that, if any vertical line intersects a graph at more than one point, the graph is not the graph of a function.

E X A M P L E   7   *Identifying the Graph of a Function*

Which of the graphs in Figure 4 are graphs of functions?

FIGURE 4

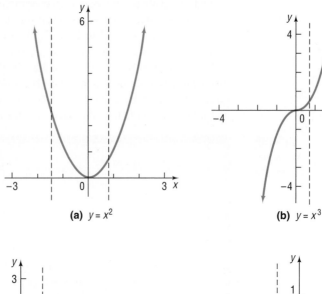

(a) $y = x^2$ (b) $y = x^3$

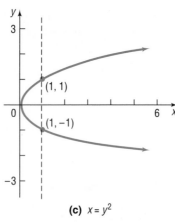   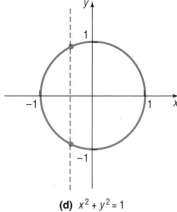

(c) $x = y^2$ (d) $x^2 + y^2 = 1$

Solution   The graphs in Figures 4(a) and 4(b) are graphs of functions, because a vertical line intersects each graph in at most one point. The graphs in Figures 4(c) and 4(d) are not graphs of functions, because some vertical line intersects each graph in more than one point.   ■

Ordered Pairs

The preceding discussion provides an alternative way to think of a function. We may consider a function f as a set of **ordered pairs** (x, y) or $(x, f(x))$, in which no two pairs have the same first element. The set of all first elements is the domain of the function, and the set of all second elements is its range. Thus, there is associated with each element x in the domain a unique element y in the range. An example is the set of all ordered pairs (x, y) such that $y = x^2$. Some of the pairs in this set are

$$(2, 2^2) = (2, 4) \qquad (0, 0^2) = (0, 0)$$

$$(-2, (-2)^2) = (-2, 4) \qquad \left(\frac{1}{2}, \left(\frac{1}{2}\right)^2\right) = \left(\frac{1}{2}, \frac{1}{4}\right)$$

In this set, no two pairs have the same *first* element (although there are pairs that have the same *second* element). This set is the *square function,* which associates with each real number x the number x^2. Look again at Figure 4(a).

On the other hand, the ordered pairs (x, y) for which $y^2 = x$ do not represent a function, because there are ordered pairs with the same first element but different second elements. For example, $(1, 1)$ and $(1, -1)$ are ordered pairs obeying the relationship $y^2 = x$ with the same first element but different second elements. Look again at Figure 4(c).

The next example illustrates how to determine the domain and range of a function if its graph is given.

E X A M P L E 8   *Obtaining Information from the Graph of a Function*

Let f be the function whose graph is given in Figure 5. Some points on the graph are labeled.

(a) What is the value of the function when $x = -6$, $x = -4$, $x = 0$, and $x = 6$?
(b) What is the domain of f?
(c) What is the range of f?
(d) List the intercepts. (Recall that these are the points, if any, where the graph crosses or touches the coordinate axes.)

FIGURE 5

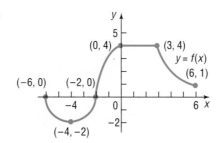

Solution   (a) Since $(-6, 0)$ is on the graph of f, the y-coordinate 0 must be the value of f at the x-coordinate -6; that is, $f(-6) = 0$. In a similar way, we find that when $x = -4$ then $y = -2$, or $f(-4) = -2$; when $x = 0$, then $y = 4$, or $f(0) = 4$; and when $x = 6$, then $y = 1$, or $f(6) = 1$.

(b) To determine the domain of f, we notice that the points on the graph of f all have x-coordinates between -6 and 6, inclusive; and, for each number x between -6 and 6, there is a point $(x, f(x))$ on the graph. Thus, the domain of f is $\{x \mid -6 \le x \le 6\}$, or the interval $[-6, 6]$.

(c) The points on the graph all have y-coordinates between -2 and 4, inclusive; and, for each such number y, there is at least one number x in the domain. Hence, the range of f is $\{y \mid -2 \le y \le 4\}$, or the interval $[-2, 4]$.

(d) The intercepts are $(-6, 0)$, $(-2, 0)$, and $(0, 4)$. ∎

When the graph of a function is given, its domain may be viewed as the shadow created by the graph on the x-axis by vertical beams of light. Its range can be viewed as the shadow created by the graph on the y-axis by horizontal beams of light. Try this technique with the graph given in Figure 5.

■ Now work Problems 25 and 27.

E X A M P L E   9   *Getting from an Island to Town*

An island is 2 miles from the nearest point P on a straight shoreline. A town is 12 miles down the shore from P.

(a) If a person can row a boat at an average speed of 3 miles per hour and the same person can walk 5 miles per hour, express the time T it takes to go from the island to town as a function of the distance x from P to where the person lands the boat. See Figure 6.

(b) How long will it take to travel from the island to town if the person lands the boat 4 miles from P?

(c) How long will it take if the person lands the boat 8 miles from P?

Solution   (a) Figure 6 illustrates the situation. The distance d_1 from the island to the landing point satisfies the equation

$$d_1^2 = 4 + x^2$$

FIGURE 6

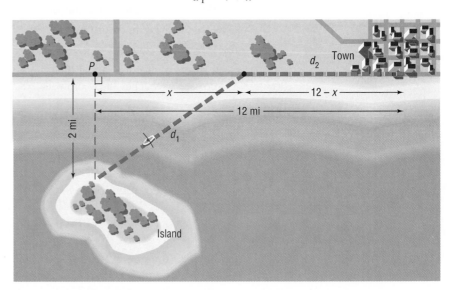

Since the average speed of the boat is 3 miles per hour, the time t_1 it takes to cover the distance d_1 is

$$d_1 = 3t_1$$

Thus,

$$t_1 = \frac{d_1}{3} = \frac{\sqrt{4 + x^2}}{3}$$

The distance d_2 from the landing point to town is $12 - x$, and the time t_2 it takes to cover this distance at an average walking speed of 5 miles per hour is

$$d_2 = 5t_2$$

Thus,

$$t_2 = \frac{d_2}{5} = \frac{12 - x}{5}$$

The total time T of the trip is $t_1 + t_2$. Thus,

$$T(x) = \frac{\sqrt{4 + x^2}}{3} + \frac{12 - x}{5}$$

(b) If the boat is landed 4 miles from P, then $x = 4$. The time T the trip takes is

$$T(4) = \frac{\sqrt{20}}{3} + \frac{8}{5} \approx 3.09 \text{ hours}$$

(c) If the boat is landed 8 miles from P, then $x = 8$. The time T the trip takes is

$$T(8) = \frac{\sqrt{68}}{3} + \frac{4}{5} \approx 3.55 \text{ hours} \qquad ■$$

■ Now work Problem 63.

Summary

We list here some of the important vocabulary introduced in this section, with a brief description of each term.

Function	A rule or correspondence between two sets of real numbers so that each number x in the first set, the domain, has corresponding to it exactly one number y in the second set.
	A set of ordered pairs (x, y) or $(x, f(x))$ in which no two pairs have the same first element.
	The range is the set of y values of the function for the x values in the domain.
	A function f may be defined implicitly by an equation involving x and y or explicitly by writing $y = f(x)$.
Unspecified domain	If a function f is defined by an equation and no domain is specified, then the domain will be taken to be the largest set of real numbers for which the rule defines a real number.
Function notation	$y = f(x)$
	f is a symbol for the rule that defines the function.
	x is the independent variable.
	y is the dependent variable.
	$f(x)$ is the value of the function at x, or the image of x.
Graph of a function	The collection of points (x, y) that satisfies the equation $y = f(x)$.
	A collection of points is the graph of a function provided vertical lines intersect the graph in at most one point (vertical-line test).

2.1

Exercise 2.1

In Problems 1–8, find the following values for each function:

(a) $f(0)$ (b) $f(1)$ (c) $f(-1)$ (d) $f(3)$

1. $f(x) = -3x^2 + 2x - 4$ **2.** $f(x) = 2x^2 + x - 1$ **3.** $f(x) = \dfrac{x}{x^2 + 1}$

4. $f(x) = \dfrac{x^2 - 1}{x + 4}$

5. $f(x) = |x| + 4$

6. $f(x) = \sqrt{x^2 + x}$

7. $f(x) = \dfrac{2x + 1}{3x - 5}$

8. $f(x) = 1 - \dfrac{1}{(x + 2)^2}$

In Problems 9–20, use the graph of the function f *given in the figure.*

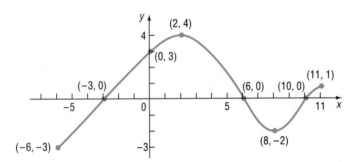

9. Find $f(0)$ and $f(-6)$.

10. Find $f(6)$ and $f(11)$.

11. Is $f(2)$ positive or negative?

12. Is $f(8)$ positive or negative?

13. For what numbers x is $f(x) = 0$?

14. For what numbers x is $f(x) > 0$?

15. What is the domain of f?

16. What is the range of f?

17. What are the x-intercepts?

18. What are the y-intercepts?

19. How often does the line $y = \frac{1}{2}$ intersect the graph?

20. How often does the line $y = 3$ intersect the graph?

In Problems 21–24, answer the questions about the given function.

21. $f(x) = \dfrac{x + 2}{x - 6}$

    (a) Is the point $(3, 14)$ on the graph of f?

    (b) If $x = 4$, what is $f(x)$?

    (c) If $f(x) = 2$, what is x?

    (d) What is the domain of f?

22. $f(x) = \dfrac{x^2 + 2}{x + 4}$

    (a) Is the point $(1, \frac{3}{5})$ on the graph of f?

    (b) If $x = 0$, what is $f(x)$?

    (c) If $f(x) = \frac{1}{2}$, what is x?

    (d) What is the domain of f?

23. $f(x) = \dfrac{2x^2}{x^4 + 1}$

    (a) Is the point $(-1, 1)$ on the graph of f?

    (b) If $x = 2$, what is $f(x)$?

    (c) If $f(x) = 1$, what is x?

    (d) What is the domain of f?

24. $f(x) = \dfrac{2x}{x - 2}$

    (a) Is the point $(\frac{1}{2}, -\frac{2}{3})$ on the graph of f?

    (b) If $x = 4$, what is $f(x)$?

    (c) If $f(x) = 1$, what is x?

    (d) What is the domain of f?

In Problems 25–36, determine whether the graph is that of a function by using the vertical-line test. If it is, use the graph to find:

(a)   Its domain and range

(b)   The intercepts, if any

(c)   Any symmetry with respect to the x-axis, y-axis, or origin

25.

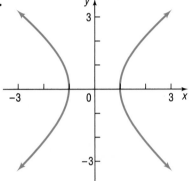

26.

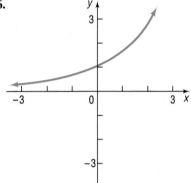

27.

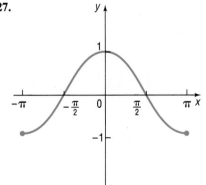

28.

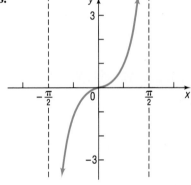

29.

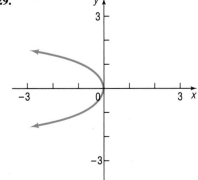

30.

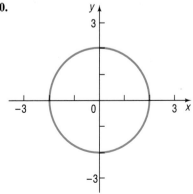

31.

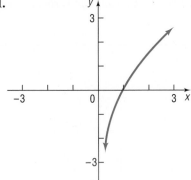

32.

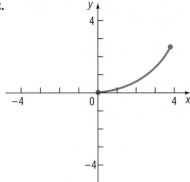

33.

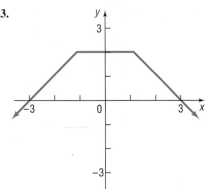

34.

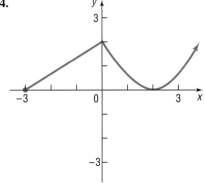

35.

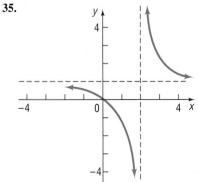

36.

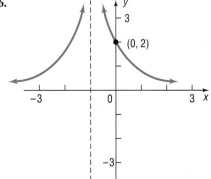

In Problems 37–50, find the domain of each function.

37. $f(x) = 3x + 4$

38. $f(x) = 5x^2 + 2$

39. $f(x) = \dfrac{x}{x^2 + 1}$

40. $f(x) = \dfrac{x^2}{x^2 + 1}$

41. $g(x) = \dfrac{x}{x^2 - 1}$

42. $h(x) = \dfrac{x}{x - 1}$

43. $F(x) = \dfrac{x - 2}{x^3 + x}$

44. $G(x) = \dfrac{x + 4}{x^3 - 4x}$

45. $h(x) = \sqrt{3x - 12}$

46. $G(x) = \sqrt{1 - x}$

47. $f(x) = \sqrt{x^2 - 9}$

48. $f(x) = \dfrac{1}{\sqrt{x^2 - 4}}$

49. $p(x) = \sqrt{\dfrac{x - 2}{x - 1}}$

50. $q(x) = \sqrt{x^2 - x - 2}$

51. If $f(x) = 2x^3 + Ax^2 + 4x - 5$ and $f(2) = 5$, what is the value of A?

52. If $f(x) = 3x^2 - Bx + 4$ and $f(-1) = 12$, what is the value of B?

53. If $f(x) = (3x + 8)/(2x - A)$ and $f(0) = 2$, what is the value of A?

54. If $f(x) = (2x - B)/(3x + 4)$ and $f(2) = \frac{1}{2}$, what is the value of B?

55. If $f(x) = (2x - A)/(x - 3)$ and $f(4) = 0$, what is the value of A? Where is f not defined?

56. If $f(x) = (x - B)/(x - A)$, $f(2) = 0$, and $f(1)$ is undefined, what are the values of A and B?

57. *Effect of Gravity on Earth* If a rock falls from a height of 20 meters on Earth, the height H (in meters) after x seconds is approximately

$$H(x) = 20 - 4.9x^2$$

    (a) What is the height of the rock when $x = 1$ second? $x = 1.1$ seconds? $x = 1.2$ seconds? $x = 1.3$ seconds?

    (b) When does the rock strike the ground?

58. *Effect of Gravity on Jupiter* If a rock falls from a height of 20 meters on the planet Jupiter, its height H (in meters) after x seconds is approximately

$$H(x) = 20 - 13x^2$$

    (a) What is the height of the rock when $x = 1$ second? $x = 1.1$ seconds? $x = 1.2$ seconds?

    (b) When does the rock strike the ground?

59. *Geometry* Express the area A of a rectangle as a function of the length x if the length is twice the width of the rectangle.

60. *Geometry* Express the area A of an isosceles right triangle as a function of the length x of one of the two equal sides.

61. Express the gross salary G of a person who earns \$5 per hour as a function of the number x of hours worked.

62. A commissioned salesperson earns \$100 base pay plus \$10 per item sold. Express the gross salary G as a function of the number x of items sold.

63. Refer to Example 9. Is there a place to land the boat so that the travel time is least? Do you think this place is closer to town or closer to P? Discuss the possibilities. Give reasons.

64. Refer to Example 9. Graph the function $T = T(x)$. Use TRACE to see how the time T varies as x changes from 0 to 12. What value of x results in the least time?

65. A cable TV company is asked to provide service to a customer whose house is located 2 miles from the road along which the cable is buried. The nearest connection box for the cable is located 5 miles down the road (see the figure).

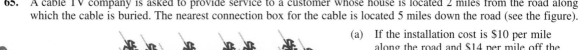

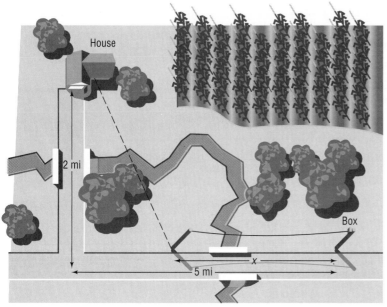

    (a) If the installation cost is \$10 per mile along the road and \$14 per mile off the road, express the total cost C of installation as a function of the distance x (in miles) from the connection box to the point where the cable installation turns off the road. Give the domain.

    (b) Compute the cost if $x = 1$ mile.

    (c) Compute the cost if $x = 3$ miles.

    (d) Graph the function $C = C(x)$. Use TRACE to see how the cost C varies as x changes from 0 to 5. What value of x results in the least cost?

66. An island is 3 miles from the nearest point P on a straight shoreline. A town is located 20 miles down the shore from P. (Refer to Figure 6 for a similar situation.)

(a) If a person has a boat that averages 12 miles per hour and the same person can run 5 miles per hour, express the time T it takes to go from the island to town as a function of x, where x is the distance from P to where the person lands the boat.

(b) How long will it take to travel from the island to town if you land the boat 8 miles from P?

(c) How long will it take if you land the boat 12 miles from P?

(d) Graph the function $T = T(x)$. Use TRACE to see how the time T varies as x changes from 0 to 20. What value of x results in the least time?

67. A page with dimensions of $8\frac{1}{2}$ inches by 11 inches has a border of uniform width x surrounding the printed matter of the page, as shown in the figure.

(a) Write a formula for the area A of the printed part of the page as a function of the width x of the border.

(b) Give the domain and range of A.

(c) Find the area of the printed page for borders of widths 1 inch, 1.2 inches, and 1.5 inches.

(d) Graph the function $A = A(x)$. Use TRACE to determine what margin should be used to obtain an area of 70 square inches and of 50 square inches.

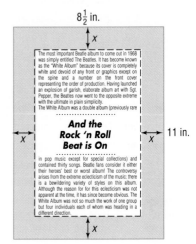

68. *Cost of Trans-Atlantic Travel* An airplane crosses the Atlantic Ocean (3000 miles) with an airspeed of 500 miles per hour. The cost C (in dollars) per passenger is given by

$$C(x) = 100 + \frac{x}{10} + \frac{36{,}000}{x}$$

where x is the ground speed (airspeed $\pm$ wind).

(a) What is the cost per passenger for quiescent (no wind) conditions?

(b) What is the cost per passenger with a head wind of 50 miles per hour?

(c) What is the cost per passenger with a tail wind of 100 miles per hour?

(d) What is the cost per passenger with a head wind of 100 miles per hour?

(e) Graph the function $C = C(x)$. As x varies from 400 to 600 miles per hour, how does the cost vary?

69. *Period of a Pendulum* The period T (in seconds) of a simple pendulum is a function of its length l (in feet) defined by the equation

$$T(l) = 2\pi\sqrt{\frac{l}{g}}$$

where $g \approx 32.2$ feet per second per second is the acceleration of gravity. Use a calculator to determine the period of a pendulum whose length is 1 foot. By how much does the period increase if the length is increased to 2 feet?

70. *Effect of Elevation on Weight* If an object weighs m pounds at sea level, then its weight W (in pounds) at a height of h miles above sea level is given approximately by

$$W(h) = m\left(\frac{4000}{4000 + h}\right)^2$$

If a woman weighs 120 pounds at sea level, how much will she weigh on Pike's Peak, which is 14,110 feet above sea level?

In Problems 71–78, tell whether the set of ordered pairs (x, y) defined by each equation is a function.

71. $y = x^2 + 2x$ **72.** $y = x^3 - 3x$ **73.** $y = \dfrac{2}{x}$ **74.** $y = \dfrac{3}{x} - 3$

75. $y^2 = 1 - x^2$ **76.** $y = \pm\sqrt{1 - 2x}$ **77.** $x^2 + y = 1$ **78.** $x + 2y^2 = 1$

79. Some functions f have the property that $f(a + b) = f(a) + f(b)$ for all real numbers a and b. Which of the following functions have this property?

  (a) $h(x) = 2x$ (b) $g(x) = x^2$ (c) $F(x) = 5x - 2$ (d) $G(x) = 1/x$

··80. Draw the graph of a function whose domain is $\{x \mid -3 \le x \le 8,\ x \ne 5\}$ and whose range is $\{y \mid -1 \le y \le 2,\ y \ne 0\}$. What point(s) in the rectangle $-3 \le x \le 8,\ -1 \le y \le 2$ cannot be on the graph? Compare your graph with those of other students. What differences do you see?

81. Are the functions $f(x) = x - 1$ and $g(x) = (x^2 - 1)/(x + 1)$ the same? Explain.

82. Describe how you would proceed to find the domain and range of a function if you were given its graph. How would your strategy change if, instead, you were given the equation defining the function?

83. How many x-intercepts can the graph of a function have? How many y-intercepts can it have?

84. Is a graph that consists of a single point the graph of a function? Can you write the equation of such a function?

85. Is there a function whose graph is symmetric with respect to the x-axis?

86. Investigate when, historically, the use of function notation $y = f(x)$ first appeared. Write a brief essay on the early uses of this notation.

2.2

More about Functions

Function Notation

The independent variable of a function is sometimes called the **argument** of the function. Thinking of the independent variable as an argument sometimes can make it easier to apply the rule of the function. For example, if f is the function defined by $f(x) = x^3$, then f is the rule that tells us to cube the argument. Thus, $f(2)$ means to cube 2, $f(a)$ means to cube the number a, and $f(x + h)$ means to cube the quantity $x + h$.

E X A M P L E   1

Finding Values of a Function

For the function G defined by $G(x) = 2x^2 - 3x$, evaluate:

  (a) $G(3)$ (b) $G(x) + G(3)$ (c) $G(-x)$
  (d) $-G(x)$ (e) $G(x + 3)$

Solution   (a) We replace x by 3 in the rule for G to get

$$G(3) = 2(3)^2 - 3(3) = 18 - 9 = 9$$

  (b) $G(x) + G(3) = (2x^2 - 3x) + (9) = 2x^2 - 3x + 9$
  (c) We replace x by $-x$ in the rule for G:

$$G(-x) = 2(-x)^2 - 3(-x) = 2x^2 + 3x$$

  (d) $-G(x) = -(2x^2 - 3x) = -2x^2 + 3x$
  (e) $G(x + 3) = 2(x + 3)^2 - 3(x + 3)$ Notice the use of parentheses here.
$$= 2(x^2 + 6x + 9) - 3x - 9$$
$$= 2x^2 + 12x + 18 - 3x - 9$$
$$= 2x^2 + 9x + 9$$

■

Notice in this example that $G(x + 3) \neq G(x) + G(3)$.

■ Now work Problem 29.

Example 1 illustrates certain uses of **function notation.** Let's look at another use.

E X A M P L E   2

Using Function Notation

For the function $f(x) = x^2 + 1$, find: $\dfrac{f(x) - f(1)}{x - 1}$, $x \neq 1$.

Solution   First, we find $f(1)$:

$$f(1) = (1)^2 + 1 = 2$$

Then

$$\frac{f(x) - f(1)}{x - 1} = \frac{(x^2 + 1) - (2)}{x - 1} = \frac{x^2 - 1}{x - 1} = \frac{(x + 1)(x - 1)}{x - 1} = x + 1 \quad ■$$

Expressions like the one we worked with in Example 2 occur frequently in calculus.

Difference Quotient

For a number c in the domain of a function f, the expression

$$\frac{f(x) - f(c)}{x - c} \qquad x \neq c \qquad (1)$$

is called the **difference quotient** of f at c.

The difference quotient of a function has an important geometric interpretation. Look at the graph of $y = f(x)$ in Figure 7. We have labeled two points on the graph: $(c, f(c))$ and $(x, f(x))$. The slope of the line containing these two points is

$$\frac{f(x) - f(c)}{x - c}$$

This line is called a **secant line.** Thus, the difference quotient of a function equals the slope of a secant line containing two points on its graph.

FIGURE 7

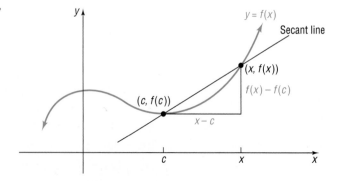

E X A M P L E   3   *Finding a Difference Quotient*

Find the difference quotient of $f(x) = 2x^2 - x + 1$ at $c = 2$.

Solution   From expression (1), we seek

$$\frac{f(x) - f(2)}{x - 2} \qquad x \neq 2$$

We begin by finding $f(2)$:

$$f(2) = 2(2)^2 - (2) + 1 = 8 - 2 + 1 = 7$$

Then the difference quotient of f at 2 is

$$\frac{f(x) - f(2)}{x - 2} = \frac{2x^2 - x + 1 - 7}{x - 2} = \frac{2x^2 - x - 6}{x - 2} = \frac{(2x + 3)(x - 2)}{x - 2} = 2x + 3 \qquad \blacksquare$$

■   Now work Problem 41.

Increasing and Decreasing Functions

Consider the graph given in Figure 8. If you look from left to right along the graph of this function, you will notice that parts of the graph are rising, parts are falling, and parts are horizontal. In such cases, the function is described as *increasing, decreasing,* and *constant,* respectively.

FIGURE 8

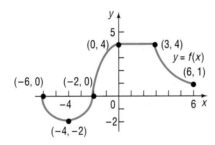

E X A M P L E   4   *Determining Where a Function Is Increasing, Decreasing, or Constant*

Where is the function in Figure 8 increasing? Where is it decreasing? Where is it constant?

Solution   To answer the question of where a function is increasing, where it is decreasing, and where it is constant, we use inequalities involving the independent variable x or we use intervals of x-coordinates. The graph in Figure 8 is rising (increasing) from the point $(-4, -2)$ to the point $(0, 4)$, so we conclude that it is increasing on the interval $[-4, 0]$ (or for $-4 \leq x \leq 0$). The graph is falling (decreasing) from the point $(-6, 0)$ to the point $(-4, -2)$ and from the point $(3, 4)$ to the point $(6, 1)$. We conclude that the graph is decreasing on the intervals $[-6, -4]$ and $[3, 6]$ (or for $-6 \leq x \leq -4$ and $3 \leq x \leq 6$). The graph is constant on the interval $[0, 3]$ (or for $0 \leq x \leq 3$).   ■

More precise definitions follow.

Increasing Function	A function f is **increasing** on an interval I if, for any choice of x_1 and x_2 in I, with $x_1 < x_2$, we have $f(x_1) < f(x_2)$.

Decreasing Function	A function f is **decreasing** on an interval I if, for any choice of x_1 and x_2 in I, with $x_1 < x_2$, we have $f(x_1) > f(x_2)$.

Constant Function	A function f is **constant** on an interval I if, for all choices of x in I, the values $f(x)$ are equal.

Thus, the graph of an increasing function goes up from left to right, the graph of a decreasing function goes down from left to right, and the graph of a constant function remains at a fixed height. Figure 9 illustrates the definitions.

FIGURE 9

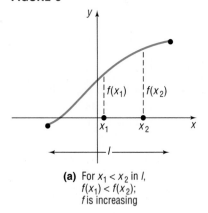

(a) For $x_1 < x_2$ in I,
$f(x_1) < f(x_2)$;
f is increasing

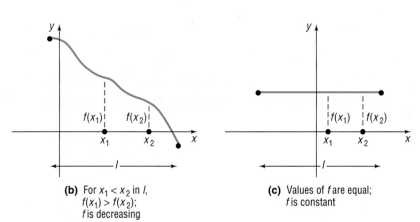

(b) For $x_1 < x_2$ in I,
$f(x_1) > f(x_2)$;
f is decreasing

(c) Values of f are equal;
f is constant

Even and Odd Functions

Even Function	A function f is **even** if for every number x in its domain the number $-x$ is also in the domain and

$$f(-x) = f(x)$$

Odd Function	A function f is **odd** if for every number x in its domain the number $-x$ is also in the domain and

$$f(-x) = -f(x)$$

Refer to Section 1.6, where the tests for symmetry are listed. The following results are then evident:

Theorem   A function is even if and only if its graph is symmetric with respect to the *y*-axis. A function is odd if and only if its graph is symmetric with respect to the origin.   ■

E X A M P L E   5   *Determining Even and Odd Functions from the Graph*

Determine whether each graph given in Figure 10 is the graph of an even function, an odd function, or a function that is neither even nor odd.

FIGURE 10

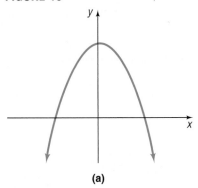

(a)

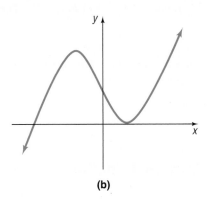

(b)

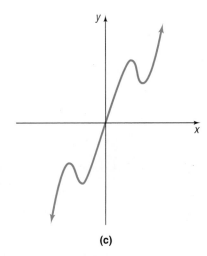

(c)

Solution   The graph in Figure 10(a) is that of an even function, because the graph is symmetric with respect to the *y*-axis. The function whose graph is given in Figure 10(b) is neither even nor odd, because the graph is neither symmetric with respect to the *y*-axis nor symmetric with respect to the origin. The function whose graph is given in Figure 10(c) is odd, because its graph is symmetric with respect to the origin.   ■

■   Now work Problem 9.

E X A M P L E   6   *Identifying Even and Odd Functions*

Determine whether each of the following functions is even, odd, or neither. Then, without graphing, determine whether the graph is symmetric with respect to the *y*-axis or with respect to the origin.

(a) $f(x) = x^2 - 5$ (b) $g(x) = x^3 - 1$
(c) $h(x) = 5x^3 - x$ (d) $F(x) = |x|$

Solution   (a) We replace *x* by $-x$ in $f(x) = x^2 - 5$. Then

$$f(-x) = (-x)^2 - 5 = x^2 - 5$$

Since $f(-x) = f(x)$, we conclude that f is an even function, and the graph is symmetric with respect to the *y*-axis.

(b) We replace *x* by $-x$. Then

$$g(-x) = (-x)^3 - 1 = -x^3 - 1$$

Since $g(-x) \neq g(x)$ and $g(-x) \neq -g(x)$, we conclude that g is neither even nor odd. The graph is not symmetric with respect to the y-axis nor with respect to the origin.

(c) We replace x by $-x$ in $h(x) = 5x^3 - x$. Then

$$h(-x) = 5(-x)^3 - (-x) = -5x^3 + x$$

Since $h(-x), = -h(x)$, h is an odd function, and the graph of h is symmetric with respect to the origin.

(d) We replace x by $-x$ in $F(x) = |x|$. Then

$$F(-x) = |-x| = |x|$$

Since $F(-x) = F(x)$, F is an even function, and the graph of F is symmetric with respect to the y-axis.   ■

■   Now work Problem 51.

Important Functions

We now give names to some of the functions we have encountered. In going through this list, pay special attention to the characteristics of each function, particularly to the shape of each graph.

Linear Function

$$f(x) = mx + b \qquad m \text{ and } b \text{ are real numbers}$$

The domain of the **linear function** f consists of all real numbers. The graph of this function is a nonvertical straight line with slope m and y-intercept b. A linear function is increasing if $m > 0$, decreasing if $m < 0$, and constant if $m = 0$.

Constant Function

$$f(x) = b \qquad b \text{ a real number}$$

See Figure 11.

FIGURE 11

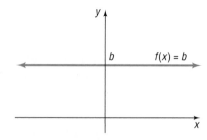

A **constant function** is a special linear function ($m = 0$). Its domain is the set of all real numbers; its range is the set consisting of a single number b. Its graph is a horizontal line whose y-intercept is b. The constant function is an even function whose graph is constant over its domain.

Identity Function

$$f(x) = x$$

See Figure 12.

FIGURE 12

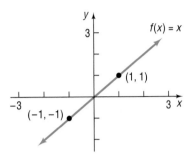

The **identity function** is also a special linear function. Its domain and its range are the set of all real numbers. Its graph is a line whose slope is $m = 1$ and whose y-intercept is 0. The line consists of all points for which the x-coordinate equals the y-coordinate. The identity function is an odd function that is increasing over its domain. Note that the graph bisects quadrants I and III.

Square Function

$$f(x) = x^2$$

See Figure 13.

FIGURE 13

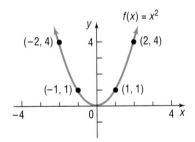

The domain of the **square function** f is the set of all real numbers; its range is the set of nonnegative real numbers. The graph of this function is a parabola, whose intercept is at $(0, 0)$. The square function is an even function that is decreasing on the interval $(-\infty, 0]$ and increasing on the interval $[0, \infty)$.

Cube Function

$$f(x) = x^3$$

See Figure 14.

FIGURE 14

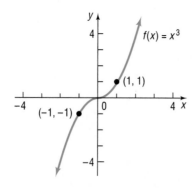

The domain and range of the **cube function** are the set of all real numbers. The intercept of the graph is at $(0, 0)$. The cube function is odd and is increasing on the interval $(-\infty, \infty)$.

Square Root Function

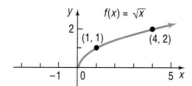
$$f(x) = \sqrt{x}$$

See Figure 15.

FIGURE 15

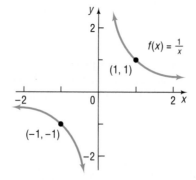

The domain and range of the **square root function** are the set of nonnegative real numbers. The intercept of the graph is at $(0, 0)$. The square root function is neither even nor odd and is increasing on the interval $[0, \infty)$.

Reciprocal Function

$$f(x) = \frac{1}{x}$$

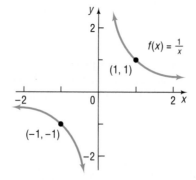

Refer to Example 8, p. 62, for a discussion of the equation $y = 1/x$. See Figure 16.

FIGURE 16

The domain and range of the **reciprocal function** are the set of all nonzero real numbers. The graph has no intercepts. The reciprocal function is decreasing on the intervals $(-\infty, 0)$ and $(0, \infty)$ and is an odd function.

Absolute Value Function

$$f(x) = |x|$$

See Figure 17.

FIGURE 17

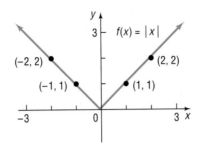

The domain of the **absolute value function** is the set of all real numbers; its range is the set of nonnegative real numbers. The intercept of the graph is at $(0, 0)$. If $x \geq 0$, then $f(x) = x$ and the graph of f is part of the line $y = x$; if $x < 0$, then $f(x) = -x$ and the graph of f is part of the line $y = -x$. The absolute value function is an even function; it is decreasing on the interval $(-\infty, 0]$ and increasing on the interval $[0, \infty)$.

Check: Graph $y = |x|$ on a square screen and compare what you see with Figure 17. Note that some graphing calculators use the symbol abs (x) for absolute value. If your utility has no built-in absolute value function, you can still graph $y = |x|$ by using the fact that $|x| = \sqrt{x^2}$. ∎

The symbol $[[x]]$, read as **"bracket x,"** stands for the largest integer less than or equal to x. For example,

$$[[1]] = 1 \qquad [[2.5]] = 2 \qquad [[\tfrac{1}{2}]] = 0 \qquad [[-\tfrac{3}{4}]] = -1 \qquad [[\pi]] = 3$$

This type of correspondence occurs frequently enough in mathematics that we give it a name.

Greatest-integer Function

$$f(x) = [[x]] = \text{Greatest integer less than or equal to } x$$

We obtain the graph of $f(x) = [[x]]$ by plotting several points. See Table 1. For values of x, $-1 \leq x < 0$, the value of $f(x) = [[x]]$ is -1; for values of x, $0 \leq x < 1$, the value of f is 0. See Figure 18 for the graph.

TABLE 1

x	$y = f(x)$ $= [[x]]$	(x, y)
-1	-1	$(-1, -1)$
$-\frac{1}{2}$	-1	$(-\frac{1}{2}, -1)$
$-\frac{1}{4}$	-1	$(-\frac{1}{4}, -1)$
0	0	$(0, 0)$
$\frac{1}{4}$	0	$(\frac{1}{4}, 0)$
$\frac{1}{2}$	0	$(\frac{1}{2}, 0)$
$\frac{3}{4}$	0	$(\frac{3}{4}, 0)$

FIGURE 18

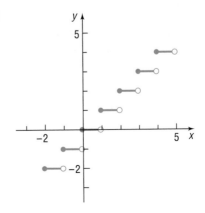

The domain of the **greatest-integer function** is the set of all real numbers; its range is the set of integers. The y-intercept of the graph is at 0. The x-intercepts lie in the interval $[0, 1)$. The greatest-integer function is neither even nor odd. It is constant on every interval of the form $[k, k + 1)$, for k an integer. In Figure 18, we use a solid dot to indicate, for example, that at $x = 1$ the value of f is $f(1) = 1$; we use an open circle to illustrate that the function does not assume the value of 0 at $x = 1$.

From the graph of the greatest-integer function, we can see why it is also called a **step function.** At $x = 0$, $x = \pm 1$, $x = \pm 2$, and so on, this function exhibits what is called a *discontinuity;* that is, at integer values, the graph suddenly "steps" from one value to another without taking on any of the intermediate values. For example, to the immediate left of $x = 3$, the y-coordinates are 2, and to the immediate right of $x = 3$, the y-coordinates are 3.

The functions we have discussed so far are basic. Whenever you encounter one of them, you should see a mental picture of its graph. For example, if you encounter the function $f(x) = x^2$, you should see in your mind's eye a picture like Figure 13.

■ Now work Problem 67.

Piecewise-defined Functions

Sometimes, a function is defined by a rule consisting of two or more equations. The choice of which equation to use depends on the value of the independent variable x. For example, the absolute value function $f(x) = |x|$ is actually defined by two equations: $f(x) = x$ if $x \geq 0$ and $f(x) = -x$ if $x < 0$. For convenience, we generally combine these equations into one expression as

$$f(x) = |x| = \begin{cases} x & \text{if } x \geq 0 \\ -x & \text{if } x < 0 \end{cases}$$

When functions are defined by more than one equation, they are called **piecewise-defined** functions.

Let's look at another example of a piecewise-defined function.

E X A M P L E   7   *Analyzing a Piecewise-defined Function*

For the following function f,

$$f(x) = \begin{cases} -x + 1 & \text{if } -1 \leq x < 1 \\ 2 & \text{if } x = 1 \\ x^2 & \text{if } x > 1 \end{cases}$$

(a) Find $f(0)$, $f(1)$, and $f(2)$.   (b) Determine the domain of f.

(c) Graph f.   (d) Use the graph to find the range of f.

Solution   (a) To find $f(0)$, we observe that when $x = 0$ the equation for f is given by $f(x) = -x + 1$. So we have

$$f(0) = -0 + 1 = 1$$

When $x = 1$, the equation for f is $f(x) = 2$. Thus

$$f(1) = 2$$

FIGURE 19
$y = f(x)$

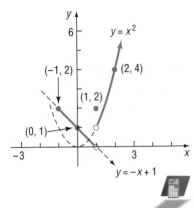

When $x = 2$, the equation for f is $f(x) = x^2$. So
$$f(2) = 2^2 = 4$$

(b) To find the domain of f, we look at its definition. We conclude that the domain of f is $\{x | x \geq -1\}$, or $[-1, \infty)$.

(c) To graph f, we graph "each piece." Thus, we first graph the line $y = -x + 1$ and keep only the part for which $-1 \leq x < 1$. Then we plot the point $(1, 2)$, because when $x = 1$, $f(x) = 2$. Finally, we graph the parabola $y = x^2$ and keep only the part for which $x > 1$. See Figure 19.

(d) From the graph, we conclude that the range of f is $\{y | y > 0\}$, or $(0, \infty)$.

Check: Graph the function f given in Example 7 and compare what you see with Figure 19. ∎

■ Now work Problem 81.

E X A M P L E   8

Cost of Electricity

In the winter, Commonwealth Edison Company supplies electricity to residences for a monthly customer charge of $9.06 plus 10.819¢ per kilowatt-hour (kWhr) for the first 400 kWhr supplied in the month, and 7.093¢ per kWhr for all usage over 400 kWhr in the month.*

(a) What is the charge for using 300 kWhr in a month?
(b) What is the charge for using 700 kWhr in a month?
(c) If C is the monthly charge for x kWhr, express C as a function of x.

Solution

(a) For 300 kWhr, the charge is $9.06 plus 10.819¢ = $0.10819 per kWhr. Thus,
$$\text{Charge} = \$9.06 + \$0.10819(300) = \$41.52$$

(b) For 700 kWhr, the charge is $9.06 plus 10.819¢ for the first 400 kWhr plus 7.093¢ for the 300 kWhr in excess of 400. Thus,
$$\text{Charge} = \$9.06 + \$0.10819(400) + \$0.07093(300) = \$73.62$$

(c) If $0 \leq x \leq 400$, the monthly charge C (in dollars) can be found by multiplying x times $0.10819 and adding the monthly customer charge of $9.06. Thus, if $0 \leq x \leq 400$, then $C(x) = 0.10819x + 9.06$. For $x > 400$, the charge is $0.10819(400) + 9.06 + 0.07093(x - 400)$, since $x - 400$ equals the usage in excess of 400 kWhr, which costs $0.07093 per kWhr. Thus, if $x > 400$, then

$$C(x) = 0.10819(400) + 9.06 + 0.07093(x - 400)$$
$$= 52.336 + 0.07093(x - 400)$$
$$= 0.07093x + 23.964$$

*Source: Commonwealth Edison Co., Chicago, Illinois, 1991.

The rule for computing C follows two rules:

$$C(x) = \begin{cases} 0.10819x + 9.06 & \text{if } 0 \leq x \leq 400 \\ 0.07093x + 23.964 & \text{if } x > 400. \end{cases}$$

See Figure 20 for the graph. ■

FIGURE 20

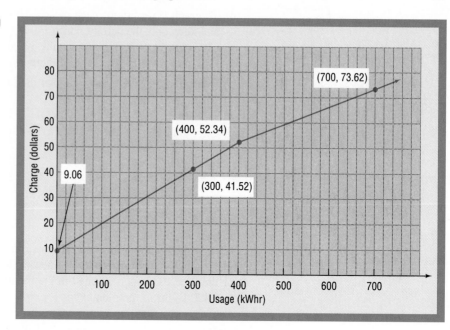

2.2

Exercise 2.2

In Problems 1–8, match each graph to the function listed whose graph most resembles the one given.

A. *Constant function*   B. *Linear function*
C. *Square function*   D. *Cube function*
E. *Square root function*   F. *Reciprocal function*
G. *Absolute value function*   H. *Greatest-integer function*

1.   **2.**   **3.**    **4.**

5.   **6.**   **7.** 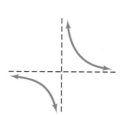   **8.**

In Problems 9–22, the graph of a function is given. Use the graph to find:

(a) Its domain and range
(b) The intervals on which it is increasing, decreasing, or constant
(c) Whether it is even, odd, or neither
(d) The intercepts, if any

9.
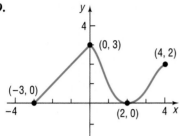

10.
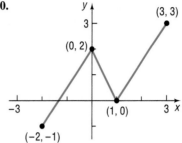

11.
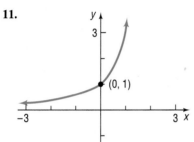

12.
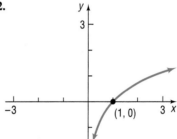

13.
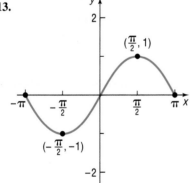

14.
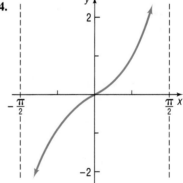

15.
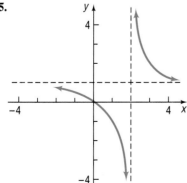

16.
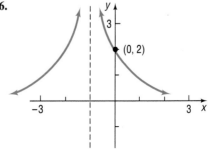

17.

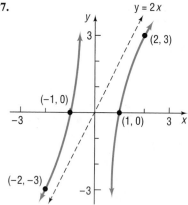

18.

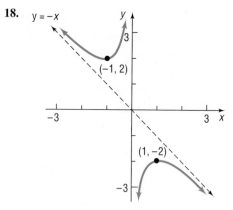

19.

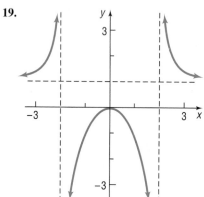

20.

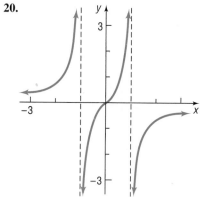

For Problems 21 and 22, assume that the graph shown is complete.

21.

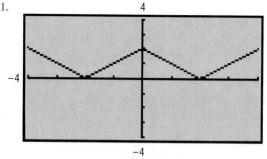

22.

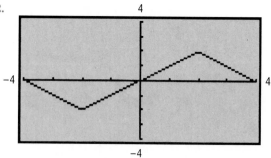

23. If $f(x) = [[2x]]$, find: (a) $f(1.2)$ (b) $f(1.6)$ (c) $f(-1.8)$

24. If $f(x) = [[x/2]]$, find: (a) $f(1.2)$ (b) $f(1.6)$ (c) $f(-1.8)$

25. If

$$f(x) = \begin{cases} x^2 & \text{if } x < 0 \\ 2 & \text{if } x = 0 \\ 2x + 1 & \text{if } x > 0 \end{cases}$$

find: (a) $f(-2)$ (b) $f(0)$ (c) $f(2)$

26. If
$$f(x) = \begin{cases} x^3 & \text{if } x < 0 \\ 3x + 2 & \text{if } x \geq 0 \end{cases}$$

find: (a) $f(-1)$ (b) $f(0)$ (c) $f(1)$

In Problems 27–38, find the following for each function:

(a) $f(-x)$ (b) $-f(x)$ (c) $f(2x)$ (d) $f(x-3)$ (e) $f(1/x)$ (f) $1/f(x)$

27. $f(x) = 2x + 5$ **28.** $f(x) = 3 - x$ **29.** $f(x) = 2x^2 - 4$ **30.** $f(x) = x^3 + 1$

31. $f(x) = x^3 - 3x$ **32.** $f(x) = x^2 + x$ **33.** $f(x) = \dfrac{x}{x^2 + 1}$ **34.** $f(x) = \dfrac{x^2}{x^2 + 1}$

35. $f(x) = |x|$ **36.** $f(x) = \dfrac{1}{x}$ **37.** $f(x) = 1 + \dfrac{1}{x}$ **38.** $f(x) = 4 + \dfrac{2}{x}$

In Problems 39–50, find the difference quotient,

$$\frac{f(x) - f(1)}{x - 1} \qquad x \neq 1$$

for each function. Be sure to simplify.

39. $f(x) = 3x$ **40.** $f(x) = -2x$ **41.** $f(x) = 1 - 3x$ **42.** $f(x) = x^2 + 1$

43. $f(x) = 3x^2 - 2x$ **44.** $f(x) = 4x - 2x^2$ **45.** $f(x) = x^3 - x$ **46.** $f(x) = x^3 + x$

47. $f(x) = \dfrac{2}{x + 1}$ **48.** $f(x) = \dfrac{1}{x^2}$ **49.** $f(x) = \sqrt{x}$ **50.** $f(x) = \sqrt{x + 3}$

In Problems 51–62, tell whether each function is even, odd, or neither without drawing a graph.

51. $f(x) = 4x^3$ **52.** $f(x) = 2x^4 - x^2$ **53.** $g(x) = 2x^2 - 5$ **54.** $h(x) = 3x^3 + 2$

55. $F(x) = \sqrt[3]{x}$ **56.** $G(x) = \sqrt{x}$ **57.** $f(x) = x + |x|$ **58.** $f(x) = \sqrt[3]{2x^2 + 1}$

59. $g(x) = \dfrac{1}{x^2}$ **60.** $h(x) = \dfrac{x}{x^2 - 1}$ **61.** $h(x) = \dfrac{x^3}{3x^2 - 9}$ **62.** $F(x) = \dfrac{x}{|x|}$

63. How many x-intercepts can a function defined on an interval have if it is increasing on that interval? Explain.

64. How many y-intercepts can a function have? Explain.

In Problems 65–90:

(a) *Find the domain of each function.*   (b) *Locate any intercepts.*
(c) *Graph each function.*   (d) *Based on the graph, find the range.*

65. $f(x) = 3x - 3$ **66.** $f(x) = 4 - 2x$ **67.** $g(x) = x^2 - 4$

68. $g(x) = x^2 + 4$ **69.** $h(x) = -x^2$ **70.** $F(x) = 2x^2$

71. $f(x) = \sqrt{x - 2}$ **72.** $g(x) = \sqrt{x} + 2$ **73.** $h(x) = \sqrt{2 - x}$

74. $F(x) = -\sqrt{x}$ **75.** $f(x) = |x| + 3$ **76.** $g(x) = |x + 3|$

77. $h(x) = -|x|$ **78.** $F(x) = |3 - x|$

79. $f(x) = \begin{cases} 2x & \text{if } x \neq 0 \\ 0 & \text{if } x = 0 \end{cases}$ **80.** $f(x) = \begin{cases} 3x & \text{if } x \neq 0 \\ 4 & \text{if } x = 0 \end{cases}$

81. $f(x) = \begin{cases} 1 + x & \text{if } x < 0 \\ x^2 & \text{if } x \geq 0 \end{cases}$ **82.** $f(x) = \begin{cases} 1/x & \text{if } x < 0 \\ \sqrt{x} & \text{if } x \geq 0 \end{cases}$

83. $f(x) = \begin{cases} |x| & \text{if } -2 \le x < 0 \\ 1 & \text{if } x = 0 \\ x^3 & \text{if } x > 0 \end{cases}$

84. $f(x) = \begin{cases} 3 + x & \text{if } -3 \le x < 0 \\ 3 & \text{if } x = 0 \\ \sqrt{x} & \text{if } x > 0 \end{cases}$

85. $g(x) = \begin{cases} 1 & \text{if } x \text{ is an integer} \\ -1 & \text{if } x \text{ is not an integer} \end{cases}$

86. $g(x) = \begin{cases} x & \text{if } x \ge 1 \\ 1 & \text{if } x < 1 \end{cases}$

87. $h(x) = 2[[x]]$

88. $f(x) = [[2x]]$

89. $F(x) = \begin{cases} 4 - x^2 & \text{if } |x| \le 2 \\ x^2 - 4 & \text{if } |x| > 2 \end{cases}$

90. $G(x) = |x^2 - 4|$

Problems 91–94 require the following definition. Secant Line *The slope of the secant line containing the two points $(x, f(x))$ and $(x + h, f(x + h))$ on the graph of a function $y = f(x)$ may be given as*

$$\frac{f(x + h) - f(x)}{h}$$

In Problems 91–94, express the slope of the secant line of each function in terms of x and h. Be sure to simplify your answer.

91. $f(x) = 2x + 5$

92. $f(x) = -3x + 2$

93. $f(x) = x^2 + 2x$

94. $f(x) = 1/x$

In Problems 95–98, the graph of a piecewise-defined function is given. Write a definition for each function.

95.

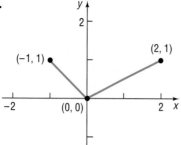

96.

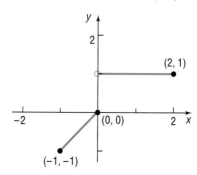

97.

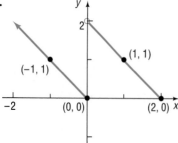

98.

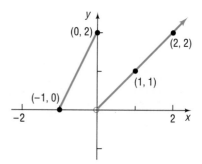

In Problems 99 and 100, decide whether each function is even. Give a reason.

99. $f(x) = \begin{cases} x^2 + 4 & \text{if } x \ne 2 \\ 6 & \text{if } x = 2 \end{cases}$

100. $f(x) = \begin{cases} x^2 + 4 & \text{if } x \ne 2 \\ 5 & \text{if } x = 2 \end{cases}$

101. Graph $y = x^2$. Then graph $y = x^2 + 2$, followed by $y = x^2 + 4$, followed by $y = x^2 - 2$. What pattern do you observe? Can you predict the graph of $y = x^2 - 4$? Of $y = x^2 + 5$?

102. Graph $y = x^2$. Then graph $y = (x - 2)^2$, followed by $y = (x - 4)^2$, followed by $y = (x + 2)^2$. What pattern do you observe? Can you predict the graph of $y = (x + 4)^2$? Of $y = (x - 5)^2$?

103. Graph $y = |x|$. Then graph $y = 2|x|$, followed by $y = 4|x|$, followed by $y = \frac{1}{2}|x|$. What pattern do you observe? Can you predict the graph of $y = \frac{1}{4}|x|$? Of $y = 5|x|$?

104. Graph $y = x^2$. Then graph $y = -x^2$. What pattern do you observe? Now try $y = |x|$ and $y = -|x|$. What do you conclude?

105. Graph $y = \sqrt{x}$. Then graph $y = \sqrt{-x}$. What pattern do you observe? Now try $y = 2x + 1$ and $y = 2(-x) + 1$. What do you conclude?

106. Graph $y = x^3$. Then graph $y = (x - 1)^3 + 2$. Could you have predicted the result?

107. *Cost of Natural Gas* On November 12, 1991, the Peoples Gas Light and Coke Company had the following rate schedule* for natural gas usage in single-family residences:

Monthly service charge	$7.00
Per therm service charge, 1st 90 therms	$0.21054/therm
Over 90 therms	$0.11242/therm
Gas charge	$0.26341/therm

(a) What is the charge for using 50 therms in a month?
(b) What is the charge for using 500 therms in a month?
(c) Construct a function that relates the monthly charge C for x therms of gas.
(d) Graph this function.

108. *Cost of Natural Gas* On November 19, 1991, Northern Illinois Gas Company had the following rate schedule† for natural gas usage in single-family residences:

Monthly customer charge	$4.00
Distribution charge, 1st 50 therms	$0.1402/therm
Over 50 therms	$0.0547/therm
Gas supply charge	$0.2406/therm

(a) What is the charge for using 40 therms in a month?
(b) What is the charge for using 202 therms in a month?
(c) Construct a function that gives the monthly charge C for x therms of gas.
(d) Graph this function.

109. Let f denote any function with the property that, whenever x is in its domain, then so is $-x$. Define the functions $E(x)$ and $O(x)$ to be

$$E(x) = \frac{1}{2}[f(x) + f(-x)] \qquad O(x) = \frac{1}{2}[f(x) - f(-x)]$$

(a) Show that $E(x)$ is an even function. (b) Show that $O(x)$ is an odd function.
(c) Show that $f(x) = E(x) + O(x)$.
(d) Draw the conclusion that any such function f can be written as the sum of an even function and an odd function.

110. Let f and g be two functions defined on the same interval $[a, b]$. Suppose that we define two functions min (f, g) and max (f, g) as follows:

$$\min(f, g)(x) = \begin{cases} f(x) & \text{if } f(x) \le g(x) \\ g(x) & \text{if } f(x) > g(x) \end{cases} \qquad \max(f, g)(x) = \begin{cases} g(x) & \text{if } f(x) \le g(x) \\ f(x) & \text{if } f(x) > g(x) \end{cases}$$

*Source: The Peoples Gas Light and Coke Company, Chicago, Illinois.
†Source: Northern Illinois Gas Company, Naperville, Illinois.

Show that

$$\min(f, g)(x) = \frac{f(x) + g(x)}{2} - \frac{|f(x) - g(x)|}{2}$$

Develop a similar formula for max (f, g).

111. Consider the equation

$$y = \begin{cases} 1 & \text{if } x \text{ is rational} \\ 0 & \text{if } x \text{ is irrational} \end{cases}$$

Is this a function? What is its domain? What is its range? What is its y-intercept, if any? What are its x-intercepts, if any? Is it even, odd, or neither? How would you describe its graph?

112. Define some functions that pass through (0, 0) and (1, 1) and are increasing for $x \geq 0$. Begin your list with $y = \sqrt{x}$, $y = x$, and $y = x^2$. Can you propose a general result about such functions?

113. Can you think of a function that is both even and odd?

2.3

Graphing Techniques

At this stage, if you were asked to graph any of the functions defined by $y = x^2$, $y = x^3$, $y = x$, $y = \sqrt{x}$, $y = |x|$, or $y = 1/x$, your response should be, "Yes, I recognize these functions and know the general shapes of their graphs." (If this is not your answer, review the previous section and Figures 12 through 17.)

Sometimes, we are asked to graph a function that is "almost" like one we already know how to graph. In this section, we look at some of these functions and develop techniques for graphing them.

Vertical Shifts

EXAMPLE 1

Vertical Shift Up

Use the graph of $f(x) = x^2$ to obtain the graph of $g(x) = x^2 + 3$.

Solution

We begin by obtaining some points on the graphs of f and g. For example, when $x = 0$, then $y = f(0) = 0$ and $y = g(0) = 3$. When $x = 1$, then $y = f(1) = 1$ and $y = g(1) = 4$. Table 2 lists these and a few other points on each graph. We conclude that the graph of g is identical to that of f, except that it is shifted vertically up 3 units. See Figure 21.

TABLE 2

x	$y = f(x)$ $= x^2$	$y = g(x)$ $= x^2 + 3$
-2	4	7
-1	1	4
0	0	3
1	1	4
2	4	7

FIGURE 21

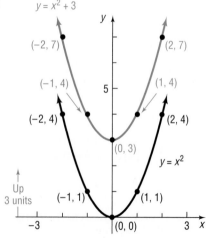

If a real number c is added to the right side of a function $y = f(x)$, the graph of the new function $y = f(x) + c$ is the graph of f **shifted vertically** up (if $c > 0$) or down (if $c < 0$). Let's look at another example.

E X A M P L E   2   *Vertical Shift Down*

Use the graph of $f(x) = x^2$ to obtain the graph of $h(x) = x^2 - 4$.

Solution   Table 3 lists some points on the graphs of f and h. The graph of h is identical to that of f, except that it is shifted down 4 units. See Figure 22.   ■

TABLE 3

x	$y = f(x)$ $= x^2$	$y = h(x)$ $= x^2 - 4$
-2	4	0
-1	1	-3
0	0	-4
1	1	-3
2	4	0

FIGURE 22

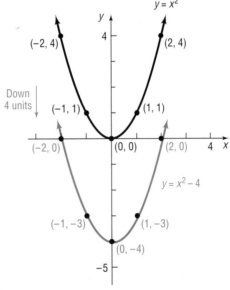

Seeing the Concept On the same screen, graph each of the following functions:

$$y = x^2, \quad y = x^2 + 1, \quad y = x^2 + 2, \quad y = x^2 - 1, \quad y = x^2 - 2$$

See Figure 23.

FIGURE 23

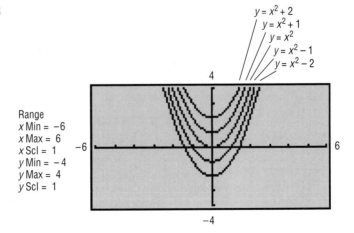

Range
x Min = -6
x Max = 6
x Scl = 1
y Min = -4
y Max = 4
y Scl = 1

■   Now work Problem 21.

Horizontal Shifts

E X A M P L E 3   *Horizontal Shift to the Right*

Use the graph of $f(x) = x^2$ to obtain the graph of $g(x) = (x - 2)^2$.

Solution   The function $g(x) = (x - 2)^2$ is basically a square function. Table 4 lists some points on the graphs of f and g. Note that when $f(x) = 0$ then $x = 0$, and when $g(x) = 0$, then $x = 2$. Also, when $f(x) = 4$, then $x = -2$ or 2, and when $g(x) = 4$, then $x = 0$ or 4. We conclude that the graph of g is identical to that of f, except that it is shifted 2 units to the right. See Figure 24.   ■

TABLE 4

x	$y = f(x)$ $= x^2$	$y = g(x)$ $= (x - 2)^2$
-2	4	16
0	0	4
2	4	0
4	16	4

FIGURE 24

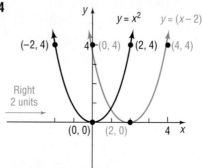

If a real number c is added to the argument x of a function f, the graph of the new function $g(x) = f(x + c)$ is the graph of f **shifted horizontally** left (if $c > 0$) or right (if $c < 0$). Let's look at another example.

E X A M P L E 4   *Horizontal Shift to the Left*

Use the graph $f(x) = x^2$ to obtain the graph of $h(x) = (x + 4)^2$.

Solution   Again, the function $h(x) = (x + 4)^2$ is basically a square function. Thus, its graph is the same as that of f, except it is shifted 4 units to the left. (Do you see why?) See Figure 25.

FIGURE 25

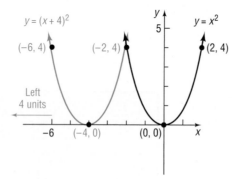

    Seeing the Concept On the same screen, graph each of the following functions:

$$y = x^2, \quad y = (x - 1)^2, \quad y = (x - 3)^2, \quad y = (x + 2)^2$$

See Figure 26.

FIGURE 26

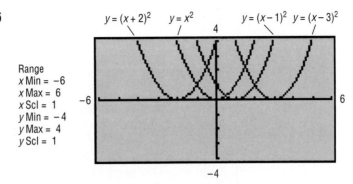

Range
x Min = -6
x Max = 6
x Scl = 1
y Min = -4
y Max = 4
y Scl = 1

■ Now work Problem 27.

Vertical and horizontal shifts are sometimes combined.

EXAMPLE 5   *Combining Vertical and Horizontal Shifts*

Graph the function: $f(x) = (x - 1)^3 + 3$

Solution   We graph f in steps. First, we note that the rule for f is basically a cube function. Thus, we begin with the graph of $y = x^3$. See Figure 27(a). Next, to get the graph of $y = (x - 1)^3$, we shift the graph of $y = x^3$ horizontally 1 unit to the right. See Figure 27(b). Finally, to get the graph of $y = (x - 1)^3 + 3$, we shift the graph of $y = (x - 1)^3$ vertically up 3 units. See Figure 27(c). Note the three points that have been plotted on each graph. Using key points such as these can be helpful in keeping track of just what is taking place.

FIGURE 27

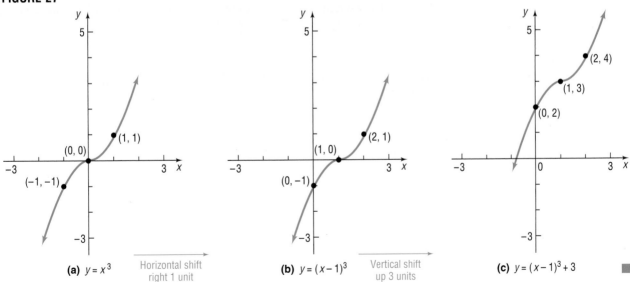

In Example 5, if the vertical shift had been done first, followed by the horizontal shift, the result would have been the same. (Try it for yourself.)

■ Now work Problem 39.

Compressions and Stretches

EXAMPLE 6

Vertical Stretch

Use the graph of $f(x) = |x|$ to obtain the graph of $g(x) = 3|x|$.

Solution

To see the relationship between the graphs of f and g, we form Table 5, listing points on each graph. For each x, the y-coordinate of a point on the graph of g is 3 times as large as the corresponding y-coordinate on the graph of f. The graph of $f(x) = |x|$ is vertically stretched by a factor of 3 [for example, from $(1, 1)$ to $(1, 3)$] to obtain the graph of $g(x) = 3|x|$. See Figure 28.

TABLE 5

x	$y = f(x)$ $= |x|$	$y = g(x)$ $= 3|x|$
-2	2	6
-1	1	3
0	0	0
1	1	3
2	2	6

FIGURE 28

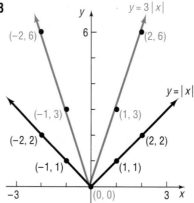

When the right side of a function $y = f(x)$ is multiplied by a positive number k, the graph of the new function $y = kf(x)$ is a vertically *compressed* (if $0 < k < 1$) or *stretched* (if $k > 1$) version of the graph of $y = f(x)$.

EXAMPLE 7

Vertical Compression

Use the graph $f(x) = |x|$ to obtain the graph of $h(x) = \frac{1}{2}|x|$.

Solution

For each x, the y-coordinate of a point on the graph of h is $\frac{1}{2}$ as large as the corresponding y-coordinate on the graph of f. The graph of $f(x) = |x|$ is vertically compressed by a factor of $\frac{1}{2}$ [for example, from $(2, 2)$ to $(2, 1)$] to obtain the graph of $h(x) = \frac{1}{2}|x|$. See Table 6 and Figure 29.

TABLE 6

x	$y = f(x)$ $= |x|$	$y = h(x)$ $= \frac{1}{2}|x|$
-2	2	1
-1	1	$\frac{1}{2}$
0	0	0
1	1	$\frac{1}{2}$
2	2	1

FIGURE 29

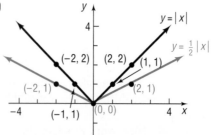

Compare Figures 28 and 29. Notice how the graph of $f(x) = |x|$ has been vertically stretched to obtain the graph $g(x) = 3|x|$, while it has been vertically compressed to obtain the graph of $h(x) = \frac{1}{2}|x|$. To put it another way, for a given x, the graph of $g(x) = 3|x|$ has larger y values than the graph of $f(x) = |x|$, while the graph of $h(x) = \frac{1}{2}|x|$ has smaller y values than that of f.

■ Now work Problem 29.

If the argument x of a function $y = f(x)$ is multiplied by a positive number k, the graph of the new function $y = f(kx)$ is also a compressed or stretched version of the graph of $y = f(x)$, but, in this case, it occurs horizontally rather than vertically. To see why, we look at the following example.

E X A M P L E   8

Horizontal Compression

Use the graph of $f(x) = \sqrt{x}$ to obtain the graph of $g(x) = \sqrt{2x}$.

Solution

The graph of f is familiar to us and is shown in Figure 30. Now we list some points on the graphs of f and g in Table 7. As the table indicates, the graph of g increases more rapidly than that of f. That is, the graph of g is compressed horizontally toward the y-axis. See Figure 30.

TABLE 7

x	$y = f(x)$ $= \sqrt{x}$	$y = g(x)$ $= \sqrt{2x}$
0	0	0
1	1	$\sqrt{2}$
2	$\sqrt{2}$	2
4	2	$2\sqrt{2}$

FIGURE 30

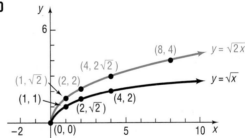

Note that since $\sqrt{2x} = \sqrt{2}\sqrt{x}$ the graph of $g(x) = \sqrt{2x}$ can also be viewed as a vertical stretch of the graph of $f(x) = \sqrt{x}$. ■

Reflections about the *x*-Axis and the *y*-Axis

E X A M P L E   9

Reflection about the x-Axis

Graph the function: $f(x) = -x^2$

Solution

We begin with the graph of $y = x^2$, as shown in Figure 31. For each point (x, y) on the graph of $y = x^2$, the point $(x, -y)$ is on the graph of $y = -x^2$, as indicated in Table 8. Thus, we can draw the graph of $y = -x^2$ by reflecting the graph of $y = x^2$ about the x-axis. See Figure 31.

TABLE 8

x	$y = x^2$	$y = -x^2$
-2	4	-4
-1	1	-1
0	0	0
1	1	-1
2	4	-4

FIGURE 31

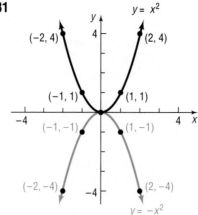

When the right side of the equation $y = f(x)$ is multiplied by -1, the graph of the new function $y = -f(x)$ is the **reflection about the x-axis** of the graph of the function $y = f(x)$.

■ Now work Problem 33.

E X A M P L E   1 0   *Reflection about the y-Axis*

Graph the function: $f(x) = \sqrt{-x}$

Solution   First, notice that the domain of f consists of all real numbers x for which $-x \geq 0$, or for which $x \leq 0$. To get the graph of $f(x) = \sqrt{-x}$, we begin with the graph of $y = \sqrt{x}$, as shown in Figure 32. For each point (x, y) on the graph of $y = \sqrt{x}$, the point $(-x, y)$ is on the graph of $y = \sqrt{-x}$. Thus, we get the graph of $y = \sqrt{-x}$ by reflecting the graph of $y = \sqrt{x}$ about the y-axis. See Figure 32.

FIGURE 32

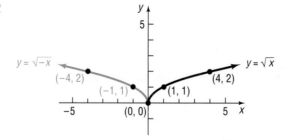

When the graph of the function $y = f(x)$ is known, the graph of the new function $y = f(-x)$ is the **reflection about the y-axis** of the graph of the function $y = f(x)$.

Summary of Graphing Techniques

Table 9 summarizes the graphing procedures we have just discussed.

TABLE 9	To Graph:	Draw the Graph of f and:
	Vertical shifts	
	$y = f(x) + c, \quad c > 0$	Raise the graph of f by c units.
	$y = f(x) - c, \quad c > 0$	Lower the graph of f by c units.
	Horizontal shifts	
	$y = f(x + c), \quad c > 0$	Shift the graph of f to the left c units.
	$y = f(x - c), \quad c > 0$	Shift the graph of f to the right c units.
	Compressing or stretching	
	$y = kf(x), \quad k > 0$	Compress or stretch the graph of f by a factor of k.
	$y = f(kx), \quad k > 0$	
	Reflection about the x-axis	
	$y = -f(x)$	Reflect the graph of f about the x-axis.
	Reflection about the y-axis	
	$y = f(-x)$	Reflect the graph of f about the y-axis.

The examples that follow combine some of the procedures outlined in this section to get the required graph.

E X A M P L E   1 1   *Combining Graphing Procedures*

Graph the function: $f(x) = \dfrac{3}{x - 2} + 1$

Solution   We use the following steps to obtain the graph of f:

STEP 1: $y = \dfrac{1}{x}$ Reciprocal function

STEP 2: $y = \dfrac{3}{x}$ Vertical stretch of the graph of $y = \dfrac{1}{x}$ by a factor of 3

STEP 3: $y = \dfrac{3}{x - 2}$ Horizontal shift to the right 2 units; replace x by $x - 2$

STEP 4: $y = \dfrac{3}{x - 2} + 1$ Vertical shift up 1 unit; add 1

See Figure 33.

FIGURE 33

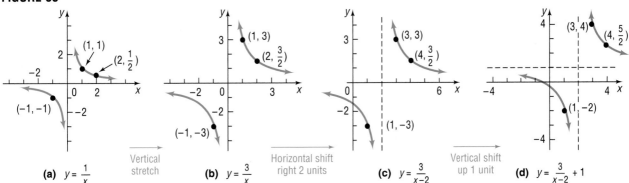

(a) $y = \dfrac{1}{x}$ Vertical stretch   (b) $y = \dfrac{3}{x}$ Horizontal shift right 2 units   (c) $y = \dfrac{3}{x-2}$ Vertical shift up 1 unit   (d) $y = \dfrac{3}{x-2} + 1$

There are other orderings of the steps shown in Example 11 that would also result in the graph of f. For example, try this one;

STEP 1: $y = \dfrac{1}{x}$ Reciprocal function

STEP 2: $y = \dfrac{1}{x - 2}$ Horizontal shift to the right 2 units; replace x by $x - 2$

STEP 3: $y = \dfrac{3}{x - 2}$ Vertical stretch of the graph of $y = \dfrac{1}{x - 2}$ by factor of 3

STEP 4: $y = \dfrac{3}{x - 2} + 1$ Vertical shift up 1 unit; add 1

■ Now work Problem 41.

E X A M P L E   1 2   *Combining Graphing Procedures*

Graph the function: $f(x) = \sqrt{1 - x} + 2$

Solution   We use the following steps to get the graph of $y = \sqrt{1 - x} + 2$:

STEP 1: $y = \sqrt{x}$ Square root function
STEP 2: $y = \sqrt{x + 1}$ Replace x by $x + 1$; horizontal shift left 1 unit
STEP 3: $y = \sqrt{-x + 1} = \sqrt{1 - x}$ Replace x by $-x$; reflect about y-axis
STEP 4: $y = \sqrt{1 - x} + 2$ Vertical shift up 2 units

See Figure 34.

FIGURE 34

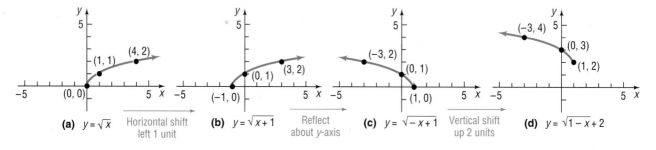

(a) $y = \sqrt{x}$ Horizontal shift left 1 unit   (b) $y = \sqrt{x + 1}$ Reflect about y-axis   (c) $y = \sqrt{-x + 1}$ Vertical shift up 2 units   (d) $y = \sqrt{1 - x} + 2$

2.3

Exercise 2.3

In Problems 1–12, match each graph to one of the following functions:

A. $y = x^2 + 2$ B. $y = -x^2 + 2$ C. $y = |x| + 2$ D. $y = -|x| + 2$

E. $y = (x - 2)^2$ F. $y = -(x + 2)^2$ G. $y = |x - 2|$ H. $y = -|x + 2|$

I. $y = 2x^2$ J. $y = -2x^2$ K. $y = 2|x|$ L. $y = -2|x|$

1.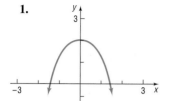

2.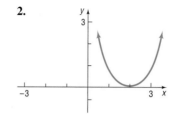

3.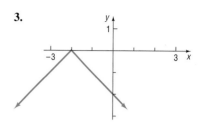

4.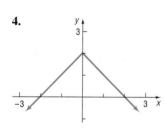

5.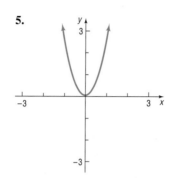

6.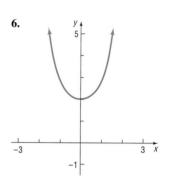

7.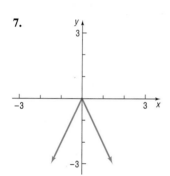

8.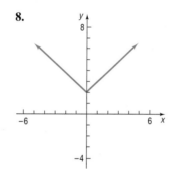

9.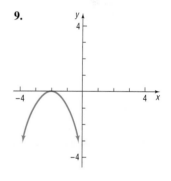

10.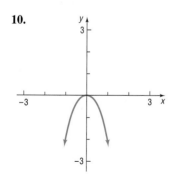

11.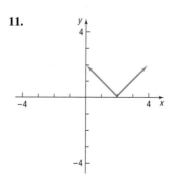

12.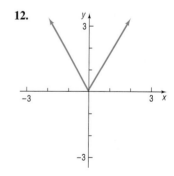

In Problems 13–20, use the function $f(x) = x^3$. Write the function whose graph is the graph of $y = x^3$, but is:

13. Shifted to the right 4 units

14. Shifted to the left 4 units

15. Shifted up 4 units

16. Shifted down 4 units.

17. Reflected about the y-axis

18. Reflected about the x-axis

19. Vertically stretched by a factor of 4

20. Horizontally stretched by a factor of 4

In Problems 21–50, graph each function using the techniques of shifting, compressing, stretching, and/or reflecting. Start with the graph of the basic function (for example, $y = x^2$) and show all stages.

21. $f(x) = x^2 - 1$

22. $f(x) = x^2 + 4$

23. $g(x) = x^3 + 1$

24. $g(x) = x^3 - 1$

25. $h(x) = \sqrt{x - 2}$

26. $h(x) = \sqrt{x + 1}$

27. $f(x) = (x - 1)^3$

28. $f(x) = (x + 2)^3$

29. $g(x) = 4\sqrt{x}$

30. $g(x) = \frac{1}{2}\sqrt{x}$

31. $h(x) = \dfrac{1}{2x}$

32. $h(x) = \dfrac{4}{x}$

33. $f(x) = -|x|$

34. $f(x) = -\sqrt{x}$

35. $g(x) = -\dfrac{1}{x}$

36. $g(x) = -x^3$

37. $h(x) = [[-x]]$

38. $h(x) = \dfrac{1}{-x}$

39. $f(x) = (x + 1)^2 - 3$

40. $f(x) = (x - 2)^2 + 1$

41. $g(x) = \sqrt{x - 2} + 1$

42. $g(x) = |x + 1| - 3$

43. $h(x) = \sqrt{-x} - 2$

44. $h(x) = \dfrac{4}{x} + 2$

45. $f(x) = (x + 1)^3 - 1$

46. $f(x) = 4\sqrt{x - 1}$

47. $g(x) = 2|1 - x|$

48. $g(x) = 4\sqrt{2 - x}$

49. $h(x) = 2[[x - 1]]$

50. $h(x) = -x^3 + 2$

In Problems 51–56, the graph of a function f is illustrated. Use the graph of f as the first step toward graphing each of the following functions:

(a) $F(x) = f(x) + 3$

(b) $G(x) = f(x + 2)$

(c) $P(x) = -f(x)$

(d) $Q(x) = \frac{1}{2}f(x)$

(e) $g(x) = f(-x)$

(f) $h(x) = 3f(x)$

51.
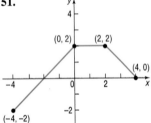

52.
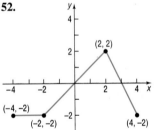

53.
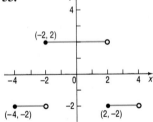

54.
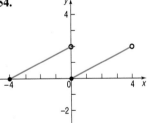

55.
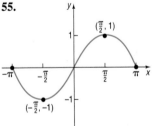

56.
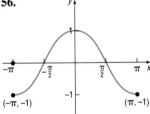

In Problems 57–62, complete the square of each quadratic expression. Then graph each function using the technique of shifting.

57. $f(x) = x^2 + 2x$

58. $f(x) = x^2 - 6x$

59. $f(x) = x^2 - 8x + 1$

60. $f(x) = x^2 + 4x + 2$

61. $f(x) = x^2 + x + 1$

62. $f(x) = x^2 - x + 1$

63. The equation $y = (x - c)^2$ defines a *family of parabolas,* one parabola for each value of c. On one set of coordinate axes, graph the members of the family for $c = 0$, $c = 3$, $c = -2$.

64. Repeat Problem 63 for the family of parabolas $y = x^2 + c$.

65. *Temperature Measurements* The relationship between Celsius (°C) and Fahrenheit (°F) for measuring temperature is given by the equation

$$F = \frac{9}{5}C + 32$$

The relationship between Celsius (°C) and Kelvin (K) is $K = C + 273$. Graph the equation $F = \frac{9}{5}C + 32$, using degrees Fahrenheit on the y-axis and degrees Celsius on the x-axis. Use the techniques introduced in this section to obtain the graph showing the relationship between Kelvin and Fahrenheit temperatures.

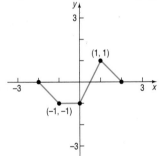

66. *Period of a Pendulum* The period T (in seconds) of a simple pendulum is a function of its length l (in feet) defined by the equation

$$T = 2\pi\sqrt{\frac{l}{g}}$$

where $g \approx 32.2$ feet per second per second is the acceleration of gravity. Graph this function using T on the y-axis and l on the x-axis. On the same coordinate axes, graph the following functions:

(a) $T = 2\pi\sqrt{\dfrac{l+2}{g}}$ (b) $T = 2\pi\sqrt{\dfrac{4l}{g}}$

Discuss how changes in the length l affect the period T of the pendulum.

67. The graph of a function f is illustrated in the figure.

(a) Draw the entire graph of $y = |f(x)|$.

(b) Draw the entire graph of $y = f(|x|)$.

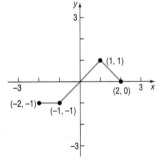

68. Repeat Problem 67 for the graph shown.

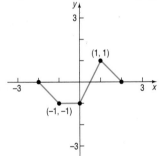

69. (a) Graph $y = x + 1$ and $y = |x + 1|$.

(b) Graph $y = 4 - x^2$ and $y = |4 - x^2|$.

(c) Graph $y = x^3 + x$ and $y = |x^3 + x|$.

(d) What do you conclude about the relationship between the graphs of $y = f(x)$ and $y = |f(x)|$?

70. (a) Graph $y = x + 1$ and $y = |x| + 1$.

(b) Graph $y = 4 - x^2$ and $y = 4 - |x|^2$.

(c) Graph $y = x^3 + x$ and $y = |x|^3 + |x|$.

(d) What do you conclude about the relationship between the graphs of $y = f(x)$ and $y = f(|x|)$?

2.4

Operations on Functions; Composite Functions

In this section, we introduce some operations on functions. We shall see that functions, like numbers, can be added, subtracted, multiplied, and divided. For example, if $f(x) = x^2 + 9$ and $g(x) = 3x + 5$, then

$$f(x) + g(x) = (x^2 + 9) + (3x + 5) = x^2 + 3x + 14$$

The new function $y = x^2 + 3x + 14$ is called the *sum function* $f + g$. Similarly,

$$f(x) \cdot g(x) = (x^2 + 9)(3x + 5) = 3x^3 + 5x^2 + 27x + 45$$

The new function $y = 3x^3 + 5x^2 + 27x + 45$ is called the *product function* $f \cdot g$. The general definitions are given next.

Sum Function

If f and g are functions:
Their **sum f + g** is the function defined by

$$(f + g)(x) = f(x) + g(x)$$

Difference Function

Their **difference f − g** is the function defined by

$$(f - g)(x) = f(x) - g(x)$$

Product Function

Their **product f · g** is the function defined by

$$(f \cdot g)(x) = f(x) \cdot g(x)$$

Quotient Function

Their **quotient f/g** is the function defined by

$$\left(\frac{f}{g} \right)(x) = \frac{f(x)}{g(x)}$$

In each case, the domain of the resulting function consists of the numbers x that are common to the domains of f and g, but the numbers x for which $g(x) = 0$ must be excluded from the domain of the quotient f/g.

Thus, the sum function, $f + g$, is defined as the sum of the values of the functions f and g, and so on.

E X A M P L E   1   *Operations on Functions*

Let f and g be two functions defined as

$$f(x) = \sqrt{x + 2} \quad \text{and} \quad g(x) = \sqrt{x - 3}$$

Find the following, and determine the domain in each case:

(a) $(f + g)(x)$ (b) $(f - g)(x)$ (c) $(f \cdot g)(x)$ (d) $(f/g)(x)$

Solution

(a) $(f + g)(x) = f(x) + g(x) = \sqrt{x + 2} + \sqrt{x - 3}$

(b) $(f - g)(x) = f(x) - g(x) = \sqrt{x + 2} - \sqrt{x - 3}$

(c) $(f \cdot g)(x) = f(x) \cdot g(x) = (\sqrt{x + 2})(\sqrt{x - 3}) = \sqrt{(x + 2)(x - 3)}$

(d) $\left(\dfrac{f}{g}\right)(x) = \dfrac{f(x)}{g(x)} = \dfrac{\sqrt{x + 2}}{\sqrt{x - 3}} = \sqrt{\dfrac{x + 2}{x - 3}}$

The domain of f consists of all numbers x for which $x \geq -2$; the domain of g consists of all numbers x for which $x \geq 3$. The numbers x common to both these domains are those for which $x \geq 3$. As a result, the numbers x for which $x \geq 3$ comprise the domain of the sum function $f + g$, the difference function $f - g$, and the product function $f \cdot g$. For the quotient function f/g, we must exclude from this set the number 3, because the denominator, g, has the value 0 when $x = 3$. Thus, the domain of f/g consists of all x for which $x > 3$.   ∎

■   Now work Problem 1.

It is sometimes helpful to view a complicated function as the sum, difference, product, or quotient of simpler functions. For example,

$F(x) = x^2 + \sqrt{x}$ is the sum of $f(x) = x^2$ and $g(x) = \sqrt{x}$.

$H(x) = (x^2 - 1)/(x^2 + 1)$ is the quotient of $f(x) = x^2 - 1$ and $g(x) = x^2 + 1$.

One use of this view of functions is to obtain a graph. The next example illustrates this graphing technique when the function to be graphed is the sum of two simpler functions. In this instance, the method used is called **adding y-coordinates.**

E X A M P L E   2

Graphing by Adding y-Coordinates

Graph the function: $F(x) = x + \sqrt{x}$

Solution

First, we notice that the domain of F is $x \geq 0$. Next, we graph the two functions $f(x) = x$ and $g(x) = \sqrt{x}$ for $x \geq 0$. See Figures 35(a) and 35(b). To plot a point $(x, F(x))$ on the graph of F, we select a nonnegative number x and add the y-coordinates $f(x)$ and $g(x)$ to get the y-coordinate $F(x) = f(x) + g(x)$. For example, when $x = 1$, then $f(1) = 1$, $g(1) = 1$, and $F(1) = f(1) + g(1) = 1 + 1 = 2$. When $x = 4$, then $f(4) = 4$, $g(4) = 2$, and $F(4) = f(4) + g(4) = 4 + 2 = 6$, and so on. See Table 10. Figure 35(c) illustrates the graph of F.

FIGURE 35

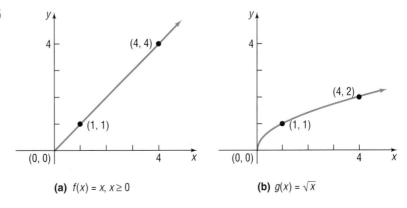

(a) $f(x) = x, x \geq 0$ (b) $g(x) = \sqrt{x}$

TABLE 10

x	$f(x) = x$	$g(x) = \sqrt{x}$	$F(x) = x + \sqrt{x}$
0	0	0	0
1	1	1	2
4	4	2	6

FIGURE 35 (continued)

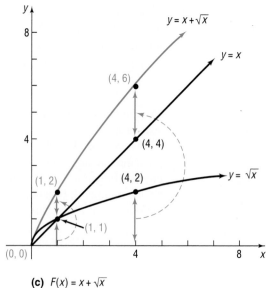

(c) $F(x) = x + \sqrt{x}$

Check: Graph the functions $y = x$, $y = \sqrt{x}$, and $F(x) = x + \sqrt{x}$ and compare what you see with Figure 35(c).   ■

Composite Functions

Consider the function $y = (2x + 3)^2$. If we write $y = f(u) = u^2$ and $u = g(x) = 2x + 3$, then, by a substitution process, we can obtain the original function: $y = f(u) = f(g(x)) = (2x + 3)^2$. This process is called **composition.** In general, suppose that f and g are two functions, and suppose that x is a number in the domain of g. By evaluating g at x, we get $g(x)$. If $g(x)$ is in the domain of f, then we may evaluate f at $g(x)$ and thereby obtain the expression $f(g(x))$. If we do this for all x such that x is in the domain of g and $g(x)$ is in the domain of f, the resulting correspondence from x to $f(g(x))$ is called a *composite function*. See Figure 36.

FIGURE 36

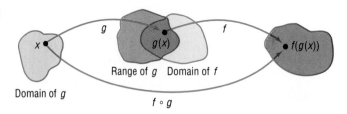

Composite Function

Given the two functions f and g, the **composite function,** denoted by $f \circ g$ (read as "f composed with g"), is defined by

$$(f \circ g)(x) = f(g(x))$$

where the domain of $f \circ g$ is the set of all numbers x in the domain of g such that $g(x)$ is in the domain of f.

ISSION POSSIBLE

Chapter 2

Consulting for the Silver Satellite & Cable TV Co.

Your team works for the Silver Satellite & Cable TV Company in the Research & Development Department. You've been asked to come up with a formula to determine the cost of running cable from a connection box to a new cable household. The first example you are working with involves the Stevens family who own a rural home with a driveway two miles long extending to the house from a nearby highway. The nearest connection box is along the highway but 5 miles from the driveway.

It costs the company $10 per mile to install cable along the highway and $14 per mile to install cable off the highway. Because the Steven house is surrounded by farmland which they own, it would be possible to run the cable overland to the house directly from the connection box or from any point between the connection box to the driveway.

1. Draw a sketch of this problem situation, assuming the highway is a straight road and the driveway is also a straight road perpendicular to the highway. Include two or more possible routes for the cable.
2. Suppose x represents the distance in miles the cable runs along the highway from the connection box before turning off toward the house. Express the total cost of installation as a function of x. (You may choose to answer #3 before #2 if you would like to examine concrete instances before creating the equation.)
3. Make a table of the possible integral values of x and the corresponding cost in each instance. Is there one choice which appears to cost the least?
4. If you charge the Stevens $80 for installation, would you be willing to let them choose which way the cable would go? Explain.
5. If you have a graphing calculator, graph the function from #2 and determine if there is a non-integral choice for x that would make the installation cost even cheaper. Zoom and trace until you have the lowest possible cost.
6. Before proceeding further with the installation, you check the local regulations for cable companies and find that there is a pending state legislation that says the cable cannot turn off the highway more than .5 miles from the Stevens' driveway. If this legislation passes, what will be the ultimate cost of installing the Stevens' cable?
7. If the cable company wishes to install cable in 5000 homes in this area, and assuming the figures for the Stevens installation are typical, how much will the new legislation cost the company over all if they cannot use the cheapest installation cost, but instead have to follow the new state regulations?

Figure 37 provides a second illustration of the definition. Notice that the "in-side" function g in $f(g(x))$ is done first.

FIGURE 37

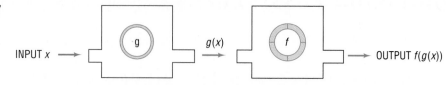

Let's look at some examples.

E X A M P L E   3

Evaluating a Composite Function

Suppose that $f(x) = 2x^2 - 3$ and $g(x) = 4x$. Find:

(a) $(f \circ g)(1)$ (b) $(g \circ f)(1)$ (c) $(f \circ f)(-2)$ (d) $(g \circ g)(-1)$

Solution

(a) $(f \circ g)(1) = f(g(1)) = f(4) = 2 \cdot 16 - 3 = 29$

$$\underset{\underset{g(1) = 4}{g(x) = 4x}}{\uparrow} \quad \underset{f(x) = 2x^2 - 3}{\uparrow}$$

(b) $(g \circ f)(1) = g(f(1)) = g(-1) = 4 \cdot (-1) = -4$

$$\underset{\underset{f(1) = -1}{f(x) = 2x^2 - 3}}{\uparrow} \quad \underset{g(x) = 4x}{\uparrow}$$

(c) $(f \circ f)(-2) = f(f(-2)) = f(5) = 2 \cdot 25 - 3 = 47$

$$\underset{f(-2) = 5}{\uparrow}$$

(d) $(g \circ g)(-1) = g(g(-1)) = g(-4) = 4 \cdot (-4) = -16$

$$\underset{g(-1) = -4}{\uparrow}$$

■ Now work Problem 17.

E X A M P L E   4

Finding a Composite Function

Suppose that $f(x) = \sqrt{x}$ and $g(x) = x^3 - 1$. Find the following composite functions, and then find the domain of each composite function:

(a) $f \circ g$ (b) $g \circ f$ (c) $f \circ f$ (d) $g \circ g$

Solution

(a) $(f \circ g)(x) = f(g(x)) = f(x^3 - 1) = \sqrt{x^3 - 1}$

The domain of $f \circ g$ is the interval $[1, \infty)$, which is found by determining those x in the domain of g for which $x^3 - 1 \geq 0$.

(b) $(g \circ f)(x) = g(f(x)) = g(\sqrt{x}) = (\sqrt{x})^3 - 1 = x^{3/2} - 1$

The domain of $g \circ f$ is $[0, \infty)$.

(c) $(f \circ f)(x) = f(f(x)) = f(\sqrt{x}) = \sqrt{\sqrt{x}} = \sqrt[4]{x}$

The domain of $f \circ f$ is $[0, \infty)$.

(d) $(g \circ g)(x) = g(g(x)) = g(x^3 - 1) = (x^3 - 1)^3 - 1$

The domain of $g \circ g$ is the set of all real numbers.

■ Now work Problem 27.

Examples 4(a) and 4(b) illustrate that, in general, $f \circ g \neq g \circ f$. However, sometimes $f \circ g$ does equal $g \circ f$, as shown in the next example.

EXAMPLE 5

Showing Two Composite Functions Equal

If $f(x) = 3x - 4$ and $g(x) = \frac{1}{3}(x + 4)$, show that $(f \circ g)(x) = (g \circ f)(x) = x$ for every x.

Solution

$$(f \circ g)(x) = f(g(x))$$

$$= f\left(\frac{x + 4}{3}\right) \qquad g(x) = \frac{1}{3}(x + 4) = \frac{x + 4}{3}$$

$$= 3\left(\frac{x + 4}{3}\right) - 4 \qquad \text{Substitute } g(x) \text{ into the rule for } f, \ f(x) = 3x - 4.$$

$$= x + 4 - 4 = x$$

$$(g \circ f)(x) = g(f(x))$$

$$= g(3x - 4) \qquad f(x) = 3x - 4$$

$$= \frac{1}{3}[(3x - 4) + 4)] \qquad \text{Substitute } f(x) \text{ into the rule for } g, \ g(x) = \frac{1}{3}(x + 4).$$

$$= \frac{1}{3}(3x) = x$$

Thus, $(f \circ g)(x) = (g \circ f)(x) = x$. ∎

In the next section, we shall see that there is an important relationship between functions f and g for which $(f \circ g)(x) = (g \circ f)(x) = x$.

◼ Now work Problem 41.

Calculus Application

Some techniques in calculus require that we be able to determine the components of a composite function. For example, the function $H(x) = \sqrt{x + 1}$ is the composition of the functions f and g, where $f(x) = \sqrt{x}$ and $g(x) = x + 1$, because $H(x) = (f \circ g)(x) = f(g(x)) = f(x + 1) = \sqrt{x + 1}$.

EXAMPLE 6

Finding the Components of a Composite Function

Find functions f and g such that $f \circ g = H$ if $H(x) = (x^2 + 1)^{50}$.

Solution

The function H takes $x^2 + 1$ and raises it to the power 50. A natural way to decompose H is to raise the function $g(x) = x^2 + 1$ to the power 50. Thus, if we let $f(x) = x^{50}$ and $g(x) = x^2 + 1$, then

FIGURE 38

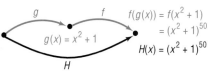

$$(f \circ g)(x) = f(g(x))$$

$$= f(x^2 + 1)$$

$$= (x^2 + 1)^{50} = H(x)$$

See Figure 38. ∎

Other functions f and g may be found for which $f \circ g = H$ in Example 6. For example, if $f(x) = x^2$ and $g(x) = (x^2 + 1)^{25}$, then

$$(f \circ g)(x) = f(g(x)) = f((x^2 + 1)^{25}) = [(x^2 + 1)^{25}]^2 = (x^2 + 1)^{50}$$

Thus, although the functions f and g found as a solution to Example 6 are not unique, there is usually a "natural" selection for f and g that comes to mind first. This natural selection usually will enable you to use your calculator most efficiently. Let's look again at Example 6. To calculate the value of H at, say, 2, we proceed as follows:

Keystrokes: | 2 | | x^2 | | + | 1 | | = | | x^y | 50 | | = |

Display: | | 2 | | 4 | | | 1 | | 5 | | 50 | | 8.8818   E34 |

EXAMPLE 7 *Finding the Components of a Composite Function*

Find functions f and g such that $f \circ g = H$ if $H(x) = 1/(x + 1)$.

Solution   Here, H is the reciprocal of $g(x) = x + 1$. Thus, if we let $f(x) = 1/x$ and $g(x) = x + 1$, we find that

$$(f \circ g)(x) = f(g(x)) = f(x + 1) = \frac{1}{x + 1} = H(x)$$ ∎

2.4

Exercise 2.4

In Problems 1–10, for the given functions f and g, find the following functions and state the domain of each:

(a) $f + g$ (b) $f - g$ (c) $f \cdot g$ (d) f/g

1. $f(x) = 3x + 4;$ $g(x) = 2x - 3$ 2. $f(x) = 2x + 1;$ $g(x) = 3x - 2$

3. $f(x) = x - 1;$ $g(x) = 2x^2$ 4. $f(x) = 2x^2 + 3;$ $g(x) = 4x^3 + 1$

5. $f(x) = \sqrt{x};$ $g(x) = 3x - 5$ 6. $f(x) = |x|;$ $g(x) = x$

7. $f(x) = 1 + \dfrac{1}{x};$ $g(x) = \dfrac{1}{x}$ 8. $f(x) = 2x^2 - x;$ $g(x) = 2x^2 + x$

9. $f(x) = \dfrac{2x + 3}{3x - 2};$ $g(x) = \dfrac{4x}{3x - 2}$ 10. $f(x) = \sqrt{x + 1};$ $g(x) = \dfrac{2}{x}$

11.   Given $f(x) = 3x + 1$ and $(f + g)(x) = 6 - \frac{1}{2}x$, find the function g.

12.   Given $f(x) = 1/x$ and $(f/g)(x) = (x + 1) / (x^2 - x)$, find the function g.

In Problems 13–16, use the method of adding y-coordinates to graph each function on the interval [0, 2].

13. $f(x) = |x| + x^2$ 14. $f(x) = |x| + \sqrt{x}$ 15. $f(x) = x^3 + x$ 16. $f(x) = x^3 + x^2$

In Problems 17–26, for the given functions f and g, find:

(a) $(f \circ g)(4)$ (b) $(g \circ f)(2)$ (c) $(f \circ f)(1)$ (d) $(g \circ g)(0)$

17. $f(x) = 2x;$ $g(x) = 3x^2 + 1$ 18. $f(x) = 3x + 2;$ $g(x) = 2x^2 - 1$

19. $f(x) = 4x^2 - 3;$ $g(x) = 3 - \frac{1}{2}x^2$ 20. $f(x) = 2x^2;$ $g(x) = 1 - 3x^2$

21. $f(x) = \sqrt{x};$ $g(x) = 2x$ 22. $f(x) = \sqrt{x + 1};$ $g(x) = 3x$

23. $f(x) = |x|;$ $g(x) = \dfrac{1}{x^2 + 1}$ 24. $f(x) = |x - 2|;$ $g(x) = \dfrac{3}{x^2 + 2}$

25. $f(x) = \dfrac{3}{x^2 + 1};$ $g(x) = \sqrt{x}$ 26. $f(x) = x^3;$ $g(x) = \dfrac{2}{x^2 + 1}$

In Problems 27–40, for the given functions f and g, find;

(a) $f \circ g$ (b) $g \circ f$ (c) $f \circ f$ (d) $g \circ g$

27. $f(x) = 2x + 3;$ $g(x) = 3x$

28. $f(x) = -x;$ $g(x) = 2x - 4$

29. $f(x) = 3x + 1;$ $g(x) = x^2$

30. $f(x) = \sqrt{x + 1};$ $g(x) = x + 4$

31. $f(x) = \sqrt{x};$ $g(x) = x^2 - 1$

32. $f(x) = \sqrt{x + 1};$ $g(x) = \dfrac{1}{x^2}$

33. $f(x) = \dfrac{x - 1}{x + 1};$ $g(x) = \dfrac{1}{x}$

34. $f(x) = x + \dfrac{1}{x};$ $g(x) = x^2$

35. $f(x) = x^2;$ $g(x) = \sqrt{x}$

36. $f(x) = 2x + 4;$ $g(x) = \frac{1}{2}x - 2$

37. $f(x) = \dfrac{1}{2x + 3};$ $g(x) = 2x + 3$

38. $f(x) = \dfrac{x + 1}{x - 1};$ $g(x) = \dfrac{x - 1}{x + 1}$

39. $f(x) = ax + b;$ $g(x) = cx + d$

40. $f(x) = \dfrac{ax + b}{cx + d};$ $g(x) = mx$

In Problems 41–48, show that $(f \circ g)(x) = (g \circ f)(x) = x$.

41. $f(x) = 2x;$ $g(x) = \frac{1}{2}x$

42. $f(x) = 4x;$ $g(x) = \frac{1}{4}x$

43. $f(x) = x^3;$ $g(x) = \sqrt[3]{x}$

44. $f(x) = x + 5;$ $g(x) = x - 5$

45. $f(x) = 2x - 6;$ $g(x) = \frac{1}{2}(x + 6)$

46. $f(x) = 4 - 3x;$ $g(x) = \frac{1}{3}(4 - x)$

47. $f(x) = ax + b;$ $g(x) = \dfrac{1}{a}(x - b), \, a \neq 0$

48. $f(x) = \dfrac{1}{x};$ $g(x) = \dfrac{1}{x}$

49. If $f(x) = 2x^3 - 3x^2 + 4x - 1$ and $g(x) = 2$, find $(f \circ g)(x)$ and $(g \circ f)(x)$.

50. If $f(x) = x/(x - 1)$, find $(f \circ f)(x)$.

In Problems 51–54, use $f(x) = x^2$, $g(x) = \sqrt{x} + 2$, and $h(x) = 1 - 3x$ to find the indicated composite function.

51. $f \circ (g \circ h)$ **52.** $(f \circ g) \circ h$ **53.** $(f + g) \circ h$ **54.** $(f \circ h) + (g \circ h)$

In Problems 55–62, let $f(x) = x^2$, $g(x) = 3x$, and $h(x) = \sqrt{x} + 1$. Express each function as a composite of f, g, and/or h.

55. $F(x) = 9x^2$ **56.** $G(x) = 3x^2$ **57.** $H(x) = |x| + 1$

58. $p(x) = 3\sqrt{x} + 3$ **59.** $q(x) = x + 2\sqrt{x} + 1$ **60.** $R(x) = 9x$

61. $P(x) = x^4$ **62.** $Q(x) = \sqrt{\sqrt{x} + 1} + 1$

In Problems 63–70, find functions f and g so that $f \circ g = H$.

63. $H(x) = (2x + 3)^4$ **64.** $H(x) = (1 + x^2)^{3/2}$ **65.** $H(x) = \sqrt{x^2 + x + 1}$

66. $H(x) = \dfrac{1}{1 + x^2}$ **67.** $H(x) = \left(1 - \dfrac{1}{x^2}\right)^2$ **68.** $H(x) = |2x^2 + 3|$

69. $H(x) = [[x^2 + 1]]$ **70.** $H(x) = (4 - x^2)^{-4}$

71. If $f(x) = 2x^2 + 5$ and $g(x) = 3x + a$, find a so that the graph of $f \circ g$ crosses the y-axis at 23.

72. If $f(x) = 3x^2 - 7$ and $g(x) = 2x + a$, find a so that the graph of $f \circ g$ crosses the y-axis at 68.

73. The surface area S (in square meters) of a hot air balloon is given by

$$S(r) = 4\pi r^2$$

where r is the radius of the balloon (in meters). If the radius r is increasing with time t (in seconds) according to the formula $r(t) = \frac{2}{3}t^3$, $t \geq 0$, find the surface area S of the balloon as a function of the time t.

74. The volume V (in cubic meters) of the hot air balloon described in Problem 73 is given by $V(r) = \frac{4}{3}\pi r^3$. If the radius r is the same function of t as in Problem 73, find the volume V as a function of the time t.

75. *Automobile Production* The number N of cars produced at a certain factory in 1 day after t hours of operation is given by $N(t) = 100t - 5t^2$, $0 \le t \le 10$. If the cost C (in dollars) of producing x cars is $C(x) = 15,000 + 8000x$, find the cost C as a function of the time t of operation of the factory.

76. *Environmental Concerns* The spread of oil leaking from a tanker is in the shape of a circle. If the radius r (in feet) of the spread after t hours is $r(t) = 200\sqrt{t}$, find the area A of the oil slick as a function of the time t.

77. *Production Cost* The price p of a certain product and the quantity x sold obey the demand equation

$$p = -\frac{1}{4}x + 100 \qquad 0 \le x \le 400$$

Suppose that the cost C of producing x units is

$$C = \frac{\sqrt{x}}{25} + 600$$

Assuming that all items produced are sold, find the cost C as a function of the price p. [*Hint:* Solve for x in the demand equation, and then form the composite.]

78. *Cost of a Commodity* The price p of a certain commodity and the quantity x sold obey the demand equation

$$p = -\frac{1}{5}x + 200 \qquad 0 \le x \le 1000$$

Suppose that the cost C of producing x units is

$$C = \frac{\sqrt{x}}{10} + 400$$

Assuming that all items produced are sold, find the cost C as a function of the price p.

79. If f and g are odd functions, show that the composite function $f \circ g$ is also odd.

80. If f is an odd function and g is an even function, show that the composite functions $f \circ g$ and $g \circ f$ are also even.

2.5

One-to-One Functions; Inverse Functions

Suppose that (x_1, y_1) and (x_2, y_2) are any two *distinct* points on the graph of a function $y = f(x)$. Then it follows that $x_1 \ne x_2$. For some functions, it also happens that the y-coordinates of distinct points are always unequal. Such functions are called *one-to-one* functions. See Figure 39.

FIGURE 39

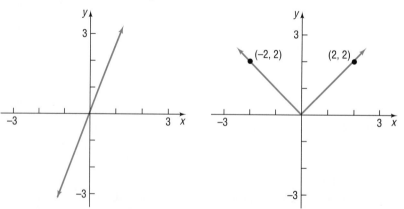

(a) $f(x) = 2x$
One-to-one:
Every distinct point has
a different y-coordinate

(b) $g(x) = |x|$
Not one-to-one:
The distinct points $(-2, 2)$ and $(2, 2)$
have the same y-coordinate

| One-to-One Function | A function *f* is said to be **one-to-one** if, for any choice of numbers x_1 and x_2, $x_1 \neq x_2$, in the domain of *f*, then $f(x_1) \neq f(x_2)$. |

In other words, if *f* is a one-to-one function, then for each *x* in the domain of *f*, there is exactly one *y* in the range, and no *y* in the range is the image of more than one *x* in the domain. See Figure 40.

FIGURE 40

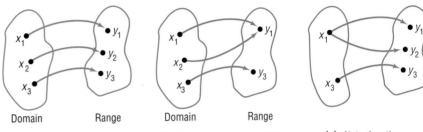

(a) One-to-one function: Each *x* in the domain has one and only one image in the range

(b) Not a one-to-one function: y_1 is the image of both x_1 and x_2

(c) Not a function: x_1 has two images, y_1 and y_2

As Figure 41 illustrates, if the graph of a function *f* is known, there is a simple test, called the **horizontal-line test,** to determine whether *f* is one-to-one.

Theorem
Horizontal-Line Test

If horizontal lines intersect the graph of a function *f* in at most one point, then *f* is one-to-one.   ■

The reason this test works can be seen in Figure 41, where the horizontal line $y = h$ intersects the graph at two distinct points, (x_1, h) and (x_2, h), with the same second element. Thus, *f* is not one-to-one.

FIGURE 41

$f(x_1) = f(x_2) = h$, but $x_1 \neq x_2$; *f* is not a one-to-one function

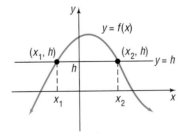

E X A M P L E   1   *Using the Horizontal-line Test*

For each given function, use the graph to determine whether the function is one-to-one.

(a) $f(x) = x^2$

(b) $g(x) = x^3$

FIGURE 42

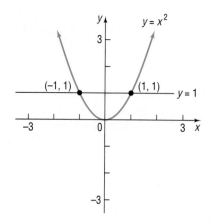

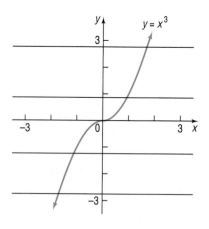

(a) A horizontal line intersects the graph twice; thus, f is not one-to-one

(b) Horizontal lines intersect the graph exactly once; thus, g is one-to-one

Solution   (a) Figure 42(a) illustrates the horizontal-line test for $f(x) = x^2$. The horizontal line $y = 1$ meets the graph of f twice, at $(1, 1)$ and at $(-1, 1)$, so f is not one-to-one.

(b) Figure 42(b) illustrates the horizontal-line test for $g(x) = x^3$. Because each horizontal line will intersect the graph of g exactly once, it follows that g is one-to-one.

■   Now work Problem 1.

Let's look more closely at the one-to-one function $g(x) = x^3$. This function is an increasing function. Because an increasing (or decreasing) function will always have different y values for unequal x values, it follows that a function that is increasing (or decreasing) on its domain is also a one-to-one function.

Theorem   An increasing (decreasing) function is a one-to-one function.   ■

Inverse of a Function

We mentioned earlier that a function $y = f(x)$ can be thought of as a rule that tells us to do something to the argument x. For example, the function $f(x) = 2x$ multiplies the argument by 2. An *inverse function* of f undoes whatever f does. For example, the function $g(x) = \frac{1}{2}x$, which divides the argument by 2, is an inverse of $f(x) = 2x$. See Figure 43.

FIGURE 43

$$g(f(x)) = g(2x) = \tfrac{1}{2}(2x) = x$$

x $f(x) = 2x$ $g(2x) = \tfrac{1}{2}(2x) = x$

For a function $y = f(x)$ to have an inverse function, f must be one-to-one. Then for each x in its domain there is exactly one y in its range; furthermore, to each y in the range, there corresponds exactly one x in the domain. The correspondence from the range of f onto the domain of f is, therefore, also a function. It is this function that is the *inverse of f*. A definition is given next.

Inverse of f

Let f denote a one-to-one function $y = f(x)$. The **inverse of f,** denoted by f^{-1}, is a function such that $f^{-1}(f(x)) = x$ for every x in the domain of f and $f(f^{-1}(x)) = x$ for every x in the domain of f^{-1}.

Warning: Be careful! The -1 used in f^{-1} is not an exponent. Thus, f^{-1} does *not* mean the reciprocal of f; it means the inverse of f.

Figure 44 illustrates the definition.
Two facts are now apparent about a function f and its inverse f^{-1}.

FIGURE 44

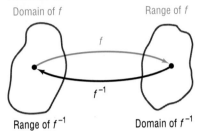

Domain of f Range of f

Range of f^{-1} Domain of f^{-1}

Domain of f = Range of f^{-1} Range of f = Domain of f^{-1}

Look again at Figure 44 to visualize the relationship. If we start with x, apply f, and then apply f^{-1}, we get x back again. If we start with x, apply f^{-1}, and then apply f, we get the number x back again. To put it simply, what f does, f^{-1} undoes, and vice versa:

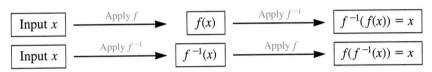

In other words,

$$f^{-1}(f(x)) = x \quad \text{and} \quad f(f^{-1}(x)) = x$$

The preceding conditions can be used to verify that a function is, in fact, the inverse of f, as Example 2 demonstrates.

EXAMPLE 2   *Verifying Inverse Functions*

(a) We verify that the inverse of $g(x) = x^3$ is $g^{-1}(x) = \sqrt[3]{x}$ by showing that

$$g^{-1}(g(x)) = g^{-1}(x^3) = \sqrt[3]{x^3} = x$$

and

$$g(g^{-1}(x)) = g(\sqrt[3]{x}) = (\sqrt[3]{x})^3 = x$$

(b) We verify that the inverse of $h(x) = 3x$ is $h^{-1}(x) = \frac{1}{3}x$ by showing that

$$h^{-1}(h(x)) = h^{-1}(3x) = \frac{1}{3}(3x) = x$$

and

$$h(h^{-1}(x)) = h(\tfrac{1}{3}x) = 3(\tfrac{1}{3}x) = x$$

(c) We verify that the inverse of $f(x) = 2x + 3$ is $f^{-1}(x) = \frac{1}{2}(x - 3)$ by showing that

$$f^{-1}(f(x)) = f^{-1}(2x + 3) = \tfrac{1}{2}[(2x + 3) - 3] + 3 = \tfrac{1}{2}(2x) = x$$

and

$$f(f^{-1}(x)) = f(\tfrac{1}{2}(x - 3)) = 2[\tfrac{1}{2}(x - 3)] = (x - 3) + 3 = x. \quad ■$$

■ Now work Problem 13.

Exploration: Graph $y = x$ on a square screen, using the viewing rectangle $-3 \le x \le 7$, $-2 \le y \le 2$. Then graph $y = x^3$ followed by its inverse, $y = \sqrt[3]{x}$. What do you observe about the graphs of $y = x^3$, its inverse $y = \sqrt[3]{x}$, and the line $y = x$?
    Do the same experiment for the functions given in Problem 13. Do you see the symmetry of the graph of f and its inverse with respect to the line $y = x$?

Geometric Interpretation

Suppose that (a, b) is a point on the graph of the one-to-one function f defined by $y = f(x)$. Then $b = f(a)$. This means that $a = f^{-1}(b)$, so (b, a) is a point on the graph of the inverse function f^{-1}. The relationship between the point (a, b) on f and the point (b, a) on f^{-1} is shown in Figure 45. The line joining (a, b) and (b, a) is perpendicular to the line $y = x$ and is bisected by the line $y = x$. (Do you see why?) It follows that the point (b, a) on f^{-1} is the reflection about the line $y = x$ of the point (a, b) on f.

FIGURE 45

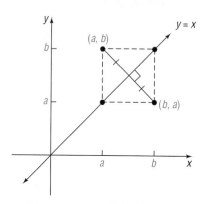

Theorem

The graph of a function f and the graph of its inverse f^{-1} are symmetric with respect to the line $y = x$. ∎

Figure 46 illustrates this result. Notice that, once the graph of f is known, the graph of f^{-1} may be obtained by folding the paper along the line $y = x$.

FIGURE 46

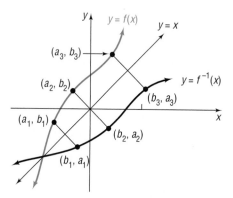

E X A M P L E   3   *Graphing the Inverse Function*

The graph in Figure 47(a) is that of a one-to-one function $y = f(x)$. Draw the graph of its inverse.

Solution   We begin by adding the graph of $y = x$ to Figure 47(a). Since the points $(-2, -1)$, $(-1, 0)$, and $(2, 1)$ are on the graph of f, we know that the points $(-1, -2)$, $(0, -1)$, and $(1, 2)$ must be on the graph of f^{-1}. Keeping in mind that the graph of f^{-1} is the reflection about the line $y = x$ of the graph of f, we can draw f^{-1}. See Figure 47(b). ∎

FIGURE 47

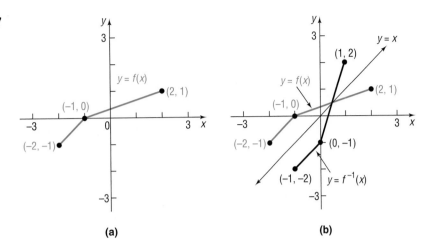

(a)

(b)

■ Now work Problem 7.

Finding the Inverse Function

The fact that the graph of a one-to-one function f and its inverse are symmetric with respect to the line $y = x$ tells us more. Look again at Figure 46. It says that we can obtain f^{-1} by interchanging the roles of x and y in f. That is, if f is defined by the equation

$$y = f(x)$$

then f^{-1} is defined by the equation

$$x = f(y)$$

Be careful! The equation $x = f(y)$ defines f^{-1} implicitly. If we can solve this equation for y, we will have the explicit form of f^{-1}, that is,

$$y = f^{-1}(x)$$

Let's use this procedure to find the inverse of $f(x) = 2x + 3$. (Since f is a linear function and is increasing, we know that f is one-to-one.)

E X A M P L E   4

Finding the Inverse Function

Find the inverse of $f(x) = 2x + 3$. Also find the domain and range of f and f^{-1}. Graph f and f^{-1} on the same coordinate axes.

Solution   In the equation $y = 2x + 3$, interchange the variables x and y. The result,

$$x = 2y + 3$$

is an equation that defines the inverse f^{-1} implicitly. Solving for y, we obtain

$$2y + 3 = x$$
$$2y = x - 3$$
$$y = \tfrac{1}{2}(x - 3)$$

The explicit form of the inverse f^{-1} is therefore

$$f^{-1}(x) = \tfrac{1}{2}(x - 3)$$

which we verified in Example 2(c).

Then we find

$$\text{Domain } f = \text{Range } f^{-1} = (-\infty, \infty)$$
$$\text{Range } f = \text{Domain } f^{-1} = (-\infty, \infty)$$

The graphs of $f(x) = 2x + 3$ and its inverse $f^{-1}(x) = \tfrac{1}{2}(x - 3)$ are shown in Figure 48. Note the symmetry of the graphs with respect to the line $y = x$. ■

FIGURE 48

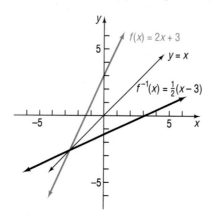

We outline next the steps to follow for finding the inverse of a one-to-one function.

Procedure for Finding the Inverse of a One-to-One Function

STEP 1:   In $y = f(x)$, interchange the variables x and y to obtain

$$x = f(y)$$

This equation defines the inverse function f^{-1} implicitly.

STEP 2:   If possible, solve the implicit equation for y in terms of x to obtain the explicit form of f^{-1}:

$$y = f^{-1}(x)$$

STEP 3:   Check the result by showing that

$$f^{-1}(f(x)) = x \quad \text{and} \quad f(f^{-1}(x)) = x$$

E X A M P L E   5

Finding the Inverse Function

The function

$$f(x) = \frac{2x + 1}{x - 1}$$

is one-to-one. Find its inverse and check the result.

Solution   STEP 1:   Interchange the variables x and y in

$$y = \frac{2x + 1}{x - 1}$$

to obtain

$$x = \frac{2y + 1}{y - 1}$$

STEP 2:   Solve for y:

$$x = \frac{2y + 1}{y - 1}$$
$$x(y - 1) = 2y + 1$$
$$xy - x = 2y + 1$$
$$xy - 2y = x + 1$$
$$(x - 2)y = x + 1$$
$$y = \frac{x + 1}{x - 2}$$

The inverse is

$$f^{-1}(x) = \frac{x + 1}{x - 2}$$

STEP 3:   *Check:*

$$f^{-1}(f(x)) = f^{-1}\left(\frac{2x + 1}{x - 1}\right) = \frac{\dfrac{2x + 1}{x - 1} + 1}{\dfrac{2x + 1}{x - 1} - 2} = \frac{2x + 1 + x - 1}{2x + 1 - 2(x - 1)} = \frac{3x}{3} = x$$

$$f(f^{-1}(x)) = f\left(\frac{x + 1}{x - 2}\right) = \frac{2\left(\dfrac{x + 1)}{x - 2}\right) + 1}{\dfrac{x + 1}{x - 2} - 1} = \frac{2(x + 1) + x - 2}{x + 1 - (x - 2)} = \frac{3x}{3} = x$$

 Check: We found that if $f(x) = (2x + 1) / (x - 1)$ then $f^{-1}(x) = (x + 1) / (x - 2)$. Graph $y = f(f^{-1}(x))$ on a square screen. What do you see? Are you surprised?   ■

■   Now work Problem 25.

If a function is not one-to-one, then it will have no inverse. Sometimes, though, an appropriate restriction on the domain of such a function will yield a new function that is one-to-one. Let's look at an example of this common practice.

E X A M P L E   6   *Finding the Inverse Function*

Find the inverse of $y = f(x) = x^2$ if $x \geq 0$.

Solution   The function $f(x) = x^2$ is not one-to-one. [Refer to Example 1(a).] However, if we restrict f to only that part of its domain for which $x \geq 0$, as indicated, we have a new function that is increasing and therefore is one-to-one. As a result, the function defined by $y = x^2$, $x \geq 0$, has an inverse, f^{-1}.

We follow the steps given previously to find f^{-1}:

STEP 1:   In the equation $y = x^2$, $x \geq 0$, interchange the variables x and y. The result is

$$x = y^2 \qquad y \geq 0$$

This equation defines (implicitly) the inverse function.

STEP 2:   We solve for y to get the explicit form of the inverse. Since $y \geq 0$, only one solution for y is obtained:

$$y = \sqrt{x}$$

so that $f^{-1}(x) = \sqrt{x}$.

STEP 3:   *Check:* $f^{-1}(f(x)) = f^{-1}(x^2) = \sqrt{x^2} = |x| = x$,   since $x \geq 0$

$$f(f^{-1})(x)) = f(\sqrt{x}) = (\sqrt{x})^2 = x$$

Figure 49 illustrates the graphs of $f(x) = x^2$, $x \geq 0$, and $f^{-1}(x) = \sqrt{x}$.   ■

FIGURE 49

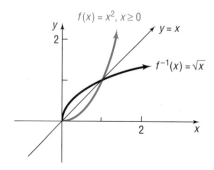

Calculators

We noted earlier that many calculators have keys that allow you to find the value of a function. These same calculators usually have a key labeled $\boxed{\text{inv}}$, $\boxed{\text{inverse}}$, $\boxed{\text{2nd}}$, or $\boxed{\text{shift}}$ that enables you to calculate the value of the corresponding inverse function. (If the actual inverse is present as a function key, such as $\boxed{\sqrt{x}}$ and $\boxed{x^2}$, the inverse key is usually disengaged for such functions.)

Try the following experiments on your calculator:

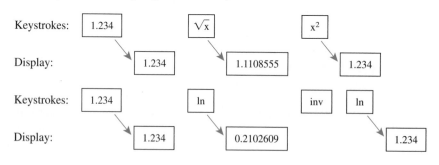

Summary

1. If a function f is one-to-one, then it has an inverse f^{-1}.
2. Domain f = Range f^{-1}; Range f = Domain f^{-1}.
3. To verify that f^{-1} is the inverse of f, show that $f^{-1}(f(x)) = x$ and $f(f^{-1}(x)) = x$.
4. The graphs of f and f^{-1} are symmetric with respect to the line $y = x$.

2.5

Exercise 2.5

In Problems 1–6, the graph of a function f is given. Use the horizontal-line test to determine whether f is one-to-one.

1.
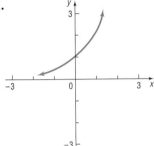

2.
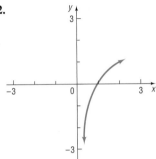

3.
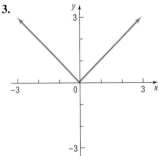

4.
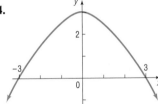

5.
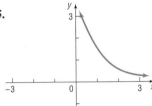

6.

In Problems 7–12, the graph of a one-to-one function f is given. Find the graph of the inverse function f^{-1}. For convenience (and as a hint), the graph of $y = x$ is also given.

7.
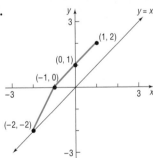

8.
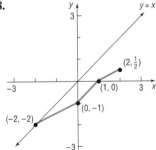

9.
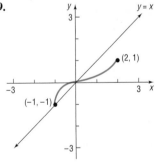

10.
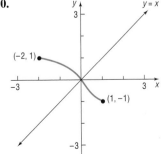

11.
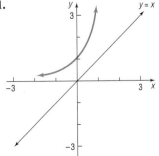

12.
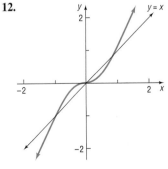

In Problems 13–22, verify that the functions f and g are inverses of each other by showing that f(g(x)) = x and g(f(x)) = x.

13. $f(x) = 3x + 4$; $g(x) = \frac{1}{3}(x - 4)$

14. $f(x) = 3 - 2x$; $g(x) = -\frac{1}{2}(x - 3)$

15. $f(x) = 4x - 8$; $g(x) = \frac{x}{4} + 2$

16. $f(x) = 2x + 6$; $g(x) = \frac{1}{2}x - 3$

17. $f(x) = x^3 - 8$; $g(x) = \sqrt[3]{x + 8}$

18. $f(x) = (x - 2)^2$, $x \geq 2$; $g(x) = \sqrt{x} + 2$, $x \geq 0$

19. $f(x) = \frac{1}{x}$; $g(x) = \frac{1}{x}$

20. $f(x) = x$; $g(x) = x$

21. $f(x) = \frac{2x + 3}{x + 4}$; $g(x) = \frac{4x - 3}{2 - x}$

22. $f(x) = \frac{x - 5}{2x + 3}$; $g(x) = \frac{3x + 5}{1 - 2x}$

In Problems 23–34, the function f is one-to-one. Find its inverse and check your answer. State the domain and range of f and f^{-1}. Graph f and f^{-1} on the same coordinate axes.

23. $f(x) = 3x$

24. $f(x) = -4x$

25. $f(x) = 4x + 2$

26. $f(x) = 1 - 3x$

27. $f(x) = x^3 - 1$

28. $f(x) = x^3 + 1$

29. $f(x) = x^2 + 4$, $x \geq 0$

30. $f(x) = x^2 + 9$, $x \geq 0$

31. $f(x) = \frac{4}{x}$

32. $f(x) = -\frac{3}{x}$

33. $f(x) = \frac{1}{x - 2}$

34. $f(x) = \frac{4}{x + 2}$

In Problems 35–46, the function f is one-to-one. Find its inverse and check your answer. State the domain and range of f and f^{-1}.

35. $f(x) = \frac{2}{3 + x}$

36. $f(x) = \frac{4}{2 - x}$

37. $f(x) = (x + 2)^2$, $x \geq -2$

38. $f(x) = (x - 1)^2$, $x \geq 1$

39. $f(x) = \frac{2x}{x - 1}$

40. $f(x) = \frac{3x + 1}{x}$

41. $f(x) = \frac{3x + 4}{2x - 3}$

42. $f(x) = \frac{2x - 3}{x + 4}$

43. $f(x) = \frac{2x + 3}{x + 2}$

44. $f(x) = \frac{-3x - 4}{x - 2}$

45. $f(x) = 2\sqrt[3]{x}$

46. $f(x) = \frac{4}{\sqrt{x}}$

47. Find the inverse of the linear function $f(x) = mx + b$, $m \neq 0$.

48. Find the inverse of the function $f(x) = \sqrt{r^2 - x^2}$, $0 \leq x \leq r$.

49. Can an even function be one-to-one? Explain.

50. Is every odd function one-to-one? Explain.

51. A function f has an inverse. If the graph of f lies in quadrant I, in which quadrant does the graph of f^{-1} lie?

52. A function f has an inverse. If the graph of f lies in quadrant II, in which quadrant does the graph of f^{-1} lie?

53. The function $f(x) = |x|$ is not one-to-one. Find a suitable restriction on the domain of f so that the new function that results is one-to-one. Then find the inverse of f.

54. The function $f(x) = x^4$ is not one-to-one. Find a suitable restriction on the domain of f so that the new function that results is one-to-one. Then find the inverse of f.

55. *Temperature Conversion* To convert from x degrees Celsius to y degrees Fahrenheit, we use the formula $y = f(x) = \frac{9}{5}x + 32$. To convert from x degrees Fahrenheit to y degrees Celsius, we use the formula $y = g(x) = \frac{5}{9}(x - 32)$. Show that f and g are inverse functions.

56. *Demand for Corn* The demand for corn obeys the equation $p(x) = 300 - 50x$, where p is the price per bushel (in dollars) and x is the number of bushels produced, in millions. Express the production amount x as a function of the price p.

57. *Period of a Pendulum* The period T (in seconds) of a simple pendulum is a function of its length l (in feet), given by $T(l) = 2\pi\sqrt{l/g}$, where $g \approx 32.2$ feet per second per second is the acceleration of gravity. Express the length l as a function of the period T.

58. Give an example of a function whose domain is the set of real numbers and that is neither increasing nor decreasing on its domain, but is one-to-one. [*Hint:* Use a piecewise-defined function.]

59. Given

$$f(x) = \frac{ax + b}{cx + d}$$

find $f^{-1}(x)$. If $c \neq 0$, under what conditions on a, b, c, and d is $f = f^{-1}$?

60. We said earlier that finding the range of a function f is not easy. However, if f is one-to-one, we can find its range by finding the domain of the inverse function f^{-1}. Use this technique to find the range of each of the following one-to-one functions:

(a) $f(x) = \dfrac{2x + 5}{x - 3}$ \qquad (b) $g(x) = 4 - \dfrac{2}{x}$ \qquad (c) $F(x) = \dfrac{3}{4 - x}$

For Problems 61–66, write a program that will graph the inverse of a function $y = f(x)$. Then graph the function f and its inverse on the same screen. Compare your answers with those of Problems 23–28.

61. $f(x) = 3x$ \qquad\qquad 62. $f(x) = -x$ \qquad\qquad 63. $f(x) = 4x + 2$

64. $f(x) = 1 - 3x$ \qquad\qquad 65. $f(x) = x^3 - 1$ \qquad\qquad 66. $f(x) = x^3 + 1$

67. If the graph of a function and its inverse intersect, must this occur on $y = x$? Can they intersect anywhere else? Must they intersect?

68. Can a one-to-one function and its inverse be equal? What must be true about the graph of f for this to happen? Give some examples to support your conclusion.

69. Draw the graph of a one-to-one function that contains the points $(-2, -3)$, $(0, 0)$, and $(1, 5)$. Now draw the graph of its inverse. Compare your graph to those of other students. Discuss any similarities. What differences do you see?

2.6

Mathematical Models: Constructing Functions

Real-world problems often result in mathematical models that involve functions. These functions need to be constructed or built based on the information given. In constructing functions, we must be able to translate the verbal description into the language of mathematics. We do this by assigning symbols to represent the independent and dependent variables and then finding the function or rule that relates these variables.

E X A M P L E   1 *Area of a Rectangle with Fixed Perimeter*

The perimeter of a rectangle is 50 feet. Express its area A as a function of the length x of a side.

Solution Consult Figure 50. If the length of the rectangle is x and if w is its width, then the sum of the lengths of the sides is the perimeter, 50.

FIGURE 50

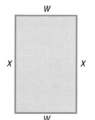

$$x + w + x + w = 50$$
$$2x + 2w = 50$$
$$x + w = 25$$
$$w = 25 - x$$

The area A is length times width, so

$$A = xw = x(25 - x)$$

The area A as a function of x is

$$A(x) = x(25 - x)$$

    ■

Note that we use the symbol A as the dependent variable and also as the name of the function that relates the length x to the area A. As we mentioned earlier, this double usage is common in applications and should cause no difficulties.

E X A M P L E   2

Economics: Demand Equations

In economics, revenue R is defined as the amount of money derived from the sale of a product and is equal to the unit selling price p of the product times the number x of units actually sold. That is,

$$R = xp$$

Usually, p and x are related: as one increases, the other decreases. Suppose that p and x are related by the following **demand equation:**

$$p = -\tfrac{1}{10}x + 20 \qquad 0 \le x \le 200$$

Express the revenue R as a function of the number x of units sold.

Solution   Since $R = xp$ and $p = -\tfrac{1}{10}x + 20$, it follows that

$$R(x) = xp = x\left(-\tfrac{1}{10}x + 20\right) = -\tfrac{1}{10}x^2 + 20x$$

    ■

■ Now work Problem 3.

E X A M P L E   3

Finding the Distance from the Origin to a Point on a Graph

Let $P = (x, y)$ be a point on the graph of $y = x^2 - 1$.

(a) Express the distance d from P to the origin O as a function of x.

(b) What is d if $x = 0$?   (c) What is d if $x = 1$?

(d) What is d if $x = \sqrt{2}/2$?

Solution   (a) Figure 51 illustrates the graph. The distance d from P to O is

$$d = \sqrt{x^2 + y^2}$$

Since P is a point on the graph of $y = x^2 - 1$, we have

$$d(x) = \sqrt{x^2 + (x^2 - 1)^2} = \sqrt{x^4 - x^2 + 1}$$

Thus, we have expressed the distance d as a function of x.

(b) If $x = 0$, the distance d is

$$d(0) = \sqrt{1} = 1$$

(c) If $x = 1$, the distance d is

$$d(1) = \sqrt{1 - 1 + 1} = 1$$

FIGURE 51
$y = x^2 - 1$

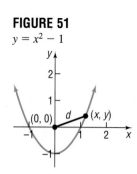

(d) If $x = \sqrt{2}/2$, the distance d is

$$d\left(\frac{\sqrt{2}}{2}\right) = \sqrt{\left(\frac{\sqrt{2}}{2}\right)^4 - \left(\frac{\sqrt{2}}{2}\right)^2 + 1} = \sqrt{\frac{1}{4} - \frac{1}{2} + 1} = \frac{\sqrt{3}}{2}$$ ∎

  ■ Now work Problem 13.

E X A M P L E   4

Filling a Swimming Pool

A rectangular swimming pool 20 meters long and 10 meters wide is 4 meters deep at one end and 1 meter deep at the other. Figure 52 illustrates a cross-sectional view of the pool. Water is being pumped into the pool at the deep end.

(a) Find a function that expresses the volume V of water in the pool as a function of the height x of the water at the deep end.

(b) Find the volume when the height is 1 meter.

(c) Find the volume when the height is 2 meters.

FIGURE 52

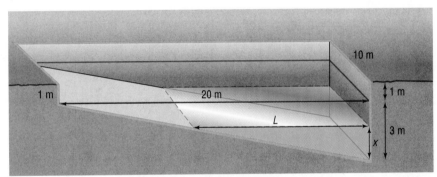

Solution (a) Let L denote the distance (in meters) measured at water level from the deep end to the short end. Notice that L and x form the sides of a triangle that is similar to the triangle whose sides are 20 meters by 3 meters. Thus, L and x are related by the equation

$$\frac{L}{x} = \frac{20}{3} \quad \text{or} \quad L = \frac{20x}{3} \qquad 0 \le x \le 3$$

The volume V of water in the pool at any time is

$$V = \left(\begin{array}{c}\text{Cross-sectional} \\ \text{triangular area}\end{array}\right)(\text{Width}) = \left(\tfrac{1}{2}Lx\right)(10) \text{ cubic meters}$$

Since $L = 20x/3$, we have

$$V(x) = \left(\frac{1}{2} \cdot \frac{20x}{3} \cdot x\right)(10) = \frac{100}{3}x^2 \text{ cubic meters}$$

(b) When the height x of the water is 1 meter, the volume $V = V(x)$ is

$$V(1) = \frac{100}{3} \cdot 1^2 = 33.3 \text{ cubic meters}$$

(c) When the height x of the water is 2 meters, the volume $V = V(x)$ is

$$V(2) = \frac{100}{3} \cdot 2^2 = \frac{400}{3} = 133.3 \text{ cubic meters}$$ ∎

E X A M P L E   5   *Area of an Isosceles Triangle*

Consider an isosceles triangle of fixed perimeter p.

(a) If x equals the length of one of the two equal sides, express the area A as a function of x.

(b) What is the domain of A?

Solution   (a) Look at Figure 53. Since the equal sides are of length x, the third side must be of length $p - 2x$. (Do you see why?) We know that the area A is

$$A = \tfrac{1}{2}(\text{Base})(\text{Height})$$

FIGURE 53

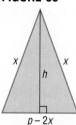

To find the height h, we drop the perpendicular to the base of length $p - 2x$ and use the fact that the perpendicular bisects the base. Then, by the Pythagorean Theorem, we have

$$h^2 = x^2 - \left(\frac{p - 2x}{2}\right)^2 = x^2 - \frac{1}{4}(p^2 - 4px + 4x^2)$$

$$= px - \frac{1}{4}p^2 = \frac{4px - p^2}{4}$$

$$h = \sqrt{\frac{4px - p^2}{4}} = \frac{\sqrt{p}}{2}\sqrt{4x - p}$$

The area A is given by

$$A = \frac{1}{2} \cdot (p - 2x)\frac{\sqrt{p}}{2}\sqrt{4x - p} = \frac{\sqrt{p}}{4}(p - 2x)\sqrt{4x - p}$$

(b) The domain of A is found as follows. Because of the expression $\sqrt{4x - p}$, we require that

$$4x - p > 0$$

$$x > \frac{p}{4}$$

Since $p - 2x$ is a side of the triangle, we also require that

$$p - 2x > 0$$

$$-2x > -p$$

$$x < \frac{p}{2}$$

Thus, the domain of A is $p/4 < x < p/2$, or $(p/4, p/2)$, and we state the functions as

$$A(x) = \frac{\sqrt{p}}{4}(p - 2x)\sqrt{4x - p} \qquad \frac{p}{4} < x < \frac{p}{2}$$ ■

■ Now work Problem 9.

E X A M P L E   6   *Minimum Payments and Interest Charged for Credit Cards*

(a) Holders of credit cards issued by banks, department stores, oil companies, and so on, receive bills each month that state minimum amounts that must be paid by a certain due date. The minimum due depends on the total amount owed. One such credit card company uses the following rules: For a bill of less than $10, the entire amount is due. For a bill of at least $10 but less than $500, the minimum due is $10. There is a minimum of $30 due on a bill of at least $500 but less than $1000, a minimum of $50 due on a bill of at least $1000 but less than $1500, and a minimum of $70 due on bills of $1500 or more. Find the function f that describes the minimum payment due on a bill of x dollars. Graph f.

(b) The card holder may pay any amount between the minimum due and the total owed. The organization issuing the card charges the card holder interest of 1.5% per month for the first $1000 owed and 1% per month on any unpaid balance over $1000. Find the function g that gives the amount of interest charged per month on a balance of x dollars. Graph g.

Solution   (a) The function f that describes the minimum payment due on a bill of x dollars is

$$f(x) = \begin{cases} x & \text{if} \quad 0 \leq x < 10 \\ 10 & \text{if} \quad 10 \leq x < 500 \\ 30 & \text{if} \quad 500 \leq x < 1000 \\ 50 & \text{if} \ 1000 \leq x < 1500 \\ 70 & \text{if} \ 1500 \leq x \end{cases}$$

To graph this function f, we proceed as follows: For $0 \leq x < 10$, draw the graph of $y = x$; for $10 \leq x < 500$, draw the graph of the constant function $y = 10$; for $500 \leq x < 1000$, draw the graph of the constant function $y = 30$; and so on. The graph of f is given in Figure 54.

FIGURE 54

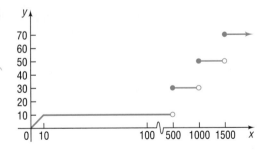

FIGURE 55

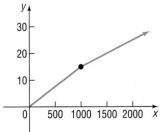

(b) If $g(x)$ is the amount of interest charged per month on a balance of x, then $g(x) = 0.015x$ for $0 \leq x \leq 1000$. The amount of the unpaid balance above $1000 is $x - 1000$. If the balance due is $x > 1000$, then the interest is $0.015(1000) + 0.01(x - 1000) = 15 + 0.01x - 10 = 5 + 0.01x$, so

$$g(x) = \begin{cases} 0.015x & \text{if } 0 \leq x \leq 1000 \\ 5 + 0.01x & \text{if } x > 1000 \end{cases}$$

See Figure 55.   ■

EXAMPLE 7   *Finding the Cost of a Can*

A company that manufactures aluminum cans requires a cylindrical container with a capacity of 500 cubic centimeters $\left(\frac{1}{2} \text{ liter}\right)$. The top and bottom of the can will be made of a special aluminum alloy that costs \$0.05 per square centimeter. The sides of the can are to be made of material that costs \$0.02 per square centimeter. Express the cost of material for the can as a function of the radius r of the can.

Solution   Figure 56 illustrates the situation. Notice that the material required to produce a cylindrical can of height h and radius r consists of a rectangle of area $2\pi rh$ and two circles, each of area πr^2. The total cost C (in cents) of manufacturing the can is therefore

FIGURE 56

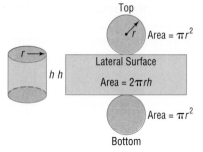

Top
Area = πr^2

Lateral Surface
Area = $2\pi rh$

Area = πr^2

Bottom

$$C = \text{Cost of top and bottom} + \text{Cost of side}$$

$$= \underbrace{2(\pi r^2)}_{\substack{\text{Total area} \\ \text{of top and} \\ \text{bottom}}} \underbrace{(5)}_{\substack{\text{Cost/unit} \\ \text{area}}} + \underbrace{(2\pi rh)}_{\substack{\text{Total} \\ \text{area of} \\ \text{side}}} \underbrace{(2)}_{\substack{\text{Cost/unit} \\ \text{area}}}$$

$$= 10\pi r^2 + 4\pi rh$$

But we have the additional restriction that the height h and radius r must be chosen so that the volume V of the can is 500 cubic centimeters. Since $V = \pi r^2 h$, we have

$$500 = \pi r^2 h \quad \text{or} \quad h = \frac{500}{\pi r^2}$$

Thus, the cost C, in cents, as a function of the radius r is

$$C(r) = 10\pi r^2 + 4\pi r \left(\frac{500}{\pi r^2}\right) = 10\pi r^2 + \frac{2000}{r}$$ ∎

EXAMPLE 8   *Making a Playpen*

A manufacturer of children's playpens makes a square model that can be opened at one corner and attached at right angles to a wall, or, perhaps, the side of a house. If each side is 3 feet in length, the open configuration doubles the available area in which the child can play from 9 square feet to 18 square feet. See Figure 57.

FIGURE 57

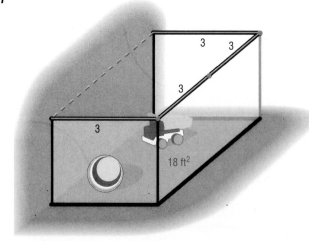

Now, suppose we place hinges at the outer corners to allow for a configuration like the one shown in Figure 58.

(a) Express the area A of this configuration as a function of the distance x between the two parallel sides.

(b) Find the domain of A.   (c) Find A if $x = 5$.

(d) Graph $A = A(x)$. For what value of x is the area largest?*

FIGURE 58

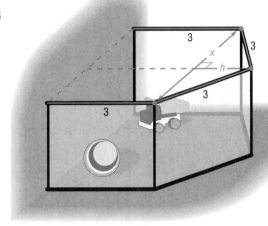

Solution

(a) Refer to Figure 58. The area A that we seek consists of the area of a rectangle (with width 3 and length x) and the area of an isosceles triangle (with base x and two equal sides of length 3). The height h of the triangle may be found using the Pythagorean Theorem:

$$h^2 = 3^2 - \left(\frac{x}{2}\right)^2 = 9 - \frac{x^2}{4} = \frac{36 - x^2}{4}$$

$$h = \tfrac{1}{2}\sqrt{36 - x^2}$$

The area A enclosed by the playpen is

$$A = \text{Area of rectangle} + \text{Area of triangle} = 3x + \tfrac{1}{2}x\left(\tfrac{1}{2}\sqrt{36 - x^2}\right)$$

$$A(x) = 3x + \frac{x\sqrt{36 - x^2}}{4}$$

Thus, we have expressed the area A as a function of x.

(b) To find the domain of A, we note first that $x > 0$, since x is a length. Also, the expression under the radical must be positive, so

$$36 - x^2 > 0$$

$$x^2 < 36$$

$$-6 < x < 6$$

*Adapted from *Proceedings, Summer Conference for College Teachers on Applied Mathematics* (University of Missouri, Rolla), 1971.

Combining these restrictions, we find that the domain of A is $0 < x < 6$, or $(0, 6)$.

(c) If $x = 5$, the area is

$$A(5) = 3(5) + \frac{5}{4}\sqrt{36 - (5)^2} \approx 19.15 \text{ square feet}$$

Thus, if the width of the playpen is 5 feet, its area is 19.15 square feet.

(d) The maximum area is about 19.81 square feet, obtained when x is about 5.58 feet. See Figure 59.   ■

FIGURE 59

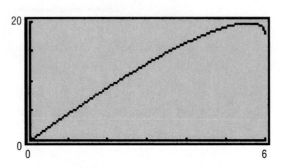

2.6

Exercise 2.6

1. *Volume of a Cylinder* The volume V of a right circular cylinder of height h and radius r is $V = \pi r^2 h$. If the height is twice the radius, express the volume V as a function of r.

2. *Volume of a Cone* The volume V of a right circular cone is $V = \frac{1}{3}\pi r^2 h$. If the height is twice the radius, express the volume V as a function of r.

3. *Demand Equation* The price p and the quantity x sold of a certain product obey the demand equation

$$p = -\frac{1}{6}x + 100 \qquad 0 \le x \le 600$$

Express the revenue R as a function of x. (Remember, $R = xp$.)

4. *Demand Equation* The price p and the quantity x sold of a certain product obey the demand equation

$$p = -\frac{1}{3}x + 100 \qquad 0 \le x \le 300$$

Express the revenue R as a function of x.

5. *Demand Equation* The price p and the quantity x sold of a certain product obey the demand equation

$$x = -5p + 100 \qquad 0 \le p \le 20$$

Express the revenue R as a function of x.

6. *Demand Equation* The price p and the quantity x sold of a certain product obey the demand equation

$$x = -20p + 500 \qquad 0 \le p \le 25$$

Express the revenue R as a function of x.

7. *Enclosing a Rectangular Field* A farmer has available 400 yards of fencing and wishes to enclose a rectangular area.
(a) Express the area A of the rectangle as a function of the width x of the rectangle.
(b) What is the domain of A?
(c) Graph $A = A(x)$. For what value of x is the area largest?

8. *Enclosing a Rectangular Field along a River* A farmer has 3000 feet of fencing available to enclose a rectangular field. One side of the field lies along a river, so only three sides require fencing.
(a) Express the area A of the rectangle as a function of x, where x is the length of the side parallel to the river.
(b) Graph $A = A(x)$. For what value of x is the area largest?

9. A wire of length x is bent into the shape of a circle.
(a) Express the circumference of the circle as a function of x.
(b) Express the area of the circle as a function of x.

10. A wire of length x is bent into the shape of a square.
(a) Express the perimeter of the square as a function of x.
(b) Express the area of the square as a function of x.

11. A right triangle has one vertex on the graph of $y = x^3$, $x > 0$, at (x, y), another at the origin, and the third on the positive y-axis at $(0, y)$, as shown in the figure on the right. Express the area A of the triangle as a function of x.

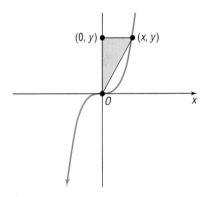

12. A right triangle has one vertex on the graph of $y = 9 - x^2$, $x > 0$, at (x, y), another at the origin, and the third on the positive x-axis at $(x, 0)$. Express the area A of the triangle as a function of x.

13. Let $P = (x, y)$ be a point on the graph of $y = x^2 - 8$.
(a) Express the distance d from P to the origin as a function of x.
(b) What is d if $x = 0$?   (c) What is d if $x = 1$?

14. Let $P = (x, y)$ be a point on the graph of $y = x^2 - 8$.
(a) Express the distance d from P to the point $(0, -1)$ as a function of x.
(b) What is d if $x = 0$?   (c) What is d if $x = -1$?

15. Let $P = (x, y)$ be a point on the graph of $y = \sqrt{x}$. Express the distance d from P to the point $(1, 0)$ as a function of x.

16. Let $P = (x, y)$ be a point on the graph of $y = 1/x$. Express the distance d from P to the origin as a function of x.

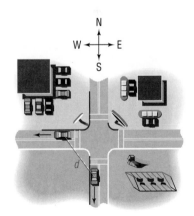

17. Two cars leave an intersection at the same time. One is headed south at a constant speed of 30 miles per hour; the other is headed west at a constant speed of 40 miles per hour (see the figure). Express the distance d between the cars as a function of the time t. [*Hint:* At $t = 0$, the cars leave the intersection.]

18. Two cars are approaching an intersection. One is 2 miles south of the intersection and is moving at a constant speed of 30 miles per hour. At the same time, the other car is 3 miles east of the intersection and is moving at a constant speed of 40 miles per hour.
(a) Express the distance d between the cars as a function of time t. [Hint: At $t = 0$, the cars are 2 miles south and 3 miles east of the intersection, respectively.]
(b) Graph $d = d(t)$. For what value of t is d smallest?

19. *Constructing an Open Box* An open box with a square base is to be made from a square piece of cardboard 24 inches on a side by cutting out a square from each corner and turning up the sides (see the figure).

(a) Express the volume V of the box as a function of the length x of the side of the square cut from each corner.

(b) Graph $V = V(x)$. For what value of x is V largest?

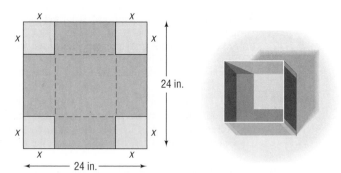

20. *Constructing an Open Box* An open box with a square base is required to have a volume of 10 cubic feet.

(a) Express the amount A of material used to make such a box as a function of the length x of a side of the square base.

(b) Graph $A = A(x)$. For what value of x is A smallest?

21. *Constructing a Closed Box* A closed box with a square base is required to have a volume of 10 cubic feet.

(a) Express the amount A of material used to make such a box as a function of the length x of a side of the square base.

(b) Graph $A = A(x)$. For what value of x is A smallest?

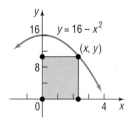

22. *Spheres* The volume V of a sphere of radius r is $V = \frac{4}{3}\pi r^3$; the surface area S of this sphere is $S = 4\pi r^2$. Express the volume V as a function of the surface area S. If the surface area doubles, how does the volume change?

23. A rectangle has one corner on the graph of $y = 16 - x^2$, another at the origin, a third on the positive y-axis, and the fourth on the positive x-axis (see the figure).

(a) Express the area A of the rectangle as a function of x.

(b) What is the domain of A?

(c) Graph $A = A(x)$. For what value of x is A largest?

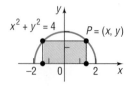

24. A rectangle is inscribed in a semicircle of radius 2 (see the figure). Let $P = (x, y)$ be the point in quadrant I that is a vertex of the rectangle and is on the circle.

(a) Express the area A of the rectangle as a function of x.

(b) Express the perimeter p of the rectangle as a function of x.

(c) Graph $A = A(x)$. For what value of x is A largest?

(d) Graph $p = p(x)$. For what value of x is p largest?

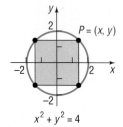

25. A rectangle is inscribed in a circle of radius 2 (see the figure). Let $P = (x, y)$ be the point in quadrant I that is a vertex of the rectangle and is on the circle.

(a) Express the area A of the rectangle as a function of x.

(b) Express the perimeter p of the rectangle as a function of x.

(c) Graph $A = A(x)$. For which value of x is A largest?

(d) Graph $p = p(x)$. For what value of x is p largest?

26. A circle of radius r is inscribed in a square (see the figure).

(a) Express the area A of the square as a function of the radius r of the circle.

(b) Express the perimeter p of the square as a function of r.

27. *Cost of a Can* A can in the shape of a right circular cylinder is required to have a volume of 500 cubic centimeters. The top and bottom are made of material that costs 6¢ per square centimeter, while the sides are made of material that costs 4¢ per square centimeter.

   (a) Express the total cost C of the material as a function of the radius r of the cylinder. (Refer to Figure 56.)
   (b) Graph $C = C(r)$. For what value of r is the cost C least?

28. *Material Needed to Make a Drum* A steel drum in the shape of a right circular cylinder is required to have a volume of 100 cubic feet.
   (a) Express the amount A of material required to make the drum as a function of the radius r of the cylinder.
   (b) How much material is required if the drum is of radius 3 feet?
   (c) Of radius 4 feet?
   (d) Of radius 5 feet?
   (e) Graph $A = A(r)$. For what value of r is A smallest?

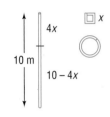

29. A wire 10 meters long is to be cut into two pieces. One piece will be shaped as a square, and the other piece will be shaped as a circle (see the figure).
   (a) Express the total area A enclosed by the pieces of wire as a function of the length x of a side of the square.
   (b) What is the domain of A?
   (c) Graph $A = A(x)$. For what value of x is A smallest?

30. A wire 10 meters long is to be cut into two pieces. One piece will be shaped as an equilateral triangle, and the other piece will be shaped as a circle.
   (a) Express the total area A enclosed by the pieces of wire as a function of the length x of a side of the equilateral triangle.
   (b) What is the domain of A?
   (c) Graph $A = A(x)$. For what value of x is A smallest?

31. A semicircle of radius r is inscribed in a rectangle so that the diameter of the semicircle is the length of the rectangle (see the figure).
   (a) Express the area A of the rectangle as a function of the radius r of the semicircle.
   (b) Express the perimeter p of the rectangle as a function of r.

32. An equilateral triangle is inscribed in a circle of radius r. See the figure. Express the circumference C of the circle as a function of the length x of a side of the triangle. [*Hint:*   First show that $r^2 = x^2/3$.]

33. An equilateral triangle is inscribed in a circle of radius r. See the figure. Express the area A within the circle, but outside the triangle, as a function of the length x of a side of the triangle.

34. *Cost of Transporting Goods* A trucking company transports goods between Chicago and New York, a distance of 960 miles. The company's policy is to charge, for each pound, $0.50 per mile for the first 100 miles, $0.40 per mile for the next 300 miles, $0.25 per mile for the next 400 miles, and no charge for the remaining 160 miles.
   (a) Graph the relationship between the cost of transportation in dollars and mileage over the entire 960 mile route.
   (b) Find the cost as a function of mileage for hauls between 100 and 400 miles from Chicago.
   (c) Find the cost as a function of mileage for hauls between 400 and 800 miles from Chicago.

35. *Car Rental Costs* An economy car rented in Florida from National Car Rental® on a weekly basis costs $95 per week.* Extra days cost $24 per day until the daily rate exceeds the weekly rate, in which case the weekly rate applies. Find the cost C of renting an economy car as a piecewise-defined function of the number x of days used, where $7 \leq x \leq 14$. Graph this function. [*Note:*   Any part of a day counts as a full day.]

36. Rework Problem 35 for a luxury car, which costs $219 on a weekly basis with extra days at $45 per day.

Source: National Car Rental®, 1995.

37. Water is poured into a container in the shape of a right circular cone with radius 4 feet and height 16 feet (see the figure). Express the volume V of water in the cone as a function of the height h of the water. [*Hint:* The volume V of a cone of radius r and height h is $V = \frac{1}{3}\pi r^2 h$.]

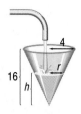

38. *Federal Income Tax* Two 1994 Tax Rate Schedules are given in the accompanying table. If x equals the amount on Form 1040, line 37, and y equals the tax due, construct a function f for each schedule.

1994 TAX RATE SCHEDULES

SCHEDULE X—USE IF YOUR FILING STATUS IS SINGLE				SCHEDULE Y-1—USE IF YOUR FILING STATUS IS MARRIED FILING JOINTLY OR QUALIFYING WIDOW(ER)			
If the amount on Form 1040, line 37, is: Over—	But not over—	Enter on Form 1040, line 38	of the amount over—	If the amount on Form 1040, line 37, is: Over—	But not over—	Enter on Form 1040, line 38	of the amount over—
$0	$22,750	---- 15%	$0	$0	$38,000	---- 15%	$0
22,750	55,100	$3,412.50 + 28%	22,750	38,000	91,850	$5,700.00 + 28%	38,000
55,100	115,000	12,470.50 + 31%	55,100	91,850	140,000	20,778.00 + 31%	91,850
115,000	250,000	31,039.50 + 36%	115,000	140,000	250,000	35,704.50 + 36%	140,000
250,000	----	79,639.50 + 39.6%	250,000	250,000	----	75,304.50 + 39.6%	250,000

Chapter Review

THINGS TO KNOW

Function

A rule or correspondence between two sets of real numbers so that each number x in the first set, the domain, has corresponding to it exactly one number y in the second set. The range is the set of y values of the function for the x values in the domain.
x is the independent variable; y is the dependent variable.
A function f may be defined implicitly by an equation involving x and y or explicitly by writing $y = f(x)$.
A function can also be characterized as a set of ordered pairs (x, y) or $(x, f(x))$ in which no two pairs have the same first element.

Function notation

$y = f(x)$
f is a symbol for the function or rule that defines the function.
x is the argument, or independent variable.
y is the dependent variable.
$f(x)$ is the value of the function at x, or the image of x.

Domain

If unspecified, the domain of a function f is the largest set of real numbers for which the rule defines a real number.

Vertical-line test

A set of points in the plane is the graph of a function if and only if every vertical line intersects the graph in at most one point.

Even function f

$f(-x) = f(x)$ for every x in the domain ($-x$ must also be in the domain).

Odd function f

$f(-x) = -f(x)$ for every x in the domain ($-x$ must also be in the domain).

One-to-one function f

If $x_1 \neq x_2$, then $f(x_1) \neq f(x_2)$ for any choice of x_1 and x_2 in the domain.

Horizontal-line test

If horizontal lines intersect the graph of a function f in at most one point, then f is one-to-one.

Inverse function f^{-1} of f

Domain of f = Range of f^{-1}; Range of f = Domain of f^{-1}
$f^{-1}(f(x)) = x$ and $f(f^{-1}(x)) = x$.
Graphs of f and f^{-1} are symmetric with respect to the line $y = x$.

IMPORTANT FUNCTIONS

Linear function

$f(x) = mx + b$ Graph is a straight line with slope m and y-intercept b.

Constant function

$f(x) = b$ Graph is a horizontal line with y-intercept b (see Figure 11).

Identity function

$f(x) = x$ Graph is a straight line with slope 1 and y-intercept 0 (see Figure 12).

Square function

$f(x) = x^2$ Graph is a parabola with intercept at $(0, 0)$ (see Figure 13).

Cube function

$f(x) = x^3$ See Figure 14.

Square root function

$f(x) = \sqrt{x}$ See Figure 15.

Reciprocal function

$f(x) = 1/x$ See Figure 16.

Absolute value function

$f(x) = |x|$ See Figure 17.

HOW TO:

Find the domain and range of a function from its graph

Find the domain of a function given its equation

Determine whether a function is even or odd without graphing it

Graph certain functions by shifting, compressing, stretching, and/or reflecting (see Table 9)

Find the composite of two functions

Find the inverse of certain one-to-one functions (see the procedure given on page 154)

Graph f^{-1} given the graph of f

Construct functions in applications, including piecewise-defined functions

FILL-IN-THE-BLANK ITEMS

1. If f is a function defined by the equation $y = f(x)$, then x is called the _____ variable and y is the _____ variable.

2. A set of points in the xy-plane is the graph of a function if and only if no _____ line contains more than one point of the set.

3. A(n) _____ function f is one for which $f(-x) = f(x)$ for every x in the domain of f; a(n) _____ function f is one for which $f(-x) = -f(x)$ for every x in the domain of f.

4. Suppose that the graph of a function f is known. Then the graph of $y = f(x - 2)$ may be obtained by a(n) _____ shift of the graph of f to the _____ a distance of 2 units.

5. If $f(x) = x + 1$ and $g(x) = x^3$, then _____ $= (x + 1)^3$.

6. If every horizontal line intersects the graph of a function f at no more than one point, then f is a(n) _____ function.

7. If f^{-1} denotes the inverse of a function f, then the graphs of f and f^{-1} are symmetric with respect to the line _____.

TRUE/FALSE ITEMS

T   F   **1.** Vertical lines intersect the graph of a function in no more than one point.

T   F   **2.** The y-intercept of the graph of the function $y = f(x)$ whose domain is all real numbers is $f(0)$.

T   F   **3.** Even functions have graphs that are symmetric with respect to the origin.

T   F   **4.** The graph of $y = f(-x)$ is the reflection about the y-axis of the graph of $y = f(x)$.

T   F   **5.** $f(g(x)) = f(x) \cdot g(x)$

T   F   **6.** If f and g are inverse functions, then the domain of f is the same as the domain of g.

T   F   **7.** If f and g are inverse functions, then their graphs are symmetric with respect to the line $y = x$.

REVIEW EXERCISES

1. Given that f is a linear function, $f(4) = -5$, and $f(0) = 3$, write the equation that defines f.

2. Given that g is a linear function with slope $= -4$ and $g(-2) = 2$, write the equation that defines g.

3. A function f is defined by

$$f(x) = \frac{Ax + 5}{6x - 2}$$

If $f(1) = 4$, find A.

4. A function g is defined by

$$g(x) = \frac{A}{x} + \frac{8}{x^2}$$

If $g(-1) = 0$, find A.

5. (a) Tell which of the following graphs are graphs of functions.
   (b) Tell which are graphs of one-to-one functions.

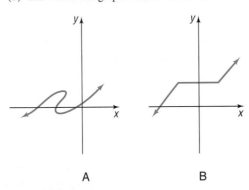   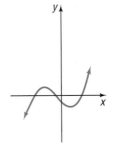   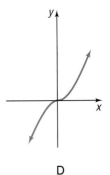

  A   B   C   D

6. Use the graph of the function f shown to find:
   (a) The domain and range of f
   (b) The intervals on which f is increasing
   (c) The intervals on which f is constant
   (d) The intercepts of f

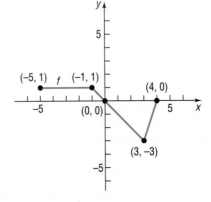

In Problems 7–12, find the following for each function:

(a) $f(-x)$ (b) $-f(x)$ (c) $f(x + 2)$ (d) $f(x - 2)$

7. $f(x) = \dfrac{3x}{x^2 - 4}$

8. $f(x) = \dfrac{x^2}{x + 2}$

9. $f(x) = \sqrt{x^2 - 4}$

10. $f(x) = |x^2 - 4|$

11. $f(x) = \dfrac{x^2 - 4}{x^2}$

12. $f(x) = \dfrac{x^3}{x^2 - 4}$

In Problems 13–18, determine whether the given function is even, odd, or neither without drawing a graph.

13. $f(x) = x^3 - 4x$

14. $g(x) = \dfrac{4 + x^2}{1 + x^4}$

15. $h(x) = \dfrac{1}{x^4} + \dfrac{1}{x^2} + 1$

16. $F(x) = \sqrt{1 - x^3}$

17. $G(x) = 1 - x + x^3$

18. $H(x) = 1 + x + x^2$

In Problems 19–30, find the domain of each function.

19. $f(x) = \dfrac{x}{x^2 - 9}$

20. $f(x) = \dfrac{3x^2}{x - 2}$

21. $f(x) = \sqrt{2 - x}$

22. $f(x) = \sqrt{x + 2}$

23. $h(x) = \dfrac{\sqrt{x}}{|x|}$

24. $g(x) = \dfrac{|x|}{x}$

25. $f(x) = \dfrac{x}{x^2 + 2x - 3}$

26. $F(x) = \dfrac{1}{x^2 - 3x - 4}$

27. $G(x) = \begin{cases} |x| & \text{if } -1 \le x \le 1 \\ 1/x & \text{if } x > 1 \end{cases}$

28. $H(x) = \begin{cases} 1/x & \text{if } 0 < x < 4 \\ x - 4 & \text{if } 4 \le x \le 8 \end{cases}$

29. $f(x) = \begin{cases} 1/(x - 2) & \text{if } x > 2 \\ 0 & \text{if } x = 2 \\ 3x & \text{if } 0 \le x < 2 \end{cases}$

30. $g(x) = \begin{cases} |1 - x| & \text{if } x < 1 \\ 3 & \text{if } x = 1 \\ x + 1 & \text{if } 1 < x \le 3 \end{cases}$

In Problems 31–50:

(a) Find the domain of each function.   (b) Locate any intercepts.

(c) Graph each function.   (d) Based on the graph, find the range.

31. $F(x) = |x| - 4$

32. $f(x) = |x| + 4$

33. $g(x) = -|x|$

34. $g(x) = \frac{1}{2}|x|$

35. $h(x) = \sqrt{x - 1}$

36. $h(x) = \sqrt{x} - 1$

37. $f(x) = \sqrt{1 - x}$

38. $f(x) = -\sqrt{x}$

39. $F(x) = \begin{cases} x^2 + 4 & \text{if } x < 0 \\ 4 - x^2 & \text{if } x \ge 0 \end{cases}$

40. $H(x) = \begin{cases} |1 - x| & \text{if } 0 \le x \le 2 \\ |x - 1| & \text{if } x > 2 \end{cases}$

41. $h(x) = (x - 1)^2 + 2$

42. $h(x) = (x + 2)^2 - 3$

43. $g(x) = (x - 1)^3 + 1$

44. $g(x) = (x + 2)^3 - 8$

45. $f(x) = \begin{cases} 2\sqrt{x} & \text{if } x \ge 4 \\ x & \text{if } 0 < x < 4 \end{cases}$

46. $f(x) = \begin{cases} 3|x| & \text{if } x < 0 \\ \sqrt{1 - x} & \text{if } 0 \le x \le 1 \end{cases}$

47. $g(x) = \dfrac{1}{x - 1} + 1$

48. $g(x) = \dfrac{1}{x + 2} - 2$

49. $h(x) = [\![-x]\!]$

50. $h(x) = -[\![x]\!]$

In Problems 51–56, the function f is one-to-one. Find the inverse of each function and check your answer. Find the domain and range of f and f^{-1}.

51. $f(x) = \dfrac{2x + 3}{5x - 2}$

52. $f(x) = \dfrac{2 - x}{3 + x}$

53. $f(x) = \dfrac{1}{x - 1}$

54. $f(x) = \sqrt{x - 2}$

55. $f(x) = \dfrac{3}{x^{1/3}}$

56. $f(x) = x^{1/3} + 1$

In Problems 57–62, for the given functions f and g, find:

(a) $(f \circ g)(2)$ (b) $(g \circ f)(-2)$ (c) $(f \circ f)(4)$ (d) $(g \circ g)(-1)$

57. $f(x) = 3x - 5$; $g(x) = 1 - 2x^2$

58. $f(x) = 4 - x$; $g(x) = 1 + x^2$

59. $f(x) = \sqrt{x + 2}$; $g(x) = 2x^2 + 1$

60. $f(x) = 1 - 3x^2$; $g(x) = \sqrt{4 - x}$

61. $f(x) = \dfrac{1}{x^2 + 4}$; $g(x) = 3x - 2$

62. $f(x) = \dfrac{2}{1 + 2x^2}$; $g(x) = 3x$

In Problems 63–68, find f ∘ g, g ∘ f, f ∘ f, and g ∘ g for each pair of functions.

63. $f(x) = \dfrac{2 - x}{x}$; $g(x) = 3x + 1$

64. $f(x) = \dfrac{2x}{x + 1}$; $g(x) = \dfrac{2x}{x - 1}$

65. $f(x) = 3x^2 + x + 1$; $g(x) = |3x|$

66. $f(x) = \sqrt{3x}$; $g(x) = 1 + x + x^2$

67. $f(x) = \dfrac{x + 1}{x - 1}$; $g(x) = \dfrac{1}{x}$

68. $f(x) = \sqrt{x^2 - 3}$; $g(x) = \sqrt{3 - x^2}$

69. For the graph of the function f shown:
  (a) Draw the graph of $y = f(-x)$
  (b) Draw the graph of $y = -f(x)$.
  (c) Draw the graph of $y = f(x + 2)$.
  (d) Draw the graph of $y = f(x) + 2$.
  (e) Draw the graph of $y = f(2 - x)$.
  (f) Draw the graph of f^{-1}.

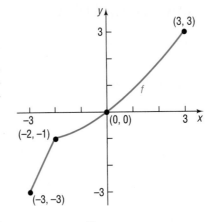

70. Repeat Problem 69 for the graph of the function g shown here.

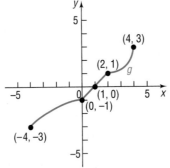

71. *Temperature Conversion* The temperature T of the air is approximately a linear function of the altitude h for altitudes within 10,000 meters of the surface of Earth. If the surface temperature is 30°C and the temperature at 10,000 meters is 5°C, find the function $T = T(h)$.

72. *Speed as a Function of Time* The speed v (in feet per second) of a car is a linear function of the time t (in seconds) for $10 \le t \le 30$. If after each second the speed of the car has increased by 5 feet per second, and if after 20 seconds the speed is 80 feet per second, how fast is the car going after 30 seconds? Find the function $v = v(t)$.

73. *Strength of a Beam* The strength of a rectangular wooden beam is proportional to the product of the width and the cube of its depth (see the figure). If the beam is to be cut from a log in the shape of a cylinder of radius 3 feet, express the strength S of the beam as a function of the width x. What is the domain of S?

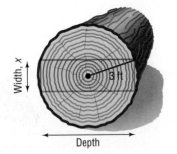

Chapter 3

PREPARING FOR THIS CHAPTER

Before getting started on this chapter, review the following concepts:

Completing the square (Appendix A, p. 821)
The discriminant of a quadratic equation (p. 20)
The intercepts of an equation (p. 59)
Graphs of certain functions: (Example 5, p. 58;
    Example 6, p. 60; Example 8, p. 62)
Solving inequalities (Section 1.4; pp. 35–43)
Division of polynomials (Appendix A, p. 805)
Complex numbers (Section 1.5; pp. 45–52)

POLYNOMIAL AND RATIONAL FUNCTIONS

Preview   The Golden Gate Bridge

The Golden Gate Bridge, a suspension bridge, spans the entrance to San Francisco Bay. Its 746-foot-tall towers are 4200 feet apart. The bridge is suspended from two huge cables more than 3 feet in diameter; the 90-foot-wide roadway is 220 feet above the water. The cables are parabolic in shape and touch the road surface at the center of the bridge. Find the height of the cable at a distance of 1000 feet from the center.

[Example 9 in Section 3.1] ■

n Chapters 1 and 2, we graphed linear functions $f(x) = ax + b$, $a \neq 0$; the square function $f(x) = x^2$; and the cube function $f(x) = x^3$. Each of these functions belongs to the class of functions called *polynomial functions,* which we discuss further in this chapter. We will also discuss *rational functions,* which are ratios of polynomial functions. In this chapter, we place special emphasis on the graphs of polynomial and rational functions. This emphasis will demonstrate the importance of evaluating polynomials (Section 3.4) and solving polynomial equations (Sections 3.5 and 3.6). Section 3.7 deals with polynomials having coefficients that are complex numbers.

3.1

Quadratic Functions

A **quadratic function** is a function of the form

$$f(x) = ax^2 + bx + c \tag{1}$$

where a, b, and c are real numbers and $a \neq 0$. The domain of a quadratic function consists of all real numbers.

Many applications require a knowledge of quadratic functions. For example, suppose that a retailer determines that the equation that relates the number x of calculators sold at the price p per calculator is given by

$$x = 15{,}000 - 750p$$

Then the revenue R derived from selling x calculators at the price p per calculator is

$$\begin{aligned} R &= xp \\ &= (15{,}000 - 750p)p \\ &= -750p^2 + 15{,}000p \end{aligned}$$

Figure 1 illustrates the graph of this revenue function, whose domain is $0 \leq p \leq 20$, since both x and p must be nonnegative.

FIGURE 1

Graph of a revenue function:
$R = -750p^2 + 15{,}000p$

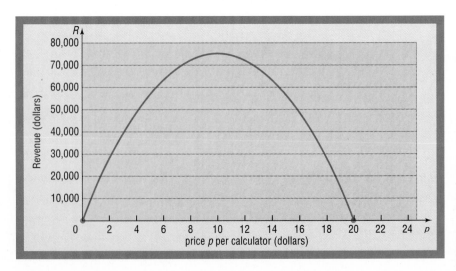

A second situation in which a quadratic function appears involves the motion of a projectile. Based on Newton's Second Law of Motion (force equals mass times acceleration, $F = ma$), it can be shown that, ignoring air resistance, the path of a projectile propelled upward at an inclination to the horizontal is the graph of a quadratic function. See Figure 2 for an illustration.

FIGURE 2
Path of a cannonball

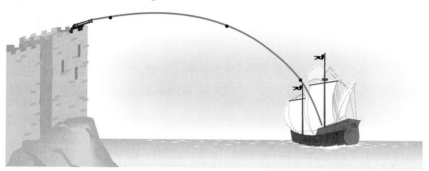

Graphing Quadratic Functions

We already know how to graph quadratic functions. For example, based on the discussion in Section 2.3, we know how to graph quadratic functions of the form $f(x) = ax^2$, $a \neq 0$. Figure 3 illustrates the graphs of $f(x) = ax^2$ for $a = 1$, $a = 3$, and $a = \frac{1}{2}$, drawn on the same set of coordinate axes. Observe that the choice of a larger value of a in $f(x) = ax^2$ results in a "thinner," or "narrower," graph.

FIGURE 3   **FIGURE 4**

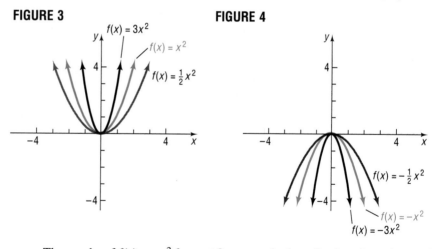

FIGURE 5

Graphs of a quadratic function,
$f(x) = ax^2 + bx + c$, $a \neq 0$

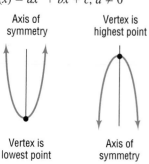

Axis of symmetry   Vertex is highest point

Vertex is lowest point   Axis of symmetry

(a) Opens up   **(b)** Opens down

The graphs of $f(x) = ax^2$ for $a < 0$ are merely the reflection about the x-axis of the corresponding graphs of $f(x) = |a|x^2$. See Figure 4.

The graphs in Figures 3 and 4 are typical of the graphs of all quadratic functions, which we call **parabolas.*** Refer to Figure 5, where two parabolas are pictured. The one on the left **opens up** and has a lowest point; the one on the right **opens down** and has a highest point. The lowest or highest point of a parabola is called the **vertex.** The vertical line passing through the vertex in each parabola in Figure 5 is called the **axis of symmetry** (usually abbreviated to **axis**) of the parabola. Because the parabola is symmetric about its axis, the axis of symmetry of a parabola can be used to advantage in graphing the parabola.

*We shall study parabolas using a geometric definition in Section 9.2.

The parabolas shown in Figure 5 are the graphs of a quadratic function $f(x) = ax^2 + bx + c$, $a \neq 0$. Notice that the coordinate axes are not included in the figure. Depending on the values of a, b, and c, the axes could be placed anywhere. The important fact is that, except possibly for compression or stretching, the shape of the graph of a quadratic function will look like one of the parabolas in Figure 5.

In the following example, we use techniques from Section 2.3 to graph a quadratic function $f(x) = ax^2 + bx + c$, $a \neq 0$. In so doing, we shall complete the square (reviewed in Appendix A.3) and write the function f in the form $f(x) = a(x - h)^2 + k$.

E X A M P L E   1   *Graphing a Quadratic Function Using Shifting, Reflecting, and the Like*

Graph the function: $f(x) = 2x^2 + 8x + 5$

Solution   We begin by completing the square on the right side:

$$f(x) = 2x^2 + 8x + 5$$
$$= 2(x^2 + 4x) + 5 \qquad \text{Factor out the 2 from } 2x^2 + 8x.$$
$$= 2(x^2 + 4x + 4) + 5 - 8 \qquad \begin{array}{l}\text{Complete the square of } 2(x^2 + 4x).\\ \text{Notice that the factor of 2 requires}\end{array}$$
$$= 2(x + 2)^2 - 3 \qquad \begin{array}{l}\text{that 8 be added and subtracted.}\end{array} \qquad (2)$$

The graph of f can be obtained in three stages, as shown in Figure 6. Now compare this graph to the graph in Figure 5(a). The graph of $f(x) = 2x^2 + 8x + 5$ is a parabola that opens up and has its vertex (lowest point) at $(-2, -3)$. Its axis of symmetry is the line $x = -2$.

FIGURE 6

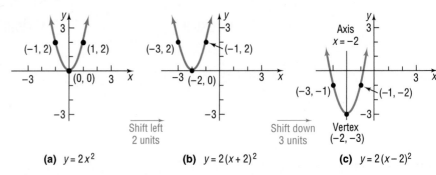

(a) $y = 2x^2$ (b) $y = 2(x + 2)^2$ (c) $y = 2(x - 2)^2$

Check: Graph $f(x) = 2x^2 + 8x + 5$ and TRACE to locate its vertex.   ∎

■   Now work Problem 17.

The method used in Example 1 can be used to graph any quadratic function $f(x) = ax^2 + bx + c$, $a \neq 0$, as follows:

$$f(x) = ax^2 + bx + c$$
$$= a\left(x^2 + \frac{b}{a}x\right) + c \qquad \text{Factor out } a \text{ from } ax^2 + bx.$$
$$= a\left(x^2 + \frac{b}{ax} + \frac{b^2}{4a^2}\right) + c - a\left(\frac{b^2}{4a^2}\right) \qquad \begin{array}{l}\text{Complete the square by adding and}\\ \text{subtracting } a(b^2/4a^2). \text{ Look closely at this}\\ \text{step!}\end{array}$$
$$= a\left(x + \frac{b}{2a}\right)^2 + c - \frac{b^2}{4a}$$
$$= a\left(x + \frac{b}{2a}\right)^2 + \frac{4ac - b^2}{4a}$$

If we let $h = -b/2a$ and let $k = (4ac - b^2)/4a$, this last equation can be rewritten in the form

$$f(x) = a(x - h)^2 + k \qquad (3)$$

The graph of f is the parabola $y = ax^2$ shifted horizontally h units and vertically k units. As a result, the vertex is at (h, k), and the graph opens up if $a > 0$ and down if $a < 0$. The axis is the vertical line $x = h$.

For example, compare equation (3) with equation (2) of Example 1.

$$f(x) = 2(x + 2)^2 - 3$$
$$f(x) = a(x - h)^2 + k$$

We conclude that $a = 2$, so the graph opens up. Also, we find that $h = -2$ and $k = -3$, so its vertex is at $(-2, -3)$.

It is not required to complete the square to obtain the vertex. In almost every case, it is easier to obtain the vertex of a quadratic function f by remembering that its x-coordinate is $h = -b/2a$. The y-coordinate can then be found by evaluating f at $-b/2a$.

These results are summarized next:

Characteristics of the Graph of a Quadratic Function

$$f(x) = ax^2 + bx + c$$

$$\text{Vertex} = \left(\frac{-b}{2a}, f\left(\frac{-b}{2a}\right) \right) \qquad \text{Axis: The line } x = \frac{-b}{2a} \qquad (4)$$

Parabola opens up if $a > 0$. Parabola opens down if $a < 0$.

E X A M P L E   2   *Locating the Vertex without Graphing*

Without graphing, locate the vertex and axis of the parabola defined by $f(x) = -3x^2 + 6x + 1$. Does it open up or down?

Solution   For this quadratic function, $a = -3$, $b = 6$, and $c = 1$. The x-coordinate of the vertex is

$$\frac{-b}{2a} = \frac{-6}{-6} = 1$$

The y-coordinate of the vertex is therefore

$$f\left(\frac{-b}{2a}\right) = f(1) = -3 + 6 + 1 = 4$$

The vertex is located at the point $(1, 4)$. The axis of symmetry is the line $x = 1$. Finally, because $a = -3 < 0$, the parabola opens down.   ∎

The information we gathered in Example 2, together with the location of the intercepts, usually provides enough information to graph $f(x) = ax^2 + bx + c$, $a \neq 0$. The y-intercept is the value of f at $x = 0$, that is, $f(0) = c$. The x-intercepts, if there are any, are found by solving the equation

$$f(x) = ax^2 + bx + c = 0$$

This equation has two, one, or no real solutions, depending on whether the discriminant $b^2 - 4ac$ is positive, 0, or negative. Thus, it has corresponding x-intercepts, as follows:

The x-Intercepts of a Quadratic Function

1. If the discriminant $b^2 - 4ac > 0$, the graph of $f(x) = ax^2 + bx + c$ has two distinct x-intercepts and so will cross the x-axis in two places.
2. If the discriminant $b^2 - 4ac = 0$, the graph of $f(x) = ax^2 + bx + c$ has one x-intercept and touches the x-axis at its vertex.
3. If the discriminant $b^2 - 4ac < 0$, the graph of $f(x) = ax^2 + bx + c$ has no x-intercept and so will not cross or touch the x-axis.

Figure 7 illustrates these possibilities for parabolas that open up.

FIGURE 7
$f(x) = ax^2 + bx + c, a > 0$

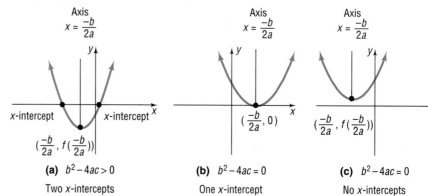

(a) $b^2 - 4ac > 0$
Two x-intercepts

(b) $b^2 - 4ac = 0$
One x-intercept

(c) $b^2 - 4ac = 0$
No x-intercepts

E X A M P L E   3   *Graphing a Quadratic Function Using Its Vertex, Axis, and Intercepts*

Use the information from Example 2 and the locations of the intercepts to graph $f(x) = -3x^2 + 6x + 1$.

Solution   In Example 2, we found the vertex to be at $(1, 4)$ and the axis to be $x = 1$. The y-intercept is found by letting $x = 0$. Thus, the y-intercept is $f(0) = 1$. The x-intercepts are found by letting $f(x) = 0$. This results in the equation

$$-3x^2 + 6x + 1 = 0$$

FIGURE 8
$f(x) = -3x^2 + 6x + 1$

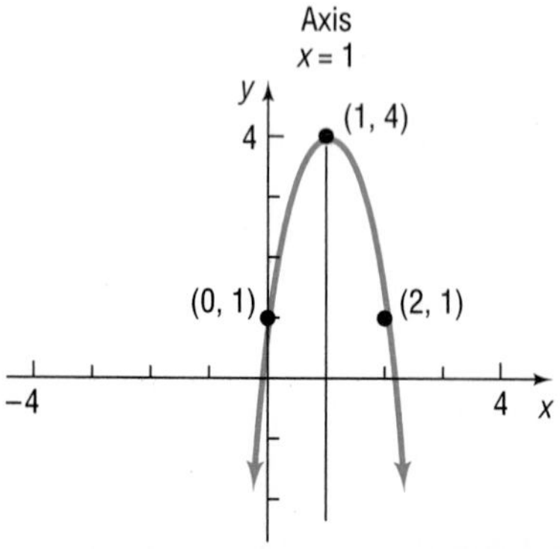

The discriminant $b^2 - 4ac = (6)^2 - 4(-3)(1) = 36 + 12 = 48 > 0$, so the equation has two real solutions and the graph has two x-intercepts. Using the quadratic formula, we find

$$x = \frac{-b + \sqrt{b^2 - 4ac}}{2a} = \frac{-6 + \sqrt{48}}{-6} = \frac{-6 + 4\sqrt{3}}{-6} = -0.15$$

and

$$x = \frac{-b - \sqrt{b^2 - 4ac}}{2a} = \frac{-6 - \sqrt{48}}{-6} = \frac{-6 - 4\sqrt{3}}{-6} = 2.15$$

The x-intercepts are approximately -0.15 and 2.15.

The graph is illustrated in Figure 8. Notice how we used the y-intercept and the axis of symmetry, $x = 1$, to obtain the additional point $(2, 1)$ on the graph.

Check: Graph $f(x) = -3x^2 + 6x + 1$. TRACE to locate the two x-intercepts and the vertex.   ■

■   Now work Problem 25.

If the graph of a quadratic function has one x-intercept or none, it may be necessary to plot some additional points to obtain the graph.

E X A M P L E   4   *Graphing a Quadratic Function Using Its Vertex, Axis, and Intercepts*

Graph $f(x) = x^2 - 6x + 9$ by determining whether its graph opens up or down and by finding its vertex, axis of symmetry, y-intercept, and x-intercepts, if any.

Solution   For $f(x) = x^2 - 6x + 9$, we have $a = 1$, $b = -6$, and $c = 9$. Since $a = 1 > 0$, the parabola opens up. The x-coordinate of the vertex is

$$\frac{-b}{2a} = \frac{-(-6)}{2 \cdot 1} = 3$$

The y-coordinate of the vertex is

$$f(3) = 9 - 6 \cdot 3 + 9 = 0$$

So the vertex is at $(3, 0)$. The axis of symmetry is the line $x = 3$. The y-intercept is $f(0) = 9$. Since the vertex $(3, 0)$ lies on the x-axis, the graph will touch the x-axis at the x-intercept. By using the axis of symmetry and the y-intercept at $(0, 9)$, we can locate the point $(6, 9)$ on the graph. See Figure 9.   ■

FIGURE 9
$f(x) = x^2 - 6x + 9$

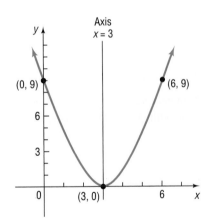

E X A M P L E   5   *Graphing a Quadratic Function Using Its Vertex, Axis, and Intercepts*

Graph $f(x) = 2x^2 + x + 1$ by determining whether its graph opens up or down and by finding its vertex, axis of symmetry, y-intercept, and x-intercepts, if any.

Solution   For $f(x) = 2x^2 + x + 1$, we have $a = 2$, $b = 1$, and $c = 1$. Since $a = 2 > 0$, the parabola opens up. The x-coordinate of the vertex is

$$\frac{-b}{2a} = -\frac{1}{4}$$

The y-coordinate of the vertex is

$$f\left(-\tfrac{1}{4}\right) = 2\left(\tfrac{1}{16}\right) + \left(-\tfrac{1}{4}\right) + 1 = \tfrac{7}{8}$$

So, the vertex is at $\left(-\tfrac{1}{4}, \tfrac{7}{8}\right)$. The axis of symmetry is the line $x = -\tfrac{1}{4}$. The y-intercept is $f(0) = 1$. The x-intercept(s), if any, obey the equation

$$2x^2 + x + 1 = 0$$

Since the discriminant $b^2 - 4ac = 1 - 8 = -7 < 0$, this equation has no real solution, and so the graph has no x-intercepts. We use the point $(0, 1)$ and the axis of symmetry $x = -\frac{1}{4}$ to locate the point $(-\frac{1}{2}, 1)$ on the graph. See Figure 10. ∎

FIGURE 10

$f(x) = 2x^2 + x + 1$

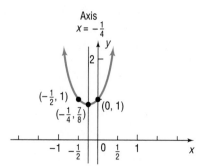

Summary

Two ways to graph a quadratic function:

1. Complete the square and apply shifting techniques (Example 1).
2. Use the results given in display (4) to find the vertex and axis of symmetry and to determine whether the graph opens up or down. Then locate the y-intercept and the x-intercepts, if there are any (Examples 2 to 5).

■   Now work Problem 31.

Applications

We have already seen that the graph of a quadratic function $f(x) = ax^2 + bx + c$ is a parabola with vertex at $(-b/2a, f(-b/2a))$. This vertex is the highest point on the graph if $a < 0$ and the lowest point on the graph if $a > 0$. If the vertex is the highest point ($a < 0$), then $f(-b/2a)$ is the **maximum value** of f. If the vertex is the lowest point ($a > 0$), then $f(-b/2a)$ is the **minimum value** of f. These ideas give rise to many applications.

E X A M P L E   6

Maximizing Revenue

A store selling calculators has found that, when the calculators are sold at a price of p dollars per unit, the revenue R (in dollars) as a function of the price p is

$$R(p) = -750p^2 + 15,000p$$

What unit price should be established in order to maximize revenue? If this price is charged, what is the maximum revenue?

Solution   The revenue R is

$$R(p) = -750p^2 + 15,000p = ap^2 + bp + c$$

The function R is a quadratic function with $a = -750$, $b = 15,000$, and $c = 0$. Because $a < 0$, the vertex is the highest point of the parabola. The revenue R is therefore a maximum when the price p is

$$p = \frac{-b}{2a} = \frac{-15,000}{2(-750)} = \frac{-15,000}{-1500} = \$10$$

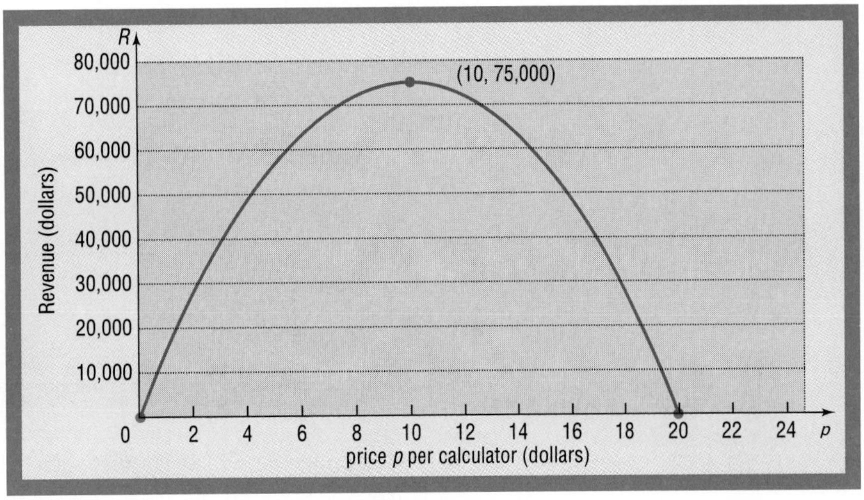

FIGURE 11
$R(p) = -750p^2 + 15,000p$

The maximum revenue R is

$$R(10) = -750(10)^2 + 15,000(10) = \$75,000$$

See Figure 11 for an illustration. ■

E X A M P L E   7

Analyzing the Motion of a Projectile

A projectile is fired from a cliff 500 feet above the water at an inclination of $45°$ to the horizontal, with a muzzle velocity of 400 feet per second. The height h of the projectile above the water is given by

$$h(x) = \frac{-32x^2}{(400)^2} + x + 500$$

where x is the horizontal distance of the projectile from the base of the cliff.

(a) Find the maximum height of the projectile.
(b) How far from the base of the cliff will the projectile strike the water?

Figure 12 on page 184 illustrates the situation.

Solution (a) The height of the projectile is given by a quadratic function:

$$h(x) = \frac{-32x^2}{(400)^2} + x + 500 = \frac{-1}{5000}x^2 + x + 500$$

We are looking for the maximum value of h. Since the maximum value is obtained at the vertex, we compute

$$x = \frac{-b}{2a} = \frac{-1}{2(-1/5000)} = \frac{5000}{2} = 2500$$

The maximum height of the projectile is

$$h(2500) = \frac{-1}{5000}(2500)^2 + 2500 + 500 = -1250 + 2500 + 500 = 1750 \text{ ft}$$

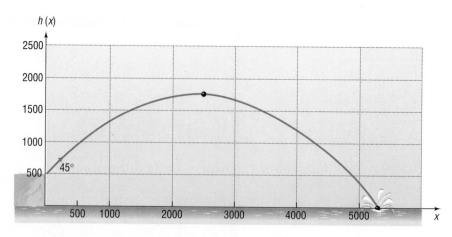

FIGURE 12

(b)   The projectile will strike the water when its height is zero. To find the distance x traversed, we need to solve the equation

$$h(x) = \frac{-1}{5000}x^2 + x + 500 = 0$$

We use the quadratic formula with

$$b^2 - 4ac = 1 - 4\left(\frac{-1}{5000}\right)(500) = 1.4$$

$$x = \frac{-1 \pm \sqrt{1.4}}{2(-1/5000)} = \begin{cases} -458 \\ 5458 \end{cases}$$

We discard the negative solution and find that the projectile will strike the water a distance of 5458 feet from the base of the cliff.   ■

Exploration Graph:

$$h(x) = \frac{-1}{5000}x^2 + x + 500 \qquad 0 \le x \le 5500$$

TRACE the path of the projectile, noting its maximum height and the distance from the base of the cliff at which it strikes the water. Compare your results with those obtained in the text. How far from the base of the cliff is the projectile when its height is 1000 ft? 1500 ft? 2000 ft?   ■

■ Now work Problem 55.

E X A M P L E   8   *Navigation*

A cruise ship leaves the Port of Miami heading due east at a constant speed of 5 knots (1 knot = 1 nautical mile per hour). At 5:00 PM, the cruise ship is 5 nautical miles due south of a cabin cruiser that is moving south at a constant speed of 10 knots. At what time are the two ships closest?

Solution   We begin with an illustration depicting the relative position of each ship at 5:00 PM. See Figure 13(a). After a time t (in hours) has passed, the cruise ship has moved east $5t$ nautical miles, and the cabin cruiser has moved south $10t$ nautical miles.

FIGURE 13

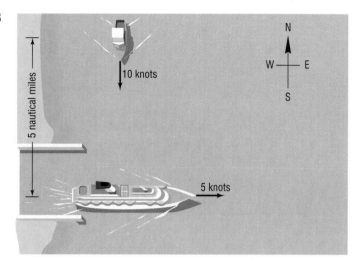

(a) Position at 5:00 PM

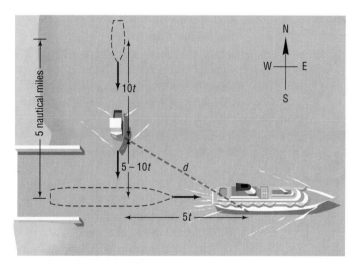

(b) Position at time t

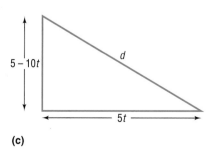

(c)

Figure 13(b) illustrates the relative position of each ship after t hours. Figure 13(c) shows a right triangle extracted from Figure 13(b). By the Pythagorean Theorem, the square of the distance d between the ships after time t is

$$d^2 = (5 - 10t)^2 + (5t)^2$$
$$= 125t^2 - 100t + 25$$

Now, the distance d is a minimum when d^2 is a minimum. Because d^2 is a quadratic function of t, it follows that d^2, and hence d, is a minimum when

$$t = \frac{-b}{2a} = \frac{100}{2(125)} = \frac{2}{5} \text{ hour}$$

Thus, the ships are closest after $\frac{2}{5}(60) = 24$ minutes, that is, at 5:24 PM. ∎

In a suspension bridge, the main cables are of parabolic shape because, if the total weight of a bridge is uniformly distributed along its length, the only cable shape that will bear the load evenly is that of a parabola. The Golden Gate Bridge in San Francisco is an example of a suspension bridge.

E X A M P L E   9   *The Golden Gate Bridge*

The Golden Gate Bridge, a suspension bridge, spans the entrance to San Francisco Bay. Its 746-foot-tall towers are 4200 feet apart. The bridge is suspended from two huge cables more than 3 feet in diameter; the 90-foot-wide roadway is 220 feet above the water. The cables are parabolic in shape and touch the road surface at the center of the bridge. Find the height of the cable at a distance of 1000 feet from the center.

Solution   We begin by choosing the placement of the coordinate axes so that the x-axis coincides with the road surface and the origin coincides with the center of the bridge. As a result, the twin towers will be vertical (height $746 - 220 = 526$ feet above the road) and located 2100 feet from the center. Also, the cable, which has the shape of a parabola, will extend from the towers, open up, and have its vertex at $(0, 0)$. As illustrated in Figure 14, the choice of placement of the axes enables us to identify the equation of the parabola as $y = ax^2$, $a > 0$. We also can see that the points $(-2100, 526)$ and $(2100, 526)$ are on the graph.

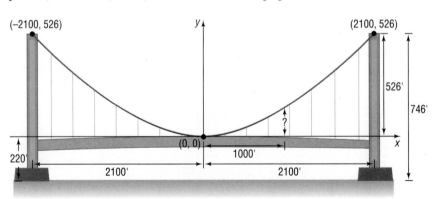

FIGURE 14

Based on these facts, we can find the value of a in $y = ax^2$:

$$y = ax^2$$
$$526 = a(2100)^2$$
$$a = \frac{526}{(2100)^2}$$

The equation of the parabola is therefore

$$y = \frac{526}{(2100)^2}x^2$$

The height of the cable when $x = 1000$ is

$$y = \frac{526}{(2100)^2}(1000)^2 \approx 119.3 \text{ feet}$$

Thus, the cable is 119.3 feet high at a distance of 1000 feet from the center of the bridge.   ■

3.1

Exercise 3.1

In Problems 1–8, match each graph to one of the following functions:

A. $y = x^2 - 1$ B. $y = -x^2 - 1$ C. $y = x^2 - 2x + 1$ D. $y = x^2 + 2x + 1$

E. $y = x^2 - 2x + 2$ F. $y = x^2 + 2x$ G. $y = x^2 - 2x$ H. $y = x^2 + 2x + 2$

1.

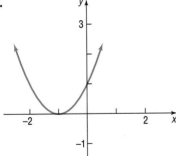

2.

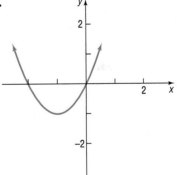

3.

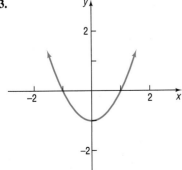

4.

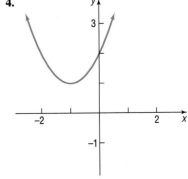

5.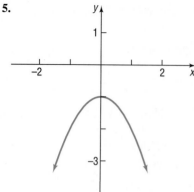

6.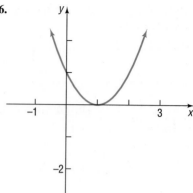

7.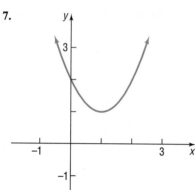

8.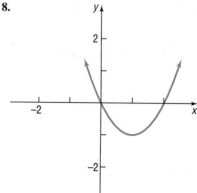

In Problems 9–24, graph the function f by starting with the graph of y = x² and using shifting, compressing, stretching, and/or reflection.

9. $f(x) = \frac{1}{4}x^2$

10. $f(x) = 2x^2$

11. $f(x) = \frac{1}{4}x^2 - 2$

12. $f(x) = 2x^2 - 3$

13. $f(x) = \frac{1}{4}x^2 + 2$

14. $f(x) = 2x^2 + 4$

15. $f(x) = -\frac{1}{4}x^2 + 1$

16. $f(x) = -2x^2 - 2$

17. $f(x) = x^2 + 4x + 2$

18. $f(x) = x^2 - 6x - 1$

19. $f(x) = 2x^2 - 4x + 1$

20. $f(x) = 3x^2 + 6x$

21. $f(x) = -x^2 - 2x$

22. $f(x) = -2x^2 + 6x + 2$

23. $f(x) = \frac{1}{2}x^2 + x - 1$

24. $f(x) = \frac{2}{3}x^2 + \frac{4}{3}x - 1$

In Problems 25–38, graph each quadratic function by determining whether its graph opens up or down and by finding its vertex, axis of symmetry, y-intercept, and x-intercepts, if any.

25. $f(x) = x^2 + 2x - 8$

26. $f(x) = x^2 - 2x - 3$

27. $f(x) = -x^2 - 3x + 4$

28. $f(x) = -x^2 + x + 2$

29. $f(x) = x^2 + 2x + 1$

30. $f(x) = -x^2 + 4x - 4$

31. $f(x) = 2x^2 - x + 2$

32. $f(x) = 4x^2 - 2x + 1$

33. $f(x) = -2x^2 + 2x - 3$

34. $f(x) = -3x^2 + 3x - 2$

35. $f(x) = 3x^2 + 6x + 2$

36. $f(x) = 2x^2 + 5x + 3$

37. $f(x) = -4x^2 - 6x + 2$

38. $f(x) = 3x^2 - 8x + 2$

In Problems 39–44, determine whether the given quadratic function has a maximum value or a minimum value and then find the value.

39. $f(x) = 2x^2 + 12x - 3$ **40.** $f(x) = 4x^2 - 8x + 3$ **41.** $f(x) = -x^2 + 10x - 4$

42. $f(x) = -2x^2 + 8x + 3$ **43.** $f(x) = -3x^2 + 12x + 1$ **44.** $f(x) = 4x^2 - 4x$

45. On one set of coordinate axes, graph the family of parabolas $f(x) = x^2 + 2x + c$ for $c = -3$, $c = 0$, and $c = 1$. Describe the characteristics of a member of this family.

46. On one set of coordinate axes, graph the family of parabolas $f(x) = x^2 + cx + 1$ for $c = -4$, $c = 0$, and $c = 4$. Describe the general characteristics of this family.

47. Graph $y = x^2 + 1$. Then graph $y = x^2 + x + 1$, followed by $y = x^2 + 2x + 1$, followed by $y = x^2 + 3x + 1$. what is happening? Do you see any pattern?

48. Graph $y = x^2 + x + 1$. Then graph $y = 2x^2 + x + 1$, followed by $y = 3x^2 + x + 1$, followed by $y = 4x^2 + x + 1$. What is happening? Do you see any pattern?

49. *Maximizing Revenue* Suppose that the manufacturer of a gas clothes dryer has found that when the unit price is p dollars the revenue R (in dollars) is

$$R = -4p^2 + 4000p$$

What unit price should be established for the dryer to maximize revenue? What is the maximum revenue?

50. *Maximizing Revenue* A tractor company has found that the revenue from sales of heavy-duty tractors is a function of the unit price p it charges. If the revenue R is

$$R = -\frac{1}{2}p^2 + 1900p$$

what unit price p should be charged to maximize revenue? What is the maximum revenue?

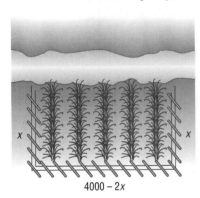

x x

$4000 - 2x$

51. *Rectangles with Fixed Perimeter* What is the largest rectangular area that can be enclosed with 400 feet of fencing? What are the dimensions of the rectangle?

52. *Rectangles with Fixed Perimeter* What are the dimensions of a rectangle of a fixed perimeter P that result in the largest area?

53. *Enclosing the Most Area with a Fence* A farmer with 4000 meters of fencing wants to enclose a rectangular plot that borders on a river. If the farmer does not fence the side along the river, what is the largest area that can be enclosed? (See the figure.)

54. *Enclosing the Most Area with a Fence* A farmer with 2000 meters of fencing wants to enclose a rectangular plot that borders on a straight highway. If the farmer does not fence the side along the highway, what is the largest area that can be enclosed?

55. *Enclosing the Most Area with a Fence* A farmer with 10,000 meters of fencing wants to enclose a rectangular field and then divide it into two plots with a fence parallel to one of the sides (see the figure). What is the largest area that can be enclosed?

56. *Enclosing the Most Area with a Fence* A farmer with 10,000 meters of fencing wants to enclose a rectangular field and then divide it into three plots with two fences parallel to one of the sides. What is the largest area that can be enclosed?

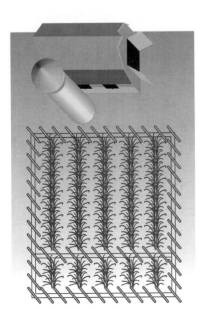

57. *Analyzing the Motion of a Projectile* A projectile is fired from a cliff 200 feet above the water at an inclination of 45° to the horizontal, with a muzzle velocity of 50 feet per second. The height h of the projectile above the water is given by

$$h(x) = \frac{-32x^2}{(50)^2} + x + 200$$

where x is the horizontal distance of the projectile from the base of the cliff.

(a) Find the maximum height of the projectile.

(b) How far from the base of the cliff will the projectile strike the water?

(c) TRACE the path of the projectile, noting its maximum height and the distance from the base of the cliff at which it strikes the water. Compare your results with those obtained in parts (a) and (b). When the height of the projectile is 100 feet above the water, how far is it from the cliff?

58. *Analyzing the Motion of a Projectile* A projectile is fired at an inclination of 45° to the horizontal, with a muzzle velocity of 100 feet per second. The height *h* of the projectile is given by

$$h(x) = \frac{-32x^2}{(100)^2} + x$$

where *x* is the horizontal distance of the projectile from the firing point.

(a) Find the maximum height of the projectile.

(b) How far from the firing point will the projectile strike the ground?

(c) TRACE the path of the projectile, noting its maximum height and the distance from the firing point to the point at which it strikes the ground. Compare your results with those obtained in parts (a) and (b). When the height of the projectile is 50 feet above the ground, how far has it traveled horizontally?

59. *Navigation* An aircraft carrier maintains a constant speed of 10 knots heading due north. At 4:00 PM, the ship's radar detects a destroyer 100 nautical miles due east of the carrier. If the destroyer is heading due west at 20 knots, when will the two ships be the closest? (1 knot = 1 nautical mile per hour)

60. *Air Traffic Control* An air traffic controller sees two aircraft flying at the same altitude on his screen. One, a Piper Cub, is headed due west at 150 miles per hour. The other, a Lear jet, is 15 miles due north of the Piper and is headed due south at 400 miles per hour. How close will the two aircraft come to each other?

61. *Suspension Bridge* A suspension bridge with weight uniformly distributed along its length has twin towers that extend 75 meters above the road surface and are 400 meters apart. The cables are parabolic in shape and are suspended from the tops of the towers. The cables touch the road surface at the center of the bridge. Find the height of the cables at a point 100 meters from the center. (Assume that the road is level.)

62. *Architecture* A parabolic arch has a span of 120 feet and a maximum height of 25 feet. Choose suitable rectangular coordinate axes and find the equation of the parabola. Then calculate the height of the arch at points 10 feet, 20 feet, and 40 feet from the center.

63. *Constructing Rain Gutters* A rain gutter is to be made of aluminum sheets that are 12 inches wide by turning up the edges 90°. What depth will provide maximum cross-sectional area and hence allow the most water to flow?

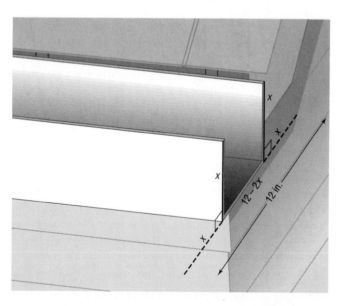

64. *Navigation* At 4 PM a cruise ship leaves the Port of Miami heading due east at a constant speed of 15 knots. At the same time, a pleasure boat located 100 nautical miles northeast of the Port of Miami is headed due south at a constant speed of 12 knots. When are the two ships closest? How close do they get to each other? (Express your answer in nautical miles; 1 knot = 1 nautical mile per hour.)

65. *Norman Windows* A Norman window has the shape of a rectangle sur-mounted by a semicircle of diameter equal to the width of the rectangle (see the figure). If the perimeter of the window is 20 feet, what dimensions will ad-mit the most light (maximize the area)? [*Hint:*   Circumference of circle = $2\pi r$; Area of circle = πr^2, where r is the radius of the circle.]

66. *Constructing a Stadium* A track and field playing area is in the shape of a rectangle with semicircles at each end (see the figure). The inside perimeter of the track is to be 1500 meters. What should the dimensions of the rectangle be so that the area of the rectangle is a maximum?

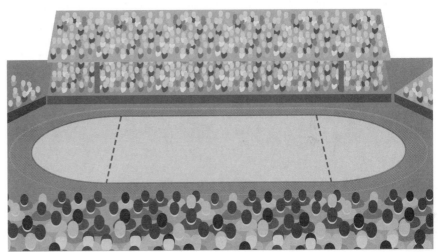

67. *Architecture* A special window has the shape of a rectangle surmounted by an equilateral triangle (see the figure). If the perimeter of the window is 16 feet, what dimensions will admit the most light? [*Hint:*   Area of an equilateral tri-angle = $(\sqrt{3}/4)x^2$, where x is the length of a side of the triangle.]

68. *Analyzing the Motion of a Projectile* A projectile is fired at an inclination of 45° to the horizontal with an initial velocity of v_0 feet per second. If the starting point is the origin, the x-axis is horizontal, and the y-axis is vertical, then the height y (in feet) after a horizontal distance x has been traversed is ap-proximately

$$y = \frac{-32x^2}{v_0^2} + x$$

(a)   Find the maximum height in terms of the initial velocity v_0.
(b)   If the initial velocity is doubled, what happens to the maximum height?
(c)   Assuming the ground is flat, how far from the starting point will the pro-jectile land if the initial velocity is 64 feet per second?

69. A charter flight club charges its members $400 per year. But, for each new member in excess of 60, the charge for every member is reduced by $5. What number of members leads to a maximum revenue?

70. A car rental agency has 24 identical cars. The owner of the agency finds that all the cars can be rented at a price of $10 per day. However, for each $2 in-crease in rental, one of the cars is not rented. What should be charged to max-imize income?

71. The graph of the function $f(x) = ax^2 + bx + c$ has vertex at $x = 0$ and passes through the points $(0, 2)$ and $(1, 8)$. Find a, b, and c.

72. The graph of the function $f(x) = ax^2 + bx + c$ has vertex at $x = 1$ and passes through the points $(0, 1)$ and $(-1, -8)$. Find a, b, and c.

73. *Chemical Reactions* A self-catalytic chemical reaction results in the formation of a compound that causes the formation ratio to increase. If the reaction rate V is given by

$$V(x) = kx(a - x) \qquad 0 \le x \le a$$

where k is a positive constant, a is the initial amount of the compound, and x is the variable amount of the compound, for what value of x is the reaction rate a maximum?

74. A rectangle has one vertex on the line $y = 10 - x$, $x > 0$, another at the origin, one on the positive x-axis, and one on the positive y-axis. Find the largest area A that can be enclosed by the rectangle.

75. *Calculus: Simpson's Rule* The figure shows the graph of $y = ax^2 + bx + c$. Suppose the points $(-h, y_0)$, $(0, y_1)$, and (h, y_2) are on the graph. It can be shown that the area enclosed by the parabola, the x-axis and the lines $x = -h$ and $x = h$ is

$$\text{Area} = \frac{h}{3}(2ah^2 + 6c)$$

Show that this area may also be given by

$$\text{Area} = \frac{h}{3}(y_0 + 4y_1 + y_2)$$

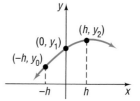

76. Let $f(x) = ax^2 + bx + c$, where a, b, and c are odd integers. If x is an integer, show that $f(x)$ must be an odd integer. [*Hint:* x is either an even integer or an odd integer.]

77. Make up a quadratic function that opens down and has only one x-intercept. Compare yours with others in the class. What are the similarities? What are the differences?

78. The figure illustrates actual data on enrollment in all public schools (both elementary and high schools) for the academic years 1980–1981 to 1988–89. Assume that the points are the points on a parabola. Further assume that a minimum enrollment of 39 million occurred between 1984–1985 and 1985–1986. Find an equation for this parabola. Assume that the trend continues and use the equation to predict public school enrollment in 1991–1992. Determine what the actual enrollment in 1991–1992 was. How does your projection compare?

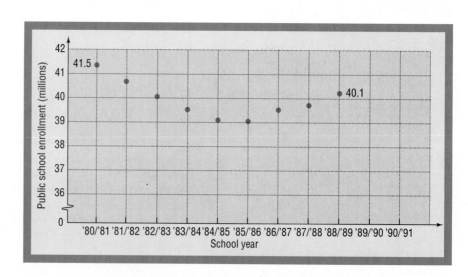

3.2

Polynomial Functions

Polynomial Function

A **polynomial function** is a function of the form

$$f(x) = a_n x^n + a_{n-1} x^{n-1} + \ldots + a_1 x + a_0 \qquad (1)$$

where $a_n, a_{n-1}, \ldots, a_1, a_0$ are real numbers and n is a non-negative integer. The domain consists of all real numbers.

Thus, a polynomial function is a function whose rule is given by a polynomial in one variable (refer to Section 1.1). The **degree** of a polynomial function is the degree of the polynomial in one variable.

EXAMPLE 1

Identifying Polynomial Functions

Determine which of the following are polynomial functions. For those that are, state the degree; for those that are not, tell why not.

(a) $f(x) = 2 - 3x^4$ (b) $g(x) = \sqrt{x}$ (c) $h(x) = \dfrac{x^2 - 2}{x^3 - 1}$

(d) $F(x) = 0$ (e) $G(x) = 8$

Solution

(a) f is a polynomial function of degree 4.

(b) g is not a polynomial function. The variable x is raised to the $\frac{1}{2}$ power, which is not a nonnegative integer.

(c) h is not a polynomial function. It is the ratio of two polynomials, and the polynomial in the denominator is of positive degree.

(d) F is the zero polynomial function; it is not assigned a degree.

(e) G is a nonzero constant function, a polynomial function of degree 0.   ■

■ Now work Problems 1 and 5.

We have already discussed in detail polynomial functions of degrees 0, 1, and 2. See Table 1 for a summary of the characteristics of the graphs of these polynomial functions.

TABLE 1

DEGREE	FORM	NAME	GRAPH
No degree	$f(x) = 0$	Zero function	The x-axis
0	$f(x) = a_0, \quad a_0 \neq 0$	Constant function	Horizontal line with y-intercept a_0
1	$f(x) = a_1 x + a_0, \quad a_1 \neq 0$	Linear function	Nonvertical, nonhorizontal line with slope a_1 and y-intercept a_0
2	$f(x) = a_2 x^2 + a_1 x + a_0, \quad a_2 \neq 0$	Quadratic function	Parabola: graph opens up if $a_2 > 0$; graph opens down if $a_2 < 0$

Power Functions

First, we consider a special kind of polynomial function called a *power function*.

Power Function of Degree n

A **power function of degree n** is a function of the form

$$f(x) = ax^n \qquad (2)$$

where a is a real number, $a \neq 0$, and $n > 0$ is an integer.

The graph of a power function of degree 1, $f(x) = ax$, is a straight line, with slope a, that passes through the origin. The graph of a power function of degree 2, $f(x) = ax^2$, is a parabola, with vertex at the origin, that opens up if $a > 0$ and down if $a < 0$.

If we know how to graph a power function of the form $f(x) = x^n$, then a compression or stretch and, perhaps, a reflection about the x-axis will enable us to obtain the graph of $g(x) = ax^n$. Consequently, we shall concentrate on graphing power functions of the form $f(x) = x^n$.

We begin with power functions of even degree of the form $f(x) = x^n$, $n \geq 2$ and n even. The domain of f is the set of all real numbers, and the range is the set of nonnegative real numbers. Such a power function is an even function (do you see why?), and hence its graph is symmetric with respect to the y-axis. Its graph always contains the origin and the points $(-1, 1)$ and $(1, 1)$.

If $n = 2$, the graph is the familiar parabola $y = x^2$ that opens up, with vertex at the origin. If $n \geq 4$, the graph of $f(x) = x^n$, n even, will be closer to the x-axis than the parabola $y = x^2$ if $-1 < x < 1$ and farther from the x-axis than the parabola $y = x^2$ if $x < -1$ or if $x > 1$. Figure 15(a) illustrates this conclusion. Figure 15(b) shows the graphs of $y = x^4$ and $y = x^8$ for further comparison.

From Figure 15, we can see that as n increases the graph of $f(x) = x^n$, $n \geq 2$ and n even, tends to flatten out near the origin and to increase very rapidly when x is far from 0. For large n, it may appear that the graph coincides with the x-axis near the origin, but it does not; the graph actually touches the x-axis only at the

FIGURE 15

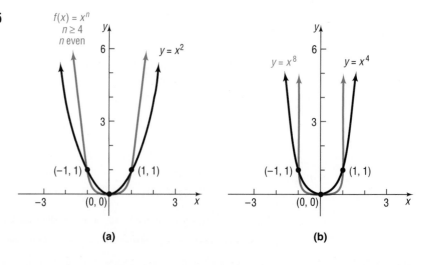

(a)   (b)

origin (see Table 2). Also, for large n, it may appear that for $x < -1$ or for $x > 1$ the graph is vertical, but it is not; it is only increasing very rapidly in those intervals. If the graphs were enlarged many times, these distinctions would be clear.

TABLE 2

	$x = 0.1$	$x = 0.3$	$x = 0.5$
$f(x) = x^8$	10^{-8}	0.0000656	0.0039063
$f(x) = x^{20}$	10^{-20}	$3.487 \cdot 10^{-11}$	0.000001
$f(x) = x^{40}$	10^{-40}	$1.216 \cdot 10^{-21}$	$9.095 \cdot 10^{-13}$

Seeing the Concept Graph $y = x^4$, $y = x^8$, and $y = x^{12}$ using the viewing rectangle $-2 \le x \le 2$, $0 \le y \le 16$. Then graph each one again using the viewing rectangle $0 \le x \le 1$, $0 \le y \le 1$. See Figure 16.

FIGURE 16

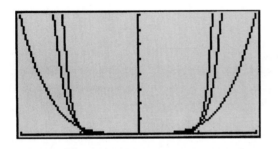
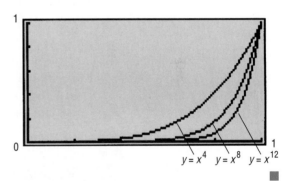

Now we consider power functions of odd degree of the form $f(x) = x^n$, $n \ge 3$ and n odd. The domain and range of f are the set of real numbers. Such a power function is an odd function (do you see why?), and hence its graph is symmetric with respect to the origin. Its graph always contains the origin and the points $(-1, -1)$ and $(1, 1)$.

The graph of $f(x) = x^n$ when $n = 3$ has been shown several times and is repeated in Figure 17. If $n \ge 5$, the graph of $f(x) = x^n$, n odd, will be closer to the x-axis than that of $y = x^3$ if $-1 < x < 1$ and farther from the x-axis than that of $y = x^3$ if $x < -1$ or if $x > 1$. Figure 17 also illustrates this conclusion.

FIGURE 17

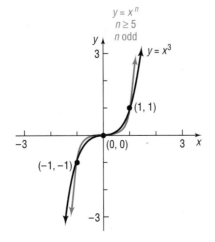

Figure 18 shows the graph of $y = x^5$ and the graph of $y = x^9$ for further comparison.

FIGURE 18

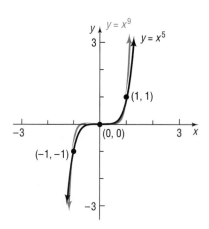

From Figures 17 and 18, we can see that as n increases the graph of $f(x) = x^n$, $n \geq 3$ and n odd, tends to flatten out near the origin and to become nearly vertical when x is far from 0.

 Seeing the Concept Graph $y = x^3$, $y = x^7$, and $y = x^{11}$ using the viewing rectangle $-2 \leq x \leq 2$, $-8 \leq y \leq 8$. Then graph each one again using the viewing rectangle $0 \leq x \leq 1$, $0 \leq y \leq 1$. See Figure 19.

FIGURE 19

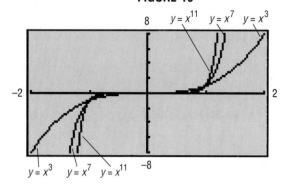
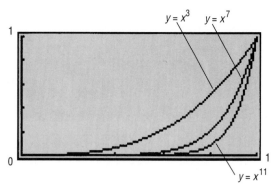

The methods of shifting, compression, stretching, and reflection studied in Section 2.3, when used in conjunction with the facts just presented, enable us to graph a variety of polynomials.

E X A M P L E   2   *Graphing Polynomial Functions Using Shifts, Reflections, and the Like*

Graph: $f(x) = 1 - x^5$

Solution   Figure 20 shows the required stages.   ■

FIGURE 20

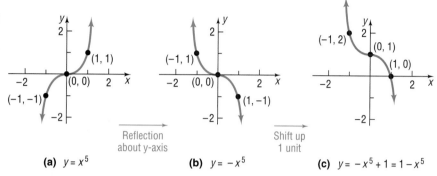

(a) $y = x^5$ (b) $y = -x^5$ (c) $y = -x^5 + 1 = 1 - x^5$

E X A M P L E   3

Graphing Polynomial Functions Using Shifts, Reflections, and the Like
Graph: $f(x) = \frac{1}{2}(x - 1)^4$

Solution   Figure 21 shows the required stages.   ■

FIGURE 21

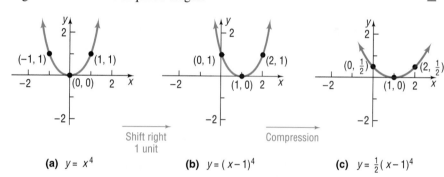

(a) $y = x^4$ (b) $y = (x - 1)^4$ (c) $y = \frac{1}{2}(x - 1)^4$

FIGURE 22

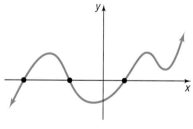

(a) Graph of a polynomial function: smooth, continuous

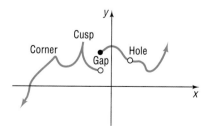

(b) Cannot be the graph of a polynomial function

■ Now work Problem 15.

Graphing Other Polynomials

To graph most polynomial functions of degree 3 or higher requires techniques beyond the scope of this text. If you take a course in calculus you will learn that the graph of every polynomial function is both smooth and continuous. By **smooth,** we mean that the graph contains no sharp corners or cusps; by **continuous,** we mean that the graph has no gaps or holes and can be drawn without lifting pencil from paper. See Figures 22(a) and 22(b).

Figure 23 shows the graph of a polynomial function with four x-intercepts. Notice that at the x-intercepts the graph must either cross the x-axis or touch the x-axis. Consequently, between consecutive x-intercepts the graph is either above the x-axis or below the x-axis. We will make use of this characteristic of the graph of a polynomial shortly.

If a polynomial function f is factored completely, it is easy to solve the equation $f(x) = 0$ and locate the x-intercepts of the graph. For example, if $f(x) = (x - 1)^2(x + 3)$, then the solutions of the equation

$$f(x) = (x - 1)^2(x + 3) = 0$$

are easily identified as 1 and -3. In general, if f is a polynomial function and r is a real number for which $f(r) = 0$, then r is called a (real) **zero of f,** or **root of f.**

FIGURE 23

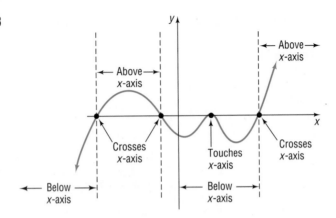

Thus, the real zeros of a polynomial function are the x-intercepts of its graph. Also, if $x - r$ is a factor of a polynomial f, then $f(r) = 0$ and so r is a zero of f. If the same factor $x - r$ occurs more than once, then r is called a **repeated,** or **multiple, zero of f.** More precisely, we have the following definition.

Zero of Multiplicity m	If $(x - r)^m$ is a factor of a polynomial f and $(x - r)^{m+1}$ is not a factor of f, then r is called a **zero of multiplicity m of f.**

E X A M P L E   4   *Identifying Zeros and Their Multiplicities*

For the polynomial

$$f(x) = 5(x - 2)(x + 3)^2\left(x - \frac{1}{2}\right)^4$$

2 is a zero of multiplicity 1

-3 is a zero of multiplicity 2

$\frac{1}{2}$ is a zero of multiplicity 4   ∎

Suppose it is possible to factor completely a polynomial function and, as a result, locate all the x-intercepts of its graph (the real zeros of the function). As mentioned earlier, these x-intercepts then divide the x-axis into open intervals, and on each such interval, the graph of the polynomial will be either above or below the x-axis. Let's look at an example.

E X A M P L E   5   *Graphing a Polynomial Using Its x-intercepts*

For the polynomial: $f(x) = x^2(x - 2)$

(a) Find the x- and y-intercepts of the graph of f.

(b) Use the x-intercepts to find the intervals on which the graph of f is above the x-axis and the intervals on which the graph of f is below the x-axis.

(c) Locate other points on the graph and connect all the points plotted with a smooth curve.

Solution   (a) The y-intercept is $f(0) = 0^2(0 - 2) = 0$. The x-intercepts satisfy the equation

$$f(x) = x^2(x - 2) = 0$$

from which we find

$$x^2 = 0 \quad \text{or} \quad x - 2 = 0$$
$$x = 0 \qquad \qquad x = 2$$

The x-intercepts are 0 and 2.

(b) The two x-intercepts divide the x-axis into three intervals:

$$-\infty < x < 0 \qquad 0 < x < 2 \qquad 2 < x < \infty$$

Since the graph of f crosses (or touches) the x-axis only at $x = 0$ and $x = 2$, it follows that the graph of f is either above the x-axis [$f(x) > 0$] or below the x-axis [$f(x) < 0$] on each of these three intervals. Thus, we need to solve the following two inequalities:

$$f(x) = x^2(x - 2) > 0 \quad \text{and} \quad f(x) = x^2(x - 2) < 0$$

Recalling the procedure discussed in Section 1.4, we set up Figure 24(a). Thus, the graph of f is above the x-axis for $2 < x < \infty$ and below the x-axis for $-\infty < x < 0$ and $0 < x < 2$.

(c) Because we used test numbers, we know three additional points on the graph: $(-1, -3)$, $(1, -1)$, and $(3, 9)$. Figure 24(b) illustrates these points, the intercepts, and a smooth, continuous curve (the graph of f) connecting them.

FIGURE 24

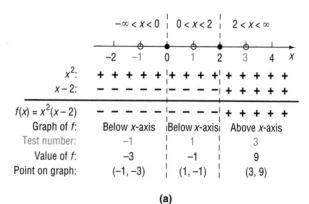

(a)

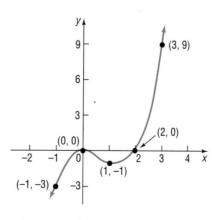

(b)

Notice that the graph of f in Figure 24(b) *crosses* the x-axis at $x = 2$, *a zero of multiplicity 1.* Notice also that the graph of f *touches* the x-axis at $x = 0$, a *zero of multiplicity 2,* since the graph is below the x-axis on both sides of $(0, 0)$.

This suggests the following result:

If r Is a Zero of Even Multiplicity	Sign of $f(x)$ does not change from one side to the other side of r.	Graph **touches** x-axis at r.

If r Is a Zero of Odd Multiplicity	Sign of $f(x)$ changes from one side to the other side of r.	Graph **crosses** x-axis at r.

■ Now work Problem 19.

Look again at Figure 24(b). We cannot be sure just how low the graph actually goes in the interval $0 < x < 2$. But we do know that somewhere in the interval $0 < x < 2$ the graph of f must change direction (from decreasing to increasing). The points at which a graph changes direction are called **turning points.** In calculus, such points are called **local maxima** or **local mimima,** and techniques for locating them are given. So we shall not ask for the location of turning points in our graphs. Instead, we will use the following result from calculus that tells us the maximum number of turning points the graph of a polynomial function can have.

Theorem   If f is a polynomial function of degree n, then f has at most $n - 1$ turning points.

■

For example, the graph of $f(x) = x^2(x - 2)$ shown in Figure 24(b) is the graph of a polynomial of degree 3 and has $3 - 1 = 2$ turning points: one at $(0, 0)$ and the other somewhere in the interval $0 < x < 2$.

 Exploration A graphing utility can be used to locate the turning points of a graph. Graph $y = x^2(x - 2)$. Use TRACE to approximate the location of the turning point in the interval $0 < x < 2$. See Figure 25.

FIGURE 25

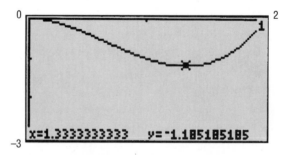

One last remark about Figure 24(b). Notice that the graph of $f(x) = x^2(x - 2)$ looks somewhat like the graph of $y = x^3$. In fact, for very large values of x, either positive or negative, there is little difference. To see for yourself, use your calculator to compare the values of $f(x) = x^2(x - 2)$ and $y = x^3$ for $x = -100,000$ and $x = 100,000$.

Theorem   For large values of x, either positive or negative, the graph of the polynomial

$$f(x) = a_n x^n + a_{n-1} x^{n-1} + \cdots + a_1 x + a_0$$

resembles the graph of the power function

$$y = a_n x^n$$

∎

The following box summarizes some features of the graph of a polynomial function.

Summary: Graph of a Polynomial Function $f(x) = a_n x^n + a_{n-1} x^{n-1} + \cdots + a_1 x + a_0,$ $a_n \neq 0$	Degree of the polynomial f: n Maximum number of turning points: $n - 1$ At zero of even multiplicity: graph of f touches x-axis At zero of odd multiplicity: graph of f crosses x-axis Between zeros, graph of f is either above the x-axis or below it. For large x, graph of f behaves like graph of $y = a_n x^n$.

E X A M P L E   6

Analyzing the Graph of a Polynomial Function

For the polynomial: $f(x) = x^3 + x^2 - 12x$

(a) Find the x- and y-intercepts of the graph of f.

(b) Determine whether the graph crosses or touches the x-axis at each x-intercept.

(c) Find the power function that the graph of f resembles for large values of x.

(d) Determine the maximum number of turning points on the graph of f.

(e) Use the x-intercepts and test numbers to find the intervals on which the graph of f is above the x-axis and the intervals on which the graph is below the x-axis.

(f) Put all the information together, and connect the points with a smooth, continuous curve to obtain the graph of f.

Solution   (a) The y-intercept is $f(0) = 0$. To find the x-intercepts, if any, we factor f:

$$f(x) = x^3 + x^2 - 12x = x(x^2 + x - 12) = x(x + 4)(x - 3)$$

Solving the equation $f(x) = x(x + 4)(x - 3) = 0$, we find that the x-intercepts, or zeros of f, are -4, 0, and 3.

(b) Since each zero of f is of multiplicity 1, the graph of f will cross the x-axis at each x-intercept.

(c) The graph of f resembles that of the power function $y = x^3$ for large values of x.

(d) The graph of f will contain at most two turning points.

(e) The three x-intercepts divide the x-axis into four intervals:

$$-\infty < x < -4 \quad -4 < x < 0 \quad 0 < x < 3 \quad 3 < x < \infty$$

To determine the sign of $f(x)$ in each interval, we choose test numbers and construct Figure 26(a).

(f) The graph of f is given in Figure 26(b).

FIGURE 26
$f(x) = x^3 + x^2 - 12x$

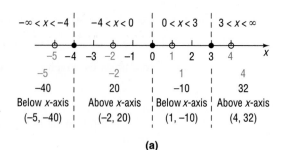

(a)

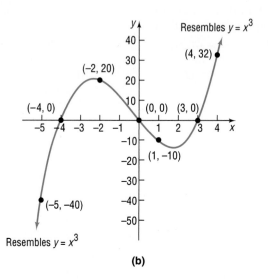

(b)

 Check: Graph $y = x^3 + x^2 - 12x$. Compare what you see with Figure 26. Use TRACE to locate the two turning points. ■

E X A M P L E   7

Analyzing the Graph of a Polynomial Function

Follow the instructions for Example 6 for the following polynomial:

$$f(x) = x^2(x - 4)(x + 1)$$

Solution   (a) The y-intercept is $f(0) = 0$. The x-intercepts satisfy the equation

$$f(x) = x^2(x - 4)(x + 1) = 0$$

So

$$x^2 = 0 \quad \text{or} \quad x - 4 = 0 \quad \text{or} \quad x + 1 = 0$$
$$x = 0 \qquad\qquad x = 4 \qquad\qquad x = -1$$

The x-intercepts are -1, 0, and 4.

(b) The intercept 0 is a zero of multiplicity 2, so the graph of f will touch the x-axis at 0; 4 and -1 are zeros of multiplicity 1, so the graph of f will cross the x-axis at 4 and -1.

(c) The graph of f resembles that of the power function $y = x^4$ for large values of x.

(d) The graph of f will contain at most three turning points.

(e) The three x-intercepts divide the x-axis into four intervals:

$$-\infty < x < -1 \qquad -1 < x < 0 \qquad 0 < x < 4 \qquad 4 < x < \infty$$

To determine the sign of $f(x)$ in each interval, we choose test numbers and construct Figure 27(a).

(f) The graph of f is given in Figure 27(b).

FIGURE 27

$f(x) = x^2(x - 4)(x + 1)$

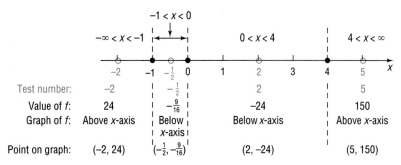

	$-\infty < x < -1$	$-1 < x < 0$	$0 < x < 4$	$4 < x < \infty$
Test number:	-2	$-\frac{1}{2}$	2	5
Value of f:	24	$-\frac{9}{16}$	-24	150
Graph of f:	Above x-axis	Below x-axis	Below x-axis	Above x-axis
Point on graph:	$(-2, 24)$	$\left(-\frac{1}{2}, -\frac{9}{16}\right)$	$(2, -24)$	$(5, 150)$

(a)

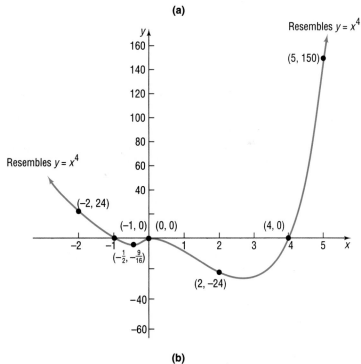

(b)

Check: Graph $y = x^2 (x - 4)(x + 1)$. Compare what you see with Figure 27. TRACE to locate the two turning points besides $(0,0)$. ∎

■ Now work Problem 37.

E X A M P L E 8 *Using a Graphing Utility to Graph a Polynomial Function*

Graph: $f(x) = \pi x^4 - \sqrt{7}x^3 + 5x - 1$

Solution   We begin by making the following observations:

1. The y-intercept is $f(0) = -1$
2. The graph will behave like $y = x^4$ for large values of x.
3. The graph will have no more than four x-intercepts and no more than three turning points.

These observations help us set the viewing rectangle for our first attempt at a complete graph. See Figure 28(a). Figure 28(b) shows the complete graph.

FIGURE 28

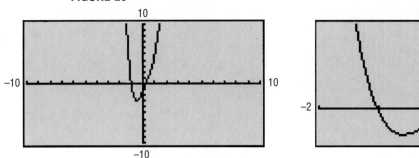

Using TRACE, we find the x-intercepts to be -1.01 and 0.20, correct to two decimal places. (Refer to the Appendix B, Section B.5, Approximations.) The only turning point is located at $(-0.57, -3.02)$, correct to two decimal places. ■

Summary

To sketch the graph of a polynomial function $y = f(x)$, follow these steps:

Steps for Graphing
a Polynomial

STEP 1: (a) Find the x-intercepts, if any, by solving the equation $f(x) = 0$.
(b) Find the y-intercept by letting $x = 0$ and finding the value of $f(0)$.

STEP 2: Determine whether the graph of f crosses or touches the x-axis at each x-intercept.

STEP 3: Find the power function that the graph of f resembles for large values of x.

STEP 4: Determine the maximum number of turning points on the graph of f.

STEP 5: Use the x-intercept(s) and test numbers to find the intervals on which the graph of f is above the x-axis and the intervals on which the graph is below the x-axis.

STEP 6: Plot the points obtained in Steps 1 and 5, and use the remaining information to connect them with a smooth, continuous curve.

3.2

Exercise 3.2

In Problems 1–10, determine which functions are polynomial functions. For those that are, state the degree. For those that are not, tell why not.

1. $f(x) = 4x + x^3$

2. $f(x) = 5x^2 + 4x^4$

3. $g(x) = \dfrac{1 - x^2}{2}$

4. $h(x) = 3 - \dfrac{1}{2}x$

5. $f(x) = 1 - \dfrac{1}{x}$

6. $f(x) = x(x - 1)$

7. $g(x) = x^{3/2} - x^2 + 2$

8. $h(x) = \sqrt{x}(\sqrt{x} - 1)$

9. $F(x) = 5x^4 - \pi x^3 + \dfrac{1}{2}$

10. $F(x) = \dfrac{x^2 - 5}{x^3}$

In Problems 11–18, use the graph of y = x⁴ to graph each function.

11. $f(x) = (x + 1)^4$

12. $f(x) = x^4 + 2$

13. $f(x) = \dfrac{1}{2}x^4$

14. $f(x) = -x^4$

15. $f(x) = 2(x + 1)^4 + 1$

16. $f(x) = 3 - (x + 2)^4$

17. $f(x) = -\dfrac{1}{2}(x - 2)^4 - 1$

18. $f(x) = 1 - 2(x + 1)^4$

In Problems 19–28, for each polynomial function, list each real zero and its multiplicity. Determine whether the graph crosses or touches the x-axis at each x-intercept.

19. $f(x) = 3(x - 7)(x + 3)^2$

20. $f(x) = 4(x + 4)(x + 3)^3$

21. $f(x) = 4(x^2 + 1)(x - 2)^3$

22. $f(x) = 2(x - 3)(x + 4)^3$

23. $f(x) = -2\left(x + \dfrac{1}{2}\right)^2(x^2 + 4)^2$

24. $f(x) = \left(x - \dfrac{1}{3}\right)^2(x - 1)^3$

25. $f(x) = (x - 5)^3(x + 4)^2$

26. $f(x) = (x + \sqrt{3})^2(x - 2)^4$

27. $f(x) = 3(x^2 + 8)(x^2 + 9)^2$

28. $f(x) = -2(x^2 + 3)^3$

In Problems 29–50, for each polynomial function f:

(a) Find the x- and y-intercepts of f.

(b) Determine whether the graph of f crosses or touches the x-axis at each x-intercept.

(c) Find the power function that the graph of f resembles for large values of x.

(d) Determine the maximum number of turning points on the graph of f.

(e) Use the x-intercept(s) and test numbers to find the intervals on which the graph of f is above and below the x-axis.

(f) Plot the points obtained in parts (a) and (e), and use the remaining information to connect them with a smooth, continuous curve.

29. $f(x) = (x - 1)^2$

30. $f(x) = (x - 2)^3$

31. $f(x) = x^2(x - 3)$

32. $f(x) = x(x + 2)^2$

33. $f(x) = 6x^3(x + 4)$

34. $f(x) = 5x(x - 1)^3$

35. $f(x) = -4x^2(x + 2)$

36. $f(x) = -\dfrac{1}{2}x^3(x + 4)$

37. $f(x) = x(x - 2)(x + 4)$

38. $f(x) = x(x + 4)(x - 3)$

39. $f(x) = 4x - x^3$

40. $f(x) = x - x^3$

41. $f(x) = x^2(x - 2)(x + 2)$

42. $f(x) = x^2(x - 3)(x + 4)$

43. $f(x) = x^2(x - 2)^2$

44. $f(x) = x^3(x - 3)$

45. $f(x) = x^2(x - 3)(x + 1)$

46. $f(x) = x^2(x - 3)(x - 1)$

47. $f(x) = x(x + 2)(x - 4)(x - 6)$

48. $f(x) = x(x - 2)(x + 2)(x + 4)$

49. $f(x) = x^2(x - 2)(x^2 + 3)$

50. $f(x) = x^2(x^2 + 1)(x + 4)$

51. Consult the illustration. Which of the following polynomial functions might have this graph? (More than one answer may be possible.)

(a) $y = -4x(x - 1)(x - 2)$

(b) $y = x^2(x - 1)^2(x - 2)$

(c) $y = 3x(x - 1)(x - 2)$

(d) $y = x(x - 1)^2(x - 2)^2$

(e) $y = x^3(x - 1)(x - 2)$

(f) $y = -x(1 - x)(x - 2)$

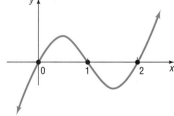

52. Consult the illustration. Which of the following polynomial functions might have this graph? (More than one answer may be possible.)

(a) $y = 2x^3(x - 1)(x - 2)^2$

(b) $y = x^2(x - 1)(x - 2)$

(c) $y = x^3(x - 1)^2(x - 2)$

(d) $y = x^2(x - 1)^2(x - 2)^2$

(e) $y = 5x(x - 1)^2(x - 2)$

(f) $y = -2x(x - 1)^2(2 - x)$

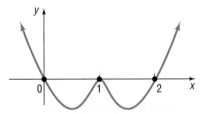

53. Consult the illustration. Which of the following polynomial functions might have this graph? (More than one answer may be possible.)

(a) $y = \dfrac{1}{2}(x^2 - 1)(x - 2)$

(b) $y = -\dfrac{1}{2}(x^2 + 1)(x - 2)$

(c) $y = (x^2 - 1)\left(1 - \dfrac{x}{2}\right)$

(d) $y = -\dfrac{1}{2}(x^2 - 1)^2(x - 2)$

(e) $y = \left(x^2 + \dfrac{1}{2}\right)(x^2 - 1)(2 - x)$

(f) $y = -(x - 1)(x - 2)(x + 1)$

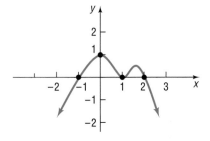

54. Consult the illustration. Which of the following polynomial functions might have this graph? (More than one answer may be possible.)

(a) $y = -\dfrac{1}{2}(x^2 - 1)(x - 2)(x + 1)$

(b) $y = -\dfrac{1}{2}(x^2 + 1)(x - 2)(x + 1)$

(c) $y = -\dfrac{1}{2}(x + 1)^2(x - 1)(x - 2)$

(d) $y = (x - 1)^2(x + 1)\left(1 - \dfrac{x}{2}\right)$

(e) $y = -(x - 1)^2(x - 2)(x + 1)$

(f) $y = -\left(x^2 + \dfrac{1}{2}\right)(x - 1)^2(x + 1)(x - 2)$

In Problems 55–60, graph each polynomial function. Approximate the x-intercepts, if any, and the turning points correct to two decimal places.

55. $y = \pi x^3 + \sqrt{2}x^2 - x - 2$

56. $y = -2x^3 + \pi x^2 + \sqrt{3}x + 1$

57. $y = 2x^4 - \pi x^3 + \sqrt{5}x - 4$

58. $y = -1.2x^4 + 0.5x^2 - \sqrt{3}x + 2$

59. $y = -2x^5 - \sqrt{2}x^2 - x - \sqrt{2}$

60. $y = \pi x^5 + \pi x^4 + \sqrt{3}x + 1$

61. Can the graph of a polynomial function have no y-intercept? Can it have no x-intercepts? Explain.

62. Write a few paragraphs that provide a general strategy for graphing a polynomial function. Be sure to mention the following: degree, intercepts, and turning points.

63. Make up a polynomial that has the following characteristics: crosses the x-axis at -1 and 4, touches the x-axis at 0 and 2, and is above the x-axis between 0 and 2. Give your polynomial to a fellow classmate and ask for a written critique of your polynomial.

64. Make up two polynomials, not of the same degree, with the following characteristics: crosses the x-axis at -2, touches the x-axis at 1, and is above the x-axis between -2 and 1. Give your polynomials to a fellow classmate and ask for a written critique of your polynomials.

65. The graph of a polynomial function is always smooth and continuous. Name a function studied earlier that is smooth and not continuous. Name one that is continuous, but not smooth.

66. Which of the following statements are true regarding the graph of the polynomial $f(x) = x^3 + bx^2 + cx + d$? (Give reasons for your conclusions.)

(a) It intersects the y-axis in one and only one point.

(b) It intersects the x-axis in at most three points.

(c) It intersects the x-axis at least once.

(d) For x very large, it behaves like the graph of $y = x^3$.

(e) It is symmetric with respect to the origin.

(f) It contains the origin.

3.3

Rational Functions

Ratios of integers are called *rational numbers*. Similarly, ratios of polynomial functions are called *rational functions*.

Rational Function

A **rational function** is a function of the form

$$R(x) = \frac{p(x)}{q(x)}$$

where p and q are polynomial functions and q is not the zero polynomial. The domain consists of all real numbers except those for which the denominator q is 0.

E X A M P L E   1   *Finding the Domain of a Rational Function*

(a) The domain of $R(x) = \dfrac{2x^2 - 4}{x + 5}$ consists of all real numbers x except -5.

(b) The domain of $R(x) = \dfrac{1}{x^2 - 4}$ consists of all real numbers x except -2 and 2.

(c) The domain of $R(x) = \dfrac{x^3}{x^2 + 1}$ consists of all real numbers.

(d) The domain of $R(x) = \dfrac{-x^2 + 2}{3}$ consists of all real numbers.

(e) The domain of $R(x) = \dfrac{x^2 - 1}{x - 1}$ consists of all real numbers x except 1.

It is important to observe that the functions

$$R(x) = \frac{x^2 - 1}{x - 1} \quad \text{and} \quad f(x) = x + 1$$

are not equal, since the domain of R is $\{x | x \neq 1\}$ and the domain of f is all real numbers.

■ Now work Problem 3.

If $R(x) = p(x)/q(x)$ is a rational function and if p and q have no common factors, then the rational function R is said to be in **lowest terms.** We shall assume throughout this section that all rational functions are in lowest terms.

For a rational function $R(x) = p(x)/q(x)$ in lowest terms, the zeros, if any, of the numerator are the x-intercepts of the graph of R and so will play a major role in graphing R. The zeros of the denominator of R [that is, the numbers x, if any, for which $q(x) = 0$], although not in the domain of R, also play a major role in the graph of R. We will discuss this role shortly.

We have already discussed the characteristics of the rational function $f(x) = 1/x$. (Refer to Example 8, page 62.) The next rational function we take up is $H(x) = 1/x^2$.

E X A M P L E   2   *Graphing* $y = \dfrac{1}{x^2}$

Graph: $H(x) = \dfrac{1}{x^2}$

Solution   The domain of $H(x) = 1/x^2$ consists of all real numbers x except 0. Thus, the graph has no y-intercept, because x can never equal 0. The graph has no x-intercept, because the equation $H(x) = 0$ has no solution. Therefore, the graph of H will not cross either of the coordinate axes. Because

$$H(-x) = \frac{1}{(-x)^2} = \frac{1}{x^2} = H(x)$$

TABLE 3

x	$H(x) = 1/x^2$
$\frac{1}{100}$	10,000
$\frac{1}{10}$	100
$\frac{1}{2}$	4
1	1
2	$\frac{1}{4}$
10	$\frac{1}{100}$
100	$\frac{1}{10,000}$

H is an even function, so its graph is symmetric with respect to the y-axis. Table 3 shows the behavior of $H(x) = 1/x^2$ for selected positive numbers x (we will use symmetry to obtain the graph of H when $x < 0$). From Table 3, it is apparent that, as the values of x approach (get closer to) 0, the values of $H(x)$ become larger and larger positive numbers. When this happens, we say that H is **unbounded in the positive direction.** We symbolize this by writing $H \to \infty$ (read as "H **approaches infinity**"). In calculus, **limits** are used to convey these ideas. There we use the symbolism $\lim\limits_{x \to 0} H(x) = \infty$, read the limit of $H(x)$ as x approaches zero equals infinity, to mean that $H(x) \to \infty$ as $x \to 0$. Look again at Table 3. As $x \to \infty$, the values of $H(x)$ approach 0. In calculus, this is symbolized by writing $\lim\limits_{x \to \infty} H = 0$. Figure 29 illustrates the graph.   ■

Sometimes the techniques of shifting, compressing, stretching, and reflection can be used to graph a rational function.

FIGURE 29

$$H(x) = \frac{1}{x^2}$$

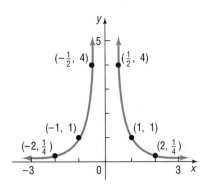

E X A M P L E   3   *Using Shifts, Reflections, and the Like, to Graph a Rational Function*

Graph the rational function: $R(x) = \dfrac{1}{(x-2)^2} + 1$

Solution   First, we take note of the fact that the domain of R consists of all real numbers except $x = 2$. To graph R, we start with the graph of $y = 1/x^2$. See Figure 30 for the steps.

FIGURE 30

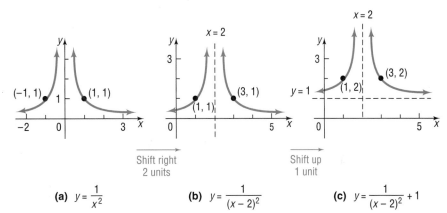

(a) $y = \dfrac{1}{x^2}$ (b) $y = \dfrac{1}{(x-2)^2}$ (c) $y = \dfrac{1}{(x-2)^2} + 1$

■ Now work Problem 19.   ■

Asymptotes

In Figure 30(c), notice that as the values of x become more negative, that is, as x becomes **unbounded in the negative direction** ($x \rightarrow -\infty$, read as "x **approaches negative infinity**"), the values $R(x)$ approach 1. In fact, we can conclude the following from Figure 30(c):

1. As $x \rightarrow -\infty$, the values $R(x)$ approach 1. $[\lim_{x \to -\infty} R(x) = 1]$

2. As x approaches 2, the values $R(x) \rightarrow \infty$. $[\lim_{x \to 2} R(x) = \infty]$

3. As $x \rightarrow \infty$, the values $R(x)$ approach 1. $[\lim_{x \to \infty} R(x) = 1]$

This behavior of the graph is depicted by the vertical line $x = 2$ and the horizontal line $y = 1$. These lines are called *asymptotes* of the graph, which we define as follows:

Let R denote a function:

Horizontal Asymptote	If, as $x \to -\infty$ or as $x \to \infty$, the values of $R(x)$ approach some fixed number L, then the line $y = L$ is a **horizontal asymptote** of the graph of R.

Vertical Asymptote	If, as x approaches some number c, the values $|R(x)| \to \infty$, then the line $x = c$ is a **vertical asymptote** of the graph of R.

Even though asymptotes of a function are not part of the graph of the function, they provide information about the way the graph looks. Figure 31 illustrates some of the possibilities.

FIGURE 31

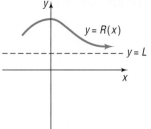

(a) As $x \to \infty$, the values of $R(x)$ approach L. That is, the points on the graph of R are getting closer to the line $y = L$; $y = L$ is a horizontal asymptote.

(b) As $x \to -\infty$, the values of $R(x)$ approach L. That is, the points on the graph of R are getting closer to the line $y = L$; $y = L$ is a horizontal asymptote.

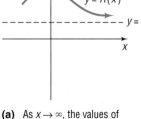

(c) As x approaches c, the values of $|R(x)| \to \infty$, That is, the points on the graph of R are getting closer to the line $x = c$; $x = c$ is a vertical asymptote.

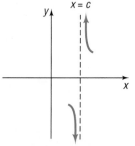

(d) As x approaches c, the values of $|R(x)| \to \infty$, That is, the points on the graph of R are getting closer to the line $x = c$; $x = c$ is a vertical asymptote.

Thus, an asymptote is a line that a certain part of the graph of a function gets closer and closer to, but never touches. However, other parts of the graph of the function may intersect a nonvertical asymptote. The graph of the function will never intersect a vertical asymptote. Notice that a horizontal asymptote, when it occurs, describes a certain behavior of the graph as $x \to \infty$ or as $x \to -\infty$, while a vertical asymptote, when it occurs, describes a certain behavior of the graph when x is close to some number c.

y
x

FIGURE 32
Oblique asymptote

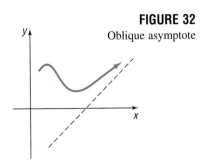

If an asymptote is neither horizontal nor vertical, it is called **oblique.** Figure 32 shows an oblique asymptote.

Finding Asymptotes

The vertical asymptotes, if any, of a rational function $R(x) = p(x)/q(x)$ are found by factoring the denominator $q(x)$. Suppose that $x - r$ is a factor of the denominator. Now, as x approaches r, symbolized as $x \rightarrow r$, the values of $x - r$ approach 0, causing the ratio to become unbounded, that is, causing $|R(x)| \rightarrow \infty$. Based on the definition, we conclude that the line $x = r$ is a vertical asymptote.

Theorem
Locating Vertical Asymptotes

A rational function $R(x) = p(x)/q(x)$ in lowest terms will have a vertical asymptote $x = r$ if $x - r$ is a factor of the denominator q. ∎

Thus, if r is a zero of the denominator of a rational function $R(x) = p(x)/q(x)$ in lowest terms, then R will have the vertical asymptote $x = r$.

Warning: If a rational function is not in lowest terms, an application of this theorem may result in an incorrect listing of vertical asymptotes.

E X A M P L E 4

Finding Vertical Asymptotes

Find the vertical asymptotes, if any, of the graph of each rational function.

(a) $R(x) = \dfrac{x}{x^2 - 4}$ (b) $F(x) = \dfrac{x + 3}{x - 1}$ (c) $H(x) = \dfrac{x^2}{x^2 + 1}$

Solution

(a) The zeros of the denominator $x^2 - 4$ are -2 and 2. Hence, the lines $x = -2$ and $x = 2$ are the vertical asymptotes of the graph of R.

(b) The only zero of the denominator is 1. Hence, the line $x = 1$ is the only vertical asymptote of the graph of F.

(c) The denominator has no zeros. Hence, the graph of H has no vertical asymptotes. ∎

As Example 4 points out, rational functions can have no vertical asymptotes, one vertical asymptote, or more than one vertical asymptote. However, the graph of a rational function will never intersect any of its vertical asymptotes. (Do you know why?)

Exploration: Graph each of the following rational functions:

$$y = \frac{1}{x - 1} \qquad y = \frac{x}{x - 1} \qquad y = \frac{x^2}{x - 1} \qquad y = \frac{x^3}{x - 1}$$

Each has the vertical asymptote $x = 1$. Use TRACE to see what happens as x approaches 1. Be sure to look at both sides of $x = 1$. ∎

The procedure for finding horizontal and oblique asymptotes is somewhat more involved. To find such asymptotes, we need to know how the values of a function behave as $x \rightarrow -\infty$ or as $x \rightarrow \infty$.

If a rational function $R(x)$ is **proper,** that is, if the degree of the numerator is less than the degree of the denominator, then as $x \rightarrow -\infty$ or as $x \rightarrow \infty$, the values of $R(x)$ approach 0. Consequently, the line $y = 0$ (the x-axis) is a horizontal asymptote of the graph.

Theorem If a rational function is proper, the line $y = 0$ is a horizontal asymptote of its graph. ∎

E X A M P L E   5

Finding Horizontal Asymptotes

Find the horizontal asymptotes, if any, of the graph of

$$R(x) = \frac{x - 12}{4x^2 + x + 1}$$

Solution The rational function R is proper, since the degree of the numerator (1) is less than the degree of the denominator (2). We conclude that the line $y = 0$ is a horizontal asymptote of the graph of R. ∎

To see why $y = 0$ is a horizontal asymptote of the function R in Example 5, we need to investigate the behavior of R for x unbounded. When x is unbounded, the numerator of R, which is $x - 12$, can be approximated by the power function $y = x$, while the denominator of R, which is $4x^2 + x + 1$, can be approximated by the power function $y = 4x^2$.

Thus,

$$R(x) = \underset{\substack{\uparrow \\ \text{For } x \text{ unbounded}}}{\frac{x - 12}{4x^2 + x + 1}} \approx \frac{x}{4x^2} = \underset{\substack{\uparrow \\ \text{As } x \to -\infty \text{ or } x \to \infty}}{\frac{1}{4x} \to 0}$$

This shows that the line $y = 0$ is a horizontal asymptote of the graph of R.

If a rational function $R(x)$ is **improper,** that is, if the degree of the numerator is greater than or equal to the degree of the denominator, we must use long division to write the rational function as the sum of a polynomial $f(x)$ plus a proper rational function $r(x)$. That is, we write

$$R(x) = \frac{p(x)}{q(x)} = f(x) + r(x)$$

where $f(x)$ is a polynomial and $r(x)$ is a proper rational function. Since $r(x)$ is proper, then $r(x) \to 0$ as $x \to -\infty$ or as $x \to \infty$. Thus,

$$R(x) = \frac{p(x)}{q(x)} \to f(x) \qquad \text{as} \qquad x \to -\infty \text{ or as } x \to \infty$$

The possibilities are listed next:

1. If $f(x) = b$, a constant, then the line $y = b$ is a horizontal asymptote of the graph of R.
2. If $f(x) = ax + b$, $a \neq 0$, then the line $y = ax + b$ is an oblique asymptote of the graph of R.
3. In all other cases, the graph of R approaches the graph of f, and there are no horizontal or oblique asymptotes.

The following examples demonstrate these conclusions.

E X A M P L E   6

Finding Horizontal or Oblique Asymptotes

Find the horizontal or oblique asymptotes, if any, of the graph of

$$H(x) = \frac{3x^4 - x^2}{x^3 - x^2 + 1}$$

Solution The rational function H is improper, since the degree of the numerator (4) is larger than the degree of the denominator (3). To find any horizontal or oblique asymptotes, we use long division:

$$
\begin{array}{r}
3x + 3 \\
x^3 - x^2 + 1 \overline{) 3x^4 - x^2 } \\
\underline{3x^4 - 3x^3 + 3x} \\
3x^3 - x^2 - 3x \\
\underline{3x^3 - 3x^2 + 3} \\
2x^2 - 3x - 3
\end{array}
$$

Thus,

$$
H(x) = \frac{3x^4 - x^2}{x^3 - x^2 + 1} = 3x + 3 + \frac{2x^2 - 3x - 3}{x^3 - x^2 + 1}
$$

Then, as $x \to -\infty$ or as $x \to \infty$, the remainder will behave as follows:

$$
\frac{2x^2 - 3x - 3}{x^3 - x^3 + 1} \approx \frac{2x^2}{x^3} = \frac{2}{x} \to 0
$$

Thus, as $x \to -\infty$ or as $x \to \infty$, we have $H(x) \to 3x + 3$. We conclude that the graph of the rational function H has an oblique asymptote $y = 3x + 3$. ∎

EXAMPLE 7

Finding Horizontal or Oblique Asymptotes

Find the horizontal or oblique asymptotes, if any, of the graph of

$$
R(x) = \frac{8x^2 - x + 2}{4x^2 - 1}
$$

Solution The rational function R is improper, since the degree of the numerator (2) equals the degree of the denominator (2). To find any horizontal or oblique asymptotes, we use long division:

$$
\begin{array}{r}
2 \\
4x^2 - 1 \overline{) 8x^2 - x + 2} \\
\underline{8x^2 - 2} \\
-x + 4
\end{array}
$$

Thus,

$$
R(x) = \frac{8x^2 - x + 2}{4x^2 - 1} = 2 + \frac{-x + 4}{4x^2 - 1}
$$

Then, as $x \to -\infty$ or as $x \to \infty$, the remainder will behave as follows:

$$
\frac{-x + 4}{4x^2 - 1} \approx \frac{-x}{4x^2} = \frac{-1}{4x} \to 0
$$

Thus, as $x \to -\infty$ or as $x \to \infty$, we have $R(x) \to 2$. We conclude that $y = 2$ is a horizontal asymptote of the graph. ∎

In Example 7, note that the quotient 2 obtained by long division is the quotient of the leading coefficients of the numerator polynomial and the denominator polynomial $\left(\frac{8}{4}\right)$. This means we can avoid the long division process for rational functions whose numerator and denominator *are of the same degree* and conclude that the quotient of the leading coefficients will give us the horizontal asymptote.

■   Now work Problem 35.

E X A M P L E   8   *Finding Horizontal or Oblique Asymptotes*

Find the horizontal or oblique asymptotes, if any, of the graph of

$$G(x) = \frac{2x^5 - x^3 + 2}{x^3 - 1}$$

Solution   The rational function G is improper, since the degree of the numerator (5) is larger than the degree of the denominator (3). To find any horizontal or oblique asymptotes, we use long division:

$$
\begin{array}{r}
2x^2 - 1 \\
x^3 - 1 \overline{)\,2x^5 - x^3 \qquad\quad + 2\,} \\
\underline{2x^5 \qquad\quad - 2x^2} \\
-x^3 + 2x^2 + 2 \\
\underline{-x^3 \qquad\quad + 1} \\
2x^2 + 1
\end{array}
$$

Thus,

$$G(x) = \frac{2x^5 - x^3 + 2}{x^3 - 1} = 2x^2 - 1 + \frac{2x^2 + 1}{x^3 - 1}$$

Then, as $x \to -\infty$ or as $x \to \infty$, the remainder will behave as follows:

$$\frac{2x^2 + 1}{x^3 - 1} \approx \frac{2x^2}{x^3} = \frac{2}{x} \to 0$$

Thus, as $x \to -\infty$ or as $x \to \infty$, we have $G(x) \to 2x^2 - 1$. We conclude that, for large values of x, the graph of G approaches the graph of $y = 2x^2 - 1$.   ■

We summarize next the procedure for finding horizontal and oblique asymptotes.

Finding Horizontal and Oblique Asymptotes of a Rational Function *R*   Consider the rational function

$$R(x) = \frac{p(x)}{q(x)} = \frac{a_n x^n + a_{n-1}x^{n-1} + \cdots + a_1 x + a_0}{b_m x^m + b_{m-1}x^{m-1} + \cdots + b_1 x + b_0}$$

in which the degree of the numerator is n and the degree of the denominator is m.

1. If R is a proper rational function ($n < m$), the graph will have the horizontal asymptote $y = 0$ (the x-axis).
2. If R is improper ($n \geq m$), use long division.
   (a) If $n = m$, the quotient obtained will be a number L ($= a_n/b_m$) and the line $y = L$ ($= a_n/b_m$) is a horizontal asymptote.
   (b) If $n = m + 1$, the quotient obtained is of the form $ax + b$ (a polynomial of degree 1), then the line $y = ax + b$ is an oblique asymptote.
   (c) If $n > m + 1$, the quotient obtained is a polynomial of degree 2 or higher, then R has neither a horizontal nor an oblique asymptote. In this case, for x unbounded, the graph of R will behave like the graph of the quotient.

Note: The graph of a rational function either has one horizontal or one oblique asymptote or else has no horizontal and no oblique asymptote.

Now we are ready to graph rational functions.

Graphing Rational Functions

We commented earlier that calculus provides the tools required to graph a polynomial function accurately. The same holds true for rational functions. However, we can gather together quite a bit of information about their graphs to get an idea of the general shape and position of the graph.

In the examples that follow, we will discuss the graph of a rational function $R(x) = p(x)/q(x)$ by applying the following steps:

Steps for Graphing a Rational Function

STEP 1: Locate the intercepts, if any, of the graph. The x-intercepts, if any, of $R(x) = p(x)/q(x)$ satisfy the equation $p(x) = 0$. The y-intercept, if there is one, is $R(0)$.

STEP 2: Test for symmetry. Replace x by $-x$ in $R(x)$. If $R(-x) = R(x)$, there is symmetry with respect to the y-axis; if $R(-x) = -R(x)$, there is symmetry with respect to the origin.

STEP 3: Locate the vertical asymptotes, if any, by factoring the denominator $q(x)$ of $R(x)$ and identifying its zeros.

STEP 4: Locate the horizontal or oblique asymptotes, if any, using the procedure given earlier. Determine points, if any, at which the graph of R intersects these asymptotes.

STEP 5: Determine where the graph is above the x-axis and where the graph is below the x-axis.

STEP 6: Graph the asymptotes, if any, found in Steps 3 and 4. Plot the points found in Steps 1, 4, and 5. Use all the information to connect the points with a smooth curve.

EXAMPLE 9   *Analyzing the Graph of a Rational Function*

Discuss the graph of the rational function: $R(x) = \dfrac{x-1}{x^2-4}$

Solution   First, we factor both the numerator and the denominator of R:

$$R(x) = \frac{x-1}{(x+2)(x-2)}$$

STEP 1: We locate the x-intercepts by finding the zeros of the numerator. By inspection, 1 is the only x-intercept. The y-intercept is $R(0) = \frac{1}{4}$.

STEP 2: Because

$$R(-x) = \frac{-x-1}{x^2-4}$$

we conclude that R is neither even nor odd. Thus, there is no symmetry with respect to the y-axis or the origin.

STEP 3: We locate the vertical asymptotes by factoring the denominator: $x^2 - 4 = (x+2)(x-2)$. The graph of R thus has two vertical asymptotes: the lines $x = -2$ and $x = 2$.

STEP 4: The degree of the numerator is less than the degree of the denominator, so R is proper and the line $y = 0$ (the x-axis) is a horizontal asymptote of the graph. To determine if the graph of R intersects the horizontal asymptote, we solve the equation $R(x) = 0$. The only solution is $x = 1$, so the graph of R intersects the horizontal asymptote at $(1, 0)$.

STEP 5: The zero of the numerator, 1, and the zeros of the denominator, -2 and 2, divide the x-axis into four intervals:

$$-\infty < x < -2 \qquad -2 < x < 1 \qquad 1 < x < 2 \qquad 2 < x < \infty$$

Now we construct Figure 33.

FIGURE 33

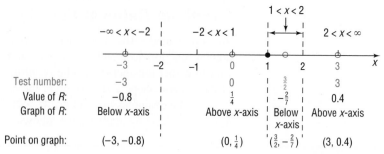

	$-\infty < x < -2$	$-2 < x < 1$	$1 < x < 2$	$2 < x < \infty$
Test number:	-3	0	$\frac{3}{2}$	3
Value of R:	-0.8	$\frac{1}{4}$	$-\frac{2}{7}$	0.4
Graph of R:	Below x-axis	Above x-axis	Below x-axis	Above x-axis
Point on graph:	$(-3, -0.8)$	$(0, \frac{1}{4})$	$(\frac{3}{2}, -\frac{2}{7})$	$(3, 0.4)$

STEP 6: We begin by graphing the asymptotes and plotting the points found in Steps 1, 4, and 5. See Figure 34(a). Next, we determine the behavior of the graph near the asymptotes. Since the x-axis is a horizontal asymptote and the graph lies below the x-axis for $-\infty < x < -2$, we can sketch a portion of the graph by placing a small arrow to the far left and under the x-axis. Since the line $x = -2$ is a vertical asymptote and the graph lies below the x-axis for $-\infty < x < -2$, we continue the sketch with an arrow placed well below the x-axis and approaching the line $x = -2$ on the left. Similar explanations account for the positions of the other portions of the graph. In particular, note how we use the facts that the graph lies above the x-axis for $-2 < x < 1$ and below the x-axis for $1 < x < 2$ and $(1, 0)$ is an x-intercept to draw the conclusion that the graph crosses the x-axis at $(1, 0)$. Figure 34(b) shows the complete sketch.

FIGURE 34

$$R(x) = \frac{x - 1}{x^2 - 4}$$

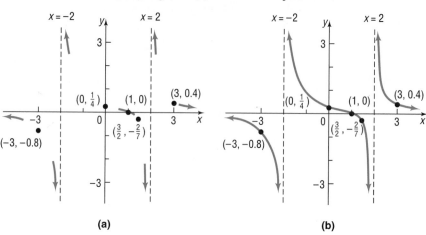

(a)   (b)

Check: Graph $R(x) = (x - 1)/(x^2 - 4)$ and compare what you see with Figure 34(b).

■ Now work Problem 37.

E X A M P L E   1 0

Analyzing the Graph of a Rational Function

Discuss the graph of the rational function: $R(x) = \dfrac{x^2 - 1}{x}$

Solution   **STEP 1:** The graph has two x-intercepts: -1 and 1. There is no y-intercept.
STEP 2: Since $R(-x) = -R(x)$, the function is odd and the graph is symmetric with respect to the origin.
STEP 3: The graph of $R(x)$ has the line $x = 0$ (the y-axis) as a vertical asymptote.

STEP 4: The rational function R is improper since the degree of the numerator (2) is larger than the degree of the denominator (1). To find any horizontal or oblique asymptotes, we use long division:

$$x\overline{)x^2 - 1}$$
$$\overline{x^2}$$
$$-1$$

The quotient is x, so the line $y = x$ is an oblique asymptote of the graph. To determine whether the graph of R intersects the asymptote $y = x$, we solve the equation $R(x) = x$:

$$R(x) = \frac{x^2 - 1}{x} = x$$
$$x^2 - 1 = x^2$$
$$-1 = 0 \quad \text{Impossible}$$

We conclude that the equation $(x^2 - 1)/x = x$ has no solution, so the graph of $R(x)$ does not intersect the line $y = x$.

STEP 5: The zeros of the numerator are -1 and 1; the denominator has the zero 0. Thus, we divide the x-axis into four intervals:

$$-\infty < x < -1 \qquad -1 < x < 0 \qquad 0 < x < 1 \qquad 1 < x < \infty$$

Now we construct Figure 35:

FIGURE 35

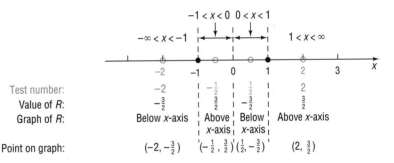

	$-\infty < x < -1$	$-1 < x < 0$	$0 < x < 1$	$1 < x < \infty$
Test number:	-2	$-\frac{1}{2}$	$\frac{1}{2}$	2
Value of R:	$-\frac{3}{2}$	$\frac{3}{2}$	$-\frac{3}{2}$	$\frac{3}{2}$
Graph of R:	Below x-axis	Above x-axis	Below x-axis	Above x-axis
Point on graph:	$(-2, -\frac{3}{2})$	$(-\frac{1}{2}, \frac{3}{2})$	$(\frac{1}{2}, -\frac{3}{2})$	$(2, \frac{3}{2})$

STEP 6: Figure 36(a) shows a partial graph using the facts we have gathered. The complete graph is given in Figure 36(b).

FIGURE 36

$$R(x) = \frac{x^2 - 1}{x}$$

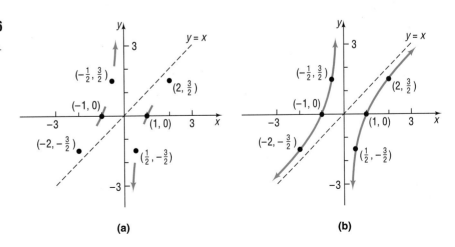

(a)   (b)

Check: Graph $R(x) = (x^2 - 1)/x$ and compare what you see with Figure 36. Could you have predicted that $y = x$ is an oblique asymptote? Graph $y = x$ and ZOOM-OUT. What do you observe? ∎

■ Now work Problem 45.

E X A M P L E   1 1   *Analyzing the Graph of a Rational Function*

Discuss the graph of the rational function: $R(x) = \dfrac{x^4 + 1}{x^2}$

Solution   **STEP 1:** The graph has no x-intercepts and no y-intercepts.
STEP 2: Since $R(-x) = R(x)$, the function is even and the graph is symmetric with respect to the y-axis.
STEP 3: The graph of $R(x)$ has the line $x = 0$ (the y-axis) as a vertical asymptote.
STEP 4: The rational function R is improper. To find any horizontal or oblique asymptotes, we use long division:

$$x^2\overline{\smash{\big)}x^4 + 1} \quad \begin{array}{c} x^2 \\ \hline \end{array}$$

$$\begin{array}{r} x^4 \\ \hline 1 \end{array}$$

The quotient is x^2, so the graph has no horizontal or oblique asymptotes. However, the graph of R will approach the graph of $y = x^2$ as $x \to -\infty$ and as $x \to \infty$.
STEP 5: The numerator has no zeros, and the denominator has one zero at 0. Thus, we divide the x-axis into the following two intervals: $-\infty < x < 0$ and $0 < x < \infty$. Now we set up Figure 37:

FIGURE 37

	$-\infty < x < 0$	$0 < x < \infty$
Test number:	-1	1
Value of R:	2	2
Graph of R:	Above x-axis	Above x-axis
Point on graph:	$(-1, 2)$	$(1, 2)$

STEP 6:   Figure 38 shows the graph.

FIGURE 38

$R(x) = \dfrac{x^4 + 1}{x^2}$

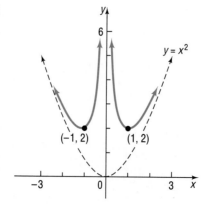

Check: Graph $R(x) = (x^4 + 1)/x^2$ and compare what you see with Figure 38. Use TRACE to find the two turning points. Enter $y = x^2$ and ZOOM-OUT. What do you see?   ■

■ Now work Problem 43.

E X A M P L E   1 2   *Analyzing the Graph of a Rational Function*

Discuss the graph of the rational function: $R(x) = \dfrac{3x^2 - 3x}{x^2 + x - 12}$

Solution   We factor R to get

$$R(x) = \frac{3x(x - 1)}{(x + 4)(x - 3)}$$

STEP 1: The graph has two x-intercepts: 0 and 1. The y-intercept is $R(0) = 0$.
STEP 2: There is no symmetry with respect to the y-axis or the origin.
STEP 3: The graph of R has two vertical asymptotes: $x = -4$ and $x = 3$.
STEP 4: Since the degree of the numerator equals the degree of the denominator, the graph has a horizontal asymptote. To find it, we form the quotient of the leading coefficient of the numerator (3) and the leading coefficient of the denominator (1). Thus, the graph of R has the horizontal asymptote $y = 3$. To find out whether the graph of R intersects the asymptote, we solve the equation $R(x) = 3$.

$$R(x) = \frac{3x^2 - 3x}{x^2 + x - 12} = 3$$
$$3x^2 - 3x = 3x^2 + 3x - 36$$
$$-6x = -36$$
$$x = 6$$

Thus, the graph intersects the line $y = 3$ only at $x = 6$, and $(6, 3)$ is a point on the graph of R.
STEP 5: The zeros of the numerator, 0 and 1, and the zeros of the denominator, -4 and 3, divide the x-axis into five intervals:

$$-\infty < x < -4 \qquad -4 < x < 0 \qquad 0 < x < 1 \quad 1 < x < 3 \qquad 3 < x < \infty$$

Now we can construct Figure 39:

FIGURE 39

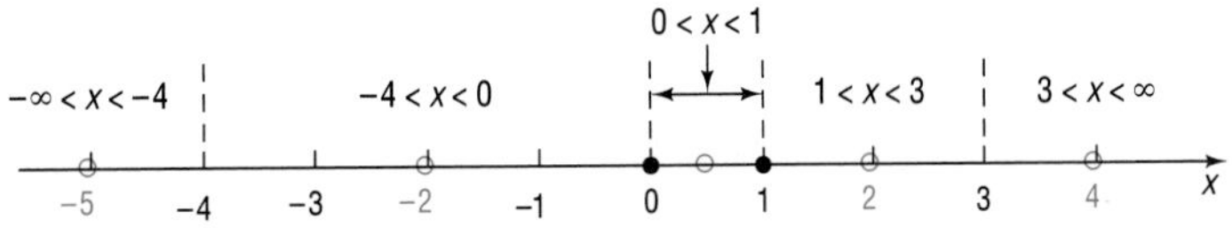

	$-\infty < x < -4$	$-4 < x < 0$	$0 < x < 1$	$1 < x < 3$	$3 < x < \infty$
Test number:	-5	-2	$\frac{1}{2}$	2	4
Value of R:	11.25	-1.8	$\frac{1}{15}$	-1	4.5
Graph of R:	Above x-axis	Below x-axis	Above x-axis	Below x-axis	Above x-axis
Point on graph:	$(-5, 11.25)$	$(-2, -1.8)$	$(\frac{1}{2}, \frac{1}{15})$	$(2, -1)$	$(4, 4.5)$

STEP 6: Figure 40(a) shows a partial graph. Notice that we have not yet used the fact that the line $y = 3$ is a horizontal asymptote, because we do not know yet whether the graph of R crosses or touches the line $y = 3$ at $(6, 3)$. To see whether the graph, in fact, crosses or touches the line $y = 3$, we plot an additional point to the right of $(6, 3)$. We use $x = 7$ to find $R(7) = \frac{63}{22} < 3$. Thus, the graph crosses $y = 3$ at $x = 6$. Because $y = 3$ is an asymptote of the graph, the

graph approaches the line $y = 3$ from above as $x \to -\infty$ and approaches the line $y = 3$ from below as $x \to \infty$. See Figure 40(b). The completed graph is shown in Figure 40(c).

FIGURE 40

$$R(x) = \frac{3x^2 - 3x}{x^2 + x - 12}$$

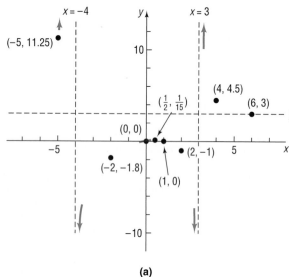

(a)

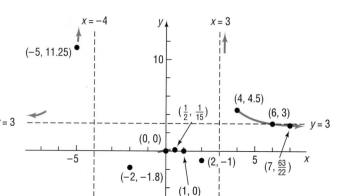

(b)

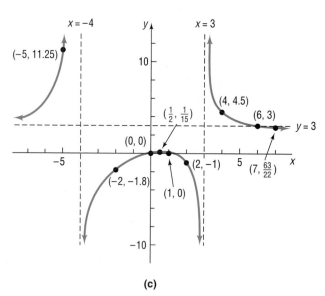

(c)

Exploration: Figure 41(a) shows the graph of

$$R(x) = \frac{3x^2 - 3x}{x^2 + x - 12}$$

Figure 41(b) shows the turning point located between 0 and 1, that is, (0.52, 0.06), correct to two decimal places. Figure 41(c) shows the turning point located to the right of $x = 6$, that is, (11.47, 12.74), correct to two decimal places. Also shown here is a view of the graph and its horizontal asymptote $y = 3$.

FIGURE 41

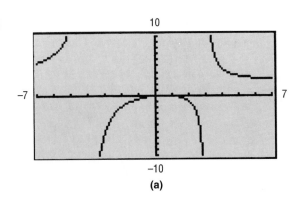

(a)

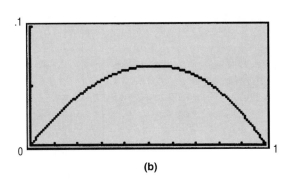

(b)

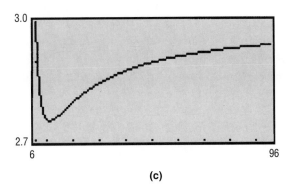

(c)

3.3

Exercise 3.3

In Problems 1–10, find the domain of each rational function.

1. $R(x) = \dfrac{4x}{x-3}$

2. $R(x) = \dfrac{5x^2}{3+x}$

3. $H(x) = \dfrac{-4x^2}{(x-2)(x+4)}$

4. $G(x) = \dfrac{6}{(x+3)(4-x)}$

5. $F(x) = \dfrac{3x(x-1)}{2x^2-5x-3}$

6. $Q(x) = \dfrac{-x(1-x)}{3x^2+5x-2}$

7. $R(x) = \dfrac{x}{x^3-8}$

8. $R(x) = \dfrac{x}{x^4-1}$

9. $H(x) = \dfrac{3x^2+x}{x^2+4}$

10. $G(x) = \dfrac{x-3}{x^4+1}$

In Problems 11–16, use the graph shown to find:

(a) The domain and range of each function

(b) The intercepts, if any

(c) Horizontal asymptotes, if any

(d) Vertical asymptotes, if any

(e) Oblique asymptotes, if any

11.

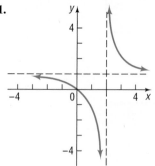

12.

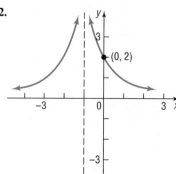

13.

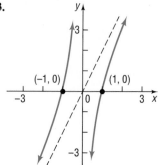

14.

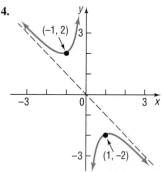

15.

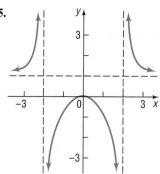

16.

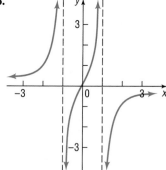

In Problems 17–26, graph each rational function using the methods of shifting, compression, stretching, and reflection, or the Steps 1 through 6 on page 215.

17. $R(x) = \dfrac{1}{(x-1)^2}$

18. $R(x) = \dfrac{3}{x}$

19. $H(x) = \dfrac{-2}{x+1}$

20. $G(x) = \dfrac{2}{(x+2)^2}$

21. $R(x) = \dfrac{1}{x^2 + 4x + 4}$

22. $R(x) = \dfrac{1}{x-1} + 1$

23. $F(x) = 1 - \dfrac{1}{x}$

24. $Q(x) = 1 + \dfrac{1}{x}$

25. $R(x) = \dfrac{x^2 - 4}{x^2}$

26. $R(x) = \dfrac{x-4}{x}$

In Problems 27–36, find the vertical, horizontal, and oblique asymptotes, if any, of each rational function. Do not graph.

27. $R(x) = \dfrac{3x}{x+4}$

28. $R(x) = \dfrac{3x+5}{x-6}$

29. $H(x) = \dfrac{x^4 + 2x^2 + 1}{x^2 - x + 1}$

30. $G(x) = \dfrac{-x^2 + 1}{x+5}$

31. $T(x) = \dfrac{x^3}{x^4 - 1}$

32. $P(x) = \dfrac{4x^5}{x^3 - 1}$

33. $Q(x) = \dfrac{5 - x^2}{3x^4}$

34. $F(x) = \dfrac{-2x^2 + 1}{2x^3 + 4x^2}$

35. $R(x) = \dfrac{3x^4 + 4}{x^3 + 3x}$

36. $R(x) = \dfrac{6x^2 + x + 12}{3x^2 - 5x - 2}$

In Problems 37–62, follow Steps 1 through 6 on page 215 to graph each rational function.

37. $R(x) = \dfrac{x + 1}{x(x + 4)}$

38. $R(x) = \dfrac{x}{(x - 1)(x + 2)}$

39. $R(x) = \dfrac{3x + 3}{2x + 4}$

40. $R(x) = \dfrac{2x + 4}{x - 1}$

41. $R(x) = \dfrac{3}{x^2 - 4}$

42. $R(x) = \dfrac{6}{x^2 - x - 6}$

43. $P(x) = \dfrac{x^4 + x^2 + 1}{x^2 - 1}$

44. $Q(x) = \dfrac{x^4 - 1}{x^2 - 4}$

45. $H(x) = \dfrac{x^3 - 1}{x^2 - 9}$

46. $G(x) = \dfrac{x^3 + 1}{x^2 + 2x}$

47. $R(x) = \dfrac{x^2}{x^2 + x - 6}$

48. $R(x) = \dfrac{x^2 + x - 12}{x^2 - 4}$

49. $G(x) = \dfrac{x}{x^2 - 4}$

50. $G(x) = \dfrac{3x}{x^2 - 1}$

51. $R(x) = \dfrac{3}{(x - 1)(x^2 - 4)}$

52. $R(x) = \dfrac{-4}{(x + 1)(x^2 - 9)}$

53. $H(x) = 4\dfrac{x^2 - 1}{x^4 - 16}$

54. $H(x) = \dfrac{x^2 + 4}{x^4 - 1}$

55. $F(x) = \dfrac{x^2 - 3x - 4}{x + 2}$

56. $F(x) = \dfrac{x^2 + 3x + 2}{x - 1}$

57. $R(x) = \dfrac{x^2 + x - 12}{x - 4}$

58. $R(x) = \dfrac{x^2 - x - 12}{x + 5}$

59. $F(x) = \dfrac{x^2 + x - 12}{x + 2}$

60. $G(x) = \dfrac{x^2 - x - 12}{x + 1}$

61. $R(x) = \dfrac{x(x - 1)^2}{(x + 3)^3}$

62. $R(x) = \dfrac{(x - 1)(x + 2)(x - 3)}{x(x - 4)^2}$

63. If the graph of a rational function R has the vertical asymptote $x = 4$, then the factor $x - 4$ must be present in the denominator of R. Explain why.

64. If the graph of a rational function R has the horizontal asymptote $y = 2$, then the degree of the numerator of R equals the degree of the denominator of R. Explain why.

65. Graph each of the following functions:

$$y = \dfrac{x^2 - 1}{x - 1} \qquad y = \dfrac{x^3 - 1}{x - 1} \qquad y = \dfrac{x^4 - 1}{x - 1} \qquad y = \dfrac{x^5 - 1}{x - 1}$$

Is $x = 1$ a vertical asymptote? Why not? What is happening for $x = 1$? What do you conjecture about $y = \dfrac{x^n - 1}{x - 1}$, $n \ge 1$ an integer, for $x = 1$?

66. Graph each of the following functions:

$$y = \dfrac{x^2}{x - 1} \qquad y = \dfrac{x^4}{x - 1} \qquad y = \dfrac{x^6}{x - 1} \qquad y = \dfrac{x^8}{x - 1}$$

What similarities do you see? What differences?

In Problems 67–72, graph each function and use TRACE to approximate the minimum value, correct to two decimal places.

67. $f(x) = x + \dfrac{1}{x}, x > 0$

68. $f(x) = 2x + \dfrac{9}{x}, x > 0$

69. $f(x) = x^2 + \dfrac{1}{x}, x > 0$

70. $f(x) = 2x^2 + \dfrac{9}{x}, x > 0$

71. $f(x) = x + \dfrac{1}{x^3}, x > 0$

72. $f(x) = 2x + \dfrac{9}{x^3}, x > 0$

73. Consult the illustration. Which of the following rational functions might have this graph? (More than one answer might be possible.)

(a) $y = \dfrac{4x^2}{x^2 - 4}$

(b) $y = \dfrac{x}{x^2 - 4}$

(c) $y = \dfrac{x^2}{x^2 - 4}$

(d) $y = \dfrac{x^2(x^2 + 1)}{(x^2 + 4)(x^2 - 4)}$

(e) $y = \dfrac{x^3}{x^2 - 4}$

(f) $y = \dfrac{x^2 - 4}{x^2}$

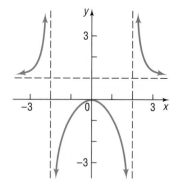

74. Consult the illustration. Which of the following rational functions might have this graph? (More than one answer may be possible.)

(a) $y = \dfrac{2x}{x^2 - 1}$

(b) $y = \dfrac{-3x}{x^2 - 1}$

(c) $y = \dfrac{x^3}{x^2 - 1}$

(d) $y = \dfrac{x^2 - 1}{-3x}$

(e) $y = \dfrac{-x^3}{(x^2 + 1)(x^2 - 1)}$

(f) $y = \dfrac{-x^2}{(x^2 + 1)(x^2 - 1)}$

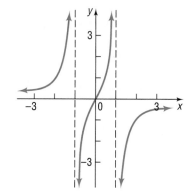

75. Consult the illustration. Make up a rational function that might have this graph. Compare yours with a friend's. What similarities do you see? What differences?

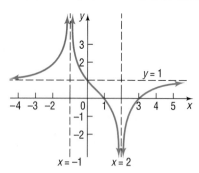

76. Can the graph of a rational function have both a horizontal and an oblique asymptote? Explain.

77. Write a few paragraphs that provide a general strategy for graphing a rational function. Be sure to mention the following: proper, improper, intercepts, and asymptotes.

78. Make up a rational function that has the following characteristics: crosses the x-axis at 3; touches the x-axis at -2; one vertical asymptote, $x = 1$; and one horizontal asymptote, $y = 2$. Give your rational function to a fellow classmate and ask for a written critique of your rational function.

79. Make up a rational function that has $y = 2x + 1$ as an oblique asymptote. Explain the methodology you used.

ISSION POSSIBLE

Chapter 3

RESPONDING TO A CHALLENGE BY THE CARDASSIANS.

While playing with virtual reality in your school's computer lab, you are suddenly faced with some very real and very nasty-looking Cardassians who have burst into your school through your computer network. As usual, they are very scornful of the intelligence of Earthlings, and when you try to protest, they throw this challenge at you. "Find a rational function defined by this graph," they say, "and we promise to support your reputation for intellectual achievement in every inter-galactic community this side of the Macklin Nebula. Otherwise, consider yourselves the laughing stock of the universe." And they departed back through the computer screen, laughing at what they perceived to be their extreme cleverness.

As you look at the computer screen now, all you can see is this graph:

You decide to tackle this as a team project to save the reputation of all the humans on this planet! The Cardassians mentioned "rational function," so you know the solution will have the form

$$R(x) = \frac{a_n x^n + a_{n-1}x^{n-1} + \ldots + a_0}{b_m x^m + b_{m-1}x^{m-1} + \ldots + b_0}$$

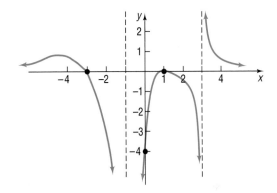

Sometimes it is easier to think of it in factored form.

1. You notice some things right away. There are two vertical asymptotes. How do they fit into the solution? Put them in.
2. There are two x-intercepts. Where do they show up in the solution? Put them in.
3. Use a graphing utility to check what you have so far. Does it look like the original graph? Notice that 1 and −3 appear to be the only real zeros. What can you tell about the multiplicity of each one? Do you need to change the multiplicity of either zero in your solution?
4. How can you account for the fact that the graph of the function goes to −∞ on both sides of one vertical asymptote and at the other vertical asymptote it goes to −∞ on one side and to ∞ on the other? Can you change the denominator in some way to account for this?
5. Is the information you have so far consistent with the horizontal asymptote $y = 0$? If not, how should you adjust your function so that it is consistent?
6. Is the information you have so far consistent with the y-intercept? If not, how should you adjust your function so that it is consistent?
7. Are you finished? Have you checked your solution on a graphing utility? Are you satisfied you have it right?
8. Are there other functions whose graph might look like the one given? If so, what common characteristics do they have? Will the Cardassians be more impressed if more than one correct answer is sent?
9. Compare your group's solution to those of other groups before typing the solution into the computer. Then send all the correct solutions off to cardass@slime.uni.

3.4

Remainder and Factor Theorems; Synthetic Division

Recall that when we divide one polynomial (the dividend) by another (the divisor) we obtain a quotient polynomial and a remainder, the remainder being either the zero polynomial or a polynomial whose degree is less than the degree of the divisor. To check our work, we verify that

$$(\text{Divisor})(\text{Quotient}) + \text{Remainder} = \text{Dividend}$$

This checking routine is the basis for a famous theorem called the **division algorithm* for polynomials,** which we now state without proof.

Theorem
Division Algorithm for Polynomials

If $f(x)$ and $g(x)$ denote polynomial functions and if $g(x)$ is not the zero polynomial, then there are unique polynomial functions $q(x)$ and $r(x)$ such that

$$\frac{f(x)}{g(x)} = q(x) + \frac{r(x)}{g(x)} \quad \text{or} \quad f(x) = g(x)q(x) + r(x) \qquad (1)$$

$$\underset{\text{dividend}}{\uparrow} \quad \underset{\text{divisor}}{\uparrow}\,\underset{\text{quotient}}{\uparrow} \quad \underset{\text{remainder}}{\uparrow}$$

where $r(x)$ is either the zero polynomial or a polynomial of degree less than that of $g(x)$. ∎

In equation (1), $f(x)$ is the **dividend,** $g(x)$ is the **divisor,** $q(x)$ is the **quotient,** and $r(x)$ is the **remainder.**

If the divisor $g(x)$ is a first-degree polynomial of the form

$$g(x) = x - c \qquad c \text{ a real number}$$

then the remainder $r(x)$ is either the zero polynomial or a polynomial of degree 0. Thus, for such divisors, the remainder is some number, say, R, and we may write

$$f(x) = (x - c)q(x) + R \qquad (2)$$

This equation is an identity in x and is true for all real numbers x. In particular, it is true when $x = c$. Thus, if $x = c$, then equation (2) becomes

$$f(c) = (c - c)q(c) + R$$
$$f(c) = R$$

and equation (2) takes the form

$$f(x) = (x - c)q(x) + f(c) \qquad (3)$$

We have now proved the following result, called the **Remainder Theorem:**

Remainder Theorem

Let f be a polynomial function. If $f(x)$ is divided by $x - c$, then the remainder is $f(c)$. ∎

---

*A systematic process in which certain steps are repeated a finite number of times is called an **algorithm.** Thus, long division is an algorithm.

E X A M P L E   1

Using the Remainder Theorem Find the remainder if $f(x) = x^3 - 4x^2 + 2x - 5$ is divided by:

(a) $x - 3$ (b) $x + 2$

Solution   (a)   We could use long division. However, it is much easier here to use the Remainder Theorem, which says the remainder is

$$f(3) = (3)^3 - 4(3)^2 + 2(3) - 5 = 27 - 36 + 6 - 5 = -8$$

(b)   To find the remainder when $f(x)$ is divided by $x + 2 = x - (-2)$, we evaluate

$$f(-2) = (-2)^3 - 4(-2)^2 + 2(-2) - 5 = -8 - 16 - 4 - 5 = -33$$

Thus, the remainder is -33.   ■

■   Now work Problem 1.

An important and useful consequence of the Remainder Theorem is the **Factor Theorem.**

Factor Theorem   Let f be a polynomial function. Then $x - c$ is a factor of $f(x)$ if and only if $f(c) = 0$.   ■

The Factor Theorem actually consists of two separate statements:

1. If $f(c) = 0$, then $x - c$ is a factor of $f(x)$.
2. If $x - c$ is a factor of $f(x)$, then $f(c) = 0$.

Thus, the proof requires two parts.

Proof   1. Suppose that $f(c) = 0$. Then, by equation (3), we have

$$f(x) = (x - c)q(x)$$

for some polynomial $q(x)$. That is, $x - c$ is a factor of $f(x)$.
2. Suppose that $x - c$ is a factor of $f(x)$. Then there is a polynomial function q such that

$$f(x) = (x - c)q(x)$$

Replacing x by c, we find that

$$f(c) = (c - c)q(c) = 0 \cdot q(c) = 0$$

This completes the proof.   ■

One use of the Factor Theorem is to determine whether a polynomial has a particular factor.

E X A M P L E   2

Using the Factor Theorem

Use the Factor Theorem to determine whether the function $f(x) = 2x^3 - x^2 + 2x - 3$ has the factor:

(a) $x - 1$ (b) $x + 3$

Solution   (a) Because $x - 1$ is of the form $x - c$ with $c = 1$, we find the value of $f(1)$:

$$f(1) = 2(1)^3 - (1)^2 + 2(1) - 3 = 2 - 1 + 2 - 3 = 0$$

By the Factor Theorem, $x - 1$ is a factor of $f(x)$.

(b) To test the factor $x + 3$, we first need to write it in the form $x - c$. Since $x + 3 = x - (-3)$, we find the value of $f(-3)$:

$$f(-3) = 2(-3)^3 - (-3)^2 + 2(-3) - 3 = -54 - 9 - 6 - 3 = -72$$

Because $f(-3) \neq 0$, we conclude from the Factor Theorem that $x - (-3) = x + 3$ is not a factor of $f(x)$. ∎

Synthetic Division

To find the quotient as well as the remainder when a polynomial function f of degree 1 or higher is divided by $g(x) = x - c$, a shortened version of long division, called **synthetic division,** makes the task simpler.

To see how synthetic division works, we will use long division to divide the polynomial $f(x) = 2x^3 - x^2 + 3$ by $g(x) = x - 3$.

$$
\begin{array}{r}
2x^2 + 5x + 15 \\
x - 3 \overline{\smash{)}\; 2x^3 - x^2 + 3} \\
\underline{2x^3 - 6x^2} \\
5x^2 \\
\underline{5x^2 - 15x} \\
15x + 3 \\
\underline{15x - 45} \\
48
\end{array}
$$

The process of synthetic division arises from rewriting the long division in a more compact form, using simpler notation. For example, in the long division above, the terms in color are not really necessary because they are identical to the terms directly above them. With these terms removed, we have

$$
\begin{array}{r}
2x^2 + 5x + 15 \\
x - 3 \overline{\smash{)}\; 2x^3 - x^2 + 3} \\
\underline{- 6x^2} \\
5x^2 \\
\underline{- 15x} \\
15x \\
\underline{- 45} \\
48
\end{array}
$$

Most of the x's that appear in this process can also be removed, provided we are careful about positioning each coefficient. In this regard, we will need to use 0 as the coefficient of x in the dividend, because that power of x is missing. Now we have

$$
\begin{array}{r}
2x^2 + 5x + 15 \\
x - 3 \overline{\smash{)}\; 2 -1 0 3} \\
\underline{-6} \\
5 \\
\underline{-15} \\
15 \\
\underline{-45} \\
48
\end{array}
$$

We can make this display more compact by moving the lines up until the numbers in color align horizontally:

$$
\begin{array}{r}
2x^2 + 5x + 1 \quad\ 5 \quad \text{Row 1} \\
x - 3\overline{)2 \quad -1 \quad\ 0 \quad\ 3} \quad \text{Row 2} \\
\underline{-6 \ -15 \ -45} \quad \text{Row 3} \\
\bigcirc \quad 5 \quad 15 \quad 48 \quad \text{Row 4}
\end{array}
$$

Now, if we place the leading coefficient of the quotient (2) in the circled position, the first three numbers in Row 4 are precisely the coefficients of the quotient, and the last number in Row 4 is the remainder. Thus, Row 1 is not really needed, so we can compress the process to three rows, where the bottom row contains the coefficients of both the quotient and the remainder:

$$
\begin{array}{r}
x - 3\overline{)2 \quad -1 \quad\ 0 \quad\ 3} \quad \text{Row 1} \\
\underline{-6 \ -15 \ -45} \quad \text{Row 2 (subtract)} \\
2 \quad 5 \quad 15 \quad 48 \quad \text{Row 3}
\end{array}
$$

Recall that the entries in Row 3 are obtained by subtracting the entries in Row 2 from those in Row 1. Rather than subtracting the entries in Row 2, we can change the sign of each entry and add. With this modification, our display will look like this:

$$
\begin{array}{r}
x - 3\overline{)2 \quad -1 \quad\ 0 \quad\ 3} \quad \text{Row 1} \\
\underline{6 \quad 15 \quad 45} \quad \text{Row 2 (add)} \\
2 \quad 5 \quad 15 \quad 48 \quad \text{Row 3}
\end{array}
$$

Notice that the entries in Row 2 are three times the prior entries in Row 3. Our last modification to the display replaces the $x - 3$ by 3. The entries in Row 3 give the quotient and the remainder as shown next.

$$
\begin{array}{r}
3\overline{)2 \quad -1 \quad\ 0 \quad\ 3} \quad \text{Row 1} \\
6 \quad 15 \quad 45 \quad \text{Row 2 (add)} \\
2 \quad 5 \quad 15 \quad 48 \quad \text{Row 3}
\end{array}
$$

Quotient   Remainder

$$2x^2 + 5x + 15 \qquad R = 48$$

Let's go through another example step by step. ■

E X A M P L E   3   *Using Synthetic Division to Find the Quotient and Remainder*

Use synthetic division to find the quotient and remainder when

$$f(x) = 3x^4 + 8x^2 - 7x + 4 \quad \text{is divided by} \quad g(x) = x - 1$$

Solution   **STEP 1:**   Write the dividend in descending powers of x. Then copy the coefficients, remembering to insert a 0 for any missing powers of x:

$$3 \quad 0 \quad 8 \quad -7 \quad 4 \quad \text{Row 1}$$

STEP 2:   Insert the usual division symbol. Since the divisor is $x - 1$, we insert 1 to the left of the division symbol

$$1\overline{)3 \quad 0 \quad 8 \quad -7 \quad 4} \quad \text{Row 1}$$

STEP 3:   Bring the 3 down two rows, and enter it in Row 3:

$$1\overline{)3 \quad 0 \quad 8 \quad -7 \quad 4} \quad \text{Row 1}$$

$$\downarrow \qquad\qquad\qquad\qquad \text{Row 2}$$

$$3 \qquad\qquad\qquad\qquad \text{Row 3}$$

STEP 4:   Multiply the latest entry in Row 3 by 1 and place the result in Row 2, but one column over to the right:

$$1\overline{)3 \quad 0 \quad 8 \quad -7 \quad 4} \quad \text{Row 1}$$

$$3 \qquad\qquad\qquad \text{Row 2}$$

$$3 \qquad\qquad\qquad\qquad \text{Row 3}$$

STEP 5:   Add the entry in Row 2 to the entry above it in Row 1, and enter the sum in Row 3:

$$1\overline{)3 \quad 0 \quad 8 \quad -7 \quad 4} \quad \text{Row 1}$$

$$3 \qquad\qquad\qquad \text{Row 2}$$

$$3 \quad 3 \qquad\qquad\qquad \text{Row 3}$$

STEP 6:   Repeat Steps 4 and 5 until no more entries are available in Row 1:

$$1\overline{)3 \quad 0 \quad 8 \quad -7 \quad 4} \quad \text{Row 1}$$

$$3 \quad 3 \quad 11 \quad 4 \quad \text{Row 2 (add)}$$

$$3 \quad 3 \quad 11 \quad 4 \quad 8 \quad \text{Row 3}$$

STEP 7:   The final entry in Row 3, an 8, is the remainder; the other entries in Row 3(3, 3, 11, and 4) are the coefficients (in descending order) of a polynomial whose degree is 1 less than that of the dividend; this is the quotient. Thus,

$$\text{Quotient} = 3x^3 + 3x^2 + 11x + 4 \qquad \text{Remainder} = 8$$

Check: (Divisor)(Quotient) + Remainder

$$= (x - 1)(3x^3 + 3x^2 + 11x + 4) + 8$$
$$= 3x^4 + 3x^3 + 11x^2 + 4x - 3x^3 - 3x^2 - 11x - 4 + 8$$
$$= 3x^4 + 8x^2 - 7x + 4 = \text{Dividend} \qquad \blacksquare$$

Let's do an example in which all seven steps are combined.

E X A M P L E   4   *Using Synthetic Division to Verify a Factor*

Use synthetic division to show that $g(x) = x + 3$ is a factor of

$$f(x) = 2x^5 + 5x^4 - 2x^3 + 2x^2 - 2x + 3$$

Solution   The divisor is $x + 3 = x - (-3)$, so the Row 3 entries will be multiplied by -3, entered in Row 2, and added to Row 1:

$$-3\overline{)2 \quad 5 \quad -2 \quad 2 \quad -2 \quad 3} \quad \text{Row 1}$$

$$-6 \quad 3 \quad -3 \quad 3 \quad -3 \quad \text{Row 2}$$

$$2 \quad -1 \quad 1 \quad -1 \quad 1 \quad 0 \quad \text{Row 3}$$

Because the remainder is 0, it follows that $f(-3) = 0$. Hence, by the Factor Theorem, $x - (-3) = x + 3$ is a factor of $f(x)$. $\qquad \blacksquare$

■ Now work Problem 13.

One important use of synthetic division is to find the value of a polynomial.

E X A M P L E   5   *Using Synthetic Division to Find the Value of a Polynomial*

Use synthetic division to find the value of $f(x) = -3x^4 + 2x^3 - x + 1$ at $x = -2$; that is, find $f(-2)$.

Solution   The Remainder Theorem tells us that the value of a polynomial function at c equals the remainder when the polynomial is divided by $x - c$. This remainder is the final entry of the third row in the process of synthetic division. We want $f(-2)$, so we divide by $x - (-2)$:

$$
\begin{array}{r}
-2\overline{)\begin{array}{rrrrr} -3 & 2 & 0 & -1 & 1 \end{array}} \\
\begin{array}{rrrrr} & 6 & -16 & 32 & -62 \end{array} \\
\hline
\begin{array}{rrrrr} -3 & 8 & -16 & 31 & -61 \end{array}
\end{array}
$$

The quotient is $q(x) = -3x^3 + 8x^2 - 16x + 31$; the remainder is $R = -61$. Because the remainder was found to be -61, it follows from the Remainder Theorem that $f(-2) = -61$.   ■

■   Now work Problem 33.

As Example 5 illustrates, we can use the process of synthetic division to find the value of a polynomial function at a number c as an alternative to merely substituting c for x. Compare the work required in Example 5 with the arithmetic involved in substituting:

$$
\begin{aligned}
f(-2) &= -3(-2)^4 + 2(-2)^3 - (-2) + 1 \\
&= -3(16) + 2(-8) + 2 + 1 \\
&= -48 - 16 + 2 + 1 = -61
\end{aligned}
$$

As you can see, finding $f(-2)$ may be easier using synthetic division.

Sometimes, neither substitution nor synthetic division avoids the need for messy calculations. Consider the problem of evaluating $f(x) = 3x^5 - 5x^4 + 0.2x^3 - 1.5x^2 + 2x - 6$ at $x = 1.2$. Here, a third method, using the *nested form* of a polynomial, is more helpful.

Nested Form of a Polynomial

Consider the polynomial $f(x) = 3x^3 - 5x^2 + 2x - 7$. We can factor $f(x)$ as follows:

$$
\begin{aligned}
f(x) &= 3x^3 - 5x^2 + 2x - 7 \\
&= (3x^3 - 5x^2 + 2x) - 7 \quad \text{Group terms containing } x. \\
&= (3x^2 - 5x + 2)x - 7 \quad \text{Factor } x. \\
&= [(3x^2 - 5x) + 2]x - 7 \quad \text{Regroup.} \\
&= [(3x - 5)x + 2]x - 7 \quad \text{Factor } x \text{ from parentheses.}
\end{aligned}
$$

Notice that this form of the polynomial contains only linear expressions. A polynomial function written in this way is said to be in **nested form.**

Let's look at some other examples.

E X A M P L E   6   *Writing a Polynomial in Nested Form*

Write each polynomial in nested form:

(a) $f(x) = 2x^2 - 3x + 5$ (b) $f(x) = 5x^3 - 6x^2 + 2$

(c) $f(x) = -5x^4 + 3x^3 - 2x^2 + 10x - 8$

Solution   (a)   We proceed as follows:

$$f(x) = (2x^2 - 3x) + 5 = (2x - 3)x + 5$$

The expression $(2x - 3)x + 5$ is the nested form of $2x^2 - 3x + 5$.

(b) $f(x) = (5x^3 - 6x^2) + 2 = (5x^2 - 6x) x + 2 = [(5x - 6)x]x + 2$

The expression $[(5x - 6)x]x + 2$ is the nested form of the polynomial $5x^3 - 6x^2 + 2$.

(c) $f(x) = (-5x^4 + 3x^3 - 2x^2 + 10x) - 8$

$= (-5x^3 + 3x^2 - 2x + 10)x - 8$

$= [(-5x^2 + 3x - 2)x + 10]x - 8$

$= \{[(-5x + 3)x - 2]x + 10\}x - 8$ ■

■ Now work Problem 39.

The advantage of evaluating a polynomial in nested form is that this method avoids the need to raise a number to a power, which on a calculator or computer can cause serious round-off errors. Furthermore, computers can perform the operation of addition much faster than the operation of multiplication, and the nested form requires fewer multiplications than the ordinary form of a polynomial. In Example 6(b), to evaluate $f(x) = 5x^3 - 6x^2 + 2$ in its ordinary form requires five multiplications and two additions:

$$\underset{\text{Addition}}{\underbrace{5 \cdot \overset{\text{Multiplication}}{\overbrace{x \cdot x \cdot x}} - 6 \cdot x \cdot x + 2}}$$

In nested form, it requires three multiplications and two additions:

$$\underset{\text{Addition}}{\underbrace{[(5 \cdot \overset{\text{Multiplication}}{\overbrace{x - 6) \cdot x}}] \cdot x + 2}}$$

Thus, to avoid errors and to speed up calculations, many computers evaluate polynomials by using the nested form.

E X A M P L E   7   *Using the Nested Form to Find the Value of a Polynomial*

Use the nested form and a calculator to evaluate the following polynomial at $x = 1.3$.

$$f(x) = 0.5x^3 - 1.2x^2 + 5.1x - 6.2$$

Solution   We write f in nested form as

$$f(x) = [(0.5x - 1.2)x + 5.1]x - 6.2$$

We start inside the parentheses by multiplying 0.5 by $x = 1.3$. Then subtract 1.2. Multiply the result by $x = 1.3$ and add 5.1. Multiply this result by $x = 1.3$ and subtract 6.2. The value is

$$f(1.3) = -0.4995$$

On a calculator, you would proceed as follows:

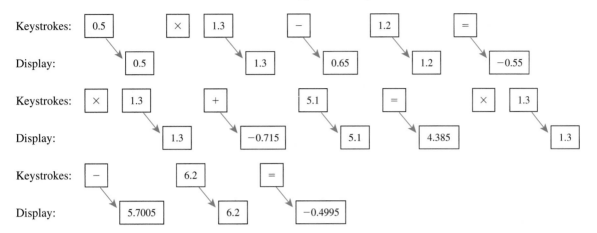

Notice that no memory key was used in this process.

Summary

Three ways to find the value of a polynomial function $f(x)$ at a number c:

1. Replace x by the number c to find $f(c)$.
2. Use synthetic division to divide $f(x)$ by $x - c$. The remainder is $f(c)$.
3. Write $f(x)$ in nested form and use a calculator to find $f(c)$.

3.4

Exercise 3.4

In Problems 1–10, use the Remainder Theorem to find the remainder when $f(x)$ is divided by $g(x)$.

1. $f(x) = x^3 - x^2 + 2x + 4$; $g(x) = x - 2$

2. $f(x) = x^3 + 2x^2 - 3x + 1$; $g(x) = x + 1$

3. $f(x) = 3x^3 + 2x^2 - x + 3$; $g(x) = x - 3$

4. $f(x) = -4x^3 + 2x^2 - x + 1$; $g(x) = x + 2$

5. $f(x) = x^5 - 4x^3 + x$; $g(x) = x + 3$

6. $f(x) = x^4 + x^2 + 2$; $g(x) = x - 2$

7. $f(x) = 4x^6 - 3x^4 + x^2 + 5$; $g(x) = x - 1$

8. $f(x) = x^5 + 5x^3 - 10$; $g(x) = x + 1$

9. $f(x) = 0.1x^3 + 0.2x$; $g(x) = x + 1.1$

10. $f(x) = 0.1x^2 - 0.2$; $g(x) = x + 2.1$

In Problems 11–22, use synthetic division to find the quotient $q(x)$ and remainder R when $f(x)$ is divided by $g(x)$.

11. $f(x) = x^3 - x^2 + 2x + 4$; $g(x) = x - 2$

12. $f(x) = x^3 + 2x^2 - 3x + 1$; $g(x) = x + 1$

13. $f(x) = 3x^3 + 2x^2 - x + 3$; $g(x) = x - 3$

14. $f(x) = -4x^3 + 2x^2 - x + 1$; $g(x) = x + 2$

15. $f(x) = x^5 - 4x^3 + x$; $g(x) = x + 3$

16. $f(x) = x^4 + x^2 + 2$; $g(x) = x - 2$

17. $f(x) = 4x^6 - 3x^4 + x^2 + 5$; $g(x) = x - 1$

18. $f(x) = x^5 + 5x^3 - 10$; $g(x) = x + 1$

19. $f(x) = 0.1x^3 + 0.2x$; $g(x) = x + 1.1$

20. $f(x) = 0.1x^2 - 0.2$; $g(x) = x + 2.1$

21. $f(x) = x^5 - 1$; $g(x) = x - 1$

22. $f(x) = x^5 + 1$; $g(x) = x + 1$

In Problems 23–32, use synthetic division to determine whether $x - c$ is a factor of $f(x)$.

23. $f(x) = 4x^3 - 3x^2 - 8x + 4$; $c = 2$

24. $f(x) = -4x^3 + 5x^2 + 8$; $c = -3$

25. $f(x) = 3x^4 - 6x^3 - 5x + 10$; $c = 2$

26. $f(x) = 4x^4 - 15x^2 - 4$; $c = 2$

27. $f(x) = 3x^6 + 82x^3 + 27$; $c = -3$

28. $f(x) = 2x^6 - 18x^4 + x^2 - 9$; $c = -3$

29. $f(x) = 4x^6 - 64x^4 + x^2 - 15$; $c = -4$

30. $f(x) = x^6 - 16x^4 + x^2 - 16$; $c = -4$

31. $f(x) = 2x^4 - x^3 + 2x - 1$; $c = \dfrac{1}{2}$

32. $f(x) = 3x^4 + x^3 - 3x + 1$; $c = -\dfrac{1}{3}$

In Problems 33–38, use synthetic division to find $f(c)$.

33. $f(x) = 5x^4 - 3x^2 + 1$; $c = 2$

34. $f(x) = -2x^3 + 3x^2 + 5$; $c = -2$

35. $f(x) = 4x^5 - 3x^3 + 2x - 1$; $c = -1$

36. $f(x) = -3x^4 + 3x^3 - 2x^2 + 5$; $c = -1$

37. $f(x) = 9x^{17} - 8x^{10} + 9x^8 + 5$; $c = 1$

38. $f(x) = 10x^{15} + 4x^{12} - 2x^5 + x^2$; $c = -1$

In Problems 39–48, write each polynomial in nested form.

39. $f(x) = 3x^3 - 2x^2 - 5x + 8$

40. $f(x) = -4x^3 + 5x^2 + 6$

41. $f(x) = 3x^4 - 6x^3 - 5x + 10$

42. $f(x) = 4x^4 - 15x^2 - 4$

43. $f(x) = 3x^6 - 82x^3 + 27$

44. $f(x) = 2x^6 - 18x^4 + x^2 - 9$

45. $f(x) = 4x^6 - 64x^4 + x^2 - 15$

46. $f(x) = x^6 - 16x^4 + x^2 - 16$

47. $f(x) = 2x^4 - x^3 + 2x - 1$

48. $f(x) = 3x^4 + x^3 - 3x + 1$

In Problems 49–58, use the nested form and a calculator to evaluate each polynomial at $x = 1.2$. Avoid using any memory key.

49. $f(x) = 3x^3 + 2x^2 - 5x + 8$

50. $f(x) = -4x^3 + 5x^2 + 6$

51. $f(x) = 3x^4 - 6x^3 - 5x + 10$

52. $f(x) = 4x^4 - 15x^2 - 4$

53. $f(x) = 3x^6 - 82x^3 + 27$

54. $f(x) = 2x^6 - 18x^4 + x^2 - 9$

55. $f(x) = 4x^6 - 64x^4 + x^2 - 15$

56. $f(x) = x^6 - 16x^4 + x^2 - 16$

57. $f(x) = 2x^4 - x^3 + 2x - 1$

58. $f(x) = 3x^4 + x^3 - 3x + 1$

59. Find k such that $f(x) = x^3 - kx^2 + kx + 2$ has the factor $x - 2$.

60. Find k such that $f(x) = x^4 - kx^3 + kx^2 + 1$ has the factor $x + 2$.

61. What is the remainder when $f(x) = 2x^{20} - 8x^{10} + x - 2$ is divided by $x - 1$?

62. What is the remainder when $f(x) = -3x^{17} + x^9 - x^5 + 2x$ is divided by $x + 1$?

63. Use the Factor Theorem to prove that $x - c$ is a factor of $x^n - c^n$ for any positive integer n.

64. Use the Factor Theorem to prove that $x + c$ is a factor of $x^n + c^n$ if $n \geq 1$ is an odd integer.

65. An IBM-AT microcomputer finds powers by multiplication. Suppose each multiplication of two numbers requires 33,333 nanoseconds and each addition or subtraction requires 500 nanoseconds. (Note that 1 nanosecond = 10^{-9} second.) If all other times are disregarded, how long will it take the computer to find the value of $f(x) = 2x^3 - 6x^2 + 4x - 10$ at $x = 2.013$ by:

(a) Replacing x by 2.013 in the expression for $f(x)$?

(b) Replacing x by 2.013 in the nested form for $f(x)$?

66. Using the microcomputer described in Problem 65, how long would it take by method (a) and by method (b) to find the value of $f(x) = ax^3 + bx^2 + cx + d$ for 5000 values of x?

67. *Programming Exercise* Write a program that simulates synthetic division and divides a polynomial by $x - c$. Your input should consist of the coefficients of the polynomial, in order from highest to lowest, followed by the number c. Your output should consist of the numbers that would appear in the third row of the process of synthetic division.

68. *Programming Exercise* Write a program that will evaluate a polynomial by using the nested form. Your input should consist of the coefficients of the polynomial, in order from highest to lowest, followed by the number at which the polynomial is to be evaluated. Test your program on the polynomials given in Problems 49–58.

 69. When dividing a polynomial by $x - c$, do you prefer to use long division or synthetic division? Does the value of c make a difference to you in choosing? Give reasons.

3.5
The Zeros of a Polynomial Function

The real zeros of a polynomial function f are the real solutions, if any, of the equation $f(x) = 0$. They are also the x-intercepts of the graph of f. For polynomial and rational functions, we have seen the importance of locating the zeros for graphing. In most cases, however, the zeros of a polynomial function are difficult to find. There are no nice formulas like the quadratic formula available to help us for polynomials of degree higher than 2. Although formulas do exist for solving any third- or fourth-degree polynomial equation, they are somewhat complicated. (If you are interested in learning about them, see Problems 75-83 for solving cubic equations; for 4th degree polynomial equations, consult a book on the theory of equations.) It has been proved that no general formulas exist for polynomial equations of degree 5 or higher. In this section, we shall learn some ways of detecting information about the character of the zeros, which, in turn, may help us find them or, at least, isolate them.

Our first theorem concerns the number of zeros a polynomial function may have. In counting the zeros of a polynomial, we count each zero as many times as its multiplicity.

Theorem
Number of Zeros

A polynomial function cannot have more zeros than its degree.

■

Proof

The proof is based on the Factor Theorem. If r is a zero of a polynomial function f, then $f(r) = 0$ and, hence, $x - r$ is a factor of $f(x)$. Thus, each zero corresponds to a factor of degree 1. Becuse f cannot have more first-degree factors than its degree, the result follows.

■

The next theorem, called **Descartes' Rule of Signs,** provides information about the number and location of the zeros of a polynomial function, so we know where to look for zeros. Descartes' Rule of Signs assumes that the polynomial is written in descending powers of x and requires that we count the number of variations in sign of the coefficients of $f(x)$ and $f(-x)$.

For example, the following polynomial function has two variations in the signs of coefficients:

$$f(x) = -3x^7 + 4x^4 + 3x^2 - 2x - 1$$
$$= \underbrace{-3x^7 + 0x^6 + 0x^5 + 4x^4}_{- \text{ to } +} + 0x^3 + \underbrace{3x^2 - 2x}_{+ \text{ to } -} - 1$$

Notice that we ignored the zero coefficients in $0x^6$, $0x^5$ and $0x^3$ in counting the number of variations in sign of $f(x)$. Replacing x by $-x$, we get

$$f(-x) = 3x^7 + 4x^4 + 3x^2 \underbrace{+ 2x}_{+ \text{ to } -} - 1$$

which has one variation in sign.

Theorem
Descartes' Rule of Signs

Let f denote a polynomial function.

The number of positive zeros of f either equals the number of variations in sign of the coefficients of $f(x)$ or else equals that number less an even integer.
The number of negative zeros of f either equals the number of variations in sign of the coefficients of $f(-x)$ or else equals that number less an even integer. ■

We shall not prove Descartes' Rule of Signs. Let's see how it is used.

E X A M P L E   1   *Using Descartes' Rule of Signs to Locate Zeros*

Discuss the zeros of: $f(x) = 3x^6 - 4x^4 + 3x^3 + 2x^2 - x - 3$

Solution   There are at most six zeros, because the polynomial is of degree 6. Since there are three variations in sign of the coefficients of $f(x)$, by Descartes' Rule of Signs we expect either three or one positive zero(s). To continue, we look at $f(-x)$:

$$f(-x) = 3x^6 - 4x^4 - 3x^3 + 2x^2 + x - 3$$

There are three variations in sign, so we expect either three or one negative zero(s). Equivalently, we now know that the graph of f has either three or one positive x-intercept(s) and three or one negative x-intercept(s). ■

■   Now work Problem 1.

Although we have not actually found the zeros in Example 1, we know something about the number of zeros and how many might be positive or negative. The next result, which you are asked to prove in Exercise 3.5 (Problem 74), is called the **Rational Zeros Theorem.** It provides information about the rational zeros of a polynomial with integer coefficients.

Theorem
Rational Zeros Theorem

Let f be a polynomial function of degree 1 or higher of the form

$$f(x) = a_n x^n + a_{n-1} x^{n-1} + \cdots + a_1 x + a_0 \qquad a_n \neq 0, a_0 \neq 0$$

where each coefficient is an integer. If p/q, in lowest terms, is a rational zero of f, then p must be a factor of a_0 and q must be a factor of a_n. ■

E X A M P L E   2   *Listing Potential Rational Zeros*

List the potential rational zeros of

$$f(x) = 2x^3 + 11x^2 - 7x - 6$$

Solution   Because f has integer coefficients, we may use the Rational Zeros Theorem. First, we list all the integers p that are factors of $a_0 = -6$ and all the integers q that are factors of $a_3 = 2$:

$$p: \quad \pm 1, \pm 2, \pm 3, \pm 6$$
$$q: \quad \pm 1, \pm 2$$

Now we form all possible ratios p/q:

$$\frac{p}{q}: \quad \pm 1,\ \pm 2,\ \pm 3,\ \pm 6,\ \pm\frac{1}{2},\ \pm\frac{3}{2}$$

If f has a rational zero, it will be found in this list, which contains 12 possibilities.

■

■ Now work Problem 13.

Be sure you understand what the Rational Zeros Theorem says: For a polynomial with integer coefficients, *if* there is a rational zero, it is one of those listed. There may not be any rational zeros. Synthetic division may be used to test each potential rational zero to determine whether it is indeed a zero. To make the work easier, the integers are usually tested first. Let's continue this example.

EXAMPLE 3   *Finding the Real Zeros of a Polynomial Function*

Continue working with Example 2 to find the zeros of

$$f(x) = 2x^3 + 11x^2 - 7x - 6$$

Solution   We gather all the information we can about the zeros:

STEP 1: There are at most three zeros.
STEP 2: By Descartes' Rule of Signs, there is one positive zero. Also, because

$$f(-x) = -2x^3 + 11x^2 + 7x - 6$$

there are two or no negative zeros.
STEP 3: Now we use the list of potential rational zeros obtained in Example 2: ± 1, ± 2, ± 3, ± 6, $\pm\frac{1}{2}$, $\pm\frac{3}{2}$. We choose to test the potential rational zero 1 using synthetic division:

$$
\begin{array}{r|rrr}
1) & 2 & 11 & -7 & -6 \\
   &   & 2 & 13 & 6 \\
\hline
   & 2 & 13 & 6 & 0
\end{array}
$$

The remainder is 0. Thus, 1 is a zero and $x - 1$ is a factor of f. The entries in the bottom row of this synthetic division can be used to factor f:

$$
\begin{aligned}
f(x) &= 2x^3 + 11x^2 - 7x - 6 \\
     &= (x - 1)(2x^2 + 13x + 6)
\end{aligned}
$$

Now any solution of the equation $2x^2 + 13x + 6 = 0$ will be a zero of f. Because of this, we call the equation $2x^2 + 13x + 6 = 0$ a **depressed equation** of f. Since the degree of the depressed equation of f is less than that of the original equation, we work with the depressed equation to find the zeros of f.
STEP 4: The depressed equation $2x^2 + 13x + 6 = 0$ is a quadratic equation with discriminant $b^2 - 4ac = 169 - 48 = 121 > 0$. It thus has two real solutions, which can be found by factoring:

$$2x^2 + 13x + 6 = (2x + 1)(x + 6) = 0$$
$$2x + 1 = 0 \qquad \text{or} \quad x + 6 = 0$$
$$x = -\frac{1}{2} \qquad\qquad x = -6$$

The zeros of f are -6, $-\frac{1}{2}$, and 1.   ■

Notice that the three zeros of f found in Example 3 are among those given in the list of potential rational zeros in Example 2. Also, notice that we can write f in factored form as

$$f(x) = 2x^3 + 11x^2 - 7x - 6 = (x - 1)(2x + 1)(x + 6) \tag{1}$$

E X A M P L E   4   *Finding the Real Zeros of a Polynomial Function*

Find the zeros of: $f(x) = 3x^5 - 2x^4 - 15x^3 + 10x^2 + 12x - 8$

Solution   We gather all the information we can about the zeros:

STEP 1:   There are at most five zeros.

STEP 2:   By Descartes' Rule of Signs, there are three or one positive zero(s). Also, because

$$f(-x) = -3x^5 - 2x^4 + 15x^3 + 10x^2 - 12x - 8$$

there are two or no negative zeros.

STEP 3:   To obtain the list of potential rational zeros, we write the factors p of $a_0 = -8$ and the factors q of $a_5 = 3$:

$$p: \quad \pm 1, \pm 2, \pm 4, \pm 8$$
$$q: \quad \pm 1, \pm 3$$

The potential rational zeros consist of all possible quotients p/q:

$$\frac{p}{q}: \quad \pm 1, \pm 2, \pm 4, \pm 8, \pm \frac{1}{3}, \pm \frac{2}{3}, \pm \frac{4}{3}, \pm \frac{8}{3}$$

We can test the potential rational zero 1 using synthetic division:

$1\overline{)}3$	-2	-15	10	12	-8
	3	1	-14	-4	8
3	1	-14	-4	8	0

The remainder is 0. Thus, 1 is a zero and $x - 1$ is a factor. As before, we use the entries in the bottom row of this synthetic division to factor f:

$$f(x) = 3x^5 - 2x^4 - 15x^3 + 10x^2 + 12x - 8$$
$$= (x - 1)(3x^4 + x^3 - 14x^2 - 4x + 8)$$

We now work with the first depressed equation of f:

$$3x^4 + x^3 - 14x^2 - 4x + 8 = 0$$

STEP 4:   Let $q_1(x) = 3x^4 + x^3 - 14x^2 - 4x + 8$. By Descartes' Rule of Signs, q_1 has two or no positive zeros. Also, because

$$q_1(-x) = 3x^4 - x^3 - 14x^2 + 4x + 8$$

q_1 has two or no negative zeros.

STEP 5:   The potential rational zeros of q_1 are the same as those listed earlier for f. We choose to test 1 again because it may be a repeated root:

$1\overline{)}3$	1	-14	-4	8
	3	4	-10	-14
3	4	-10	-14	-6

The remainder tells us that 1 is not a zero of q_1. Now we test -1:

$-1\overline{)}3$	1	-14	-4	8
	-3	2	12	-8
3	-2	-12	8	0

We find that -1 is a zero of q_1 and therefore $x - (-1) = x + 1$ is a factor of q_1. Thus, we have

$$f(x) = (x - 1)(x + 1)(3x^3 - 2x^2 - 12x + 8)$$

STEP 6:   We work now with the second depressed equation of f:

$$3x^3 - 2x^2 - 12x + 8 = 0$$

Let $q_2(x) = 3x^3 - 2x^2 - 12x + 8$. By Descartes' Rule of Signs, q_2 has two or no positive zeros. Also, because

$$q_2(-x) = -3x^3 - 2x^2 + 12x + 8$$

q_2 has one negative zero. The list of potential rational zeros of q_2 is the same as that of f. However, because 1 was not a zero of q_1, it cannot be a zero of q_2. Also, the fact that -1 is a zero of q_1 does not mean it cannot also be a zero of q_2 (that is, it could be a repeated root of q_1). We know there is one negative zero (which may not be rational), so we test -1 once more to determine whether it is a root of q_2:

$$
\begin{array}{r|rrrr}
-1) & 3 & -2 & -12 & 8 \\
   &   & -3 & 5 & 7 \\
\hline
   & 3 & -5 & -7 & 15
\end{array}
$$

It is not. Next, we choose to test -2:

$$
\begin{array}{r|rrrr}
-2) & 3 & -2 & -12 & 8 \\
   &   & -6 & 16 & -8 \\
\hline
   & 3 & -8 & 4 & 0
\end{array}
$$

We find that -2 is a zero, so $x - (-2) = x + 2$ is a factor. Thus, we have

$$f(x) = (x - 1)(x + 1)(x + 2)(3x^2 - 8x + 4) \tag{2}$$

STEP 7: The new depressed equation of f, $3x^2 - 8x + 4 = 0$, is a quadratic equation with a discriminant of $b^2 - 4ac = (-8)^2 - 4(3)(4) = 16$. Therefore, this equation has two real solutions, and, in this case, we find them by factoring:

$$3x^2 - 8x + 4 = 0$$
$$(3x - 2)(x - 2) = 0$$
$$3x - 2 = 0 \quad \text{or} \quad x - 2 = 0$$
$$x = \frac{2}{3} \qquad\qquad x = 2$$

The zeros of f are -2, -1, $\frac{2}{3}$, 1, and 2. ∎

■ Now work Problem 25.

The procedure outlined in Example 4 for finding the zeros of a polynomial can also be used to solve polynomial equations.

E X A M P L E 5 *Solving a Polynomial Equation*

Solve the equation: $x^5 - 5x^4 + 12x^3 - 24x^2 + 32x - 16 = 0$

Solution The solutions of this equation are the zeros of the polynomial function

$$f(x) = x^5 - 5x^4 + 12x^3 - 24x^2 + 32x - 16$$

STEP 1: There are at most five real solutions.
STEP 2: By Descartes' Rule of Signs, there are five, three, or one positive solution(s). Because

$$f(-x) = -x^5 - 5x^4 - 12x^3 - 24x^2 - 32x - 16$$

there are no negative solutions.
STEP 3: Because $a_5 = 1$ and there are no negative solutions, the potential rational solutions are the positive integers 1, 2, 4, 8, and 16. We test the potential rational solution 1 first, using synthetic division:

$$\begin{array}{r|rrrrrr} 1) & 1 & -5 & 12 & -24 & 32 & -16 \\ & & 1 & -4 & 8 & -16 & 16 \\ \hline & 1 & -4 & 8 & -16 & 16 & 0 \end{array}$$

Thus, 1 is a solution and

$$x^5 - 5x^4 + 12x^3 - 24x^2 + 32x - 16 = (x - 1)(x^4 - 4x^3 + 8x^2 - 16x + 16)$$

The remaining solutions satisfy the depressed equation

$$x^4 - 4x^3 + 8x^2 - 16x + 16 = 0$$

STEP 4: The potential rational solutions are still 1, 2, 4, 8 and 16. We test 1 first, since it may be a repeated solution:

$$\begin{array}{r|rrrrr} 1) & 1 & -4 & 8 & -16 & 16 \\ & & 1 & -3 & 5 & -11 \\ \hline & 1 & -3 & 5 & -11 & 5 \end{array}$$

Thus, 1 is not a solution of the depressed equation. We try 2 next:

$$\begin{array}{r|rrrrr} 2) & 1 & -4 & 8 & -16 & 16 \\ & & 2 & -4 & 8 & -16 \\ \hline & 1 & -2 & 4 & -8 & 0 \end{array}$$

Thus, 2 is a solution of the depressed equation and

$$x^5 - 5x^4 + 12x^3 - 24x^2 + 32x - 16 = (x - 1)(x - 2)(x^3 - 2x^2 + 4x - 8)$$

The remaining solutions satisfy the new depressed equation

$$x^3 - 2x^2 + 4x - 8 = 0$$

STEP 5: The potential rational solutions are now 1, 2, 4, and 8. We know 1 is not a solution (why?), so we start with 2:

$$\begin{array}{r|rrrr} 2) & 1 & -2 & 4 & -8 \\ & & 2 & 0 & 8 \\ \hline & 1 & 0 & 4 & 0 \end{array}$$

Thus, 2 is a solution of the new depressed equation and is a repeated solution of the original equation, so

$$x^5 - 5x^4 + 12x^3 - 24x^2 + 32x - 16 = (x - 1)(x - 2)^2(x^2 + 4)$$

The remaining solutions satisfy the depressed equation

$$x^2 + 4 = 0$$

which has no real solutions.

STEP 6: Thus, the real solutions are 1 and 2 (the latter being a repeated solution). ∎

■ Now work Problem 37.

E X A M P L E   6   *Finding the Real Zeros of a Polynomial*

Use Descartes' Rule of Signs and the Rational Zeros Theorem to find the real zeros of the polynomial function

$$g(x) = x^5 - x^4 - x^3 + x^2 - 2x + 2$$

Use the zeros to factor g over the real numbers. Then graph g.

Solution   **STEP 1:** There are at most five zeros.
STEP 2: There are four, two, or no positive zeros. Also, because

$$g(-x) = -x^5 - x^4 + x^3 + x^2 + 2x + 2$$

there is one negative zero.

STEP 3: The potential rational zeros of g are ± 1, ± 2. We test 1:

$$
\begin{array}{r|rrrrrr}
1) & 1 & -1 & -1 & 1 & -2 & 2 \\
 & & 1 & 0 & -1 & 0 & -2 \\
\hline
 & 1 & 0 & -1 & 0 & -2 & 0
\end{array}
$$

Thus, 1 is a zero, so $x - 1$ is a factor and

$$g(x) = (x - 1)(x^4 - x^2 - 2)$$

STEP 4: The depressed equation $x^4 - x^2 - 2 = 0$ is quadratic in form and can be factored as follows:

$$x^4 - x^2 - 2 = 0$$
$$(x^2 - 2)(x^2 + 1) = 0$$
$$x^2 - 2 = 0 \quad \text{or} \quad x^2 + 1 = 0$$
$$x = \pm\sqrt{2}$$

Because $x^2 + 1 = 0$ has no real solution, the depressed equation has only the two solutions $\sqrt{2}$ and $-\sqrt{2}$.

Thus, the zeros of g are $-\sqrt{2}$, 1, and $\sqrt{2}$. The factored form of g over the real numbers is

$$
\begin{aligned}
g(x) &= x^5 - x^4 - x^3 + x^2 - 2x + 2 \\
&= (x - 1)(x^4 - x^2 + 2) \\
&= (x - 1)(x^2 - 2)(x^2 + 1) \\
&= (x - 1)(x - \sqrt{2})(x + \sqrt{2})(x^2 + 1)
\end{aligned}
$$

Now we construct Figure 42:

FIGURE 42

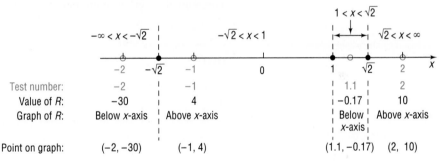

The graph of g has at most four turning points. For large values of x, the graph will behave like the graph of $y = x^5$. Figure 43 illustrates the graph of g.

FIGURE 43

$g(x) = x^5 - x^4 - x^3 + x^2 - 2x + 2$

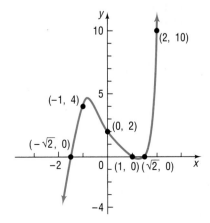

Check: Graph $g(x) = x^5 - x^4 - x^3 + x^2 - 2x + 2$. Use TRACE to locate the turning points and to verify the x-intercepts. ■

■ Now work Problem 47.

In Example 6, the quadratic factor $x^2 + 1$ that appears in the factored form of $g(x)$ is called *irreducible,* because the polynomial $x^2 + 1$ cannot be factored over the real numbers. In general, we say that a quadratic factor $ax^2 + bx + c$ is **irreducible** if it cannot be factored over the real numbers, that is, if it is prime over the real numbers.

Refer back to the polynomial function f of Example 4. We found that f has five real zeros, so, by the Factor Theorem, its factored form will contain five linear factors. The polynomial function g of Example 6 has three real zeros, and its factored form contains three linear factors and one irreducible quadratic factor. The next result tells us what to expect when we factor a polynomial.

Theorem   Every polynomial function (with real coefficients) can be uniquely factored into a product of linear factors and/or irreducible quadratic factors. ■

We shall prove this result in Section 3.7, and, in fact, we shall draw several additional conclusions about the zeros of a polynomial function. One conclusion is worth noting now. If a polynomial (with real coefficients) is of odd degree, then it must contain at least one linear factor. (Do you see why?) Therefore, it will have at least one real zero.

Corollary   A polynomial function (with real coefficients) of odd degree has at least one real zero. ■

Summary

To obtain information about the real zeros of a polynomial function, follow these steps:

Procedure for Finding
the Real Zeros
of a Polynomial
Function

STEP 1:   Use the degree of the polynomial to determine the maximum number of zeros.

STEP 2:   Use Descartes' Rule of Signs to determine the possible number of positive zeros and negative zeros.

STEP 3:   (a) If the polynomial has integer coefficients, use the Rational Zeros Theorem to identify those rational numbers that potentially can be zeros.
(b) Use synthetic division to test each potential rational zero.
(c) Each time a zero (and thus a factor) is found, repeat Steps 2 and 3 on the depressed equation.

STEP 4:   In attempting to find the zeros, remember to use (if possible) the factoring techniques you already know (special products, factoring by grouping, and so on).

If these procedures fail to locate all the zeros, you may have to be satisfied with "estimating" or "approximating" the zeros, the subject of the next section.

HISTORICAL FEATURE   ■ Formulas for the solution of third- and fourth-degree polynomial equations exist, and, while not very practical, they do have an interesting history.

In the 1500's in Italy, mathematical contests were a popular pastime, and persons possessing methods for solving problems kept them secret. (Solutions that were published were already common knowledge.) Niccolo of Brescia (1499–1557), commonly referred to as Tartaglia ("the stammerer"), had the secret for solving cubic (third-degree) equations, which gave him a decided advantage in the contests. Girolamo Cardano (1501–1576) found out that Tartaglia had the secret, and, being interested in cubics, he requested it from Tartaglia. The reluctant Tartaglia hesitated for some time, but finally, swearing Cardano to secrecy with midnight oaths by candlelight, told him the secret. Cardano then published the solution in his book *Ars Magna* (1545), giving Tartaglia the credit but rather compromising the secrecy. Tartaglia exploded into bitter recriminations, and each wrote pamphlets that reflected on the other's mathematics, moral character, and ancestry. Tartaglia's method is discussed in Problems 75–83 in Exercise 3.5.

The quartic (fourth-degree) equation was solved by Cardano's student Lodovico Ferrari, and this solution also was included, with credit and this time with permission, in the *Ars Magna*.

Attempts were made to solve the fifth-degree equation in similar ways, all of which failed. In the early 1800's, P. Ruffini, Niels Abel, and Evariste Galois all found ways to show that it is not possible to solve fifth-degree equations by formula, but the proofs required the introduction of new methods. Galois' methods eventually developed into a large part of modern algebra.   ■

3.5

Exercise 3.5

In Problems 1–12, tell the maximum number of zeros each polynomial function may have. Then, use Descartes' Rule of Signs to determine how many positive and how many negative zeros each polynomial function may have. Do not attempt to find the zeros.

1. $f(x) = -4x^7 + x^3 - x^2 + 2$

2. $f(x) = 5x^4 + 2x^2 - 6x - 5$

3. $f(x) = 2x^6 - 3x^2 - x + 1$

4. $f(x) = -3x^5 + 4x^4 + 2$

5. $f(x) = 3x^3 - 2x^2 + x + 2$

6. $f(x) = -x^3 - x^2 + x + 1$

7. $f(x) = -x^4 + x^2 - 1$

8. $f(x) = x^4 + 5x^3 - 2$

9. $f(x) = x^5 + x^4 + x^2 + x + 1$

10. $f(x) = x^5 - x^4 + x^3 - x^2 + x - 1$

11. $f(x) = x^6 - 1$

12. $f(x) = x^6 + 1$

In Problems 13–24, list the potential rational zeros of each polynomial function. Do not attempt to find the zeros.

13. $f(x) = 3x^4 - 3x^3 + x^2 - x + 1$

14. $f(x) = x^5 - x^4 + 2x^2 + 3$

15. $f(x) = x^5 - 6x^2 + 9x - 3$

16. $f(x) = 2x^5 - x^4 - x^2 + 1$

17. $f(x) = -4x^3 - x^2 + x + 2$

18. $f(x) = 6x^4 - x^2 + 2$

19. $f(x) = 3x^4 - x^2 + 2$

20. $f(x) = -4x^3 + x^2 + x + 2$

21. $f(x) = 2x^5 - x^3 + 2x^2 + 4$

22. $f(x) = 3x^5 - x^2 + 2x + 3$

23. $f(x) = 6x^4 + 2x^3 - x^2 + 2$

24. $f(x) = -6x^3 - x^2 + x + 3$

In Problems 25–36, use Descartes' Rule of Signs and the Rational Zeros Theorem to find all the real zeros of each polynomial function. Use the zeros to factor f over the real numbers.

25. $f(x) = x^3 + 2x^2 - 5x - 6$

26. $f(x) = x^3 + 8x^2 + 11x - 20$

27. $f(x) = 2x^3 - x^2 + 2x - 1$

28. $f(x) = 2x^3 + x^2 + 2x + 1$

29. $f(x) = x^4 + x^2 - 2$

30. $f(x) = x^4 - 3x^2 - 4$

31. $f(x) = 4x^4 + 7x^2 - 2$

32. $f(x) = 4x^4 + 15x^2 - 4$

33. $f(x) = x^4 + x^3 - 3x^2 - x + 2$

34. $f(x) = x^4 - x^3 - 6x^2 + 4x + 8$

35. $f(x) = 4x^5 - 8x^4 - x + 2$

36. $f(x) = 4x^5 + 12x^4 - x - 3$

In Problems 37–46, solve each equation in the real number system.

37. $x^4 - x^3 + 2x^2 - 4x - 8 = 0$

38. $2x^3 + 3x^2 + 2x + 3 = 0$

39. $3x^3 + 4x^2 - 7x + 2 = 0$

40. $2x^3 - 3x^2 - 3x - 5 = 0$

41. $3x^3 - x^2 - 15x + 5 = 0$

42. $2x^3 - 11x^2 + 10x + 8 = 0$

43. $x^4 + 4x^3 + 2x^2 - x + 6 = 0$

44. $x^4 - 2x^3 + 10x^2 - 18x + 9 = 0$

45. $x^3 - \frac{2}{3}x^2 + \frac{8}{3}x + 1 = 0$

46. $x^3 + \frac{3}{2}x^2 + 3x - 2 = 0$

In Problems 47–56, find the intercepts of each polynomial function $f(x)$. Find the intervals x for which the graph of f is above the x-axis and below the x-axis. Obtain several other points on the graph, and connect them with a smooth curve. [Hint: Use the factored form of f (see Problems 27–36).]

47. $f(x) = 2x^3 - x^2 + 2x - 1$

48. $f(x) = 2x^3 + x^2 + 2x + 1$

49. $f(x) = x^4 + x^2 - 2$

50. $f(x) = x^4 - 3x^2 - 4$

51. $f(x) = 4x^4 + 7x^2 - 2$

52. $f(x) = 4x^4 + 15x^2 - 4$

53. $f(x) = x^4 + x^3 - 3x^2 - x + 2$

54. $f(x) = x^4 - x^3 - 6x^2 + 4x + 8$

55. $f(x) = 4x^5 - 8x^4 - x + 2$

56. $f(x) = 4x^5 + 12x^4 - x - 3$

In Problems 57–64, solve each equation in the complex number system.

57. $x^4 - 3x^2 - 4 = 0$

58. $x^4 - 5x^2 - 36 = 0$

59. $x^4 + x^3 - x - 1 = 0$

60. $x^4 - x^3 + x - 1 = 0$

61. $x^4 + 3x^3 - x^2 - 12x - 12 = 0$

62. $x^4 - 3x^3 - 5x^2 + 27x - 36 = 0$

63. $x^5 - x^4 + 2x^3 - 2x^2 + x - 1 = 0$

64. $x^5 + x^4 + x^3 + x^2 - 2x - 2 = 0$

65. One solution of the equation $x^3 - 8x^2 + 16x - 3 = 0$ is 3. Find the sum of the remaining solutions.

66. One solution of the equation $x^3 + 5x^2 + 5x - 2 = 0$ is -2. Find the sum of the remaining solutions.

67. Is $\frac{1}{3}$ a zero of $f(x) = 2x^3 + 3x^2 - 6x + 7$? Explain.

68. Is $\frac{1}{3}$ a zero of $f(x) = 4x^3 - 5x^2 - 3x + 1$? Explain.

69. Is $\frac{3}{5}$ a zero of $f(x) = 2x^6 - 5x^4 + x^3 - x + 1$? Explain.

70. Is $\frac{2}{3}$ a zero of $f(x) = x^7 + 6x^5 - x^4 + x + 2$? Explain.

71. What is the length of the edge of a cube if, after a slice 1 inch thick is cut from one side, the volume remaining is 294 cubic inches?

72. What is the length of the edge of a cube if its volume could be doubled by an increase of 6 centimeters in one edge, an increase of 12 centimeters in a second edge, and a decrease of 4 centimeters in the third edge?

73. Let $f(x)$ be a polynomial function whose coefficients are integers. Suppose that r is a real zero of f and that the leading coefficient of f is 1. Use the Rational Zeros Theorem to show that r is either an integer or an irrational number.

74. Prove the Rational Zeros Theorem. [*Hint:* Let p/q, where p and q have no common factors except 1 and -1, be a solution of the polynomial $f(x) = a_n x^n + a_{n-1} x^{n-1} + \cdots + a_1 x + a_0$, whose coefficients are all integers. Show that $a_n p^n + a_{n-1} p^{n-1} q + \cdots + a_1 p q^{n-1} + a_0 q^n = 0$. Now, because p is a factor of the first n terms of this equation, p must also be a factor of the term $a_0 q^n$. Since p is not a factor of q (why?), p must be a factor of a_0. Similarly, q must be a factor of a_n.]

Problems 75–83 develop the Tartaglia–Cardano solution of the cubic equation and show why it is not altogether practical.

75. Show that the general cubic equation $y^3 + by^2 + cy + d = 0$ can be transformed into an equation of the form $x^3 + px + q = 0$ by using the substitution $y = x - b/3$.

76. In the equation $x^3 + px + q = 0$, replace x by $H + K$. Let $3HK = -p$, and show that $H^3 + K^3 = -q$. [*Hint:* $3H^2K + 3HK^2 = 3HKx$.]

77. Based on Problem 76, we have the two equations

$$3HK = -p \quad \text{and} \quad H^3 + K^3 = -q$$

Solve for K in $3HK = -p$ and substitute into $H^3 + K^3 = -q$. Then show that

$$H = \sqrt[3]{\frac{-q}{2} + \sqrt{\frac{q^2}{4} + \frac{p^3}{27}}}$$

[*Hint:* Look for an equation that is quadratic in form.]

78. Use the solution for H from Problem 77 and the equation $H^3 + K^3 = -q$ to show that

$$K = \sqrt[3]{\frac{-q}{2} - \sqrt{\frac{q^2}{4} + \frac{p^3}{27}}}$$

79. Use the results from Problems 76–78 to show that the solution of $x^3 + px + q = 0$ is

$$x = \sqrt[3]{\frac{-q}{2} + \sqrt{\frac{q^2}{4} + \frac{p^3}{27}}} + \sqrt[3]{\frac{-q}{2} - \sqrt{\frac{q^2}{4} + \frac{p^3}{27}}}$$

80. Use the result of Problem 79 to solve the equation $x^3 - 6x - 9 = 0$.

81. Use a calculator and the result of Problem 79 to solve the equation $x^3 + 3x - 14 = 0$.

82. Use the methods of this chapter to solve the equation $x^3 + 3x - 14 = 0$.

83. *Requires Complex Numbers* Show that the formula derived in Problem 79 leads to the cube root of a complex number when applied to the equation $x^3 - 6x + 4 = 0$. Use the methods of this chapter to solve the equation.

84. A function f has the property that $f(2 + x) = f(2 - x)$ for all x. If f has exactly four real zeros, find their sum.

3.6

Approximating the Real Zeros of a Polynomial Function

Sometimes the procedures we discussed in Section 3.5 yield limited information about the zeros of a polynomial. Let's look at an example.

EXAMPLE 1   *Finding the Real Zeros of a Polynomial Function*

Discuss the zeros of: $f(x) = x^5 - x^3 - 1$

Solution   **STEP 1:** f has at most five zeros.

STEP 2: f has one positive zero, and because $f(-x) = -x^5 + x^3 - 1$, f has two or no negative zeros.

STEP 3: The potential rational zeros are ± 1, neither of which is an actual zero. We conclude that f has one positive irrational zero and perhaps two negative irrational zeros. ∎

To obtain more information about the zeros of the polynomial of Example 1, we need some additional results.

Upper and Lower Bounds

The search for the zeros of a polynomial function can be reduced somewhat if upper and lower bounds to the zeros can be found. A number M is an **upper bound** to the zeros of a polynomial f if no zero of f exceeds M. The number m is a **lower bound** if no zero of f is less than m.

Thus, if m is a lower bound and M is an upper bound to the zeros of a polynomial f, then

$$m \leq \text{Any zero of } f \leq M$$

One immediate advantage of knowing the values of a lower bound m and an upper bound M is that, for polynomials with integer coefficients, it may allow you to eliminate some potential rational zeros, that is, any that lie outside the interval $[m, M]$. The next result tells us how to locate lower and upper bounds.

Theorem
Bounds on Zeros

Let f denote a polynomial function whose leading coefficient is positive.

If $M > 0$ is a real number and if the third row in the process of synthetic division of f by $x - M$ contains only numbers that are positive or 0, then M is an upper bound to the zeros of f.

If $m < 0$ is a real number and if the third row in the process of synthetic division of f by $x - m$ contains numbers that are alternately positive (or 0) and negative (or 0), then m is a lower bound to the zeros of f.   ∎

Proof (Outline)

We shall give only an outline of the proof of the first part of the theorem. Suppose that M is a positive real number, and the third row in the process of synthetic division of the polynomial f by $x - M$ contains only numbers that are positive or 0. Then there is a quotient q and a remainder R so that

$$f(x) = (x - M)q(x) + R$$

where the coefficients of $q(x)$ are positive or 0 and the remainder $R \geq 0$. Then, for any $x > M$, we must have $x - M > 0$, $q(x) > 0$, and $R \geq 0$, so that $f(x) > 0$. That is, there is no zero of f larger than M.   ∎

E X A M P L E   2

Finding Upper and Lower Bounds to the Zeros

Find upper and lower bounds to the zeros of: $f(x) = x^5 - x^3 - 1$

Solution

To get an upper bound to the zeros, the usual practice is to start with 1 and continue with 2, 3, ... , until the third row of the process of synthetic division yields only numbers that are positive or 0. Thus, we start with 1:

1) 1	0	−1	0	0	−1	2) 1	0	−1	0	0	−1
	1	1	0	0	0		2	4	6	12	24
1	1	0	0	0	−1	1	2	3	6	12	23

With 2, the third row has only positive numbers; thus, 2 is an upper bound.

To get a lower bound to the zeros, we start with −1 and continue with −2, −3, ... , until the third row of the process of synthetic division yields numbers that alternate in sign:

−1) 1	0	−1	0	0	−1
	−1	1	0	0	0
1	−1	0	0	0	−1

Count as   Count as   Count as
positive   negative   positive

Because the entries alternate in sign, -1 is a lower bound. Thus, the zeros of f lie between -1 and 2. ∎

■ Now work Problem 1.

Note: In determining lower bounds, a 0 in the bottom row following a nonzero entry may be counted as positive or negative, as needed. If the next entry is also a 0, it must be counted opposite to the way the preceding 0 was counted (refer to Example 2).

In Example 2, we found that the zeros of $f(x) = x^5 - x^3 - 1$ lie in the interval $[-1, 2]$. However, remember that in finding the lower bound -1 and the upper bound 2 we tested only integers. Were we to test other positive numbers less than 2 and other negative numbers greater than -1, we might be able to fine-tune the bounds and find a smaller interval containing the zeros of f. However, the effort required to do this is usually not worth it, since more efficient methods are available. We look at one such method next.

Intermediate Value Theorem

The next result, called the **Intermediate Value Theorem,** is based on the fact that the graph of a polynomial function is continuous; that is, it contains no "jumps" or "gaps."

Theorem
Intermediate Value Theorem

Let f denote a polynomial function. If $a < b$ and if $f(a)$ and $f(b)$ are of opposite sign, then there is at least one zero of f between a and b. ∎

Although the proof of this result requires advanced methods in calculus, it is easy to "see" why the result is true. Look at Figure 44.

FIGURE 44
If $f(a) < 0$ and $f(b) > 0$, there is a zero between a and b.

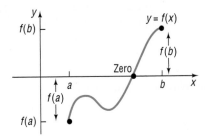

EXAMPLE   3

Using the Intermediate Value Theorem to Locate Zeros
Show that $f(x) = x^5 - x^3 - 1$ has a zero between 1 and 2.

Solution

We know from Example 1 that f has exactly one positive zero. Now, look back at the solution to Example 2, where we used synthetic division to divide f by $x - 1$ and then by $x - 2$. There, we see that

$$f(1) = -1 \quad \text{and} \quad f(2) = 23$$

Because $f(1) < 0$ and $f(2) > 0$, it follows from the Intermediate Value Theorem that f has a zero between 1 and 2. ∎

Note that the zero we now know to lie between 1 and 2 is irrational, because we found in Example 1 that the only possible rational zeros are -1 and 1.

■ Now work Problem 7.

We can use the Intermediate Value Theorem to get a better approximation of the zero of a function f as follows:

Approximating the Zeros of a Polynomial Function

STEP 1: Find two consecutive integers a and $a + 1$ such that f has a zero between them.
STEP 2: Divide the interval $[a, a + 1]$ into 10 equal subintervals.
STEP 3: Evaluate f at each endpoint of the subintervals until the Intermediate Value Theorem applies; that interval then contains a zero.
STEP 4: Repeat the process starting at Step 2 until the desired accuracy is achieved.

E X A M P L E   4

Approximating the Zeros of a Polynomial Function

Find the positive zero of $f(x) = x^5 - x^3 - 1$ correct to two decimal places.

Solution   From Example 3 we know that the positive zero is between 1 and 2. We divide the interval $[1, 2]$ into 10 equal subintervals: $[1, 1.1], [1.1, 1.2], [1.2, 1.3],$ $[1.3, 1.4], [1.4, 1.5], [1.5, 1.6], [1.6, 1.7], [1.7, 1.8], [1.8, 1.9], [1.9, 2]$. Now, we find the value of f at each endpoint until the Intermediate Value Theorem applies. The easiest method is to write $f(x)$ in nested form and use a calculator. Thus, we write

$$f(x) = x^5 - x^3 - 1$$
$$= (x^2 - 1) \cdot x \cdot x \cdot x - 1 = (x \cdot x - 1) \cdot x \cdot x \cdot x - 1$$

$$f(1.0) = -1 \qquad\qquad f(1.2) = -0.23968$$
$$f(1.1) = -0.72049 \qquad f(1.3) = 0.51593$$

We can stop here and conclude that the zero is between 1.2 and 1.3. Now we divide the interval $[1.2, 1.3]$ into 10 equal subintervals and proceed to evaluate f at each endpoint:

$$f(1.20) = -0.23968 \qquad f(1.23) = -0.0455613$$
$$f(1.21) = -0.1778185 \qquad f(1.24) = 0.025001$$
$$f(1.22) = -0.1131398$$

We conclude that the zero lies between 1.23 and 1.24, and so, correct to two decimal places, the zero is 1.23.   ∎

Comment: Use the graph of $f(x) = x^5 - x^3 - 1$ to conclude that f has a positive zero. Then use ZOOM and TRACE to approximate it correct to two decimal places.   ∎

There are many other numerical techniques for approximating the zeros of a polynomial. The one outlined in Example 4 (a variation of the *bisection method*) has the advantages that it will always work, that it can be programmed rather easily on a computer, and each time it is used another decimal place of accuracy is achieved. See Problem 33 for the bisection method, which places the zero in a succession of intervals, with each new interval being half the length of the preceding one.

3.6

Exercise 3.6

In Problems 1–6, find integer-valued upper and lower bounds to the zeros of each polynomial function.

1. $f(x) = 2x^3 + x^2 - 1$

2. $f(x) = 3x^3 - 2x^2 + x + 4$

3. $f(x) = x^3 - 5x^2 - 11x + 11$

4. $f(x) = 2x^3 - x^2 - 11x - 6$

5. $f(x) = x^4 + 3x^3 - 5x^2 + 9$

6. $f(x) = 4x^4 - 12x^3 + 27x^2 - 54x + 81$

In Problems 7–12, use the Intermediate Value Theorem to show that each polynomial function has a zero in the given interval.

7. $f(x) = 8x^4 - 2x^2 + 5x - 1;$ $[0, 1]$

8. $f(x) = x^4 + 8x^3 - x^2 + 2;$ $[-1, 0]$

9. $f(x) = 2x^3 + 6x^2 - 8x + 2;$ $[-5, -4]$

10. $f(x) = 3x^3 - 10x + 9;$ $[-3, -2]$

11. $f(x) = x^5 - x^4 + 7x^3 - 7x^2 - 18x + 18;$ $[1.4, 1.5]$

12. $f(x) = x^5 - 3x^4 - 2x^3 + 6x^2 + x + 2;$ $[1.7, 1.8]$

In Problems 13–16, each polynomial function has exactly one positive zero. Use the method of Example 4 to approximate the zero correct to two decimal places.

13. $f(x) = x^3 + x^2 + x - 4$

14. $f(x) = 2x^4 + x^2 - 1$

15. $f(x) = 2x^4 - 3x^3 - 4x^2 - 8$

16. $f(x) = 3x^3 - 2x^2 - 20$

In Problems 17–20, each equation has a solution r in the interval indicated. Use the method of Example 4 to approximate this solution correct to two decimal places.

17. $8x^4 - 2x^2 + 5x - 1 = 0;$ $0 \le r \le 1$

18. $x^4 + 8x^3 - x^2 + 2 = 0;$ $-1 \le r \le 0$

19. $2x^3 + 6x^2 - 8x + 2 = 0;$ $-5 \le r \le -4$

20. $3x^3 - 10x + 9 = 0;$ $-3 \le r \le -2$

In Problems 21–24, approximate the positive zero correct to two decimal places.

21. $f(x) = x^3 + x^2 + x - 4$

22. $f(x) = 2x^4 + x^2 - 1$

23. $f(x) = 2x^4 - 3x^3 - 4x^2 - 8$

24. $f(x) = 3x^3 - 2x^2 - 20$

In Problems 25–28, approximate the solutions correct to two decimal places.

25. $8x^4 - 2x^2 + 5x - 1 = 0$

26. $x^4 + 8x^3 - x^2 + 2 = 0$

27. $2x^3 + 6x^2 - 8x + 2 = 0$

28. $3x^3 - 10x + 9 = 0$

29. You are given a polynomial equation to solve. Write a brief paragraph outlining your strategy.

30. Suppose the potential rational zeros of a polynomial function are ± 3 and ± 7. Explain why the Upper and Lower Bounds Theorem makes it a good idea to test ± 3 first, before ± 7.

31. *Programming Exercise* Write a computer program that will estimate the positive zero of a polynomial function to any desired degree of accuracy. Input should consist of the coefficients of the polynomial, in order from highest to lowest, followed by the degree N of accuracy wanted (that is, the zero is to be estimated to within 10^{-N}), followed by two consecutive integers between which the zero lies. Output will consist of two decimal numbers between which the zero lies. The program should contain a subroutine that writes the polynomial in nested form.

32. *Programming Exercise* Modify the program in Problem 31 to include a subroutine that will locate the two consecutive integers between which the zero lies.

33. *Bisection Method for Approximating Zeros of a Function f* We begin with two consecutive integers, a and $a + 1$, such that $f(a)$ and $f(a + 1)$ are of oppposite sign. Evaluate f at the midpoint m_1 of a and $a + 1$. If $f(m_1) = 0$, then m_1 is the zero of f, and we are finished. Otherwise, $f(m_1)$ is of opposite sign to either $f(a)$ or $f(a + 1)$. Suppose it is $f(a)$ and $f(m_1)$ that are of opposite sign. Now evaluate f at the midpoint m_2 of a and m_1. Repeat this process until the desired degree of accuracy is obtained. Note that each iteration places the zero in an interval whose length is half that of the previous interval. Use the bisection method to solve Problems 13–20.

3.7
Complex Polynomials; Fundamental Theorem of Algebra

A variable z in the complex number system is referred to as a **complex variable.** A **complex polynomial function** f of degree n is a complex function of the form

$$f(z) = a_n z^n + a_{n-1} z^{n-1} + \cdots + a_1 z + a_0 \qquad (1)$$

where $a_n, a_{n-1}, \ldots, a_1, a_0$ are complex numbers, $a_n \neq 0$, and n is a nonnegative integer. Here, a_n is called the **leading coefficient** of f. A complex number r is called a (complex) **zero** of a complex function f if $f(r) = 0$.

In Chapter 2, we discovered that some quadratic equations have no real solutions, but that in the complex number system every quadratic equation has a solution, either real or complex. The next result, proved by Karl Friedrich Gauss (1777–1855) when he was 22 years of age,* gives an extension to complex polynomials. In fact, this result is so important and useful it has become known as the **Fundamental Theorem of Algebra.**

Fundamental Theorem
of Algebra

Every complex polynomial function $f(z)$ of degree $n \geq 1$ has at least one complex zero. ∎

We shall not prove this result, as the proof is beyond the scope of this book. However, using the Fundamental Theorem of Algebra and the Factor Theorem, we can prove the following result:

Theorem

Every complex polynomial function $f(z)$ of degree $n \geq 1$ can be factored into n linear factors (not necessarily distinct) of the form

$$f(z) = a_n(z - r_1)(z - r_2) \cdot \ldots \cdot (z - r_n) \tag{2}$$

where $a_n, r_1, r_2, \ldots, r_n$ are complex numbers.

Proof   Let

$$f(z) = a_n z^n + a_{n-1} z^{n-1} + \cdots + a_1 z + a_0$$

By the Fundamental Theorem of Algebra, f has at least one zero, say, r_1. Then, by the Factor Theorem, $z - r_1$ is a factor, and

$$f(z) = (z - r_1)q_1(z)$$

where $q_1(z)$ is a complex polynomial of degree $n - 1$ whose leading coefficient is a_n. Again, by the Fundamental Theorem of Algebra, the complex polynomial $q_1(z)$ has at least one zero, say, r_2. By the Factor Theorem, $q_1(z)$ has the factor $z - r_2$, so

$$q_1(z) = (z - r_2)q_2(z)$$

where $q_2(z)$ is a complex polynomial of degree $n - 2$ whose leading coefficient is a_n. Consequently,

$$f(z) = (z - r_1)(z - r_2)q_2(z)$$

Repeating this argument n times, we finally arrive at

$$f(z) = (z - r_1)(z - r_2) \cdot \ldots \cdot (z - r_n)q_n(z)$$

*In all, Gauss gave four different proofs of this theorem, the first one in 1799 being the subject of his doctoral dissertation.

where $q_n(z)$ is a complex polynomial of degree $n - n = 0$ whose leading coefficient is a_n. Thus, $q_n(z) = a_n z^0 = a_n$, and so

$$f(z) = a_n(z - r_1)(z - r_2) \cdot \ldots \cdot (z - r_n)$$ ■

Complex Polynomials with Real Coefficients

We can use the Fundamental Theorem of Algebra to obtain valuable information about the zeros of complex polynomials whose coefficients are real numbers.

Theorem
Conjugate Pairs

Let $f(z)$ be a complex polynomial whose coefficients are real numbers. If $r = a + bi$ is a zero of f, then the complex conjugate $\bar{r} = a - bi$ is also a zero of f. ■

In other words, for complex polynomials whose coefficients are real numbers, the zeros occur in conjugate pairs.

Proof Let

$$f(z) = a_n z^n + a_{n-1} z^{n-1} + \cdots + a_1 z + a_0$$

where $a_n, a_{n-1}, \ldots, a_1, a_0$ are real numbers and $a_n \neq 0$. If r is a zero of f, then $f(r) = 0$, so

$$a_n r^n + a_{n-1} r^{n-1} + \cdots + a_1 r + a_0 = 0$$

We take the conjugate of both sides to get

$\overline{a_n r^n + a_{n-1} r^{n-1} + \cdots + a_1 r + a_0} = \bar{0}$	
$\overline{a_n r^n} + \overline{a_{n-1} r^{n-1}} + \cdots + \overline{a_1 r} + \overline{a_0} = \bar{0}$	The conjugate of a sum equals the sum of the conjugates (see Section 1.5).
$\overline{a_n}(\bar{r})^n + \overline{a_{n-1}}(\bar{r})^{n-1} + \cdots + \overline{a_1}\bar{r} + \overline{a_0} = \bar{0}$	The conjugate of a product equals the product of the conjugates.
$a_n(\bar{r})^n + a_{n-1}(\bar{r})^{n-1} + \cdots + a_1\bar{r} + a_0 = 0$	The conjugate of a real number equals the real number.

This last equation states that $f(\bar{r}) = 0$; that is, $\bar{r}$ is a zero of f. ■

The value of this result should be clear. Once we know that, say, $3 + 4i$ is a zero of a polynomial with real coefficients, then we know that $3 - 4i$ is also a zero. This result has an important corollary.

Corollary A complex polynomial f of odd degree with real coefficients has at least one real zero. ■

Proof Because complex zeros occur as conjugate pairs in a complex polynomial with real coefficients, there will always be an even number of zeros that are not real numbers. Consequently, since f is of odd degree, one of its zeros has to be a real number. ■

For example, the polynomial $f(z) = z^5 - 3z^4 + 4z^3 - 5$ has at least one zero that is a real number, since f is of degree 5 (odd) and has real coefficients.

Now we can prove the theorem we conjectured earlier in Section 3.5.

Theorem   Every polynomial function with real coefficients can be uniquely factored over the real numbers into a product of linear factors and/or irreducible quadratic factors.

■

Proof   Every complex polynomial f of degree n has exactly n zeros and can be factored into a product of n linear factors. If its coefficients are real, then those zeros that are complex numbers will always occur as conjugate pairs. As a result, if $r = a + bi$ is a complex zero, then so is $\bar{r} = a - bi$. Consequently, when the linear factors $z - r$ and $z - \bar{r}$ of f are multiplied, we have

$$(z - r)(z - \bar{r}) = z^2 - (r + \bar{r})z + r\bar{r} = z^2 - 2az + a^2 + b^2$$

This second-degree polynomial has real coefficients and is irreducible (over the real numbers). Thus, the factors of f are either linear or irreducible quadratic factors.   ■

E X A M P L E   1   *Using the Conjugate Pairs Theorem*

A polynomial f of degree 5 whose coefficients are real numbers has the zeros 1, $5i$, and $1 + i$. Find the remaining two zeros.

Solution   Since complex zeros appear as conjugate pairs, it follows that $-5i$, the conjugate of $5i$, and $1 - i$, the conjugate of $1 + i$, are the two remaining zeros.   ■

■ Now work Problem 1.

Polynomials with Complex Coefficients

The division algorithm for polynomials (see Section 3.4) is true for polynomials with complex coefficients. As a result, the Remainder Theorem and Factor Theorem are also true. In fact, the process of synthetic division also works for polynomials with complex coefficients.

E X A M P L E   2   *Verifying a Complex Zero*

Use synthetic division and the Factor Theorem to show that $1 + 2i$ is a zero of

$$f(z) = (1 + i)z^2 + (2 - i)z + (3 - 4i)$$

Solution   We use synthetic division and divide $f(z)$ by $z - (1 + 2i)$:

$$
\begin{array}{r|ccc}
1 + 2i & 1 + i & 2 - i & 3 - 4i \\
 & & -1 + 3i & -3 + 4i \\
\hline
 & 1 + i & 1 + 2i & 0
\end{array}
$$

Thus, $z - (1 + 2i)$ is a factor, and $1 + 2i$ is a zero.   ■

Of course, we could have shown that $1 + 2i$ is a zero of the polynomial $f(z)$ in Example 2 by using substitution as follows:

$$f(1 + 2i) = (1 + i)(1 + 2i)^2 + (2 - i)(1 + 2i) + (3 - 4i)$$
$$= -7 + i + 4 + 3i + 3 - 4i = 0$$

■ Now work Problem 23.

Based on equation (2), a complex polynomial f of degree n has n linear factors. These n linear factors do not have to be distinct; some may be repeated more than once. When a linear factor $z - r$ appears exactly m times in the factored form of f, then r is called a **zero of multiplicity** m of f. This leads us to the following conclusion.

Theorem   If a zero of multiplicity m of a complex polynomial f is counted m times, then a complex polynomial f of degree $n \geq 1$ will have exactly n zeros.   ■

E X A M P L E   3   *Listing all the Zeros of a Polynomial Function*

The complex polynomial function

$$f(z) = (2 + i)(z - 5)^3(z + i)^2[z - (3 + i)]^4(z - i)$$

of degree 10 has $2 + i$ as leading coefficient. Its zeros are listed next.

$$
\begin{aligned}
5: &\quad \text{Multiplicity} \quad 3 \\
-i: &\quad \text{Multiplicity} \quad 2 \\
3 + i: &\quad \text{Multiplicity} \quad 4 \\
i: &\quad \text{Multiplicity} \quad 1 \\
\text{Degree:} &\quad\quad\quad\quad\quad 10
\end{aligned}
$$
■

E X A M P L E   4   *Using the Zeros to Form a Polynomial*

Form a polynomial $f(z)$ with complex coefficients of degree 3 and with the following zeros:

$$
\begin{aligned}
1 + i: &\quad \text{Multiplicity 1} \\
-i: &\quad \text{Multiplicity 2}
\end{aligned}
$$

Solution   Since $1 + i$ is a zero of multiplicity 1 and $-i$ is a zero of multiplicity 2, then $z - (1 + i)$ and $(z + i)^2$ are factors of f. Thus, $f(z)$ is of the form

$$
\begin{aligned}
f(z) &= [z - (1 + i)](z + i)^2 \\
&= [z - (1 + i)](z^2 + 2iz - 1) \\
&= z^3 + (-1 + i)z^2 + (1 - 2i)z + 1 + i
\end{aligned}
$$

Although other polynomials with complex coefficients have the three required zeros, the only ones of degree 3 will be $f(z)$ or $kf(z)$, where $k \neq 0$ is some complex number.   ■

3.7

Exercise 3.7

In Problems 1–10, information is given about a complex polynomial $f(z)$ whose coefficients are real numbers. Find the remaining zeros of f.

1. Degree 3; zeros: $3, 4 - i$

2. Degree 3; zeros: $4, 3 + i$

3. Degree 4; zeros: $i, 1 + i$

4. Degree 4; zeros: $1, 2, 2 + i$

5. Degree 5; zeros: $1, i, 2i$

6. Degree 5; zeros: $0, 1, 2, i$

7. Degree 4; zeros: $i, 2, -2$

8. Degree 4; zeros: $2 - i, -i$

9. Degree 6; zeros: $2, 2 + i, -3 -i, 0$

10. Degree 6; zeros: $i, 3 - 2i, -2 + i$

In Problems 11 and 12, tell why the facts given are contradictory.

11. $f(z)$ is a complex polynomial of degree 3 whose coefficients are real numbers; its zeros are $4 + i, 4 - i$, and $2 + i$.

12. $f(z)$ is a complex polynomial of degree 3 whose coefficients are real numbers; its zeros are $2, i$, and $3 + i$.

13. $f(z)$ is a complex polynomial of degree 4 whose coefficients are real numbers; three of its zeros are $2, 1 + 2i$, and $1 - 2i$. Explain why the remaining zero must be a real number.

14. $f(z)$ is a complex polynomial of degree 4 whose coefficients are real numbers; two of its zeros are -3 and $4 - i$. Explain why one of the remaining zeros must be a real number. Write down one of the missing zeros.

15. Find all the zeros of $f(z) = z^3 - 1$.

16. Find all the zeros of $f(z) = z^4 - 1$.

In Problems 17–22, evaluate each complex polynomial function f at $z = 1 + i$.

17. $f(z) = iz - 3$

18. $f(z) = 3z + i$

19. $f(z) = 3z^2 - z$

20. $f(z) = (4 + i)z^2 + 5 - 2i$

21. $f(z) = z^3 + iz - 1 + i$

22. $f(z) = iz^3 - 2z^2 + 1$

In Problems 23–28, use synthetic division to find the value of f(r).

23. $f(z) = 5z^5 - iz^4 + 2;\quad r = 1 + i$

24. $f(z) = iz^4 + (2 + i)z^2 - z;\quad r = 1 - i$

25. $f(z) = (1 + i)z^4 - z^3 + iz;\quad r = 2 - i$

26. $f(z) = 2iz^3 + 8z^2 - 4iz + 1;\quad r = 2 + i$

27. $f(z) = iz^5 + iz^3 + iz;\quad r = 1 + 2i$

28. $f(z) = z^4 + z^2 + 1;\quad r = 1 - 2i$

In Problems 29–34, form a polynomial f(z) with complex coefficients having the given degree and zeros.

29. Degree 3; zeros: $3 + 2i$, multiplicity 1; 4, multiplicity 2

30. Degree 3; zeros: i, multiplicity 2; $1 + 2i$, multiplicity 1

31. Degree 3; zeros: 2, multiplicity 1; $-i$, multiplicity 1; $1 + i$, multiplicity 1

32. Degree 3; zeros: i, multiplicity 1; $4 - i$, multiplicity 1; $2 + i$, multiplicity 1

33. Degree 4; zeros: 3, multiplicity 2; $-i$, multiplicity 2

34. Degree 4; zeros: 1, multiplicity 3; $1 + i$, multiplicity 1

Chapter Review

THINGS TO KNOW

Quadratic function	$f(x) = ax^2 + bx + c, a \neq 0$	Vertex: $(-b/2a, f(-b/2a))$ Axis: The line $x = -b/2a$ Parabola opens up if $a > 0$. Parabola opens down if $a < 0$.
Power function	$f(x) = x^n, n \geq 2$ even	Even function: Passes through $(-1, 1), (0, 0), (1, 1)$ Opens up
	$f(x) = x^n, n \geq 3$ odd	Odd function: Passes through $(-1, -1), (0, 0), (1, 1)$ Increasing
Polynomial function	$f(x) = a_n x^n + a_{n-1} x^{n-1} +$ $\cdots + a_1 x + a_0, a_n \neq 0$	At most $n - 1$ turning points; behaves like $y = a_n x^n$ for large x.

Rational function

$$R(x) = \frac{p(x)}{q(x)},$$

p, q are polynomial functions in lowest terms   See Steps 1 through 6 on page 215.

Zeros of a polynomial f	Numbers for which $f(x) = 0$; these are the x-intercepts of the graph of f.
Remainder Theorem	If a polynomial $f(x)$ is divided by $x - c$, then the remainder is $f(c)$.
Factor Theorem	$x - c$ is a factor of a polynomial $f(x)$ if and only if $f(c) = 0$.
Descartes' Rule of Signs	Let f denote a polynomial function. The number of positive zeros of f either equals the number of variations in sign of the coefficients of $f(x)$ or else equals that number less some even integer. The number of negative zeros of f either equals the number of variations in sign of the coefficients of $f(-x)$ or else equals that number less some even integer.
Rational Zeros Theorem	Let f be a polynomial function of degree 1 or higher of the form

$$f(x) = a_n x^n + a_{n-1} x^{n-1} + \cdots + a_1 x + a_0, \ a_n \neq 0, \ a_0 \neq 0$$

	where each coefficient is an integer. If p/q, in lowest terms, is a rational zero of f, then p must be a factor of a_0 and q must be a factor of a_n.
Intermediate Value Theorem	If $a < b$ and if $f(a)$ and $f(b)$ are of opposite sign, then there is at least one zero of f between a and b.
Fundamental Theorem of Algebra	Every complex polynomial function $f(z)$ of degree $n \geq 1$ has at least one complex zero.
Conjugate Pairs Theorem	Let $f(z)$ be a complex polynomial whose coefficients are real numbers. If $r = a + bi$ is a zero of f, then the complex conjugate $\bar{r} = a - bi$ is also a zero of f.

How To:

Graph quadratic functions

Graph polynomial functions

Graph rational functions (see Steps 1 through 6, page 215)

Use synthetic division to divide a polynomial by $x - c$

Write a polynomial in nested form

Find the zeros of a polynomial by using Descartes' Rule of Signs, the Rational Zeros Theorem, and depressed equations

Solve polynomial equations using Descartes' Rule of Signs, the Rational Zeros Theorem, and depressed equations

Approximate the zeros of a polynomial

Fill-In-The-Blank Items

1. The graph of a quadratic function is called a(n) _____. Its lowest or highest point is called the _____.

2. In the process of long division,

$$(\text{Divisor})(\text{Quotient}) + \underline{\hspace{1.5cm}} = \underline{\hspace{1.5cm}}$$

3. When a polynomial function f is divided by $x - c$, the remainder is _____.

4. A polynomial function f has the factor $x - c$ if and only if _____.

5. A number r for which $f(r) = 0$ is called a(n) _____ of the function f.

6. The polynomial function $f(x) = x^5 - 2x^3 + x^2 + x - 1$ has either _____ or _____ positive zeros; it has _____ or _____ negative zeros.

7. The possible rational zeros of $f(x) = 2x^5 - x^3 + x^2 - x + 1$ are _____.

8. The line _____ is a horizontal asymptote of: $R(x) = \dfrac{x^3 - 1}{x^3 + 1}$

9. The line _____ is a vertical asymptote of: $R(x) = \dfrac{x^3 - 1}{x^3 + 1}$

10. If $3 + 4i$ is a zero of a polynomial of degree 5 with real coefficients, then so is _____.

TRUE/FALSE ITEMS

T   F   **1.** Every polynomial of degree 3 with real coefficients has exactly three real zeros.

T   F   **2.** If $2 - 3i$ is a zero of a polynomial with real coefficients, then so is $-2 + 3i$.

T   F   **3.** The graph of $R(x) = \dfrac{x^2}{x - 1}$ has exactly one vertical asymptote.

T   F   **4.** The graph of $f(x) = x^2(x - 3)(x + 4)$ has exactly three x-intercepts.

T   F   **5.** If f is a polynomial function of degree 4 and if $f(2) = 5$, then

$$\frac{f(x)}{x - 2} = p(x) + \frac{5}{x - 2}$$

where $p(x)$ is a polynomial of degree 3.

REVIEW EXERCISES

In Problems 1–10, graph each quadratic function by determining whether its graph opens up or down and by finding its vertex, axis of symmetry, y-intercept, and x-intercepts, if any.

1. $f(x) = (x - 2)^2 + 2$ **2.** $f(x) = (x + 1)^2 - 4$ **3.** $f(x) = \frac{1}{4}x^2 - 16$

4. $f(x) = -\frac{1}{2}x^2 + 2$ **5.** $f(x) = -4x^2 + 4x$ **6.** $f(x) = 9x^2 - 6x + 3$

7. $f(x) = \frac{9}{2}x^2 + 3x + 1$ **8.** $f(x) = -x^2 + x + \frac{1}{2}$ **9.** $f(x) = 3x^2 + 4x - 1$

10. $f(x) = -2x^2 - x + 4$

In Problems 11–16, graph each function using the techniques of shifting, compressing, stretching, and reflection.

11. $f(x) = (x + 2)^3$ **12.** $f(x) = -x^3 + 3$ **13.** $f(x) = -(x - 1)^4$

14. $f(x) = (x - 1)^4 - 2$ **15.** $f(x) = (x - 1)^4 + 2$ **16.** $f(x) = (1 - x)^3$

In Problems 17–22, determine whether the given quadratic function has a maximum value or a minimum value, and then find the value.

17. $f(x) = 3x^2 - 6x + 4$ **18.** $f(x) = 2x^2 + 8x + 5$ **19.** $f(x) = -x^2 + 8x - 4$

20. $f(x) = -x^2 - 10x - 3$ **21.** $f(x) = -3x^2 + 12x + 4$ **22.** $f(x) = -2x^2 + 4$

In Problems 23–30:

(a) Find the x- and y-intercepts of each polynomial function f.

(b) Determine whether the graph of f touches or crosses the x-axis at each x-intercept.

(c) Find the power function that the graph of f resembles for large values of x.

(d) Determine the maximum number of turning points on the graph of f.

(e) Use the x-intercept(s) and test numbers to find the intervals on which the graph of f is above and below the x-axis.

(f) Plot the points obtained in parts (a) and (e), and use the remaining information to connect them with a smooth curve.

23. $f(x) = x(x + 2)(x + 4)$ **24.** $f(x) = x(x - 2)(x - 4)$ **25.** $f(x) = (x - 2)^2(x + 4)$

26. $f(x) = (x - 2)(x + 4)^2$ **27.** $f(x) = x^3 - 4x^2$ **28.** $f(x) = x^3 + 4x$

29. $f(x) = (x - 1)^2(x + 3)(x + 1)$ **30.** $f(x) = (x - 4)(x + 2)^2(x - 2)$

In Problems 31–40, discuss each rational function following the six steps outlined in Section 3.3.

31. $R(x) = \dfrac{2x - 6}{x}$

32. $R(x) = \dfrac{4 - x}{x}$

33. $H(x) = \dfrac{x + 2}{x(x - 2)}$

34. $H(x) = \dfrac{x}{x^2 - 1}$

35. $R(x) = \dfrac{x^2 + x - 6}{x^2 - x - 6}$

36. $R(x) = \dfrac{x^2 - 6x + 9}{x^2}$

37. $F(x) = \dfrac{x^3}{x^2 - 4}$

38. $F(x) = \dfrac{3x^3}{(x - 1)^2}$

39. $R(x) = \dfrac{2x^4}{(x - 1)^2}$

40. $R(x) = \dfrac{x^4}{x^2 - 9}$

In Problems 41–44, use synthetic division to find the quotient $q(x)$ and remainder R when $f(x)$ is divided by $g(x)$.

41. $f(x) = 8x^3 - 3x^2 + x + 4$; $g(x) = x - 1$

42. $f(x) = 2x^3 + 8x^2 - 5x + 5$; $g(x) = x - 2$

43. $f(x) = x^4 - 2x^3 + x - 1$; $g(x) = x + 2$

44. $f(x) = x^4 - x^2 + 3x$; $g(x) = x + 1$

45. Find the value of $f(x) = 12x^6 - 8x^4 + 1$ at $x = 4$.

46. Find the value of $f(x) = -16x^3 + 18x^2 - x + 2$ at $x = -2$.

In Problems 47 and 48 use Descartes' Rule of Signs to demonstrate how many positive and negative zeros each polynomial function may have. Do not attempt to find the zeros.

47. $f(x) = 12x^8 - x^7 + 8x^4 - 2x^3 + x + 3$

48. $f(x) = -6x^5 + x^4 + 5x^3 + x + 1$

49. List all the potential rational zeros of: $f(x) = 12x^8 - x^7 + 6x^4 - x^3 + x - 3$

50. List all the potential rational zeros of: $f(x) = -6x^5 + x^4 + 2x^3 - x + 1$

In Problems 51–56, use Descartes' Rule of Signs and the Rational Zeros Theorem to find all the real zeros of each polynomial function. Use the zeros to factor f over the real numbers.

51. $f(x) = x^3 - 3x^2 - 6x + 8$

52. $f(x) = x^3 - x^2 - 10x - 8$

53. $f(x) = 4x^3 + 4x^2 - 7x + 2$

54. $f(x) = 4x^3 - 4x^2 - 7x - 2$

55. $f(x) = x^4 - 4x^3 + 9x^2 - 20x + 20$

56. $f(x) = x^4 + 6x^3 + 11x^2 + 12x + 18$

In Problems 57–60, solve each equation in the real number system.

57. $2x^4 + 2x^3 - 11x^2 + x - 6 = 0$

58. $3x^4 + 3x^3 - 17x^2 + x - 6 = 0$

59. $2x^4 + 7x^3 + x^2 - 7x - 3 = 0$

60. $2x^4 + 7x^3 - 5x^2 - 28x - 12 = 0$

In Problems 61–70, find the intercepts of each polynomial $f(x)$. Find the numbers x for which the graph of f is above and below the x-axis. Obtain several other points on the graph and connect them with a smooth curve.

61. $f(x) = x^3 - 3x^2 - 6x + 8$

62. $f(x) = x^3 - x^2 - 10x - 8$

63. $f(x) = 4x^3 + 4x^2 - 7x + 2$

64. $f(x) = 4x^3 - 4x^2 - 7x - 2$

65. $f(x) = x^4 - 4x^3 + 9x^2 - 20x + 20$

66. $f(x) = x^4 + 6x^3 + 11x^2 + 12x + 18$

67. $f(x) = 2x^4 + 2x^3 - 11x^2 + x - 6$

68. $f(x) = 3x^4 + 3x^3 - 17x^2 + x - 6$

69. $f(x) = 2x^4 + 7x^3 + x^2 - 7x - 3$

70. $f(x) = 2x^4 + 7x^3 - 5x^2 - 28x - 12$

In Problems 71–74, use the Intermediate Value Theorem to show that each polynomial has a zero in the given interval.

71. $f(x) = 3x^3 - x - 1$;   [0, 1]

72. $f(x) = 2x^3 - x^2 - 3$;   [1, 2]

73. $f(x) = 8x^4 - 4x^3 - 2x - 1$;   [0, 1]

74. $f(x) = 3x^4 + 4x^3 - 8x - 2$;   [1, 2]

In Problems 75–78, find integer-valued upper and lower bounds to the zeros of each polynomial function.

75. $f(x) = 2x^3 - x^2 - 4x + 2$

76. $f(x) = 2x^3 + x^2 - 10x - 5$

77. $f(x) = 2x^3 - 7x^2 - 10x + 35$

78. $f(x) = 3x^3 - 7x^2 - 6x + 14$

In Problems 79–82, each polynomial has exactly one positive zero. Approximate the zero correct to two decimal places.

79. $f(x) = x^3 - x - 2$

80. $f(x) = 2x^3 - x^2 - 3$

81. $f(x) = 8x^4 - 4x^3 - 2x - 1$

82. $f(x) = 3x^4 + 4x^3 - 8x - 2$

In Problems 83–86, information is given about a complex polynomial f(z) whose coefficients are real numbers. Find the remaining zeros of f.

83. Degree 3; zeros: $4 + i$, 6

84. Degree 3; zeros: $3 + 4i$, 5

85. Degree 4; zeros: i, $1 + i$

86. Degree 4; zeros: 1, 2, $1 + i$

In Problems 87–90, form a polynomial f(z) with complex coefficients having the given degree and zeros.

87. Degree 4;   zeros: 1, multiplicity 2; i, multiplicity 1; 3, multiplicity 1

88. Degree 4;   zeros: i, multiplicity 2; 2, multiplicity 2

89. Degree 3;   zeros: $1 + i$, 2, 3, each of multiplicity 1

90. Degree 3;   zeros: 1, $1 + i$, $1 + 2i$, each of multiplicity 1

91. Find the quotient and remainder if $x^4 + 2x^3 - 7x^2 - 8x + 12$ is divided by $(x - 2)(x - 1)$.

92. Find the quotient and remainder if $x^4 + 2x^3 - 4x^2 - 5x - 6$ is divided by $(x - 2)(x + 3)$.

In Problems 93–96, solve each equation in the complex number system.

93. $x^3 - x^2 - 8x + 12 = 0$

94. $x^3 - 3x^2 - 4x + 12 = 0$

95. $3x^4 - 4x^3 + 4x^2 - 4x + 1 = 0$

96. $x^4 + 4x^3 + 2x^2 - 8x - 8 = 0$

In Problems 97–100, write each polynomial in nested form. Then use a calculator to evaluate each polynomial at x = 1.5. Avoid using any memory key.

97. $f(x) = 8x^3 - 3x^2 + x - 6$

98. $f(x) = 5x^3 + 4x^2 - 6x + 8$

99. $f(x) = x^4 - 2x^3 + x - 1$

100. $f(x) = x^4 - x^2 + 3x$

101. Find integer-valued upper and lower bounds to the zeros of

$$f(x) = 4x^5 - 3x^4 + 8x^2 + x + 2$$

102. Find integer-valued upper and lower bounds to the zeros of

$$f(x) = 8x^6 - x^4 + 6x^2 + 24x + 15$$

103. Find the point on the line $y = x$ that is closest to the point (3, 1). [*Hint:*   Find the minimum value of the function $f(x) = d^2$, where d is the distance from (3, 1) to a point on the line.]

104. Find the point on the line $y = x + 1$ that is closest to the point (4, 1).

105. A horizontal bridge is in the shape of a parabolic arch. Given the information shown in the figure, what is the height *h* of the arch 2 feet from shore?

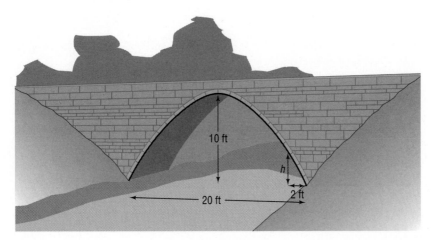

106. Find the length and width of a rectangle whose perimeter is 20 feet and whose area is 16 square feet.

107. Design a polynomial function with the following characteristics: degree 6; four real zeros, one of multiplicity 3; *y*-intercept 3; behaves like $y = -5x^6$ for large values of *x*. Is this polynomial unique? Compare your polynomial with those of other students. What terms will be the same as everyone else's? Add some more characteristics such as symmetry or naming the real zeros. How does this modify the polynomial?

108. Design a rational function with the following characteristics: three real zeros, one of multiplicity 2; *y*-intercept 1; vertical asymptotes $x = -2$ and $x = 3$; oblique asymptote $y = 2x + 1$. Is this rational function unique? Compare yours with those of other students. What will be the same as everyone else's? Add some more characteristics such as symmetry or naming the real zeros. How does this modify the rational function?

109. The illustration shows the graph of a polynomial function.

(a) Is the degree of the polynomial even or odd?

(b) Is the leading coefficient positive or negative?

(c) Is the function even, odd, or neither?

(d) Why is x^2 necessarily a factor of the polynomial?

(e) What is the minimum degree of the polynomial?

(f) Formulate five different polynomials whose graphs could look like the one shown. Compare yours to those of other students. What similarities do you see? What differences?

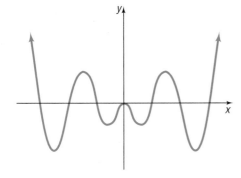

PREPARING FOR THIS CHAPTER

Before getting started on this chapter, review the following concepts:

Exponents (pp. 6–9 and Appendix A, Section A.2)
Functions (Sections 2.1 and 2.2)
Inverse functions (pp. 150–156)
Simple interest (p. 25)

EXPONENTIAL AND LOGARITHMIC FUNCTIONS

Preview   Alcohol and Driving

The concentration of alcohol in a person's blood is measurable. Recent medical research suggests that the risk R (given as a percent) of having an accident while driving a car can be modeled by the equation

$$R = 6e^{kx}$$

where x is the variable concentration of alcohol in the blood and k is a constant.

(a) *Suppose that a concentration of alcohol in the blood of 0.04 results in a 10% risk (R = 10) of an accident. Find the constant k in the equation.*
(b) *Using this value of k, what is the risk if the concentration is 0.17?*
(c) *Using the same value of k, what concentration of alcohol corresponds to a risk of 100%?*
(d) *If the law asserts that anyone with a risk of having an accident of 20% or more should not have driving privileges, at what concentration of alcohol in the blood should a driver be arrested and charged with DUI (Driving Under the Influence)? [See Example 9 in Section 4.2.]*

$\mathcal{U}$ ntil now, our study of functions has concentrated primarily on polynomial and rational functions. These functions belong to the class of **algebraic functions,** that is, functions that can be expressed in terms of sums, differences, products, quotients, powers, or roots of polynomials. Functions that are not algebraic are termed **transcendental** (they transcend, or go beyond, algebraic functions).

In this chapter, we study two transcendental functions: the *exponential* and *logarithmic functions.* These functions occur frequently in a wide variety of applications.

4.1

Exponential Functions

In Chapter 1, we gave a definition for raising a real number a to a rational power. Based on that discussion, we gave meaning to expressions of the form

$$a^r$$

where the base a is a positive real number and the exponent r is a rational number.

But what is the meaning of a^x, where the base a is a positive real number and the exponent x is an irrational number? Although a rigorous definition requires methods discussed in calculus, the basis for the definition is easy to follow: Select a rational number r that is formed by truncating (removing) all but a finite number of digits from the irrational number x. Then it is reasonable to expect that

$$a^x \approx a^r$$

For example, take the irrational number $\pi = 3.14159. \ldots$ Then, an approximation to a^π is

$$a^\pi \approx a^{3.14}$$

where the digits after the hundredths position have been removed from the value for π. A better approximation would be

$$a^\pi \approx a^{3.14159}$$

where the digits after the hundred-thousandths position have been removed. Continuing in this way, we can obtain approximations to a^π to any desired degree of accuracy.

Most scientific calculators have an $\boxed{x^y}$ key (or a $\boxed{y^x}$ key) for working with exponents. To use this key, first enter the base x, then press the $\boxed{x^y}$ key, enter y, and press the $\boxed{=}$ key.

EXAMPLE 1   *Using a Calculator to Evaluate Powers of 2*

Using a calculator with an $\boxed{x^y}$ key, evaluate:

(a) $2^{1.4}$ (b) $2^{1.41}$ (c) $2^{1.414}$ (d) $2^{1.4142}$ (e) $2^{\sqrt{2}}$

Solution   (a) $2^{1.4} \approx 2.6390158$ (b) $2^{1.41} \approx 2.6573716$
(c) $2^{1.414} \approx 2.6647497$ (d) $2^{1.4142} \approx 2.6651191$
(e) $2^{\sqrt{2}} \approx 2.6651441$ ■

■   Now work Problem 1.

It can be shown that the familiar laws of rational exponents hold for real exponents.

Theorem

Laws of Exponents

If s, t, a, and b are real numbers with $a > 0$ and $b > 0$, then

$$a^s \cdot a^t = a^{s+t} \qquad (a^s)^t = a^{st} \qquad (ab)^s = a^s \cdot b^s$$

$$1^s = 1 \qquad a^{-s} = \frac{1}{a^s} = \left(\frac{1}{a}\right)^s \qquad a^0 = 1 \qquad (1)$$

We are now ready for the following definition.

Exponential Function

An **exponential function** is a function of the form

$$f(x) = a^x$$

where a is a positive real number and $a \neq 1$. The domain of f is the set of all real numbers.

We exclude the base $a = 1$, because this function is simply the constant function $f(x) = 1^x = 1$. We also need to exclude bases that are negative, because, otherwise, we would have to exclude many values of x from the domain, such as $x = \frac{1}{2}$, $x = \frac{3}{4}$, and so on. [Recall that $(-2)^{1/2}$, $(-3)^{3/4}$, and so on, are not defined in the system of real numbers.]

Graphs of Exponential Functions

First, we graph the exponential function $y = 2^x$.

E X A M P L E   2

Graphing an Exponential Function

Graph the exponential function: $f(x) = 2^x$

Solution

The domain of $f(x) = 2^x$ consists of all real numbers. We begin by locating some points on the graph of $f(x) = 2^x$, as listed in Table 1.

Since $2^x > 0$ for all x, the range of f is $(0, \infty)$. From this, we conclude that the graph has no x-intercepts, and, in fact, the graph will lie above the x-axis. As Table 1 indicates, the y-intercept is 1. Table 1 also indicates that as $x \to -\infty$ the value of $f(x) = 2^x$ gets closer and closer to 0. Thus, the x-axis is a horizontal asymptote to the graph as $x \to -\infty$. Look again at Table 1. As $x \to \infty$, $f(x) = 2^x$ grows very quickly, causing the graph of $f(x) = 2^x$ to rise very rapidly. Thus, it is apparent that f is an increasing function and, hence, is one-to-one. Using all this information, we plot some of the points from Table 1 and connect them with a smooth, continuous curve, as shown in Figure 1.

TABLE 1

x	$f(x) = 2^x$
-10	$2^{-10} \approx 0.00098$
-3	$2^{-3} = \frac{1}{8}$
-2	$2^{-2} = \frac{1}{4}$
-1	$2^{-1} = \frac{1}{2}$
0	$2^0 = 1$
1	$2^1 = 2$
2	$2^2 = 4$
3	$2^3 = 8$
10	$2^{10} = 1024$

FIGURE 1

$y = 2^x$

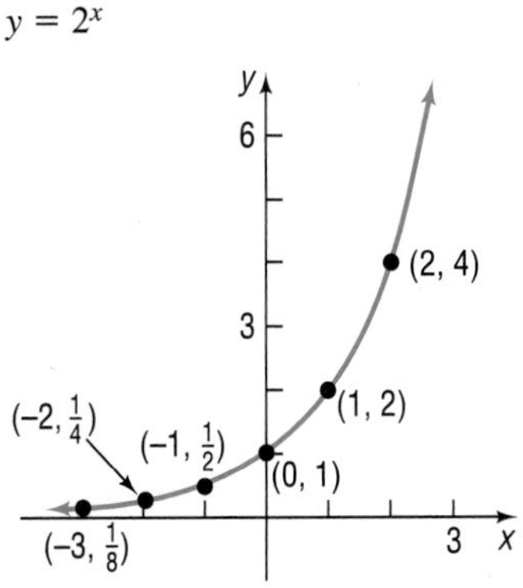

As we shall see, graphs that look like the one in Figure 1 occur very frequently in a variety of situations. For example, look at the graph

FIGURE 2

Ethiopia's population is growing at one of the fastest paces in the world, exacerbating continuing shortfalls in food needed to feed its millions of people. The nation, which is only about three-quarters the size of Alaska, grows with more than 2 million births each year.

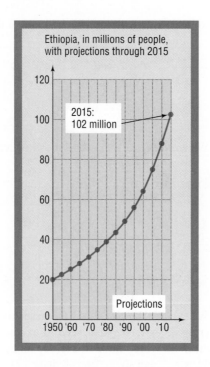

Ethiopia, in millions of people, with projections through 2015

2015:
102 million

Projections

Sources: News reports, World Bank, U.S. Agency for International Development, UN Food and Agriculture Organization, Population Reference Bureau, UNICEF, U.S. Department of Agriculture, Human Nutrition Information Service.

FIGURE 3

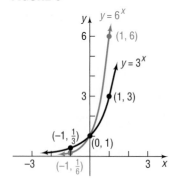

in Figure 2, which illustrates the population in Ethiopia. Researchers might conclude from this graph that the population in Ethiopia is "behaving exponentially"; that is, the graph exhibits "rapid, or exponential, growth." We shall have more to say about situations that lead to exponential growth later in this chapter. For now, we continue to seek properties of the exponential functions.

The graph of $f(x) = 2^x$ in Figure 1 is typical of all exponential functions that have a base larger than 1. Such functions are increasing functions and hence are one-to-one. Their graphs lie above the x-axis, pass through the point $(0, 1)$, and thereafter rise rapidly as $x \to \infty$. As $x \to -\infty$, the x-axis is a horizontal asymptote. There are no vertical asymptotes. Finally, the graphs are smooth and continuous, with no corners or gaps. Figure 3 illustrates the graphs of two more exponential functions whose bases are larger than 1. Notice that for the larger base the graph is steeper when $x > 0$ and is closer to the x-axis when $x < 0$.

The display summarizes the information we have about $f(x) = a^x$, $a > 1$:

$$f(x) = a^x \qquad a > 1$$

Domain: $(-\infty, \infty)$ Range: $(0, \infty)$

x-intercepts: None y-intercept: 1

Horizontal asymptote: x-axis, as $x \to -\infty$

f is an increasing function

f is one-to-one passing through $(0, 1)$ and $(1, a)$

Seeing the Concept: Graph $y = 2^x$ and compare what you see to Figure 1. Clear the screen and graph $y = 3^x$ and $y = 6^x$ and compare what you see to Figure 3. Clear the screen and graph $y = 10^x$ and $y = 100^x$. Which viewing rectangle seems to work best? ■

Now we consider $f(x) = a^x$ when $0 < a < 1$.

E X A M P L E   3

Graphing an Exponential Function

Graph the exponential function: $f(x) = \left(\frac{1}{2}\right)^x$

Solution   The domain of $f(x) = \left(\frac{1}{2}\right)^x$ consists of all real numbers. As before, we locate some points on the graph, as listed in Table 2. Since $\left(\frac{1}{2}\right)^x > 0$ for all x, the range of f is $(0, \infty)$. Thus, the graph lies above the x-axis and so has no x-intercepts. The y-intercept is 1. As $x \to -\infty$, $f(x) = \left(\frac{1}{2}\right)^x$ grows very quickly. As $x \to \infty$, the values of $f(x)$ approach 0. Thus, the x-axis $(y = 0)$ is a horizontal asymptote as $x \to \infty$. It is apparent that f is a decreasing function and, hence, is one-to-one. Figure 4 illustrates the graph.

TABLE 2

x	$f(x) = \left(\frac{1}{2}\right)^x$
-10	$\left(\frac{1}{2}\right)^{-10} = 1024$
-3	$\left(\frac{1}{2}\right)^{-3} = 8$
-2	$\left(\frac{1}{2}\right)^{-2} = 4$
-1	$\left(\frac{1}{2}\right)^{-1} = 2$
0	$\left(\frac{1}{2}\right)^{0} = 1$
1	$\left(\frac{1}{2}\right)^{1} = \frac{1}{2}$
2	$\left(\frac{1}{2}\right)^{2} = \frac{1}{4}$
3	$\left(\frac{1}{2}\right)^{3} = \frac{1}{8}$
10	$\left(\frac{1}{2}\right)^{10} \approx 0.00098$

FIGURE 4

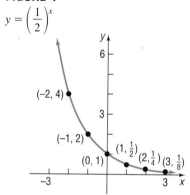

FIGURE 5

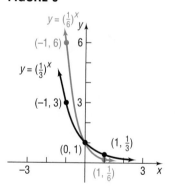

Note that we could have obtained the graph of $y = \left(\frac{1}{2}\right)^x$ from the graph of $y = 2^x$. If $f(x) = 2^x$, then $f(-x) = 2^{-x} = \frac{1}{2^x} = \left(\frac{1}{2}\right)^x$. Thus, the graph of $y = \left(\frac{1}{2}\right)^x = 2^{-x}$ is a reflection about the y-axis of the graph of $y = 2^x$. Compare Figures 1 and 4.

The graph of $f(x) = \left(\frac{1}{2}\right)^x$ in Figure 4 is typical of all exponential functions that have a base between 0 and 1. Such functions are decreasing, one-to-one functions. Their graphs lie above the x-axis and pass through the point $(0, 1)$. The graphs rise rapidly as $x \to -\infty$. As $x \to \infty$, the x-axis is a horizontal asymptote. There are no vertical asymptotes. Finally, the graphs are smooth and continuous, with no corners or gaps. Figure 5 illustrates the graphs of two more exponential functions whose bases are between 0 and 1. Notice that the choice of a base closer to 0 results in a graph that is steeper when $x < 0$ and closer to the x-axis when $x > 0$.

The display summarizes the information we have about $f(x) = a^x$, $0 < a < 1$:

$$f(x) = a^x \qquad 0 < a < 1$$

Domain: $(-\infty, \infty)$ Range: $(0, \infty)$

x-intercepts: None y-intercept: 1

Horizontal asymptote: x-axis, as $x \to \infty$

f is a decreasing function

f is one-to-one passing through $(0, 1)$ and $(1, a)$

Seeing the Concept: Graph $y = \left(\frac{1}{2}\right)^x$ and compare what you see to Figure 4. Clear the screen and graph $y = \left(\frac{1}{3}\right)^x$ and $y = \left(\frac{1}{6}\right)^x$ and compare what you see to Figure 5. Clear the screen and graph $y = \left(\frac{1}{10}\right)^x$ and $y = \left(\frac{1}{100}\right)^x$. Which viewing rectangle seems to work best? ■

The techniques of shifting, compression, stretching, and reflection may be used to graph many functions that are basically exponential functions.

EXAMPLE 4　*Graphing Functions That Are Basically Exponential Using Shifts, Reflections, and the Like*

Graph: $f(x) = 2^{-x} - 3$

Solution　Figure 6 shows the various steps.

FIGURE 6

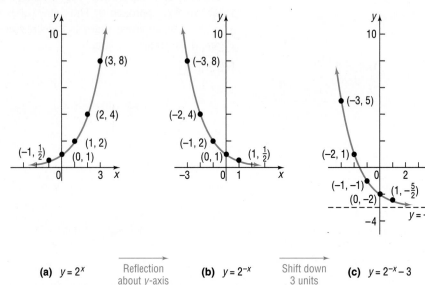

(a) $y = 2^x$ ――Reflection about y-axis→ **(b)** $y = 2^{-x}$ ――Shift down 3 units→ **(c)** $y = 2^{-x} - 3$

Note that the horizontal asymptote of $f(x) = 2^{-x} - 3$ is the line $y = -3$, as Figure 6(c) illustrates.　■

EXAMPLE 5　*Graphing Functions That Are Basically Exponential Using Shifts, Reflections, and the Like*

Graph: $f(x) = -(2^{x-3})$

Solution　Figure 7 shows the various steps.

FIGURE 7

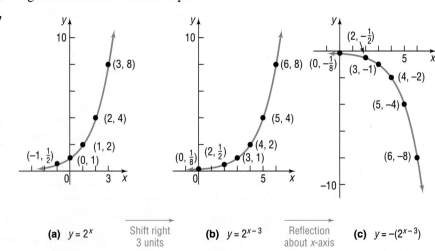

(a) $y = 2^x$ ――Shift right 3 units→ **(b)** $y = 2^{x-3}$ ――Reflection about x-axis→ **(c)** $y = -(2^{x-3})$

Check: Graph $y = -(2^{x-3})$ and compare what you see to Figure 7. Use the TRACE feature to verify the points on the graph shown in Figure 7.   ■

■   Now work Problems 11 and 13.

The Base *e*

As we shall see shortly, many problems that occur in nature require the use of an exponential function whose base is a certain irrational number, symbolized by the letter *e*.

Let's look now at one way of arriving at this important number *e*.

The Number *e*	The **number e** is defined as the number that the expression

$$\left(1 + \frac{1}{n}\right)^n \qquad (2)$$

approaches as $n \to \infty$. In calculus, this is expressed using limit notation as

$$e = \lim_{n \to \infty} \left(1 + \frac{1}{n}\right)^n$$

TABLE 3

n	$\dfrac{1}{n}$	$1 + \dfrac{1}{n}$	$\left(1 + \dfrac{1}{n}\right)^n$
1	1	2	2
2	0.5	1.5	2.25
5	0.2	1.2	2.48832
10	0.1	1.1	2.59374246
100	0.01	1.01	2.704813829
1,000	0.001	1.001	2.716923932
10,000	0.0001	1.0001	2.718145926
100,000	0.00001	1.00001	2.718268237
1,000,000	0.000001	1.000001	2.718280469
1,000,000,000	10^{-9}	$1 + 10^{-9}$	2.718281828

Table 3 illustrates what happens to the defining expression (2) as *n* takes on increasingly large values. The last number in the last column in the table is correct to nine decimal places and is the same as the entry given for *e* on your calculator (if expressed correct to nine decimal places).

The exponential function $f(x) = e^x$, whose base is the number *e*, occurs with such frequency in applications that it is usually referred to as *the* exponential function. Indeed, many calculators have the key $\boxed{e^x}$ or $\boxed{exp(x),}$ which may be used to evaluate the exponential function for a given value of *x*.* Now use your calculator to find e^x for $x = -2$,

*If your calculator does not have this key but does have a $\boxed{\text{SHIFT}}$ key and an $\boxed{\text{ln}}$ key, you can display the number *e* as follows:

Keystrokes: $\boxed{1}$ $\boxed{\text{SHIFT}}$ $\boxed{\text{ln}}$

Display: ↘ $\boxed{1}$ ↘ $\boxed{2.7182818}$

The reason this works will become clear in Section 4.2.

$x = -1$, $x = 0$, $x = 1$, and $x = 2$, as we have done to create Table 4 (after rounding). The graph of the exponential function $f(x) = e^x$ is given in Figure 8. Since $2 < e < 3$, the graph of $y = e^x$ lies between the graphs of $y = 2^x$ and $y = 3^x$. (Refer to Figures 1 and 3.)

TABLE 4

x	e^x
-2	0.14
-1	0.37
0	1
1	2.72
2	7.39

FIGURE 8

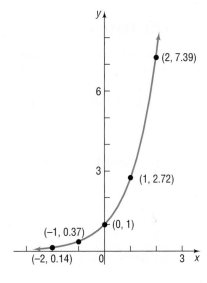

$y = e^x$

Seeing the Concept: Graph $y = e^x$ and compare what you see to Figure 8. Use TRACE to verify the points on the graph shown in Figure 8. Now graph $y = 2^x$ and $y = 3^x$. Notice that the graph of $y = e^x$ lies between these two graphs. ■

■ Now work Problem 21.

There are many applications involving the exponential function. Let's look at one.

E X A M P L E 6

Response to Advertising

Suppose that the percent R of people who respond to a newspaper advertisement for a new product and purchase the item advertised after t days is found using the formula

$$R = 50 - 100e^{-0.3t}$$

(a) What percent has responded and purchased after 5 days?

(b) What percent has responded and purchased after 10 days?

(c) What is the highest percent of people expected to respond and purchase?

(d) Graph $R = 50 - 100e^{-0.3t}$, $t > 0$. TRACE and compare the values of R for $t = 5$ and $t = 10$ to the ones obtained in Example 6. How many days are required for R to exceed 40%?

Solution   (a) After 5 days, we have $t = 5$. The corresponding percent R of people responding and purchasing is

$$R = 50 - 100e^{(-0.3)(5)} = 50 - 100e^{-1.5}$$

We use a calculator to evaluate this expression:

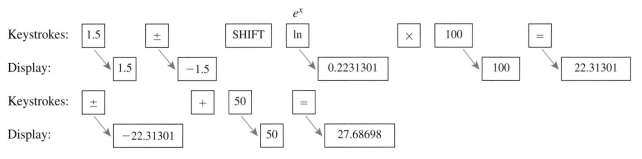

Thus, about 28% will have responded after 5 days.

(b) After 10 days, we have $t = 10$. The corresponding percent R of people responding and purchasing is

$$R = 50 - 100e^{(-0.3)(10)} = 50 - 100e^{-3} \approx 45.021$$

About 45% will have responded and purchased after 10 days.

(c) As time passes, more people are expected to respond and purchase. The highest percent expected is therefore found for the value of R as $t \to \infty$. Since $e^{-0.3t} = 1/e^{0.3t}$, it follows that $e^{-0.3t} \to 0$ as $t \to \infty$. Thus, the highest percent expected is 50%.

(d) See Figure 9.

FIGURE 9

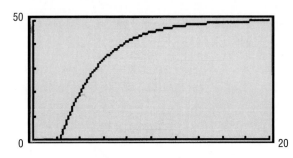

It will require 7 days to exceed 40%.

■ Now work Problem 33.

Summary

PROPERTIES OF THE EXPONENTIAL FUNCTION

$f(x) = a^x, a > 1$ Domain: $(-\infty, \infty)$; Range: $(0, \infty)$; x-intercepts: none; y-intercept: 1; horizontal asymptote: x-axis as $x \to -\infty$; increasing; one-to-one See Figure 3 for a typical graph.

$f(x) = a^x, 0 < a < 1$ Domain: $(-\infty, \infty)$; Range: $(0, \infty)$; x-intercepts: none, y-intercept: 1; horizontal asymptote: x-axis as $x \to \infty$; decreasing; one-to-one See Figure 5 for a typical graph.

4.1

Exercise 4.1

In Problems 1–10, approximate each number using a calculator. Express your answer rounded to three decimal places.

1. (a) $3^{2.2}$ (b) $3^{2.23}$ (c) $3^{2.236}$ (d) $3^{\sqrt{5}}$
2. (a) $5^{1.7}$ (b) $5^{1.73}$ (c) $5^{1.732}$ (d) $5^{\sqrt{3}}$
3. (a) $2^{3.14}$ (b) $2^{3.141}$ (c) $2^{3.1415}$ (d) 2^{π}
4. (a) $2^{2.7}$ (b) $2^{2.71}$ (c) $2^{2.718}$ (d) 2^{e}
5. (a) $3.1^{2.7}$ (b) $3.14^{2.71}$ (c) $3.141^{2.718}$ (d) π^{e}
6. (a) $2.7^{3.1}$ (b) $2.71^{3.14}$ (c) $2.718^{3.141}$ (d) e^{π}
7. $e^{1.2}$ 8. $e^{-1.3}$ 9. $e^{-0.85}$ 10. $e^{2.1}$

In Problems 11–18, the graph of an exponential function is given. Match each graph to one of the following functions:

A. $y = 3^{x}$ B. $y = 3^{-x}$ C. $y = -3^{x}$ D. $y = -3^{-x}$
E. $y = 3^{x} - 1$ F. $y = 3^{x-1}$ G. $y = 3^{1-x}$ H. $y = 1 - 3^{x}$

11.
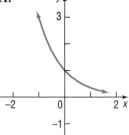

12.
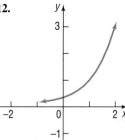

13.
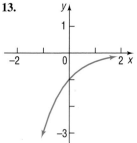

14.
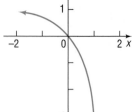

15.
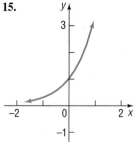

16.
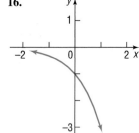

17.
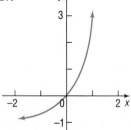

18.
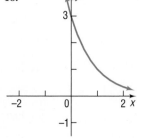

In Problems 19–26, use the graph of $y = e^x$ (Figure 8), along with the techniques of shifting, compression, stretching, and reflection, to graph each function.

19. $y = e^{-x}$ **20.** $y = -e^x$ **21.** $y = e^{x+2}$ **22.** $y = e^x - 1$

23. $y = 5 - e^{-x}$ **24.** $y = 9 - 3e^{-x}$ **25.** $y = 2 - e^{-x/2}$ **26.** $y = 7 - 3e^{-2x}$

27. If $4^x = 7$, what does 4^{-2x} equal?   **28.** If $2^x = 3$, what does 4^{-x} equal?

29. If $3^{-x} = 2$, what does 3^{2x} equal?   **30.** If $5^{-x} = 3$, what does 5^{3x} equal?

31. *Optics* If a single pane of glass obliterates 3% of the light passing through it, then the percent p of light that passes through n successive panes is given approximately by the equation

$$p = 100e^{-0.03n}$$

(a) What percent of light will pass through 10 panes?

(b) What percent of light will pass through 25 panes?

32. *Atmospheric Pressure* The atmospheric pressure p on a balloon or plane decreases with increasing height. This pressure, measured in millimeters of mercury, is related to the number of kilometers h above sea level by the formula

$$p = 760e^{-0.145h}$$

(a) Find the atmospheric pressure at a height of 2 kilometers (over a mile).

(b) What is it at a height of 10 kilometers (over 30,000 feet)?

33. *Space Satellites* The number of watts w provided by a space satellite's power supply over a period of d days is given by the formula

$$w = 50e^{-0.004d}$$

(a) How much power will be available after 30 days?

(b) How much power will be available after 1 year (365 days)?

34. *Healing of Wounds* The normal healing of wounds can be modeled by an exponential function. If A_0 represents the original area of the wound and if A equals the area of the wound after n days, then the formula

$$A = A_0 e^{-0.35n}$$

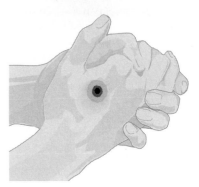

describes the area of a wound on the nth day following an injury when no infection is present to retard the healing. Suppose a wound initially had an area of 100 square centimeters.

(a) If healing is taking place, how large should the area of the wound be after 3 days?

(b) How large should it be after 10 days?

35. *Drug Medication* The formula

$$D = 5e^{-0.4h}$$

can be used to find the number of milligrams D of a certain drug that is in a patient's bloodstream h hours after the drug has been administered. How many milligrams will be present after 1 hour? After 6 hours?

36. *Spreading of Rumors* A model for the number of people N in a college community who have heard a certain rumor is

$$N = P(1 - e^{-0.15d})$$

where P is the total population of the community and d is the number of days that have elapsed since the rumor began. In a community of 1000 students, how many students will have heard the rumor after 3 days?

37. *Response to TV Advertising* The percent R of viewers who respond to a television commercial for a new product after t days is found by using the formula

$$R = 70 - 100e^{-0.2t}$$

(a) What percent is expected to respond after 10 days?

(b) What percent has responded after 20 days?

(c) What is the highest percent of people expected to respond?

 (d) Graph $R = 70 - 100e^{-0.2t}$, $t > 0$. TRACE and compare the values of R for $t = 10$ and $t = 20$ to the ones obtained in parts (a) and (b). How many days are required for R to exceed 40%?

38. *Profit* The annual profit P of a company due to the sales of a particular item after it has been on the market x years is determined to be

$$P = \$100{,}000 - \$60{,}000\left(\tfrac{1}{2}\right)^x$$

(a) What is the profit after 5 years?

(b) What is the profit after 10 years?

(c) What is the most profit the company can expect from this product?

(d) Graph the profit function. TRACE and compare the values of P for $x = 5$ and $x = 10$ to the one obtained in parts (a) and (b). How many years does it take before a profit of \$65,000 is obtained?

39. *Alternating Current in a RL Circuit* The equation governing the amount of current I (in amperes) after time t (in seconds) in a single RL circuit consisting of a resistance R (in ohms), an inductance L (in henrys), and an electromotive force E (in volts) is

$$I = \frac{E}{R}[1 - e^{-(R/L)t}]$$

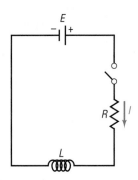

(a) If $E = 120$ volts, $R = 10$ ohms, and $L = 5$ henrys, how much current I_1 is available after 0.3 second? After 0.5 second? After 1 second?

(b) What is the maximum current?

(c) Graph this function $I = I_1(t)$, measuring I along the y-axis and t along the x-axis.

(d) If $E = 120$ volts, $R = 5$ ohms, and $L = 10$ henrys, how much current I_2 is available after 0.3 second? After 0.5 second? After 1 second?

(e) What is the maximum current?

(f) Graph this function $I = I_2(t)$ on the same coordinate axes as $I_1(t)$.

40. *Alternating Current in a RC Circuit* The equation governing the amount of current I (in milliamperes) after time t (in milliseconds) in a single RC circuit consisting of a resistance R (in ohms), a capacitance C (in microfarads), and an electromotive force E (in volts) is

$$I = \frac{E}{R}e^{-t/(RC)}$$

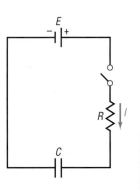

(a) If $E = 120$ volts, $R = 2000$ ohms, and $C = 1.0$ microfarad, how much current I_1 is available initially ($t = 0$)? After 1000 milliseconds? After 3000 milliseconds?

(b) What is the maximum current?

(c) Graph this function $I = I_1(t)$, measuring I along the y-axis and t along the x-axis.

(d) If $E = 120$ volts, $R = 1000$ ohms, and $C = 2.0$ microfarads, how much current I_2 is available initially? After 1000 milliseconds? After 3000 milliseconds?

(e) What is the maximum current?

(f) Graph this function $I = I_2(t)$ on the same coordinate axes as $I_1(t)$.

41. *The Challenger Disaster** After the *Challenger* disaster in 1986, a study of the 23 launches that preceded the fatal flight was made. A mathematical model was developed involving the relationship between the Fahrenheit temperature x around the O-rings and the number y of eroded or leaky primary O-rings. The model stated that

$$y = 6\,[1 + e^{-(5.085-0.1156x)}]^{-1}$$

where the number 6 indicates the 6 primary O-rings on the spacecraft.

(a) What is the predicted number of eroded or leaky primary O-rings at a temperature of 100°F?

(b) What is the predicted number of eroded or leaky primary O-rings at a temperature of 60°F?

(c) What is the predicted number of eroded or leaky primary O-rings at a temperature of 30°F?

(d) Graph the equation and TRACE. At what temperature is the predicted number of eroded or leaky O-rings 1? 3? 5?

42. *Postage Stamps*† The cumulative number y of different postage stamps (regular and commemorative only) issued by the U.S. Post Office can be approximated (modeled) by the exponential function

$$y = 78e^{0.025x}$$

where x is the number of years since 1848.

(a) What is the predicted cumulative number of stamps that will have been issued by the year 1995? Check with the Postal Service and comment on the accuracy of using the function.

(b) What is the predicted cumulative number of stamps that will have been issued by the year 1998?

(c) The cumulative number of stamps actually issued by the United States was 2 in 1848, 88 in 1868, 218 in 1888, and 341 in 1908. What conclusion can you draw about using the given function as a model over the first few decades in which stamps were issued?

43. *Another Formula for e* Use a calculator to compute the value of

$$2 + \frac{1}{2!} + \frac{1}{3!} + \ \ldots \ + \frac{1}{n!}$$

for $n = 4, 6, 8$, and 10. Compare each result with e.
[*Hint:* $1! = 1, 2! = 2 \cdot 1, 3! = 3 \cdot 2 \cdot 1, n! = n(n-1) \cdot \ldots \cdot (3)(2)(1)$].

44. *Another Formula for e* Use a calculator to compute the various values of the expression

$$2 + \cfrac{1}{1 + \cfrac{1}{2 + \cfrac{2}{3 + \cfrac{3}{4 + \cfrac{4}{\text{etc.}}}}}}$$

Compare the values to e.

45. If $f(x) = a^x$, show that: $\dfrac{f(x+h) - f(x)}{h} = a^x \left(\dfrac{a^h - 1}{h} \right)$

46. If $f(x) = a^x$, show that: $f(A + B) = f(A) \cdot f(B)$

*Linda Tappin, "Analyzing Data Relating to the *Challenger* Disaster," *Mathematics Teacher,* Vol. 87, No. 6, September 1994, pp. 423–426.
†David Kullman, "Patterns of Postage-stamp Production," *Mathematics Teacher,* Vol. 85, No. 3, March 1992, pp. 188–189.

47. If $f(x) = a^x$, show that: $f(-x) = \dfrac{1}{f(x)}$

48. If $f(x) = a^x$, show that: $f(\alpha x) = [f(x)]^\alpha$

Problems 49 and 50 provide definitions for two other transcendental functions.

49. The **hyperbolic sine function,** designated by sinh x, is defined as

$$\sinh x = \frac{1}{2}(e^x - e^{-x})$$

(a) Show that $f(x) = \sinh x$ is an odd function.

(b) Graph $y = e^x$ and $y = e^{-x}$ on the same set of coordinate axes, and use the method of subtracting y-coordinates to obtain a graph of $f(x) = \sinh x$.

50. The **hyperbolic cosine function,** designated by cosh x, is defined as

$$\cosh x = \frac{1}{2}(e^x + e^{-x})$$

(a) Show that $f(x) = \cosh x$ is an even function.

(b) Graph $y = e^x$ and $y = e^{-x}$ on the same set of coordinate axes, and use the method of adding y-coordinates to obtain a graph of $f(x) = \cosh x$.

(c) Refer to Problem 49. Show that, for every x,

$$(\cosh x)^2 - (\sinh x)^2 = 1$$

51. *Historical problem* Pierre de Fermat (1601-1665) conjectured that the function

$$f(x) = 2^{(2^x)} + 1$$

for $x = 1, 2, 3, \ldots$, would always have a value equal to a prime number. But Leonhard Euler (1707–1783) showed that this formula fails for $x = 5$. Use a calculator to determine the prime numbers produced by f for $x = 1, 2, 3, 4$. Then show that $f(5) = 641 \times 6,700,417$, which is not prime.

52. The bacteria in a 4 liter container double every minute. After 60 minutes the container is full. How long did it take to fill half the container?

53. Explain in your own words what the number e is. Provide at least two applications that require the use of this number.

54. Do you think there is a power function that increases more rapidly than an exponential function whose base is greater than 1? Explain.

4.2

Logarithmic Functions

Recall that a one-to-one function $y = f(x)$ has an inverse that is defined (implicitly) by the equation $x = f(y)$. In particular, the exponential function $y = f(x) = a^x$, $a > 0$, $a \neq 1$, is one-to-one and, hence, has an inverse that is defined implicitly by the equation

$$x = a^y \qquad a > 0, a \neq 1$$

This inverse is so important that it is given a name, the *logarithmic function.*

Logarithmic Function

> The **logarithmic function to the base a,** where $a > 0$ and $a \neq 1$, is denoted by $y = \log_a x$ (read as "y is the logarithm to the base a of x") and is defined by
>
> $$y = \log_a x \quad \text{if and only if} \quad x = a^y$$

EXAMPLE 1   *Relating Logarithms to Exponents*

(a) If $y = \log_3 x$, then $x = 3^y$. Thus, if $x = 9$, then $y = 2$, so $9 = 3^2$ is equivalent to $2 = \log_3 9$.

(b) If $y = \log_5 x$, then $x = 5^y$. Thus, if $x = \frac{1}{5} = 5^{-1}$, then $y = -1$, so $\frac{1}{5} = 5^{-1}$ is equivalent to $-1 = \log_5 \left(\frac{1}{5}\right)$. ∎

EXAMPLE 2   *Changing Exponential Expressions to Logarithmic Expressions*

Change each exponential expression to an equivalent expression involving a logarithm.

(a) $1.2^3 = m$ (b) $e^b = 9$ (c) $a^4 = 24$

Solution   We use the fact that $y = \log_a x$ and $x = a^y$, $a > 0$, $a \neq 1$, are equivalent.

(a) If $1.2^3 = m$, then $3 = \log_{1.2} m$.

(b) If $e^b = 9$, then $b = \log_e 9$.

(c) If $a^4 = 24$, then $4 = \log_a 24$. ∎

■ Now work Problem 1.

EXAMPLE 3   *Changing Logarithmic Expressions to Exponential Expressions*

Change each logarithmic expression to an equivalent expression involving an exponent.

(a) $\log_a 4 = 5$ (b) $\log_e b = -3$ (c) $\log_3 5 = c$

Solution   (a) If $\log_a 4 = 5$, then $a^5 = 4$.

(b) If $\log_e b = -3$, then $e^{-3} = b$.

(c) If $\log_3 5 = c$, then $3^c = 5$. ∎

■ Now work Problem 13.

To find the exact value of a logarithm, we write the logarithm in exponential notation and use the following fact:

$$\text{If } a^u = a^v, \quad \text{then} \quad u = v. \tag{1}$$

The result (1) is a consequence of the fact that exponential functions are one-to-one.

EXAMPLE 4   *Finding the Exact Value of a Logarithmic Function*

Find the exact value of:

(a) $\log_2 8$ (b) $\log_3 \frac{1}{3}$ (c) $\log_5 25$

Solution   (a) For $y = \log_2 8$, we have the equivalent exponential equation $2^y = 8 = 2^3$, so, by (1), $y = 3$. Thus, $\log_2 8 = 3$.

(b) For $y = \log_3 \frac{1}{3}$, we have $3^y = \frac{1}{3} = 3^{-1}$, so $y = -1$. Thus, $\log_3 \frac{1}{3} = -1$.

(c) For $y = \log_5 25$, we have $5^y = 25 = 5^2$, so $y = 2$. Thus, $\log_5 25 = 2$.   ■

■ Now work Problem 25.

Domain of a Logarithmic Function

The logarithmic function $y = \log_a x$ has been defined as the inverse of the exponential function $y = a^x$. That is, if $f(x) = a^x$, then $f^{-1}(x) = \log_a x$. Based on the discussion given in Section 4.5 on inverse functions, we know that for a function f and its inverse f^{-1}

$$\text{Domain } f^{-1} = \text{Range } f \quad \text{and} \quad \text{Range } f^{-1} = \text{Domain } f$$

Consequently, it follows that:

Domain of logarithmic function = Range of exponential function = $(0, \infty)$
Range of logarithmic function = Domain of exponential function = $(-\infty, \infty)$

In the next box, we summarize some properties of the logarithmic function:

$$y = \log_a x \quad \text{(defining equation: } x = a^y)$$
$$\text{Domain: } 0 < x < \infty \quad \text{Range: } -\infty < y < \infty$$

Notice that the domain of a logarithmic function consists of the *positive* real numbers.

E X A M P L E   5   *Finding the Domain of a Logarithmic Function*
Find the domain of each logarithmic function:

(a) $F(x) = \log_2 (1 - x)$ (b) $g(x) = \log_5 \left(\dfrac{1+x}{1-x} \right)$ (c) $h(x) = \log_{1/2} |x|$

Solution   (a) The domain of F consists of all x for which $(1 - x) > 0$; that is, all $x < 1$, or $(-\infty, 1)$.

(b) The domain of g is restricted to

$$\frac{1 + x}{1 - x} > 0$$

Solving this inequality, we find that the domain of g consists of all x between -1 and 1, that is, $-1 < x < 1$, or $(-1, 1)$.

(c) Since $|x| > 0$ provided $x \neq 0$, the domain of h consists of all nonzero real numbers.   ■

■ Now work Problem 39.

Graphs of Logarithmic Functions

Since exponential functions and logarithmic functions are inverses of each other, the graph of a logarithmic function $y = \log_a x$ is the reflection about the line $y = x$ of the graph of the exponential function $y = a^x$, as shown in Figure 10.

FIGURE 10

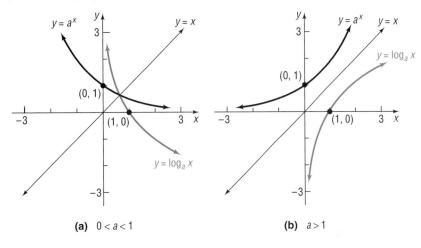

(a) $0 < a < 1$ **(b)** $a > 1$

Facts about the Graph
of a Logarithmic Function
$f(x) = \log_a x$

1. The x-intercept of the graph is 1. There is no y-intercept.
2. The y-axis is a vertical asymptote of the graph.
3. A logarithmic function is decreasing if $0 < a < 1$ and increasing if $a > 1$.
4. The graph is smooth and continuous, with no corners or gaps.

If the base of a logarithmic function is the number e, then we have the **natural logarithm function.** This function occurs so frequently in applications that it is given a special symbol, **ln** (from the Latin, *logarithmus naturalis*). Thus,

$$y = \ln x \quad \text{if and only if} \quad x = e^y$$

Since $y = \ln x$ and the exponential function $y = e^x$ are inverse functions, we can obtain the graph of $y = \ln x$ by reflecting the graph of $y = e^x$ about the line $y = x$. See Figure 11.

FIGURE 11

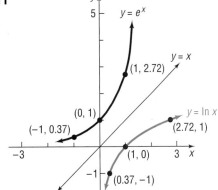

TABLE 5

x	$\ln x$
$\frac{1}{2}$	-0.69
2	0.69
3	1.10

Using a calculator with an $\boxed{ln}$ key, we can obtain other points on the graph of $f(x) = \ln x$. See Table 5.

Check: Graph $y = e^x$ and $y = \ln x$ on the same square screen. Use TRACE to verify the points on the graph given in Figure 11. Do you see the symmetry of the two graphs with respect to the line $y = x$?   ■

E X A M P L E   6   *Graphing Functions That Are Basically Logarithmic Using Shifting, Reflection, and the Like*

Graph $y = -\ln x$ by starting with the graph of $y = \ln x$.

Solution   The graph of $y = -\ln x$ is obtained by a reflection about the *x*-axis of the graph of $y = \ln x$. See Figure 12.

FIGURE 12

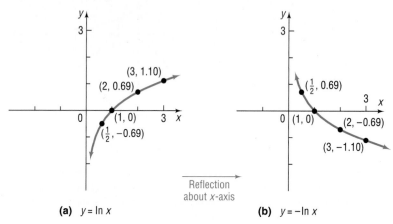

(a) $y = \ln x$ (b) $y = -\ln x$ ■

E X A M P L E   7   *Graphing Functions That Are Basically Logarithmic Using Shifting, Reflection, and the Like*

Graph: $y = \ln(x + 2)$

Solution   The domain consists of all *x* for which

$$x + 2 > 0 \quad \text{or} \quad x > -2$$

The graph is obtained by applying a horizontal shift to the left 2 units, as shown in Figure 13. Notice that the line $x = -2$ is a vertical asymptote.

FIGURE 13

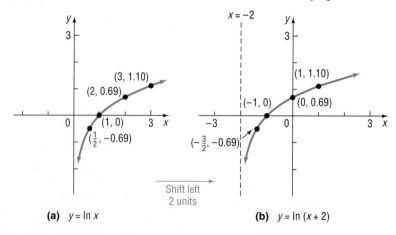

(a) $y = \ln x$ (b) $y = \ln(x + 2)$ ■

E X A M P L E   8   *Graphing Functions That Are Basically Logarithmic Using Shifting, Reflection, and the Like*

Graph: $y = \ln(1 - x)$

Solution   The domain consists of all x for which
$$1 - x > 0 \quad \text{or} \quad x < 1$$

To obtain the graph of $y = \ln(1 - x)$, we use the steps illustrated in Figure 14. Note that the vertical asymptote of $y = \ln(1 - x)$ is $x = 1$.

FIGURE 14

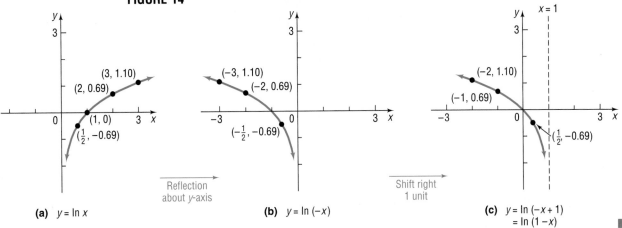

(a) $y = \ln x$ **(b)** $y = \ln(-x)$ **(c)** $y = \ln(-x + 1)$
                                                                    $= \ln(1 - x)$

■   Now work Problem 61.

E X A M P L E   9   *Alcohol and Driving*

The concentration of alcohol in a person's blood is measurable. Recent medical research suggests that the risk R (given as a percent) of having an accident while driving a car can be modeled by the equation

$$R = 6e^{kx}$$

where x is the variable concentration of alcohol in the blood and k is a constant.

(a) Suppose that a concentration of alcohol in the blood of 0.04 results in a 10% risk ($R = 10$) of an accident. Find the constant k in the equation.

(b) Using this value of k, what is the risk if the concentration is 0.17?

(c) Using the same value of k, what concentration of alcohol corresponds to a risk of 100%?

(d) If the law asserts that anyone with a risk of having an accident of 20% or more should not have driving privileges, at what concentration of alcohol in the blood should a driver be arrested and charged with a DUI—Driving Under the Influence?

Solution   (a) For a concentration of alcohol in the blood of 0.04 and a risk of 10%, we let $x = 0.04$ and $R = 10$ in the equation and solve for k.

$$R = 6e^{kx}$$
$$10 = 6e^{k(0.04)}$$
$$\frac{10}{6} = e^{0.04k} \qquad \text{Change to a logarithmic expression.}$$
$$0.04k = \ln \frac{10}{6} = 0.5108256$$
$$k = 12.77$$

(b) Using $k = 12.77$ and $x = 0.17$ in the equation, we find the risk R to be

$$R = 6e^{kx} = 6e^{(12.77)(0.17)} = 52.6$$

For a concentration of alcohol in the blood of 0.17, the risk of an accident is about 52.6%.

(c) Using $k = 12.77$ and $R = 100$ in the equation, we find the concentration x of alcohol in the blood to be

$$R = 6e^{kx}$$

$$100 = 6e^{12.77x}$$

$$\frac{100}{6} = e^{12.77x} \qquad \text{Change to a logarithmic expression.}$$

$$12.77x = \ln \frac{100}{6} = 2.8134$$

$$x = 0.22$$

For a concentration of alcohol in the blood of 0.22, the risk of an accident is 100%.

(d) Using $k = 12.77$ and $R = 20$ in the equation, we find the concentration x of alcohol in the blood to be

$$R = 6e^{kx}$$

$$20 = 6e^{12.77x}$$

$$\frac{20}{6} = e^{12.77x}$$

$$12.77x = \ln \frac{20}{6} = 1.204$$

$$x = 0.094$$

A driver with a concentration of alcohol in the blood of 0.094 or more should be arrested and charged with DUI.   ∎

Note: Most states use 0.10 as the blood alcohol content at which a DUI citation is given. A few states use 0.08.

Summary

PROPERTIES OF THE LOGARITHMIC FUNCTION

$f(x) = \log_a x, \quad a > 1$

($y = \log_a x$ means $x = a^y$)

Domain: $(0, \infty)$; Range: $(-\infty, \infty)$; x-intercept: 1; y-intercept: none; vertical asymptote: y-axis; increasing; one-to-one
See Figure 10(b) for a typical graph.

$f(x) = \log_a x, \quad 0 < a < 1$

($y = \log_a x$ means $x = a^y$)

Domain: $(0, \infty)$; Range: $(-\infty, \infty)$; x-intercept: 1; y-intercept: none; vertical asymptote: y-axis; decreasing; one-to-one
See Figure 10(a) for a typical graph.

4.2

Exercise 4.2

In Problems 1–12, change each exponential expression to an equivalent expression involving a logarithm.

1. $9 = 3^2$ **2.** $16 = 4^2$ **3.** $a^2 = 1.6$ **4.** $a^3 = 2.1$ **5.** $1.1^2 = M$ **6.** $2.2^3 = N$

7. $2^x = 7.2$ **8.** $3^x = 4.6$ **9.** $x^{\sqrt{2}} = \pi$ **10.** $x^\pi = e$ **11.** $e^x = 8$ **12.** $e^{2.2} = M$

In Problems 13–24, change each logarithmic expression to an equivalent expression involving an exponent.

13. $\log_2 8 = 3$ **14.** $\log_3(\frac{1}{9}) = -2$ **15.** $\log_a 3 = 6$ **16.** $\log_b 4 = 2$

17. $\log_3 2 = x$ **18.** $\log_2 6 = x$ **19.** $\log_2 M = 1.3$ **20.** $\log_3 N = 2.1$

21. $\log_{\sqrt{2}} \pi = x$ **22.** $\log_\pi x = \frac{1}{2}$ **23.** $\ln 4 = x$ **24.** $\ln x = 4$

In Problems 25–36, find the exact value of each logarithm without using a calculator or a table.

25. $\log_2 1$ **26.** $\log_8 8$ **27.** $\log_5 25$ **28.** $\log_3(\frac{1}{9})$ **29.** $\log_{1/2} 16$ **30.** $\log_{1/3} 9$

31. $\log_{10} \sqrt{10}$ **32.** $\log_5 \sqrt[3]{25}$ **33.** $\log_{\sqrt{2}} 4$ **34.** $\log_{\sqrt{3}} 9$ **35.** $\ln \sqrt{e}$ **36.** $\ln e^3$

In Problems 37–46, find the domain of each function.

37. $f(x) = \ln(3 - x)$ **38.** $g(x) = \ln(x^2 - 1)$ **39.** $F(x) = \log_2 x^2$

40. $H(x) = \log_5 x^3$ **41.** $h(x) = \log_{1/2}(x^2 - x - 6)$ **42.** $G(x) = \log_{1/2}\left(\dfrac{1}{x}\right)$

43. $f(x) = \dfrac{1}{\ln x}$ **44.** $g(x) = \ln(x - 5)$ **45.** $g(x) = \log_5\left(\dfrac{x + 1}{x}\right)$ **46.** $h(x) = \log_3\left(\dfrac{x^2}{x - 1}\right)$

In Problems 47–50, use a calculator to evaluate each expression. Round your answer to three decimal places.

47. $\ln \dfrac{5}{3}$ **48.** $\dfrac{\ln 5}{3}$ **49.** $\dfrac{\ln 10/3}{0.04}$ **50.** $\dfrac{\ln 2/3}{-0.1}$

51. Find a such that the graph of $f(x) = \log_a x$ contains the point $(2, 2)$.

52. Find a such that the graph of $f(x) = \log_a x$ contains the point $(\frac{1}{2}, -4)$.

In Problems 53–60, the graph of a logarithmic function is given. Match each graph to one of the following functions:

A. $y = \log_3 x$ **B.** $y = \log_3(-x)$ **C.** $y = -\log_3 x$ **D.** $y = -\log_3(-x)$

E. $y = \log_3 x - 1$ **F.** $y = \log_3(x - 1)$ **G.** $y = \log_3(1 - x)$ **H.** $y = 1 - \log_3 x$

53.

54.

55.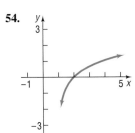

56.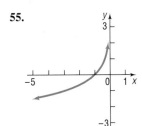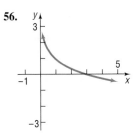

57.

58.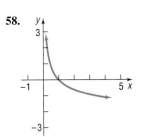

59.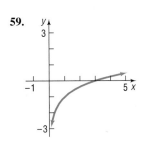

60.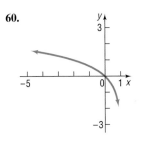

In Problems 61–70, use the graph of $y = \ln x$, along with the techniques of shifting, compression, stretching, and reflection, to graph each function.

61. $f(x) = \ln(x + 4)$ **62.** $f(x) = \ln(x - 3)$ **63.** $f(x) = \ln(-x)$ **64.** $f(x) = -\ln(-x)$

65. $g(x) = \ln 2x$ **66.** $h(x) = \ln \frac{1}{2}x$ **67.** $f(x) = 3 \ln x$ **68.** $f(x) = -2 \ln x$

69. $g(x) = \ln(3 - x)$ **70.** $h(x) = \ln(4 - x)$

71. *Optics* If a single pane of glass obliterates 10% of the light passing through it, then the percent P of light that passes through n successive panes is given approximately by the equation

$$P = 100e^{-0.1n}$$

(a) How many panes are necessary to block at least 50% of the light?

(b) How many panes are necessary to block at least 75% of the light?

72. *Chemistry* The pH of a chemical solution is given by the formula

$$\text{pH} = -\log_{10} [\text{H}^+]$$

where $[\text{H}^+]$ is the concentration of hydrogen ions in moles per liter. Values of pH range from 0 (acidic) to 14 (alkaline).

(a) Find the pH of a 1 liter container of water with 0.0000001 mole of hydrogen ion.

(b) Find the hydrogen ion concentration of a mildly acidic solution with a pH of 4.2.

73. *Space Satellites* The number of watts w provided by a space satellite's power supply of d days is given by the formula

$$w = 50e^{-0.004d}$$

(a) How long will it take for the available power to drop to 30 watts?

(b) How long will it take for the available power to drop to only 5 watts?

74. *Healing of Wounds* The normal healing of wounds can be modeled by an exponential function. If A_0 represents the original area of the wound and if A equals the area of the wound after n days, then the formula

$$A = A_0 e^{-0.35n}$$

describes the area of a wound on the nth day following an injury when no infection is present to retard the healing. Suppose a wound initially had an area of 100 square centimeters.

(a) If healing is taking place, how many days should pass before the wound is one-half its original size?

(b) How long before the wound is 10% of its original size?

75. *Drug Medication* The formula

$$D = 5e^{-0.4h}$$

can be used to find the number of milligrams D of a certain drug that is in a patient's bloodstream h hours after the drug has been administered. When the number of milligrams reaches 2, the drug is to be administered again. What is the time between injections?

76. *Spreading of Rumors* A model for the number of people N in a college community who have heard a certain rumor is

$$N = P(1 - e^{-0.15d})$$

where P is the total population of the community and d is the number of days that have elapsed since the rumor began. In a community of 1000 students, how many days will elapse before 450 students have heard the rumor?

77. *Current in a RL Circuit* The equation governing the amount of current I (in amperes) after time t (in seconds) in a simple RL circuit consisting of a resistance R (in ohms), an inductance L (in henrys), and an electromotive force E (in volts) is

$$I = \frac{E}{R}[1 - e^{-(R/L)t}]$$

If $E = 12$ volts, $R = 10$ ohms, and $L = 5$ henrys, how long does it take to obtain a current of 0.5 ampere? Of 1.0 ampere? Graph the equation.

78. *Learning Curve* Psychologists sometimes use the function

$$L(t) = A(1 - e^{-kt})$$

to measure the amount L learned at time t. The number A represents the amount to be learned, and the number k measures the rate of learning. Suppose that a student has an amount A of 200 vocabulary words to learn. A psychologist determines that the student learned 20 vocabulary words after 5 minutes.

(a) Determine the rate of learning k.
(b) Approximately how many words will the student have learned after 10 minutes?
(c) After 15 minutes?
(d) How long does it take for the student to learn 180 words?

79. *Alcohol and Driving* The concentration of alcohol in a person's blood is measurable. Suppose the risk R (given as a percent) of having an accident while driving a car can be modeled by the equation

$$R = 3e^{kx}$$

where x is the variable concentration of alcohol in the blood and k is a constant.

(a) Suppose a concentration of alcohol in the blood of 0.06 results in a 10% risk ($R = 10$) of an accident. Find the constant k in the equation.
(b) Using this value of k, what is the risk if the concentration is 0.17?
(c) Using the same value of k, what concentration of alcohol corresponds to a risk of 100%?
(d) If the law asserts that anyone with a risk of having an accident of 15% or more should not have driving privileges, at what concentration of alcohol in the blood should a driver be arrested and charged with a DUI?

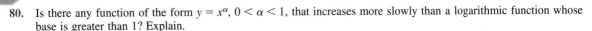

(e) Compare this situation with that of Example 9. If you were a lawmaker, which situation would you support? Give your reasons.

80. Is there any function of the form $y = x^\alpha$, $0 < \alpha < 1$, that increases more slowly than a logarithmic function whose base is greater than 1? Explain.

81. *Constructing a Function* Look back at Figure 2 on page 264. Assuming that the points (1950, 20) and (1990, 50) are on the graph, find an exponential equation $y = Ae^{bt}$, that fits the data. [*Hint:*   Let $t = 0$ correspond to the year 1950. Then show that $A = 20$. Now find b.] Is the projection of 102 million in 2015 confirmed by your model? Try to obtain similar data about the United States birthrate and construct a function to fit those data.

82. *Critical Thinking* In buying a new car, one consideration might be how well the price of the car holds up over time. Different makes of cars have different depreciation rates. One way to compute a depreciation rate for a car is given here. Suppose the current prices of a certain Mercedes automobile are as follows:

NEW	1 YEAR OLD	2 YEARS OLD	3 YEARS OLD	4 YEARS OLD	5 YEARS OLD
$38,000	$36,600	$32,400	$28,750	$25,400	$21,200

Use the formula New = Old(e^{Rt}) to find R, the annual depreciation rate, for a specific time t. When might be the best time to trade the car in? Consult the NADA ("blue") book and compare two like models that you are interested in. Which has the better depreciation rate?

4.3

Properties of Logarithms

Logarithms have some very useful properties that can be derived directly from the definition and the laws of exponents.

E X A M P L E   1   *Establishing Properties of Logarithms*

(a) Show that $\log_a 1 = 0$.   (b) Show that $\log_a a = 1$.

Solution   (a) This fact was established when we graphed $y = \log_a x$ (see Figure 10). Algebraically, for $y = \log_a 1$, we have $a^y = 1 = a^0$, so $y = 0$.

(b) For $y = \log_a a$, we have $a^y = a = a^1$, so $y = 1$.

$$\log_a 1 = 0 \qquad \log_a a = 1$$

∎

Theorem   In the properties given next, M and a are positive real numbers, with $a \neq 1$, and
Properties of Logarithms r is any real number.

The number $\log_a M$ is the exponent to which a must be raised to obtain M. That is,

$$a^{\log_a M} = M \tag{1}$$

The logarithm to the base a of a raised to a power equals that power. That is,

$$\log_a a^r = r \tag{2}$$

∎

Proof of Property (1)   Let $x = \log_a M$. Change this logarithmic expression to the equivalent exponential expression:

$$a^x = M$$

But $x = \log_a M$, so

$$a^{\log_a M} = M$$ ∎

Proof of Property (2)   Let $x = a^r$. Change this exponential expression to the equivalent logarithmic expression:

$$\log_a x = r$$

But $x = a^r$, so

$$\log_a a^r = r$$ ∎

E X A M P L E   2   *Using Properties (1) and (2)*

(a) $2^{\log_2 \pi} = \pi$ (b) $\log_{0.2} 0.2^{-\sqrt{2}} = -\sqrt{2}$ (c) $\ln e^{kt} = kt$ ∎

Other useful properties of logarithms are given now.

Theorem	In the following properties, M, N, and a are positive real numbers, with $a \neq 1$, and r is any real number.

The Log of a Product Equals
the Sum of the Logs

$$\log_a MN = \log_a M + \log_a N \qquad (3)$$

The Log of a Quotient Equals
the Difference of the Logs

$$\log_a\left(\frac{M}{N}\right) = \log_a M - \log_a N \qquad (4)$$

$$\log_a\left(\frac{1}{N}\right) = -\log_a N \qquad (5)$$

$$\log_a M^r = r \log_a M \qquad (6)$$

We shall derive properties (3) and (6) and leave the derivations of properties (4) and (5) as exercises (see Problems 63 and 64).

Proof of Property (3) Let $A = \log_a M$ and let $B = \log_a N$. These expressions are equivalent to the exponential expressions

$$a^A = M \quad \text{and} \quad a^B = N$$

Now

$$\log_a MN = \log_a a^A a^B = \log_a a^{A+B} \qquad \text{Law of exponents}$$
$$= A + B \qquad \text{Property (2) of logarithms}$$
$$= \log_a M + \log_a N$$

Proof of Property (6) Let $A = \log_a M$. This expression is equivalent to

$$a^A = M$$

Now

$$\log_a M^r = \log_a(a^A)^r = \log_a a^{rA} \qquad \text{Law of exponents}$$
$$= rA \qquad \text{Property (2) of logarithms}$$
$$= r \log_a M$$

Logarithms can be used to transform products into sums, quotients into differences, and powers into factors. Such transformations prove useful in certain types of calculus problems.

E X A M P L E   3   *Writing a Logarithmic Expression as a Sum of Logarithms*

Write $\log_a(x\sqrt{x^2 + 1})$ as a sum of logarithms. Express all powers as factors.

Solution

$$\log_a(x\sqrt{x^2 + 1}) = \log_a x + \log_a \sqrt{x^2 + 1} \qquad \text{Property (3)}$$
$$= \log_a x + \log_a(x^2 + 1)^{1/2}$$
$$= \log_a x + \tfrac{1}{2} \log_a(x^2 + 1) \qquad \text{Property (6)}$$

EXAMPLE 4   *Writing a Logarithmic Expression as a Difference of Logarithms*

Write

$$\log_a \frac{x^2}{(x-1)^3}$$

as a difference of logarithms. Express all powers as factors.

Solution

$$\log_a \frac{x^2}{(x-1)^3} \underset{\substack{\uparrow \\ \text{Property (4)}}}{=} \log_a x^2 - \log_a(x-1)^3 \underset{\substack{\uparrow \\ \text{Property (6)}}}{=} 2\log_a x - 3\log_a(x-1)$$ ■

■ Now work Problem 13.

EXAMPLE 5   *Writing a Logarithmic Expression as a Sum and Difference of Logarithms*

Write

$$\log_a \frac{x^3\sqrt{x^2+1}}{(x+1)^4}$$

as a sum and difference of logarithms. Express all powers as factors.

Solution

$$\log_a \frac{x^3\sqrt{x^2+1}}{(x+1)^4} = \log_a(x^3\sqrt{x^2+1}) - \log_a(x+1)^4$$

$$= \log_a x^3 + \log_a \sqrt{x^2+1} - \log_a(x+1)^4$$

$$= \log_a x^3 + \log_a(x^2+1)^{1/2} - \log_a(x+1)^4$$

$$= 3\log_a x + \tfrac{1}{2}\log_a(x^2+1) - 4\log_a(x+1)$$ ■

Another use of properties (3) through (6) is to write sums and/or differences of logarithms with the same base as a single logarithm.

EXAMPLE 6   *Writing Expressions as a Single Logarithm*

Write each of the following as a single logarithm:

(a) $\log_a 7 + 4\log_a 3$
(b) $\tfrac{2}{3}\log_a 8 - \log_a(3^4 - 8)$
(c) $\log_a x + \log_a 9 + \log_a(x^2+1) - \log_a 5$

Solution   (a) $\log_a 7 + 4\log_a 3 = \log_a 7 + \log_a 3^4$ Property (6)

$$= \log_a 7 + \log_a 81$$

$$= \log_a(7 \cdot 81)$$ Property (3)

$$= \log_a 567$$

(b) $\tfrac{2}{3}\log_a 8 - \log_a(3^4 - 8) = \log_a 8^{2/3} - \log_a(81 - 8)$ Property (6)

$$= \log_a 4 - \log_a 73$$

$$= \log_a\left(\tfrac{4}{73}\right)$$ Property (4)

(c) $\log_a x + \log_a 9 + \log_a(x^2+1) - \log_a 5 = \log_a 9x + \log_a(x^2+1) - \log_a 5$

$$= \log_a[9x(x^2+1)] - \log_a 5$$

$$= \log_a\left[\frac{9x(x^2+1)}{5}\right]$$ ■

Warning: A common error made by some students is to express the logarithm of a sum as the sum of logarithms:

$$\log_a(M + N) \quad \text{is not equal to} \quad \log_a M + \log_a N$$

Correct Statement $\log_a MN = \log_a M + \log_a N$ Property (3)

Another common error is to express the difference of logarithms as the quotient of logarithms:

$$\log_a M - \log_a N \quad \text{is not equal to} \quad \frac{\log_a M}{\log_a N}$$

Correct Statement $\log_a M - \log_a N = \log_a\!\left(\dfrac{M}{N}\right)$ Property (4)

■   Now work Problem 23.

There remain two other properties of logarithms we need to know. They are a consequence of the fact that the logarithmic function $y = \log_a x$ is one-to-one.

Theorem   In the following properties, *M, N,* and *a* are positive real numbers, with $a \neq 1$:

$$
\begin{array}{ll}
\text{If} \quad M = N, \quad \text{then} \quad \log_a M = \log_a N. & (7)\\[4pt]
\text{If} \quad \log_a M = \log_a N, \quad \text{then} \quad M = N. & (8)
\end{array}
$$

■

Properties (7) and (8) are useful for solving *logarithmic equations,* a topic discussed in the next section.

Using a Calculator to Evaluate Logarithms with Bases other Than *e* or 10

Logarithms to the base 10, called **common logarithms,** were used to facilitate arithmetic computations before the widespread use of calculators. (See the Historic Feature at the end of this section.) Natural logarithms, that is, logarithms whose base is the number *e,* remain very important because they arise frequently in the study of natural phenomena.

Common logarithms are usually abbreviated by writing **log,** with the base understood to be 10, just as natural logarithms are abbreviated by **ln,** with the base understood to be *e.*

Most calculators have both $\boxed{\log}$ and $\boxed{\ln}$ keys to calculate the common logarithm and natural logarithm of a number. Let's look at an example to see how to calculate logarithms having a base other than 10 or *e.*

E X A M P L E   7   *Evaluating Logarithms Whose Base Is Neither 10 nor e*

Evaluate: $\log_2 7$

Solution   Let $y = \log_2 7$. Then $2^y = 7$, so

$$
\begin{array}{ll}
2^y = 7 & \\[4pt]
\ln 2^y = \ln 7 & \text{Property (7)}\\[4pt]
y \ln 2 = \ln 7 & \text{Property (6)}\\[4pt]
y = \dfrac{\ln 7}{\ln 2} & \text{Solve for } y.\\[8pt]
\quad = 2.8074 & \text{Use calculator (}\boxed{\ln}\text{ key).}
\end{array}
$$

■

Example 7 shows how to change the base from 2 to e. In general, to change from the base b to the base a, we use the **change-of-base formula.**

Theorem
Change-of-Base Formula

If $a \neq 1$, $b \neq 1$, and M are positive real numbers, then

$$\log_a M = \frac{\log_b M}{\log_b a} \qquad (9)$$

Proof   We derive this formula as follows: Let $y = \log_a M$. Then $a^y = M$, so

$$\log_b a^y = \log_b M \qquad \text{Property (7)}$$
$$y \log_b a = \log_b M \qquad \text{Property (6)}$$
$$y = \frac{\log_b M}{\log_b a} \qquad \text{Solve for } y.$$
$$\log_a M = \frac{\log_b M}{\log_b a} \qquad y = \log_a M$$

Since calculators have only keys for $\boxed{\log}$ and $\boxed{\ln}$, in practice, the change-of-base formula uses either $b = 10$ or $b = e$. Thus,

$$\log_a M = \frac{\log M}{\log a} \quad \text{and} \quad \log_a M = \frac{\ln M}{\ln a} \qquad (10)$$

Comment: To graph logarithmic functions when the base is different from e or 10 requires the change-of-base formula. Thus, for example, to graph $y = \log_2 x$, we would instead graph $y = (\ln x)/(\ln 2)$. Try it.

E X A M P L E   8

Using the Change-of-Base Formula
Calculate:

(a) $\log_5 89$ (b) $\log_{\sqrt{2}} \sqrt{5}$

Solution   (a) $\log_5 89 = \dfrac{\log 89}{\log 5} \approx \dfrac{1.94939}{0.69897} = 2.7889$

or

$$\log_5 89 = \frac{\ln 89}{\ln 5} \approx \frac{4.4886}{1.6094} = 2.7889$$

(b) $\log_{\sqrt{2}} \sqrt{5} = \dfrac{\log \sqrt{5}}{\log \sqrt{2}} = \dfrac{\frac{1}{2} \log 5}{\frac{1}{2} \log 2} \approx \dfrac{0.69897}{0.30103} = 2.3219$

or

$$\log_{\sqrt{2}} \sqrt{5} = \frac{\ln \sqrt{5}}{\ln \sqrt{2}} = \frac{\frac{1}{2} \ln 5}{\frac{1}{2} \ln 2} \approx \frac{1.6094}{0.6931} = 2.3219$$

■   Now work Problem 41.

Summary of Properties of Logarithms

In the list that follows, $a > 0$, $a \neq 1$, and $b > 0$, $b \neq 1$; also, $M > 0$ and $N > 0$.

Definition	$y = \log_a x$ means $x = a^y$
Properties of logarithms	$\log_a 1 = 0; \quad \log_a a = 1$
	$a^{\log_a M} = M; \quad \log_a a^r = r$
	$\log_a MN = \log_a M + \log_a N$
	$\log_a\left(\dfrac{M}{N}\right) = \log_a M - \log_a N$
	$\log_a\left(\dfrac{1}{N}\right) = -\log_a N$
	$\log_a M^r = r \log_a M$
Change-of-base formula	$\log_a M = \dfrac{\log_b M}{\log_b a}$

HISTORICAL FEATURE   ■ Logarithms were invented about 1590 by John Napier (1550–1617) and Jobst Bürgi (1552–1632), working independently. Napier, whose work had the greater influence, was a Scottish lord, a secretive man whose neighbors were inclined to believe him to be in league with the devil. His approach to logarithms was quite different from ours; it was based on the relationship between arithmetic and geometric sequences (see Chapter 11), and not on the inverse function relationship of logarithms to exponential functions (described in Section 4.2). Napier's tables, published in 1614, listed what would now be called *natural logarithms* of sines and were rather difficult to use. A London professor, Henry Briggs, became interested in the tables and visited Napier. In their conversations, they developed the idea of common logarithms, and Briggs then converted Napier's tables into tables of common logarithms, which were published in 1617. Their importance for calculation was immediately recognized, and by 1650 they were being printed as far away as China. They remained an important calculation tool until the advent of the inexpensive handheld calculator about 1972, which has decreased their calculational, but not their theoretical, importance.

A side effect of the invention of logarithms was the popularization of the decimal system of notation for real numbers.   ■

4.3

Exercise 4.3

In Problems 1–12, suppose $\ln 2 = a$ *and* $\ln 3 = b$. *Use properties of logarithms to write each logarithm in terms of a and b.*

1. $\ln 6$

2. $\ln \frac{2}{3}$

3. $\ln 1.5$

4. $\ln 0.5$

5. $\ln 2e$

6. $\ln\left(\dfrac{3}{e}\right)$

7. $\ln 12$

8. $\ln 24$

9. $\ln \sqrt[5]{18}$

10. $\ln \sqrt[4]{48}$

11. $\log_2 3$

12. $\log_3 2$

In Problems 13–22, write each expression as a sum and/or difference of logarithms. Express powers as factors.

13. $\ln(x^2\sqrt{1-x})$

14. $\ln(x\sqrt{1+x^2})$

15. $\log_2\left(\dfrac{x^3}{x-3}\right)$

16. $\log_5\left(\dfrac{\sqrt[3]{x^2+1}}{x^2-1}\right)$

17. $\log\left[\dfrac{x(x+2)}{(x+3)^2}\right]$

18. $\log\dfrac{x^3\sqrt{x+1}}{(x-2)^2}$

19. $\ln\left[\dfrac{x^2-x-2}{(x+4)^2}\right]^{1/3}$

20. $\ln\left[\dfrac{(x-4)^2}{x^2-1}\right]^{2/3}$

21. $\ln\dfrac{5x\sqrt{1-3x}}{(x-4)^3}$

22. $\ln\left[\dfrac{5x^2\sqrt[3]{1-x}}{4(x+1)^2}\right]$

In Problems 23–32, write each expression as a single logarithm.

23. $3\log_5 u + 4\log_5 v$

24. $\log_3 u^2 - \log_3 v$

25. $\log_{1/2}\sqrt{x} - \log_{1/2} x^3$

26. $\log_2\left(\dfrac{1}{x}\right) + \log_2\left(\dfrac{1}{x^2}\right)$

27. $\ln\left(\dfrac{x}{x-1}\right) + \ln\left(\dfrac{x+1}{x}\right) - \ln(x^2-1)$

28. $\log\left(\dfrac{x^2+2x-3}{x^2-4}\right) - \log\left(\dfrac{x^2+7x+6}{x+2}\right)$

29. $8\log_2\sqrt{3x-2} - \log_2\left(\dfrac{4}{x}\right) + \log_2 4$

30. $21\log_3\sqrt[3]{x} + \log_3 9x^2 - \log_5 25$

31. $2\log_a 5x^3 - \tfrac{1}{2}\log_a(2x+3)$

32. $\tfrac{1}{3}\log(x^3+1) + \tfrac{1}{2}\log(x^2+1)$

In Problems 33–40, use the change of base formula and a calculator to evaluate each logarithm. Round your answer to three decimal places.

33. $\log_3 21$

34. $\log_5 18$

35. $\log_{1/3} 71$

36. $\log_{1/2} 15$

37. $\log_{\sqrt{2}} 7$

38. $\log_{\sqrt{5}} 8$

39. $\log_\pi e$

40. $\log_\pi \sqrt{2}$

41. Show that: $\log_a(x+\sqrt{x^2-1}) + \log_a(x-\sqrt{x^2-1}) = 0$

42. Show that: $\log_a(\sqrt{x}+\sqrt{x-1}) + \log_a(\sqrt{x}-\sqrt{x-1}) = 0$

43. Show that: $\ln(1+e^{2x}) = 2x + \ln(1+e^{-2x})$

44. If $f(x) = \log_a x$, show that: $\dfrac{f(x+h)-f(x)}{h} = \log_a\left(1+\dfrac{h}{x}\right)^{1/h}$, $h \neq 0$

45. If $f(x) = \log_a x$, show that: $-f(x) = \log_{1/a} x$

46. If $f(x) = \log_a x$, show that: $f(1/x) = -f(x)$

47. If $f(x) = \log_a x$, show that: $f(AB) = f(A) + f(B)$

48. If $f(x) = \log_a x$, show that: $f(x^\alpha) = \alpha f(x)$

In Problems 49–58, express y as a function of x. The constant C is a positive number.

49. $\ln y = \ln x + \ln C$

50. $\ln y = \ln(x+C)$

51. $\ln y = \ln x + \ln(x+1) + \ln C$

52. $\ln y = 2\ln x - \ln(x+1) + \ln C$

53. $\ln y = 3x + \ln C$

54. $\ln y = -2x + \ln C$

55. $\ln(y-3) = -4x + \ln C$

56. $\ln(y+4) = 5x + \ln C$

57. $3\ln y = \tfrac{1}{2}\ln(2x+1) - \tfrac{1}{3}\ln(x+4) + \ln C$

58. $2\ln y = -\tfrac{1}{2}\ln x + \tfrac{1}{3}\ln(x^2+1) + \ln C$

59. Find the value of $\log_2 3 \cdot \log_3 4 \cdot \log_4 5 \cdot \log_5 6 \cdot \log_6 7 \cdot \log_7 8$.

60. Find the value of $\log_2 4 \cdot \log_4 6 \cdot \log_6 8$.

61. Find the value of $\log_2 3 \cdot \log_3 4 \cdot \ldots \cdot \log_n(n+1) \cdot \log_{n+1} 2$.

62. Find the value of $\log_2 2 \cdot \log_2 4 \cdot \ldots \cdot \log_2 2^n$.

63. Show that $\log_a(M/N) = \log_a M - \log_a N$, where a, M, and N are positive real numbers, with $a \neq 1$.

64. Show that $\log_a(1/N) = -\log_a N$, where a and N are positive real numbers, with $a \neq 1$.

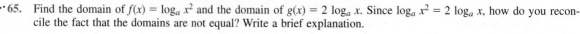

65. Find the domain of $f(x) = \log_a x^2$ and the domain of $g(x) = 2\log_a x$. Since $\log_a x^2 = 2\log_a x$, how do you reconcile the fact that the domains are not equal? Write a brief explanation.

ISSION POSSIBLE

Chapter 4

McNEWTON'S COFFEE

Your team has been called in to solve a problem encountered by a fast food restaurant. They believe that their coffee should be brewed at 170° Fahrenheit; however, at that temperature it is too hot to drink, and a customer who accidentally spills the coffee on himself might receive third degree burns.

What they need is a special container that will heat the water to 170°, brew the coffee at that temperature, then cool it quickly to a drinkable temperature, say 140°F, and hold it there, or at least keep it at or above 120°F for a reasonable period of time without further cooking. To cool down the coffee, three companies have submitted proposals with these specifications:

(a) The CentiKeeper Company has a container that will reduce the temperature of a liquid from 200°F to 100°F in 90 minutes by maintaining a constant temperature of 70°F.

(b) The TempControl Company has a container that will reduce the temperature of a liquid from 200°F to 110°F in 60 minutes by maintaining a constant temperature of 60°F.

(c) The Hot'n'Cold, Inc., has a container that will reduce the temperature of a liquid from 210°F to 90°F in 30 minutes by maintaining a constant temperature of 50°F.

Your job is to make a recommendation as to which container to purchase. For this you will need Newton's Law of Cooling which follows:

$$u = T + (u_0 - T)e^{kt}, \quad k < 0$$

In this formula, T represents the temperature of the surrounding medium, u_0 is the initial temperature of the heated object, t is the length of time in minutes, k is a negative constant, and u represents the temperature at time t.

1. Use Newton's Law of Cooling to find the constant k of the formula for each container.
2. How long does it take each container to lower the coffee temperature from 170°F to 140°F?
3. How long will the coffee temperature remain between 120°F and 140°F?
4. On the basis of this information, which company should get the contract with McNewton's? What are your reasons?
5. What are "capital cost" and "operating cost"? How might they affect your choice?

4.4

Logarithmic and Exponential Equations

Logarithmic Equations

Equations that contain terms of the form $\log_a x$, where a is a positive real number, with $a \neq 1$, are often called **logarithmic equations.**

Let's see how we can use the properties of logarithms to solve logarithmic equations.

E X A M P L E   1 Solve: $\log_3(4x - 7) = 2$

Solution We change the expression to exponential form to solve:

$$\log_3(4x - 7) = 2$$
$$4x - 7 = 3^2$$
$$4x - 7 = 9$$
$$4x = 16$$
$$x = 4$$

∎

E X A M P L E   2 Solve: $2 \log_5 x = \log_5 9$

Solution

$$2 \log_5 x = \log_5 9$$
$$\log_5 x^2 = \log_5 9 \qquad \text{Property (6), Section 4.3}$$
$$x^2 = 9 \qquad \text{Property (8), Section 4.3}$$
$$x = 3 \quad \text{or} \quad x = -3 \qquad$$

Recall that logarithms of negative numbers are not defined, so, in the expression $2 \log_5 x$, x must be positive. Therefore, -3 is extraneous and we discard it.

The equation has only one solution, 3.

∎

◼ Now work Problem 1.

E X A M P L E   3 Solve: $\log_4(x + 3) + \log_4(2 - x) = 1$

Solution We need to express the left side as a single logarithm. Then we will change the expression to exponential form.

$$\log_4(x + 3) + \log_4(2 - x) = 1$$
$$\log_4[(x + 3)(2 - x)] = 1 \qquad \text{Property (3), Section 4.3}$$
$$(x + 3)(2 - x) = 4^1 = 4$$
$$-x^2 - x + 6 = 4$$
$$x^2 + x - 2 = 0$$
$$(x + 2)(x - 1) = 0$$
$$x = -2 \quad \text{or} \quad x = 1$$

You should verify that both of these are solutions.

∎

◼ Now work Problem 11.

Care must be taken when solving logarithmic equations. Be sure to check each apparent solution in the original equation and discard any that are extraneous. In the expression $\log_a M$, remember that a and M are positive and $a \neq 1$.

Exponential Equations

Equations that involve terms of the form a^x, $a > 0$, $a \neq 1$, are often referred to as **exponential equations.** Such equations sometimes can be solved by appropriately applying the laws of exponents and equation (1), p. 401, namely,

$$\text{If } a^u = a^v, \quad \text{then } u = v. \tag{1}$$

To use equation (1), each side of the equality must be written with the same base.

E X A M P L E   4   Solve the equation: $3^{x+1} = 81$

Solution   Since $81 = 3^4$, we can write the equation as

$$3^{x+1} = 81 = 3^4$$

Now we have the same base, 3, on each side, so we can apply (1) to obtain

$$x + 1 = 4$$
$$x = 3$$

  ■ Now work Problem 19.

E X A M P L E   5   Solve the equation: $e^{-x^2} = (e^x)^2 \cdot \dfrac{1}{e^3}$

Solution   We use some laws of exponents first to get the same base e on each side:

$$e^{-x^2} = (e^x)^2 \cdot \frac{1}{e^3} = e^{2x} \cdot e^{-3} = e^{2x-3}$$

Now apply (1) to get

$$-x^2 = 2x - 3$$
$$x^2 + 2x - 3 = 0$$
$$(x + 3)(x - 1) = 0$$
$$x = -3 \quad \text{or} \quad x = 1$$

E X A M P L E   6   Solve the equation: $4^x - 2^x - 12 = 0$

Solution   We note that $4^x = (2^2)^x = 2^{2x} = (2^x)^2$, so the equation is actually quadratic in form, and we can rewrite it as

$$(2^x)^2 - 2^x - 12 = 0$$

Now we can factor as usual:

$$(2^x - 4)(2^x + 3) = 0$$
$$2^x - 4 = 0 \quad \text{or} \quad 2^x + 3 = 0$$
$$2^x = 4 \qquad \qquad 2^x = -3$$

The equation on the left has the solution $x = 2$, since $2^x = 4 = 2^2$; the equation on the right has no solution, since $2^x > 0$ for all x. ∎

In each of the preceding three examples, we were able to write each exponential expression using the same base, obtaining exact solutions to the equation. When this is not possible, logarithms can sometimes be used to obtain the solution.

E X A M P L E   7   Solve for x: $2^x = 5$

Solution   We write the exponential equation as the equivalent logarithmic equation:

$$2^x = 5$$

$$x = \log_2 5 = \underset{\uparrow}{\dfrac{\ln 5}{\ln 2}}$$

Change-of-base formula (10)

Alternatively, we can solve the equation $2^x = 5$ by taking the natural logarithm (or common logarithm) of each side. Taking the natural logarithm,

$$2^x = 5$$
$$\ln 2^x = \ln 5$$
$$x \ln 2 = \ln 5$$
$$x = \dfrac{\ln 5}{\ln 2}$$

Using a calculator, the solution, rounded to three decimal places, is:

$$x = \dfrac{\ln 5}{\ln 2} = 2.322$$

∎

∎   Now work Problem 33.

E X A M P L E   8   Solve for x: $8 \cdot 3^x = 5$

Solution

$$8 \cdot 3^x = 5$$ Isolate 3^x on the left side.

$$3^x = \tfrac{5}{8}$$ Proceed as in Example 7.

$$x = \log_3\left(\tfrac{5}{8}\right) = \dfrac{\ln \tfrac{5}{8}}{\ln 3}$$

Using a calculator, the solution, rounded to three decimal places, is:

$$x = \dfrac{\ln\left(\tfrac{5}{8}\right)}{\ln 3} = -0.428$$

∎

E X A M P L E   9   Solve for x: $5^{x-2} = 3^{3x+2}$

Solution   Because the bases are different, we take the natural logarithm of each side and apply appropriate properties of logarithms. The result is an equation in x that we can solve.

$$5^{x-2} = 3^{3x+2}$$
$$\ln 5^{x-2} = \ln 3^{3x+2} \qquad \text{Property (7)}$$
$$(x - 2)\ln 5 = (3x + 2)\ln 3 \qquad \text{Property (6)}$$
$$(\ln 5)x - 2 \ln 5 = (3 \ln 3)x + 2 \ln 3$$
$$(\ln 5 - 3 \ln 3)x = 2 \ln 3 + 2 \ln 5$$
$$x = \frac{2(\ln 3 + \ln 5)}{\ln 5 - 3 \ln 3} = -3.212 \qquad \blacksquare$$

■   Now work Problem 41.

Graphing Utility Solutions

The techniques introduced in this section only apply to certain types of logarithmic and exponential equations. Solutions for other types are usually studied in calculus, using numerical methods. In the next example, we show how a graphing utility may be used to obtain solutions.

E X A M P L E   1 0   *Solving Equations Using a Graphing Utility*

Solve: $x + e^x = 2$
Express the solution(s) correct to two decimal places.

Solution A   This type of exponential equation cannot be solved by previous methods. A graphing utility, though, can be used here. We begin by noting that the equation to be solved is equivalent to

$$x + e^x = 2$$
$$x + e^x - 2 = 0$$

The solution(s) of this equation is the same as the x-intercept(s) of the graph of the function $f(x) = x + e^x - 2$. This function f is increasing (do you see why?) and so has at most one x-intercept. Since $f(0) = -1 < 0$ and $f(1) = 1 + e - 2 > 0$, it follows from the Intermediate Value Theorem (Section 3.6, p. 247), that there is one x-intercept and it is between 0 and 1. So we graph the function, with $0 \le x \le 1$. See Figure 15.

FIGURE 15

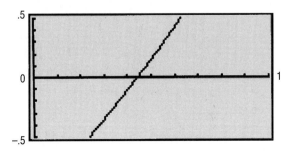

By using TRACE, ZOOM-IN, and/or BOX, we obtain the solution $x = 0.44$, correct to two decimal places.

Solution B   The equation to be solved is equivalent to

$$x + e^x = 2$$
$$e^x = 2 - x$$

We graph the two equations $y = e^x$ and $y = 2 - x$. See Figure 16. The x-coordinate of their point of intersection is the solution we seek. Using TRACE, ZOOM-IN, and/or BOX, we obtain the solution $x = 0.44$, correct to two decimal places.

FIGURE 16

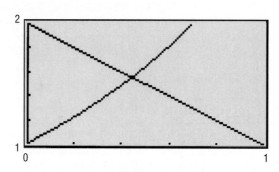

4.4

Exercise 4.4

In Problems 1–58, solve each equation.

1. $\log_2(2x + 1) = 3$
2. $\log_3(3x - 2) = 2$
3. $\log_3(x^2 + 1) = 2$
4. $\log_5(x^2 + x + 4) = 2$
5. $\frac{1}{2}\log_3 x = 2 \log_3 2$
6. $-2 \log_4 x = \log_4 9$
7. $2 \log_5 x = 3 \log_5 4$
8. $3 \log_2 x = -\log_2 27$
9. $3 \log_2(x - 1) + \log_2 4 = 5$
10. $2 \log_3(x + 4) - \log_3 9 = 2$
11. $\log_{10} x + \log_{10}(x + 15) = 2$
12. $\log_4 x + \log_4(x - 3) = 1$
13. $\log_x 4 = 2$
14. $\log_x\left(\frac{1}{8}\right) = 3$
15. $\log_3(x - 1)^2 = 2$
16. $\log_2(x + 4)^3 = 6$
17. $\log_{1/2}(3x + 1)^{1/3} = -2$
18. $\log_{1/3}(1 - 2x)^{1/2} = -1$
19. $2^{2x+1} = 4$
20. $5^{1-2x} = \frac{1}{5}$
21. $3^{x^3} = 9^x$
22. $4^{x^2} = 2^x$
23. $8^{x^2-2x} = \frac{1}{2}$
24. $9^{-x} = \frac{1}{3}$
25. $2^x \cdot 8^{-x} = 4^x$
26. $\left(\frac{1}{2}\right)^{1-x} = 4$
27. $2^{2x} - 2^x - 12 = 0$
28. $3^{2x} + 3^x - 2 = 0$
29. $3^{2x} + 3^{x+1} - 4 = 0$
30. $4^x - 2^x = 0$
31. $4^x = 8$
32. $9^{2x} = 27$
33. $2^x = 10$
34. $3^x = 14$
35. $8^{-x} = 1.2$
36. $2^{-x} = 1.5$
37. $3^{1-2x} = 4^x$
38. $2^{x+1} = 5^{1-2x}$
39. $\left(\frac{3}{5}\right)^x = 7^{1-x}$
40. $\left(\frac{4}{3}\right)^{1-x} = 5^x$
41. $1.2^x = (0.5)^{-x}$
42. $(0.3)^{1+x} = 1.7^{2x-1}$
43. $\pi^{1-x} = e^x$
44. $e^{x+3} = \pi^x$
45. $5(2^{3x}) = 8$
46. $0.3(4^{0.2x}) = 0.2$
47. $400e^{0.2x} = 600$
48. $500e^{0.3x} = 600$
49. $\log_a(x - 1) - \log_a(x + 6) = \log_a(x - 2) - \log_a(x + 3)$
50. $\log_a x + \log_a(x - 2) = \log_a(x + 4)$
51. $\log_{1/3}(x^2 + x) - \log_{1/3}(x^2 - x) = -1$
52. $\log_4(x^2 - 9) - \log_4(x + 3) = 3$
53. $\log_2 8^x = -3$
54. $\log_3 3^x = -1$
55. $\log_2(x^2 + 1) - \log_4 x^2 = 1$
    [*Hint:* Change $\log_4 x^2$ to base 2.]
56. $\log_2(3x + 2) - \log_4 x = 3$
57. $\log_{16} x + \log_4 x + \log_2 x = 7$
58. $\log_9 x + 3 \log_3 x = 14$

In Problems 59–70, use a graphing utility to solve each equation. Express your answer correct to two decimal places.

59. $e^x = -x$

60. $e^{2x} = x + 2$

61. $e^x = x^2$

62. $e^x = x^3$

63. $\ln x = -x$

64. $\ln 2x = -x + 2$

65. $\ln x = x^3 - 1$

66. $\ln x = -x^2$

67. $e^x + \ln x = 4$

68. $e^x - \ln x = 4$

69. $e^{-x} = \ln x$

70. $e^{-x} = -\ln x$

4.5

Compound Interest

Interest is money paid for the use of money. The total amount borrowed (whether by an individual from a bank in the form of a loan or by a bank from an individual in the form of a savings account) is called the **principal.** The **rate of interest,** expressed as a percent, is the amount charged for the use of the principal for a given period of time, usually on a yearly (that is, per annum) basis.

If a principal of P dollars is borrowed for a period of t years at a per annum interest rate r, expressed as a decimal, the interest I charged is

Simple Interest Formula

$$I = Prt \tag{1}$$

Interest charged according to formula (1) is called **simple interest.**

In working with problems involving interest we use the term **payment period** as follows:

Annually	Once per year
Semiannually	Twice per year
Quarterly	4 times per year
Monthly	12 times per year
Daily	365 times per year*

When the interest due at the end of a payment period is added to the principal so that the interest computed at the end of the next payment period is based on this new principal amount (old principal + interest), the interest is said to have been **compounded.** Thus, **compound interest** is interest paid on previously earned interest.

E X A M P L E   1

Computing Compound Interest

A credit union pays interest of 8% per annum compounded quarterly on a certain savings plan. If $1000 is deposited in such a plan and the interest is left to accumulate, how much is in the account after 1 year?

Solution   We use the simple interest formula, $I = Prt$. The principal P is $1000 and the rate of interest is 8% = 0.08. After the first quarter of a year, the time t is $\frac{1}{4}$ year, so the interest earned is

$$I = Prt = (\$1000)(0.08)(\tfrac{1}{4}) = \$20$$

The new principal is $P + I = \$1000 + \$20 = \$1020$. At the end of the second quarter, the interest on this principal is

$$I = (\$1020)(0.08)(\tfrac{1}{4}) = \$20.40$$

*Some banks use a 360 day "year."

At the end of the third quarter, the interest on the new principal of $1020 + $20.40 = $1040.40 is

$$I = (\$1040.40)(0.08)\left(\tfrac{1}{4}\right) = \$20.81$$

Finally, after the fourth quarter, the interest is

$$I = (\$1061.21)(0.08)\left(\tfrac{1}{4}\right) = \$21.22$$

Thus, after 1 year the account contains $1082.43.   ■

The pattern of the calculations performed in Example 1 leads to a general formula for compound interest. To fix our ideas, let P represent the principal to be invested at a per annum interest rate r, which is compounded n times per year. (For computing purposes, r is expressed as a decimal.) The interest earned after each compounding period is the principal times r/n. Thus, the amount A after one compounding period is

$$A = P + P\left(\frac{r}{n}\right) = P\left(1 + \frac{r}{n}\right)$$

After two compounding periods, the amount A, based on the new principal $P(1 + r/n)$, is

$$A = \underbrace{P\left(1 + \frac{r}{n}\right)}_{\substack{\text{New} \\ \text{principal}}} + \underbrace{P\left(1 + \frac{r}{n}\right)\left(\frac{r}{n}\right)}_{\substack{\text{Interest on} \\ \text{new principal}}} = P\left(1 + \frac{r}{n}\right)\left(1 + \frac{r}{n}\right) = P\left(1 + \frac{r}{n}\right)^2$$

After three compounding periods,

$$A = P\left(1 + \frac{r}{n}\right)^2 + P\left(1 + \frac{r}{n}\right)^2\left(\frac{r}{n}\right) = P\left(1 + \frac{r}{n}\right)^2\left(1 + \frac{r}{n}\right) = P\left(1 + \frac{r}{n}\right)^3$$

Continuing in this way, after n compounding periods (1 year),

$$A = P\left(1 + \frac{r}{n}\right)^n$$

Because t years will contain $n \cdot t$ compounding periods, after t years we have

$$A = P\left(1 + \frac{r}{n}\right)^{nt}$$

Theorem   The amount A after t years due to a principal P invested at an annual interest rate r compounded n times per year is

Compound Interest Formula

$$A = P\left(1 + \frac{r}{n}\right)^{nt} \qquad (2)$$

■

E X A M P L E   2   *Comparing Investments Using Different Compounding Periods*

Investing $1000 at an annual rate of 10% compounded annually, quarterly, monthly, and daily will yield the following amounts after 1 year:

Annual compounding: $A = P(1 + r)$

$$= (\$1000)(1 + 0.10) = \$1100.00$$

Quarterly compounding: $A = P\left(1 + \dfrac{r}{4}\right)^4$

$$= (\$1000)(1 + 0.025)^4 = \$1103.81$$

Monthly compounding: $A = P\left(1 + \dfrac{r}{12}\right)^{12}$

$$= (\$1000)(1 + 0.00833)^{12} = \$1104.71$$

Daily compounding: $A = P\left(1 + \dfrac{r}{365}\right)^{365}$

$$= (\$1000)(1 + 0.000274)^{365} = \$1105.16 \qquad \blacksquare$$

■ Now work Problem 1.

From Example 2, we can see that the effect of compounding more frequently is that the account after 1 year is higher: $\$1000$ compounded 4 times a year at 10% results in $\$1103.81$; $\$1000$ compounded 12 times a year at 10% results in $\$1104.71$; and $\$1000$ compounded 365 times a year at 10% results in $\$1105.16$. This leads to the following question: What would happen to the amount after 1 year if the number of times the interest is compounded were increased without bound?

Let's find the answer. Suppose P is the principal, r is the per annum interest rate, and n is the number of times the interest is compounded each year. The amount after 1 year is

$$A = P\left(1 + \frac{r}{n}\right)^n$$

Now suppose that the number n of times the interest is compounded per year gets larger and larger; that is, suppose that $n \to \infty$. Then,

$$A = \left(1 + \frac{r}{n}\right)^n = \left[1 + \frac{1}{n/r}\right]^n = \left(\left[1 + \frac{1}{n/r}\right]^{n/r}\right)^r = \left[\left(1 + \frac{1}{h}\right)^h\right]^r \tag{3}$$

$$h = \frac{n}{r}$$

Thus, in (3), as $n \to \infty$, then $h = n/r \to \infty$, and the expression in brackets equals e, [Refer to (2) on p. 267] so $A \to e^r$. Table 6 compares $(1 + r/n)^n$, for large values of n, to e^r for $r = 0.05$, $r = 0.10$, $r = 0.15$, and $r = 1$. The larger n gets, the closer $(1 + r/n)^n$ gets to e^r. Thus, no matter how frequent the compounding, the amount after 1 year has the definite ceiling Pe^r.

TABLE 6

	$\left(1 + \dfrac{r}{n}\right)^n$			
	$n = 100$	$n = 1000$	$n = 10,000$	e^r
$r = 0.05$	1.0512579	1.05127	1.051271	1.0512711
$r = 0.10$	1.1051157	1.1051654	1.1051703	1.1051709
$r = 0.15$	1.1617037	1.1618212	1.1618329	1.1618342
$r = 1$	2.7048138	2.7169239	2.7181459	2.7182818

When interest is compounded so that the amount after 1 year is Pe^r, we say the interest is **compounded continuously.**

Theorem

The amount A after t years due to a principal P invested at an annual interest rate r compounded continuously is

Continuous Compounding

$$A = Pe^{rt} \qquad (4)$$

■

E X A M P L E   3

Using Continuous Compounding

The amount A that results from investing a principal P of $1000 at an annual rate r of 10% compounded continuously for a time t of 1 year is

$$A = \$1000e^{0.10} = (\$1000)(1.10517) = \$1105.17$$

■

■ Now work Problem 9.

The **effective rate of interest** is the equivalent annual simple rate of interest that would yield the same amount as compounding after 1 year. For example, based on Example 3, a principal of $1000 will result in $1105.17 at a rate of 10% compounded continuously. To get this same amount using a simple rate of interest would require that interest of $1105.17 − $1000.00 = $105.17 be earned on the principal. Since $105.17 is 10.517% of $1000, a simple rate of interest of 10.517% is needed to equal 10% compounded continuously. Thus, the effective rate of interest of 10% compounded continuously is 10.517%.

Based on the results of Examples 2 and 3, we find the following comparisons:

	ANNUAL RATE	EFFECTIVE RATE
Annual compounding	10%	10%
Quarterly compounding	10%	10.381%
Monthly compounding	10%	10.471%
Daily compounding	10%	10.516%
Continuous compounding	10%	10.517%

■ Now work Problem 21.

E X A M P L E   4

Computing the Value of an IRA

On January 2, 1996, $2000 is placed in an Individual Retirement Account (IRA) that will pay interest of 10% per annum compounded continuously. What will the IRA be worth on January 1, 2016?

Solution

The amount A after 20 years is

$$A = Pe^{rt} = \$2000e^{(0.10)(20)} = \$14{,}778.11$$

Check: Graph $y = 2000e^{0.1x}$ and use TRACE to verify that when $x = 20$ then $y = \$14{,}778.11$.

■

Exploration: How long will it be until $y = \$40{,}000$?

■

When people engaged in finance speak of the "time value of money," they are usually referring to the **present value** of money. The present value of A dollars to be received at a future date is the principal you would need to invest now so that

FIGURE 17

Time is Money

it would grow to A dollars in the specified time period. Thus, the present value of money to be received at a future date is always less than the amount to be received, since the amount to be received will equal the present value (money invested now) *plus* the interest accrued over the time period.

We use the compound interest formula (2) to get a formula for present value. If P is the present value of A dollars to be received after t years at a per annum interest rate r compounded n times per year, then, by formula (2),

$$A = P\left(1 + \frac{r}{n}\right)^{nt}$$

To solve for P, we divide both sides by $(1 + r/n)^{nt}$, and the result is

$$\frac{A}{(1 + r/n)^{nt}} = P \quad \text{or} \quad P = A\left(1 + \frac{r}{n}\right)^{-nt}$$

Theorem   The present value P of A dollars to be received after t years, assuming a per annum interest rate r compounded n times per year, is

Present Value Formulas

$$P = A\left(1 + \frac{r}{n}\right)^{-nt} \tag{5}$$

If the interest is compounded continuously, then

$$P = Ae^{-rt} \tag{6}$$

∎

To prove (6), solve formula (4) for P.

E X A M P L E   5   *Computing the Value of a Zero-Coupon Bond*

A zero-coupon (noninterest-bearing) bond can be redeemed in 10 years for $1000. How much should you be willing to pay for it now if you want a return of:

(a)   8% compounded monthly?

(b)   7% compounded continuously?

Solution   (a)   We are seeking the present value of $1000. Thus, we use formula (5) with $A = \$1000$, $n = 12$, $r = 0.08$, and $t = 10$:

$$P = A\left(1 + \frac{r}{n}\right)^{-nt}$$

$$= \$1000\left(1 + \frac{0.08}{12}\right)^{-12(10)}$$

$$= \$450.52$$

For a return of 8% compounded monthly, you should pay $450.52 for the bond.

(b)   Here, we use formula (6) with $A = \$1000$, $r = 0.07$, and $t = 10$:

$$P = Ae^{-rt}$$
$$= \$1000e^{-(0.07)(10)}$$
$$= \$496.59$$

For a return of 7% compounded continuously, you should pay $496.56 for the bond.   ■

■   Now work Problem 11.

E X A M P L E   6

Rate of Interest Required to Double an Investment

What annual rate of interest compounded annually should you seek if you want to double your investment in 5 years?

Solution   If P is the principal and we want P to double, the amount A will be $2P$. We use the compound interest formula with $n = 1$ and $t = 5$ to find r:

$$2P = P(1 + r)^5$$
$$2 = (1 + r)^5$$
$$1 + r = \sqrt[5]{2}$$
$$r = \sqrt[5]{2} - 1 = 1.148698 - 1 = 0.148698$$

The annual rate of interest needed to double the principal in 5 years is 14.87%.   ■

■   Now work Problem 23.

E X A M P L E   7

Doubling and Tripling Time for an Investment

(a)   How long will it take for an investment to double in value if it earns 5% compounded continuously?

(b)   How long will it take to triple at this rate?

Solution   (a)   If P is the initial investment and we want P to double, the amount A will be $2P$. We use formula (4) for continuously compounded interest with $r = 0.05$. Then

$$A = Pe^{rt}$$
$$2P = Pe^{0.05t}$$
$$2 = e^{0.05t}$$
$$0.05t = \ln 2$$
$$t = \frac{\ln 2}{0.05} = 13.86$$

It will take about 14 years to double the investment.

(b)   To triple the investment, we set $A = 3P$ in formula (4).

$$A = Pe^{rt}$$
$$3P = Pe^{0.05t}$$
$$3 = e^{0.05t}$$
$$0.05t = \ln 3$$
$$t = \frac{\ln 3}{0.05} = 21.97$$

It will take about 22 years to triple the investment.

■   Now work Problem 29.

4.5

Exercise 4.5

In Problems 1–10, find the amount that results from each investment.

1. $100 invested at 4% compounded quarterly after a period of 2 years
2. $50 invested at 6% compounded monthly after a period of 3 years
3. $500 invested at 8% compounded quarterly after a period of $2\frac{1}{2}$ years
4. $300 invested at 12% compounded monthly after a period of $1\frac{1}{2}$ years
5. $600 invested at 5% compounded daily after a period of 3 years
6. $700 invested at 6% compounded daily after a period of 2 years
7. $10 invested at 11% compounded continuously after a period of 2 years
8. $40 invested at 7% compounded continuously after a period of 3 years
9. $100 invested at 10% compounded continuously after a period of $2\frac{1}{4}$ years
10. $100 invested at 12% compounded continuously after a period of $3\frac{3}{4}$ years

In Problems 11–20, find the principal needed now to get each amount; that is, find the present value.

11.   To get $100 after 2 years at 6% compounded monthly
12.   To get $75 after 3 years at 8% compounded quarterly
13.   To get $1000 after $2\frac{1}{2}$ years at 6% compounded daily
14.   To get $800 after $3\frac{1}{2}$ years at 7% compounded monthly
15.   To get $600 after 2 years at 4% compounded quarterly
16.   To get $300 after 4 years at 3% compounded daily
17.   To get $80 after $3\frac{1}{4}$ years at 9% compounded continuously
18.   To get $800 after $2\frac{1}{2}$ years at 8% compounded continuously
19.   To get $400 after 1 year at 10% compounded continuously
20.   To get $1000 after 1 year at 12% compounded continuously
21.   Find the effective rate of interest for $5\frac{1}{4}$% compounded quarterly.
22.   What interest rate compounded quarterly will give an effective interest rate of 7%?
23.   What annual rate of interest is required to double an investment in 3 years?
24.   What annual rate of interest is required to double an investment in 10 years?

In Problems 25–28, which of the two rates would yield the larger amount in 1 year? [Hint: Start with a principal of $10,000 in each instance.]

25. 6% compounded quarterly or $6\frac{1}{4}$% compounded annually?

26. 9% compounded quarterly or $9\frac{1}{4}$% compounded annually?

27. 9% compounded monthly or 8.8% compounded daily?

28. 8% compounded semiannually or 7.9% compounded daily?

29. How long does it take for an investment to double in value if it is invested at 8% per annum compounded monthly? Compounded continuously?

30. How long does it take for an investment to double in value if it is invested at 10% per annum compounded monthly? Compounded continuously?

31. If you have $100 to invest at 8% per annum compounded monthly, how long will it be before the amount is $150? If the compounding is continuous, how long will it be?

32. If you have $100 to invest at 10% per annum compounded monthly, how long will it be before the amount is $175? If the compounding is continuous, how long will it be?

33. How many years will it take for an initial investment of $10,000 to grow to $25,000? Assume a rate of interest of 6% compounded continuously.

34. How many years will it take for an initial investment of $25,000 to grow to $80,000? Assume a rate of interest of 7% compounded continuously.

35. What will a $90,000 house cost 5 years from now if the inflation rate over that period averages 3% compounded annually?

36. A department store charges 1.25% per month on the unpaid balance for customers with charge accounts (interest is compounded monthly). A customer charges $200 and does not pay her bill for 6 months. What is the bill at that time?

37. You will be buying a new car for $15,000 in 3 years. How much money should you ask your parents for now so that, if you invest it at 5% compounded continuously, you will have enough to buy the car?

38. You will require $3000 in 6 months to pay off a loan that has no prepayment privileges. If you have the $3000 now, how much of it should you save in an account paying 3% compounded monthly so that in 6 months you will have exactly $3000?

39. You are contemplating the purchase of 100 shares of a stock selling for $15 per share. The stock pays no dividends. The history of the stock indicates that it should grow at an annual rate of 15% per year. How much will the 100 shares of stock be worth in 5 years?

40. You are contemplating the purchase of 100 shares of a stock selling for $15 per share. The stock pays no dividends. Your broker says the stock will be worth $20 per share in 2 years. What is the annual rate of return on this investment?

41. A business purchased for $650,000 in 1994 is sold in 1997 for $850,000. What is the annual rate of return for this investment?

42. You have just inherited a diamond ring appraised at $5000. If diamonds have appreciated in value at an annual rate of 8%, what was the value of the ring 10 years ago when the ring was purchased?

43. You place $1000 in a bank account that pays 5.6% compounded continuously. After 1 year, will you have enough money to buy a computer system that costs $1060? If another bank will pay you 5.9% compounded monthly, is this a better deal?

44. On January 1, you place $1000 in a Certificate of Deposit that pays 6.8% compounded continuously and matures in 3 months. Then you place the $1000 and the interest in a passbook account that pays 5.25% compounded monthly. How much do you have in the passbook account on May 1?

45. You invest $2000 in a bond trust that pays 9% interest compounded semiannually. Your friend invests $2000 in a Certificate of Deposit (CD) that pays $8\frac{1}{2}$% compounded continuously. Who has more money after 20 years, you or your friend?

46. Suppose that you have access to an investment that will pay 10% interest compounded continuously. Which is better: To be given $1000 now so that you can take advantage of this investment opportunity or to be given $1325 after 3 years?

47. You have just purchased a house for $150,000, with the seller holding a second mortgage of $50,000. You promise to pay the seller $50,000 plus all accrued interest 5 years from now. The seller offers you three interest options on the second mortgage:

(a) Simple interest at 12% per annum   (b) $11\frac{1}{2}$% interest compounded monthly

(c) $11\frac{1}{4}$% interest compounded continuously

Which option is best; that is, which results in the least interest on the loan?

48. A bank advertises that it pays interest on saving accounts at the rate of 4.25% compounded daily. Find the effective rate if the bank uses (a) 360 days or (b) 365 days in determining the daily rate.

Problems 49–52 involve zero-coupon bonds. A zero-coupon bond *is a bond that is sold now at a discount and will pay its face value at some time in the future when it matures; no interest payments are made.*

49. A zero-coupon bond can be redeemed in 20 years for $10,000. How much should you be willing to pay for it now if you want a return of:

(a) 10% compounded monthly?   (b) 10% compounded continuously?

50. A child's grandparents are considering buying a $40,000 face value zero-coupon bond at birth so that she will have enough money for her college education 17 years later. If money is worth 8% compounded annually, what should they pay for the bond?

51. How much should a $10,000 face value zero-coupon bond, maturing in 10 years, be sold for now if its rate of return is to be 8% compounded annually?

52. If you pay $12,485.52 for a $25,000 face value zero-coupon bond that matures in 8 years, what is your annual rate of return?

53. Explain in your own words what the term *compound interest* means. What does *continuous compounding* mean?

54. Explain in your own words the meaning of present value.

55. Write a program that will calculate the amount after n years if a principal P is invested at r% per annum compounded quarterly. Use it to verify your answers to Problem 1 and 3.

56. Write a program that will calculate the principal needed now to get the amount A in n years at r% per annum compounded daily. Use it to verify your answer to Problem 13.

57. Write a program that will calculate the annual rate of interest required to double an investment in n years. Use it to verify your answers to Problems 23 and 24.

58. Write a program that will calculate the number of months required for an initial investment of x dollars to grow to y dollars at r% per annum compounded continuously. Use it to verify your answer to Problems 33 and 34.

59. *Time to Double or Triple an Investment* The formula

$$y = \frac{\ln m}{n \ln\left(1 + \dfrac{r}{n}\right)}$$

can be used to find the number of years y required to multiply an investment m times when r is the per annum interest rate compounded n times a year.

(a) How many years will it take to double the value of an IRA that compounds annually at the rate of 12%?

(b) How many years will it take to triple the value of a savings account that compounds quarterly at an annual rate of 6%?

(c) Give a derivation of this formula.

60. *Time to Reach an Investment Goal* The formula

$$y = \frac{\ln A - \ln P}{r}$$

can be used to find the number of years y required for an investment P to grow to a value A when compounded continuously at an annual rate r.

(a) How long will it take to increase an initial investment of $1000 to $8000 at an annual rate of 10%?

(b) What annual rate is required to increase the value of a $2000 IRA to $30,000 in 35 years?

(c) Give a derivation of this formula.

61. *Critical Thinking* You have just contracted to buy a house and will seek financing in the amount of $100,000. You go to several banks. Bank 1 will lend you $100,000 at the rate of 8.75% amortized over 30 years with a loan origination fee of 1.75%. Bank 2 will lend you $100,000 at the rate of 8.375% amortized over 15 years with a loan origination fee of 1.5%. Bank 3 will lend you $100,000 at the rate of 9.125% amortized over 30 years with no loan origination fee. Bank 4 will lend you $100,000 at the rate of 8.625% amortized over 15 years with no loan origination fee. Which loan would you take? Why? Be sure to have sound reasons for your choice. If the amount of the monthly payment does not matter to you, which loan would you take? Again, have sound reasons for your choice. Use the information in the table to assist you. Compare your final decision with others in the class. Discuss.

	MONTHLY PAYMENT	LOAN ORIGINATION FEE
Bank 1	$786.70	$1750.00
Bank 2	$977.42	$1500.00
Bank 3	$813.63	$0.00
Bank 4	$990.68	$0.00

4.6

Growth and Decay

Many natural phenomena have been found to follow the law that an amount A varies with time t according to

$$A = A_0 e^{kt} \tag{1}$$

where A_0 is the original amount ($t = 0$) and $k \neq 0$ is a constant.

If $k > 0$, then equation (1) states that the amount A is increasing over time; if $k < 0$, the amount A is decreasing over time. In either case, when an amount A varies over time according to equation (1), it is said to follow the **exponential law** or the **law of uninhibited growth** ($k > 0$) **or decay** ($k < 0$). See Figure 18.

FIGURE 18

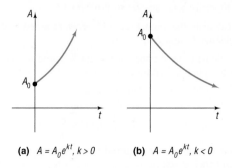

(a) $A = A_0 e^{kt}, k > 0$ (b) $A = A_0 e^{kt}, k < 0$

For example, we saw in Section 4.5 that continuously compounded interest follows the law of uninhibited growth. In this section we shall look at three additional phenomena that follow the exponential law.

Biology

Mitosis, or division of cells, is a universal process in the growth of living organisms such as amebas, plants, human skin cells, and many others. Based on an ideal situation in which no cells die and no by-products are produced, the number of cells present at a given time follows the law of uninhibited growth. Actually, however, after enough time has passed, growth at an exponential rate will cease due to the influence of factors such as lack of living space, dwindling food supply, and so on. The law of uninhibited growth accurately reflects only the early stages of the mitosis process.

The mitosis process begins with a culture containing N_0 cells. Each cell in the culture grows for a certain period of time and then divides into two identical cells. We assume that the time needed for each cell to divide in two is constant and does not change as the number of cells increases. These cells then grow, and eventually each divides in two; and so on.

A formula that gives the number N of cells in the culture after a time t has passed (in the early stages of growth) is

Uninhibited Growth of Cells

$$N(t) = N_0 e^{kt} \qquad k > 0 \qquad\qquad\qquad (2)$$

where k is a positive constant.

In using equation (2) to model the growth of cells, we are using a function that yields positive real numbers, even though we are counting the number of cells, which must be an integer. This is a common practice in many applications.

E X A M P L E   1   *Bacterial Growth*

A colony of bacteria increases according to the law of uninhibited growth. If the number of bacteria doubles in 3 hours, how long will it take for the size of the colony to triple?

Solution   Using formula (2), the number N of cells at a time t is

$$N(t) = N_0 e^{kt}$$

where N_0 is the initial number of bacteria present and k is a positive number. We first seek the number k. The number of cells doubles in 3 hours; thus, we have

$$N(3) = 2N_0$$

But $N(3) = N_0 e^{k(3)}$, so

$$N_0 e^{k(3)} = 2N_0$$
$$e^{3k} = 2$$
$$3k = \ln 2 \quad \text{\footnotesize Write the exponential equation as a logarithm.}$$
$$k = \tfrac{1}{3}\ln 2 \approx \tfrac{1}{3}(0.6931) = 0.2310$$

Formula (2) for this growth process is therefore

$$N(t) = N_0 e^{0.2310t}$$

The time t needed for the size of the colony to triple requires that $N = 3N_0$. Thus, we substitute $3N_0$ for N to get

$$3N_0 = N_0 e^{0.2310t}$$
$$3 = e^{0.2310t}$$
$$0.2310t = \ln 3$$
$$t = \frac{1}{0.2310}\ln 3 \approx \frac{1.0986}{0.2310} = 4.756 \text{ hours}$$

It will take about 4.756 hours for the size of the colony to triple.   ■

■ Now work Problem 1.

Radioactive Decay

Radioactive materials follow the law of uninhibited decay. Thus, the amount A of a radioactive material present at time t is given by the formula

<table>
<tr><td>Uninhibited Radioactive Decay</td><td>$$A = A_0 e^{kt} \qquad k < 0 \qquad\qquad (3)$$</td></tr>
</table>

where A_0 is the original amount of radioactive material and k is a negative number.

All radioactive substances have a specific **half-life,** which is the time required for half of the radioactive substance to decay. In **carbon dating,** we use the fact that all living organisms contain two kinds of carbon, carbon-12 (a stable carbon) and carbon-14 (a radioactive carbon, with a half-life of 5600 years). While an organism is living, the ratio of carbon-12 to carbon-14 is constant. But when an organism dies, the original amount of carbon-12 present remains unchanged, whereas the amount of carbon-14 begins to decrease. This change in the amount of carbon-14 present relative to the amount of carbon-12 present makes it possible to calculate

E X A M P L E   2   *Estimating the Age of Ancient Tools*

Traces of burned wood found along with ancient stone tools in an archaeological dig in Chile were found to contain approximately 1.67% of the original amount of carbon-14. If the half-life of carbon-14 is 5600 years, approximately when was the tree cut and burned?

Solution   Using equation (3), the amount A of carbon-14 present at time t is

$$A = A_0 e^{kt}$$

where A_0 is the original amount of carbon-14 present and k is a negative number. We first seek the number k. To find it, we use the fact that after 5600 years half of the original amount of carbon-14 remains. Thus,

$$\frac{1}{2} A_0 = A_0 e^{k(5600)}$$

$$\frac{1}{2} = e^{5600k}$$

$$5600k = \ln \frac{1}{2}$$

$$k = \frac{1}{5600} \ln \frac{1}{2} \approx -0.000124$$

Formula (3) therefore becomes

$$A = A_0 e^{-0.000124t}$$

If the amount A of carbon-14 now present is 1.67% of the original amount, it follows that

$$0.0167 A_0 = A_0 e^{-0.000124t}$$

$$0.0167 = e^{-0.000124t}$$

$$-0.000124t = \ln 0.0167$$

$$t = \frac{1}{-0.000124} \ln 0.0167 \approx 33{,}000 \text{ years}$$

The tree was cut and burned about 33,000 years ago. Some archaeologists use this conclusion to argue that humans lived in the Americas 33,000 years ago, much earlier than is generally accepted.   ■

■   Now work Problem 3.

Newton's Law of Cooling

Newton's Law of Cooling* states that the temperature of a heated object decreases exponentially over time toward the temperature of the surrounding medium. That is, the temperature u of a heated object at a given time t satisfies the equation

Newton's Law of Cooling

$$u = T + (u_0 - T)e^{kt} \qquad k < 0 \tag{4}$$

where T is the constant temperature of the surrounding medium, u_0 is the initial temperature of the heated object, and k is a negative number.

E X A M P L E   3   *Using Newton's Law of Cooling*

An object is heated to 100°C and is then allowed to cool in a room whose air temperature is 30°C. If the temperature of the object is 80°C after 5 minutes, when will its temperature be 50°C?

Solution   Using equation (4) with $T = 30$ and $u_0 = 100$, the temperature (in degrees Celsius) of the object at time t (in minutes) is

$$u = 30 + (100 - 30)e^{kt} = 30 + 70e^{kt} \tag{5}$$

where k is a negative number. To find k, we use the fact that $u = 80$ when $t = 5$. Then

$$80 = 30 + 70e^{k(5)}$$
$$50 = 70e^{5k}$$
$$e^{5k} = \frac{50}{70}$$
$$5k = \ln \frac{5}{7}$$
$$k = \frac{1}{5} \ln \frac{5}{7} \approx -0.0673$$

Formula (5) therefore becomes

$$u = 30 + 70e^{-0.0673t}$$

---

*Named after Sir Isaac Newton (1642–1727), one of the cofounders of calculus.

Now, we want to find t when $u = 50°C$, so

$$50 = 30 + 70e^{-0.0673t}$$

$$20 = 70e^{-0.0673t}$$

$$e^{-0.0673t} = \frac{20}{70}$$

$$-0.0673t = \ln\frac{2}{7}$$

$$t = \frac{1}{-0.0673}\ln\frac{2}{7} \approx 18.6 \text{ minutes}$$

Thus, the temperature of the object will be 50°C after about 18.6 minutes.

Check: Graph $y = 30 + 70e^{-0.0673x}$ and use TRACE to verify that when $x = 18.6$ then $y = 50$. ■

Exploration: What is y when $x = 30$ minutes? When is $y = 35°C$? When is $y = 30°C$? ■

4.6

Exercise 4.6

1. *Growth of an Insect Population* The size P of a certain insect population at time t (in days) obeys the equation $P = 500e^{0.02t}$. After how many days will the population reach 1000? 2000?

2. *Growth of Bacteria* The number N of bacteria present in a culture at time t (in hours) obeys the equation $N = 1000e^{0.01t}$. After how many hours will the population equal 1500? 2000?

3. *Radioactive Decay* Strontium-90 is a radioactive material that decays according to the law $A = A_0 e^{-0.0244t}$, where A_0 is the initial amount present and A is the amount present at time t (in years). What is the half-life of strontium-90?

4. *Radioactive Decay* Iodine-131 is a radioactive material that decays according to the law $A = A_0 e^{-0.087t}$, where A_0 is the initial amount present and A is the amount present at time t (in days). What is the half-life of iodine-131?

5. Use the information in Problem 3 to determine how long it takes for 100 grams of strontium-90 to decay to 10 grams.

6. Use the information in Problem 4 to determine how long it takes for 100 grams of iodine-131 to decay to 10 grams.

7. *Growth of a Colony of Mosquitoes* The population of a colony of mosquitoes obeys the law of uninhibited growth. If there are 1000 mosquitoes initially, and there are 1800 after 1 day, what is the size of the colony after 3 days? How long is it until there are 10,000 mosquitoes?

8. *Bacterial Growth* A culture of bacteria obeys the law of uninhibited growth. If 500 bacteria are present initially, and there are 800 after 1 hour, how many will be present in the culture after 5 hours? How long is it until there are 20,000 bacteria?

9. *Population Growth* The population of a southern city follows the exponential law. If the population doubled in size over an 18 month period and the current population is 10,000, what will the population be 2 years from now?

10. *Population Growth* The population of a midwestern city follows the exponential law. If the population decreased from 900,000 to 800,000 from 1993 to 1995, what will the population be in 1997?

11. *Radioactive Decay* The half-life of radium is 1690 years. If 10 grams are present now, how much will be present in 50 years?

12. *Radioactive Decay* The half-life of radioactive potassium is 1.3 billion years. If 10 grams are present now, how much will be present in 100 years? In 1000 years?

13. *Estimating the Age of A Tree* A piece of charcoal is found to contain 30% of the carbon-14 it originally had. When did the tree from which the charcoal came die? Use 5600 years as the half-life of carbon-14.

14. *Estimating the Age of a Fossil* A fossilized leaf contains 70% of its normal amount of carbon-14. How old is the fossil?

15. *Cooling Time of a Pizza* A pizza baked at 450°F is removed from the oven at 5:00 PM into a room that is a constant 70°F. After 5 minutes, the pizza is at 300°F. At what time can you begin eating the pizza if you want its temperature to be 135°F?

16. A thermometer reading 72°F is placed in a refrigerator where the temperature is a constant 38°F. If the thermometer reads 60°F after 2 minutes, what will it read after 7 minutes? How long will it take before the thermometer reads 39°F?

17. A thermometer reading 8°C is brought into a room with a constant temperature of 35°C. If the thermometer reads 15°C after 3 minutes, what will it read after being in the room for 5 minutes? For 10 minutes? [*Hint:* You need to construct a formula similar to equation (4).]

18. *Thawing Time of a Steak* A frozen steak has a temperature of 28°F. It is placed in a room with a constant temperature of 70°F. After 10 minutes, the temperature of the steak has risen to 35°F. What will the temperature of the steak be after 30 minutes? How long will it take the steak to thaw to a temperature of 45°F? [See the hint given for Problem 17.]

19. *Decomposition of Salt in Water* Salt (NaCl) decomposes in water into sodium (NA^+) and chloride (Cl^-) ions according to the law of uninhibited decay. If the initial amount of salt is 25 kilograms and, after 10 hours, 15 kilograms of salt are left, how much salt is left after 1 day? How long does it take until $\frac{1}{2}$ kilogram of salt is left?

20. *Voltage of a Condenser* The voltage of a certain condenser decreases over time according to the law of uninhibited decay. If the initial voltage is 40 volts, and 2 seconds later it is 10 volts, what is the voltage after 5 seconds?

21. *Radioactivity from Chernobyl* After the release of radioactive material into the atmosphere from a nuclear power plant at Chernobyl (Ukraine) in 1986, the hay in Austria was contaminated by iodine-131 (see Problem 4). If it is all right to feed the hay to cows when 10% of the iodine-131 remains, how long do the farmers need to wait to use this hay?

22. *Pig Roasts* The hotel Bora-Bora is having a pig roast. At noon, the chef put the pig in a large earthen oven. The pig's original temperature was 75°F. At 2:00 PM, the chef checked the pig's temperature and was upset because it had reached only 100°F. If the oven's temperature remains a constant 325°F, at what time may the hotel serve its guests, assuming that pork is done when it reaches 175°F?

4.7

Logarithmic Scales

Common logarithms often appear in the measurement of quantities, because they provide a way to scale down positive numbers that vary from very small to very large. For example, if a certain quantity can take on values from $0.0000000001 = 10^{-10}$ to $10,000,000,000 = 10^{10}$, the common logarithms of such numbers would be between -10 and 10.

Loudness of Sound

Our first application utilizes a logarithmic scale to measure the loudness of a sound. Physicists define the **intensity of a sound wave** as the amount of energy the wave transmits through a given area. For example, the least intense sound that a human ear can detect at a frequency of 100 hertz is about 10^{-12} watt per square meter. The **loudness** $L(x)$, measured in **decibels** (named in honor of Alexander Graham Bell), of a sound of intensity x (measured in watts per square meter) is defined as

Loudness

$$L(x) = 10 \log \frac{x}{I_0} \qquad (1)$$

where $I_0 = 10^{-12}$ watt per square meter is the least intense sound that a human ear can detect. If we let $x = I_0$ in equation (1), we get

$$L(I_0) = 10 \log \frac{I_0}{I_0} = 10 \log 1 = 0$$

Thus, at the threshold of human hearing, the loudness is 0 decibels. Figure 19 gives the loudness of some common sounds.

FIGURE 19

Loudness of common sounds (in decibels)

DECIBELS

140	Shotgun blast, jet 100 feet away at takeoff	Pain
130	Motor test chamber	Human ear pain threshold
120	Firecrackers, severe thunder, pneumatic jackhammer, hockey crowd	Uncomfortably loud
110	Amplified rock music	
100	Textile loom, subway train, elevated train, farm tractor, power lawn mower, newspaper press	Loud
90	Heavy city traffic, noisy factory	
80	Diesel truck going 40 mi/hr 50 feet away, crowded restaurant, garbage disposal, average factory, vacuum cleaner	Moderately loud
70	Passenger car going 50 mi/hr 50 feet away	
60	Quiet typewriter, singing birds, window air conditioner, quiet automobile	Quiet
50	Normal conversation, average office	
40	Household refrigerator, quiet office	Very quiet
30	Average home, dripping faucet, whisper 5 feet away	
20	Light rainfall, rustle of leaves	Average person's threshold of hearing
10	Whisper across room	Just audible
0		Threshold for acute hearing

Note that a decibel is not a linear unit like the meter. For example, a noise level of 10 decibels is 10 times as great as a noise level of 0 decibels. [If $L(x) = 10$, then $x = 10I_0$.] A noise level of 20 decibels is 100 times as great as a noise level of 0 decibels. [If $L(x) = 20$, then $x = 100I_0$.] A noise level of 30 decibels is 1000 times as great as a noise level of 0 decibels, and so on.

EXAMPLE 1   *Finding the Intensity of a Sound*

Use Figure 19 to find the intensity of the sound of a dripping faucet.

Solution   From Figure 19, we see that the loudness of the sound of dripping water is 30 decibels. Thus, by equation (1), its intensity x may be found as follows:

$$30 = 10 \log\left(\frac{x}{I_0}\right)$$

$$3 = \log\left(\frac{x}{I_0}\right) \qquad \text{Divide by 10.}$$

$$\frac{x}{I_0} = 10^3 \qquad \text{Write in exponential form.}$$

$$x = 1000I_0$$

where $I_0 = 10^{-12}$ watt per square meter. Thus, the intensity of the sound of a dripping faucet is 1000 times as great as a noise level of 0 decibels; that is, such a sound has an intensity of $1000 \cdot 10^{-12} = 10^{-9}$ watt per square meter.   ■

■ Now work Problem 5.

EXAMPLE 2   *Finding the Loudness of a Sound*

Use Figure 19 to determine the loudness of a subway train if it is known that this sound is 10 times as intense as the sound due to heavy city traffic.

Solution   The sound due to heavy city traffic has a loudness of 90 decibels. Its intensity, therefore, is the value of x in the equation

$$90 = 10 \log\left(\frac{x}{I_0}\right)$$

A sound 10 times as intense as x has loudness $L(10x)$. Thus, the loudness of the subway train is

$$\begin{aligned}
L(10x) &= 10 \log\left(\frac{10x}{I_0}\right) & \text{Replace } x \text{ by } 10x. \\
&= 10 \log\left(10 \cdot \frac{x}{I_0}\right) & \\
&= 10 \left[\log 10 + \log\left(\frac{x}{I_0}\right)\right] & \text{Log of product = Sum of logs} \\
&= 10 \log 10 + 10 \log\left(\frac{x}{I_0}\right) & \log 10 = 1 \\
&= 10 + 90 = 100 \text{ decibels} & \blacksquare
\end{aligned}$$

Magnitude of an Earthquake

Our second application uses a logarithmic scale to measure the magnitude of an earthquake.

The **Richter scale*** is one way of converting seismographic readings into numbers that provide an easy reference for measuring the magnitude M of an earthquake. All earthquakes are compared to a **zero-level earthquake** whose seismographic reading measures 0.001 millimeter at a distance of 100 kilometers from the epicenter. An earthquake whose seismographic reading measures x millimeters has **magnitude** $M(x)$ given by

Magnitude of an Earthquake

$$M(x) = \log\left(\frac{x}{x_0}\right) \tag{2}$$

where $x_0 = 10^{-3}$ is the reading of a zero-level earthquake the same distance from its epicenter.

E X A M P L E   3   *Finding the Magnitude of an Earthquake*

What is the magnitude of an earthquake whose seismographic reading is 0.1 millimeter at a distance of 100 kilometers from its epicenter?

Solution   If $x = 0.1$, the magnitude $M(x)$ of this earthquake is

$$M(0.1) = \log\left(\frac{x}{x_0}\right) = \log\left(\frac{0.1}{0.001}\right) = \log\left(\frac{10^{-1}}{10^{-3}}\right) = \log 10^2 = 2$$

This earthquake thus measures 2.0 on the Richter scale. $\blacksquare$

■   Now work Problem 7.

*Named after the American scientist, C.F. Richter, who devised it in 1935.

Based on formula (2), we define the **intensity of an earthquake** as the ratio of x to x_0. For example, the intensity of the earthquake described in Example 3 is $\frac{0.1}{0.001} = 10^2 = 100$. That is, it is 100 times as intense as a zero-level earthquake.

E X A M P L E   4   *Comparing the Intensity of Two Earthquakes*

The devastating San Francisco earthquake of 1906 measured 8.9 on the Richter scale. How did the intensity of that earthquake compare to the Papua New Guinea earthquake of 1988, which measured 6.7 on the Richter scale?

FIGURE 20

Solution   Let x_1 and x_2 denote the seismographic readings, respectively, of the 1906 San Francisco earthquake and the Papua New Guinea earthquake. Then, based on formula (2),

$$8.9 = \log\left(\frac{x_1}{x_0}\right) \qquad 6.7 = \log\left(\frac{x_2}{x_0}\right)$$

Consequently,

$$\frac{x_1}{x_0} = 10^{8.9} \qquad \frac{x_2}{x_0} = 10^{6.7}$$

The 1906 San Francisco earthquake was $10^{8.9}$ times as intense as a zero-level earthquake. The Papua New Guinea earthquake was $10^{6.7}$ times as intense as a zero-level earthquake. Thus,

$$\frac{x_1}{x_2} = \frac{10^{8.9}x_0}{10^{6.7}x_0} = 10^{2.2} \approx 158$$

$$x_1 \approx 158x_2$$

Hence, the San Francisco earthquake was 158 times as intense as the Papua New Guinea earthquake.   ■

Example 4 demonstrates that the relative intensity of two earthquakes can be found by raising 10 to a power equal to the difference of their readings on the Richter scale.

4.7

Exercise 4.7

1. *Loudness of a Dishwasher* Find the loudness of a dishwasher that operates at an intensity of 10^{-5} watt per square meter. Express your answer in decibels.

2. *Loudness of a Diesel Engine* Find the loudness of a diesel engine that operates at an intensity of 10^{-3} watt per square meter. Express your answer in decibels.

3. *Loudness of a Jet Engine* With engines at full throttle, a Boeing 727 jetliner produces noise at an intensity of 0.15 watt per square meter. Find the loudness of the engines in decibels.

4. *Loudness of a Whisper* A whisper produces noise at an intensity of $10^{-9.8}$ watt per square meter. What is the loudness of a whisper in decibels?

5. *Intensity of a Sound at Threshold of Pain* For humans, the threshold of pain due to sound averages 130 decibels. What is the intensity of such a sound in watts per square meter?

6. If one sound is 50 times as intense as another, what is the difference in the loudness of the two sounds? Express your answer in decibels.

7. *Magnitude of an Earthquake* Find the magnitude of an earthquake whose seismographic reading is 10.0 millimeters at a distance of 100 kilometers from its epicenter.

8. *Magnitude of an Earthquake* Find the magnitude of an earthquake whose seismographic reading is 1210 millimeters at a distance of 100 kilometers from its epicenter.

9. *Comparing Earthquakes* The Mexico City earthquake of 1978 registered 7.85 on the Richter scale. What would a seismograph 100 kilometers from the epicenter have measured for this earthquake? How does this earthquake compare in intensity to the 1906 San Francisco earthquake, which registered 8.9 on the Richter scale?

10. *Comparing Earthquakes* Two earthquakes differ by 1.0 when measured on the Richter scale. How would the seismographic readings differ at a distance of 100 kilometers from the epicenter? How do their intensities compare?

Chapter Review

THINGS TO KNOW

Properties of the exponential function

$f(x) = a^x, a > 1$ Domain: $(-\infty, \infty)$; Range: $(0, \infty)$; x-intercepts: none; y-intercept: 1; horizontal asymptote: x-axis as $x \to -\infty$; increasing; one-to-one
See Figure 3 for a typical graph.

$f(x) = a^x, 0 < a < 1$ Domain: $(-\infty, \infty)$; Range: $(0, \infty)$; x-intercepts: none, y-intercept: 1; horizontal asymptote: x-axis as $x \to \infty$; decreasing; one-to-one
See Figure 5 for a typical graph.

Properties of the logarithmic function

$f(x) = \log_a x, a > 1$ Domain: $(0, \infty)$; Range: $(-\infty, \infty)$; x-intercept: 1; y-intercept: none; vertical asymptote: y-axis; increasing;
($y = \log_a x$ means $x = a^y$)   one-to-one
See Figure 10(b) for a typical graph.

$f(x) = \log_a x, 0 < a < 1$ Domain: $(0, \infty)$; Range: $(-\infty, \infty)$; x-intercept: 1; y-intercept: none; vertical asymptote: y-axis; decreasing;
($y = \log_a x$ means $x = a^y$)   one-to-one
See Figure 10(a) for a typical graph.

Number e

Value approached by the expression $\left(1 + \dfrac{1}{n}\right)^n$ as $n \to \infty$; that is, $\lim\limits_{n \to \infty} \left(1 + \dfrac{1}{n}\right)^n = e$

Natural logarithm

$y = \ln x$ means $x = e^y$

Properties of logarithms

$\log_a 1 = 0$ $\log_a a = 1$ $a^{\log_a M} = M$ $\log_a a^r = r$

$\log_a MN = \log_a M + \log_a N$ $\log_a\!\left(\dfrac{M}{N}\right) = \log_a M - \log_a N$ $\log_a\!\left(\dfrac{1}{N}\right) = -\log_a N$ $\log_a M^r = r \log_a M$

FORMULAS

Change-of-base formula	$\log_a M = \dfrac{\log_b M}{\log_b a}$
Compound interest	$A = P\left(1 + \dfrac{r}{n}\right)^{nt}$
Continuous compounding	$A = Pe^{rt}$
Present value	$P = A\left(1 + \dfrac{r}{n}\right)^{-nt}$ or $P = Ae^{-rt}$
Growth and decay	$A = A_0 e^{kt}$

HOW TO:

Graph exponential and logarithmic functions

Solve certain exponential equations

Solve certain logarithmic equations

Solve problems involving compound interest

Solve problems involving growth and decay

Solve problems involving intensity of sound and intensity of earthquakes

FILL-IN-THE-BLANK ITEMS

1. The graph of every exponential function $f(x) = a^x$, $a > 0$, $a \neq 1$, passes through the two points _____.

2. If the graph of an exponential function $f(x) = a^x$, $a > 0$, $a \neq 1$, is decreasing, then its base must be less than _____.

3. If $3^x = 3^4$, then $x =$ _____.

4. The logarithm of a product equals the _____ of the logarithms.

5. For every base, the logarithm of _____ equals 0.

6. If $\log_8 M = \log_5 7 / \log_5 8$, then $M =$ _____.

7. The domain of the logarithmic function $f(x) = \log_a x$ consists of _____.

8. The graph of every logarithmic function $f(x) = \log_a x$, $a > 0$, $a \neq 1$, passes through the two points _____.

9. If the graph of a logarithmic function $f(x) = \log_a x$, $a > 0$, $a \neq 1$, is increasing, then its base must be larger than _____.

10. If $\log_3 x = \log_3 7$, then $x =$ _____.

TRUE/FALSE ITEMS

T F **1.** The graph of every exponential function $f(x) = a^x$, $a > 0$, $a \neq 1$, will contain the points $(0, 1)$ and $(1, a)$.

T F **2.** The graphs of $y = 3^{-x}$ and $y = \left(\frac{1}{3}\right)^x$ are identical.

T F **3.** The present value of $1000 to be received after 2 years at 10% per annum compounded continuously is approximately $1205.

T F **4.** If $y = \log_a x$, then $y = a^x$.

T F **5.** The graph of every logarithmic function $f(x) = \log_a x$, $a > 0$, $a \neq 1$, will contain the points $(1, 0)$ and $(a, 1)$.

T   F   **6.** $a^{\log_M a} = M$, where $a > 0$, $a \neq 1$, $M > 0$

T   F   **7.** $\log_a(M + N) = \log_a M + \log_a N$, where $a > 0$, $a \neq 1$, $M > 0$, $N > 0$

T   F   **8.** $\log_a M - \log_a N = \log_a(M/N)$, where $a > 0$, $a \neq 1$, $M > 0$, $N > 0$

REVIEW EXERCISES

In Problems 1–6, evaluate each expression.

1. $\log_2(\frac{1}{8})$ **2.** $\log_3 81$ **3.** $\ln e^{\sqrt{2}}$ **4.** $e^{\ln 0.1}$ **5.** $2^{\log_2 0.4}$ **6.** $\log_2 2^{\sqrt{3}}$

In Problems 7–12, write each expression as a single logarithm.

7. $3 \log_4 x^2 + \frac{1}{2} \log_4 \sqrt{x}$ **8.** $-2 \log_3\left(\frac{1}{x}\right) + \frac{1}{3} \log_3 \sqrt{x}$

9. $\ln\left(\dfrac{x - 1}{x}\right) + \ln\left(\dfrac{x}{x + 1}\right) - \ln(x^2 - 1)$ **10.** $\log(x^2 - 9) - \log(x^2 + 7x + 12)$

11. $2 \log 2 + 3 \log x - \frac{1}{2} [\log(x + 3) + \log(x - 2)]$ **12.** $\frac{1}{2} \ln(x^2 + 1) - 4 \ln\frac{1}{2} - \frac{1}{2}[\ln(x - 4) + \ln x]$

In Problems 13–20, find y as a function of x. The constant C is a positive number.

13. $\ln y = 2x^2 + \ln C$ **14.** $\ln(y - 3) = \ln 2x^2 + \ln C$

15. $\frac{1}{2} \ln y = 3x^2 + \ln C$ **16.** $\ln 2y = \ln(x + 1) + \ln(x + 2) + \ln C$

17. $\ln(y - 3) + \ln(y + 3) = x + C$ **18.** $\ln(y - 1) + \ln(y + 1) = -x + C$

19. $e^{y + C} = x^2 + 4$ **20.** $e^{3y - C} = (x + 4)^2$

In Problems 21–30, graph each function. Begin each problem either with the graph of $y = e^x$ or with the graph of $y = \ln x$.

21. $f(x) = e^{-x}$ **22.** $f(x) = \ln(-x)$ **23.** $f(x) = 1 - e^x$ **24.** $f(x) = 3 + \ln x$

25. $f(x) = 3e^x$ **26.** $f(x) = \frac{1}{2} \ln x$ **27.** $f(x) = e^{|x|}$ **28.** $f(x) = \ln|x|$

29. $f(x) = 3 - e^{-x}$ **30.** $f(x) = 4 - \ln(-x)$

In Problems 31–50, solve each equation.

31. $4^{1 - 2x} = 2$ **32.** $8^{6 + 3x} = 4$ **33.** $3^{x^2 + x} = \sqrt{3}$

34. $4^{x - x^2} = \frac{1}{2}$ **35.** $\log_x 64 = -3$ **36.** $\log_{\sqrt{2}} x = -6$

37. $5^x = 3^{x + 2}$ **38.** $5^{x + 2} = 7^{x - 2}$ **39.** $9^{2x} = 27^{3x - 4}$

40. $25^{2x} = 5^{x^2 - 12}$ **41.** $\log_3 \sqrt{x - 2} = 2$ **42.** $2^{x + 1} \cdot 8^{-x} = 4$

43. $8 = 4^{x^2} \cdot 2^{5x}$ **44.** $2^x \cdot 5 = 10^x$ **45.** $\log_6(x + 3) + \log_6(x + 4) = 1$

46. $\log_{10}(7x - 12) = 2 \log_{10} x$ **47.** $e^{1 - x} = 5$ **48.** $e^{1 - 2x} = 4$

49. $2^{3x} = 3^{2x + 1}$ **50.** $2^{x^3} = 3^{x^2}$

In Problems 51–54, use the following result: If x is the atmospheric pressure (measured in millimeters of mercury), then the formula for the altitude h(x) (measured in meters above sea level) is

$$h(x) = (30T + 8000)\log\left(\frac{P_0}{x}\right)$$

where T is the temperature (in degrees Celsius) and P_0 is the atmospheric pressure at sea level, which is approximately 760 millimeters of mercury.

51. *Finding the Altitude of an Airplane* At what height is an aircraft whose instruments record an outside temperature of 0°C and a barometric pressure of 300 millimeters of mercury?

52. *Finding the Height of a Mountain* How high is a mountain if instruments placed on its peak record a temperature of 5°C and a barometric pressure of 500 millimeters of mercury?

53. *Atmospheric Pressure Outside an Airplane* What is the atmospheric pressure outside an aircraft flying at an altitude of 10,000 meters if the outside air temperature is $-100°C$?

54. *Atmospheric Pressure at High Altitudes* What is the atmospheric pressure (in millimeters of mercury) on Mt. Everest, which has an altitude of approximately 8900 meters, if the air temperature is 5°C?

55. *Amplifying Sound* An amplifier's power output P (in watts) is related to its decibel voltage gain d by the formula $P = 25e^{0.1d}$.

(a) Find the power output for a decibel voltage gain of 4 decibels.

(b) For a power output of 50 watts, what is the decibel voltage gain?

56. *Limiting Magnitude of a Telescope* A telescope is limited in its usefulness by the brightness of the star it is aimed at and by the diameter of its lens. One measure of a star's brightness is its *magnitude:* the dimmer the star, the larger its magnitude. A formula for the limiting magnitude L of a telescope, that is, the magnitude of the dimmest star it can be used to view, is given by

$$L = 9 + 5.1 \log d$$

where d is the diameter (in inches) of the lens.

(a) What is the limiting magnitude of a 3.5 inch telescope?

(b) What diameter is required to view a star of magnitude 14?

57. *Product Demand* The demand for a new product increases rapidly at first and then levels off. The percent P of actual purchases of this product after it has been on the market t months is

$$P = 90 - 80\left(\frac{3}{4}\right)^t$$

(a) What is the percent of purchases of the product after 5 months?

(b) What is the percent of purchases of the product after 10 months?

(c) What is the maximum percent of purchases of the product?

(d) How many months does it take before 40% of purchases occurs?

(e) How many months before 70% of purchases occurs?

58. *Disseminating Information* A survey of a certain community of 10,000 residents shows that the number of residents N who have heard a piece of information after m months is given by the formula

$$m = 55.3 - 6 \ln(10,000 - N)$$

How many months will it take for half of the citizens to learn about a community program of free blood pressure readings?

59. *Salvage Value* The number of years n for a piece of machinery to depreciate to a known salvage value can be found using the formula

$$n = \frac{\log_{10} s - \log_{10} i}{\log_{10}(1 - d)}$$

where s is the salvage value of the machinery, i is its initial value, and d is the annual rate of depreciation.

(a) How many years will it take for a piece of machinery to decline in value from $90,000 to $10,000 if the annual rate of depreciation is 0.20 (20%)?

(b) How many years will it take for a piece of machinery to lose half of its value if the annual rate of depreciation is 15%?

60. *Funding a College Education* A child's grandparents purchase a $10,000 bond fund that matures in 18 years to be used for her college education. The bond fund pays 4% interest compounded semiannually. How much will the bond fund be worth at maturity?

61. *Funding a College Education* A child's grandparents wish to purchase a bond fund that matures in 18 years to be used for her college education. The bond fund pays 4% interest compounded semiannually. How much should they purchase so that the bond fund will be worth $85,000 at maturity?

62. *Funding an IRA* First Colonial Bankshares Corporation advertised the following IRA investment plans.

(a) Assuming continuous compounding, what was the annual rate of interest they offered?

(b) First Colonial Bankshares claims that $4000 invested today will have a value of over $32,000 in 20 years. Use the answer found in part (a) to find the actual value of $4000 in 20 years. Assume continuous compounding.

TARGET IRA PLANS

FOR EACH $5000 MATURITY VALUE DESIRED	
DEPOSIT:	**AT A TERM OF:**
$620.17	20 years
$1045.02	15 years
$1760.92	10 years
$2967.26	5 years

63. *Loudness of a Garbage Disposal* Find the loudness of a garbage disposal unit that operates at an intensity of 10^{-4} watt per square meter. Express your answer in decibels.

64. *Comparing Earthquakes* On September 9, 1985, the western suburbs of Chicago experienced a mild earthquake that registered 3.0 on the Richter scale. How did this earthquake compare in intensity to the great San Francisco earthquake of 1906, which registered 8.9 on the Richter scale?

65. *Estimating the Date a Prehistoric Man Died* The bones of a prehistoric man found in the desert of New Mexico contain approximately 5% of the original amount of carbon-14. If the half-life of carbon-14 is 5600 years, approximately how long ago did the man die?

66. *Temperature of a Skillet* A skillet is removed from an oven whose temperature is 450°F and placed in a room whose temperature is 70°F. After 5 minutes, the temperature of the skillet is 400°F. How long will it be until its temperature is 150°F?

67. In a room whose constant temperature is 70°F, will a pizza heated to 450°F ever reach exactly 70°F? Newton's Law of Cooling seems to say no! What really happens? What assumptions are made in using Newton's law? Write a brief paragraph explaining your conclusions.

Chapter 5

PREPARING FOR THIS CHAPTER

Before getting started on this chapter, review the following concepts:

Pythagorean Theorem (p. 10)
Unit circle (p. 63)
Functions (p. 96)
Domain of a function (p. 99)
Even and odd functions (pp. 114–116)

TRIGONOMETRIC FUNCTIONS

Preview Gibb's Hill Lighthouse, Southampton, Bermuda

In operation since 1846, it stands 117 feet high on a hill 245 feet high, so its beam of light is 362 feet above sea level. A brochure states that the light itself can be seen on the horizon about 26 miles distant. Verify the accuracy of this statement. The brochure further states that ships 40 miles away can see the light and planes flying at 10,000 feet can see it 120 miles away. Verify the accuracy of these statements. What assumption did the brochure make about the height of the ship?

[See Example 9 and Problem 40 in Section 5.5] ■

Trigonometry was developed by Greek astronomers who regarded the sky as the inside of a sphere, so it was natural that triangles on a sphere were investigated early (by Menelaus of Alexandria about AD 100) and that triangles in the plane were studied much later. The first book containing a systematic treatment of plane and spherical trigonometry was written by the Persian astronomer Nasîr ed-dîn (about AD 1250).

Regiomontanus (1436–1476) is the man most responsible for moving trigonometry from astronomy into mathematics. His work was improved by Copernicus (1473–1543) and Copernicus's student Rhaeticus (1514–1576). Rhaeticus's book was the first to define the six trigonometric functions as ratios of sides of triangles, although he did not give the functions their present names. Credit for this is due to Thomas Fincke (1583), but Fincke's notation was by no means universally accepted at the time. The notation was finally stabilized by the textbooks of Leonhard Euler (1707–1783).

Trigonometry has since evolved from its use by surveyors, navigators, and engineers to present applications involving ocean tides, the rise and fall of food supplies in certain ecologies, brain wave patterns, and many other phenomena.

There are two widely accepted approaches to the development of the *trigonometric functions:* One uses circles, especially the *unit circle;* the other uses *right triangles.* In this book, we introduce trigonometric functions using the unit circle. In Section 5.4, we shall show that right triangle trigonometry is a special case of the unit circle approach.

5.1

Angles and Their Measure

A **ray,** or **half-line,** is that portion of a line that starts at a point V on the line and extends indefinitely in one direction. The starting point V of a ray is called its **vertex.** See Figure 1.

If two rays are drawn with a common vertex, they form an **angle.** We call one of the rays of an angle the **initial side** and the other the **terminal side.** The angle that is formed is identified by showing the direction and amount of rotation from the initial side to the terminal side. If the rotation is in the counterclockwise direction, the angle is **positive;** if the rotation is clockwise, the angle is **negative.** See Figure 2. Lowercase Greek letters, such as α (alpha), β (beta), γ (gamma), θ (theta), and so on, will be used to denote angles. Notice in Figure 2(a) that the angle α is positive because the direction of the rotation from the initial side to the terminal side is counterclockwise. The angle β in Figure 2(b) is negative because the rotation is clockwise. The angle γ in Figure 2(c) is positive. Notice that the angle α in Figure 2(a) and the angle γ have the same initial side and the same terminal side. However, α and γ are unequal because the amount of rotation required to go from the initial side to the terminal side is greater for angle γ than for angle α.

FIGURE 1

FIGURE 2

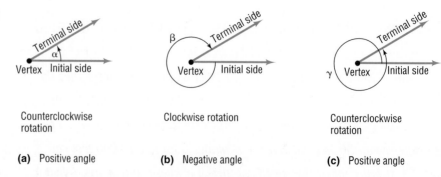

(a) Positive angle

(b) Negative angle

(c) Positive angle

An angle θ is said to be in **standard position** if its vertex is at the origin of a rectangular coordinate system and its initial side coincides with the positive x-axis. See Figure 3.

FIGURE 3

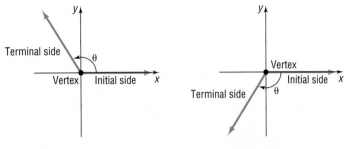

(a) θ in standard position;
     θ positive

(b) θ in standard position;
     θ negative

When an angle θ is in standard position, the terminal side either will lie in a quadrant, in which case we say θ **lies in that quadrant,** or it will lie on the x-axis or the y-axis, in which case we say θ is a **quadrantal angle.** For example, the angle θ in Figure 4(a) lies in quadrant II, the angle θ in Figure 4(b) lies in quadrant IV, and the angle θ in Figure 4(c) is a quadrantal angle.

FIGURE 4

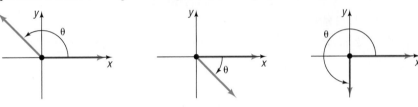

(a) θ lies in quadrant II

(b) θ lies in quadrant IV

(c) θ is a quadrantal angle

We measure angles by determining the amount of rotation needed for the initial side to become coincident with the terminal side. There are two commonly used measures for angles: *degrees* and *radians.*

Degrees

The angle formed by rotating the initial side exactly once in the counterclockwise direction until it coincides with itself (1 revolution) is said to measure 360 degrees, abbreviated 360°. Thus **one degree, 1°,** is $\frac{1}{360}$ revolution. A **right angle** is an angle of 90°, or $\frac{1}{4}$ revolution; a **straight angle** is an angle of 180°, or $\frac{1}{2}$ revolution. See Figure 5. As Figure 5(b) shows, it is customary to indicate a right angle by using the symbol $\llcorner$.

FIGURE 5

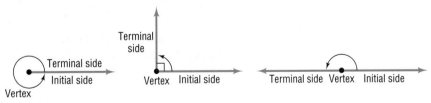

(a) 1 revolution
     counterclockwise, 360°

(b) $\frac{1}{4}$ revolution
     counterclockwise, 90°

(c) $\frac{1}{2}$ revolution
     counterclockwise, 180°

E X A M P L E   1   *Drawing an Angle*

Draw each angle:

(a) 45°   (b) −90°   (c) 225°   (d) 405°

Solution   (a) An angle of 45° is $\frac{1}{2}$ of a right angle. See Figure 6.

(b) An angle of $-90°$ is $\frac{1}{4}$ revolution in the clockwise direction. See Figure 7.

FIGURE 6

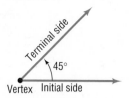

FIGURE 7

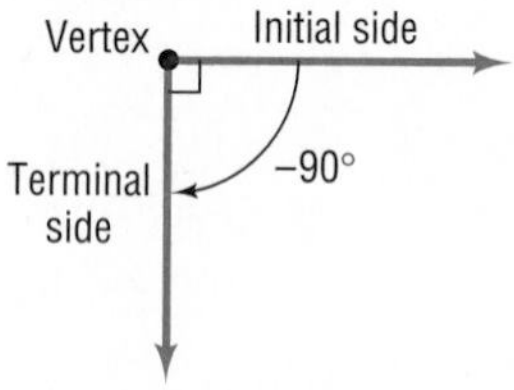

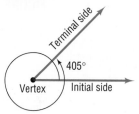

(c) An angle of 225° consists of a rotation through 180° followed by a rotation through 45°. See Figure 8.

(d) An angle of 405° consists of 1 revolution (360°) followed by a rotation through 45°. See Figure 9.

FIGURE 8

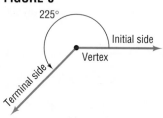

FIGURE 9

■

■ Now work Problem 1.

Although subdivisions of a degree may be obtained by using decimals, we also may use the notion of *minutes* and *seconds*. **One minute,** denoted by **1′,** is defined as $\frac{1}{60}$ degree. **One second,** denoted by **1″,** is defined as $\frac{1}{60}$ minute or, equivalently, $\frac{1}{3600}$ degree. An angle of, say, 30 degrees, 40 minutes, 10 seconds is written compactly as 30°40′10″. To summarize:

$$1 \text{ counterclockwise revolution} = 360°$$
$$60' = 1° \qquad 60'' = 1' \tag{1}$$

Because calculators use decimals, it is important to be able to convert from the degree, minute, second notation (D°M′S″) to a decimal form, and vice versa. Check your calculator; it may have a special key that does the conversion for you. If you do not have this key on your calculator, you can perform the conversion as described next.

Before getting started, though, you must set the calculator to receive degrees, because there are two common ways to measure angles. Many calculators show which mode is currently in use by displaying ⎡DEG⎤ for degree mode or ⎡RAD⎤ for radian mode. (We will define radians shortly.) Usually, a key is used to change from one mode to another. Check your instruction manual to find out how your particular calculator works.

Now let's see how to convert from the degree, minute, second notation (D°M′S″) to a decimal form, and vice versa, by looking at some examples: $15°30′ = 15.5°$, because $30′ = \frac{1}{2}° = 0.5°$, and $32.25° = 32°15′$, because $0.25° = \frac{1}{4}° = 15′$. For most conversions, a calculator will be helpful.

E X A M P L E   2   *Converting between Degrees, Minutes, Seconds, and Decimal Forms*

(a) Convert $50°6′21″$ to a decimal in degrees.

(b) Convert $21.256°$ to the D°M′S″ form. Round off your answer to the nearest second.

Solution   (a) Because $1′ = \frac{1}{60}°$ and $1″ = \frac{1}{60}′ = \left(\frac{1}{60} \cdot \frac{1}{60}\right)°$, we convert as follows:

$$50°6′21″ = \left(50 + 6 \cdot \frac{1}{60} + 21 \cdot \frac{1}{60} \cdot \frac{1}{60}\right)°$$
$$\approx (50 + 0.1 + 0.005833)°$$
$$= 50.105833°$$

(b) We start with the decimal part of $21.256°$, that is, $0.256°$:

$$0.256° = (0.256)(1°) = (0.256)(60′) = 15.36′$$
$$\underset{1° = 60′}{\uparrow}$$

Now we work with the decimal part of $15.36′$, that is, $0.36′$:

$$0.36′ = (0.36)(1′) = (0.36)(60″) = 21.6″ \approx 22″$$

Thus,

$$21.256° = 21° + 0.256° = 21° + 15.36′ = 21° + 15′ + 0.36′$$
$$= 21° + 15′ + 21.6″ \approx 21°15′22″$$

◼ Now work Problems 57 and 63.

In many applications, such as describing the exact location of a star or the precise position of a boat at sea, angles measured in degrees, minutes, and even seconds are used. For calculation purposes, these are transformed to decimal form. In many other applications, especially those in calculus, angles are measured using *radians*.

Radians

Consider a circle of radius r. Construct an angle whose vertex is at the center of this circle, called a **central angle,** and whose rays subtend an arc on the circle whose length equals r. See Figure 10. The measure of such an angle is **1 radian.**

Now consider a circle and two central angles, θ and θ_1. Suppose that these central angles subtend arcs of lengths s and s_1, respectively, as shown in Figure 11. From geometry, we know that the ratio of the measures of the angles equals the ratio of the corresponding lengths of the arcs subtended by those angles; that is,

$$\frac{\theta}{\theta_1} = \frac{s}{s_1} \tag{2}$$

FIGURE 10

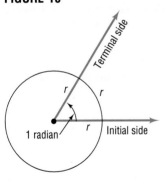

FIGURE 11

$$\frac{\theta}{\theta_1} = \frac{s}{s_1}$$

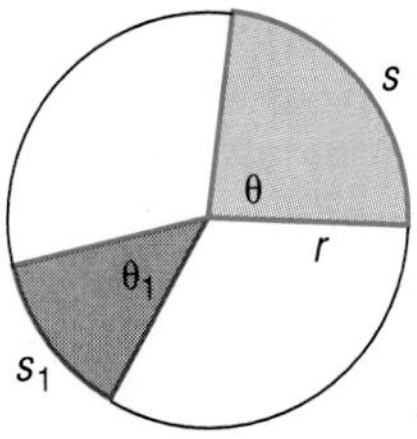

Suppose that θ and θ_1 are measured in radians, and suppose that $\theta_1 = 1$ radian. Refer again to Figure 11. Then the amount of arc s_1 subtended by the central angle θ_1 equals the radius r of the circle. Thus, $s_1 = r$, so formula (2) reduces to

$$\frac{\theta}{1} = \frac{s}{r} \quad \text{or} \quad s = r\theta \tag{3}$$

Theorem

Arc Length

For a circle of radius r, a central angle of θ radians subtends an arc whose length s is

$$s = r\theta \tag{4}$$

∎

Note: Formulas must be consistent with regard to the units used. In equation (4), we write

$$s = r\theta$$

To see the units, however, we must go back to equation (3) and write

$$\frac{\theta \text{ radians}}{1 \text{ radian}} = \frac{s \text{ length units}}{r \text{ length units}}$$

$$s \text{ length units} = (r \text{ length units})\frac{\theta \text{ radians}}{1 \text{ radian}}$$

Since the radians cancel, we are left with

$$s \text{ length units} = (r \text{ length units})\theta \quad \text{\scriptsize $s = r\theta$}$$

where θ appears to be "dimensionless" but, in fact, is measured in radians. Thus, in using the formula $s = r\theta$, the dimension of radians for θ is usually omitted, and any convenient unit of length (such as inches or meters) may be used for s and r.

E X A M P L E   3

Finding the Length of Arc of a Circle

Find the length of the arc of a circle of radius 2 meters subtended by a central angle of 0.25 radian.

Solution

We use equation (4) with $r = 2$ meters and $\theta = 0.25$. The length s of the arc is

$$s = r\theta = 2(0.25) = 0.5 \text{ meter}$$

∎

■   Now work Problem 33.

FIGURE 12

1 revolution $= 2\pi$ radians

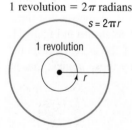

Relationship between Degrees and Radians

Consider a circle of radius r. A central angle of 1 revolution will subtend an arc equal to the circumference of the circle (Figure 12). Because the circumference of a circle equals $2\pi r$, we use $s = 2\pi r$ in equation (4) to find that, for an angle θ of 1 revolution,

$$s = r\theta$$

$$2\pi r = r\theta$$

$$\theta = 2\pi \text{ radians}$$

Thus,

$$1 \text{ revolution} = 2\pi \text{ radians} \qquad (5)$$

so that

$$360° = 2\pi \text{ radians}$$

or

$$180° = \pi \text{ radians} \qquad (6)$$

Divide both sides of equation (6) by 180. Then

$$1 \text{ degree} = \frac{\pi}{180} \text{ radian}$$

Divide both sides of (6) by π. Then

$$\frac{180}{\pi} \text{ degrees} = 1 \text{ radian}$$

Thus, we have the following two conversion formulas:

$$1 \text{ degree} = \frac{\pi}{180} \text{ radian} \qquad 1 \text{ radian} = \frac{180}{\pi} \text{ degrees} \qquad (7)$$

EXAMPLE 4   *Converting from Degrees to Radians*

Convert each angle in degrees to radians:

(a) 60°   (b) 150°   (c) −45°   (d) 90°

Solution   (a) $60° = 60 \cdot 1 \text{ degree} = 60 \cdot \dfrac{\pi}{180} \text{ radian} = \dfrac{\pi}{3} \text{ radians}$

(b) $150° = 150 \cdot \dfrac{\pi}{180} \text{ radian} = \dfrac{5\pi}{6} \text{ radians}$

(c) $-45° = -45 \cdot \dfrac{\pi}{180} \text{ radian} = -\dfrac{\pi}{4} \text{ radian}$

(d) $90° = 90 \cdot \dfrac{\pi}{180} \text{ radian} = \dfrac{\pi}{2} \text{ radians}$ ∎

■   Now work Problem 13.

Example 4 illustrates that angles that are fractions of a revolution are expressed in radian measure as fractional multiples of π, rather than as decimals. Thus, a right angle, as in Example 4(d), is left in the form $\pi/2$ radians, which is exact, rather than using the approximation $\pi/2 \approx 3.1416/2 = 1.5708$ radians.

E X A M P L E   5   *Converting Radians to Degrees*

Convert each angle in radians to degrees.

(a) $\dfrac{\pi}{6}$ radian   (b) $\dfrac{3\pi}{2}$ radians   (c) $-\dfrac{3\pi}{4}$ radians   (d) $\dfrac{7\pi}{3}$ radians

Solution   (a) $\dfrac{\pi}{6}$ radian $= \dfrac{\pi}{6} \cdot 1$ radian $= \dfrac{\pi}{6} \cdot \dfrac{180}{\pi}$ degrees $= 30°$

(b) $\dfrac{3\pi}{2}$ radians $= \dfrac{3\pi}{2} \cdot \dfrac{180}{\pi}$ degrees $= 270°$

(c) $-\dfrac{3\pi}{4}$ radians $= -\dfrac{3\pi}{4} \cdot \dfrac{180}{\pi}$ degrees $= -135°$

(d) $\dfrac{7\pi}{3}$ radians $= \dfrac{7\pi}{3} \cdot \dfrac{180}{\pi}$ degrees $= 420°$ ■

■ Now work Problem 23.

Table 1 lists the degree and radian measures of some commonly encountered angles. You should learn to feel equally comfortable using degree or radian measure for these angles.

TABLE 1

DEGREES	0°	30°	45°	60°	90°	120°	135°	150°	180°
RADIANS	0	$\dfrac{\pi}{6}$	$\dfrac{\pi}{4}$	$\dfrac{\pi}{3}$	$\dfrac{\pi}{2}$	$\dfrac{2\pi}{3}$	$\dfrac{3\pi}{4}$	$\dfrac{5\pi}{6}$	π
DEGREES	210°	225°	240°	270°	300°	315°	330°	360°	
RADIANS	$\dfrac{7\pi}{6}$	$\dfrac{5\pi}{4}$	$\dfrac{4\pi}{3}$	$\dfrac{3\pi}{2}$	$\dfrac{5\pi}{3}$	$\dfrac{7\pi}{4}$	$\dfrac{11\pi}{6}$	2π	

E X A M P L E   6   *Finding the Length of Arc of a Circle*

Find the length of the arc of a circle of radius $r = 3$ feet subtended by a central angle of 30°.

Solution   We use equation (4), but first we must convert the central angle of 30° to radians. Since $30° = \pi/6$ radian, we use $\theta = \pi/6$ and $r = 3$ feet in equation (4). The length of the arc is

$$s = r\theta = 3 \cdot \dfrac{\pi}{6} = \dfrac{\pi}{2} \approx \dfrac{3.14}{2} = 1.57 \text{ feet}$$ ■

When an angle is measured in degrees, the degree symbol always will be shown. However, when an angle is measured in radians, we will follow the usual practice and omit the word *radians*. Thus, if the measure of an angle is given as $\pi/6$, it is understood to mean $\pi/6$ radian.

Circular Motion

We have already defined the average speed of an object as the distance traveled divided by the elapsed time. Suppose that an object moves along a circle of radius r at a constant speed. If s is the distance traveled in time t along this circle, then the **linear speed** v of the object is defined as

FIGURE 13

$v = \dfrac{s}{t}$,

$\omega = \dfrac{\theta}{t}$

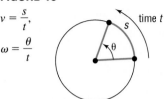

$$v = \frac{s}{t} \qquad (8)$$

As this object travels along the circle, suppose that θ (measured in radians) is the central angle swept out in time t (See Figure 13.). Then the **angular speed** ω (the Greek letter omega) of this object is the angle (measured in radians) swept out divided by the elapsed time; that is,

$$\omega = \frac{\theta}{t} \qquad (9)$$

Angular speed is the way the speed of a phonograph record is described. For example, a 45 rpm (revolutions per minute) record is one that rotates at an angular speed of

$$\frac{45 \text{ revolutions}}{\text{Minute}} = \frac{45 \text{ revolutions}}{\text{Minute}} \cdot \frac{2\pi \text{ radians}}{\text{Revolution}} = \frac{90\pi \text{ radians}}{\text{Minute}}$$

There is an important relationship between linear speed and angular speed. In the formula $s = r\theta$, divide each side by t:

$$\frac{s}{t} = r\frac{\theta}{t}$$

Then, using equations (8) and (9), we obtain

$$v = r\omega \qquad (10)$$

When using equation (10), remember that $v = s/t$ (the linear speed) has the dimensions of length per unit of time (such as feet per second or miles per hour), r (the radius of the circular motion) has the same length dimension as s, and ω (the angular speed) has the dimensions of radians per unit of time. As noted earlier, we leave the radian dimension off the numerical value of the angular speed ω so that both sides of the equation will be dimensionally consistent (with "length per unit of time"). If the angular speed is given in terms of *revolutions* per unit of time (as is often the case), be sure to convert it to *radians* per unit of time before attempting to use equation (10).

E X A M P L E 7   *Finding Linear Speed*

Find the linear speed of a $33\frac{1}{3}$ rpm record at the point where the needle is 3 inches from the spindle (center of the record).

Solution   Look at Figure 14. The point P is traveling along a circle of radius $r = 3$ inches. The angular speed ω of the record is

FIGURE 14

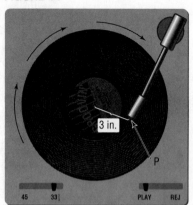

3 in.

P

45   33⅓   PLAY   REJ

$$\omega = \frac{33\frac{1}{3} \text{ revolutions}}{\text{Minute}} = \frac{100 \text{ revolutions}}{3 \text{ minutes}} \cdot \frac{2\pi \text{ radians}}{\text{Revolution}}$$

$$= \frac{200\pi \text{ radians}}{3 \text{ minutes}}$$

From equation (10), the linear speed v of the point P is

$$v = r\omega = 3 \text{ inches} \cdot \frac{200\pi \text{ radians}}{3 \text{ minutes}} = \frac{200\pi \text{ inches}}{\text{Minute}} \approx \frac{628 \text{ inches}}{\text{Minute}}$$ ■

■   Now work Problem 71.

5.1

Exercise 5.1

In Problems 1–12, draw each angle.

1. 30°	**2.** 60°	**3.** 135°	**4.** −120°	**5.** 450°	**6.** 540°
7. $3\pi/4$	**8.** $4\pi/3$	**9.** $-\pi/6$	**10.** $-2\pi/3$	**11.** $16\pi/3$	**12.** $21\pi/4$

In Problems 13–22, convert each angle in degrees to radians. Express your answer as a multiple of π.

13. 30°	**14.** 120°	**15.** 240°	**16.** 330°	**17.** −60°
18. −30°	**19.** 180°	**20.** 270°	**21.** 135°	**22.** −225°

In Problems 23–32, convert each angle in radians to degrees.

23. $\pi/3$	**24.** $5\pi/6$	**25.** $-5\pi/4$	**26.** $-2\pi/3$	**27.** $\pi/2$
28. 4π	**29.** $\pi/12$	**30.** $5\pi/12$	**31.** $2\pi/3$	**32.** $5\pi/4$

In Problems 33–40, s denotes the length of arc of a circle of radius r subtended by the central angle θ. Find the missing quantity.

33. $r = 10$ meters, $\theta = \frac{1}{2}$ radian, $s = ?$ **34.** $r = 6$ feet, $\theta = 2$ radians, $s = ?$

35. $\theta = \frac{1}{3}$ radian, $s = 2$ feet, $r = ?$ **36.** $\theta = \frac{1}{4}$ radian, $s = 6$ centimeters, $r = ?$

37. $r = 5$ miles, $s = 3$ miles, $\theta = ?$ **38.** $r = 6$ meters, $s = 8$ meters, $\theta = ?$

39. $r = 2$ inches, $\theta = 30°$, $s = ?$ **40.** $r = 3$ meters, $\theta = 120°$, $s = ?$

In Problems 41–48, convert each angle in degrees to radians. Express your answer in decimal form, rounded to two decimal places.

41. 17°	**42.** 73°	**43.** −40°	**44.** −51°	**45.** 125°	**46.** 200°	**47.** 340°	**48.** 350°

In Problems 49–56, convert each angle in radians to degrees. Express your answer in decimal form, rounded to two decimal places.

49. 3.14	**50.** π	**51.** 10.25	**52.** 0.75	**53.** 2	**54.** 3	**55.** 6.32	**56.** $\sqrt{2}$

In Problems 57–62, convert each angle to a decimal in degrees. Round off your answer to two decimal places.

57. 40°10′25″   **58.** 61°42′21″   **59.** 1°2′3″   **60.** 73°40′40″   **61.** 9°9′9″   **62.** 98°22′45″

In Problems 63–68, convert each angle to D°M′S″ *form. Round off your answer to the nearest second.*

63. 40.32°   **64.** 61.24°   **65.** 18.255°   **66.** 29.411°   **67.** 19.99°   **68.** 44.01°

69. *Minute Hand of a Clock* The minute hand of a clock is 6 inches long. How far does the tip of the minute hand move in 15 minutes? How far does it move in 25 minutes?

70. *Movement of a Pendulum* A pendulum swings through an angle of 20° each second. If the pendulum is 40 inches long, how far does its tip move each second?

71. An object is traveling around a circle with a radius of 5 centimeters. If in 20 seconds a central angle of $\frac{1}{3}$ radian is swept out, what is the angular speed of the object? What is its linear speed?

72. An object is traveling around a circle with a radius of 2 meters. If in 20 seconds the object travels 5 meters, what is its angular speed? What is its linear speed?

73. *Bicycle Wheels* The diameter of each wheel of a bicycle is 26 inches. If you are traveling at a speed of 35 miles per hour on this bicycle, through how many revolutions per minute are the wheels turning?

74. *Car Wheels* The radius of each wheel of a car is 15″. If the wheels are turning at the rate of 3 revolutions per second, how fast is the car moving? Express your answer in inches per second and in miles per hour.

75. *Windshield Wipers* The windshield wiper of a car is 18 inches long. How many inches will the tip of the wiper trace out in $\frac{1}{3}$ revolution?

76. *Windshield Wipers* The windshield wiper of a car is 18 inches long. If it takes 1 second to trace out $\frac{1}{3}$ revolution, how fast is the tip of the wiper moving?

77. *Speed of the Moon* The mean distance of the Moon from Earth is 2.39×10^5 miles. Assuming that the orbit of the Moon around Earth is circular and that 1 revolution takes 27.3 days, find the linear speed of the Moon. Express your answer in miles per hour.

78. *Speed of the Earth* The mean distance of Earth from the Sun is 9.29×10^7 miles. Assuming that the orbit of Earth around the Sun is circular and that 1 revolution takes 365 days, find the linear speed of Earth. Express your answer in miles per hour.

79. *Pulleys* Two pulleys, one with radius 2 inches and the other with radius 8 inches, are connected by a belt. (See the figure.) If the 2 inch pulley is caused to rotate at 3 revolutions per minute, determine the revolutions per minute of the 8 inch pulley. [*Hint:* The linear speeds of the pulleys, that is, the speed of the belt, are the same.]

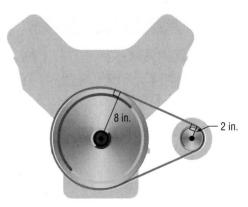

80. *Pulleys* Two pulleys, one with radius r_1 and the other with radius r_2, are connected by a belt. The pulley with radius r_1 rotates at ω_1 revolutions per minute, whereas the pulley with radius r_2 rotates at ω_2 revolutions per minute. Show that $r_1/r_2 = \omega_2/\omega_1$.

81. *Computing the Speed of River Current* To approximate the speed of the current of a river, a circular paddle wheel with radius 4 feet is lowered into the water. If the current causes the wheel to rotate at a speed of 10 revolutions per minute, what is the speed of the current? Express your answer in miles per hour.

82. *Spin Balancing Tires* A spin balancer rotates the wheel of a car at 480 revolutions per minute. If the diameter of the wheel is 26 inches, what road speed is being tested? Express your answer in miles per hour. At how many revolutions per minute should the balancer be set to test a road speed of 80 miles per hour?

83. *Nautical Miles* A **nautical mile** equals the length of arc subtended by a central angle of 1 minute on a great circle* on the surface of Earth. (See the figure.) If the radius of Earth is taken as 3960 miles, express 1 nautical mile in terms of ordinary, or **statute,** miles (5280 feet).

84. *Difference in Time of Sun Rise* Naples, Florida, is approximately 90 miles due west of Ft. Lauderdale. How much sooner would a person in Ft. Lauderdale first see the rising Sun than a person in Naples? [*Hint:* Consult the figure. When a person at Q sees the first rays of the Sun, a person at P is still in the dark. The person at P sees the first rays after Earth has rotated so that P is at the location Q. Now use the fact that in 24 hours a length of arc of $2\pi(3960)$ miles is subtended.]

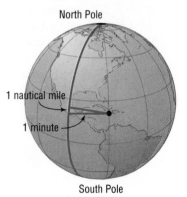

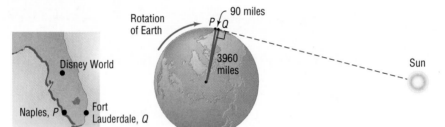

85. *Keeping up with the Sun* How fast would you have to travel on the surface of Earth to keep up with the Sun (that is, so that the Sun would appear to remain in the same position in the sky)?

86. Do you prefer to measure angles using degrees or radians? Provide justification and a rationale for your choice.

87. Discuss why ships and airplanes use nautical miles to measure distance. Explain the difference between a nautical mile and a statute mile.

88. Investigate the way speed bicycles work. In particular, explain the differences and similarities between 10-speed and 18-speed derailleurs. Be sure to include a discussion of linear speed and angular speed.

5.2

Trigonometric Functions: Unit Circle Approach

We are now ready to introduce the trigonometric functions. As we said earlier, the approach taken uses the unit circle.

The Unit Circle

Recall that the **unit circle** is a circle whose radius is 1 and whose center is at the origin of a rectangular coordinate system. Because the radius r of the unit circle

---

*Any circle drawn on the surface of Earth that divides Earth into two equal hemispheres.

is 1, we see from the formula $s = r\theta$ that on the unit circle a central angle of θ radians subtends an arc whose length s is

$$s = \theta$$

FIGURE 15
Unit circle: $x^2 + y^2 = 1$

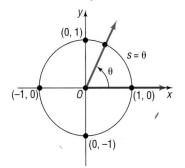

See Figure 15. Thus, on the unit circle, the length measure of the arc s equals the radian measure of the central angle θ. In other words, on the unit circle, the real number used to measure an angle θ in radians corresponds exactly with the real number used to measure the length of the arc subtended by that angle.

For example suppose that $r = 1$ foot. Then, if $\theta = 3$ radians, $s = 3$ feet; if $\theta = 8.2$ radians, then $s = 8.2$ feet; and so on.

Now, let t be any real number and let θ be the angle, in standard position, equal to t radians. Let P be the point on the unit circle that is also on the terminal side of θ. If $t \geq 0$, this point P is reached by moving *counterclockwise* along the unit circle, starting at $(1, 0)$, for a length of arc equal to t units. See Figure 16(a). If $t < 0$, this point P is reached by moving *clockwise* along the unit circle, starting at $(1, 0)$, for a length of arc equal to $|t|$ units. See Figure 16(b).

FIGURE 16

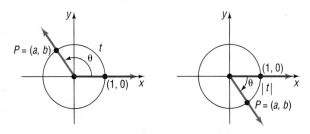

(a) $\theta = t$ radians; length of arc from $(1, 0)$ to P is t units, $t \geq 0$

(b) $\theta = t$ radians; length of arc from $(1, 0)$ to P is $|t|$ units, $t < 0$

Thus, to each real number t there corresponds a unique point $P = (a, b)$ on the unit circle. This is the important idea here. No matter what real number t is chosen, there corresponds a unique point P on the unit circle. We use the coordinates of the point $P = (a, b)$ on the unit circle corresponding to the real number t to define the **six trigonometric functions.**

Trigonometric Functions

Let t be a real number and let $P = (a, b)$ be the point on the unit circle that corresponds to t.

Sine Function

The **sine function** associates with t the y-coordinate of P and is denoted by

$$\sin t = b$$

Cosine Function	The **cosine function** associates with t the x-coordinate of P and is denoted by

$$\cos t = a$$

Tangent Function	If $a \neq 0$, the **tangent function** is defined as

$$\tan t = \frac{b}{a}$$

Cosecant Function	If $b \neq 0$, the **cosecant function** is defined as

$$\csc t = \frac{1}{b}$$

Secant Function	If $a \neq 0$, the **secant function** is defined as

$$\sec t = \frac{1}{a}$$

Cotangent Function	If $b \neq 0$, the **cotangent function** is defined as

$$\cot t = \frac{a}{b}$$

Notice in these definitions that if $a = 0$, that is, if the point $P = (0, b)$ is on the y-axis, then the tangent function and the secant function are undefined. Also, if $b = 0$, that is, if the point $P = (a, 0)$ is on the x-axis, then the cosecant function and the cotangent function are undefined.

Because we use the unit circle in these definitions of the trigonometric functions, they are also sometimes referred to as **circular functions.**

E X A M P L E   1   *Finding the Value of the Six Trigonometric Functions*

Let t be a real number and let $P = (-\frac{1}{2}, \sqrt{3}/2)$ be the point on the unit circle that corresponds to t. See Figure 17. Then

FIGURE 17

$\theta = t$ radians

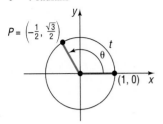

$$\sin t = \frac{\sqrt{3}}{2} \qquad \cos t = -\frac{1}{2} \qquad \tan t = \frac{\sqrt{3}/2}{-\frac{1}{2}} = -\sqrt{3}$$

$$\csc t = \frac{1}{\sqrt{3}/2} = \frac{2\sqrt{3}}{3} \qquad \sec t = \frac{1}{-\frac{1}{2}} = -2 \qquad \cot t = \frac{-\frac{1}{2}}{\sqrt{3}/2} = \frac{-\sqrt{3}}{3}$$ ∎

Trigonometric Functions of Angles

Let P be the point on the unit circle corresponding to the real number t. Then the angle θ, in standard position and measured in radians, whose terminal side is the ray from the origin through P is

$$\theta = t \text{ radians}$$

See Figure 18.

FIGURE 18

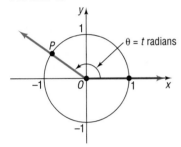

Thus, on the unit circle, the measure of the angle θ in radians equals the value of the real number t. As a result, we can say that

$$\underset{\underset{\text{Real number}}{\uparrow}}{\sin t} = \underset{\underset{\theta = t \text{ radians}}{\uparrow}}{\sin \theta}$$

and so on. We now can define the trigonometric functions of the angle θ.

> If $\theta = t$ radians, the **six trigonometric functions of the angle** θ are defined as
>
> | | | |
> |---|---|---|
> | $\sin \theta = \sin t$ | $\cos \theta = \cos t$ | $\tan \theta = \tan t$ |
> | $\csc \theta = \csc t$ | $\sec \theta = \sec t$ | $\cot \theta = \cot t$ |

Even though the distinction between trigonometric functions of real numbers and trigonometric functions of angles is important, it is customary to refer to trigonometric functions of real numbers and trigonometric functions of angles collectively as *the trigonometric functions*. We shall follow this practice from now on.

If an angle θ is measured in degrees, we shall use the degree symbol when writing a trigonometric function of θ, as, for example, in $\sin 30°$ and $\tan 45°$. If an angle θ is measured in radians, then no symbol is used when writing a trigonometric function of θ, as, for example, in $\cos \pi$ and $\sec \pi/3$.

Finally, since the values of the trigonometric functions of an angle θ are determined by the coordinates of the point $P = (a, b)$ on the unit circle corresponding to θ, the units used to measure the angle θ are irrelevant. For example, it does not matter whether we write $\theta = \pi/2$ radians or $\theta = 90°$. The point on the unit circle corresponding to this angle is $P = (0, 1)$. Hence,

$$\sin \frac{\pi}{2} = \sin 90° = 1 \quad \text{and} \quad \cos \frac{\pi}{2} = \cos 90° = 0$$

Evaluating the Trigonometric Functions

To find the exact value of a trigonometric function of an angle θ requires that we locate the corresponding point P on the unit circle. In fact, though, any circle whose center is at the origin can be used.

Let θ be any nonquadrantal angle placed in standard position. Let $P^* = (a^*, b^*)$ be the point where the terminal side of θ intersects the unit circle. See Figure 19.

FIGURE 19

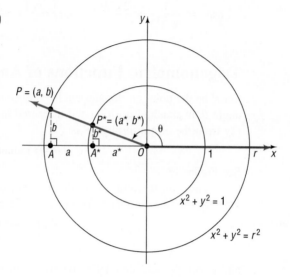

Let $P = (a, b)$ be any point on the terminal side of θ. Suppose that r is the distance from the origin to P. Then P is on the circle $x^2 + y^2 = r^2$. Refer again to Figure 19. Notice that the triangles OA^*P^* and OAP are similar; thus, ratios of corresponding sides are equal:

$$\frac{b^*}{1} = \frac{b}{r} \qquad \frac{a^*}{1} = \frac{a}{r} \qquad \frac{b^*}{a^*} = \frac{b}{a}$$

$$\frac{1}{b^*} = \frac{r}{b} \qquad \frac{1}{a^*} = \frac{r}{a} \qquad \frac{a^*}{b^*} = \frac{a}{b}$$

These results lead us to formulate the following theorem:

Theorem   For an angle θ in standard position, let $P = (a, b)$ be any point on the terminal side of θ. Let r equal the distance from the origin to P. Then

$$\sin \theta = \frac{b}{r} \qquad\qquad \cos \theta = \frac{a}{r} \qquad\qquad \tan \theta = \frac{b}{a}, \quad a \neq 0$$

$$\csc \theta = \frac{r}{b}, \quad b \neq 0 \qquad \sec \theta = \frac{r}{a}, \quad a \neq 0 \qquad \cot \theta = \frac{a}{b}, \quad b \neq 0$$

■

E X A M P L E   2   *Finding the Exact Value of the Six Trigonometric Functions*

Find the exact value of each of the six trigonometric functions of an angle θ if $(4, -3)$ is a point on its terminal side.

FIGURE 20

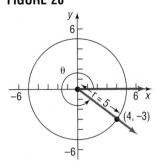

Solution   Figure 20 illustrates the situation for θ a positive angle. For the point $(a, b) = (4, -3)$, we have $a = 4$ and $b = -3$. Then $r = \sqrt{a^2 + b^2} = \sqrt{16 + 9} = 5$. Thus,

$$\sin\theta = \frac{b}{r} = -\frac{3}{5} \qquad \cos\theta = \frac{a}{r} = \frac{4}{5} \qquad \tan\theta = \frac{b}{a} = -\frac{3}{4}$$

$$\csc\theta = \frac{r}{b} = -\frac{5}{3} \qquad \sec\theta = \frac{r}{a} = \frac{5}{4} \qquad \cot\theta = \frac{a}{b} = -\frac{4}{3}$$ ■

■   Now work Problem 1.

To evaluate the trigonometric functions of a given angle θ in standard position, we need to find the coordinates of any point on the terminal side of that angle. This is not always so easy to do. In the examples that follow, we will evaluate the trigonometric functions of certain angles for which this process is relatively easy. A calculator will need to be used to evaluate trigonometric functions of most angles.

E X A M P L E   3   *Finding the Exact Value of the Six Trigonometric Functions of Quadrantal Angles*

Find the exact value of each of the trigonometric functions at

(a) $\theta = 0 = 0°$ (b) $\theta = \pi/2 = 90°$ (c) $\theta = \pi = 180°$

(d) $\theta = 3\pi/2 = 270°$

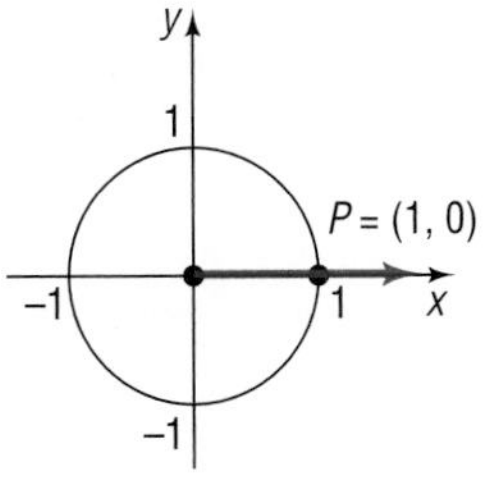

Solution   (a) The point $P = (1, 0)$ is on the terminal side of $\theta = 0 = 0°$ and is a distance of 1 unit from the origin. See Figure 21. Thus,

FIGURE 21
$\theta = 0 = 0°$

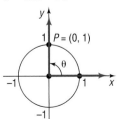

$$\sin 0 = \sin 0° = \frac{0}{1} = 0 \qquad \cos 0 = \cos 0° = \frac{1}{1} = 1$$

$$\tan 0 = \tan 0° = \frac{0}{1} = 0 \qquad \sec 0 = \sec 0° = \frac{1}{1} = 1$$

Since the y-coordinate of P is 0, $\csc 0$ and $\cot 0$ are not defined.

FIGURE 22
$\theta = \pi/2 = 90°$

(b) The point $P = (0, 1)$ is on the terminal side of $\theta = \pi/2 = 90°$ and is a distance of 1 unit from the origin. See Figure 22. Thus,

$$\sin\frac{\pi}{2} = \sin 90° = \frac{1}{1} = 1 \qquad \cos\frac{\pi}{2} = \cos 90° = \frac{0}{1} = 0$$

$$\csc\frac{\pi}{2} = \csc 90° = \frac{1}{1} = 1 \qquad \cot\frac{\pi}{2} = \cot 90° = \frac{0}{1} = 0$$

Since the x-coordinate of P is 0, $\tan \pi/2$ and $\sec \pi/2$ are not defined.

FIGURE 23

$\theta = \pi = 180°$

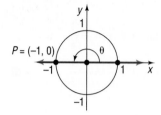

(c) The point $P = (-1, 0)$ is on the terminal side of $\theta = \pi = 180°$ and is a distance of 1 unit from the origin. See Figure 23. Thus,

$$\sin \pi = \sin 180° = \frac{0}{1} = 0 \qquad \cos \pi = \cos 180° = \frac{-1}{1} = -1$$

$$\tan \pi = \tan 180° = \frac{0}{1} = 0 \qquad \sec \pi = \sec 180° = \frac{1}{-1} = -1$$

Since the y-coordinate of P is 0, $\csc \pi$ and $\cot \pi$ are not defined.

FIGURE 24

$\theta = 3\pi/2 = 270°$

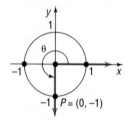

(d) The point $P = (0, -1)$ is on the terminal side of $\theta = 3\pi/2 = 270°$ and is a distance of 1 unit from the origin. See Figure 24. Thus,

$$\sin \frac{3\pi}{2} = \sin 270° = \frac{-1}{1} = -1 \qquad \cos \frac{3\pi}{2} = \cos 270° = \frac{0}{1} = 0$$

$$\csc \frac{3\pi}{2} = \csc 270° = \frac{1}{-1} = -1 \qquad \cot \frac{3\pi}{2} = \cot 270° = \frac{0}{-1} = 0$$

Since the x-coordinate of P is 0, $\tan 3\pi/2$ and $\sec 3\pi/2$ are not defined. ■

Note: The results obtained in Example 3 are the same whether we pick a point on the unit circle or a point on any circle whose center is at the origin. For example, $P = (0, 5)$ is a point on the terminal side of $\theta = \pi/2 = 90°$ and is a distance of 5 units from the origin. Using this point, $\sin \theta = \frac{5}{5} = 1$, $\cos \theta = \frac{0}{5} = 0$, and so on, as in Example 3(b).

Table 2 summarizes the values of the trigonometric functions found in Example 3.

TABLE 2   QUADRANTAL ANGLES

θ (RADIANS)	θ (DEGREES)	$\sin \theta$	$\cos \theta$	$\tan \theta$	$\csc \theta$	$\sec \theta$	$\cot \theta$
0	0°	0	1	0	Not defined	1	Not defined
$\pi/2$	90°	1	0	Not defined	1	Not defined	0
π	180°	0	-1	0	Not defined	-1	Not defined
$3\pi/2$	270°	-1	0	Not defined	-1	Not defined	0

■ Now work Problems 19 and 39.

E X A M P L E   4   *Finding Exact Values of Trigonometric Functions ($\theta = 45°$, $\theta = -45°$)*

Find the exact value of each of the trigonometric functions at:

(a) $\theta = \pi/4 = 45°$ (b) $\theta = -\pi/4 = -45°$

Solution   (a) We seek the coordinates of a point $P = (a, b)$ on the terminal side of $\theta = \pi/4 = 45°$. See Figure 25. First, we observe that P lies on the line $y = x$. (Do you see why? Since $\theta = 45° = \frac{1}{2} \cdot 90°$, P must lie on the line that bisects quadrant I.) Suppose that P also lies on the unit circle so that P is a distance of 1 unit from the origin. Then it follows that

FIGURE 25
$\theta = \pi/4 = 45°$

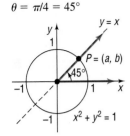

$$a^2 + b^2 = 1 \quad a = b, a > 0, b > 0$$
$$a^2 + a^2 = 1$$
$$2a^2 = 1$$
$$a = \frac{1}{\sqrt{2}} = \frac{\sqrt{2}}{2}, \qquad b = \frac{\sqrt{2}}{2}$$

Thus,

$$\sin\frac{\pi}{4} = \sin 45° = \frac{\sqrt{2}}{2} \qquad \cos\frac{\pi}{4} = \cos 45° = \frac{\sqrt{2}}{2} \qquad \tan\frac{\pi}{4} = \tan 45° = \frac{\sqrt{2}/2}{\sqrt{2}/2} = 1$$

$$\csc\frac{\pi}{4} = \csc 45° = \frac{1}{\sqrt{2}/2} = \sqrt{2} \qquad \sec\frac{\pi}{4} = \sec 45° = \frac{1}{\sqrt{2}/2} = \sqrt{2} \qquad \cot\frac{\pi}{4} = \cot 45° = \frac{\sqrt{2}/2}{\sqrt{2}/2} = 1$$

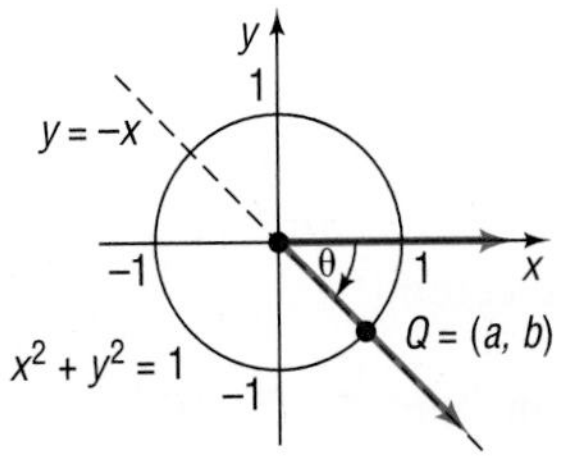

(b) We seek the coordinates of a point $Q = (a, b)$ on the terminal side of $\theta = -\pi/4 = -45°$. See Figure 26. First, we notice that Q lies on the line $y = -x$. If Q also lies on the unit circle $x^2 + y^2 = 1$, then

FIGURE 26
$\theta = -\pi/4 = -45°$

$$a^2 + b^2 = 1 \quad b = -a, a > 0$$
$$a^2 + (-a)^2 = 1$$
$$2a^2 = 1$$
$$a = \frac{1}{\sqrt{2}} = \frac{\sqrt{2}}{2}, \qquad b = -a = -\frac{\sqrt{2}}{2}$$

Thus,

$$\sin\left(-\frac{\pi}{4}\right) = \sin(-45°) \qquad \cos\left(-\frac{\pi}{4}\right) = \cos(-45°) \qquad \tan\left(-\frac{\pi}{4}\right) = \tan(-45°)$$

$$= -\frac{\sqrt{2}}{2} \qquad\qquad = \frac{\sqrt{2}}{2} \qquad\qquad = \frac{-\sqrt{2}/2}{\sqrt{2}/2} = -1$$

$$\csc\left(-\frac{\pi}{4}\right) = \csc(-45°) \qquad \sec\left(-\frac{\pi}{4}\right) = \sec(-45°) \qquad \cot\left(-\frac{\pi}{4}\right) = \cot(-45°)$$

$$= \frac{1}{-\sqrt{2}/2} = -\sqrt{2} \qquad\qquad = \frac{1}{\sqrt{2}/2} = \sqrt{2} \qquad\qquad = \frac{\sqrt{2}/2}{-\sqrt{2}/2} = -1$$

∎

In solving Example 4(b), we could have located the point Q by using the coordinates of $P = (\sqrt{2}/2, \sqrt{2}/2)$ from Example 4(a), noting the symmetry of Q and P with respect to the x-axis.

E X A M P L E   5   *Finding the Exact Value of a Trigonometric Expression*

Find the exact value of each expression:

(a) $\sin 45° \sin 180°$ (b) $\tan\frac{\pi}{4} - \sin\frac{3\pi}{2}$

FIGURE 27

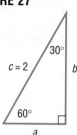

(a)

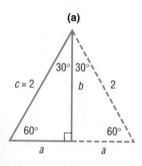

(b)

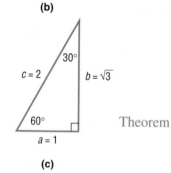

(c)

Solution

(a) $\sin 45° \cos 180° = \dfrac{\sqrt{2}}{2} \cdot (-1) = \dfrac{-\sqrt{2}}{2}$

From Example 4(a) ⤴   ⤴ From Table 2

(b) $\tan \dfrac{\pi}{4} - \sin \dfrac{3\pi}{2} = 1 - (-1) = 2$

From Example 4(a) ⤴   ⤴ From Table 2

■

■ Now work Problems 15 and 37.

Trigonometric Functions of 30° and 60°

Consider a right triangle in which one of the angles is 30°. It then follows that the other angle is 60°. Figure 27(a) illustrates such a triangle with hypotenuse of length 2. Our problem is to determine a and b.

We begin by placing next to this triangle another triangle congruent to the first, as shown in Figure 27(b). Notice that we now have a triangle whose angles are each 60°. This triangle is therefore equilateral, so each side is of length 2. In particular, the base is $2a = 2$, and so $a = 1$. By the Pythagorean Theorem, b satisfies the equation $a^2 + b^2 = c^2$, so we have

$$a^2 + b^2 = c^2$$
$$1^2 + b^2 = 2^2$$
$$b^2 = 4 - 1 = 3$$
$$b = \sqrt{3}$$

This results in Figure 27(c) and leads to the following theorem:

Theorem

In a 30, 60, 90 degree right triangle, the length of the side opposite the 30° angle is half the length of the hypotenuse. The length of the adjacent side is $\sqrt{3}/2$ times the length of the hypotenuse. ■

E X A M P L E   6

Finding Exact Values of the Trigonometric Functions ($\theta = 60°$)

Find the exact value of each of the trigonometric functions of $\pi/3 = 60°$.

Solution

We reposition the triangle in Figure 27(c) in a rectangular coordinate system. We seek the coordinates of the point $P = (a, b)$ on the terminal side of $\theta = \pi/3 = 60°$, a distance of 2 units from the origin. See Figure 28.

FIGURE 28

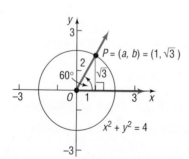

Based on the preceding theorem, it follows that $a = 1$ and $b = \sqrt{3}$. Since $r = 2$, we have

$$\sin \frac{\pi}{3} = \sin 60° = \frac{\sqrt{3}}{2} \qquad \cos \frac{\pi}{3} = \cos 60° = \frac{1}{2} \qquad \tan \frac{\pi}{3} = \tan 60° = \frac{\sqrt{3}}{1} = \sqrt{3}$$

$$\csc \frac{\pi}{3} = \csc 60° = \frac{2}{\sqrt{3}} = \frac{2\sqrt{3}}{3} \qquad \sec \frac{\pi}{3} = \sec 60° = \frac{2}{1} = 2 \qquad \cot \frac{\pi}{3} = \cot 60° = \frac{1}{\sqrt{3}} = \frac{\sqrt{3}}{3} \quad \blacksquare$$

E X A M P L E   7 *Finding Exact Values of the Trigonometric Functions ($\theta = 30°$)*

Find the exact value of each of the trigonometric functions of $\pi/6 = 30°$.

Solution Again, we reposition the triangle in Figure 27(c) in a rectangular coordinate system. We seek the coordinates of the point $P = (a, b)$ on the terminal side of $\theta = \pi/6 = 30°$, a distance of 2 units from the origin. See Figure 29. Based on the preceding theorem, it follows that $a = \sqrt{3}$ and $b = 1$. Since $r = 2$, we have

$$\sin \frac{\pi}{6} = \sin 30° = \frac{1}{2} \qquad \cos \frac{\pi}{6} = \cos 30° = \frac{\sqrt{3}}{2} \qquad \tan \frac{\pi}{6} = \tan 30° = \frac{1}{\sqrt{3}} = \frac{\sqrt{3}}{3}$$

$$\csc \frac{\pi}{6} = \csc 30° = \frac{2}{1} = 2 \qquad \sec \frac{\pi}{6} = \sec 30° = \frac{2}{\sqrt{3}} = \frac{2\sqrt{3}}{3} \qquad \cot \frac{\pi}{6} = \cot 30° = \frac{\sqrt{3}}{1} = \sqrt{3}$$

FIGURE 29

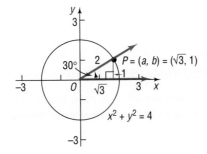

Table 3 summarizes the information just derived for $\pi/6$ (30°), $\pi/4$ (45°), and $\pi/3$ (60°). Until you memorize the entries in Table 3, you should draw an appropriate diagram to determine the values given in the table.

TABLE 3

θ (RADIANS)	θ (DEGREES)	$\sin \theta$	$\cos \theta$	$\tan \theta$	$\csc \theta$	$\sec \theta$	$\cot \theta$
$\pi/6$	30°	$\frac{1}{2}$	$\sqrt{3}/2$	$\sqrt{3}/3$	2	$2\sqrt{3}/3$	$\sqrt{3}$
$\pi/4$	45°	$\sqrt{2}/2$	$\sqrt{2}/2$	1	$\sqrt{2}$	$\sqrt{2}$	1
$\pi/3$	60°	$\sqrt{3}/2$	$\frac{1}{2}$	$\sqrt{3}$	$2\sqrt{3}/3$	2	$\sqrt{3}/3$

■ Now work Problems 21 and 31.

E X A M P L E   8 *Constructing a Rain Gutter*

A rain gutter is to be constructed of aluminum sheets 12 inches wide. After marking off a length of 4 inches from each edge, this length is bent up at an angle θ. See Figure 30. The area A of the opening can be expressed as a function of θ:

$$A(\theta) = 16 \sin \theta \, (\cos \theta + 1)$$

Find the area A of the opening for $\theta = 30°$, $\theta = 45°$, and $\theta = 60°$.

FIGURE 30

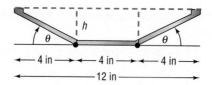

Solution

For $\theta = 30°$: $A(30°) = 16 \sin 30° (\cos 30° + 1)$

$$= 16\left(\frac{1}{2}\right)\left(\frac{\sqrt{3}}{2} + 1\right) = 4\sqrt{3} + 8$$

The area of the opening for $\theta = 30°$ is about 14.9 square inches.

For $\theta = 45°$: $A(45°) = 16 \sin 45° (\cos 45° + 1)$

$$= 16\left(\frac{\sqrt{2}}{2}\right)\left(\frac{\sqrt{2}}{2} + 1\right) = 8 + 8\sqrt{2}$$

The area of the opening for $\theta = 45°$ is about 19.3 square inches.

For $\theta = 60°$: $A(60°) = 16 \sin 60° (\cos 60° + 1)$

$$= 16\left(\frac{\sqrt{3}}{2}\right)(\tfrac{1}{2} + 1) = 12\sqrt{3}$$

The area of the opening for $\theta = 60°$ is about 20.8 square inches. ■

Using a Calculator to Find Values of Trigonometric Functions

Before getting started, you must first decide whether to enter the angle in the calculator using radians or degrees and then set the calculator to the correct mode. If your calculator does not display the mode, you can determine the current mode by entering 30 and then pressing the key marked $\boxed{\sin}$. If you are in the degree mode, the display will show $\boxed{0.5}$ (sin 30° = 0.5). If you are in the radian mode, the display will show $\boxed{-0.9880316}$. Most calculators have a key that allows you to change from one mode to the other. (Check your instruction manual to find out how your calculator handles degrees and radians.)

Your calculator probably only has the keys marked $\boxed{\sin}$, $\boxed{\cos}$, and $\boxed{\tan}$. To find the values of the remaining three trigonometric functions, secant, cosecant, and cotangent, we use the facts that

$$\sec \theta = \frac{1}{\cos \theta} \qquad \csc \theta = \frac{1}{\sin \theta} \qquad \cot \theta = \frac{1}{\tan \theta}$$

These facts are a direct consequence of the theorem stated on page 336.

E X A M P L E   9

Using a Calculator to Approximate the Value of a Trigonometric Function

Use a calculator to find the approximate value of

(a) $\cos 48°$ (b) $\csc 21°$ (c) $\tan \dfrac{\pi}{12}$

Round your answer to two decimal places.

Solution   (a) First, we set the mode to receive degrees.

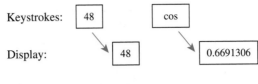

Thus,

$$\cos 48° \approx 0.67$$

rounded to two decimal places.

(b) Most calculators do not have a $\boxed{\text{CSC}}$ key. The manufacturers assume that the user knows some trigonometry. Thus, to find the value of csc 21°, we use the fact that csc 21° = 1/(sin 21°) and proceed as follows:

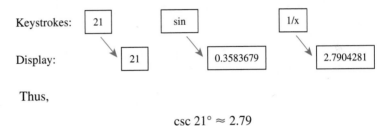

Thus,

$$\csc 21° \approx 2.79$$

rounded to two decimal places.

(c) Set the mode to receive radians. Then, to find tan(π/12),

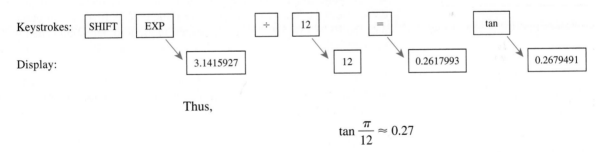

Thus,

$$\tan \frac{\pi}{12} \approx 0.27$$

rounded to two decimal places.   ■

■   Now work Problem 47.

Summary

For the quadrantal angles and the angles 30°, 45°, 60°, and their integral multiples, we can find the exact value of each trigonometric function by using the geometric features of these angles and symmetry.

Figure 31 shows points on the unit circle that are on the terminal sides of some angles that are integral multiples of π/6 (30°), π/4 (45°), and π/3 (60°).

Notice the relationship between integral multiples of an angle and symmetry. For example, in Figure 31(a), the point on the unit circle corresponding to 7π/4

FIGURE 31

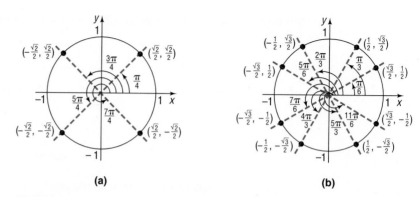

(a)   (b)

is the reflection about the x-axis of the point corresponding to $\pi/4$. The point on the unit circle corresponding to $5\pi/4$ is the reflection about the origin of the point corresponding to $\pi/4$. Thus, using Figure 31(a), we see that

$$\sin\frac{7\pi}{4} = -\frac{\sqrt{2}}{2} \qquad \cos\left(-\frac{\pi}{4}\right) = \frac{\sqrt{2}}{2} \qquad \tan\frac{5\pi}{4} = \frac{-\sqrt{2}/2}{-\sqrt{2}/2} = 1$$

Using Figure 31(b), we see that

$$\sin\frac{11\pi}{6} = -\frac{1}{2} \qquad \cos\frac{7\pi}{6} = -\frac{\sqrt{3}}{2} \qquad \tan\frac{4\pi}{3} = \frac{-\sqrt{3}/2}{-\frac{1}{2}} = \sqrt{3}$$

For most other angles, besides the quadrantal angles and those listed in Figure 31, we can only approximate the value of each trigonometric function using a calculator.

HISTORICAL FEATURE

FIGURE 32

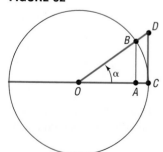

■ The name *sine* for the sine function is due to a medieval confusion. The name comes from the Sanskrit word *jīva* (meaning chord), first used in India by Aryabhata the Elder (AD 510). He really meant half-chord, but abbreviated it. This was brought into Arabic as *jība,* which was meaningless. Because the proper Arabic word *jaib* would be written the same way (short vowels are not written out in Arabic), *jība* was pronounced as *jaib,* which meant bosom or hollow, and *jaib* remains as the Arabic word for sine to this day. Scholars translating the Arabic works into Latin found that the word *sinus* also meant bosom or hollow, and from *sinus* we get the word *sine.*

The name *tangent,* due to Thomas Fincke (1583), can be understood by looking at Figure 32. The line segment $\overline{DC}$ is tangent to the circle at C. If $d(O, B) = d(O, C) = 1$, then the length of the line segment $\overline{DC}$ is

$$d(D, C) = \frac{d(D, C)}{1} = \frac{d(D, C)}{d(O, C)} = \tan\alpha$$

The old name for the tangent is *umbra versa* (meaning turned shadow), referring to the use of the tangent in solving height problems with shadows.

The names of the remaining functions came about as follows. If α and β are complementary angles, then $\cos\alpha = \sin\beta$. Because β is the complement of α, it was natural to write the cosine of α as *sin co α*. Probably for reasons involving ease of pronunciation, the *co* migrated to the front, and then cosine received a three-letter abbreviation to match sin, sec, and tan. The two other cofunctions were similarly treated, except that the long forms *cotan* and *cosec* survive to this day in some countries. ■

5.2

Exercise 5.2

In Problems 1–10, a point on the terminal side of an angle θ is given. Find the exact value of each of the six trigonometric functions of θ.

1. $(-3, 4)$ **2.** $(5, -12)$ **3.** $(2, -3)$ **4.** $(-1, -2)$ **5.** $(-2, -2)$

6. $(1, -1)$ **7.** $(-3, -2)$ **8.** $(2, 2)$ **9.** $(\frac{1}{3}, -\frac{1}{4})$ **10.** $(-0.3, -0.4)$

In Problems 11–30, find the exact value of each expression. Do not use a calculator.

11. $\sin 45° + \cos 60°$

12. $\sin 30° - \cos 45°$

13. $\sin 90° + \tan 45°$

14. $\cos 180° - \sin 180°$

15. $\sin 45° \cos 45°$

16. $\tan 45° \cos 30°$

17. $\csc 45° \tan 60°$

18. $\sec 30° \cot 45°$

19. $4 \sin 90° - 3 \tan 180°$

20. $5 \cos 90° - 8 \sin 270°$

21. $2 \sin \dfrac{\pi}{3} - 3 \tan \dfrac{\pi}{6}$

22. $2 \sin \dfrac{\pi}{4} + 3 \tan \dfrac{\pi}{4}$

23. $\sin \dfrac{\pi}{4} - \cos \dfrac{\pi}{4}$

24. $\tan \dfrac{\pi}{3} + \cos \dfrac{\pi}{3}$

25. $2 \sec \dfrac{\pi}{4} + 4 \cot \dfrac{\pi}{3}$

26. $3 \csc \dfrac{\pi}{3} + \cot \dfrac{\pi}{4}$

27. $\tan \pi - \cos 0$

28. $\sin \dfrac{3\pi}{2} + \tan \pi$

29. $\csc \dfrac{\pi}{2} + \cot \dfrac{\pi}{2}$

30. $\sec \pi - \csc \dfrac{\pi}{2}$

In Problems 31–46, find the exact value of each of the six trigonometric functions of the given angle. If any are not defined, say "not defined." Do not use a calculator.

31. $2\pi/3$ **32.** $3\pi/4$ **33.** $150°$ **34.** $330°$

35. $-\pi/6$ **36.** $-\pi/3$ **37.** $225°$ **38.** $210°$

39. $5\pi/2$ **40.** 3π **41.** $-180°$ **42.** $-270°$

43. $3\pi/2$ **44.** $-\pi$ **45.** $450°$ **46.** $-90°$

In Problems 47–70, use a calculator to find the approximate value of each expression rounded to two decimal places.

47. $\sin 28°$ **48.** $\cos 14°$ **49.** $\tan 21°$ **50.** $\sin 15°$

51. $\sec 41°$ **52.** $\csc 55°$ **53.** $\cot 70°$ **54.** $\tan 80°$

55. $\sin \dfrac{\pi}{10}$ **56.** $\cos \dfrac{\pi}{8}$ **57.** $\tan \dfrac{5\pi}{12}$ **58.** $\sin \dfrac{3\pi}{10}$

59. $\sec \dfrac{\pi}{12}$ **60.** $\csc \dfrac{5\pi}{13}$ **61.** $\cot \dfrac{\pi}{18}$ **62.** $\sin \dfrac{\pi}{18}$

63. $\sin 1$ **64.** $\tan 1$ **65.** $\sin 1°$ **66.** $\tan 1°$

67. $\cos 21.5°$ **68.** $\cos 35.2°$ **69.** $\tan 0.3$ **70.** $\tan 0.1$

In Problems 71–82, find the exact value of each expression if θ = 60°. Do not use a calculator.

71. $\sin \theta$ **72.** $\cos \theta$ **73.** $\sin \dfrac{\theta}{2}$ **74.** $\cos \dfrac{\theta}{2}$

75. $(\sin \theta)^2$ **76.** $(\cos \theta)^2$ **77.** $\sin 2\theta$ **78.** $\cos 2\theta$

79. $2 \sin \theta$ **80.** $2 \cos \theta$ **81.** $\dfrac{\sin \theta}{2}$ **82.** $\dfrac{\cos \theta}{2}$

83. Find the exact value of $\sin 45° + \sin 135° + \sin 225° + \sin 315°$.

84. Find the exact value of $\tan 60° + \tan 150°$.

85. If $\sin \theta = 0.1$, find $\sin(\theta + \pi)$.

86. If $\cos \theta = 0.3$, find $\cos(\theta + \pi)$.

87. If $\tan \theta = 3$, find $\tan(\theta + \pi)$.

88. If $\cot \theta = -2$, find $\cot(\theta + \pi)$.

89. If $\sin \theta = \frac{1}{5}$, find $\csc \theta$.

90. If $\cos \theta = \frac{2}{3}$, find $\sec \theta$.

The path of a projectile fired at an inclination θ to the horizontal with initial speed v_0 is a parabola (see the figure). The range R of the projectile, that is, the horizontal distance that the projectile travels, is found by using the formula

$$R = \frac{v_0^2 \sin 2\theta}{g}$$

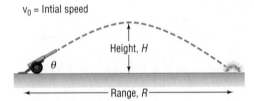

$v_0 =$ Intial speed

Height, H

θ

Range, R

where $g \approx 32.2$ feet per second per second ≈ 9.8 meters per second per second is the acceleration due to gravity. The maximum height H of the projectile is

$$H = \frac{v_0^2 \sin^2\theta}{2g}$$

In Problems 91–94, find the range R and maximum height H.

91. The projectile is fired at an angle of $45°$ to the horizontal with an initial speed of 100 feet per second.

92. The projectile is fired at an angle of $30°$ to the horizontal with an initial speed of 150 meters per second.

93. The projectile is fired at an angle of $25°$ to the horizontal with an initial speed of 500 meters per second.

94. The projectile is fired at an angle of $50°$ to the horizontal with an initial speed of 200 feet per second.

95. If friction is ignored, the time t (in seconds) required for a block to slide down an inclined plane (see the figure) is given by the formula

$$t = \sqrt{\frac{2a}{g \sin \theta \cos \theta}}$$

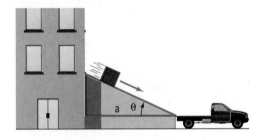

where a is the length (in feet) of the base and $g \approx 32$ feet per second per second is the acceleration of gravity. How long does it take a block to slide down an inclined plane with base $a = 10$ feet when
(a) $\theta = 30°$?   (b) $\theta = 45°$?   (c) $\theta = 60°$?

96. In a certain piston engine, the distance x (in meters) from the center of the drive shaft to the head of the piston is given by

$$x = \cos \theta + \sqrt{16 + 0.5 \cos 2\theta}$$

where θ is the angle between the crank and the path of the piston head (see the figure). Find x when $\theta = 30°$ and when $\theta = 45°$.

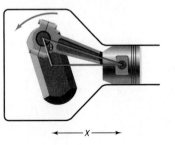

x

97. *Calculating the Time of a Trip* Two oceanfront homes are located 8 miles apart on a straight stretch of beach, each a distance of 1 mile from a paved road that parallels the ocean. Sally can walk 8 miles per hour along the paved road, but only 3 miles per hour in the sand that separates the road from the ocean. Because of a river directly between the two houses, it is necessary to walk in the sand to the road, continue on the road, and then walk directly back in the sand to get from one house to the other. See the illustration. The time T to get from one house to the other as a function of the angle θ shown in the illustration is

$$T(\theta) = 1 + \frac{2}{3 \sin \theta} - \frac{1}{4 \tan \theta}, \; 0° < \theta < 90°$$

(a) Calculate the time T for $\theta = 30°$. How long is Sally on the paved road?
(b) Calculate the time T for $\theta = 45°$. How long is Sally on the paved road?
(c) Calculate the time T for $\theta = 60°$. How long is Sally on the paved road?
(d) Calculate the time T for $\theta = 90°$. Describe the path taken. Why can't the formula for T be used?

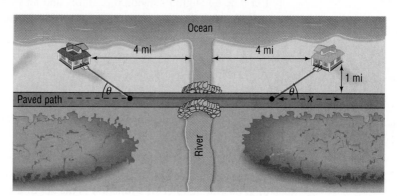

98. *Designing Fine Decorative Pieces* A designer of decorative art plans to market solid gold spheres encased in clear crystal cones. Each sphere is of fixed radius R and will be enclosed in a cone of height h and radius r. See the illustration. Many cones can be used to enclose the sphere, each having a different slant angle θ. The volume V of the cone can be expressed as a function of the slant angle θ of the cone:

$$V(\theta) = \frac{1}{3}\pi R^3 \frac{(1 + \sec \theta)^3}{\tan^2 \theta}, \; 0° < \theta < 90°$$

What volume V is required to enclose a sphere of radius 2 centimeters in a cone whose slant angle θ is 30°? 45°? 60°?

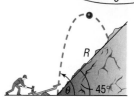

99. *Projectile Motion* An object is propelled upward at an angle θ, $45° < \theta < 90°$, to the horizontal with an initial velocity of v_0 feet per second from the base of a plane that makes an angle of 45° with the horizontal. See the illustration. If air resistance is ignored, the distance R it travels up the inclined plane is given by

$$R = \frac{v_0^2 \sqrt{2}}{32}(\sin 2\theta - \cos 2\theta - 1)$$

Find the distance R that the object travels along the inclined plane if the initial velocity is 32 feet per second and $\theta = 60°$.

100. If θ ($0 < \theta < \pi$) is the angle between a horizontal ray directed to the right (say, the positive x-axis) and a nonhorizontal, nonvertical line L, show that the slope m of L equals $\tan \theta$. The angle θ is called the **inclination** of L. [*Hint:* See the illustration, where we have drawn the line $L*$ parallel to L and passing through the origin. Use the fact that $L*$ intersects the unit circle at the point $(\cos \theta, \sin \theta)$.]

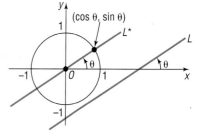

101. Write a brief paragraph that explains how to quickly compute the trigonometric functions of 30°, 45°, and 60°.

102. Write a brief paragraph that explains how to quickly compute the trigonometric functions of 0°, 90°, 180°, and 270°.

103. How would you explain the meaning of the sine function to a fellow student who has just completed college algebra?

5.3

Properties of the Trigonometric Functions

FIGURE 33

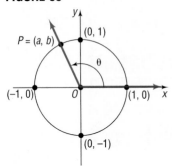

Domain and Range of the Trigonometric Functions

Let θ be an angle in standard position, and let $P = (a, b)$ be a point on the terminal side of θ. Suppose, for convenience, that P also lies on the unit circle. See Figure 33. Then, by definition,

$$\sin \theta = b \qquad \cos \theta = a \qquad \tan \theta = \frac{b}{a}, \quad a \neq 0$$

$$\csc \theta = \frac{1}{b}, \quad b \neq 0 \qquad \sec \theta = \frac{1}{a}, \quad a \neq 0 \qquad \cot \theta = \frac{a}{b}, \quad b \neq 0$$

For $\sin \theta$ and $\cos \theta$, θ can be any angle, so it follows that the domain of the sine function and cosine function is the set of all real numbers.

> The domain of the sine function is the set of all real numbers.
>
> The domain of the cosine function is the set of all real numbers.

If $a = 0$, then the tangent function and the secant function are not defined. Thus, for the tangent function and secant function the x-coordinate of $P = (a, b)$ cannot be 0. On the unit circle, there are two such points, $(0, 1)$ and $(0, -1)$. These two points correspond to the angles $\pi/2$ (90°) and $3\pi/2$ (270°) or, more generally, to any angle that is an odd multiple of $\pi/2$ (90°), such as $\pi/2$ (90°), $3\pi/2$ (270°), $5\pi/2$ (450°), $-\pi/2$ (−90°), $-3\pi/2$ (−270°), and so on. Such angles must therefore be excluded from the domain of the tangent function and secant function.

> The domain of the tangent function is the set of all real numbers, except odd multiples of $\pi/2$ (90°).
>
> The domain of the secant function is the set of all real numbers, except odd multiples of $\pi/2$ (90°).

If $b = 0$, then the cotangent function and the cosecant function are not defined. Thus, for the cotangent function and cosecant function the y-coordinate of $P = (a, b)$ cannot be 0. On the unit circle, there are two such points, $(1, 0)$ and $(-1, 0)$. These two points correspond to the angles 0 (0°) and π (180°) or, more generally, to any angle that is an integral multiple of π (180°), such as 0 (0°), π (180°), 2π (360°), 3π (540°), $-\pi$ (−180°), and so on. Such angles must therefore be excluded from the domain of the cotangent function and cosecant function.

> The domain of the cotangent function is the set of all real numbers, except integral multiples of π (180°).
>
> The domain of the cosecant function is the set of all real numbers, except integral multiples of π (180°).

Next, we determine the range of each of the six trigonometric functions. Refer again to Figure 33. Let $P = (a, b)$ be the point on the unit circle that corresponds to the angle θ. It follows that $-1 \leq a \leq 1$ and $-1 \leq b \leq 1$. Consequently, since $\sin \theta = b$ and $\cos \theta = a$, we have

$$-1 \leq \sin \theta \leq 1 \qquad -1 \leq \cos \theta \leq 1$$

Thus, the range of both the sine function and the cosine function consists of all real numbers between -1 and 1, inclusive. In terms of absolute value notation, we have $|\sin \theta| \leq 1$ and $|\cos \theta| \leq 1$.

Similarly, if θ is not a multiple of π (180°), then $\csc \theta = 1/b$. Since $b = \sin \theta$ and $|b| = |\sin \theta| \leq 1$, it follows that $|\csc \theta| = 1/|b| \geq 1$. Thus, the range of the cosecant function consists of all real numbers less than or equal to -1 or greater than or equal to 1. That is,

$$\csc \theta \leq -1 \quad \text{or} \quad \csc \theta \geq 1$$

If θ is not an odd multiple of $\pi/2$ (90°), then, by definition, $\sec \theta = 1/a$. Since $a = \cos \theta$ and $|a| = |\cos \theta| \leq 1$, it follows that $|\sec \theta| = 1/|a| \geq 1$. Thus, the range of the secant function consists of all real numbers less than or equal to -1 or greater than or equal to 1.

$$\sec \theta \leq -1 \quad \text{or} \quad \sec \theta \geq 1$$

The range of both the tangent function and the cotangent function consists of all real numbers. You are asked to prove this in Problems 91 and 92.

$$-\infty < \tan \theta < \infty \qquad -\infty < \cot \theta < \infty$$

Table 4 summarizes the preceding results.

TABLE 4

FUNCTION	SYMBOL	DOMAIN	RANGE
sine	$f(\theta) = \sin \theta$	All real numbers	All real numbers from -1 to 1, inclusive
cosine	$f(\theta) = \cos \theta$	All real numbers	All real numbers from -1 to 1, inclusive
tangent	$f(\theta) = \tan \theta$	All real numbers, except odd multiples of $\pi/2$ (90°)	All real numbers
cosecant	$f(\theta) = \csc \theta$	All real numbers, except integral multiples of π (180°)	All real numbers greater than or equal to 1 or less than or equal to -1
secant	$f(\theta) = \sec \theta$	All real numbers, except odd multiples of $\pi/2$ (90°)	All real numbers greater than or equal to 1 or less than or equal to -1
cotangent	$f(\theta) = \cot \theta$	All real numbers, except integral multiples of π (180°)	All real numbers

■   Now work Problem 85.

Period of the Trigonometric Functions

FIGURE 34

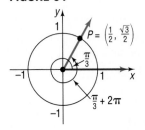

Look at Figure 34. This figure shows that, for an angle of $\pi/3$ radian, the corresponding point P on the unit circle is $(1/2, \sqrt{3}/2)$. Notice that, for an angle of $\pi/3 + 2\pi$ radians, the corresponding point P on the unit circle is also $(1/2, \sqrt{3}/2)$. Thus,

$$\sin \frac{\pi}{3} = \frac{\sqrt{3}}{2} \quad \text{and} \quad \sin\left(\frac{\pi}{3} + 2\pi\right) = \frac{\sqrt{3}}{2}$$

$$\cos \frac{\pi}{3} = \frac{1}{2} \quad \text{and} \quad \cos\left(\frac{\pi}{3} + 2\pi\right) = \frac{1}{2}$$

This example illustrates a more general situation. For a given angle θ, measured in radians, suppose we know that the corresponding point $P = (a, b)$ on the unit circle. Now add 2π to θ. The point on the unit circle corresponding to $\theta + 2\pi$ is identical to the point P on the unit circle corresponding to θ. See Figure 35. Thus, the values of the trigonometric functions of $\theta + 2\pi$ are equal to the values of the corresponding trigonometric functions of θ.

If we add (or subtract) integral multiples of 2π to θ, the trigonometric values remain unchanged. That is, for all θ,

FIGURE 35

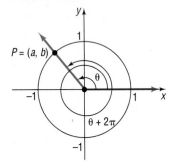

$$\sin(\theta + 2\pi k) = \sin \theta \qquad \cos(\theta + 2\pi k) = \cos \theta \tag{1}$$

where k is any integer.

Functions that exhibit this kind of behavior are called *periodic functions*.

Seeing the Concept To see the periodic behavior of the sine function, graph $y = \sin x$, $y = \sin(x + 2\pi)$, $y = \sin(x - 2\pi)$, and $y = \sin(x + 4\pi)$ on the same screen.

Periodic Function	A function *f* is called **periodic** if there is a positive number *p* such that, whenever θ is in the domain of *f*, so is $\theta + p$, and

$$f(\theta + p) = f(\theta)$$

Period	If there is a smallest such number *p*, this smallest value is called the **(fundamental) period** of *f*.

Thus, based on equation (1), the sine and cosine functions are periodic. In fact, the sine and cosine functions have period 2π. You are asked to prove this fact in Problems 93 and 94. The secant and cosecant functions are also periodic with period 2π; the tangent and cotangent functions are periodic with period π. You are asked to prove these statements in Problems 95 through 98.

Periodic Properties

$$\sin(\theta + 2\pi k) = \sin\theta \qquad \cos(\theta + 2\pi k) = \cos\theta \qquad \tan(\theta + \pi k) = \tan\theta$$

$$\csc(\theta + 2\pi k) = \csc\theta \qquad \sec(\theta + 2\pi k) = \sec\theta \qquad \cot(\theta + \pi k) = \cot\theta$$

where k is any integer.

Because the sine, cosine, secant, and cosecant functions have period 2π, once we know their values for $0 \le \theta < 2\pi$, we know all their values; similarly, since the tangent and cotangent functions have period π, once we know their values for $0 \le \theta < \pi$, we know all their values.

E X A M P L E   1

FIGURE 36

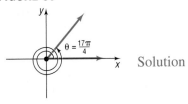

(a)

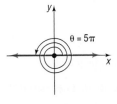

(b)

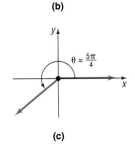

(c)

Finding Exact Values Using Periodic Properties

Find the exact value of

(a) $\sin\dfrac{17\pi}{4}$ (b) $\cos 5\pi$ (c) $\tan\dfrac{5\pi}{4}$

Solution   (a) It is best to sketch the angle first, as shown in Figure 36(a). Since the period of the sine function is 2π, each full revolution can be ignored. This leaves the angle $\pi/4$. Thus,

$$\sin\frac{17\pi}{4} = \sin\!\left(\frac{\pi}{4} + 4\pi\right) = \sin\frac{\pi}{4} = \frac{\sqrt{2}}{2}$$

(b) See Figure 36(b). Since the period of the cosine function is 2π, each full revolution can be ignored. This leaves the angle π. Thus,

$$\cos 5\pi = \cos(\pi + 4\pi) = \cos\pi = -1$$

(c) See Figure 36(c). Since the period of the tangent function is π, each half revolution can be ignored. This leaves the angle $\pi/4$. Thus,

$$\tan\frac{5\pi}{4} = \tan\!\left(\frac{\pi}{4} + \pi\right) = \tan\frac{\pi}{4} = 1 \qquad \blacksquare$$

The periodic properties of the trigonometric functions will be very helpful to us when we study their graphs in the next chapter.

■   Now work Problem 1.

The Signs of the Trigonometric Functions

FIGURE 37

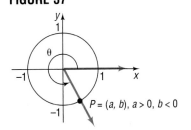

Let $P = (a, b)$ be the point on the unit circle that corresponds to the angle θ. If we know in which quadrant the point P lies, then we can determine the signs of the trigonometric functions of θ. For example, if $P = (a, b)$ lies in quadrant IV, as shown in Figure 37, then we know that $a > 0$ and $b < 0$. Consequently,

$$\sin\theta = b < 0 \qquad \cos\theta = a > 0 \qquad \tan\theta = \frac{b}{a} < 0$$

$$\csc\theta = \frac{1}{b} < 0 \qquad \sec\theta = \frac{1}{a} > 0 \qquad \cot\theta = \frac{a}{b} < 0$$

Table 5 lists the signs of the six trigonometric functions for each quadrant. See also Figure 38.

TABLE 5

QUADRANT OF P	sin θ, csc θ	cos θ, sec θ	tan θ, cot θ
I	Positive	Positive	Positive
II	Positive	Negative	Negative
III	Negative	Negative	Positive
IV	Negative	Positive	Negative

FIGURE 38

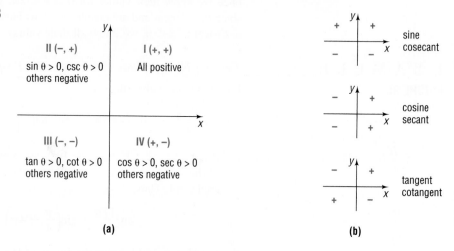

(a)   (b)

E X A M P L E   2

Finding the Quadrant an Angle θ Lies In

If sin $\theta < 0$ and cos $\theta < 0$, name the quadrant in which the angle θ lies.

Solution   Let $P = (a, b)$ be the point on the unit circle corresponding to θ. Then sin $\theta = b < 0$ and cos $\theta = a < 0$. Thus, $P = (a, b)$ must be in quadrant III, so θ lies in quadrant III.   ■

■   Now work Problem 17.

Fundamental Identities

If $P = (a, b)$ is a point on the unit circle corresponding to θ, then

$$\sin \theta = b \qquad\qquad \cos \theta = a \qquad\qquad \tan \theta = \frac{b}{a}, \quad \text{if } a \neq 0$$

$$\csc \theta = \frac{1}{b}, \quad \text{if } b \neq 0 \qquad \sec \theta = \frac{1}{a}, \quad \text{if } a \neq 0 \qquad \cot \theta = \frac{a}{b}, \quad \text{if } b \neq 0$$

Thus, we have the **reciprocal identities:**

Reciprocal Identities

$$\csc \theta = \frac{1}{\sin \theta} \qquad \sec \theta = \frac{1}{\cos \theta} \qquad \cot \theta = \frac{1}{\tan \theta} \qquad (2)$$

Two other fundamental identities that are easy to see are the **quotient identities:**

Quotient Identities

$$\tan \theta = \frac{\sin \theta}{\cos \theta} \qquad \cot = \frac{\cos \theta}{\sin \theta} \tag{3}$$

The proofs of formulas (2) and (3) follow from the definitions of the trigonometric functions. (See Problems 99 and 100.)

Seeing the Concept To see the identity $\tan \theta = (\sin \theta)/(\cos \theta)$, graph $y = \tan x$ and $y = (\sin x)/(\cos x)$ on the same screen.

If $\sin \theta$ and $\cos \theta$ are known, formulas (2) and (3) make it easy to find the values of the remaining trigonometric functions.

E X A M P L E   3   *Finding Exact Values Using Identities When Sine and Cosine Are Given*

Given $\sin \theta = 1/\sqrt{5}$ and $\cos \theta = 2/\sqrt{5}$, find the exact values of the four remaining trigonometric functions of θ.

Solution   Based on a quotient identity from formula (3), we have

$$\tan \theta = \frac{\sin \theta}{\cos \theta} = \frac{1/\sqrt{5}}{2/\sqrt{5}} = \frac{1}{2}$$

Then we use the reciprocal identities from formula (2) to get

$$\csc \theta = \frac{1}{\sin \theta} = \frac{1}{1/\sqrt{5}} = \sqrt{5} \qquad \sec \theta = \frac{1}{\cos \theta} = \frac{1}{2/\sqrt{5}} = \frac{\sqrt{5}}{2} \qquad \cot \theta = \frac{1}{\tan \theta} = \frac{1}{\frac{1}{2}} = 2$$

■

■ Now work Problem 25.

The equation of the unit circle is $x^2 + y^2 = 1$. Thus, if $P = (a, b)$ is the point on the terminal side of an angle θ and if P lies on the unit circle, then

$$b^2 + a^2 = 1$$

But $b = \sin \theta$ and $a = \cos \theta$. Thus,

$$(\sin \theta)^2 + (\cos \theta)^2 = 1 \tag{4}$$

It is customary to write $\sin^2 \theta$ instead of $(\sin \theta)^2$, $\cos^2 \theta$ instead of $(\cos \theta)^2$, and so on. With this notation, we can rewrite equation (4) as

$$\sin^2 \theta + \cos^2 \theta = 1 \tag{5}$$

If $\cos \theta \neq 0$, we can divide each side of equation (5) by $\cos^2 \theta$:

$$\frac{\sin^2 \theta}{\cos^2 \theta} + 1 = \frac{1}{\cos^2 \theta}$$

$$\left(\frac{\sin \theta}{\cos \theta}\right)^2 + 1 = \left(\frac{1}{\cos \theta}\right)^2$$

Now use formulas (2) and (3) to get

$$\tan^2 \theta + 1 = \sec^2 \theta \qquad (6)$$

Similarly, if $\sin \theta \neq 0$, we can divide equation (5) by $\sin^2 \theta$ and use formulas (2) and (3) to get the result:

$$1 + \cot^2 \theta = \csc^2 \theta \qquad (7)$$

Collectively, the identities in equations (5), (6), and (7) are referred to as the **Pythagorean identities.**

Let's pause here to summarize the fundamental identities.

Fundamental Identities

$$\tan \theta = \frac{\sin \theta}{\cos \theta} \qquad \cot \theta = \frac{\cos \theta}{\sin \theta}$$

$$\cot \theta = \frac{1}{\tan \theta} \qquad \sec \theta = \frac{1}{\cos \theta} \qquad \csc \theta = \frac{1}{\sin \theta}$$

$$\sin^2 \theta + \cos^2 \theta = 1 \qquad \tan^2 \theta + 1 = \sec^2 \theta \qquad 1 + \cot^2 \theta = \csc^2 \theta$$

The Pythagorean identity

$$\sin^2 \theta + \cos^2 \theta = 1$$

can be solved for $\sin \theta$ in terms of $\cos \theta$ (or vice versa) as follows:

$$\sin^2 \theta = 1 - \cos^2 \theta$$
$$\sin \theta = \pm\sqrt{1 - \cos^2 \theta}$$

where the $+$ sign is used if $\sin \theta > 0$ and the $-$ sign is used if $\sin \theta < 0$.

E X A M P L E   4   *Finding Exact Values Given One Value and the Sign of Another*

Given that $\sin \theta = \frac{1}{3}$ and $\cos \theta < 0$, find the exact value of each of the remaining five trigonometric functions.

Solution   We solve this problem in two ways: the first way uses the definition of the trigonometric functions; the second method uses the fundamental identities.

Solution 1   *Using the Definition*

Suppose that $P = (a, b)$ is a point on the terminal side of θ that lies a distance of $r = 3$ units from the origin. Since $\sin \theta > 0$ and $\cos \theta < 0$, the point P lies in quadrant II. See Figure 39. (Do you see why we chose 3 units? Notice that $\sin \theta = \frac{1}{3} = b/r$. The choice $r = 3$ will make our calculations easy.) With this choice, $b = 1$ and $r = 3$. Since $\cos \theta = a/r < 0$, it follows that $a < 0$. Thus,

FIGURE 39

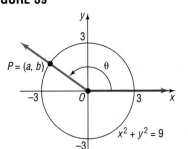

$$a^2 + b^2 = r^2 \qquad b = 1, r = 3, a < 0$$
$$a^2 + 1^2 = 3^2$$
$$a^2 = 8$$
$$a = -2\sqrt{2}$$

Thus,

$$\cos\theta = \frac{a}{r} = \frac{-2\sqrt{2}}{3} \qquad \tan\theta = \frac{b}{a} = \frac{1}{-2\sqrt{2}} = \frac{-\sqrt{2}}{4}$$

$$\csc\theta = \frac{r}{b} = \frac{3}{1} = 3 \qquad \sec\theta = \frac{r}{a} = \frac{3}{-2\sqrt{2}} = \frac{-3\sqrt{2}}{4} \qquad \cot\theta = \frac{a}{b} = \frac{-2\sqrt{2}}{1} = -2\sqrt{2}$$

■

Solution 2   *Using Identities*

First, we solve equation (5) for cos θ:

$$\sin^2\theta + \cos^2\theta = 1$$
$$\cos^2\theta = 1 - \sin^2\theta$$
$$\cos\theta = \pm\sqrt{1 - \sin^2\theta}$$

Because cos $\theta < 0$, we choose the minus sign:

$$\cos\theta = -\sqrt{1 - \sin^2\theta} = -\sqrt{1 - \frac{1}{9}} = -\sqrt{\frac{8}{9}} = -\frac{2\sqrt{2}}{3}$$
$$\underset{\sin\theta = \frac{1}{3}}{\uparrow}$$

Now we know the values of sin θ and cos θ, so we can use formulas (2) and (3) to get

$$\tan\theta = \frac{\sin\theta}{\cos\theta} = \frac{\frac{1}{3}}{-2\sqrt{2}/3} = \frac{1}{-2\sqrt{2}} = \frac{-\sqrt{2}}{4} \qquad \cot\theta = \frac{1}{\tan\theta} = -2\sqrt{2}$$

$$\sec\theta = \frac{1}{\cos\theta} = \frac{1}{-2\sqrt{2}/3} = \frac{-3}{2\sqrt{2}} = \frac{-3\sqrt{2}}{4} \qquad \csc\theta = \frac{1}{\sin\theta} = \frac{1}{\frac{1}{3}} = 3$$

■

■ Now work Problem 33.

Even–Odd Properties

Recall that a function f is even if $f(-\theta) = f(\theta)$ for all θ in the domain of f; a function f is odd if $f(-\theta) = -f(\theta)$ for all θ in the domain of f. We will now show that the trigonometric functions sine, tangent, cotangent, and cosecant are odd functions, whereas the functions cosine and secant are even functions.

Theorem
Even–Odd Properties

$$\sin(-\theta) = -\sin\theta \qquad \cos(-\theta) = \cos\theta \qquad \tan(-\theta) = -\tan\theta$$
$$\csc(-\theta) = -\csc\theta \qquad \sec(-\theta) = \sec\theta \qquad \cot(-\theta) = -\cot\theta$$

FIGURE 40

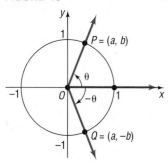

Proof Let $P = (a, b)$ be the point on the terminal side of the angle θ that is on the unit circle. (See Figure 40.) The point Q on the terminal side of the angle $-\theta$ that is on the unit circle will have coordinates $(a, -b)$. Using the definition for the trigonometric functions, we have

$$\sin \theta = b \qquad \cos \theta = a \qquad \sin(-\theta) = -b \qquad \cos(-\theta) = a$$

so that

$$\sin(-\theta) = -\sin \theta \qquad \cos(-\theta) = \cos \theta$$

Now, using these results and some of the fundamental identities, we have

$$\tan(-\theta) = \frac{\sin(-\theta)}{\cos(-\theta)} = \frac{-\sin \theta}{\cos \theta} = -\tan \theta \qquad \cot(-\theta) = \frac{1}{\tan(-\theta)} = \frac{1}{-\tan \theta} = -\cot \theta$$

$$\sec(-\theta) = \frac{1}{\cos(-\theta)} = \frac{1}{\cos \theta} = \sec \theta \qquad \csc(-\theta) = \frac{1}{\sin(-\theta)} = \frac{1}{-\sin \theta} = -\csc \theta$$

∎

Seeing the Concept To see that the cosine function is even, graph $y = \cos x$ and then $y = \cos(-x)$. Clear the screen. Now graph $y = -\sin x$ and $y = \sin(-x)$.

E X A M P L E 5

Finding Exact Values Using Even–Odd Properties

Find the exact value of

(a) $\sin(-45°)$ (b) $\cos(-\pi)$ (c) $\cot(-3\pi/2)$ (d) $\tan(-37\pi/4)$

Solution   (a) $\sin(-45°) = \underset{\substack{\uparrow \\ \text{Odd function}}}{-\sin 45°} = -\frac{\sqrt{2}}{2}$ (b) $\cos(-\pi) = \underset{\substack{\uparrow \\ \text{Even function}}}{\cos \pi} = -1$

(c) $\cot\left(-\frac{3\pi}{2}\right) = \underset{\substack{\uparrow \\ \text{Odd function}}}{-\cot \frac{3\pi}{2}} = 0$

(d) $\tan\left(-\frac{37\pi}{4}\right) = \underset{\substack{\uparrow \\ \text{Odd function}}}{-\tan \frac{37\pi}{4}} = -\tan\left(\frac{\pi}{4} + 9\pi\right) = \underset{\substack{\uparrow \\ \text{Period is } \pi}}{-\tan \frac{\pi}{4}} = -1$

∎

■ Now work Problem 49.

5.3

Exercise 5.3

In Problems 1–16, use the fact that the trigonometric functions are periodic to find the exact value of each expression. Do not use a calculator.

1. $\sin 405°$ **2.** $\cos 420°$ **3.** $\tan 405°$ **4.** $\sin 390°$ **5.** $\csc 450°$ **6.** $\sec 540°$

7. $\cot 390°$ **8.** $\sec 420°$ **9.** $\cos \frac{33\pi}{4}$ **10.** $\sin \frac{9\pi}{4}$ **11.** $\tan 21\pi$ **12.** $\csc \frac{9\pi}{2}$

13. $\sec \frac{17\pi}{4}$ **14.** $\cot \frac{17\pi}{4}$ **15.** $\tan \frac{19\pi}{6}$ **16.** $\sec \frac{25\pi}{6}$

In Problems 17–24, name the quadrant in which the angle θ lies.

17. $\sin\theta > 0,\quad \cos\theta < 0$

18. $\sin\theta < 0,\quad \cos\theta > 0$

19. $\sin\theta < 0,\quad \tan\theta < 0$

20. $\cos\theta > 0,\quad \tan\theta > 0$

21. $\cos\theta > 0,\quad \tan\theta < 0$

22. $\cos\theta < 0,\quad \tan\theta > 0$

23. $\sec\theta < 0,\quad \sin\theta > 0$

24. $\csc\theta > 0,\quad \cos\theta < 0$

In Problems 25–32, sin θ and cos θ are given. Find the exact value of each of the four remaining trigonometric functions.

25. $\sin\theta = 2/\sqrt{5},\quad \cos\theta = 1/\sqrt{5}$

26. $\sin\theta = -1/\sqrt{5},\quad \cos\theta = -2/\sqrt{5}$

27. $\sin\theta = \frac{1}{2},\quad \cos\theta = \sqrt{3}/2$

28. $\sin\theta = \sqrt{3}/2,\quad \cos\theta = \frac{1}{2}$

29. $\sin\theta = -\frac{1}{3},\quad \cos\theta = 2\sqrt{2}/3$

30. $\sin\theta = 2\sqrt{2}/3,\quad \cos\theta = -\frac{1}{3}$

31. $\sin\theta = 0.2588,\quad \cos\theta = 0.9659$

32. $\sin\theta = 0.6428,\quad \cos\theta = 0.7660$

In Problems 33–48, find the exact value of each of the remaining trigonometric functions of θ.

33. $\sin\theta = \frac{12}{13},\quad 90° < \theta < 180°$

34. $\cos\theta = \frac{3}{5},\quad 270° < \theta < 360°$

35. $\cos\theta = -\frac{4}{5},\quad \pi < \theta < 3\pi/2$

36. $\sin\theta = -\frac{5}{13},\quad \pi < \theta < 3\pi/2$

37. $\sin\theta = \frac{5}{13},\quad \cos\theta < 0$

38. $\cos\theta = \frac{4}{5},\quad \sin\theta < 0$

39. $\cos\theta = -\frac{1}{3},\quad \csc\theta > 0$

40. $\sin\theta = -\frac{2}{3},\quad \sec\theta > 0$

41. $\sin\theta = \frac{2}{3},\quad \tan\theta < 0$

42. $\cos\theta = -\frac{1}{4},\quad \tan\theta > 0$

43. $\sec\theta = 2,\quad \sin\theta < 0$

44. $\csc\theta = 3,\quad \cot\theta < 0$

45. $\tan\theta = \frac{3}{4},\quad \sin\theta < 0$

46. $\cot\theta = \frac{4}{3},\quad \cos\theta < 0$

47. $\tan\theta = -\frac{1}{3},\quad \sin\theta > 0$

48. $\sec\theta = -2,\quad \tan\theta > 0$

In Problems 49–66, use the even–odd properties to find the exact value of each expression. Do not use a calculator.

49. $\sin(-60°)$

50. $\cos(-30°)$

51. $\tan(-30°)$

52. $\sin(-135°)$

53. $\sec(-60°)$

54. $\csc(-30°)$

55. $\sin(-90°)$

56. $\cos(-270°)$

57. $\tan\left(-\dfrac{\pi}{4}\right)$

58. $\sin(-\pi)$

59. $\cos\left(-\dfrac{\pi}{4}\right)$

60. $\sin\left(-\dfrac{\pi}{3}\right)$

61. $\tan(-\pi)$

62. $\sin\left(-\dfrac{3\pi}{2}\right)$

63. $\csc\left(-\dfrac{\pi}{4}\right)$

64. $\sec(-\pi)$

65. $\sec\left(-\dfrac{\pi}{6}\right)$

66. $\csc\left(-\dfrac{\pi}{3}\right)$

In Problems 67–78, find the exact value of each expression. Do not use a calculator.

67. $\sin(-\pi) + \cos 5\pi$

68. $\tan\left(-\dfrac{5\pi}{4}\right) - \cot\dfrac{7\pi}{2}$

69. $\sec(-\pi) + \csc\left(-\dfrac{\pi}{2}\right)$

70. $\tan(-6\pi) + \cos\dfrac{9\pi}{4}$

71. $\sin\left(-\dfrac{9\pi}{4}\right) - \tan\left(-\dfrac{9\pi}{4}\right)$

72. $\cos\left(-\dfrac{17\pi}{4}\right) - \sin\left(-\dfrac{3\pi}{2}\right)$

73. $\sin^2 40° + \cos^2 40°$

74. $\sec^2 18° - \tan^2 18°$

75. $\sin 80° \csc 80°$

76. $\tan 10° \cot 10°$

77. $\tan 40° - \dfrac{\sin 40°}{\cos 40°}$

78. $\cot 20° - \dfrac{\cos 20°}{\sin 20°}$

79. If $\sin\theta = 0.3$, find the value of $\sin\theta + \sin(\theta + 2\pi) + \sin(\theta + 4\pi)$.

80. If $\cos\theta = 0.2$, find the value of $\cos\theta + \cos(\theta + 2\pi) + \cos(\theta + 4\pi)$.

81. If $\tan\theta = 3$, find the value of $\tan\theta + \tan(\theta + \pi) + \tan(\theta + 2\pi)$.

82. If $\cot \theta = -2$, find the value of $\cot \theta + \cot(\theta - \pi) + \cot(\theta - 2\pi)$.

83. *Calculating the Time of a Trip* From a parking lot, you want to walk to a house on the ocean. The house is located 1500 feet down a paved path that parallels the ocean, which is 500 feet away. See the illustration. Along the path you can walk 300 feet per minute, but in the sand on the beach you can only walk 100 feet per minute.

The time T to get from the parking lot to the beachhouse can be expressed as a function of the angle θ shown in the illustration is

$$T(\theta) = 5 - \frac{5}{3 \tan \theta} + \frac{5}{\sin \theta}, \quad 0 < \theta < \frac{\pi}{2}$$

Calculate the time T if you walk directly from the parking lot to the house. [*Hint:* $\tan \theta = 500/1500$.]

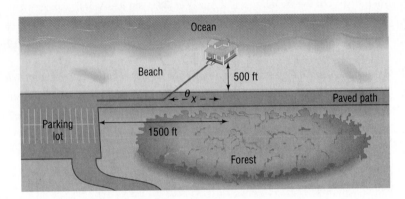

84. *Calculating the Time of a Trip* Two oceanfront homes are located 8 miles apart on a straight stretch of beach, each a distance of 1 mile from a paved road that parallels the ocean. Sally can walk 8 miles per hour along the paved road, but only 3 miles per hour in the sand that separates the road from the ocean. Because of a river directly between the two houses, it is necessary to walk in the sand to the road, continue on the road, and then walk directly back in the sand to get from one house to the other. See the illustration. The time T to get from one house to the other as a function of the angle θ shown in the illustration is

$$T(\theta) = 1 + \frac{2}{3 \sin \theta} - \frac{1}{4 \tan \theta}, \quad 0 < \theta < \frac{\pi}{2}$$

(a) Calculate the time T for $\tan \theta = 1/4$.
(b) Describe the path taken.
(c) Explain why θ must be larger than 14°.

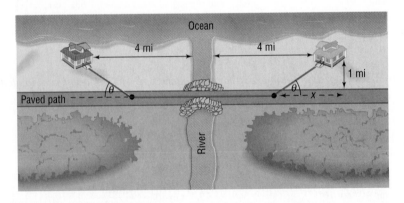

85. For what numbers θ is $f(\theta) = \tan \theta$ not defined?

86. For what numbers θ is $f(\theta) = \cot \theta$ not defined?

87. For what numbers θ is $f(\theta) = \sec \theta$ not defined?

88. For what numbers θ is $f(\theta) = \csc \theta$ not defined?

89. What is the value of $\sin k\pi$, where k is any integer?

90. What is the value of $\cos k\pi$, where k is any integer?

91. Show that the range of the tangent function is the set of all real numbers.

92. Show that the range of the cotangent function is the set of all real numbers.

93. Show that the period of $f(\theta) = \sin \theta$ is 2π. [*Hint:* Assume that $0 < p < 2\pi$ exists so that $\sin(\theta + p) = \sin \theta$ for all θ. Let $\theta = 0$ to find p. Then let $\theta = \pi/2$ to obtain a contradiction.]

94. Show that the period of $f(\theta) = \cos \theta$ is 2π.

95. Show that the period of $f(\theta) = \sec \theta$ is 2π.

96. Show that the period of $f(\theta) = \csc \theta$ is 2π.

97. Show that the period of $f(\theta) = \tan \theta$ is π.

98. Show that the period of $f(\theta) = \cot \theta$ is π.

99. Prove the reciprocal identities given in formula (2).

100. Prove the quotient identities given in formula (3).

101. Establish the identity: $(\sin \theta \cos \phi)^2 + (\sin \theta \sin \phi)^2 + \cos^2 \theta = 1$

102. Write down five characteristics of the tangent function. Explain the meaning of each.

103. Describe your understanding of the meaning of a periodic function.

5.4

Right Triangle Trigonometry

A triangle in which one angle is a right angle (90°) is called a **right triangle.** Recall that the side opposite the right angle is called the **hypotenuse,** and the remaining two sides are called the **legs** of the triangle. In Figure 41(a), we have labeled the hypotenuse as c, to indicate its length is c units, and, in a like manner, we have labeled the legs as a and b. Because the triangle is a right triangle, the Pythagorean Theorem tells us that

$$a^2 + b^2 = c^2$$

FIGURE 41

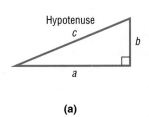

(a)

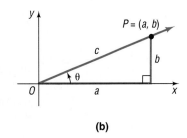

(b)

Now, suppose that θ is an **acute angle;** that is, $0° < \theta < 90°$ (if θ is measured in degrees) or $0 < \theta < \pi/2$ (if θ is measured in radians). Place θ in standard position, and let $P = (a, b)$ be any point except the origin O on the terminal side of θ. Form a right triangle by dropping the perpendicular from P to the x-axis, as shown in Figure 41(b).

By referring to the lengths of the sides of the triangle by the names hypotenuse (c), opposite (b), and adjacent (a), as indicated in Figure 42, we can express the trigonometric functions of θ as ratios of the sides of a right triangle:

FIGURE 42

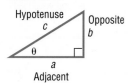

$$\sin \theta = \frac{\text{Opposite}}{\text{Hypotenuse}} = \frac{b}{c} \qquad \cos \theta = \frac{\text{Adjacent}}{\text{Hypotenuse}} = \frac{a}{c}$$

$$\tan \theta = \frac{\text{Opposite}}{\text{Adjacent}} = \frac{b}{a} \qquad \csc \theta = \frac{\text{Hypotenuse}}{\text{Opposite}} = \frac{c}{b} \qquad (1)$$

$$\sec \theta = \frac{\text{Hypotenuse}}{\text{Adjacent}} = \frac{c}{a} \qquad \cot \theta = \frac{\text{Adjacent}}{\text{Opposite}} = \frac{a}{b}$$

Notice that each of the trigonometric functions of the acute angle θ is positive.

E X A M P L E 1   *Finding the Value of Trigonometric Functions from a Right Triangle*

Find the exact value of each of the six trigonometric functions of the angle θ in Figure 43.

Solution We see in Figure 43 that the two given sides of the triangle are

$$c = \text{Hypotenuse} = 5 \qquad a = \text{Adjacent} = 3$$

FIGURE 43

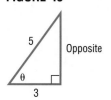

To find the length of the opposite side, we use the Pythagorean Theorem:

$$(\text{Adjacent})^2 + (\text{Opposite})^2 = (\text{Hypotenuse})^2$$
$$3^2 + (\text{Opposite})^2 = 5^2$$
$$(\text{Opposite})^2 = 25 - 9 = 16$$
$$\text{Opposite} = 4$$

Now that we know the lengths of the three sides, we use the ratios in (1) to find the value of each of the six trigonometric functions:

$$\sin \theta = \frac{\text{Opposite}}{\text{Hypotenuse}} = \frac{4}{5} \qquad \cos \theta = \frac{\text{Adjacent}}{\text{Hypotenuse}} = \frac{3}{5} \qquad \tan \theta = \frac{\text{Opposite}}{\text{Adjacent}} = \frac{4}{3}$$

$$\csc \theta = \frac{\text{Hypotenuse}}{\text{Opposite}} = \frac{5}{4} \qquad \sec \theta = \frac{\text{Hypotenuse}}{\text{Adjacent}} = \frac{5}{3} \qquad \cot \theta = \frac{\text{Adjacent}}{\text{Opposite}} = \frac{3}{4}$$

■

■ Now work Problem 1.

Thus, the values of the trigonometric functions of an acute angle are ratios of the lengths of the sides of a right triangle. This way of viewing the trigonometric functions leads to many applications and, in fact, was the point of view used by early mathematicians (before calculus) in studying the subject of trigonometry.

Complementary Angles: Cofunctions

FIGURE 44

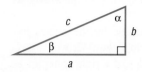

Two acute angles are called **complementary** if their sum is a right angle. Because the sum of the angles of any triangle is $180°$, it follows that, for a right triangle, the two acute angles are complementary.

Refer now to Figure 44; we have labeled the angle opposite side b as β and the angle opposite side a as α. Notice that side b is adjacent to angle α and side a is adjacent to angle β. As a result,

$$\sin \beta = \frac{b}{c} = \cos \alpha \qquad \cos \beta = \frac{a}{c} = \sin \alpha \qquad \tan \beta = \frac{b}{a} = \cot \alpha$$

$$\csc \beta = \frac{c}{b} = \sec \alpha \qquad \sec \beta = \frac{c}{a} = \csc \alpha \qquad \cot \beta = \frac{a}{b} = \tan \alpha$$

$$(2)$$

Because of these relationships, the functions sine and cosine, tangent and cotangent, and secant and cosecant are called **cofunctions** of each other. The identities (2) may be expressed in words as follows:

Theorem   Cofunctions of complementary angles are equal.   ■

Examples of this theorem are given next:

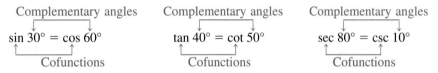

If θ is an acute angle measured in degrees, the angle $90° - \theta$ (or $\pi/2 - \theta$, if θ is in radians) is the angle complementary to θ. Table 6 restates the preceding theorem on cofunctions.

TABLE 6

θ (DEGREES)	θ (RADIANS)
$\sin \theta = \cos(90° - \theta)$	$\sin \theta = \cos(\pi/2 - \theta)$
$\cos \theta = \sin(90° - \theta)$	$\cos \theta = \sin(\pi/2 - \theta)$
$\tan \theta = \cot(90° - \theta)$	$\tan \theta = \cot(\pi/2 - \theta)$
$\csc \theta = \sec(90° - \theta)$	$\csc \theta = \sec(\pi/2 - \theta)$
$\sec \theta = \csc(90° - \theta)$	$\sec \theta = \csc(\pi/2 - \theta)$
$\cot \theta = \tan(90° - \theta)$	$\cot \theta = \tan(\pi/2 - \theta)$

Although the angle θ in Table 6 is acute, we will see later that these results are valid for any angle θ.

 Seeing the Concept Graph $y = \sin x$ and $y = \cos(90° - x)$. Be sure that the mode is set to degrees.

E X A M P L E   2

Using the Theorem on Cofunctions

(a) $\sin 62° = \cos(90° - 62°) = \cos 28°$ (b) $\tan \dfrac{\pi}{12} = \cot\left(\dfrac{\pi}{2} - \dfrac{\pi}{12}\right) = \cot \dfrac{5\pi}{12}$

(c) $\cos \dfrac{\pi}{4} = \sin\left(\dfrac{\pi}{2} - \dfrac{\pi}{4}\right) = \sin \dfrac{\pi}{4}$ (d) $\csc \dfrac{\pi}{6} = \sec\left(\dfrac{\pi}{2} - \dfrac{\pi}{6}\right) = \sec \dfrac{\pi}{3}$

■

■   Now work Problem 57(a).

Reference Angle

Next, we concentrate on angles that lie in a quadrant. Once we know in which quadrant an angle lies, we know the sign of each value of the trigonometric functions of that angle. The use of a certain reference angle may help us to evaluate the trigonometric functions of such an angle.

Reference Angle

> Let θ denote a nonacute angle that lies in a quadrant. The acute angle formed by the terminal side of θ and either the positive x-axis or the negative x-axis is called the **reference angle** for θ.

Figure 45 illustrates the reference angle for some general angles θ. Note that a reference angle is always an acute angle, that is, an angle whose measure is between 0° and 90°.

FIGURE 45

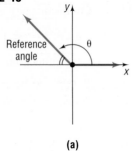

(a)

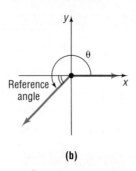

(b)

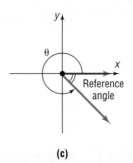

(c)

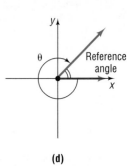

(d)

Although formulas can be given for calculating reference angles, usually it is easier to find the reference angle for a given angle by making a quick sketch of the angle.

EXAMPLE 3

Finding Reference Angles

Find the reference angle for each of the following angles:

(a) 150°   (b) −45°   (c) $9\pi/4$ (d) $-5\pi/6$

Solution

(a) Refer to Figure 46. The reference angle for 150° is 30°.

(b) Refer to Figure 47. The reference angle for −45° is 45°.

FIGURE 46

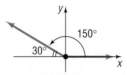

FIGURE 47

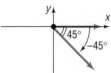

(c) Refer to Figure 48. The reference angle for $9\pi/4$ is $\pi/4$.

(d) Refer to Figure 49. The reference angle for $-5\pi/6$ is $\pi/6$.

FIGURE 48

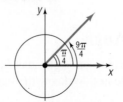

FIGURE 49

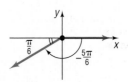

■ Now work Problem 11.

The advantage of using reference angles is that, except for the correct sign, the values of the trigonometric functions of a general angle θ equal the values of the trigonometric functions of its reference angle.

Theorem

Reference Angles

If θ is an angle that lies in a quadrant and if α is its reference angle, then

$$
\begin{array}{lll}
\sin \theta = \pm \sin \alpha & \cos \theta = \pm \cos \alpha & \tan \theta = \pm \tan \alpha \\
\csc \theta = \pm \csc \alpha & \sec \theta = \pm \sec \alpha & \cot \theta = \pm \cot \alpha
\end{array}
\tag{3}
$$

FIGURE 50

$\sin \theta = b/c$, $\sin \alpha = b/c$;

$\cos \theta = a/c$, $\cos \alpha = |a|/c$

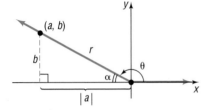

where the $+$ or $-$ sign depends on the quadrant in which θ lies.   ■

For example, suppose that θ lies in quadrant II and α is its reference angle. See Figure 50. If (a, b) is a point on the terminal side of θ and if $c = \sqrt{a^2 + b^2}$, we have

$$
\sin \theta = \frac{b}{c} = \sin \alpha \qquad \cos \theta = \frac{a}{c} = \frac{-|a|}{c} = -\cos \alpha
$$

and so on.

The next example illustrates how the theorem on reference angles is used.

E X A M P L E   4

Using Reference Angles to Find the Value of Trigonometric Functions

Find the exact value of each of the following trigonometric functions using reference angles:

(a) $\sin 135°$ (b) $\cos 240°$ (c) $\cos \dfrac{5\pi}{6}$ (d) $\tan\left(-\dfrac{\pi}{3}\right)$

Solution

(a) Refer to Figure 51. The angle $135°$ is in quadrant II, where the sine function is positive. The reference angle for $135°$ is $45°$. Thus,

$$
\sin 135° = \sin 45° = \frac{\sqrt{2}}{2}
$$

FIGURE 51

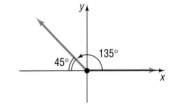

(b) Refer to Figure 52. The angle $240°$ is in quadrant III, where the cosine function is negative. The reference angle for $240°$ is $60°$. Thus,

$$
\cos 240° = -\cos 60° = -\tfrac{1}{2}
$$

FIGURE 52

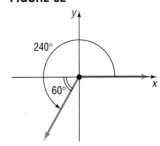

(c) Refer to Figure 53. The angle $5\pi/6$ is in quadrant II, where the cosine function is negative. The reference angle for $5\pi/6$ is $\pi/6$. Thus,

$$\cos\frac{5\pi}{6} = -\cos\frac{\pi}{6} = -\frac{\sqrt{3}}{2}$$

FIGURE 53

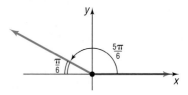

(d) Refer to Figure 54. The angle $-\pi/3$ is in quadrant IV, where the tangent function is negative. The reference angle for $-\pi/3$ is $\pi/3$. Thus,

$$\tan\left(-\frac{\pi}{3}\right) = -\tan\frac{\pi}{3} = -\sqrt{3}$$

FIGURE 54

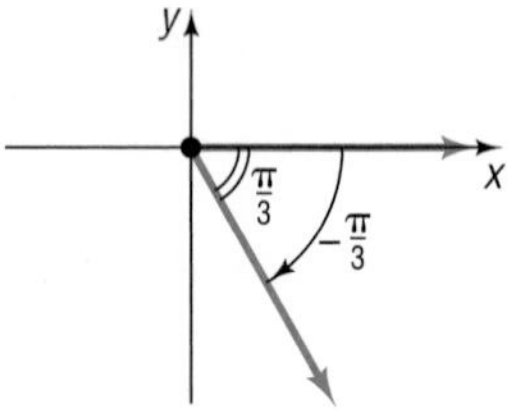

■

■ Now work Problems 27 and 45.

E X A M P L E   5

Using Right Triangles to Find Values When One Value Is Known

Given that $\cos\theta = -\frac{2}{3}$, $\pi/2 < \theta < \pi$, find the exact value of each of the remaining trigonometric functions.

Solution

Using Right Triangles We have already discussed two ways of solving this type of problem: using the definition of the trigonometric functions and using identities. (Refer to Example 4 in Section 5.3). Here we present a third method: using right triangles.

FIGURE 55

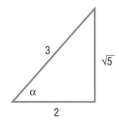

The angle θ lies in quadrant II, so we know that $\sin\theta$ and $\csc\theta$ are positive, whereas the other trigonometric functions are negative. If α is the reference angle for θ, then $\cos\alpha = \frac{2}{3}$. The values of the remaining trigonometric functions of the angle α can be found by drawing the appropriate triangle. We use Figure 55 to obtain

$$\sin\alpha = \frac{\sqrt{5}}{3} \qquad \cos\alpha = \frac{2}{3} \qquad \tan\alpha = \frac{\sqrt{5}}{2}$$

$$\csc\alpha = \frac{3}{\sqrt{5}} = \frac{3\sqrt{5}}{5} \qquad \sec\alpha = \frac{3}{2} \qquad \cot\alpha = \frac{2}{\sqrt{5}} = \frac{2\sqrt{5}}{5}$$

Now, we assign the appropriate sign to each of these values to find the values of the trigonometric functions of θ:

$$\sin\theta = \frac{\sqrt{5}}{3} \qquad \cos\theta = -\frac{2}{3} \qquad \tan\theta = -\frac{\sqrt{5}}{2}$$

$$\csc\theta = \frac{3\sqrt{5}}{5} \qquad \sec\theta = -\frac{3}{2} \qquad \cot\theta = -\frac{2\sqrt{5}}{5}$$

■

E X A M P L E   6

Constructing a Rain Gutter

A rain gutter is to be constructed of aluminum sheets 12 inches wide. After marking off a length of 4 inches from each edge, this length is bent up at an angle θ. See Figure 56.

(a) Express the area A of the opening as a function of θ.

(b) Graph $A = A(\theta)$. Find the angle θ that makes A largest. (This bend will allow the most water to flow through the gutter).

FIGURE 56

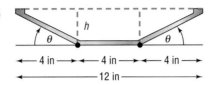

Solution   (a) The area A of the opening is the sum of the areas of two right triangles and one rectangle:

$$A = 2 \cdot \tfrac{1}{2} \cdot h\sqrt{16 - h^2} + 4h = (4 \sin \theta)(4 \cos \theta) + 4(4 \sin \theta)$$
$$A(\theta) = 16 \sin \theta(\cos \theta + 1)$$

(b) See Figure 57. The angle θ that makes A largest is 60°.   ■

FIGURE 57

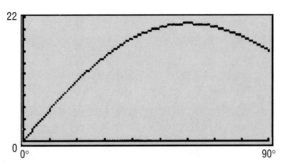

5.4

Exercise 5.4

In Problems 1–10, find the exact value of each of the six trigonometric functions of the angle θ in each figure.

1.

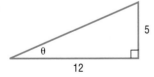

2.

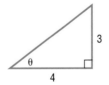

3.

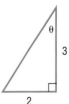

4.

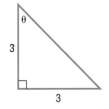

5.

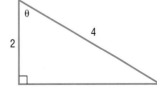

6.

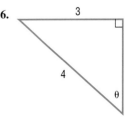

7.

8.
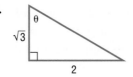

9.
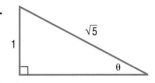

10.
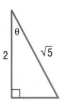

In Problems 11–26, find the reference angle of each angle.

11. $-30°$ **12.** $60°$ **13.** $120°$ **14.** $300°$ **15.** $210°$

16. $330°$ **17.** $5\pi/4$ **18.** $5\pi/6$ **19.** $8\pi/3$ **20.** $7\pi/4$

21. $-135°$ **22.** $-240°$ **23.** $-2\pi/3$ **24.** $-7\pi/6$ **25.** $420°$

26. $480°$

In Problems 27–56, find the exact value of each expression. Do not use a calculator.

27. $\sin 150°$ **28.** $\cos 210°$ **29.** $\cos 315°$ **30.** $\sin 120°$

31. $\sec 240°$ **32.** $\csc 300°$ **33.** $\cot 330°$ **34.** $\tan 225°$

35. $\sin \dfrac{3\pi}{4}$ **36.** $\cos \dfrac{2\pi}{3}$ **37.** $\cot \dfrac{7\pi}{6}$ **38.** $\csc \dfrac{7\pi}{4}$

39. $\cos(-60°)$ **40.** $\tan(-120°)$ **41.** $\sin\left(-\dfrac{2\pi}{3}\right)$ **42.** $\cot\left(-\dfrac{\pi}{6}\right)$

43. $\tan \dfrac{14\pi}{3}$ **44.** $\sec \dfrac{11\pi}{4}$ **45.** $\csc(-315°)$ **46.** $\sec(-225°)$

47. $\sin 38° - \cos 52°$ **48.** $\tan 12° - \cot 78°$ **49.** $\dfrac{\cos 10°}{\sin 80°}$ **50.** $\dfrac{\cos 40°}{\sin 50°}$

51. $1 - \cos^2 20° - \cos^2 70°$ **52.** $1 + \tan^2 5° - \csc^2 85°$ **53.** $\tan 20° - \dfrac{\cos 70°}{\cos 20°}$

54. $\cot 40° - \dfrac{\sin 50°}{\sin 40°}$ **55.** $\cos 35° \sin 55° + \sin 35° \cos 55°$ **56.** $\sec 35° \csc 55° - \tan 35° \cot 55°$

57. If $\sin \theta = \frac{1}{3}$, find the exact value of: (a) $\cos(90° - \theta)$ (b) $\cos^2 \theta$ (c) $\csc \theta$ (d) $\sec\left(\dfrac{\pi}{2} - \theta\right)$

58. If $\sin \theta = 0.2$, find the exact value of: (a) $\cos\left(\dfrac{\pi}{2} - \theta\right)$ (b) $\cos^2 \theta$ (c) $\sec(90° - \theta)$ (d) $\csc \theta$

59. If $\tan \theta = 4$, find the exact value of: (a) $\sec^2 \theta$ (b) $\cot \theta$ (c) $\cot\left(\dfrac{\pi}{2} - \theta\right)$ (d) $\csc^2 \theta$

60. If $\sec \theta = 3$, find the exact value of: (a) $\cos \theta$ (b) $\tan^2 \theta$ (c) $\csc(90° - \theta)$ (d) $\sin^2 \theta$

61. If $\csc \theta = 4$, find the exact value of: (a) $\sin \theta$ (b) $\cot^2 \theta$ (c) $\sec(90° - \theta)$ (d) $\sec^2 \theta$

62. If $\cot \theta = 2$, find the exact value of: (a) $\tan \theta$ (b) $\csc^2 \theta$ (c) $\tan\left(\dfrac{\pi}{2} - \theta\right)$ (d) $\sec^2 \theta$

63. If $\sin \theta = 0.3$, find the exact value of $\sin \theta + \cos\left(\dfrac{\pi}{2} - \theta\right)$.

64. If $\tan \theta = 4$ find the exact value of $\tan \theta + \tan\left(\dfrac{\pi}{2} - \theta\right)$.

65. Find the exact value of: $\sin 1° + \sin 2° + \sin 3° + \cdots + \sin 358° + \sin 359°$.

66. Find the exact value of: $\cos 1° + \cos 2° + \cos 3° + \cdots + \cos 358° + \cos 359°$.

67. Find the acute angle θ that satisfies the equation: $\sin \theta = \cos(2\theta + 30°)$

68. Find the acute angle θ that satisfies the equation: $\tan \theta = \cot(\theta + 45°)$

69. *Calculating the Time of a Trip* Two oceanfront homes are located 8 miles apart on a straight stretch of beach, each a distance of 1 mile from a paved road that parallels the ocean. Sally can walk 8 miles per hour along the paved road, but only 3 miles per hour in the sand that separates the road from the ocean. Because of a river directly between the two houses, it is necessary to walk in the sand to the road, continue on the road, and then walk directly back in the sand to get from one house to the other. See the illustration.

(a) Express the time T to get from one house to the other as a function of the angle θ shown in the illustration.

 (b) Graph $T = T(\theta)$. What angle θ results in the least time? What is the least time? How long is Sally on the paved road?

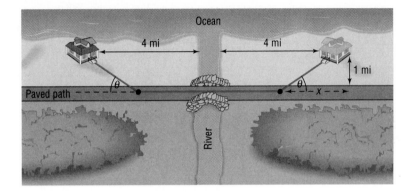

70. *Designing Fine Decorative Pieces* A designer of decorative art plans to market solid gold spheres encased in clear crystal cones. Each sphere is of fixed radius R and will be enclosed in a cone of height h and radius r. See the illustration. Many cones can be used to enclose the sphere, each having a different slant angle θ.

(a) Express the volume V of the cone as a function of the slant angle θ of the cone. [*Hint:* The volume V of a cone of height h and radius r is $V = \frac{1}{3}\pi r^2 h$.]

 (b) What slant angle θ should be used for the volume V of the cone to be a minimum? (This choice minimizes the amount of crystal required and gives maximum emphasis to the gold sphere).

71. *Calculating the Time of a Trip* From a parking lot, you want to walk to a house on the ocean. The house is located 1500 feet down a paved path that parallels the ocean, which is 500 feet away. See the illustration. Along the path you can walk 300 feet per minute, but in the sand on the beach you can only walk 100 feet per minute.

(a) Calculate the time T if you walk 1500 feet along the paved path and then 500 feet in the sand to the house.

(b) Calculate the time T if you walk in the sand first for 500 feet and then walk along the beach for 1500 feet to the house.

(c) Express the time T to get from the parking lot to the beachhouse as a function of the angle θ shown in the illustration.

(d) Calculate the time T if you walk 1000 feet along the paved path and then walk directly to the house.

 (e) Graph $T = T(\theta)$. For what angle θ is T least? What is the least time? What is x for this angle?

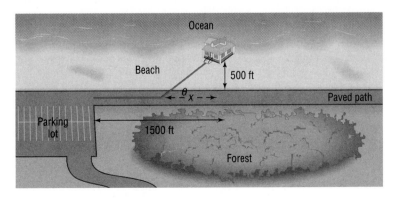

72. *Carrying a Ladder around a Corner* A ladder of length L is carried horizontally around a corner from a hall 3 feet wide into a hall 4 feet wide. See the illustration. Find the length L of the ladder as a function of the angle θ shown in the illustration.

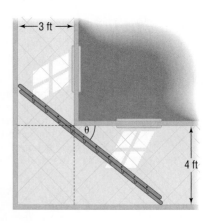

 73. Give three examples that will show a fellow student how to use the Theorem on Reference Angles. Give them to the fellow student and ask for a critique.

74. Refer to Example 4, Section 5.3, and Example 5, Section 5.4. Which of the three methods of solution do you prefer? Which is your least favorite? Are there situations in which one method is sometimes best and others in which another method is best? Give reasons.

75. Use a calculator set in radian mode to complete the following table. What can you conclude about the ratio $(\sin \theta)/\theta$ as θ approaches 0?

θ	0.5	0.4	0.2	0.1	0.01	0.001	0.0001	0.00001
$\sin \theta$								
$\dfrac{\sin \theta}{\theta}$								

76. Use a calculator set in radian mode to complete the following table. What can you conclude about the ratio $(\cos \theta - 1)/\theta$ as θ approaches 0?

θ	0.5	0.4	0.2	0.1	0.01	0.001	0.0001	0.00001
$\cos \theta - 1$								
$\dfrac{\cos \theta - 1}{\theta}$								

77. Suppose that the angle θ is a central angle of a circle of radius 1 (see the figure). Show that

(a) Angle $OAC = \dfrac{\theta}{2}$ (b) $|CD| = \sin \theta$ and $|OD| = \cos \theta$

(c) $\tan \dfrac{\theta}{2} = \dfrac{\sin \theta}{1 + \cos \theta}$

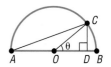

78. Show that the area of an isosceles triangle is $A = a^2 \sin \theta \cos \theta$, where a is the length of one of the two equal sides and θ is the measure of one of the two equal angles (see the figure).

79. Let $n > 0$ be any real number, and let θ be any angle for which $0 < \theta < \pi/(1 + n)$. Then we can construct a triangle with the angles θ and $n\theta$ and included side of length 1 (do you see why?) and place it on the unit circle as illustrated. Now, drop the perpendicular from C to $D = (x, 0)$, and show that

$$x = \frac{\tan n\theta}{\tan \theta + \tan n\theta}$$

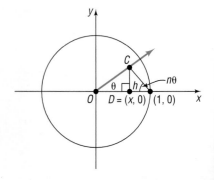

80. Refer to the accompanying figure. The smaller circle, whose radius is a, is tangent to the larger circle, whose radius is b. The ray OA contains a diameter of each circle, and the ray OB is tangent to each circle. Show that

$$\cos \theta = \frac{\sqrt{ab}}{\dfrac{a+b}{2}}$$

(that is, $\cos \theta$ equals the ratio of the geometric mean of a and b to the arithmetic mean of a and b). [*Hint:* First show that $\sin \theta = (b-a)/(b+a)$.]

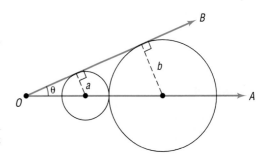

81. Refer to the figure in the margin. If $|OA| = 1$, show that

(a) Area $\triangle OAC = \frac{1}{2} \sin \alpha \cos \alpha$ (b) area $\triangle OCB = \frac{1}{2}|OB|^2 \sin \beta \cos \beta$

(c) Area $\triangle OAB = \frac{1}{2}|OB| \sin(\alpha + \beta)$ (d) $|OB| = \dfrac{\cos \alpha}{\cos \beta}$

(e) $\sin(\alpha + \beta) = \sin \alpha \cos \beta + \cos \alpha \sin \beta$

[*Hint:* Area $\triangle OAB =$ Area $\triangle OAC +$ Area $\triangle OCB$.]

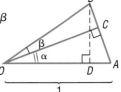

82. Refer to the accompanying figure, where a unit circle is drawn. The line DB is tangent to the circle.

(a) Express the area of $\triangle OBC$ in terms of $\sin \theta$ and $\cos \theta$.

(b) Express the area of $\triangle OBD$ in terms of $\sin \theta$ and $\cos \theta$.

(c) The area of the sector of the circle $\overset{\frown}{OBC}$ is $\frac{1}{2}\theta$, where θ is measured in radians. Use the results of parts (a) and (b) and the fact that

$$\text{Area } \triangle OBC < \text{Area } \overset{\frown}{OBC} < \text{Area } \triangle OBD$$

to show that

$$1 < \frac{\theta}{\sin \theta} < \frac{1}{\cos \theta}$$

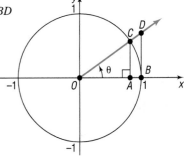

83. If $\cos \alpha = \tan \beta$ and $\cos \beta = \tan \alpha$, where α and β are acute angles, show that

$$\sin \alpha = \sin \beta = \sqrt{\frac{3 - \sqrt{5}}{2}}$$

84. If θ is an acute angle, explain why $\sec \theta > 1$.

85. If θ is an acute angle, explain why $0 < \sin \theta < 1$.

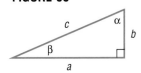

5.5

Applications

FIGURE 58

Solving Right Triangles

In the discussion that follows, we shall always label a right triangle so that side a is opposite angle α, side b is opposite angle β, and side c is the hypotenuse, as shown in Figure 58. **To solve a right triangle** means to find the missing lengths of its sides and the measurements of its angles. We shall follow the practice of expressing the lengths of the sides rounded to two decimal places and of expressing angles in degrees rounded to one decimal place.

To solve a right triangle, we need to know either an angle and a side or else two sides. Then we make use of the Pythagorean Theorem and the fact that the sum of the angles of a triangle is $180°$. The sum of the unknown angles in a right triangle is therefore $90°$. Thus, for the right triangle shown in Figure 58, we have

$$c^2 = a^2 + b^2 \qquad \alpha + \beta = 90°$$

EXAMPLE 1 *Solving a Right Triangle*

Use Figure 59. If $b = 2$ and $\alpha = 40°$, find a, c, and β.

FIGURE 59

40°
c
2
β
a

Solution Since $\alpha = 40°$ and $\alpha + \beta = 90°$, we find that $\beta = 50°$. To find the sides a and c, we use the following relationships:

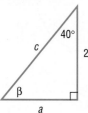

$$\tan 40° = \frac{a}{2} \qquad \text{and} \qquad \cos 40° = \frac{2}{c}$$

So

$$a = 2 \tan 40° \approx 1.68 \quad \text{and} \qquad c = \frac{2}{\cos 40°} \approx 2.61$$

■ Now work Problem 1.

EXAMPLE 2 *Solving a Right Triangle*

Use Figure 60. If $a = 3$ and $b = 2$, find c, α, and β.

FIGURE 60

α
c
2
β
3

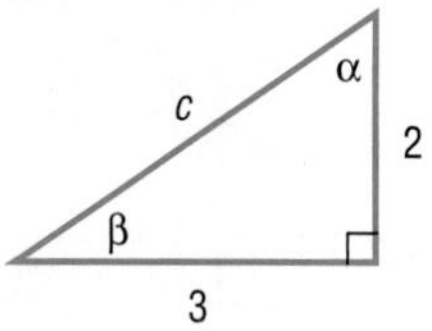

Solution Since $a = 3$ and $b = 2$, then by the Pythagorean Theorem we have

$$c^2 = a^2 + b^2 = 9 + 4 = 13$$
$$c = \sqrt{13} \approx 3.61$$

To find angle α, we use the relationship

$$\tan \alpha = \frac{3}{2}$$

Set the mode on your calculator to degrees. On your calculator there should be a key marked [tan⁻¹] or a key marked [inv] or [SHIFT].* Use the key [tan⁻¹] or the keys [SHIFT] [tan] as follows to find α:

Keystrokes:
3
÷
2
=
SHIFT
tan
Display:
3
2
1.5
56.309932

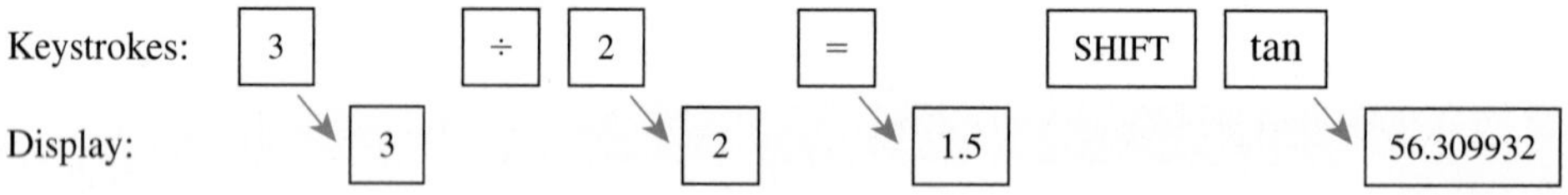

Thus, $\alpha = 56.3°$ rounded to one decimal place. Since $\alpha + \beta = 90°$, we find that $\beta = 33.7°$. ■

■ Now work Problem 11.

One common use for trigonometry is to measure heights and distances that are either awkward or impossible to measure by ordinary means.

*If not, consult your owner's manual.

MISSION POSSIBLE

Chapter 5

IDENTIFYING THE MOUNTAINS OF THE HAWAIIAN ISLANDS SEEN FROM OAHU

A wealthy tourist with a strong desire for the perfect scrapbook has called in your consulting firm for help in labeling a photograph he took on a clear day from the southeast shore of Oahu. He shows you the photograph in which you see mostly sea and sky. But on the horizon are three mountain peaks, equally spaced and apparently all of the same height. The tourist tells you that the T-shirt vendor at the beach informed him that the three mountain peaks are the volcanoes of Lanai, Maui, and the Big Island (Hawaii). The vendor even added, "You almost never see the Big Island from here."

The tourist thinks his photograph may have some special value. He has some misgivings, however. He is wondering if, in fact, it is ever really possible to see the Big Island from Oahu. He wants accurate labels on his photos. So he asks you for help.

You realize that some trigonometry is needed as well as a good atlas and encyclopedia. After doing some research you discover the following facts:

The heights of the peaks vary. Lanaihale, the tallest peak on Lanai, is only about 3370 feet above sea level. Maui's Haleakala is 10,023 feet above sea level. Mauna Kea on the Big Island is the highest of all, 13,796 feet above sea level. (If measured from its base on the ocean floor, it would qualify as the tallest mountain on Earth.)

Measuring from the southeast corner of Oahu, the distance to Lanai is about 65 miles, to Maui it's about 110 miles, and to the Big Island it is about 190 miles. These distances, measured along the surface of the Earth, represent the length of arc that originates at sea level on Oahu and terminates in a imaginary location directly below the peak of each mountain at what would be sea level.

1. These volcanic peaks are all different heights. If your wealthy client took a photo of three mountain peaks that all look the same height, how could they possibly represent these three volcanoes?
2. The radius of the earth is approximately 3960 miles. Based on that figure, what is the circumference of the earth? How is the distance between islands related to the circumference of the earth?
3. To determine which of these mountain peaks would actually be visible from Oahu, you need to consider that the tourist standing on the shore and looking "straight out" would have a line of sight tangent to the surface of the earth at that point. Make a rough sketch of the right triangle formed by the tourist's line of sight, the radius from the center of the earth to the tourist, and a line from the center of the earth that passes through Mauna Kea. Can you determine the angle formed at the center of the earth?
4. What would the length of the hypotenuse of the triangle be? Can you tell from that whether or not Mauna Kea would be visible from Oahu?
5. Repeat this procedure with the information about Maui and Lanai. Will they be visible from Oahu?
6. There is another island off Oahu that the T-shirt vendor did not mention. Molokai is about 40 miles from Oahu, and its highest peak, "Kamakou," is 4961 feet above sea level. Would Kamakou be visible from Oahu?
7. Your team should now be prepared to name the three volcanic peaks in the photograph. How do you decide which one is which?

E X A M P L E   3   *Finding the Width of a River*

A surveyor can measure the width of a river by setting up a transit* at a point C on one side of the river and taking a sighting of a point A on the other side. Refer to Figure 61. After turning through an angle of 90° at C, a distance of 200 meters is walked off to point B. Using the transit at B, the angle β is measured and found to be 20°. What is the width of the river?

Solution   We seek the length of side b. We know a and β, so we use the relationship

$$\tan \beta = \frac{b}{a}$$

FIGURE 61

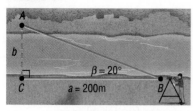

to get

$$\tan 20° = \frac{b}{200}$$
$$b = 200 \tan 20° \approx 72.79 \text{ meters}$$

Thus, the width of the river is approximately 73 meters, rounded to the nearest meter.   ∎

■   Now work Problem 21.

E X A M P L E   4   *Finding the Inclination of a Mountain Trail*

A straight trail with a uniform inclination leads from a hotel, elevation 8000 feet, to a scenic overlook, elevation 11,100 feet. The length of the trail is 14,100 feet. What is the inclination (grade) of the trail?

Solution   Figure 62 illustrates the situation. We seek the angle β. As the illustration shows,

FIGURE 62

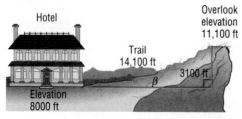

$$\sin \beta = \frac{3100}{14{,}100}$$

Using a calculator,

$$\beta \approx 12.7°$$

The inclination (grade) of the trail is approximately 12.7°.   ∎

Vertical heights can sometimes be measured using either the angle of elevation or the angle of depression. If a person is looking up at an object, the acute angle measured from the horizontal to a line-of-sight observation of the object is called the **angle of elevation.** See Figure 63(a).

If a person is standing on a cliff looking down at an object, the acute angle made by the line of sight observation of the object and the horizontal is called the **angle of depression.** See Figure 63(b).

---

*An instrument used in surveying to measure angles.

FIGURE 63

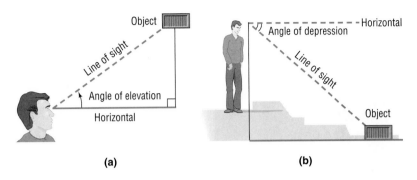

(a)   (b)

EXAMPLE 5

Finding Heights Using the Angle of Elevation

To determine the height of a radio transmission tower, a surveyor walks off a distance of 300 meters from the base of the tower. See Figure 64(a). The angle of elevation is then measured and found to be 40°. If the transit is 2 meters off the ground when the sighting is taken, how high is the radio tower?

FIGURE 64

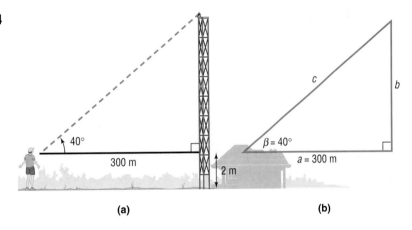

(a)   (b)

Solution   Figure 64(b) shows a triangle that replicates the illustration in Figure 64(a). To find the length b, we use the fact that $\tan \beta = b/a$. Then

$$b = a \tan \beta = 300 \tan 40° = 251.73 \text{ meters}$$

Because the transit is 2 meters high, the actual height of the tower is approximately 254 meters, rounded to the nearest meter.   ■

■ Now work Problem 23.

The idea behind Example 5 can also be used to find the height of an object with a base that is not accessible to the horizontal.

EXAMPLE 6

Finding the Height of a Statue on a Building

Adorning the top of the Board of Trade building in Chicago is a statue of the Greek goddess Ceres, goddess of wheat. From street level, two observations are taken 400 feet from the center of the building. The angle of elevation to the base of the statue is found to be 45.0°; the angle of elevation to the top of the statue is 47.2°. See Figure 65(a). What is the height of the statue?

FIGURE 65

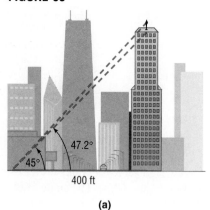

47.2°
45°
400 ft

(a)

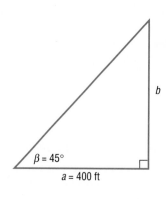

b

$\beta = 45°$
$a = 400$ ft

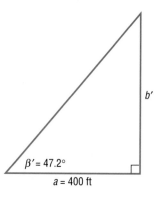

b'

$\beta' = 47.2°$
$a = 400$ ft

(b)

Solution   Figure 65(b) shows two triangles that replicate Figure 65(a). The height of the statue of Ceres will be $b' - b$. To find b and b', we refer to Figure 65(b):

$$\tan 45° = \frac{b}{400} \qquad\qquad \tan 47.2° = \frac{b'}{400}$$

$$b = 400 \tan 45° = 400 \qquad\qquad b' = 400 \tan 47.2° = 431.96$$

The height of the statue is approximately 32 feet.   ∎

When it is not possible to walk off a distance from the base of the object whose height we seek, a more imaginative solution is required.

E X A M P L E   7   *Finding the Height of a Mountain*

To measure the height of a mountain, a surveyor takes two sightings of the peak at a distance 900 meters apart on a direct line to the mountain.* See Figure 66(a). The first observation results in an angle of elevation of 47°, whereas the second results in an angle of elevation of 35°. If the transit is 2 meters high, what is the height h of the mountain?

FIGURE 66

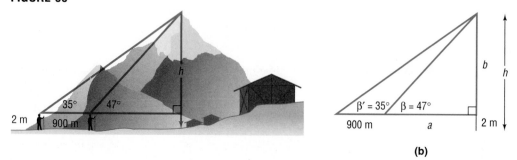

h

35°   47°
2 m   900 m

b
h

$\beta' = 35°$ $\beta = 47°$
900 m a 2 m

(b)

Solution   Figure 66(b) shows two triangles that replicate the illustration in Figure 66(a). From the two right triangles shown, we find

*For simplicity, we assume that these sightings are at the same level.

$$\tan \beta' = \frac{b}{a+900} \qquad \tan \beta = \frac{b}{a}$$

$$\tan 35° = \frac{b}{a+900} \qquad \tan 47° = \frac{b}{a}$$

This is a system of two equations involving two variables, a and b. Because we seek b, we choose to solve the right-hand equation for a and substitute the result, $a = b/\tan 47° = b \cot 47°$, in the left-hand equation. The result is

$$\tan 35° = \frac{b}{b \cot 47° + 900}$$

$$b = (b \cot 47° + 900) \tan 35°$$

$$b = b \cot 47° \tan 35° + 900 \tan 35°$$

$$b(1 - \cot 47° \tan 35°) = 900 \tan 35°$$

$$b = \frac{900 \tan 35°}{1 - \cot 47° \tan 35°} = \frac{900 \tan 35°}{1 - \dfrac{\tan 35°}{\tan 47°}} = 1816$$

The height of the peak from ground level is therefore approximately $1816 + 2 = 1818$ meters. ∎

■ Now work Problem 29.

In navigation, the **direction** or **bearing** from O of an object at P means the positive angle measured clockwise* from the north (N) to the ray OP. See Figure 67. Based on Figure 67, we would say that the bearing of P from O is θ degrees.

FIGURE 67
The bearing of P from O is θ

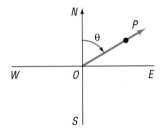

E X A M P L E   8   *Finding the Bearing of an Airplane*

A Boeing 777 aircraft takes off from O'Hare Airport on a runway having a bearing of 22°. After flying for 1 mile, the pilot of the aircraft requests permission to turn 90° and head toward the northwest. The request is granted.

(a) What is the new bearing?

(b) After the plane goes 2 miles in this direction, what bearing should the control tower use to locate the aircraft?

*Notice that this convention is just the opposite of what we are used to.

Solution   (a) Figure 68 illustrates the situation. After flying 1 mile from the airport O (the control tower), the aircraft is at P. After turning 90° toward the northwest, we see that

$$\text{Angle } NOP = 22° \qquad \text{Angle } RQN = 90° - 22° = 68°$$

FIGURE 68

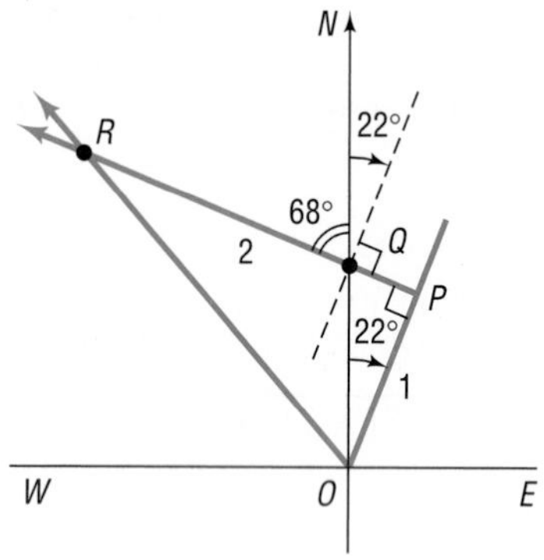

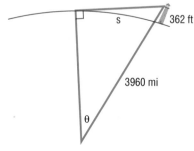

The bearing of this aircraft is therefore

$$360° - \text{Angle } RQN = 360° - 68° = 292°$$

(b) After flying 2 miles at a bearing of 292°, the aircraft is at R. If $\theta = $ Angle ROP, then

$$\tan \theta = \frac{2}{1} = 2 \quad \text{so} \quad \theta = 63.4°$$

As a result,

$$\text{Angle } RON = \theta - 22° = 41.4°$$

The bearing of the plane from the control tower at O is

$$360° - \text{Angle } RON = 360° - 41.4° = 318.6°$$ ■

E X A M P L E   9

The Gibb's Hill Lighthouse, Southampton, Bermuda

In operation since 1846, it stands 117 feet high on a hill 245 feet high, so its beam of light is 362 feet above sea level. A brochure states that the light can be seen on the horizon about 26 miles distant. Verify the accuracy of this statement.

Solution   Figure 69 illustrates the situation. The central angle θ obeys the equation

FIGURE 69

![Figure 69: right triangle with sides labeled s, 362 ft, 3960 mi, and angle θ]

$$\cos \theta = \frac{3960}{3960 + \dfrac{362}{5280}} = 0.999982687$$

so

$$\theta = 0.00588 \text{ radian} = 0.33715° = 20.23'$$

The brochure does not indicate whether the distance is measured in nautical miles or in statute miles. The distance s in nautical miles is 20.23, the measurement of θ in minutes. (Refer to Problem 83, Section 5.1). The distance s in statute miles is

$$s = r\theta = 3960(0.00588) = 23.3 \text{ miles}$$

In either case, it would seem that the brochure overstated the distance somewhat. ■

5.5

Exercise 5.5

In Problems 1–14, use the right triangle shown in the margin. Then, using the given information, solve the triangle.

1. $b = 5$, $\beta = 20°$;   find a, c, and α
2. $b = 4$, $\beta = 10°$;   find a, c, and α
3. $a = 6$, $\beta = 40°$;   find b, c, and α
4. $a = 7$, $\beta = 50°$;   find b, c, and α
5. $b = 4$, $\alpha = 10°$;   find a, c, and β
6. $b = 6$, $\alpha = 20°$;   find a, c, and β

7. $a = 5$, $\alpha = 25°$; find b, c, and β

8. $a = 6$, $\alpha = 40°$; find b, c, and β

9. $c = 9$, $\beta = 20°$; find b, a, and α

10. $c = 10$, $\alpha = 40°$; find b, a, and β

11. $a = 5$, $b = 3$; find c, α, and β

12. $a = 2$, $b = 8$; find c, α, and β

13. $a = 2$, $c = 5$; find b, α, and β

14. $b = 4$, $c = 6$; find a, α, and β

15. A right triangle has a hypotenuse of length 3 inches. If one angle is 35°, find the length of each leg.

16. A right triangle has a hypotenuse of length 2 centimeters. If one angle is 40°, find the length of each leg.

17. A right triangle contains a 35° angle. If one leg is of length 5 inches, what is the length of the hypotenuse? [*Hint:* Two answers are possible.]

18. A right triangle contains an angle of $\pi/10$ radian. If one leg is of length 3 meters, what is the length of the hypotenuse? [*Hint:* Two answers are possible.]

19. The hypotenuse of a right triangle is 5 inches. If one leg is 2 inches, find the degree measure of each angle.

20. The hypotenuse of a right triangle is 3 feet. If one leg is 1 foot, find the degree measure of each angle.

21. *Finding the Width of a Gorge* Find the distance from A to C across the gorge illustrated in the figure.

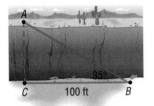

22. *Finding the Distance Across a Pond* Find the distance from A to C across the pond illustrated in the figure.

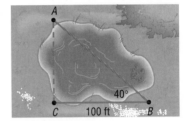

23. *The Eiffel Tower* The tallest tower built before the era of television masts, the Eiffel Tower was completed on March 31, 1889. Find the height of the Eiffel Tower (before a television mast was added to the top) using the information given in the illustration.

24. *Finding the Distance of a Ship from Shore* A ship, offshore from a vertical cliff known to be 100 feet in height, takes a sighting of the top of the cliff. If the angle of elevation is found to be 25°, how far offshore is the ship?

25. Suppose that you are headed toward a plateau 50 meters high. If the angle of elevation to the top of the plateau is 20°, how far are you from the base of the plateau?

26. *Statue of Liberty* A ship is just offshore of New York City. A sighting is taken of the Statue of Liberty, which is about 305 feet tall. If the angle of elevation to the top of the statue is 20°, how far is the ship from the base of the statue?

27. A 22 foot extension ladder leaning against a building makes a 70° angle with the ground. How far up the building does the ladder touch?

28. *Finding the Height of a Building* To measure the height of a building, two sightings are taken a distance of 50 feet apart. If the first angle of elevation is 40° and the second is 32°, what is the height of the building?

29. *Great Pyramid of Cheops* One of the original Seven Wonders of the World, the Great Pyramid of Cheops was built about 2580 BC. Its original height was 480 feet 11 inches, but due to the loss of its topmost stones, it is now shorter.* Find the current height of the Great Pyramid, using the information given in the illustration.

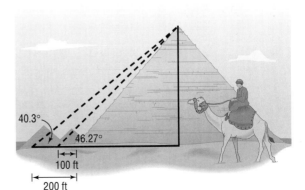

**Source: Guinness Book of World Records.*

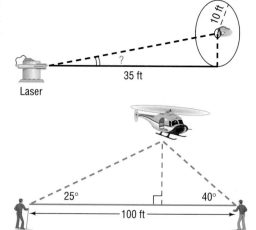

30. A laser beam is to be directed through a small hole in the center of a circle of radius 10 feet. The origin of the beam is 35 feet from the circle (see the figure). At what angle of elevation should the beam be aimed to ensure that it goes through the hole?

31. *Finding the Angle of Elevation of the Sun* At 10 AM on April 26, 1996, a building 300 feet high casts a shadow 50 feet long. What is the angle of elevation of the Sun?

32. *Mt. Rushmore* To measure the height of Lincoln's caricature on Mt. Rushmore, two sightings 800 feet from the base of the mountain are taken. If the angle of elevation to the bottom of Lincoln's face is 32° and the angle of elevation to the top is 35°, what is the height of Lincoln's face?

33. *Finding the Height of a Helicopter* Two observers simultaneously measure the angle of elevation of a helicopter. One angle is measured as 25°, the other as 40° (see the figure). If the observers are 100 feet apart and the helicopter lies over the line joining them, how high is the helicopter?

34. *Finding the Distance between Two Objects* A blimp, suspended in the air at a height of 500 feet, lies directly over a line from Soldier Field to the Adler Planetarium on Lake Michigan (see the figure). If the angle of depression from the blimp to the stadium is 32° and from the blimp to the planetarium is 23°, find the distance between Soldier Field and the Adler Planetarium.

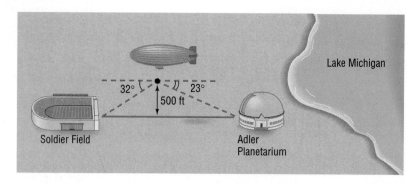

35. *Finding the Length of a Guy Wire* A radio transmission tower is 200 feet high. How long should a guy wire be if it is to be attached to the tower 10 feet from the top and is to make an angle of 21° with the ground?

36. *Finding the Height of a Tower* A guy wire 80 feet long is attached to the top of a radio transmission tower, making an angle of 25° with the ground. How high is the tower?

37. *Washington Monument* The angle of elevation of the Washington Monument is 35.1° at the instant it casts a shadow 789 feet long. Use this information to calculate the height of the monument.

38. *Finding the Length of a Mountain Trail* A straight trail with a uniform inclination of 17° leads from a hotel at an elevation of 9000 feet to a mountain lake at an elevation of 11,200 feet. What is the length of the trail?

39. *Finding the Speed of a Truck* A state trooper is hidden 30 feet from a highway. One second after a truck passes, the angle θ between the highway and the line of observation from the patrol car to the truck is measured. See the illustration.
  (a) If the angle measures 15°, how fast is the truck traveling? Express the answer in feet per second and in miles per hour.
  (b) If the angle measures 20°, how fast is the truck traveling? Express the answer in feet per second and in miles per hour.
  (c) If the speed limit is 55 miles per hour and a speeding ticket is issued for speeds of 5 miles per hour or more over the limit, for what angles should the trooper issue a ticket?

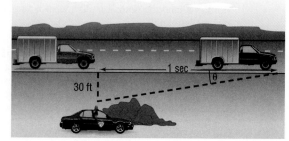

40. *The Gibb's Hill Lighthouse, Southampton, Bermuda.* In operation since 1846, it stands 117 feet high on a hill 245 feet high, so its beam of light is 362 feet above sea level. A brochure states that ships 40 miles away can see the light and planes flying at 10,000 feet can see it 120 miles away. Verify the accuracy of these statements. What assumption did the brochure make about the height of the ship? (See illustration, Section 1.1, Problem 115.)

41. *Finding the Bearing of an Aircraft* A DC-9 aircraft leaves Midway Airport from a runway whose bearing is 40°. After flying for 1/2 mile, the pilot requests permission to turn 90° and head toward the southeast. The permission is granted.
(a) What is the new bearing?
(b) After the airplane goes 1 mile in this direction, what bearing should the control tower use to locate the aircraft?

42. *Finding the Bearing of a Ship* A ship leaves the port of Miami with a bearing of 100° and a speed of 15 knots. After 1 hour, the ship turns 90° toward the south.
(a) What is the new bearing?
(b) After 2 hours, maintaining the same speed, what is the bearing of the ship from port?

43. *Shooting Free Throws in Basketball* The eyes of a basketball player are 6 feet above the floor. The player is at the free-throw line, which is 15 feet from the center of the basket rim (see the figure). What is the angle of elevation from the player's eyes to the center of the rim? [*Hint:* The rim is 10 feet above the floor.]

44. *Finding the Pitch of a Roof* A carpenter is preparing to put a roof on a garage that is 20 feet by 40 feet by 20 feet. A steel support beam 46 feet in length is positioned in the center of the garage. To support the roof, another beam will be attached to the top of the center beam (see the figure). At what angle of elevation is the new beam? In other words, what is the pitch of the roof?

45. *Determining Distances At Sea* The navigator of a ship at sea spots two lighthouses she recognizes as ones 3 miles apart along a straight seashore. She determines that the angles formed between two line of sight observations of the lighthouses and the line from the ship directly to shore are 15° and 35°. See the illustration.
(a) How far is the ship from shore?
(b) How far is the ship from lighthouse *A*?
(c) How far is the ship from lighthouse *B*?

46. *Constructing a Highway* A highway whose primary directions are North-South is being constructed along the west coast of Florida. Near Naples, a bay obstructs the straight path of the road. Since the cost of a bridge is prohibitive, engineers decide to go around the bay. The illustration shows the path they decide upon and the measurements taken. What is the length of highway needed to go around the bay?

43.

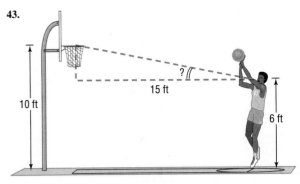

44.

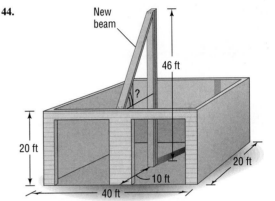

45.

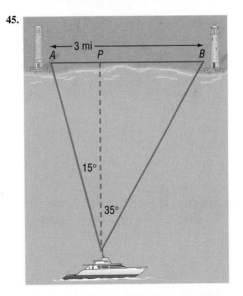

46.

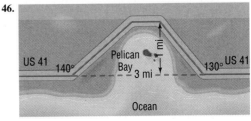

47. *Surveillance Satellites* A surveillance satellite circles Earth at a height of h miles above the surface. Suppose d is the distance, in miles, on the surface of the Earth that can be observed from the satellite. See the illustration.
(a) Find an equation that relates the central angle θ to the height h.
(b) Find an equation that relates the observable distance d and θ.
(c) Find an equation that relates d and h.
(d) If d is to be 2500 miles, how high must the satellite orbit above Earth?
(e) If the satellite orbits at a height of 300 miles, what distance d on the surface can be observed?

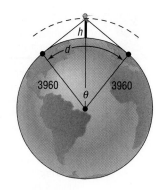

 48. Compare Problem 115 in Section 1.1 with Example 9 and Problem 40 of this section. One uses the Pythagorean Theorem in its solution, while the others use trigonometry. Write a short paper that contrasts the two methods of solution. Be sure to point out the benefits as well as the shortcomings of each method. Then decide which solution you prefer. Be sure to have reasons.

Chapter Review

THINGS TO KNOW

Definitions

Standard position of an angle	Vertex is at the origin; initial side is along the positive x-axis
Degree (1°)	$1° = \frac{1}{360}$ revolution
Radian (1)	The measure of a central angle whose rays subtend an arc whose length is the radius of the circle
Unit circle	Center at origin; radius = 1 Equation: $x^2 + y^2 = 1$
Trigonometric functions	$P = (a, b)$ is the point on the unit circle corresponding to $\theta = t$ radians;

$$\sin t = \sin \theta = b \qquad\qquad \cos t = \cos \theta = a$$

$$\tan t = \tan \theta = \frac{b}{a}, \quad a \neq 0 \qquad \cot t = \cot \theta = \frac{a}{b}, \quad b \neq 0$$

$$\csc t = \csc \theta = \frac{1}{b}, \quad b \neq 0 \qquad \sec t = \sec \theta = \frac{1}{a}, \quad a \neq 0$$

Periodic function	$f(\theta + p) = f(\theta)$, for all θ, $p > 0$, where the smallest such p is the fundamental period
Acute angle	An angle θ whose measure is $0° < \theta < 90°$ (or $0 < \theta < \pi/2$)
Complementary angles	Two acute angles whose sum is $90°$ ($\pi/2$)
Cofunction	The following pairs of functions are cofunctions of each other: sine and cosine; tangent and cotangent; secant and cosecant
Reference angle of θ	The acute angle formed by the terminal side of θ and either the positive or negative x-axis

Formulas

1 revolution = 360°
$\qquad\quad = 2\pi$ radians

$s = r\theta$	θ is measured in radians; s is the length of arc subtended by the central angle θ of the circle of radius r.
$v = r\omega$	v is the linear speed along the circle of radius r; ω is the angular speed (measured in radians per unit time)

Table of Values

θ (RADIANS)	θ (DEGREES)	$\sin \theta$	$\cos \theta$	$\tan \theta$	$\csc \theta$	$\sec \theta$	$\cot \theta$
0	0°	0	1	0	Not defined	1	Not defined
$\pi/6$	30°	$\frac{1}{2}$	$\sqrt{3}/2$	$\sqrt{3}/3$	2	$2\sqrt{3}/3$	$\sqrt{3}$
$\pi/4$	45°	$\sqrt{2}/2$	$\sqrt{2}/2$	1	$\sqrt{2}$	$\sqrt{2}$	1
$\pi/3$	60°	$\sqrt{3}/2$	$\frac{1}{2}$	$\sqrt{3}$	$2\sqrt{3}/3$	2	$\sqrt{3}/3$
$\pi/2$	90°	1	0	Not defined	1	Not defined	0
π	180°	0	-1	0	Not defined	-1	Not defined
$3\pi/2$	270°	-1	0	Not defined	-1	Not defined	0

Properties of the Trigonometric Functions

$f(\theta) = \sin \theta$

Domain: all real numbers
Range: all real numbers from -1 to 1, inclusive
Periodic: period $= 2\pi$ (360°)
Odd function

$f(\theta) = \cos \theta$

Domain: all real numbers
Range: all real numbers from -1 to 1, inclusive
Periodic: period $= 2\pi$ (360°)
Even function

$f(\theta) = \tan \theta$

Domain: all real numbers, except odd multiples of $\pi/2$ (90°)
Range: all real numbers
Periodic: period $= \pi$ (180°)
Odd function

$f(\theta) = \csc \theta$

Domain: all real numbers, except integral multiples of π (180°)
Range: all real numbers greater than or equal to 1 or less than or equal to -1
Periodic: period $= 2\pi$ (360°)
Odd function

$f(\theta) = \sec \theta$

Domain: all real numbers, except odd multiples of $\pi/2$ (90°)
Range: all real numbers greater than or equal to 1 or less than or equal to -1
Periodic: period $= 2\pi$ (360°)
Even function

$f(\theta) = \cot \theta$

Domain: all real numbers, except integral multiples of π (180°)
Range: all real numbers
Periodic: period $= \pi$ (180°)
Odd function

Identities

$$\tan \theta = \frac{\sin \theta}{\cos \theta}, \quad \cot \theta = \frac{\cos \theta}{\sin \theta}$$

$$\cot \theta = \frac{1}{\tan \theta}, \quad \sec \theta = \frac{1}{\cos \theta}, \quad \csc \theta = \frac{1}{\sin \theta}$$

$$\sin^2 \theta + \cos^2 \theta = 1, \quad \tan^2 \theta + 1 = \sec^2 \theta, \quad 1 + \cot^2 \theta = \csc^2 \theta$$

How To

Convert an angle from radian measure to degree measure

Convert an angle from degree measure to radian measure

Find the value of each of the remaining trigonometric functions if the value of one function and the quadrant of the angle are given

Use the theorem on cofunctions of complementary angles

Use reference angles to find the value of a trigonometric function

Use a calculator to find the value of a trigonometric function

Use right triangle trigonometry to solve right triangles and applied problems

Fill-in-the-Blank Items

1. Two rays drawn with a common vertex form a(n) _____. One of the rays is called the _____ _____; the other is called the _____ _____.

2. In the formula $s = r\theta$ for measuring the length s of arc along a circle of radius r, the angle θ must be measured in _____.

3. 180 degrees = _____ radians.

4. Two acute angles whose sum is a right angle are called _____.

5. The sine and _____ functions are cofunctions.

6. An angle is in _____ _____ if its vertex is at the origin and its initial side coincides with the positive x-axis.

7. The reference angle of 135° is _____.

8. The sine, cosine, cosecant, and secant functions have period _____; the tangent and cotangent functions have period _____.

True/False Items

T   F   1. In the formula $s = r\theta$, r is the radius of a circle and s is the arc subtended by a central angle θ, where θ is measured in degrees.

T   F   2. $|\sin \theta| \le 1$

T   F   3. $1 + \tan^2 \theta = \csc^2 \theta$

T   F   4. The only even trigonometric functions are the cosine and secant functions.

T   F   5. $\tan 62° = \cot 38°$

T   F   6. $\sin 182° = \cos 2°$

Review Exercises

In Problems 1–4, convert each angle in degrees to radians. Express your answer as a multiple of π.

1. 135°   2. 210°   3. 18°   4. 15°

In Problems 5–8, convert each angle in radians to degrees.

5. $3\pi/4$ 6. $2\pi/3$ 7. $-5\pi/2$ 8. $-3\pi/2$

In Problems 9–30, find the exact value of each expression. Do not use a calculator.

9. $\tan \dfrac{\pi}{4} - \sin \dfrac{\pi}{6}$

10. $\cos \dfrac{\pi}{3} + \sin \dfrac{\pi}{2}$

11. $3 \sin 45° - 4 \tan \dfrac{\pi}{6}$

12. $4 \cos 60° + 3 \tan \dfrac{\pi}{3}$

13. $6 \cos \dfrac{3\pi}{4} + 2 \tan\left(-\dfrac{\pi}{3}\right)$

14. $3 \sin \dfrac{2\pi}{3} - 4 \cos \dfrac{5\pi}{2}$

15. $\sec\left(-\dfrac{\pi}{3}\right) - \cot\left(-\dfrac{5\pi}{4}\right)$

16. $4 \csc \dfrac{3\pi}{4} - \cot\left(-\dfrac{\pi}{4}\right)$

17. $\tan \pi + \sin \pi$

18. $\cos \dfrac{\pi}{2} - \csc\left(-\dfrac{\pi}{2}\right)$

19. $\cos 180° - \tan(-45°)$

20. $\sin 270° + \cos(-180°)$

21. $\sin^2 20° + \dfrac{1}{\sec^2 20°}$

22. $\dfrac{1}{\cos^2 40°} - \dfrac{1}{\cot^2 40°}$

23. $\sec 50° \cos 50°$

24. $\tan 10° \cot 10°$

25. $\dfrac{\sin 50°}{\cos 40°}$

26. $\dfrac{\tan 20°}{\cot 70°}$

27. $\dfrac{\sin(-40°)}{\cos 50°}$

28. $\tan(-20°) \cot 20°$

29. $\sin 400° \sec(-50°)$

30. $\cot 200° \cot(-70°)$

In Problems 31–46, find the exact value of each of the remaining trigonometric functions.

31. $\sin \theta = -\frac{4}{5}, \quad \cos \theta > 0$

32. $\cos \theta = -\frac{3}{5}, \quad \sin \theta < 0$

33. $\tan \theta = \frac{12}{5}, \quad \sin \theta < 0$

34. $\cot \theta = \frac{12}{5}, \quad \cos \theta < 0$

35. $\sec \theta = -\frac{5}{4}, \quad \tan \theta < 0$

36. $\csc \theta = -\frac{5}{3}, \quad \cot \theta < 0$

37. $\sin \theta = \frac{12}{13}, \quad \theta$ in quadrant II

38. $\cos \theta = -\frac{3}{5}, \quad \theta$ in quadrant III

39. $\sin \theta = -\frac{5}{13}, \quad 3\pi/2 < \theta < 2\pi$

40. $\cos \theta = \frac{12}{13}, \quad 3\pi/2 < \theta < 2\pi$

41. $\tan \theta = \frac{1}{3}, \quad 180° < \theta < 270°$

42. $\tan \theta = -\frac{2}{3}, \quad 90° < \theta < 180°$

43. $\sec \theta = 3, \quad 3\pi/2 < \theta < 2\pi$

44. $\csc \theta = -4, \quad \pi < \theta < 3\pi/2$

45. $\cot \theta = -2, \quad \pi/2 < \theta < \pi$

46. $\tan \theta = -2, \quad 3\pi/2 < \theta < 2\pi$

In Problems 47–50, solve each triangle.

47.

48.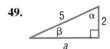

49.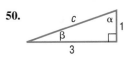

50.

51. Find the length of arc subtended by a central angle of $30°$ on a circle of radius 2 feet.

52. The minute hand of a clock is 8 inches long. How far does the tip of the minute hand move in 30 minutes? How far does it move in 20 minutes?

53. *Angular Speed of a Race Car* A race car is driven around a circular track at a constant speed of 180 miles per hour. If the diameter of the track is $\frac{1}{2}$ mile, what is the angular speed of the car? Express your answer in revolutions per hour (which is equivalent to laps per hour).

54. *Angular Speed of a Race Car* Repeat Problem 53 if the car goes only 150 miles per hour.

55. *Measuring the Length of a Lake* From a stationary hot air balloon 500 feet above the ground, two sightings of a lake are made (see the figure). How long is the lake?

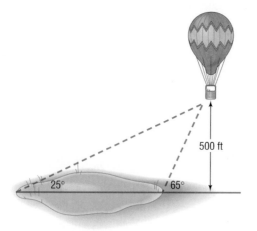

56. *Finding the Speed of a Glider* From a glider 200 feet above the ground, two sightings of a stationary object directly in front are taken 1 minute apart (see the figure). What is the speed of the glider?

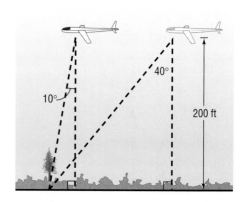

57. *Finding the Width of a River* Find the distance from *A* to *C* across the river illustrated in the figure.

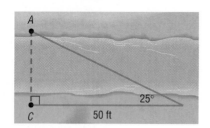

58. *Finding the Height of a Building* Find the height of the building shown in this figure.

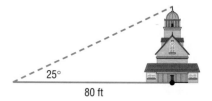

59. *Finding the Distance to Shore* The Sears Tower in Chicago is 1454 feet tall and is situated about 1 mile inland from the shore of Lake Michigan, as indicated in the figure. An observer in a pleasure boat on the lake directly in front of the Sears Tower looks at the top of the tower and measures the angle of elevation as 5°. How far offshore is the boat?

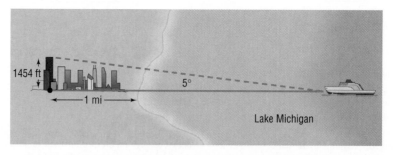

60. *Finding the Grade of a Mountain Trail* A straight trail with a uniform inclination leads from a hotel, elevation 5000 feet, to a lake in a valley, elevation 4100 feet. The length of the trail is 4100 feet. What is the inclination (grade) of the trail?

PREPARING FOR THIS CHAPTER

Before getting started on this chapter, review the following concepts:

Graph of a function (p. 101)
Values of the trigonometric functions
   of certain angles (Table 2, p. 338; Table 3, p. 341)
Periodic functions (pp. 350–351)
Graphing techniques (Section 2.3)
Inverse functions (pp. 153–156)

Preview   Touch-Tone Phones

On a Touch-Tone phone, each button produces a unique sound. The
sound produced is the sum of two tones, given by

$$y = \sin 2\pi l t \quad \text{and} \quad y = \sin 2\pi h t$$

where l and h are the low and high frequencies (cycles per second) of the
two tones. For example, if you touch 7, the low frequency is $l = 852$
cycles per second and the high frequency is $h = 1209$ cycles per second.
The sound emitted by touching 7 is

$$y = \sin 2\pi(852)t + \sin 2\pi(1209)t$$

Graph the sound emitted by touching 7.

[Problem 29 in Exercise 6.5] ■

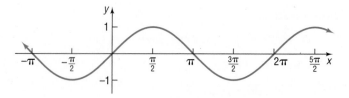

n this chapter we graph the six trigonometric functions. After discussing the
graphs of the sine and cosine functions,
applications are given to damped vibrations and
simple harmonic motion. Then we graph the remaining trigonometric functions and conclude the
chapter with a discussion of the inverse trigonometric functions.

6.1

Graphs of the Sine and Cosine Functions

In Chapter 5 we defined the trigonometric functions $f(\theta) = \sin \theta$, $f(\theta) = \cos \theta$, and so on. In this chapter, we shall use the traditional symbols x to represent the independent variable (or argument) and y for the dependent variable (or value at x) for each function. Thus, we write the six trigonometric functions as

$$y = \sin x \qquad y = \cos x \qquad y = \tan x$$
$$y = \csc x \qquad y = \sec x \qquad y = \cot x$$

Our purpose in this section is to graph each of these functions. Unless indicated otherwise, we shall use radian measure throughout for the independent variable x.

The Graph of $y = \sin x$

Since the sine function has period 2π (refer to Section 7.6), we need to graph $y = \sin x$ only on the interval $[0, 2\pi]$. The remainder of the graph will consist of repetitions of this portion of the graph.

We begin by constructing Table 1, which lists some points on the graph of $y = \sin x$, $0 \le x \le 2\pi$. As the table shows, the graph of $y = \sin x$, $0 \le x \le 2\pi$, begins at the origin. As x increases from 0 to $\pi/2$, the value of $y = \sin x$ increases from 0 to 1; as x increases from $\pi/2$ to π to $3\pi/2$, the value of y decreases from 1 to 0 to -1; as x increases from $3\pi/2$ to 2π, the value of y increases from -1 to 0. If we plot the points listed in Table 1 and connect them with a smooth curve, we obtain the graph shown in Figure 1.

TABLE 1

x	$y = \sin x$	(x, y)
0	0	$(0, 0)$
$\pi/6$	$\frac{1}{2}$	$(\pi/6, \frac{1}{2})$
$\pi/2$	1	$(\pi/2, 1)$
$5\pi/6$	$\frac{1}{2}$	$(5\pi/6, \frac{1}{2})$
π	0	$(\pi, 0)$
$7\pi/6$	$-\frac{1}{2}$	$(7\pi/6, -\frac{1}{2})$
$3\pi/2$	-1	$(3\pi/2, -1)$
$11\pi/6$	$-\frac{1}{2}$	$(11\pi/6, -\frac{1}{2})$
2π	0	$(2\pi, 0)$

FIGURE 1
$y = \sin x$,
$0 \le x \le 2\pi$

The graph in Figure 1 is one period of the graph of $y = \sin x$. To obtain a more complete graph of $y = \sin x$, we repeat this period in each direction, as shown in Figure 2.

FIGURE 2
$y = \sin x$,
$-\infty < x < \infty$

The graph of $y = \sin x$ illustrates some of the facts we already know about the sine function:

Characteristics of the Sine Function	

Characteristics
of the Sine Function

1. The domain is the set of all real numbers.
2. The range consists of all real numbers from -1 to 1, inclusive.
3. The sine function is an odd function, as the symmetry of the graph with respect to the origin indicates.
4. The sine function is periodic, with period 2π.
5. The x-intercepts are $\ldots, -2\pi, -\pi, 0, \pi, 2\pi, 3\pi, \ldots$; the y-intercept is 0.
6. The maximum value is 1 and occurs at $x = \ldots, -3\pi/2, \pi/2, 5\pi/2, 9\pi/2, \ldots$; the minimum value is -1 and occurs at $x = \ldots, -\pi/2, 3\pi/2, 7\pi/2, 11\pi/2, \ldots$.

■   Now work Problems 1, 3, and 5.

The graphing techniques introduced in Chapter 2 may be used to graph functions that are variations of the sine function.

EXAMPLE   1   *Graphing Variations of $y = \sin x$ Using Shifts, Reflections, and the Like*

Use the graph of $y = \sin x$ to graph $y = -\sin x + 2$.

Solution   Figure 3 illustrates the steps.

FIGURE 3

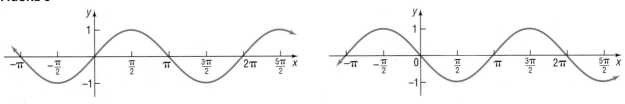

(a) $y = \sin x$

Reflection about x-axis

(b) $y = -\sin x$

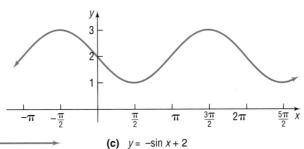

(c) $y = -\sin x + 2$

Vertical shift

Check: Graph $y = -\sin x + 2$ and compare the result with Figure 3(c).   ■

EXAMPLE   2   *Graphing Variations of $y = \sin x$ Using Shifts, Reflections, and the Like*

Use the graph of $y = \sin x$ to graph $y = \sin\left(x - \dfrac{\pi}{4}\right)$.

Solution   Figure 4 illustrates the steps.

FIGURE 4

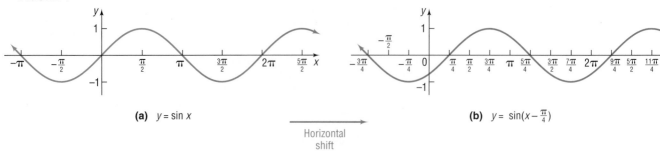

(a) $y = \sin x$

Horizontal
shift

(b) $y = \sin(x - \frac{\pi}{4})$

 Check: Graph $y = \sin\left(x - \dfrac{\pi}{4}\right)$ and compare the result with Figure 4(b). ■

■ Now work Problem 11.

The Graph of $y = \cos x$

The cosine function also has period 2π. Thus, we proceed as we did with the sine function by constructing Table 2, which lists some points on the graph of $y = \cos x$, $0 \le x \le 2\pi$. As the table shows, the graph of $y = \cos x$, $0 \le x \le 2\pi$, begins at the point $(0, 1)$. As x increases from 0 to $\pi/2$ to π, the value of y decreases from 1 to 0 to -1; as x increases from π to $3\pi/2$ to 2π, the value of y increases from -1 to 0 to 1. As before, we plot the points in Table 2 to get one period of the graph of $y = \cos x$. See Figure 5.

TABLE 2

x	$y = \cos x$	(x, y)
0	1	$(0, 1)$
$\pi/3$	$\frac{1}{2}$	$(\pi/3, \frac{1}{2})$
$\pi/2$	0	$(\pi/2, 0)$
$2\pi/3$	$-\frac{1}{2}$	$(2\pi/3, -\frac{1}{2})$
π	-1	$(\pi, -1)$
$4\pi/3$	$-\frac{1}{2}$	$(4\pi/3, -\frac{1}{2})$
$3\pi/2$	0	$(3\pi/2, 0)$
$5\pi/3$	$\frac{1}{2}$	$(5\pi/3, \frac{1}{2})$
2π	1	$(2\pi, 1)$

FIGURE 5

$y = \cos x$,
$0 \le x \le 2\pi$

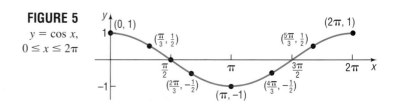

A more complete graph of $y = \cos x$ is obtained by repeating this period in each direction, as shown in Figure 6.

FIGURE 6

$y = \cos x$, $-\infty < x < \infty$

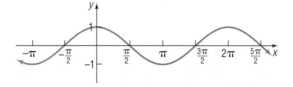

The graph of $y = \cos x$ illustrates some of the facts we already know about the cosine function:

Characteristics of the Cosine Function

1. The domain is the set of all real numbers.
2. The range consists of all real numbers from -1 to 1, inclusive.
3. The cosine function is an even function, as the symmetry of the graph with respect to the y-axis indicates.
4. The cosine function is periodic, with period 2π.
5. The x-intercepts are . . . , $-3\pi/2$, $-\pi/2$, $\pi/2$, $3\pi/2$, $5\pi/2$, . . . ; the y-intercept is 1.
6. The maximum value is 1 and occurs at $x =$ . . . , -2π, 0, 2π, 4π, 6π, . . . ; the minimum value is -1 and occurs at $x =$ . . . , $-\pi$, π, 3π, 5π, ...

Again, the graphing techniques from Chapter 2 may be used to graph variations of the cosine function.

EXAMPLE 3   *Graphing Variations of $y = \cos x$ Using Shifts, Reflections, and the Like*

Use the graph of $y = \cos x$ to graph $y = 2 \cos x$

Solution   Figure 7 illustrates the graph, which is a vertical stretch of the graph of $y = \cos x$.

FIGURE 7
$y = 2 \cos x$

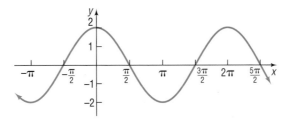

Check: Graph $y = 2 \cos x$ and compare the result with Figure 7.   ■

EXAMPLE 4   *Graphing Variations of $y = \cos x$ Using Shifts, Reflections, and the Like*

Use the graph of $y = \cos x$ to graph $y = \cos 3x$.

Solution   Figure 8 illustrates the graph, which is a horizontal compression of the graph of $y = \cos x$. Notice that, due to this compression, the period of $y = \cos 3x$ is $2\pi/3$, whereas the period of $y = \cos x$ is 2π. We shall comment more about this in the next section.

FIGURE 8
$y = \cos 3x$

Check: Graph $y = \cos 3x$. Use TRACE to verify that the period is $2\pi/3$.   ■

■   Now work Problem 13.

6.1

Exercise 6.1

In Problems 1–10, refer to the graphs of y = sin x and y = cos x to answer each question, if necessary.

1. What is the y-intercept of $y = \sin x$?

2. What is the y-intercept of $y = \cos x$?

3. For what numbers x, $-\pi \le x \le \pi$ is the graph of $y = \sin x$ increasing?

4. For what numbers x, $-\pi \le x \le \pi$ is the graph of $y = \cos x$ decreasing?

5. What is the largest value of $y = \sin x$?

6. What is the smallest value of $y = \cos x$?

7. For what numbers x, $0 \le x \le 2\pi$, does $\sin x = 0$?

8. For what numbers x, $0 \le x \le 2\pi$, does $\cos x = 0$?

9. For what numbers x, $-2\pi \le x \le 2\pi$, does $\sin x = 1$? What about $\sin x = -1$?

10. For what numbers x, $-2\pi \le x \le 2\pi$, does $\cos x = 1$? What about $\cos x = -1$?

In Problems 11–26, graph each function. Show at least one period.

11. $y = 3 \sin x$

12. $y = 4 \cos x$

13. $y = \cos\left(x + \dfrac{\pi}{4}\right)$

14. $y = \sin(x - \pi)$

15. $y = \sin x - 1$

16. $y = \cos x + 1$

17. $y = -2 \sin x$

18. $y = -3 \cos x$

19. $y = \sin \pi x$

20. $y = \cos \dfrac{\pi}{2} x$

21. $y = 2 \sin x + 2$

22. $y = 3 \cos x + 3$

23. $y = -2 \cos\left(x - \dfrac{\pi}{2}\right)$

24. $y = -3 \sin\left(x + \dfrac{\pi}{2}\right)$

25. $y = 3 \sin (\pi - x)$

26. $y = 2 \cos (\pi - x)$

 27. Graph

$$y = \sin x \quad \text{and} \quad y = \cos\left(x - \dfrac{\pi}{2}\right)$$

Do you think that $\sin x = \cos\left(x - \dfrac{\pi}{2}\right)$?

28. Graph $y = \sin x$, $y = 2 \sin x$, $y = \frac{1}{2} \sin x$, and $y = 8 \sin x$. What do you conclude about the graph of $y = A \sin x$, $A > 0$?

29. Graph $y = \sin x$, $y = \sin 2x$, $y = \sin 4x$, and $y = \sin \frac{1}{2}x$. What do you conclude about the graph of $y = \sin \omega x$?

30. Graph $y = \sin x$, $y = \sin[x - (\pi/3)]$, $y = \sin[x - (\pi/4)]$, and $y = \sin[x - (\pi/6)]$. What do you conclude about the graph of $y = \sin(x - \phi)$, $\phi > 0$?

6.2

Sinusoidal Graphs

The graph of $y = \cos x$, when compared to the graph of $y = \sin x$, suggests that the graph of $y = \sin x$ is the same as the graph of $y = \cos x$ after a horizontal shift of $\pi/2$ units to the right. See Figure 9.

FIGURE 9

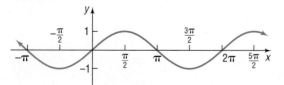

(a) $y = \sin x$

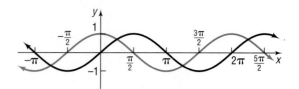

(b) $y = \cos x$ $y = \cos\left(x - \dfrac{\pi}{2}\right)$

Based on Figure 9, we conjecture that

$$\sin x = \cos\left(x - \frac{\pi}{2}\right)$$

(We shall prove this fact in Chapter 7.) Because of this similarity, the graphs of sine functions and cosine functions are referred to as **sinusoidal graphs.**

 Check: Graph $y = \sin x$ and $y = \cos\left(x - \dfrac{\pi}{2}\right)$. How many graphs do you see?

Let's look at some general characteristics of sinusoidal graphs.

In Example 3 of the previous section, we obtained the graph of $y = 2 \cos x$, which we reproduce in Figure 10. Notice that the values of $y = 2 \cos x$ lie between -2 and 2, inclusive.

FIGURE 10

$y = 2 \cos x$

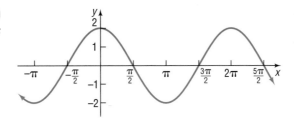

In general, the values of the functions $y = A \sin x$ and $y = A \cos x$, where $A \neq 0$, will always satisfy the inequalities.

$$-|A| \leq A \sin x \leq |A| \quad \text{and} \quad -|A| \leq A \cos x \leq |A|$$

respectively. The number $|A|$ is called the **amplitude** of $y = A \sin x$ or $y = A \cos x$. See Figure 11.

FIGURE 11

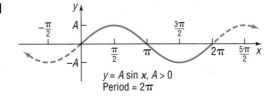

$y = A \sin x, A > 0$
Period $= 2\pi$

In Example 4 of Section 6.1, we obtained the graph of $y = \cos 3x$, which we reproduce in Figure 12. Notice that the period of this function is $2\pi/3$.

FIGURE 12

$y = \cos 3x$

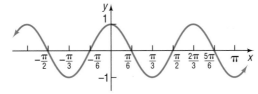

In general, if $\omega > 0$, the functions $y = \sin \omega x$ and $y = \cos \omega x$ will have period $T = 2\pi/\omega$. To see why, recall that the graph of $y = \sin \omega x$ is obtained from the graph of $y = \sin x$ by performing an appropriate horizontal compression or stretch. This horizontal compression replaces the interval $[0, 2\pi]$, which contains one period of the graph of $y = \sin x$, by the interval $[0, 2\pi/\omega]$, which contains one period of the graph of $y = \sin \omega x$. Thus, the period of the functions $y = \sin \omega x$ and $y = \cos \omega x$, $\omega > 0$, is $2\pi/\omega$. See Figure 13.

FIGURE 13

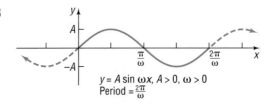

$y = A \sin \omega x, A > 0, \omega > 0$
Period $= \frac{2\pi}{\omega}$

If $\omega < 0$ in $y = \sin \omega x$ or $y = \cos \omega x$, we use the facts that

$$\sin \omega x = -\sin(-\omega x) \quad \text{and} \quad \cos \omega x = \cos(-\omega x)$$

to obtain an equivalent form in which the coefficient of x is positive.

Theorem   If $\omega > 0$, the amplitude and period of $y = A \sin \omega x$ and $y = A \cos \omega x$ are given by

$$\text{Amplitude} = |A| \qquad \text{Period} = T = \frac{2\pi}{\omega} \tag{1}$$

■

E X A M P L E   1   *Finding the Amplitude and Period of a Sinusoidal Function*

Determine the amplitude and period of $y = 3 \sin 4x$, and graph the function.

Solution   Comparing $y = 3 \sin 4x$ to $y = A \sin \omega x$, we find that $A = 3$ and $\omega = 4$. Thus, from equation (1), the amplitude is $|A| = 3$ and the period is $T = 2\pi/\omega = 2\pi/4 = \pi/2$. We can use this information to graph $y = 3 \sin 4x$ by beginning as shown in Figure 14(a). Notice that the amplitude is used to scale the y-axis, and the period is used to scale the x-axis. (This will usually result in a different scale for each axis.) Now we fill in the graph of the sine function. See Figure 14(b).

FIGURE 14

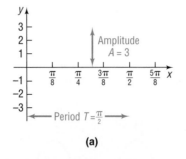

(a)

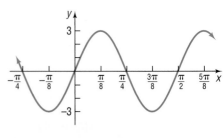

(b) $y = 3 \sin 4x$

Check: Graph $y = 3 \sin 4x$ and compare the result with Figure 14(b).   ■

■   Now work Problem 5.

E X A M P L E   2   *Finding the Amplitude and Period of a Sinusoidal Function*

Determine the amplitude and period of $y = -4 \cos \pi x$, and graph the function.

Solution   Comparing $y = -4 \cos \pi x$ with $y = A \cos \omega x$, we find that $A = -4$ and $\omega = \pi$. Thus, the amplitude is $|A| = |-4| = 4$, and the period is $T = 2\pi/\omega = 2\pi/\pi = 2$. We use the amplitude to scale the y-axis, the period to scale the x-axis, and fill in the graph of the cosine function, thus obtaining the graph of $y = 4 \cos \pi x$ shown in Figure 15(a). Now, since we want the graph of $y = -4 \cos \pi x$, we reflect the graph of $y = 4 \cos \pi x$ about the x-axis, as shown in Figure 15(b).

FIGURE 15

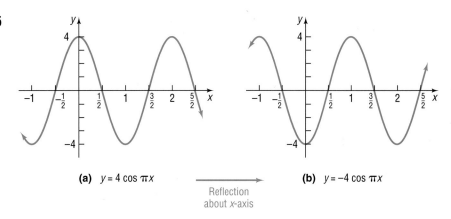

(a) $y = 4 \cos \pi x$ Reflection about x-axis   (b) $y = -4 \cos \pi x$

Check: Graph $y = -4 \cos \pi x$ and compare the result with Figure 15(b).

E X A M P L E   3   *Finding the Amplitude and Period of a Sinusoidal Function*

Determine the amplitude and period of $y = 2 \sin(-\pi x)$, and graph the function.

Solution   Since the sine function is odd, we use the equivalent form

$$y = -2 \sin \pi x$$

Comparing $y = -2 \sin \pi x$ to $y = A \sin \omega x$, we find that $A = -2$ and $\omega = \pi$. Thus, the amplitude is $|A| = 2$, and the period is $T = 2\pi/\omega = 2\pi/\pi = 2$. Again, we use the amplitude to scale the y-axis, the period to scale the x-axis, and fill in the graph of the sine function. Figure 16(a) shows the resulting graph of $y = 2 \sin \pi x$. To obtain the graph of $y = -2 \sin \pi x$, we reflect the graph of $y = 2 \sin \pi x$ about the x-axis, as shown in Figure 16(b).

FIGURE 16

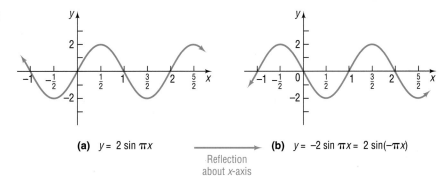

(a) $y = 2 \sin \pi x$ Reflection about x-axis   (b) $y = -2 \sin \pi x = 2 \sin(-\pi x)$

Check: Graph $y = 2 \sin(-\pi x)$ and compare the result with Figure 16(b). ■

We also can use the ideas of amplitude and period to identify a sinusoidal function when its graph is given.

E X A M P L E   4   *Finding an Equation for a Sinusoidal Graph*

Find an equation for the graph shown in Figure 17.

FIGURE 17

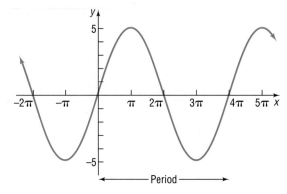

Solution This graph can be viewed as the graph of a sine function with amplitude $A = 5$.*
The period T is observed to be 4π. Thus, by equation (1),

$$T = \frac{2\pi}{\omega}$$

$$4\pi = \frac{2\pi}{\omega}$$

$$\omega = \frac{2\pi}{4\pi} = \frac{1}{2}$$

A sine function whose graph is given in Figure 17 is

$$y = A \sin \omega x = 5 \sin \frac{x}{2}$$

 Check: Graph $y = 5 \sin \dfrac{x}{2}$ and compare the result with Figure 17. ■

E X A M P L E   5   *Finding an Equation for a Sinusoidal Graph*

Find an equation for the graph shown in Figure 18.

FIGURE 18

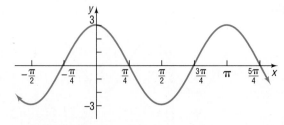

*The equation could also be viewed as a cosine function with a horizontal shift, but viewing it as a sine function is easier.

Solution   From the graph we conclude that it is easiest to view the equation as a cosine function with amplitude $A = 3$ and period $T = \pi$. Thus, $2\pi/\omega = \pi$, so $\omega = 2$. A cosine function whose graph is given in Figure 18 is

$$y = A \cos \omega x = 3 \cos 2x$$

 Check: Graph $y = 3 \cos 2x$ and compare the result with Figure 18. ■

E X A M P L E   6   *Finding an Equation for a Sinusoidal Graph*

Find an equation for the graph shown in Figure 19.

FIGURE 19

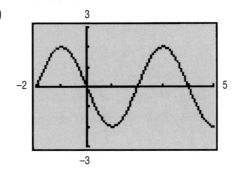

Solution   The graph is sinusoidal, with amplitude $A = 2$. The period is 4, so $2\pi/\omega = 4$ or $\omega = \pi/2$. Since the graph passes through the origin, it is easiest to view the equation as a sine function, but notice that the graph is actually the reflection of a sine function about the x-axis (since the graph is decreasing near the origin). Thus, we have

$$y = -A \sin \omega x = -2 \sin \frac{\pi}{2}x, \qquad A > 0$$

 Check: Graph $y = -2 \sin \dfrac{\pi}{2}x$ and compare the result with Figure 19. ■

■ Now work Problem 31.

Phase Shift

We have seen that the graph of $y = A \sin \omega x$, $\omega > 0$, has amplitude $|A|$ and period $T = 2\pi/\omega$. Thus, one period can be drawn as x varies from 0 to $2\pi/\omega$ or, equivalently, as ωx varies from 0 to 2π. See Figure 20.

FIGURE 20

One period of $y = A \sin \omega x$, $A > 0$, $\omega > 0$

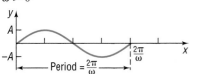

We now want to discuss the graph of

$$y = A \sin(\omega x - \phi)$$

where $\omega > 0$ and ϕ (the Greek letter phi) are real numbers. The graph will be a sine curve of amplitude $|A|$. As $\omega x - \phi$ varies from 0 to 2π, one period will be traced out. This period will begin when

$$\omega x - \phi = 0 \quad \text{or} \quad x = \frac{\phi}{\omega}$$

FIGURE 21

One period of $y = A \sin(\omega x - \phi)$, $A > 0$, $\omega > 0$, $\phi > 0$

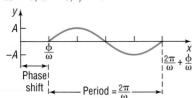

and will end when

$$\omega x - \phi = 2\pi \quad \text{or} \quad x = \frac{2\pi}{\omega} + \frac{\phi}{\omega}$$

See Figure 21.

Thus, we see the graph of $y = A \sin(\omega x - \phi)$ is the same as the graph of $y = A \sin \omega x$, except that it has been shifted ϕ/ω units (to the right if $\phi > 0$ and to the left if $\phi < 0$). This number ϕ/ω is called the **phase shift** of the graph of $y = A \sin(\omega x - \phi)$.

For the graphs of $y = A \sin(\omega x - \phi)$ or $y = A \cos(\omega x - \phi)$, $\omega > 0$,

$$\text{Amplitude} = |A| \qquad \text{Period} = T = \frac{2\pi}{\omega} \qquad \text{Phase shift} = \frac{\phi}{\omega}$$

E X A M P L E 7

Finding the Amplitude, Period, and Phase Shift of a Sinusoidal Function

Find the amplitude, period, and phase shift of $y = 3 \sin(2x - \pi)$, and graph the function.

Solution

Comparing $y = 3 \sin(2x - \pi)$ to $y = A \sin(\omega x - \phi)$, we find that $A = 3$, $\omega = 2$, and $\phi = \pi$. The graph is a sine curve with amplitude $A = 3$ and period $T = 2\pi/\omega = 2\pi/2 = \pi$. One period of the sine curve begins at $2x - \pi = 0$ or $x = \pi/2$ (this is the phase shift) and ends at $2x - \pi = 2\pi$ or $x = 3\pi/2$. See Figure 22.

FIGURE 22

$y = 3 \sin(2x - \pi)$

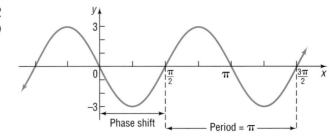

 Check: Graph $y = 3 \sin(2x - \pi)$ and compare the result with Figure 22.

■

E X A M P L E 8

Finding the Amplitude, Period, and Phase Shift of a Sinusoidal Function

Find the amplitude, period, and phase shift of $y = 3 \cos(4x + 2\pi)$, and graph the function.

Solution

Comparing $y = 3 \cos(4x + 2\pi)$ to $y = A \cos(\omega x - \phi)$, we see that $A = 3$, $\omega = 4$, and $\phi = -2\pi$. The graph is a cosine curve with amplitude $A = 3$ and period $T = 2\pi/\omega = 2\pi/4 = \pi/2$. One period of the cosine curve begins at $4x + 2\pi = 0$ or $x = -\pi/2$ (the phase shift) and ends at $4x + 2\pi = 2\pi$ or $x = 0$. See Figure 23.

FIGURE 23
$y = 3 \cos(4x + 2\pi)$

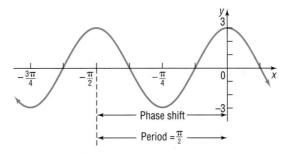

Check: Graph $y = 3 \cos(4x + 2\pi)$ and compare the result with Figure 23. ∎

■ Now work Problem 45.

6.2

Exercise 6.2

In Problems 1–10, determine the amplitude and period of each function without graphing.

1. $y = 2 \sin x$ **2.** $y = 3 \cos x$ **3.** $y = -4 \cos 2x$ **4.** $y = -\sin \frac{1}{2}x$

5. $y = 6 \sin \pi x$ **6.** $y = -3 \cos 3x$ **7.** $y = -\frac{1}{2} \cos \frac{3}{2}x$ **8.** $y = \frac{4}{3} \sin \frac{2}{3}x$

9. $y = \dfrac{5}{3} \sin\left(-\dfrac{2\pi}{3}x\right)$ **10.** $y = \dfrac{9}{5} \cos\left(-\dfrac{3\pi}{2}x\right)$

In Problems 11–20, match the given function to one of the graphs (A)–(J).

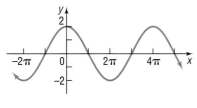

(A)

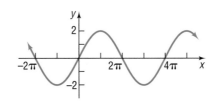

(B)

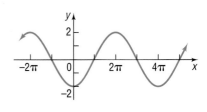

(C)

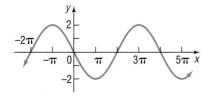

(D)

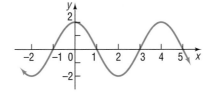

(E)

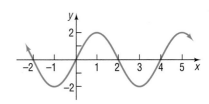

(F)

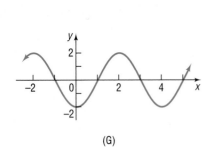

(G)

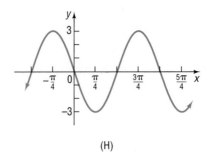

(H)

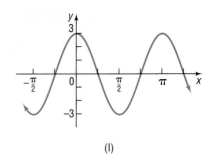

(I)

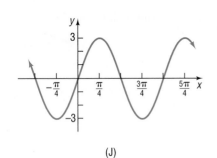

(J)

11. $y = 2 \sin \dfrac{\pi}{2} x$

12. $y = 2 \cos \dfrac{\pi}{2} x$

13. $y = 2 \cos \frac{1}{2} x$

14. $y = 3 \cos 2x$

15. $y = -3 \sin 2x$

16. $y = 2 \sin \frac{1}{2} x$

17. $y = -2 \cos \frac{1}{2} x$

18. $y = -2 \cos \dfrac{\pi}{2} x$

19. $y = 3 \sin 2x$

20. $y = -2 \sin \frac{1}{2} x$

In Problems 21–30, graph each function.

21. $y = 5 \sin 4x$

22. $y = 4 \cos 6x$

23. $y = 5 \cos \pi x$

24. $y = 2 \sin \pi x$

25. $y = -2 \cos 2\pi x$

26. $y = -5 \cos 2\pi x$

27. $y = -4 \sin \frac{1}{2} x$

28. $y = -2 \cos \frac{1}{2} x$

29. $y = \frac{3}{2} \sin(-\frac{2}{3} x)$

30. $y = \frac{4}{3} \cos(-\frac{1}{3} x)$

In Problems 31–44, find a function whose graph is given.

31.

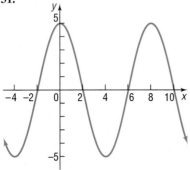

32.

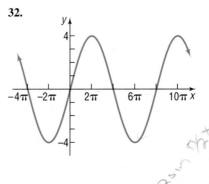

33.

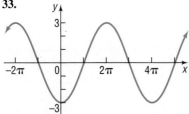

34.

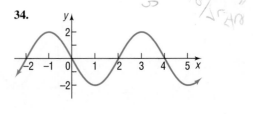

35.
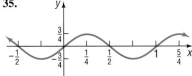

36.
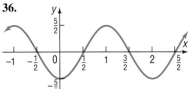

37.
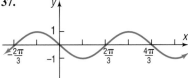

38.
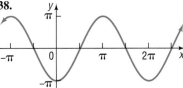

39.
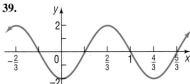

40.
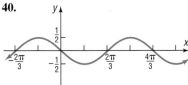

41.
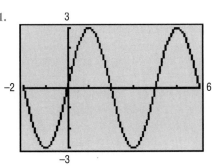

42.
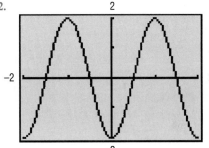

43.
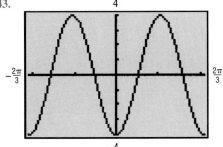

44.
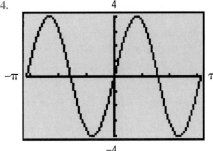

In Problems 45–52, find the amplitude, period, and phase shift of each function. Graph one period.

45. $y = 4 \sin(2x - \pi)$

46. $y = 3 \sin(3x - \pi)$

47. $y = 2 \cos\left(3x + \dfrac{\pi}{2}\right)$

48. $y = 3 \cos(2x + \pi)$

49. $y = -3 \sin\left(2x + \dfrac{\pi}{2}\right)$

50. $y = -2 \cos\left(2x - \dfrac{\pi}{2}\right)$

51. $y = 4 \sin(\pi x + 2)$

52. $y = 2 \cos(2\pi x + 4)$

53. $y = 3 \cos(\pi x - 2)$

54. $y = 2 \cos(2\pi x - 4)$

55. $y = 3 \sin\left(-2x + \dfrac{\pi}{2}\right)$

56. $y = 3 \cos\left(-2x + \dfrac{\pi}{2}\right)$

57. *Alternating Current (ac) Circuits* The current I, in amperes, flowing through an ac (alternating current) circuit at time t is

$$I = 220 \sin 60\pi t, \qquad t \geq 0$$

What is the period? What is the amplitude? Graph this function over two periods.

58. *Alternating Current (ac) Circuits* The current I, in amperes, flowing through an ac (alternating current) circuit at time t is

$$I = 120 \sin 30\pi t, \qquad t \geq 0$$

What is the period? What is the amplitude? Graph this function over two periods.

59. *Alternating Current (ac) Circuits* The current I, in amperes, flowing through an ac (alternating current) circuit at time t is

$$I = 120 \sin\left(30\pi t - \dfrac{\pi}{3}\right), \qquad t \geq 0$$

What is the period? What is the amplitude? What is the phase shift? Graph this function over two periods.

60. *Alternating Current (ac) Circuits* The current I, in amperes, flowing through an ac (alternating current) circuit at time t is

$$I = 220 \sin\left(60\pi t - \dfrac{\pi}{6}\right), \qquad t \geq 0$$

What is the period? What is the amplitude? What is the phase shift? Graph this function over two periods.

61. *Alternating Current (ac) Generators* The voltage V produced by an ac generator is

$$V = 220 \sin 120\pi t$$

(a) What is the amplitude? What is the period?
(b) Graph V over two periods, beginning at $t = 0$.
(c) If a resistance of $R = 10$ ohms is present, what is the current I? [*Hint:* Use Ohm's Law, $V = IR$].
(d) What is the amplitude and period of the current I?
(e) Graph I over two periods, beginning at $t = 0$.

62. *Alternating Current (ac) Generators* The voltage V produced by an ac generator is

$$V = 120 \sin 120\pi t$$

(a) What is the amplitude? What is the period?
(b) Graph V over two periods, beginning at $t = 0$.
(c) If a resistance of $R = 20$ ohms is present, what is the current I? [*Hint:* Use Ohm's Law, $V = IR$].
(d) What is the amplitude and period of the current I?
(e) Graph I over two periods, beginning at $t = 0$.

63. *Alternating Current (ac) Generators* The voltage V produced by an ac generator is sinusoidal. As a function of time, the voltage V is

$$V = V_0 \sin 2\pi ft$$

where f is the **frequency,** the number of complete oscillations (cycles) per second. [In the United States and Canada, f is 60 hertz (Hz)]. The **power** P delivered to a resistance R at any time t is defined as

$$P = \dfrac{V^2}{R}$$

(a) Show that $P = \dfrac{V_0^2}{R} \sin^2 2\pi ft$.
(b) The graph of P is shown in the figure. Express P as a sinusoidal function.
(c) Deduce that

$$\sin^2 2\pi ft = \dfrac{1}{2}(1 - \cos 4\pi ft)$$

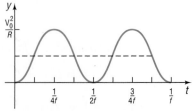

Power in an a-c generator

64. *Biorhythms* In the theory of biorhythms, a sine function of the form

$$P = 100 \sin \omega t$$

is used to measure the percent P of a person's potential at time t, where t is measured in days and $t = 0$ is the person's birthday. Three characteristics are commonly measured:

Physical potential: period of 23 days

Emotional potential: period of 28 days

Intellectual potential: period of 33 days

(a) Find ω for each characteristic.
(b) Graph all three functions.
(c) Is there a time t when all three characteristics have 100% potential? When is it?
(d) Suppose that you are 20 years old today ($t = 7305$ days). Describe your physical, emotional, and intellectual potential for the next 30 days.

65. Explain how the amplitude and period of a sinusoidal graph are used to establish the scale on each coordinate axis.

66. Find an application in your major field that leads to a sinusoidal graph. Write a paper about your findings.

6.3

Applications

Combining Waves; Amplifying and Damping; Graphing Sums and Products

Many physical and biological applications require the graphs of sums and products of functions, such as

$$f(x) = \sin x + \cos 2x \quad \text{and} \quad g(x) = e^x \sin x$$

For example, if two tones are emitted, the sound produced is the sum of the waves produced by the two tones. See Problem 29 for an explanation of Touch-Tone phones.

To graph the sum of two (or more) functions, we can use the method of adding y-coordinates described next.

E X A M P L E   1   *Graphing the Sum of Two Functions*

Use the method of adding y-coordinates to graph $h(x) = x + \sin x$.

Solution   First, we graph the component functions,

$$y = h_1(x) = x \qquad y = h_2(x) = \sin x$$

in the same coordinate system. See Figure 24(a). Now, we select several values of x, say, $x = 0$, $x = \pi/2$, $x = \pi$, $x = 3\pi/2$, and $x = 2\pi$, at which we compute $h(x) = h_1(x) + h_2(x)$. Table 3 shows the computation. We plot these points and connect them to get the graph, as shown in Figure 24(b).

FIGURE 24

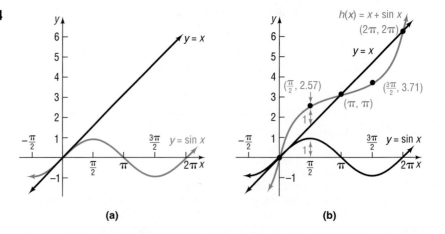

(a)   (b)

TABLE 3

x	0	$\pi/2$	π	$3\pi/2$	2π
$y = h_1(x) = x$	0	$\pi/2$	π	$3\pi/2$	2π
$y = h_2(x) = \sin x$	0	1	0	-1	0
$h(x) = x + \sin x$	0	$\pi/2 + 1 \approx 2.57$	π	$3\pi/2 - 1 \approx 3.71$	2π
POINT ON GRAPH OF h	$(0, 0)$	$(\pi/2, 2.57)$	(π, π)	$(3\pi/2, 3.71)$	$(2\pi, 2\pi)$

In Example 1, note that the graph of $h(x) = x + \sin x$ intersects the line $y = x$ whenever $\sin x = 0$. Also, notice that the graph of h is not periodic.

Check: Graph $y = x + \sin x$ and compare the result with Figure 24(b). Use TRACE to verify that the graphs intersect when $\sin x = 0$.

The next example shows a periodic graph of the sum of two functions.

E X A M P L E   2   *Graphing the Sum of Two Sinusoidal Functions*
Use the method of adding y-coordinates to graph

$$f(x) = \sin x + \cos 2x$$

Solution   Table 4 shows the steps for computing several points on the graph of f. Figure 25 illustrates the graphs of the component functions, $y = f_1(x) = \sin x$ and $y = f_2(x) = \cos 2x$, and the graph of $f(x) = \sin x + \cos 2x$, which is shown in color.

TABLE 4

x	$-\pi/2$	0	$\pi/2$	π	$3\pi/2$	2π
$y = f_1(x) = \sin x$	-1	0	1	0	-1	0
$y = f_2(x) = \cos 2x$	-1	1	-1	1	-1	1
$f(x) = \sin x + \cos 2x$	-2	1	0	1	-2	1
POINT ON GRAPH OF f	$(-\pi/2, -2)$	$(0, 1)$	$(\pi/2, 0)$	$(\pi, 1)$	$(3\pi/2, -2)$	$(2\pi, 1)$

FIGURE 25

$f(x) = \sin x + \cos 2x$

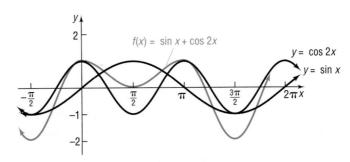

Check: Graph $y = \sin x + \cos 2x$ and compare the result with Figure 25.   ∎

If we are asked to graph a function f that is the difference of two functions g and h, that is,

$$f(x) = g(x) - h(x)$$

we may view f as

$$f(x) = g(x) + [-h(x)]$$

and use the method of adding y-coordinates, as described previously.

■ Now work Problem 1.

The amplitude of any real oscillating spring or swinging pendulum decreases with time due to air resistance, friction, and so on. See Figure 26. The graph of such phenomena is generally given by the product of two functions.

FIGURE 26

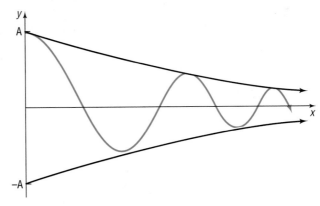

E X A M P L E   3   *Graphing the Product of Two Functions*

Graph $f(x) = x \sin x$.

Solution   Here, f is the product of $y = x$ and $y = \sin x$. Using properties of absolute value and the fact that $|\sin x| \le 1$, we find that

$$|f(x)| = |x \sin x| = |x| \, |\sin x| \le |x|$$

If $x \ge 0$, this reduces to

$$|f(x)| \le x \quad \text{or} \quad -x \le f(x) \le x, \qquad x > 0$$

If $x < 0$ we have

$$|f(x)| \leq {}^-x \quad \text{or} \quad x \leq f(x) \leq -x, \qquad x < 0$$

Thus, in every case the graph of f will lie between the lines $y = x$ and $y = -x$. Furthermore, we conclude that the graph of f will touch these lines when $\sin x = \pm 1$, that is, when $x = -\pi/2, \pi/2, 3\pi/2$, and so on. Since $y = f(x) = x \sin x = 0$ when $x = -\pi, 0, \pi, 2\pi$, and so on, we know the location of the x-intercepts of the graph of f. Finally, the function f is an even function $[f(-x) = -x \sin(-x) = x \sin x = f(x)]$, so the graph is symmetric with respect to the y-axis. See Table 5. Figure 27 illustrates the graph.

TABLE 5

x	**0**	$\pi/2$	π	$3\pi/2$	2π
$y = x$	0	$\pi/2$	π	$3\pi/2$	2π
$y = \sin x$	0	1	0	-1	0
$y = f(x) = x \sin x$	0	$\pi/2$	0	$-3\pi/2$	0
POINT ON GRAPH OF f	$(0, 0)$	$(\pi/2, \pi/2)$	$(\pi, 0)$	$(3\pi/2, -3\pi/2)$	$(2\pi, 0)$

FIGURE 27

$f(x) = x \sin x$

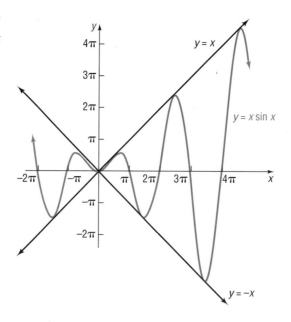

E X A M P L E   4   *Analyzing the Graph of $y = x \sin x$*

Graph $y = x \sin x$, along with $y = x$ and $y = -x$, for $0 \leq x \leq 4\pi$. Determine where $y = x \sin x$ touches $y = x$. Compare this to where $y = x \sin x$ has a turning point (local maximum). Express answers correct to two decimal places.

Solution   Figure 28 shows the graphs of $y = x \sin x$, $y = x$, and $y = -x$. Using TRACE and BOX, we find that $y = x \sin x$ touches $y = x$ at $x = 1.57$ and at $x = 7.85$. The turning points (local maxima) occur just to the right of each of these values at $x = 2.02$ and at $x = 7.97$.

FIGURE 28

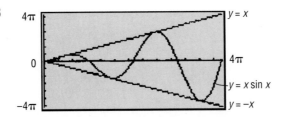

If $x \geq 0$, the graph of $f(x) = x \sin x$ is called an **amplified sine wave,** and x is called the **amplifying factor.** By using different factors, we can obtain other sinusoidal waves. The next example illustrates a **damping factor.**

E X A M P L E   5

Graphing the Product of Two Functions

The damped vibration curve

$$f(x) = e^{-x} \sin x, \qquad x \geq 0$$

is of importance in many applications, such as the motion of musical strings, the vibrations of a pendulum, and the current in an electrical circuit. Graph this function.

Solution

The function f is the product of $y = e^{-x}$ and $y = \sin x$. Using properties of absolute value and the fact that $|\sin x| \leq 1$, we find that

$$|f(x)| = |e^{-x} \sin x| = |e^{-x}| \, |\sin x| \leq |e^{-x}| = e^{-x}$$
$$\underset{e^{-x} > 0 \text{ for all } x}{\uparrow}$$

Thus,

$$-e^{-x} \leq f(x) \leq e^{-x}$$

and the graph of f will lie between the graphs of $y = e^{-x}$ and $y = -e^{-x}$. Also, the graph of f will touch these graphs when $|\sin x| = 1$, that is, when $x = -\pi/2$, $\pi/2$, $3\pi/2$, and so on. The x-intercepts of the graph of f occur at $x = -\pi$, 0, π, 2π, and so on. See Table 6. See Figure 29 for the graph.

TABLE 6

x	0	$\pi/2$	π	$3\pi/2$	2π
$y = e^{-x}$	1	$e^{-\pi/2}$	$e^{-\pi}$	$e^{-3\pi/2}$	$e^{-2\pi}$
$y = \sin x$	0	1	0	-1	0
$f(x) = e^{-x} \sin x$	0	$e^{-\pi/2}$	0	$-e^{-3\pi/2}$	0
POINT ON GRAPH OF f	$(0, 0)$	$(\pi/2, e^{-\pi/2})$	$(\pi, 0)$	$(3\pi/2, -e^{-3\pi/2})$	$(2\pi, 0)$

FIGURE 29
$f(x) = e^{-x} \sin x,$
$x \geq 0$

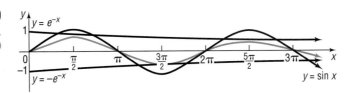

E X A M P L E 6 *Analyzing the Graph of $y = e^{-x} \sin x$*

Graph $y = e^{-x} \sin x$, along with $y = e^{-x}$ and $y = -e^{-x}$ for $0 \le x \le 3\pi$. Determine where $y = e^{-x} \sin x$ touches $y = e^{-x}$. Compare this to where $y = e^{-x} \sin x$ has a turning point (local maximum). Express answers correct to two decimal places.

Solution Figure 30 shows the graphs of $y = e^{-x} \sin x$, $y = e^{-x}$, and $y = -e^{-x}$. Using TRACE and BOX, we find that $y = e^{-x} \sin x$ touches $y = e^{-x}$ at $x = \frac{\pi}{2} \approx 1.57$. The turning point (local maximum) occurs at $x = 0.78$.

FIGURE 30

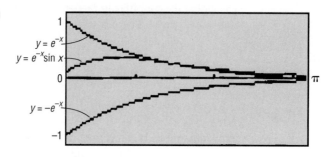

■ Now work Problem 21.

Simple Harmonic Motion

There are many physical phenomena that can be described as simple harmonic motion. Radio and television waves, light waves, sound waves, and water waves exhibit motion that is simple harmonic. Even yearly low and high temperatures at a given location can be modeled using an equation for simple harmonic motion.

The swinging of a pendulum, the vibrations of a tuning fork, and the bobbing of a weight attached to a coiled spring are also examples of vibrational motion. In this type of motion, an object swings back and forth over the same path. In each illustration in Figure 31, the point B is the **equilibrium (rest) position** of the vibrating object. The **amplitude** of vibration is the distance from the object's rest position to its point of greatest displacement (either point A or point C in Figure 31). The **period** of a vibrating object is the time required to complete one vibration— that is, the time it takes to go from, say, point A through B to C and back to A.

FIGURE 31

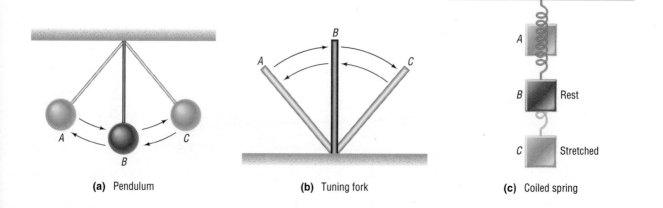

(a) Pendulum   (b) Tuning fork   (c) Coiled spring

Simple Harmonic Motion

Simple harmonic motion is a special kind of vibrational motion in which the acceleration **a** of the object is directly proportional to the negative of its displacement **d** from its rest position. That is, $a = -kd$, $k > 0$.

For example, when the mass hanging from the spring in Figure 31(c) is pulled down from its rest position B to the point C, the force of the spring tries to restore the mass to its rest position. Assuming that there is no frictional force* to retard the motion, the amplitude will remain constant. The force increases in direct proportion to the distance the mass is pulled from its rest position. Since the force increases directly, the acceleration of the mass of the object must do likewise, because (by Newton's Second Law of Motion) force is directly proportional to acceleration. Thus, the acceleration of the object varies directly with its displacement, and the motion is an example of simple harmonic motion.

Simple harmonic motion is related to circular motion. To see this relationship, consider a circle of radius a, with center at $(0, 0)$. See Figure 32. Suppose that an object initially placed at $(a, 0)$ moves counterclockwise around the circle at constant angular speed ω. Suppose further that after time t has elapsed, the object is at the point $P = (x, y)$ on the circle. The angle θ, in radians, swept out by the ray $\overrightarrow{OP}$ in this time t is

FIGURE 32

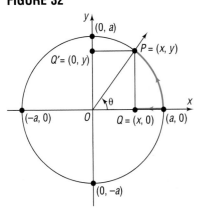

$$\theta = \omega t$$

The coordinates of the point P at time t are

$$x = a \cos \theta = a \cos \omega t$$
$$y = a \sin \theta = a \sin \omega t$$

Corresponding to each position $P = (x, y)$ of the object moving about the circle, there is the point $Q = (x, 0)$, called the **projection of P on the x-axis.** As P moves around the circle at a constant rate, the point Q moves back and forth between the points $(a, 0)$ and $(-a, 0)$ along the x-axis with a motion that is simple harmonic. Similarly, for each point P there is a point $Q' = (0, y)$, called the **projection of P on the y-axis.** As P moves around the circle, the point Q' moves back and forth between the points $(0, a)$ and $(0, -a)$ on the y-axis with a motion that is simple harmonic. Thus, simple harmonic motion can be described as the projection of constant circular motion on a coordinate axis.

Theorem
Simple Harmonic Motion

An object that moves on a coordinate axis so that its distance d from the origin at time t is given by either

$$d = a \cos \omega t \quad \text{or} \quad d = a \sin \omega t$$

where a and ω are constants, moves with simple harmonic motion. The motion has amplitude $|a|$ and period $2\pi/\omega$. ■

The **frequency** f of an object in simple harmonic motion is the number of oscillations per unit time. Since the period is the time required for one oscillation, it follows that frequency is the reciprocal of the period; that is,

*If friction is present, the amplitude will decrease with time to 0. This type of motion is an example of **damped motion.**

$$f = \frac{\omega}{2\pi}$$

E X A M P L E   7

Finding an Equation for an Object in Harmonic Motion

Suppose that the object attached to the coiled spring in Figure 31(c) is pulled down a distance of 5 inches from its rest position and then released. If the time for one oscillation is 3 seconds, write an equation that relates the distance d of the object from its rest position after time t (in seconds). Assume no friction.

Solution

The motion of the object is simple harmonic. Since the object is released at time $t = 0$ when its distance d from the rest position is 5 inches, it is easiest to use the equation

$$d = a \cos \omega t$$

to describe the motion. (Do you see why? For this equation, when $t = 0$, then $d = a = 5$.) Now the amplitude is 5 and the period is 3. Thus,

$$a = 5 \quad \text{and} \quad \frac{2\pi}{\omega} = \text{Period} = 3 \quad \text{or} \quad \omega = \frac{2\pi}{3}$$

An equation of the motion of the object is

$$d = 5 \cos \frac{2\pi}{3}t$$

∎

Note: In the solution to Example 7, we let $a = 5$, indicating that the positive direction of the motion is down. If we wanted the positive direction to be up, we would let $a = -5$.

■   Now work Problem 33.

E X A M P L E   8

Analyzing the Motion of an Object

Suppose that the distance x (in meters) an object travels in time t (in seconds) satisfies the equation

$$x = 10 \sin 5t$$

(a) Describe the motion of the object.

(b) What is the maximum displacement from its resting position?

(c) What is the time required for one oscillation?

(d) What is the frequency?

Solution

We observe that the given equation is of the form

$$d = a \sin \omega t \quad d = 10 \sin 5t$$

where $a = 10$ and $\omega = 5$.

(a) The motion is simple harmonic.

(b) The maximum displacement of the object from its resting position is the amplitude: $a = 10$ meters.

(c) The time required for one oscillation is the period:

$$\text{Period} = \frac{2\pi}{\omega} = \frac{2\pi}{5} \text{ seconds}$$

(d) The frequency is the reciprocal of the period. Thus,

$$\text{Frequency} = f = \frac{5}{2\pi} \text{ oscillation per second}$$

■

■ Now work Problem 41.

6.3

Exercise 6.3

In Problems 1–20, use the method of adding y-coordinates to graph each function.

1. $f(x) = x + \cos x$

2. $f(x) = x + \cos 2x$

3. $f(x) = x - \sin x$

4. $f(x) = x - \cos x$

5. $f(x) = \sin x + \cos x$

6. $f(x) = \sin 2x + \cos x$

7. $g(x) = \sin x + \sin 2x$

8. $g(x) = \cos 2x + \cos x$

9. $h(x) = \sqrt{x} + \sin x$

10. $h(x) = \sqrt{x} + \cos x$

11. $F(x) = 2 \sin x - \cos 2x$

12. $F(x) = 2 \cos 2x - \sin x$

13. $f(x) = 2 \sin \pi x + \cos \pi x$

14. $f(x) = 2 \cos \frac{\pi}{2}x + \sin \frac{\pi}{2}x$

15. $f(x) = \frac{x^2}{\pi^2} + \sin 2x$

16. $f(x) = \frac{x^2}{\pi^2} - \cos 2x$

17. $f(x) = |x| + \sin \frac{\pi}{2}x$

18. $f(x) = |x| + \cos \pi x$

19. $f(x) = 3 \sin 2x + 2 \cos 3x$

20. $f(x) = 2 \sin 3x + 3 \cos 2x$

In Problems 21–28, graph each function.

21. $f(x) = x \cos x$

22. $f(x) = x \sin 2x$

23. $f(x) = x^2 \sin x$

24. $f(x) = x^2 \cos x$

25. $f(x) = |x| \cos x$

26. $f(x) = |x| \sin x$

27. $f(x) = e^{-x} \cos 2x, \ x \geq 0$

28. $f(x) = e^{-x} \sin 2x, \ x \geq 0$

29. *Touch-Tone Phones* On a Touch-Tone phone, each button produces a unique sound. The sound produced is the sum of two tones, given by

$$y = \sin 2\pi l t \quad \text{and} \quad y = \sin 2\pi h t$$

where l and h are the low and high frequencies (cycles per second) shown on the illustration. For example, if you touch 7, the low frequency is $l = 852$ cycles per second and the high frequency is $h = 1209$ cycles per second. The sound emitted by touching 7 is

$$y = \sin 2\pi(852)t + \sin 2\pi(1209)t$$

Graph the sound emitted by touching 7.

30. Graph the sound emitted by the star key (*) on a Touch-Tone phone. See Problem 29.

31. *The Sawtooth Curve* An oscilloscope often displays a *sawtooth curve*. This curve can be approximated by sinusoidal curves of varying periods and amplitudes.

(a) Graph the following function, which can be used to approximate the sawtooth curve:

$$f(x) = \tfrac{1}{2} \sin 2\pi x + \tfrac{1}{4} \sin 4\pi x \qquad 0 \leq x \leq 2$$

(b) A better approximation to the sawtooth curve is given by

$$f(x) = \tfrac{1}{2} \sin 2\pi x + \tfrac{1}{4} \sin 4\pi x + \tfrac{1}{8} \sin 8\pi x$$

Graph this function for $0 \leq x \leq 4$ and compare the result to the graph obtained in part (a).

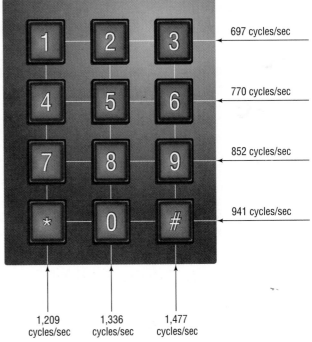

Touch Tone Phone

697 cycles/sec

770 cycles/sec

852 cycles/sec

941 cycles/sec

1,209 cycles/sec   1,336 cycles/sec   1,477 cycles/sec

(c) A third and even better approximation to the sawtooth curve is given by

$$f(x) = \tfrac{1}{2}\sin 2\pi x + \tfrac{1}{4}\sin 4\pi x + \tfrac{1}{8}\sin 8\pi x + \tfrac{1}{16}\sin 16\pi x$$

Graph this function for $0 \le x \le 4$ and compare the result to the graphs obtained in parts (a) and (b).

(d) What do you think the next approximation to the sawtooth curve is?

32. *Charging a Capacitor* If a charged capacitor is connected to a coil by closing a switch (see the figure), energy is transferred to the coil and then back to the capacitor in an oscillatory motion. The voltage V (in volts) across the capacitor will gradually diminish to 0 with time t.

(a) Graph the equation relating V and t:

$$V(t) = e^{-1.9t}\cos \pi t \qquad 0 \le t \le 3$$

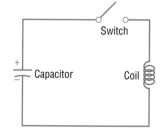

(b) At what times t will the graph of V touch the graph of $y = e^{-1.9t}$? When does V touch the graph of $y = -e^{-1.9t}$?

(c) Graph $V = V(t)$, $0 \le t \le 3$.
When will the voltage V be between -0.1 and 0.1 volt?

In Problems 33–36, an object attached to a coiled spring is pulled down a distance a from its rest position and then released. Assuming that the motion is simple harmonic with period T, write an equation that relates the distance d of the object from its rest position after t seconds. Also assume that the positive direction of the motion is down.

33. $a = 5$; $T = 2$ seconds   34. $a = 10$; $T = 3$ seconds

35. $a = 6$; $T = \pi$ seconds   36. $a = 4$; $T = \pi/2$ seconds

37. Rework Problem 33 under the same conditions except that, at time $t = 0$, the object is at its resting position and moving down.

38. Rework Problem 34 under the same conditions except that, at time $t = 0$, the object is at its resting position and moving down.

39. Rework Problem 35 under the same conditions except that, at time $t = 0$, the object is at its resting position and moving down.

40. Rework Problem 36 under the same conditions except that, at time $t = 0$, the object is at its resting position and moving down.

In Problems 41–48, the distance d (in meters) an object travels in time t (in seconds) is given.
(a) Describe the motion of the object.
(b) What is the maximum displacement from its resting position?
(c) What is the time required for one oscillation?
(d) What is the frequency?

41. $d = 5\sin 3t$ 42. $d = 4\sin 2t$ 43. $d = 6\cos \pi t$ 44. $d = 5\cos \dfrac{\pi}{2}t$

45. $d = -3\sin \tfrac{1}{2}t$ 46. $d = -2\cos 2t$ 47. $d = 6 + 2\cos 2\pi t$ 48. $d = 4 + 3\sin \pi t$

49. Graph the function $f(x) = (\sin x)/x$, $x > 0$. Based on the graph, what do you conjecture about the value of $(\sin x)/x$ for x close to 0?

50. Graph $y = x\sin x$, $y = x^2\sin x$, and $y = x^3\sin x$ for $x > 0$. What patterns do you observe?

51. Graph $y = \dfrac{1}{x}\sin x$, $y = \dfrac{1}{x^2}\sin x$, and $y = \dfrac{1}{x^3}\sin x$ for $x > 0$. What patterns do you observe?

52. How would you explain to a friend what simple harmonic motion is?

ISSION POSSIBLE

Chapter 6

DETERMINING HIGH TIDE AT ROCK HARBOR

Your consulting firm has gone off to vacation together on Cape Cod. (Even a consulting firm can take a vacation.) While there, you choose to go deep sea fishing on Cape Cod Bay. The Captain of the Madame B tells you to be on the dock at high tide Thursday morning. After you hang up the phone, you find yourselves wondering when high tide will be, but rather than call the Captain back, you decide to use your all-powerful trig brains to figure out the answer.

1. You know that the length of time between a high tide and low tide is about 6 1/4 hours. And you know that the high tide today (Monday) was at 3 A.M. What time will the high tide occur on Thursday morning?
2. Can you think of any reasons why a fishing boat would want to leave the harbor at high tide?
3. When you visited the harbor at low tide you saw markings on the pier that indicated the height of the water. The depth is 13.5 ft. at low tide and 15 ft. at high tide. You suddenly realize that you have enough information to rough out a graph of the tides for the Captain. Draw a sinusoidal graph that shows the height of the water at the pier starting with the high tide on Monday at 3 A.M. and continuing past the high tide Thursday morning. How many high tides will occur between 3 A.M. Monday and 3 A.M. Thursday?
4. What would be the equation of this graph, taking into account the amplitude, frequency, and any phase shifts. Would it be easier to use a sine or a cosine function?
5. High and low tides are not as regular as such a formula would indicate. What are some factors you can think of that would affect the height of the tides?
6. When the Captain returns to the deck, he leaves a certain amount of slack in the ropes used to fasten the boat to the pier. Why? What could go wrong if he didn't do this?

6.4

The Graphs of $y = \tan x$, $y = \csc x$, $y = \sec x$, and $y = \cot x$

The Graph of $y = \tan x$

Because the tangent function has period π, we only need to determine the graph over some interval of length π. The rest of the graph will consist of repetitions of that graph. Because the tangent function is not defined at $\ldots, -3\pi/2, -\pi/2, \pi/2, 3\pi/2, \ldots$, we shall concentrate on the interval $(-\pi/2, \pi/2)$, of length π and construct Table 7, which lists some points on the graph of $y = \tan x$, $-\pi/2 < x < \pi/2$. As before, we plot the points in the table and connect them with a smooth curve. See Figure 33 for a partial graph of $y = \tan x$, where $-\pi/3 \le x \le \pi/3$.

TABLE 7

x	$y = \tan x$	(x, y)
$-\pi/3$	$-\sqrt{3} \approx -1.73$	$(-\pi/3, -\sqrt{3})$
$-\pi/4$	-1	$(-\pi/4, -1)$
$-\pi/6$	$-\sqrt{3}/3 \approx -0.58$	$(-\pi/6, -\sqrt{3}/3)$
0	0	$(0, 0)$
$\pi/6$	$\sqrt{3}/3 \approx 0.58$	$(\pi/6, \sqrt{3}/3)$
$\pi/4$	1	$(\pi/4, 1)$
$\pi/3$	$\sqrt{3} \approx 1.73$	$(\pi/3, \sqrt{3})$

FIGURE 33
$y = \tan x$,
$-\pi/3 \le x \le \pi/3$

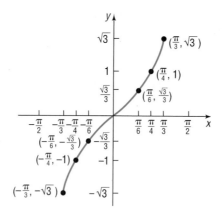

To complete the graph of $y = \tan x$, we need to investigate the behavior of the function as x approaches $-\pi/2$ and $\pi/2$. We must be careful, though, because $y = \tan x$ is not defined at these numbers. To determine this behavior, we use the identity

$$\tan x = \frac{\sin x}{\cos x}$$

If x is close to $\pi/2$, but remains less than $\pi/2$, then $\sin x$ will be close to 1 and $\cos x$ will be positive and close to 0. (Refer back to the graphs of the sine function and the cosine function.) Hence, the ratio $(\sin x)/(\cos x)$ will be positive and large. In fact, the closer x gets to $\pi/2$, the closer $\sin x$ gets to 1 and $\cos x$ gets to 0, so $\tan x$ approaches ∞. In other words, the vertical line $x = \pi/2$ is a vertical asymptote to the graph of $y = \tan x$.

If x is close to $-\pi/2$, but remains greater than $-\pi/2$, then $\sin x$ will be close to -1 and $\cos x$ will be positive and close to 0. Hence, the ratio $(\sin x)/(\cos x)$ approaches $-\infty$. In other words, the vertical line $x = -\pi/2$ is also a vertical asymptote to the graph.

With these observations, we can complete one period of the graph, and we obtain the graph of $y = \tan x$ by repeating this period, as shown in Figure 34.

FIGURE 34

$y = \tan x$, $-\infty < x < \infty$,
x not equal to odd multiples of $\pi/2$

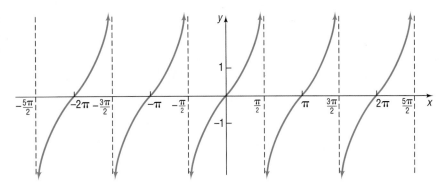

Check: Graph $y = \tan x$ and compare the result with Figure 34. Use TRACE to see what happens as x gets close to $\pi/2$, but is less than $\pi/2$. Be sure to set the RANGE accordingly. ■

The graph of $y = \tan x$ illustrates some of the facts we already know about the tangent function:

Characteristics of the Tangent Function

1. The domain is the set of all real numbers, except odd multiples of $\pi/2$.
2. The range consists of all real numbers.
3. The tangent function is an odd function, as the symmetry of the graph with respect to the origin indicates.
4. The tangent function is periodic, with period π.
5. The x-intercepts are . . . , -2π, $-\pi$, 0, π, 2π, 3π, . . . ; the y-intercept is 0.
6. Vertical asymptotes occur at $x = $ . . . , $-3\pi/2$, $-\pi/2$, $\pi/2$, $3\pi/2$, ...

■ Now work Problems 1 and 9.

E X A M P L E   1

Graphing Variations of $y = \tan x$ Using Shifts, Reflections, and the Like

Graph $y = \tan\left(x + \dfrac{\pi}{4}\right)$.

Solution   We start with the graph of $y = \tan x$ and shift it horizontally to the left $\pi/4$ unit. See Figure 35.

FIGURE 35

$y = \tan\left(x + \dfrac{\pi}{4}\right)$

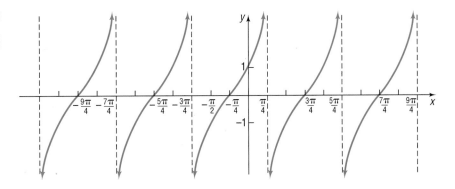

Check: Graph $y = \tan\left(x + \dfrac{\pi}{4}\right)$ and compare the result with Figure 35.

■ Now work Problem 11.

The Graphs of $y = \csc x$, $y = \sec x$, and $y = \cot x$

The cosecant and secant functions, sometimes referred to as **reciprocal functions,** are graphed by making use of the reciprocal identities

$$\csc x = \frac{1}{\sin x} \quad \text{and} \quad \sec x = \frac{1}{\cos x}$$

For example, the value of the cosecant function $y = \csc x$ at a given number x equals the reciprocal of the corresponding value of the sine function, provided the value of the sine function is not 0. If the value of $\sin x$ is 0, then, at such numbers x, the cosecant function is not defined. In fact, the graph of the cosecant function has vertical asymptotes at integral multiples of π. See Figure 36 for the graph.

FIGURE 36
$y = \csc x$, $-\infty < x < \infty$, x not equal to integral multiples of π, $|y| \geq 1$

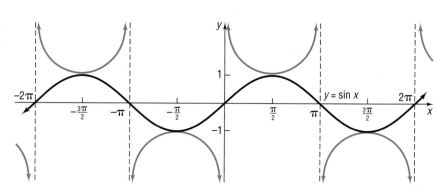

Check: Graph $y = \csc x$ and compare the result with Figure 36. Use TRACE to see what happens when x is close to 0.

E X A M P L E   2

Graphing Variations of $y = \csc x$ Using Shifts, Reflections, and the Like

Graph $y = 2 \csc(x - \pi/2)$, $-\pi \leq x \leq \pi$.

Solution

Figure 37 shows the required steps.

Check: Graph $y = x$ and compare the result with Figure 37.

Using the idea of reciprocals, we can similarly obtain the graph of $y = \sec x$. See Figure 38.

FIGURE 37

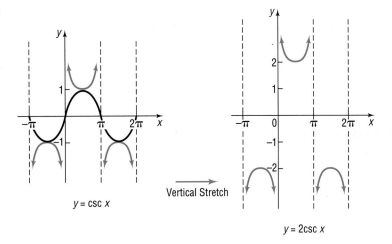

$y = \csc x$

Vertical Stretch

$y = 2\csc x$

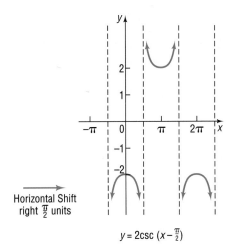

Horizontal Shift
right $\frac{\pi}{2}$ units

$y = 2\csc \left(x - \frac{\pi}{2}\right)$

FIGURE 38

$y = \sec x$, $-\infty < x < \infty$, x not equal to
odd multiples of $\pi/2$, $|y| \geq 1$

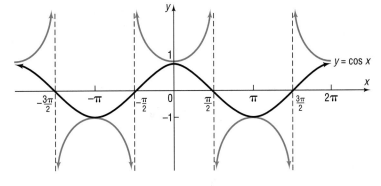

$y = \cos x$

TABLE 8

x	$y = \cot x$	(x, y)
$\pi/6$	$\sqrt{3}$	$(\pi/6, \sqrt{3})$
$\pi/4$	1	$(\pi/4, 1)$
$\pi/3$	$\sqrt{3}/3$	$(\pi/3, \sqrt{3}/3)$
$\pi/2$	0	$(\pi/2, 0)$
$2\pi/3$	$-\sqrt{3}/3$	$(2\pi/3, -\sqrt{3}/3)$
$3\pi/4$	-1	$(3\pi/4, -1)$
$5\pi/6$	$-\sqrt{3}$	$(5\pi/6, -\sqrt{3})$

We obtain the graph of $y = \cot x$ as we did the graph of $y = \tan x$. The period
of $y = \cot x$ is π. Because the cotangent function is not defined for integral multiples of π, we shall concentrate on the interval $(0, \pi)$. Table 8 lists some points
on the graph of $y = \cot x$, $0 < x < \pi$. As x approaches 0, but remains greater than
0, the value of $\cos x$ will be close to 1 and the value of $\sin x$ will be positive and

close to 0. Hence, the ratio $(\cos x)/(\sin x) = \cot x$ will be positive and large, so as x approaches 0, $\cot x$ approaches ∞. Similarly, as x approaches but remains less than π, the value of $\cos x$ will be close to -1 and the value of $\sin x$ will be positive and close to 0. Hence, the ratio $(\cos x)/(\sin x) = \cot x$ will be negative and will approach $-\infty$ as x approaches π. Figure 39 shows the graph.

FIGURE 39
$y = \cot x$, $-\infty < x < \infty$, x not equal to integral multiples of π, $-\infty < y < \infty$

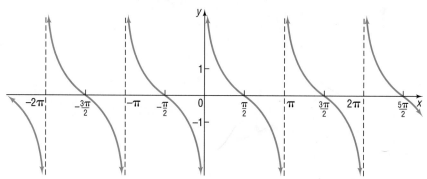

   Check: Graph $y = \cot x$ and compare the result with Figure 39. Use TRACE to see what happens when x is close to 0. ∎

6.4

Exercise 6.4

1. What is the y-intercept of $y = \tan x$?
2. What is the y-intercept of $y = \cot x$?
3. What is the y-intercept of $y = \sec x$?
4. What is the y-intercept of $y = \csc x$?
5. For what numbers x, $-2\pi \le x \le 2\pi$, does $\sec x = 1$? What about $\sec x = -1$?
6. For what numbers x, $-2\pi \le x \le 2\pi$, does $\csc x = 1$? What about $\csc x = -1$?
7. For what numbers x, $-2\pi \le x \le 2\pi$, does the graph of $y = \sec x$ have vertical asymptotes?
8. For what numbers x, $-2\pi \le x \le 2\pi$, does the graph of $y = \csc x$ have vertical asymptotes?
9. For what numbers x, $-2\pi \le x \le 2\pi$, does the graph of $y = \tan x$ have vertical asymptotes?
10. For what numbers x, $-2\pi \le x \le 2\pi$, does the graph of $y = \cot x$ have vertical asymptotes?

In Problems 11–14, match each graph to a function.

A. $y = -\tan x$ B. $y = \tan\left(x + \dfrac{\pi}{2}\right)$ C. $y = \tan(x + \pi)$ D. $y = -\tan\left(x - \dfrac{\pi}{2}\right)$

11.

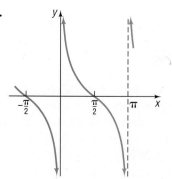

12.

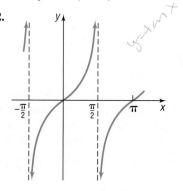

13.

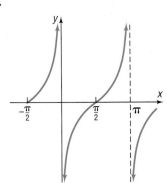

14.

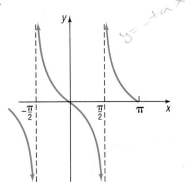

In Problems 15–30, graph each function.

15. $y = -\sec x$
16. $y = -\cot x$
17. $y = \sec\left(x - \dfrac{\pi}{2}\right)$
18. $y = \csc(x - \pi)$

19. $y = \tan(x - \pi)$
20. $y = \cot(x - \pi)$
21. $y = 3\tan 2x$
22. $y = 4\tan \frac{1}{2}x$

23. $y = \sec 2x$
24. $y = \csc \frac{1}{2}x$
25. $y = \cot \pi x$
26. $y = \cot 2x$

27. $y = -3\tan 4x$
28. $y = -3\tan 2x$
29. $y = 2\sec \frac{1}{2}x$
30. $y = 2\sec 3x$

31.   *Carrying a Ladder Around a Corner* A ladder of length L is carried horizontally around a corner from a hall 3 feet wide into a hall 4 feet wide. See the illustration.

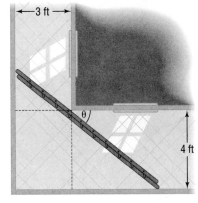

   (a) Show that the length L of the ladder as a function of the angle θ is
$$L = 3\sec \theta + 4\csc \theta$$

   (b) Graph L, $0 < \theta < \dfrac{\pi}{2}$

   (c) Graph L, $0 < \theta < \dfrac{\pi}{2}$. Where is L the least?

   (d) What is the length of the largest ladder that can be carried around the corner? Why is this also the least value of L?

32.   Graph
$$y = \tan x \quad \text{and} \quad y = -\cot\left(x + \frac{\pi}{2}\right)$$

Do you think that $\tan x = -\cot\left(x + \dfrac{\pi}{2}\right)$?

6.5

The Inverse Trigonometric Functions

In Section 2.5 we discussed inverse functions, and we noted that if a function is one-to-one it will have an inverse. We also observed that if a function is not one-to-one, it may be possible to restrict its domain in some suitable manner such that the restricted function is one-to-one. In this section, we use these ideas to define inverse trigonometric functions. (You may wish to review Section 2.5 at this time.) We begin with the inverse of the sine function.

The Inverse Sine Function

In Figure 40, we reproduce the graph of $y = \sin x$. Because every horizontal line $y = b$, where b is between -1 and 1, intersects the graph of $y = \sin x$ infinitely many times, it follows from the horizontal-line test that the function $y = \sin x$ is not one-to-one.

FIGURE 40

$y = \sin x, \ -\infty < x < \infty, \ -1 \leq y \leq 1$

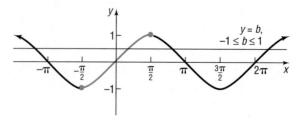

However, if we restrict the domain of $y = \sin x$ to the interval $[-\pi/2, \pi/2]$, the restricted function

$$y = \sin x, \qquad -\frac{\pi}{2} \leq x \leq \frac{\pi}{2}$$

is one-to-one and hence, will have an inverse.* See Figure 41.

FIGURE 41

$y = \sin x, \ -\pi/2 \leq x \leq \pi/2, \ -1 \leq y \leq 1$

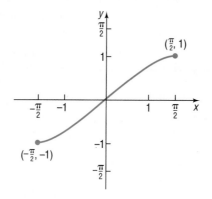

The inverse function is called the **inverse sine** of x and is symbolized by $y = \sin^{-1} x$. Thus,

Inverse Sine Function

$$y = \sin^{-1} x \quad \text{means} \quad x = \sin y \tag{1}$$

$$\text{where} \quad -\frac{\pi}{2} \leq y \leq \frac{\pi}{2} \quad \text{and} \quad -1 \leq x \leq 1$$

Because $y = \sin^{-1} x$ means $x = \sin y$, we read $y = \sin^{-1} x$ as "y is the angle whose sine equals x." Alternatively, we can say that "y is the inverse sine of x." Be careful about the notation used. The superscript -1 that appears in $y = \sin^{-1} x$ is not an exponent, but is reminiscent of the symbolism f^{-1} used to denote the inverse of a function f. (To avoid this notation, some books use the notation $y = \arcsin x$ instead of $y = \sin^{-1} x$.)

Based on the general discussion of functions and their inverses (Section 2.5), we have the following results:

*Although there are many other ways to restrict the domain and obtain a one-to-one function, mathematicians have agreed on a consistent use of the interval $[-\pi/2, \pi/2]$ in order to define the inverse of $y = \sin x$.

$$\sin^{-1}(\sin u) = u \qquad \text{where} \qquad -\frac{\pi}{2} \le u \le \frac{\pi}{2}$$

$$\sin(\sin^{-1} v) = v \qquad \text{where} \qquad -1 \le v \le 1$$

Let's examine the function $y = \sin^{-1} x$. Its domain is $-1 \le x \le 1$, and its range is $-\pi/2 \le y \le \pi/2$. Its graph can be obtained by reflecting the restricted portion of the graph of $y = \sin x$ about the line $y = x$, as shown in Figure 42.

FIGURE 42
$y = \sin^{-1} x, \; -1 \le x \le 1,$
$\quad -\pi/2 \le y \le \pi/2$

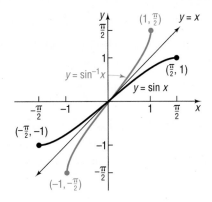

Check: Graph $y = \sin^{-1} x$ and compare the result with Figure 42. ■

For some numbers x it is possible to find the exact value of $y = \sin^{-1} x$.

E X A M P L E   1   *Finding the Exact Value of an Inverse Sine Function*
Find the exact value of $\sin^{-1} 1$.

Solution   Let $\theta = \sin^{-1} 1$. Then we seek the angle θ, $-\pi/2 \le \theta \le \pi/2$, whose sine equals 1:

$$\theta = \sin^{-1} 1, \qquad -\frac{\pi}{2} \le \theta \le \frac{\pi}{2}$$

$$\sin \theta = 1, \qquad -\frac{\pi}{2} \le \theta \le \frac{\pi}{2} \qquad \text{By definition}$$

From Figure 43 we see that the only angle θ within the interval $[-\pi/2, \pi/2]$ whose sine is 1 is $\pi/2$. (Note that $\sin(5\pi/2)$ also equals 1, but $5\pi/2$ lies outside the interval $[-\pi/2, \pi/2]$ and hence is not admissible.)

FIGURE 43

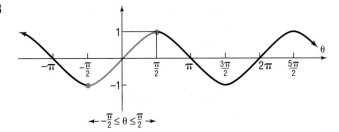

We conclude that

$$\theta = \frac{\pi}{2}$$

So

$$\sin^{-1} 1 = \frac{\pi}{2}$$ ◼

◼   Now work Problem 1.

E X A M P L E   2   *Finding the Exact Value of an Inverse Sine Function*

Find the exact value of $\sin^{-1}(\sqrt{3}/2)$.

Solution   Let $\theta = \sin^{-1}(\sqrt{3}/2)$. Then, we seek the angle θ, $-\pi/2 \le \theta \le \pi/2$, whose sine equals $\sqrt{3}/2$:

$$\theta = \sin^{-1} \frac{\sqrt{3}}{2}, \qquad -\frac{\pi}{2} \le \theta \le \frac{\pi}{2}$$

$$\sin \theta = \frac{\sqrt{3}}{2}, \qquad -\frac{\pi}{2} \le \theta \le \frac{\pi}{2} \quad \text{By definition}$$

From Figure 44, we see that the only angle θ within the interval $[-\pi/2, \pi/2]$ whose sine is $\sqrt{3}/2$ is $\pi/3$. (Note that $\sin(2\pi/3)$ also equals $\sqrt{3}/2$, but $2\pi/3$ lies outside the interval $[-\pi/2, \pi/2]$ and hence is not admissible.)

FIGURE 44

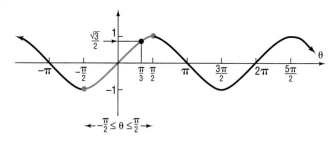

We conclude that

$$\theta = \frac{\pi}{3}$$

So

$$\sin^{-1} \frac{\sqrt{3}}{2} = \frac{\pi}{3}$$ ◼

E X A M P L E   3   *Finding the Exact Value of an Inverse Sine Function*

Find the exact value of $\sin^{-1}(-\frac{1}{2})$.

Solution   Let $\theta = \sin^{-1}(-\frac{1}{2})$. Then we seek the angle θ, $-\pi/2 \le \theta \le \pi/2$, whose sine equals $-\frac{1}{2}$:

$$\theta = \sin^{-1}\left(-\frac{1}{2}\right), \qquad -\frac{\pi}{2} \le \theta \le \frac{\pi}{2}$$

$$\sin \theta = -\frac{1}{2}, \qquad -\frac{\pi}{2} \le \theta \le \frac{\pi}{2}$$

(Refer to Figure 44, if necessary.) The only angle within the interval $[-\pi/2, \pi/2]$ whose sine is $-\frac{1}{2}$ is $-\pi/6$. Thus,

$$\theta = -\frac{\pi}{6}$$

So

$$\sin^{-1}\left(-\frac{1}{2}\right) = -\frac{\pi}{6}$$ ■

■ Now work Problem 3.

For most numbers x, the value $y = \sin^{-1} x$ must be approximated on a calculator. Some calculators have a single key labeled $\boxed{\sin^{-1}}$ or $\boxed{\text{arcsin}}$. For others, you need to press the $\boxed{\text{inv}}$ key or the $\boxed{\text{SHIFT}}$ key, followed by the $\boxed{\sin}$ key, to evaluate the inverse sine function.

E X A M P L E   4   *Finding an Approximate Value of an Inverse Sine Function*

Find the approximate value of

(a) $\sin^{-1} \frac{1}{3}$ (b) $\sin^{-1}(-\frac{1}{4})$

Solution   Because we want the angle measured in radians, we first set the mode of the calculator to radians.

(a) Keystrokes:*

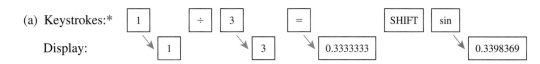

Thus, $\sin^{-1} \frac{1}{3} \approx 0.34$.

(b) Keystrokes:

Display:

$\boxed{0.25}$ $\boxed{+/-}$ $\boxed{\text{SHIFT}}$ $\boxed{\sin}$

$\boxed{0.25}$ $\boxed{-0.25}$ $\boxed{-0.2526802}$

Thus, $\sin^{-1}(-\frac{1}{4}) \approx -0.25$.   ■

■ Now work Problem 13.

The Inverse Cosine Function

In Figure 45 we reproduce the graph of $y = \cos x$. Because every horizontal line $y = b$, where b is between -1 and 1, intersects the graph of $y = \cos x$ infinitely many times, it follows that the cosine function is not one-to-one.

*On some calculators, $\boxed{\sin^{-1}}$ is pressed first; then $\boxed{1/3}$ is entered. Consult your owner's manual for the correct sequence.

FIGURE 45
$y = \cos x, \; -\infty < x < \infty, \; -1 \le y \le 1$

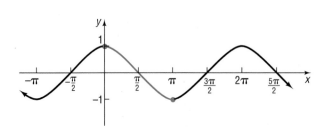

FIGURE 46
$y = \cos x, \; 0 \le x \le \pi, \; -1 \le y \le 1$

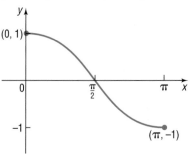

However, if we restrict the domain of $y = \cos x$ to the interval $[0, \pi]$, the restricted function

$$y = \cos x, \qquad 0 \le x \le \pi$$

is one-to-one and hence will have an inverse.* See Figure 46.

The inverse function is called the **inverse cosine** of x and is symbolized by $y = \cos^{-1} x$ (or by $y = \arccos x$). Thus,

Inverse Cosine Function

$$y = \cos^{-1} x \quad \text{means} \quad x = \cos y \tag{2}$$
$$\text{where} \quad 0 \le y \le \pi \quad \text{and} \quad -1 \le x \le 1$$

Here, y is the angle whose cosine is x. The domain of the function $y = \cos^{-1} x$ is $-1 \le x \le 1$, and its range is $0 \le y \le \pi$. The graph of $y = \cos^{-1} x$ can be obtained by reflecting the restricted portion of the graph of $y = \cos x$ about the line $y = x$, as shown in Figure 47.

FIGURE 47
$y = \cos^{-1} x, \; -1 \le x \le 1, \; 0 \le y \le \pi$

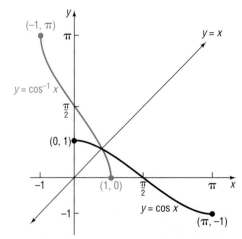

Check: Graph $y = \cos^{-1} x$ and compare the result with Figure 47.   ■

*This is the generally accepted restriction to define the inverse.

In general,

$$\cos^{-1}(\cos u) = u, \qquad \text{where} \quad 0 \le u \le \pi$$
$$\cos(\cos^{-1} v) = v, \qquad \text{where} \quad -1 \le v \le 1$$

E X A M P L E   5

Finding the Exact Value of an Inverse Cosine Function
Find the exact value of $\cos^{-1} 0$.

Solution   Let $\theta = \cos^{-1} 0$. Then we seek the angle θ, $0 \le \theta \le \pi$, whose cosine equals 0:

$$\theta = \cos^{-1} 0, \qquad 0 \le \theta \le \pi$$
$$\cos \theta = 0, \qquad 0 \le \theta \le \pi$$

From Figure 48, we see that the only angle θ within the interval $[0, \pi]$ whose cosine is 0 is $\pi/2$. (Note that $\cos(3\pi/2)$ also equals 0, but $3\pi/2$ lies outside the interval $[0, \pi]$ and hence is not admissible.)

FIGURE 48

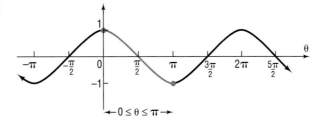

We conclude that

$$\theta = \frac{\pi}{2}$$

So

$$\cos^{-1} 0 = \frac{\pi}{2}$$

■

E X A M P L E   6

Finding the Exact Value of an Inverse Cosine Function
Find the exact value of $\cos^{-1}(\sqrt{2}/2)$.

Solution   Let $\theta = \cos^{-1}(\sqrt{2}/2)$. Then we seek the angle θ, $0 \le \theta \le \pi$, whose cosine equals $\sqrt{2}/2$:

$$\theta = \cos^{-1} \frac{\sqrt{2}}{2}, \qquad 0 \le \theta \le \pi$$
$$\cos \theta = \frac{\sqrt{2}}{2}, \qquad 0 \le \theta \le \pi$$

From Figure 49, we see that the only angle θ within the interval $[0, \pi]$ whose cosine is $\sqrt{2}/2$ is $\pi/4$.

FIGURE 49

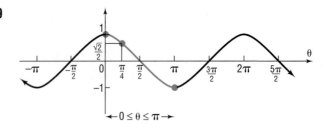

We conclude that

$$\theta = \frac{\pi}{4}$$

So

$$\cos^{-1} \frac{\sqrt{2}}{2} = \frac{\pi}{4}$$ ∎

E X A M P L E   7   *Finding the Exact Value of an Inverse Cosine Function*
Find the exact value of $\cos^{-1}(-\frac{1}{2})$.

Solution   Let $\theta = \cos^{-1}(-\frac{1}{2})$. Then we seek the angle θ, $0 \le \theta \le \pi$, whose cosine equals $-\frac{1}{2}$:

$$\theta = \cos^{-1}(-\tfrac{1}{2}), \qquad 0 \le \theta \le \pi$$
$$\cos \theta = -\tfrac{1}{2}, \qquad 0 \le \theta \le \pi$$

(Refer to Figure 49, if necessary.) The only angle within the interval $[0, \pi]$ whose cosine is $-\frac{1}{2}$ is $2\pi/3$. Thus,

$$\theta = \frac{2\pi}{3}$$

So

$$\cos^{-1}\left(-\frac{1}{2}\right) = \frac{2\pi}{3}$$ ∎

The Inverse Tangent Function

In Figure 50 we reproduce the graph of $y = \tan x$. Because every horizontal line intersects the graph infinitely many times, it follows that the tangent function is not one-to-one.

FIGURE 50
$y = \tan x$, $-\infty < x < \infty$, x not equal to
odd multiples of $\pi/2$, $-\infty < y < \infty$

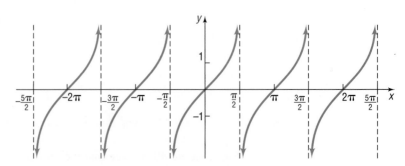

However, if we restrict the domain of $y = \tan x$ to the interval $(-\pi/2, \pi/2)$,* the restricted function

$$y = \tan x, \qquad -\frac{\pi}{2} < x < \frac{\pi}{2}$$

is one-to-one and hence has an inverse. See Figure 51.

FIGURE 51
$y = \tan x, -\pi/2 < x < \pi/2,$
$-\infty < y < \infty$

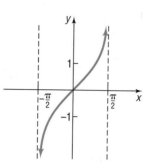

The inverse function is called the **inverse tangent** of x and is symbolized by $y = \tan^{-1} x$ (or by $y = \arctan x$). Thus,

Inverse Tangent Function

$$y = \tan^{-1} x \quad \text{means} \quad x = \tan y \tag{3}$$

$$\text{where} \quad -\frac{\pi}{2} < y < \frac{\pi}{2} \quad \text{and} \quad -\infty < x < \infty$$

Here, y is the angle whose tangent is x. The domain of the function $y = \tan^{-1} x$ is $-\infty < x < \infty$, and its range is $-\pi/2 < y < \pi/2$. The graph of $y = \tan^{-1} x$ can be obtained by reflecting the restricted portion of the graph of $y = \tan x$ about the line $y = x$, as shown in Figure 52.

FIGURE 52
$y = \tan^{-1} x, -\infty < x < \infty,$
$-\pi/2 < y < \pi/2$

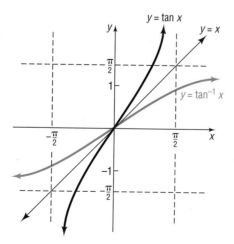

Check: Graph $y = \tan^{-1} x$ and compare the results with Figure 52. ∎

*This is the generally accepted restriction.

In general,

$$
\tan^{-1}(\tan u) = u, \qquad \text{where} \quad -\frac{\pi}{2} < u < \frac{\pi}{2}
$$

$$
\tan(\tan^{-1} v) = v, \qquad \text{where} \quad -\infty < v < \infty
$$

E X A M P L E   8   *Finding the Exact Value of an Inverse Tangent Function*

Find the exact value of $\tan^{-1} 1$.

Solution   Let $\theta = \tan^{-1} 1$. Then we seek the angle θ, $-\pi/2 < \theta < \pi/2$, whose tangent equals 1:

$$
\theta = \tan^{-1} 1, \qquad -\frac{\pi}{2} < \theta < \frac{\pi}{2}
$$

$$
\tan \theta = 1, \qquad -\frac{\pi}{2} < \theta < \frac{\pi}{2}
$$

Refer to Figure 51. The only angle θ within the interval $(-\pi/2, \pi/2)$ whose tangent is 1 is $\pi/4$.

We conclude that

$$
\theta = \frac{\pi}{4}
$$

So

$$
\tan^{-1} 1 = \frac{\pi}{4} \qquad \blacksquare
$$

E X A M P L E   9   *Finding the Exact Value of an Inverse Tangent Function*

Find the exact value of $\tan^{-1}(-\sqrt{3})$.

Solution   Let $\theta = \tan^{-1}(-\sqrt{3})$. Then we seek the angle θ, $-\pi/2 < \theta < \pi/2$, whose tangent equals $-\sqrt{3}$:

$$
\theta = \tan^{-1}(-\sqrt{3}), \qquad -\frac{\pi}{2} < \theta < \frac{\pi}{2}
$$

$$
\tan \theta = -\sqrt{3} \qquad -\frac{\pi}{2} < \theta < \frac{\pi}{2}
$$

Refer again to Figure 51. The only angle θ within the interval $(-\pi/2, \pi/2)$ whose tangent is $-\sqrt{3}$ is $-\pi/3$. Thus,

$$
\theta = -\frac{\pi}{3}
$$

So

$$
\tan^{-1}(-\sqrt{3}) = -\frac{\pi}{3} \qquad \blacksquare
$$

■   Now work Problem 5.

E X A M P L E   1 0   *Finding the Exact Value of Expressions Involving Inverse Trigonometric Functions*

Find the exact value of $\sin[\cos^{-1}(\sqrt{3}/2)]$.

Solution   Let $\theta = \cos^{-1}(\sqrt{3}/2)$. We seek the angle θ, $0 \leq \theta \leq \pi$, whose cosine equals $\sqrt{3}/2$:

$$\cos \theta = \frac{\sqrt{3}}{2}, \qquad 0 \leq \theta \leq \pi$$

$$\theta = \frac{\pi}{6}$$

Now

$$\sin\left(\cos^{-1}\frac{\sqrt{3}}{2}\right) = \sin \theta = \sin \frac{\pi}{6} = \frac{1}{2}$$ ∎

E X A M P L E   1 1   *Finding the Exact Value of Expressions Involving Inverse Trigonometric Functions*

Find the exact value of $\cos[\tan^{-1}(-1)]$.

Solution   Let $\theta = \tan^{-1}(-1)$. We seek the angle θ, $-\pi/2 < \theta < \pi/2$, whose tangent equals -1:

$$\tan \theta = -1, \qquad -\frac{\pi}{2} < \theta < \frac{\pi}{2}$$

$$\theta = -\frac{\pi}{4},$$

Now

$$\cos[\tan^{-1}(-1)] = \cos \theta = \cos\left(-\frac{\pi}{4}\right) = \frac{\sqrt{2}}{2}$$ ∎

E X A M P L E   1 2   *Finding the Exact Value of Expressions Involving Inverse Trigonometric Functions*

Find the exact value of $\sec(\sin^{-1}\frac{1}{2})$.

Solution   Let $\theta = \sin^{-1}\frac{1}{2}$. We seek the angle θ, $-\pi/2 \leq \theta \leq \pi/2$, whose sine equals $\frac{1}{2}$.

$$\sin \theta = \frac{1}{2}, \qquad -\frac{\pi}{2} \leq \theta \leq \frac{\pi}{2}$$

$$\theta = \frac{\pi}{6}$$

Now

$$\sec\left(\sin^{-1}\frac{1}{2}\right) = \sec \theta = \sec \frac{\pi}{6} = \frac{2}{\sqrt{3}} = \frac{2\sqrt{3}}{3}$$ ∎

■ Now work Problem 25.

It is not necessary to be able to find the angle in order to solve problems like those given in Examples 10–12.

E X A M P L E   1 3   *Finding the Exact Value of Expressions Involving Inverse Trigonometric Functions*

Find the exact value of $\sin(\tan^{-1}\frac{1}{2})$.

Solution   Let $\theta = \tan^{-1}\frac{1}{2}$. Then $\tan\theta = \frac{1}{2}$, where $-\pi/2 < \theta < \pi/2$. Because $\tan\theta > 0$, it follows that $0 < \theta < \pi/2$. Now, in Figure 53, we draw a triangle in the appropriate quadrant depicting $\tan\theta = \frac{1}{2}$. The hypotenuse of this triangle is easily found to be of length $\sqrt{5}$. Hence, the sine of θ is $1/\sqrt{5}$, so

$$\sin\left(\tan^{-1}\frac{1}{2}\right) = \sin\theta = \frac{1}{\sqrt{5}} = \frac{\sqrt{5}}{5} \qquad\blacksquare$$

FIGURE 53
$\tan\theta = \frac{1}{2}$

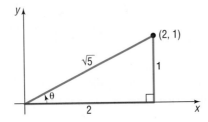

E X A M P L E   1 4   *Finding the Exact Value of Expressions Involving Inverse Trigonometric Functions*

Find the exact value of $\tan(\cos^{-1}\frac{1}{3})$

Solution   Let $\theta = \cos^{-1}\frac{1}{3}$. Then $\cos\theta = \frac{1}{3}$, where $0 \le \theta \le \pi$. Because $\cos\theta > 0$, it follows that $0 \le \theta \le \pi/2$. Look at the triangle in Figure 54. The opposite side to angle θ is $\sqrt{8} = 2\sqrt{2}$. Thus,

$$\tan(\cos^{-1}\tfrac{1}{3}) = \tan\theta = 2\sqrt{2}$$

FIGURE 54
$\cos\theta = \frac{1}{3}$

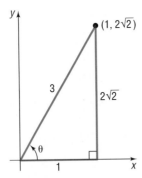

E X A M P L E   1 5   *Finding the Exact Value of Expressions Involving Inverse Trigonometric Functions*

Find the exact value of $\cos[\sin^{-1}(-\frac{1}{3})]$.

Solution   Let $\theta = \sin^{-1}(-\frac{1}{3})$. Then $\sin\theta = -\frac{1}{3}$ and $-\pi/2 \le \theta \le \pi/2$. Because $\sin\theta < 0$, it follows that $-\pi/2 \le \theta \le 0$. Based on Figure 55, we conclude that

$$\cos\left[\sin^{-1}\left(-\frac{1}{3}\right)\right] = \cos\theta = \frac{2\sqrt{2}}{3}$$

FIGURE 55
$\sin \theta = -\frac{1}{3}$

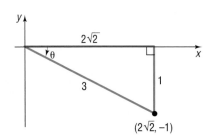

E X A M P L E   1 6

Finding the Exact Value of Expressions Involving Inverse Trigonometric Functions

FIGURE 56
$\cos \theta = -\frac{1}{3}$

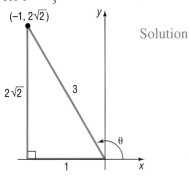

Find the exact value of $\tan[\cos^{-1}(-\frac{1}{3})]$.

Solution   Let $\theta = \cos^{-1}(-\frac{1}{3})$. Then $\cos \theta = -\frac{1}{3}$ and $0 \leq \theta \leq \pi$. Because $\cos \theta < 0$, it follows that $\pi/2 \leq \theta \leq \pi$. Based on Figure 56, we conclude that

$$\tan\left[\cos^{-1}\left(-\frac{1}{3}\right)\right] = \tan \theta = \frac{2\sqrt{2}}{-1} = -2\sqrt{2}$$

■   Now work Problem 39.

E X A M P L E   1 7

Establishing an Identity Involving Inverse Trigonometric Functions

Show that $\sin(\tan^{-1} v) = \dfrac{v}{\sqrt{1 + v^2}}$

Solution   Let $\theta = \tan^{-1} v$ so that $\tan \theta = v$, $-\pi/2 < \theta < \pi/2$. There are two possibilities: either $-\pi/2 < \theta < 0$ or $0 \leq \theta < \pi/2$. If $0 \leq \theta < \pi/2$, then $\tan \theta = v \geq 0$. Based on Figure 57, we conclude that

$$\sin(\tan^{-1} v) = \sin \theta = \frac{v}{\sqrt{1 + v^2}}$$

If $-\pi/2 < \theta < 0$, then $\tan \theta = v < 0$. Based on Figure 58, we conclude that

$$\sin(\tan^{-1} v) = \sin \theta = \frac{v}{\sqrt{1 + v^2}}$$

FIGURE 57
$v > 0$

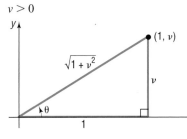

FIGURE 58
$v < 0$

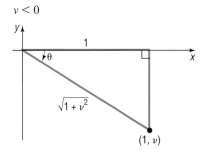

An alternative solution to Example 17 uses the fundamental identities. If $\theta = \tan^{-1} v$, then $\tan \theta = v$ and $-\pi/2 < \theta < \pi/2$, as before. As a result, we know that $\sec \theta > 0$. Thus,

$$\sin(\tan^{-1} v) = \sin \theta = \cos \theta \tan \theta = \frac{\tan \theta}{\sec \theta} = \frac{\tan \theta}{\sqrt{1 + \tan^2 \theta}} = \frac{v}{\sqrt{1 + v^2}}$$

$$\uparrow \qquad\qquad\qquad \uparrow$$
$$\tan \theta = \frac{\sin \theta}{\cos \theta} \qquad \sec^2 \theta = 1 + \tan^2 \theta$$
$$\sec \theta > 0$$

■ Now work Problem 57.

The Remaining Inverse Trigonometric Functions

The inverse cotangent, inverse secant, and inverse cosecant functions are defined as follows:

$$y = \cot^{-1} x \quad \text{means} \quad x = \cot y \tag{4}$$
$$\text{where} \quad -\infty < x < \infty \quad \text{and} \quad 0 < y < \pi$$

$$y = \sec^{-1} x \quad \text{means} \quad x = \sec y \tag{5}$$
$$\text{where} \quad |x| \geq 1 \quad \text{and} \quad 0 \leq y \leq \pi, \ y \neq \frac{\pi}{2}$$

$$y = \csc^{-1} x \quad \text{means} \quad x = \csc y \tag{6}$$
$$\text{where} \quad |x| \geq 1 \quad \text{and} \quad \frac{-\pi}{2} \leq y \leq \frac{\pi}{2}, \ y \neq 0$$

Most calculators do not have keys for evaluating these inverse trigonometric functions. The easiest way to evaluate them is to convert to an inverse trigonometric function whose range is the same as the one to be evaluated. In this regard, notice that $y = \cot^{-1} x$ and $y = \sec^{-1} x$ (except where undefined) each have the same range as $y = \cos^{-1} x$, while $y = \csc^{-1} x$ (except where undefined) has the same range as $y = \sin^{-1} x$.

EXAMPLE 18   *Approximating the Value of Inverse Trigonometric Functions*

Use a calculator to approximate to two decimal places:

(a) $\sec^{-1} 3$ (b) $\csc^{-1}(-4)$ (c) $\cot^{-1} \frac{1}{2}$ (d) $\cot^{-1}(-2)$

Solution   First, set your calculator to radian mode.

(a) Let $\theta = \sec^{-1} 3$. Then $\sec \theta = 3$ and $0 \leq \theta \leq \pi$, $\theta \neq \pi/2$. Thus, $\cos \theta = \frac{1}{3}$ and

$$\sec^{-1} 3 = \theta = \cos^{-1} \frac{1}{3} \approx 1.23$$
$$\uparrow$$
$$\text{Use a calculator.}$$

FIGURE 59
$\cot \theta = \frac{1}{2}, 0 < \theta < \pi$

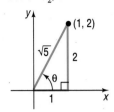

FIGURE 60
$\cot \theta = -2, 0 < \theta < \pi$

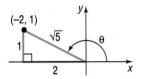

(b) Let $\theta = \csc^{-1}(-4)$. Then $\csc \theta = -4$, $-\pi/2 \leq \theta \leq \pi/2$, $\theta \neq 0$. Thus, $\sin \theta = -\frac{1}{4}$ and

$$\csc^{-1}(-4) = \theta = \sin^{-1}(-\tfrac{1}{4}) \approx -0.25$$

(c) Let $\theta = \cot^{-1} \frac{1}{2}$. Then $\cot \theta = \frac{1}{2}$, $0 < \theta < \pi$. From these facts we know that θ is in quadrant I. We draw Figure 59 to help us to find $\cos \theta$. Thus, $\cos \theta = 1/\sqrt{5}$, $0 < \theta < \pi/2$, and

$$\cot^{-1} \frac{1}{2} = \theta = \cos^{-1} \frac{1}{\sqrt{5}} \approx 1.11$$

(d) Let $\theta = \cot^{-1}(-2)$. Then $\cot \theta = -2$, $0 < \theta < \pi$. From these facts we know that θ lies in quadrant II. We draw Figure 60 to help us to find $\cos \theta$. Thus, $\cos \theta = -2/\sqrt{5}$, $\pi/2 < \theta < \pi$, and

$$\cot^{-1}(-2) = \theta = \cos^{-1}\left(-\frac{2}{\sqrt{5}}\right) \approx 2.68$$ ■

■   Now work Problem 67.

6.5

Exercise 6.5

In Problems 1–12, find the exact value of each expression.

1. $\sin^{-1} 0$

2. $\cos^{-1} 1$

3. $\sin^{-1}(-1)$

4. $\cos^{-1}(-1)$

5. $\tan^{-1} 0$

6. $\tan^{-1}(-1)$

7. $\sin^{-1} \dfrac{\sqrt{2}}{2}$

8. $\tan^{-1} \dfrac{\sqrt{3}}{3}$

9. $\tan^{-1} \sqrt{3}$

10. $\sin^{-1}\left(-\dfrac{\sqrt{3}}{2}\right)$

11. $\cos^{-1}\left(-\dfrac{\sqrt{3}}{2}\right)$

12. $\sin^{-1}\left(-\dfrac{\sqrt{2}}{2}\right)$

In Problems 13–24, use a calculator to find the approximate value of each expression rounded to two decimal places.

13. $\sin^{-1} 0.1$

14. $\cos^{-1} 0.6$

15. $\tan^{-1} 5$

16. $\tan^{-1} 0.2$

17. $\cos^{-1} \frac{7}{8}$

18. $\sin^{-1} \frac{1}{8}$

19. $\tan^{-1}(-0.4)$

20. $\tan^{-1}(-3)$

21. $\sin^{-1}(^-0.12)$

22. $\cos^{-1}(-0.44)$

23. $\cos^{-1} \dfrac{\sqrt{2}}{3}$

24. $\sin^{-1} \dfrac{\sqrt{3}}{5}$

In Problems 25–46, find the exact value of each expression.

25. $\cos\left(\sin^{-1} \dfrac{\sqrt{2}}{2}\right)$

26. $\sin\left(\cos^{-1} \dfrac{1}{2}\right)$

27. $\tan\left[\cos^{-1}\left(-\dfrac{\sqrt{3}}{2}\right)\right]$

28. $\tan\left[\sin^{-1}\left(-\dfrac{1}{2}\right)\right]$

29. $\sec\left(\cos^{-1} \dfrac{1}{2}\right)$

30. $\cot\left[\sin^{-1}\left(-\dfrac{1}{2}\right)\right]$

31. $\csc(\tan^{-1} 1)$

32. $\sec(\tan^{-1} \sqrt{3})$

33. $\sin[\tan^{-1}(-1)]$

34. $\cos\left[\sin^{-1}\left(-\dfrac{\sqrt{3}}{2}\right)\right]$

35. $\sec\left[\sin^{-1}\left(-\dfrac{1}{2}\right)\right]$

36. $\csc\left[\cos^{-1}\left(-\dfrac{\sqrt{3}}{2}\right)\right]$

37. $\tan\left(\sin^{-1} \dfrac{1}{3}\right)$

38. $\tan\left(\cos^{-1} \dfrac{1}{3}\right)$

39. $\sec\left(\tan^{-1} \dfrac{1}{2}\right)$

40. $\cos\left(\sin^{-1} \dfrac{\sqrt{2}}{3}\right)$

41. $\cot\left[\sin^{-1}\left(-\dfrac{\sqrt{2}}{3}\right)\right]$

42. $\csc[\tan^{-1}(-2)]$

43. $\sin[\tan^{-1}(-3)]$

44. $\cot\left[\cos^{-1}\left(-\dfrac{\sqrt{3}}{3}\right)\right]$

45. $\sec\left(\sin^{-1}\dfrac{2\sqrt{5}}{5}\right)$ **46.** $\csc\left(\tan^{-1}\dfrac{1}{2}\right)$

In Problems 47–56, use a calculator to approximate the value of each expression rounded to two decimal places.

47. $\sin^{-1}(\tan 0.5)$ **48.** $\cos^{-1}(\tan 0.4)$ **49.** $\tan^{-1}(\sin 0.1)$ **50.** $\tan^{-1}(\cos 0.2)$

51. $\cos^{-1}(\sin 1)$ **52.** $\tan^{-1}(\cos 1)$ **53.** $\sin^{-1}\left(\tan\dfrac{\pi}{8}\right)$ **54.** $\cos^{-1}\left(\sin\dfrac{\pi}{8}\right)$

55. $\tan^{-1}\left(\sin\dfrac{\pi}{8}\right)$ **56.** $\tan^{-1}\left(\cos\dfrac{\pi}{8}\right)$

57. Show that $\sec(\tan^{-1} v) = \sqrt{1 + v^2}$.   **58.** Show that $\tan(\sin^{-1} v) = v/\sqrt{1 - v^2}$.

59. Show that $\tan(\cos^{-1} v) = \sqrt{1 - v^2}/v$.   **60.** Show that $\sin(\cos^{-1} v) = \sqrt{1 - v^2}$.

61. Show that $\cos(\sin^{-1} v) = \sqrt{1 - v^2}$.   **62.** Show that $\cos(\tan^{-1} v) = 1/\sqrt{1 + v^2}$.

63. Show that $\sin^{-1} v + \cos^{-1} v = \pi/2$.   **64.** Show that $\tan^{-1} v + \cot^{-1} v = \pi/2$.

65. Show that $\tan^{-1} 1/v = \pi/2 - \tan^{-1} v$.   **66.** Show that $\cot^{-1} e^v = \tan^{-1} e^{-v}$.

In Problems 67–78, use a calculator to approximate each expression rounded to two decimal places.

67. $\sec^{-1} 4$ **68.** $\csc^{-1} 5$ **69.** $\cot^{-1} 2$ **70.** $\sec^{-1}(-3)$

71. $\csc^{-1}(-3)$ **72.** $\cot^{-1}(-\frac{1}{2})$ **73.** $\cot^{-1}(-\sqrt{5})$ **74.** $\cot^{-1}(-8.1)$

75. $\csc^{-1}(-\frac{3}{2})$ **76.** $\sec^{-1}(-\frac{4}{3})$ **77.** $\cot^{-1}(-\frac{3}{2})$ **78.** $\cot^{-1}(-\sqrt{10})$

79. The drive wheel of an engine is 13 inches in diameter, and the pulley on the rotary pump is 5 inches in diameter. If the shafts of the drive wheel and the pulley are 2 feet apart, what length of belt is required to join them as shown in figure below?

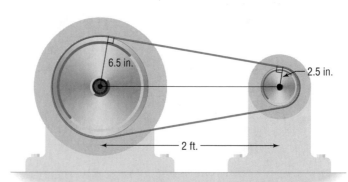

80. Rework Problem 79 if the belt is crossed, as shown in the figure below.

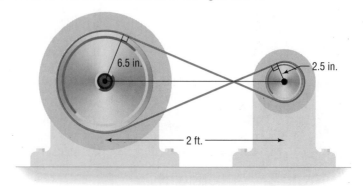

81. For what numbers x does $\sin(\sin^{-1} x) = x$?

82. For what numbers x does $\cos(\cos^{-1} x) = x$?

83. For what numbers x does $\sin^{-1}(\sin x) = x$?

84. For what numbers x does $\cos^{-1}(\cos x) = x$?

85. Draw the graph of $y = \cot^{-1} x$.

86. Draw the graph of $y = \sec^{-1} x$.

87. Draw the graph of $y = \csc^{-1} x$.

88. Cadillac Mountain, elevation 1530 feet, is located in Acadia National Park, Maine, and is the highest peak on the east coast of the United States. It is said that a person standing on the summit will be the first person in the United States to see the rays of the rising Sun. How much sooner would a person atop Cadillac Mountain see the first rays than a person standing below, at sea level? [*Hint:* Consult the figure. When the person at D sees the first rays of the Sun, the person at P does not. The person at P sees the first rays of the Sun only after Earth has rotated so that P is at location Q. Compute the length of arc s subtended by the central angle θ. Then use the fact that in 24 hours a length of $2\pi(3960)$ miles is subtended, and find the time it takes to subtend the length s.]

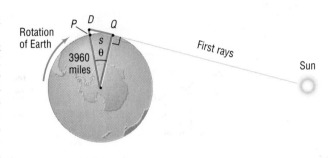

 89. Explain in your own words how you would use your calculator to find the value of $\cot^{-1} 10$.

90. Consult three books on calculus and write down the definition in each of $y = \sec^{-1} x$ and $y = \csc^{-1} x$. Compare these with the definitions given in this book.

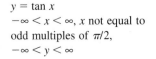

THINGS TO KNOW

The graphs of the six trigonometric functions	$y = \sin x$ $-\infty < x < \infty$ $-1 \le y \le 1$	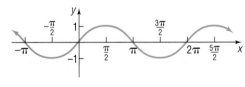
	$y = \cos x$ $-\infty < x < \infty$ $-1 \le y \le 1$	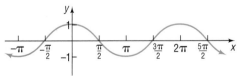
	$y = \tan x$ $-\infty < x < \infty$, x not equal to odd multiples of $\pi/2$, $-\infty < y < \infty$	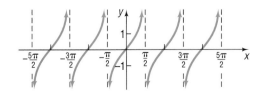

$y = \csc x$
$-\infty < x < \infty$, x not equal to
integral multiples of π,
$|y| \geq 1$

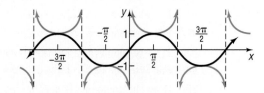

$y = \sec x$
$-\infty < x < \infty$, x not equal to
odd multiples of $\pi/2$,
$|y| \geq 1$

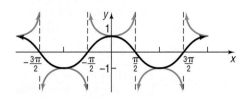

$y = \cot x$
$-\infty < x < \infty$, x not equal to
integral multiples of π,
$-\infty < y < \infty$

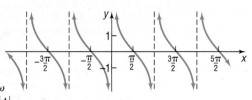

Sinusoidal graphs	$y = A \sin \omega x$, $\omega > 0$	Period $= 2\pi/\omega$
	$y = A \cos \omega x$, $\omega > 0$	Amplitude $= |A|$
	$y = A \sin(\omega x - \phi)$	Phase shift $= \phi/\omega$
	$y = A \cos(\omega x - \phi)$	

Definitions of the six
inverse trigonometric functions

$y = \sin^{-1} x$ means $x = \sin y$ where $-1 \leq x \leq 1$, $-\pi/2 \leq y \leq \pi/2$
$y = \cos^{-1} x$ means $x = \cos y$ where $-1 \leq x \leq 1$, $0 \leq y \leq \pi$
$y = \tan^{-1} x$ means $x = \tan y$ where $-\infty < x < \infty$, $-\pi/2 < y < \pi/2$
$y = \cot^{-1} x$ means $x = \cot y$ where $-\infty < x < \infty$, $0 < y < \pi$
$y = \sec^{-1} x$ means $x = \sec y$ where $|x| \geq 1$, $0 \leq y \leq \pi$, $y \neq \pi/2$
$y = \csc^{-1} x$ means $x = \csc y$ where $|x| \geq 1$, $-\pi/2 \leq y \leq \pi/2$, $y \neq 0$

How To:

Graph the trigonometric functions, including variations

Find the period and amplitude of a sinusoidal function and use them to graph the function

Find a function whose sinusoidal graph is given

Graph certain sums of functions by adding y-coordinates

Graph certain products of functions

Find the exact value of certain inverse trigonometric functions

Use a calculator to find the approximate values of inverse trigonometric functions

Fill-in-the-Blank Items

1. The following function has amplitude 3 and period 2:

$$y = \underline{\hspace{1cm}} \sin \underline{\hspace{1cm}} x$$

2. The function $y = 3 \sin 6x$ has amplitude _____ and period _____.

3. The amplifying factor of $f(x) = 5x \cos 3\pi x$ is _____.

4. The function $y = \sin^{-1} x$ has domain _____ and range _____.

5. The value of $\sin^{-1}[\cos(\pi/2)]$ is _____.

6. The motion of an object obeys the equation $x = 4 \cos 6t$. Such motion is described as _____ _____ _____.

7. Which of the trigonometric functions have graphs that are symmetric with respect to the y-axis?

8. Which of the trigonometric functions have graphs that are symmetric with respect to the origin?

TRUE/FALSE ITEMS

T   F   **1.** The graphs of $y = \sin x$ and $y = \cos x$ are identical except for a horizontal shift.

T   F   **2.** The graphs of $y = \tan x$, $y = \cot x$, $y = \sec x$, and $y = \csc x$ have infinitely many vertical asymptotes.

T   F   **3.** For $y = -2 \sin \pi x$, the amplitude is 2 and the period is $\pi/2$.

T   F   **4.** The domain of $y = \sin^{-1} x$ is $-\pi/2 \le x \le \pi/2$.

T   F   **5.** $\cos(\sin^{-1} 0) = 1$ and $\sin(\cos^{-1} 0) = 1$.

REVIEW EXERCISES

In Problems 1–4, determine the amplitude and period of each function without graphing.

1. $y = 4 \cos x$ **2.** $y = \sin 2x$ **3.** $y = -8 \sin \dfrac{\pi}{2}x$ **4.** $y = -2 \cos 3\pi x$

In Problems 5–12, find the amplitude, period, and phase shift of each function. Graph one period.

5. $y = 4 \sin 3x$ **6.** $y = 2 \cos \frac{1}{3}x$ **7.** $y = -2 \sin\left(\dfrac{\pi}{2}x + \dfrac{1}{2}\right)$ **8.** $y = -6 \sin(2\pi x - 2)$

9. $y = \frac{1}{2} \sin(\frac{3}{2}x - \pi)$ **10.** $y = \frac{3}{2} \cos(6x + 3\pi)$ **11.** $y = -\frac{2}{3} \cos(\pi x - 6)$ **12.** $y = -7 \sin\left(\dfrac{\pi}{3}x + \dfrac{4}{3}\right)$

In Problems 13–16, find a function whose graph is given.

13.

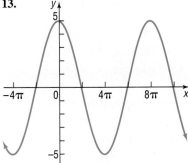

14.

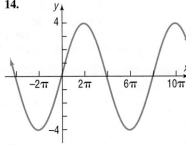

15.

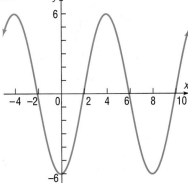

16.

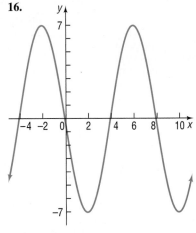

In Problems 17–20, graph each function. Each graph should contain at least one period.

17. $y = \tan(x + \pi)$ **18.** $y = -\tan\left(x - \dfrac{\pi}{2}\right)$ **19.** $y = -2 \tan 3x$ **20.** $y = 4 \tan 2x$

In Problems 21–26, use the method of adding y-coordinates to graph each function.

21. $f(x) = 2x + \sin 2x$ **22.** $f(x) = 2x + \cos 2x$ **23.** $f(x) = \sin \pi x + \cos \dfrac{\pi}{2}x$

24. $f(x) = \sin \dfrac{\pi}{2}x + \cos \pi x$ **25.** $f(x) = 3 \sin \pi x + 2 \cos \pi x$ **26.** $f(x) = 3 \cos 2\pi x + 4 \sin 2\pi x$

In Problems 27–32, graph each function.

27. $f(x) = x \cos 2x$ **28.** $f(x) = x \sin \pi x$ **29.** $f(x) = e^{-x} \sin \pi x$

30. $f(x) = e^{-x} \cos \pi x$ **31.** $f(x) = e^{x} \sin \pi x$ **32.** $f(x) = e^{x} \cos \pi x$

In Problems 33–48, find the exact value of each expression.

33. $\sin^{-1} 1$ **34.** $\cos^{-1} 0$ **35.** $\tan^{-1} 1$ **36.** $\sin^{-1}\left(-\tfrac{1}{2}\right)$

37. $\cos^{-1}\left(-\dfrac{\sqrt{3}}{2}\right)$ **38.** $\tan^{-1}(-\sqrt{3})$ **39.** $\sin\left(\cos^{-1}\dfrac{\sqrt{2}}{2}\right)$ **40.** $\cos(\sin^{-1} 0)$

41. $\tan\left[\sin^{-1}\left(-\dfrac{\sqrt{3}}{2}\right)\right]$ **42.** $\tan[\cos^{-1}(-\tfrac{1}{2})]$ **43.** $\sec\left(\tan^{-1}\dfrac{\sqrt{3}}{3}\right)$ **44.** $\csc\left(\sin^{-1}\dfrac{\sqrt{3}}{2}\right)$

45. $\sin(\tan^{-1}\tfrac{3}{4})$ **46.** $\cos(\sin^{-1}\tfrac{3}{5})$ **47.** $\tan[\sin^{-1}(-\tfrac{4}{5})]$ **48.** $\tan[\cos^{-1}(-\tfrac{3}{5})]$

In Problems 49–52, the distance d (in feet) an object travels in time t (in seconds) is given.
(a) Describe the motion of the object.
(b) What is the maximum displacement from its resting position?
(c) What is the time required for one oscillation?
(d) What is the frequency?

49. $d = 6 \sin 2t$ **50.** $d = 2 \cos 4t$ **51.** $d = -2 \cos \pi t$ **52.** $d = -3 \sin \dfrac{\pi}{2}t$

53. *Alternating Voltage* The electromotive force E, in volts in a certain ac circuit obeys the equation

$$E = 120 \sin 120 \pi t, \ t \geq 0$$

where t is measured in seconds.
(a) What is the maximum value of E?
(b) What is the period?
(c) Graph this function over two periods.

54. *Alternating Current* The current I, in amperes, flowing through an ac (alternating current) circuit at time t is

$$I = 220 \sin\left(30 \pi t + \dfrac{\pi}{6}\right), \qquad t \geq 0$$

(a) What is the period?
(b) What is the amplitude?
(c) What is the phase shift?
(d) Graph this function over two periods.

Chapter 7

PREPARING FOR THIS CHAPTER

Before getting started on this chapter, review the following concepts:

Fundamental identities (p. 354)
Even–odd properties (p. 355)
Values of the trigonometric functions
    of certain angles (Table 2, p. 338; Table 3, p. 341)
Solving equations (Section 1.2)

ANALYTIC TRIGONOMETRY

Preview   Projectile Motion

*An object is propelled upward at an angle θ to the horizontal with an initial velocity of v_0 feet per second. If air resistance is ignored, the horizontal distance R that it travels, the **range**, is given by*

$$R = \frac{1}{16} v_0^2 \sin\theta \cos\theta$$

(a) Show that $R = \dfrac{1}{32} v_0^2 \sin 2\theta$.

(b) Find the angle θ for which R is a maximum.

[Example 4, Section 7.3] ∎

In this chapter we continue the derivation of identities involving the trigonometric functions. These identities play an important role in calculus, the physical and life sciences, and economics, where they are used to simplify complicated expressions.

The last section of this chapter provides techniques for solving equations that contain trigonometric functions.

7.1

Trigonometric Identities

First, we review a fundamental definition:

Two functions f and g are said to be **identically equal** if

$$f(x) = g(x)$$

for every value of x for which both functions are defined. Such an equation is referred to as an **identity.** An equation that is not an identity is called a **conditional equation.**

For example, the following expressions are identities:

$$(x + 1)^2 = x^2 + 2x + 1 \qquad \sin^2 x + \cos^2 x = 1 \qquad \csc x = \frac{1}{\sin x}$$

The following expressions are conditional equations:

$2x + 5 = 0$ True only if $x = -\dfrac{5}{2}$

$\sin x = 0$ True only if $x = k\pi$, k an integer

$\sin x = \cos x$ True only if $x = \dfrac{\pi}{4} + 2k\pi$ or $x = \dfrac{5\pi}{4} + 2k\pi$, k an integer

The following box summarizes the trigonometric identities we have established thus far:

Quotient Identities

$$\tan \theta = \frac{\sin \theta}{\cos \theta} \qquad \cot \theta = \frac{\cos \theta}{\sin \theta}$$

Reciprocal Identities

$$\csc \theta = \frac{1}{\sin \theta} \qquad \sec \theta = \frac{1}{\cos \theta} \qquad \cot \theta = \frac{1}{\tan \theta}$$

Pythagorean Identities

$$\sin^2 \theta + \cos^2 \theta = 1 \qquad \tan^2 \theta + 1 = \sec^2 \theta$$
$$1 + \cot^2 \theta = \csc^2 \theta$$

Even–Odd Identities

$$\sin(-\theta) = -\sin \theta \qquad \cos(-\theta) = \cos \theta \qquad \tan(-\theta) = -\tan \theta$$
$$\csc(-\theta) = -\csc \theta \qquad \sec(-\theta) = \sec \theta \qquad \cot(-\theta) = -\cot \theta$$

This list of identities comprises what we shall refer to as the **basic trigonometric identities.** These identities should not merely be memorized, but should be *known* (just as you know your name rather than have it memorized). In fact, the use made of a basic identity is often a minor variation of the form listed here. For example, we might want to use $\sin^2\theta = 1 - \cos^2\theta$ instead of $\sin^2\theta + \cos^2\theta = 1$. For this reason, among others, you need to know these relationships and be quite comfortable with variations of them.

In the examples that follow, the directions will read "Establish the identity. . . . " As you will see, this is accomplished by starting with one side of the given equation (usually the one containing the more complicated expression) and, using appropriate basic identities and algebraic manipulations, arriving at the other side. The selection of an appropriate basic identity to obtain the desired result is learned only through experience and lots of practice.

EXAMPLE 1   *Establishing an Identity*

Establish the identity: $\sec\theta \cdot \sin\theta = \tan\theta$

Solution   We start with the left side, because it contains the more complicated expression, and apply a reciprocal identity:

$$\sec\theta \cdot \sin\theta = \frac{1}{\cos\theta} \cdot \sin\theta = \frac{\sin\theta}{\cos\theta} = \tan\theta$$

Having arrived at the right side, the identity is established.   ■

■   Now work Problem 1.

EXAMPLE 2   *Establishing an Identity*

Establish the identity: $\sin^2(-\theta) + \cos^2(-\theta) = 1$

Solution   We begin with the left side and apply even–odd identities:

$$\begin{aligned}
\sin^2(-\theta) + \cos^2(-\theta) &= [\sin(-\theta)]^2 + [\cos(-\theta)]^2 \\
&= (-\sin\theta)^2 + (\cos\theta)^2 \\
&= (\sin\theta)^2 + (\cos\theta)^2 \\
&= 1
\end{aligned}$$ ■

EXAMPLE 3   *Establishing an Identity*

Establish the identity: $\dfrac{\sin^2(-\theta) - \cos^2(-\theta)}{\sin(-\theta) - \cos(-\theta)} = \cos\theta - \sin\theta$

Solution   We begin with two observations: The left side appears to contain the more complicated expression. Also, the left side contains expressions with the argument $-\theta$, whereas the right side contains expressions with the argument θ. We decide, therefore, to start with the left side and apply even–odd identities:

$$\begin{aligned}
\frac{\sin^2(-\theta) - \cos^2(-\theta)}{\sin(-\theta) - \cos(-\theta)} &= \frac{[\sin(-\theta)]^2 - [\cos(-\theta)]^2}{\sin(-\theta) - \cos(-\theta)} \\
&= \frac{(-\sin\theta)^2 - (\cos\theta)^2}{-\sin\theta - \cos\theta} \qquad \text{Even–odd identities} \\
&= \frac{(\sin\theta)^2 - (\cos\theta)^2}{-\sin\theta - \cos\theta} \qquad \text{Simplify.} \\
&= \frac{(\sin\theta - \cos\theta)(\sin\theta + \cos\theta)}{-(\sin\theta + \cos\theta)} \qquad \text{Factor.} \\
&= \cos\theta - \sin\theta \qquad \text{Cancel and simplify.} \quad ■
\end{aligned}$$

E X A M P L E   4   *Establishing an Identity*

Establish the identity: $\dfrac{1 + \tan \theta}{1 + \cot \theta} = \tan \theta$

Solution $\dfrac{1 + \tan \theta}{1 + \cot \theta} = \dfrac{1 + \tan \theta}{1 + \dfrac{1}{\tan \theta}} = \dfrac{1 + \tan \theta}{\dfrac{\tan \theta + 1}{\tan \theta}} = \dfrac{\tan \theta (1 + \tan \theta)}{\tan \theta + 1} = \tan \theta$ ∎

∎   Now work Problem 9.

When sums or differences of quotients appear, it is usually best to rewrite them as a single quotient, especially if the other side of the identity consists of only one term.

E X A M P L E   5   *Establishing an Identity*

Establish the identity: $\dfrac{\sin \theta}{1 + \cos \theta} + \dfrac{1 + \cos \theta}{\sin \theta} = 2 \csc \theta$

Solution   The left side is more complicated, so we start with it and proceed to add:

$$\dfrac{\sin \theta}{1 + \cos \theta} + \dfrac{1 + \cos \theta}{\sin \theta} = \dfrac{\sin^2 \theta + (1 + \cos \theta)^2}{(1 + \cos \theta)(\sin \theta)}$$

$$= \dfrac{\sin^2 \theta + 1 + 2 \cos \theta + \cos^2 \theta}{(1 + \cos \theta)(\sin \theta)}$$

$$= \dfrac{(\sin^2 \theta + \cos^2 \theta) + 1 + 2 \cos \theta}{(1 + \cos \theta)(\sin \theta)}$$

$$= \dfrac{2 + 2 \cos \theta}{(1 + \cos \theta)(\sin \theta)}$$

$$= \dfrac{2(1 + \cos \theta)}{(1 + \cos \theta)(\sin \theta)}$$

$$= \dfrac{2}{\sin \theta}$$

$$= 2 \csc \theta \qquad\qquad ∎$$

E X A M P L E   6   *Establishing an Identity*

Establish the identity: $\dfrac{1}{\cos \theta} - \dfrac{\cos \theta}{1 + \sin \theta} = \tan \theta$

Solution $\dfrac{1}{\cos \theta} - \dfrac{\cos \theta}{1 + \sin \theta} = \dfrac{1 + \sin \theta - \cos^2 \theta}{\cos \theta (1 + \sin \theta)}$

$$= \dfrac{\sin \theta + (1 - \cos^2 \theta)}{\cos \theta (1 + \sin \theta)}$$

$$= \dfrac{\sin \theta + \sin^2 \theta}{\cos \theta (1 + \sin \theta)} \qquad 1 - \cos^2 \theta = \sin^2 \theta$$

$$= \frac{\sin \theta (1 + \sin \theta)}{\cos \theta (1 + \sin \theta)}$$

$$= \tan \theta \qquad \blacksquare$$

■ Now work Problem 23.

Sometimes, multiplying the numerator and denominator by an appropriate factor will result in a simplification.

E X A M P L E   7   *Establishing an Identity*

Establish the identity: $\dfrac{1 - \sin \theta}{\cos \theta} = \dfrac{\cos \theta}{1 + \sin \theta}$

Solution   We start with the left side and multiply the numerator and the denominator by $1 + \sin \theta$. (Alternatively, we could multiply the numerator and denominator of the right side by $1 - \sin \theta$.)

$$\frac{1 - \sin \theta}{\cos \theta} = \frac{1 - \sin \theta}{\cos \theta} \cdot \frac{1 + \sin \theta}{1 + \sin \theta}$$

$$= \frac{1 - \sin^2 \theta}{\cos \theta (1 + \sin \theta)}$$

$$= \frac{\cos^2 \theta}{\cos \theta (1 + \sin \theta)}$$

$$= \frac{\cos \theta}{1 + \sin \theta} \qquad \blacksquare$$

■ Now work Problem 35.

Although a lot of practice is the only real way to learn how to establish identities, the following guidelines should prove helpful:

Guidelines
for Establishing
Identities

1. It is almost always preferable to start with the side containing the more complicated expression.
2. Rewrite sums or differences of quotients as a single quotient.
3. Sometimes rewriting one side in terms of sines and cosines only will help.
4. Always keep your goal in mind. As you manipulate one side of the expression, you must keep in mind the form of the expression on the other side.

Warning: Be careful not to handle identities to be established as if they were equations. You *cannot* establish an identity by such methods as adding the same expression to each side and obtaining a true statement. This practice is not allowed, because the original statement is precisely the one that you are trying to establish. You do not know until it has been established that it is, in fact, true.

7.1

Exercise 7.1

In Problems 1–80, establish each identity.

1. $\csc \theta \cdot \cos \theta = \cot \theta$

2. $\csc \theta \cdot \tan \theta = \sec \theta$

3. $1 + \tan^2(-\theta) = \sec^2 \theta$

4. $1 + \cot^2(-\theta) = \csc^2 \theta$

5. $\cos \theta(\tan \theta + \cot \theta) = \csc \theta$

6. $\sin \theta(\cot \theta + \tan \theta) = \sec \theta$

7. $\tan \theta \cot \theta - \cos^2 \theta = \sin^2 \theta$

8. $\sin \theta \csc \theta - \cos^2 \theta = \sin^2 \theta$

9. $(\sec \theta - 1)(\sec \theta + 1) = \tan^2 \theta$

10. $(\csc \theta - 1)(\csc \theta + 1) = \cot^2 \theta$

11. $(\sec \theta + \tan \theta)(\sec \theta - \tan \theta) = 1$

12. $(\csc \theta + \cot \theta)(\csc \theta - \cot \theta) = 1$

13. $\sin^2 \theta(1 + \cot^2 \theta) = 1$

14. $(1 - \sin^2 \theta)(1 + \tan^2 \theta) = 1$

15. $(\sin \theta + \cos \theta)^2 + (\sin \theta - \cos \theta)^2 = 2$

16. $\tan^2 \theta \cos^2 \theta + \cot^2 \theta \sin^2 \theta = 1$

17. $\sec^4 \theta - \sec^2 \theta = \tan^4 \theta + \tan^2 \theta$

18. $\csc^4 \theta - \csc^2 \theta = \cot^4 \theta + \cot^2 \theta$

19. $\sec \theta - \tan \theta = \dfrac{\cos \theta}{1 + \sin \theta}$

20. $\csc \theta - \cot \theta = \dfrac{\sin \theta}{1 + \cos \theta}$

21. $3 \sin^2 \theta + 4 \cos^2 \theta = 3 + \cos^2 \theta$

22. $9 \sec^2 \theta - 5 \tan^2 \theta = 5 + 4 \sec^2 \theta$

23. $1 - \dfrac{\cos^2 \theta}{1 + \sin \theta} = \sin \theta$

24. $1 - \dfrac{\sin^2 \theta}{1 - \cos \theta} = -\cos \theta$

25. $\dfrac{1 + \tan \theta}{1 - \tan \theta} = \dfrac{\cot \theta + 1}{\cot \theta - 1}$

26. $\dfrac{\csc \theta - 1}{\csc \theta + 1} = \dfrac{1 - \sin \theta}{1 + \sin \theta}$

27. $\dfrac{\sec \theta}{\csc \theta} + \dfrac{\sin \theta}{\cos \theta} = 2 \tan \theta$

28. $\dfrac{\csc \theta - 1}{\cot \theta} = \dfrac{\cot \theta}{\csc \theta + 1}.$

29. $\dfrac{1 + \sin \theta}{1 - \sin \theta} = \dfrac{\csc \theta + 1}{\csc \theta - 1}$

30. $\dfrac{\cos \theta + 1}{\cos \theta - 1} = \dfrac{1 + \sec \theta}{1 - \sec \theta}$

31. $\dfrac{1 - \sin \theta}{\cos \theta} + \dfrac{\cos \theta}{1 - \sin \theta} = 2 \sec \theta$

32. $\dfrac{\cos \theta}{1 + \sin \theta} + \dfrac{1 + \sin \theta}{\cos \theta} = 2 \sec \theta$

33. $\dfrac{\sin \theta}{\sin \theta - \cos \theta} = \dfrac{1}{1 - \cot \theta}$

34. $1 - \dfrac{\sin^2 \theta}{1 + \cos \theta} = \cos \theta$

35. $\dfrac{1 - \sin \theta}{1 + \sin \theta} = (\sec \theta - \tan \theta)^2$

36. $\dfrac{1 - \cos \theta}{1 + \cos \theta} = (\csc \theta - \cot \theta)^2$

37. $\dfrac{\cos \theta}{1 - \tan \theta} + \dfrac{\sin \theta}{1 - \cot \theta} = \sin \theta + \cos \theta$

38. $\dfrac{\cot \theta}{1 - \tan \theta} + \dfrac{\tan \theta}{1 - \cot \theta} = 1 + \tan \theta + \cot \theta$

39. $\tan \theta + \dfrac{\cos \theta}{1 + \sin \theta} = \sec \theta$

40. $\dfrac{\sin \theta \cos \theta}{\cos^2 \theta - \sin^2 \theta} = \dfrac{\tan \theta}{1 - \tan^2 \theta}$

41. $\dfrac{\tan \theta + \sec \theta - 1}{\tan \theta - \sec \theta + 1} = \tan \theta + \sec \theta$

42. $\dfrac{\sin \theta - \cos \theta + 1}{\sin \theta + \cos \theta - 1} = \dfrac{\sin \theta + 1}{\cos \theta}$

43. $\dfrac{\tan \theta - \cot \theta}{\tan \theta + \cot \theta} = \sin^2 \theta - \cos^2 \theta$

44. $\dfrac{\sec \theta - \cos \theta}{\sec \theta + \cos \theta} = \dfrac{\sin^2 \theta}{1 + \cos^2 \theta}$

45. $\dfrac{\tan \theta - \cot \theta}{\tan \theta + \cot \theta} = 2 \sin^2 \theta - 1$

46. $\dfrac{\tan \theta - \cot \theta}{\tan \theta + \cot \theta} = 1 - 2 \cos^2 \theta$

47. $\dfrac{\sec \theta + \tan \theta}{\cot \theta + \cos \theta} = \tan \theta \sec \theta$

48. $\dfrac{\sec \theta}{1 + \sec \theta} = \dfrac{1 - \cos \theta}{\sin^2 \theta}$

49. $\dfrac{1 - \tan^2 \theta}{1 + \tan^2 \theta} = 2 \cos^2 \theta - 1$

50. $\dfrac{1 - \cot^2 \theta}{1 + \cot^2 \theta} = 1 - 2 \cos^2 \theta$

51. $\dfrac{\sec\theta - \csc\theta}{\sec\theta\csc\theta} = \sin\theta - \cos\theta$

52. $\dfrac{\sin^2\theta - \tan\theta}{\cos^2\theta - \cot\theta} = \tan^2\theta$

53. $\sec\theta - \cos\theta = \sin\theta\tan\theta$

54. $\tan\theta + \cot\theta = \sec\theta\csc\theta$

55. $\dfrac{1}{1 - \sin\theta} + \dfrac{1}{1 + \sin\theta} = 2\sec^2\theta$

56. $\dfrac{1 + \sin\theta}{1 - \sin\theta} - \dfrac{1 - \sin\theta}{1 + \sin\theta} = 4\tan\theta\sec\theta$

57. $\dfrac{\sec\theta}{1 - \sin\theta} = \dfrac{1 + \sin\theta}{\cos^3\theta}$

58. $\dfrac{1 - \sin\theta}{1 + \sin\theta} = (\sec\theta - \tan\theta)^2$

59. $\dfrac{(\sec\theta - \tan\theta)^2 + 1}{\csc\theta(\sec\theta - \tan\theta)} = 2\tan\theta$

60. $\dfrac{\sec^2\theta - \tan^2\theta + \tan\theta}{\sec\theta} = \sin\theta + \cos\theta$

61. $\dfrac{\sin\theta + \cos\theta}{\cos\theta} - \dfrac{\sin\theta - \cos\theta}{\sin\theta} = \sec\theta\csc\theta$

62. $\dfrac{\sin\theta + \cos\theta}{\sin\theta} - \dfrac{\cos\theta - \sin\theta}{\cos\theta} = \sec\theta\csc\theta$

63. $\dfrac{\sin^3\theta + \cos^3\theta}{\sin\theta + \cos\theta} = 1 - \sin\theta\cos\theta$

64. $\dfrac{\sin^3\theta + \cos^3\theta}{1 - 2\cos^2\theta} = \dfrac{\sec\theta - \sin\theta}{\tan\theta - 1}$

65. $\dfrac{\cos^2\theta - \sin^2\theta}{1 - \tan^2\theta} = \cos^2\theta$

66. $\dfrac{\cos\theta + \sin\theta - \sin^3\theta}{\sin\theta} = \cot\theta + \cos^2\theta$

67. $\dfrac{(2\cos^2\theta - 1)^2}{\cos^4\theta - \sin^4\theta} = 1 - 2\sin^2\theta$

68. $\dfrac{1 - 2\cos^2\theta}{\sin\theta\cos\theta} = \tan\theta - \cot\theta$

69. $\dfrac{1 + \sin\theta + \cos\theta}{1 + \sin\theta - \cos\theta} = \dfrac{1 + \cos\theta}{\sin\theta}$

70. $\dfrac{1 + \cos\theta + \sin\theta}{1 + \cos\theta - \sin\theta} = \sec\theta + \tan\theta$

71. $(a\sin\theta + b\cos\theta)^2 + (a\cos\theta - b\sin\theta)^2 = a^2 + b^2$

72. $(2a\sin\theta\cos\theta)^2 + a^2(\cos^2\theta - \sin^2\theta)^2 = a^2$

73. $\dfrac{\tan\alpha + \tan\beta}{\cot\alpha + \cot\beta} = \tan\alpha\tan\beta$

74. $(\tan\alpha + \tan\beta)(1 - \cot\alpha\cot\beta) + (\cot\alpha + \cot\beta)(1 - \tan\alpha\tan\beta) = 0$

75. $(\sin\alpha + \cos\beta)^2 + (\cos\beta + \sin\alpha)(\cos\beta - \sin\alpha) = 2\cos\beta(\sin\alpha + \cos\beta)$

76. $(\sin\alpha - \cos\beta)^2 + (\cos\beta + \sin\alpha)(\cos\beta - \sin\alpha) = -2\cos\beta(\sin\alpha - \cos\beta)$

77. $\ln|\sec\theta| = -\ln|\cos\theta|$

78. $\ln|\tan\theta| = \ln|\sin\theta| - \ln|\cos\theta|$

79. $\ln|1 + \cos\theta| + \ln|1 - \cos\theta| = 2\ln|\sin\theta|$

80. $\ln|\sec\theta + \tan\theta| + \ln|\sec\theta - \tan\theta| = 0$

 81. Write a few paragraphs outlining your strategy for establishing identities.

7.2

Sum and Difference Formulas

In this section, we continue our derivation of trigonometric identities by obtaining formulas that involve the sum or difference of two angles, such as $\cos(\alpha + \beta)$, $\cos(\alpha - \beta)$, or $\sin(\alpha + \beta)$. These formulas are referred to as the **sum and difference formulas**. We begin with the formulas for $\cos(\alpha + \beta)$ and $\cos(\alpha - \beta)$.

Theorem
Sum and Difference Formulas
for Cosines

$$\cos(\alpha + \beta) = \cos\alpha\cos\beta - \sin\alpha\sin\beta \qquad (1)$$

$$\cos(\alpha - \beta) = \cos\alpha\cos\beta + \sin\alpha\sin\beta \qquad (2)$$

In words, formula (1) states that the cosine of the sum of two angles equals the cosine of the first times the cosine of the second minus the sine of the first times the sine of the second.

Proof   We shall prove formula (2) first. Although this formula is true for all numbers α and β, we shall assume in our proof that $0 < \beta < \alpha < 2\pi$. We begin with a circle with center at the origin (0, 0) and radius of 1 unit (the unit circle), and we place the angles α and β in standard position, as shown in Figure 1(a). The point $P_1 = (x_1, y_1)$ lies on the terminal side of β, and the point $P_2 = (x_2, y_2)$ lies on the terminal side of α.

Now, place the angle $\alpha - \beta$ in standard position, as shown in Figure 1(b), where the point A has coordinates (1, 0) and the point $P_3 = (x_3, y_3)$ is on the terminal side of the angle $\alpha - \beta$.

Looking at triangle OP_1P_2 in Figure 1(a) and triangle OAP_3 in Figure 1(b), we see that these triangles are congruent. (Do you see why? Two sides and the included angle, $\alpha - \beta$, are equal.) Hence, the unknown side of each triangle must be equal; that is,

$$d(A, P_3) = d(P_1, P_2)$$

Using the distance formula, we find that

$$\sqrt{(x_3 - 1)^2 + y_3^2} = \sqrt{(x_2 - x_1)^2 + (y_2 - y_1)^2}$$

$$(x_3 - 1)^2 + y_3^2 = (x_2 - x_1)^2 + (y_2 - y_1)^2 \quad \text{Square each side.}$$

$$x_3^2 - 2x_3 + 1 + y_3^2 = x_2^2 - 2x_1x_2 + x_1^2 + y_2^2 - 2y_1y_2 + y_1^2 \quad (3)$$

Since $P_1 = (x_1, y_1)$, $P_2 = (x_2, y_2)$, and $P_3 = (x_3, y_3)$ are points on the unit circle $x^2 + y^2 = 1$, it follows that

$$x_1^2 + y_1^2 = 1 \qquad x_2^2 + y_2^2 = 1 \qquad x_3^2 + y_3^2 = 1$$

Consequently, equation (3) simplifies to

$$x_3^2 + y_3^2 - 2x_3 + 1 = (x_2^2 + y_2^2) + (x_1^2 + y_1^2) - 2x_1x_2 - 2y_1y_2$$

$$2 - 2x_3 = 2 - 2x_1x_2 - 2y_1y_2$$

$$x_3 = x_1x_2 + y_1y_2 \quad (4)$$

But $P_1 = (x_1, y_1)$ is on the terminal side of angle β and is a distance of 1 unit from the origin. Thus,

$$\sin \beta = \frac{y_1}{1} = y_1 \qquad \cos \beta = \frac{x_1}{1} = x_1 \quad (5)$$

Similarly,

$$\sin \alpha = \frac{y_2}{1} = y_2 \qquad \cos \alpha = \frac{x_2}{1} = x_2 \qquad \cos(\alpha - \beta) = \frac{x_3}{1} = x_3 \quad (6)$$

Using equations (5) and (6) in equation (4), we get

$$\cos(\alpha - \beta) = \cos \alpha \cos \beta + \sin \alpha \sin \beta$$

which is formula (2).

The proof of formula (1) follows from formula (2). We use the fact that $\alpha + \beta = \alpha - (-\beta)$. Then

FIGURE 1

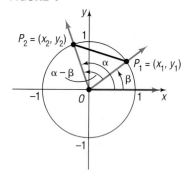

(a)

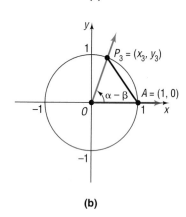

(b)

$$\cos(\alpha + \beta) = \cos[\alpha - (-\beta)]$$
$$= \cos \alpha \cos(-\beta) + \sin \alpha \sin(-\beta) \quad \text{Use formula (2).}$$
$$= \cos \alpha \cos \beta - \sin \alpha \sin \beta \quad \text{Even–odd identities} \quad \blacksquare$$

One use of formulas (1) and (2) is to obtain the exact value of the cosine of an angle that can be expressed as the sum or difference of angles whose sine and cosine are known exactly.

E X A M P L E   1

Using the Sum Formula to Find Exact Values

Find the exact value of cos 75°.

Solution   Since $75° = 45° + 30°$, we use formula (2) to obtain

$$\cos 75° = \cos(45° + 30°) = \cos 45° \cos 30° - \sin 45° \sin 30°$$

$$\uparrow$$
$$\text{Formula (1)}$$

$$= \frac{\sqrt{2}}{2} \cdot \frac{\sqrt{3}}{2} - \frac{\sqrt{2}}{2} \cdot \frac{1}{2} = \frac{1}{4}(\sqrt{6} - \sqrt{2}) \qquad \blacksquare$$

E X A M P L E   2

Using the Difference Formula to Find Exact Values

Find the exact value of $\cos(\pi/12)$.

Solution

$$\cos \frac{\pi}{12} = \cos\left(\frac{3\pi}{12} - \frac{2\pi}{12}\right) = \cos\left(\frac{\pi}{4} - \frac{\pi}{6}\right)$$

$$= \cos \frac{\pi}{4} \cos \frac{\pi}{6} + \sin \frac{\pi}{4} \sin \frac{\pi}{6} \quad \text{Use formula (2).}$$

$$= \frac{\sqrt{2}}{2} \cdot \frac{\sqrt{3}}{2} + \frac{\sqrt{2}}{2} \cdot \frac{1}{2} = \frac{1}{4}(\sqrt{6} + \sqrt{2}) \qquad \blacksquare$$

■   Now work Problem 3.

Another use of formulas (1) and (2) is to establish other identities. One important pair of identities is given next:

$$\cos\left(\frac{\pi}{2} - \theta\right) = \sin \theta \qquad \qquad (7a)$$

$$\sin\left(\frac{\pi}{2} - \theta\right) = \cos \theta \qquad \qquad (7b)$$

Proof   To prove formula (7a), we use the formula for $\cos(\alpha - \beta)$ with $\alpha = \pi/2$ and $\beta = \theta$:

$$\cos\left(\frac{\pi}{2} - \theta\right) = \cos \frac{\pi}{2} \cos \theta + \sin \frac{\pi}{2} \sin \theta$$

$$= 0 \cdot \cos \theta + 1 \cdot \sin \theta$$

$$= \sin \theta$$

To prove formula (7b), we make use of the identity (7a) just established:

$$\sin\left(\frac{\pi}{2} - \theta\right) = \cos\left[\frac{\pi}{2} - \left(\ \ - \theta\right)\right] = \cos \theta$$

Use (7a).

Formulas (7a) and (7b) should look familiar. They are the basis for the theorem stated in Chapter 5: Cofunctions of complementary angles are equal.

Furthermore, since $\cos(\pi/2 - \theta) = \cos(\theta - \pi/2)$, it follows from formula (7a) that $\cos(\theta - \pi/2) = \sin \theta$. Thus, the graphs of $y = \sin x$ and $y = \cos(x - \pi/2)$ are identical, a fact we used earlier in Section 6.2.

■ Now work Problem 29.

Formulas for $\sin(\alpha + \beta)$ and $\sin(\alpha - \beta)$

Having established the identities in formulas (7a) and (7b), we now can derive the sum and difference formulas for $\sin(\alpha + \beta)$ and $\sin(\alpha - \beta)$.

Proof

$$\sin(\alpha + \beta) = \cos\left[\frac{\pi}{2} - (\alpha + \beta)\right] \qquad \text{Formula (7a)}$$

$$= \cos\left[\left(\frac{\pi}{2} - \alpha\right) - \beta\right]$$

$$= \cos\left(\frac{\pi}{2} - \alpha\right)\cos \beta + \sin\left(\frac{\pi}{2} - \alpha\right)\sin \beta \quad \text{Formula (2)}$$

$$= \sin \alpha \cos \beta + \cos \alpha \sin \beta \qquad \text{Formulas (7a) and (7b)}$$

$$\sin(\alpha - \beta) = \sin[\alpha + (-\beta)]$$

$$= \sin \alpha \cos(-\beta) + \cos \alpha \sin(-\beta)$$

$$= \sin \alpha \cos \beta + \cos \alpha(-\sin \beta) \qquad \text{Even–odd identities}$$

$$= \sin \alpha \cos \beta - \cos \alpha \sin \beta$$

Thus,

Theorem
Sum and Difference Formulas
for Sines

$$\sin(\alpha + \beta) = \sin \alpha \cos \beta + \cos \alpha \sin \beta \qquad (8)$$
$$\sin(\alpha - \beta) = \sin \alpha \cos \beta - \cos \alpha \sin \beta \qquad (9)$$

In words, formula (8) states that the sine of the sum of two angles equals the sine of the first times the cosine of the second plus the cosine of the first times the sine of the second.

E X A M P L E   3

Using the Sum Formula to Find Exact Values

Find the exact value of $\sin(7\pi/12)$.

Solution

$$\sin\frac{7\pi}{12} = \sin\left(\frac{3\pi}{12} + \frac{4\pi}{12}\right) = \sin\left(\frac{\pi}{4} + \frac{\pi}{3}\right)$$

$$= \sin \frac{\pi}{4} \cos \frac{\pi}{3} + \cos \frac{\pi}{4} \sin \frac{\pi}{3} \quad \text{Formula (8)}$$

$$= \frac{\sqrt{2}}{2} \cdot \frac{1}{2} + \frac{\sqrt{2}}{2} \cdot \frac{\sqrt{3}}{2} = \frac{1}{4}(\sqrt{2} + \sqrt{6}) \qquad ■$$

E X A M P L E   4   *Using the Difference Formula to Find Exact Values*

Find the exact value of $\sin 165°$.

Solution
$$\sin 165° = \sin(225° - 60°)$$
$$= \sin 225° \cos 60° - \cos 225° \sin 60° \quad \text{Formula (9)}$$
$$= \frac{-\sqrt{2}}{2} \cdot \frac{1}{2} - \frac{-\sqrt{2}}{2} \cdot \frac{\sqrt{3}}{2} = \frac{1}{4}(\sqrt{6} - \sqrt{2}) \qquad ■$$

■ Now work Problem 9.

E X A M P L E   5   *Using the Difference Formula to Find Exact Values*

Find the exact value of $\cos 80° \cos 20° + \sin 80° \sin 20°$.

Solution The form of the expression $\cos 80° \cos 20° + \sin 80° \sin 20°$ is that of the right side of the formula for $\cos(\alpha - \beta)$ with $\alpha = 80°$ and $\beta = 20°$. Thus,

$$\cos 80° \cos 20° + \sin 80° \sin 20° = \cos(80° - 20°) = \cos 60° = \frac{1}{2} \qquad ■$$

E X A M P L E   6   *Using the Sum Formula to Find Exact Values*

Find the exact value of $\sin \dfrac{\pi}{9} \cos \dfrac{\pi}{18} + \cos \dfrac{\pi}{9} \sin \dfrac{\pi}{18}$.

Solution We observe that the form of the given expression is that of the right side of the formula for $\sin(\alpha + \beta)$ with $\alpha = \pi/9$ and $\beta = \pi/18$. Thus,

$$\sin \frac{\pi}{9} \cos \frac{\pi}{18} + \cos \frac{\pi}{9} \sin \frac{\pi}{18} = \sin\left(\frac{\pi}{9} + \frac{\pi}{18}\right) = \sin \frac{3\pi}{18} = \sin \frac{\pi}{6} = \frac{1}{2} \quad ■$$

■ Now work Problem 19.

E X A M P L E   7   *Finding Exact Values*

If it is known that $\sin \alpha = \frac{4}{5}$, $\pi/2 < \alpha < \pi$, and that $\sin \beta = -2/\sqrt{5} = -2\sqrt{5}/5$, $\pi < \beta < 3\pi/2$, find the exact value of:

(a) $\cos \alpha$ (b) $\cos \beta$ (c) $\cos(\alpha + \beta)$ (d) $\sin(\alpha + \beta)$

Solution (a) See Figure 2. Since $(x, 4)$ is on the circle $x^2 + y^2 = 25$ and is in Quadrant II, we have

$$x^2 + 4^2 = 25, \quad x < 0$$
$$x^2 = 25 - 16 = 9$$
$$x = -3$$

Thus,

$$\cos \alpha = \frac{x}{r} = -\frac{3}{5}$$

FIGURE 2
Given $\sin \alpha = \frac{4}{5}$, $\pi/2 < \alpha < \pi$

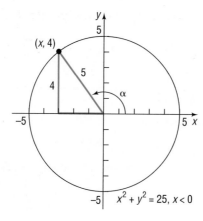

FIGURE 3
Given $\sin \beta = -2/\sqrt{5}$, $\pi < \beta < 3\pi/2$

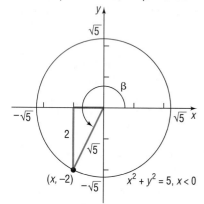

(b) See Figure 3. Since $(x, -2)$ is on the circle $x^2 + y^2 = 5$ and is in Quadrant III, we have

$$x^2 + (-2)^2 = 5, \qquad x < 0$$
$$x^2 = 5 - 4 = 1$$
$$x = -1$$

Thus,

$$\cos \beta = \frac{x}{r} = -\frac{1}{\sqrt{5}} = -\frac{\sqrt{5}}{5}$$

(c) Using the results found in parts (a) and (b) and formula (1), we have

$$\cos(\alpha + \beta) = \cos \alpha \cos \beta - \sin \alpha \sin \beta$$
$$= -\frac{3}{5}\left(-\frac{\sqrt{5}}{5}\right) - \frac{4}{5}\left(-\frac{2\sqrt{5}}{5}\right) = \frac{11\sqrt{5}}{25}$$

(d)
$$\sin(\alpha + \beta) = \sin \alpha \cos \beta + \cos \alpha \sin \beta$$
$$= \frac{4}{5}\left(-\frac{\sqrt{5}}{5}\right) + \left(-\frac{3}{5}\right)\left(-\frac{2\sqrt{5}}{5}\right) = \frac{2\sqrt{5}}{25}$$

■ Now work Problem 23.

Formulas for $\tan(\alpha + \beta)$ and $\tan(\alpha - \beta)$

We use the identity $\tan \theta = (\sin \theta)/(\cos \theta)$ and the sum formulas for $\sin(\alpha + \beta)$ and $\cos(\alpha + \beta)$ to derive a formula for $\tan(\alpha + \beta)$.

Proof
$$\tan(\alpha + \beta) = \frac{\sin(\alpha + \beta)}{\cos(\alpha + \beta)} = \frac{\sin \alpha \cos \beta + \cos \alpha \sin \beta}{\cos \alpha \cos \beta - \sin \alpha \sin \beta}$$

Now we divide the numerator and denominator by $\cos \alpha \cos \beta$:

$$\tan(\alpha + \beta) = \frac{\dfrac{\sin \alpha \cos \beta + \cos \alpha \sin \beta}{\cos \alpha \cos \beta}}{\dfrac{\cos \alpha \cos \beta - \sin \alpha \sin \beta}{\cos \alpha \cos \beta}} = \frac{\dfrac{\sin \alpha \cos \beta}{\cos \alpha \cos \beta} + \dfrac{\cos \alpha \sin \beta}{\cos \alpha \cos \beta}}{\dfrac{\cos \alpha \cos \beta}{\cos \alpha \cos \beta} - \dfrac{\sin \alpha \sin \beta}{\cos \alpha \cos \beta}}$$

$$= \frac{\dfrac{\sin \alpha}{\cos \alpha} + \dfrac{\sin \beta}{\cos \beta}}{1 - \dfrac{\sin \alpha \sin \beta}{\cos \alpha \cos \beta}} = \frac{\tan \alpha + \tan \beta}{1 - \tan \alpha \tan \beta}$$

We use the sum formula for $\tan(\alpha + \beta)$ to get the difference formula:

$$\tan(\alpha - \beta) = \tan[\alpha + (-\beta)] = \frac{\tan \alpha + \tan(-\beta)}{1 - \tan \alpha \tan(-\beta)} = \frac{\tan \alpha - \tan \beta}{1 + \tan \alpha \tan \beta} \qquad \blacksquare$$

Thus, we have proved the following results:

Theorem
Sum and Difference Formulas
for Tangents

$$\tan(\alpha + \beta) = \frac{\tan \alpha + \tan \beta}{1 - \tan \alpha \tan \beta} \qquad (10)$$

$$\tan(\alpha - \beta) = \frac{\tan \alpha - \tan \beta}{1 + \tan \alpha \tan \beta} \qquad (11)$$

$\blacksquare$

In words, formula (10) states that the tangent of the sum of two angles equals the tangent of the first plus the tangent of the second divided by 1 minus their product.

EXAMPLE 8   *Establishing an Identity*

Prove the identity: $\tan(\theta + \pi) = \tan \theta$

Solution

$$\tan(\theta + \pi) = \frac{\tan \theta + \tan \pi}{1 - \tan \theta \tan \pi} = \frac{\tan \theta + 0}{1 - \tan \theta \cdot 0} = \tan \theta \qquad \blacksquare$$

The result obtained in Example 8 verifies that the tangent function is periodic with period π, a fact we mentioned earlier.

Warning: Be careful when using formulas (10) and (11). These formulas can be used only for angles α and β for which $\tan \alpha$ and $\tan \beta$ are defined, that is, all angles except odd multiples of $\pi/2$.

EXAMPLE 9   *Establishing an Identity*

Prove the identity: $\tan\left(\theta + \dfrac{\pi}{2}\right) = -\cot \theta$

Solution   We cannot use formula (10), since $\tan(\pi/2)$ is not defined. Instead, we proceed as follows:

$$\tan\left(\theta + \frac{\pi}{2}\right) = \frac{\sin\left(\theta + \dfrac{\pi}{2}\right)}{\cos\left(\theta + \dfrac{\pi}{2}\right)} = \frac{\sin \theta \cos \dfrac{\pi}{2} + \cos \theta \sin \dfrac{\pi}{2}}{\cos \theta \cos \dfrac{\pi}{2} - \sin \theta \sin \dfrac{\pi}{2}}.$$

$$= \frac{(\sin \theta)(0) + (\cos \theta)(1)}{(\cos \theta)(0) - (\sin \theta)(1)} = \frac{\cos \theta}{-\sin \theta} = -\cot \theta$$

$\blacksquare$

EXAMPLE 10   *Finding Exact Values*

Find the exact value of $\sin(\cos^{-1} \frac{1}{2} + \sin^{-1} \frac{3}{5})$.

Solution   Let $\alpha = \cos^{-1} \frac{1}{2}$ and $\beta = \sin^{-1} \frac{3}{5}$. Then

$$\cos \alpha = \tfrac{1}{2}, \quad 0 \le \alpha \le \pi, \quad \text{and} \quad \sin \beta = \tfrac{3}{5}, \quad -\frac{\pi}{2} \le \beta \le \frac{\pi}{2}$$

FIGURE 4

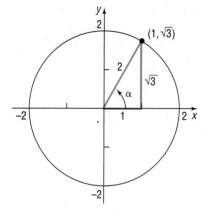

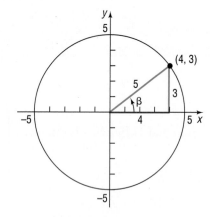

(a) $\cos \alpha = \frac{1}{2}, 0 \le \alpha \le \pi$ (b) $\sin \beta = \frac{3}{5}, -\pi/2 \le \beta \le \pi/2$

Based on Figure 4, we obtain $\sin \alpha = \sqrt{3}/2$ and $\cos \beta = \frac{4}{5}$. Thus,

$$\sin\left(\cos^{-1}\frac{1}{2} + \sin^{-1}\frac{3}{5}\right) = \sin(\alpha + \beta) = \sin \alpha \cos \beta + \cos \alpha \sin \beta$$

$$= \frac{\sqrt{3}}{2} \cdot \frac{4}{5} + \frac{1}{2} \cdot \frac{3}{5} = \frac{4\sqrt{3} + 3}{10}$$

■

■ Now work Problem 57.

E X A M P L E   1 1   *Writing a Trigonometric Expression as an Algebraic Expression*

Write $\sin(\sin^{-1} u + \cos^{-1} v)$ as an algebraic expression containing u and v (that is, without any trigonometric functions).

Solution   Let $\alpha = \sin^{-1} u$ and $\beta = \cos^{-1} v$. Then

$$\sin \alpha = u, \quad -\frac{\pi}{2} \le \alpha \le \frac{\pi}{2} \quad \text{and} \quad \cos \beta = v, \quad 0 \le \beta \le \pi$$

Since $-\pi/2 \le \alpha \le \pi/2$, we know that $\cos \alpha \ge 0$. As a result,

$$\cos \alpha = \sqrt{1 - \sin^2 \alpha} = \sqrt{1 - u^2}$$

Similarly, since $0 \le \beta \le \pi$, we know that $\sin \beta \ge 0$. Thus,

$$\sin \beta = \sqrt{1 - \cos^2 \beta} = \sqrt{1 - v^2}$$

Now,

$$\sin(\sin^{-1} u + \cos^{-1} v) = \sin(\alpha + \beta) = \sin \alpha \cos \beta + \cos \alpha \sin \beta$$

$$= uv + \sqrt{1 - u^2} \sqrt{1 - v^2}$$ ■

Summary

The following box summarizes the sum and difference formulas:

Sum and Difference Formulas

$$\cos(\alpha + \beta) = \cos \alpha \cos \beta - \sin \alpha \sin \beta$$
$$\cos(\alpha - \beta) = \cos \alpha \cos \beta + \sin \alpha \sin \beta$$

$$\sin(\alpha + \beta) = \sin \alpha \cos \beta + \cos \alpha \sin \beta$$
$$\sin(\alpha - \beta) = \sin \alpha \cos \beta - \cos \alpha \sin \beta$$
$$\tan(\alpha + \beta) = \frac{\tan \alpha + \tan \beta}{1 - \tan \alpha \tan \beta}$$
$$\tan(\alpha - \beta) = \frac{\tan \alpha - \tan \beta}{1 + \tan \alpha \tan \beta}$$

7.2

Exercise 7.2

In Problems 1–12, find the exact value of each trigonometric function.

1. $\sin \dfrac{5\pi}{12}$ **2.** $\sin \dfrac{\pi}{12}$ **3.** $\cos \dfrac{7\pi}{12}$ **4.** $\tan \dfrac{7\pi}{12}$ **5.** $\cos 165°$ **6.** $\sin 105°$

7. $\tan 15°$ **8.** $\tan 195°$ **9.** $\sin \dfrac{17\pi}{12}$ **10.** $\tan \dfrac{19\pi}{12}$ **11.** $\sec\left(-\dfrac{\pi}{12}\right)$ **12.** $\cot\left(-\dfrac{5\pi}{12}\right)$

In Problems 13–22, find the exact value of each expression.

13. $\sin 20° \cos 10° + \cos 20° \sin 10°$ **14.** $\sin 20° \cos 80° - \cos 20° \sin 80°$

15. $\cos 70° \cos 20° - \sin 70° \sin 20°$ **16.** $\cos 40° \cos 10° + \sin 40° \sin 10°$

17. $\dfrac{\tan 20° + \tan 25°}{1 - \tan 20° \tan 25°}$ **18.** $\dfrac{\tan 40° - \tan 10°}{1 + \tan 40° \tan 10°}$

19. $\sin \dfrac{\pi}{12} \cos \dfrac{7\pi}{12} - \cos \dfrac{\pi}{12} \sin \dfrac{7\pi}{12}$ **20.** $\cos \dfrac{5\pi}{12} \cos \dfrac{7\pi}{12} - \sin \dfrac{5\pi}{12} \sin \dfrac{7\pi}{12}$

21. $\sin \dfrac{\pi}{12} \cos \dfrac{5\pi}{12} - \sin \dfrac{5\pi}{12} \cos \dfrac{\pi}{12}$ **22.** $\sin \dfrac{\pi}{18} \cos \dfrac{5\pi}{18} + \cos \dfrac{\pi}{18} \sin \dfrac{5\pi}{18}$

In Problems 23–28, find the exact value of each of the following under the given conditions:

(a) $\sin(\alpha + \beta)$ (b) $\cos(\alpha + \beta)$ (c) $\sin(\alpha - \beta)$ (d) $\tan(\alpha - \beta)$

23. $\sin \alpha = \frac{3}{5}, 0 < \alpha < \pi/2;$ $\cos \beta = 2/\sqrt{5}, -\pi/2 < \beta < 0$

24. $\cos \alpha = 1/\sqrt{5}, 0 < \alpha < \pi/2;$ $\sin \beta = -\frac{4}{5}, -\pi/2 < \beta < 0$

25. $\tan \alpha = -\frac{4}{3}, \pi/2 < \alpha < \pi;$ $\cos \beta = \frac{1}{2}, 0 < \beta < \pi/2$

26. $\tan \alpha = \frac{5}{12}, \pi < \alpha < 3\pi/2;$ $\sin \beta = -\frac{1}{2}, \pi < \beta < 3\pi/2$

27. $\sin \alpha = \frac{5}{13}, -3\pi/2 < \alpha < -\pi;$ $\tan \beta = -\sqrt{3}, \pi/2 < \beta < \pi$

28. $\cos \alpha = \frac{1}{2}, -\pi/2 < \alpha < 0;$ $\sin \beta = \frac{1}{3}, 0 < \beta < \pi/2$

In Problems 29–54, establish each identity.

29. $\sin\left(\dfrac{\pi}{2} + \theta\right) = \cos \theta$ **30.** $\cos\left(\dfrac{\pi}{2} + \theta\right) = -\sin \theta$ **31.** $\sin(\pi - \theta) = \sin \theta$

32. $\cos(\pi - \theta) = -\cos \theta$ **33.** $\sin(\pi + \theta) = -\sin \theta$ **34.** $\cos(\pi + \theta) = -\cos \theta$

35. $\tan(\pi - \theta) = -\tan \theta$ **36.** $\tan(2\pi - \theta) = -\tan \theta$ **37.** $\sin\left(\dfrac{3\pi}{2} + \theta\right) = -\cos \theta$

38. $\cos\left(\dfrac{3\pi}{2} + \theta\right) = \sin \theta$ **39.** $\sin(\alpha + \beta) + \sin(\alpha - \beta) = 2 \sin \alpha \cos \beta$

40. $\cos(\alpha + \beta) + \cos(\alpha - \beta) = 2 \cos \alpha \cos \beta$ **41.** $\dfrac{\sin(\alpha + \beta)}{\sin \alpha \cos \beta} = 1 + \cot \alpha \tan \beta$

42. $\dfrac{\sin(\alpha + \beta)}{\cos \alpha \cos \beta} = \tan \alpha + \tan \beta$

43. $\dfrac{\cos(\alpha + \beta)}{\cos \alpha \cos \beta} = 1 - \tan \alpha \tan \beta$

44. $\dfrac{\cos(\alpha - \beta)}{\sin \alpha \cos \beta} = \cot \alpha + \tan \beta$

45. $\dfrac{\sin(\alpha + \beta)}{\sin(\alpha - \beta)} = \dfrac{\tan \alpha + \tan \beta}{\tan \alpha - \tan \beta}$

46. $\dfrac{\cos(\alpha + \beta)}{\cos(\alpha - \beta)} = \dfrac{1 - \tan \alpha \tan \beta}{1 + \tan \alpha \tan \beta}$

47. $\cot(\alpha + \beta) = \dfrac{\cot \alpha \cot \beta - 1}{\cot \beta + \cot \alpha}$

48. $\cot(\alpha - \beta) = \dfrac{\cot \alpha \cot \beta + 1}{\cot \beta - \cot \alpha}$

49. $\sec(\alpha + \beta) = \dfrac{\csc \alpha \csc \beta}{\cot \alpha \cot \beta - 1}$

50. $\sec(\alpha - \beta) = \dfrac{\sec \alpha \sec \beta}{1 + \tan \alpha \tan \beta}$

51. $\sin(\alpha - \beta) \sin(\alpha + \beta) = \sin^2 \alpha - \sin^2 \beta$

52. $\cos(\alpha - \beta) \cos(\alpha + \beta) = \cos^2 \alpha - \sin^2 \beta$

53. $\sin(\theta + k\pi) = (-1)^k \cdot \sin \theta, k$ any integer

54. $\cos(\theta + k\pi) = (-1)^k \cdot \cos \theta, k$ any integer

In Problems 55–64, find the exact value for each expression.

55. $\sin(\sin^{-1} \frac{1}{2} + \cos^{-1} 0)$

56. $\sin\left(\sin^{-1} \dfrac{\sqrt{3}}{2} + \cos^{-1} 1\right)$

57. $\sin[\sin^{-1} \frac{3}{5} - \cos^{-1}(-\frac{4}{5})]$

58. $\sin[\sin^{-1}(-\frac{4}{5}) - \tan^{-1} \frac{3}{4}]$

59. $\cos(\tan^{-1} \frac{4}{3} + \cos^{-1} \frac{5}{13})$

60. $\sin[\tan^{-1} \frac{5}{12} - \sin^{-1}(-\frac{3}{5})]$

61. $\sec(\sin^{-1} \frac{5}{13} - \tan^{-1} \frac{3}{4})$

62. $\sec(\tan^{-1} \frac{4}{3} + \cot^{-1} \frac{5}{12})$

63. $\cot\left(\sec^{-1} \dfrac{5}{3} + \dfrac{\pi}{6}\right)$

64. $\cos\left(\dfrac{\pi}{4} - \csc^{-1} \dfrac{5}{3}\right)$

In Problems 65–70, write each trigonometric expression as an algebraic expression containing u and v.

65. $\cos(\cos^{-1} u + \sin^{-1} v)$

66. $\sin(\sin^{-1} u - \cos^{-1} v)$

67. $\sin(\tan^{-1} u - \sin^{-1} v)$

68. $\cos(\tan^{-1} u + \tan^{-1} v)$

69. $\tan(\sin^{-1} u - \cos^{-1} v)$

70. $\sec(\tan^{-1} u + \cos^{-1} v)$

71. *Calculus* Show that the difference quotient for $f(x) = \sin x$ is given by

$$\dfrac{f(x + h) - f(x)}{h} = \dfrac{\sin(x + h) - \sin x}{h} = \cos x \cdot \dfrac{\sin h}{h} - \sin x \cdot \dfrac{1 - \cos h}{h}$$

72. *Calculus* Show that the difference quotient for $f(x) = \cos x$ is given by

$$\dfrac{f(x + h) - f(x)}{h} = \dfrac{\cos(x + h) - \cos x}{h} = -\sin x \cdot \dfrac{\sin h}{h} - \cos x \cdot \dfrac{1 - \cos h}{h}$$

73. Show that $\sin(\sin^{-1} u + \cos^{-1} u) = 1$.

74. Show that $\cos(\sin^{-1} u + \cos^{-1} u) = 0$.

75. Explain why formula (11) cannot be used to show that

$$\tan\left(\dfrac{\pi}{2} - \theta\right) = \cot \theta$$

Establish this identity by using formulas (7a) and (7b).

76. If $\tan \alpha = x + 1$ and $\tan \beta = x - 1$, show that $2 \cot(\alpha - \beta) = x^2$.

77. *Geometry: Angle between Two Lines* Let L_1 and L_2 denote two nonvertical intersecting lines, and let θ denote the acute angle between L_1 and L_2 (see the figure). Show that

$$\tan \theta = \dfrac{m_2 - m_1}{1 + m_1 m_2}$$

where m_1 and m_2 are the slopes of L_1 and L_2, respectively. [*Hint:* Use the facts that $\tan \theta_1 = m_1$ and $\tan \theta_2 = m_2$.]

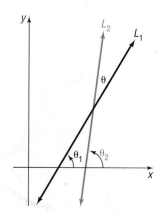

78. If $\alpha + \beta + \gamma = 180°$ and $\cot \theta = \cot \alpha + \cot \beta + \cot \gamma$, $0 < \theta < 90°$, show that

$$\sin^3 \theta = \sin(\alpha - \theta) \sin(\beta - \theta) \sin(\gamma - \theta)$$

79. Discuss the following derivation:

$$\tan(\theta + \frac{\pi}{2}) = \frac{\tan \theta + \tan(\pi/2)}{1 - \tan \theta \tan(\pi/2)} = \frac{\dfrac{\tan \theta}{\tan(\pi/2)} + 1}{\dfrac{1}{\tan(\pi/2)} - \tan \theta} = \frac{0 + 1}{0 - \tan \theta} = \frac{1}{-\tan \theta} = -\cot \theta$$

Can you justify each step?

7.3

Double-angle and Half-angle Formulas

In this section we derive formulas for $\sin 2\theta$, $\cos 2\theta$, $\sin \frac{1}{2}\theta$, and $\cos \frac{1}{2}\theta$ in terms of $\sin \theta$ and $\cos \theta$. They are easily derived using the sum formulas.

Double-angle Formulas

In the sum formulas for $\sin(\alpha + \beta)$ and $\cos(\alpha + \beta)$, let $\alpha = \beta = \theta$. Then

$$\sin(\alpha + \beta) = \sin \alpha \cos \beta + \cos \alpha \sin \beta$$
$$\sin(\theta + \theta) = \sin \theta \cos \theta + \cos \theta \sin \theta$$
$$\sin 2\theta = 2 \sin \theta \cos \theta \tag{1}$$

and

$$\cos(\alpha + \beta) = \cos \alpha \cos \beta - \sin \alpha \sin \beta$$
$$\cos(\theta + \theta) = \cos \theta \cos \theta - \sin \theta \sin \theta$$
$$\cos 2\theta = \cos^2 \theta - \sin^2 \theta \tag{2}$$

An application of the Pythagorean identity $\sin^2 \theta + \cos^2 \theta = 1$ results in two other ways to write formula (2) for $\cos 2\theta$:

$$\cos 2\theta = \cos^2 \theta - \sin^2 \theta = (1 - \sin^2 \theta) - \sin^2 \theta = 1 - 2 \sin^2 \theta$$

and

$$\cos 2\theta = \cos^2 \theta - \sin^2 \theta = \cos^2 \theta - (1 - \cos^2 \theta) = 2 \cos^2 \theta - 1$$

Thus, we have established the following **double-angle formulas:**

Theorem

Double-angle Formulas

$$
\begin{array}{lr}
\sin 2\theta = 2 \sin \theta \cos \theta & (3) \\
\cos 2\theta = \cos^2 \theta - \sin^2 \theta & (4a) \\
\cos 2\theta = 1 - 2 \sin^2 \theta & (4b) \\
\cos 2\theta = 2 \cos^2 \theta - 1 & (4c)
\end{array}
$$

E X A M P L E   1

Finding Exact Values Using the Double-angle Formula

If $\sin \theta = \frac{3}{5}$, $\pi/2 < \theta < \pi$, find the exact value of:

(a) $\sin 2\theta$ (b) $\cos 2\theta$

Solution   (a) Because $\sin 2\theta = 2 \sin \theta \cos \theta$ and we already know that $\sin \theta = \frac{3}{5}$, we only need to find $\cos \theta$ (see Figure 5). Since $\pi/2 < \theta < \pi$, it follows that $(x, 3)$ is on the circle $x^2 + y^2 = 25$ and is in Quadrant II. Thus,

FIGURE 5

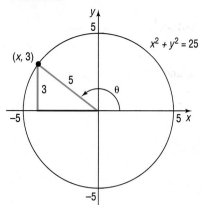

$$x^2 + 3^2 = 25, \qquad x < 0$$
$$x^2 = 25 - 9 = 16$$
$$x = -4$$

Thus, $\cos \theta = -\frac{4}{5}$. Now we use formula (3) to obtain

$$\sin 2\theta = 2 \sin \theta \cos \theta = 2(\tfrac{3}{5})(-\tfrac{4}{5}) = -\tfrac{24}{25}$$

(b) Because we are given $\sin \theta = \frac{3}{5}$, it is easiest to use formula (4b) to get $\cos 2\theta$:

$$\cos 2\theta = 1 - 2 \sin^2 \theta = 1 - 2(\tfrac{9}{25}) = 1 - \tfrac{18}{25} = \tfrac{7}{25} \qquad \blacksquare$$

Warning: In finding $\cos 2\theta$ in Example 1(b), we chose to use a version of the double-angle formula, formula (4b). Note that we are unable to use the Pythagorean identity $\cos 2\theta = \pm\sqrt{1 - \sin^2 2\theta}$, with $\sin 2\theta = -\frac{24}{25}$, because we have no way of knowing which sign to choose.

   ■   Now work Problems 1(a) and (b).

E X A M P L E   2   *Establishing Identities*

(a) Develop a formula for $\tan 2\theta$ in terms of $\tan \theta$.

(b) Develop a formula for $\sin 3\theta$ in terms of $\sin \theta$ and $\cos \theta$.

Solution   (a) In the sum formula for $\tan(\alpha + \beta)$, let $\alpha = \beta = \theta$. Then

$$\tan(\alpha + \beta) = \frac{\tan \alpha + \tan \beta}{1 - \tan \alpha \tan \beta}$$

$$\tan(\theta + \theta) = \frac{\tan \theta + \tan \theta}{1 - \tan \theta \tan \theta}$$

$$\tan 2\theta = \frac{2 \tan \theta}{1 - \tan^2 \theta} \tag{5}$$

(b) To get a formula for $\sin 3\theta$, we use the sum formula and write 3θ as $2\theta + \theta$.

$$\sin 3\theta = \sin(2\theta + \theta) = \sin 2\theta \cos \theta + \cos 2\theta \sin \theta$$

Now use the double-angle formulas to get

$$\sin 3\theta = (2 \sin \theta \cos \theta)(\cos \theta) + (\cos^2 \theta - \sin^2 \theta)(\sin \theta)$$
$$= 2 \sin \theta \cos^2 \theta + \sin \theta \cos^2 \theta - \sin^3 \theta$$
$$= 3 \sin \theta \cos^2 \theta - \sin^3 \theta \qquad \blacksquare$$

The formula obtained in Example 2(b) can also be written as

$$\sin 3\theta = 3 \sin \theta \cos^2 \theta - \sin^3 \theta = 3 \sin \theta(1 - \sin^2 \theta) - \sin^3 \theta$$
$$= 3 \sin \theta - 4 \sin^3 \theta$$

That is, $\sin 3\theta$ is a third-degree polynomial in the variable $\sin \theta$. In fact, $\sin n\theta$, n a positive odd integer, can always be written as a polynomial of degree n in the variable $\sin \theta$.[*]

_____

[*]Due to the work done by P. L. Chebyshëv, these polynomials are sometimes called *Chebyshëv polynomials*.

■ Now work Problem 47.

Other Variations of the Double-angle Formulas

By rearranging the double-angle formulas (4b) and (4c), we obtain other formulas that we will use a little later in this section.

If we solve formula (4b) for $\sin^2 \theta$, we get

$$\cos 2\theta = 1 - 2 \sin^2 \theta$$

$$2 \sin^2 \theta = 1 - \cos 2\theta$$

$$\sin^2 \theta = \frac{1 - \cos 2\theta}{2} \tag{6}$$

Similarly, we can solve for $\cos^2 \theta$ in formula (4c):

$$\cos 2\theta = 2 \cos^2 \theta - 1$$

$$2 \cos^2 \theta = 1 + \cos 2\theta$$

$$\cos^2 \theta = \frac{1 + \cos 2\theta}{2} \tag{7}$$

Formulas (6) and (7) can be used to develop a formula for $\tan^2 \theta$:

$$\tan^2 \theta = \frac{\sin^2 \theta}{\cos^2 \theta} = \frac{\dfrac{1 - \cos 2\theta}{2}}{\dfrac{1 + \cos 2\theta}{2}}$$

$$\tan^2 \theta = \frac{1 - \cos 2\theta}{1 + \cos 2\theta} \tag{8}$$

Formulas (6) through (8) do not have to be memorized since their derivations are so straightforward.

Formulas (6) and (7) are important in calculus. The next example illustrates a problem that arises in calculus requiring the use of formula (7).

EXAMPLE 3   *Establishing an Identity*

Write an equivalent expression for $\cos^4 \theta$ that does not involve any powers of sine or cosine greater than 1.

Solution   The idea here is to apply formula (7) twice:

$$\cos^4 \theta = (\cos^2 \theta)^2 = \left(\frac{1 + \cos 2\theta}{2}\right)^2 \qquad \text{Formula (7)}$$

$$= \frac{1}{4}(1 + 2\cos 2\theta + \cos^2 2\theta)$$

$$= \frac{1}{4} + \frac{1}{2}\cos 2\theta + \frac{1}{4}\cos^2 2\theta$$

$$= \frac{1}{4} + \frac{1}{2}\cos 2\theta + \frac{1}{4}\left[\frac{1 + \cos 2(2\theta)}{2}\right] \quad \text{Formula (7)}$$

$$= \frac{1}{4} + \frac{1}{2}\cos 2\theta + \frac{1}{8}(1 + \cos 4\theta)$$

$$= \frac{3}{8} + \frac{1}{2}\cos 2\theta + \frac{1}{8}\cos 4\theta \qquad \blacksquare$$

■   Now work Problem 21.

Identities, such as the double-angle formulas, can sometimes be used to rewrite expressions in a more suitable form. Let's look at an example.

EXAMPLE 4

FIGURE 6

Projectile Motion

An object is propelled upward at an angle θ to the horizontal with an initial velocity of v_0 feet per second. See Figure 6. If air resistance is ignored, the horizontal distance R it travels, the **range,** is given by

$$R = \frac{1}{16} v_0^2 \sin\theta \cos\theta$$

(a) Show that $R = \frac{1}{32} v_0^2 \sin 2\theta$.

(b) Find the angle θ for which R is a maximum.

Solution

(a) We rewrite the given expression for the range using the double-angle formula $\sin 2\theta = 2\sin\theta \cos\theta$. Then

$$R = \frac{1}{16} v_0^2 \sin\theta \cos\theta = \frac{1}{16} v_0^2 \frac{\sin 2\theta}{2} = \frac{1}{32} v_0^2 \sin 2\theta$$

(b) In this form, the largest value for the range R can be easily found. For a fixed initial speed v_0, the angle θ of inclination to the horizontal determines the value of R. Since the largest value of a sine function is 1, occurring when the argument is $90°$, it follows that for maximum R we must have

$$2\theta = 90°$$

$$\theta = 45°$$

An inclination to the horizontal of $45°$ results in maximum range.   ■

Half-angle Formulas

Another important use of formulas (6) through (8) is to prove the **half-angle formulas.** In formulas (6) through (8), let $\theta = \alpha/2$. Then:

$$\sin^2\frac{\alpha}{2} = \frac{1 - \cos\alpha}{2} \qquad \cos^2\frac{\alpha}{2} = \frac{1 + \cos\alpha}{2}$$

$$\tan^2\frac{\alpha}{2} = \frac{1 - \cos\alpha}{1 + \cos\alpha} \qquad (9)$$

If we solve for the trigonometric functions on the left sides of equations (9), we obtain the half-angle formulas:

Theorem
Half-angle Formulas

$$\sin \frac{\alpha}{2} = \pm \sqrt{\frac{1 - \cos \alpha}{2}} \tag{10a}$$

$$\cos \frac{\alpha}{2} = \pm \sqrt{\frac{1 + \cos \alpha}{2}} \tag{10b}$$

$$\tan \frac{\alpha}{2} = \pm \sqrt{\frac{1 - \cos \alpha}{1 + \cos \alpha}} \tag{10c}$$

where the $+$ or $-$ sign is determined by the quadrant of the angle $\alpha/2$.

We use the half-angle formulas in the next example.

E X A M P L E   5   *Finding Exact Values Using Half-angle Formulas*

Find the exact value of:

(a) $\cos 15°$ (b) $\sin(-15°)$

Solution   (a) Because $15° = 30°/2$, we can use the half-angle formula for $\cos(\alpha/2)$ with $\alpha = 30°$. Also, because $15°$ is in quadrant I, $\cos 15° > 0$, and we choose the $+$ sign in using formula (10b):

$$\cos 15° = \cos \frac{30°}{2} = \sqrt{\frac{1 + \cos 30°}{2}}$$

$$= \sqrt{\frac{1 + \sqrt{3}/2}{2}} = \sqrt{\frac{2 + \sqrt{3}}{4}} = \frac{\sqrt{2 + \sqrt{3}}}{2}$$

(b) We use the fact that $\sin(-15°) = -\sin 15°$ and then apply formula (10a):

$$\sin(-15°) = -\sin \frac{30°}{2} = -\sqrt{\frac{1 - \cos 30°}{2}}$$

$$= -\sqrt{\frac{1 - \sqrt{3}/2}{2}} = -\sqrt{\frac{2 - \sqrt{3}}{4}} = -\frac{\sqrt{2 - \sqrt{3}}}{2}$$

It is interesting to compare the answer found in Example 5(a) with the answer to Example 2 of Section 7.2. There we calculated

$$\cos \frac{\pi}{12} = \cos 15° = \frac{1}{4} (\sqrt{6} + \sqrt{2})$$

Based on these results, we conclude that

$$\frac{1}{4} (\sqrt{6} + \sqrt{2}) \quad \text{and} \quad \frac{\sqrt{2 + \sqrt{3}}}{2}$$

are equal. (Since each expression is positive, you can verify this equality by squaring each expression.) Thus, two very different looking, yet correct, answers can be obtained, depending on the approach taken to solve a problem.

■   Now work Problem 11.

E X A M P L E   6

Finding Exact Values Using Half-angle Formulas

If $\cos \alpha = -\frac{3}{5}$, $\pi < \alpha < 3\pi/2$, find the exact value of:

(a) $\sin \dfrac{\alpha}{2}$ (b) $\cos \dfrac{\alpha}{2}$ (c) $\tan \dfrac{\alpha}{2}$

Solution   First, we observe that if $\pi < \alpha < 3\pi/2$ then $\pi/2 < \alpha/2 < 3\pi/4$. As a result, $\alpha/2$ lies in quadrant II.

(a) Because $\alpha/2$ lies in quadrant II, $\sin(\alpha/2) > 0$. Thus, we use the $+$ sign in formula (10a) to get

$$\sin \frac{\alpha}{2} = \sqrt{\frac{1 - \cos \alpha}{2}} = \sqrt{\frac{1 - (-\frac{3}{5})}{2}}$$

$$= \sqrt{\frac{\frac{8}{5}}{2}} = \sqrt{\frac{4}{5}} = \frac{2}{\sqrt{5}} = \frac{2\sqrt{5}}{5}$$

(b) Because $\alpha/2$ lies in quadrant II, $\cos(\alpha/2) < 0$. Thus, we use the $-$ sign in formula (10b) to get

$$\cos \frac{\alpha}{2} = -\sqrt{\frac{1 + \cos \alpha}{2}} = -\sqrt{\frac{1 + (-\frac{3}{5})}{2}}$$

$$= -\sqrt{\frac{\frac{2}{5}}{2}} = -\frac{1}{\sqrt{5}} = -\frac{\sqrt{5}}{5}$$

(c) Because $\alpha/2$ lies in quadrant II, $\tan(\alpha/2) < 0$. Thus, we use the $-$ sign in formula (10c) to get

$$\tan \frac{\alpha}{2} = -\sqrt{\frac{1 - \cos \alpha}{1 + \cos \alpha}} = -\sqrt{\frac{1 - (-\frac{3}{5})}{1 + (-\frac{3}{5})}} = -\sqrt{\frac{\frac{8}{5}}{\frac{2}{5}}} = -2$$ ■

Another way to solve Example 6(c) is to use the solutions found in parts (a) and (b):

$$\tan \frac{\alpha}{2} = \frac{\sin(\alpha/2)}{\cos(\alpha/2)} = \frac{2\sqrt{5}/5}{-\sqrt{5}/5} = -2$$

■   Now work Problems 1(c) and (d).

There is a formula for $\tan(\alpha/2)$ that does contain $+$ and $-$ signs, making it more useful than Formula 10(c). Because

$$1 - \cos \alpha = 2 \sin^2 \left(\frac{\alpha}{2}\right) \quad \text{Formula 9}$$

and

$$\sin \alpha = \sin 2 \left(\frac{\alpha}{2}\right) = 2 \sin \left(\frac{\alpha}{2}\right) \cos \left(\frac{\alpha}{2}\right) \quad \text{Double–angle formula}$$

we have

$$\frac{1 - \cos \alpha}{\sin \alpha} = \frac{2 \sin^2\left(\dfrac{\alpha}{2}\right)}{2 \sin\left(\dfrac{\alpha}{2}\right) \cos\left(\dfrac{\alpha}{2}\right)} = \frac{\sin\left(\dfrac{\alpha}{2}\right)}{\cos\left(\dfrac{\alpha}{2}\right)} = \tan\left(\dfrac{\alpha}{2}\right)$$

Since it also can be shown that

$$\frac{1 - \cos \alpha}{\sin \alpha} = \frac{\sin \alpha}{1 + \cos \alpha}$$

we have the following two half-angle formulas:

Half-angle Formulas for tan α/2

$$\tan\left(\frac{\alpha}{2}\right) = \frac{1 - \cos \alpha}{\sin \alpha} = \frac{\sin \alpha}{1 + \cos \alpha} \tag{11}$$

With this formula, the solution to Example 6(c) can be given as

$$\cos \alpha = -\tfrac{3}{5}$$

$$\sin \alpha = -\sqrt{1 - \cos^2 \alpha} = -\sqrt{1 - \tfrac{9}{25}} = -\sqrt{\tfrac{16}{25}} = -\tfrac{4}{5}$$

Thus, by equation (11),

$$\tan \frac{\alpha}{2} = \frac{1 - \cos \alpha}{\sin \alpha} = \frac{1 - (-\tfrac{3}{5})}{-\tfrac{4}{5}} = \frac{\tfrac{8}{5}}{-\tfrac{4}{5}} = -2$$

The next example illustrates a problem that arises in calculus.

E X A M P L E   7

Using Half-angle Formulas in Calculus

If $z = \tan(\alpha/2)$, show that:

(a) $\sin \alpha = \dfrac{2z}{1 + z^2}$ (b) $\cos \alpha = \dfrac{1 - z^2}{1 + z^2}$

Solution   (a)

$$\frac{2z}{1 + z^2} = \frac{2 \tan(\alpha/2)}{1 + \tan^2(\alpha/2)} = \frac{2 \tan(\alpha/2)}{\sec^2(\alpha/2)} = \frac{2 \cdot \dfrac{\sin(\alpha/2)}{\cos(\alpha/2)}}{\dfrac{1}{\cos^2(\alpha/2)}}$$

$$= \frac{2 \sin(\alpha/2)}{\cos(\alpha/2)} \cdot \cos^2 \frac{\alpha}{2} = 2 \sin \frac{\alpha}{2} \cos \frac{\alpha}{2}$$

$$= \sin 2\left(\frac{\alpha}{2}\right) = \sin \alpha$$

    ↑

Double-angle formula (3)

(b)

$$\frac{1 - z^2}{1 + z^2} = \frac{1 - \tan^2(\alpha/2)}{1 + \tan^2(\alpha/2)} = \frac{1 - \dfrac{\sin^2(\alpha/2)}{\cos^2(\alpha/2)}}{1 + \dfrac{\sin^2(\alpha/2)}{\cos^2(\alpha/2)}}$$

$$= \frac{\cos^2(\alpha/2) - \sin^2(\alpha/2)}{\cos^2(\alpha/2) + \sin^2(\alpha/2)} = \frac{\cos 2(\alpha/2)}{1} = \cos \alpha$$

    ↑

Double-angle formula (4a)
and Pythagorean identity

7.3

Exercise 7.3

In Problems 1–10, use the information given about the angle θ to find the exact value of:

(a) $\sin 2\theta$ (b) $\cos 2\theta$ (c) $\sin \dfrac{\theta}{2}$ (d) $\cos \dfrac{\theta}{2}$

1. $\sin \theta = \frac{3}{5}, \quad 0 < \theta < \pi/2$ 2. $\cos \theta = \frac{3}{5}, \quad 0 < \theta < \pi/2$ 3. $\tan \theta = \frac{4}{3}, \quad \pi < \theta < 3\pi/2$

4. $\tan \theta = \frac{1}{2}, \quad \pi < \theta < 3\pi/2$ 5. $\cos \theta = -\sqrt{2}/\sqrt{3}, \quad \pi/2 < \theta < \pi$ 6. $\sin \theta = -1/\sqrt{3}, \quad 3\pi/2 < \theta < 2\pi$

7. $\sec \theta = 3, \quad \sin \theta > 0$ 8. $\csc \theta = -\sqrt{5}, \quad \cos \theta < 0$ 9. $\cot \theta = -2, \quad \sec \theta < 0$

10. $\sec \theta = 2, \quad \csc \theta < 0$ 11. $\tan \theta = -3, \quad \sin \theta < 0$ 12. $\cot \theta = 3, \quad \cos \theta < 0$

In Problems 13–22, use the half-angle formulas to find the exact value of each trigonometric function.

13. $\sin 22.5°$ 14. $\cos 22.5°$ 15. $\tan \dfrac{7\pi}{8}$ 16. $\tan \dfrac{9\pi}{8}$ 17. $\cos 165°$

18. $\sin 195°$ 19. $\sec \dfrac{15\pi}{8}$ 20. $\csc \dfrac{7\pi}{8}$ 21. $\sin\left(-\dfrac{\pi}{8}\right)$ 22. $\cos\left(-\dfrac{3\pi}{8}\right)$

23. Show that $\sin^4 \theta = \frac{3}{8} - \frac{1}{2}\cos 2\theta + \frac{1}{8}\cos 4\theta$.
24. Develop a formula for $\cos 3\theta$ as a third-degree polynomial in the variable $\cos \theta$.
25. Show that $\sin 4\theta = (\cos \theta)(4 \sin \theta - 8 \sin^3 \theta)$.
26. Develop a formula for $\cos 4\theta$ as a fourth-degree polynomial in the variable $\cos \theta$.
27. Find an expression for $\sin 5\theta$ as a fifth-degree polynomial in the variable $\sin \theta$.
28. Find an expression for $\cos 5\theta$ as a fifth-degree polynomial in the variable $\cos \theta$.

In Problems 29–48, establish each identity.

29. $\cos^4 \theta - \sin^4 \theta = \cos 2\theta$ 30. $\dfrac{\cot \theta - \tan \theta}{\cot \theta + \tan \theta} = \cos 2\theta$

31. $\cot 2\theta = \dfrac{\cot^2 \theta - 1}{2 \cot \theta}$ 32. $\cot 2\theta = \frac{1}{2}(\cot \theta - \tan \theta)$

33. $\sec 2\theta = \dfrac{\sec^2 \theta}{2 - \sec^2 \theta}$ 34. $\csc 2\theta = \frac{1}{2}\sec \theta \csc \theta$

35. $\cos^2 2\theta - \sin^2 2\theta = \cos 4\theta$ 36. $(4 \sin \theta \cos \theta)(1 - 2 \sin^2 \theta) = \sin 4\theta$

37. $\dfrac{\cos 2\theta}{1 + \sin 2\theta} = \dfrac{\cot \theta - 1}{\cot \theta + 1}$ 38. $\sin^2 \theta \cos^2 \theta = \frac{1}{8}(1 - \cos 4\theta)$

39. $\sec^2 \dfrac{\theta}{2} = \dfrac{2}{1 + \cos \theta}$ 40. $\csc^2 \dfrac{\theta}{2} = \dfrac{2}{1 - \cos \theta}$

41. $\cot^2 \dfrac{\theta}{2} = \dfrac{\sec \theta + 1}{\sec \theta - 1}$ 42. $\tan \dfrac{\theta}{2} = \csc \theta - \cot \theta$

43. $\cos \theta = \dfrac{1 - \tan^2(\theta/2)}{1 + \tan^2(\theta/2)}$ 44. $1 - \frac{1}{2}\sin 2\theta = \dfrac{\sin^3 \theta + \cos^3 \theta}{\sin \theta + \cos \theta}$

45. $\dfrac{\sin 3\theta}{\sin \theta} - \dfrac{\cos 3\theta}{\cos \theta} = 2$ 46. $\dfrac{\cos \theta + \sin \theta}{\cos \theta - \sin \theta} - \dfrac{\cos \theta - \sin \theta}{\cos \theta + \sin \theta} = 2 \tan 2\theta$

47. $\tan 3\theta = \dfrac{3 \tan \theta - \tan^3 \theta}{1 - 3 \tan^2 \theta}$ 48. $\tan \theta + \tan(\theta + 120°) + \tan(\theta + 240°) = 3 \tan 3\theta$

In Problems 49–62, find the exact value of each expression.

49. $\sin(2 \sin^{-1} \frac{1}{2})$

50. $\sin[2 \sin^{-1}(\sqrt{3}/2)]$

51. $\cos(2 \sin^{-1} \frac{3}{5})$

52. $\cos(2 \cos^{-1} \frac{4}{5})$

53. $\tan[2 \cos^{-1}(-\frac{3}{5})]$

54. $\tan(2 \tan^{-1} \frac{3}{4})$

55. $\sin(2 \cos^{-1} \frac{4}{5})$

56. $\cos[2 \tan^{-1}(-\frac{4}{3})]$

57. $\sin^2(\frac{1}{2} \cos^{-1} \frac{3}{5})$

58. $\cos^2(\frac{1}{2} \sin^{-1} \frac{3}{5})$

59. $\sec(2 \tan^{-1} \frac{3}{4})$

60. $\csc[2 \sin^{-1}(-\frac{3}{5})]$

61. $\cot^2(\frac{1}{2} \tan^{-1} \frac{4}{3})$

62. $\cot^2(\frac{1}{2} \cos^{-1} \frac{5}{13})$

63. Find the value of the number C: $\frac{1}{2} \sin^2 x + C = -\frac{1}{4} \cos 2x$

64. Find the value of the number C: $\frac{1}{2} \cos^2 x + C = \frac{1}{4} \cos 2x$

65. *Area of an Isosceles Triangle* Show that the area A of an isosceles triangle whose equal sides are of length s and the angle between them is θ is

$$A = \frac{1}{2} s^2 \sin \theta$$

[*Hint:* See the illustration. The height h bisects the angle θ and is the perpendicular bisector of the base.]

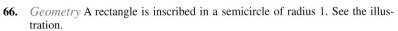

66. *Geometry* A rectangle is inscribed in a semicircle of radius 1. See the illustration.

(a) Express the area A of the rectangle as a function of the angle θ shown in the illustration.

(b) Show that $A = \sin 2\theta$.

(c) Find the angle θ that results in the largest area A.

(d) Find the dimensions of this largest rectangle.

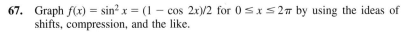

67. Graph $f(x) = \sin^2 x = (1 - \cos 2x)/2$ for $0 \le x \le 2\pi$ by using the ideas of shifts, compression, and the like.

68. Repeat Problem 67 for $g(x) = \cos^2 x$.

69. Use the fact that

$$\cos \frac{\pi}{12} = \frac{1}{4}(\sqrt{6} + \sqrt{2})$$

to find $\sin(\pi/24)$ and $\cos(\pi/24)$.

70. Show that

$$\sin \frac{\pi}{8} = \frac{\sqrt{2 - \sqrt{2}}}{2}$$

and use it to find $\sin(\pi/16)$ and $\cos(\pi/16)$.

71. Show that $\sin^3 \theta + \sin^3(\theta + 120°) + \sin^3(\theta + 240°) = -\frac{3}{4} \sin 3\theta$.

72. If $\tan \theta = a \tan(\theta/3)$, express $\tan(\theta/3)$ in terms of a.

In Problems 73 and 74, establish each identity.

73. $\ln|\sin \theta| = \frac{1}{2}(\ln|1 - \cos 2\theta| - \ln 2)$

74. $\ln|\cos \theta| = \frac{1}{2}(\ln|1 + \cos 2\theta| - \ln 2)$

75. *Projectile Motion* An object is propelled upward at an angle θ, $45° < \theta < 90°$, to the horizontal with an initial velocity of v_0 feet per second from the base of a plane that makes an angle of $45°$ with the horizontal. See the illustration. If air resistance is ignored, the distance R it travels up the inclined plane is given by

$$R = \frac{v_0^2 \sqrt{2}}{16} \cos \theta(\sin \theta - \cos \theta)$$

(a) Show that

$$R = \frac{v_0^2 \sqrt{2}}{32}(\sin 2\theta - \cos 2\theta - 1)$$

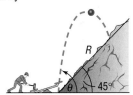

(b) Graph $R = R(\theta)$. What value of θ makes R the largest?

76. *Sawtooth Curve* An oscilloscope often displays a sawtooth curve. This curve can be approximated by sinusoidal curves of varying periods and amplitudes. A first approximation to the sawtooth curve is given by

$$y = \frac{1}{2}\sin 2\pi x + \frac{1}{4}\sin 4\pi x.$$

Show that $y = \sin 2\pi x \cos^2 \pi x.$

77. Go to the library and research Chebyshëv polynomials. Write a report on your findings.

7.4

Product-to-Sum and Sum-to-Product Formulas

The sum and difference formulas can be used to derive formulas for writing the products of sines and/or cosines as sums or differences. These identities are usually called the **product-to-sum formulas.**

Theorem
Product-to-Sum Formulas

$$\sin \alpha \sin \beta = \tfrac{1}{2}[\cos(\alpha - \beta) - \cos(\alpha + \beta)] \qquad (1)$$

$$\cos \alpha \cos \beta = \tfrac{1}{2}[\cos(\alpha - \beta) + \cos(\alpha + \beta)] \qquad (2)$$

$$\sin \alpha \cos \beta = \tfrac{1}{2}[\sin(\alpha + \beta) + \sin(\alpha - \beta)] \qquad (3)$$

These formulas do not have to be memorized. Instead, you should remember how they are derived. Then, when you want to use them, either look them up or derive them, as needed.

To derive formulas (1) and (2), write down the sum and difference formulas for the cosine:

$$\cos(\alpha - \beta) = \cos \alpha \cos \beta + \sin \alpha \sin \beta \qquad (4)$$

$$\cos(\alpha + \beta) = \cos \alpha \cos \beta - \sin \alpha \sin \beta \qquad (5)$$

Subtract equation (5) from equation (4) to get

$$\cos(\alpha - \beta) - \cos(\alpha + \beta) = 2 \sin \alpha \sin \beta$$

from which

$$\sin \alpha \sin \beta = \tfrac{1}{2}[\cos(\alpha - \beta) - \cos(\alpha + \beta)]$$

Now, add equations (4) and (5) to get

$$\cos(\alpha - \beta) + \cos(\alpha + \beta) = 2 \cos \alpha \cos \beta$$

from which

$$\cos \alpha \cos \beta = \tfrac{1}{2}[\cos(\alpha - \beta) + \cos(\alpha + \beta)]$$

To derive product-to-sum formula (3), use the sum and difference formulas for sine in a similar way. (You are asked to do this in Problem 41 at the end of this section.)

$\mathcal{M}$ISSION POSSIBLE

Chapter 7

HOW FAR AND HOW HIGH DOES A BASEBALL NEED TO GO FOR AN OUT-OF-THE-PARK HOME RUN?

The sportscasters for the Cleveland Indians decided they had better be prepared for any eventuality when the new ballpark, Jacob's Field, opened in 1994. One eventuality they worried about was Albert Belle hitting a home run that went so high and so far that it left the ball park. What would they tell their listeners about the height and distance the ball went?

So they called in the Mission Possible team. The sportscasters wanted to know the distance from homeplate to the highest point of the stadium and the distance from ground level to the highest point of the stadium for every 5° from the 3rd base line around to the 1st base line. The problem was complicated by the fact that the distance from homeplate to the outfield fence varied from 325 feet to 410 feet. Furthermore, the height that would have to be cleared also varied, depending on where the ball was hit.

1. First, how many distances and heights do the sportscasters want?
2. To see one solution, use the distance from homeplate across 2nd base to the outfield fence, 410 feet. Using a transit, you find the angle of elevation from homeplate to the highest point of the stadium is 10° and the angle of elevation from the base of the outfield fence to the highest point of the stadium is 32.5°. Use this information to find the minimum distance from homeplate to the highest point of the stadium and the distance from ground level to the highest point of the stadium.
3. You tell the sportscasters that if they provide you the two angles of elevation for each 5° movement around the stadium, you can solve their problem. After receiving the two angles of elevation for each 5° movement around the stadium, you decide that it would be faster to develop a general formula, so that by inputting the two angles and the distance from homeplate to the fence, the distance from homeplate to the top of the roof and the height of the roof would be computed. Let α and β denote the angles of elevation of the top of the roof from homeplate and from the base of the outfield fence and let L be the distance from homeplate to the outfield fence. What are the correct formulas?
4. Compare your formulas with those of other groups. Are they all the same? Are they all equivalent? Which ones are simplest?
5. Now write each correct formula in the form shown below:

$$\text{height} = \frac{L \sin \alpha \sin \beta}{\sin(\beta - \alpha)} \qquad \text{distance} = \frac{L \sin \beta}{\sin(\beta - \alpha)}$$

6. What is the probable trajectory of a hit baseball on a windless day? How might the wind and other factors affect the path of the baseball?
7. If a player actually did hit an out-of-the-park homerun, how would the distance the ball traveled compare to the figures you have been developing for the sportscasters? What other factors will add to the distance?
8. Could there be a longest homerun? How would you measure it?

EXAMPLE 1   *Expressing Products as Sums*

Express each of the following products as a sum containing only sines or cosines:

(a) $\sin 6\theta \sin 4\theta$ (b) $\cos 3\theta \cos \theta$ (c) $\sin 3\theta \cos 5\theta$

Solution   (a) We use formula (1) to get

$$\sin 6\theta \sin 4\theta = \tfrac{1}{2}\left[\cos(6\theta - 4\theta) - \cos(6\theta + 4\theta)\right]$$
$$= \tfrac{1}{2}(\cos 2\theta - \cos 10\theta)$$

(b) We use formula (2) to get

$$\cos 3\theta \cos \theta = \tfrac{1}{2}\left[\cos(3\theta - \theta) + \cos(3\theta + \theta)\right]$$
$$= \tfrac{1}{2}(\cos 2\theta + \cos 4\theta)$$

(c) We use formula (3) to get

$$\sin 3\theta \cos 5\theta = \tfrac{1}{2}\left[\sin(3\theta + 5\theta) + \sin(3\theta - 5\theta)\right]$$
$$= \tfrac{1}{2}\left[\sin 8\theta + \sin(-2\theta)\right] = \tfrac{1}{2}(\sin 8\theta - \sin 2\theta) \qquad \blacksquare$$

■   Now work Problem 1.

The **sum-to-product formulas** are given next.

Theorem
Sum-to-Product Formulas

$$\sin \alpha + \sin \beta = 2 \sin \frac{\alpha + \beta}{2} \cos \frac{\alpha - \beta}{2} \qquad (6)$$

$$\sin \alpha - \sin \beta = 2 \sin \frac{\alpha - \beta}{2} \cos \frac{\alpha + \beta}{2} \qquad (7)$$

$$\cos \alpha + \cos \beta = 2 \cos \frac{\alpha + \beta}{2} \cos \frac{\alpha - \beta}{2} \qquad (8)$$

$$\cos \alpha - \cos \beta = -2 \sin \frac{\alpha + \beta}{2} \sin \frac{\alpha - \beta}{2} \qquad (9)$$

■

We shall derive formula (6) and leave the derivations of formulas (7) through (9) as exercises (see Problems 42 through 44).

Proof $$2 \sin \frac{\alpha + \beta}{2} \cos \frac{\alpha - \beta}{2} = 2 \cdot \frac{1}{2}\left[\sin\left(\frac{\alpha + \beta}{2} + \frac{\alpha - \beta}{2}\right) + \sin\left(\frac{\alpha + \beta}{2} - \frac{\alpha - \beta}{2}\right)\right]$$

↑
Product-to-sum formula (3)

$$= \sin \frac{2\alpha}{2} + \sin \frac{2\beta}{2} = \sin \alpha + \sin \beta \qquad \blacksquare$$

EXAMPLE 2   *Expressing Sums (or Differences) as a Product*

Express each sum or difference as a product of sines and/or cosines:

(a) $\sin 5\theta - \sin 3\theta$ (b) $\cos 3\theta + \cos 2\theta$

Solution   (a) We use formula (7) to get

$$\sin 5\theta - \sin 3\theta = 2 \sin \frac{5\theta - 3\theta}{2} \cos \frac{5\theta + 3\theta}{2}$$

$$= 2 \sin \theta \cos 4\theta$$

(b) $$\cos 3\theta + \cos 2\theta = 2 \cos \frac{3\theta + 2\theta}{2} \cos \frac{3\theta - 2\theta}{2} \quad \text{Formula (8)}$$

$$= 2 \cos \frac{5\theta}{2} \cos \frac{\theta}{2}$$ ■

■   Now work Problem 11.

7.4

Exercise 7.4

In Problems 1–10, express each product as a sum containing only sines or cosines

1. $\sin 4\theta \sin 2\theta$ **2.** $\cos 4\theta \cos 2\theta$ **3.** $\sin 4\theta \cos 2\theta$ **4.** $\sin 3\theta \sin 5\theta$

5. $\cos 3\theta \cos 5\theta$ **6.** $\sin 4\theta \cos 6\theta$ **7.** $\sin \theta \sin 2\theta$ **8.** $\cos 3\theta \cos 4\theta$

9. $\sin \dfrac{3\theta}{2} \cos \dfrac{\theta}{2}$ **10.** $\sin \dfrac{\theta}{2} \cos \dfrac{5\theta}{2}$

In Problems 11–18, express each sum or difference as a product of sines and/or cosines.

11. $\sin 4\theta - \sin 2\theta$ **12.** $\sin 4\theta + \sin 2\theta$ **13.** $\cos 2\theta + \cos 4\theta$ **14.** $\cos 5\theta - \cos 3\theta$

15. $\sin \theta + \sin 3\theta$ **16.** $\cos \theta + \cos 3\theta$ **17.** $\cos \dfrac{\theta}{2} - \cos \dfrac{3\theta}{2}$ **18.** $\sin \dfrac{\theta}{2} - \sin \dfrac{3\theta}{2}$

In Problems 19–36, establish each identity.

19. $\dfrac{\sin \theta + \sin 3\theta}{2 \sin 2\theta} = \cos \theta$ **20.** $\dfrac{\cos \theta + \cos 3\theta}{2 \cos 2\theta} = \cos \theta$ **21.** $\dfrac{\sin 4\theta + \sin 2\theta}{\cos 4\theta + \cos 2\theta} = \tan 3\theta$

22. $\dfrac{\cos \theta - \cos 3\theta}{\sin 3\theta - \sin \theta} = \tan 2\theta$ **23.** $\dfrac{\cos \theta - \cos 3\theta}{\sin \theta + \sin 3\theta} = \tan \theta$ **24.** $\dfrac{\cos \theta - \cos 5\theta}{\sin \theta + \sin 5\theta} = \tan 2\theta$

25. $\sin \theta(\sin \theta + \sin 3\theta) = \cos \theta(\cos \theta - \cos 3\theta)$ **26.** $\sin \theta(\sin 3\theta + \sin 5\theta) = \cos \theta(\cos 3\theta - \cos 5\theta)$

27. $\dfrac{\sin 4\theta + \sin 8\theta}{\cos 4\theta + \cos 8\theta} = \tan 6\theta$ **28.** $\dfrac{\sin 4\theta - \sin 8\theta}{\cos 4\theta - \cos 8\theta} = -\cot 6\theta$

29. $\dfrac{\sin 4\theta + \sin 8\theta}{\sin 4\theta - \sin 8\theta} = -\dfrac{\tan 6\theta}{\tan 2\theta}$ **30.** $\dfrac{\cos 4\theta - \cos 8\theta}{\cos 4\theta + \cos 8\theta} = \tan 2\theta \tan 6\theta$

31. $\dfrac{\sin \alpha + \sin \beta}{\sin \alpha - \sin \beta} = \tan \dfrac{\alpha + \beta}{2} \cot \dfrac{\alpha - \beta}{2}$ **32.** $\dfrac{\cos \alpha + \cos \beta}{\cos \alpha - \cos \beta} = -\cot \dfrac{\alpha + \beta}{2} \cot \dfrac{\alpha - \beta}{2}$

33. $\dfrac{\sin \alpha + \sin \beta}{\cos \alpha + \cos \beta} = \tan \dfrac{\alpha + \beta}{2}$ **34.** $\dfrac{\sin \alpha - \sin \beta}{\cos \alpha - \cos \beta} = -\cot \dfrac{\alpha + \beta}{2}$

35. $1 + \cos 2\theta + \cos 4\theta + \cos 6\theta = 4 \cos \theta \cos 2\theta \cos 3\theta$

36. $1 - \cos 2\theta + \cos 4\theta - \cos 6\theta = 4 \sin \theta \cos 2\theta \sin 3\theta$

37. *Touch-Tone Phones* On a Touch-Tone phone, each button produces a unique sound. The sound produced is the sum of two tones, given by

$$y = \sin 2\pi l t \quad \text{and} \quad y = \sin 2\pi h t$$

where l and h are the low and high frequencies (cycles per second) shown on the illustration. For example, if you touch 7, the low frequency is $l = 852$ cycles per second and the high frequency is $h = 1209$ cycles per second. The sound emitted by touching 7 is

$$y = \sin 2\pi(852)t + \sin 2\pi(1209)t$$

Write this sound as a product of sines and/or cosines.

38. Write the sound emitted by touching the # key as a product of sines and/or cosines.

39. If $\alpha + \beta + \gamma = \pi$, show that
$\sin 2\alpha + \sin 2\beta + \sin 2\gamma = 4 \sin \alpha \sin \beta \sin \gamma$.

40. If $\alpha + \beta + \gamma = \pi$, show that
$\tan \alpha + \tan \beta + \tan \gamma = \tan \alpha \tan \beta \tan \gamma$.

41. Derive formula (3).   **42.** Derive formula (7).

43. Derive formula (8).   **44.** Derive formula (9).

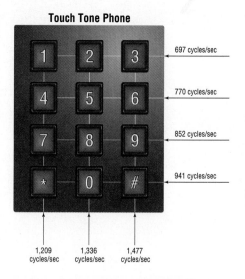

Touch Tone Phone

697 cycles/sec
770 cycles/sec
852 cycles/sec
941 cycles/sec

1,209 cycles/sec   1,336 cycles/sec   1,477 cycles/sec

7.5

Trigonometric Equations

The previous sections of this chapter were devoted to trigonometric identities—that is, equations involving trigonometric functions that are satisfied by every value in the domain of the variable. In this section, we discuss **trigonometric equations**—that is, equations involving trigonometric functions that are satisfied only by some values of the variable (or, possibly, are not satisfied by any values of the variable). The values that satisfy the equation are called **solutions** of the equation.

E X A M P L E   1   *Checking Whether a Given Number is a Solution of a Trigonometric Equation*

Determine whether $\theta = \pi/4$ is a solution of the equation $\sin \theta = \frac{1}{2}$. Is $\theta = \pi/6$ a solution?

Solution   Replace θ by $\pi/4$ in the given equation. The result is

$$\sin \frac{\pi}{4} = \frac{\sqrt{2}}{2} \neq \frac{1}{2}$$

We conclude that $\pi/4$ is not a solution.
    Next, replace θ by $\pi/6$ in the equation. The result is

$$\sin \frac{\pi}{6} = \frac{1}{2}$$

Thus, $\pi/6$ is a solution of the given equation.   ∎

The equation given in Example 1 has other solutions besides $\theta = \pi/6$. For example, $\theta = 5\pi/6$ is also a solution, as is $\theta = 13\pi/6$. (You should check this for

FIGURE 7
$y = \sin x$

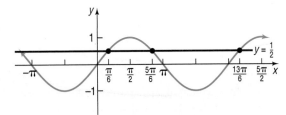

yourself.) In fact, the equation has an infinite number of solutions due to the periodicity of the sine function, as can be seen in Figure 7.

Unless otherwise indicated, in solving trigonometric equations, we need to find *all* solutions. As the next example illustrates, finding all the solutions can be accomplished by first finding solutions over an interval whose length equals the period of the function and then adding multiples of that period to the solutions found. Let's look at some examples.

E X A M P L E   2   *Solving a Trigonometric Equation*

Solve the equation: $\cos \theta = \frac{1}{2}$

Solution   The period of the cosine function is 2π. In the interval $[0, 2\pi)$, there are two angles θ for which $\cos \theta = \frac{1}{2}$: $\theta = \pi/3$ and $\theta = 5\pi/3$. See Figure 8. Because the cosine function has period 2π, all the solutions of $\cos \theta = \frac{1}{2}$ may be given by

$$\theta = \frac{\pi}{3} + 2k\pi \quad \text{or} \quad \theta = \frac{5\pi}{3} + 2k\pi, \quad k \text{ any integer}$$

FIGURE 8
$\cos \theta = \frac{1}{2}$

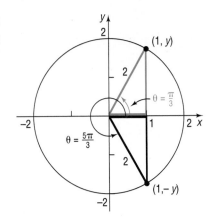

In most of our work, we shall be interested only in finding solutions of trigonometric equations for $0 \le \theta < 2\pi$.

■   Now work Problem 1.

E X A M P L E   3   *Solving a Trigonometric Equation*

Solve the equation: $\sin 2\theta = 1, \ 0 \le \theta < 2\pi$

Solution   The period of the sine function is 2π. In the interval $[0, 2\pi)$, the sine function has the value 1 only at $\pi/2$. Because the argument is 2θ in the given equation, we have

$$2\theta = \frac{\pi}{2} + 2k\pi, \qquad k \text{ any integer}$$

$$\theta = \frac{\pi}{4} + k\pi$$

In the interval $[0, 2\pi)$, the solutions of $\sin 2\theta = 1$ are $\pi/4$ ($k = 0$) and $\pi/4 + \pi = 5\pi/4$ ($k = 1$). Note that $k = -1$ gives $\theta = -3\pi/4$ and $k = 2$ gives $\theta = 9\pi/4$, both of which are outside $[0, 2\pi)$. ■

Warning: In solving a trigonometric equation for θ, $0 \le \theta < 2\pi$, in which the argument is not θ (as in Example 3), you must write down all the solutions, first, and then list those that are in the interval $[0, 2\pi)$. Otherwise, solutions may be lost. For example, in solving $\sin 2\theta = 1$, if you merely write the solution $2\theta = \pi/2$, you will find only $\theta = \pi/4$ and miss the solution $\theta = 5\pi/4$.

■ Now work Problem 7.

E X A M P L E 4   *Solving a Trigonometric Equation*

Solve the equation: $\sin 2\theta = \frac{1}{2}$, $0 \le \theta < 2\pi$

Solution   The period of the sine function is 2π. In the interval $[0, 2\pi)$, the sine function has the value $\frac{1}{2}$ at $\pi/6$ and $5\pi/6$. See Figure 9. Consequently, because the argument is 2θ in the equation $\sin 2\theta = \frac{1}{2}$, we have

$$2\theta = \frac{\pi}{6} + 2k\pi \quad \text{or} \quad 2\theta = \frac{5\pi}{6} + 2k\pi, \qquad k \text{ any integer}$$

$$\theta = \frac{\pi}{12} + k\pi \qquad\qquad \theta = \frac{5\pi}{12} + k\pi$$

FIGURE 9
$\sin 2\theta = \frac{1}{2}$, $0 \le \theta < 2\pi$

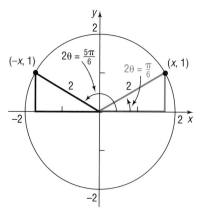

In the interval $[0, 2\pi)$, the solutions of $\sin 2\theta = \frac{1}{2}$ are $\pi/12$, $\pi/12 + \pi = 13\pi/12$, $5\pi/12$, and $5\pi/12 + \pi = 17\pi/12$. ■

E X A M P L E 5   *Solving a Trigonometric Equation*

Solve the equation: $\tan \dfrac{\theta}{4} = 1$, $0 \le \theta < 2\pi$

Solution   The period of the tangent function is π. In the interval $[0, \pi)$, the tangent function has the value 1 only at $\pi/4$. Because the argument is $\theta/4$ in the given equation, we have

$$\frac{\theta}{4} = \frac{\pi}{4} + k\pi, \qquad k \text{ any integer}$$

$$\theta = \pi + 4k\pi$$

In the interval $[0, 2\pi)$, $\theta = \pi$ is the only solution.   ■

E X A M P L E   6

Solving a Trigonometric Equation with a Calculator

Use a calculator to solve the equation: $\sin \theta = 0.3, 0 \le \theta < 2\pi$

Solution

To solve $\sin \theta = 0.3$ on a calculator, we first choose the mode. If we set it to radians, we find

Keystrokes: `0.3`   `SHIFT` $\overset{\sin^{-1}}{\boxed{\text{sin}}}$

Display: `0.3`   `0.3046927`

The angle 0.3046927 radian is the angle $-\pi/2 \le \theta \le \pi/2$ for which $\sin \theta = 0.3$. Another angle for which $\sin \theta = 0.3$ is $\pi - 0.3046927$. See Figure 10. The angle $\pi - 0.3046927$ is the angle in quadrant II, where $\sin \theta = 0.3$. Thus, the solutions for $\sin \theta = 0.3, 0 \le \theta < 2\pi$, are

$$\theta = 0.3046927 \quad \text{or} \quad \pi - 0.3046927 \approx 2.8369$$ ■

FIGURE 10
$\sin \theta = 0.3$

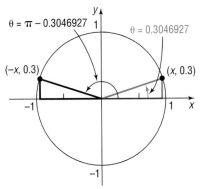

Warning: Example 6 illustrates that caution must be exercised when solving trigonometric equations on a calculator. Remember that the calculator supplies an angle only within the restrictions of the definition of the inverse trigonometric function. To find the remaining solutions, you must identify other quadrants, if any, in which the angle may be located.

■   Now work Problem 13.

Many trigonometric equations can be solved by applying techniques that we already know, such as applying the quadratic formula (if the equation is a second-degree polynomial) or factoring.

E X A M P L E   7

Solving a Trigonometric Equation Quadratic in Form

Solve the equation: $2 \sin^2 \theta - 3 \sin \theta + 1 = 0, 0 \le \theta < 2\pi$

Solution

The equation we wish to solve is a quadratic equation (in $\sin \theta$) that can be factored:

$$2 \sin^2 \theta - 3 \sin \theta + 1 = 0 \quad {\scriptstyle 2x^2 - 3x + 1 = 0, \quad x = \sin \theta}$$
$$(2 \sin \theta - 1)(\sin \theta - 1) = 0 \quad {\scriptstyle (2x - 1)(x - 1) = 0}$$
$$2 \sin \theta - 1 = 0 \quad \text{or} \quad \sin \theta - 1 = 0$$
$$\sin \theta = \tfrac{1}{2} \qquad\qquad \sin \theta = 1$$

Thus,

$$\theta = \frac{\pi}{6} \qquad \theta = \frac{5\pi}{6} \qquad \theta = \frac{\pi}{2}$$ ■

■   Now work Problem 23.

When a trigonometric equation contains more than one trigonometric function, identities sometimes can be used to obtain an equivalent equation that contains only one trigonometric function.

EXAMPLE 8

Solving a Trigonometric Equation Using Identities

Solve the equation: $3 \cos \theta + 3 = 2 \sin^2 \theta, 0 \le \theta < 2\pi$

Solution   The equation in its present form contains sines and cosines. However, a form of the Pythagorean identity can be used to transform the equation into an equivalent expression containing only cosines:

$$3 \cos \theta + 3 = 2 \sin^2 \theta$$
$$3 \cos \theta + 3 = 2(1 - \cos^2 \theta) \quad \text{\small sin}^2 \theta = 1 - \cos^2 \theta$$
$$3 \cos \theta + 3 = 2 - 2 \cos^2 \theta$$
$$2 \cos^2 \theta + 3 \cos \theta + 1 = 0 \qquad \qquad \text{\small Quadratic in cos } \theta$$
$$(2 \cos \theta + 1)(\cos \theta + 1) = 0 \qquad \qquad \text{\small Factor}$$
$$2 \cos \theta + 1 = 0 \quad \text{or} \quad \cos \theta + 1 = 0$$
$$\cos \theta = -\tfrac{1}{2} \qquad \qquad \cos \theta = -1$$

Thus,

$$\theta = \frac{2\pi}{3} \qquad \theta = \frac{4\pi}{3} \qquad \theta = \pi$$

Check: Graph $y = 3 \cos x + 3$ and $y = 2 \sin^2 x$ and use TRACE to find their point(s) of intersection. How close are your approximate solutions to the exact ones found in this example?   ∎

EXAMPLE 9

Solving a Trigonometric Equation Using Identities

Solve the equation: $\cos 2\theta + 3 = 5 \cos \theta, 0 \le \theta < 2\pi$

Solution   First, we observe that the given equation contains two cosine functions, but with different arguments, θ and 2θ. We use the double-angle formula $\cos 2\theta = 2 \cos^2 \theta - 1$ to obtain an equivalent equation containing only $\cos \theta$:

$$\cos 2\theta + 3 = 5 \cos \theta$$
$$(2 \cos^2 \theta - 1) + 3 = 5 \cos \theta$$
$$2 \cos^2 \theta - 5 \cos \theta + 2 = 0$$
$$(\cos \theta - 2)(2 \cos \theta - 1) = 0$$
$$\cos \theta = 2 \quad \text{or} \quad \cos \theta = \tfrac{1}{2}$$

For any angle θ, $-1 \le \cos \theta \le 1$; thus, the equation $\cos \theta = 2$ has no solution. The solutions of $\cos \theta = \tfrac{1}{2}$ are

$$\theta = \frac{\pi}{3} \qquad \theta = \frac{5\pi}{3}$$

∎

■   Now work Problem 33.

EXAMPLE 10

Solving a Trigonometric Equation Using Identities

Solve the equation: $\cos^2 \theta + \sin \theta = 2, 0 \le \theta < 2\pi$

Solution   We use a form of the Pythagorean identity:

$$\cos^2 \theta + \sin \theta = 2$$
$$(1 - \sin^2 \theta) + \sin \theta = 2$$
$$\sin^2 \theta - \sin \theta + 1 = 0$$

This is a quadratic equation in $\sin \theta$. The discriminant is $b^2 - 4ac = 1 - 4 = -3 < 0$. Therefore, the equation has no real solution.

Check: Graph $y = \cos^2 x + \sin x$ and $y = 2$ to see that the two graphs do not intersect anywhere.   ∎

EXAMPLE   1 1   *Solving a Trigonometric Equation Using Identities*

Solve the equation: $\sin \theta \cos \theta = -\frac{1}{2}, 0 \le \theta < 2\pi$

Solution   The left side of the given equation is in the form of the double-angle formula $2 \sin \theta \cos \theta = \sin 2\theta$, except for a factor of 2. Thus, we multiply each side by 2:

$$\sin \theta \cos \theta = -\frac{1}{2}$$
$$2 \sin \theta \cos \theta = -1$$
$$\sin 2\theta = -1$$

The argument here is 2θ. Thus, we need to write all the solutions of this equation and then list those that are in the interval $[0, 2\pi)$.

$$2\theta = \frac{3\pi}{2} + 2k\pi, \qquad k \text{ any integer}$$

$$\theta = \frac{3\pi}{4} + k\pi$$

The solutions in the interval $[0, 2\pi)$ are

$$\theta = \frac{3\pi}{4} \qquad \theta = \frac{7\pi}{4}$$ ∎

Sometimes it is necessary to square both sides of an equation in order to obtain expressions that allow the use of identities. Remember, however, that when squaring both sides extraneous solutions may be introduced. As a result, apparent solutions must be checked.

EXAMPLE   1 2   *Other Methods for Solving a Trigonometric Equation*

Solve the equation: $\sin \theta + \cos \theta = 1, 0 \le \theta < 2\pi$

Solution A   Attempts to use available identities do not lead to equations that are easy to solve. (Try it yourself.) Thus, given the form of this equation, we decide to square each side:

$$\sin \theta + \cos \theta = 1$$
$$(\sin \theta + \cos \theta)^2 = 1$$
$$\sin^2 \theta + 2 \sin \theta \cos \theta + \cos^2 \theta = 1$$
$$2 \sin \theta \cos \theta = 0 \quad \sin^2 \theta + \cos^2 \theta = 1$$
$$\sin \theta \cos \theta = 0$$

Thus,

$$\sin \theta = 0 \quad \text{or} \quad \cos \theta = 0$$

and the apparent solutions are

$$\theta = 0 \qquad \theta = \pi \qquad \theta = \frac{\pi}{2} \qquad \theta = \frac{3\pi}{2}$$

Because we squared both sides of the original equation, we must check these apparent solutions to see if any are extraneous:

$$\theta = 0: \quad \sin 0 + \cos 0 = 0 + 1 = 1 \qquad \text{A solution}$$

$$\theta = \pi: \quad \sin \pi + \cos \pi = 0 + (-1) = -1 \qquad \text{Not a solution}$$

$$\theta = \frac{\pi}{2}: \quad \sin \frac{\pi}{2} + \cos \frac{\pi}{2} = 1 + 0 = 1 \qquad \text{A solution}$$

$$\theta = \frac{3\pi}{2}: \quad \sin \frac{3\pi}{2} + \cos \frac{3\pi}{2} = -1 + 0 = -1 \qquad \text{Not a solution}$$

Thus, $\theta = 3\pi/2$ and $\theta = \pi$ are extraneous. The actual solutions are $\theta = 0$ and $\theta = \pi/2$. ∎

We can solve the equation given in Example 12 another way.

Solution B   We start with the equation

$$\sin \theta + \cos \theta = 1$$

and divide each side by $\sqrt{2}$. (The reason for this choice will become apparent shortly.) Then

$$\frac{1}{\sqrt{2}} \sin \theta + \frac{1}{\sqrt{2}} \cos \theta = \frac{1}{\sqrt{2}}$$

The left side now resembles the formula for the sine of the sum of two angles, one of which is θ. The other angle is unknown (call it ϕ.) Then

$$\sin(\theta + \phi) = \sin \theta \cos \phi + \cos \theta \sin \phi = \frac{1}{\sqrt{2}} \qquad (1)$$

where

$$\cos \phi = \frac{1}{\sqrt{2}} \qquad \sin \phi = \frac{1}{\sqrt{2}}, \qquad 0 \le \phi < 2\pi$$

The angle ϕ is therefore $\pi/4$. As a result, equation (1) becomes

$$\sin \left(\theta + \frac{\pi}{4} \right) = \frac{1}{\sqrt{2}}$$

We solve this equation to get

$$\theta + \frac{\pi}{4} = \frac{\pi}{4} \quad \text{or} \quad \theta + \frac{\pi}{4} = \frac{3\pi}{4}$$

$$\theta = 0 \qquad \qquad \theta = \frac{\pi}{2}$$

These solutions agree with the solutions found earlier. ∎

This second method of solution can be used to solve any linear equation in the variables $\sin\theta$ and $\cos\theta$.

E X A M P L E   1 3

Solving a Trigonometric Equation Linear in sin θ and cos θ

Solve

$$a \sin\theta + b \cos\theta = c, \qquad 0 \le \theta < 2\pi \qquad (2)$$

where a, b, and c are constants and either $a \ne 0$ or $b \ne 0$.

Solution

We divide each side of equation (2) by $\sqrt{a^2 + b^2}$. Then

$$\frac{a}{\sqrt{a^2 + b^2}} \sin\theta + \frac{b}{\sqrt{a^2 + b^2}} \cos\theta = \frac{c}{\sqrt{a^2 + b^2}} \qquad (3)$$

There is a unique angle ϕ, $0 \le \phi < 2\pi$, for which

$$\cos\phi = \frac{a}{\sqrt{a^2 + b^2}} \quad \text{and} \quad \sin\phi = \frac{b}{\sqrt{a^2 + b^2}} \qquad (4)$$

FIGURE 11

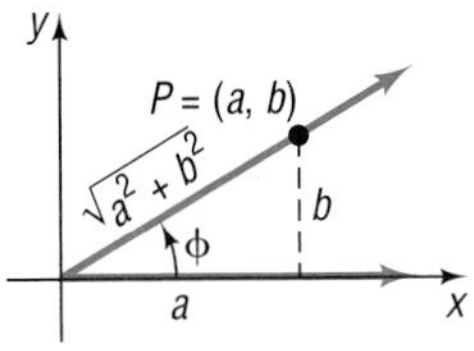

(see Figure 11). Thus, equation (3) may be written as

$$\sin\theta \cos\phi + \cos\theta \sin\phi = \frac{c}{\sqrt{a^2 + b^2}}$$

or, equivalently,

$$\sin(\theta + \phi) = \frac{c}{\sqrt{a^2 + b^2}} \qquad (5)$$

where ϕ satisfies equations (4).

If $|c| > \sqrt{a^2 + b^2}$, then $\sin(\theta + \phi) > 1$ or $\sin(\theta + \phi) < -1$, and equation (5) has no solution.

If $|c| \le \sqrt{a^2 + b^2}$, then the solutions of equation (5) are

$$\theta + \phi = \sin^{-1}\frac{c}{\sqrt{a^2 + b^2}} \quad \text{or} \quad \theta + \phi = \pi - \sin^{-1}\frac{c}{\sqrt{a^2 + b^2}}$$

Because the angle ϕ is determined by equations (4), these are the solutions to equation (2). ■

Graphing Utility Solutions

The techniques introduced in this section only apply to certain types of trigonometric equations. Solutions for other types are usually studied in calculus, using numerical methods. In the next example, we show how a graphing utility may be used to obtain solutions.

E X A M P L E   1 4

Solving Trigonometric Equations Using a Graphing Utility

Solve: $5 \sin x + x = 3$

Express the solution(s) correct to two decimal places.

Solution

This type of trigonometric equation cannot be solved by previous methods. A graphing utility, though, can be used here. We begin by noting that the given equation is equivalent to

$$5 \sin x = 3 - x$$

FIGURE 12

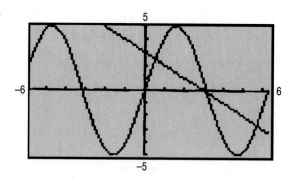

The solution(s) of this equation is the same as the points of intersection of the graphs of $y = 5 \sin x$ and $y = 3 - x$. See Figure 12. There are three points of intersection, whose x-coordinates are the solutions we seek. Using TRACE, ZOOM-IN, and/or BOX, we find

$$x = 0.51 \qquad x = 3.17 \qquad x = 5.71$$

correct to two decimal places.   ■

Note: The solutions of the equation

$$5 \sin x + x = 3$$

can be found in other ways.

1. Graph $y = 5 \sin x + x$ and $y = 3$. Then find the x-coordinates of the points of intersection.
2. Graph the function $f(x) = 5 \sin x + x - 3$. Then find the x-intercepts.

Try these two ways. Decide which of the three you prefer.

7.5

Exercise 7.5

In Problems 1–12, solve each equation on the interval $0 \le \theta < 2\pi$.

1. $\sin \theta = \dfrac{1}{2}$

2. $\tan \theta = 1$

3. $\tan \theta = -\dfrac{1}{\sqrt{3}}$

4. $\cos \theta = -\dfrac{\sqrt{3}}{2}$

5. $\cos \theta = 0$

6. $\sin \theta = \dfrac{\sqrt{2}}{2}$

7. $\sin 3\theta = -1$

8. $\tan \dfrac{\theta}{2} = \sqrt{3}$

9. $\cos\left(2\theta - \dfrac{\pi}{2}\right) = -1$

10. $\sin\left(3\theta + \dfrac{\pi}{18}\right) = 1$

11. $\sec \dfrac{3\theta}{2} = -2$

12. $\cot \dfrac{2\theta}{3} = -\sqrt{3}$

In Problems 13–20, solve each equation on the interval $0 \le \theta < 2\pi$. Round your answer to two decimal places.

13. $\sin \theta = 0.4$

14. $\cos \theta = 0.6$

15. $\tan \theta = 5$

16. $\cot \theta = 2$

17. $\cos \theta = -0.9$

18. $\sin \theta = -0.2$

19. $\sec \theta = -4$

20. $\csc \theta = -3$

In Problems 21–50, solve each equation on the interval $0 \le \theta < 2\pi$.

21. $2 \cos^2 \theta + \cos \theta = 0$

22. $\sin^2 \theta - 1 = 0$

23. $2 \sin^2 \theta - \sin \theta - 1 = 0$

24. $2 \cos^2 \theta + \cos \theta - 1 = 0$

25. $(\tan \theta - 1)(\sec \theta - 1) = 0$

26. $(\cot \theta + 1)(\csc \theta - \tfrac{1}{2}) = 0$

27. $\cos \theta = \sin \theta$

28. $\cos \theta + \sin \theta = 0$

29. $\tan \theta = 2 \sin \theta$

30. $\sin 2\theta = \cos \theta$

31. $\sin \theta = \csc \theta$

32. $\tan \theta = \cot \theta$

33. $\cos 2\theta = \cos \theta$

34. $\sin 2\theta \sin \theta = \cos \theta$

35. $\sin 2\theta + \sin 4\theta = 0$

36. $\cos 2\theta + \cos 4\theta = 0$

37. $\cos 4\theta - \cos 6\theta = 0$

38. $\sin 4\theta - \sin 6\theta = 0$

39. $1 + \sin \theta = 2 \cos^2 \theta$

40. $\sin^2 \theta = 2 \cos \theta + 2$

41. $\tan^2 \theta = \frac{3}{2} \sec \theta$

42. $\csc^2 \theta = \cot \theta + 1$

43. $3 - \sin \theta = \cos 2\theta$

44. $\cos 2\theta + 5 \cos \theta + 3 = 0$

45. $\sec^2 \theta + \tan \theta = 0$

46. $\sec \theta = \tan \theta + \cot \theta$

47. $\sin \theta - \sqrt{3} \cos \theta = 1$

48. $\sqrt{3} \sin \theta + \cos \theta = 1$

49. $\tan 2\theta + 2 \sin \theta = 0$

50. $\tan 2\theta + 2 \cos \theta = 0$

In Problems 51–56, solve each equation for $-\pi \le x \le \pi$. *Express the solution(s) correct to two decimal places.*

51. Solve the equation $\cos x = e^x$ by graphing $y = \cos x$ and $y = e^x$ and finding their point(s) of intersection.

52. Solve the equation $\cos x = e^x$ by graphing $y = \cos x - e^x$ and finding the x-intercept(s).

53. Solve the equation $2 \sin x = 0.7x$ by graphing $y = 2 \sin x$ and $y = 0.7x$ and finding their point(s) of intersection.

54. Solve the equation $2 \sin x = 0.7x$ by graphing $y = 2 \sin x - 0.7x$ and finding the x-intercept(s).

55. Solve the equation $\cos x = x^2$ by graphing $y = \cos x$ and $y = x^2$ and finding their point(s) of intersection.

56. Solve the equation $\cos x = x^2$ by graphing $y = \cos x - x^2$ and finding the x-intercept(s).

In Problems 57–68, use a graphing utility to solve each equation. Express the solution(s) correct to two decimal places.

57. $x + 5 \cos x = 0$

58. $x - 4 \sin x = 0$

59. $22x - 17 \sin x = 3$

60. $19x + 8 \cos x = 2$

61. $\sin x + \cos x = x$

62. $\sin x - \cos x = x$

63. $x^2 - 2 \cos x = 0$

64. $x^2 + 3 \sin x = 0$

65. $x^2 - 2 \sin 2x = 3x$

66. $x^2 = x + 3 \cos 2x$

67. $6 \sin x - e^x = 2,\ x > 0$

68. $4 \cos 3x - e^x = 1,\ x > 0$

69. *Constructing a Rain Gutter* A rain gutter is to be constructed of aluminum sheets 12 inches wide. After marking off a length of 4 inches from each edge, this length is bent up at an angle θ. See the illustration. The area A of the opening as a function of θ is given by

$$A = 16 \sin \theta(\cos \theta + 1),\ 0° \le \theta \le 90°$$

(a) In calculus, you will be asked to find the angle θ that maximizes A by solving the equation

$$\cos 2\theta + \cos \theta = 0,\ 0° \le \theta \le 90°$$

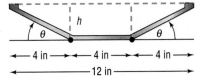

Solve this equation for θ by using the double-angle formula.

(b) Solve the equation for θ by writing the sum of the two cosines as a product.

(c) What is the maximum area A of the opening?

70. *Projectile Motion* An object is propelled upward at an angle θ, $45° < \theta < 90°$, to the horizontal with an initial velocity of v_0 feet per second from the base of a plane that makes an angle of $45°$ with the horizontal. See the illustration. If air resistance is ignored, the distance R it travels up the inclined plane is given by

$$R = \frac{v_0^2 \sqrt{2}}{32} (\sin 2\theta - \cos 2\theta - 1)$$

(a) In calculus, you will be asked to find the angle θ that maximizes R by solving the equation

$$\sin 2\theta + \cos 2\theta = 0$$

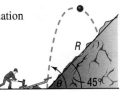

Solve this equation for θ using the method of Example 13.

(b) Solve this equation for θ by dividing each side by $\cos 2\theta$.

(c) What is the maximum distance R if $v_0 = 32$ feet per second?

71. *Heat Transfer* In the study of heat transfer, the equation $x + \tan x = 0$ occurs. Graph $y = -x$ and $y = \tan x$ for $x \geq 0$. Conclude that there are an infinite number of points of intersection of these two graphs. Now find the first two positive solutions of $x + \tan x = 0$ correct to two decimal places.

72. *Carrying a Ladder around a Corner* A ladder of length L is carried horizontally around a corner from a hall 3 feet wide into a hall 4 feet wide. See the illustration.

(a) Express L as a function of θ.

(b) In calculus, you will be asked to find the length of the longest ladder that can turn the corner by solving the equation

$$3 \sec \theta \tan \theta - 4 \csc \theta \cot \theta = 0, \quad 0° < \theta < 90°$$

Solve this equation for θ.

(c) What is the length of the largest ladder that can be carried around the corner?

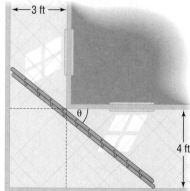

The following discussion of **Snell's Law of Refraction** (named after Willebrod Snell, 1591–1626) is needed for Problems 73–79: Light, sound, and other waves travel at different speeds, depending on the media (air, water, wood, and so on) through which they pass. Suppose that light travels from a point A in one medium, where its speed is v_1, to a point B in another medium, where its speed is v_2. Refer to the figure, where the angle θ_1 is called the **angle of incidence** and the angle θ_2 is the **angle of refraction**. Snell's Law,* which can be proved using calculus, states that

$$\frac{\sin \theta_1}{\sin \theta_2} = \frac{v_1}{v_2}$$

The ratio v_1/v_2 is called the **index of refraction.** Some values are given in the following table.

SOME INDEXES OF REFRACTION

MEDIUM	INDEX OF REFRACTION
Water	1.33
Ethyl alcohol	1.36
Carbon bisulfide	1.63
Air (1 atm and 20°C)	1.0003
Methylene iodide	1.74
Fused quartz	1.46
Glass, crown	1.52
Glass, dense flint	1.66
Sodium chloride	1.53

For light of wavelength 589 nanometers, measured with respect to a vacuum. The index with respect to air is negligibly different in most cases.

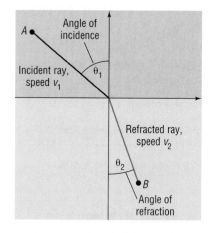

θ_1	θ_2
10°	7°45'
20°	15°30'
30°	22°30'
40°	29°0'
50°	35°0'
60°	40°30'
70°	45°30'
80°	50°0'

73. The index of refraction of light in passing from a vacuum into water is 1.33. If the angle of incidence is 40°, determine the angle of refraction.

74. The index of refraction of light in passing from a vacuum into dense glass is 1.66. If the angle of incidence is 50°, determine the angle of refraction.

75. Ptolemy, who lived in the city of Alexandria in Egypt during the second century AD, gave the measured values in the table in the margin for the angle of incidence θ_1 and the angle of refraction θ_2 for a light beam passing from air into water. Do these values agree with Snell's Law? If so, what index of refraction results? (These data are interesting as the oldest recorded physical measurements.)[†]

*Because this law was also deduced by René Descartes, in France it is known as Descartes' Law.

[†]Adapted from Halliday and Resnick, *Physics, Parts 1 & 2*, 3rd ed. New York: Wiley, 1978, p. 953.

76. The speed of yellow sodium light (wavelength of 589 nanometers) in a certain liquid is measured to be 1.92×10^8 meters per second. What is the index of refraction of this liquid, with respect to air, for sodium light?[*]

77. A beam of light with a wavelength of 589 nanometers traveling in air makes an angle of incidence of 40° upon a slab of transparent material, and the refracted beam makes an angle of refraction of 26°. Find the index of refraction of the material.[†]

78. A light ray with a wavelength of 589 nanometers (produced by a sodium lamp) traveling through air makes an angle of incidence of 30° on a smooth, flat slab of crown glass. Find the angle of refraction.[‡]

79. A light beam passes from one medium to another through a thick slab of material whose index of refraction is n_2. Show that the emerging beam is parallel to the incident beam.[‡]

80. Explain in your own words how you would use your calculator to solve the equation $\sin x = 0.3$, $0 \le x < 2\pi$. How would you modify your approach in order to solve the equation $\cot x = 5$, $0 < x < 2\pi$?

Chapter Review

THINGS TO KNOW

Formulas

Sum and difference formulas	$\cos(\alpha + \beta) = \cos \alpha \cos \beta - \sin \alpha \sin \beta$
	$\cos(\alpha - \beta) = \cos \alpha \cos \beta + \sin \alpha \sin \beta$
	$\sin(\alpha + \beta) = \sin \alpha \cos \beta + \cos \alpha \sin \beta$
	$\sin(\alpha - \beta) = \sin \alpha \cos \beta - \cos \alpha \sin \beta$
	$\tan(\alpha + \beta) = \dfrac{\tan \alpha + \tan \beta}{1 - \tan \alpha \tan \beta}$
	$\tan(\alpha - \beta) = \dfrac{\tan \alpha - \tan \beta}{1 + \tan \alpha \tan \beta}$
Double-angle formulas	$\sin 2\theta = 2 \sin \theta \cos \theta$
	$\cos 2\theta = \cos^2 \theta - \sin^2 \theta$
	$\cos 2\theta = 1 - 2 \sin^2 \theta$
	$\cos 2\theta = 2 \cos^2 \theta - 1$
	$\tan 2\theta = \dfrac{2 \tan \theta}{1 - \tan^2 \theta}$
Half-angle formulas	$\sin^2 \dfrac{\alpha}{2} = \dfrac{1 - \cos \alpha}{2}$
	$\cos^2 \dfrac{\alpha}{2} = \dfrac{1 + \cos \alpha}{2}$
	$\tan^2 \dfrac{\alpha}{2} = \dfrac{1 - \cos \alpha}{1 + \cos \alpha}$

$$\left. \begin{array}{l} \sin \dfrac{\alpha}{2} = \pm \sqrt{\dfrac{1 - \cos \alpha}{2}} \\[2mm] \cos \dfrac{\alpha}{2} = \pm \sqrt{\dfrac{1 + \cos \alpha}{2}} \\[2mm] \tan \dfrac{\alpha}{2} = \pm \sqrt{\dfrac{1 - \cos \alpha}{1 + \cos \alpha}} \end{array} \right\}$$

where the + or − sign is determined by the quadrant of the angle $\alpha/2$

$$= \frac{1 - \cos \alpha}{\sin \alpha} = \frac{\sin \alpha}{1 + \cos \alpha}$$

---

[*]Adapted from Serway, *Physics,* 3rd ed. Philadelphia: W.B. Saunders, p. 805.
[†]Adapted from Serway, *Physics,* 3rd ed. Philadelphia: W.B. Saunders, p. 805.
[‡]Ibid.

Product-to-sum formulas

$$\sin \alpha \sin \beta = \tfrac{1}{2} \left[\cos(\alpha - \beta) - \cos(\alpha + \beta) \right]$$

$$\cos \alpha \cos \beta = \tfrac{1}{2} \left[\cos(\alpha - \beta) + \cos(\alpha + \beta) \right]$$

$$\sin \alpha \cos \beta = \tfrac{1}{2} \left[\sin(\alpha + \beta) + \sin(\alpha - \beta) \right]$$

Sum-to-product formulas

$$\sin \alpha + \sin \beta = 2 \sin \frac{\alpha + \beta}{2} \cos \frac{\alpha - \beta}{2}$$

$$\sin \alpha - \sin \beta = 2 \sin \frac{\alpha - \beta}{2} \cos \frac{\alpha + \beta}{2}$$

$$\cos \alpha + \cos \beta = 2 \cos \frac{\alpha + \beta}{2} \cos \frac{\alpha - \beta}{2}$$

$$\cos \alpha - \cos \beta = -2 \sin \frac{\alpha + \beta}{2} \sin \frac{\alpha - \beta}{2}$$

How To

Establish identities
Solve a trigonometric equation

Fill-in-the-Blank Items

1. Suppose that f and g are two functions with the same domain. If $f(x) = g(x)$ for every x in the domain, the equation is called a(n) _____. Otherwise, it is called a(n) _____ equation.

2. $\cos(\alpha + \beta) = \cos \alpha \cos \beta$ _____ $\sin \alpha \sin \beta$

3. $\sin(\alpha + \beta) = \sin \alpha \cos \beta$ _____ $\cos \alpha \sin \beta$

4. $\cos 2\theta = \cos^2 \theta -$ _____ $=$ _____ $- 1 = 1 -$ _____

5. $\sin^2 \dfrac{\alpha}{2} = \dfrac{}{2}$

True/False Items

T   F   1. $\sin(-\theta) + \sin \theta = 0$ for all θ.

T   F   2. $\sin(\alpha + \beta) = \sin \alpha + \sin \beta + 2 \sin \alpha \sin \beta$.

T   F   3. $\cos 2\theta$ has three equivalent forms: $\cos^2 \theta - \sin^2 \theta$, $1 - 2 \sin^2 \theta$, and $2 \cos^2 \theta - 1$.

T   F   4. $\cos \dfrac{\alpha}{2} = \pm \dfrac{\sqrt{1 + \cos \alpha}}{2}$, where the $+$ or $-$ sign depends on the angle $\alpha/2$.

T   F   5. Most trigonometric equations have unique solutions.

T   F   6. The equation $\tan \theta = \pi/2$ has no solution.

Review Exercises

In Problems 1–32, establish each identity.

1. $\tan \theta \cot \theta - \sin^2 \theta = \cos^2 \theta$

2. $\sin \theta \csc \theta - \sin^2 \theta = \cos^2 \theta$

3. $\cos^2 \theta (1 + \tan^2 \theta) = 1$

4. $(1 - \cos^2 \theta)(1 + \cot^2 \theta) = 1$

5. $4 \cos^2 \theta + 3 \sin^2 \theta = 3 + \cos^2 \theta$

6. $4 \sin^2 \theta + 2 \cos^2 \theta = 4 - 2 \cos^2 \theta$

7. $\dfrac{1 - \cos \theta}{\sin \theta} + \dfrac{\sin \theta}{1 - \cos \theta} = 2 \csc \theta$

8. $\dfrac{\sin \theta}{1 + \cos \theta} + \dfrac{1 + \cos \theta}{\sin \theta} = 2 \csc \theta$

9. $\dfrac{\cos \theta}{\cos \theta - \sin \theta} = \dfrac{1}{1 - \tan \theta}$

10. $1 - \dfrac{\cos^2 \theta}{1 + \sin \theta} = \sin \theta$

11. $\dfrac{\csc \theta}{1 + \csc \theta} = \dfrac{1 - \sin \theta}{\cos^2 \theta}$

12. $\dfrac{1 + \sec \theta}{\sec \theta} = \dfrac{\sin^2 \theta}{1 - \cos \theta}$

13. $\csc \theta - \sin \theta = \cos \theta \cot \theta$

14. $\dfrac{\csc \theta}{1 - \cos \theta} = \dfrac{1 + \cos \theta}{\sin^3 \theta}$

15. $\dfrac{1 - \sin \theta}{\sec \theta} = \dfrac{\cos^3 \theta}{1 + \sin \theta}$

16. $\dfrac{1 - \cos \theta}{1 + \cos \theta} = (\csc \theta - \cot \theta)^2$

17. $\dfrac{1 - 2 \sin^2 \theta}{\sin \theta \cos \theta} = \cot \theta - \tan \theta$

18. $\dfrac{(2 \sin^2 \theta - 1)^2}{\sin^4 \theta - \cos^4 \theta} = 1 - 2 \cos^2 \theta$

19. $\dfrac{\cos(\alpha + \beta)}{\cos \alpha \sin \beta} = \cot \beta - \tan \alpha$

20. $\dfrac{\sin(\alpha - \beta)}{\sin \alpha \cos \beta} = 1 - \cot \alpha \tan \beta$

21. $\dfrac{\cos(\alpha - \beta)}{\cos \alpha \cos \beta} = 1 + \tan \alpha \tan \beta$

22. $\dfrac{\cos(\alpha + \beta)}{\sin \alpha \cos \beta} = \cot \alpha - \tan \beta$

23. $(1 + \cos \theta)\left(\tan \dfrac{\theta}{2}\right) = \sin \theta$

24. $\sin \theta \tan \dfrac{\theta}{2} = 1 - \cos \theta$

25. $2 \cot \theta \cot 2\theta = \cot^2 \theta - 1$

26. $2 \sin 2\theta(1 - 2 \sin^2 \theta) = \sin 4\theta$

27. $1 - 8 \sin^2 \theta \cos^2 \theta = \cos 4\theta$

28. $\dfrac{\sin 3\theta \cos \theta - \sin \theta \cos 3\theta}{\sin 2\theta} = 1$

29. $\dfrac{\sin 2\theta + \sin 4\theta}{\cos 2\theta + \cos 4\theta} = \tan 3\theta$

30. $\dfrac{\sin 2\theta + \sin 4\theta}{\sin 2\theta - \sin 4\theta} + \dfrac{\tan 3\theta}{\tan \theta} = 0$

31. $\dfrac{\cos 2\theta - \cos 4\theta}{\cos 2\theta + \cos 4\theta} - \tan \theta \tan 3\theta = 0$

32. $\cos 2\theta - \cos 10\theta = (\tan 4\theta)(\sin 2\theta + \sin 10\theta)$

In Problems 33–40, find the exact value of each expression.

33. $\sin 165°$

34. $\tan 105°$

35. $\cos \dfrac{5\pi}{12}$

36. $\sin\left(-\dfrac{\pi}{12}\right)$

37. $\cos 80° \cos 20° + \sin 80° \sin 20°$

38. $\sin 70° \cos 40° - \cos 70° \sin 40°$

39. $\tan \dfrac{\pi}{8}$

40. $\sin \dfrac{5\pi}{8}$

In Problems 41–50, use the information given about the angles α and β to find the exact value of:

(a) $\sin(\alpha + \beta)$ (b) $\cos(\alpha + \beta)$ (c) $\sin(\alpha - \beta)$ (d) $\tan(\alpha + \beta)$

(e) $\sin 2\alpha$ (f) $\cos 2\beta$ (g) $\sin \dfrac{\beta}{2}$ (h) $\cos \dfrac{\alpha}{2}$

41. $\sin \alpha = \frac{4}{5}$, $0 < \alpha < \pi/2$; $\sin \beta = \frac{5}{13}$, $\pi/2 < \beta < \pi$

42. $\cos \alpha = \frac{4}{5}$, $0 < \alpha < \pi/2$; $\cos \beta = \frac{5}{13}$, $-\pi/2 < \beta < 0$

43. $\sin \alpha = -\frac{3}{5}$, $\pi < \alpha < 3\pi/2$; $\cos \beta = \frac{12}{13}$, $3\pi/2 < \beta < 2\pi$

44. $\sin \alpha = -\frac{4}{5}$, $-\pi/2 < \alpha < 0$; $\cos \beta = -\frac{5}{13}$, $\pi/2 < \beta < \pi$

45. $\tan \alpha = \frac{3}{4}$, $\pi < \alpha < 3\pi/2$; $\tan \beta = \frac{12}{5}$, $0 < \beta < \pi/2$

46. $\tan \alpha = -\frac{4}{3}$, $\pi/2 < \alpha < \pi$; $\cot \beta = \frac{12}{5}$, $\pi < \beta < 3\pi/2$

47. $\sec \alpha = 2$, $-\pi/2 < \alpha < 0$; $\sec \beta = 3$, $3\pi/2 < \beta < 2\pi$

48. $\csc \alpha = 2$, $\pi/2 < \alpha < \pi$; $\sec \beta = -3$, $\pi/2 < \beta < \pi$

49. $\sin \alpha = -\frac{2}{3}$, $\pi < \alpha < 3\pi/2$; $\cos \beta = -\frac{2}{3}$, $\pi < \beta < 3\pi/2$

50. $\tan \alpha = -2$, $\pi/2 < \alpha < \pi$; $\cot \beta = -2$, $\pi/2 < \beta < \pi$

In Problems 51–70, solve each equation on the interval $0 \leq \theta < 2\pi$.

51. $\cos \theta = \frac{1}{2}$

52. $\sin \theta = -\sqrt{3}/2$

53. $\cos \theta = -\sqrt{2}/2$

54. $\tan \theta = -\sqrt{3}$

55. $\sin 2\theta = -1$

56. $\cos 2\theta = 0$

57. $\tan 2\theta = 0$

58. $\sin 3\theta = 1$

59. $\sin \theta = 0.9$

60. $\tan \theta = 25$

61. $\sin \theta = \tan \theta$

62. $\cos \theta = \sec \theta$

63. $\sin \theta + \sin 2\theta = 0$

64. $\cos 2\theta = \sin \theta$

65. $\sin 2\theta - \cos \theta - 2 \sin \theta + 1 = 0$

66. $\sin 2\theta - \sin \theta - 2 \cos \theta + 1 = 0$

67. $2 \sin^2 \theta - 3 \sin \theta + 1 = 0$

68. $2 \cos^2 \theta + \cos \theta - 1 = 0$

69. $\sin \theta - \cos \theta = 1$

70. $\sin \theta + 2 \cos \theta = 1$

Chapter 8

PREPARING FOR THIS CHAPTER

Before getting started on this chapter, review the following concepts:

Solving right triangles (p. 369)
Rectangular coordinates; Graphing equations (Section 1.6)
Complex numbers (Section 1.5)

Preview   Correcting a Navigational Error

A motorized sailboat leaves Naples, Florida, bound for Key West, 150 miles away. Maintaining a constant speed of 15 miles per hour, but encountering heavy crosswinds and strong currents, the crew finds, after 4 hours, that the sailboat is off course by 20°.

(a) How far is the sailboat from Key West at this time?

(b) Through what angle should the sailboat turn to correct its course?

(c) How much time has been added to the trip because of this?

(Assume that the speed remains at 15 miles per hour.)

[Example 3 in Section 8.2]   ■

J n Chapter 5, we used the trigonometric functions to solve right triangles—that is, triangles with a 90° angle. In this chapter, we will use the trigonometric functions to solve oblique triangles—that is, triangles that do not have a 90° angle. To solve such triangles, we shall develop the Law of Sines (in Section 8.1) and the Law of Cosines (in Section 8.2). In addition to finding the sides and angles of such triangles, we shall derive formulas for finding their areas (in Section 8.3).

The final sections of this chapter deal with polar coordinates (an alternative to rectangular coordinates for plotting points), graphing in polar coordinates, and finding roots of complex numbers (Demoivre's Theorem).

The Law of Sines

If none of the angles of a triangle is a right angle, the triangle is called **oblique.** Thus, an oblique triangle will have either three acute angles or two acute angles and one obtuse angle (an angle between 90° and 180°). See Figure 1.

FIGURE 1
Oblique triangles

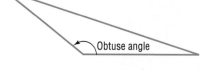

(a) All angles are acute   **(b)** Two acute angles and one obtuse angle

FIGURE 2

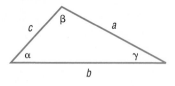

In the discussion that follows, we shall always label an oblique triangle so that side a is opposite angle α, side b is opposite angle β, and side c is opposite angle γ, as shown in Figure 2.

To **solve an oblique triangle** means to find the lengths of its sides and the measurements of its angles. To do this, we shall need to know the length of one side along with two other facts: either two angles, or the other two sides, or one angle and one other side.* Thus, there are four possibilities to consider:

CASE 1: One side and two angles are known (SAA or ASA).
CASE 2: Two sides and the angle opposite one of them are known (SSA).
CASE 3: Two sides and the included angle are known (SAS).
CASE 4: Three sides are known (SSS).

Figure 3 illustrates the four cases.

FIGURE 3

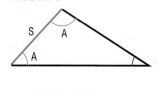
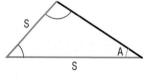
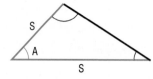
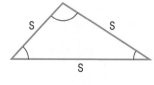

Case 1: SAA or ASA   Case 2: SSA   Case 3: SAS   Case 4: SSS

The **Law of Sines** is used to solve triangles for which Case 1 or 2 holds.

---

*Recall from plane geometry the fact that knowing three angles of a triangle determines a family of *similar triangles*—that is, triangles that have the same shape but different sizes.

Theorem
Law of Sines

For a triangle with sides a, b, c and opposite angles α, β, γ, respectively,

$$\frac{\sin \alpha}{a} = \frac{\sin \beta}{b} = \frac{\sin \gamma}{c} \qquad (1)$$

Proof

To prove the Law of Sines, we construct an altitude of length h from one of the vertices of such a triangle. Figure 4(a) shows h for a triangle with three acute angles, and Figure 4(b) shows h for a triangle with an obtuse angle. In each case, the altitude is drawn from the vertex at β. Using either illustration, we have

$$\sin \gamma = \frac{h}{a}$$

from which

$$h = a \sin \gamma \qquad (2)$$

From Figure 4(a), it also follows that

$$\sin \alpha = \frac{h}{c}$$

from which

$$h = c \sin \alpha \qquad (3)$$

From Figure 4(b), it follows that

$$\sin \alpha = \sin(180° - \alpha) = \frac{h}{c}$$

which again gives

$$h = c \sin \alpha$$

Thus, whether the triangle has three acute angles or has two acute angles and one obtuse angle, equations (2) and (3) hold. As a result, we may equate the expressions for h in equations (2) and (3) to get

$$a \sin \gamma = c \sin \alpha$$

from which

$$\frac{\sin \alpha}{a} = \frac{\sin \gamma}{c} \qquad (4)$$

In a similar manner, by constructing the altitude h' from the vertex of angle α as shown in Figure 5, we can show that

$$\sin \beta = \frac{h'}{c} \quad \text{and} \quad \sin \gamma = \frac{h'}{b}$$

Thus,

$$h' = c \sin \beta = b \sin \gamma$$

and

$$\frac{\sin \beta}{b} = \frac{\sin \gamma}{c} \qquad (5)$$

FIGURE 4

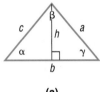

(a)

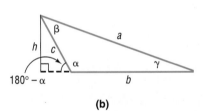

(b)

FIGURE 5

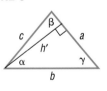

(a)

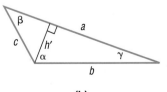

(b)

When equations (4) and (5) are combined, we have equation (1), the Law of Sines.   ■

In applying the Law of Sines to solve triangles, we use the fact that the sum of the angles of any triangle equals 180°; that is,

$$\alpha + \beta + \gamma = 180° \tag{6}$$

Our first two examples show how to solve a triangle when one side and two angles are known (Case 1: SAA or ASA).

E X A M P L E   1   *Using the Law of Sines to Solve a SAA Triangle*

Solve the triangle: $\alpha = 40°$, $\beta = 60°$, $a = 4$

Solution   Figure 6 shows the triangle that we want to solve. The third angle γ is easily found using equation (6):

$$\alpha + \beta + \gamma = 180°$$
$$40° + 60° + \gamma = 180°$$
$$\gamma = 80°$$

FIGURE 6

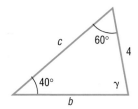

Now we use the Law of Sines (twice) to find the unknown sides b and c:

$$\frac{\sin \alpha}{a} = \frac{\sin \beta}{b} \qquad \frac{\sin \alpha}{a} = \frac{\sin \gamma}{c}$$

Because $a = 4$, $\alpha = 40°$, $\beta = 60°$, and $\gamma = 80°$, we have

$$\frac{\sin 40°}{4} = \frac{\sin 60°}{b} \qquad \frac{\sin 40°}{4} = \frac{\sin 80°}{c}$$

Thus,

$$b = \frac{4 \sin 60°}{\sin 40°} \approx 5.39 \qquad c = \frac{4 \sin 80°}{\sin 40°} \approx 6.13$$

From a calculator   From a calculator   ■

Notice that in Example 1 we found b and c by working with the given side a. This is better than finding b first and working with a rounded value of b to find c.

E X A M P L E   2   *Using the Law of Sines to Solve an ASA Triangle*

Solve the triangle: $\alpha = 35°$, $\beta = 15°$, $c = 5$

Solution   Figure 7 illustrates the triangle that we want to solve. Because we know two angles ($\alpha = 35°$ and $\beta = 15°$), it is easy to find the third angle using equation (6):

$$\alpha + \beta + \gamma = 180°$$
$$35° + 15° + \gamma = 180°$$
$$\gamma = 130°$$

FIGURE 7

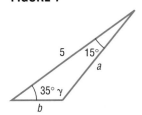

Now we know the three angles and one side ($c = 5$) of the triangle. To find the remaining two sides a and b, we use the Law of Sines (twice):

$$\frac{\sin \alpha}{a} = \frac{\sin \gamma}{c} \qquad\qquad \frac{\sin \beta}{b} = \frac{\sin \gamma}{c}$$

$$\frac{\sin 35°}{a} = \frac{\sin 130°}{5} \qquad\qquad \frac{\sin 15°}{b} = \frac{\sin 130°}{5}$$

$$a = \frac{5 \sin 35°}{\sin 130°} \approx 3.74 \qquad\qquad b = \frac{5 \sin 15°}{\sin 130°} \approx 1.69 \qquad \blacksquare$$

Note: In subsequent examples and in the exercises that follow, unless otherwise indicated, we shall measure angles in degrees and round off to one decimal place; we shall round off all sides to two decimal places. To avoid round-off errors when using a calculator, we will store unrounded values in memory for use in subsequent calculations.

■ Now work Problem 1.

The Ambiguous Case

FIGURE 8

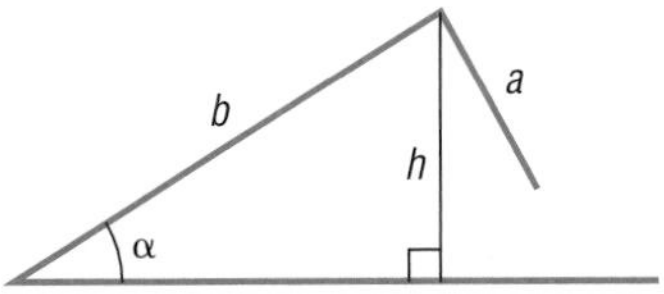

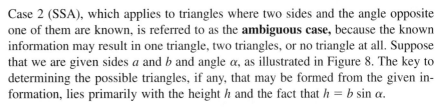

Case 2 (SSA), which applies to triangles where two sides and the angle opposite one of them are known, is referred to as the **ambiguous case,** because the known information may result in one triangle, two triangles, or no triangle at all. Suppose that we are given sides a and b and angle α, as illustrated in Figure 8. The key to determining the possible triangles, if any, that may be formed from the given information, lies primarily with the height h and the fact that $h = b \sin \alpha$.

No Triangle: If $a < b \sin \alpha = h$, then clearly side a is not sufficiently long to form a triangle. See Figure 9.

One Right Triangle: If $a = b \sin \alpha = h$, then side a is just long enough to form a right triangle. See Figure 10.

FIGURE 9
$a < b \sin \alpha$

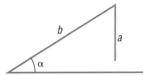

FIGURE 10
$a = b \sin \alpha$

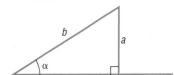

Two Triangles: If $a < b$ and $h = b \sin \alpha < a$, then two distinct triangles can be formed from the given information. See Figure 11.

One Triangle: If $a \geq b$, then only one triangle can be formed. See Figure 12.

FIGURE 11
$b \sin \alpha < a$ and $a < b$

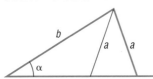

FIGURE 12
$a \geq b$

Fortunately, we do not have to rely on an illustration to draw the correct conclusion in the ambiguous case. The Law of Sines will lead us to the correct determination. Let's see how.

E X A M P L E   3   *Using the Law of Sines to Solve a SSA Triangle (One Solution)*

Solve the triangle: $a = 3$, $b = 2$, $\alpha = 40°$

Solution   See Figure 13(a). Because $a = 3$, $b = 2$, and $\alpha = 40°$ are known, we use the Law of Sines to find the angle β:

$$\frac{\sin \alpha}{a} = \frac{\sin \beta}{b}$$

FIGURE 13(a)

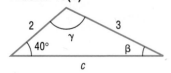

Then

$$\frac{\sin 40°}{3} = \frac{\sin \beta}{2}$$

$$\sin \beta = \frac{2 \sin 40°}{3} \approx 0.43$$

There are two angles β, $0° < \beta < 180°$, for which $\sin \beta \approx 0.43$:

$$\beta \approx 25.4° \quad \text{and} \quad \beta \approx 154.6°$$

[**Note:** Here we computed β using the stored value of $\sin \beta$. If you use the rounded value, $\sin \beta \approx 0.43$, you will obtain slightly different results.]

The second possibility is ruled out, because $\alpha = 40°$, making $\alpha + \beta \approx 194.6° > 180°$. Now, using $\beta \approx 25.4°$, we find

$$\gamma = 180° - \alpha - \beta \approx 180° - 40° - 25.4° = 114.6°$$

The third side c may now be determined using the Law of Sines:

$$\frac{\sin \gamma}{c} = \frac{\sin \alpha}{a}$$

$$\frac{\sin 114.6°}{c} = \frac{\sin 40°}{3}$$

$$c = \frac{3 \sin 114.6°}{\sin 40°} \approx 4.24$$

FIGURE 13(b)

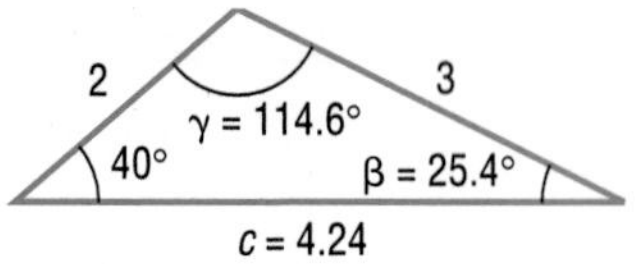

Figure 13(b) illustrates the solved triangle.   ■

E X A M P L E   4   *Using the Law of Sines to Solve a SSA Triangle (Two Solutions)*

Solve the triangle: $a = 6$, $b = 8$, $\alpha = 35°$

Solution   Because $a = 6$, $b = 8$, and $\alpha = 35°$ are known, we use the Law of Sines to find the angle β:

$$\frac{\sin \alpha}{a} = \frac{\sin \beta}{b}$$

Then

$$\frac{\sin 35°}{6} = \frac{\sin \beta}{8}$$

$$\sin \beta = \frac{8 \sin 35°}{6} \approx 0.76$$

$$\beta_1 \approx 49.9° \quad \text{or} \quad \beta_2 \approx 130.1°$$

For both possibilities we have $\alpha + \beta < 180°$. Hence, there are two triangles—one containing the angle $\beta = \beta_1 \approx 49.9°$ and the other containing the angle $\beta = \beta_2 \approx 130.1°$. The third angle γ is either

$$\gamma_1 = 180° - \alpha - \beta_1 \approx 95.1° \quad \text{or} \quad \gamma_2 = 180° - \alpha - \beta_2 \approx 14.9°$$

$$\begin{array}{cc} \uparrow & \uparrow \\ \alpha = 35° & \alpha = 35° \\ \beta_1 = 49.9° & \beta_2 = 130.1° \end{array}$$

The third side c obeys the Law of Sines, so we have

FIGURE 14

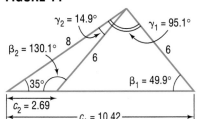

$$\frac{\sin \gamma}{c} = \frac{\sin \alpha}{a}$$

$$\frac{\sin 95.1°}{c_1} = \frac{\sin 35°}{6} \qquad \text{or} \qquad \frac{\sin 14.9°}{c_2} = \frac{\sin 35°}{6}$$

$$c_1 = \frac{6 \sin 95.1°}{\sin 35°} \approx 10.42 \qquad c_2 = \frac{6 \sin 14.9°}{\sin 35°} \approx 2.69$$

The two solved triangles are illustrated in Figure 14. ∎

EXAMPLE 5   *Using the Law of Sines to Solve a SSA Triangle (No Solution)*

Solve the triangle: $a = 2, c = 1, \gamma = 50°$

Solution   Because $a = 2$, $c = 1$, and $\gamma = 50°$ are known, we use the Law of Sines to find the angle α:

$$\frac{\sin \alpha}{a} = \frac{\sin \gamma}{c}$$

FIGURE 15

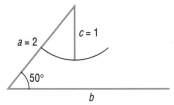

$$\frac{\sin \alpha}{2} = \frac{\sin 50°}{1}$$

$$\sin \alpha = 2 \sin 50° \approx 1.53$$

There is no angle α for which $\sin \alpha > 1$. Hence, there can be no triangle with the given measurements. Figure 15 illustrates the measurements given. Notice that, no matter how we attempt to position side c, it will never intersect side b to form a triangle. ∎

■ Now work Problem 17.

Applied Problems

The Law of Sines is particularly useful for solving certain applied problems.

E X A M P L E 6

Rescue at Sea

Coast Guard Station Zulu is located 120 miles due west of Station X-ray. A ship at sea sends an SOS call that is received by each station. The call to Station Zulu indicates that the location of the ship is 40° east of north; the call to Station X-ray indicates that the location of the ship is 30° west of north.

(a) How far is each station from the ship?

(b) If a helicopter capable of flying 200 miles per hour is dispatched from the nearest station to the ship, how long will it take to reach the ship?

Solution (a) Figure 16 illustrates the situation. The angle γ is found to be

$$\gamma = 180° - 50° - 60° = 70°$$

FIGURE 16

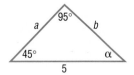

The Law of Sines can now be used to find the two distances a and b that we seek:

$$\frac{\sin 50°}{a} = \frac{\sin 70°}{120}$$

$$a = \frac{120 \sin 50°}{\sin 70°} \approx 97.82 \text{ miles}$$

$$\frac{\sin 60°}{b} = \frac{\sin 70°}{120}$$

$$b = \frac{120 \sin 60°}{\sin 70°} \approx 110.59 \text{ miles}$$

Thus, Station Zulu is about 111 miles from the ship, and Station X-ray is about 98 miles from the ship.

(b) The time t needed for the helicopter to reach the ship from Station X-ray is found by using the formula

$$(\text{Velocity, } v)(\text{Time, } t) = \text{Distance, } a$$

Then

$$t = \frac{a}{v} = \frac{97.82}{200} \approx 0.49 \text{ hour} \approx 29 \text{ minutes}$$

It will take about 29 minutes for the helicopter to reach the ship. ■

■ Now work Problem 29.

8.1

Exercise 8.1

In Problems 1–8, solve each triangle.

1.
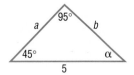

2.
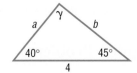

3.
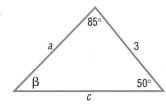

4.

5.

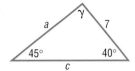

6.

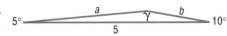

7.

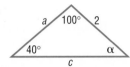

8.

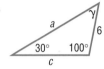

In Problems 9–16, solve each triangle.

9. $\alpha = 40°$, $\beta = 20°$, $a = 2$

10. $\alpha = 50°$, $\gamma = 20°$, $a = 3$

11. $\beta = 70°$, $\gamma = 10°$, $b = 5$

12. $\alpha = 70°$, $\beta = 60°$, $c = 4$

13. $\alpha = 110°$, $\gamma = 30°$, $c = 3$

14. $\beta = 10°$, $\gamma = 100°$, $b = 2$

15. $\alpha = 40°$, $\beta = 40°$, $c = 2$

16. $\beta = 20°$, $\gamma = 70°$, $a = 1$

In Problems 17–28, two sides and an angle are given. Determine whether the given information results in one triangle, two triangles, or no triangle at all. Solve any triangle(s) that results.

17. $a = 3$, $b = 2$, $\alpha = 50°$

18. $b = 4$, $c = 3$, $\beta = 40°$

19. $b = 5$, $c = 3$, $\beta = 100°$

20. $a = 2$, $c = 1$, $\alpha = 120°$

21. $a = 4$, $b = 5$, $\alpha = 60°$

22. $b = 2$, $c = 3$, $\beta = 40°$

23. $b = 4$, $c = 6$, $\beta = 20°$

24. $a = 3$, $b = 7$, $\alpha = 70°$

25. $a = 2$, $c = 1$, $\gamma = 100°$

26. $b = 4$, $c = 5$, $\beta = 95°$

27. $a = 2$, $c = 1$, $\gamma = 25°$

28. $b = 4$, $c = 5$, $\beta = 40°$

29. *Rescue at Sea* Coast Guard Station Able is located 150 miles due south of Station Baker. A ship at sea sends an SOS call that is received by each station. The call to Station Able indicates that the ship is located 35° north of east; the call to Station Baker indicates that the ship is located 30° south of east.
(a) How far is each station from the ship?
(b) If a helicopter capable of flying 200 miles per hour is dispatched from the nearest station to the ship, how long will it take to reach the ship?

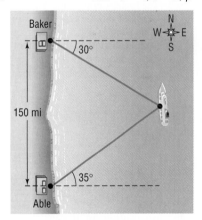

30. *Surveying* Consult the figure. To find the distance from the house at A to the house at B, a surveyor measures the angle BAC to be 40°, then walks off a distance of 100 feet to C, and measures the angle ACB to be 50°. What is the distance from A to B?

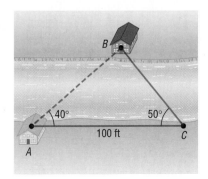

31. *Finding the Length of a Ski Lift* Consult the figure. To find the length of the span of a proposed ski lift from *A* to *B*, a surveyor measures the angle *DAB* to be 25°, then walks off a distance of 1000 feet to *C*, and measures the angle *ACB* to be 15°. What is the distance from *A* to *B*?

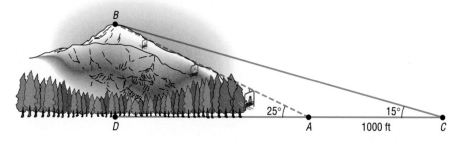

32. *Finding the Height of a Mountain* Use the illustration in Problem 31 to find the height *BD* of the mountain at *B*.

33. *Finding the Height of an Airplane* An aircraft is spotted by two observers who are 1000 feet apart. As the airplane passes over the line joining them, each observer takes a sighting of the angle of elevation to the plane, as indicated in the figure. How high is the airplane?

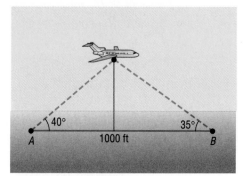

34. *Finding the Height of the Bridge Over the Royal Gorge* The highest bridge in the world is the bridge over the Royal Gorge of the Arkansas River in Colorado.* Sightings to the same point at water level directly under the bridge are taken from each side of the 880-foot-long bridge, as indicated in the figure. How high is the bridge?

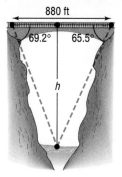

35. *Navigation* An airplane flies from city *A* to city *B*, a distance of 150 miles, then turns through an angle of 40° and heads toward city *C*, as shown in the figure.
(a) If the distance between cities *A* and *C* is 300 miles, how far is it from city *B* to city *C*?
(b) Through what angle should the pilot turn at city *C* to return to City *A*?

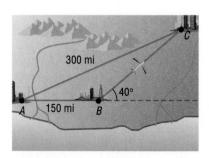

Source: Guinness Book of World Records.

36. *Time Lost due to a Navigation Error* In attempting to fly from city *A* to city *B*, an aircraft followed a course that was 10° in error, as indicated in the figure. After flying a distance of 50 miles, the pilot corrected the course by turning at point *C* and flying 70 miles farther. If the constant speed of the aircraft was 250 miles per hour, how much time was lost due to the error?

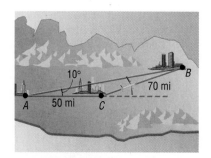

37. *Finding the Lean of the Leaning Tower of Pisa* The famous Leaning Tower of Pisa was originally 184.5 feet high.* After walking 123 feet from the base of the tower, the angle of elevation to the top of the tower is found to be 60°. Find the angle *CAB* indicated in the figure. Also, find the perpendicular distance from *C* to *AB*.

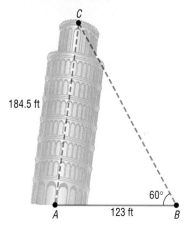

38. *Crankshafts on Cars* On a certain automobile, the crankshaft is 3 inches long and the connecting rod is 9 inches long (see the figure). At the time when the angle *OPA* is 15°, how far is the piston (*P*) from the center (*O*) of the crankshaft?

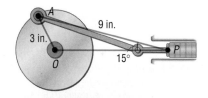

39. *Constructing a Highway* U.S. 41, a highway whose primary directions are north–south, is being constructed along the west coast of Florida. Near Naples, a bay obstructs the straight path of the road. Since the cost of a bridge is prohibitive, engineers decide to go around the bay. The illustration shows the path that they decide on and the measurements taken. What is the length of highway needed to go around the bay?

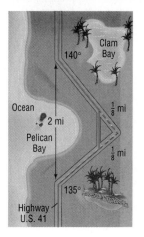

*In their 1986 report on the fragile 7-century-old bell tower, scientists in Pisa, Italy, said the Leaning Tower of Pisa had increased its famous lean by 1 millimeter, or 0.04 inch. That is about the annual average, although the tilting has slowed to about half that much in the previous 2 years. (*Source:* United Press International, June 29, 1986.)

   Update PISA, ITALY. September, 1995. The Leaning Tower of Pisa has suddenly shifted, jeopardizing years of preservation work to stabilize it, Italian newspapers said Sunday. The tower, built on shifting subsoil between 1174 and 1350 as a belfry for the nearby cathedral, recently moved .07 inches in one night. The tower has been closed to tourists since 1990 but officials hope to partially reopen it next year.

40. *Determining Distances at Sea* The navigator of a ship at sea spots two lighthouses that she knows to be 3 miles apart along a straight seashore. She determines that the angles formed between two line-of-sight observations of the light-houses and the line from the ship directly to shore are 15° and 35°. See the illustration.
(a) How far is the ship from lighthouse *A*?
(b) How far is the ship from lighthouse *B*?
(c) How far is the ship from shore?

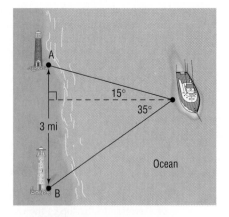

41. *Calculating Distances at Sea* The navigator of a ship at sea has the harbor in sight at which the ship is to dock. She spots a lighthouse she knows is 1 mile down the coast from the mouth of the harbor, and she measures the angle between the line-of-sight observations of the harbor and lighthouse to be 20°. With the ship heading directly toward the harbor, she repeats this measurement after 5 minutes of traveling at 12 miles per hour. If the new angle is 30°, how far is the ship from the harbor?

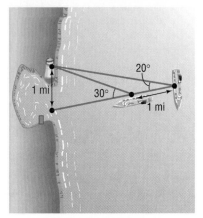

42. *Finding Distances* A forest ranger is walking on a path inclined at 5° to the horizontal directly toward a 100-foot-tall fire observation tower. The angle of elevation of the top of the tower is 40°. How far is the ranger from the tower at this time?

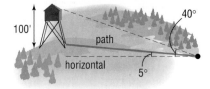

43. *Mollweide's Formula* For any triangle, **Mollweide's Formula** (named after Karl Mollweide, 1774–1825) states that

$$\frac{a + b}{c} = \frac{\cos \frac{1}{2}(\alpha - \beta)}{\sin \frac{1}{2}\gamma}$$

Derive it. [*Hint:* Use the Law of Sines and then a sum-to-product formula.] Notice that this formula involves all six parts of a triangle. As a result, it is sometimes used to check the solution of a triangle.

44. *Mollweide's Formula* Another form of Mollweide's Formula is

$$\frac{a - b}{c} = \frac{\sin \frac{1}{2}(\alpha - \beta)}{\cos \frac{1}{2}\gamma}$$

Derive it.

45. For any triangle, derive the formula

$$a = b \cos \gamma + c \cos \beta$$

[*Hint:* Use the fact that $\sin \alpha = \sin(180° - \beta - \gamma)$.]

46. *Law of Tangents* For any triangle, derive the **Law of Tangents:**

$$\frac{a - b}{a + b} = \frac{\tan \frac{1}{2}(\alpha - \beta)}{\tan \frac{1}{2}(\alpha + \beta)}$$

[*Hint:* Use Mollweide's Formula.]

47. *Circumscribing a Triangle* Show that

$$\frac{\sin \alpha}{a} = \frac{\sin \beta}{b} = \frac{\sin \gamma}{c} = \frac{1}{2r}$$

where r is the radius of the circle circumscribing the triangle ABC whose sides are a, b, c, as shown in the figure in the margin. [*Hint:* Draw the diameter AB'. Then $\beta =$ angle $ABC =$ angle $AB'C$ and angle $ACB' = 90°$.]

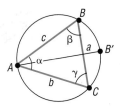

 48. Make up three problems involving oblique triangles. One should result in one triangle, the second in two triangles, and the third in no triangle.

8.2

The Law of Cosines

In Section 8.1, we used the Law of Sines to solve Case 1 (SAA or ASA) and Case 2 (SSA) of an oblique triangle. In this section, we derive the Law of Cosines and use it to solve the remaining cases, 3 and 4.

CASE 3: Two sides and the included angle are known (SAS).
CASE 4: Three sides are known (SSS).

Theorem
Law of Cosines

For a triangle with sides a, b, c and opposite angles α, β, γ, respectively.

$$c^2 = a^2 + b^2 - 2ab \cos \gamma \qquad (1)$$
$$b^2 = a^2 + c^2 - 2ac \cos \beta \qquad (2)$$
$$a^2 = b^2 + c^2 - 2bc \cos \alpha \qquad (3)$$

Proof

We shall prove only formula (1) here. Formulas (2) and (3) may be proved using the same argument.

We begin by strategically placing a triangle on a rectangular coordinate system so that the vertex of angle γ is at the origin and side b lies along the positive x-axis. Regardless of whether γ is acute, as in Figure 17(a), or obtuse, as in Figure 17(b), the vertex B has coordinates $(a \cos \gamma, a \sin \gamma)$. Vertex A has coordinates $(b, 0)$.

FIGURE 17

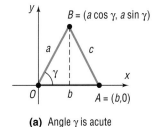

(a) Angle γ is acute

(b) Angle γ is obtuse

We can now use the distance formula to compute c^2:

$$\begin{aligned} c^2 &= (b - a\cos\gamma)^2 + (0 - a\sin\gamma)^2 \\ &= b^2 - 2ab\cos\gamma + a^2\cos^2\gamma + a^2\sin^2\gamma \\ &= a^2(\cos^2\gamma + \sin^2\gamma) + b^2 - 2ab\cos\gamma \\ &= a^2 + b^2 - 2ab\cos\gamma \end{aligned}$$

■

Each of formulas (1), (2), and (3) may be stated in words as follows:

Theorem Law of Cosines The square of one side of a triangle equals the sum of the squares of the other two sides minus twice their product times the cosine of their included angle. ■

Observe that if the triangle is a right triangle (so that, say, $\gamma = 90°$), then formula (1) becomes the familiar Pythagorean Theorem: $c^2 = a^2 + b^2$. Thus, the Pythagorean Theorem is a special case of the Law of Cosines.

Let's see how to use the Law of Cosines to solve Case 3 (SAS), which applies to triangles for which two sides and the included angle are known.

EXAMPLE 1

Using the Law of Cosines to Solve a SAS Triangle

Solve the triangle: $a = 2,\ b = 3,\ \gamma = 60°$

FIGURE 18

Solution See Figure 18. The Law of Cosines makes it easy to find the third side, c:

$$\begin{aligned} c^2 &= a^2 + b^2 - 2ab\cos\gamma \\ &= 4 + 9 - 2\cdot 2\cdot 3\cdot \cos 60° \\ &= 13 - (12\cdot \tfrac{1}{2}) = 7 \\ c &= \sqrt{7} \end{aligned}$$

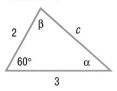

Side c is of length $\sqrt{7}$. To find the angles α and β, we may use either the Law of Sines or the Law of Cosines. It is preferable to use the Law of Cosines, since it will lead to an equation with one solution. Using the Law of Sines would lead to an equation with two solutions that would need to be checked to determine which solution fits the given data. Thus, we choose to use formulas (2) and (3) of the Law of Cosines.

For α:

$$\begin{aligned} a^2 &= b^2 + c^2 - 2bc\cos\alpha \\ 2bc\cos\alpha &= b^2 + c^2 - a^2 \\ \cos\alpha &= \frac{b^2 + c^2 - a^2}{2bc} = \frac{9 + 7 - 4}{2\cdot 3\sqrt{7}} = \frac{12}{6\sqrt{7}} = \frac{2\sqrt{7}}{7} \\ \alpha &\approx 40.9° \end{aligned}$$

For β:

$$\begin{aligned} b^2 &= a^2 + c^2 - 2ac\cos\beta \\ \cos\beta &= \frac{a^2 + c^2 - b^2}{2ac} = \frac{4 + 7 - 9}{4\sqrt{7}} = \frac{1}{2\sqrt{7}} = \frac{\sqrt{7}}{14} \\ \beta &\approx 79.1° \end{aligned}$$

Notice that $\alpha + \beta + \gamma = 40.9° + 79.1° + 60° = 180°$, as required. ■

■ Now work Problem 1.

The next example illustrates how the Law of Cosines is used when three sides of a triangle are known, Case 4 (SSS).

E X A M P L E   2   *Using the Law of Cosines to Solve a SSS Triangle*

Solve the triangle: $a = 4, b = 3, c = 6$

Solution   See Figure 19. To find the angles α, β, and γ, we proceed as we did in the latter part of the solution to Example 1.

FIGURE 19

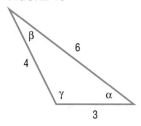

For α:

$$\cos \alpha = \frac{b^2 + c^2 - a^2}{2bc} = \frac{9 + 36 - 16}{2 \cdot 3 \cdot 6} = \frac{29}{36}$$

$$\alpha \approx 36.3°$$

For β:

$$\cos \beta = \frac{a^2 + c^2 - b^2}{2ac} = \frac{16 + 36 - 9}{2 \cdot 4 \cdot 6} = \frac{43}{48}$$

$$\beta \approx 26.4°$$

Since we know α and β,

$$\gamma = 180° - \alpha - \beta \approx 180° - 36.3° - 26.4° = 117.3°$$ ■

■   Now work Problem 17.

E X A M P L E   3   *Correcting a Navigational Error*

A motorized sailboat leaves Naples, Florida, bound for Key West, 150 miles away. Maintaining a constant speed of 15 miles per hour, but encountering heavy crosswinds and strong currents, the crew finds, after 4 hours, that the sailboat is off course by 20°.

(a) How far is the sailboat from Key West at this time?

(b) Through what angle should the sailboat turn to correct its course?

(c) How much time has been added to the trip because of this? (Assume that the speed remains at 15 miles per hour.)

Solution   See Figure 20. With a speed of 15 miles per hour, the sailboat has gone 60 miles after 4 hours. We seek the distance x of the sailboat from Key West. We also seek the angle θ that the sailboat should turn through to correct its course.

FIGURE 20

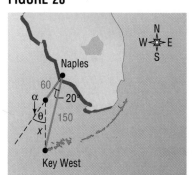

(a) To find x, we use the Law of Cosines, since we know two sides and the included angle.

$$x^2 = 150^2 + 60^2 - 2(150)(60) \cos 20° = 9186$$

$$x = 95.8$$

The sailboat is about 96 miles from Key West.

(b) We now know three sides of the triangle, so we can use the Law of Cosines again to find the angle α opposite the side of length 150 miles.

$$150^2 = 96^2 + 60^2 - 2(96)(60) \cos \alpha$$

$$9684 = -11,520 \cos \alpha$$

$$\cos \alpha \approx -0.8406$$

$$\alpha \approx 147.2°$$

The sailboat should turn through an angle of

$$\theta = 180° - \alpha \approx 180° - 147.2° = 32.8°$$

The sailboat should turn through an angle of about 33° to correct its course.

(c) The total length of the trip is now 60 + 96 = 156 miles. The extra 6 miles will only require about 0.4 hours or 24 minutes more if the speed of 15 miles per hour is maintained. ■

■ Now work Problem 25.

HISTORICAL FEATURE   ■ The Law of Sines was known vaguely long before it was explicitly stated by Nasîr ed-dîn (about AD 1250). Ptolemy (about AD 150) was aware of it in a form using a chord function instead of the sine function. But it was first clearly stated in Europe by Regiomontanus, writing in 1464.

The Law of Cosines appears first in Euclid's *Elements* (Book II), but in a well-disguised form in which squares built on the sides of triangles are added and a rectangle representing the cosine term is subtracted. It was thus known to all mathematicians because of their familiarity with Euclid's work. An early modern form of the Law of Cosines—that for finding the angle when the sides are known—was stated by François Vièta (in 1593).

The Law of Tangents (see Problem 46 of Exercise 8.1) has become obsolete. In the past it was used in place of the Law of Cosines, because the Law of Cosines was very inconvenient for calculation with logarithms or slide rules. Mixing of addition and multiplication is now quite easy on a calculator, however, and the Law of Tangents has been shelved along with the slide rule. ■

8.2

Exercise 8.2

In Problems 1–8, solve each triangle.

1.
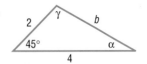

2.
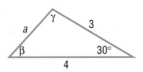

3.
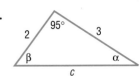

4.
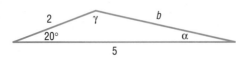

5.
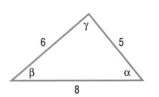

6.
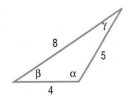

7.
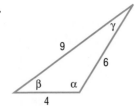

8.
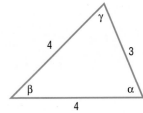

In Problems 9–24, solve each triangle.

9. $a = 3, b = 4, \gamma = 40°$

10. $a = 2, c = 1, \beta = 10°$

11. $b = 1, c = 3, \alpha = 80°$

12. $a = 6, b = 4, \gamma = 60°$

13. $a = 3, c = 2, \beta = 110°$

14. $b = 4, c = 1, \alpha = 120°$

15. $a = 2, b = 2, \gamma = 50°$

16. $a = 3, c = 2, \beta = 90°$

17. $a = 12, b = 13, c = 5$

18. $a = 4, b = 5, c = 3$

19. $a = 2, b = 2, c = 2$

20. $a = 3, b = 3, c = 2$

21. $a = 5, b = 8, c = 9$

22. $a = 4, b = 3, c = 6$

23. $a = 10, b = 8, c = 5$

24. $a = 9, b = 7, c = 10$

25. *Surveying* Consult the figure. To find the distance from the house at A to the house at B, a surveyor measures the angle ACB, which is found to be 70°, then walks off the distance to each house, 50 feet and 70 feet, respectively. How far apart are the houses?

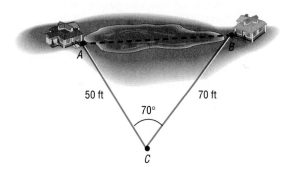

26. *Navigation* An airplane flies from city A to city B, a distance of 150 miles, then turns through an angle of 50° and flies to city C, a distance of 100 miles (see the figure).
(a) How far is it from city A to city C?
(b) Through what angle should the pilot turn at city C to return to city A?

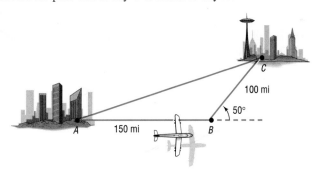

27. *Revising a Flight Plan* In attempting to fly from city A to city B, a distance of 330 miles, a pilot inadvertently took a course that was 10° in error, as indicated in the figure.
(a) If the aircraft maintains an average speed of 220 miles per hour and if the error in direction is discovered after 15 minutes, through what angle should the pilot turn to head toward city B?
(b) What new average speed should the pilot maintain so that the total time of the trip is 90 minutes?

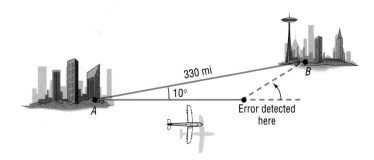

28. *Avoiding a Tropical Storm* A cruise ship maintains an average speed of 15 knots in going from San Juan, Puerto Rico, to Barbados, West Indies, a distance of 600 nautical miles. To avoid a tropical storm, the captain heads out of San Juan in a direction of 20° off a direct heading to Barbados. The captain maintains the 15 knot speed for 10 hours, after which time the path to Barbados becomes clear of storms.
(a) Through what angle should the captain turn to head directly to Barbados?
(b) How long will it be before the ship reaches Barbados if the same 15 knot speed is maintained?

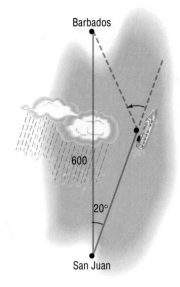

29. *Major League Baseball Field* A Major League baseball diamond is actually a square 90 feet on a side. The pitching rubber is located 60.5 feet from home plate on a line joining home plate and second base.
(a) How far is it from the pitching rubber to first base?
(b) How far is it from the pitching rubber to second base?
(c) If a pitcher faces home plate, through what angle does he need to turn to face first base?

30. *Little League Baseball Field* Aceording to Little League baseball official regulations, a diamond is a square 60 feet on a side. The pitching rubber is located 46 feet from home plate on a line joining home plate and second base.
(a) How far is it from the pitching rubber to first base?
(b) How far is it from the pitching rubber to second base?
(c) If a pitcher faces home plate, through what angle does he need to turn to face first base?

31. *Finding the Length of a Guy Wire* The height of a radio tower is 500 feet, and the ground on one side of the tower slopes upward at an angle of 10° (see the figure).
(a) How long should a guy wire be if it is to connect to the top of the tower and be secured at a point on the slope 100 feet from the base of the tower?
(b) How long should a second guy wire be if it is to connect to the middle of the tower and be secured at a point 100 feet from the base?

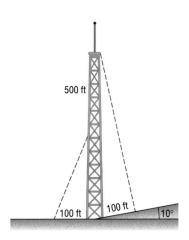

32. *Finding the Length of a Guy Wire* A radio tower 500 feet high is located on the side of a hill with an inclination to the horizontal of 5° (see the figure). How long should two guy wires be if they are to connect to the top of the tower and be secured at two points 100 feet directly above and directly below the base of the tower?

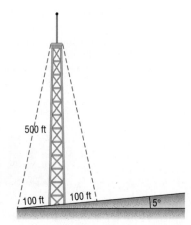

33. *Wrigley Field, Home of the Chicago Cubs* The distance from home plate to dead center in Wrigley Field is 400 feet (see the figure). How far is it from dead center to third base?

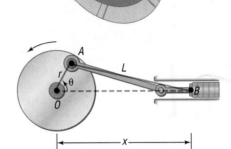

34. *Little League Baseball* The distance from home plate to dead center at the Oak Lawn Little League field is 280 feet. How far is it from dead center to third base? [*Hint:* The distance between the bases in Little League is 60 feet.]

35. *Rods and Pistons* Rod *OA* (see the figure) rotates about the fixed point *O* so that point *A* travels on a circle of radius *r*. Connected to point *A* is another rod *AB* of length $L > r$, and point *B* is connected to a piston. Show that the distance *x* between point *O* and point *B* is given by

$$x = r \cos \theta + \sqrt{r^2 \cos^2 \theta + L^2 - r^2}$$

where θ is the angle of rotation of rod *OA*.

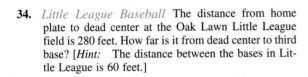

36. *Geometry* Show that the length *d* of a chord of a circle of radius *r* is given by the formula

$$d = 2r \sin \frac{\theta}{2}$$

where θ is the central angle formed by the radii to the ends of the chord (see the figure). Use this result to derive the fact that $\sin \theta < \theta$, where $\theta > 0$ is measured in radians.

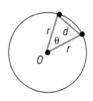

37. For any triangle, show that

$$\cos \frac{\gamma}{2} = \sqrt{\frac{s(s - c)}{ab}}$$

where $s = \frac{1}{2}(a + b + c)$. [*Hint:* Use a half-angle formula and the Law of Cosines.]

38. For any triangle, show that

$$\sin \frac{\gamma}{2} = \sqrt{\frac{(s - a)(s - b)}{ab}}$$

where $s = \frac{1}{2}(a + b + c)$.

39. Use the Law of Cosines to prove the identity

$$\frac{\cos \alpha}{a} + \frac{\cos \beta}{b} + \frac{\cos \gamma}{c} = \frac{a^2 + b^2 + c^2}{2abc}$$

40. Write down your strategy for solving an oblique triangle.

8.3

The Area of a Triangle

In this section, we shall derive several formulas for calculating the area A of a triangle. The most familiar of these is the following:

Theorem The area A of a triangle is

$$A = \tfrac{1}{2}bh \qquad (1)$$

where b is the base and h is an altitude drawn to that base. ∎

FIGURE 21
$A = \tfrac{1}{2}bh$

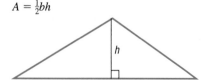

FIGURE 22

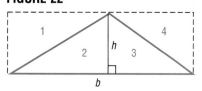

Proof The derivation of this formula is rather easy once a rectangle of base b and height h is constructed around the triangle. See Figures 21 and 22.

Triangles 1 and 2 in Figure 22 are equal in area, as are triangles 3 and 4. Consequently, the area of the triangle with base b and altitude h is exactly half the area of the rectangle, which is bh. ∎

If the base b and altitude h to that base are known, then we can easily find the area of such a triangle using formula (1). Usually, though, the information required to use formula (1) is not given. Suppose, for example, that we know two sides a and b and the included angle γ (see Figure 23). Then the altitude h can be found by noting that

$$\frac{h}{a} = \sin \gamma$$

so that

$$h = a \sin \gamma$$

Using this fact in formula (1) produces

$$A = \tfrac{1}{2}bh = \tfrac{1}{2}b(a \sin \gamma) = \tfrac{1}{2}ab \sin \gamma$$

Thus, we have the formula

$$A = \tfrac{1}{2}ab \sin \gamma \qquad (2)$$

FIGURE 23
$h = a \sin \gamma$

By dropping altitudes from the other two vertices of the triangle, we obtain the following corresponding formulas:

$$A = \tfrac{1}{2}bc \sin \alpha \qquad (3)$$
$$A = \tfrac{1}{2}ac \sin \beta \qquad (4)$$

It is easiest to remember these formulas using the following wording:

Theorem The area A of a triangle equals one-half the product of two of its sides times the sine of their included angle. ∎

E X A M P L E   1   *Finding the Area of a SAS Triangle*

Find the area A of the triangle for which $a = 8$, $b = 6$, and $\gamma = 30°$.

Solution   We use formula (2) to get

$$A = \tfrac{1}{2}ab \sin \gamma = \tfrac{1}{2} \cdot 8 \cdot 6 \sin 30° = 12$$

◼

◼   Now work Problem 1.

If the three sides of a triangle are known, another formula, called **Heron's Formula** (named after Heron of Alexandria), can be used to find the area of a triangle.

Theorem   The area A of a triangle with sides a, b, and c is
Heron's Formula

$$A = \sqrt{s(s-a)(s-b)(s-c)} \qquad (5)$$

where $s = \tfrac{1}{2}(a + b + c)$.   ◼

Proof   The proof we shall give uses the Law of Cosines and is quite different from the proof given by Heron.

From the Law of Cosines,

$$c^2 = a^2 + b^2 - 2ab \cos \gamma$$

and the two half-angle formulas,

$$\cos^2 \frac{\gamma}{2} = \frac{1 + \cos \gamma}{2} \qquad \sin^2 \frac{\gamma}{2} = \frac{1 - \cos \gamma}{2}$$

we find

$$\cos^2 \frac{\gamma}{2} = \frac{1 + \cos \gamma}{2} = \frac{1 + \dfrac{a^2 + b^2 - c^2}{2ab}}{2}$$

$$= \frac{a^2 + 2ab + b^2 - c^2}{4ab} = \frac{(a+b)^2 - c^2}{4ab}$$

$$= \frac{(a + b - c)(a + b + c)}{4ab} = \frac{2(s-c) \cdot 2s}{4ab} = \frac{s(s-c)}{ab} \qquad (6)$$

$$\underset{\substack{\uparrow \\ a+b-c = a+b+c-2c \\ = 2s - 2c}}{}$$

Similarly,

$$\sin^2 \frac{\gamma}{2} = \frac{(s-a)(s-b)}{ab} \qquad (7)$$

Now we use formula (2) for the area:

$$A = \frac{1}{2}ab \sin \gamma$$

$$= \frac{1}{2}ab \cdot 2 \sin \frac{\gamma}{2} \cos \frac{\gamma}{2} \qquad \sin \gamma = \sin 2\left(\frac{\gamma}{2}\right) = 2 \sin \frac{\gamma}{2} \cos \frac{\gamma}{2}$$

$$= ab \sqrt{\frac{(s-a)(s-b)}{ab}} \sqrt{\frac{s(s-c)}{ab}} \qquad \text{Use equations (6) and (7).}$$

$$= \sqrt{s(s-a)(s-b)(s-c)} \qquad \blacksquare$$

EXAMPLE 2   *Finding the Area of a SSS Triangle*

Find the area of a triangle whose sides are 4, 5, and 7.

Solution   We let $a = 4$, $b = 5$, and $c = 7$. Then

$$s = \tfrac{1}{2}(a + b + c) = \tfrac{1}{2}(4 + 5 + 7) = 8$$

Heron's Formula then gives the area A as

$$A = \sqrt{s(s-a)(s-b)(s-c)} = \sqrt{8 \cdot 4 \cdot 3 \cdot 1} = \sqrt{96} = 4\sqrt{6} \qquad \blacksquare$$

■ Now work Problem 17.

HISTORICAL FEATURE   ■ Heron's Formula (also known as *Hero's Formula*) is due to Heron of Alexandria (about AD 75), who had, besides his mathematical talents, a good deal of engineering skill. In various temples his mechanical devices produced effects that seemed supernatural, and visitors presumably were thus influenced to generosity. Heron's book *Metrica,* on making such devices, has survived and was discovered in 1896 in the city of Constantinople.

Heron's Formula for the area of a triangle caused some mild discomfort in Greek mathematics, because a product with two factors was an area and with three factors was a volume, but four factors seemed contradictory in Heron's time.

Karl Mollweide (1774–1875), a mathematician and astronomer, discovered the formulas named for him (see Problems 43 and 44 of Exercise 8.1). These formulas are not too important in themselves but often simplify the derivation of other formulas, as demonstrated in Historical Problems 1 and 2, which follow.   ■

HISTORICAL PROBLEMS   ■ 1. This derivation of Heron's formula uses Mollweide's Formula.

(a) Show that

$$\frac{s}{c} = \frac{\cos(\alpha/2)\cos(\beta/2)}{\sin(\gamma/2)}$$

where $s = \tfrac{1}{2}(a + b + c)$. *Hint:* Use Mollweide's Formula (see Problem 43 in Exercise 8.1) and add 1 to both sides. Then use the fact that

$$\sin \frac{\gamma}{2} = \sin \frac{180° - (\alpha + \beta)}{2} = \cos \frac{\alpha + \beta}{2}$$

(b) Similarly, show that

$$\frac{s}{a} = \frac{\cos(\beta/2)\cos(\gamma/2)}{\sin(\alpha/2)} \quad \text{and} \quad \frac{s}{b} = \frac{\cos(\alpha/2)\cos(\gamma/2)}{\sin(\beta/2)}$$

(c) Use the results of parts (a) and (b) to show that

$$\frac{s-a}{a} = \frac{\sin(\beta/2)\sin(\gamma/2)}{\sin(\alpha/2)} \qquad \frac{s-b}{b} = \frac{\sin(\alpha/2)\sin(\gamma/2)}{\sin(\beta/2)}$$

$$\frac{s-c}{c} = \frac{\sin(\alpha/2)\sin(\beta/2)}{\sin(\gamma/2)}$$

(d) Now form the product

$$\frac{s}{c} \cdot \frac{s-a}{a} \cdot \frac{s-b}{b} \cdot \frac{s-c}{c}$$

After cancellations, multiply each side by $cabc$, use the double-angle formulas, and use the fact that $A = \frac{1}{2}bc \sin \alpha = \frac{1}{2}ac \sin \beta$. Heron's Formula will then follow.

2. We again use Mollweide's Formula to derive some other interesting formulas. (These were derived in another way in Problems 37 and 38 of Exercise 8.2.)

(a) Add 1 to both sides of the second form of Mollweide's Formula (see Problem 44 in Exercise 8.1) and simplify to obtain

$$\frac{s-b}{c} = \frac{\sin(\alpha/2)\cos(\beta/2)}{\cos(\gamma/2)}$$

(b) Similarly, show that

$$\frac{s-c}{b} = \frac{\sin(\alpha/2)\cos(\gamma/2)}{\cos(\beta/2)}$$

(c) Use the results of part (b) and Problem 1(b) and (c) to show that

$$\cos^2\frac{\gamma}{2} = \frac{s(s-c)}{ab} \quad \text{and} \quad \sin^2\frac{\gamma}{2} = \frac{(s-a)(s-b)}{ab}$$

(d) Show that

$$\tan^2\frac{\gamma}{2} = \frac{(s-a)(s-b)}{s(s-c)}$$

3. (a) If h_1, h_2, and h_3 are the altitudes dropped from A, B, and C, respectively, in a triangle (see the figure), show that

$$\frac{1}{h_1} + \frac{1}{h_2} + \frac{1}{h_3} = \frac{s}{K}$$

where K is the area of the triangle and $s = \frac{1}{2}(a + b + c)$. [*Hint:* $h_1 = 2K/a$.]

(b) Show that a formula for the altitude h from a vertex to the opposite side a of a triangle is

$$h = \frac{a \sin \beta \sin \gamma}{\sin \alpha}$$

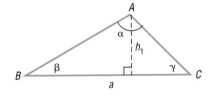

4. *Inscribed Circle.* The lines that bisect each angle of a triangle meet in a single point O, and the perpendicular distance r from O to each side of the triangle is the same. The circle with center at O and radius r is called the *inscribed circle* of the triangle (see the figure).

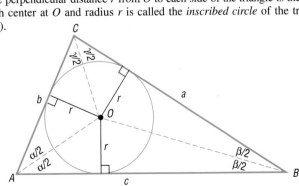

(a) Apply Problem 3(b) to triangle OAB to show that

$$r = \frac{c \sin(\alpha/2) \sin(\beta/2)}{\cos(\gamma/2)}$$

(b) Use the results of part (a) and Problem 1(c) to show that

$$r = (s - c) \tan \frac{\gamma}{2}$$

(c) Show that

$$\cot \frac{\alpha}{2} + \cot \frac{\beta}{2} + \cot \frac{\gamma}{2} = \frac{s}{r}$$

(d) Show that the area K of triangle ABC is $K = rs$. Then show that

$$r = \sqrt{\frac{(s - a)(s - b)(s - c)}{s}}$$

where $s = \frac{1}{2}(a + b + c)$.

5. Find the coordinates of the center O of the inscribed circle in terms of a, b, c, α, β, and γ. Then write the general equation of the inscribed circle. [*Hint:* Use a system of rectangular coordinates with the vertex A at the origin and side c along the positive x-axis.] ■

8.3

Exercise 8.3

In Problems 1–8, find the area of each triangle. Round off answers to two decimal places.

1.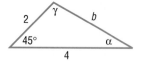

2.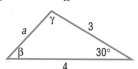

3.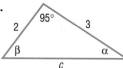

4.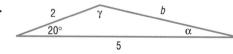

5.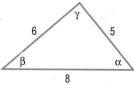

6.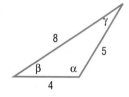

7.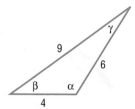

8.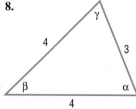

In Problems 9–24, find the area of each triangle. Round off answers to two decimal places.

9. $a = 3$, $b = 4$, $\gamma = 40°$

10. $a = 2$, $c = 1$, $\beta = 10°$

11. $b = 1$, $c = 3$, $\alpha = 80°$

12. $a = 6$, $b = 4$, $\gamma = 60°$

13. $a = 3$, $c = 2$, $\beta = 110°$

14. $b = 4$, $c = 1$, $\alpha = 120°$

15. $a = 2$, $b = 2$, $\gamma = 50°$

16. $a = 3$, $c = 2$, $\beta = 90°$

17. $a = 12$, $b = 13$, $c = 5$

18. $a = 4$, $b = 5$, $c = 3$

19. $a = 2$, $b = 2$, $c = 2$

20. $a = 3$, $b = 3$, $c = 2$

21. $a = 5$, $b = 8$, $c = 9$

22. $a = 4$, $b = 3$, $c = 6$

23. $a = 10$, $b = 8$, $c = 5$

24. $a = 9$, $b = 7$, $c = 10$

25. *Cost of a Triangular Lot* The dimensions of a triangular lot are 100 feet by 50 feet by 75 feet. If the price of such land is $3 per square foot, how much does the lot cost?

26. *Approximating the Area of a Lake* To approximate the area of a lake, a surveyor walks around the perimeter of the lake, taking the measurements shown in the illustration. Using this technique, what is the approximate area of the lake? [*Hint:*   Use the Law of Cosines on the three triangles shown and then find the sum of their areas.]

27. *Area of a Triangle* Prove that the area A of a triangle is given by the formula

$$A = \frac{a^2 \sin \beta \sin \gamma}{2 \sin \alpha}$$

28. *Area of a Triangle* Prove the two other forms of the formula given in Problem 27,

$$A = \frac{b^2 \sin \alpha \sin \gamma}{2 \sin \beta} \quad \text{and} \quad A = \frac{c^2 \sin \alpha \sin \beta}{2 \sin \gamma}$$

In Problems 29–36, use the results of Problem 27 or 28 to find the area of each triangle. Round off answers to two decimal places.

29. $\alpha = 40°, \beta = 20°, a = 2$ 30. $\alpha = 50°, \gamma = 20°, a = 3$ 31. $\beta = 70°, \gamma = 10°, b = 5$

32. $\alpha = 70°, \beta = 60°, c = 4$ 33. $\alpha = 110°, \gamma = 30°, c = 3$ 34. $\beta = 10°, \gamma = 100°, b = 2$

35. $\alpha = 40°, \beta = 40°, c = 2$ 36. $\beta = 20°, \gamma = 70°, a = 1$

37. *Geometry* Consult the figure in the margin, which shows a circle of radius r with center at O. Find the area A of the shaded region as a function of the central angle θ.

8.4

Polar Coordinates

So far, we have always used a system of rectangular coordinates to plot points in the plane. Now we are ready to describe another system called *polar coordinates.* As we shall soon see, in many instances polar coordinates offer certain advantages over rectangular coordinates.

FIGURE 24

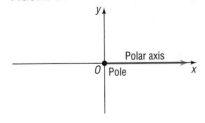

In a rectangular coordinate system, you will recall, a point in the plane is represented by an ordered pair of numbers (x, y), where x and y equal the signed distance of the point from the y-axis and x-axis, respectively. In a polar coordinate system, we select a point, called the **pole,** and then a ray with vertex at the pole, called the **polar axis.** Comparing the rectangular and polar coordinate systems, we see (in Figure 24) that the origin in rectangular coordinates coincides with the pole in polar coordinates, and the positive x-axis in rectangular coordinates coincides with the polar axis in polar coordinates.

A point P in a polar coordinate system is represented by an ordered pair of numbers (r, θ). The number r is the distance of the point from the pole, and θ is an angle (in degrees or radians) formed by the polar axis and a ray from the pole through the point. We call the ordered pair (r, θ) the **polar coordinates** of the point. See Figure 25.

FIGURE 25

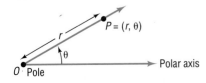

FIGURE 26

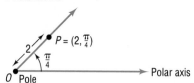

As an example, suppose that the polar coordinates of a point P are $(2, \pi/4)$. We locate P by first drawing an angle of $\pi/4$ radian, placing its vertex at the pole and its initial side along the polar axis. Then we go out a distance of 2 units along the terminal side of the angle to reach the point P. See Figure 26.

■   Now work Problem 1.

Recall that an angle measured counterclockwise is positive, whereas one measured clockwise is negative. This convention has some interesting consequences relating to polar coordinates. Let's see what these consequences are.

E X A M P L E   1   *Finding Several Polar Coordinates of a Single Point*

Consider again a point P with polar coordinates $(2, \pi/4)$, as shown in Figure 27(a). Because $\pi/4$, $9\pi/4$, and $-7\pi/4$ all have the same terminal side, we also could have located this point P by using the polar coordinates $(2, 9\pi/4)$ or $(2, -7\pi/4)$, as shown in Figures 27(b) and (c).

FIGURE 27

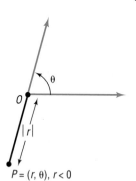

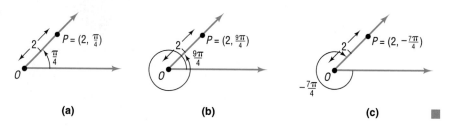

(a)   (b)   (c)   ■

P = (r, θ), r < 0

FIGURE 28

In using polar coordinates (r, θ), it is possible for the first entry r to be negative. When this happens, we follow the convention that the location of the point, instead of being on the terminal side of θ, is on the ray from the pole extending in the direction *opposite* the terminal side of θ at a distance $|r|$ from the pole. See Figure 28 for an illustration.

E X A M P L E   2   *Polar Coordinates (r, θ), $r < 0$.*

Consider again the point P with polar coordinates $(2, \pi/4)$, as shown in Figure 29(a). This same point P can be assigned the polar coordinates $(-2, 5\pi/4)$, as indicated in Figure 29(b). To locate the point $(-2, 5\pi/4)$, we use the ray in the opposite direction of $5\pi/4$ and go out 2 units along that ray to find the point P.

FIGURE 29

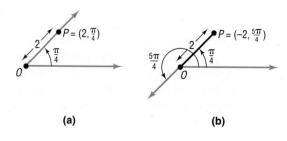

(a)   (b)   ■

MISSION POSSIBLE

Chapter 8

LOCATING LOST TREASURE

While scuba diving off Wreck Hill in Bermuda, a group of 5 entrepreneurs discovered a treasure map in a small water-tight cask on a pirate schooner that had sunk in 1747. The map directed them to an area of Bermuda now known as The Flatts, but when they got there, they realized that the most important landmark on the map was gone. They called in the Mission Possible team to help them recreate the map. They promised you 25% of whatever treasure was found.

The directions on the map read as follows:

1. From the tallest palm tree, sight the highest hill. Drop your eyes vertically until you sight the base of the hill.
2. Turn 40° clockwise from that line and walk 70 paces to the big red rock.
3. From the red rock walk 50 paces back to the sight line between the palm tree and the hill. Dig there.

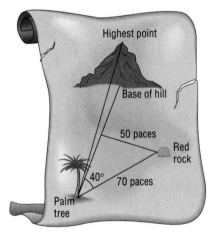

The 5 entrepreneurs said they believed they had found the red rock and the highest hill in the vicinity, but the "tallest palm tree" had long since fallen and disintegrated. It had occurred to them that the treasure must be located on a circle with radius 50 "paces" centered around the red rock, but they had decided against digging a trench 470 feet in circumference, especially since they had no assurance that the treasure was still there. (They had decided that a "pace" must be about a yard.)

1. Determine a plan to locate the position of the lost palm tree, and write out an explanation of your procedure for the entrepreneurs.
2. Unfortunately, it turns out that the entrepreneurs had more in common with the 18th century pirates than you had bargained for. Once you told them the location of the lost palm tree, they tied you all to the red rock, saying they could take it from there. From the location of the palm tree, they sighted 40° counterclockwise from the rock to the hill, then ran about 30 yards to the circle they had traced about the rock and began to dig frantically. Nothing. After about an hour, they drove off shouting back at you, "25% of nothing is nothing!"
3. Fortunately, the entrepreneurs had left the shovels. After you managed to untie yourselves, you went to the correct location and found the treasure. Where was it? How far from the palm tree? Explain.
4. People who scuba dive for sunken treasure have certain legal obligations. What are they? Should you share the treasure with a lawyer, just to make sure you get to keep the rest?

These examples show a major difference between rectangular coordinates and polar coordinates. In the former, each point has exactly one pair of rectangular coordinates; in the latter, a point can have infinitely many pairs of polar coordinates.

Summary   A point with polar coordinates (r, θ) also can be represented by any of the following:

$$(r, \theta + 2k\pi) \quad \text{or} \quad (-r, \theta + \pi + 2k\pi), \qquad k \text{ any integer}$$

The polar coordinates of the pole are $(0, \theta)$, where θ can be any angle.

E X A M P L E   3   *Plotting Points Using Polar Coordinates*

Plot the points with the following polar coordinates:

(a) $(3, 5\pi/3)$ (b) $(2, -\pi/4)$ (c) $(3, 0)$ (d) $(-2, \pi/4)$

Solution   Figure 30 shows the points.

FIGURE 30

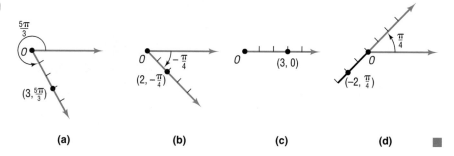

(a)   (b)   (c)   (d)

■   Now work Problem 9.

E X A M P L E   4   *Finding Other Polar Coordinates of a Given Point*

Plot the point P with polar coordinates $(3, \pi/6)$, and find other polar coordinates (r, θ) of this same point for which:

(a) $r > 0$, $2\pi \le \theta \le 4\pi$ (b) $r < 0$, $0 \le \theta \le 2\pi$
(c) $r > 0$, $-2\pi \le \theta \le 0$

Solution   The point $(3, \pi/6)$ is plotted in Figure 31.

(a) We add 1 revolution (2π radians) to the angle $\pi/6$ to get $P = (3, \pi/6 + 2\pi) = (3, 13\pi/6)$. See Figure 32.

FIGURE 31

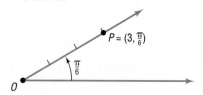

FIGURE 32

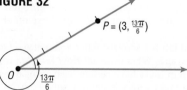

(b) We add $\frac{1}{2}$ revolution (π radians) to the angle and replace r by $-r$ to get $P = (-3, \pi/6 + \pi) = (-3, 7\pi/6)$. See Figure 33.

FIGURE 33

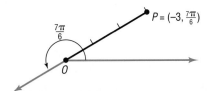

(c) We subtract 2π from the angle $\pi/6$ to get $P = (3, \pi/6 - 2\pi) = (3, -11\pi/6)$. See Figure 34.

FIGURE 34

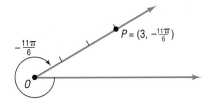

■ Now work Problem 13.

Conversion between Polar Coordinates and Rectangular Coordinates

It is sometimes convenient and, indeed, necessary to be able to convert coordinates or equations in rectangular form to polar form, and vice versa. To do this, we recall that the origin in rectangular coordinates is the pole in polar coordinates and that the positive x-axis in rectangular coordinates is the polar axis in polar coordinates.

Theorem
Conversion from Polar
Coordinates to Rectangular
Coordinates

If P is a point with polar coordinates (r, θ), the rectangular coordinates (x, y) of P are given by

$$x = r \cos \theta \qquad y = r \sin \theta \qquad (1)$$

■

Proof   Suppose that P has the polar coordinates (r, θ). We seek the rectangular coordinates (x, y) of P. Refer to Figure 35.

FIGURE 35

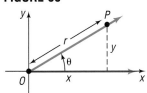

If $r = 0$, then, regardless of θ, the point P is the pole, for which the rectangular coordinates are $(0, 0)$. Thus, formula (1) is valid for $r = 0$.

If $r > 0$, the point P is on the terminal side of θ and $r = d(O, P)$. Thus,

$$\cos \theta = \frac{x}{r} \qquad \sin \theta = \frac{y}{r}$$

so

$$x = r \cos \theta \qquad y = r \sin \theta$$

If $r < 0$, then the point $P = (r, \theta)$ can be represented as $(-r, \pi + \theta)$, where $-r > 0$. Thus,

$$\cos(\pi + \theta) = -\cos\theta = \frac{x}{-r} \qquad \sin(\pi + \theta) = -\sin\theta = \frac{y}{-r}$$

so

$$x = r\cos\theta \qquad y = r\sin\theta \qquad\blacksquare$$

E X A M P L E   5   *Converting from Polar Coordinates to Rectangular Coordinates*

Find the rectangular coordinates of the points with the following polar coordinates:

(a) $(6, \pi/6)$ (b) $(-2, 5\pi/4)$ (c) $(-4, -\pi/4)$

Solution   We use formula (1): $x = r\cos\theta$ and $y = r\sin\theta$.

(a) See Figure 36(a).

$$x = r\cos\theta = 6\cos\frac{\pi}{6} = 6\cdot\frac{\sqrt{3}}{2} = 3\sqrt{3}$$

$$y = r\sin\theta = 6\sin\frac{\pi}{6} = 6\cdot\frac{1}{2} = 3$$

The rectangular coordinates of the point $(6, \pi/6)$ are $(3\sqrt{3}, 3)$.

(b) See Figure 36(b).

$$x = r\cos\theta = -2\cos\frac{5\pi}{4} = -2\left(-\frac{\sqrt{2}}{2}\right) = \sqrt{2}$$

$$y = r\sin\theta = -2\sin\frac{5\pi}{4} = -2\left(-\frac{\sqrt{2}}{2}\right) = \sqrt{2}$$

The rectangular coordinates of the point $(-2, 5\pi/4)$ are $(\sqrt{2}, \sqrt{2})$.

(c) See Figure 36(c).

$$x = r\cos\theta = -4\cos\left(-\frac{\pi}{4}\right) = -4\cdot\frac{\sqrt{2}}{2} = -2\sqrt{2}$$

$$y = r\sin\theta = -4\sin\left(-\frac{\pi}{4}\right) = -4\left(-\frac{\sqrt{2}}{2}\right) = 2\sqrt{2}$$

The rectangular coordinates of the given point $(-4, -\pi/4)$ are $(-2\sqrt{2}, 2\sqrt{2})$.

FIGURE 36

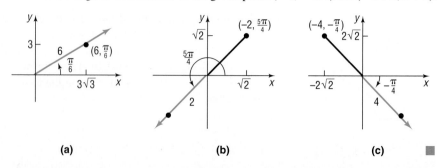

(a)   (b)   (c)   ■

Note: Most calculators have the capability of converting from polar coordinates to rectangular coordinates. Consult your owner's manual to learn the proper key strokes. Since in most cases this procedure is tedious, you will find that using formula (1) is faster.

■ Now work Problems 21 and 33.

To convert from rectangular coordinates (x, y) to polar coordinates (r, θ) is a little more complicated. Let's look at some examples.

E X A M P L E   6   *Converting from Rectangular Coordinates to Polar Coordinates*

FIGURE 37

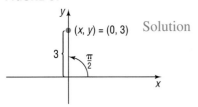

Find polar coordinates of a point whose rectangular coordinates are $(0, 3)$.

Solution   See Figure 37. The point $(0, 3)$ lies on the y-axis a distance of 3 units from the origin (pole). Polar coordinates for this point can be given by $(3, \pi/2)$. ■

Figure 38 shows polar coordinates of points that lie on either the x-axis or the y-axis.

FIGURE 38

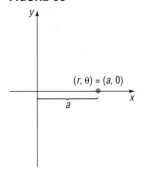

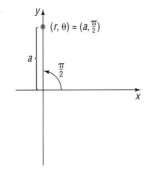

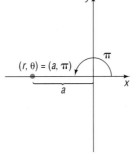

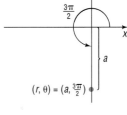

(a) $(x, y) = (a, 0), a > 0$ (b) $(x, y) = (0, a), a > 0$ (c) $(x, y) = (-a, 0), a > 0$ (d) $(x, y) = (0, -a), a > 0$

■ Now work Problem 37.

E X A M P L E   7   *Converting from Rectangular Coordinates to Polar Coordinates*

Find polar coordinates of a point whose rectangular coordinates are $(2, -2)$.

Solution   See Figure 39. The distance r from the origin to the point $(2, -2)$ is
$$r = \sqrt{x^2 + y^2} = \sqrt{(2)^2 + (-2)^2} = \sqrt{8} = 2\sqrt{2}$$

FIGURE 39

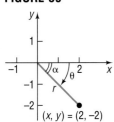

To find θ, we use the reference angle α. Then
$$\alpha = \tan^{-1}\left|\frac{y}{x}\right| = \tan^{-1}\left|\frac{-2}{2}\right| = \tan^{-1} 1 = \frac{\pi}{4}$$

Thus, $\theta = -\pi/4$, and a set of polar coordinates for this point is $(2\sqrt{2}, -\pi/4)$. Other possible representations include $(2\sqrt{2}, 7\pi/4)$ and $(-2\sqrt{2}, 3\pi/4)$. ■

E X A M P L E   8   *Converting from Rectangular Coordinates to Polar Coordinates*

Find polar coordinates of a point whose rectangular coordinates are $(-1, -\sqrt{3})$.

Solution   See Figure 40. The distance r from the origin to the point $(-1, -\sqrt{3})$ is

FIGURE 40

$$r = \sqrt{x^2 + y^2} = \sqrt{(-1)^2 + (-\sqrt{3})^2} = \sqrt{4} = 2$$

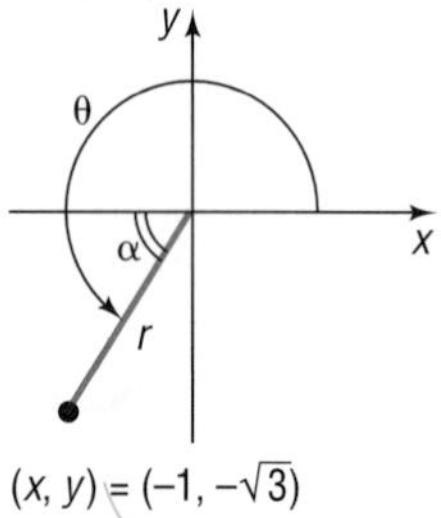

To find θ, we use the reference angle α. Then

$$\alpha = \tan^{-1}\left|\frac{y}{x}\right| = \tan^{-1}\left|\frac{-\sqrt{3}}{-1}\right| = \tan^{-1}\sqrt{3} = \frac{\pi}{3}$$

Thus,

$$\theta = \pi + \alpha = \pi + \frac{\pi}{3} = \frac{4\pi}{3}$$

and a set of polar coordinates is $(2, 4\pi/3)$. Other possible representations include $(-2, \pi/3)$ and $(2, -2\pi/3)$. ■

Figure 41 shows how to find polar coordinates of a point that lies in a quadrant when its rectangular coordinates (x, y) are given.

FIGURE 41

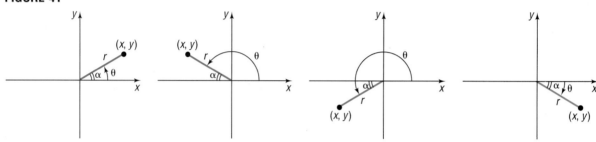

(a) $r = \sqrt{x^2 + y^2}$
$\theta = \alpha = \tan^{-1}\left|\frac{y}{x}\right|$

(b) $r = \sqrt{x^2 + y^2}$
$\theta = \pi - \alpha = \pi - \tan^{-1}\left|\frac{y}{x}\right|$

(c) $r = \sqrt{x^2 + y^2}$
$\theta = \pi + \alpha = \pi + \tan^{-1}\left|\frac{y}{x}\right|$

(d) $r = \sqrt{x^2 + y^2}$
$\theta = -\alpha = -\tan^{-1}\left|\frac{y}{x}\right|$

■   Now work Problem 41.

Based on the preceding discussion, we have the formulas

$$r^2 = x^2 + y^2 \qquad \tan\theta = \frac{y}{x} \qquad \text{if } x \neq 0 \qquad\qquad (2)$$

Warning: Be careful when using formula (2). Always plot the point (x, y) first, as we did in Examples 7 and 8, so that r and θ are chosen to reflect the correct quadrant. Look again at Figure 41.

Note: Most calculators have the capability of converting from rectangular coordinates to polar coordinates. Consult your owner's manual to learn the proper key strokes. Since in most cases this procedure is tedious, you will find that using formula (2) is faster.

Formulas (1) and (2) may also be used to transform equations.

E X A M P L E   9   *Transforming Equations from Polar to Rectangular Form*

Transform the equation $r = 4\sin\theta$ from polar coordinates to rectangular coordinates, and identify the graph.

Solution   If we multiply each side by r, it will be easier to apply formulas (1) and (2):

$$r = 4 \sin \theta$$
$$r^2 = 4r \sin \theta \quad \text{Multiply each side by } r.$$
$$x^2 + y^2 = 4y \quad \text{Apply formulas (1) and (2).}$$

This is the equation of a circle:

$$x^2 + (y^2 - 4y) = 0$$
$$x^2 + (y^2 - 4y + 4) = 4$$
$$x^2 + (y - 2)^2 = 4$$

Its center is at (0, 2), and its radius is 2.   ■

■ Now work Problem 57.

E X A M P L E   1 0   *Transforming an Equation from Rectangular to Polar Form*

Transform the equation $4xy = 9$ from rectangular coordinates to polar coordinates.

Solution   We use formula (1):

$$4xy = 9$$
$$4(r \cos \theta)(r \sin \theta) = 9 \quad \text{Formula (1)}$$
$$4r^2 \cos \theta \sin \theta = 9$$
$$2r^2 \sin 2\theta = 9 \quad \text{Double-angle formula}$$

■

8.4

Exercise 8.4

In Problems 1–12, plot each point given in polar coordinates.

1. $(3, 90°)$	**2.** $(4, 270°)$	**3.** $(-2, 0)$	**4.** $(-3, \pi)$
5. $(6, \pi/6)$	**6.** $(5, 5\pi/3)$	**7.** $(-2, 135°)$	**8.** $(-3, 120°)$
9. $(-1, -\pi/3)$	**10.** $(-3, -3\pi/4)$	**11.** $(-2, -\pi)$	**12.** $(-3, -\pi/2)$

In Problems 13–20, plot each point given in polar coordinates, and find other polar coordinates (r, θ) of the point for which:

(a) $r > 0, \quad -2\pi \le \theta < 0$ (b) $r < 0, \quad 0 \le \theta < 2\pi$ (c) $r > 0, \quad 2\pi \le \theta < 4\pi$

13. $(5, 2\pi/3)$	**14.** $(4, 3\pi/4)$	**15.** $(-2, 3\pi)$	**16.** $(-3, 4\pi)$
17. $(1, \pi/2)$	**18.** $(2, \pi)$	**19.** $(-3, -\pi/4)$	**20.** $(-2, -2\pi/3)$

In Problems 21–36, polar coordinates of a point are given. Find the rectangular coordinates of each point.

21. $(3, \pi/2)$	**22.** $(4, 3\pi/2)$	**23.** $(-2, 0)$	**24.** $(-3, \pi)$
25. $(6, 150°)$	**26.** $(5, 300°)$	**27.** $(-2, 3\pi/4)$	**28.** $(-3, 2\pi/3)$
29. $(-1, -\pi/3)$	**30.** $(-3, -3\pi/4)$	**31.** $(-2, -180°)$	**32.** $(-3, -90°)$
33. $(7.5, 110°)$	**34.** $(-3.1, 182°)$	**35.** $(6.3, 3.8)$	**36.** $(8.1, 5.2)$

In Problems 37–48, the rectangular coordinates of a point are given. Find polar coordinates for each point.

37. $(3, 0)$ **38.** $(0, 2)$ **39.** $(-1, 0)$ **40.** $(0, -2)$

41. $(1, -1)$ **42.** $(-3, 3)$ **43.** $(\sqrt{3}, 1)$ **44.** $(-2, -2\sqrt{3})$

45. $(1.3, -2.1)$ **46.** $(-0.8, -2.1)$ **47.** $(8.3, 4.2)$ **48.** $(-2.3, 0.2)$

In Problems 49–56, the letters x and y represent rectangular coordinates. Write each equation using polar coordinates (r, θ).

49. $2x^2 + 2y^2 = 3$ **50.** $x^2 + y^2 = x$ **51.** $x^2 = 4y$ **52.** $y^2 = 2x$

53. $2xy = 1$ **54.** $4x^2y = 1$ **55.** $x = 4$ **56.** $y = -3$

In Problems 57–64, the letters r and θ represent polar coordinates. Write each equation using rectangular coordinates (x, y).

57. $r = \cos \theta$ **58.** $r = \sin \theta + 1$ **59.** $r^2 = \cos \theta$ **60.** $r = \sin \theta - \cos \theta$

61. $r = 2$ **62.** $r = 4$ **63.** $r = \dfrac{4}{1 - \cos \theta}$ **64.** $r = \dfrac{3}{3 - \cos \theta}$

65. Show that the formula for the distance d between two points $P_1 = (r_1, \theta_1)$ and $P_2 = (r_2, \theta_2)$ is
$$d = \sqrt{r_1^2 + r_2^2 - 2r_1r_2 \cos(\theta_2 - \theta_1)}$$

8.5

Polar Equations and Graphs

Polar Equation   An equation whose variables are polar coordinates is called a **polar equation.** The **graph of a polar equation** is the set of all points whose polar coordinates satisfy the equation.

Just as a rectangular grid may be used to plot points given by rectangular coordinates, as in Figure 42(a), we can use a grid consisting of concentric circles

FIGURE 42

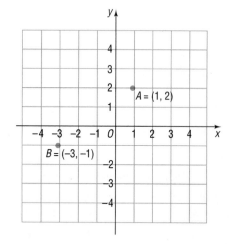

(a) Rectangular grid

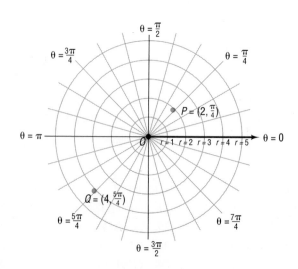

(b) Polar grid

(with centers at the pole) and rays (with vertices at the pole) to plot points given by polar coordinates, as shown in Figure 42(b). We shall use such **polar grids** to graph polar equations.

One method we can use to graph a polar equation is to convert the equation to rectangular coordinates. In the discussion that follows, (x, y) represent the rectangular coordinates of a point P and (r, θ) represent polar coordinates of the point P.

E X A M P L E   1   *Identifying and Graphing a Polar Equation (Circle)*

Identify and graph the equation: $r = 3$

Solution   We convert the polar equation to a rectangular equation:

$$r = 3$$
$$r^2 = 9$$
$$x^2 + y^2 = 9$$

Thus, the graph of $r = 3$ is a circle, with center at the pole and radius 3. See Figure 43.

FIGURE 43
$r = 3$ or $x^2 + y^2 = 9$

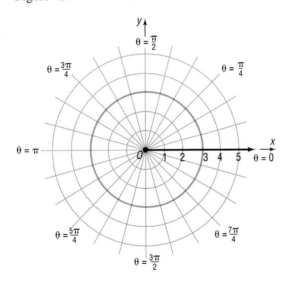

■   Now work Problem 1.

E X A M P L E   2   ·   *Identifying and Graphing a Polar Equation (Line)*

Identify and graph the equation: $\theta = \pi/4$

Solution   We convert the polar equation to a rectangular equation:

$$\theta = \frac{\pi}{4}$$

$$\tan \theta = \tan \frac{\pi}{4} = 1$$

$$\frac{y}{x} = 1$$

$$y = x$$

The graph of $\theta = \pi/4$ is a line passing through the pole making an angle of $\pi/4$ with the polar axis. See Figure 44. ∎

FIGURE 44

$\theta = \dfrac{\pi}{4}$ or $y = x$

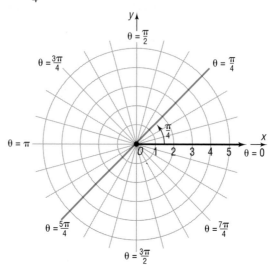

EXAMPLE 3   *Identifying and Graphing a Polar Equation (Line)*

Identify and graph the equation: $\theta = -\pi/6$

Solution   We convert the polar equation to a rectangular equation:

$$\theta = -\frac{\pi}{6}$$

$$\tan \theta = \tan\left(-\frac{\pi}{6}\right) = -\frac{\sqrt{3}}{3}$$

$$\frac{y}{x} = -\frac{\sqrt{3}}{3}$$

$$y = -\frac{\sqrt{3}}{3}x$$

The graph of $\theta = -\pi/6$ is a line passing through the pole making an angle of $-\pi/6$ with the polar axis. See Figure 45. ∎

FIGURE 45

$\theta = -\dfrac{\pi}{6}$ or $y = -\dfrac{\sqrt{3}}{3}x$

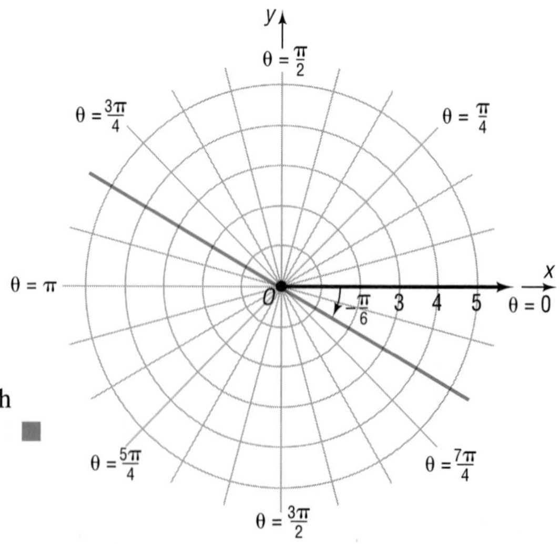

∎   Now work Problem 3.

EXAMPLE 4   *Identifying and Graphing a Polar Equation (Horizontal Line)*

Identify and graph the equation: $r \sin \theta = 2$

Solution   Since $y = r \sin \theta$, we can simply write the equation as

$$y = 2$$

We conclude that the graph of $r \sin \theta = 2$ is a horizontal line 2 units above the pole. See Figure 46.

FIGURE 46
$r \sin \theta = 2$ or $y = 2$

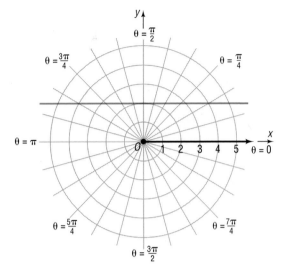

Comment: To graph an equation in polar coordinates, the format $r = f(\theta)$ must be used. Also, RANGE values must be set. It is best to use a square screen and radian measure.

Check: Verify the graph of $r \sin \theta = 2$ by graphing $r = 2/(\sin \theta)$.

E X A M P L E   5   *Identifying and Graphing a Polar Equation (Vertical Line)*

Identify and graph the equation: $r \cos \theta = -3$

Solution   Since $x = r \cos \theta$, we can simply write the equation as

$$x = -3$$

We conclude that the graph of $r \cos \theta = -3$ is a vertical line 3 units to the left of the pole. See Figure 47.

FIGURE 47
$r \cos \theta = -3$ or $x = -3$

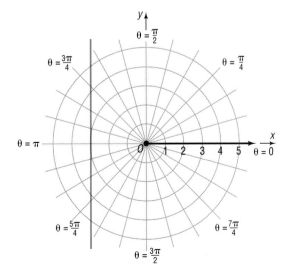

Based on Examples 4 and 5, we are led to the following results. (The proofs are left as exercises.)

Theorem If a is a nonzero real number:

The graph of the equation

$$r \sin \theta = a$$

is a horizontal line a units above the pole if $a > 0$ and $|a|$ units below the pole if $a < 0$.

The graph of the equation

$$r \cos \theta = a$$

is a vertical line a units to the right of the pole if $a > 0$ and $|a|$ units to the left of the pole if $a < 0$. ■

■ Now work Problem 7.

E X A M P L E   6 *Identifying and Graphing a Polar Equation (Circle)*

Identify and graph the equation: $r = 4 \sin \theta$

Solution To transform the equation to rectangular coordinates, we multiply each side by r:

$$r^2 = 4r \sin \theta$$

Now we use the facts that $r^2 = x^2 + y^2$ and $y = r \sin \theta$. Then

$$x^2 + y^2 = 4y$$
$$x^2 + (y^2 - 4y) = 0$$
$$x^2 + (y - 2)^2 = 4$$

This is the equation of a circle with center at $(0, 2)$ in rectangular coordinates and radius 2. See Figure 48. ■

FIGURE 48
$r = 4 \sin \theta$ or $x^2 + (y - 2)^2 = 4$

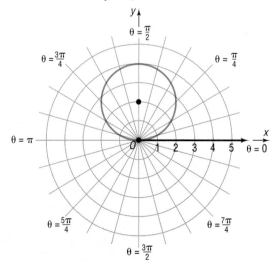

E X A M P L E   7   *Identifying and Graphing a Polar Equation (Circle)*

Identify and graph the equation: $r = -2 \cos \theta$

Solution   We proceed as in Example 6:

$$r^2 = -2r \cos \theta$$

$$x^2 + y^2 = -2x$$

$$x^2 + 2x + y^2 = 0$$

$$(x + 1)^2 + y^2 = 1$$

This is the equation of a circle with center at $(-1, 0)$ in rectangular coordinates and radius 1. See Figure 49.

Check: Graph $r = 4 \sin \theta$ and compare the result with Figure 48. Clear the screen and do the same for $r = -2 \cos \theta$ and compare with Figure 49. ■

FIGURE 49

$r = -2 \cos \theta$ or $(x + 1)^2 + y^2 = 1$

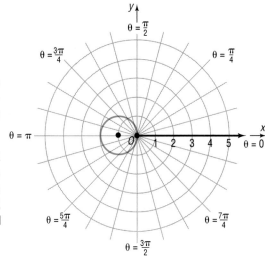

Exploration: Graph $r = \sin \theta$, $r = 2 \sin \theta$, and $r = 3 \sin \theta$. Do you see the pattern? Clear the screen and graph $r = -\sin \theta$, $r = -2 \sin \theta$, and $r = -3 \sin \theta$. Do you see the pattern? Clear the screen and graph $r = \cos \theta$, $r = 2 \cos \theta$, and $r = 3 \cos \theta$. Do you see the pattern? Clear the screen and graph $r = -\cos \theta$, $r = -2 \cos \theta$, and $r = -3 \cos \theta$. Do you see the pattern?

Based on Examples 6 and 7, we are led to the following results. (The proofs are left as exercises.)

Theorem   Let a be a positive real number. Then:

	EQUATION	DESCRIPTION
(a)	$r = 2a \sin \theta$	Circle: radius a; center at $(0, a)$ in rectangular coordinates
(b)	$r = -2a \sin \theta$	Circle: radius a; center at $(0, -a)$ in rectangular coordinates
(c)	$r = 2a \cos \theta$	Circle: radius a; center at $(a, 0)$ in rectangular coordinates
(d)	$r = -2a \cos \theta$	Circle: radius a; center at $(-a, 0)$ in rectangular coordinates

■

The method of converting a polar equation to an identifiable rectangular equation in order to graph it is not always helpful, nor is it always necessary. Usually, we set up a table that lists several points on the graph. By checking for symmetry, it may be possible to reduce the number of points needed to draw the graph.

Symmetry

In polar coordinates the points (r, θ) and $(r, -\theta)$ are symmetric with respect to the polar axis (and to the x-axis). See Figure 50(a). The points (r, θ) and $(r, \pi - \theta)$ are symmetric with respect to the line $\theta = \pi/2$ (y-axis). See Figure 50(b). The points (r, θ) and $(-r, \theta)$ are symmetric with respect to the pole (origin). See Figure 50(c).

FIGURE 50

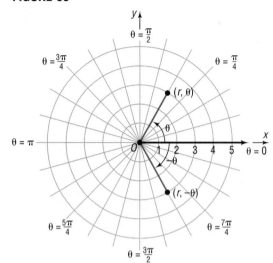

(a) Points symmetric with respect to the polar axis

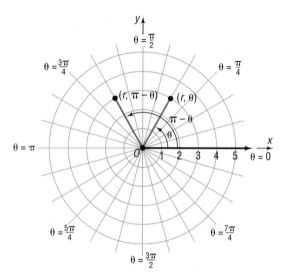

(b) Points symmetric with respect to the line $\theta = \frac{\pi}{2}$

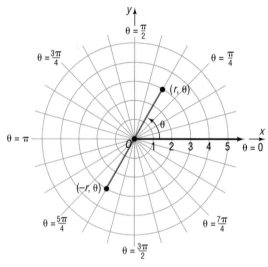

(c) Points symmetric with respect to the pole

The following tests are a consequence of these observations.

Theorem
Tests for Symmetry

Symmetry with Respect to the Polar Axis (*x*-Axis):
In a polar equation, replace θ by $-\theta$. If an equivalent equation results, the graph is symmetric with respect to the polar axis.

Symmetry with Respect to the Line $\theta = \pi/2$ (*y*-Axis):
In a polar equation, replace θ by $\pi - \theta$. If an equivalent equation results, the graph is symmetric with respect to the line $\theta = \pi/2$.

Symmetry with Respect to the Pole (Origin):
In a polar equation, replace r by $-r$. If an equivalent equation results, the graph is symmetric with respect to the pole.

The three tests for symmetry given here are *sufficient* conditions for symmetry, but they are not *necessary* conditions. That is, an equation may fail these tests and still have a graph that is symmetric with respect to the polar axis, the line $\theta = \pi/2$, or the pole. For example, the graph of $r = \sin 2\theta$ turns out to be symmetric with respect to the polar axis, the line $\theta = \pi/2$, and the pole, but all three tests given here fail.

E X A M P L E   8   *Graphing a Polar Equation (Cardioid)*

Graph the equation: $r = 1 - \sin \theta$

Solution   We check for symmetry first:

Polar Axis: Replace θ by $-\theta$. The result is

$$r = 1 - \sin(-\theta) = 1 + \sin \theta$$

The test fails, so the graph may or may not be symmetric with respect to the polar axis.

The Line $\theta = \pi/2$: Replace θ by $\pi - \theta$. The result is

$$r = 1 - \sin(\pi - \theta) = 1 - (\sin \pi \cos \theta - \cos \pi \sin \theta)$$
$$= 1 - [0 \cdot \cos \theta - (-1) \sin \theta] = 1 - \sin \theta$$

Thus, the graph is symmetric with respect to the line $\theta = \pi/2$.

The Pole: Replace r by $-r$. Then the result is $-r = 1 - \sin \theta$, so $r = -1 + \sin \theta$. The test fails, so the graph may or may not be symmetric with respect to the pole.

Next, we identify points on the graph by assigning values to the angle θ and calculating the corresponding values of r. Due to the symmetry with respect to the line $\theta = \pi/2$, we only need to assign values to θ from $-\pi/2$ to $\pi/2$, as given in Table 1.

Now we plot the points (r, θ) from Table 1 and trace out the graph, beginning at the point $(2, -\pi/2)$ and ending at the point $(0, \pi/2)$. Then we reflect this portion of the graph about the line $\theta = \pi/2$ (y-axis) to obtain the complete graph. See Figure 51.

TABLE 1

θ	$r = 1 - \sin \theta$
$-\pi/2$	$1 + 1 = 2$
$-\pi/3$	$1 + \sqrt{3}/2 \approx 1.87$
$-\pi/6$	$1 + \frac{1}{2} = \frac{3}{2}$
0	1
$\pi/6$	$1 - \frac{1}{2} = \frac{1}{2}$
$\pi/3$	$1 - \sqrt{3}/2 \approx 0.13$
$\pi/2$	0

FIGURE 51

$r = 1 - \sin \theta$

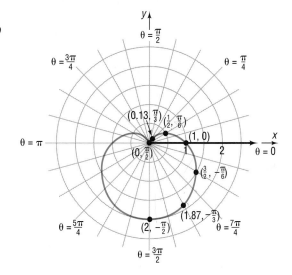

 Check: Graph $r = 1 - \sin \theta$ and compare the result with Figure 51. ∎

Exploration: Graph $r = 1 + \sin \theta$. Clear the screen and graph $r = 1 - \cos \theta$. Clear the screen and graph $r = 1 + \cos \theta$. Do you see a pattern?

The curve in Figure 51 is an example of a *cardioid* (a heart-shaped curve).

Cardioid

Cardioids are characterized by equations of the form

$$r = a(1 + \cos \theta) \qquad r = a(1 + \sin \theta)$$
$$r = a(1 - \cos \theta) \qquad r = a(1 - \sin \theta)$$

where $a > 0$. The graph of a cardioid contains the pole.

■ Now work Problem 25.

E X A M P L E   9 *Graphing a Polar Equation (Limaçon)*

Graph the equation: $r = 3 + 2 \cos \theta$

Solution   We check for symmetry first:

Polar Axis: Replace θ by $-\theta$. The result is

$$r = 3 + 2 \cos(-\theta) = 3 + 2 \cos \theta$$

Thus, the graph is symmetric with respect to the polar axis.

The Line $\theta = \pi/2$: Replace θ by $\pi - \theta$. The result is

$$r = 3 + 2 \cos(\pi - \theta) = 3 + 2(\cos \pi \cos \theta + \sin \pi \sin \theta)$$
$$= 3 - 2 \cos \theta$$

The test fails, so the graph may or may not be symmetric with respect to the line $\theta = \pi/2$.

The Pole: Replace r by $-r$. The test fails, so the graph may or may not be symmetric with respect to the pole.

Next, we identify points on the graph by assigning values to the angle θ and calculating the corresponding values of r. Due to the symmetry with respect to the polar axis, we only need to assign values to θ from 0 to π, as given in Table 2.

Now we plot the points (r, θ) from Table 2 and trace out the graph, beginning at the point $(5, 0)$ and ending at the point $(\pi, 1)$. Then we reflect this portion of the graph about the pole (x-axis) to obtain the complete graph. See Figure 52.

TABLE 2

θ	$r = 3 + 2 \cos \theta$
0	5
$\pi/6$	$3 + \sqrt{3} \approx 4.73$
$\pi/3$	4
$\pi/2$	3
$2\pi/3$	2
$5\pi/6$	$3 - \sqrt{3} \approx 1.27$
π	1

FIGURE 52

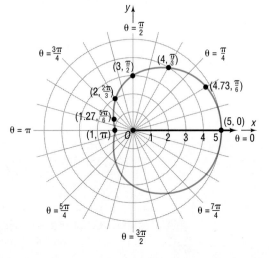
$r = 3 + 2 \cos \theta$

Check: **Graph** $r = 3 + 2 \cos \theta$ and compare the result with Figure 52.   ■

Exploration: Graph $r = 3 - 2 \cos \theta$. Clear the screen and graph $r = 3 + 2 \sin \theta$. Clear the screen and graph $r = 3 - 2 \sin \theta$. Do you see a pattern?

The curve in Figure 52 is an example of a *limaçon* (the French word for *snail*) without an inner loop.

Limaçon without an
Inner Loop

Limaçons without an inner loop are characterized by equations of the form

$$r = a + b \cos \theta \qquad r = a + b \sin \theta$$
$$r = a - b \cos \theta \qquad r = a - b \sin \theta$$

where $a > 0$, $b > 0$, and $a > b$. The graph of a limaçon without an inner loop does not contain the pole.

■ Now work Problem 31.

E X A M P L E   1 0   *Graphing a Polar Equation (Limaçon with Inner Loop)*

Graph the equation: $r = 1 + 2 \cos \theta$

Solution   First, we check for symmetry:

Polar Axis: Replace θ by $-\theta$. The result is

$$r = 1 + 2 \cos(-\theta) = 1 + 2 \cos \theta$$

Thus, the graph is symmetric with respect to the polar axis.

The Line $\theta = \pi/2$: Replace θ by $\pi - \theta$. The result is

$$r = 1 + 2 \cos(\pi - \theta) = 1 + 2(\cos \pi \cos \theta + \sin \pi \sin \theta)$$
$$= 1 - 2 \cos \theta$$

The test fails, so the graph may or may not be symmetric with respect to the line $\theta = \pi/2$.

The Pole: Replace r by $-r$. The test fails, so the graph may or may not be symmetric with respect to the pole.

TABLE 3

θ	$r = 1 + 2 \cos \theta$
0	3
$\pi/6$	$1 + \sqrt{3} \approx 2.73$
$\pi/3$	2
$\pi/2$	1
$2\pi/3$	0
$5\pi/6$	$1 - \sqrt{3} \approx -0.73$
π	-1

Next, we identify points on the graph of $r = 1 + 2 \cos \theta$ by assigning values to the angle θ and calculating the corresponding values of r. Due to the symmetry with respect to the polar axis, we only need to assign values to θ from 0 to π, as given in Table 3.

Now we plot the points (r, θ) from Table 3, beginning at $(3, 0)$ and ending at $(-1, \pi)$. See Figure 53(a). Finally, we reflect this portion of the graph about the polar axis (x-axis) to obtain the complete graph. See Figure 53(b).

FIGURE 53

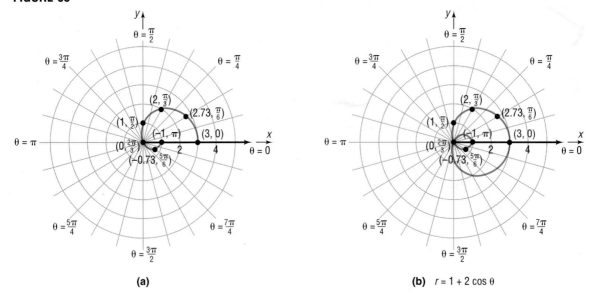

(a)   **(b)** $r = 1 + 2 \cos \theta$

    Check: Graph $r = 1 + 2 \cos \theta$ and compare the result with Figure 53(b).   ■

    Exploration: Graph $r = 1 - 2 \cos \theta$. Clear the screen and graph $r = 1 + 2 \sin \theta$. Clear the screen and graph $r = 1 - 2 \sin \theta$. Do you see a pattern?

The curve in Figure 53(b) is an example of a limaçon with an inner loop.

Limaçon with an Inner Loop

> **Limaçons with an inner loop** are characterized by equations of the form
>
> $$r = a + b \cos \theta \qquad r = a + b \sin \theta$$
> $$r = a - b \cos \theta \qquad r = a - b \sin \theta$$
>
> where $a > 0$, $b > 0$, and $a < b$. The graph of a limaçon with an inner loop will pass through the pole twice.

■ Now work Problem 33.

E X A M P L E   1 1   *Graphing a Polar Equation (Rose)*

Graph the equation: $r = 2 \cos 2\theta$

Solution   We check for symmetry:

Polar Axis: If we replace θ by $-\theta$, the result is

$$r = 2 \cos 2(-\theta) = 2 \cos 2\theta$$

Thus, the graph is symmetric with respect to the polar axis.

The Line $\theta = \pi/2$: If we replace θ by $\pi - \theta$, we obtain

$$r = 2 \cos 2(\pi - \theta) = 2 \cos(2\pi - 2\theta) = 2 \cos(-2\theta) = 2 \cos 2\theta$$

Thus, the graph is symmetric with respect to the line $\theta = \pi/2$.

The Pole: Since the graph is symmetric with respect to both the polar axis and the line $\theta = \pi/2$, it must be symmetric with respect to the pole.

TABLE 4

θ	$r = 2 \cos 2\theta$
0	2
$\pi/6$	1
$\pi/4$	0
$\pi/3$	-1
$\pi/2$	-2

Next, we construct Table 4. Due to the symmetry with respect to the polar axis, the line $\theta = \pi/2$, and the pole, we consider only values of θ from 0 to $\pi/2$.

We plot and connect these points in Figure 54(a). Finally, because of symmetry, we reflect this portion of the graph first about the polar axis (*x*-axis) and then about the line $\theta = \pi/2$ (*y*-axis) to obtain the complete graph. See Figure 54(b).

FIGURE 54

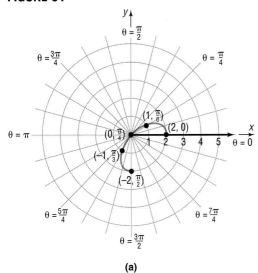

(a)

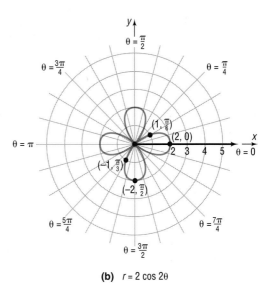

(b) $r = 2 \cos 2\theta$

Check: Graph $r = 2 \cos 2\theta$ and compare the results with Figure 54(b). Notice how the graph begins the same as Figure 54(a). ∎

The curve in Figure 54(b) is called a *rose* with four petals.

Rose

> **Rose** curves are characterized by equations of the form
> $$r = a \cos n\theta \qquad r = a \sin n\theta, \qquad a > 0$$
> and have graphs that are rose-shaped. If n is even, the rose has $2n$ petals; if n is odd, the rose has n petals.

■ Now work Problem 37.

E X A M P L E   1 2   *Graphing a Polar Equation (Lemniscate)*

Graph the equation: $r^2 = 4 \sin 2\theta$

Solution

TABLE 5

θ	$r^2 = 4 \sin 2\theta$	r
0	0	0
$\pi/6$	$2\sqrt{3}$	± 1.9
$\pi/4$	4	± 2
$\pi/3$	$2\sqrt{3}$	± 1.9
$\pi/2$	0	0

We leave it to you to verify that the graph is symmetric with respect to the pole. Table 5 lists points on the graph for values of $\theta = 0$ through $\theta = \pi/2$. Note that there are no points on the graph for $\pi/2 < \theta < \pi$ (quadrant II), since $\sin 2\theta < 0$ for such values. The points from Table 5 where $r \geq 0$ are plotted in Figure 55(a). The remaining points on the graph may be obtained by using symmetry. Figure 55(b) shows the final graph.

FIGURE 55

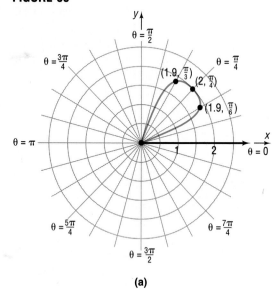

(a)

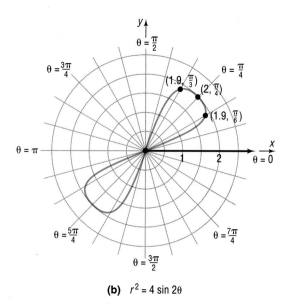

(b) $r^2 = 4 \sin 2\theta$

 Check: Graph $r^2 = 4 \sin 2\theta$ by entering $r = 2\sqrt{\sin 2\theta}$. Compare the result with Figure 55(b). ∎

The curve in Figure 55(b) is an example of a *lemniscate.*

Lemniscate

Lemniscates are characterized by equations of the form
$$r^2 = a^2 \sin 2\theta \qquad r^2 = a^2 \cos 2\theta$$
where $a \neq 0$, and have graphs that are propeller shaped.

■ Now work Problem 41.

E X A M P L E   1 3   *Graphing a Polar Equation (Spiral)*
Sketch the graph of: $r = e^{\theta/5}$

Solution   The tests for symmetry with respect to the pole, the polar axis, and the line $\theta = \pi/2$ fail. Furthermore, there is no number θ for which $r = 0$. Hence, the graph does not pass through the pole. We observe that r is positive for all θ, r increases as θ increases, $r \to 0$ as $\theta \to -\infty$, and $r \to \infty$ as $\theta \to \infty$. With the help of a calculator, we obtain the values in Table 6. See Figure 56 for the graph. ∎

The curve in Figure 56 is called a **logarithmic spiral,** since its equation may be written as $\theta = 5 \ln r$ and it spirals infinitely both toward the pole and away from it.

TABLE 6

θ	$r = e^{\theta/5}$
$-3\pi/2$	0.39
$-\pi$	0.53
$-\pi/2$	0.73
$-\pi/4$	0.85
0	1
$\pi/4$	1.17
$\pi/2$	1.37
π	1.87
$3\pi/2$	2.57
2π	3.51

FIGURE 56
$r = e^{\theta/5}$

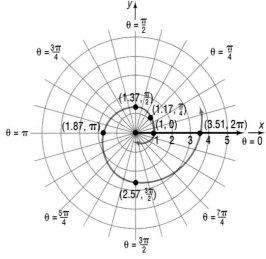

Classification of Polar Equations

The equations of some lines and circles in polar coordinates and their corresponding equations in rectangular coordinates are given in Table 7. Also included are the names and the graphs of a few of the more frequently encountered polar equations.

Calculus Comment

For those of you who are planning to study calculus, a comment about one important role of polar equations is in order.

In rectangular coordinates, the equation $x^2 + y^2 = 1$, whose graph is the unit circle, does not define a function. In fact, on the interval $[-1, 1]$ it defines two functions,

$$y = \sqrt{1 - x^2} \quad \text{Upper semicircle} \qquad y = -\sqrt{1 - x^2} \quad \text{Lower semicircle}$$

In polar coordinates, the equation $r = 1$, whose graph is also the unit circle, does define a function. That is, for each choice of θ there is only one corresponding value of r, that is, $r = 1$. Since many uses of calculus require that functions be used, the opportunity to express nonfunctions in rectangular coordinates as functions in polar coordinates becomes extremely useful.

Note also that the vertical line test for functions is valid only for equations in rectangular coordinates.

HISTORICAL FEATURE ■ Polar coordinates seem to have been invented by Jacob Bernoulli (1654–1705) about 1691, although, as with most such ideas, earlier traces of the notion exist. Early users of calculus remained committed to rectangular coordinates, and polar coordinates did not become widely used until the early 1800's. Even then, it was mostly geometers who used them for describing odd curves. Finally, about the mid-1800's, applied mathematicians realized the tremendous simplification polar coordinates make possible in the description of objects with circular or cylindrical symmetry. From then on their use became widespread. ■

TABLE 7

LINES

Description	Line passing through the pole making an angle α with the polar axis	Vertical line	Horizontal line
Rectangular equation	$y = (\tan \alpha)x$	$x = a$	$y = b$
Polar equation	$\theta = \alpha$	$r \cos \theta = a$	$r \sin \theta = b$
Typical graph	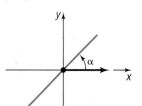		

CIRCLES

Description	Center at the pole, radius a	Passing through the pole, tangent to the line $\theta = \pi/2$, center on the polar axis, radius a	Passing through the pole, tangent to the polar axis, center on the line $\theta = \pi/2$, radius a
Rectangular equation	$x^2 + y^2 = a^2, a > 0$	$x^2 + y^2 = \pm 2ax, a > 0$	$x^2 + y^2 = \pm 2ay, a > 0$
Polar equation	$r = a, a > 0$	$r = \pm 2a \cos \theta, a > 0$	$r = \pm 2a \sin \theta, a > 0$
Typical graph	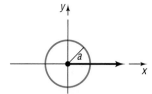	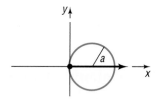	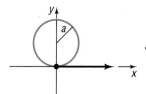

OTHER EQUATIONS

Name	Cardioid	Limaçon without inner loop	Limaçon with inner loop
Polar equations	$r = a \pm a \cos \theta, a > 0$ $r = a \pm a \sin \theta, a > 0$	$r = a \pm b \cos \theta, 0 < b < a$ $r = a \pm b \sin \theta, 0 < b < a$	$r = a \pm b \cos \theta, 0 < a < b$ $r = a \pm b \sin \theta, 0 < a < b$
Typical graph	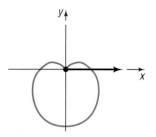	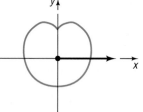	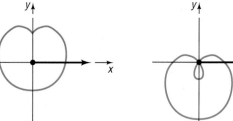

Name	Lemniscate	Rose with three petals	Rose with four petals
Polar equations	$r^2 = a^2 \cos 2\theta, a > 0$ $r^2 = a^2 \sin 2\theta, a > 0$	$r = a \sin 3\theta, a > 0$ $r = a \cos 3\theta, a > 0$	$r = a \sin 2\theta, a > 0$ $r = a \cos 2\theta, a > 0$
Typical graph	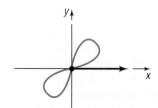	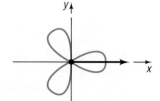	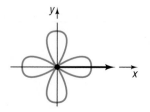

8.5

Exercise 8.5

In Problems 1–16, identify and graph each polar equation.

1. $r = 4$ **2.** $r = 2$ **3.** $\theta = \pi/3$ **4.** $\theta = -\pi/4$

5. $r \sin \theta = 4$ **6.** $r \cos \theta = 4$ **7.** $r \cos \theta = -2$ **8.** $r \sin \theta = -2$

9. $r = 2 \cos \theta$ **10.** $r = 2 \sin \theta$ **11.** $r = -4 \sin \theta$ **12.** $r = -4 \cos \theta$

13. $r \sec \theta = 4$ **14.** $r \csc \theta = 8$ **15.** $r \csc \theta = -2$ **16.** $r \sec \theta = -4$

In Problems 17–24, match each of the graphs (A) through (H) to one of the following polar equations.

17. $r = 2$ **18.** $\theta = \pi/4$ **19.** $r = 2 \cos \theta$ **20.** $r \cos \theta = 2$

21. $r = 1 + \cos \theta$ **22.** $r = 2 \sin \theta$ **23.** $\theta = 3\pi/4$ **24.** $r \sin \theta = 2$

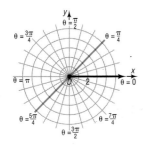

(A)

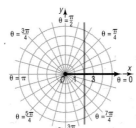

(B)

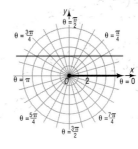

(C)

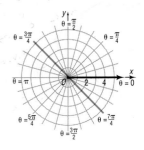

(D)

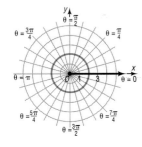

(E)

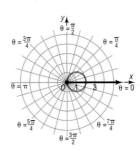

(F)

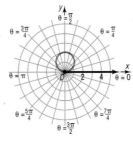

(G)

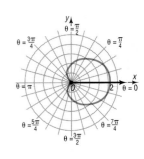

(H)

In Problems 25–48, identify and graph each polar equation. Be sure to test for symmetry.

25. $r = 2 + 2 \cos \theta$ **26.** $r = 1 + \sin \theta$ **27.** $r = 3 - 3 \sin \theta$ **28.** $r = 2 - 2 \cos \theta$

29. $r = 2 + \sin \theta$ **30.** $r = 2 - \cos \theta$ **31.** $r = 4 - 2 \cos \theta$ **32.** $r = 4 + 2 \sin \theta$

33. $r = 1 + 2 \sin \theta$ **34.** $r = 1 - 2 \sin \theta$ **35.** $r = 2 - 3 \cos \theta$ **36.** $r = 2 + 4 \cos \theta$

37. $r = 3 \cos 2\theta$ **38.** $r = 2 \sin 2\theta$ **39.** $r = 4 \sin 3\theta$ **40.** $r = 3 \cos 4\theta$

41. $r^2 = 9 \cos 2\theta$ **42.** $r^2 = \sin 2\theta$ **43.** $r = 2^\theta$ **44.** $r = 3^\theta$

45. $r = 1 - \cos \theta$ **46.** $r = 3 + \cos \theta$ **47.** $r = 1 - 3 \cos \theta$ **48.** $r = 4 \cos 3\theta$

In Problems 49–58, graph each polar equation.

49. $r = \dfrac{2}{1 - \cos \theta}$ (parabola)   **50.** $r = \dfrac{2}{1 - 2 \cos \theta}$ (hyperbola)

51. $r = \dfrac{1}{3 - 2 \cos \theta}$ (ellipse)   **52.** $r = \dfrac{1}{1 - \cos \theta}$ (parabola)

53. $r = \theta$, $\theta \geq 0$ (spiral of Archimedes)
54. $r = \dfrac{3}{\theta}$ (reciprocal spiral)

55. $r = \csc \theta - 2$, $0 < \theta < \pi$ (conchoid)
56. $r = \sin \theta \tan \theta$ (cissoid)

57. $r = \tan \theta$ (kappa curve)
58. $r = \cos \dfrac{\theta}{2}$

59. Show that the graph of the equation $r \sin \theta = a$ is a horizontal line a units above the pole if $a > 0$ and $|a|$ units below the pole if $a < 0$.

60. Show that the graph of the equation $r \cos \theta = a$ is a vertical line a units to the right of the pole if $a > 0$ and $|a|$ units to the left of the pole if $a < 0$.

61. Show that the graph of the equation $r = 2a \sin \theta$, $a > 0$, is a circle of radius a with center at $(0, a)$ in rectangular coordinates.

62. Show that the graph of the equation $r = -2a \sin \theta$, $a > 0$, is a circle of radius a with center at $(0, -a)$ in rectangular coordinates.

63. Show that the graph of the equation $r = 2a \cos \theta$, $a > 0$, is a circle of radius a with center at $(a, 0)$ in rectangular coordinates.

64. Show that the graph of the equation $r = -2a \cos \theta$, $a > 0$, is a circle of radius a with center at $(-a, 0)$ in rectangular coordinates.

65. Explain why the following test for symmetry is valid: Replace r by $-r$ and θ by $-\theta$ in a polar equation. If an equivalent equation results, the graph is symmetric with respect to the line $\theta = \pi/2$ (y-axis).
(a) Show that the test on page 520 fails for $r^2 = \cos \theta$, but this new test works.
(b) Show that the test on page 520 works for $r^2 = \sin \theta$, yet this new test fails.

66. Develop a new test for symmetry with respect to the pole.
(a) Find a polar equation for which this new test fails, yet the test on page 520 works.
(b) Find a polar equation for which the test on page 520 fails, yet the new test works.

67. Write down two different tests for symmetry with respect to the polar axis. Find examples in which one works and the other fails. Which test do you prefer to use? Justify your position.

8.6

The Complex Plane; Demoivre's Theorem

When we first introduced complex numbers, we were not prepared to give a geometric interpretation of a complex number. Now we are ready. Although there are several such interpretations we could give, the one that follows is the easiest to understand.

A complex number $z = x + yi$ can be interpreted geometrically as the point (x, y) in the xy-plane. Thus, each point in the plane corresponds to a complex number and, conversely, each complex number corresponds to a point in the plane. We shall refer to the collection of such points as the **complex plane.** The x-axis will be referred to as the **real axis,** because any point that lies on the real axis is of the form $z = x + 0i = x$, a real number. The y-axis is called the **imaginary axis,** because any point that lies on it is of the form $z = 0 + yi = yi$, a pure imaginary number. See Figure 57.

FIGURE 57
Complex plane

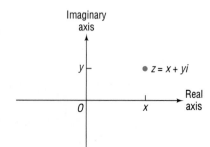

Magnitude

FIGURE 58

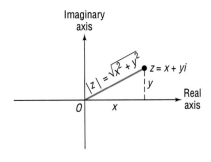

Let $z = x + yi$ be a complex number. The **magnitude** or **modulus** of z, denoted by $|z|$, is defined as the distance from the origin to the point (x, y). Thus,

$$|z| = \sqrt{x^2 + y^2} \qquad (1)$$

See Figure 58 for an illustration.

This definition for $|z|$ is consistent with the definition for the absolute value of a real number: If $z = x + yi$ is real, then $z = x + 0i$ and

$$|z| = \sqrt{x^2 + 0^2} = \sqrt{x^2} = |x|$$

Recall (Section 1.5) that if $z = x + yi$, then its **conjugate,** denoted by $\bar{z}$, is $\bar{z} = x - yi$. Because $z\bar{z} = x^2 + y^2$, it follows from equation (1) that the magnitude of z can be written as

$$|z| = \sqrt{z\bar{z}} \qquad (2)$$

Polar Form of a Complex Number

When a complex number is written in the form $z = x + yi$, we say that it is in **rectangular,** or **Cartesian, form,** because (x, y) are the rectangular coordinates of the corresponding point in the complex plane. Suppose that (r, θ) are the polar coordinates of this point. Then

$$x = r \cos \theta \qquad y = r \sin \theta \qquad (3)$$

If $r \geq 0$ and $0 \leq \theta < 2\pi$, the complex number $z = x + yi$ may be written in **polar form** as

Polar Form of z

$$z = x + yi = (r \cos \theta) + (r \sin \theta)i = r(\cos \theta + i \sin \theta) \qquad (4)$$

FIGURE 59

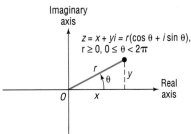

See Figure 59.

If $z = r(\cos \theta + i \sin \theta)$ is the polar form of a complex number, the angle θ, $0 \leq \theta < 2\pi$, is called the **argument of z.** Also, because $r \geq 0$, from equation (3) we have $r = \sqrt{x^2 + y^2}$. Thus, from equation (1) it follows that the magnitude of $z = r(\cos \theta + i \sin \theta)$ is

$$|z| = r$$

E X A M P L E 1

Plotting a Point in the Complex Plane and Writing a Complex Number in Polar Form

Plot the point corresponding to $z = \sqrt{3} - i$ in the complex plane, and write an expression for z in polar form.

Solution The point corresponding to $z = \sqrt{3} - i$ has the rectangular coordinates $(\sqrt{3}, -1)$. The point, located in quadrant IV, is plotted in Figure 60. Because $x = \sqrt{3}$ and $y = -1$, it follows that

$$r = \sqrt{x^2 + y^2} = \sqrt{(\sqrt{3})^2 + (-1)^2} = \sqrt{4} = 2$$

FIGURE 60

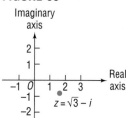

and

$$\sin \theta = \frac{y}{r} = \frac{-1}{2} \qquad \cos \theta = \frac{x}{r} = \frac{\sqrt{3}}{2}, \qquad 0 \le \theta < 2\pi$$

Thus, $\theta = 11\pi/6$ and $r = 2$, so the polar form of $z = \sqrt{3} - i$ is

$$z = r(\cos \theta + i \sin \theta) = 2\left(\cos \frac{11\pi}{6} + i \sin \frac{11\pi}{6}\right)$$ ■

■ Now work Problem 1.

E X A M P L E 2 *Plotting a Point in the Complex Plane and Converting from Polar to Rectangular Form*

Plot the point corresponding to $z = 2(\cos 30° + i \sin 30°)$ in the complex plane, and write an expression for z in rectangular form.

Solution To plot the complex number $z = 2(\cos 30° + i \sin 30°)$, we plot the point whose polar coordinates are $(r, \theta) = (2, 30°)$, as shown in Figure 61. In rectangular form,

FIGURE 61

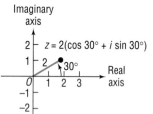

$$z = 2(\cos 30° + i \sin 30°) = 2\left(\frac{\sqrt{3}}{2} + \frac{1}{2}i\right) = \sqrt{3} + i$$ ■

■ Now work Problem 13.

The polar form of a complex number provides an alternative for finding products and quotients of complex numbers.

Theorem Let $z_1 = r_1(\cos \theta_1 + i \sin \theta_1)$ and $z_2 = r_2(\cos \theta_2 + i \sin \theta_2)$ be two complex numbers. Then

$$z_1 z_2 = r_1 r_2 [\cos(\theta_1 + \theta_2) + i \sin(\theta_1 + \theta_2)] \qquad (5)$$

If $z_2 \ne 0$, then

$$\frac{z_1}{z_2} = \frac{r_1}{r_2}[\cos(\theta_1 - \theta_2) + i \sin(\theta_1 - \theta_2)] \qquad (6)$$

■

Proof We shall prove formula (5). The proof of formula (6) is left as an exercise (see Problem 56).

$$z_1 z_2 = [r_1(\cos \theta_1 + i \sin \theta_1)][r_2(\cos \theta_2 + i \sin \theta_2)]$$
$$= r_1 r_2 [(\cos \theta_1 + i \sin \theta_1)(\cos \theta_2 + i \sin \theta_2)]$$
$$= r_1 r_2 [(\cos \theta_1 \cos \theta_2 - \sin \theta_1 \sin \theta_2) + i(\sin \theta_1 \cos \theta_2 + \cos \theta_1 \sin \theta_2)]$$
$$= r_1 r_2 [(\cos(\theta_1 + \theta_2) + i \sin(\theta_1 + \theta_2)]$$ ■

Because the magnitude of a complex number z is r and its argument is θ, when $z = r(\cos\theta + i\sin\theta)$, we can restate this theorem as follows:

Theorem   The magnitude of the product (quotient) of two complex numbers equals the product (quotient) of their magnitudes; the argument of the product (quotient) of two complex numbers equals the sum (difference) of their arguments.   ■

Let's look at an example of how this theorem can be used.

E X A M P L E   3   *Finding Products and Quotients of Complex Numbers in Polar Form*

If $z = 3(\cos 20° + i\sin 20°)$ and $w = 5(\cos 100° + \sin 100°)$, find the following (leave your answers in polar form):

(a) zw (b) z/w

Solution   (a) $$zw = [3(\cos 20° + i\sin 20°)][5(\cos 100° + i\sin 100°)]$$
$$= (3\cdot 5)[\cos(20° + 100°) + i\sin(20° + 100°)]$$
$$= 15(\cos 120° + i\sin 120°)$$

(b) $$\frac{z}{w} = \frac{3(\cos 20° + i\sin 20°)}{5(\cos 100° + i\sin 100°)}$$
$$= \tfrac{3}{5}[\cos(20° - 100°) + i\sin(20° - 100°)]$$
$$= \tfrac{3}{5}[\cos(-80°) + i\sin(-80°)]$$
$$= \tfrac{3}{5}(\cos 280° + i\sin 280°) \quad \text{Argument must lie between 0° and 360°}$$ ■

■ Now work Problem 23.

Demoivre's Theorem

Demoivre's Theorem, stated by Abraham Demoivre (1667–1754) in 1730, but already known to many people by 1710, is important for the following reason: The fundamental processes of algebra are the four operations of addition, subtraction, multiplication, and division, together with powers and the extraction of roots. Demoivre's Theorem allows these latter fundamental algebraic operations to be applied to complex numbers.

Demoivre's Theorem, in its most basic form, is a formula for raising a complex number z to the power n, where $n \geq 1$ is a positive integer. Let's see if we can guess the form of the result.

Let $z = r(\cos\theta + i\sin\theta)$ be a complex number. Then, based on equation (5), we have

$$n = 2: \quad z^2 = r^2(\cos 2\theta + i\sin 2\theta)$$
$$n = 3: \quad z^3 = z^2 \cdot z$$
$$= [r^2(\cos 2\theta + i\sin 2\theta)][r(\cos\theta + i\sin\theta)]$$
$$= r^3(\cos 3\theta + i\sin 3\theta) \quad \text{Equation (5)}$$
$$n = 4: \quad z^4 = z^3 \cdot z$$
$$= [r^3(\cos 3\theta + i\sin 3\theta)][r(\cos\theta + i\sin\theta)]$$
$$= r^4(\cos 4\theta + i\sin 4\theta) \quad \text{Equation (5)}$$

The pattern should now be clear.

Theorem
Demoivre's Theorem

If $z = r(\cos \theta + i \sin \theta)$ is a complex number, then

$$z^n = r^n(\cos n\theta + i \sin n\theta) \tag{7}$$

where $n \geq 1$ is a positive integer. ∎

We shall not prove Demoivre's Theorem because it requires mathematical induction (which is not discussed until Section 11.4).

Let's look at some examples.

E X A M P L E 4   *Using Demoivre's Theorem*

$$[2(\cos 20° + i \sin 20°)]^3 = 2^3[\cos(3 \cdot 20°) + i \sin(3 \cdot 20°)]$$
$$= 8(\cos 60° + i \sin 60°)$$
$$= 8\left(\frac{1}{2} + \frac{\sqrt{3}}{2}i\right) = 4 + 4\sqrt{3}\, i$$ ∎

∎ Now work Problem 31.

E X A M P L E 5   *Using Demoivre's Theorem*

Write $(1 + i)^5$ in the standard form $a + bi$.

Solution   To apply Demoivre's Theorem, we must first write the complex number in polar form. Thus, since the magnitude of $1 + i$ is $\sqrt{1^2 + 1^2} = \sqrt{2}$, we begin by writing

$$1 + i = \sqrt{2}\left(\frac{1}{\sqrt{2}} + \frac{1}{\sqrt{2}}i\right) = \sqrt{2}\left(\cos\frac{\pi}{4} + i \sin\frac{\pi}{4}\right)$$

Now

$$(1 + i)^5 = \left[\sqrt{2}\left(\cos\frac{\pi}{4} + i \sin\frac{\pi}{4}\right)\right]^5$$
$$= (\sqrt{2})^5\left[\cos\left(5 \cdot \frac{\pi}{4}\right) + i \sin\left(5 \cdot \frac{\pi}{4}\right)\right]$$
$$= 4\sqrt{2}\left(\cos\frac{5\pi}{4} + i \sin\frac{5\pi}{4}\right)$$
$$= 4\sqrt{2}\left[-\frac{1}{\sqrt{2}} + \left(-\frac{1}{\sqrt{2}}\right)i\right] = -4 - 4i$$ ∎

E X A M P L E 6   *Using a Calculator with Demoivre's Theorem*

Write $(3 + 4i)^3$ in the standard form $a + bi$.

Solution   Again, we start by writing $3 + 4i$ in polar form. This time we will use degrees for the argument. The magnitude of $3 + 4i$ is $\sqrt{3^2 + 4^2} = \sqrt{25} = 5$, so we write

$$3 + 4i = 5\left(\frac{3}{5} + \frac{4}{5}i\right) \approx 5(\cos 53.1° + i \sin 53.1°)$$

Although we have written the angle rounded to one decimal place (53.1°), we keep the actual value of the angle in memory. Now

$$
\begin{aligned}
(3 + 4i)^3 &\approx [5(\cos 53.1° + i \sin 53.1°)]^3 \\
&= 5^3[\cos(3 \cdot 53.1°) + i \sin(3 \cdot 53.1°)] \\
&= 125(\cos 159.3° + i \sin 159.3°) \\
&\approx 125[-0.935 + i(0.353)] = -117 + 44i
\end{aligned}
$$

In this computation, we used the actual values stored in memory, not the rounded values shown. The final answer, $-117 + 44i$, is exact as you can verify by cubing $3 + 4i$. ■

Complex Roots

Let w be a given complex number, and let $n \geq 2$ denote a positive integer. Any complex number z that satisfies the equation

$$
z^n = w
$$

is called a **complex *n*th root** of w. In keeping with previous usage, if $n = 2$, the solutions of the equation $z^2 = w$ are called **complex square roots** of w, and if $n = 3$, the solutions of the equation $z^3 = w$ are called **complex cube roots** of w.

Theorem
Finding Complex Roots

Let $w = r(\cos \theta + i \sin \theta)$ be a complex number. If $w \neq 0$, there are n distinct complex nth roots of w, given by the formula

$$
z_k = \sqrt[n]{r}\left[\cos\left(\frac{\theta}{n} + \frac{2k\pi}{n}\right) + i \sin\left(\frac{\theta}{n} + \frac{2k\pi}{n}\right)\right] \tag{8}
$$

where $k = 0, 1, 2, \ldots, n - 1$.

Proof (Outline)

We shall not prove this result in its entirety. Instead, we shall show only that each z_k in equation (8) obeys the equation $z_k^n = w$ and, hence, each z_k is a complex nth root of w.

$$
\begin{aligned}
z_k^n &= \left\{\sqrt[n]{r}\left[\cos\left(\frac{\theta}{n} + \frac{2k\pi}{n}\right) + i \sin\left(\frac{\theta}{n} + \frac{2k\pi}{n}\right)\right]\right\}^n \\
&= (\sqrt[n]{r})^n\left[\cos n\left(\frac{\theta}{n} + \frac{2k\pi}{n}\right) + i \sin n\left(\frac{\theta}{n} + \frac{2k\pi}{n}\right)\right] \quad \text{Demoivre's Theorem} \\
&= r[\cos(\theta + 2k\pi) + i \sin(\theta + 2k\pi)] \\
&= r(\cos \theta + i \sin \theta) = w
\end{aligned}
$$

Thus, each z_k, $k = 0, 1, \ldots, n - 1$, is a complex nth root of w. To complete the proof, we would need to show that each z_k, $k = 0, 1, 2, \ldots, n - 1$, is, in fact, distinct and that there are no complex nth roots of w other than those given by equation (8). ■

E X A M P L E   7

Finding Complex Cube Roots

Find the complex cube roots of $-1 + \sqrt{3}\, i$. Leave your answers in polar form, with θ in degrees.

Solution   First, we express $-1 + \sqrt{3}\,i$ in polar form using degrees:

$$-1 + \sqrt{3}\,i = 2\left(-\frac{1}{2} + \frac{\sqrt{3}}{2}i\right) = 2(\cos 120° + i \sin 120°)$$

The three complex cube roots of $-1 + \sqrt{3}\,i = 2(\cos 120° + i \sin 120°)$ are

$$z_k = \sqrt[3]{2}\left[\cos\left(\frac{120°}{3} + \frac{360°k}{3}\right) + i \sin\left(\frac{120°}{3} + \frac{360°k}{3}\right)\right], \qquad k = 0, 1, 2$$

Thus,

$$z_0 = \sqrt[3]{2}(\cos 40° + i \sin 40°)$$
$$z_1 = \sqrt[3]{2}(\cos 160° + i \sin 160°)$$
$$z_2 = \sqrt[3]{2}(\cos 280° + i \sin 280°)$$ ■

Notice that each of the three complex cube roots of $-1 + \sqrt{3}\,i$ has the same magnitude, $\sqrt[3]{2}$. This means that the points corresponding to each cube root lie the same distance from the origin; hence, the three points lie on a circle with center at the origin and radius $\sqrt[3]{2}$. Furthermore, the arguments of these cube roots are 40°, 160°, and 280°, the difference of consecutive pairs being 120°. This means that the three points are equally spaced on the circle, as shown in Figure 62. These results are not coincidental. In fact, you are asked to show that these results hold for complex nth roots in Problems 53 through 55.

FIGURE 62

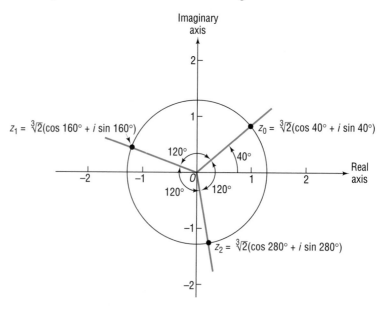

■   Now work Problem 43.

HISTORICAL FEATURE   ■   The Babylonians, Greeks, and Arabs considered square roots of negative quantities to be impossible and equations with complex solutions to be unsolvable. The first hint that there was some connection between real solutions of equations and complex numbers came when Girolamo Cardano (1501–1576) and Tartaglia (1499–1557) found *real* roots of cubic equations by taking cube roots of *complex* quantities. For centuries thereafter, mathematicians worked with complex numbers without much belief in their actual existence. In 1673, John Wallis appears to have been the first to suggest the graphical representation of complex numbers, a truly significant idea that was not pursued further until about 1800. Several people, including Karl Friedrich Gauss (1777–1855), then rediscovered the idea, and the graphical representation helped to establish complex numbers as equal members of the number family. In practical applications, complex numbers have found their greatest uses in the area of alternating current, where they are a commonplace tool, and subatomic physics. ■

HISTORICAL PROBLEMS

■ 1. In Problem 83, Exercise 3.5, we found $x = 2$ to be a solution of the cubic equation $x^3 - 6x + 4 = 0$. Recall that we could not apply the Tartaglia–Cardano method, because it led to the cube root of a complex number. Use Demoivre's Theorem to find the cube root and finish the problem.

2. The quadratic formula will work perfectly well if the coefficients are complex numbers. Solve the following, using Demoivre's Theorem where necessary. [*Hint:*   The answers are "nice."]

   (a) $z^2 - (2 + 5i)z - 3 + 5i = 0$ (b) $z^2 - (1 + i)\, z - 2 - i = 0$ ■

8.6

Exercise 8.6

In Problems 1–12, plot each complex number in the complex plane and write it in polar form. Express the argument in degrees.

1. $1 + i$
2. $-1 + i$
3. $\sqrt{3} - i$
4. $1 - \sqrt{3}\, i$
5. $-3i$
6. -2
7. $4 - 4i$
8. $9\sqrt{3} + 9i$
9. $3 - 4i$
10. $2 + \sqrt{3}\, i$
11. $-2 + 3i$
12. $\sqrt{5} - i$

In Problems 13–22, write each complex number in rectangular form.

13. $2(\cos 120° + i \sin 120°)$
14. $3(\cos 210° + i \sin 210°)$
15. $4\left(\cos \dfrac{7\pi}{4} + i \sin \dfrac{7\pi}{4}\right)$

16. $2\left(\cos \dfrac{5\pi}{6} + i \sin \dfrac{5\pi}{6}\right)$
17. $3\left(\cos \dfrac{3\pi}{2} + i \sin \dfrac{3\pi}{2}\right)$
18. $4\left(\cos \dfrac{\pi}{2} + i \sin \dfrac{\pi}{2}\right)$

19. $0.2(\cos 100° + i \sin 100°)$
20. $0.4(\cos 200° + i \sin 200°)$
21. $2\left(\cos \dfrac{\pi}{18} + i \sin \dfrac{\pi}{18}\right)$

22. $3\left(\cos \dfrac{\pi}{10} + i \sin \dfrac{\pi}{10}\right)$

In Problems 23–30, find zw and z/w. Leave your answer in polar form.

23. $z = 2(\cos 40° + i \sin 40°)$
    $w = 4(\cos 20° + i \sin 20°)$
24. $z = \cos 120° + i \sin 120°$
    $w = \cos 100° + i \sin 100°$
25. $z = 3(\cos 130° + i \sin 130°)$
    $w = 4(\cos 270° + i \sin 270°)$

26. $z = 2(\cos 80° + i \sin 80°)$
    $w = 6(\cos 200° + i \sin 200°)$
27. $z = 2\left(\cos \dfrac{\pi}{8} + i \sin \dfrac{\pi}{8}\right)$
    $w = 2\left(\cos \dfrac{\pi}{10} + i \sin \dfrac{\pi}{10}\right)$
28. $z = 4\left(\cos \dfrac{3\pi}{8} + i \sin \dfrac{3\pi}{8}\right)$
    $w = 2\left(\cos \dfrac{9\pi}{16} + i \sin \dfrac{9\pi}{16}\right)$

29. $z = 2 + 2i$
    $w = \sqrt{3} - i$
30. $z = 1 - i$
    $w = 1 - \sqrt{3}\, i$

In Problems 31–42, write each expression in the standard form a + bi.

31. $[4(\cos 40° + i \sin 40°)]^3$
32. $[3(\cos 80° + i \sin 80°)]^3$
33. $\left[2\left(\cos \dfrac{\pi}{10} + i \sin \dfrac{\pi}{10}\right)\right]^5$

34. $\left[\sqrt{2}\left(\cos \dfrac{5\pi}{16} + i \sin \dfrac{5\pi}{16}\right)\right]^4$
35. $[\sqrt{3}(\cos 10° + i \sin 10°)]^6$
36. $[\tfrac{1}{2}(\cos 72° + i \sin 72°)]^5$

37. $\left[\sqrt{5}\left(\cos \dfrac{3\pi}{16} + i \sin \dfrac{3\pi}{16}\right)\right]^4$
38. $\left[\sqrt{3}\left(\cos \dfrac{5\pi}{18} + i \sin \dfrac{5\pi}{18}\right)\right]^6$
39. $(1 - i)^5$

40. $(\sqrt{3} - i)^6$
41. $(\sqrt{2} - i)^6$
42. $(1 - \sqrt{5}\, i)^8$

In Problems 43–50, find all the complex roots. Leave your answers in polar form with θ in degrees.

43. The complex cube roots of $1 + i$

44. The complex fourth roots of $\sqrt{3} - i$

45. The complex fourth roots of $4 - 4\sqrt{3}\,i$

46. The complex cube roots of $-8 - 8i$

47. The complex fourth roots of $-16i$

48. The complex cube roots of -8

49. The complex fifth roots of i

50. The complex fifth roots of $-i$

51. Find the four complex roots of unity (1). Plot each.

52. Find the six complex roots of unity (1). Plot each.

53. Show that each complex nth root of a nonzero complex number w has the same magnitude.

54. Use the result of Problem 53 to draw the conclusion that each complex nth root lies on a circle with center at the origin. What is the radius of this circle?

55. Refer to Problem 54. Show that the complex nth roots of a nonzero complex number w are equally spaced on the circle.

56. Prove formula (6).

Chapter Review

THINGS TO KNOW

Formulas

Law of Sines	$\dfrac{\sin \alpha}{a} = \dfrac{\sin \beta}{b} = \dfrac{\sin \gamma}{c}$
Law of Cosines	$c^2 = a^2 + b^2 - 2ab \cos \gamma$
	$b^2 = a^2 + c^2 - 2ac \cos \beta$
	$a^2 = b^2 + c^2 - 2bc \cos \alpha$
Area of a triangle	$A = \frac{1}{2}bh$
	$A = \frac{1}{2}ab \sin \gamma$
	$A = \frac{1}{2}bc \sin \alpha$
	$A = \frac{1}{2}ac \sin \beta$
	$A = \sqrt{s(s-a)(s-b)(s-c)}$, where $s = \frac{1}{2}(a+b+c)$
Relationship between polar coordinates (r, θ) and rectangular coordinates (x, y)	$x = r \cos \theta,\ y = r \sin \theta$
	$x^2 + y^2 = r^2,\ \tan \theta = \dfrac{y}{x},\ x \neq 0$
Polar form of a complex number	If $z = x + iy$, then $z = r(\cos \theta + i \sin \theta)$,
	where $r = \|z\| = \sqrt{x^2 + y^2}$, $\sin \theta = \dfrac{y}{r}$ $\cos \theta = \dfrac{x}{r}$ $0 \leq \theta < 2\pi$
Demoivre's Theorem	If $z = r(\cos \theta + i \sin \theta)$, then
	$z^n = r^n(\cos n\theta + i \sin n\theta)$, where $n \geq 1$ is a positive integer
nth root of a complex number	$\sqrt[n]{z} = \sqrt[n]{r}\left[\cos\left(\dfrac{\theta}{n} + \dfrac{2k\pi}{n}\right) + i \sin\left(\dfrac{\theta}{n} + \dfrac{2k\pi}{n}\right)\right]$, $k = 0, \ldots, n-1$

HOW TO:

Use the Law of Sines to solve a SAA, ASA, or SSA triangle

Use the Law of Cosines to solve a SAS or SSS triangle

Find the area of a triangle

Plot polar coordinates

Convert from polar to rectangular coordinates

Convert from rectangular to polar coordinates

Graph polar equations (see Table 7)

Write a complex number in polar form, $z = r(\cos \theta + i \sin \theta)$, $0° \leq \theta < 360°$

Use Demoivre's Theorem to find powers of complex numbers

Use the complex roots theorem to find complex roots

FILL-IN-THE-BLANK ITEMS

1. If two sides and the angle opposite one of them are known, the Law of _____ is used to determine whether the known information results in no triangle, one triangle, or two triangles.

2. If three sides of a triangle are given, the Law of _____ is used to solve the triangle.

3. If three sides of a triangle are given, _____ Formula is used to find the area of the triangle.

4. In polar coordinates, the origin is called the _____, and the positive x-axis is referred to as the _____.

5. Another representation in polar coordinates for the point $(2, \pi/3)$ is (_____, $4\pi/3$).

6. Using polar coordinates (r, θ), the circle $x^2 + y^2 = 2x$ takes the form _____.

7. In a polar equation, replace θ by $-\theta$. If an equivalent equation results, the graph is symmetric with respect to _____.

8. When a complex number z is written in the polar form $z = r(\cos\theta + i\sin\theta)$, the nonnegative number r is the _____ _____ of z, and the angle θ, $0 \le \theta < 2\pi$, is the _____ of z.

TRUE/FALSE ITEMS

T   F   **1.** An oblique triangle in which two sides and an angle are given always results in at least one triangle.

T   F   **2.** Given three sides of a triangle, there is a formula for finding its area.

T   F   **3.** The polar coordinates of a point are unique.

T   F   **4.** The rectangular coordinates of a point are unique.

T   F   **5.** The tests for symmetry in polar coordinates are conclusive.

T   F   **6.** Demoivre's Theorem is useful for raising a complex number to a positive integer power.

REVIEW EXERCISES

In Problems 1–20, find the remaining angle(s) and side(s) of each triangle, if it exists. If no triangle exists, say "No triangle."

1. $\alpha = 50°$, $\beta = 30°$, $a = 1$	**2.** $\alpha = 10°$, $\gamma = 40°$, $c = 2$	**3.** $\alpha = 100°$, $a = 5$, $c = 2$
4. $a = 2$, $c = 5$, $\alpha = 60°$	**5.** $a = 3$, $c = 1$, $\gamma = 110°$	**6.** $a = 3$, $c = 1$, $\gamma = 20°$
7. $a = 3$, $c = 1$, $\beta = 100°$	**8.** $a = 3$, $b = 5$, $\beta = 80°$	**9.** $a = 2$, $b = 3$, $c = 1$
10. $a = 10$, $b = 7$, $c = 8$	**11.** $a = 1$, $b = 3$, $\gamma = 40°$	**12.** $a = 4$, $b = 1$, $\gamma = 100°$
13. $a = 5$, $b = 3$, $\alpha = 80°$	**14.** $a = 2$, $b = 3$, $\alpha = 20°$	**15.** $a = 1$, $b = \frac{1}{2}$, $c = \frac{4}{3}$
16. $a = 3$, $b = 2$, $c = 2$	**17.** $a = 3$, $\alpha = 10°$, $b = 4$	**18.** $a = 4$, $\alpha = 20°$, $\beta = 100°$
19. $c = 5$, $b = 4$, $\alpha = 70°$	**20.** $a = 1$, $b = 2$, $\gamma = 60°$	

In Problems 21–30, find the area of each triangle.

21. $a = 2$, $b = 3$, $\gamma = 40°$	**22.** $b = 5$, $c = 4$, $\alpha = 20°$	**23.** $b = 4$, $c = 10$, $\alpha = 70°$
24. $a = 2$, $b = 1$, $\gamma = 100°$	**25.** $a = 4$, $b = 3$, $c = 5$	**26.** $a = 10$, $b = 7$, $c = 8$
27. $a = 4$, $b = 2$, $c = 5$	**28.** $a = 3$, $b = 2$, $c = 2$	**29.** $\alpha = 50°$, $\beta = 30°$, $a = 1$
30. $\alpha = 10°$, $\gamma = 40°$, $c = 3$		

In Problems 31–36, plot each point given in polar coordinates, and find its rectangular coordinates.

31. $(3, \pi/6)$	**32.** $(4, 2\pi/3)$	**33.** $(-2, 4\pi/3)$
34. $(-1, 5\pi/4)$	**35.** $(-3, -\pi/2)$	**36.** $(-4, -\pi/4)$

In Problems 37–42, the rectangular coordinates of a point are given. Find two pairs of polar coordinates (r, θ) for each point, one with r > 0 and the other with r < 0. Express θ in radians.

37. $(-3, 3)$ **38.** $(1, -1)$ **39.** $(0, -2)$

40. $(2, 0)$ **41.** $(3, 4)$ **42.** $(-5, 12)$

In Problems 43–48, the letters x and y represent rectangular coordinates. Write each equation using polar coordinates (r, θ).

43. $3x^2 + 3y^2 = 6y$ **44.** $2x^2 - 2y^2 = 5y$ **45.** $2x^2 - y^2 = \dfrac{y}{x}$

46. $x^2 + 2y^2 = \dfrac{y}{x}$ **47.** $x(x^2 + y^2) = 4$ **48.** $y(x^2 - y^2) = 3$

In Problems 49–54, write each polar equation as an equation in rectangular coordinates (x, y).

49. $r = 2 \sin \theta$ **50.** $3r = \sin \theta$ **51.** $r = 5$

52. $\theta = \pi/4$ **53.** $r \cos \theta + 3r \sin \theta = 6$ **54.** $r^2 \tan \theta = 1$

In Problems 55–60, sketch the graph of each polar equation. Be sure to test for symmetry.

55. $r = 4 \cos \theta$ **56.** $r = 3 \sin \theta$ **57.** $r = 3 - 3 \sin \theta$

58. $r = 2 + \cos \theta$ **59.** $r = 4 - \cos \theta$ **60.** $r = 1 - 2 \sin \theta$

In Problems 61–64, write each complex number in polar form. Express each argument in degrees.

61. $-1 - i$ **62.** $-\sqrt{3} + i$ **63.** $4 - 3i$ **64.** $3 - 2i$

In Problems 65–70, write each complex number in the standard form a + bi.

65. $2(\cos 150° + i \sin 150°)$ **66.** $3(\cos 60° + i \sin 60°)$ **67.** $3\left(\cos \dfrac{2\pi}{3} + i \sin \dfrac{2\pi}{3}\right)$

68. $4\left(\cos \dfrac{3\pi}{4} + i \sin \dfrac{3\pi}{4}\right)$ **69.** $0.1(\cos 350° + i \sin 350°)$ **70.** $0.5(\cos 160° + i \sin 160°)$

In Problems 71–76, find zw and z/w. Leave your answers in polar form.

71. $z = \cos 80° + i \sin 80°$
    $w = \cos 50° + i \sin 50°$

72. $z = \cos 205° + i \sin 205°$
    $w = \cos 85° + i \sin 85°$

73. $z = 3\left(\cos \dfrac{9\pi}{5} + i \sin \dfrac{9\pi}{5}\right)$
    $w = 2\left(\cos \dfrac{\pi}{5} + i \sin \dfrac{\pi}{5}\right)$

74. $z = 2\left(\cos \dfrac{5\pi}{3} + i \sin \dfrac{5\pi}{3}\right)$
    $w = 3\left(\cos \dfrac{\pi}{3} + i \sin \dfrac{\pi}{3}\right)$

75. $z = 5(\cos 10° + i \sin 10°)$
    $w = \cos 355° + i \sin 355°$

76. $z = 4(\cos 50° + i \sin 50°)$
    $w = \cos 340° + i \sin 340°$

In Problems 77–84, write each expression in the standard form a + bi.

77. $[3(\cos 20° + i \sin 20°)]^3$ **78.** $[2(\cos 50° + i \sin 50°)]^3$ **79.** $\left[\sqrt{2}\left(\cos \dfrac{5\pi}{8} + i \sin \dfrac{5\pi}{8}\right)\right]^4$

80. $\left[2\left(\cos \dfrac{5\pi}{16} + i \sin \dfrac{5\pi}{16}\right)\right]^4$ **81.** $(1 - \sqrt{3}\,i)^6$ **82.** $(2 - 2i)^8$

83. $(3 + 4i)^4$ **84.** $(1 - 2i)^4$

85. Find all the complex cube roots of 27.

86. Find all the complex fourth roots of -16.

87. *Navigation* An airplane flies from city *A* to city *B*, a distance of 100 miles, then turns through an angle of 20° and heads toward city *C*, as indicated in the figure. If the distance from *A* to *C* is 300 miles, how far is it from city *B* to city *C*?

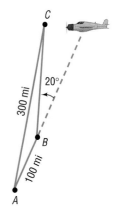

88. *Correcting a Navigation Error* Two cities *A* and *B* are 300 miles apart. In flying from city *A* to city *B*, a pilot inadvertently took a course that was 5° in error.

(a) If the error was discovered after flying 10 minutes at a constant speed of 420 miles per hour, through what angle should the pilot turn to correct the course? (Consult the figure.)

(b) What new constant speed should be maintained so that no time is lost due to the error? (Assume that the speed would have been a constant 420 miles per hour if no error had occurred.)

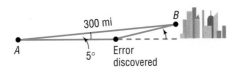

89. *Determining Distances at Sea* The navigator of a ship at sea spots two lighthouses that she knows to be 2 miles apart along a straight seashore. She determines that the angles formed between two line-of-sight observations of the lighthouses and the line from the ship directly to shore are 12° and 30°. See the illustration.

(a) How far is the ship from lighthouse *A*?

(b) How far is the ship from lighthouse *B*?

(c) How far is the ship from shore?

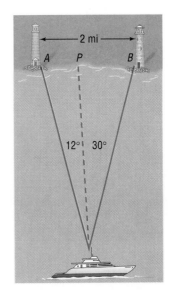

90. *Constructing a Highway* A highway whose primary directions are north–south is being constructed along the west coast of Florida. Near Naples, a bay obstructs the straight path of the road. Since the cost of a bridge is prohibitive, engineers decide to go around the bay. The illustration shows the path they decide on and the measurements taken. What is the length of highway needed to go around the bay?

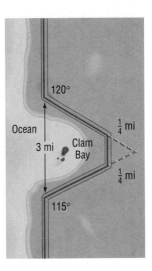

91. *Correcting a Navigational Error* A yacht leaves St. Thomas bound for an island in the British West Indies, 200 miles away. Maintaining a constant speed of 18 miles per hour, but encountering heavy crosswinds and strong currents, the crew finds after 4 hours that the sailboat is off course by 15°.

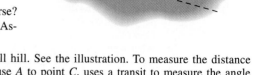

(a) How far is the sailboat from the island at this time?

(b) Through what angle should the sailboat turn to correct its course?

(c) How much time has been added to the trip because of this? (Assume that the speed remains at 18 miles per hour.)

92. *Surveying* Two homes are located on opposite sides of a small hill. See the illustration. To measure the distance between them, a surveyor walks a distance of 50 feet from house A to point C, uses a transit to measure the angle ACB, which is found to be 80°, and then walks to house B, a distance of 60 feet. How far apart are the houses?

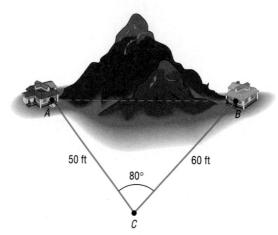

93. *Approximating the Area of a Lake* To approximate the area of a lake, a surveyor walks around the perimeter of the lake, taking the measurements shown in the illustration. Using this technique, what is the approximate area of the lake? [*Hint:* Use the Law of Cosines on the three triangles shown and then find the sum of their areas.]

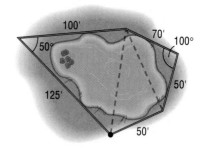

94. *Calculating the Cost of Land* The irregular parcel of land shown in the figure is being sold for $100 per square foot. What is the cost of this parcel?

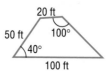

Chapter 9

ANALYTIC GEOMETRY

PREPARING FOR THIS CHAPTER

Before getting started on this chapter, review the following concepts:

Distance formula (p. 54)
Completing the square (Appendix A, Section A.3)
Intercepts (p. 59)
Symmetry (pp. 59–60)
Circles (pp. 63–66)
Double-angle and half-angle formulas (Section 7.3)
Polar coordinates (Section 8.4)
Amplitude and period of sinusoidal graphs (p. 392)

Preview   Satellite Dish

*A satellite dish is shaped like a **paraboloid of revolution,** a surface formed by rotating a parabola about its axis of symmetry. The signals that emanate from a satellite strike the surface of the dish and are reflected to a single point, where the receiver is located. If the dish is 8 feet across at its opening and is 3 feet deep at its center, at what position should the receiver be placed? [Example 8 in Section 9.2].*

Historically, Apollonius (200 BC) was among the first to study *conics* and discover some of their interesting properties. Today, conics are still studied because of their many uses. *Paraboloids of revolution* (parabolas rotated about their axes of symmetry) are used as signal collectors (the satellite dishes used with radar and cable TV, for example), as solar energy collectors, and as reflectors (telescopes, light projection, and so on). The planets circle the Sun in approximately *elliptical* orbits. Elliptical surfaces can be used to reflect signals such as light and sound from one place to another. And *hyperbolas* can be used to determine the positions of ships at sea.

The Greeks used the methods of Euclidean geometry to study conics. We shall use the more powerful methods of analytic geometry, bringing to bear both algebra and geometry, for our study of conics. Thus, we shall give a geometric description of each conic, and then, using rectangular coordinates and the distance formula, we shall find equations that represent conics. We used this same development, you may recall, when we first defined a circle in Section 1.6.

The chapter concludes with a section on equations of conics in polar coordinates, followed by a discussion of plane curves and parametric equations.

Preliminaries

FIGURE 1

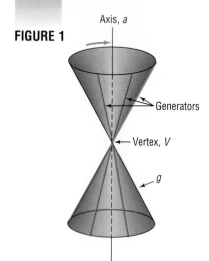

The word *conic* derives from the word *cone,* which is a geometric figure that can be constructed in the following way: Let a and g be two distinct lines that intersect at a point V. Keep the line a fixed. Now rotate the line g about a while maintaining the same angle between a and g. The collection of points swept out (generated) by the line g is called a **(right circular) cone.** See Figure 1. The fixed line a is called the **axis** of the cone; the point V is called its **vertex;** the lines that pass through V and make the same angle with a as g are called **generators** of the cone. Thus, each generator is a line that lies entirely on the cone. The cone consists of two parts, called **nappes,** that intersect at the vertex.

Conics, an abbreviation for **conic sections,** are curves that result from the intersection of a (right circular) cone and a plane. The conics we shall study arise when the plane does not contain the vertex, as shown in Figure 2. These conics are **circles** when the plane is perpendicular to the axis of the cone and intersects each generator; **ellipses** when the plane is tilted slightly so that it intersects each generator, but intersects only one nappe of the cone; **parabolas** when the plane is tilted further so that it is parallel to one (and only one) generator and intersects only one nappe of the cone; and **hyperbolas** when the plane intersects both nappes.

FIGURE 2

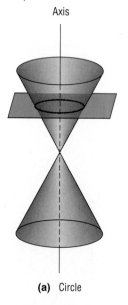

(a) Circle

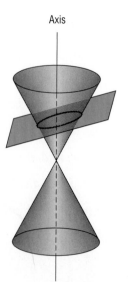

(b) Ellipse

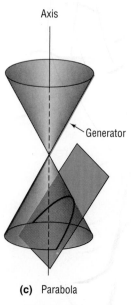

(c) Parabola

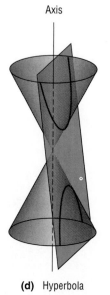

(d) Hyperbola

If the plane does contain the vertex, the intersection of the plane and the cone is a point, a line, or a pair of intersecting lines. These are usually called **degenerate conics.**

9.2

The Parabola

We stated earlier (Section 3.1) that the graph of a quadratic function is a parabola. In this section, we begin with a geometric definition of parabola and use it to obtain an equation.

Parabola

> A **parabola** is defined as the collection of all points P in the plane that are the same distance from a fixed point F as they are from a fixed line D. The point F is called the **focus** of the parabola, and the line D is its **directrix.** As a result, a parabola is the set of points P for which

$$d(F, P) = d(P, D) \qquad (1)$$

FIGURE 3

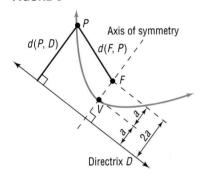

Figure 3 shows a parabola. The line through the focus F and perpendicular to the directrix D is called the **axis of symmetry** of the parabola. The point of intersection of the parabola with its axis of symmetry is called the **vertex** V.

Because the vertex V lies on the parabola, it must satisfy equation (1): $d(F, V) = d(V, D)$. Thus, the vertex is midway between the focus and the directrix. We shall let a equal the distance $d(F, V)$ from F to V. Now we are ready to derive an equation for a parabola. To do this, we use a rectangular system of coordinates, positioned so that the vertex V, focus F, and directrix D of the parabola are conveniently located. If we choose to locate the vertex V at the origin $(0, 0)$, then we can conveniently position the focus F on either the x-axis or the y-axis.

First, we consider the case where the focus F is on the positive x-axis, as shown in Figure 4. Because the distance from F to V is a, the coordinates of F will be $(a, 0)$ with $a > 0$. Similarly, because the distance from V to the directrix D is also a and because D must be perpendicular to the x-axis (since the x-axis is the axis of symmetry), the equation of the directrix D must be $x = -a$. Now, if $P = (x, y)$ is any point on the parabola, then P must obey equation (1):

$$d(F, P) = d(P, D)$$

FIGURE 4

$y^2 = 4ax$

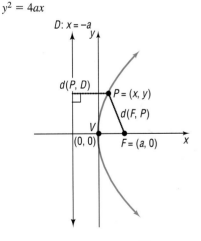

So, we have

$$\sqrt{(x - a)^2 + y^2} = |x + a| \qquad \text{Use the distance formula.}$$
$$(x - a)^2 + y^2 = (x + a)^2 \qquad \text{Square both sides.}$$
$$x^2 - 2ax + a^2 + y^2 = x^2 + 2ax + a^2$$
$$y^2 = 4ax$$

Theorem

The equation of a parabola with vertex at $(0, 0)$, focus at $(a, 0)$, and directrix $x = -a$, $a > 0$, is

Equation of a Parabola;
Vertex at $(0, 0)$,
Focus at $(a, 0)$, $a > 0$

$$y^2 = 4ax \qquad (2)$$

E X A M P L E   1   *Finding the Equation of a Parabola*

Find an equation of the parabola with vertex at (0, 0) and focus at (3, 0). Graph the equation.

FIGURE 5
$y^2 = 12x$

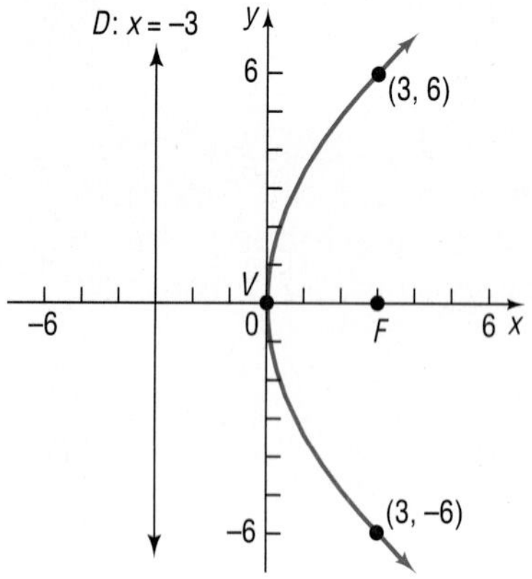

Solution   The distance from the vertex (0, 0) to the focus (3, 0) is $a = 3$. Based on equation (2), the equation of this parabola is

$$y^2 = 4ax$$
$$y^2 = 12x \qquad a = 3$$

To graph this parabola, it is helpful to plot the two points on the graph above and below the focus. To locate them, we let $x = 3$. Then

$$y^2 = 12x = 36$$
$$y = \pm 6$$

The points on the parabola above and below the focus are (3, −6) and (3, 6). See Figure 5.   ∎

In general, the points on a parabola $y^2 = 4ax$ that lie above and below the focus $(a, 0)$ are each at a distance $2a$ from the focus. This follows from the fact that if $x = a$ then $y^2 = 4ax = 4a^2$, or $y = \pm 2a$. The line segment joining these two points is called the **latus rectum**; its length is $4a$.

Comment: To graph the parabola $y^2 = 12x$ discussed in Example 1, we need to graph the two functions $y = \sqrt{12x}$ and $y = -\sqrt{12x}$. Do this and compare what you see with Figure 5.   ∎

■   Now work Problem 9.

By reversing the steps we used to obtain equation (2), it follows that the graph of an equation of the form of equation (2) is a parabola; its vertex is at (0, 0), its focus is at $(a, 0)$, its directrix is the line $x = -a$, and its axis of symmetry is the x-axis.

For the remainder of this section, the direction "Discuss the equation" will mean to find the vertex, focus, and directrix of the parabola and graph it.

E X A M P L E   2   *Discussing the Equation of a Parabola*

Discuss the equation: $y^2 = 8x$

FIGURE 6
$y^2 = 8x$

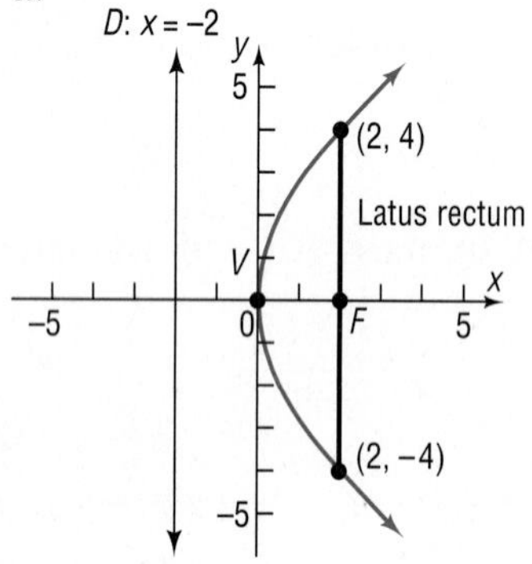

Solution   The equation $y^2 = 8x$ is of the form $y^2 = 4ax$, where $4a = 8$. Thus, $a = 2$. Consequently, the graph of the equation is a parabola with vertex at (0, 0) and focus on the positive x-axis at (2, 0). The directrix is the vertical line $x = -2$. The two points defining the latus rectum are obtained by letting $x = 2$. Then $y^2 = 16$, or $y = \pm 4$. These points help in graphing the parabola, since they determine the *opening* of the graph. See Figure 6.   ∎

Recall that we arrived at equation (2) after placing the focus on the positive x-axis. If the focus is placed on the negative x-axis, positive y-axis, or negative y-axis, a different form of the equation for the parabola results. The four forms of the equation of a parabola with vertex at (0, 0) and focus on a coordinate axis a distance a from (0, 0) are given in Table 1, and their graphs are given in Figure 7. Notice that each graph is symmetric with respect to its axis of symmetry.

TABLE 1   **EQUATIONS OF A PARABOLA: VERTEX AT (0, 0); FOCUS ON AXIS; $a > 0$**

VERTEX	FOCUS	DIRECTRIX	EQUATION	DESCRIPTION
(0, 0)	(a, 0)	$x = -a$	$y^2 = 4ax$	Parabola, axis of symmetry is the x-axis, opens to right
(0, 0)	(−a, 0)	$x = a$	$y^2 = -4ax$	Parabola, axis of symmetry is the x-axis, opens to left
(0, 0)	(0, a)	$y = -a$	$x^2 = 4ay$	Parabola, axis of symmetry is the y-axis, opens up
(0, 0)	(0, −a)	$y = a$	$x^2 = -4ay$	Parabola, axis of symmetry is the y-axis, opens down

FIGURE 7

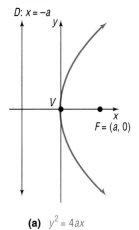 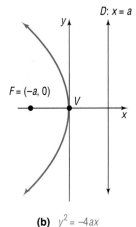 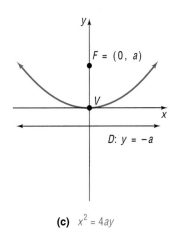 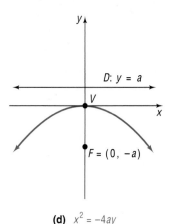

(a) $y^2 = 4ax$ (b) $y^2 = -4ax$ (c) $x^2 = 4ay$ (d) $x^2 = -4ay$

E X A M P L E   3

Discussing the Equation of a Parabola

Discuss the equation: $x^2 = -12y$

Solution   The equation $x^2 = -12y$ is of the form $x^2 = -4ay$, with $a = 3$. Consequently, the graph of the equation is a parabola with vertex at (0, 0), focus at (0, − 3), and directrix the line $y = 3$. The parabola opens down, and its axis of symmetry is the y-axis. To obtain the points defining the latus rectum, let $y = -3$. Then $x^2 = 36$, or $x = \pm 6$. See Figure 8.

FIGURE 8
$x^2 = -12y$

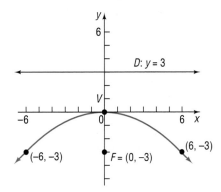

■   Now work Problem 27.

E X A M P L E   4

Finding the Equation of a Parabola

Find the equation of the parabola with focus at (0, 4) and directrix the line $y = -4$. Graph the equation.

Solution   A parabola whose focus is at (0, 4) and whose directrix is the horizontal line $y = -4$ will have its vertex at (0, 0). (Do you see why? The vertex is midway between the focus and the directrix.) Thus, the equation of this parabola is of the form $x^2 = 4ay$, with $a = 4$; that is,

$$x^2 = 16y$$

Figure 9 shows the graph.

FIGURE 9
$x^2 = 16y$

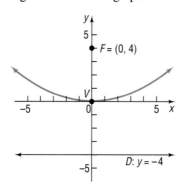

EXAMPLE 5

Finding the Equation of a Parabola

Find the equation of a parabola with vertex at (0, 0) if its axis of symmetry is the x-axis and its graph contains the point $(-\frac{1}{2}, 2)$. Find its focus and directrix, and graph the equation.

Solution   Because the vertex is at the origin and the axis of symmetry is the x-axis, we see from Table 1 that the form of the equation is

$$y^2 = kx$$

FIGURE 10
$y^2 = -8x$

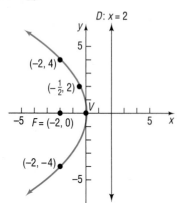

Because the point $(-\frac{1}{2}, 2)$ is on the parabola, the coordinates $x = -\frac{1}{2}$, $y = 2$ must satisfy the equation. Putting $x = -\frac{1}{2}$ and $y = 2$ into the equation, we find

$$4 = k(-\tfrac{1}{2})$$
$$k = -8$$

Thus, the equation of the parabola is

$$y^2 = -8x$$

Comparing this equation to $y^2 = -4ax$, we find that $a = 2$. The focus is therefore at $(-2, 0)$, and the directrix is the line $x = 2$. Letting $x = -2$, we find $y^2 = 16$ or $y = \pm 4$. The points $(-2, 4)$ and $(-2, -4)$ define the latus rectum. See Figure 10. ■

■   Now work Problem 19.

Vertex at *(h, k)*

If a parabola with vertex at the origin and axis of symmetry along a coordinate axis is shifted horizontally h units and then vertically k units, the result is a parabola with vertex at *(h, k)* and axis of symmetry parallel to a coordinate axis. The equations of such parabolas have the same forms as those in Table 1, but with x replaced by $x - h$ and y replaced by $y - k$. Table 2 gives the forms of the equations of such parabolas. Figure 11(a)–(d) illustrates the graphs for $h > 0$, $k > 0$.

TABLE 2   PARABOLAS WITH VERTEX AT (h, k), AXIS OF SYMMETRY PARALLEL TO A COORDINATE AXIS, $a > 0$

VERTEX	FOCUS	DIRECTRIX	EQUATION	DESCRIPTION
(h, k)	$(h + a, k)$	$x = -a + h$	$(y-k)^2 = 4a(x - h)$	Parabola, axis of symmetry parallel to x-axis, opens to right
(h, k)	$(h - a, k)$	$x = a + h$	$(y - k)^2 = -4a(x - h)$	Parabola, axis of symmetry parallel to x-axis, opens to left
(h, k)	$(h, k + a)$	$y = -a + k$	$(x - h)^2 = 4a(y - k)$	Parabola, axis of symmetry parallel to y-axis, opens up
(h, k)	$(h, k - a)$	$y = a + k$	$(x - h)^2 = -4a(y - k)$	Parabola, axis of symmetry parallel to y-axis, opens down

FIGURE 11

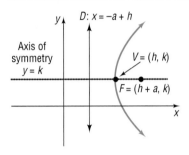

(a) $(y - k)^2 = 4a(x - h)$

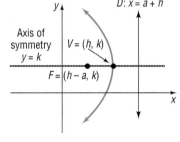

(b) $(y - k)^2 = -4a(x - h)$

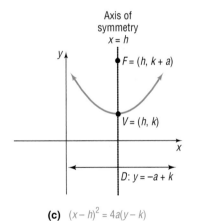

(c) $(x - h)^2 = 4a(y - k)$

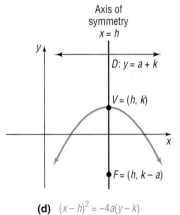

(d) $(x - h)^2 = -4a(y - k)$

E X A M P L E   6   *Finding the Equation of a Parabola, Vertex Not at Origin*

Find an equation of the parabola with vertex at $(-2, 3)$ and focus at $(0, 3)$. Graph the equation.

Solution   The vertex $(-2, 3)$ and focus $(0, 3)$ both lie on the horizontal line $y = 3$ (the axis of symmetry). The distance a from $(-2, 3)$ to $(0, 3)$x is $a = 2$. Also, because the focus lies to the right of the vertex, we know that the parabola opens to the right. Consequently, the form of the equation is

$$(y - k)^2 = 4a(x - h)$$

FIGURE 12
$(y - 3)^2 = 8(x + 2)$
$D: x = -4$

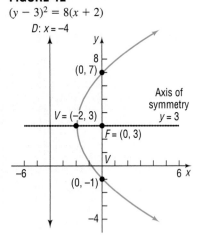

where $(h, k) = (-2, 3)$ and $a = 2$. Therefore, the equation is

$$(y - 3)^2 = 4 \cdot 2[x - (-2)]$$
$$(y - 3)^2 = 8(x + 2)$$

If $x = 0$, then $(y - 3)^2 = 16$. Thus, $y - 3 = \pm 4$ and $y = -1, y = 7$. The points $(0, -1)$ and $(0, 7)$ define the latus rectum; the line $x = -4$ is the directrix. See Figure 12. ∎

■ Now work Problem 17.

Polynomial equations define parabolas whenever they involve two variables that are quadratic in one variable and linear in the other. To discuss this type of equation, we first complete the square of the quadratic variable.

E X A M P L E 7   *Discussing the Equation of a Parabola*

Discuss the equation: $x^2 + 4x - 4y = 0$

Solution   To discuss the equation $x^2 + 4x - 4y = 0$, we complete the square involving the variable x. Thus,

FIGURE 13
$x^2 + 4x - 4y = 0$

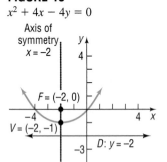

$$x^2 + 4x - 4y = 0$$
$$x^2 + 4x = 4y \qquad \text{Isolate the terms involving } x \text{ on the left side.}$$
$$x^2 + 4x + 4 = 4y + 4 \qquad \text{Complete the square on the left side.}$$
$$(x + 2)^2 = 4(y + 1)$$

This equation is of the form $(x - h)^2 = 4a(y - k)$, with $h = -2$, $k = -1$, and $a = 1$. The graph is a parabola with vertex at $(h, k) = (-2, -1)$ that opens up. The focus is at $(-2, 0)$, and the directrix is the line $y = -2$. See Figure 13. ∎

Parabolas find their way into many applications. For example, as we discussed in Section 5.1, suspension bridges have cables in the shape of a parabola. Another property of parabolas that is used in applications is their reflecting property.

FIGURE 14
Searchlight

Reflecting Property

Suppose that a mirror is shaped like a **paraboloid of revolution,** a surface formed by rotating a parabola about its axis of symmetry. If a light (or any other emitting source) is placed at the focus of the parabola, all the rays emanating from the light will reflect off the mirror in lines parallel to the axis of symmetry. This principle is used in the design of searchlights, flashlights, certain automobile headlights, and other such devices. See Figure 14.

Conversely, suppose that rays of light (or other signals) emanate from a distant source so that they are essentially parallel. When these rays strike the surface of a parabolic mirror whose axis of symmetry is parallel to these rays, they are reflected to a single point at the focus. This principle is used in the design of some solar energy devices, satellite dishes, and the mirrors used in some types of telescopes. See Figure 15.

FIGURE 15

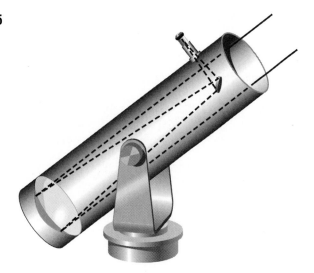

E X A M P L E   8   *Satellite Dish*

A satellite dish is shaped like a paraboloid of revolution. The signals that emanate from a satellite strike the surface of the dish and are reflected to a single point, where the receiver is located. If the dish is 8 feet across at its opening and is 3 feet deep at its center, at what position should the receiver be placed?

FIGURE 16

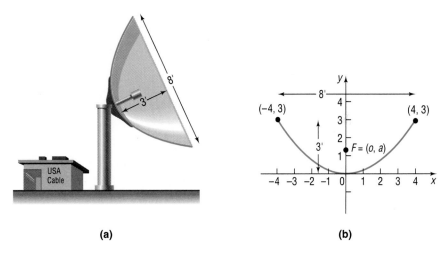

(a)   (b)

Solution   Figure 16(a) shows the satellite dish. We draw the parabola used to form the dish on a rectangular coordinate system so that the vertex of the parabola is at the origin and its focus is on the positive y-axis. See Figure 16(b). The form of the equation of the parabola is

$$x^2 = 4ay$$

and its focus is at $(0, a)$. Since $(4, 3)$ is a point on the graph, we have

$$4^2 = 4a(3)$$

$$a = \frac{4}{3}$$

The receiver should be located $1\frac{1}{3}$ feet from the base of the dish, along its axis of symmetry.   ■

9.2

Exercise 9.2

In Problems 1–8, the graph of a parabola is given. Match each graph to its equation.

A. $y^2 = 4x$ B. $x^2 = 4y$ C. $y^2 = -4x$

D. $x^2 = -4y$ E. $(y-1)^2 = 4(x-1)$ F. $(x+1)^2 = 4(y+1)$

G. $(y-1)^2 = -4(x-1)$ H. $(x+1)^2 = -4(y+1)$

1.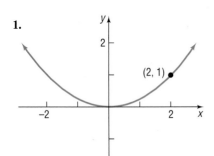

2.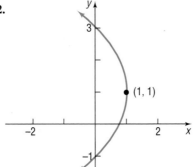

3.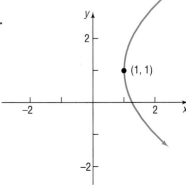

4.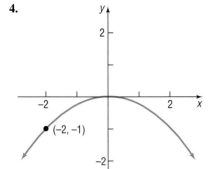

5.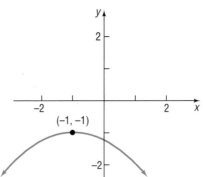

6.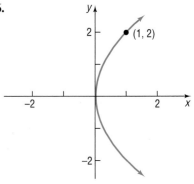

7.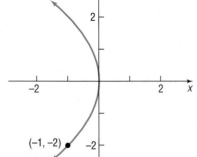

8.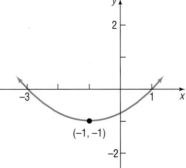

In Problems 9–24, find the equation of the parabola described. Find the two points that define the latus rectum, and graph the equation.

9. Focus at (4, 0); vertex at (0, 0)

10. Focus at (0, 2); vertex at (0, 0)

11. Focus at (0, −3); vertex at (0, 0)

12. Focus at (−4, 0); vertex at (0, 0)

13. Focus at (−2, 0); directrix the line $x = 2$

14. Focus at (0, −1); directrix the line $y = 1$

15. Directrix the line $y = -\frac{1}{2}$; vertex at (0, 0)

16. Directrix the line $x = -\frac{1}{2}$; vertex at (0, 0)

17. Vertex at (2, −3); focus at (2, −5)

18. Vertex at (4, −2); focus at (6, −2)

19. Vertex at (0, 0); axis of symmetry the y-axis; containing the point (2, 3)

20. Vertex at (0, 0); axis of symmetry the x-axis; containing the point (2, 3)

21. Focus at (−3, 4); directrix the line $y = 2$

22. Focus at (2, 4); directrix the line $x = -4$

23. Focus at (−3, −2); directrix the line $x = 1$

24. Focus at (−4, 4); directrix the line $y = -2$

In Problems 25–42, find the vertex, focus, and directrix of each parabola. Graph the equation.

25. $x^2 = 4y$

26. $y^2 = 8x$

27. $y^2 = -16x$

28. $x^2 = -4y$

29. $(y - 2)^2 = 8(x + 1)$

30. $(x + 4)^2 = 16(y + 2)$

31. $(x - 3)^2 = -(y + 1)$

32. $(y + 1)^2 = -4(x - 2)$

33. $(y + 3)^2 = 8(x - 2)$

34. $(x - 2)^2 = 4(y - 3)$

35. $y^2 - 4y + 4x + 4 = 0$

36. $x^2 + 6x - 4y + 1 = 0$

37. $x^2 + 8x = 4y - 8$

38. $y^2 - 2y = 8x - 1$

39. $y^2 + 2y - x = 0$

40. $x^2 - 4x = 2y$

41. $x^2 - 4x = y + 4$

42. $y^2 + 12y = -x + 1$

In Problems 43–50, write an equation for each parabola.

43.

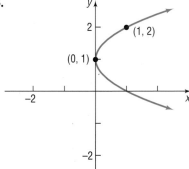

44.

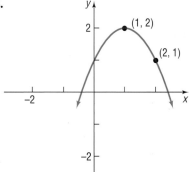

45.

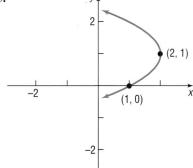

46.

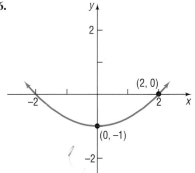

47.

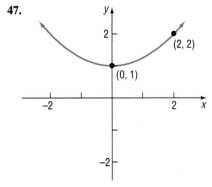

48.

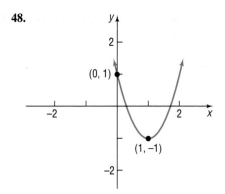

49.

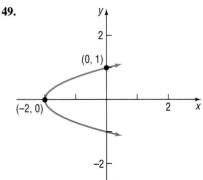

50.

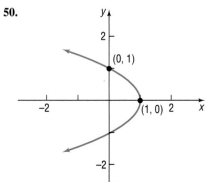

51. *Satellite Dish* A satellite dish is shaped like a paraboloid of revolution. The signals that emanate from a satellite strike the surface of the dish and are reflected to a single point, where the receiver is located. If the dish is 10 feet across at its opening and is 4 feet deep at its center, at what position should the receiver be placed?

52. *Constructing a TV Dish* A cable TV receiving dish is in the shape of a paraboloid of revolution. Find the location of the receiver, which is placed at the focus, if the dish is 6 feet across at its opening and 2 feet deep.

53. *Constructing a Flashlight* The reflector of a flashlight is in the shape of a paraboloid of revolution. Its diameter is 4 inches and its depth is 1 inch. How far from the vertex should the light bulb be placed so that the rays will be reflected parallel to the axis?

54. *Constructing a Headlight* A sealed-beam headlight is in the shape of a paraboloid of revolution. The bulb, which is placed at the focus, is 1 inch from the vertex. If the depth is to be 2 inches, what is the diameter of the headlight at its opening?

55. *Suspension Bridges* The cables of a suspension bridge are in the shape of a parabola, as shown in the figure. The towers supporting the cable are 600 feet apart and 80 feet high. If the cables touch the road surface midway between the towers, what is the height of the cable at a point 150 feet from a tower?

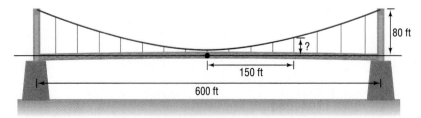

56. *Suspension Bridges* The cables of a suspension bridge are in the shape of a parabola. The towers supporting the cable are 400 feet apart and 100 feet high. If the cables are at a height of 10 feet midway between the towers, what is the height of the cable at a point 50 feet from a tower?

57. *Searchlights* A searchlight is shaped like a paraboloid of revolution. If the light source is located 2 feet from the base along the axis of symmetry and the opening is 5 feet across, how deep should the searchlight be?

58. *Searchlights* A searchlight is shaped like a paraboloid of revolution. If the light source is located 2 feet from the base along the axis of symmetry and the depth of the searchlight is 4 feet, what should the width of the opening be?

59. *Solar Heat* A mirror is shaped like a paraboloid of revolution and will be used to concentrate the rays of the sun at its focus, creating a heat source. If the mirror is 20 feet across at its opening and is 6 feet deep, where will the heat source be concentrated?

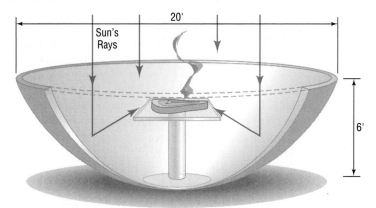

60. *Reflecting Telescopes* A reflecting telescope contains a mirror shaped like a paraboloid of revolution. If the mirror is 4 inches across at its opening and is 3 feet deep, where will the light collected be concentrated?

61. *Parabolic Arch Bridge* A bridge is built in the shape of a parabolic arch. The bridge has a span of 120 feet and a maximum height of 25 feet. See the illustration. Choose a suitable rectangular coordinate system and find the height of the arch at distances of 10, 30, and 50 feet from the center.

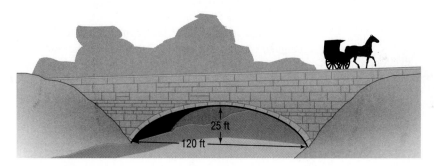

62. *Parabolic Arch Bridge* A bridge is built in the shape of a parabolic arch and is to have a span of 100 feet. The height of the arch a distance of 40 feet from the center is to be 10 feet. Find the height of the arch at its center.

63. Show that an equation of the form

$$Ax^2 + Ey = 0 \qquad A \neq 0, E \neq 0$$

is the equation of a parabola with vertex at (0, 0) and axis of symmetry the y-axis. Find its focus and directrix.

64. Show that an equation of the form

$$Cy^2 + Dx = 0 \qquad C \neq 0, D \neq 0$$

is the equation of a parabola with vertex at (0, 0) and axis of symmetry the x-axis. Find its focus and directrix.

65. Show that the graph of an equation of the form

$$Ax^2 + Dx + Ey + F = 0 \qquad A \neq 0$$

(a) Is a parabola if $E \neq 0$.
(b) Is a vertical line if $E = 0$ and $D^2 - 4AF = 0$.
(c) Is two vertical lines if $E = 0$ and $D^2 - 4AF > 0$.
(d) Contains no points if $E = 0$ and $D^2 - 4AF < 0$.

66. Show that the graph of an equation of the form

$$Cy^2 + Dx + Ey + F = 0 \qquad C \neq 0$$

(a) Is a parabola if $D \neq 0$.
(b) Is a horizontal line if $D = 0$ and $E^2 - 4CF = 0$.
(c) Is two horizontal lines if $D = 0$ and $E^2 - 4CF > 0$.
(d) Contains no points if $D = 0$ and $E^2 - 4CF < 0$.

9.3

The Ellipse

Ellipse

FIGURE 17

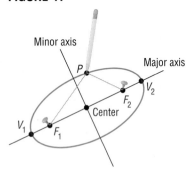

> An **ellipse** is the collection of all points in the plane the sum of whose distances from two fixed points, called the **foci,** is a constant.

The definition actually contains within it a physical means for drawing an ellipse. Find a piece of string (the length of this string is the constant referred to in the definition). Then take two thumbtacks (the foci) and stick them on a piece of cardboard so that the distance between them is less than the length of the string. Now attach the ends of the string to the thumbtacks and, using the point of a pencil, pull the string taut. Keeping the string taut, rotate the pencil around the two thumbtacks. The pencil traces out an ellipse, as shown in Figure 17.

In Figure 17, the foci are labeled F_1 and F_2. The line containing the foci is called the **major axis.** The midpoint of the line segment joining the foci is called the **center** of the ellipse. The line through the center and perpendicular to the major axis is called the **minor axis.**

FIGURE 18

$d(F_1, P) + d(F_2, P) = 2a$

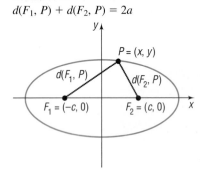

The two points of intersection of the ellipse and the major axis are the **vertices,** V_1 and V_2, of the ellipse. The distance from one vertex to the other is called the **length of the major axis.** The ellipse is symmetric with respect to its major axis and with respect to its minor axis.

With these ideas in mind, we are now ready to find the equation of an ellipse in a rectangular coordinate system. First, we place the center of the ellipse at the origin. Second, we position the ellipse so that its major axis coincides with a coordinate axis. Suppose that the major axis coincides with the x-axis, as shown in Figure 18. If c is the distance from the center to a focus, then one focus will be at $F_1 = (-c, 0)$ and the other at $F_2 = (c, 0)$. As we shall see, it is convenient to let $2a$ denote the constant distance referred to in the definition. Thus, if $P = (x, y)$ is any point on the ellipse, we have

$$d(F_1, P) + d(F_2, P) = 2a \qquad \text{Sum of the distances from } P \text{ to the foci equals a constant}$$

$$\sqrt{(x + c)^2 + y^2} + \sqrt{(x - c)^2 + y^2} = 2a \qquad \text{Use the distance formula.}$$

$$\sqrt{(x + c)^2 + y^2} = 2a - \sqrt{(x - c)^2 + y^2} \qquad \text{Isolate one radical.}$$

$$(x + c)^2 + y^2 = 4a^2 - 4a\sqrt{(x - c)^2 + y^2} \qquad \text{Square both sides.}$$
$$\qquad\qquad\qquad + (x - c)^2 + y^2$$

$$x^2 + 2cx + c^2 + y^2 = 4a^2 - 4a\sqrt{(x - c)^2 + y^2} \qquad \text{Simplify}$$
$$\qquad\qquad\qquad + x^2 - 2cx + c^2 + y^2$$

$$4cx - 4a^2 = -4a\sqrt{(x - c)^2 + y^2} \qquad \text{Isolate the radical.}$$

$$cx - a^2 = -a\sqrt{(x - c)^2 + y^2} \qquad \text{Divide each side by 4.}$$

$$c^2x^2 - 2a^2cx + a^4 = a^2[(x - c)^2 + y^2] \qquad \text{Square both sides again.}$$

$$c^2x^2 - 2a^2cx + a^4 = a^2(x^2 - 2cx + c^2 + y^2)$$

$$(c^2 - a^2)x^2 - a^2y^2 = a^2c^2 - a^4$$

$$(a^2 - c^2)x^2 + a^2y^2 = a^2(a^2 - c^2) \qquad \text{Multiply each side by } -1;$$
$$\text{factor } a^2 \text{ on the right side. (1)}$$

MISSION POSSIBLE

Chapter 9

BUILDING A BRIDGE OVER THE EAST RIVER

Your team is working for the transportation authority in New York City. You have been asked to study the construction plans for a new bridge over the East River in New York City. The space between supports needs to be 1050 feet; the height at the center of the arch needs to be 350 feet. One company has suggested that the support be in the shape of a parabola; another company suggests a semiellipse. The engineering team will determine the relative strengths of the two plans; your job is to find out if there are any differences in the channel widths.

An empty tanker needs a 280 foot clearance to pass beneath the bridge. You need to find the width of the channel for each of the two different plans.

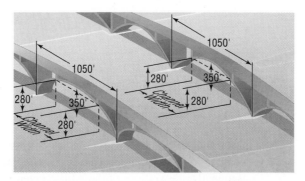

1. To determine the equation of a parabola with these characteristics, first place the parabola on coordinate axes in a convenient location and sketch it.
2. What is the equation of the parabola? (If using a decimal in the equation, you may want to carry six decimal places. If using a fraction, your answer will be more exact.)
3. How wide is the channel that the tanker will pass through if the shape of the support is parabolic?
4. To determine the equation of a semiellipse with these characteristics, place the semiellipse on coordinate axes in a convenient location and sketch it.
5. What is the equation of the ellipse? How wide is the channel that the tanker will pass through?
6. Now that you know which of the two provides the wider channel, consider some other factors. Your department is also in charge of channel depth, amount of traffic on the river, and various other factors. For example, if there were flooding in the river and the water level rose by 10 feet, how would the clearances be affected? Make a decision about which plan you think would be the better one as far as your department is concerned, and explain why you think so.

To obtain points on the ellipse off the x-axis, it must be that $a > c$. To see why, look again at Figure 18:

$$d(F_1, P) + d(F_2, P) > d(F_1, F_2)$$

The sum of the lengths of two sides of a triangle is greater than the length of the third side.

$$2a > 2c$$

$d(F_1, P) + d(F_2, P) = 2a$; $d(F_1, F_2) = 2c$

$$a > c$$

Since $a > c$, we also have $a^2 > c^2$, so $a^2 - c^2 > 0$. Let $b^2 = a^2 - c^2$, $b > 0$. Then $a > b$ and equation (1) can be written as

$$b^2 x^2 + a^2 y^2 = a^2 b^2$$

$$\frac{x^2}{a^2} + \frac{y^2}{b^2} = 1 \quad \text{Divide each side by } a^2 b^2.$$

Theorem   An equation of the ellipse with center at $(0, 0)$ and foci at $(-c, 0)$ and $(c, 0)$ is

Equation of an Ellipse;
Center at (0, 0);
Foci at ($\pm c$, 0);
Major Axis along
the x-Axis

$$\frac{x^2}{a^2} + \frac{y^2}{b^2} = 1 \qquad \text{where } a > b > 0 \text{ and } b^2 = a^2 - c^2 \qquad (2)$$

The major axis is the x-axis.   ■

As you can verify, the ellipse defined by equation (2) is symmetric with respect to the x-axis, y-axis, and origin.

To find the vertices of the ellipse defined by equation (2), let $y = 0$. The vertices satisfy the equation $x^2/a^2 = 1$, the solutions of which are $x = \pm a$. Consequently, the vertices of the ellipse given by equation (2) are $V_1 = (-a, 0)$ and $V_2 = (a, 0)$. The y-intercepts of the ellipse, found by letting $x = 0$, have co-ordinates $(0, -b)$ and $(0, b)$. These four intercepts, $(a, 0)$, $(-a, 0)$, $(0, b)$, and $(0, -b)$, are used to graph the ellipse. See Figure 19.

Notice in Figure 19 the right triangle formed with the points $(0, 0)$, $(c, 0)$, and $(0, b)$. Because $b^2 = a^2 - c^2$ (or $b^2 + c^2 = a^2$), the distance from the focus at $(c, 0)$ to the point $(0, b)$ is a.

FIGURE 19

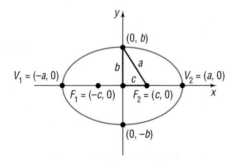

EXAMPLE 1   *Finding an Equation of an Ellipse*

Find an equation of the ellipse with center at the origin, one focus at $(3, 0)$, and a vertex at $(-4, 0)$. Graph the equation.

Solution   The ellipse has its center at the origin, and the major axis coincides with the x-axis. One focus is at $(c, 0) = (3, 0)$, so $c = 3$. One vertex is at $(-a, 0) = (-4, 0)$, so $a = 4$. From equation (2), it follows that

FIGURE 20

$$\frac{x^2}{16} + \frac{y^2}{7} = 1$$

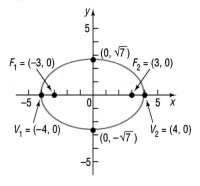

$$b^2 = a^2 - c^2 = 16 - 9 = 7$$

so an equation of the ellipse is

$$\frac{x^2}{16} + \frac{y^2}{7} = 1$$

Figure 20 shows the graph.   ■

An equation of the form of equation (2), with $a > b$, is the equation of an ellipse with center at the origin, foci on the x-axis at $(-c, 0)$ and $(c, 0)$, where $c^2 = a^2 - b^2$, and major axis along the x-axis.

For the remainder of this section, the direction "Discuss the equation" will mean to find the center, major axis, foci, and vertices of the ellipse and graph it.

E X A M P L E   2   *Discussing the Equation of an Ellipse*

Discuss the equation:

$$\frac{x^2}{25} + \frac{y^2}{9} = 1$$

Solution   The given equation is of the form of equation (2), with $a^2 = 25$ and $b^2 = 9$. The equation is that of an ellipse with center $(0, 0)$ and major axis along the x-axis. The vertices are at $(\pm a, 0) = (\pm 5, 0)$. Because $b^2 = a^2 - c^2$, we find

$$c^2 = a^2 - b^2 = 25 - 9 = 16$$

The foci are at $(\pm c, 0) = (\pm 4, 0)$. Figure 21 shows the graph.

FIGURE 21

$$\frac{x^2}{25} + \frac{y^2}{9} = 1$$

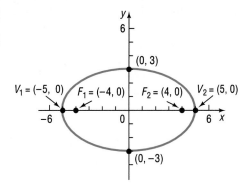

Comment: To graph the ellipse $(x^2/25) + (y^2/9) = 1$ discussed in Example 2, we need to graph the two functions $y = 3\sqrt{1 - (x^2/25)}$ and $y = -3\sqrt{1 - (x^2/25)}$. Do this and compare what you see with Figure 21. Be sure to set the viewing rectangle for a square screen.   ■

■   Now work Problems 5 and 15.

Notice in Figures 20 and 21 how we used the intercepts of the equation to graph each ellipse. Following this practice will make it easier for you to obtain an accurate graph of an ellipse.

If the major axis of an ellipse with center at (0, 0) coincides with the y-axis, then the foci are at $(0, -c)$ and $(0, c)$. Using the same steps as before, the definition of an ellipse leads to the following result:

Theorem

Equation of an Ellipse;
Center at (0, 0);
Foci at (0, ± c);
Major Axis
along the y-Axis

An equation of the ellipse with center at (0, 0) and foci at $(0, -c)$ and $(0, c)$ is

$$\frac{x^2}{b^2} + \frac{y^2}{a^2} = 1 \qquad \text{where } a > b > 0 \text{ and } b^2 = a^2 - c^2 \qquad (3)$$

The major axis is the y-axis; the vertices are at $(0, -a)$ and $(0, a)$. ∎

Figure 22 illustrates the graph of such an ellipse. Again, notice the right triangle with the points at (0, 0), $(b, 0)$, and $(0, c)$.

FIGURE 22

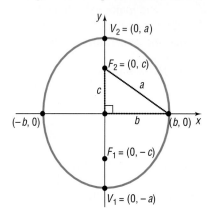

Look closely at equations (2) and (3). Although they may look alike, there is a difference! In equation (2), the larger number, a^2, is in the denominator of the x^2-term, so the major axis of the ellipse is along the x-axis. In equation (3), the larger number, a^2, is in the denominator of the y^2-term, so the major axis is along the y-axis.

E X A M P L E 3

Discussing the Equation of an Ellipse

Discuss the equation:

$$9x^2 + y^2 = 9$$

Solution

To put the equation in proper form, we divide each side by 9:

$$x^2 + \frac{y^2}{9} = 1$$

The larger number, 9, is in the denominator of the y^2-term so, based on equation (3), this is the equation of an ellipse with center at the origin and major axis along the y-axis. Also, we conclude that $a^2 = 9$, $b^2 = 1$, and $c^2 = a^2 - b^2 = 9 - 1 = 8$. The vertices are at $(0, \pm a) = (0, \pm 3)$, and the foci are at $(0, \pm c) = (0, \pm 2\sqrt{2})$. The graph is given in Figure 23. ∎

E X A M P L E 4

Finding an Equation of an Ellipse

Find an equation of the ellipse having one focus at $(0, 2)$ and vertices at $(0, -3)$ and $(0, 3)$. Graph the equation.

FIGURE 23

$$x^2 + \frac{y^2}{9} = 1$$

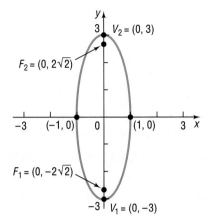

Solution   Because the vertices are at $(0, -3)$ and $(0, 3)$, the center of this ellipse is at the origin. Also, its major axis coincides with the y-axis. The given information also reveals that $c = 2$ and $a = 3$, so $b^2 = a^2 - c^2 = 9 - 4 = 5$. The form of the equation of this ellipse is given by equation (3):

$$\frac{x^2}{b^2} + \frac{y^2}{a^2} = 1$$

$$\frac{x^2}{5} + \frac{y^2}{9} = 1$$

Figure 24 shows the graph.

FIGURE 24

$$\frac{x^2}{5} + \frac{y^2}{9} = 1$$

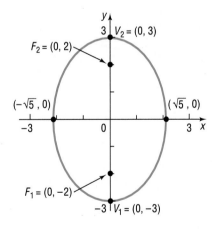

■   Now work Problems 9 and 17.

The circle may be considered a special kind of ellipse. To see why, let $a = b$ in equation (2) or in equation (3). Then

$$\frac{x^2}{a^2} + \frac{y^2}{a^2} = 1$$

$$x^2 + y^2 = a^2$$

This is the equation of a circle with center at the origin and radius a. The value of c is

$$c^2 = a^2 - b^2 = 0$$

We conclude that the closer the two foci of an ellipse are, the more the ellipse will look like a circle.

Center at (*h, k*)

If an ellipse with center at the origin and major axis coinciding with a coordinate axis is shifted horizontally *h* units and then vertically *k* units, the result is an ellipse with center at (*h, k*) and major axis parallel to a coordinate axis. Table 3 gives the forms of the equations of such ellipses, and Figure 25 shows their graphs.

TABLE 3 ELLIPSES WITH CENTER AT (*h, k*) AND MAJOR AXIS PARALLEL TO A COORDINATE AXIS

CENTER	MAJOR AXIS	FOCI	VERTICES	EQUATION
(*h, k*)	Parallel to *x*-axis	(*h* ± *c, k*)	(*h* ± *a, k*)	$\dfrac{(x-h)^2}{a^2} + \dfrac{(y-k)^2}{b^2} = 1,$ $a > b$ and $b^2 = a^2 - c^2$
(*h, k*)	Parallel to *y*-axis	(*h, k* ± *c*)	(*h, k* ± *a*)	$\dfrac{(x-h)^2}{b^2} + \dfrac{(y-k)^2}{a^2} = 1,$ $a > b$ and $b^2 = a^2 - c^2$

FIGURE 25

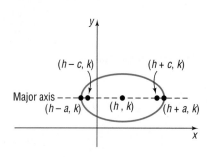

(a) $\dfrac{(x-h)^2}{a^2} + \dfrac{(y-k)^2}{b^2} = 1$

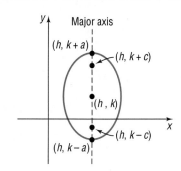

(b) $\dfrac{(x-h)^2}{b^2} + \dfrac{(y-k)^2}{a^2} = 1$

E X A M P L E   5   *Finding an Equation of an Ellipse, Center Not at the Origin*

Find an equation for the ellipse with center at (2, −3), one focus at (3, −3), and one vertex at (5, −3). Graph the equation.

Solution   The center is at (*h, k*) = (2, −3), so *h* = 2 and *k* = −3. The major axis is parallel to the *x*-axis. The distance from the center (2, −3) to a focus (3, −3) is *c* = 1; the distance from the center (2, −3) to a vertex (5, −3) is *a* = 3. Thus, $b^2 = a^2 - c^2 = 9 - 1 = 8$. The form of the equation is

$$\frac{(x-h)^2}{a^2} + \frac{(y-k)^2}{b^2} = 1 \quad \text{where } h = 2, k = -3, a = 3, b = 2\sqrt{2}$$

$$\frac{(x-2)^2}{9} + \frac{(y+3)^2}{8} = 1$$

Figure 26 shows the graph. ∎

■ Now work Problem 41.

FIGURE 26

$$\frac{(x-2)^2}{9} + \frac{(y+3)^2}{8} = 1$$

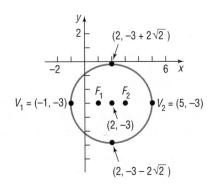

E X A M P L E   6

Discussing the Equation of an Ellipse

Discuss the equation: $4x^2 + y^2 - 8x + 4y + 4 = 0$

Solution

FIGURE 27

$(x-1)^2 + \dfrac{(y+2)^2}{4} = 1$

We proceed to complete the square in x and in y:

$$4x^2 + y^2 - 8x + 4y + 4 = 0$$
$$4x^2 - 8x + y^2 + 4y = -4$$
$$4(x^2 - 2x) + (y^2 + 4y) = -4$$
$$4(x^2 - 2x + 1) + (y^2 + 4y + 4) = -4 + 4 + 4 \quad \text{Complete each square.}$$
$$4(x-1)^2 + (y+2)^2 = 4$$
$$(x-1)^2 + \frac{(y+2)^2}{4} = 1 \quad\quad\quad \text{Divide each side by 4.}$$

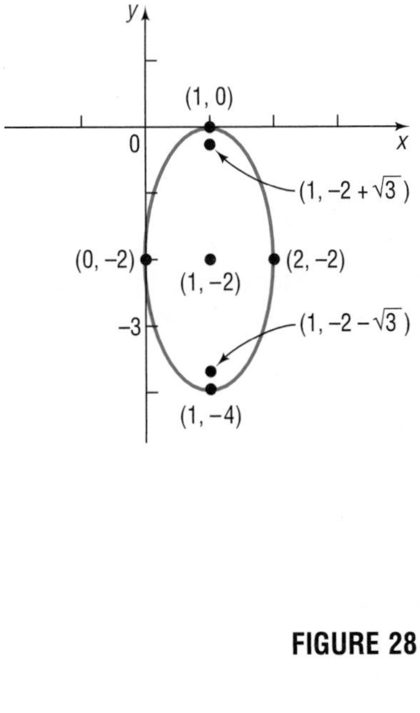

This is the equation of an ellipse with center at $(1, -2)$ and major axis parallel to the y-axis. Since $a^2 = 4$ and $b^2 = 1$, we have $c^2 = a^2 - b^2 = 4 - 1 = 3$. The vertices are at $(h, k \pm a) = (1, -2 \pm 2)$ or $(1, 0)$ and $(1, -4)$. The foci are at $(h, k \pm c) = (1, -2 \pm \sqrt{3})$ or $(1, -2 - \sqrt{3})$ and $(1, -2 + \sqrt{3})$. Figure 27 shows the graph. ∎

Applications

Ellipses are found in many applications in science and engineering. For example, the orbits of the planets around the Sun are elliptical, with the Sun's position at a focus. See Figure 28.

FIGURE 28

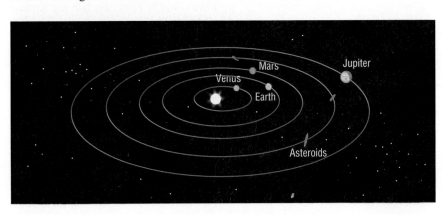

Stone and concrete bridges are often shaped as semielliptical arches. Elliptical gears are used in machinery when a variable rate of motion is required.

Ellipses also have an interesting reflection property. If a source of light (or sound) is placed at one focus, the waves transmitted by the source will reflect off the ellipse and concentrate at the other focus. This is the principal behind "whispering galleries," which are rooms designed with elliptical ceilings. A person standing at one focus of the ellipse can whisper and be heard by a person standing at the other focus, because all the sound waves that reach the ceiling are reflected to the other person.

E X A M P L E   7   *Whispering Galleries*

Figure 29 shows the specifications for an elliptical ceiling in a hall designed to be a whispering gallery. In a whispering gallery, a person standing at one focus of the ellipse can whisper and be heard by another person standing at the other focus, because all the sound waves that reach the ceiling from one focus are reflected to the other focus. Where in the hall are the foci located?

FIGURE 29

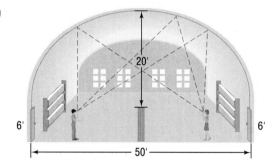

Solution   We set up a rectangular coordinate system so that the center of the ellipse is at the origin and the major axis is along the x-axis. See Figure 30. The equation of the ellipse is

$$\frac{x^2}{a^2} + \frac{y^2}{b^2} = 1$$

where $a = 25$ and $b = 20$. Since

$$c^2 = a^2 - b^2 = 25^2 - 20^2 = 625 - 400 = 225$$

we have $c = 15$. Thus, the foci are located 15 feet from the center of the ellipse along the major axis. ∎

FIGURE 30

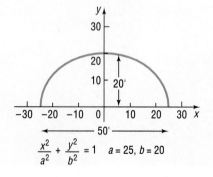

$\dfrac{x^2}{a^2} + \dfrac{y^2}{b^2} = 1$ $a = 25, b = 20$

9.3

Exercise 9.3

In Problems 1–4, the graph of an ellipse is given. Match each graph to its equation.

A. $\dfrac{x^2}{4} + y^2 = 1$ B. $x^2 + \dfrac{y^2}{4} = 1$ C. $\dfrac{x^2}{16} + \dfrac{y^2}{4} = 1$ D. $\dfrac{x^2}{4} + \dfrac{y^2}{16} = 1$

1.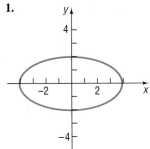

2.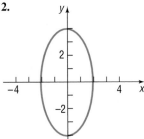

3.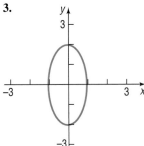

4.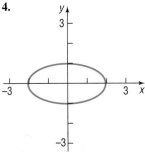

In Problems 5–14, find the vertices and foci of each ellipse. Graph each equation.

5. $\dfrac{x^2}{25} + \dfrac{y^2}{4} = 1$ **6.** $\dfrac{x^2}{9} + \dfrac{y^2}{4} = 1$ **7.** $\dfrac{x^2}{9} + \dfrac{y^2}{25} = 1$

8. $\dfrac{x^2}{4} + \dfrac{y^2}{16} = 1$ **9.** $4x^2 + y^2 = 16$ **10.** $x^2 + 9y^2 = 18$

11. $4y^2 + x^2 = 8$ **12.** $4y^2 + 9x^2 = 36$ **13.** $x^2 + y^2 = 16$

14. $x^2 + y^2 = 4$

In Problems 15–24, find an equation for each ellipse. Graph the equation.

15. Center at $(0, 0)$; focus at $(3, 0)$; vertex at $(5, 0)$

16. Center at $(0, 0)$; focus at $(-1, 0)$; vertex at $(3, 0)$

17. Center at $(0, 0)$; focus at $(0, -4)$; vertex at $(0, 5)$

18. Center at $(0, 0)$; focus at $(0, 1)$; vertex at $(0, -2)$

19. Foci at $(\pm 2, 0)$; length of the major axis is 6

20. Focus at $(0, -4)$; vertices at $(0, \pm 8)$

21. Foci at $(0, \pm 3)$; x-intercepts are ± 2

22. Foci at $(0, \pm 2)$; length of the major axis is 8

23. Center at $(0, 0)$; vertex at $(0, 4)$; $b = 1$

24. Vertices at $(\pm 5, 0)$; $c = 2$

In Problems 25–28, write an equation for each ellipse.

25.

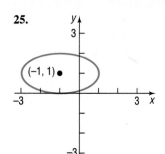

26.

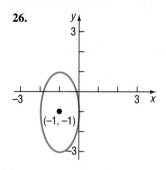

27.

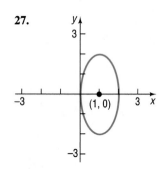

28.

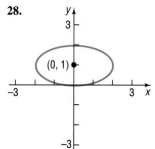

In Problems 29–40, find the center, foci, and vertices of each ellipse. Graph each equation.

29. $\dfrac{(x-3)^2}{4} + \dfrac{(y+1)^2}{9} = 1$

30. $\dfrac{(x+4)^2}{9} + \dfrac{(y+2)^2}{4} = 1$

31. $(x+5)^2 + 4(y-4)^2 = 16$

32. $9(x-3)^2 + (y+2)^2 = 18$

33. $x^2 + 4x + 4y^2 - 8y + 4 = 0$

34. $x^2 + 3y^2 - 12y + 9 = 0$

35. $2x^2 + 3y^2 - 8x + 6y + 5 = 0$

36. $4x^2 + 3y^2 + 8x - 6y = 5$

37. $9x^2 + 4y^2 - 18x + 16y - 11 = 0$

38. $x^2 + 9y^2 + 6x - 18y + 9 = 0$

39. $4x^2 + y^2 + 4y = 0$

40. $9x^2 + y^2 - 18x = 0$

In Problems 41–50, find an equation for each ellipse. Graph the equation.

41. Center at $(2, -2)$; vertex at $(7, -2)$; focus at $(4, -2)$

42. Center at $(-3, 1)$; vertex at $(-3, 3)$; focus at $(-3, 0)$

43. Vertices at $(4, 3)$ and $(4, 9)$; focus at $(4, 8)$

44. Foci at $(1, 2)$ and $(-3, 2)$; vertex at $(-4, 2)$

45. Foci at $(5, 1)$ and $(-1, 1)$; length of the major axis is 8

46. Vertices at $(2, 5)$ and $(2, -1)$; $c = 2$

47. Center at $(1, 2)$; focus at $(4, 2)$; contains the point $(1, 3)$

48. Center at $(1, 2)$; focus at $(1, 4)$; contains the point $(2, 2)$

49. Center at $(1, 2)$; vertex at $(4, 2)$; contains the point $(1, 3)$

50. Center at $(1, 2)$; vertex at $(1, 4)$; contains the point $(2, 2)$

In Problems 51–54, graph each function. [Hint: Notice that each function is half an ellipse.]

51. $f(x) = \sqrt{16 - 4x^2}$

52. $f(x) = \sqrt{9 - 9x^2}$

53. $f(x) = -\sqrt{64 - 16x^2}$

54. $f(x) = -\sqrt{4 - 4x^2}$

55. *Semielliptical Arch Bridge* An arch in the shape of the upper half of an ellipse is used to support a bridge that is to span a river 20 meters wide. The center of the arch is 6 meters above the center of the river (see the figure). Write an equation for the ellipse in which the *x*-axis coincides with the water level and the *y*-axis passes through the center of the arch.

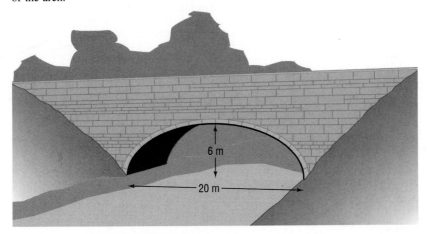

56. *Semielliptical Arch Bridge* The arch of a bridge is a semiellipse with a horizontal major axis. The span is 30 feet, and the top of the arch is 10 feet above the major axis. The roadway is horizontal and is 2 feet above the top of the arch. Find the vertical distance from the roadway to the arch at 5 foot intervals along the roadway.

57. *Whispering Galleries* A hall 100 feet in length is to be designed as a whispering gallery. If the foci are located 25 feet from the center, how high will the ceiling be at the center?

58. *Whispering Galleries* A person, standing at one focus of a whispering gallery, is 6 feet from the nearest wall. Her friend is standing at the other focus, 100 feet away. What is the length of this whispering gallery? How high is its elliptical ceiling at the center?

59. *Semielliptical Arch Bridge* A bridge is built in the shape of a semielliptical arch. The bridge has a span of 120 feet and a maximum height of 25 feet. Choose a suitable rectangular coordinate system and find the height of the arch at distances of 10, 30, and 50 feet from the center.

60. *Semielliptical Arch Bridge* A bridge is built in the shape of a semielliptical arch and is to have a span of 100 feet. The height of the arch, a distance of 40 feet from the center, is to be 10 feet. Find the height of the arch at its center.

61. *Semielliptical Arch* An arch in the form of half an ellipse is 40 feet wide and 15 feet high at the center. Find the height of the arch at intervals of 10 feet along its width.

62. *Semielliptical Arch Bridge* An arch for a bridge over a highway is in the form of half an ellipse. The top of the arch is 20 feet above the ground level (the major axis). The highway has four lanes, each 12 feet wide; a center safety strip 8 feet wide; and two side strips, each 4 feet wide. What should the span of the bridge be (the length of its major axis) if the height 28 feet from the center is to be 13 feet?

*In Problems 63–66, use the fact that the orbit of a planet about the Sun is an ellipse, with the Sun at one focus. The **aphelion** of a planet is its greatest distance from the Sun and the **perihelion** is its shortest distance. The **mean distance** of a planet from the Sun is the length of the semimajor axis of the elliptical orbit. See the illustration.*

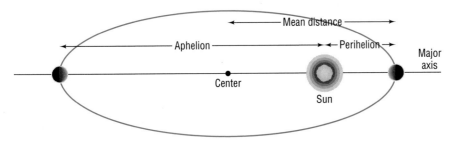

63. *Earth* The mean distance of Earth from the Sun is 93 million miles. If the aphelion of Earth is 94.5 million miles, what is the perihelion? Write an equation for the orbit of Earth around the Sun.

64. *Mars* The mean distance of Mars from the Sun is 142 million miles. If the perihelion of Mars is 128.5 million miles, what is the aphelion? Write an equation for the orbit of Mars about the Sun.

65. *Jupiter* The aphelion of Jupiter is 507 million miles. If the distance from the Sun to the center of its elliptical orbit is 23.2 million miles, what is the perihelion? What is the mean distance? Write an equation for the orbit of Jupiter around the Sun.

66. *Pluto* The perihelion of Pluto is 4551 million miles and the distance of the Sun from the center of its elliptical orbit is 897.5 million miles. Find the aphelion of Pluto. What is the mean distance of Pluto from the Sun? Write an equation for the orbit of Pluto about the Sun.

67. Consult the figure. A racetrack is in the shape of an ellipse, 100 feet long and 50 feet wide. What is the width 10 feet from the side?

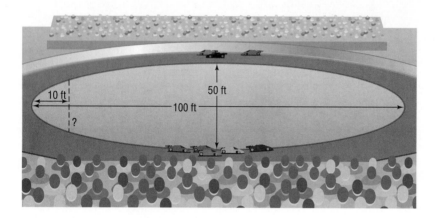

68. A racetrack is in the shape of an ellipse 80 feet long and 40 feet wide. What is the width 10 feet from the side?

69. Show that an equation of the form

$$Ax^2 + Cy^2 + F = 0 \qquad A \neq 0, C \neq 0, F \neq 0$$

where A and C are of the same sign and F is of opposite sign:

(a) Is the equation of an ellipse with center at $(0, 0)$ if $A \neq C$.
(b) Is the equation of a circle with center at $(0, 0)$ if $A = C$.

70. Show that the graph of an equation of the form

$$Ax^2 + Cy^2 + Dx + Ey + F = 0 \qquad A \neq 0, C \neq 0$$

where A and C are of the same sign:

(a) Is an ellipse if $(D^2/4A) + (E^2/4C) - F$ is the same sign as A.
(b) Is a point if $(D^2/4A) + (E^2/4C) - F = 0$.
(c) Contains no points if $(D^2/4A) + (E^2/4C) - F$ is of opposite sign to A.

71. The **eccentricity** e of an ellipse is defined as the number c/a, where a and c are the numbers given in equation (2). Because $a > c$, it follows that $e < 1$. Write a brief paragraph about the general shape of each of the following ellipses. Be sure to justify your conclusions.

(a) Eccentricity close to 0   (b) Eccentricity = 0.5   (c) Eccentricity close to 1

9.4

The Hyperbola

Hyperbola

A **hyperbola** is the collection of all points in the plane the difference of whose distances from two fixed points, called the **foci,** is a constant.

Figure 31 illustrates a hyperbola with foci F_1 and F_2. The line containing the foci is called the **transverse axis.** The midpoint of the line segment joining the foci is called the **center** of the hyperbola. The line through the center and perpendicular to the transverse axis is called the **conjugate axis.** The hyperbola consists of two separate curves, called **branches,** that are symmetric with respect to the transverse axis, conjugate axis, and center. The two points of intersection of the hyperbola and the transverse axis are the **vertices,** V_1 and V_2, of the hyperbola.

FIGURE 31

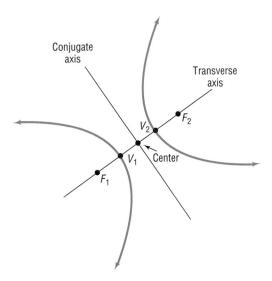

With these ideas in mind, we are now ready to find the equation of a hyperbola in a rectangular coordinate system. First, we place the center at the origin. Next, we position the hyperbola so that its transverse axis coincides with a coordinate axis. Suppose that the transverse axis coincides with the x-axis, as shown in Figure 32. If c is the distance from the center to a focus, then one focus will be at $F_1 = (-c, 0)$ and the other at $F_2 = (c, 0)$. Now we let the constant difference of the distances from any point $P = (x, y)$ on the hyperbola to the foci F_1 and F_2 be denoted by $\pm 2a$. (If P is on the right branch, the $+$ sign is used; if P is on the left branch, the $-$ sign is used.) The coordinates of P must satisfy the equation

$$d(F_1, P) - d(F_2, P) = \pm 2a \qquad \text{Difference of the distances from } P \text{ to the foci equals } \pm 2a.$$

$$\sqrt{(x + c)^2 + y^2} - \sqrt{(x - c)^2 + y^2} = \pm 2a \qquad \text{Use the distance formula.}$$

FIGURE 32

$d(F_1, P) - d(F_2, P) = \pm 2a$

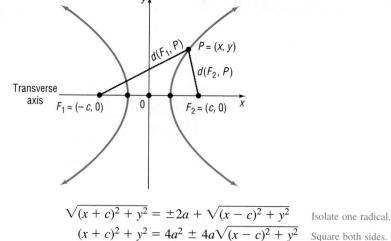

$$\sqrt{(x + c)^2 + y^2} = \pm 2a + \sqrt{(x - c)^2 + y^2} \qquad \text{Isolate one radical.}$$

$$(x + c)^2 + y^2 = 4a^2 \pm 4a\sqrt{(x - c)^2 + y^2} \qquad \text{Square both sides.}$$
$$+ (x - c)^2 + y^2$$

Next, we remove the parentheses:

$$x^2 + 2cx + c^2 + y^2 = 4a^2 \pm 4a\sqrt{(x - c)^2 + y^2} + x^2 - 2cx + c^2 + y^2$$

$$4cx - 4a^2 = \pm 4a\sqrt{(x - c)^2 + y^2} \qquad \text{Isolate the radical.}$$

$$cx - a^2 = \pm a\sqrt{(x - c)^2 + y^2} \qquad \text{Divide each side by 4.}$$

$$(cx - a^2)^2 = a^2[(x - c)^2 + y^2] \qquad \text{Square both sides.}$$

$$c^2x^2 - 2ca^2x + a^4 = a^2(x^2 - 2cx + c^2 + y^2)$$

$$c^2x^2 + a^4 = a^2x^2 + a^2c^2 + a^2y^2$$

$$(c^2 - a^2)x^2 - a^2y^2 = a^2c^2 - a^4$$

$$(c^2 - a^2)x^2 - a^2y^2 = a^2(c^2 - a^2) \qquad\qquad\qquad\qquad (1)$$

To obtain points on the hyperbola off the *x*-axis, it must be that $a < c$. To see why, look again at Figure 32.

$$d(F_1, P) < d(F_2, P) + d(F_1, F_2) \qquad \text{Use triangle } F_1PF_2.$$

$$d(F_1, P) - d(F_2, P) < d(F_1, F_2) \qquad\qquad \begin{array}{l}P \text{ is on the right branch,}\\ \text{so } d(F_1, P) - d(F_2, P) = 2a.\end{array}$$

$$2a < 2c$$

$$a < c$$

Since $a < c$, we also have $a^2 < c^2$, so $c^2 - a^2 > 0$. Let $b^2 = c^2 - a^2$, $b > 0$. Then equation (1) can be written as

$$b^2x^2 - a^2y^2 = a^2b^2$$

$$\frac{x^2}{a^2} - \frac{y^2}{b^2} = 1$$

To find the vertices of the hyperbola defined by this equation, let $y = 0$. The vertices satisfy the equation $x^2/a^2 = 1$, the solutions of which are $x = \pm a$. Consequently, the vertices of the hyperbola are $V_1 = (-a, 0)$ and $V_2 = (a, 0)$.

Theorem   An equation of the hyperbola with center at $(0, 0)$, foci at $(-c, 0)$ and $(c, 0)$, and vertices at $(-a, 0)$ and $(a, 0)$ is

Equation of a Hyperbola;
Center at (0, 0); Foci at
($\pm c$, 0); Vertices at ($\pm a$, 0);
Transverse Axis along x-Axis

$$\frac{x^2}{a^2} - \frac{y^2}{b^2} = 1 \qquad \text{where } b^2 = c^2 - a^2 \qquad (2)$$

The transverse axis is the x-axis. ■

As you can verify, the hyperbola defined by equation (2) is symmetric with respect to the x-axis, y-axis, and origin. To find the y-intercepts, if any, let $x = 0$ in equation (2). This results in the equation $y^2/b^2 = -1$, which has no solution. We conclude that the hyperbola defined by equation (2) has no y-intercepts. In fact, since $x^2/a^2 - 1 = y^2/b^2 \geq 0$, it follows that $x^2/a^2 \geq 1$. Thus, there are no points on the graph for $-a < x < a$. See Figure 33.

FIGURE 33
$$\frac{x^2}{a^2} - \frac{y^2}{b^2} = 1,$$
$$b^2 = c^2 - a^2$$

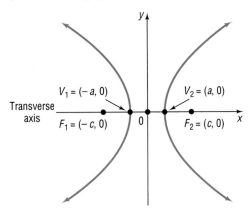

E X A M P L E   1

Finding an Equation of a Hyperbola

Find an equation of the hyperbola with center at the origin, one focus at (3, 0), and one vertex at (−2, 0). Graph the equation.

Solution

The hyperbola has its center at the origin, and the transverse axis coincides with the x-axis. One focus is at $(c, 0) = (3, 0)$, so $c = 3$. One vertex is at $(-a, 0) = (-2, 0)$, so $a = 2$. From equation (2), it follows that $b^2 = c^2 - a^2 = 9 - 4 = 5$, so an equation of the hyperbola is

$$\frac{x^2}{4} - \frac{y^2}{5} = 1$$

FIGURE 34
$$\frac{x^2}{4} - \frac{y^2}{5} = 1$$

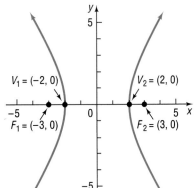

See Figure 34. ■

Comment: To graph the hyperbola $(x^2/4) - (y^2/5) = 1$ discussed in Example 1, we need to graph the two functions $y = \sqrt{5}\sqrt{(x^2/4) - 1}$ and $y = -\sqrt{5}\sqrt{(x^2/4) - 1}$. Do this and compare what you see with Figure 34. ■

■ Now work Problem 5.

An equation of the form of equation (2) is the equation of a hyperbola with center at the origin, foci on the x-axis at $(-c, 0)$ and $(c, 0)$, where $c^2 = a^2 + b^2$, and transverse axis along the x-axis.

For the remainder of this section, the direction "Discuss the equation" will mean to find the center, transverse axis, vertices, and foci of the hyperbola and graph it.

E X A M P L E 2   *Discussing the Equation of a Hyperbola*

Discuss the equation: $\dfrac{x^2}{16} - \dfrac{y^2}{4} = 1$

Solution   The given equation is of the form of equation (2), with $a^2 = 16$ and $b^2 = 4$. Thus, the graph of the equation is a hyperbola with center at (0, 0) and transverse axis along the x-axis. Also, we know that $c^2 = a^2 + b^2 = 16 + 4 = 20$. The vertices are at $(\pm a, 0) = (\pm 4, 0)$, and the foci are at $(\pm c, 0) = (\pm 2\sqrt{5}, 0)$. Figure 35 shows the graph.

FIGURE 35

$\dfrac{x^2}{16} - \dfrac{y^2}{4} = 1$

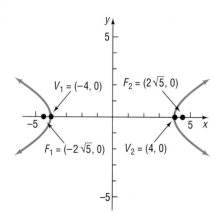

The next result gives the form of the equation of a hyperbola with center at the origin and transverse axis along the y-axis.

Theorem   An equation of the hyperbola with center at (0, 0), foci at $(0, -c)$ and $(0, c)$, and vertices at $(0, -a)$ and $(0, a)$ is

Equation of a Hyperbola;
Center at (0, 0); Foci at
(0, ±c); Vertices at (0, ±a);
Transverse Axis along y-Axis

$$\frac{y^2}{a^2} - \frac{x^2}{b^2} = 1 \qquad \text{where } b^2 = c^2 - a^2 \qquad (3)$$

The transverse axis is the y-axis. ∎

Figure 36 shows the graph of a typical hyperbola defined by equation (3).

FIGURE 36

$\dfrac{y^2}{a^2} - \dfrac{x^2}{b^2} = 1,$

$b^2 = c^2 - a^2$

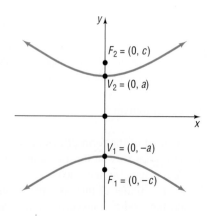

Notice the difference in the form of equations (2) and (3). When the y^2-term is subtracted from the x^2-term, the transverse axis is the x-axis. When the x^2-term is subtracted from the y^2-term, the transverse axis is the y-axis.

E X A M P L E   3   *Discussing the Equation of a Hyperbola*

Discuss the equation: $y^2 - 4x^2 = 4$

Solution   To put the equation in proper form, we divide each side by 4:

$$\frac{y^2}{4} - x^2 = 1$$

Since the x^2-term is subtracted from the y^2-term, the equation is that of a hyperbola with center at the origin and transverse axis along the y-axis. Also, comparing the above equation to equation (3), we find $a^2 = 4$, $b^2 = 1$, and $c^2 = a^2 + b^2 = 5$. The vertices are at $(0, \pm a) = (0, \pm 2)$, and the foci are at $(0, \pm c) = (0, \pm\sqrt{5})$. The graph is given in Figure 37.

FIGURE 37
$$\frac{y^2}{4} - x^2 = 1$$

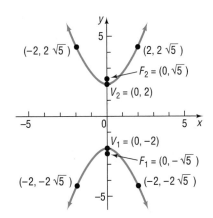

E X A M P L E   4   *Finding an Equation of a Hyperbola*

Find an equation of the hyperbola having one vertex at $(0, 2)$ and foci at $(0, -3)$ and $(0, 3)$. Graph the equation.

Solution   Since the foci are at $(0, -3)$ and $(0, 3)$, the center of the hyperbola is at the origin. Also, the transverse axis is along the y-axis. The given information also reveals that $c = 3$, $a = 2$, and $b^2 = c^2 - a^2 = 9 - 4 = 5$. The form of the equation of the hyperbola is given by equation (3):

$$\frac{y^2}{a^2} - \frac{x^2}{b^2} = 1$$

$$\frac{y^2}{4} - \frac{x^2}{5} = 1$$

See Figure 38.

■   Now work Problem 7.

FIGURE 38

$$\frac{y^2}{4} - \frac{x^2}{5} = 1$$

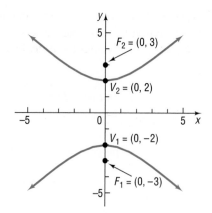

Look at the equations of the hyperbolas in Examples 3 and 4. For the hyperbola in Example 3, $a^2 = 4$ and $b^2 = 1$, so $a > b$; for the hyperbola in Example 4, $a^2 = 4$ and $b^2 = 5$, so $a < b$. We conclude that, for hyperbolas, there are no requirements involving the relative size of a and b. Contrast this situation to the case of an ellipse, in which the relative sizes of a and b dictate which axis is the major axis. Hyperbolas have another feature to distinguish them from ellipses and parabolas: Hyperbolas have asymptotes.

Asymptotes

Recall from Section 5.3 that a horizontal or oblique asymptote of a graph is a line with the property that the distance from the line to points on the graph approaches 0 as $x \to -\infty$ or as $x \to \infty$.

Theorem   The hyperbola $\dfrac{x^2}{a^2} - \dfrac{y^2}{b^2} = 1$ has the two oblique asymptotes

Asymptotes of a Hyperbola

$$y = \frac{b}{a}x \quad \text{and} \quad y = -\frac{b}{a}x$$

Proof   We begin by solving for y in the equation of the hyperbola:

$$\frac{x^2}{a^2} - \frac{y^2}{b^2} = 1$$

$$\frac{y^2}{b^2} = \frac{x^2}{a^2} - 1$$

$$y^2 = b^2\left(\frac{x^2}{a^2} - 1\right)$$

If $x \neq 0$, we can rearrange the right side in the form

$$y^2 = \frac{b^2 x^2}{a^2}\left(1 - \frac{a^2}{x^2}\right)$$

$$y = \pm\frac{bx}{a}\sqrt{1 - \frac{a^2}{x^2}}$$

Now, as $x \to -\infty$ or as $x \to \infty$, the term a^2/x^2 approaches 0, so the expression under the radical approaches 1. Thus, as $x \to -\infty$ or as $x \to \infty$, the value of y approaches $\pm bx/a$; that is, the graph of the hyperbola approaches the lines

$$y = -\frac{b}{a}x \quad \text{and} \quad y = \frac{b}{a}x$$

Thus, these lines are oblique asymptotes of the hyperbola. ∎

The asymptotes of a hyperbola are not part of the hyperbola, but they do serve as a guide for graphing a hyperbola. For example, suppose that we want to graph the equation

$$\frac{x^2}{a^2} - \frac{y^2}{b^2} = 1$$

We begin by plotting the vertices $(-a, 0)$ and $(a, 0)$. Then we plot the points $(0, -b)$ and $(0, b)$ and use these four points to construct a rectangle, as shown in Figure 39. The diagonals of this rectangle have slopes b/a and $-b/a$, and their extensions are the asymptotes $y = (b/a)x$ and $y = -(b/a)x$ of the hyperbola.

FIGURE 39

$$\frac{x^2}{a^2} - \frac{y^2}{b^2} = 1$$

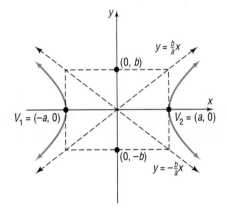

Theorem   The hyperbola $\dfrac{y^2}{a^2} - \dfrac{x^2}{b^2} = 1$ has the two oblique asymptotes

Asymptotes of a Hyperbola

$$y = \frac{a}{b}x \quad \text{and} \quad y = -\frac{a}{b}x$$

∎

You are asked to prove this result in Problem 60.

E X A M P L E   5   *Discussing the Equation of a Hyperbola*

Discuss the equation: $9x^2 - 4y^2 = 36$

FIGURE 40

$$\frac{x^2}{4} - \frac{y^2}{9} = 1$$

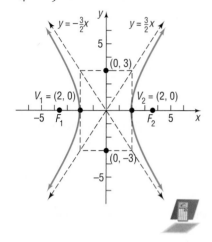

Solution   First, we divide each side by 36 to put the equation in proper form:

$$\frac{x^2}{4} - \frac{y^2}{9} = 1$$

This is the equation of a hyperbola with center at the origin and transverse axis along the x-axis. Using $a^2 = 4$ and $b^2 = 9$, we find $c^2 = a^2 + b^2 = 13$. The vertices are at $(\pm a, 0) = (\pm 2, 0)$, the foci are at $(\pm c, 0) = (\pm\sqrt{13}, 0)$, and the asymptotes have the equations

$$y = \frac{3}{2}x \quad \text{and} \quad y = -\frac{3}{2}x$$

Now form the rectangle containing the points $(\pm a, 0)$ and $(0, \pm b)$, that is, $(-2, 0)$ $(2, 0)$, $(0, -3)$, and $(0, 3)$. The extensions of the diagonals of this rectangle are the asymptotes. See Figure 40 for the graph. ■

Exploration: Graph the upper portion of the hyperbola $9x^2 - 4y^2 = 36$ discussed in Example 5 and its asymptotes $y = \frac{3}{2}x$ and $y = -\frac{3}{2}x$. Now use ZOOM and TRACE to see what happens as x becomes unbounded in the positive direction. What happens as x becomes unbounded in the negative direction? ■

■ Now work Problem 17.

Center at (h, k)

If a hyperbola with center at the origin and transverse axis coinciding with a coordinate axis is shifted horizontally h units and then vertically k units, the result is a hyperbola with center at (h, k) and transverse axis parallel to a coordinate axis. Table 4 gives the forms of the equations of such hyperbolas. See Figure 41 for the graphs.

TABLE 4   HYPERBOLAS WITH CENTER AT (h, k) AND TRANSVERSE AXIS PARALLEL TO A COORDINATE AXIS

CENTER	TRANSVERSE AXIS	FOCI	VERTICES	EQUATION	ASYMPTOTES
(h, k)	Parallel to x-axis	$(h \pm c, k)$	$(h \pm a, k)$	$\dfrac{(x-h)^2}{a^2} - \dfrac{(y-k)^2}{b^2} = 1,$ $b^2 = c^2 - a^2$	$y - k = \pm\dfrac{b}{a}(x - h)$
(h, k)	Parallel to y-axis	$(h, k \pm c)$	$(h, k \pm a)$	$\dfrac{(y-k)^2}{a^2} - \dfrac{(x-h)^2}{b^2} = 1,$ $b^2 = c^2 - a^2$	$y - k = \pm\dfrac{a}{b}(x - h)$

FIGURE 41

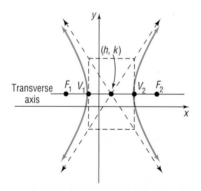

(a) $\dfrac{(x-h)^2}{a^2} - \dfrac{(y-k)^2}{b^2} = 1$

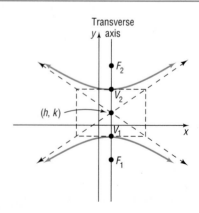

(b) $\dfrac{(y-k)^2}{a^2} - \dfrac{(x-h)^2}{b^2} = 1$

E X A M P L E   6

Finding an Equation of a Hyperbola, Center Not at the Origin

Find an equation for the hyperbola with center at $(1, -2)$, one focus at $(4, -2)$, and one vertex at $(3, -2)$. Graph the equation.

Solution
The center is at $(h, k) = (1, -2)$, so $h = 1$ and $k = -2$. The transverse axis is parallel to the x-axis. The distance from the center $(1, -2)$ to the focus $(4, -2)$ is $c = 3$; the distance from the center $(1, -2)$ to the vertex $(3, -2)$ is $a = 2$. Thus, $b^2 = c^2 - a^2 = 9 - 4 = 5$. The equation is

$$\frac{(x - h)^2}{a^2} - \frac{(y - k)^2}{b^2} = 1$$

$$\frac{(x - 1)^2}{4} - \frac{(y + 2)^2}{5} = 1$$

See Figure 42.

FIGURE 42

$$\frac{(x - 1)^2}{4} - \frac{(y + 2)^2}{5} = 1$$

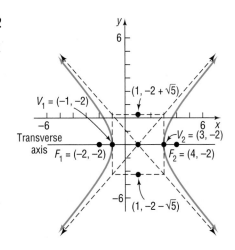

■ Now work Problem 27.

E X A M P L E   7

Discussing the Equation of a Hyperbola

Discuss the equation: $-x^2 + 4y^2 - 2x - 16y + 11 = 0$

Solution
We complete the squares in x and in y:

$$-x^2 + 4y^2 - 2x - 16y + 11 = 0$$

$$-(x^2 + 2x) + 4(y^2 - 4y) = -11 \qquad \text{Group terms.}$$

$$-(x^2 + 2x + 1) + 4(y^2 - 4y + 4) = -1 + 16 - 11 \qquad \text{Complete each square.}$$

$$-(x + 1)^2 + 4(y - 2)^2 = 4$$

$$(y - 2)^2 - \frac{(x + 1)^2}{4} = 1 \qquad \text{Divide by 4.}$$

This is the equation of a hyperbola with center at $(-1, 2)$ and transverse axis parallel to the y-axis. Also, $a^2 = 1$ and $b^2 = 4$, so $c^2 = a^2 + b^2 = 5$. The vertices are at $(h, k \pm a) = (-1, 2 \pm 1)$, or $(-1, 1)$ and $(-1, 3)$. The foci are at $(h, k \pm c) = (-1, 2 \pm \sqrt{5})$. The asymptotes are $y - 2 = \frac{1}{2}(x + 1)$ and $y - 2 = -\frac{1}{2}(x + 1)$. Figure 43 shows the graph.

FIGURE 43

$$(y - 2)^2 - \frac{(x + 1)^2}{4} = 1$$

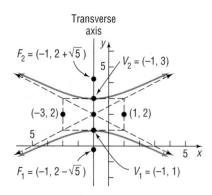

Applications

Suppose that a gun is fired from an unknown source S. An observer at O_1 hears the report (sound of gun shot) 1 second after another observer at O_2. Because sound travels at about 1100 feet per second, it follows that the point S must be 1100 feet closer to O_2 than to O_1. Thus, S lies on one branch of a hyperbola with foci at O_1 and O_2. (Do you see why? The difference of the distances from S to O_1 and from S to O_2 is the constant 1100.) If a third observer at O_3 hears the same report 2 seconds after O_1 hears it, then S will lie on a branch of a second hyperbola with foci at O_1 and O_3. The intersection of the two hyperbolas will pinpoint the location of S. See Figure 44 for an illustration.

FIGURE 44

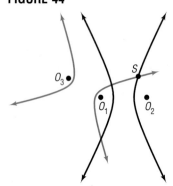

LORAN

In the LOng RAnge Navigation system (LORAN), a master radio sending station and a secondary sending station emit signals that can be received by a ship at sea. (See Figure 45.) Because a ship monitoring the two signals will usually be nearer to one of the two stations, there will be a difference in the distances that the two signals travel, which will register as a slight time difference between the signals. As

FIGURE 45

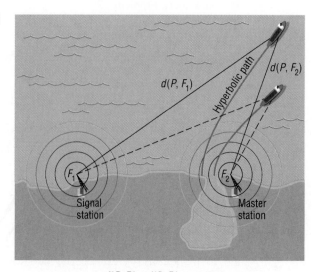

$$d(P, F_1) - d(P, F_2) = \text{constant}$$

long as the time difference remains constant, the difference of the two distances will also be constant. If the ship follows a path corresponding to the fixed time difference, it will follow the path of a hyperbola whose foci are located at the positions of the two sending stations. So for each time difference a different hyperbolic path results, each bringing the ship to a different shore location. Navigation charts show the various hyperbolic paths corresponding to different time differences.

E X A M P L E   8

LORAN

Two LORAN stations are positioned 250 miles apart along a straight shore.

(a) A ship records a time difference of 0.00086 second between the LORAN signals. Set up an appropriate rectangular coordinate system to determine where the ship would reach shore if it were to follow the hyperbola corresponding to this time difference.

(b) If the ship wants to enter a harbor located between the two stations 25 miles from the master station, what time difference should it be looking for?

(c) If the ship is 80 miles offshore when the desired time difference is obtained, what is the exact location of the ship? [*Note:* The speed of each radio signal is 186,000 miles per second.]

Solution

(a) We set up a rectangular coordinate system so that the two stations lie on the *x*-axis and the origin is midway between them. See Figure 46.

FIGURE 46

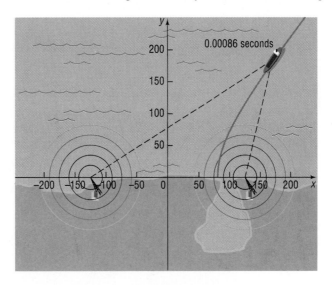

The ship lies on a hyperbola whose foci are the locations of the two stations. The reason for this is that the constant time difference of the signals from each station results in a constant difference in the distance of the ship from each station. Since the time difference is 0.00086 second and the speed of the signal is 186,000 miles per second, the difference of the distances from the ship to each station (foci) is

$$\text{Distance} = \text{Speed} \times \text{Time} = 186{,}000 \times 0.00086 = 160 \text{ miles}$$

The difference of the distances from the ship to each station, 160, equals $2a$, so $a = 80$ and the vertex of the corresponding hyperbola is at $(80, 0)$. Since the focus is at (125, 0), following this hyperbola the ship would reach shore 45 miles from the master station.

(b) To reach shore 25 miles from the master station, the ship should follow a hyperbola with vertex at (100, 0). For this hyperbola, $a = 100$, so the constant difference of the distances from the ship to each station is 200. The time difference the ship should look for is

$$\text{Time} = \frac{\text{Distance}}{\text{Speed}} = \frac{200}{186{,}000} = 0.001075 \text{ second}$$

(c) To find the exact location of the ship, we need to find the equation of the hyperbola with vertex at (100, 0) and a focus at (125, 0). The form of the equation of this hyperbola is

$$\frac{x^2}{a^2} - \frac{y^2}{b^2} = 1$$

where $a = 100$. Since $c = 125$, we have

$$b^2 = c^2 - a^2 = 125^2 - 100^2 = 5625$$

The equation of the hyperbola is

$$\frac{x^2}{100^2} - \frac{y^2}{5625} = 1$$

Since the ship is 80 miles from shore, we use $y = 80$ in the equation and solve for x.

$$\frac{x^2}{100^2} - \frac{80^2}{5625} = 1$$

$$\frac{x^2}{100^2} = 1 + \frac{80^2}{5625} = 2.14$$

$$x^2 = 100^2(2.14)$$

$$x = 146$$

The ship is at the position (146, 80). See Figure 47.

FIGURE 47

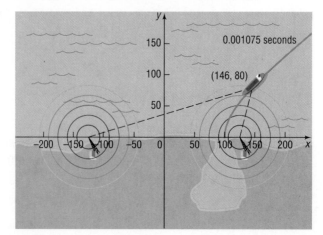

9.4

Exercise 9.4

In Problems 1–4, the graph of a hyperbola is given. Match each graph to its equation.

A. $\dfrac{x^2}{4} - y^2 = 1$ B. $x^2 - \dfrac{y^2}{4} = 1$ C. $\dfrac{y^2}{4} - x^2 = 1$ D. $y^2 - \dfrac{x^2}{4} = 1$

1.
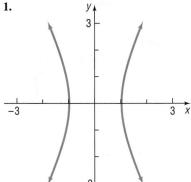

2.
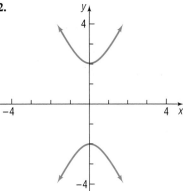

3.
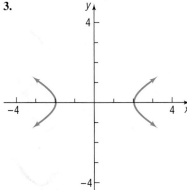

4.
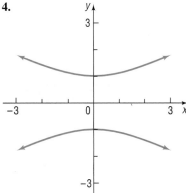

In Problems 5–14, find an equation for the hyperbola described. Graph the equation.

5. Center at $(0, 0)$; focus at $(3, 0)$; vertex at $(1, 0)$

6. Center at $(0, 0)$; focus at $(0, 5)$; vertex at $(0, 3)$

7. Center at $(0, 0)$; focus at $(0, -6)$; vertex at $(0, 4)$

8. Center at $(0, 0)$; focus at $(-3, 0)$; vertex at $(2, 0)$

9. Foci at $(-5, 0)$ and $(5, 0)$; vertex at $(3, 0)$

10. Focus at $(0, 6)$; vertices at $(0, -2)$ and $(0, 2)$

11. Vertices at $(0, -6)$ and $(0, 6)$; asymptote the line $y = 2x$

12. Vertices at $(-4, 0)$ and $(4, 0)$; asymptote the line $y = 2x$

13. Foci at $(-4, 0)$ and $(4, 0)$; asymptote the line $y = -x$

14. Foci at $(0, -2)$ and $(0, 2)$; asymptote the line $y = -x$

In Problems 15–22, find the center, transverse axis, vertices, foci, and asymptotes. Graph each equation.

15. $\dfrac{x^2}{16} - \dfrac{y^2}{4} = 1$ **16.** $\dfrac{y^2}{16} - \dfrac{x^2}{4} = 1$ **17.** $4x^2 - y^2 = 16$ **18.** $y^2 - 4x^2 = 16$

19. $y^2 - 9x^2 = 9$ **20.** $x^2 - y^2 = 4$ **21.** $y^2 - x^2 = 25$ **22.** $2x^2 - y^2 = 4$

In Problems 23–26, write an equation for each hyperbola.

23.

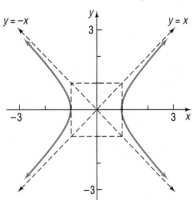

24.

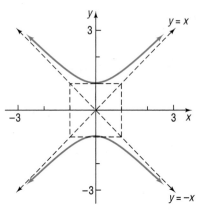

25.

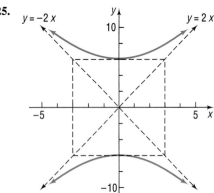

26.

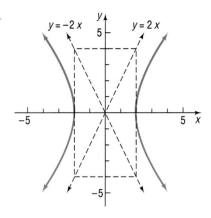

In Problems 27–34, find an equation for the hyperbola described. Graph the equation.

27. Center at $(4, -1)$; focus at $(7, -1)$; vertex at $(6, -1)$

28. Center at $(-3, 1)$; focus at $(-3, 6)$; vertex at $(-3, 4)$

29. Center at $(-3, -4)$; focus at $(-3, -8)$; vertex at $(-3, -2)$

30. Center at $(1, 4)$; focus at $(-2, 4)$; vertex at $(0, 4)$

31. Foci at $(3, 7)$ and $(7, 7)$; vertex at $(6, 7)$

32. Focus at $(-4, 0)$; vertices at $(-4, 4)$ and $(-4, 2)$

33. Vertices at $(-1, -1)$ and $(3, -1)$; asymptote the line $(x - 1)/2 = (y + 1)/3$

34. Vertices at $(1, -3)$ and $(1, 1)$; asymptote the line $(x - 1)/2 = (y + 1)/3$

In Problems 35–48, find the center, transverse axis, vertices, foci, and asymptotes. Graph each equation.

35. $\dfrac{(x - 2)^2}{4} - \dfrac{(y + 3)^2}{9} = 1$ **36.** $\dfrac{(y + 3)^2}{4} - \dfrac{(x - 2)^2}{9} = 1$

37. $(y - 2)^2 - 4(x + 2)^2 = 4$

38. $(x + 4)^2 - 9(y - 3)^2 = 9$

39. $(x + 1)^2 - (y + 2)^2 = 4$

40. $(y - 3)^2 - (x + 2)^2 = 4$

41. $x^2 - y^2 - 2x - 2y - 1 = 0$

42. $y^2 - x^2 - 4y + 4x - 1 = 0$

43. $y^2 - 4x^2 - 4y - 8x - 4 = 0$

44. $2x^2 - y^2 + 4x + 4y - 4 = 0$

45. $4x^2 - y^2 - 24x - 4y + 16 = 0$

46. $2y^2 - x^2 + 2x + 8y + 3 = 0$

47. $y^2 - 4x^2 - 16x - 2y - 19 = 0$

48. $x^2 - 3y^2 + 8x - 6y + 4 = 0$

In Problems 49–52, graph each function. [Hint: Notice that each function is half a hyperbola.]

49. $f(x) = \sqrt{16 + 4x^2}$

50. $f(x) = -\sqrt{9 + 9x^2}$

51. $f(x) = -\sqrt{-25 + x^2}$

52. $f(x) = \sqrt{-1 + x^2}$

53. *LORAN* Two LORAN stations are positioned 200 miles apart along a straight shore.
  (a) A ship records a time difference of 0.00038 second between the LORAN signals. Set up an appropriate rectangular coordinate system to determine where the ship would reach shore if it were to follow the hyperbola corresponding to this time difference.
  (b) If the ship wants to enter a harbor located between the two stations 20 miles from the master station, what time difference should it be looking for?
  (c) If the ship is 50 miles offshore when the desired time difference is obtained, what is the exact location of the ship? [*Note:* The speed of each radio signal is 186,000 miles per second.]

54. *LORAN* Two LORAN stations are positioned 100 miles apart along a straight shore.
  (a) A ship records a time difference of 0.00032 second between the LORAN signals. Set up an appropriate rectangular coordinate system to determine where the ship would reach shore if it were to follow the hyperbola corresponding to this time difference.
  (b) If the ship wants to enter a harbor located between the two stations 10 miles from the master station, what time difference should it be looking for?
  (c) If the ship is 20 miles offshore when the desired time difference is obtained, what is the exact location of the ship? [*Note:* The speed of each radio signal is 186,000 miles per second.]

55. *Calibrating Instruments* In a test of their recording devices a team of seismologists positioned two of the devices 2000 feet apart, with the device at point A to the west of the device at point B. At a point between the devices and 200 feet from point B, a small amount of explosive was detonated and a note made of the time at which the sound reached each device. A second explosion is to be carried out at a point directly north of point B.
  (a) How far north should the site of the second explosion be chosen so that the measured time difference recorded by the devices for the second detonation is the same as that recorded for the first detonation?
  (b) Explain why this experiment can be used to calibrate the instruments.

56. Explain in your own words the LORAN system of navigation.

57. The **eccentricity** e of a hyperbola is defined as the number c/a, where a and c are the numbers given in equation (2). Because $c > a$, it follows that $e > 1$. Describe the general shape of a hyperbola whose eccentricity is close to 1. What is the shape if e is very large?

58. A hyperbola for which $a = b$ is called an **equilateral hyperbola.** Find the eccentricity e of an equilateral hyperbola. [*Note:* The eccentricity of a hyperbola is defined in Problem 57.]

59. Two hyperbolas that have the same set of asymptotes are called **conjugate.** Show that the hyperbolas

$$\frac{x^2}{4} - y^2 = 1 \quad \text{and} \quad y^2 - \frac{x^2}{4} = 1$$

are conjugate. Graph each hyperbola on the same set of coordinate axes.

60. Prove that the hyperbola

$$\frac{y^2}{a^2} - \frac{x^2}{b^2} = 1$$

has the two oblique asymptotes

$$y = \frac{a}{b}x \quad \text{and} \quad y = -\frac{a}{b}x$$

61. Show that the graph of an equation of the form

$$Ax^2 + Cy^2 + F = 0 \qquad A \neq 0,\ C \neq 0,\ F \neq 0$$

where A and C are of opposite sign, is a hyperbola with center at $(0, 0)$.

62. Show that the graph of an equation of the form

$$Ax^2 + Cy^2 + Dx + Ey + F = 0 \qquad A \neq 0,\ C \neq 0$$

where A and C are of opposite sign:
(a) Is a hyperbola if $(D^2/4A) + (E^2/4C) - F \neq 0$.
(b) Is two intersecting lines if $(D^2/4A) + (E^2/4C) - F = 0$.

9.5

Rotation of Axes; General Form of a Conic

In this section, we show that the graph of a general second-degree polynomial containing two variables x and y, that is, an equation of the form

$$Ax^2 + Bxy + Cy^2 + Dx + Ey + F = 0 \tag{1}$$

where A, B, and C are not simultaneously 0, is a conic. We shall not concern ourselves here with the degenerate cases of equation (1), such as $x^2 + y^2 = 0$, whose graph is a single point $(0, 0)$; or $x^2 + 3y^2 + 3 = 0$, whose graph contains no points; or $x^2 - 4y^2 = 0$, whose graph is two lines, $x - 2y = 0$ and $x + 2y = 0$.

We begin with the case where $B = 0$. In this case, the term containing xy is not present, so equation (1) has the form

$$Ax^2 + Cy^2 + Dx + Ey + F = 0$$

where either $A \neq 0$ or $C \neq 0$.

We have already discussed the procedure for identifying the graph of this kind of equation; we complete the squares of the quadratic expressions in x or y, or both. Once this has been done, the conic can be identified by comparing it to one of the forms studied in Sections 9.2 through 9.4.

In fact, though, we can identify the conic directly from the equation without completing the squares.

Theorem
Identifying Conics without Completing the Squares

Excluding degenerate cases, the equation

$$Ax^2 + Cy^2 + Dx + Ey + F = 0 \tag{2}$$

where either $A \neq 0$ or $C \neq 0$:

(a) Defines a parabola if $AC = 0$.

(b) Defines an ellipse (or a circle) if $AC > 0$.

(c) Defines a hyperbola if $AC < 0$.   ∎

Proof (a) If $AC = 0$, then either $A = 0$ or $C = 0$, but not both, so the form of equation (2) is either

$$Ax^2 + Dx + Ey + F = 0, \qquad A \neq 0$$

or

$$Cy^2 + Dx + Ey + F = 0, \qquad C \neq 0$$

Using the results of Problems 65 and 66 in Exercise 9.2, it follows that, except for the degenerate cases, the equation is a parabola.

(b) If $AC > 0$, then A and C are of the same sign. Using the results of Problems 69 and 70 in Exercise 9.3, except for the degenerate cases, the equation is an ellipse if $A \neq C$ or a circle if $A = C$.

(c) If $AC < 0$, then A and C are of opposite sign. Using the results of Problems 61 and 62 in Exercise 9.4, except for the degenerate cases, the equation is a hyperbola. ∎

We shall not be concerned with the degenerate cases of equation (2). However, in practice, you should be alert to the possibility of degeneracy.

E X A M P L E   1   *Identifying a Conic without Completing the Squares*

Identify each equation without completing the squares.

(a) $3x^2 + 6y^2 + 6x - 12y = 0$ (b) $2x^2 - 3y^2 + 6y + 4 = 0$

(c) $y^2 - 2x + 4 = 0$

Solution   (a) We compare the given equation to equation (2) and conclude that $A = 3$ and $C = 6$. Since $AC = 18 > 0$, the equation is an ellipse.

(b) Here, $A = 2$ and $C = -3$, so $AC = -6 < 0$. The equation is a hyperbola.

(c) Here, $A = 0$ and $C = 1$, so $AC = 0$. The equation is a parabola. ∎

■ Now work Problem 1.

Although we can now identify the type of conic represented by any equation of the form of equation (2) without completing the squares, we will still need to complete the squares if we desire additional information about a conic.

Now we turn our attention to equations of the form of equation (1), where $B \neq 0$. To discuss this case, we first need to investigate a new procedure: *rotation of axes*.

Rotation of Axes

In a **rotation of axes,** the origin remains fixed while the x-axis and y-axis are rotated through an angle θ to a new position; the new positions of the x- and y-axes are denoted by x' and y', respectively, as shown in Figure 48(a).

Now look at Figure 48(b). There the point P has the coordinates (x, y) relative to the xy-plane, while the same point P has coordinates (x', y') relative to the $x'y'$-plane. We seek relationships that will enable us to express x and y in terms of x', y', and θ.

As Figure 48(b) shows, r denotes the distance from the origin O to the point P, and α denotes the angle between the positive x'-axis and the ray from O through P. Then, using the definitions of sine and cosine, we have

$$x' = r \cos \alpha \qquad y' = r \sin \alpha \qquad (3)$$
$$x = r \cos(\theta + \alpha) \qquad y = r \sin(\theta + \alpha) \qquad (4)$$

Now

$$x = r \cos(\theta + \alpha)$$
$$= r(\cos \theta \cos \alpha - \sin \theta \sin \alpha) \qquad \text{Sum formula}$$
$$= (r \cos \alpha)(\cos \theta) - (r \sin \alpha)(\sin \theta)$$
$$= x' \cos \theta - y' \sin \theta \qquad \text{By equation (3)}$$

FIGURE 48

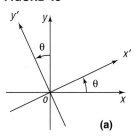

(a)

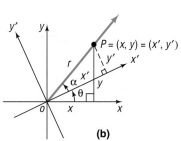

(b)

Similarly,

$$y = r \sin(\theta + \alpha)$$
$$= r(\sin \theta \cos \alpha + \cos \theta \sin \alpha)$$
$$= x' \sin \theta + y' \cos \theta$$

Theorem
Rotation Formulas

If the x- and y-axes are rotated through an angle θ, the coordinates (x, y) of a point P relative to the xy-plane and the coordinates (x', y') of the same point relative to the new x'- and y'-axes are related by the formulas

$$x = x' \cos \theta - y' \sin \theta \qquad y = x' \sin \theta + y' \cos \theta \qquad (5)$$

■

E X A M P L E   2

Rotating Axes

Express the equation $xy = 1$ in terms of new $x'y'$-coordinates by rotating the axes through a 45° angle. Discuss the new equation.

Solution

Let $\theta = 45°$ in equation (5). Then

$$x = x' \cos 45° - y' \sin 45° = x'\frac{\sqrt{2}}{2} - y'\frac{\sqrt{2}}{2} = \frac{\sqrt{2}}{2}(x' - y')$$

$$y = x' \sin 45° + y' \cos 45° = x'\frac{\sqrt{2}}{2} + y'\frac{\sqrt{2}}{2} = \frac{\sqrt{2}}{2}(x' + y')$$

FIGURE 49

$xy = 1$ or $\frac{x'^2}{2} - \frac{y'^2}{2} = 1$

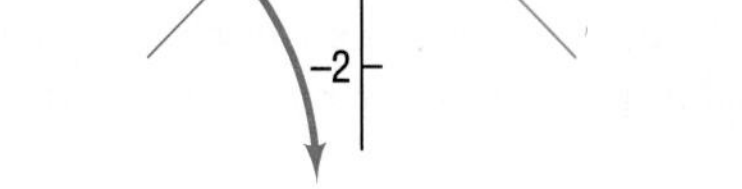

Substituting these expressions for x and y in $xy = 1$ gives

$$\left[\frac{\sqrt{2}}{2}(x' - y')\right]\left[\frac{\sqrt{2}}{2}(x' + y')\right] = 1$$

$$\frac{1}{2}(x'^2 - y'^2) = 1$$

$$\frac{x'^2}{2} - \frac{y'^2}{2} = 1$$

This is the equation of a hyperbola with center at $(0, 0)$ and transverse axis along the x'-axis. The vertices are at $(\pm\sqrt{2}, 0)$ on the x'-axis; the asymptotes are $y' = x'$ and $y' = -x'$ (which correspond to the original x- and y-axes). See Figure 49 for the graph.   ■

As Example 2 illustrates, a rotation of axes through an appropriate angle can transform a second-degree equation in x and y containing an xy-term into one in x' and y' in which no $x'y'$-term appears. In fact, we will show that a rotation of axes through an appropriate angle will transform any equation of the form of equation (1) into an equation in x' and y' without an $x'y'$-term.

To find the formula for choosing an appropriate angle θ through which to rotate the axes, we begin with equation (1),

$$Ax^2 + Bxy + Cy^2 + Dx + Ey + F = 0, \qquad B \neq 0$$

Next we rotate through an angle θ using rotation formulas (5):

$$A(x' \cos \theta - y' \sin \theta)^2 + B(x' \cos \theta - y' \sin \theta)(x' \sin \theta + y' \cos \theta)$$
$$+ C(x' \sin \theta + y' \cos \theta)^2 + D(x' \cos \theta - y' \sin \theta)$$
$$+ E(x' \sin \theta + y' \cos \theta) + F = 0$$

By expanding and collecting like terms, we obtain

$$(A \cos^2 \theta + B \sin \theta \cos \theta + C \sin^2 \theta)x'^2 + [B(\cos^2 \theta - \sin^2 \theta) + 2(C - A)(\sin \theta \cos \theta)]x'y'$$
$$+ (A \sin^2 \theta - B \sin \theta \cos \theta + C \cos^2 \theta)y'^2 \qquad (6)$$
$$+ (D \cos \theta + E \sin \theta)x'$$
$$+ (-D \sin \theta + E \cos \theta)y' + F = 0$$

In equation (6), the coefficient of $x'y'$ is

$$B' = 2(C - A)(\sin \theta \cos \theta) + B(\cos^2 \theta - \sin^2 \theta)$$

Since we want to eliminate the $x'y'$-term, we select an angle θ so that $B' = 0$. Thus,

$$2(C - A)(\sin \theta \cos \theta) + B(\cos^2 \theta - \sin^2 \theta) = 0$$
$$(C - A)(\sin 2\theta) + B \cos 2\theta = 0 \quad \text{Double-angle formulas}$$
$$B \cos 2\theta = (A - C)(\sin 2\theta)$$
$$\cot 2\theta = \frac{A - C}{B}, \qquad B \neq 0$$

Theorem To transform the equation

$$Ax^2 + Bxy + Cy^2 + Dx + Ey + F = 0, \qquad B \neq 0$$

into an equation in x' and y' without an $x'y'$-term, rotate the axes through an angle θ that satisfies the equation

$$\cot 2\theta = \frac{A - C}{B} \qquad (7)$$

∎

Equation (7) has an infinite number of solutions for θ. We shall adopt the convention of choosing the acute angle θ that satisfies (7). Then we have the following two possibilities:

If $\cot 2\theta > 0$, then $0 < 2\theta \leq \pi/2$ so that $0 < \theta \leq \pi/4$.

If $\cot 2\theta < 0$, then $\pi/2 < 2\theta < \pi$ so that $\pi/4 < \theta < \pi/2$.

Each of these results in a counterclockwise rotation of the axes through an acute angle θ.*

---

*Any rotation (clockwise or counterclockwise) through an angle θ that satisfies $\cot 2\theta = (A - C)/B$ will eliminate the $x'y'$-term. However, the final forms of the transformed equation may be different (but be equivalent), depending on the angle chosen.

Warning: Be careful if you use a calculator to solve equation (7).

1. If $\cot 2\theta = 0$, then $2\theta = \pi/2$ and $\theta = \pi/4$.
2. If $\cot 2\theta \neq 0$, first find $\cos 2\theta$. Then use the inverse cosine function key(s) to obtain 2θ, $0 < 2\theta < \pi$. Finally, divide by 2 to obtain the correct acute angle θ.

E X A M P L E   3   *Discussing an Equation Using a Rotation of Axes*

Discuss the equation: $x^2 + \sqrt{3}xy + 2y^2 - 10 = 0$

Solution   Since an xy-term is present, we must rotate the axes. Using $A = 1$, $B = \sqrt{3}$, and $C = 2$ in equation (7), the appropriate acute angle θ through which to rotate the axes satisfies the equation

$$\cot 2\theta = \frac{A - C}{B} = \frac{-1}{\sqrt{3}} = \frac{-\sqrt{3}}{3}, \qquad 0° < 2\theta < 180°$$

Since $\cot 2\theta = -\sqrt{3}/3$, we find $2\theta = 120°$, so $\theta = 60°$. Using $\theta = 60°$ in rotation formulas (5), we find

$$x = \frac{1}{2}x' - \frac{\sqrt{3}}{2}y' = \frac{1}{2}(x' - \sqrt{3}y')$$

$$y = \frac{\sqrt{3}}{2}x' + \frac{1}{2}y' = \frac{1}{2}(\sqrt{3}x' + y')$$

Substituting these values into the original equation and simplifying, we have

$$x^2 + \sqrt{3}xy + 2y^2 - 10 = 0$$

$$\frac{1}{4}(x' - \sqrt{3}y')^2 + \sqrt{3}\left[\frac{1}{2}(x' - \sqrt{3}y')\right]\left[\frac{1}{2}(\sqrt{3}x' + y')\right] + 2\left[\frac{1}{4}(\sqrt{3}x' + y')^2\right] = 10$$

Multiply both sides by 4 and expand to obtain

$$x'^2 - 2\sqrt{3}x'y' + 3y'^2 + \sqrt{3}(\sqrt{3}x'^2 - 2x'y' - \sqrt{3}y'^2) + 2(3x'^2 + 2\sqrt{3}x'y' + y'^2) = 40$$

$$10x'^2 + 2y'^2 = 40$$

$$\frac{x'^2}{4} + \frac{y'^2}{20} = 1$$

This is the equation of an ellipse with center at $(0, 0)$ and major axis along the y'-axis. The vertices are at $(0, \pm 2\sqrt{5})$ on the y'-axis. See Figure 50 for the graph.

FIGURE 50

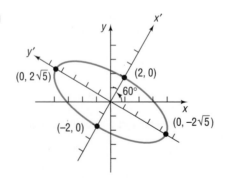

■ Now work Problem 21.

In Example 3, the acute angle θ through which to rotate the axes was easy to find because of the numbers we used in the given equation. In general, the equation $\cot 2\theta = (A - C)/B$ will not have such a "nice" solution. As the next example shows, we can still find the appropriate rotation formulas without using a calculator approximation by applying half-angle formulas.

E X A M P L E   4   *Discussing an Equation Using a Rotation of Axes*

Discuss the equation: $4x^2 - 4xy + y^2 + 5\sqrt{5}x + 5 = 0$

Solution   Letting $A = 4$, $B = -4$, and $C = 1$ in equation (7), the appropriate angle θ through which to rotate the axes satisfies

$$\cot 2\theta = \frac{A - C}{B} = \frac{3}{-4}$$

In order to use rotation formulas (5), we need to know the values of $\sin \theta$ and $\cos \theta$. Since we seek an acute angle θ, we know that $\sin \theta > 0$ and $\cos \theta > 0$. Thus, we use the half-angle formulas in the form

$$\sin \theta = \sqrt{\frac{1 - \cos 2\theta}{2}} \qquad \cos \theta = \sqrt{\frac{1 + \cos 2\theta}{2}}$$

Now we need to find the value of $\cos 2\theta$. Since $\cot 2\theta = -\frac{3}{4}$ and $\pi/2 < 2\theta < \pi$, it follows that $\cos 2\theta = -\frac{3}{5}$. Thus,

$$\sin \theta = \sqrt{\frac{1 - \cos 2\theta}{2}} = \sqrt{\frac{1 - (-\frac{3}{5})}{2}} = \sqrt{\frac{4}{5}} = \frac{2}{\sqrt{5}} = \frac{2\sqrt{5}}{5}$$

$$\cos \theta = \sqrt{\frac{1 + \cos 2\theta}{2}} = \sqrt{\frac{1 + (-\frac{3}{5})}{2}} = \sqrt{\frac{1}{5}} = \frac{1}{\sqrt{5}} = \frac{\sqrt{5}}{5}$$

With these values, rotation formulas (5) give us

$$x = \frac{\sqrt{5}}{5}x' - \frac{2\sqrt{5}}{5}y' = \frac{\sqrt{5}}{5}(x' - 2y')$$

$$y = \frac{2\sqrt{5}}{5}x' + \frac{\sqrt{5}}{5}y' = \frac{\sqrt{5}}{5}(2x' + y')$$

Substituting these values in the original equation and simplifying, we obtain

$$4x^2 - 4xy + y^2 + 5\sqrt{5}x + 5 = 0$$

$$4\left[\frac{\sqrt{5}}{5}(x' - 2y')\right]^2 - 4\left[\frac{\sqrt{5}}{5}(x' - 2y')\right]\left[\frac{\sqrt{5}}{5}(2x' + y')\right]$$
$$+ \left[\frac{\sqrt{5}}{5}(2x' + y')\right]^2 + 5\sqrt{5}\left[\frac{\sqrt{5}}{5}(x' - 2y')\right] = -5$$

Multiply both sides by 5 and expand to obtain

$$4(x'^2 - 4x'y' + 4y'^2) - 4(2x'^2 - 3x'y' - 2y'^2)$$
$$+ 4x'^2 + 4x'y' + y'^2 + 25(x' - 2y') = -25$$
$$25y'^2 - 50y' + 25x' = -25$$
$$y'^2 - 2y' + x' = -1$$
$$y'^2 - 2y' + 1 = -x' \quad \text{\small Complete the}$$
$$(y' - 1)^2 = -x' \quad \text{\small square in } y'.$$

FIGURE 51

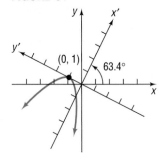

This is the equation of a parabola with vertex at $(0, 1)$ in the $x'y'$-plane. The axis of symmetry is parallel to the x'-axis. Using a calculator to solve $\sin \theta = 2\sqrt{5}/5$, we find that $\theta \approx 63.4°$. See Figure 51 for the graph. ∎

■ Now work Problem 27.

Identifying Conics without a Rotation of Axes

Suppose that we are required only to identify (rather than discuss) an equation of the form

$$Ax^2 + Bxy + Cy^2 + Dx + Ey + F = 0, \qquad B \neq 0 \qquad (8)$$

If we apply rotation formulas (5) to this equation, we obtain an equation of the form

$$A'x'^2 + B'x'y' + C'y'^2 + D'x' + E'y' + F' = 0 \qquad (9)$$

where A', B', C', D', E', and F' can be expressed in terms of A, B, C, D, E, F, and the angle θ of rotation (see Problem 43 at the end of this section). It can be shown that the value of $B^2 - 4AC$ in equation (8) and the value of $B'^2 - 4A'C'$ in equation (9) are equal no matter what angle θ of rotation is chosen (see Problem 45). In particular, if the angle θ of rotation satisfies equation (7), then $B' = 0$ in equation (9), and $B^2 - 4AC = -4A'C'$. Since equation (9) then has the form of equation (2),

$$A'x'^2 + C'y'^2 + D'x' + E'y' + F' = 0$$

we can identify it without completing the squares, as we did in the beginning of this section. In fact, now we can identify the conic described by any equation of the form of equation (8) without a rotation of axes.

Theorem

Identifying Conics without a Rotation of Axes

Except for degenerate cases, the equation

$$Ax^2 + Bxy + Cy^2 + Dx + Ey + F = 0$$

(a) Defines a parabola if $B^2 - 4AC = 0$.
(b) Defines an ellipse (or a circle) if $B^2 - 4AC < 0$.
(c) Defines a hyperbola if $B^2 - 4AC > 0$. ∎

You are asked to prove this theorem in Problem 46.

EXAMPLE 5   *Identifying a Conic without a Rotation of Axes*

Identify the equation: $8x^2 - 12xy + 17y^2 - 4\sqrt{5}x - 2\sqrt{5}y - 15 = 0$

Solution   Here, $A = 8$, $B = -12$, and $C = 17$, so that $B^2 - 4AC = -400$. Since $B^2 - 4AC < 0$, the equation defines an ellipse. ∎

■ Now work Problem 33.

9.5

Exercise 9.5

In Problems 1–10, identify each equation without completing the squares.

1. $x^2 + 4x + y + 3 = 0$ para
2. $2y^2 - 3y + 3x = 0$ parabala
3. $6x^2 + 3y^2 - 12x + 6y = 0$ ellip
4. $2x^2 + y^2 - 8x + 4y + 2 = 0$ ellip
5. $3x^2 - 2y^2 + 6x + 4 = 0$ hyp
6. $4x^2 - 3y^2 - 8x + 6y + 1 = 0$ hyp
7. $2y^2 - x^2 - y + x = 0$ hyp
8. $y^2 - 8x^2 - 2x - y = 0$ hyp
9. $x^2 + y^2 - 8x + 4y = 0$ circle
10. $2x^2 + 2y^2 - 8x + 8y = 0$ circle

In Problems 11–20, determine the appropriate rotation formulas to use so that the new equation contains no xy-term.

11. $x^2 + 4xy + y^2 - 3 = 0$
12. $x^2 - 4xy + y^2 - 3 = 0$
13. $5x^2 + 6xy + 5y^2 - 8 = 0$
14. $3x^2 - 10xy + 3y^2 - 32 = 0$
15. $13x^2 - 6\sqrt{3}xy + 7y^2 - 16 = 0$
16. $11x^2 + 10\sqrt{3}xy + y^2 - 4 = 0$
17. $4x^2 - 4xy + y^2 - 8\sqrt{5}x - 16\sqrt{5}y = 0$
18. $x^2 + 4xy + 4y^2 + 5\sqrt{5}y + 5 = 0$
19. $25x^2 - 36xy + 40y^2 - 12\sqrt{13}x - 8\sqrt{13}y = 0$
20. $34x^2 - 24xy + 41y^2 - 25 = 0$

In Problems 21–32, rotate the axes so that the new equation contains no xy-term. Discuss and graph the new equation. (Refer to Problems 11–20 for Problems 21–30.)

21. $x^2 + 4xy + y^2 - 3 = 0$
22. $x^2 - 4xy + y^2 - 3 = 0$
23. $5x^2 + 6xy + 5y^2 - 8 = 0$
24. $3x^2 - 10xy + 3y^2 - 32 = 0$
25. $13x^2 - 6\sqrt{3}xy + 7y^2 - 16 = 0$
26. $11x^2 + 10\sqrt{3}xy + y^2 - 4 = 0$
27. $4x^2 - 4xy + y^2 - 8\sqrt{5}x - 16\sqrt{5}y = 0$
28. $x^2 + 4xy + 4y^2 + 5\sqrt{5}y + 5 = 0$
29. $25x^2 - 36xy + 40y^2 - 12\sqrt{13}x - 8\sqrt{13}y = 0$
30. $34x^2 - 24xy + 41y^2 - 25 = 0$
31. $16x^2 + 24xy + 9y^2 - 130x + 90y = 0$
32. $16x^2 + 24xy + 9y^2 - 60x + 80y = 0$

In Problems 33–42, identify each equation without applying a rotation of axes.

33. $x^2 + 3xy - 2y^2 + 3x + 2y + 5 = 0$
34. $2x^2 - 3xy + 4y^2 + 2x + 3y - 5 = 0$
35. $x^2 - 7xy + 3y^2 - y - 10 = 0$
36. $2x^2 - 3xy + 2y^2 - 4x - 2 = 0$
37. $9x^2 + 12xy + 4y^2 - x - y = 0$
38. $10x^2 + 12xy + 4y^2 - x - y + 10 = 0$
39. $10x^2 - 12xy + 4y^2 - x - y - 10 = 0$
40. $4x^2 + 12xy + 9y^2 - x - y = 0$
41. $3x^2 - 2xy + y^2 + 4x + 2y - 1 = 0$
42. $3x^2 + 2xy + y^2 + 4x - 2y + 10 = 0$

In Problems 43–46, apply rotation formulas (5) to

$$Ax^2 + Bxy + Cy^2 + Dx + Ey + F = 0$$

to obtain the equation

$$A'x'^2 + B'x'y + C'y'^2 + D'x' + E'y' + F' = 0$$

43. Express A', B', C', D', E', and F' in terms of A, B, C, D, E, F, and the angle θ of rotation.

44. Show that $A + C = A' + C'$, and thus show that $A + C$ is **invariant;** that is, its value does not change under a rotation of axes.

45. Refer to Problem 44. Show that $B^2 - 4AC$ is invariant.

46. Prove that, except for degenerate cases, the equation

$$Ax^2 + Bxy + Cy^2 + Dx + Ey + F = 0$$

(a) Defines a parabola if $B^2 - 4AC = 0$.

(b) Defines an ellipse (or a circle) if $B^2 - 4AC < 0$.

(c) Defines a hyperbola if $B^2 - 4AC > 0$.

47. Use rotation formulas (5) to show that distance is invariant under a rotation of axes. That is, show that the distance from $P_1 = (x_1, y_1)$ to $P_2 = (x_2, y_2)$ in the xy-plane equals the distance from $P_1 = (x_1', y_1')$ to $P_2 = (x_2', y_2')$ in the $x'y'$-plane.

48. Show that the graph of the equation $x^{1/2} + y^{1/2} = a^{1/2}$ is part of the graph of a parabola.

 49. Formulate a strategy for discussing and graphing an equation of the form $Ax^2 + Cy^2 + Dx + Ey + F = 0$. How does your strategy change if the equation is of the form $Ax^2 + Bxy + Cy^2 + Dx + Ey + F = 0$?

9.6

Polar Equations of Conics

In Sections 9.2, 9.3, and 9.4, we gave separate definitions for the parabola, ellipse, and hyperbola based on geometric properties and the distance formula. In this section, we present an alternative definition that simultaneously defines all these conics. As we shall see, this approach is well suited to polar coordinate representation. (Refer to Section 8.4.)

Conic

Let D denote a fixed line called the **directrix;** let F denote a fixed point called the **focus,** which is not on D; and let e be a fixed positive number called the **eccentricity.** A **conic** is the set of points P in the plane such that the ratio of the distance from F to P to the distance from P to D equals e. Thus, a conic is the collection of points P for which

$$\frac{d(F, P)}{d(P, D)} = e \tag{1}$$

If $e = 1$, the conic is a **parabola.**

If $e < 1$, the conic is an **ellipse.**

If $e > 1$, the conic is a **hyperbola.**

Observe that if $e = 1$ the definition of a parabola in equation (1) is exactly the same as the definition used earlier in Section 11.2.

In the case of an ellipse, the **major axis** is a line through the focus perpendicular to the directrix. In the case of a hyperbola, the **transverse axis** is a line through the focus perpendicular to the directrix. For both an ellipse and a hyperbola, the eccentricity e satisfies

$$e = \frac{c}{a} \tag{2}$$

where c is the distance from the center to the focus and a is the distance from the center to a vertex.

Just as we did earlier using rectangular coordinates, we derive equations for the conics in polar coordinates by choosing a convenient position for the focus F and the directrix D. The focus F is positioned at the pole, and the directrix D is either parallel to the polar axis or perpendicular to it.

Suppose that we start with the directrix D perpendicular to the polar axis at a distance p units to the left of the pole (the focus F). See Figure 52.

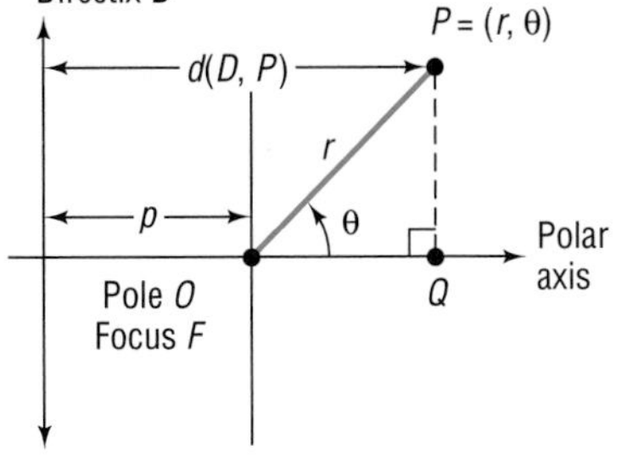

If $P = (r, \theta)$ is any point on the conic, then, by equation (1),

FIGURE 52

Directix D

$$\frac{d(F, P)}{d(D, P)} = e \quad \text{or} \quad d(F, P) = e \cdot d(D, P) \tag{3}$$

Now we use the point Q obtained by dropping the perpendicular from P to the polar axis to calculate $d(D, P)$:

$$d(D, P) = p + d(O, Q) = p + r \cos \theta$$

Using this expression and the fact that $d(F, P) = d(O, P) = r$ in equation (3), we get

$$d(F, P) = e \cdot d(D, P)$$
$$r = e(p + r \cos \theta)$$
$$r = ep + er \cos \theta$$
$$r - er \cos \theta = ep$$
$$r(1 - e \cos \theta) = ep$$
$$r = \frac{ep}{1 - e \cos \theta}$$

Theorem

Polar Equation of a Conic; Focus at Pole; Directrix Perpendicular to Polar Axis a Distance p to the Left of the Pole

The polar equation of a conic with focus at the pole and directrix perpendicular to the polar axis at a distance p to the left of the pole is

$$r = \frac{ep}{1 - e \cos \theta} \tag{4}$$

where e is the eccentricity of the conic.   ∎

E X A M P L E   1

Identifying and Graphing the Polar Equation of a Conic

Identify and graph the equation: $r = \dfrac{4}{2 - \cos \theta}$

Solution

The given equation is not quite in the form of equation (4), since the first term in the denominator is 2 instead of 1. Thus, we divide the numerator and denominator by 2 to obtain

$$r = \frac{2}{1 - \frac{1}{2} \cos \theta}$$

This equation is in the form of equation (4), with

$$e = \tfrac{1}{2} \quad \text{and} \quad ep = \tfrac{1}{2}p = 2$$

Thus, $e = \tfrac{1}{2}$ and $p = 4$. We conclude that the conic is an ellipse, since $e = \tfrac{1}{2} < 1$. One focus is at the pole, and the directrix is perpendicular to the polar axis, a dis-

FIGURE 53

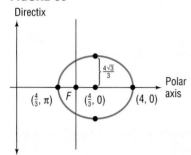

tance of 4 units to the left of the pole. It follows that the major axis is along the polar axis. To find the vertices, we let $\theta = 0$ and $\theta = \pi$. Thus, the vertices of the ellipse are $(4, 0)$ and $(\frac{4}{3}, \pi)$. At the midpoint of the vertices, we locate the center of the ellipse at $(\frac{4}{3}, 0)$. [Do you see why? The vertices $(4, 0)$ and $(\frac{4}{3}, \pi)$ in polar coordinates are $(4, 0)$ and $(-\frac{4}{3}, 0)$ in rectangular coordinates. The midpoint in rectangular coordinates is $(\frac{4}{3}, 0)$, which is also $(\frac{4}{3}, 0)$ in polar coordinates.] Thus, $a =$ distance from the center to a vertex $= \frac{8}{3}$. Using $a = \frac{8}{3}$ and $e = \frac{1}{2}$ in equation (2), $e = c/a$, we find $c = \frac{4}{3}$. Finally, using $a = \frac{8}{3}$ and $c = \frac{4}{3}$ in $b^2 = a^2 - c^2$, we have

$$b^2 = a^2 - c^2 = \frac{64}{9} - \frac{16}{9} = \frac{48}{9}$$

$$b = \frac{4\sqrt{3}}{3}$$

Figure 53 shows the graph.

Check: Graph $r = 4/(2 - \cos \theta)$ and compare the result with Figure 53. ■

Exploration: Graph $r = 4/(2 + \cos \theta)$ and compare the result with Figure 53. What do you conclude? Clear the screen and graph $r = 4/(2 - \sin \theta)$ and then $r = 4/(2 + \sin \theta)$. Compare each of these graphs with Figure 53. What do you conclude?

■ Now work Problem 5.

Equation (4) was obtained under the assumption that the directrix was perpendicular to the polar axis at a distance p units to the left of the pole. A similar derivation (see Problem 37), in which the directrix is perpendicular to the polar axis at a distance p units to the right of the pole, results in the equation

$$r = \frac{ep}{1 + e \cos \theta}$$

In Problems 38 and 39 you are asked to derive the polar equations of conics with focus at the pole and directrix parallel to the polar axis. Table 5 summarizes the polar equations of conics.

TABLE 5 **POLAR EQUATIONS OF CONICS (FOCUS AT THE POLE, ECCENTRICITY e)**

EQUATION	DESCRIPTION
(a) $r = \dfrac{ep}{1 - e \cos \theta}$	Directrix is perpendicular to the polar axis at a distance p units to the left of the pole.
(b) $r = \dfrac{ep}{1 + e \cos \theta}$	Directrix is perpendicular to the polar axis at a distance p units to the right of the pole.
(c) $r = \dfrac{ep}{1 + e \sin \theta}$	Directrix is parallel to the polar axis at a distance p units above the pole.
(d) $r = \dfrac{ep}{1 - e \sin \theta}$	Directrix is parallel to the polar axis at a distance p units below the pole.

ECCENTRICITY
If $e = 1$, the conic is a parabola; the axis of symmetry is perpendicular to the directrix.
If $e < 1$, the conic is an ellipse; the major axis is perpendicular to the directrix.
If $e > 1$, the conic is a hyperbola; the transverse axis is perpendicular to the directrix.

E X A M P L E   2

Identifying and Graphing the Polar Equation of a Conic

Identify and graph the equation: $r = \dfrac{6}{3 + 3 \sin \theta}$

Solution

To place the equation in proper form, we divide the numerator and denominator by 3 to get

$$r = \frac{2}{1 + \sin \theta}$$

Referring to Table 5, we conclude that this equation is in the form of equation (c) with

$$e = 1 \quad \text{and} \quad ep = 2$$

FIGURE 54

$(1, \frac{\pi}{2})$

Directix

$(2, \pi)$ $(2, 0)$ Polar axis

F

Thus, $e = 1$ and $p = 2$. The conic is a parabola with focus at the pole. The directrix is parallel to the polar axis at a distance 2 units above the pole; the axis of symmetry is perpendicular to the polar axis. The vertex of the parabola is at $(1, \pi/2)$. (Do you see why?) See Figure 54 for the graph. Notice that we plotted two additional points, $(2, 0)$ and $(2, \pi)$, to assist in graphing.

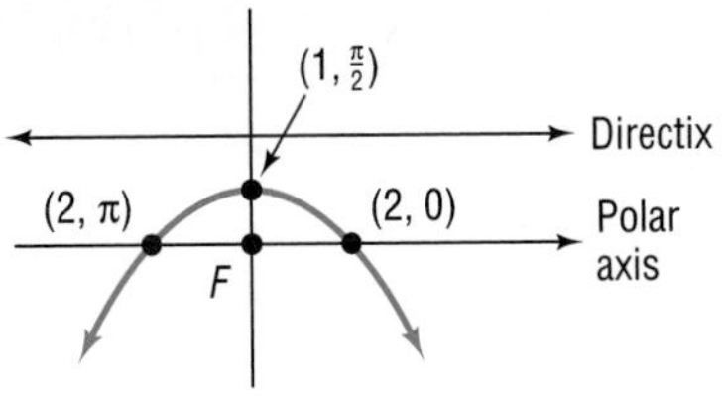

Check: Graph $r = 6/(3 + 3 \sin \theta)$ and compare the result with Figure 54.   ■

■   Now work Problem 7.

E X A M P L E   3

Identifying and Graphing the Polar Equation of a Conic

Identify and graph the equation: $r = \dfrac{3}{1 + 3 \cos \theta}$

Solution

This equation is in the form of equation (b) in Table 5. We conclude that

$$e = 3 \quad \text{and} \quad ep = 3p = 3$$

FIGURE 55

$(3, \frac{\pi}{2})$

$(\frac{3}{4}, 0)$ $(\frac{9}{8}, 0)$ $(-\frac{3}{2}, \pi)$

0

$b = \frac{3\sqrt{2}}{4}$

Polar axis

$(3, \frac{3\pi}{2})$

Thus, $e = 3$ and $p = 1$. This is the equation of a hyperbola with a focus at the pole. The directrix is perpendicular to the polar axis, 1 unit to the right of the pole. The transverse axis is along the polar axis. To find the vertices, we let $\theta = 0$ and $\theta = \pi$. Thus, the vertices are $(\frac{3}{4}, 0)$ and $(-\frac{3}{2}, \pi)$. The center is at the midpoint of $(\frac{3}{4}, 0)$ and $(-\frac{3}{2}, \pi)$, which is $(\frac{9}{8}, 0)$. Thus, $c = $ distance from the center to a focus $= \frac{9}{8}$. Since $e = 3$, it follows from equation (2), $e = c/a$, that $a = \frac{3}{8}$. Finally, using $a = \frac{3}{8}$ and $c = \frac{9}{8}$ in $b^2 = c^2 - a^2$, we find

$$b^2 = c^2 - a^2 = \frac{81}{64} - \frac{9}{64} = \frac{72}{64} = \frac{9}{8}$$

$$b = \frac{3}{2\sqrt{2}} = \frac{3\sqrt{2}}{4}$$

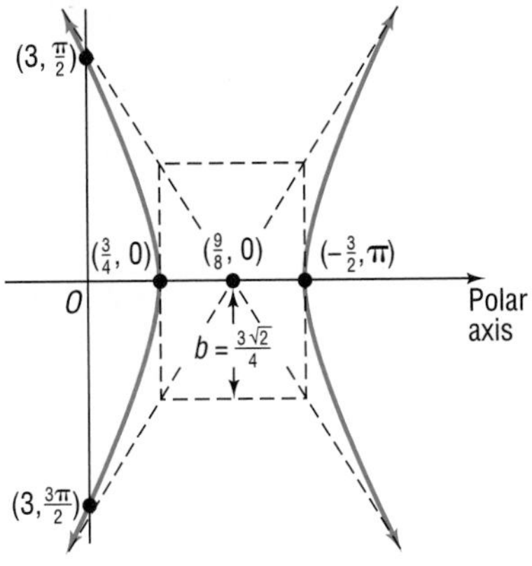

Figure 55 shows the graph. Notice that we plotted two additional points, $(3, \pi/2)$ and $(3, 3\pi/2)$, on the left branch and used symmetry to obtain the right branch. The asymptotes of this hyperbola were found in the usual way by constructing the rectangle shown.

Check: Graph $r = 3/(1 + 3 \cos \theta)$ and compare the result with Figure 55.   ■

■   Now work Problem 11.

E X A M P L E   4   *Converting a Polar Equation to a Rectangular Equation*

Convert the polar equation

$$r = \frac{1}{3 - 3 \cos \theta}$$

to a rectangular equation.

Solution   The strategy here is first to rearrange the equation and square each side, before using the transformation equations:

$$r = \frac{1}{3 - 3 \cos \theta}$$

$$3r - 3r \cos \theta = 1$$

$$3r = 1 + 3r \cos \theta \qquad \text{Rearrange the equation.}$$

$$9r^2 = (1 + 3r \cos \theta)^2 \qquad \text{Square each side.}$$

$$9(x^2 + y^2) = (1 + 3x)^2 \qquad \text{Use the transformation equations.}$$

$$9x^2 + 9y^2 = 9x^2 + 6x + 1$$

$$9y^2 = 6x + 1$$

This is the equation of a parabola in rectangular coordinates.   ■

■   Now work Problem 19.

9.6

Exercise 9.6

In Problems 1–6, identify the conic that each polar equation represents. Also, give the position of the directrix.

1. $r = \dfrac{1}{1 + \cos \theta}$

2. $r = \dfrac{3}{1 - \sin \theta}$

3. $r = \dfrac{4}{2 - 3 \sin \theta}$

4. $r = \dfrac{2}{1 + 2 \cos \theta}$

5. $r = \dfrac{3}{4 - 2 \cos \theta}$

6. $r = \dfrac{6}{8 + 2 \sin \theta}$

In Problems 7–18, identify and graph each equation.

7. $r = \dfrac{1}{1 + \cos \theta}$

8. $r = \dfrac{3}{1 - \sin \theta}$

9. $r = \dfrac{8}{4 + 3 \sin \theta}$

10. $r = \dfrac{10}{5 + 4 \cos \theta}$

11. $r = \dfrac{9}{3 - 6 \cos \theta}$

12. $r = \dfrac{12}{4 + 8 \sin \theta}$

13. $r = \dfrac{8}{2 - \sin \theta}$

14. $r = \dfrac{8}{2 + 4 \cos \theta}$

15. $r(3 - 2 \sin \theta) = 6$

16. $r(2 - \cos \theta) = 2$

17. $r = \dfrac{6 \sec \theta}{2 \sec \theta - 1}$

18. $r = \dfrac{3 \csc \theta}{\csc \theta - 1}$

In Problems 19–30, convert each polar equation to a rectangular equation.

19. $r = \dfrac{1}{1 + \cos \theta}$

20. $r = \dfrac{3}{1 - \sin \theta}$

21. $r = \dfrac{8}{4 + 3 \sin \theta}$

22. $r = \dfrac{10}{5 + 4 \cos \theta}$

23. $r = \dfrac{9}{3 - 6 \cos \theta}$

24. $r = \dfrac{12}{4 + 8 \sin \theta}$

25. $r = \dfrac{8}{2 - \sin \theta}$ **26.** $r = \dfrac{8}{2 + 4 \cos \theta}$ **27.** $r(3 - 2 \sin \theta) = 6$

28. $r(2 - \cos \theta) = 2$ **29.** $r = \dfrac{6 \sec \theta}{2 \sec \theta - 1}$ **30.** $r = \dfrac{3 \csc \theta}{\csc \theta - 1}$

In Problems 31–36, find a polar equation for each conic. For each, a focus is at the pole.

31. $e = 1$; directrix is parallel to the polar axis 1 unit above the pole

32. $e = 1$; directrix is parallel to the polar axis 2 units below the pole

33. $e = \frac{4}{5}$; directrix is perpendicular to the polar axis 3 units to the left of the pole

34. $e = \frac{2}{3}$; directrix is parallel to the polar axis 3 units above the pole

35. $e = 6$; directrix is parallel to the polar axis 2 units below the pole

36. $e = 5$; directrix is perpendicular to the polar axis 5 units to the right of the pole

37. Derive equation (b) in Table 5: $r = \dfrac{ep}{1 + e \cos \theta}$

38. Derive equation (c) in Table 5: $r = \dfrac{ep}{1 + e \sin \theta}$

39. Derive equation (d) in Table 5: $r = \dfrac{ep}{1 - e \sin \theta}$

40. The planet Mercury travels around the Sun in an elliptical orbit given approximately by

$$r = \frac{(3.442)10^7}{1 - 0.206 \cos \theta}$$

where r is measured in miles and the Sun is at the pole. Find the distance from Mercury to the Sun at *aphelion* (greatest distance from the Sun) and at *perihelion* (shortest distance from the Sun). See the figure in the margin.

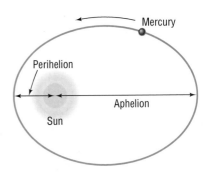

9.7

Plane Curves and Parametric Equations

Equations of the form $y = f(x)$, where f is a function, have graphs that are intersected no more than once by any vertical line. The graphs of many of the conics and certain other, more complicated graphs do not have this characteristic. Yet each graph, like the graph of a function, is a collection of points (x, y) in the xy-plane; that is, each is a *plane curve*. In this section, we discuss another way of representing such graphs.

Plane Curve

Let $x = f(t)$ and $y = g(t)$, where f and g are two functions whose common domain is some interval I. The collection of points defined by

$$(x, y) = (f(t), g(t))$$

is called a **plane curve.** The equations

$$x = f(t) \qquad y = g(t)$$

where t is in I, are called **parametric equations** of the curve. The variable t is called a **parameter.**

Parametric equations are particularly useful in describing movement along a curve. Suppose that a curve is defined by the parametric equations

$$x = f(t) \qquad y = g(t)$$

where f and g are each defined over some interval I. For a given value of t in I, we can find the value of $x = f(t)$ and $y = g(t)$, thus obtaining a point (x, y) on the curve. In fact, as t varies over the interval I in some order, say from left to right, successive values of t give rise to a directed movement along the curve. That is, as t varies over I in some order, the curve is traced out in a certain direction by the corresponding succession of points (x, y). Let's look at an example.

E X A M P L E   1

Discussing a Curve Defined by Parametric Equations

Discuss the curve defined by the parametric equations

$$x = 3t^2 \qquad y = 2t, \qquad -2 \le t \le 2 \tag{1}$$

Solution   For each number t, $-2 \le t \le 2$, there corresponds a number x and a number y. For example, when $t = -2$, then $x = 12$ and $y = -4$. When $t = 0$, then $x = 0$ and $y = 0$. Indeed, we can set up a table listing various choices of the parameter t and the corresponding values for x and y, as shown in Table 6. Plotting these points and connecting them with a smooth curve leads to Figure 56.

TABLE 6

t	x	y	(x, y)
-2	12	-4	$(12, -4)$
-1	3	-2	$(3, -2)$
0	0	0	$(0, 0)$
1	3	2	$(3, 2)$
2	12	4	$(12, 4)$

FIGURE 56

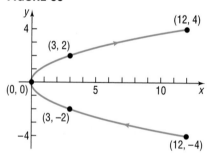

Notice the arrows on the curve in Figure 56. They indicate the direction, or **orientation,** of the curve for increasing values of the parameter t.

■   Now work Problem 1.

Comment: Most graphing utilities have the capability of graphing parametric equations. Check your owner's manual to see how it is done.

Check: Graph $x = 3t^2$, $y = 2t$, $-2 \le t \le 2$, and compare the result with Figure 56.

Parametric equations defining a curve are not unique. You can verify that the curve in Figure 56 may also be defined by any of the following parametric equations:

$$x = \frac{3t^2}{4} \qquad y = t, \qquad -4 \le t \le 4$$

or

$$x = 3t^2 + 12t + 12 \qquad y = 2t + 4, \qquad -4 \le t \le 0$$

or

$$x = 3t^{2/3} \qquad y = 2\sqrt[3]{t}, \qquad -8 \le t \le 8$$

The curve in Figure 56 should be familiar. To identify it accurately, we find the corresponding rectangular equation by eliminating the parameter t from the parametric equations (1) given in Example 1,

$$x = 3t^2 \qquad y = 2t, \qquad -2 \le t \le 2$$

Noting that we can readily solve for t in $y = 2t$, obtaining $t = y/2$, we substitute this expression in the other equation:

$$x = 3t^2 = 3\left(\frac{y}{2}\right)^2 = \frac{3y^2}{4}$$
$$\underset{t = \frac{y}{2}}{\uparrow}$$

This equation, $x = 3y^2/4$, is the equation of a parabola with vertex at $(0, 0)$ and axis along the x-axis.

Note that the parameterized curve defined by equation (1) and shown in Figure 56 is only a part of the parabola $x = 3y^2/4$. Thus, the graph of the rectangular equation obtained by eliminating the parameter will, in general, contain more points than the original parameterized curve. Care must therefore be taken when a parameterized curve is sketched after eliminating the parameter. Even so, the process of eliminating the parameter t of a parameterized curve in order to identify it accurately is sometimes a better approach than merely plotting points. However, the elimination process sometimes requires a little ingenuity.

E X A M P L E   2

Finding the Rectangular Equation of a Curve Defined Parametrically

Find the rectangular equation of the curve whose parametric equations are

$$x = a \cos t \qquad y = a \sin t$$

where $a > 0$ is a constant. Graph this curve, indicating its orientation.

Solution

The presence of sines and cosines in the parametric equations suggests that we use a Pythagorean identity. In fact, since

$$\cos t = \frac{x}{a} \qquad \sin t = \frac{y}{a}$$

we find that

$$\cos^2 t + \sin^2 t = 1$$
$$\left(\frac{x}{a}\right)^2 + \left(\frac{y}{a}\right)^2 = 1$$
$$x^2 + y^2 = a^2$$

FIGURE 57

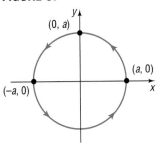

Thus, the curve is a circle with center at $(0, 0)$ and radius a. As the parameter t increases, say from $t = 0$ [the point $(a, 0)$] to $t = \pi/2$ [the point $(0, a)$] to $t = \pi$ [the point $(-a, 0)$], we see that the corresponding points are traced in a counterclockwise direction around the circle. Hence, the orientation is as indicated in Figure 57. ∎

■   Now work Problem 13.

Let's discuss the curve in Example 2 further. The domain of each parametric equation is $-\infty < t < \infty$. Thus, the graph in Figure 57 is actually being repeated each time t increases by 2π. If we wanted the curve to consist of exactly 1 revolution in the counterclockwise direction, we could write

$$x = a \cos t \qquad y = a \cos t, \qquad 0 \le t \le 2\pi$$

This curve starts at $t = 0$ [the point $(a, 0)$] and, proceeding counterclockwise around the circle, ends at $t = 2\pi$ [also the point $(a, 0)$].

If we wanted the curve to consist of exactly three revolutions in the counterclockwise direction, we could write

$$x = a \cos t \qquad y = a \sin t, \qquad -2\pi \le t \le 4\pi$$

FIGURE 58

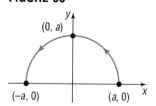

or

$$x = a \cos t \qquad y = a \sin t, \qquad 0 \le t \le 6\pi$$

or

$$x = a \cos t \qquad y = a \sin t, \qquad 2\pi \le t \le 8\pi$$

If we wanted the curve to consist of the upper semicircle of radius a with a counterclockwise orientation, we could write

$$x = a \cos t \qquad y = a \sin t, \qquad 0 \le t \le \pi$$

See Figure 58.

If we wanted the curve to consist of the left semicircle of radius a with a clockwise orientation, we could write

$$x = -a \sin t \qquad y = -a \cos t, \qquad 0 \le t \le \pi$$

See Figure 59.

FIGURE 59

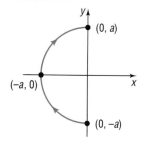

Time as a Parameter

If we think of the parameter t as time, then the parametric equations $x = f(t)$ and $y = g(t)$ of a curve C specify how the x- and y-coordinates of a moving point vary with time.

For example, we can use parametric equations to describe the motion of an object, sometimes referred to as **curvilinear motion.** Using parametric equations, we can specify not only where the object travels, that is, its location (x, y), but also when it gets there, that is, the time t.

E X A M P L E 3   *Describing the Motion of an Object*

Describe the motion of an object that moves along the curve

$$x = 4 \sin t \qquad y = 3 \cos t, \qquad 0 \le t \le 2\pi$$

Solution   We eliminate the parameter t by using the Pythagorean identity $\sin^2 t + \cos^2 t = 1$, obtaining

$$\left(\frac{x}{4}\right)^2 + \left(\frac{y}{3}\right)^2 = 1$$

$$\frac{x^2}{16} + \frac{y^2}{9} = 1$$

The curve is an ellipse, the center is at $(0, 0)$, the major axis is along the x-axis, and the vertices are at $(\pm 4, 0)$. As t varies from $t = 0$ to $t = 2\pi$, the object moves around the ellipse in a clockwise direction starting at $(0, 3)$, reaching $(4, 0)$ when $t = \pi/2$ and $(0, -3)$ when $t = \pi$, and ending at $(0, 3)$ when $t = 2\pi$. See Figure 60.

FIGURE 60

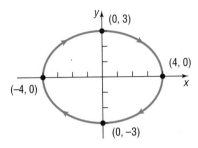

Check: Graph $x = 4 \sin t$, $y = 3 \cos t$, $0 \le t \le 2\pi$, and compare the result with Figure 60. ■

Finding Parametric Equations

We now take up the question of how to find parametric equations of a given curve.

If the curve is defined by the equation $y = f(x)$, where f is a function, one way of finding parametric equations is simply to let $x = t$. Then $y = f(t)$. Thus,

$$x = t \qquad y = f(t), \qquad t \text{ in the domain of } f$$

are parametric equations of the curve.

E X A M P L E   4

Finding Parametric Equations for a Curve Defined by a Rectangular Equation

Find parametric equations for the equation: $y = x^2 - 4$

Solution   Let $x = t$. Then the parametric equations are

$$x = t \qquad y = t^2 - 4, \qquad -\infty < t < \infty$$ ■

Another less obvious approach to Example 4 is to let $x = t^3$. Then the parametric equations become

$$x = t^3 \qquad y = t^6 - 4, \qquad -\infty < t < \infty$$

Care must be taken when using this approach, since the substitution for x must be a function that allows x to take on all the values stipulated by the domain of f. Thus, for example, letting $x = t^2$ so that $y = t^4 - 4$ does not result in equivalent parametric equations for $y = x^2 - 4$, since only points for which $x \ge 0$ are obtained.

E X A M P L E   5

Finding Parametric Equations for an Object in Motion

Find parametric equations for the ellipse

$$x^2 + \frac{y^2}{9} = 1$$

where the parameter t is time (in seconds) and

(a) The motion around the ellipse is clockwise, begins at the point $(0, 3)$, and requires 1 second for a complete revolution

(b) The motion around the ellipse is counterclockwise, begins at the point $(1, 0)$, and requires 2 seconds for a complete revolution

Solution   (a) Since the motion begins at the point $(0, 3)$, we want $x = 0$ and $y = 3$ when $t = 0$. Furthermore, since the given equation is an ellipse, we begin by letting

$$x = \sin \omega t \qquad \frac{y}{3} = \cos \omega t$$

for some constant ω. These parametric equations clearly satisfy the equation. Furthermore, with this choice, when $t = 0$, we have $x = 0$ and $y = 3$. For the motion to be clockwise, the motion will have to begin with the value of x increasing and y decreasing as t increases. Thus, $\omega > 0$. Finally, since 1 revolution requires 1 second, the period is $2\pi/\omega = 1$, so $\omega = 2\pi$. Thus, parametric equations that satisfy the conditions stipulated are

$$x = \sin 2\pi t \qquad y = 3 \cos 2\pi t, \qquad 0 \le t \le 1 \tag{2}$$

(b) Since the motion begins at the point $(1, 0)$, we want $x = 1$ and $y = 0$ when $t = 0$. Furthermore, since the given equation is an ellipse, we begin by letting

$$x = \cos \omega t \qquad \frac{y}{3} = \sin \omega t$$

for some constant ω. These parametric equations clearly satisfy the equation. Furthermore, with this choice, when $t = 0$, we have $x = 1$ and $y = 0$. For the motion to be counterclockwise, the motion will have to begin with the value of x decreasing and y increasing as t increases. Thus, $\omega > 0$. Finally, since 1 revolution requires 2 seconds, the period is $2\pi/\omega = 2$, so $\omega = \pi$. Thus, the parametric equations that satisfy the conditions stipulated are

$$x = \cos \pi t \qquad y = 3 \sin \pi t, \qquad 0 \le t \le 2 \tag{3}$$

∎

Either of equations (2) or (3) can serve as parametric equations for the ellipse $x^2 + (y^2/9) = 1$ given in Example 5. The direction of the motion, the beginning point, and the time for 1 revolution merely serve to help arrive at a particular parametric representation.

∎ Now work Problem 25.

The Cycloid

Suppose that a circle of radius a rolls along a horizontal line without slipping. As the circle rolls along the line, a point P on the circle will trace out a curve called a **cycloid** (see Figure 61). We now seek parametric equations* for a cycloid.

We begin with a circle of radius a and take the fixed line on which the circle rolls as the x-axis. Let the origin be one of the points at which the point P

---

*Any attempt to derive the rectangular equation of a cycloid would soon demonstrate how complicated the task is.

FIGURE 61
Cycloid

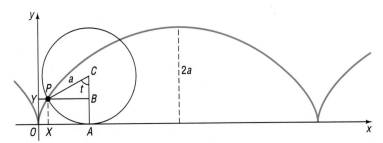

comes in contact with the x-axis. Figure 61 illustrates the position of this point P after the circle has rolled somewhat. The angle t (in radians) measures the angle through which the circle has rolled.

Since we require no slippage, it follows that

$$\text{Arc } AP = d(O, A)$$

Therefore,

$$at = d(O, A)$$

The x-coordinate of the point P is

$$d(O, X) = d(O, A) - d(X, A) = at - a \sin t = a(t - \sin t)$$

The y-coordinate of the point P is equal to

$$d(O, Y) = d(A, C) - d(B, C) = a - a \cos t = a(1 - \cos t)$$

Thus, the parametric equations of the cycloid are

$$x = a(t - \sin t) \qquad y = a(1 - \cos t) \qquad (4)$$

Check: Graph $x = t - \sin t$, $y = 1 - \cos t$, $0 \le t \le 2\pi$, and compare the result with Figure 61.

Applications to Mechanics

If a is negative in equations (4), we obtain an inverted cycloid, as shown in Figure 62(a). The inverted cycloid occurs as a result of some remarkable applications in the field of mechanics. We shall mention two of them: the *brachistochrone* and the *tautochrone.**

FIGURE 62

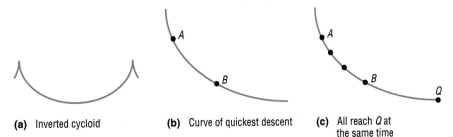

(a) Inverted cycloid   (b) Curve of quickest descent   (c) All reach Q at the same time

*In Greek, *brachistochrone* means "the shortest time," and *tautochrone* means "equal time."

FIGURE 63

A flexible pendulum constrained by cycloids swings in a cycloid

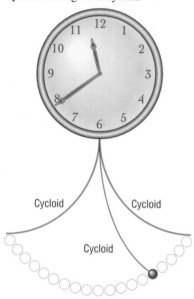

Cycloid   Cycloid

Cycloid

The **brachistochrone** is the curve of quickest descent. If a particle is constrained to follow some path from one point A to a lower point B (not on the same vertical line) and is acted on only by gravity, the time needed to make the descent is least if the path is an inverted cycloid. See Figure 62(b). This remarkable discovery, which is attributed to many famous mathematicians (including Johann Bernoulli and Blaise Pascal), was a significant step in creating the branch of mathematics known as the *calculus of variations*.

To define the **tautochrone**, let Q be the lowest point on an inverted cycloid. If several particles placed at various positions on an inverted cycloid simultaneously begin to slide down the cycloid, they will reach the point Q at the same time, as indicated in Figure 62(c). The tautochrone property of the cycloid was used by Christian Huygens (1629–1695), the Dutch mathematician, physicist, and astronomer, to construct a pendulum clock with a bob that swings along a cycloid (see Figure 63). In Huygens' clock, the bob was made to swing along a cycloid by suspending the bob on a thin wire constrained by two plates shaped like cycloids. In a clock of this design, the period of the pendulum is independent of its amplitude.

9.7

Exercise 9.7

In Problems 1–20, graph the curve whose parametric equations are given, and show its orientation. Find the rectangular equation of each curve.

1. $x = 3t + 2$, $y = t + 1$; $0 \le t \le 4$

2. $x = t - 3$, $y = 2t + 4$; $0 \le t \le 2$

3. $x = t + 2$, $y = \sqrt{t}$; $t \ge 0$

4. $x = \sqrt{2t}$, $y = 4t$; $t \ge 0$

5. $x = t^2 + 4$, $y = t^2 - 4$; $-\infty < t < \infty$

6. $x = \sqrt{t} + 4$, $y = \sqrt{t} - 4$; $t \ge 0$

7. $x = 3t^2$, $y = t + 1$; $-\infty < t < \infty$

8. $x = 2t - 4$, $y = 4t^2$; $-\infty < t < \infty$

9. $x = 2e^t$, $y = 1 + e^t$; $t \ge 0$

10. $x = e^t$, $y = e^{-t}$; $t \ge 0$

11. $x = \sqrt{t}$, $y = t^{3/2}$; $t \ge 0$

12. $x = t^{3/2} + 1$, $y = \sqrt{t}$; $t \ge 0$

13. $x = 2 \cos t$, $y = 3 \sin t$; $0 \le t \le 2\pi$

14. $x = 2 \cos t$, $y = 3 \sin t$; $0 \le t \le \pi$

15. $x = 2 \cos t$, $y = 3 \sin t$; $-\pi \le t \le 0$

16. $x = 2 \cos t$, $y = \sin t$; $0 \le t \le \pi/2$

17. $x = \sec t$, $y = \tan t$; $0 \le t \le \pi/4$

18. $x = \csc t$, $y = \cot t$; $\pi/4 \le t \le \pi/2$

19. $x = \sin^3 t$, $y = \cos^3 t$; $0 \le t \le 2\pi$

20. $x = t^2$, $y = \ln t$; $t > 0$

In Problems 21–24, find two different parametric equations for each rectangular equation.

21. $y = x^3$ 22. $y = x^4 + 1$ 23. $x = y^{3/2}$ 24. $x = \sqrt{y}$

In Problems 25–28, find parametric equations for an object that moves along the ellipse $(x^2/4) + (y^2/9) = 1$ with the motion described.

25. The motion begins at $(2, 0)$, is clockwise, and requires 2 seconds for a complete revolution.

26. The motion begins at $(0, 3)$, is clockwise, and requires 1 second for a complete revolution.

27. The motion begins at $(0, 3)$, is counterclockwise, and requires 1 second for a complete revolution.

28. The motion begins at $(2, 0)$, is counterclockwise, and requires 3 seconds for a complete revolution.

In Problems 29 and 30, the parametric equations of four curves are given. Graph each of them, indicating the orientation.

29. C_1: $x = t$, $y = t^2$; $-4 \leq t \leq 4$
   C_2: $x = \cos t$, $y = 1 - \sin^2 t$; $0 \leq t \leq \pi$
   C_3: $x = e^t$, $y = e^{2t}$; $0 \leq t \leq \ln 4$
   C_4: $x = \sqrt{t}$, $y = t$; $0 \leq t \leq 16$

30. C_1: $x = t$, $y = \sqrt{1 - t^2}$; $-1 \leq t \leq 1$
   C_2: $x = \sin t$, $y = \cos t$; $0 \leq t \leq 2\pi$
   C_3: $x = \cos t$, $y = \sin t$; $0 \leq t \leq 2\pi$
   C_4: $x = \sqrt{1 - t^2}$, $y = t$; $-1 \leq t \leq 1$

31. Show that the parametric equations for a line passing through the points (x_1, y_1) and (x_2, y_2) are

$$x = (x_2 - x_1)t + x_1 \qquad y = (y_2 - y_1)t + y_1, \qquad -\infty < t < \infty$$

What is the orientation of this line?

32. *Projectile Motion* The position of a projectile fired with an initial velocity v_0 feet per second and at an angle θ to the horizontal at the end of t seconds is given by the parametric equations

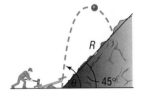

$$x = (v_0 \cos \theta)t \qquad y = (v_0 \sin \theta)t - 16t^2$$

(a) Obtain the rectangular equation of the trajectory and identify the curve.
(b) Show that the projectile hits the ground ($y = 0$) when $t = \frac{1}{16}v_0 \sin \theta$.
(c) How far has the projectile traveled (horizontally) when it strikes the ground? In other words, find the **range R.**
(d) Find the time t when $x = y$. Then find the horizontal distance x and the vertical distance y traveled by the projectile in this time. Then compute $\sqrt{x^2 + y^2}$. This is the distance R, the range, that the projectile travels up a plane inclined at $45°$ to the horizontal ($x = y$). (See also Problem 75 in Exercise 7.3.)

In Problems 33–36, graph the curve defined by the given parametric equations.

33. $x = \sin^3 t$, $y = \cos^3 t$
34. $x = \sin t + \cos t$, $y = \sin t - \cos t$
35. $x = 4 \sin t - 2 \sin 2t$,
    $y = 4 \cos t - 2 \cos 2t$
36. $x = 4 \sin t + 2 \sin 2t$,
    $y = 4 \cos t + 2 \cos 2t$

37. Look up the curves called *hypocycloid* and *epicycloid*. Write a report on what you find. Be sure to draw comparisons with the cycloid.

Chapter Review

THINGS TO KNOW

Equations

Parabola	See Tables 1 and 2.
Ellipse	See Table 3.
Hyperbola	See Table 4.

General equation of a conic $Ax^2 + Bxy + Cy^2 + Dx + Ey + F = 0$

Parabola if $B^2 - 4AC = 0$
Ellipse (or circle) if
   $B^2 - 4AC < 0$
Hyperbola if $B^2 - 4AC > 0$

Conic in polar coordinates $\dfrac{d(F, P)}{d(P, D)} = e$

Parabola if $e = 1$
Ellipse if $e < 1$
Hyperbola if $e > 1$

Plane curve $(x, y) = (f(t), g(t))$, t a parameter

Polar equations of a conic   See Table 5.

Definitions

Parabola	Set of points P in the plane for which $d(F, P) = d(P, D)$, where F is the focus and D is the directrix
Ellipse	Set of points P in the plane, the sum of whose distances from two fixed points (the foci) is a constant
Hyperbola	Set of points P in the plane, the difference of whose distances from two fixed points (the foci) is a constant

Formulas

Rotation formulas	$x = x' \cos\theta - y' \sin\theta$ $y = x' \sin\theta + y' \cos\theta$
Angle θ of rotation that eliminates the $x'y'$-term	$\cot 2\theta = \dfrac{A - C}{B}$

How To:

Find the vertex, focus, and directrix of a parabola given its equation

Graph a parabola given its equation

Find an equation of a parabola given certain information about the parabola

Find the center, foci, and vertices of an ellipse given its equation

Graph an ellipse given its equation

Find an equation of an ellipse given certain information about the ellipse

Find the center, foci, vertices, and asymptotes of a hyperbola given its equation

Graph a hyperbola given its equation

Find an equation of a hyperbola given certain information about the hyperbola

Identify conics without completing the square

Identify conics without a rotation of axes

Use rotation formulas to transform second-degree equations so that no xy-term is present

Identify and graph conics given by a polar equation

Graph parametric equations

Find the rectangular equation given the parametric equations

Fill-in-the-Blank Items

1. A(n) _____ is the collection of all points in the plane such that the distance from each point to a fixed point equals its distance to a fixed line.

2. A(n) _____ is the collection of all points in the plane the sum of whose distances from two fixed points is a constant.

3. A(n) _____ is the collection of all points in the plane the difference of whose distances from two fixed points is a constant.

4. For an ellipse, the foci lie on the _____ axis; for a hyperbola, the foci lie on the _____ axis.

5. For the ellipse $(x^2/9) + (y^2/16) = 1$, the major axis is along the _____.

6. The equations of the asymptotes of the hyperbola $(y^2/9) - (x^2/4) = 1$ are _____ and _____.

7. To transform the equation

$$Ax^2 + Bxy + Cy^2 + Dx + Ey + F = 0, \qquad B \neq 0$$

into one in x' and y' without an $x'y'$-term, rotate the axes through an acute angle θ that satisfies the equation _____.

8. The polar equation

$$r = \frac{8}{4 - 2 \sin \theta}$$

is a conic whose eccentricity is _____. It is a(n) _____ whose directrix is _____ to the polar axis at a distance _____ units _____ the pole.

9. The parametric equations $x = 2 \sin t$ and $y = 3 \cos t$ represent a(n) _____.

TRUE/FALSE ITEMS

T   F   **1.** On a parabola, the distance from any point to the focus equals the distance from that point to the directrix.

T   F   **2.** The foci of an ellipse lie on its minor axis.

T   F   **3.** The foci of a hyperbola lie on its transverse axis.

T   F   **4.** Hyperbolas always have asymptotes, and ellipses never have asymptotes.

T   F   **5.** A hyperbola never intersects its conjugate axis.

T   F   **6.** A hyperbola always intersects its transverse axis.

T   F   **7.** The equation $ax^2 + 6y^2 - 12y = 0$ defines an ellipse if $a > 0$.

T   F   **8.** The equation $3x^2 + bxy + 12y^2 = 10$ defines a parabola if $b = -12$.

T   F   **9.** If (r, θ) are polar coordinates, the equation $r = 2/(2 + 3 \sin \theta)$ defines a hyperbola.

T   F   **10.** Parametric equations defining a curve are unique.

REVIEW EXERCISES

In Problems 1–20, identify each equation. If it is a parabola, give its vertex, focus, and directrix; if it is an ellipse, give its center, vertices, and foci; if it is a hyperbola, give its center, vertices, foci, and asymptotes.

1. $y^2 = -16x$

2. $16x^2 = y$

3. $\dfrac{x^2}{25} - y^2 = 1$

4. $\dfrac{y^2}{25} - x^2 = 1$

5. $\dfrac{y^2}{25} + \dfrac{x^2}{16} = 1$

6. $\dfrac{x^2}{9} + \dfrac{y^2}{16} = 1$

7. $x^2 + 4y = 4$

8. $3y^2 - x^2 = 9$

9. $4x^2 - y^2 = 8$

10. $9x^2 + 4y^2 = 36$

11. $x^2 - 4x = 2y$

12. $2y^2 - 4y = x - 2$

13. $y^2 - 4y - 4x^2 + 8x = 4$

14. $4x^2 + y^2 + 8x - 4y + 4 = 0$

15. $4x^2 + 9y^2 - 16x - 18y = 11$

16. $4x^2 + 9y^2 - 16x + 18y = 11$

17. $4x^2 - 16x + 16y + 32 = 0$

18. $4y^2 + 3x - 16y + 19 = 0$

19. $9x^2 + 4y^2 - 18x + 8y = 23$

20. $x^2 - y^2 - 2x - 2y = 1$

In Problems 21–36, obtain an equation of the conic described. Graph the equation.

21. Parabola; focus at $(-2, 0)$; directrix the line $x = 2$

22. Ellipse; center at $(0, 0)$; focus at $(0, 3)$; vertex at $(0, 5)$

23. Hyperbola; center at $(0, 0)$; focus at $(0, 4)$; vertex at $(0, -2)$

24. Parabola; vertex at $(0, 0)$; directrix the line $y = -3$

25. Ellipse; foci at $(-3, 0)$ and $(3, 0)$; vertex at $(4, 0)$

26. Hyperbola; vertices at $(-2, 0)$ and $(2, 0)$; focus at $(4, 0)$

27. Parabola; vertex at $(2, -3)$; focus at $(2, -4)$

28. Ellipse; center at $(-1, 2)$; focus at $(0, 2)$; vertex at $(2, 2)$

29. Hyperbola; center at $(-2, -3)$; focus at $(-4, -3)$; vertex at $(-3, -3)$

30. Parabola; focus at $(3, 6)$; directrix the line $y = 8$

31. Ellipse; foci at $(-4, 2)$ and $(-4, 8)$; vertex at $(-4, 10)$

32. Hyperbola; vertices at $(-3, 3)$ and $(5, 3)$; focus at $(7, 3)$

33. Center at $(-1, 2)$; $a = 3$; $c = 4$; transverse axis parallel to the x-axis

34. Center at $(4, -2)$; $a = 1$; $c = 4$; transverse axis parallel to the y-axis

35. Vertices at $(0, 1)$ and $(6, 1)$; asymptote the line $3y + 2x - 9 = 0$

36. Vertices at $(4, 0)$ and $(4, 4)$; asymptote the line $y + 2x - 10 = 0$

In Problems 37–46, identify each conic without completing the squares and without applying a rotation of axes.

37. $y^2 + 4x + 3y - 8 = 0$

38. $2x^2 - y + 8x = 0$

39. $x^2 + 2y^2 + 4x - 8y + 2 = 0$

40. $x^2 - 8y^2 - x - 2y = 0$

41. $9x^2 - 12xy + 4y^2 + 8x + 12y = 0$

42. $4x^2 + 4xy + y^2 - 8\sqrt{5}x + 16\sqrt{5}y = 0$

43. $4x^2 + 10xy + 4y^2 - 9 = 0$

44. $4x^2 - 10xy + 4y^2 - 9 = 0$

45. $x^2 - 2xy + 3y^2 + 2x + 4y - 1 = 0$

46. $4x^2 + 12xy - 10y^2 + x + y - 10 = 0$

In Problems 47–52, rotate the axes so that the new equation contains no xy-term. Discuss and graph the new equation.

47. $2x^2 + 5xy + 2y^2 - \frac{9}{2} = 0$

48. $2x^2 - 5xy + 2y^2 - \frac{9}{2} = 0$

49. $6x^2 + 4xy + 9y^2 - 20 = 0$

50. $x^2 + 4xy + 4y^2 + 16\sqrt{5}x - 8\sqrt{5}y = 0$

51. $4x^2 - 12xy + 9y^2 + 12x + 8y = 0$

52. $9x^2 - 24xy + 16y^2 + 80x + 60y = 0$

In Problems 53–58, identify the conic that each polar equation represents, and graph it.

53. $r = \dfrac{4}{1 - \cos \theta}$

54. $r = \dfrac{6}{1 + \sin \theta}$

55. $r = \dfrac{6}{2 - \sin \theta}$

56. $r = \dfrac{2}{3 + 2 \cos \theta}$

57. $r = \dfrac{8}{4 + 8 \cos \theta}$

58. $r = \dfrac{10}{5 + 20 \sin \theta}$

In Problems 59–62, convert each polar equation to a rectangular equation.

59. $r = \dfrac{4}{1 - \cos \theta}$

60. $r = \dfrac{6}{2 - \sin \theta}$

61. $r = \dfrac{8}{4 + 8 \cos \theta}$

62. $r = \dfrac{2}{3 + 2 \cos \theta}$

In Problems 63–68, graph the curve whose parametric equations are given and show its orientation. Find the rectangular equation of each curve.

63. $x = 4t - 2$, $y = 1 - t$; $-\infty < t < \infty$

64. $x = 2t^2 + 6$, $y = 5 - t$; $-\infty < t < \infty$

65. $x = 3 \sin t$, $y = 4 \cos t + 2$; $0 \le t \le 2\pi$

66. $x = \ln t$, $y = t^3$; $t > 0$

67. $x = \sec^2 t$, $y = \tan^2 t$; $0 \le t \le \pi/4$

68. $x = t^{3/2}$, $y = 2t + 4$; $t \ge 0$

69. Find an equation of the hyperbola whose foci are the vertices of the ellipse $4x^2 + 9y^2 = 36$ and whose vertices are the foci of this ellipse.

70. Find an equation of the ellipse whose foci are the vertices of the hyperbola $x^2 - 4y^2 = 16$ and whose vertices are the foci of this hyperbola.

71. Describe the collection of points in a plane so that the distance from each point to the point $(3, 0)$ is three-fourths of its distance from the line $x = \frac{16}{3}$.

72. Describe the collection of points in a plane so that the distance from each point to the point $(5, 0)$ is five-fourths of its distance from the line $x = \frac{16}{5}$.

73. *Mirrors* A mirror is shaped like a paraboloid of revolution. If a light source is located 1 foot from the base along the axis of symmetry and the opening is 2 feet across, how deep should the mirror be?

74. *Parabolic Arch Bridge* A bridge is built in the shape of a parabolic arch. The bridge has a span of 60 feet and a maximum height of 20 feet. Find the height of the arch at distances of 5, 10, and 20 feet from the center.

75. *Semielliptical Arch Bridge* A bridge is built in the shape of a semielliptical arch. The bridge has a span of 60 feet and a maximum height of 20 feet. Find the height of the arch at distances of 5, 10, and 20 feet from the center.

76. *Whispering Galleries* The figure shows the specifications for an elliptical ceiling in a hall designed to be a whispering gallery. Where in the hall are the foci located?

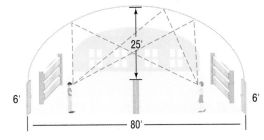

77. *LORAN* Two LORAN stations are positioned 150 miles apart along a straight shore.

(a) A ship records a time difference of 0.00032 second between the LORAN signals. Set up an appropriate rectangular coordinate system to determine where the ship would reach shore if it were to follow the hyperbola corresponding to this time difference.

(b) If the ship wants to enter a harbor located between the two stations 15 miles from the master station, what time difference should it be looking for?

(c) If the ship is 20 miles offshore when the desired time difference is obtained, what is the exact location of the ship? [*Note:* The speed of each radio signal is 186,000 miles per second.]

78. Formulate a strategy for discussing and graphing an equation of the form $Ax^2 + Bxy + Cy^2 + Dx + Ey + F = 0$.

Chapter 10

SYSTEMS OF EQUATIONS AND INEQUALITIES

PREPARING FOR THIS CHAPTER

Before getting started on this chapter, review the following concepts:

The Straight Line (p. 70)

Also, for Section 10.5: Circles (pp. 63–66)
Parabolas (Section 9.2)
Ellipses (Section 9.3)
Hyperbolas (Section 9.4)

Preview   Running a Race

In a 1 mile race, the winner crosses the finish line 10 feet ahead of the second place runner and 20 feet ahead of the third place runner. Assuming that each runner maintains a constant speed throughout the race, by how many feet does the second place runner beat the third place runner? [Problem 80 in Exercise 10.4] ■

In this chapter we take up the problem of solving equations and inequalities containing two or more variables. As the section titles suggest, there are various ways to solve such problems.

The *method of substitution* for solving equations in several unknowns goes back to ancient times.

The *method of elimination,* although it had existed for centuries, was put into systematic order by Karl Friedrich Gauss (1777–1855) and by Camille Jordan (1838–1922). This method is now used for solving large systems by computer.

The theory of *matrices* was developed in 1857 by Arthur Cayley (1821–1895), although only later were matrices used as we use them in this chapter. Matrices have become a very flexible instrument, useful in almost all areas of mathematics.

The method of *determinants* was invented by Seki Kōwa (1642–1708) in 1683 in Japan and by Gottfried Wilhelm von Leibniz (1646–1716) in 1693 in Germany. Both used them only in relation to linear equations. *Cramer's Rule* is named after Gabriel Cramer (1704–1752) of Switzerland, who popularized the use of determinants for solving linear systems.

Section 10.6 introduces *linear programming,* a modern application of linear inequalities to certain types of problems. This topic is particularly useful for students interested in operations research.

10.1

Systems of Linear Equations: Substitution; Elimination

We begin with an example.

E X A M P L E 1

Movie Theater Ticket Sales

A movie theater sells tickets for $8.00 each, with Seniors receiving a discount of $2.00. One evening the theater took in $3580 in revenue. If x represents the number of tickets sold at $8.00 and y the number of tickets sold at the discounted price of $6.00, write an equation that relates these variables.

Solution   Each nondiscounted ticket brings in $8.00, so x tickets will bring in $8x$ dollars. Similarly, y discounted tickets bring in $6y$ dollars. If the total brought in is $3580, we must have

$$8x + 6y = 3580$$ ∎

In Example 1, suppose we also know that 525 tickets were sold that evening. Then we have another equation relating the variables x and y:

$$x + y = 525$$

The two equations

$$8x + 6y = 3580$$
$$x + y = 525$$

form a *system* of equations.

In general, a **system of equations** is a collection of two or more equations, each containing one or more variables. Example 2 gives some samples of systems of equations.

E X A M P L E 2

Examples of Systems of Equations

(a) $\begin{cases} 2x + y = 5 & (1) \\ -4x + 6y = -2 & (2) \end{cases}$ (1) Two equations containing two variables, x and y

(b) $\begin{cases} x + y^2 = 5 & (1) \\ 2x + y = 4 & (2) \end{cases}$ (1) Two equations containing two variables, x and y

(c) $\begin{cases} x + y + z = 6 & (1) \\ 3x - 2y + 4z = 9 & (2) \\ x - y - z = 0 & (3) \end{cases}$ (1) Three equations containing three variables, x, y, and z

(d) $\begin{cases} x + y + z = 5 & (1) \\ x - y = 2 & (2) \end{cases}$ (1) Two equations containing three variables, x, y, and z

(e) $\begin{cases} x + y + z = 6 & (1) \\ 2x + 2z = 4 & (2) \\ y + z = 2 & (3) \\ x = 4 & (4) \end{cases}$ (1) Four equations containing three variables, x, y, and z ■

We use a brace, as shown above, to remind us that we are dealing with a system of equations. We also will find it convenient to number each equation in the system.

A **solution** of a system of equations consists of values for the variables that reduce each equation of the system to a true statement. To **solve** a system of equations means to find all solutions of the system.

For example, $x = 2$, $y = 1$ is a solution of the system in Example 2(a), because

$$2(2) + 1 = 5 \quad \text{and} \quad -4(2) + 6(1) = -2$$

A solution of the system in Example 2(b) is $x = 1$, $y = 2$, because

$$1 + 2^2 = 5 \quad \text{and} \quad 2(1) + 2 = 4$$

Another solution of the system in Example 2(b) is $x = \frac{11}{4}$, $y = -\frac{3}{2}$, which you can check for yourself. A solution of the system in Example 2(c) is $x = 3$, $y = 2$, $z = 1$, because

$$\begin{cases} 3 + 2 + 1 = 6 & (1) \; x = 3, \, y = 2, \, z = 1 \\ 3(3) - 2(2) + 4(1) = 9 & (2) \\ 3 - 2 - 1 = 0 & (3) \end{cases}$$

Note that $x = 3$, $y = 3$, $z = 0$ is not a solution of the system in Example 2(c):

$$\begin{cases} 3 + 3 + 0 = 6 & (1) \; x = 3, \, y = 3, \, z = 0 \\ 3(3) - 2(3) + 4(0) = 3 \neq 9 & (2) \\ 3 - 3 - 0 = 0 & (3) \end{cases}$$

Although these values satisfy equations (1) and (3), they do not satisfy equation (2). Any solution of the system must satisfy *each* equation of the system.

■   Now work Problem 3.

When a system of equations has at least one solution, it is said to be **consistent;** otherwise, it is called **inconsistent.**

An equation in n variables is said to be **linear** if it is equivalent to an equation of the form

$$a_1x_1 + a_2x_2 + \cdots + a_nx_n = b$$

where $x_1, x_2, \ldots, x_n$ are n distinct variables, $a_1, a_2, \ldots, a_n$, b are constants, and at least one of the a's is not 0.

Some examples of linear equations are

$$2x + 3y = 2 \qquad 5x - 2y + 3z = 10 \qquad 8x + 8y - 2z + 5w = 0$$

If each equation in a system of equations is linear, then we have a **system of linear equations.** Thus, the systems in Examples 2(a), (c), (d), and (e) are linear, whereas the system in Example 2(b) is nonlinear. We concentrate on solving linear systems in Sections 10.1–10.3, and we will take up nonlinear systems in Section 10.4.

Two Linear Equations Containing Two Variables

We can view the problem of solving a system of two linear equations containing two variables as a geometry problem. The graph of each equation in such a system is a straight line. Thus, a system of two equations containing two variables represents a pair of lines. The lines either (1) intersect or (2) are parallel or (3) are **coincident** (that is, identical).

1. If the lines intersect, then the system of equations has one solution, given by the point of intersection. The system is **consistent** and the equations are **independent.**
2. If the lines are parallel, then the system of equations has no solution, because the lines never intersect. The system is **inconsistent.**
3. If the lines are coincident, then the system of equations has infinitely many solutions, represented by the totality of points on the line. The system is **consistent** and the equations are **dependent.**

Figure 1 illustrates these conclusions.

FIGURE 1

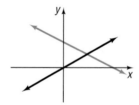
(a) Intersecting lines; system has one solution

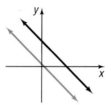
(b) Parallel lines; system has no solution

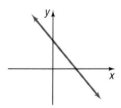
(c) Coincident lines; system has infinitely many solutions

E X A M P L E   3

Graphing a System of Equations

Graph the system $\begin{cases} 2x + y = 5 & \text{(1)} \\ -4x + 6y = 12 & \text{(2)} \end{cases}$

Solution

Equation (1) in slope–intercept form is $y = -2x + 5$, which has slope -2 and y-intercept 5. Equation (2) in slope–intercept form is $y = \frac{2}{3}x + 2$, which has slope $\frac{2}{3}$ and y-intercept 2. Figure 2 shows their graph. ∎

From the graph in Figure 2, we see that the lines intersect, so the system given in Example 3 is consistent. We can also use the graph as a means of approximating the solution. For this system, the solution appears to be close to the point (1, 3). The actual solution, which you should verify, is $\left(\frac{9}{8}, \frac{11}{4}\right)$. To obtain exact solutions we use algebraic methods. The first algebraic method we take up is the *method of substitution*.

FIGURE 2

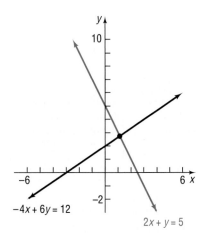

 Check: Graph the lines $2x + y = 5$ and $-4x + 6y = 12$ and compare what you see with Figure 2. Use TRACE to verify that the point of intersection is $(1.125, 2.75)$. ■

Method of Substitution

We illustrate the **method of substitution** by solving the system given in Example 3.

E X A M P L E   4

Solving a System of Equations by Substitution

Solve: $\begin{cases} 2x + y = 5 & \text{(1)} \\ -4x + 6y = 12 & \text{(2)} \end{cases}$

Solution

We solve the first equation for y, obtaining

$$y = 5 - 2x \tag{1}$$

We substitute this result for y in the second equation. The result is an equation containing just the variable x, which we can then solve for:

$$-4x + 6y = 12$$
$$-4x + 6(5 - 2x) = 12$$
$$-4x + 30 - 12x = 12$$
$$-16x = -18$$
$$x = \frac{-18}{-16} = \frac{9}{8}$$

Once we know that $x = \frac{9}{8}$, we can easily find the value of y by **back-substitution,** that is, by substituting $\frac{9}{8}$ for x in one of the original equations. We use the first one:

$$2x + y = 5$$
$$2\left(\frac{9}{8}\right) + y = 5$$
$$\frac{9}{4} + y = 5$$
$$y = 5 - \frac{9}{4} = \frac{20}{4} - \frac{9}{4} = \frac{11}{4}$$

The solution of the system is $x = \frac{9}{8}$, $y = \frac{11}{4}$.   ■

The method used to solve the system in Example 4 is called **substitution.** The steps to be used are outlined next.

Steps for Solving by Substitution

STEP 1: Pick one of the equations and solve for one of the variables in terms of the remaining variables.

STEP 2: Substitute the result in the remaining equations.

STEP 3: If one equation in one variable results, solve this equation. Otherwise, repeat Step 1 until a single equation with one variable remains.

STEP 4: Find the values of the remaining variables by back-substitution.

STEP 5: Check the solution found.

E X A M P L E   5   *Solving a System of Equations by Substitution*

Solve: $\begin{cases} 3x - 2y = 5 & (1) \\ 5x - y = 6 & (2) \end{cases}$

Solution   STEP 1: After looking at the two equations, we conclude that it is easiest to solve for the variable y in equation (2):

$$5x - y = 6$$
$$y = 5x - 6$$

STEP 2: We substitute this result into equation (1) and simplify:

$$3x - 2y = 5$$
$$3x - 2(5x - 6) = 5$$
$$-7x + 12 = 5$$
$$-7x = -7$$
$$x = 1$$

STEP 3: Because we now have one solution, $x = 1$, we proceed to Step 4.

STEP 4: Knowing $x = 1$, we can find y from the equation

$$y = 5x - 6 = 5(1) - 6 = -1$$

STEP 5: *Check:* $\begin{cases} 3(1) - 2(-1) = 3 + 2 = 5 \\ 5(1) - (-1) \ = 5 + 1 = 6 \end{cases}$

The solution of the system is $x = 1$, $y = -1$.   ■

E X A M P L E   6   *Solving a System of Equations by Substitution*

Solve: $\begin{cases} 2x - 3y = 7 & (1) \\ 4x + 5y = 3 & (2) \end{cases}$

Solution   STEP 1: In looking over the system, we conclude that there is no way to solve for one of the variables without introducing fractions. We solve for the variable x in equation (1):

$$2x - 3y = 7$$
$$2x = 3y + 7$$
$$x = \frac{3}{2}y + \frac{7}{2}$$

STEP 2: We substitute this result for x in equation (2) and simplify;

$$4x + 5y = 3$$
$$4\left(\frac{3}{2}y + \frac{7}{2}\right) + 5y = 3$$
$$6y + 14 + 5y = 3$$
$$11y + 14 = 3$$
$$11y = -11$$

STEP 3: $\qquad\qquad\qquad\qquad\qquad\qquad\qquad\qquad y = -1$

STEP 4: $\quad x = \frac{3}{2}y + \frac{7}{2} = \frac{3}{2}(-1) + \frac{7}{2} = \frac{4}{2} = 2$

STEP 5: *Check:* $\begin{cases} 2(2) - 3(-1) = 4 + 3 = 7 \\ 4(2) + 5(-1) = 8 - 5 = 3 \end{cases}$

The solution is $x = 2$, $y = -1$. ■

■ Now use substitution to work Problem 13.

Method of Elimination

A second method for solving a system of linear equations is the *method of elimination*. This method is usually preferred over substitution if substitution leads to fractions or if the system contains more than two variables. Elimination also provides the necessary motivation for solving systems using matrices (the subject of the next section).

The idea behind the method of elimination is to keep replacing the original equations in the system with equivalent equations until a system of equations with an obvious solution is reached. When we proceed in this way, we obtain **equivalent systems of equations.** The rules for obtaining equivalent equations are the same as those studied in Chapter 1. However, we may also interchange any two equations of the system and/or replace any equation in the system by the sum (or difference) of that equation and any other equation in the system.

Rules for Obtaining
an Equivalent System
of Equations

1. Interchange any two equations of the system.
2. Multiply (or divide) each side of an equation by the same nonzero constant.
3. Replace any equation in the system by the sum (or difference) of that equation and any other equation in the system.

An example will give you the idea. As you work through the example, pay particular attention to the pattern being followed.

E X A M P L E 7   *Solving a System of Equations by Elimination*

Solve: $\begin{cases} 2x + 3y = 1 & (1) \\ -x + y = -3 & (2) \end{cases}$

Solution   We multiply each side of equation (2) by 2 so that the coefficients of x in the two equations are negatives of one another. The result is the equivalent system

$$\begin{cases} 2x + 3y = 1 & (1) \\ -2x + 2y = -6 & (2) \end{cases}$$

If we now replace equation (2) of this system by the sum of the two equations, we obtain an equation containing just the variable y, which we can solve for:

$$\begin{cases} 2x + 3y = 1 \quad {\scriptstyle(1)} \\ -2x + 2y = -6 \quad {\scriptstyle(2)} \end{cases}$$

$$5y = -5$$

$$y = -1$$

We back-substitute by using this value for y in equation (1) and simplify to get

$$2x + 3(-1) = 1$$

$$2x = 4$$

$$x = 2$$

Thus, the solution of the original system is $x = 2$, $y = -1$. We leave it to you to check the solution. ◼

The procedure used in Example 7 is called the **method of elimination.** Notice the pattern of the solution. First, we eliminated the variable x from the second equation. Then we back-substituted; that is, we substituted the value found for y back into the first equation to find x.

Let's return to the movie theater example (Example 1).

E X A M P L E   8   *Movie Theater Ticket Sales*

A movie theater sells tickets for $8.00 each, with Seniors receiving a discount of $2.00. One evening the theater sold 525 tickets and took in $3580 in revenue. How many of each type of ticket was sold?

Solution   If x represents the number of tickets sold at $8.00 and y the number of tickets sold at the discounted price of $6.00, then the given information results in the system of equations

$$\begin{cases} 8x + 6y = 3580 \\ x + y = 525 \end{cases}$$

We use elimination and multiply the second equation by -6 and then add the equations.

$$\begin{cases} 8x + 6y = 3580 \\ -6x - 6y = -3150 \end{cases}$$

$$2x = 430$$

$$x = 215$$

Since $x + y = 525$, then $y = 525 - x = 525 - 215 = 310$. Thus, 215 nondiscounted tickets and 310 Senior discount tickets were sold. ◼

◼ Now use elimination to work Problem 13.

The previous examples dealt with consistent systems of equations that had a unique solution. The next two examples deal with two other possibilities that may occur, the first being a system that has no solution.

EXAMPLE 9

An Inconsistent System of Equations

Solve: $\begin{cases} 2x + y = 5 & (1) \\ 4x + 2y = 8 & (2) \end{cases}$

Solution We choose to use the method of substitution and solve equation (1) for *y*:

$$2x + y = 5$$
$$y = 5 - 2x$$

Substituting in equation (2), we get

$$4x + 2y = 8$$
$$4x + 2(5 - 2x) = 8$$
$$4x + 10 - 4x = 8$$
$$0 \cdot x = -2$$

FIGURE 3

This equation has no solution. Thus, we conclude that the system itself has no solution and is therefore inconsistent. ■

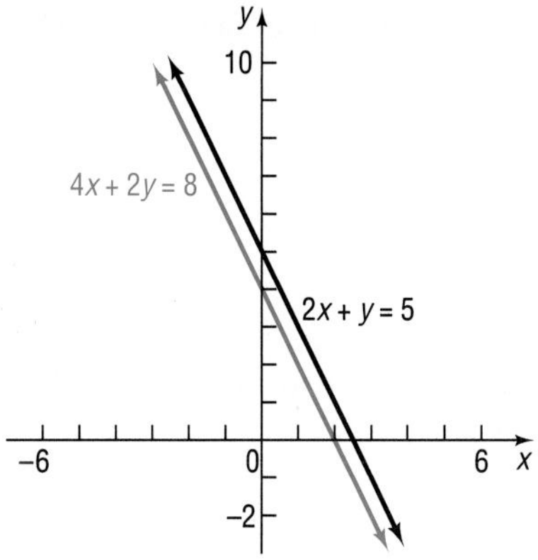

Figure 3 illustrates the pair of lines whose equations form the system in Example 9. Notice that the graphs of the two equations are lines, each with slope −2; one has a *y*-intercept of 5, the other a *y*-intercept of 4. Thus, the lines are parallel and have no point of intersection. This geometric statement is equivalent to the algebraic statement that the system has no solution.

Check: Graph the lines $2x + y = 5$ and $4x + 2y = 8$ and compare what you see with Figure 3. How can you be sure that the lines are parallel? ■

The next example is an illustration of a system with infinitely many solutions.

EXAMPLE 10

Solving a System of Equations with Infinitely Many Solutions

Solve: $\begin{cases} 2x + y = 4 & (1) \\ -6x - 3y = -12 & (2) \end{cases}$

Solution We choose to use the method of elimination:

$$\begin{cases} 2x + y = 4 & (1) \\ -6x - 3y = -12 & (2) \end{cases}$$

$$\begin{cases} 6x + 3y = 12 & (1) \text{ Multiply each side of equation (1) by 3.} \\ -6x - 3y = -12 & (2) \end{cases}$$

$$\begin{cases} 6x + 3y = 12 & (1) \text{ Replace equation (2) by the sum of} \\ 0 = 0 & (2) \text{ equations (1) and (2).} \end{cases}$$

The original system is thus equivalent to a system containing one equation, so the equations are dependent. This means that any values of *x* and *y* for which $6x + 3y = 12$ or, equivalently, $2x + y = 4$ are solutions. For example, $x = 2, y = 0$; $x = 0, y = 4$; $x = -2, y = 8$; $x = 4, y = -4$; and so on, are solutions. There are, in fact, infinitely many values of *x* and *y* for which $2x + y = 4$, so the original system has infinitely many solutions. We will write the solutions of the original systems either as

$$y = 4 - 2x$$

FIGURE 4

$$\begin{cases} 2x + y = 4 \\ -6x - 3y = -12 \end{cases}$$

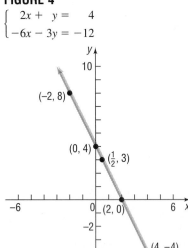

where x can be any real number, or as

$$x = 2 - \frac{1}{2}y$$

where y can be any real number. ■

Figure 4 illustrates the situation presented in Example 10. Notice that the graphs of the two equations are lines, each with slope -2 and each with y-intercept 4. Thus, the lines are coincident. Notice also that equation (2) in the original system is just -3 times equation (1), indicating that the two equations are dependent.

For the system in Example 10, we can write down some of the infinite number of solutions by assigning values to x and then finding $y = 4 - 2x$. Thus,

If $x = 4$, then $y = -4$.

If $x = 0$, then $y = 4$.

If $x = \frac{1}{2}$, then $y = 3$.

The pairs (x, y) are points on the line in Figure 4.

 Check: Graph the lines $2x + y = 4$ and $-6x - 3y = -12$ and compare what you see with Figure 4. How can you be sure that the lines are coincident? ■

■ Now work Problems 19 and 23.

Three Linear Equations Containing Three Variables

Just as with a system of two linear equations containing two variables, a system of three linear equations containing three variables also has either exactly one solution, or no solution, or infinitely many solutions.

Let's see how elimination works on a system of three equations containing three variables.

E X A M P L E 1 1

Solving a System of Three Linear Equations with Three Variables

Use the method of elimination to solve the system of equations;

$$\begin{cases} x + y - z = -1 & (1) \\ 4x - 3y + 2z = 16 & (2) \\ 2x - 2y - 3z = 5 & (3) \end{cases}$$

Solution   For a system of three equations, we attempt to eliminate one variable at a time, using pairs of equations. Our plan of attack on this system will be to first eliminate the variable x from equations (1) and (2) and then from equations (1) and (3). Next, we will eliminate the variable y from the resulting equations, leaving one equation containing only the variable z. Back-substitution can then be used to obtain the values of y and then x.

We begin by multiplying each side of equation (1) by -2, in anticipation of eliminating the variable x from equation (3) by adding equations (1) and (3):

$$\begin{cases} -2x - 2y + 2z = 2 & (1) \text{ Multiply each side of equation (1)} \\ 4x - 3y + 2z = 16 & (2) \quad \text{by } -2 \\ 2x - 2y - 3z = 5 & (3) \end{cases}$$

$$\begin{cases} -2x - 2y + 2z = 2 & (1) \\ 4x - 3y + 2z = 16 & (2) \\ -4y - z = 7 & (3) \end{cases}$$

(3) Replace equation (3) by the sum of equations (1) and (3).

We now eliminate the variable x from equation (2):

$$\begin{cases} -4x - 4y + 4z = 4 & (1) \\ 4x - 3y + 2z = 16 & (2) \\ -4y - z = 7 & (3) \end{cases}$$

(1) Multiply each side of equation (1) by 2.

$$\begin{cases} -4x - 4y + 4z = 4 & (1) \\ -7y + 6z = 20 & (2) \\ -4y - z = 7 & (3) \end{cases}$$

(2) Replace equation (2) by the sum of equations (1) and (2).

We now eliminate y from equation (3):

$$\begin{cases} -4x - 4y + 4z = 4 & (1) \\ -28y + 24z = 80 & (2) \\ 28y + 7z = -49 & (3) \end{cases}$$

(2) Multiply each side of equation (2) by 4.
(3) Multiply each side of equation (3) by -7.

$$\begin{cases} -4x - 4y + 4z = 4 & (1) \\ -28y + 24z = 80 & (2) \\ 31z = 31 & (3) \end{cases}$$

(3) Replace equation (3) by the sum of equations (2) and (3).

$$\begin{cases} -4x - 4y + 4z = 4 & (1) \\ -28y + 24z = 80 & (2) \\ z = 1 & (3) \end{cases}$$

(3) Multiply each side of equation (3) by $\frac{1}{31}$.

$$\begin{cases} -4x - 4y + 4 = 4 & (1) \\ -28y + 24 = 80 & (2) \\ z = 1 & (3) \end{cases}$$

(1) Back-substitute; replace z by 1 in equations (1) and (2).

$$\begin{cases} -4x - 4y = 0 & (1) \\ y = -2 & (2) \\ z = 1 & (3) \end{cases}$$

(2) Solve equation (2) for y.

$$\begin{cases} -4x + 8 = 0 & (1) \\ y = -2 & (2) \\ z = 1 & (3) \end{cases}$$

(1) Back-substitute; replace y by -2.

$$\begin{cases} x = 2 & (1) \\ y = -2 & (2) \\ z = 1 & (3) \end{cases}$$

The solution of the original system is $x = 2$, $y = -2$, $z = 1$. (You should check this.) ■

Look back over the solution given in Example 11. Note the pattern of making equation (3) contain only the variable z, followed by making equation (2) contain only the variable y and equation (1) contain only the variable x. Although which variables to isolate is your choice, the methodology remains the same for all systems.

10.1

Exercise 10.1

In Problems 1–10, verify that the values of the variables listed are solutions of the system of equations.

1. $\begin{cases} 2x - y = 5 \\ 5x + 2y = 8 \end{cases}$

$x = 2, y = -1$

2. $\begin{cases} 3x + 2y = 2 \\ x - 7y = -30 \end{cases}$

$x = -2, y = 4$

3. $\begin{cases} 3x - 4y = 4 \\ \frac{1}{2}x - 3y = -\frac{1}{2} \end{cases}$

$x = 2, y = \frac{1}{2}$

4. $\begin{cases} 2x + \frac{1}{2}y = 0 \\ 3x - 4y = -\frac{19}{2} \end{cases}$

$x = -\frac{1}{2}, y = 2$

5. $\begin{cases} x^2 - y^2 = 3 \\ xy = 2 \end{cases}$

$x = 2, y = 1$

6. $\begin{cases} x^2 - y^2 = 3 \\ xy = 2 \end{cases}$

$x = -2, y = -1$

7. $\begin{cases} \dfrac{x}{1 + x} + 3y = 6 \\ x + 9y^2 = 36 \end{cases}$

$x = 0, y = 2$

8. $\begin{cases} \dfrac{x}{x - 1} + y = 5 \\ 3x - y = 3 \end{cases}$

$x = 2, y = 3$

9. $\begin{cases} 3x + 3y + 2z = 4 \\ x - y - z = 0 \\ 2y - 3z = -8 \end{cases}$

$x = 1, y = -1, z = 2$

10. $\begin{cases} 4x - z = 7 \\ 8x + 5y - z = 0 \\ -x - y + 5z = 6 \end{cases}$

$x = 2, y = -3, z = 1$

In Problems 11–46, solve each system of equations. If the system has no solution, say it is inconsistent. Use either substitution or elimination.

11. $\begin{cases} x + y = 8 \\ x - y = 4 \end{cases}$

12. $\begin{cases} x + 2y = 5 \\ x + y = 3 \end{cases}$

13. $\begin{cases} 5x - y = 13 \\ 2x + 3y = 12 \end{cases}$

14. $\begin{cases} x + 3y = 5 \\ 2x - 3y = -8 \end{cases}$

15. $\begin{cases} 3x = 24 \\ x + 2y = 0 \end{cases}$

16. $\begin{cases} 4x + 5y = -3 \\ -2y = -4 \end{cases}$

17. $\begin{cases} 3x - 6y = 2 \\ 5x + 4y = 1 \end{cases}$

18. $\begin{cases} 2x + 4y = \frac{2}{3} \\ 3x - 5y = -10 \end{cases}$

19. $\begin{cases} 2x + y = 1 \\ 4x + 2y = 3 \end{cases}$

20. $\begin{cases} x - y = 5 \\ -3x + 3y = 2 \end{cases}$

21. $\begin{cases} 2x - y = 0 \\ 3x + 2y = 7 \end{cases}$

22. $\begin{cases} 3x + 3y = -1 \\ 4x + y = \frac{8}{3} \end{cases}$

23. $\begin{cases} x + 2y = 4 \\ 2x + 4y = 8 \end{cases}$

24. $\begin{cases} 3x - y = 7 \\ 9x - 3y = 21 \end{cases}$

25. $\begin{cases} 2x - 3y = -1 \\ 10x + y = 11 \end{cases}$

26. $\begin{cases} 3x - 2y = 0 \\ 5x + 10y = 4 \end{cases}$

27. $\begin{cases} 2x + 3y = 6 \\ x - y = \frac{1}{2} \end{cases}$

28. $\begin{cases} \frac{1}{2}x + y = -2 \\ x - 2y = 8 \end{cases}$

29. $\begin{cases} \frac{1}{2}x + \frac{1}{3}y = 3 \\ \frac{1}{4}x - \frac{2}{3}y = -1 \end{cases}$

30. $\begin{cases} \frac{1}{3}x - \frac{3}{2}y = -5 \\ \frac{3}{4}x + \frac{1}{3}y = 11 \end{cases}$

31. $\begin{cases} 3x - 5y = 3 \\ 15x + 5y = 21 \end{cases}$

32. $\begin{cases} 2x - y = -1 \\ x + \frac{1}{2}y = \frac{3}{2} \end{cases}$

33. $\begin{cases} x - y = 6 \\ 2x - 3z = 16 \\ 2y + z = 4 \end{cases}$

34. $\begin{cases} 2x + y = -4 \\ -2y + 4z = 0 \\ 3x - 2z = -11 \end{cases}$

35. $\begin{cases} x - 2y + 3z = 7 \\ 2x + y + z = 4 \\ -3x + 2y - 2z = -10 \end{cases}$

36. $\begin{cases} 2x + y - 3z = 0 \\ -2x + 2y + z = -7 \\ 3x - 4y - 3z = 7 \end{cases}$

37. $\begin{cases} x - y - z = 1 \\ 2x + 3y + z = 2 \\ 3x + 2y = 0 \end{cases}$

38. $\begin{cases} 2x - 3y - z = 0 \\ -x + 2y + z = 5 \\ 3x - 4y - z = 1 \end{cases}$

39. $\begin{cases} x - y - z = 1 \\ -x + 2y - 3z = -4 \\ 3x - 2y - 7z = 0 \end{cases}$

40. $\begin{cases} 2x - 3y - z = 0 \\ 3x + 2y + 2z = 2 \\ x + 5y + 3z = 2 \end{cases}$

41. $\begin{cases} 2x - 2y + 3z = 6 \\ 4x - 3y + 2z = 0 \\ -2x + 3y - 7z = 1 \end{cases}$

42. $\begin{cases} 3x - 2y + 2z = 6 \\ 7x - 3y + 2z = -1 \\ 2x - 3y + 4z = 0 \end{cases}$

43. $\begin{cases} x + y - z = 6 \\ 3x - 2y + z = -5 \\ x + 3y - 2z = 14 \end{cases}$

44. $\begin{cases} x - y + z = -4 \\ 2x - 3y + 4z = -15 \\ 5x + y - 2z = 12 \end{cases}$

45. $\begin{cases} x + 2y - z = -3 \\ 2x - 4y + z = -7 \\ -2x + 2y - 3z = 4 \end{cases}$

46. $\begin{cases} x + 4y - 3z = -8 \\ 3x - y + 3z = 12 \\ x + y + 6z = 1 \end{cases}$

47. Solve: $\begin{cases} \dfrac{1}{x} + \dfrac{1}{y} = 8 \\ \dfrac{3}{x} - \dfrac{5}{y} = 0 \end{cases}$

48. Solve: $\begin{cases} \dfrac{4}{x} - \dfrac{3}{y} = 0 \\ \dfrac{6}{x} + \dfrac{3}{2y} = 2 \end{cases}$

[Hint: Let $u = 1/x$ and $v = 1/y$, and solve for u and v. Then $x = 1/u$ and $y = 1/v$.]

In Problems 49–54, solve each system of equations. Approximate the solution correct to two decimal places.

49. $\begin{cases} y = \sqrt{2}x - 20\sqrt{7} \\ y = -0.1x + 20 \end{cases}$

50. $\begin{cases} y = -\sqrt{3}x + 100 \\ y = 0.2x + \sqrt{19} \end{cases}$

51. $\begin{cases} \sqrt{2}x + \sqrt{3}y + \sqrt{6} = 0 \\ \sqrt{3}x - \sqrt{2}y + 60 = 0 \end{cases}$

52. $\begin{cases} \sqrt{5}x - \sqrt{6}y + 60 = 0 \\ 0.2x + 0.3y + \sqrt{5} = 0 \end{cases}$

53. $\begin{cases} \sqrt{3}x + \sqrt{2}y = \sqrt{0.3} \\ 100x - 95y = 20 \end{cases}$

54. $\begin{cases} \sqrt{6}x - \sqrt{5}y + \sqrt{1.1} = 0 \\ y = -0.2x + 0.1 \end{cases}$

55. The sum of two numbers is 81. The difference of twice one and three times the other is 62. Find the two numbers.

56. The difference of two numbers is 40. Six times the smaller one less the larger one is 5. Find the two numbers.

57. The perimeter of a rectangular floor is 90 feet. Find the dimensions of the floor if the length is twice the width.

58. The length of fence required to enclose a rectangular field is 3000 meters. What are the dimensions of the field if it is known that the difference between its length and width is 50 meters?

59. *Cost of Fast Food* Four large cheeseburgers and two chocolate shakes cost a total of $7.90. Two shakes cost 15¢ more than one cheeseburger. What is the cost of a cheeseburger? A shake?

60. *Movie Theatre Tickets* A movie theater charges $9.00 for adults and $7.00 for senior citizens. On a day when 325 people paid an admission, the total receipts were $2495. How many who paid were adults? How many were seniors?

61. *Mixing Nuts* A store sells cashews for $5.00 per pound and peanuts for $1.50 per pound. The manager decides to mix 30 pounds of peanuts with some cashews and sell the mixture for $3.00 per pound. How many pounds of cashews should be mixed with the peanuts so that the mixture will produce the same revenue as would selling the nuts separately?

62. *Financial Planning* A recently retired couple need $12,000 per year to supplement their Social Security. They have $150,000 to invest to obtain this income. They have decided on two investment options: AA bonds yielding 10% per annum and a Bank Certificate yielding 5%.
(a) How much should be invested in each to realize exactly $12,000?
(b) If, after 2 years, the couple requires $14,000 per year in income, how should they reallocate their investment to achieve the new amount?

63. *Computing Wind Speed* With a tail wind, a small Piper aircraft can fly 600 miles in 3 hours. Against this same wind, the Piper can fly the same distance in 4 hours. Find the average wind speed and the average airspeed of the Piper.

64. *Computing Wind Speed* The average airspeed of a single-engine aircraft is 150 miles per hour. If the aircraft flew the same distance in 2 hours with the wind as it flew in 3 hours against the wind, what was the wind speed?

65. *Restaurant Management* A restaurant manager wants to purchase 200 sets of dishes. One design costs $25 per set, while another costs $45 per set. If she only has $7400 to spend, how many of each design should be ordered?

66. *Cost of Fast Food* One group of people purchased 10 hot dogs and 5 soft drinks at a cost of $12.50. A second bought 7 hot dogs and 4 soft drinks at a cost of $9.00. What is the cost of a single hot dog? A single soft drink?

67. *Computing a Refund* The grocery store we use does not mark prices on its goods. My wife went to this store, bought three 1 pound packages of bacon and two cartons of eggs, and paid a total of $7.45. Not knowing she went to the store, I also went to the same store, purchased two 1 pound packages of bacon and three cartons of eggs, and paid a total of $6.45. Now we want to return two 1 pound packages of bacon and two cartons of eggs. How much will be refunded?

68. *Finding the Current of a Stream* A swimmer requires 3 hours to swim 15 miles downstream. The return trip upstream takes 5 hours. Find the average speed of the swimmer in still water. How fast is the current of the stream? (Assume that the speed of the swimmer is the same in each direction.)

69. The sum of three numbers is 48. The sum of the two larger numbers is three times the smallest. The sum of the two smaller numbers is 6 more than the largest. Find the numbers.

70. A coin collection consists of 37 coins (nickels, dimes, and quarters). If the collection has a face value of $3.25 and there are 5 more dimes than there are nickels, how many of each coin are in the collection?

71. *Electricity: Kirchhoff's Rules* An application of *Kirchhoff's Rules* to the circuit shown results in the following system of equations:

$$\begin{cases} I_2 = I_1 + I_3 \\ 5 - 3I_1 - 5I_2 = 0 \\ 10 - 5I_2 - 7I_3 = 0 \end{cases}$$

Find the currents I_1, I_2, and I_3.*

*Source: Based on Raymond Serway, Physics, 3rd ed. Philadelphia: Saunders, 1990, Prob. 26, p. 790.

72. *Electricity: Kirchhoff's Rules* An application of Kirchhoff's Rules to the circuit shown results in the following system of equations:

$$\begin{cases} I_3 = I_1 + I_2 \\ 8 = 4I_3 + 6I_2 \\ 8I_1 = 4 + 6I_2 \end{cases}$$

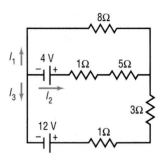

Find the currents I_1, I_2, and I_3.[†]

73. A Broadway theater has 500 seats, divided into orchestra, main, and balcony seating. Orchestra seats sell for $50, main seats for $35, and balcony seats for $25. If all the seats are sold, the gross revenue to the theater is $17,100. If all the main and balcony seats are sold, but only half the orchestra seats are sold, the gross revenue is $14,600. How many are there of each kind of seat?

74. *Laboratory Work Stations* A chemistry laboratory can be used by 38 students at one time. The laboratory has 16 work stations, some set up for 2 students each and the others set up for 3 students each. How many are there of each kind of work station?

75. Make up three systems of two linear equations containing two variables that have:

(a) No solution (b) Exactly one solution (c) Infinitely many solutions

Give them to a friend to solve and critique.

76. Write a brief paragraph outlining your strategy for solving a system of two linear equations containing two variables.

77. Do you prefer the method of substitution or the method of elimination for solving a system of two linear equations containing two variables? What about for solving a system of three linear equations containing three variables? Give reasons.

78. *Curve Fitting* Find real numbers b and c such that the parabola $y = x^2 + bx + c$ passes through the points $(1, 3)$ and $(3, 5)$.

79. *Curve Fitting* Find real numbers b and c such that the parabola $y = x^2 + bx + c$ passes through the points $(1, 2)$ and $(-1, 3)$.

80. *Curve Fitting* Find real numbers b and c such that the parabola $y = x^2 + bx + c$ passes through the points (x_1, y_1) and (x_2, y_2).

81. *Curve Fitting* Find real numbers a, b, and c such that the parabola $y = ax^2 + bx + c$ passes through the points $(-1, 4)$, $(2, 3)$, and $(0, 1)$.

82. *Curve Fitting* Find real numbers a, b, and c such that the parabola $y = ax^2 + bx + c$ passes through the points $(-1, -2)$, $(1, -4)$, and $(2, 4)$.

83. Solve: $\begin{cases} y = m_1x + b_1 \\ y = m_2x + b_2 \end{cases}$

where $m_1 \neq m_2$.

84. Solve: $\begin{cases} y = m_1x + b_1 \\ y = m_2x + b_2 \end{cases}$

where $m_1 = m_2 = m$ and $b_1 \neq b_2$.

85. Solve: $\begin{cases} y = m_1x + b_1 \\ y = m_2x + b_2 \end{cases}$

where $m_1 = m_2 = m$ and $b_1 = b_2 = b$.

[†]*Source:* Ibid., Prob. 27, p. 790.

10.2

Systems of Linear Equations: Matrices

The systematic approach of the method of elimination for solving a system of linear equations provides another method of solution that involves a simplified notation. Consider the following system of linear equations:

$$\begin{cases} x + 4y = 14 \\ 3x - 2y = 0 \end{cases}$$

If we choose not to write the symbols used for the variables, we can represent this system as

$$\left[\begin{array}{cc|c} 1 & 4 & 14 \\ 3 & -2 & 0 \end{array}\right]$$

where it is understood that the first column represents the coefficients of the variable x, the second column the coefficients of y, and the third column the constants on the right side of the equal signs. The vertical line serves as a reminder of the equal signs. The large square brackets are the traditional symbols used to denote a *matrix* in algebra.

Matrix

A **matrix** is defined as a rectangular array of numbers,

$$
\begin{array}{c}
\\
\text{Row 1} \\
\text{Row 2} \\
\vdots \\
\text{Row } i \\
\vdots \\
\text{Row } m
\end{array}
\begin{array}{c}
\begin{array}{ccccc}
\text{Column 1} & \text{Column 2} & \text{Column } j & \text{Column } n
\end{array} \\
\left[\begin{array}{ccccc}
a_{11} & a_{12} & \cdots & a_{1j} & \cdots & a_{1n} \\
a_{21} & a_{22} & \cdots & a_{2j} & \cdots & a_{2n} \\
\vdots & \vdots & & \vdots & & \vdots \\
a_{i1} & a_{i2} & \cdots & a_{ij} & \cdots & a_{in} \\
\vdots & \vdots & & \vdots & & \vdots \\
a_{m1} & a_{m2} & \cdots & a_{mj} & \cdots & a_{mn}
\end{array}\right]
\end{array} \quad (1)
$$

Each number a_{ij} of the matrix has two indices: the **row index** i and the **column index** j. The matrix shown in display (1) has m rows and n columns. The numbers a_{ij} are usually referred to as the **entries** of the matrix.

Now we will use matrix notation to represent a system of linear equations. The matrices used to represent systems of linear equations are called **augmented matrices.**

E X A M P L E   1

Writing the Augmented Matrix of a System of Equations

Write the augmented matrix of each system of equations.

(a) $\begin{cases} 3x - 4y = -6 & (1) \\ 2x - 3y = -5 & (2) \end{cases}$

(b) $\begin{cases} 2x - y + z = 0 & (1) \\ x + z - 1 = 0 & (2) \\ x + 2y - 8 = 0 & (3) \end{cases}$

Solution

(a) The augmented matrix is

$$\left[\begin{array}{cc|c} 3 & -4 & -6 \\ 2 & -3 & -5 \end{array}\right]$$

(b) Care must be taken that the system is written with the coefficients of all variables present (if any variable is missing, its coefficient is 0) and with all constants to the right of the equal signs. Thus, we need to rearrange the given system as follows:

$$\begin{cases} 2x - y + z = 0 & \text{(1)} \\ x + z - 1 = 0 & \text{(2)} \\ x + 2y - 8 = 0 & \text{(3)} \end{cases}$$

$$\begin{cases} 2x - y + z = 0 & \text{(1)} \\ x + 0 \cdot y + z = 1 & \text{(2)} \\ x + 2y + 0 \cdot z = 8 & \text{(3)} \end{cases}$$

The augmented matrix is

$$\left[\begin{array}{ccc|c} 2 & -1 & 1 & 0 \\ 1 & 0 & 1 & 1 \\ 1 & 2 & 0 & 8 \end{array}\right]$$ ∎

If we do not include the constants to the right of the equal sign, that is, to the right of the vertical bar in the augmented matrix, of a system of equations, the resulting matrix is called the **coefficient matrix** of the system. For the systems discussed in Example 1, the coefficient matrices are

$$\begin{bmatrix} 3 & -4 \\ 2 & -3 \end{bmatrix} \quad \text{and} \quad \begin{bmatrix} 2 & -1 & 1 \\ 1 & 0 & 1 \\ 1 & 2 & 0 \end{bmatrix}$$

∎   Now work Problem 3.

E X A M P L E   2

Writing the System of Linear Equations from the Augmented Matrix

Write the system of linear equations corresponding to each augmented matrix.

(a) $\left[\begin{array}{cc|c} 5 & 2 & 13 \\ -3 & 1 & -10 \end{array}\right]$ (b) $\left[\begin{array}{ccc|c} 3 & -1 & -1 & 7 \\ 2 & 0 & 2 & 8 \\ 0 & 1 & 1 & 0 \end{array}\right]$

Solution   (a) The matrix has two rows and so represents a system of two equations. The two columns to the left of the vertical bar indicate that the system has two variables. If x and y are used to denote these variables, the system of equations is

$$\begin{cases} 5x + 2y = 13 & \text{(1)} \\ -3x + y = -10 & \text{(2)} \end{cases}$$

(b) This matrix represents a system of three equations containing three variables. If x, y, and z are the three variables, this system is

$$\begin{cases} 3x - y - z = 7 & \text{(1)} \\ 2x + 2z = 8 & \text{(2)} \\ y + z = 0 & \text{(3)} \end{cases}$$ ∎

Row Operations on a Matrix

E X A M P L E   3

Solving a System of Equations

Solve the system of equations

$$\begin{cases} 4x - 3y = 11 & \text{(1)} \\ 3x + 2y = 4 & \text{(2)} \end{cases}$$

Solution We shall use a variation of the method of elimination to solve the system. First, multiply each side of equation (2) by -1 and add it to equation (1). Replace equation (1) with the result:

$$\begin{cases} x - 5y = 7 & (1) \\ 3x + 2y = 4 & (2) \end{cases}$$

Multiply each side of equation (1) by -3 and add it to equation (2). Replace equation (2) with the result:

$$\begin{cases} x - 5y = 7 & (1) \\ 0 \cdot x + 17y = -17 & (2) \end{cases}$$

Multiply each side of equation (2) by $\frac{1}{17}$:

$$\begin{cases} x - 5y = 7 & (1) \\ y = -1 & (2) \end{cases}$$

Now we back-substitute $y = -1$ into equation (1) to get

$$x - 5y = 7$$
$$x - 5(-1) = 7$$
$$x = 2$$

The solution of the system is $x = 2$, $y = -1$. ∎

The pattern of solution shown above provides a systematic way to solve any system of equations. The idea is to start with the augmented matrix of the system,

$$\begin{bmatrix} 4 & -3 & | & 11 \\ 3 & 2 & | & 4 \end{bmatrix} \qquad \begin{cases} 4x - 3y = 11 & (1) \\ 3x + 2y = 4 & (2) \end{cases}$$

and eventually arrive at the matrix,

$$\begin{bmatrix} 1 & -5 & | & 7 \\ 0 & 1 & | & -1 \end{bmatrix} \qquad \begin{cases} x - 5y = 7 & (1) \\ y = -1 & (2) \end{cases}$$

Let's go through the procedure again, this time starting with the augmented matrix and keeping the final augmented matrix given above in mind:

$$\begin{bmatrix} 4 & -3 & | & 11 \\ 3 & 2 & | & 4 \end{bmatrix} \qquad \begin{cases} 4x - 3y = 11 & (1) \\ 3x + 2y = 4 & (2) \end{cases}$$

As before, we start by multiplying each side of equation (2) by -1 and adding it to equation (1). This is equivalent to multiplying each entry in the second row of the matrix by -1, adding the result to the corresponding entries in row 1, and replacing row 1 by these entries. The result of this step is that the number 1 appears in row 1, column 1:

$$\begin{bmatrix} 1 & -5 & | & 7 \\ 3 & 2 & | & 4 \end{bmatrix} \qquad \begin{cases} x - 5y = 7 & (1) \\ 3x + 2y = 4 & (2) \end{cases}$$

Multiply each entry in the first row by -3, add the result to the entries in the second row, and replace the second row by these entries. The result of this step is that the number 0 appears in row 2, column 1:

$$\begin{bmatrix} 1 & -5 & | & 7 \\ 0 & 17 & | & -17 \end{bmatrix} \qquad \begin{cases} x - 5y = 7 & \text{(1)} \\ 0 \cdot x + 17y = -17 & \text{(2)} \end{cases}$$

Multiply each entry in the second row by $\dfrac{1}{17}$. The result of this step is that the number 1 appears in row 2, column 2:

$$\begin{bmatrix} 1 & -5 & | & 7 \\ 0 & 1 & | & -1 \end{bmatrix} \qquad \begin{cases} x - 5y = 7 & \text{(1)} \\ y = -1 & \text{(2)} \end{cases}$$

Now that we know that $y = -1$, we can back-substitute to find that $x = 2$.

The manipulations just performed on the augmented matrix are called **row operations.** There are three basic row operations:

Row Operations

1. Interchange any two rows.
2. Replace a row by a nonzero multiple of that row.
3. Replace a row by the sum of that row and a constant multiple of some other row.

These three row operations correspond to the three rules given earlier for obtaining an equivalent system of equations. Thus, when a row operation is performed on a matrix, the resulting matrix represents a system of equations equivalent to the system represented by the original matrix.

For example, consider the augmented matrix

$$\begin{bmatrix} 1 & 2 & | & 3 \\ 4 & -1 & | & 2 \end{bmatrix}$$

Suppose we want to apply a row operation to this matrix that results in a matrix whose entry in row 2, column 1 is a 0. The row operation to use is

$$\text{Multiply each entry in row 1 by } -4 \text{ and add the result} \\ \text{to the corresponding entries in row 2.} \qquad \text{(2)}$$

If we use R_2 to represent the new entries in row 2 and we use r_1 and r_2 to represent the original entries in rows 1 and 2, respectively, then we can represent the row operation in statement (2) by

$$R_2 = -4r_1 + r_2$$

Then

$$\begin{bmatrix} 1 & 2 & | & 3 \\ 4 & -1 & | & 2 \end{bmatrix} \xrightarrow[R_2 = -4r_1 + r_2]{} \begin{bmatrix} 1 & 2 & | & 3 \\ -4(1)+4 & -4(2)+(-1) & | & -4(3)+2 \end{bmatrix} = \begin{bmatrix} 1 & 2 & | & 3 \\ 0 & -9 & | & -10 \end{bmatrix}$$

As desired, we now have the entry 0 in row 2, column 1.

E X A M P L E 4   *Applying a Row Operation to an Augmented Matrix*

Apply the row operation $R_2 = -3r_1 + r_2$ to the augmented matrix

$$\begin{bmatrix} 1 & -2 & | & 2 \\ 3 & -5 & | & 9 \end{bmatrix}$$

Solution   The row operation $R_2 = -3r_1 + r_2$ tells us that the entries in row 2 are to be replaced by the entries obtained after multiplying each entry in row 1 by -3 and adding the result to the corresponding entries in row 2. Thus,

$$\begin{bmatrix} 1 & -2 & | & 2 \\ 3 & -5 & | & 9 \end{bmatrix} \xrightarrow[\substack{\uparrow \\ R_2 = -3r_1 + r_2}]{} \begin{bmatrix} 1 & -2 & | & 2 \\ -3(1) + 3 & (-3)(-2) + (-5) & | & -3(2) + 9 \end{bmatrix} = \begin{bmatrix} 1 & -2 & | & 2 \\ 0 & 1 & | & 3 \end{bmatrix}$$

■

E X A M P L E 5   *Finding a Particular Row Operation*

Using the matrix

$$\begin{bmatrix} 1 & -2 & | & 2 \\ 0 & 1 & | & 3 \end{bmatrix}$$

find a row operation that will result in a matrix with a 0 in row 1, column 2.

Solution   We want a 0 in row 1, column 2. This result can be accomplished by multiplying row 2 by 2 and adding the result to row 1. That is, we apply the row operation $R_1 = 2r_2 + r_1$:

$$\begin{bmatrix} 1 & -2 & | & 2 \\ 0 & 1 & | & 3 \end{bmatrix} \xrightarrow[\substack{\uparrow \\ R_1 = 2r_2 + r_1}]{} \begin{bmatrix} 2(0) + 1 & 2(1) + (-2) & | & 2(3) + 2 \\ 0 & 1 & | & 3 \end{bmatrix} = \begin{bmatrix} 1 & 0 & | & 8 \\ 0 & 1 & | & 3 \end{bmatrix}.$$

■

A word about the notation we have introduced. A row operation such as $R_1 = 2r_2 + r_1$ changes the entries in row 1. Note also that to change the entries in a given row we multiply the entries in some other row by an appropriate number and add the results to the original entries of the row to be changed.

■   Now work Problems 11 and 15.

Now let's see how we use row operations to solve a system of linear equations.

E X A M P L E 6   *Solving a System of Equations Using Matrices*

Solve: $\begin{cases} 4x + 3y = 11 & (1) \\ x - 3y = -1 & (2) \end{cases}$

Solution   First, we write the augmented matrix that represents this system:

$$\begin{bmatrix} 4 & 3 & | & 11 \\ 1 & -3 & | & -1 \end{bmatrix}$$

The first step requires getting the entry 1 in row 1, column 1. An interchange of rows 1 and 2 is the easiest way to do this:

$$\begin{bmatrix} 1 & -3 & | & -1 \\ 4 & 3 & | & 11 \end{bmatrix}$$

Next, we want a 0 under the entry 1 in column 1. We use the row operation $R_2 = -4r_1 + r_2$:

$$\begin{bmatrix} 1 & -3 & | & -1 \\ 4 & 3 & | & 11 \end{bmatrix} \rightarrow \begin{bmatrix} 1 & -3 & | & -1 \\ 0 & 15 & | & 15 \end{bmatrix}$$

$$\uparrow$$
$$R_2 = -4r_1 + r_2$$

Now we want the entry 1 in row 2, column 2. We use $R_2 = \frac{1}{15}r_2$:

$$\begin{bmatrix} 1 & -3 & | & -1 \\ 0 & 15 & | & 15 \end{bmatrix} \rightarrow \begin{bmatrix} 1 & -3 & | & -1 \\ 0 & 1 & | & 1 \end{bmatrix}$$

$$\uparrow$$
$$R_2 = \frac{1}{15}r_2$$

The second row of the matrix on the right represents the equation $y = 1$. Thus, using $y = 1$, we back-substitute into the equation $x - 3y = -1$ (from the first row) to get

$$x - 3(1) = -1 \quad y = 1$$
$$x = 2$$

The solution of the system is $x = 2$, $y = 1$. ■

■ Now work Problem 31.

The steps we used to solve the system of linear equations in Example 6 can be summarized as follows:

Matrix Method for Solving a System of Linear Equations

STEP 1: Write the augmented matrix that represents the system.
STEP 2: Perform row operations that place the entry 1 in row 1, column 1.
STEP 3: Perform row operations that leave the entry 1 in row 1, column 1 unchanged, while causing 0's to appear below it in column 1.
STEP 4: Perform row operations that place the entry 1 in row 2, column 2 and leave the entries in columns to the left unchanged. If it is impossible to place a 1 in row 2, column 2, then proceed to place a 1 in row 2, column 3. Once a 1 is in place, perform row operations to place 0's under it.
STEP 5: Now repeat Step 4, placing a 1 in the next row, but one column to the right. Continue until the bottom row or the vertical bar is reached.
STEP 6: If any rows are obtained that contain only 0's on the left side of the vertical bar, then place such rows at the bottom of the matrix.

After Steps 1 to 6 have been completed, the matrix is said to be in **echelon form.** A little thought should convince you that a matrix is in echelon form when:

1. The entry in row 1, column 1 is a 1, and 0's appear below it.
2. The first nonzero entry in each row after the first row is a 1, 0's appear below it, and it appears to the right of the first nonzero entry in any row above.
3. Any rows that contain all 0's to the left of the vertical bar appear at the bottom.

Two advantages of solving a system of equations by writing the augmented matrix in echelon form are the following:

1. The process is algorithmic; that is, it consists of repetitive steps so that it can be programmed on a computer.
2. The process works on any system of linear equations, no matter how many equations or variables are present.

The next example shows how to write a matrix in echelon form.

E X A M P L E 7   *Solving a System of Equations Using Matrices*

$$\text{Solve:} \begin{cases} x - y + z = 8 & (1) \\ 2x + 3y - z = -2 & (2) \\ 3x - 2y - 9z = 9 & (3) \end{cases}$$

Solution   **STEP 1:**   The augmented matrix of the system is

$$\begin{bmatrix} 1 & -1 & 1 & | & 8 \\ 2 & 3 & -1 & | & -2 \\ 3 & -2 & -9 & | & 9 \end{bmatrix}$$

STEP 2:   Because the entry 1 is already present in row 1, column 1, we can go to Step 3.

STEP 3:   Perform the row operations $R_2 = -2r_1 + r_2$ and $R_3 = -3r_1 + r_3$. Each of these leaves the entry 1 in row 1, column 1 unchanged, while causing 0's to appear under it:

$$\begin{bmatrix} 1 & -1 & 1 & | & 8 \\ 2 & 3 & -1 & | & -2 \\ 3 & -2 & -9 & | & 9 \end{bmatrix} \rightarrow \begin{bmatrix} 1 & -1 & 1 & | & 8 \\ 0 & 5 & -3 & | & -18 \\ 0 & 1 & -12 & | & -15 \end{bmatrix}$$

$$R_2 = -2r_1 + r_2$$
$$R_3 = -3r_1 + r_3$$

STEP 4:   The easiest way to obtain the entry 1 in row 2, column 2 without altering column 1 is to interchange rows 2 and 3 (another way would be to multiply row 2 by $\frac{1}{5}$, but this introduces fractions):

$$\begin{bmatrix} 1 & -1 & 1 & | & 8 \\ 0 & 1 & -12 & | & -15 \\ 0 & 5 & -3 & | & -18 \end{bmatrix}$$

To get 0's under the 1 in row 2, column 2, perform the row operation $R_3 = -5r_2 + r_3$:

$$\begin{bmatrix} 1 & -1 & 1 & | & 8 \\ 0 & 1 & -12 & | & -15 \\ 0 & 5 & -3 & | & -18 \end{bmatrix} \rightarrow \begin{bmatrix} 1 & -1 & 1 & | & 8 \\ 0 & 1 & -12 & | & -15 \\ 0 & 0 & 57 & | & 57 \end{bmatrix}$$

$$R_3 = -5r_2 + r_3$$

STEP 5:   Continuing, we place a 1 in row 3, column 3 by using $R_3 = \frac{1}{57}r_3$:

$$\begin{bmatrix} 1 & -1 & 1 & | & 8 \\ 0 & 1 & -12 & | & -15 \\ 0 & 0 & 57 & | & 57 \end{bmatrix} \rightarrow \begin{bmatrix} 1 & -1 & 1 & | & 8 \\ 0 & 1 & -12 & | & -15 \\ 0 & 0 & 1 & | & 1 \end{bmatrix}$$

$$R_3 = \frac{1}{57}r_3$$

Because we have reached the bottom row, the matrix is in echelon form and we can stop.

The system of equations represented by the matrix in echelon form is

$$\begin{cases} x - y + z = 8 \\ y - 12z = -15 \\ z = 1 \end{cases}$$

Using $z = 1$, we back-substitute to get

$$\begin{cases} x - y + 1 = 8 \quad \text{\small From row 1 of the matrix} \\ y - 12(1) = -15 \quad \text{\small From row 2 of the matrix} \end{cases}$$

Thus, we get $y = -3$, and back-substituting into $x - y = 7$, we find $x = 4$. The solution of the system is $x = 4$, $y = -3$, $z = 1$. ∎

Sometimes, it is advantageous to write a matrix in **reduced echelon form.** In this form, row operations are used to obtain entries that are 0 above (as well as below) the leading 1 in a row. For example, the echelon form obtained in the solution to Example 7 is

$$\begin{bmatrix} 1 & -1 & 1 & \bigm| & 8 \\ 0 & 1 & -12 & \bigm| & -15 \\ 0 & 0 & 1 & \bigm| & 1 \end{bmatrix}$$

To write this matrix in reduced echelon form, we proceed as follows:

$$\begin{bmatrix} 1 & -1 & 1 & \bigm| & 8 \\ 0 & 1 & -12 & \bigm| & -15 \\ 0 & 0 & 1 & \bigm| & 1 \end{bmatrix} \rightarrow \underset{\substack{\uparrow \\ R_1 = r_2 + r_1}}{\begin{bmatrix} 1 & 0 & -11 & \bigm| & -7 \\ 0 & 1 & -12 & \bigm| & -15 \\ 0 & 0 & 1 & \bigm| & 1 \end{bmatrix}} \rightarrow \underset{\substack{\uparrow \\ R_1 = 11r_3 + r_1 \\ R_2 = 12r_3 + r_2}}{\begin{bmatrix} 1 & 0 & 0 & \bigm| & 4 \\ 0 & 1 & 0 & \bigm| & -3 \\ 0 & 0 & 1 & \bigm| & 1 \end{bmatrix}}$$

The matrix is now written in reduced echelon form. The advantage of writing the matrix in this form is that the solution to the system, $x = 4$, $y = -3$, $z = 1$, is readily found, without the need to back-substitute. Another advantage will be seen in Section 12.1, where the inverse of a matrix is discussed.

■ Now work Problem 49.

The matrix method for solving a system of linear equations also identifies systems that have infinitely many solutions and systems that are inconsistent. Let's see how.

E X A M P L E 8 *Solving a System of Equations Using Matrices*

Solve: $\begin{cases} 6x - y - z = 4 \quad \text{\small (1)} \\ -12x + 2y + 2z = -8 \quad \text{\small (2)} \\ 5x + y - z = 3 \quad \text{\small (3)} \end{cases}$

Solution We start with the augmented matrix of the system:

$$\begin{bmatrix} 6 & -1 & -1 & \bigm| & 4 \\ -12 & 2 & 2 & \bigm| & -8 \\ 5 & 1 & -1 & \bigm| & 3 \end{bmatrix} \underset{\substack{\uparrow \\ R_1 = -1r_3 + r_1}}{\rightarrow} \begin{bmatrix} 1 & -2 & 0 & \bigm| & 1 \\ -12 & 2 & 2 & \bigm| & -8 \\ 5 & 1 & -1 & \bigm| & 3 \end{bmatrix} \underset{\substack{\uparrow \\ R_2 = 12r_1 + r_2. \\ R_3 = -5r_1 + r_3}}{\rightarrow} \begin{bmatrix} 1 & -2 & 0 & \bigm| & 1 \\ 0 & -22 & 2 & \bigm| & 4 \\ 0 & 11 & -1 & \bigm| & -2 \end{bmatrix}$$

Obtaining a 1 in row 2, column 2 without altering column 1 can be accomplished only by $R_2 = -\frac{1}{22}r_2$ or by $R_3 = \frac{1}{11}r_3$. (Do you see why?) We shall use the first of these:

$$\begin{bmatrix} 1 & -2 & 0 & | & 1 \\ 0 & -22 & 2 & | & 4 \\ 0 & 11 & -1 & | & -2 \end{bmatrix} \rightarrow \begin{bmatrix} 1 & -2 & 0 & | & 1 \\ 0 & 1 & -\frac{1}{11} & | & -\frac{2}{11} \\ 0 & 11 & -1 & | & -2 \end{bmatrix} \rightarrow \begin{bmatrix} 1 & -2 & 0 & | & 1 \\ 0 & 1 & -\frac{1}{11} & | & -\frac{2}{11} \\ 0 & 0 & 0 & | & 0 \end{bmatrix}$$
$$R_2 = -\frac{1}{22}r_2 \qquad\qquad R_3 = -11r_2 + r_3$$

This matrix is in echelon form. Because the bottom row consists entirely of 0's, the system actually consists of only two equations:

$$\begin{cases} x - 2y = 1 & (1) \\ y - \frac{1}{11}z = -\frac{2}{11} & (2) \end{cases}$$

We shall back-substitute the solution for y from the second equation, $y = \frac{1}{11}z - \frac{2}{11}$, into the first equation to get

$$x = 2y + 1 = 2(\tfrac{1}{11}z - \tfrac{2}{11}) + 1 = \tfrac{2}{11}z + \tfrac{7}{11}$$

Thus, the original system is equivalent to the system

$$\begin{cases} x = \frac{2}{11}z + \frac{7}{11} & (1) \\ y = \frac{1}{11}z - \frac{2}{11} & (2) \end{cases}$$

where z can be any real number.

Let's look at the situation. The original system of three equations is equivalent to a system containing two equations. This means any values of x, y, z that satisfy both

$$x = \tfrac{2}{11}z + \tfrac{7}{11} \quad \text{and} \quad y = \tfrac{1}{11}z - \tfrac{2}{11}$$

will be solutions. For example, $z = 0$, $x = \frac{7}{11}$, $y = -\frac{2}{11}$; $z = 1$, $x = \frac{9}{11}$, $y = -\frac{1}{11}$; and $z = -1$, $x = \frac{5}{11}$, $y = -\frac{3}{11}$ are some of the solutions of the original system. There are, in fact, infinitely many values of x, y, and z for which the two equations are satisfied. That is, the original system has infinitely many solutions. We will write the solution of the original system as

$$\begin{cases} x = \frac{2}{11}z + \frac{7}{11} \\ y = \frac{1}{11}z - \frac{2}{11} \end{cases}$$

where z can be any real number. ∎

We can also find the solution by writing the augmented matrix in reduced echelon form. Starting with the echelon form, we have

$$\begin{bmatrix} 1 & -2 & 0 & | & 1 \\ 0 & 1 & -\frac{1}{11} & | & -\frac{2}{11} \\ 0 & 0 & 0 & | & 0 \end{bmatrix} \rightarrow \begin{bmatrix} 1 & 0 & -\frac{2}{11} & | & \frac{7}{11} \\ 0 & 1 & -\frac{1}{11} & | & -\frac{2}{11} \\ 0 & 0 & 0 & | & 0 \end{bmatrix}$$
$$R_1 = 2r_2 + r_1$$

The matrix on the right is in reduced echelon form. The corresponding system of equations is

$$\begin{cases} x - \frac{2}{11}z = \frac{7}{11} & \text{(1)} \\ y - \frac{1}{11}z = -\frac{2}{11} & \text{(2)} \end{cases}$$

or, equivalently,

$$\begin{cases} x = \frac{2}{11}z + \frac{7}{11} & \text{(1)} \\ y = \frac{1}{11}z - \frac{2}{11} & \text{(2)} \end{cases}$$

where z can be any real number.

■ Now work Problem 53.

E X A M P L E 9 *Solving a System of Equations Using Matrices*

Solve: $\begin{cases} x + y + z = 6 \\ 2x - y - z = 3 \\ x + 2y + 2z = 0 \end{cases}$

Solution The augmented matrix is

$$\begin{bmatrix} 1 & 1 & 1 & | & 6 \\ 2 & -1 & -1 & | & 3 \\ 1 & 2 & 2 & | & 0 \end{bmatrix} \rightarrow \begin{bmatrix} 1 & 1 & 1 & | & 6 \\ 0 & -3 & -3 & | & -9 \\ 0 & 1 & 1 & | & -6 \end{bmatrix} \rightarrow \begin{bmatrix} 1 & 1 & 1 & | & 6 \\ 0 & 1 & 1 & | & -6 \\ 0 & -3 & -3 & | & -9 \end{bmatrix} \rightarrow \begin{bmatrix} 1 & 1 & 1 & | & 6 \\ 0 & 1 & 1 & | & -6 \\ 0 & 0 & 0 & | & -27 \end{bmatrix}$$

$\uparrow$
$R_2 = -2r_1 + r_2$
$R_3 = -1r_1 + r_3$

$\uparrow$
Interchange rows 2 and 3.

$\uparrow$
$R_3 = 3r_2 + r_3.$

This matrix is in echelon form. The bottom row is equivalent to the equation

$$0x + 0y + 0z = -27$$

which has no solution. Hence, the original system is inconsistent. ■

■ Now work Problems 23 and 29.

The matrix method is especially effective for systems of equations for which the number of equations and the number of variables are unequal. Here, too, such a system is either inconsistent or consistent. If it is consistent, it will have either exactly one solution or infinitely many solutions.

Let's look at a system of four equations containing three variables.

E X A M P L E 1 0 *Solving a System of Equations Using Matrices*

Solve: $\begin{cases} x - 2y + z = 0 & \text{(1)} \\ 2x + 2y - 3z = -3 & \text{(2)} \\ y - z = -1 & \text{(3)} \\ -x + 4y + 2z = 13 & \text{(4)} \end{cases}$

Solution The augmented matrix is

$$\begin{bmatrix} 1 & -2 & 1 & | & 0 \\ 2 & 2 & -3 & | & -3 \\ 0 & 1 & -1 & | & -1 \\ -1 & 4 & 2 & | & 13 \end{bmatrix} \rightarrow \begin{bmatrix} 1 & -2 & 1 & | & 0 \\ 0 & 6 & -5 & | & -3 \\ 0 & 1 & -1 & | & -1 \\ 0 & 2 & 3 & | & 13 \end{bmatrix} \rightarrow \begin{bmatrix} 1 & -2 & 1 & | & 0 \\ 0 & 1 & -1 & | & -1 \\ 0 & 6 & -5 & | & -3 \\ 0 & 2 & 3 & | & 13 \end{bmatrix}$$

$\uparrow$
$R_2 = -2r_1 + r_2$
$R_4 = r_1 + r_4$

$\uparrow$
Interchange rows 2 and 3.

$$\rightarrow \begin{bmatrix} 1 & -2 & 1 & | & 0 \\ 0 & 1 & -1 & | & -1 \\ 0 & 0 & 1 & | & 3 \\ 0 & 0 & 5 & | & 15 \end{bmatrix} \rightarrow \begin{bmatrix} 1 & -2 & 1 & | & 0 \\ 0 & 1 & -1 & | & -1 \\ 0 & 0 & 1 & | & 3 \\ 0 & 0 & 0 & | & 0 \end{bmatrix}$$

$$R_3 = -6r_2 + r_3 \qquad\qquad R_4 = -5r_3 + r_4$$
$$R_4 = -2r_2 + r_4$$

We could stop here, since the matrix is in echelon form, and back-substitute $z = 3$ to find x and y. Or we can continue to obtain the reduced echelon form:

$$\rightarrow \begin{bmatrix} 1 & 0 & -1 & | & -2 \\ 0 & 1 & -1 & | & -1 \\ 0 & 0 & 1 & | & 3 \\ 0 & 0 & 0 & | & 0 \end{bmatrix} \rightarrow \begin{bmatrix} 1 & 0 & 0 & | & 1 \\ 0 & 1 & 0 & | & 2 \\ 0 & 0 & 1 & | & 3 \\ 0 & 0 & 0 & | & 0 \end{bmatrix}$$

$$R_1 = 2r_2 + r_1 \qquad\qquad R_1 = r_3 + r_1$$
$$R_2 = r_3 + r_2$$

The matrix is now in reduced echelon form, and we can see that the solution is $x = 1$, $y = 2$, $z = 3$. ■

E X A M P L E   1 1   *Mixing Acids*

A chemistry laboratory has three containers of nitric acid, HNO_3. One container holds a solution with a concentration of 10% HNO_3, the second holds 20% HNO_3, and the third holds 40% HNO_3. How many liters of each solution should be mixed to obtain 100 liters of a solution whose concentration is 25% HNO_3?

Solution   Let x, y, and z represent the number of liters of 10%, 20%, and 40% concentrations of HNO_3, respectively. We want 100 liters in all, and the concentration of HNO_3 from each solution must sum to 25% of 100 liters. Thus, we find that

$$\begin{cases} x + y + z = 100 \\ 0.10x + 0.20y + 0.40z = 0.25(100) \end{cases}$$

Now, the augmented matrix is

$$\begin{bmatrix} 1 & 1 & 1 & | & 100 \\ 0.10 & 0.20 & 0.40 & | & 25 \end{bmatrix} \rightarrow \begin{bmatrix} 1 & 1 & 1 & | & 100 \\ 0 & 0.10 & 0.30 & | & 15 \end{bmatrix}$$

$$R_2 = -0.10r_1 + r_2$$

$$\rightarrow \begin{bmatrix} 1 & 1 & 1 & | & 100 \\ 0 & 1 & 3 & | & 150 \end{bmatrix} \rightarrow \begin{bmatrix} 1 & 0 & -2 & | & -50 \\ 0 & 1 & 3 & | & 150 \end{bmatrix}$$

$$R_2 = 10r_2 \qquad\qquad R_1 = -1r_2 + r_1$$

The matrix is now in reduced echelon form. The final matrix represents the system

$$\begin{cases} x - 2z = -50 & (1) \\ y + 3z = 150 & (2) \end{cases}$$

which has infinitely many solutions given by

$$\begin{cases} x = 2z - 50 & (1) \\ y = -3z + 150 & (2) \end{cases}$$

where z is any real number. However, the practical considerations of this problem require us to restrict the solutions to $x \geq 0$, $y \geq 0$, $z \geq 0$. Furthermore, we require $25 \leq z \leq 50$, because otherwise $x < 0$ or $y < 0$. Some of the possible solutions are given in Table 1. The final determination of what solution the laboratory will pick very likely depends on availability, cost differences, and other considerations.

TABLE 1

LITERS OF 10% SOLUTION	LITERS OF 20% SOLUTION	LITERS OF 40% SOLUTION	LITERS OF 25% SOLUTION
0	75	25	100
10	60	30	100
12	57	31	100
16	51	33	100
26	36	38	100
38	18	44	100
46	6	48	100
50	0	50	100

10.2

Exercise 10.2

In Problems 1–10, write the augmented matrix of the given system of equations.

1. $\begin{cases} x - 5y = 5 \\ 4x + 3y = 6 \end{cases}$

2. $\begin{cases} 3x + 4y = 7 \\ 4x - 2y = 5 \end{cases}$

3. $\begin{cases} 2x + 3y - 6 = 0 \\ 4x - 6y + 2 = 0 \end{cases}$

4. $\begin{cases} 9x - y = 0 \\ 3x - y - 4 = 0 \end{cases}$

5. $\begin{cases} 0.01x - 0.03y = 0.06 \\ 0.13x + 0.10y = 0.20 \end{cases}$

6. $\begin{cases} \frac{4}{3}x - \frac{3}{2}y = \frac{3}{4} \\ -\frac{1}{4}x + \frac{1}{3}y = \frac{2}{3} \end{cases}$

7. $\begin{cases} x - y + z = 10 \\ 3x + 2y = 5 \\ x + y + 2z = 2 \end{cases}$

8. $\begin{cases} 5x - y - z = 0 \\ x + y = 5 \\ 2x - 3z = 2 \end{cases}$

9. $\begin{cases} x + y - z = 2 \\ 3x - 2y = 2 \end{cases}$

10. $\begin{cases} 2x + 3y - 4z = 0 \\ x - 5z + 2 = 0 \end{cases}$

In Problems 11–20 perform, in order (a), followed by (b), followed by (c) on the given augmented matrix.

11. $\begin{bmatrix} 1 & -3 & -5 & | & -2 \\ 2 & -5 & -4 & | & 5 \\ -3 & 5 & 4 & | & 6 \end{bmatrix}$
   (a) $R_2 = -2r_1 + r_2$
   (b) $R_3 = 3r_1 + r_3$
   (c) $R_3 = 4r_2 + r_3$

12. $\begin{bmatrix} 1 & -3 & -3 & | & -3 \\ 2 & -5 & 2 & | & -4 \\ -3 & 2 & 4 & | & 6 \end{bmatrix}$
   (a) $R_2 = -2r_1 + r_2$
   (b) $R_3 = 3r_1 + r_3$
   (c) $R_3 = 7r_2 + r_3$

13. $\begin{bmatrix} 1 & -3 & 4 & | & 3 \\ 2 & -5 & 6 & | & 6 \\ -3 & 3 & 4 & | & 6 \end{bmatrix}$
   (a) $R_2 = -2r_1 + r_2$
   (b) $R_3 = 3r_1 + r_3$
   (c) $R_3 = 6r_2 + r_3$

14. $\begin{bmatrix} 1 & -3 & 3 & | & -5 \\ 2 & -5 & -3 & | & -5 \\ -3 & -2 & 4 & | & 6 \end{bmatrix}$
   (a) $R_2 = -2r_1 + r_2$
   (b) $R_3 = 3r_1 + r_3$
   (c) $R_3 = 11r_2 + r_3$

15. $\begin{bmatrix} 1 & -3 & 2 & | & -6 \\ 2 & -5 & 3 & | & -4 \\ -3 & -6 & 4 & | & 6 \end{bmatrix}$ (a) $R_2 = -2r_1 + r_2$
      (b) $R_3 = 3r_1 + r_3$
      (c) $R_3 = 15r_2 + r_3$

16. $\begin{bmatrix} 1 & -3 & -4 & | & -6 \\ 2 & -5 & 6 & | & -6 \\ -3 & 1 & 4 & | & 6 \end{bmatrix}$ (a) $R_2 = -2r_1 + r_2$
      (b) $R_3 = 3r_1 + r_3$
      (c) $R_3 = 8r_2 + r_3$

17. $\begin{bmatrix} 1 & -3 & 1 & | & -2 \\ 2 & -5 & 6 & | & -2 \\ -3 & 1 & 4 & | & 6 \end{bmatrix}$ (a) $R_2 = -2r_1 + r_2$
      (b) $R_3 = 3r_1 + r_3$
      (c) $R_3 = 8r_2 + r_3$

18. $\begin{bmatrix} 1 & -3 & -1 & | & 2 \\ 2 & -5 & 2 & | & 6 \\ -3 & -6 & 4 & | & 6 \end{bmatrix}$ (a) $R_2 = -2r_1 + r_2$
      (b) $R_3 = 3r_1 + r_3$
      (c) $R_3 = 15r_2 + r_3$

19. $\begin{bmatrix} 1 & -3 & -2 & | & 3 \\ 2 & -5 & 2 & | & -1 \\ -3 & -2 & 4 & | & 6 \end{bmatrix}$ (a) $R_2 = -2r_1 + r_2$
      (b) $R_3 = 3r_1 + r_3$
      (c) $R_3 = 11r_2 + r_3$

20. $\begin{bmatrix} 1 & -3 & 5 & | & -3 \\ 2 & -5 & 1 & | & -4 \\ -3 & 3 & 4 & | & 6 \end{bmatrix}$ (a) $R_2 = -2r_1 + r_2$
      (b) $R_3 = 3r_1 + r_3$
      (c) $R_3 = 6r_2 + r_3$

In Problems 21–30, the reduced echelon form of a system of linear equations is given. Write the system of equations corresponding to the given matrix. Use x, y, or x, y, z or x_1, x_2, x_3, x_4 as variables. Determine whether the system is consistent or inconsistent. If it is consistent, give the solution.

21. $\begin{bmatrix} 1 & 0 & | & 5 \\ 0 & 1 & | & -1 \end{bmatrix}$

22. $\begin{bmatrix} 1 & 0 & | & -4 \\ 0 & 1 & | & 0 \end{bmatrix}$

23. $\begin{bmatrix} 1 & 0 & 0 & | & 1 \\ 0 & 1 & 0 & | & 2 \\ 0 & 0 & 0 & | & 3 \end{bmatrix}$

24. $\begin{bmatrix} 1 & 0 & 0 & | & 0 \\ 0 & 1 & 0 & | & 0 \\ 0 & 0 & 0 & | & 2 \end{bmatrix}$

25. $\begin{bmatrix} 1 & 0 & 2 & | & -1 \\ 0 & 1 & -4 & | & -2 \\ 0 & 0 & 0 & | & 0 \end{bmatrix}$

26. $\begin{bmatrix} 1 & 0 & 4 & | & 4 \\ 0 & 1 & 3 & | & 2 \\ 0 & 0 & 0 & | & 0 \end{bmatrix}$

27. $\begin{bmatrix} 1 & 0 & 0 & 0 & | & 1 \\ 0 & 1 & 0 & 1 & | & 2 \\ 0 & 0 & 1 & 2 & | & 3 \end{bmatrix}$ $(1, 2-t, 3-2t, t)$

28. $\begin{bmatrix} 1 & 0 & 0 & 0 & | & 1 \\ 0 & 1 & 0 & 2 & | & 2 \\ 0 & 0 & 1 & 3 & | & 0 \end{bmatrix}$

29. $\begin{bmatrix} 1 & 0 & 0 & 4 & | & 2 \\ 0 & 1 & 1 & 3 & | & 3 \\ 0 & 0 & 0 & 0 & | & 0 \end{bmatrix}$

30. $\begin{bmatrix} 1 & 0 & 0 & 0 & | & 1 \\ 0 & 1 & 0 & 0 & | & 2 \\ 0 & 0 & 1 & 2 & | & 3 \end{bmatrix}$

In Problems 31–72, solve each system of equations using matrices (row operations). If the system has no solution, say it is inconsistent.

31. $\begin{cases} x + y = 8 \\ x - y = 4 \end{cases}$

32. $\begin{cases} x + 2y = 5 \\ x + y = 3 \end{cases}$

33. $\begin{cases} x - 5y = -13 \\ 3x + 2y = 12 \end{cases}$

34. $\begin{cases} x + 3y = 5 \\ 2x - 3y = -8 \end{cases}$

35. $\begin{cases} 3x - 6y = 24 \\ 5x + 4y = 12 \end{cases}$

36. $\begin{cases} 2x + 4y = 16 \\ 3x - 5y = -9 \end{cases}$

37. $\begin{cases} 2x + y = 1 \\ 4x + 2y = 6 \end{cases}$

38. $\begin{cases} x - y = 5 \\ -3x + 3y = 2 \end{cases}$

39. $\begin{cases} 2x - 4y = -2 \\ 3x + 2y = 3 \end{cases}$

40. $\begin{cases} 3x + 3y = 3 \\ 4x + 2y = \frac{8}{3} \end{cases}$

41. $\begin{cases} x + 2y = 4 \\ 2x + 4y = 8 \end{cases}$

42. $\begin{cases} 3x - y = 7 \\ 9x - 3y = 21 \end{cases}$

43. $\begin{cases} 2x + 3y = 6 \\ x - y = \frac{1}{2} \end{cases}$

44. $\begin{cases} \frac{1}{2}x + y = -2 \\ x - 2y = 8 \end{cases}$

45. $\begin{cases} 3x - 5y = 3 \\ 15x + 5y = 21 \end{cases}$

46. $\begin{cases} 2x - y = -1 \\ x + \frac{1}{2}y = \frac{3}{2} \end{cases}$

47. $\begin{cases} x - y = 6 \\ 2x - 3z = 16 \\ 2y + z = 4 \end{cases}$

48. $\begin{cases} 2x + y = -4 \\ -2y + 4z = 0 \\ 3x - 2z = -11 \end{cases}$

49. $\begin{cases} x - 2y + 3z = 7 \\ 2x + y + z = 4 \\ -3x + 2y - 2z = -10 \end{cases}$

50. $\begin{cases} 2x + y - 3z = 0 \\ -2x + 2y + z = -7 \\ 3x - 4y - 3z = 7 \end{cases}$

51. $\begin{cases} 2x - 2y - 2z = 2 \\ 2x + 3y + z = 2 \\ 3x + 2y = 0 \end{cases}$

52. $\begin{cases} 2x - 3y - z = 0 \\ -x + 2y + z = 5 \\ 3x - 4y - z = 1 \end{cases}$

53. $\begin{cases} -x + y + z = -1 \\ -x + 2y - 3z = -4 \\ 3x - 2y - 7z = 0 \end{cases}$

54. $\begin{cases} 2x - 3y - z = 0 \\ 3x + 2y + 2z = 2 \\ x + 5y + 3z = 2 \end{cases}$

55. $\begin{cases} 2x - 2y + 3z = 6 \\ 4x - 3y + 2z = 0 \\ -2x + 3y - 7z = 1 \end{cases}$

56. $\begin{cases} 3x - 2y + 2z = 6 \\ 7x - 3y + 2z = -1 \\ 2x - 3y + 4z = 0 \end{cases}$

57. $\begin{cases} x + y - z = 6 \\ 3x - 2y + z = -5 \\ x + 3y - 2z = 14 \end{cases}$

58. $\begin{cases} x - y + z = -4 \\ 2x - 3y + 4z = -15 \\ 5x + y - 2z = 12 \end{cases}$

59. $\begin{cases} x + 2y - z = -3 \\ 2x - 4y + z = -7 \\ -2x + 2y - 3z = 4 \end{cases}$

60. $\begin{cases} x + 4y - 3z = -8 \\ 3x - y + 3z = 12 \\ x + y + 6z = 1 \end{cases}$

61. $\begin{cases} 3x + y - z = \frac{2}{3} \\ 2x - y + z = 1 \\ 4x + 2y = \frac{8}{3} \end{cases}$

62. $\begin{cases} x + y = 1 \\ 2x - y + z = 1 \\ x + 2y + z = \frac{8}{3} \end{cases}$

63. $\begin{cases} x + y + z + w = 4 \\ 2x - y + z = 0 \\ 3x + 2y + z - w = 6 \\ x - 2y - 2z + 2w = -1 \end{cases}$

64. $\begin{cases} x + y + z + w = 4 \\ -x + 2y + z = 0 \\ 2x + 3y + z - w = 6 \\ -2x + y - 2z + 2w = -1 \end{cases}$

65. $\begin{cases} x + 2y + z = 1 \\ 2x - y + 2z = 2 \\ 3x + y + 3z = 3 \end{cases}$

66. $\begin{cases} x + 2y - z = 3 \\ 2x - y + 2z = 6 \\ x - 3y + 3z = 4 \end{cases}$

67. $\begin{cases} x - y + z = 5 \\ 3x + 2y - 2z = 0 \end{cases}$

68. $\begin{cases} 2x + y - z = 4 \\ -x + y + 3z = 1 \end{cases}$

69. $\begin{cases} 2x + 3y - z = 3 \\ x - y - z = 0 \\ -x + y + z = 0 \\ x + y + 3z = 5 \end{cases}$

70. $\begin{cases} x - 3y + z = 1 \\ 2x - y - 4z = 0 \\ x - 3y + 2z = 1 \\ x - 2y = 5 \end{cases}$

71. $\begin{cases} 4x + y + z - w = 4 \\ x - y + 2z + 3w = 3 \end{cases}$

72. $\begin{cases} -4x + y = 5 \\ 2x - y + z - w = 5 \\ z + w = 4 \end{cases}$

73. *Curve Fitting* Find the parabola $y = ax^2 + bx + c$ that passes through the points $(1, 2)$, $(-2, -7)$, and $(2, -3)$.

74. *Curve Fitting* Find the parabola $y = ax^2 + bx + c$ that passes through the points $(1, -1)$, $(3, -1)$, and $(-2, 14)$.

75. *Curve Fitting* Find the function $f(x) = ax^3 + bx^2 + cx + d$ for which $f(-3) = -112, f(-1) = -2, f(1) = 4$, and $f(2) = 13$.

76. *Curve Fitting* Find the function $f(x) = ax^3 + bx^2 + cx + d$ for which $f(-2) = -10, f(-1) = 3, f(1) = 5$, and $f(3) = 15$.

77. *Mixing Acids* A chemistry laboratory has three containers of sulfuric acid, H_2SO_4. One container holds a solution with a concentration of 15% H_2SO_4, the second holds 25% H_2SO_4, and the third holds 50% H_2SO_4. How many liters of each solution should be mixed to obtain 100 liters of a solution with a concentration of 40% H_2SO_4? Construct a table similar to Table 1 illustrating some of the possible combinations.

78. *Painting a House* Three painters, Mike, Dan, and Katy, working together can paint the exterior of a home in 10 hours. Dan and Katy together have painted a similar house in 15 hours. One day, all three worked on this same kind of house for 4 hours, after which Katy left. Mike and Dan required 8 more hours to finish. Assuming no gain or loss in efficiency, how long should it take each person to complete such a job alone?

79. *Prices of Fast Food* One group of customers bought 8 deluxe hamburgers, 6 orders of large fries, and 6 large colas for $26.10. A second group ordered 10 deluxe hamburgers, 6 large fries, and 8 large colas and paid $31.60. Is there sufficient information to determine the price of each food item? If not, construct a table showing the various possibilities. Assume that the hamburgers cost between $1.75 and $2.25, the fries between $0.75 and $1.00, and the colas between $0.60 and $0.90.

80. *Prices of Fast Food* Use the information given in Problem 79, and suppose that a third group purchased 3 deluxe hamburgers, 2 large fries, and 4 large colas for $10.95. Now is there sufficient information to determine the price of each food item?

81. *Financial Planning* Three retired couples each require an additional annual income of $2000 per year. As their financial consultant, you recommend that they invest some money in Treasury bills that yield 7%, some money in corporate bonds that yield 9%, and some money in junk bonds that yield 11%. Prepare a table for each couple showing the various ways their goals can be achieved:

(a) If the first couple has $20,000 to invest.

(b) If the second couple has $25,000 to invest.

(c) If the third couple has $30,000 to invest.

(d) What advice would you give each couple regarding the amount to invest and the choices available? [Higher yields generally carry more risk].

82. *Financial Planning* A retired couple has $25,000 to invest. As their financial consultant, you recommend that they invest some money in Treasury bills that yield 7%, some money in corporate bonds that yield 9%, and some money in junk bonds that yield 11%. Prepare a table showing the various ways this couple can achieve the following goals:

(a) The couple wants $1500 per year in income.

(b) The couple wants $2000 per year in income.

(c) The couple wants $2500 per year in income.

(d) What advice would you give this couple regarding the income they require and the choices available? [Higher yields generally carry more risk].

83. *Electricity: Kirchhoff's Rules* An application of Kirchhoff's Rules to the circuit shown results in the following system of equations:

$$\begin{cases} I_1 + I_2 = I_3 \\ 16 - 8 - 9I_3 - 3I_1 = 0 \\ 16 - 4 - 9I_3 - 9I_2 = 0 \\ 8 - 4 - 9I_2 + 3I_1 = 0 \end{cases}$$

Find the currents I_1, I_2, and I_3.*

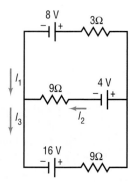

---

*Source: Based on Raymond Serway, *Physics,* 3rd ed. Philadelphia: Saunders, 1990, Prob. 31, p. 790.

84. *Electricity: Kirchhoff's Rules* An application of Kirchhoff's Rules to the circuit shown results in the following system of equations:

$$\begin{cases} -4 + 8 - 2I_2 = 0 \\ 8 = 5I_4 + I_1 \\ 4 = 3I_3 + I_1 \\ I_3 + I_4 = I_1 \end{cases}$$

Find the currents I_1, I_2, I_3, and I_4.*

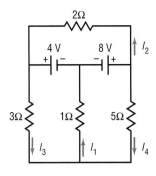

85. *Electricity: Kirchhoff's Rules* An application of Kirchhoff's Rules to the circuit shown results in the following system of equations:

$$\begin{cases} I_1 = I_3 + I_2 \\ 24 - 6I_1 - 3I_3 = 0 \\ 12 + 24 - 6I_1 - 6I_2 = 0 \end{cases}$$

Find the currents I_1, I_2, and I_3.†

86. Write a brief paragraph or two that outlines your strategy for solving a system of linear equations using matrices.

87. When solving a system of linear equations using matrices, do you prefer to place the augmented matrix in echelon form or in reduced echelon form? Give reasons for your choice.

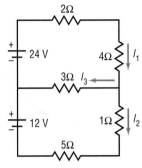

88. Make up three systems of three linear equations containing three variables that have:

(a) No solution (b) Exactly one solution (c) Infinitely many solutions

Give them to a friend to solve and critique.

89. Consider the system of equations

$$\begin{cases} a_1x + b_1y = c_1 \\ a_2x + b_2y = c_2 \end{cases}$$

If $D = a_1b_2 - a_2b_1 \neq 0$, use matrices to show that the solution is

$$x = \frac{1}{D}(c_1b_2 - c_2b_1), \qquad y = \frac{1}{D}(a_1c_2 - a_2c_1)$$

90. For the system in Problem 89, suppose that $D = a_1b_2 - a_2b_1 = 0$. Use matrices to show that the system is inconsistent if either $a_1c_2 \neq a_2c_1$ or $b_1c_2 \neq b_2c_1$ and has infinitely many solutions if both $a_1c_2 = a_2c_1$ and $b_1c_2 = b_2c_1$.

91. The graph of a linear equation containing three variables is a plane. Give a geometrical argument for what can result when solving a system of two linear equations containing three variables. [*Hint:* Two planes in a three-dimensional space are either coincident (the same), parallel, or intersect in a line.]

92. Refer to Problem 91. Give a geometrical argument for what can result when solving a system of three linear equations containing three variables.

93. Refer to Problem 91. Give a geometrical argument for what can result when solving a system of four linear equations containing three variables.

*Source: Ibid., Prob. 34, p. 791.
†Source: Ibid., Prob. 38, p. 791.

10.3

Systems of Linear Equations: Determinants

In the preceding section, we described a method of using matrices to solve any system of linear equations. This section deals with yet another method for solving systems of linear equations; however, it can be used only when the number of equations equals the number of variables. Although the method will work for any system (provided the number of equations equals the number of variables), it is most often used for systems of two equations containing two variables or three equations containing three variables. This method, called *Cramer's Rule*, is based on the concept of a *determinant*.

2 by 2 Determinants

2 by 2 Determinant

If a, b, c, and d are four real numbers, the symbol

$$D = \begin{vmatrix} a & b \\ c & d \end{vmatrix}$$

is called a **2 by 2 determinant.** Its value is the number $ad - bc$; that is,

$$D = \begin{vmatrix} a & b \\ c & d \end{vmatrix} = ad - bc \tag{1}$$

A device that may be helpful for remembering the value of a 2 by 2 determinant is the following:

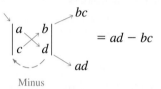

$$= ad - bc$$

Minus

E X A M P L E 1

Evaluating a 2 × 2 Determinant

$$\begin{vmatrix} 3 & -2 \\ 6 & 1 \end{vmatrix} = (3)(1) - (6)(-2) = 3 - (-12) = 15 \qquad \blacksquare$$

■ Now work Problem 3.

Let's now see the role that a 2 by 2 determinant plays in the solution of a system of two equations containing two variables. Consider the system

$$\begin{cases} ax + by = s & (1) \\ cx + dy = t & (2) \end{cases} \tag{2}$$

We shall use the method of elimination to solve this system.

Provided $d \neq 0$ and $b \neq 0$, this system is equivalent to the system

$$\begin{cases} adx + bdy = sd & (1) \quad \text{Multiply by } d. \\ bcx + bdy = tb & (2) \quad \text{Multiply by } b. \end{cases}$$

On subtracting the second equation from the first equation, we get

$$\begin{cases} (ad - bc)x + 0 \cdot y = sd - tb & (1) \\ bcx \quad + \; bdy = tb & (2) \end{cases}$$

Now, the first equation can be rewritten using determinant notation:

$$\begin{vmatrix} a & b \\ c & d \end{vmatrix} x = \begin{vmatrix} s & b \\ t & d \end{vmatrix}$$

If $D = \begin{vmatrix} a & b \\ c & d \end{vmatrix} = ad - bc \neq 0$, we can solve for x to get

$$x = \frac{\begin{vmatrix} s & b \\ t & d \end{vmatrix}}{\begin{vmatrix} a & b \\ c & d \end{vmatrix}} = \frac{\begin{vmatrix} s & b \\ t & d \end{vmatrix}}{D} \qquad (3)$$

Return now to the original system (2). Provided $a \neq 0$ and $c \neq 0$, the system is equivalent to

$$\begin{cases} acx + bcy = cs & (1) \quad \text{Multiply by } c. \\ acx + ady = at & (2) \quad \text{Multiply by } a. \end{cases}$$

On subtracting the first equation from the second equation, we get

$$\begin{cases} acx + bcy = cs & (1) \\ 0 \cdot x + (ad - bc)y = at - cs & (2) \end{cases}$$

The second equation now can be rewritten using determinant notation:

$$\begin{vmatrix} a & b \\ c & d \end{vmatrix} y = \begin{vmatrix} a & s \\ c & t \end{vmatrix}$$

If $D = \begin{vmatrix} a & b \\ c & d \end{vmatrix} = ad - bc \neq 0$, we can solve for y to get

$$y = \frac{\begin{vmatrix} a & s \\ c & t \end{vmatrix}}{\begin{vmatrix} a & b \\ c & d \end{vmatrix}} = \frac{\begin{vmatrix} a & s \\ c & t \end{vmatrix}}{D} \qquad (4)$$

Equations (3) and (4) lead us to the following result, called **Cramer's Rule:**

Theorem
Cramer's Rule for Two Equations
Containing Two Variables

The solution to the system of equations

$$\begin{cases} ax + by = s & (1) \\ cx + dy = t & (2) \end{cases} \qquad (5)$$

is given by

$$x = \frac{\begin{vmatrix} s & b \\ t & d \end{vmatrix}}{\begin{vmatrix} a & b \\ c & d \end{vmatrix}}, \qquad y = \frac{\begin{vmatrix} a & s \\ c & t \end{vmatrix}}{\begin{vmatrix} a & b \\ c & d \end{vmatrix}} \qquad (6)$$

provided that

$$D = \begin{vmatrix} a & b \\ c & d \end{vmatrix} = ad - bc \neq 0 \qquad \blacksquare$$

In the derivation given for Cramer's Rule above, we assumed that none of the numbers a, b, c, and d were 0. In Problem 58 at the end of this section you will be asked to complete the proof under the less stringent conditions that $D = ad - bc \neq 0$.

Now look carefully at the pattern in Cramer's Rule. The denominator in the solution (6) is the determinant of the coefficients of the variables:

$$\begin{cases} ax + by = s \\ cx + dy = t \end{cases} \qquad D = \begin{vmatrix} a & b \\ c & d \end{vmatrix}$$

In the solution for x, the numerator is the determinant, denoted by D_x, formed by replacing the entries in the first column (the coefficients of x) in D by the constants on the right side of the equal sign:

$$D_x = \begin{vmatrix} s & b \\ t & d \end{vmatrix}$$

In the solution for y, the numerator is the determinant, denoted by D_y, formed by replacing the entries in the second column (the coefficients of y) in D by the constants on the right side of the equal sign:

$$D_y = \begin{vmatrix} a & s \\ c & t \end{vmatrix}$$

Cramer's Rule then states that, if $D \neq 0$,

$$x = \frac{D_x}{D}, \qquad y = \frac{D_y}{D} \qquad\qquad (7)$$

E X A M P L E   2

Solving a System of Equations Using Determinants

Use Cramer's Rule, if applicable, to solve the system

$$\begin{cases} 3x - 2y = 4 & (1) \\ 6x + y = 13 & (2) \end{cases}$$

Solution The determinant D of the coefficients of the variables is

$$D = \begin{vmatrix} 3 & -2 \\ 6 & 1 \end{vmatrix} = (3)(1) - (6)(-2) = 15$$

Because $D \neq 0$, Cramer's Rule (7) can be used:

$$x = \frac{D_x}{D} = \frac{\begin{vmatrix} 4 & -2 \\ 13 & 1 \end{vmatrix}}{15} = \frac{30}{15} = 2, \qquad y = \frac{D_y}{D} = \frac{\begin{vmatrix} 3 & 4 \\ 6 & 13 \end{vmatrix}}{15} = \frac{15}{15} = 1$$

The solution is $x = 2$, $y = 1$. ∎

If, in attempting to use Cramer's Rule, the determinant of D of the coefficients of the variables is found to equal 0 (so that Cramer's Rule is not applicable), then the system either is inconsistent or has infinitely many solutions. (Refer to Problem 90 in Exercise 10.2.)

■ Now work Problem 11.

$\mathcal{M}$ISSION POSSIBLE

Chapter 10

$\mathcal{C}$ATTLE RANCHING

Your team owns a company in Wyoming that sells fencing material to ranchers. One day a rancher comes in to buy fencing with which to make two square cattle pens of equal size. She'd like to have a total of 4500 square feet fenced in altogether. She thinks if the two squares are next to each other, the fence between them could work as one side of each pen, which would allow her to use less fence. She's hoping you can help her work out what length the sides should be. She's got enough in her budget for 300 running foot of fence.

Naturally, you want to help her get the most efficient use of the fencing as possible. Even though that means she'll buy less fencing this time, it will pay off in the long run because she will become a loyal customer.

1. Make a sketch of the cattle pens. Label the lengths of the sides with variables.
2. Create equations relating the lengths of the sides of the two pens to the area and perimeter the rancher has in mind.
3. Solve this system and show the rancher the lengths of the pens she'll get with the measurements she's given you. Will 300 feet be enough fencing?
4. If the rancher's budget is really limited to 300 feet of fencing and she must have two equal-sized pens in the shape she described, then what will be the size of each pen in square footage?
5. Consider other shapes that could be made that would give the rancher at least 4500 square feet with 300 feet of fence. Don't limit yourself to the configuration the rancher mentioned. For example, what if the pens were not equal-sized; what would the dimensions be then?
6. What if she doesn't need to keep the cattle separated by age, sex, or breed? If she makes one big pen with her 300 feet of fence, she could include more square feet. Show the rancher what the other possibilities would be if she used, for example, a pen which was triangular, or rectangular, or square, or even circular. Next to each shape of pen, put the number of square feet it would enclose. Is there a best shape in terms of including the most square footage for the least fencing? What is it?

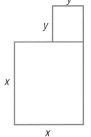

3 by 3 Determinants

In order to use Cramer's Rule to solve a system of three equations containing three variables, we need to define a 3 by 3 determinant.

A **3 by 3 determinant** is symbolized by

$$\begin{vmatrix} a_{11} & a_{12} & a_{13} \\ a_{21} & a_{22} & a_{23} \\ a_{31} & a_{32} & a_{33} \end{vmatrix} \tag{8}$$

in which $a_{11}, a_{12}, \ldots$ are real numbers.

As with matrices, we use a double subscript to identify an entry by indicating its row and column numbers. For example, the entry a_{23} is in row 2, column 3.

The value of a 3 by 3 determinant may be defined in terms of 2 by 2 determinants by the following formula:

$$\begin{vmatrix} a_{11} & a_{12} & a_{13} \\ a_{21} & a_{22} & a_{23} \\ a_{31} & a_{32} & a_{33} \end{vmatrix} = a_{11}\begin{vmatrix} a_{22} & a_{23} \\ a_{32} & a_{33} \end{vmatrix} \overset{\text{Minus}}{\underset{}{-}} a_{12}\begin{vmatrix} a_{21} & a_{23} \\ a_{31} & a_{33} \end{vmatrix} + a_{13}\begin{vmatrix} a_{21} & a_{22} \\ a_{31} & a_{32} \end{vmatrix} \tag{9}$$

|2 by 2 determinant left after removing row and column containing a_{11}|2 by 2 determinant left after removing row and column containing a_{12}|2 by 2 determinant left after removing row and column containing a_{13}|

Be sure to take note of the minus sign that appears with the second term—it's easy to forget it! Formula (9) is best remembered by noting that each entry in row 1 is multiplied by the 2 by 2 determinant that remains after the row and column containing the entry have been removed, as follows:

$$\begin{vmatrix} a_{11} & a_{12} & a_{13} \\ a_{21} & a_{22} & a_{23} \\ a_{31} & a_{32} & a_{33} \end{vmatrix} \quad a_{11}\begin{vmatrix} a_{22} & a_{23} \\ a_{32} & a_{33} \end{vmatrix} \quad \text{First term}$$

Write these entries as a 2 by 2 determinant.

$$\begin{vmatrix} a_{11} & a_{12} & a_{13} \\ a_{21} & a_{22} & a_{23} \\ a_{31} & a_{32} & a_{33} \end{vmatrix} \quad a_{12}\begin{vmatrix} a_{21} & a_{23} \\ a_{31} & a_{33} \end{vmatrix} \quad \text{Second term}$$

$$\begin{vmatrix} a_{11} & a_{12} & a_{13} \\ a_{21} & a_{22} & a_{23} \\ a_{31} & a_{32} & a_{33} \end{vmatrix} \quad a_{13}\begin{vmatrix} a_{21} & a_{22} \\ a_{31} & a_{32} \end{vmatrix} \quad \text{Third term}$$

Now insert the minus sign before the middle expression and add:

$$\begin{vmatrix} a_{11} & a_{12} & a_{13} \\ a_{21} & a_{22} & a_{23} \\ a_{31} & a_{32} & a_{33} \end{vmatrix} = a_{11}\begin{vmatrix} a_{22} & a_{23} \\ a_{32} & a_{33} \end{vmatrix} \overset{\text{Minus}}{\underset{}{-}} a_{12}\begin{vmatrix} a_{21} & a_{23} \\ a_{31} & a_{33} \end{vmatrix} + a_{13}\begin{vmatrix} a_{21} & a_{22} \\ a_{31} & a_{32} \end{vmatrix}$$

Formula (9) exhibits one way to find the value of a 3 by 3 determinant, *by expanding across row 1.* In fact, the expansion can take place across any row or down any column. The terms to be added or subtracted consist of the row (or column) entry times the value of the 2 by 2 determinant that remains after removing the row and column entry. The value of the determinant is found by adding or subtracting the terms according to the following scheme:

$$
\begin{array}{ccc}
+ & - & + \\
- & + & - \\
+ & - & +
\end{array}
$$

For example, if we choose to expand down column 2, we obtain

$$
\begin{vmatrix} a_{11} & a_{12} & a_{13} \\ a_{21} & a_{22} & a_{23} \\ a_{31} & a_{32} & a_{33} \end{vmatrix} = -a_{12} \begin{vmatrix} a_{21} & a_{23} \\ a_{31} & a_{33} \end{vmatrix} + a_{22} \begin{vmatrix} a_{11} & a_{13} \\ a_{31} & a_{33} \end{vmatrix} - a_{32} \begin{vmatrix} a_{11} & a_{13} \\ a_{21} & a_{23} \end{vmatrix}
$$

Expand down column 2 (−, +, −)

If we choose to expand across row 3, we obtain

$$
\begin{vmatrix} a_{11} & a_{12} & a_{13} \\ a_{21} & a_{22} & a_{23} \\ a_{31} & a_{32} & a_{33} \end{vmatrix} = a_{31} \begin{vmatrix} a_{12} & a_{13} \\ a_{22} & a_{23} \end{vmatrix} - a_{32} \begin{vmatrix} a_{11} & a_{13} \\ a_{21} & a_{23} \end{vmatrix} + a_{33} \begin{vmatrix} a_{11} & a_{12} \\ a_{21} & a_{22} \end{vmatrix}
$$

Expand across row 3 (+, −, +)

It can be shown that the value of a determinant does not depend on the choice of the row or column used in the expansion.

E X A M P L E 3

Evaluating a 3 × 3 Determinant

Find the value of the 3 by 3 determinant: $\begin{vmatrix} 3 & 4 & -1 \\ 4 & 6 & 2 \\ 8 & -2 & 3 \end{vmatrix}$

Solution   We choose to expand across row 1.

Remember the minus sign.
↓

$$
\begin{vmatrix} 3 & 4 & -1 \\ 4 & 6 & 2 \\ 8 & -2 & 3 \end{vmatrix} = 3 \begin{vmatrix} 6 & 2 \\ -2 & 3 \end{vmatrix} - 4 \begin{vmatrix} 4 & 2 \\ 8 & 3 \end{vmatrix} + (-1) \begin{vmatrix} 4 & 6 \\ 8 & -2 \end{vmatrix}
$$

$$
= 3(18 + 4) - 4(12 - 16) + (-1)(-8 - 48)
$$

$$
= 3(22) - 4(-4) + (-1)(-56)
$$

$$
= 66 + 16 + 56 = 138 \qquad\blacksquare
$$

We could also find the value of the 3 by 3 determinant in Example 3 by expanding down column 3 (the signs are +, −, +):

$$
\begin{vmatrix} 3 & 4 & -1 \\ 4 & 6 & 2 \\ 8 & -2 & 3 \end{vmatrix} = (-1) \begin{vmatrix} 4 & 6 \\ 8 & -2 \end{vmatrix} - 2 \begin{vmatrix} 3 & 4 \\ 8 & -2 \end{vmatrix} + 3 \begin{vmatrix} 3 & 4 \\ 4 & 6 \end{vmatrix}
$$

$$
= -1(-8 - 48) - 2(-6 - 32) + 3(18 - 16)
$$

$$
= 56 + 76 + 6 = 138
$$

■ Now work Problem 7.

Systems of Three Equations Containing Three Variables

Consider the following system of three equations containing three variables:

$$\begin{cases} a_{11}x + a_{12}y + a_{13}z = c_1 \\ a_{21}x + a_{22}y + a_{23}z = c_2 \\ a_{31}x + a_{32}y + a_{33}z = c_3 \end{cases} \qquad (10)$$

It can be shown that if the determinant D of the coefficients of the variables is not 0, that is, if

$$D = \begin{vmatrix} a_{11} & a_{12} & a_{13} \\ a_{21} & a_{22} & a_{23} \\ a_{31} & a_{32} & a_{33} \end{vmatrix} \neq 0$$

then the unique solution of system (10) is given by

Cramer's Rule for Three
Equations Containing Three
Variables

$$x = \frac{D_x}{D}, \qquad y = \frac{D_y}{D}, \qquad z = \frac{D_z}{D}$$

where

$$D_x = \begin{vmatrix} c_1 & a_{12} & a_{13} \\ c_2 & a_{22} & a_{23} \\ c_3 & a_{32} & a_{33} \end{vmatrix} \qquad D_y = \begin{vmatrix} a_{11} & c_1 & a_{13} \\ a_{21} & c_2 & a_{23} \\ a_{31} & c_3 & a_{33} \end{vmatrix} \qquad D_z = \begin{vmatrix} a_{11} & a_{12} & c_1 \\ a_{21} & a_{22} & c_2 \\ a_{31} & a_{32} & c_3 \end{vmatrix}$$

The similarity of this pattern and the pattern observed earlier for a system of two equations containing two variables should be apparent.

E X A M P L E   4   *Using Cramer's Rule*

Use Cramer's Rule, if applicable, to solve the following system:

$$\begin{cases} 2x + y - z = 3 \quad (1) \\ -x + 2y + 4z = -3 \quad (2) \\ x - 2y - 3z = 4 \quad (3) \end{cases}$$

Solution   The value of the determinant D of the coefficients of the variables is

$$D = \begin{vmatrix} 2 & 1 & -1 \\ -1 & 2 & 4 \\ 1 & -2 & -3 \end{vmatrix} = 2 \begin{vmatrix} 2 & 4 \\ -2 & -3 \end{vmatrix} - 1 \begin{vmatrix} -1 & 4 \\ 1 & -3 \end{vmatrix} + (-1) \begin{vmatrix} -1 & 2 \\ 1 & -2 \end{vmatrix}$$

$$= 2(2) - 1(-1) + (-1)(0)$$

$$= 4 + 1 = 5$$

Because $D \neq 0$, we proceed to find the value of D_x, D_y, and D_z:

$$D_x = \begin{vmatrix} 3 & 1 & -1 \\ -3 & 2 & 4 \\ 4 & -2 & -3 \end{vmatrix} = 3 \begin{vmatrix} 2 & 4 \\ -2 & -3 \end{vmatrix} - 1 \begin{vmatrix} -3 & 4 \\ 4 & -3 \end{vmatrix} + (-1) \begin{vmatrix} -3 & 2 \\ 4 & -2 \end{vmatrix}$$

$$= 3(2) - 1(-7) + (-1)(-2) = 15$$

$$D_y = \begin{vmatrix} 2 & 3 & -1 \\ -1 & -3 & 4 \\ 1 & 4 & -3 \end{vmatrix} = 2 \begin{vmatrix} -3 & 4 \\ 4 & -3 \end{vmatrix} - 3 \begin{vmatrix} -1 & 4 \\ 1 & -3 \end{vmatrix} + (-1) \begin{vmatrix} -1 & -3 \\ 1 & 4 \end{vmatrix}$$

$$= 2(-7) - 3(-1) + (-1)(-1)$$

$$= -14 + 3 + 1 = -10$$

$$D_z = \begin{vmatrix} 2 & 1 & 3 \\ -1 & 2 & -3 \\ 1 & -2 & 4 \end{vmatrix} = 2 \begin{vmatrix} 2 & -3 \\ -2 & 4 \end{vmatrix} - 1 \begin{vmatrix} -1 & -3 \\ 1 & 4 \end{vmatrix} + 3 \begin{vmatrix} -1 & 2 \\ 1 & -2 \end{vmatrix}$$

$$= 2(2) - 1(-1) + 3(0) = 5$$

As a result,

$$x = \frac{D_x}{D} = \frac{15}{5} = 3, \qquad y = \frac{D_y}{D} = \frac{-10}{5} = -2, \qquad z = \frac{D_z}{D} = \frac{5}{5} = 1$$

The solution is $x = 3$, $y = -2$, $z = 1$. ∎

If the determinant of the coefficients of the variables of a system of three linear equations containing three variables is 0, then Cramer's Rule is not applicable. In such a case, the system either is inconsistent or has infinitely many solutions.

■ Now work Problem 29.

More about Determinants

Determinants have several properties that are sometimes helpful for obtaining their value. We list some of them here.

Theorem   The value of a determinant changes sign if any two rows (or any two columns) are interchanged. (11)

Proof for 2 by 2 Determinants
$$\begin{vmatrix} a & b \\ c & d \end{vmatrix} = ad - bc \quad \text{and} \quad \begin{vmatrix} c & d \\ a & b \end{vmatrix} = bc - ad = -(ad - bc)$$

E X A M P L E   5   *Demonstrating a Theorem (11)*
$$\begin{vmatrix} 3 & 4 \\ 1 & 2 \end{vmatrix} = 6 - 4 = 2 \qquad \begin{vmatrix} 1 & 2 \\ 3 & 4 \end{vmatrix} = 4 - 6 = -2$$

Theorem   If all the entries in any row (or any column) equal 0, the value of the determinant is 0. (12)

Proof   Merely expand across the row (or down the column) containing the 0's.

Theorem   If any two rows (or any two columns) of a determinant have corresponding entries that are equal, the value of the determinant is 0. (13)

You are asked to prove this result for a 3 by 3 determinant in which the entries in column 1 equal the entries in column 3 in Problem 61 at the end of this section.

E X A M P L E 6   *Demonstrating a Theorem (13)*

$$\begin{vmatrix} 1 & 2 & 3 \\ 1 & 2 & 3 \\ 4 & 5 & 6 \end{vmatrix} = 1\begin{vmatrix} 2 & 3 \\ 5 & 6 \end{vmatrix} - 2\begin{vmatrix} 1 & 3 \\ 4 & 6 \end{vmatrix} + 3\begin{vmatrix} 1 & 2 \\ 4 & 5 \end{vmatrix}$$

$$= 1(-3) - 2(-6) + 3(-3)$$

$$= -3 + 12 - 9 = 0$$ ∎

Theorem   If any row (or any column) of a determinant is multiplied by a nonzero number k, the value of the determinant is also changed by a factor of k.   (14) ∎

You are asked to prove this result for a 3 by 3 determinant using row 2 in Problem 60 at the end of this section.

E X A M P L E 7   *Demonstrating a Theorem (14)*

$$\begin{vmatrix} 1 & 2 \\ 4 & 6 \end{vmatrix} = 6 - 8 = -2$$

$$\begin{vmatrix} k & 2k \\ 4 & 6 \end{vmatrix} = 6k - 8k = -2k = k(-2) = k\begin{vmatrix} 1 & 2 \\ 4 & 6 \end{vmatrix}$$ ∎

Theorem   If the entries of any row (or any column) of a determinant are multiplied by a nonzero number k and the result is added to the corresponding entries of another row (or column), the value of the determinant remains unchanged.   (15) ∎

In Problem 62 at the end of this section, you are asked to prove this result for a 3 by 3 determinant using rows 1 and 2.

E X A M P L E 8   *Demonstrating a Theorem (15)*

$$\begin{vmatrix} 3 & 4 \\ 5 & 2 \end{vmatrix} = \begin{vmatrix} -7 & 0 \\ 5 & 2 \end{vmatrix} = -14$$

↑ Multiply row 2 by -2 and add to row 1. ∎

10.3

Exercise 10.3

In Problems 1–10, find the value of each determinant.

1. $\begin{vmatrix} 3 & 1 \\ 4 & 2 \end{vmatrix}$ **2.** $\begin{vmatrix} 6 & 1 \\ 5 & 2 \end{vmatrix}$ **3.** $\begin{vmatrix} 6 & 4 \\ -1 & 3 \end{vmatrix}$ **4.** $\begin{vmatrix} 8 & -3 \\ 4 & 2 \end{vmatrix}$ **5.** $\begin{vmatrix} -3 & -1 \\ 4 & 2 \end{vmatrix}$

6. $\begin{vmatrix} -4 & 2 \\ -5 & 3 \end{vmatrix}$ **7.** $\begin{vmatrix} 3 & 4 & 2 \\ 1 & -1 & 5 \\ 1 & 2 & -2 \end{vmatrix}$ **8.** $\begin{vmatrix} 1 & 3 & -2 \\ 6 & 1 & -5 \\ 8 & 2 & 3 \end{vmatrix}$ **9.** $\begin{vmatrix} 4 & -1 & 2 \\ 6 & -1 & 0 \\ 1 & -3 & 4 \end{vmatrix}$ **10.** $\begin{vmatrix} 3 & -9 & 4 \\ 1 & 4 & 0 \\ 8 & -3 & 1 \end{vmatrix}$

In Problems 11–38, solve each system of equations using Cramer's Rule, if it is applicable. If it is not, say so.

11. $\begin{cases} x + y = 8 \\ x - y = 4 \end{cases}$ **12.** $\begin{cases} x + 2y = 5 \\ x + y = 3 \end{cases}$ **13.** $\begin{cases} 5x - y = 13 \\ 2x + 3y = 12 \end{cases}$ **14.** $\begin{cases} x + 3y = 5 \\ 2x - 3y = -8 \end{cases}$

15. $\begin{cases} 3x = 24 \\ x + 2y = 0 \end{cases}$

16. $\begin{cases} 4x + 5y = -3 \\ -2y = -4 \end{cases}$

17. $\begin{cases} 3x - 6y = 24 \\ 5x + 4y = 12 \end{cases}$

18. $\begin{cases} 2x + 4y = 16 \\ 3x - 5y = -9 \end{cases}$

19. $\begin{cases} 3x - 2y = 4 \\ 6x - 4y = 0 \end{cases}$

20. $\begin{cases} -x + 2y = 5 \\ 4x - 8y = 6 \end{cases}$

21. $\begin{cases} 2x - 4y = -2 \\ 3x + 2y = 3 \end{cases}$

22. $\begin{cases} 3x + 3y = 3 \\ 4x + 2y = \frac{8}{3} \end{cases}$

23. $\begin{cases} 2x - 3y = -1 \\ 10x + 10y = 5 \end{cases}$

24. $\begin{cases} 3x - 2y = 0 \\ 5x + 10y = 4 \end{cases}$

25. $\begin{cases} 2x + 3y = 6 \\ x - y = \frac{1}{2} \end{cases}$

26. $\begin{cases} \frac{1}{2}x + y = -2 \\ x - 2y = 8 \end{cases}$

27. $\begin{cases} 3x - 5y = 3 \\ 15x + 5y = 21 \end{cases}$

28. $\begin{cases} 2x - y = -1 \\ x + \frac{1}{2}y = \frac{3}{2} \end{cases}$

29. $\begin{cases} x + y - z = 6 \\ 3x - 2y + z = -5 \\ x + 3y - 2z = 14 \end{cases}$

30. $\begin{cases} x - y + z = -4 \\ 2x - 3y + 4z = -15 \\ 5x + y - 2z = 12 \end{cases}$

31. $\begin{cases} x + 2y - z = -3 \\ 2x - 4y + z = -7 \\ -2x + 2y - 3z = 4 \end{cases}$

32. $\begin{cases} x + 4y - 3z = -8 \\ 3x - y + 3z = 12 \\ x + y + 6z = 1 \end{cases}$

33. $\begin{cases} x - 2y + 3z = 1 \\ 3x + y - 2z = 0 \\ 2x - 4y + 6z = 2 \end{cases}$

34. $\begin{cases} x - y + 2z = 5 \\ 3x + 2y = 4 \\ -2x + 2y - 4z = -10 \end{cases}$

35. $\begin{cases} x + 2y - z = 0 \\ 2x - 4y + z = 0 \\ -2x + 2y - 3z = 0 \end{cases}$

36. $\begin{cases} x + 4y - 3z = 0 \\ 3x - y + 3z = 0 \\ x + y + 6z = 0 \end{cases}$

37. $\begin{cases} x - 2y + 3z = 0 \\ 3x + y - 2z = 0 \\ 2x - 4y + 6z = 0 \end{cases}$

38. $\begin{cases} x - y + 2z = 0 \\ 3x + 2y = 0 \\ -2x + 2y - 4z = 0 \end{cases}$

39. Solve: $\begin{cases} \dfrac{1}{x} + \dfrac{1}{y} = 8 \\ \dfrac{3}{x} - \dfrac{5}{y} = 0 \end{cases}$

40. Solve: $\begin{cases} \dfrac{4}{x} - \dfrac{3}{y} = 0 \\ \dfrac{6}{x} + \dfrac{3}{2y} = 2 \end{cases}$

[*Hint:*   Let $u = 1/x$ and $v = 1/y$ and solve for u and v.]

In Problems 41–46, solve for x.

41. $\begin{vmatrix} x & x \\ 4 & 3 \end{vmatrix} = 5$

42. $\begin{vmatrix} x & 1 \\ 3 & x \end{vmatrix} = -2$

43. $\begin{vmatrix} x & 1 & 1 \\ 4 & 3 & 2 \\ -1 & 2 & 5 \end{vmatrix} = 2$

44. $\begin{vmatrix} 3 & 2 & 4 \\ 1 & x & 5 \\ 0 & 1 & -2 \end{vmatrix} = 0$

45. $\begin{vmatrix} x & 2 & 3 \\ 1 & x & 0 \\ 6 & 1 & -2 \end{vmatrix} = 7$

46. $\begin{vmatrix} x & 1 & 2 \\ 1 & x & 3 \\ 0 & 1 & 2 \end{vmatrix} = -4x$

In Problems 47–54, use properties of determinants to find the value of each determinant if it is known that

$$\begin{vmatrix} x & y & z \\ u & v & w \\ 1 & 2 & 3 \end{vmatrix} = 4$$

47. $\begin{vmatrix} 1 & 2 & 3 \\ u & v & w \\ x & y & z \end{vmatrix}$

48. $\begin{vmatrix} x & y & z \\ u & v & w \\ 2 & 4 & 6 \end{vmatrix}$

49. $\begin{vmatrix} x & y & z \\ -3 & -6 & -9 \\ u & v & w \end{vmatrix}$

50. $\begin{vmatrix} 1 & 2 & 3 \\ x-u & y-v & z-w \\ u & v & w \end{vmatrix}$ **51.** $\begin{vmatrix} 1 & 2 & 3 \\ x-3 & y-6 & z-9 \\ 2u & 2v & 2w \end{vmatrix}$ **52.** $\begin{vmatrix} x & y & z-x \\ u & v & w-u \\ 1 & 2 & 2 \end{vmatrix}$

53. $\begin{vmatrix} 1 & 2 & 3 \\ 2x & 2y & 2z \\ u-1 & v-2 & w-3 \end{vmatrix}$ **54.** $\begin{vmatrix} x+3 & y+6 & z+9 \\ 3u-1 & 3v-2 & 3w-3 \\ 1 & 2 & 3 \end{vmatrix}$

55. *Geometry: Equation of a Line* An equation of the line containing the two points (x_1, y_1) and (x_2, y_2) may be expressed as the determinant

$$\begin{vmatrix} x & y & 1 \\ x_1 & y_1 & 1 \\ x_2 & y_2 & 1 \end{vmatrix} = 0$$

Prove this result by expanding the determinant and comparing the result to the two-point form of the equation of a line.

56. *Geometry: Collinear Points* Using the result obtained in Problem 55, show that three distinct points (x_1, y_1), (x_2, y_2), and (x_3, y_3) are collinear (lie on the same line) if and only if

$$\begin{vmatrix} x_1 & y_1 & 1 \\ x_2 & y_2 & 1 \\ x_3 & y_3 & 1 \end{vmatrix} = 0$$

57. Show that $\begin{vmatrix} x^2 & x & 1 \\ y^2 & y & 1 \\ z^2 & z & 1 \end{vmatrix} = (y-z)(x-y)(x-z).$

58. Complete the proof of Cramer's Rule for two equations containing two variables. [*Hint:* In system (5), page 643, if $a = 0$, then $b \neq 0$ and $c \neq 0$, since $D = -bc \neq 0$. Now show that equations (6) provide a solution of the system when $a = 0$. There are then three remaining cases: $b = 0$, $c = 0$, and $d = 0$.]

59. Interchange columns 1 and 3 of a 3 by 3 determinant. Show that the value of the new determinant is -1 times the value of the original determinant.

60. Multiply each entry in row 2 of a 3 by 3 determinant by the number k, $k \neq 0$. Show that the value of the new determinant is k times the value of the original determinant.

61. Prove that a 3 by 3 determinant in which the entries in column 1 equal those in column 3 has the value 0.

62. Prove that, if row 2 of a 3 by 3 determinant is multiplied by k, $k \neq 0$, and the result is added to the entries in row 1, then there is no change in the value of the determinant.

10.4

Systems of Nonlinear Equations

There is no general methodology for solving a system of nonlinear equations. There are times when substitution is best; other times, elimination is best; and there are times when neither of these methods works. Experience and a certain degree of imagination are your allies here.

Before we begin, two comments are in order:

1. If the system contains two variables and if the equations in the system are easy to graph, then graph them. By graphing each equation in the system, we can get an idea of how many solutions a system has and approximately where they are located.
2. Extraneous solutions can creep in when solving nonlinear systems, so it is imperative that all apparent solutions be checked.

E X A M P L E   1   *Solving a System of Nonlinear Equations Using Substitution*

Solve the following system of equations:

$$\begin{cases} 3x - y = -2 & \text{(1)} \quad \text{A line} \\ 2x^2 - y = 0 & \text{(2)} \quad \text{A parabola} \end{cases}$$

Solution   First, we notice that the system contains two variables and that we know how to graph each equation. In Figure 5, we see that the system apparently has two solutions.

We shall use substitution to solve the system. Equation (1) is easily solved for y:

FIGURE 5

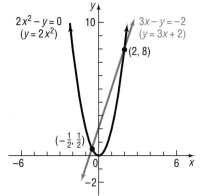

$2x^2 - y = 0$
$(y = 2x^2)$

$3x - y = -2$
$(y = 3x + 2)$

$(2, 8)$

$\left(-\frac{1}{2}, \frac{1}{2}\right)$

$$3x - y = -2$$
$$y = 3x + 2$$

We substitute this expression for y in equation (2). The result is an equation containing just the variable x, which we can then solve:

$$2x^2 - y = 0$$
$$2x^2 - (3x + 2) = 0$$
$$2x^2 - 3x - 2 = 0$$
$$(2x + 1)(x - 2) = 0$$
$$2x + 1 = 0 \quad \text{or} \quad x - 2 = 0$$
$$x = -\frac{1}{2} \qquad x = 2$$

Using these values for x in $y = 3x + 2$, we find

$$y = 3\left(-\frac{1}{2}\right) + 2 = \frac{1}{2} \quad \text{or} \quad y = 3(2) + 2 = 8$$

The apparent solutions are $x = -\frac{1}{2}$, $y = \frac{1}{2}$ and $x = 2$, $y = 8$.
Check: For $x = -\frac{1}{2}$, $y = \frac{1}{2}$:

$$\begin{cases} 3\left(-\frac{1}{2}\right) - \frac{1}{2} = -\frac{3}{2} - \frac{1}{2} = -2 & \text{(1)} \\ 2\left(-\frac{1}{2}\right)^2 - \frac{1}{2} = 2\left(\frac{1}{4}\right) - \frac{1}{2} = 0 & \text{(2)} \end{cases}$$

For $x = 2$, $y = 8$:

$$\begin{cases} 3(2) - 8 = 6 - 8 = -2 & \text{(1)} \\ 2(2)^2 - 8 = 2(4) - 8 = 0 & \text{(2)} \end{cases}$$

Each solution checks. Now we know the graphs in Figure 5 intersect at $\left(-\frac{1}{2}, \frac{1}{2}\right)$ and at $(2, 8)$.

   Check: Graph $3x - y = -2$ and $2x^2 - y = 0$ and compare what you see with Figure 5. Use ZOOM and TRACE to find the two points of intersection. ■

■   Now work Problem 3.

Our next example illustrates how the method of elimination works for nonlinear systems.

E X A M P L E   2   *Solving a System of Nonlinear Equations Using Elimination*

Solve: $\begin{cases} x^2 + y^2 = 13 & \text{(1)} \quad \text{A circle} \\ x^2 - y = 7 & \text{(2)} \quad \text{A parabola} \end{cases}$

Solution   First, we graph each equation, as shown in Figure 6. Based on the graph, we expect four solutions. By subtracting equation (2) from equation (1), the variable x is eliminated, leaving

FIGURE 6

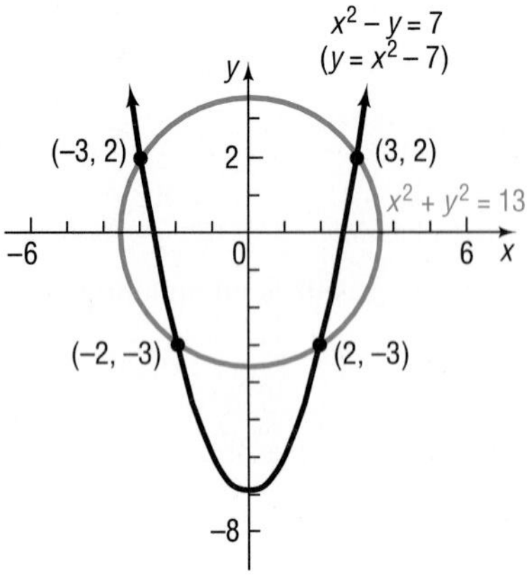

$$y^2 + y = 6$$

This quadratic equation in y is easily solved by factoring:

$$y^2 + y - 6 = 0$$
$$(y + 3)(y - 2) = 0$$
$$y = -3 \quad \text{or} \quad y = 2$$

We use these values for y in equation (2) to find x. If $y = 2$, then $x^2 = y + 7 = 9$ and $x = 3$ or -3. If $y = -3$, then $x^2 = y + 7 = 4$ and $x = 2$ or -2. Thus, we have four solutions: $x = 3, y = 2; x = -3, y = 2; x = 2, y = -3;$ and $x = -2, y = -3$. You should verify that, in fact, these four solutions also satisfy equation (1), so that all four are solutions of the system. The four points, $(3, 2), (-3, 2), (2, -3),$ and $(-2, -3)$, are the points of intersection of the graphs. Look again at Figure 6.

Check: Graph $x^2 + y^2 = 13$ and $x^2 - y = 7$. [Remember that to graph $x^2 + y^2 = 13$ requires two functions: $y = \sqrt{13 - x^2}$ and $y = -\sqrt{13 - x^2}$.] Compare what you see with Figure 6. Use ZOOM and TRACE to find the four points of intersection. ■

■ Now work Problem 1.

E X A M P L E   3   *Solving a System of Nonlinear Equations Using Elimination*

Solve: $\begin{cases} x^2 - y^2 = 1 & \text{(1)} \\ x^3 - y^2 = x & \text{(2)} \end{cases}$

Solution   Because the second equation is not so easy to graph, we omit the graphing step. We use elimination, subtracting equation (2) from equation (1), to obtain

$$x^2 - x^3 = 1 - x$$
$$x^2(1 - x) = 1 - x$$
$$x^2(1 - x) - (1 - x) = 0$$
$$(x^2 - 1)(1 - x) = 0$$
$$x^2 - 1 = 0 \quad \text{or} \quad 1 - x = 0$$
$$x = \pm 1 \qquad \qquad x = 1$$

We now use equation (1) to get y. If $x = 1$, then $1 - y^2 = 1$ and $y = 0$. If $x = -1$, then $1 - y^2 = 1$ and $y = 0$. There are two apparent solutions: $x = 1, y = 0$ and $x = -1, y = 0$. Because each of these solutions also satisfies equation (2), the system has two solutions: $x = 1, y = 0$ and $x = -1, y = 0$. The graphs of these equations will intersect at $(1, 0)$ and at $(-1, 0)$. ■

Check: Graph $x^2 - y^2 = 1$ and $x^3 - y^2 = x$. [You will need to graph four functions: $y = \sqrt{x^2 - 1}, y = -\sqrt{x^2 - 1}, y = \sqrt{x^3 - x},$ and $y = -\sqrt{x^3 - x}$.] Use ZOOM and TRACE to verify the two points of intersection. See Figure 7.

FIGURE 7

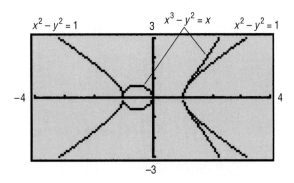

$x^2 - y^2 = 1$ 3 $x^3 - y^2 = x$ $x^2 - y^2 = 1$

-4 4

-3

E X A M P L E   4

Solving a System of Nonlinear Equations Using Elimination

Solve: $\begin{cases} x^2 + x + y^2 - 3y + 2 = 0 & (1) \\ x + 1 + \dfrac{y^2 - y}{x} = 0 & (2) \end{cases}$

Solution   First, we multiply equation (2) by x to eliminate the fraction. The result is an equivalent system because x cannot be 0 [look at equation (2) to see why]:

$$\begin{cases} x^2 + x + y^2 - 3y + 2 = 0 & (1) \\ x^2 + x + y^2 - y = 0 & (2) \end{cases}$$

Now subtract equation (2) from equation (1) to eliminate x. The result is

$$-2y + 2 = 0$$
$$y = 1$$

To find x, we back-substitute $y = 1$ in equation (1):

$$x^2 + x + 1 - 3 + 2 = 0$$
$$x^2 + x = 0$$
$$x(x + 1) = 0$$
$$x = 0 \quad \text{or} \quad x = -1$$

Because x cannot be 0, the value $x = 0$ is extraneous, and we discard it. Thus, the solution is $x = -1$, $y = 1$.

Check: We now check $x = -1$, $y = 1$:

$$\begin{cases} (-1)^2 + (-1) + 1^2 - 3(1) + 2 = 1 - 1 + 1 - 3 + 2 = 0 & (1) \\ -1 + 1 + \dfrac{1^2 - 1}{-1} = 0 + \dfrac{0}{-1} = 0 & (2) \end{cases}$$

Thus, the only solution to the system is $x = -1$, $y = 1$. The graphs of these equations will intersect at $(-1, 1)$. ∎

■ Now work Problems 17 and 41.

E X A M P L E   5

Solving a System of Nonlinear Equations

Solve: $\begin{cases} x^2 - y^2 = 4 & (1) \quad \text{A hyperbola} \\ y = x^2 & (2) \quad \text{A parabola} \end{cases}$

Solution

FIGURE 8

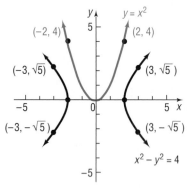

Either substitution or elimination can be used here. We use substitution and replace x^2 by y in equation (1). The result is

$$y - y^2 = 4$$
$$y^2 - y + 4 = 0$$

This is a quadratic equation whose discriminant is $1 - 4 \cdot 4 = -15 < 0$. Thus, the equation has no real solutions, and hence, the system is inconsistent. The graphs of these two equations will not intersect. See Figure 8.

 Check: Graph $x^2 - y^2 = 4$ and $y = x^2$ and compare what you see with Figure 8. Select a viewing rectangle that leaves no doubt that the curves never intersect. ∎

The following examples illustrate two of the more imaginative ways to solve systems of nonlinear equations.

E X A M P L E   6

Solving a System of Nonlinear Equations

Solve: $\begin{cases} 4x^2 - 9xy - 28y^2 = 0 & (1) \\ 16x^2 - 4xy = 16 & (2) \end{cases}$

Solution

We take note of the fact that equation (1) can be factored:

$$4x^2 - 9xy - 28y^2 = 0$$
$$(4x + 7y)(x - 4y) = 0$$

This results in the two equations

$$4x + 7y = 0 \quad \text{or} \quad x - 4y = 0$$
$$x = -\tfrac{7}{4}y \qquad\qquad x = 4y$$

We substitute each of these values for x in equation (2):

$$16x^2 - 4xy = 16 \qquad\qquad 16x^2 - 4xy = 16$$
$$16(-\tfrac{7}{4}y)^2 - 4(-\tfrac{7}{4}y)y = 16 \qquad 16(4y)^2 - 4(4y)y = 16$$
$$49y^2 + 7y^2 = 16 \qquad\qquad 16(16y^2) - 16y^2 = 16$$
$$56y^2 = 16 \qquad\qquad\qquad 15y^2 = 1$$
$$7y^2 = 2 \qquad\qquad\qquad\quad y^2 = \tfrac{1}{15}$$
$$y^2 = \tfrac{2}{7}$$

Thus, we have

$$y = \pm\sqrt{\frac{2}{7}} = \pm\frac{\sqrt{14}}{7} \qquad y = \pm\frac{\sqrt{15}}{15}$$

$$x = -\frac{7}{4}y = \mp\frac{\sqrt{14}}{4} \qquad x = 4y = \pm\frac{4\sqrt{15}}{15}$$

You should verify for yourself that, in fact, the four solutions $x = -\sqrt{14}/4$, $y = \sqrt{14}/7$; $x = \sqrt{14}/4$, $y = -\sqrt{14}/7$; $x = 4\sqrt{15}/15$, $y = \sqrt{15}/15$; and $x = -4\sqrt{15}/15$, $y = -\sqrt{15}/15$ are actually solutions of the system. ∎

E X A M P L E   7

Solving a System of Nonlinear Equations

Solve: $\begin{cases} 3xy - 2y^2 = -2 & (1) \\ 9x^2 + 4y^2 = 10 & (2) \end{cases}$

Solution We multiply equation (1) by 2 and add the result to equation (2) to eliminate the y^2-terms:

$$\begin{cases} 6xy - 4y^2 = -4 & (1) \\ 9x^2 + 4y^2 = 10 & (2) \end{cases}$$

$$9x^2 + 6xy = 6$$

$$3x^2 + 2xy = 2 \qquad \text{Divide each side by 3.}$$

Since $x \neq 0$ (do you see why?), we can solve for y in this equation to get

$$y = \frac{2 - 3x^2}{2x}, \qquad x \neq 0 \tag{1}$$

Now substitute for y in equation (2) of the system:

$$9x^2 + 4y^2 = 10$$

$$9x^2 + 4\left(\frac{2 - 3x^2}{2x}\right)^2 = 10$$

$$9x^2 + \frac{4 - 12x^2 + 9x^4}{x^2} = 10$$

$$9x^4 + 4 - 12x^2 + 9x^4 = 10x^2$$

$$18x^4 - 22x^2 + 4 = 0$$

$$9x^4 - 11x^2 + 2 = 0$$

This quadratic equation (in x^2) can be factored:

$$(9x^2 - 2)(x^2 - 1) = 0$$

$$9x^2 - 2 = 0 \qquad \text{or} \quad x^2 - 1 = 0$$

$$x^2 = \tfrac{2}{9} \qquad\qquad\qquad x^2 = 1$$

$$x = \pm \frac{\sqrt{2}}{3} \qquad\qquad\quad x = \pm 1$$

To find y, we use equation (1):

If $x = \dfrac{\sqrt{2}}{3}$: $\quad y = \dfrac{2 - 3x^2}{2x} = \dfrac{2 - \frac{2}{3}}{2(\sqrt{2}/3)} = \dfrac{4}{2\sqrt{2}} = \sqrt{2}$

If $x = -\dfrac{\sqrt{2}}{3}$: $\quad y = \dfrac{2 - 3x^2}{2x} = \dfrac{2 - \frac{2}{3}}{-2(\sqrt{2}/3)} = \dfrac{4}{-2\sqrt{2}} = -\sqrt{2}$

If $x = 1$: $\qquad\quad y = \dfrac{2 - 3x^2}{2x} = \dfrac{2 - 3}{2} = -\dfrac{1}{2}$

If $x = -1$: $\qquad y = \dfrac{2 - 3x^2}{2x} = \dfrac{2 - 3}{-2} = \dfrac{1}{2}$

The system has four solutions. Check them for yourself. ■

■ Now work Problem 37.

E X A M P L E   8

Running a Marathon

In a 50 mile marathon race, the winner crosses the finish line 1 mile ahead of the second place runner and 4 miles ahead of the third place runner. Assuming that each runner maintains a constant speed throughout the race, by how many miles does the second place runner beat the third place runner?

←——————————— 3 miles ———————————→ ←————— 1 mile —————→

Solution   Let v_1, v_2, v_3 denote the speeds of the first, second, and third place runners, respectively. Let t_1 and t_2 denote the times (in hours) required for the first place runner and second place runner to finish the race. Then we have the system of equations

$$\begin{cases} 50 = v_1 t_1 & (1) \quad \text{First place runner goes 50 miles in } t_1 \text{ hours.} \\ 49 = v_2 t_1 & (2) \quad \text{Second place runner goes 49 miles in } t_1 \text{ hours.} \\ 46 = v_3 t_1 & (3) \quad \text{Third place runner goes 46 miles in } t_1 \text{ hours.} \\ 50 = v_2 t_2 & (4) \quad \text{Second place runner goes 50 miles in } t_2 \text{ hours.} \end{cases}$$

We seek the distance of the third place runner from the finish at time t_2. That is, we seek

$$50 - v_3 t_2 = 50 - v_3\left(t_1 \cdot \frac{t_2}{t_1}\right)$$

$$= 50 - (v_3 t_1) \cdot \frac{t_2}{t_1}$$

$$= 50 - 46 \cdot \frac{50/v_2}{50/v_1} \quad \begin{cases} \text{From (3), } v_3 t_1 = 46; \\ \text{from (4), } t_2 = 50/v_2; \\ \text{from (1), } t_1 = 50/v_1. \end{cases}$$

$$= 50 - 46 \cdot \frac{v_1}{v_2}$$

$$= 50 - 46 \cdot \frac{50}{49} \qquad \text{Form the quotient of (1) and (2).}$$

$$\approx 3.06 \text{ miles} \qquad \blacksquare$$

HISTORICAL FEATURE   ■ Recall that, in the beginning of this section, we said imagination and experience are important in solving simultaneous nonlinear equations. Indeed, these kinds of problems lead into some of the deepest and most difficult parts of modern mathematics. Look again at the graphs in Examples 1 and 2 of this section (Figures 5 and 6). We see that Example 1 has two solutions, and Example 2 has four solutions. We might conjecture that the number of solutions is equal to the product of the degrees of the equations involved. This conjecture was indeed made by Etienne Bezout (1739–1783), but working out the details took about 150 years. It turns out that, to arrive at the correct number of intersections, we must count not only the complex number intersections, but also those intersections that, in a certain sense, lie at infinity. For example, a parabola and a line lying on the axis of the parabola intersect at the vertex and at infinity. This topic is part of the study of algebraic geometry. ■

HISTORICAL PROBLEM ■ 1. A papyrus dating back to 1950 BC contains the following problem: A given surface area of 100 units of area shall be represented as the sum of two squares whose sides are to each other as $1:\frac{3}{4}$. Solve for the sides by solving the system of equations

$$\begin{cases} x^2 + y^2 = 100 \\ \quad\quad x = \frac{3}{4}y \end{cases}$$

■

10.4

Exercise 10.4

In Problems 1–12, graph each equation of the system and then solve the system.

1. $\begin{cases} x^2 + y^2 = 4 \\ x^2 + 2x + y^2 = 0 \end{cases}$
2. $\begin{cases} x^2 + y^2 = 8 \\ x^2 + y^2 + 4y = 0 \end{cases}$
3. $\begin{cases} y = 3x - 5 \\ x^2 + y^2 = 5 \end{cases}$

4. $\begin{cases} x^2 + y^2 = 10 \\ y = x + 2 \end{cases}$
5. $\begin{cases} x^2 + y^2 = 4 \\ y^2 - x = 4 \end{cases}$
6. $\begin{cases} x^2 + y^2 = 16 \\ x^2 - 2y = 8 \end{cases}$

7. $\begin{cases} xy = 4 \\ x^2 + y^2 = 8 \end{cases}$
8. $\begin{cases} x^2 = y \\ xy = 1 \end{cases}$
9. $\begin{cases} x^2 + y^2 = 4 \\ y = x^2 - 9 \end{cases}$

10. $\begin{cases} xy = 1 \\ y = 2x + 1 \end{cases}$
11. $\begin{cases} y = x^2 - 4 \\ y = 6x - 13 \end{cases}$
12. $\begin{cases} x^2 + y^2 = 10 \\ xy = 3 \end{cases}$

In Problems 13–44, solve each system. Use any method you wish.

13. $\begin{cases} 2x^2 + y^2 = 18 \\ xy = 4 \end{cases}$
14. $\begin{cases} x^2 - y^2 = 21 \\ x + y = 7 \end{cases}$
15. $\begin{cases} y = 2x + 1 \\ 2x^2 + y^2 = 1 \end{cases}$

16. $\begin{cases} x^2 - 4y^2 = 16 \\ 2y - x = 2 \end{cases}$
17. $\begin{cases} x + y + 1 = 0 \\ x^2 + y^2 + 6y - x = -5 \end{cases}$
18. $\begin{cases} 2x^2 - xy + y^2 = 8 \\ xy = 4 \end{cases}$

19. $\begin{cases} 4x^2 - 3xy + 9y^2 = 15 \\ 2x + 3y = 5 \end{cases}$
20. $\begin{cases} 2y^2 - 3xy + 6y + 2x + 4 = 0 \\ 2x - 3y + 4 = 0 \end{cases}$

21. $\begin{cases} x^2 - 4y^2 + 7 = 0 \\ 3x^2 + y^2 = 31 \end{cases}$
22. $\begin{cases} 3x^2 - 2y^2 + 5 = 0 \\ 2x^2 - y^2 + 2 = 0 \end{cases}$
23. $\begin{cases} 7x^2 - 3y^2 + 5 = 0 \\ 3x^2 + 5y^2 = 12 \end{cases}$

24. $\begin{cases} x^2 - 3y^2 + 1 = 0 \\ 2x^2 - 7y^2 + 5 = 0 \end{cases}$
25. $\begin{cases} x^2 + 2xy = 10 \\ 3x^2 - xy = 2 \end{cases}$
26. $\begin{cases} 5xy + 13y^2 + 36 = 0 \\ xy + 7y^2 = 6 \end{cases}$

27. $\begin{cases} 2x^2 + y^2 = 2 \\ x^2 - 2y^2 + 8 = 0 \end{cases}$
28. $\begin{cases} y^2 - x^2 + 4 = 0 \\ 2x^2 + 3y^2 = 6 \end{cases}$
29. $\begin{cases} x^2 + 2y^2 = 16 \\ 4x^2 - y^2 = 24 \end{cases}$

30. $\begin{cases} 4x^2 + 3y^2 = 4 \\ 2x^2 - 6y^2 = -3 \end{cases}$
31. $\begin{cases} \dfrac{5}{x^2} - \dfrac{2}{y^2} + 3 = 0 \\ \dfrac{3}{x^2} + \dfrac{1}{y^2} = 7 \end{cases}$
32. $\begin{cases} \dfrac{2}{x^2} - \dfrac{3}{y^2} + 1 = 0 \\ \dfrac{6}{x^2} - \dfrac{7}{y^2} + 2 = 0 \end{cases}$

33. $\begin{cases} \dfrac{1}{x^4} + \dfrac{6}{y^4} = 6 \\ \dfrac{2}{x^4} - \dfrac{2}{y^4} = 19 \end{cases}$
34. $\begin{cases} \dfrac{1}{x^4} - \dfrac{1}{y^4} = 1 \\ \dfrac{1}{x^4} + \dfrac{1}{y^4} = 4 \end{cases}$
35. $\begin{cases} x^2 - 3xy + 2y^2 = 0 \\ x^2 + xy = 6 \end{cases}$

36. $\begin{cases} x^2 - xy - 2y^2 = 0 \\ xy + x + 6 = 0 \end{cases}$

37. $\begin{cases} xy - x^2 + 3 = 0 \\ 3xy - 4y^2 = 2 \end{cases}$

38. $\begin{cases} 5x^2 + 4xy + 3y^2 = 36 \\ x^2 + xy + y^2 = 9 \end{cases}$

39. $\begin{cases} x^3 - y^3 = 26 \\ x - y = 2 \end{cases}$

40. $\begin{cases} x^3 + y^3 = 26 \\ x + y = 2 \end{cases}$

41. $\begin{cases} y^2 + y + x^2 - x - 2 = 0 \\ y + 1 + \dfrac{x-2}{y} = 0 \end{cases}$

42. $\begin{cases} x^3 - 2x^2 + y^2 + 3y - 4 = 0 \\ x - 2 + \dfrac{y^2 - y}{x^2} = 0 \end{cases}$

43. $\begin{cases} \log_x y = 3 \\ \log_x(4y) = 5 \end{cases}$

44. $\begin{cases} \log_x(2y) = 3 \\ \log_x(4y) = 2 \end{cases}$

45. The difference of two numbers is 2 and the sum of their squares is 10. Find the numbers.

46. The sum of two numbers is 7 and the difference of their squares is 21. Find the numbers.

47. The product of two numbers is 4 and the sum of their squares is 8. Find the numbers.

48. The product of two numbers is 10 and the difference of their squares is 21. Find the numbers.

49. The difference of two numbers is the same as their product, and the sum of their reciprocals is 5. Find the numbers.

50. The sum of two numbers is the same as their product, and the difference of their reciprocals is 3. Find the numbers.

51. The ratio of a to b is $\frac{2}{3}$. The sum of a and b is 10. What is the ratio of $a + b$ to $b - a$?

52. The ratio of a to b is $\frac{4}{3}$. The sum of a and b is 14. What is the ratio of $a - b$ to $a + b$?

In Problems 53–60, graph each equation in the system. Be sure to label the points of intersection.

53. $\begin{cases} y = x^2 + 1 \\ y = x + 1 \end{cases}$

54. $\begin{cases} y = x^2 + 1 \\ y = 4x + 1 \end{cases}$

55. $\begin{cases} y = \sqrt{36 - x^2} \\ y = 8 - x \end{cases}$

56. $\begin{cases} y = \sqrt{4 - x^2} \\ y = 2x + 4 \end{cases}$

57. $\begin{cases} y = \sqrt{x} \\ y = 2 - x \end{cases}$

58. $\begin{cases} y = \sqrt{x} \\ y = 6 - x \end{cases}$

59. $\begin{cases} x = 2y \\ x = y^2 - 2y \end{cases}$

60. $\begin{cases} y = x - 1 \\ y = x^2 - 6x + 9 \end{cases}$

In Problems 61–66, graph each equation and find the point(s) of intersection, if any.

61. The line $x + 2y = 0$ and the circle $(x - 1)^2 + (y - 1)^2 = 5$

62. The line $x + 2y + 6 = 0$ and the circle $(x + 1)^2 + (y + 1)^2 = 5$

63. The circle $(x - 1)^2 + (y + 2)^2 = 4$ and the parabola $y^2 + 4y - x + 1 = 0$

64. The circle $(x + 2)^2 + (y - 1)^2 = 4$ and the parabola $y^2 - 2y - x - 5 = 0$

65. The graph of $y = \dfrac{4}{x - 3}$ and the circle $x^2 - 6x + y^2 + 1 = 0$

66. The graph of $y = \dfrac{4}{x + 2}$ and the circle $x^2 + 4x + y^2 - 4 = 0$

 In Problems 67–74, solve each of the systems of equations. Express the solution(s) correct to two decimal places.

67. $\begin{cases} y = x^{2/3} \\ y = e^{-x} \end{cases}$

68. $\begin{cases} y = x^{3/2} \\ y = e^{-x} \end{cases}$

69. $\begin{cases} x^2 + y^3 = 2 \\ x^3 y = 4 \end{cases}$

70. $\begin{cases} x^3 + y^2 = 2 \\ x^2 y = 4 \end{cases}$

71. $\begin{cases} x^4 + y^4 = 12 \\ xy^2 = 2 \end{cases}$

72. $\begin{cases} x^4 + y^4 = 6 \\ xy = 1 \end{cases}$

73. $\begin{cases} xy = 2 \\ y = \ln x \end{cases}$

74. $\begin{cases} x^2 + y^2 = 4 \\ y = \ln x \end{cases}$

75. *Geometry* The perimeter of a rectangle is 16 inches and its area is 15 square inches. What are its dimensions?

76. *Geometry* An area of 52 square feet is to be enclosed by two squares whose sides are in the ratio of 2:3. Find the sides of the squares.

77. *Geometry* Two circles have perimeters that add up to 12π centimeters and areas that add up to 20π square centimeters. Find the radius of each circle.

78. *Geometry* The altitude of an isosceles triangle drawn to its base is 3 centimeters, and its perimeter is 18 centimeters. Find the length of its base.

79. *The Tortoise and the Hare* In a 21 meter race between a tortoise and a hare, the tortoise leaves 9 minutes before the hare. The hare, by running at an average speed of 0.5 meter per hour faster than the tortoise, crosses the finish line 3 minutes before the tortoise. What are the average speeds of the tortoise and the hare?

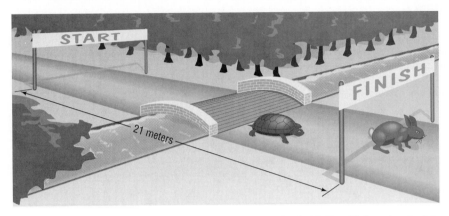

80. *Running a Race* In a 1 mile race, the winner crosses the finish line 10 feet ahead of the second place runner and 20 feet ahead of the third place runner. Assuming that each runner maintains a constant speed throughout the race, by how many feet does the second place runner beat the third place runner?

81. *Constructing a Box* A rectangular piece of cardboard, whose area is 216 square centimeters, is made into an open box by cutting a 2 centimeter square from each corner and turning up the sides. See the figure. If the box is to have a volume of 224 cubic centimeters, what size cardboard should you start with?

82. *Constructing a Cylindrical Tube* A rectangular piece of cardboard, whose area is 216 square centimeters, is made into a cylindrical tube by joining together two sides of the rectangle. (See the figure.) If the tube is to have a volume of 224 cubic centimeters, what size cardboard should you start with?

83. *Fencing* A farmer has 300 feet of fence available to enclose 4500 square feet in the shape of adjoining squares, with sides of length x and y. See the figure. Find x and y.

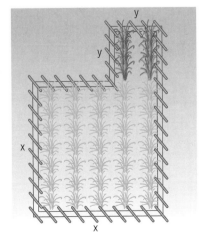

84. *Bending Wire* A wire 60 feet long is cut into two pieces. Is it possible to bend one piece into the shape of a square and the other into the shape of a circle so that the total area enclosed by the two pieces is 100 square feet? If this is possible, find the length of the side of the square and the radius of the circle.

85. *Geometry* Find formulas for the length l and width w of a rectangle in terms of its area A and perimeter P.

86. *Geometry* Find formulas for the base b and one of the equal sides l of an isosceles triangle in terms of its altitude h and perimeter P.

87. *Descartes' Method of Equal Roots* Descartes' method for finding tangents depends on the idea that, for many graphs, the tangent line at a given point is the *unique* line that intersects the graph at that point only. We will apply his method to find an equation of the tangent line to the parabola $y = x^2$ at the point (2, 4); see the figure. First, we know the equation of the tangent line must be in the form $y = mx + b$. Using the fact that the point (2, 4) is on the line, we can solve for b in terms of m and get the equation $y = mx + (4 - 2m)$. Now we want (2, 4) to be the *unique* solution to the system

$$\begin{cases} y = x^2 \\ y = mx + 4 - 2m \end{cases}$$

From this system, we get $x^2 = mx + 4 - 2m$ or $x^2 - mx + (2m - 4) = 0$. By using the quadratic formula, we get

$$x = m \pm \frac{\sqrt{m^2 - 4(2m - 4)}}{2}$$

To obtain a unique solution for x, the two roots must be equal; in other words, the expression $m^2 - 4(2m - 4)$ must be 0. Complete the work to get m, and write an equation of the tangent line.

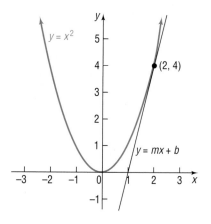

In Problems 88–94, use Descartes' method from Problem 87 to find the equation of the line tangent to each graph at the given point.

88. $x^2 + y^2 = 10$;   at (1, 3)   **89.** $y = x^2 + 2$;   at (1, 3)   **90.** $x^2 + y = 5$;   at $(-2, 1)$

91. $2x^2 + 3y^2 = 14$;   at (1, 2)   **92.** $3x^2 + y^2 = 7$;   at $(-1, 2)$ **93.** $x^2 - y^2 = 3$;   at (2, 1)

94. $2y^2 - x^2 = 14$;   at (2, 3)

95. If r_1 and r_2 are two solutions of a quadratic equation $ax^2 + bx + c = 0$, then it can be shown that

$$r_1 + r_2 = -\frac{b}{a} \quad \text{and} \quad r_1 r_2 = \frac{c}{a}$$

Solve this system of equations for r_1 and r_2.

96. A circle and a line intersect at most twice. A circle and a parabola intersect at most four times. Deduce that a circle and the graph of a polynomial of degree 3 intersect at most six times. What do you conjecture about a polynomial of degree 4? What about a polynomial of degree n? Can you explain your conclusions using an algebraic argument?

97. Suppose that you are the manager of a sheet metal shop. A customer asks you to manufacture 10,000 boxes, each box being open on top. The boxes are required to have a square base and a 9 cubic foot capacity. You construct the boxes by cutting a square out from each corner of a square piece of sheet metal and folding along the edges.
(a) What are the dimensions of the square to be cut if the area of the square piece of sheet metal is 100 square feet?
(b) Could you make the box using a smaller piece of sheet metal? Make a list of the dimensions of the box for various pieces of sheet metal.

10.5

Systems of Inequalities

In Chapter 1, we discussed inequalities in one variable. In this section, we discuss inequalities in two variables. Samples are given in Example 1.

E X A M P L E   1

Samples of Inequalities in Two Variables

(a) $3x + y - 6 < 0$ (b) $x^2 + y^2 < 4$ (c) $y^2 \leq x$ ∎

An inequality in two variables x and y is **satisfied** by an ordered pair (a, b) if, when x is replaced by a and y by b, a true statement results. A **graph of an inequality in two variables** x and y consists of all points (x, y) whose coordinates satisfy the inequality.

Let's look at an example.

E X A M P L E   2

Graphing a Linear Inequality

Graph the linear inequality: $3x + y - 6 \leq 0$

Solution   We begin with the associated problem of the graph of the linear equality

$$3x + y - 6 = 0$$

formed by replacing (for now) the $\leq$ symbol with an $=$ sign. The graph of the linear equation is a line. See Figure 9(a). This line is part of the graph of the inequality we seek because the inequality is nonstrict. (Do you see why? We are seeking points for which $3x + y - 6$ is less than *or equal to* 0.)

FIGURE 9

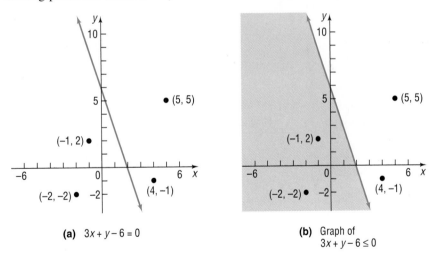

(a) $3x + y - 6 = 0$

(b)   Graph of
$3x + y - 6 \leq 0$

Now, let's test a few randomly selected points to see whether they belong to the graph of the inequality

	$3x + y - 6$	**CONCLUSION**
$(4, -1)$	$3(4) + (-1) - 6 = 5 > 0$	Does not belong to graph
$(5, 5)$	$3(5) + 5 - 6 = 14 > 0$	Does not belong to graph
$(-1, 2)$	$3(-1) + 2 - 6 = -7 < 0$	Belongs to graph
$(-2, -2)$	$3(-2) + (-2) - 6 = -14 < 0$	Belongs to graph

Look again at Figure 9(a). Notice that the two points that belong to the graph both lie on the same side of the line, and the two points that do not belong to the graph lie on the opposite side. As it turns out, this is always the case. Thus, the graph we seek consists of all points that lie on the same side of the line as do $(-1, 2)$ and $(-2, -2)$. The graph we seek is the shaded region in Figure 9(b). ■

The graph of any inequality in two variables may be obtained in a like way. First, the equation corresponding to the inequality is graphed, using dashes if the inequality is strict and solid marks if it is nonstrict. This graph, in almost every case, will separate the xy-plane into two or more regions. In each region either all points satisfy the inequality or no points satisfy the inequality. Thus, the use of a single test point in each region is all that is required to determine whether the points of that region are part of the graph or not. The steps to follow are given next.

Steps for Graphing an Inequality

STEP 1:   Replace the inequality symbol by an equal sign and graph the resulting equation. If the inequality is strict, use dashes; if it is nonstrict, use a solid mark. This graph separates the xy-plane into two or more regions.

STEP 2:   In each of the regions, select a test point P.
  (a) If the coordinates of P satisfy the inequality, then so do all the points in that region. Indicate this by shading the region.
  (b) If the coordinates of P do not satisfy the inequality, then none of the points in that region do.

E X A M P L E   3

FIGURE 10

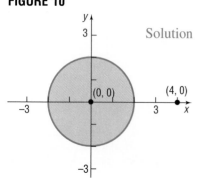

Graphing an Inequality

Graph: $x^2 + y^2 \leq 4$

Solution   First we graph the equation $x^2 + y^2 = 4$, a circle of radius 2, center at the origin. A solid color will be used because the inequality is not strict. We use two test points, one inside the circle, the other outside;

Inside $(0, 0)$: $x^2 + y^2 = 0^2 + 0^2 = 0 \leq 4$ Belongs to the graph

Outside $(4, 0)$: $x^2 + y^2 = 4^2 + 0^2 = 16 > 4$ Does not belong to graph

All the points inside and on the circle satisfy the inequality. See Figure 10. ■

■ Now work Problem 7.

Linear inequalities are inequalities in one of the forms

$$Ax + By < C, \qquad Ax + By > C, \qquad Ax + By \leq C, \qquad Ax + By \geq C$$

The graph of the corresponding equation of a linear inequality is a line, which separates the xy-plane into two regions, called **half-planes.** See Figure 11.

As shown there, if $Ax + By = C$ is the equation of the boundary line, then it divides the plane into two half-planes; one for which $Ax + By < C$ and the other for which $Ax + By > C$. Because of this, for linear inequalities, only one test point is required.

FIGURE 11

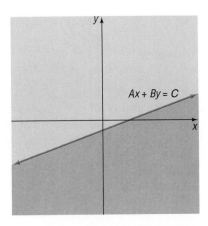

E X A M P L E   4   *Graphing Linear Inequalities*

Graph: (a) $y < 2$ (b) $y \geq 2x$

Solution   (a) The graph of the equation $y = 2$ is a horizontal line and is not part of the graph of the inequality. Since $(0, 0)$ satisfies the inequality, the graph consists of the half-plane below the line $y = 2$. See Figure 12.

FIGURE 12

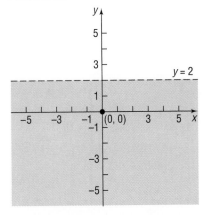

FIGURE 13

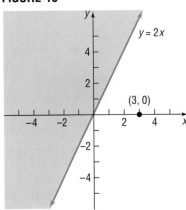

(b) The graph of the equation $y = 2x$ is a line and is part of the graph of the inequality. Using $(3, 0)$ as a test point, we find it does not satisfy the inequality $[0 < 2 \cdot 3]$. Thus, points in the half-plane on the opposite side of $y = 2x$ satisfy the inequality. See Figure 13.   ■

   ■   Now work Problem 3.

Systems of Inequalities in Two Variables

The **graph of a system of inequalities** in two variables x and y is the set of all points (x, y) that simultaneously satisfy *each* of the inequalities in the system. Thus, the graph of a system of inequalities can be obtained by graphing each inequality individually and then determining where, if at all, they intersect.

EXAMPLE 5   *Graphing a System of Linear Inequalities*

Graph the system: $\begin{cases} x + y \geq 2 \\ 2x - y \leq 4 \end{cases}$

Solution   First, we graph the inequality $x + y \geq 2$ as the shaded region in Figure 14(a). Next, we graph the inequality $2x - y \leq 4$ as the shaded region in Figure 14(b). Now, superimpose the two graphs, as shown in Figure 14(c). The points that are in both shaded regions—the overlapping, darker region in Figure 14(c)—are the solutions we seek to the system, because they simultaneously satisfy each linear inequality.

FIGURE 14

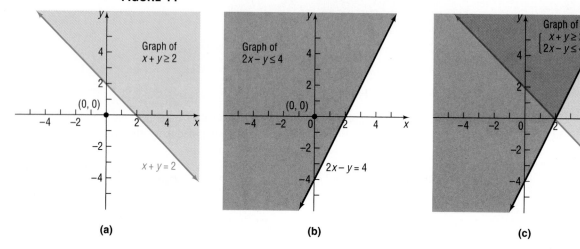

(a)   (b)   (c)

EXAMPLE 6   *Graphing a System of Linear Inequalities*

Graph the system: $\begin{cases} x + y \leq 2 \\ x + y \geq 0 \end{cases}$

Solution   See Figure 15. The overlapping, darker shaded region between the two boundary lines is the graph of the system.

FIGURE 15

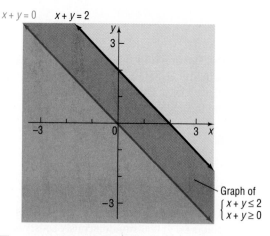

Now work Problem 25.

E X A M P L E   7   *Graphing a System of Linear Inequalities*

Graph the system: $\begin{cases} 2x - y \ge 2 \\ 2x - y \ge 0 \end{cases}$

Solution   See Figure 16. The overlapping, darker shaded region is the graph of the system. Note that the graph of the system is identical to the graph of the single inequality $2x - y \ge 2$.

FIGURE 16

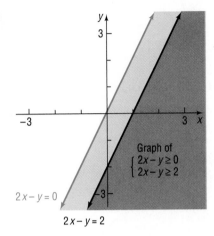

E X A M P L E   8   *Graphing a System of Linear Inequalities*

Graph the system: $\begin{cases} x + 2y \le 2 \\ x + 2y \ge 6 \end{cases}$

Solution   See Figure 17. Because no overlapping region results, there are no points in the *xy*-plane that simultaneously satisfy each inequality. Hence, the system has no solution.

FIGURE 17

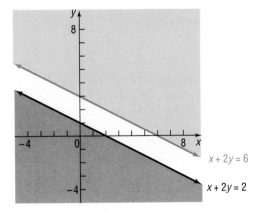

E X A M P L E   9   *Graphing a System of Inequalities*

Graph the system: $\begin{cases} y \ge x^2 - 4 \\ x + y \le 2 \end{cases}$

Solution   Figure 18 shows that the graph of the system consists of the region enclosed by the graphs of the parabola $y = x^2 - 4$ and the line $x + y = 2$. The points of intersection of the two equations are found by solving the system of equations

$$\begin{cases} y = x^2 - 4 \\ x + y = 2 \end{cases}$$

Using substitution, we find

$$x + (x^2 - 4) = 2$$
$$x^2 + x - 6 = 0$$
$$(x + 3)(x - 2) = 0$$
$$x = -3, \qquad x = 2$$

The two points of intersection are $(-3, 5)$ and $(2, 0)$.

FIGURE 18

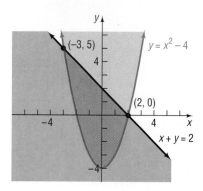

■   Now work Problem 19.

E X A M P L E   1 0   *Graphing a System of Four Linear Inequalities*

Graph the system: $\begin{cases} x + y \geq 3 \\ 2x + y \geq 4 \\ x \geq 0 \\ y \geq 0 \end{cases}$

Solution   The two inequalities $x \geq 0$ and $y \geq 0$ require that the graph be in quadrant I. Thus, we concentrate on the other two inequalities. The overlapping, darker shaded region in Figure 19 shows the graph of the system.

FIGURE 19

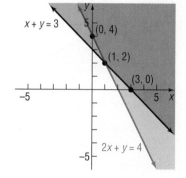

E X A M P L E   1 1   *Financial Planning*

A retired couple has up to $25,000 to invest. As their financial adviser, you recommend that they place at least $15,000 in Treasury bills yielding 6% and at most $5000 in corporate bonds yielding 9%.

(a) Using x to denote the amount of money invested in Treasury bills and y the amount invested in corporate bonds, write a system of linear inequalities that describes the possible amounts of each investment.

(b) Graph the system.

Solution   (a) The system of linear inequalities is

$$\begin{cases} x \geq 0 & \text{\small x and y are nonnegative variables since they represent money invested.} \\ y \geq 0 & \\ x + y \leq 25,000 & \text{\small The total of the two investments, $x + y$, cannot exceed \$25,000.} \\ x \geq 15,000 & \text{\small At least \$15,000 in Treasury bills} \\ y \leq 5000 & \text{\small At most \$5000 in corporate bonds} \end{cases}$$

(b) See the shaded region in Figure 20. Note that the inequalities $x \geq 0$ and $y \geq 0$ again require that the graph of the system be in quadrant I.

FIGURE 20

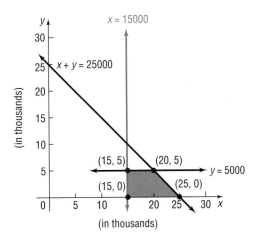

The graph of the system of linear inequalities in Figure 20 is said to be **bounded,** because it can be contained within some circle of sufficiently large radius. A graph that cannot be contained in any circle is said to be **unbounded.** For example, the graph of the system of linear inequalities in Figure 19 is unbounded, since it extends indefinitely in a particular direction.

Notice in Figures 19 and 20 that those points belonging to the graph that are also points of intersection of boundary lines have been plotted. Such points are referred to as **vertices** or **corner points** of the graph. Thus, the system graphed in Figure 19 has three corner points: (0, 4), (1, 2), and (3, 0). The system graphed in Figure 20 has four corner points: (15, 0), (25, 0), (20, 5), and (15, 5).

These ideas will be used in the next section in developing a method for solving linear programming problems, an important application of linear inequalities.

■   Now work Problem 33.

10.5

Exercise 10.5

In Problems 1–12, graph each inequality.

1. $x \geq 0$ **2.** $y \geq 0$ **3.** $x \geq 4$ **4.** $y \leq 2$

5. $2x + y \geq 6$ **6.** $3x + 2y \leq 6$ **7.** $x^2 + y^2 > 1$ **8.** $x^2 + y^2 \leq 9$

9. $y \leq x^2 - 1$ **10.** $y > x^2 + 2$ **11.** $xy \geq 4$ **12.** $xy \leq 1$

In Problems 13–30, graph each system of inequalities.

13. $\begin{cases} x + y \leq 2 \\ 2x + y \geq 4 \end{cases}$ **14.** $\begin{cases} 3x - y \geq 6 \\ x + 2y \leq 2 \end{cases}$ **15.** $\begin{cases} 2x - y \leq 4 \\ 3x + 2y \geq -6 \end{cases}$ **16.** $\begin{cases} 4x - 5y \leq 0 \\ 2x - y \geq 2 \end{cases}$

17. $\begin{cases} 2x - 3y \leq 0 \\ 3x + 2y \leq 6 \end{cases}$ **18.** $\begin{cases} 4x - y \geq 2 \\ x + 2y \geq 2 \end{cases}$ **19.** $\begin{cases} x^2 + y^2 \leq 9 \\ x + y \geq 3 \end{cases}$ **20.** $\begin{cases} x^2 + y^2 \geq 9 \\ x + y \leq 3 \end{cases}$

21. $\begin{cases} y \geq x^2 - 4 \\ y \leq x - 2 \end{cases}$ **22.** $\begin{cases} y^2 \leq x \\ y \geq x \end{cases}$ **23.** $\begin{cases} xy \geq 4 \\ y \geq x^2 + 1 \end{cases}$ **24.** $\begin{cases} y + x^2 \leq 1 \\ y \geq x^2 - 1 \end{cases}$

25. $\begin{cases} x - 2y \leq 6 \\ 2x - 4y \geq 0 \end{cases}$ **26.** $\begin{cases} x + 4y \leq 8 \\ x + 4y \geq 4 \end{cases}$ **27.** $\begin{cases} 2x + y \geq -2 \\ 2x + y \geq 2 \end{cases}$ **28.** $\begin{cases} x - 4y \leq 4 \\ x - 4y \geq 0 \end{cases}$

29. $\begin{cases} 2x + 3y \geq 6 \\ 2x + 3y \leq 0 \end{cases}$ **30.** $\begin{cases} 2x + y \geq 0 \\ 2x + y \geq 2 \end{cases}$

In Problems 31–40, graph each system of linear inequalities. Tell whether the graph is bounded or unbounded, and label the corner points.

31. $\begin{cases} x \geq 0 \\ y \geq 0 \\ 2x + y \leq 6 \\ x + 2y \leq 6 \end{cases}$ **32.** $\begin{cases} x \geq 0 \\ y \geq 0 \\ x + y \geq 4 \\ 2x + 3y \geq 6 \end{cases}$ **33.** $\begin{cases} x \geq 0 \\ y \geq 0 \\ x + y \geq 2 \\ 2x + y \geq 4 \end{cases}$ **34.** $\begin{cases} x \geq 0 \\ y \geq 0 \\ 3x + y \leq 6 \\ 2x + y \leq 2 \end{cases}$ **35.** $\begin{cases} x \geq 0 \\ y \geq 0 \\ x + y \geq 2 \\ 2x + 3y \leq 12 \\ 3x + y \leq 12 \end{cases}$

36. $\begin{cases} x \geq 0 \\ y \geq 0 \\ x + y \geq 2 \\ x + y \leq 10 \\ 2x + y \leq 3 \end{cases}$ **37.** $\begin{cases} x \geq 0 \\ y \geq 0 \\ x + y \geq 2 \\ x + y \leq 8 \\ 2x + y \leq 10 \end{cases}$ **38.** $\begin{cases} x \geq 0 \\ y \geq 0 \\ x + y \geq 2 \\ x + y \leq 8 \\ x + 2y \geq 1 \end{cases}$ **39.** $\begin{cases} x \geq 0 \\ y \geq 0 \\ x + 2y \geq 1 \\ x + 2y \leq 10 \end{cases}$ **40.** $\begin{cases} x \geq 0 \\ y \geq 0 \\ x + 2y \geq 1 \\ x + 2y \leq 10 \\ x + y \geq 2 \\ x + y \leq 8 \end{cases}$

In Problems 41–44, write a system of linear inequalities that has the given graph.

41.

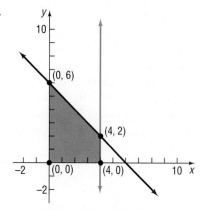

42.

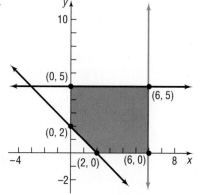

43.

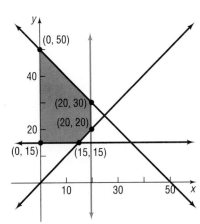

44.

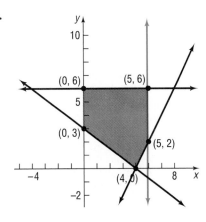

45. *Financial Planning* A retired couple has up to $50,000 to invest. As their financial adviser, you recommend they place at least $35,000 in Treasury bills yielding 7% and at most $10,000 in corporate bonds yielding 10%.
(a) Using x to denote the amount of money invested in Treasury bills and y the amount invested in corporate bonds, write a system of linear inequalities that describes the possible amounts of each investment.
(b) Graph the system and label the corner points.

46. *Manufacturing Trucks* Mike's Toy Truck Company manufactures two models of toy trucks, a standard model and a deluxe model. Each standard model requires 2 hours for painting and 3 hours for detail work; each deluxe model requires 3 hours for painting and 4 hours for detail work. Two painters and three detail workers are employed by the company, and each works 40 hours per week.
(a) Using x to denote the number of standard model trucks and y to denote the number of deluxe model trucks, write a system of linear inequalities that describes the possible number of each model of truck that can be manufactured in a week.
(b) Graph the system and label the corner points.

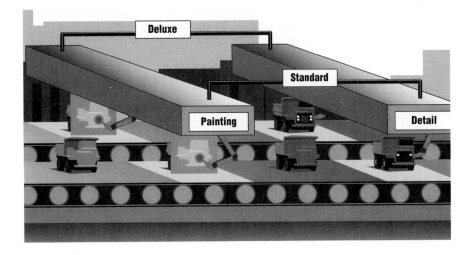

47. *Blending Coffee* A store that specializes in coffee has available 75 pounds of A grade coffee and 120 pounds of B grade coffee. These will be blended into 1 pound packages as follows: An economy blend that contains 4 ounces of A grade coffee and 12 ounces of B grade coffee and a superior blend that contains 8 ounces of A grade coffee and 8 ounces of B grade coffee.
(a) Using x to denote the number of packages of the economy blend and y to denote the number of packages of the superior blend, write a system of linear inequalities that describes the possible number of packages of each kind of blend.
(b) Graph the system and label the corner points.

48. *Mixed Nuts* A store that specializes in selling nuts has 90 pounds of cashews and 120 pounds of peanuts available. These are to be mixed in 12 ounce packages as follows: a lower priced package containing 8 ounces of peanuts and 4 ounces of cashews and a quality package containing 6 ounces of peanuts and 6 ounces of cashews.

(a) Use x to denote the number of lower priced packages and use y to denote the number of quality packages. Write a system of linear inequalities that describes the possible number of each kind of package.

(b) Graph the system and label the corner points.

49. *Transporting Goods* A small truck can carry no more than 1600 pounds of cargo or 150 cubic feet of cargo. A printer weighs 20 pounds and occupies 3 cubic feet of space. A microwave oven weighs 30 pounds and occupies 2 cubic feet of space.

(a) Using x to represent the number of microwave ovens and y to represent the number of printers, write a system of linear inequalities that describes the number of ovens and printers that can be hauled by the truck.

(b) Graph the system and label the corner points.

10.6

Linear Programming

Historically, linear programming evolved as a technique for solving problems involving resource allocation of goods and materials for the U.S. Air Force during World War II. Today, linear programming techniques are used to solve a wide variety of problems, such as optimizing airline scheduling and establishing telephone lines. Although most practical linear programming problems involve systems of several hundred linear inequalities containing several hundred variables, we shall limit our discussion to problems containing only two variables, because we can solve such problems using graphing techniques.*

We begin by returning to Example 11 of the previous section.

E X A M P L E   1

Financial Planning

A retired couple has up to $25,000 to invest. As their financial adviser, you recommend that they place at least $15,000 in Treasury bills yielding 6% and at most $5000 in corporate bonds yielding 9%. How much money should be placed in each investment so that income is maximized? ■

The problem given here is typical of a *linear programming problem.* The problem requires that a certain linear expression, the income, be maximized. If I represents income, x the amount invested in Treasury bills at 6%, and y the amount invested in corporate bonds at 9%, then

$$I = 0.06x + 0.09y$$

This linear expression is called the **objective function.** Furthermore, the problem requires that the maximum income be achieved under certain conditions or **constraints,** each of which is a linear inequality involving the variables. (See Example 11 in Section 10.5.) The linear programming problem given in Example 1 may be restated as

Maximize $I = 0.06x + 0.09y$

subject to the conditions that

---

*The **simplex method** is a way to solve linear programming problems involving many inequalities and variables. This method was developed by George Dantzig in 1946 and is particularly well suited for computerization. In 1984, Narendra Karmarkar of Bell Laboratories discovered a way of solving large linear programming problems that improves on the simplex method.

$$x \geq 0, \qquad y \geq 0$$
$$x + y \leq 25{,}000$$
$$x \geq 15{,}000$$
$$y \leq 5000$$

In general, every linear programming problem has two components:

1. A linear objective function that is to be maximized or minimized.
2. A collection of linear inequalities that must be satisfied simultaneously.

Linear Programming Problem

A **linear programming problem** in two variables x and y consists of maximizing (or minimizing) a linear objective function

$$z = Ax + By, \qquad A \text{ and } B \text{ are real numbers, not both } 0$$

subject to certain conditions, or constraints, expressible as linear inequalities in x and y.

To maximize (or minimize) the quantity $z = Ax + By$, we need to identify points (x, y) that make the expression for z the largest (or smallest) possible. But not all points (x, y) are eligible; only those that also satisfy each linear inequality (constraint) can be used. We refer to each point (x, y) that satisfies the system of linear inequalities (the constraints) as a **feasible point.** Thus, in a linear programming problem, we seek the feasible point(s) that maximizes (or minimizes) the objective function.

Let's look again at the linear programming problem in Example 1.

E X A M P L E   2

Analyzing a Linear Programming Problem

Consider the linear programming problem:

Maximize $I = 0.06x + 0.09y$

subject to the conditions that

$$x \geq 0, \qquad y \geq 0$$
$$x + y \leq 25{,}000$$
$$x \geq 15{,}000$$
$$y \leq 5000$$

Graph the constraints. Then graph the objective function for $I = 0, 0.9, 1.35, 1.65,$ and 1.8.

Solution   Figure 21 shows the graph of the constraints. We superimpose on this graph the graph of the objective function for the given values of I.

For $I = 0$, the objective function is the line $0 = 0.06x + 0.09y$.

For $I = 0.9$, the objective function is the line $0.9 = 0.06x + 0.09y$.

For $I = 1.35$, the objective function is the line $1.35 = 0.06x + 0.09y$.

For $I = 1.65$, the objective function is the line $1.65 = 0.06x + 0.09y$.

For $I = 1.8$, the objective function is the line $1.8 = 0.06x + 0.09y$.

FIGURE 21

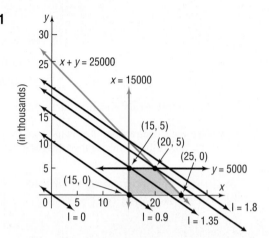

A **solution** to a linear programming problem consists of the feasible point(s) that maximizes (or minimizes) the objective function, together with the corresponding value of the objective function. One condition for a linear programming problem in two variables to have a solution is that the graph of the feasible points be bounded. (Refer to page 669.)

If none of the feasible points maximizes (or minimizes) the objective function or if there are no feasible points, then the linear programming problem has no solution.

Consider the linear programming problem stated in Example 2, and look again at Figure 21. The feasible points are the points that lie in the shaded region. For example, (20, 3) is a feasible point, as is (15, 5), (20, 5), (18, 4), etc. To find the solution of the problem requires that we find a feasible point (x, y) that makes $I = 0.06x + 0.09y$ as large as possible. Notice that as I increases in value from $I = 0$ to $I=0.9$ to $I = 1.35$ to $I = 1.65$ to $I = 1.8$, we obtain a collection of parallel lines. Furthermore, notice that the largest value of I that can be obtained while feasible points are present is $I = 1.65$, which corresponds to the line $1.65 = 0.06x + 0.09y$. Any larger value of I results in a line that does not pass through any feasible points. Finally, notice that the feasible point that yields $I = 1.65$ is the point (20, 5), a corner point. These observations form the basis of the following result, which we state without proof.

Theorem
Location of Solution of a
Linear Programming Problem

If a linear programming problem has a solution, it is located at a corner point of the graph of the feasible points.

If a linear programming problem has multiple solutions, at least one of them is located at a corner point of the graph of the feasible points.

In either case, the corresponding value of the objective function is unique. ■

We shall not consider here linear programming problems that have no solution. As a result, we can outline the procedure for solving a linear programming problem as follows:

Procedure for Solving a
Linear Programming Problem

STEP 1: Write an expression for the quantity to be maximized (or minimized). This expression is the objective function.
STEP 2: Write all the constraints as a system of linear inequalities and graph the system.
STEP 3: List the corner points of the graph of the feasible points.
STEP 4: List the corresponding values of the objective function at each corner point. The largest (or smallest) of these is the solution.

E X A M P L E   3   *Solving a Minimum Linear Programming Problem*

Minimize the expression

$$z = 2x + 3y$$

subject to the constraints

$$y \leq 5, \quad x \leq 6, \quad x + y \geq 2, \quad x \geq 0, \quad y \geq 0$$

FIGURE 22

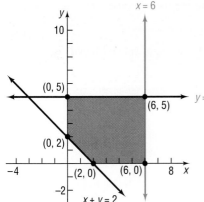

Solution   The objective function is $z = 2x + 3y$. We seek the smallest value of z that can occur if x and y are solutions of the system of linear inequalities

$$\begin{cases} y \leq 5 \\ x \leq 6 \\ x + y \geq 2 \\ x \geq 0 \\ y \geq 0 \end{cases}$$

The graph of this system (the feasible points) is shown as the shaded region in Figure 22. We have also plotted the corner points. Table 2 lists the corner points and the corresponding values of the objective function. From the table, we can see that the minimum value of z is 4, and it occurs at the point $(2, 0)$.

TABLE 2

CORNER POINT	VALUE OF THE OBJECTIVE FUNCTION
(x, y)	$z = 2x + 3y$
$(0, 2)$	$z = 2(0) + 3(2) = 6$
$(0, 5)$	$z = 2(0) + 3(5) = 15$
$(6, 5)$	$z = 2(6) + 3(5) = 27$
$(6, 0)$	$z = 2(6) + 3(0) = 12$
$(2, 0)$	$z = 2(2) + 3(0) = 4$

■   Now work Problems 3 and 9.

E X A M P L E   4   *Maximizing Profit*

At the end of every month, after filling orders for its regular customers, a coffee company has some pure Colombian coffee and some special-blend coffee remaining. The practice of the company has been to package a mixture of the two coffees into 1 pound packages as follows: a low-grade mixture containing 4 ounces of Colombian coffee and 12 ounces of special-blend coffee and a high-grade mixture containing 8 ounces of Colombian and 8 ounces of special-blend coffee. A profit of $0.30 per package is made on the low-grade mixture, whereas a profit of $0.40 per package is made on the high-grade mixture. This month, 120 pounds of special-blend coffee and 100 pounds of pure Colombian coffee remain. How many packages of each mixture should be prepared to achieve a maximum profit? Assume that all packages prepared can be sold.

Solution   We begin by assigning symbols for the two variables:

x = Number of packages of the low-grade mixture

y = Number of packages of the high-grade mixture

If P denotes the profit, then

$$P = \$0.30x + \$0.40y$$

This expression is the objective function. We seek to maximize P subject to certain constraints on x and y. Because x and y represent numbers of packages, the only meaningful values for x and y are nonnegative integers. Thus, we have the two constraints

$$x \geq 0, \qquad y \geq 0 \quad \text{Nonnegative constraints}$$

We also have only so much of each type of coffee available. For example, the total amount of Colombian coffee used in the two mixtures cannot exceed 100 pounds, or 1600 ounces. Because we use 4 ounces in each low-grade package and 8 ounces in each high-grade package, we are led to the constraint

$$4x + 8y \leq 1600 \quad \text{Colombian coffee constraint}$$

Similarly, the supply of 120 pounds, or 1920 ounces, of special-blend coffee leads to the constraint

$$12x + 8y \leq 1920 \quad \text{Special-blend coffee constraint}$$

The linear programming problem may be stated as

$$\text{Maximize} \quad P = 0.3x + 0.4y$$

subject to the constraints

$$x \geq 0, \quad y \geq 0, \quad 4x + 8y \leq 1600, \quad 12x + 8y \leq 1920$$

The graph of the constraints (the feasible points) is illustrated in Figure 23. We list the corner points and evaluate the objective function at each one. In Table 3, we can see that the maximum profit, \$84, is achieved with 40 packages of the low-grade mixture and 180 packages of the high-grade mixture.

FIGURE 23

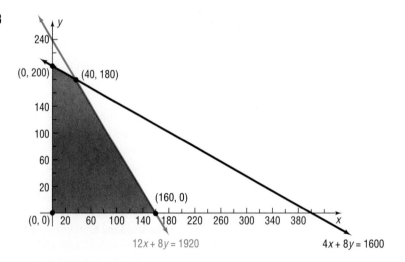

12x + 8y = 1920

4x + 8y = 1600

TABLE 3

CORNER POINT	VALUE OF PROFIT
(x, y)	$P = 0.3x + 0.4y$
$(0, 0)$	$P = 0$
$(0, 200)$	$P = 0.3(0) + 0.4(200) = \$80$
$(40, 180)$	$P = 0.3(40) + 0.4(180) = \$84$
$(160, 0)$	$P = 0.3(160) + 0.4(0) = \$48$

■   Now work Problem 17.

10.6

Exercise 10.6

In Problems 1–6, find the maximum and minimum value of the given objective function.

The figure illustrates the graph of the feasible points of a linear programming problem.

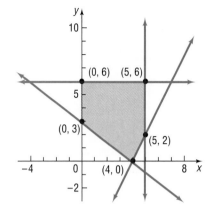

1. $z = x + y$ **2.** $z = 2x + 3y$ **3.** $z = x + 10y$

4. $z = 10x + y$ **5.** $z = 5x + 7y$ **6.** $z = 7x + 5y$

In Problems 7–16, solve each linear programming problem.

7. Maximize $z = 2x + y$ subject to $x \geq 0$, $y \geq 0$, $x + y \leq 6$, $x + y \geq 1$

8. Maximize $z = x + 3y$ subject to $x \geq 0$, $y \geq 0$, $x + y \geq 3$, $x \leq 5$, $y \leq 7$

9. Minimize $z = 2x + 5y$ subject to $x \geq 0$, $y \geq 0$, $x + y \geq 2$, $x \leq 5$, $y \leq 3$

10. Minimize $z = 3x + 4y$ subject to $x \geq 0$, $y \geq 0$, $2x + 3y \geq 6$, $x + y \leq 8$

11. Maximize $z = 3x + 5y$ subject to $x \geq 0$, $y \geq 0$, $x + y \geq 2$, $2x + 3y \leq 12$, $3x + 2y \leq 12$

12. Maximize $z = 5x + 3y$ subject to $x \geq 0$, $y \geq 0$, $x + y \geq 2$, $x + y \leq 8$, $2x + y \leq 10$

13. Minimize $z = 5x + 4y$ subject to $x \geq 0$, $y \geq 0$, $x + y \geq 2$, $2x + 3y \leq 12$, $3x + y \leq 12$

14. Minimize $z = 2x + 3y$ subject to $x \geq 0$, $y \geq 0$, $x + y \geq 3$, $x + y \leq 9$, $x + 3y \geq 6$

15. Maximize $z = 5x + 2y$ subject to $x \geq 0$, $y \geq 0$, $x + y \leq 10$, $2x + y \geq 10$, $x + 2y \geq 10$

16. Maximize $z = 2x + 4y$ subject to $x \geq 0$, $y \geq 0$, $2x + y \geq 4$, $x + y \leq 9$

17. *Maximizing Profit* A manufacturer of skis produces two types: downhill and cross-country. Use the following table to determine how many of each kind of ski should be produced to achieve a maximum profit. What is the maximum profit? What would the maximum profit be if the maximum time available for manufacturing is increased to 48 hours?

	DOWNHILL	CROSS-COUNTRY	MAXIMUM TIME AVAILABLE
Manufacturing time per ski	2 hours	1 hour	40 hours
Finishing time per ski	1 hour	1 hour	32 hours
Profit per ski	$70	$50	

18. *Farm Management* A farmer has 70 acres of land available for planting either soybeans or wheat. The cost of preparing the soil, the workdays required, and the expected profit per acre planted for each type of crop are given in the following table:

	SOYBEANS	WHEAT
Preparation cost per acre	$60	$30
Workdays required per acre	3	4
Profit per acre	$180	$100

The farmer cannot spend more than $1800 in preparation costs nor more than a total of 120 workdays. How many acres of each crop should be planted in order to maximize the profit? What is the maximum profit? What is the maximum profit if the farmer is willing to spend no more than $2400 on preparation?

19. *Farm Management* A small farm in Illinois has 100 acres of land available on which to grow corn and soybeans. The following table shows the cultivation cost per acre, the labor cost per acre, and the expected profit per acre. The column on the right shows the amount of money available for each of these expenses. Find the number of acres of each crop that should be planted in order to maximize profit.

	SOYBEANS	CORN	MONEY AVAILABLE
Cultivation cost per acre	$40	$60	$1800
Labor cost per acre	$60	$60	$2400
Profit per acre	$200	$250	

20. *Dietary Requirements* A certain diet requires at least 60 units of carbohydrates, 45 units of protein, and 30 units of fat each day. Each ounce of Supplement A provides 5 units of carbohydrates, 3 units of protein, and 4 units of fat. Each ounce of Supplement B provides 2 units of carbohydrates, 2 units of protein, and 1 unit of fat. If Supplement A costs $1.50 per ounce and Supplement B costs $1.00 per ounce, how many ounces of each supplement should be taken daily to minimize the cost of the diet?

21. *Production Scheduling* In a factory, machine 1 produces 8" plyers at the rate of 60 units per hour, and 6" plyers at the rate of 70 units per hour. Machine 2 produces 8" plyers at the rate of 40 units per hour and 6" plyers at the rate of 20 units per hour. It costs $50 per hour to operate machine 1, while machine 2 costs $30 per hour to operate. The production schedule requires that at least 240 units of 8" plyers and at least 140 units of 6" plyers must be produced during each 10 hour day. Which combination of machines will cost the least money to operate?

22. *Farm Management* An owner of a fruit orchard hires a crew of workers to prune at least 25 of his 50 fruit trees. Each newer tree requires one hour to prune, while each older tree needs one-and-a-half hours. The crew contracts to work for at least 30 hours and charges $15 for each newer tree and $20 for each older tree. To minimize his cost, how many of each kind of tree will the orchard owner have pruned? What will be the cost?

23. *Managing a Meat Market* A meat market combines ground beef and ground pork in a single package for meat loaf. The ground beef is 75% lean (75% beef, 25% fat) and costs the market $0.75 per pound. The ground pork is 60% lean and costs the market $0.45 per pound. The meat loaf must be at least 70% lean. If the market wants to use at least 50 lbs of its available pork, but no more than 200 lbs of its available groundbeef, how much ground beef should be mixed with ground pork so that the cost is minimized?

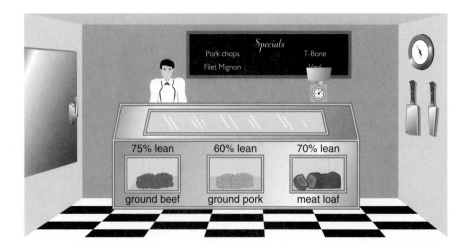

24. *Return on Investment* An investment broker is instructed by her client to invest up to $20,000, some in a junk bond yielding 9% per annum and some in Treasury bills yielding 7% per annum. The client wants to invest at least $8000 in T bills and no more than $12,000 in the junk bond.
    (a) How much should the broker recommend that the client place in each investment to maximize income if the client insists that the amount invested in T-bills must equal or exceed the amount placed in junk bonds?
    (b) How much should the broker recommend the client place in each investment to maximize income if the client insists that the amount invested in T-bills must not exceed the amount placed in junk bonds?

25. *Maximizing Profit on Ice Skates* A factory manufactures two kinds of ice skates: racing skates and figure skates. The racing skates require 6 work-hours in the fabrication department, whereas the figure skates require 4 work-hours there. The racing skates require 1 work-hour in the finishing department, whereas the figure skates require 2 work-hours there. The fabricating department has available at most 120 work-hours per day, and the finishing department has no more than 40 work-hours per day available. If the profit on each racing skate is $10 and the profit on each figure skate is $12, how many of each should be manufactured each day to maximize profit? (Assume that all skates made are sold.)

26 *Financial Planning* A retired couple has up to $50,000 to place in fixed-income securities. Their financial adviser suggests two securities to them: one is an AAA bond that yields 8% per annum; the other is a Certificate of Deposit (CD) that yields 4%. After careful consideration of the alternatives, the couple decides to place at most $20,000 in the AAA bond and at least $15,000 in the CD. They also instruct the financial adviser to place at least as much in the CD as in the AAA bond. How should the financial adviser proceed to maximize the return on their investment?

27. *Product Design* An entrepreneur is having a design group produce at least six samples of a new kind of fastener that he wants to market. It costs $9.00 to produce each metal fastener and $4.00 to produce each plastic fastener. He wants to have at least two of each version of the fastener, and needs to have all the samples 24 hours from now. It takes 4 hours to produce each metal sample and 2 hours to produce each plastic sample. To minimize the cost of the samples, how many of each kind should the entrepreneur order? What will be the cost of the samples?

28. *Animal Nutrition* Ted's dog Amadeus likes two kinds of canned dog food. "Gourmet Dog" costs 40 cents a can and has 20 units of a vitamin complex; the calorie content is 75 calories. "Chow Hound" costs 32 cents a can and has 35 units of vitamins and 50 calories. Ted likes Amadeus to have at least 1175 units of vitamins a month and at least 2375 calories during the same time period. Ted has space to store only 60 cans of dog food at a time. How much of each kind of dog food should Ted buy each month in order to minimize his cost?

29. *Airline Revenue* An airline has two classes of service: first class and coach. Management's experience has been that each aircraft should have at least 8 but not more than 16 first-class seats and at least 80 but not more than 120 coach seats.
 (a) If management decides that the ratio of first class to coach seats should never exceed 1:12, with how many of each type seat should an aircraft be configured to maximize revenue?
 (b) If management decides that the ratio of first class to coach seats should never exceed 1:8, with how many of each type seat should an aircraft be configured to maximize revenue?
 (c) If you were management, what would you do?
 {*Hint:* Assume that the airline charges $C for a coach seat and $F for a first class seat; $C > 0, F > 0$.}

30. *Minimizing Cost* A farm that specializes in raising frying chickens supplements the regular chicken feed with four vitamins. The owner wants the supplemental food to contain at least 50 units of vitamin I, 90 units of vitamin II, 60 units of vitamin III, and 100 units of vitamin IV per 100 ounces of feed. Two supplements are available: supplement A, which contains 5 units of vitamin I, 25 units of vitamin II, 10 units of vitamin III, and 35 units of vitamin IV per ounce; and supplement B, which contains 25 units of vitamin I, 10 units of vitamin II, 10 units of vitamin III, and 20 units of vitamin IV per ounce. If supplement A costs $0.06 per ounce and supplement B costs $0.08 per ounce, how much of each supplement should the manager of the farm buy to add to each 100 ounces of feed in order to keep the total cost at a minimum, while still meeting the owner's vitamin specifications?

31. Explain in your own words what a linear programming problem is and how it can be solved.

Chapter Review

THINGS TO KNOW

Systems of equations

Systems with no solutions are inconsistent.   Systems with a solution are consistent.

Consistent systems have either a unique solution or an infinite number of solutions.

Linear programming problem

Maximize (or minimize) a linear objective function, $z = Ax + By$, subject to certain conditions, or constraints, expressible as linear inequalities in x and y.

Feasible point

A point (x, y) that satisfies the constraints of a linear programming problem

Location of solution

If a linear programming problem has a solution, it is located at a corner point of the graph of the feasible points.

If a linear programming problem has multiple solutions, at least one of them is located at a corner point of the graph of the feasible points.

In either case, the corresponding value of the objective function is unique.

How To:

Solve a system of linear equations using the method of substitution

Solve a system of linear equations using the method of elimination

Solve a system of linear equations using matrices

Solve a system of linear equations using determinants

Solve a system of nonlinear equations

Graph a system of inequalities

Find the corner points of the graph of a system of linear inequalities

Solve linear programming problems

Fill-in-the-Blank Items

1. If a system of equations has no solution, it is said to be _____.

2. An m by n rectangular array of numbers is called a(n) _____.

3. Cramer's Rules uses _____ to solve a system of linear equations.

4. The matrix used to represent a system of linear equations is called a(n) _____ matrix.

5. The graph of a linear inequality is called a(n) _____.

6. A linear programming problem requires that a linear expression, called the _____ _____, be maximized or minimized.

7. Each point that satisfies the constraints of a linear programming problem is called a(n) _____ _____.

True/False Items

T F **1.** A system of two linear equations containing two unknowns always has at least one solution.

T F **2.** The augmented matrix of a system of two equations containing three variables has two rows and four columns.

T F **3.** A 3 by 3 determinant can never equal 0.

T F **4.** A consistent system of equations will have exactly one solution.

T F **5.** The graph of a linear inequality is a half-plane.

T F **6.** The graph of a system of linear inequalities is sometimes unbounded.

T F **7.** If a linear programming problem has a solution, it is located at a corner point of the graph of the feasible points.

Review Exercises

In Problems 1–20, solve each system of equations using the method of substitution or the method of elimination. If the system has no solution, say it is inconsistent.

1. $\begin{cases} 2x - y = 5 \\ 5x + 2y = 8 \end{cases}$

2. $\begin{cases} 2x + 3y = 2 \\ 7x - y = 3 \end{cases}$

3. $\begin{cases} 3x - 4y = 4 \\ x - 3y = \frac{1}{2} \end{cases}$

4. $\begin{cases} 2x + y = 0 \\ 5x - 4y = -\frac{13}{2} \end{cases}$

5. $\begin{cases} x - 2y - 4 = 0 \\ 3x + 2y - 4 = 0 \end{cases}$ **6.** $\begin{cases} x - 3y + 5 = 0 \\ 2x + 3y - 5 = 0 \end{cases}$ **7.** $\begin{cases} y = 2x - 5 \\ x = 3y + 4 \end{cases}$ **8.** $\begin{cases} x = 5y + 2 \\ y = 5x + 2 \end{cases}$

9. $\begin{cases} x - y + 4 = 0 \\ \frac{1}{2}x + \frac{1}{6}y + \frac{2}{5} = 0 \end{cases}$ **10.** $\begin{cases} x + \frac{1}{4}y = 2 \\ y + 4x + 2 = 0 \end{cases}$ **11.** $\begin{cases} x - 2y - 8 = 0 \\ 2x + 2y - 10 = 0 \end{cases}$ **12.** $\begin{cases} x - 3y + \frac{7}{2} = 0 \\ \frac{1}{2}x + 3y - 5 = 0 \end{cases}$

13. $\begin{cases} y - 2x = 11 \\ 2y - 3x = 18 \end{cases}$ **14.** $\begin{cases} 3x - 4y - 12 = 0 \\ 5x + 2y + 6 = 0 \end{cases}$ **15.** $\begin{cases} 2x + 3y - 13 = 0 \\ 3x - 2y = 0 \end{cases}$ **16.** $\begin{cases} 4x + 5y = 21 \\ 5x + 6y = 42 \end{cases}$

17. $\begin{cases} 3x - 2y = 8 \\ x - \frac{2}{3}y = 12 \end{cases}$ **18.** $\begin{cases} 2x + 5y = 10 \\ 4x + 10y = 15 \end{cases}$ **19.** $\begin{cases} x + 2y - z = 6 \\ 2x - y + 3z = -13 \\ 3x - 2y + 3z = -16 \end{cases}$ **20.** $\begin{cases} x + 5y - z = 2 \\ 2x + y + z = 7 \\ x - y + 2z = 11 \end{cases}$

In Problems 21–30, solve each system of equations using matrices. If the system has no solution, say it is inconsistent.

21. $\begin{cases} 3x - 2y = 1 \\ 10x + 10y = 5 \end{cases}$ **22.** $\begin{cases} 3x + 2y = 6 \\ x - y = -\frac{1}{2} \end{cases}$ **23.** $\begin{cases} 5x + 6y - 3z = 6 \\ 4x - 7y - 2z = -3 \\ 3x + y - 7z = 1 \end{cases}$ **24.** $\begin{cases} 2x + y + z = 5 \\ 4x - y - 3z = 1 \\ 8x + y - z = 5 \end{cases}$

25. $\begin{cases} x - 2z = 1 \\ 2x + 3y = -3 \\ 4x - 3y - 4z = 3 \end{cases}$ **26.** $\begin{cases} x + 2y - z = 2 \\ 2x - 2y + z = -1 \\ 6x + 4y + 3z = 5 \end{cases}$ **27.** $\begin{cases} x - y + z = 0 \\ x - y - 5z - 6 = 0 \\ 2x - 2y + z - 1 = 0 \end{cases}$ **28.** $\begin{cases} 4x - 3y + 5z = 0 \\ 2x + 4y - 3z = 0 \\ 6x + 2y + z = 0 \end{cases}$

29. $\begin{cases} x - y - z - t = 1 \\ 2x + y + z + 2t = 3 \\ x - 2y - 2z - 3t = 0 \\ 3x - 4y + z + 5t = -3 \end{cases}$ **30.** $\begin{cases} x - 3y + 3z - t = 4 \\ x + 2y - z = -3 \\ x + 3z + 2t = 3 \\ x + y + 5z = 6 \end{cases}$

In Problems 31–36, find the value of each determinant.

31. $\begin{vmatrix} 3 & 4 \\ 1 & 3 \end{vmatrix}$ **32.** $\begin{vmatrix} -4 & 0 \\ 1 & 3 \end{vmatrix}$ **33.** $\begin{vmatrix} 1 & 4 & 0 \\ -1 & 2 & 6 \\ 4 & 1 & 3 \end{vmatrix}$ **34.** $\begin{vmatrix} 2 & 3 & 10 \\ 0 & 1 & 5 \\ -1 & 2 & 3 \end{vmatrix}$ **35.** $\begin{vmatrix} 2 & 1 & -3 \\ 5 & 0 & 1 \\ 2 & 6 & 0 \end{vmatrix}$ **36.** $\begin{vmatrix} -2 & 1 & 0 \\ 1 & 2 & 3 \\ -1 & 4 & 2 \end{vmatrix}$

In Problems 37–42, use Cramer's Rule, if applicable, to solve each system.

37. $\begin{cases} x - 2y = 4 \\ 3x + 2y = 4 \end{cases}$ **38.** $\begin{cases} x - 3y = -5 \\ 2x + 3y = 5 \end{cases}$ **39.** $\begin{cases} 2x + 3y - 13 = 0 \\ 3x - 2y = 0 \end{cases}$ **40.** $\begin{cases} 3x - 4y - 12 = 0 \\ 5x + 2y + 6 = 0 \end{cases}$

41. $\begin{cases} x + 2y - z = 6 \\ 2x - y + 3z = -13 \\ 3x - 2y + 3z = -16 \end{cases}$ **42.** $\begin{cases} x - y + z = 8 \\ 2x + 3y - z = -2 \\ 3x - y - 9z = 9 \end{cases}$

In Problems 43–52, solve each system of equations.

43. $\begin{cases} 2x + y + 3 = 0 \\ x^2 + y^2 = 5 \end{cases}$ **44.** $\begin{cases} x^2 + y^2 = 16 \\ 2x - y^2 = -8 \end{cases}$ **45.** $\begin{cases} 2xy + y^2 = 10 \\ 3y^2 - xy = 2 \end{cases}$ **46.** $\begin{cases} 3x^2 - y^2 = 1 \\ 7x^2 - 2y^2 - 5 = 0 \end{cases}$

47. $\begin{cases} x^2 + y^2 = 64 \\ x^2 = 3y \end{cases}$ **48.** $\begin{cases} 2x^2 + y^2 = 9 \\ x^2 + y^2 = 9 \end{cases}$ **49.** $\begin{cases} 3x^2 + 4xy + 5y^2 = 8 \\ x^2 + 3xy + 2y^2 = 0 \end{cases}$ **50.** $\begin{cases} 3x^2 + 2xy - 2y^2 = 6 \\ xy - 2y^2 + 4 = 0 \end{cases}$

51. $\begin{cases} x^2 - 3x + y^2 + y = -2 \\ \dfrac{x^2 - x}{y} + y + 1 = 0 \end{cases}$ **52.** $\begin{cases} x^2 + x + y^2 = y + 2 \\ x + 1 = \dfrac{2 - y}{x} \end{cases}$

In Problems 53–58, graph each system of inequalities. Tell whether the graph is bounded or unbounded, and label the corner points.

53. $\begin{cases} -2x + y \le 2 \\ x + y \ge 2 \end{cases}$

54. $\begin{cases} x - 2y \le 6 \\ 2x + y \ge 2 \end{cases}$

55. $\begin{cases} x \ge 0 \\ y \ge 0 \\ x + y \le 4 \\ 2x + 3y \le 6 \end{cases}$

56. $\begin{cases} x \ge 0 \\ y \ge 0 \\ 3x + y \ge 6 \\ 2x + y \ge 2 \end{cases}$

57. $\begin{cases} x \ge 0 \\ y \ge 0 \\ 2x + y \le 8 \\ x + 2y \ge 2 \end{cases}$

58. $\begin{cases} x \ge 0 \\ y \ge 0 \\ 3x + y \le 9 \\ 2x + 3y \ge 6 \end{cases}$

In Problems 59–62, graph each system of inequalities.

59. $\begin{cases} x^2 + y^2 \le 16 \\ x + y \ge 2 \end{cases}$

60. $\begin{cases} y^2 \le x - 1 \\ x - y \le 3 \end{cases}$

61. $\begin{cases} y \le x^2 \\ xy \le 4 \end{cases}$

62. $\begin{cases} x^2 + y^2 \ge 1 \\ x^2 + y^2 \le 4 \end{cases}$

In Problems 63–68, solve each linear programming problem.

63. Maximize $z = 3x + 4y$ subject to $x \ge 0$, $y \ge 0$, $3x + 2y \ge 6$, $x + y \le 8$

64. Maximize $z = 2x + 4y$ subject to $x \ge 0$, $y \ge 0$, $x + y \le 6$, $x \ge 2$

65. Minimize $z = 3x + 5y$ subject to $x \ge 0$, $y \ge 0$, $x + y \ge 1$, $3x + 2y \le 12$, $x + 3y \le 12$

66. Minimize $z = 3x + y$ subject to $x \ge 0$, $y \ge 0$, $x \le 8$, $y \le 6$, $2x + y \ge 4$

67. Maximize $z = 5x + 4y$ subject to $x \ge 0$, $y \ge 0$, $x + 2y \ge 2$, $3x + 4y \le 12$, $y \ge x$

68. Maximize $z = 4x + 5y$ subject to $x \ge 0$, $y \ge 0$, $2x + 3y \ge 6$, $x \ge y$, $2x + y \le 12$

69. Find A such that the system of equations has infinitely many solutions.

$$\begin{cases} 2x + 5y = 5 \\ 4x + 10y = A \end{cases}$$

70. Find A such that the system in Problem 69 is inconsistent.

71. *Curve Fitting* Find the quadratic function $y = ax^2 + bx + c$ that passes through the three points $(0, 1)$, $(1, 0)$, and $(-2, 1)$.

72. *Curve Fitting* Find the general equation of the circle that passes through the three points $(0, 1)$, $(1, 0)$, and $(-2, 1)$. [*Hint:* The general equation of a circle is $x^2 + y^2 + Dx + Ey + F = 0$.]

73. *Blending Coffee* A coffee distributor is blending a new coffee that will cost $3.90 per pound. It will consist of a blend of $3.00 per pound coffee and $6.00 per pound coffee. What amounts of each type of coffee should be mixed to achieve the desired blend? [*Hint:* Assume that the weight of the blended coffee is 100 pounds.]

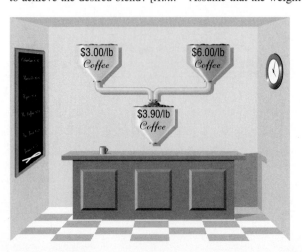

74. *Farming* A 1000 acre farm in Illinois is used to raise corn and soy beans. The cost per acre for raising corn is $65 and the cost per acre for soy beans is $45. If $54,325 has been budgeted for costs and all the acreage is to be used, how many acres should be allocated for each crop?

75. *Cookie Orders* A cookie company makes three kinds of cookies, oatmeal raisin, chocolate chip, and shortbread, packaged in small, medium, and large boxes. The small box contains 1 dozen oatmeal raisin and 1 dozen chocolate chip; the medium box has 2 dozen oatmeal raisin, 1 dozen chocolate chip, and 1 dozen shortbread; the large box contains 2 dozen oatmeal raisin, 2 dozen chocolate chip, and 3 dozen shortbread. If you require exactly 15 dozen oatmeal raisin, 10 dozen chocolate chip, and 11 dozen shortbread, how many of each size box should you buy?

76. *Mixed Nuts* A store that specializes in selling nuts has 72 pounds of cashews and 120 pounds of peanuts available. These are to be mixed in 12 ounce packages as follows: a lower-priced package containing 8 ounces of peanuts and 4 ounces of cashews and a quality package containing 6 ounces of peanuts and 6 ounces of cashews.
 (a) Use x to denote the number of lower-priced packages and use y to denote the number of quality packages. Write a system of linear inequalities that describes the possible number of each kind of package.
 (b) Graph the system and label the corner points.

77. A small rectangular lot has a perimeter of 68 feet. If its diagonal is 26 feet, what are the dimensions of the lot?

78. The area of a rectangular window is 4 square feet. If the diagonal measures $2\sqrt{2}$ feet, what are the dimensions of the window?

79. *Geometry* A certain right triangle has a perimeter of 14 inches. If the hypotenuse is 6 inches long, what are the lengths of the legs?

80. *Geometry* A certain isosceles triangle has a perimeter of 18 inches. If the altitude is 6 inches, what is the length of the base?

81. How much fence is required to enclose 5000 square feet by two squares whose sides are in the ratio of 1:2?

82. *Mixing Acids* A chemistry laboratory has three containers of hydrochloric acid, HCl. One container holds a solution with a concentration of 10% HCl, the second holds 25% HCl, and the third holds 40% HCl. How many liters of each should be mixed to obtain 100 liters of a solution with a concentration of 30% HCl? Construct a table showing some of the possible combinations.

83. *Calculating Allowances* Katy, Mike, Danny, and Colleen agreed to do yard work at home for $45 to be split among them. After they finished, their father determined that Mike deserves twice what Katy gets, Katy and Colleen deserve the same amount, and Danny deserves half of what Katy gets. How much does each one receive?

84. *Finding the Speed of the Jet Stream* On a flight between Midway Airport in Chicago and Ft. Lauderdale, Florida, a Boeing 737 jet maintains an airspeed of 475 miles per hour. If the trip from Chicago to Ft. Lauderdale takes 2 hours, 30 minutes and the return flight takes 2 hours, 50 minutes, what is the speed of the jet stream? (Assume the speed of the jet stream remains constant at the various altitudes of the plane.)

85. *Constant Rate Jobs* If Katy and Mike work together for 1 hour and 20 minutes, they will finish a certain job. If Mike and Danny work together for 1 hour and 36 minutes, the same job can be finished. If Danny and Katy work together, they can complete this job in 2 hours and 40 minutes. How long will it take each of them working alone to finish the job?

86. *Maximizing Profit on Figurines* A factory manufactures two kinds of ceramic figurines: a dancing girl and a mermaid, each requiring three processes—molding, painting, and glazing. The daily labor available for molding is no more than 90 work-hours, labor available for painting does not exceed 120 work-hours, and labor available for glazing is no more than 60 work-hours. The dancing girl requires 3 work-hours for molding, 6 work-hours for painting, and 2 work-hours for glazing. The mermaid requires 3 work-hours for molding, 4 work-hours for painting, and 3 work-hours for glazing. If the profit on each figurine is $25 for dancing girls and $30 for mermaids, how many of each should be produced each day to maximize profit? If management decides to produce the number of each figurine that maximizes profit, determine which of these processes has excess work-hours assigned to it.

87. A factory produces gasoline engines and diesel engines. Each week the factory is obligated to deliver at least 20 gasoline engines and at least 15 diesel engines. Due to physical limitations, however, the factory cannot make more than 60 gasoline engines nor more than 40 diesel engines. Finally, to prevent layoffs, a total of at least 50 engines must be produced. If gasoline engines cost $450 each to produce and diesel engines cost $550 each to produce, how many of each should be produced per week to minimize the cost? What is the excess capacity of the factory; that is, how many of each kind of engine are being produced in excess of the number the factory is obligated to deliver?

88. Describe four ways of solving a system of three linear equations containing three variables. Which method do you prefer? Why?

Chapter 11

SEQUENCES; INDUCTION; COUNTING; PROBABILITY

Preview Creating a Floor Design

A ceramic tile floor is designed in the shape of a trapezoid 20 feet wide at the base and 10 feet wide at the top. The tiles, 12 inches by 12 inches, are to be placed so that each successive row contains one less tile than the row below. How many tiles will be required? [Example 7 of Section 11.2] ■

This chapter introduces topics that are covered in more detail in courses titled *Finite Mathematics* or *Discrete Mathematics*. Applications of these topics can be found in the fields of computer science, engineering, business and economics, the social sciences, and the physical and biological sciences.

The chapter may be divided into four independent parts:

Sections 11.1–11.3, Sequences, which are functions whose domain is the set of natural numbers. Sequences form the basis for the *recursively defined* functions and *recursive procedures* used in computer programming.

Section 11.4, Mathematical Induction, a technique for proving theorems involving the natural numbers.

Section 11.5, the Binomial Theorem, a formula for the expansion of $(x + a)^n$, where n is any natural number.

Sections 11.6–11.8. The first two sections deal with techniques and formulas for counting the number of objects in a set, a part of the branch of mathematics called *combinatorics*. These formulas are used in computer science to analyze algorithms and recursive functions and to study stacks and queues. They are also used to determine *probabilities,* the likelihood that a certain outcome of a random experiment will occur.

11.1

Sequences

Sequence

A **sequence** is a function whose domain is the set of positive integers.

Because a sequence is a function, it will have a graph. In Figure 1(a), you will recognize the graph of the function $f(x) = 1/x$, $x > 0$. If all the points on this graph were removed except those whose x-coordinates are positive integers—that is, if all points were removed except $(1, 1)$, $(2, \frac{1}{2})$, $(3, \frac{1}{3})$ and so on—the remaining points would be the graph of the sequence $f(n) = 1/n$, as shown in Figure 1(b).

A sequence is usually represented by listing its values in order. For example, the sequence whose graph is given in Figure 1(b) might be represented as

$$f(1), f(2), f(3), f(4), \ldots \quad \text{or} \quad 1, \frac{1}{2}, \frac{1}{3}, \frac{1}{4}, \ldots$$

The list never ends, as the ellipsis dots indicate. The numbers in this ordered list are called the **terms** of the sequence.

In dealing with sequences, we usually use subscripted letters, for example, a_1 to represent the first term, a_2 for the second term, a_3 for the third term, and so on. Thus, for the sequence $f(n) = 1/n$, we write

$$a_1 = f(1) = 1$$

$$a_2 = f(2) = \frac{1}{2}$$

$$a_3 = f(3) = \frac{1}{3}$$

$$a_4 = f(4) = \frac{1}{4}$$

$$\vdots$$

$$a_n = f(n) = \frac{1}{n}$$

$$\vdots$$

FIGURE 1

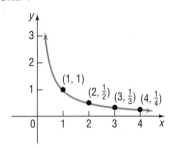

(a) $f(x) = \frac{1}{x}, x > 0$

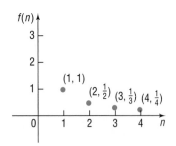

(b) $f(n) = \frac{1}{n}$

In other words, we usually do not use the traditional function notation $f(n)$ for sequences. For this particular sequence, we have a rule for the nth term, namely, $a_n = 1/n$, so it is easy to find any term of the sequence.

When a formula for the nth term of a sequence is known, rather than write out the terms of the sequence, we usually represent the entire sequence by placing braces around the formula for the nth term. For example, the sequence whose nth term is $b_n = (\frac{1}{2})^n$ may be represented as

$$\{b_n\} = \left\{\left(\frac{1}{2}\right)^n\right\}$$

or by

$$b_1 = \frac{1}{2}$$

$$b_2 = \frac{1}{4}$$

$$b_3 = \frac{1}{8}$$

$$\vdots$$

$$b_n = \left(\frac{1}{2}\right)^n$$

$$\vdots$$

E X A M P L E   1

Writing the First Several Terms of a Sequence

Write down the first six terms of the following sequence and graph it.

$$\{a_n\} = \left\{\frac{n-1}{n}\right\}$$

Solution

$$a_1 = 0$$

$$a_2 = \frac{1}{2}$$

$$a_3 = \frac{2}{3}$$

$$a_4 = \frac{3}{4}$$

$$a_5 = \frac{4}{5}$$

$$a_6 = \frac{5}{6}$$

FIGURE 2

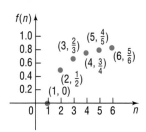

See Figure 2.

Check: Write a program that will generate the first six terms of the sequence $\{(n-1)/n\}$. Write a second program that will graph this sequence. Compare what you see with Figure 2.

E X A M P L E   2   *Writing the First Several Terms of a Sequence*

Write down the first six terms of the following sequence and graph it.

$$\{b_n\} = \left\{(-1)^{n-1}\left(\frac{2}{n}\right)\right\}$$

FIGURE 3

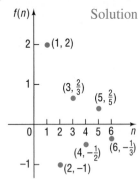

Solution

$$b_1 = 2$$

$$b_2 = -1$$

$$b_3 = \frac{2}{3}$$

$$b_4 = -\frac{1}{2}$$

$$b_5 = \frac{2}{5}$$

$$b_6 = -\frac{1}{3}$$

See Figure 3.   ■

E X A M P L E   3   *Writing the First Several Terms of a Sequence*

Write down the first six terms of the following sequence and graph it.

$$\{c_n\} = \begin{cases} n & \text{if } n \text{ is even} \\ 1/n & \text{if } n \text{ is odd} \end{cases}$$

Solution

FIGURE 4

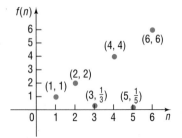

$$c_1 = 1$$

$$c_2 = 2$$

$$c_3 = \frac{1}{3}$$

$$c_4 = 4$$

$$c_5 = \frac{1}{5}$$

$$c_6 = 6$$

See Figure 4.   ■

■ Now work Problems 3 and 5.

Sometimes a sequence is indicated by an observed pattern in the first few terms that makes it possible to infer the makeup of the nth term. In the example that follows, a sufficient number of terms of the sequence is given so that a natural choice for the nth term is suggested.

E X A M P L E   4   *Determining a Sequence from a Pattern*

(a) $e, \dfrac{e^2}{2}, \dfrac{e^3}{3}, \dfrac{e^4}{4}, \ldots$ $a_n = \dfrac{e^n}{n}$

(b) $1, \dfrac{1}{3}, \dfrac{1}{9}, \dfrac{1}{27}, \ldots$ $b_n = \dfrac{1}{3^{n-1}}$

(c) 1, 3, 5, 7, . . . $c_n = 2n - 1$

(d) 1, 4, 9, 16, 25, . . . $d_n = n^2$

(e) $1, -\dfrac{1}{2}, \dfrac{1}{3}, -\dfrac{1}{4}, \dfrac{1}{5}, \ldots$ $e_n = (-1)^{n+1}\left(\dfrac{1}{n}\right)$ ■

Notice in the sequence $\{e_n\}$ in Example 4(e) that the signs of the terms **alternate.** When this occurs, we use factors such as $(-1)^{n+1}$, which equals 1 if n is odd and -1 if n is even, or $(-1)^n$, which equals -1 if n is odd and 1 if n is even.

■   Now work Problem 13.

The Factorial Symbol

Factorial Symbol, $n!$

If $n \geq 0$ is an integer, the **factorial symbol $n!$** is defined as follows:

$$0! = 1 \qquad 1! = 1$$
$$n! = n(n - 1) \cdot \ldots \cdot 3 \cdot 2 \cdot 1 \qquad \text{if } n \geq 2$$

For example, $2! = 2 \cdot 1 = 2$, $3! = 3 \cdot 2 \cdot 1 = 6$, $4! = 4 \cdot 3 \cdot 2 \cdot 1 = 24$, and so on. Table 1 lists the values of $n!$ for $0 \leq n \leq 6$.

TABLE 1

n	0	1	2	3	4	5	6
$n!$	1	1	2	6	24	120	720

Because

$$n! = \underbrace{n(n - 1)(n - 2) \cdot \ldots \cdot 3 \cdot 2 \cdot 1}_{(n-1)!}$$

we can use the formula

$$n! = n(n - 1)!$$

to find successive factorials. For example, because $6! = 720$, we have

$$7! = 7 \cdot 6! = 7(720) = 5040$$

and

$$8! = 8 \cdot 7! = 8(5040) = 40{,}320$$

Comment: Your calculator may have a factorial key. Use it to see how fast factorials increase in value. Find the value of 69!. What happens when you try to find 70!? In fact, 70! is larger than 10^{100} (a *googol*), the largest number most calculators can display.

Recursion Formulas

A second way of defining a sequence is to assign a value to the first (or the first few) terms and specify the nth term by a formula or equation that involves one or more of the terms preceding it. Sequences defined this way are said to be defined **recursively,** and the rule or formula is called a **recursive formula.**

E X A M P L E   5

Writing the Terms of a Recursively Defined Sequence

Write down the first five terms of the following recursively defined sequence.

$$s_1 = 1, \qquad s_n = 4s_{n-1}$$

Solution
The first term is given as $s_1 = 1$. To get the second term, we use $n = 2$ in the formula to get $s_2 = 4s_1 = 4 \cdot 1 = 4$. To get the third term, we use $n = 3$ in the formula to get $s_3 = 4s_2 = 4 \cdot 4 = 16$. To get a new term requires that we know the value of the preceding term. The first five terms are

$$s_1 = 1$$
$$s_2 = 4 \cdot 1 = 4$$
$$s_3 = 4 \cdot 4 = 16$$
$$s_4 = 4 \cdot 16 = 64$$
$$s_5 = 4 \cdot 64 = 256$$
∎

E X A M P L E   6

Writing the Terms of a Recursively Defined Sequence

Write down the first five terms of the following recursively defined sequence.

$$u_1 = 1, \qquad u_2 = 1, \qquad u_{n+2} = u_n + u_{n+1}$$

Solution
We are given the first two terms. To get the third term requires that we know each of the previous two terms. Thus,

$$u_1 = 1$$
$$u_2 = 1$$
$$u_3 = u_1 + u_2 = 2$$
$$u_4 = u_3 + u_2 = 2 + 1 = 3$$
$$u_5 = u_4 + u_3 = 3 + 2 = 5$$
∎

The sequence defined in Example 6 is called a **Fibonacci sequence,** and the terms of this sequence are called **Fibonacci numbers.** These numbers appear in a wide variety of applications (see Problems 55 and 56).

E X A M P L E   7

Writing the Terms of a Recursively Defined Sequence

Write down the first five terms of the following recursively defined sequence.

$$f_1 = 1, \qquad f_{n+1} = (n + 1)f_n$$

Solution   Here

$$f_1 = 1$$
$$f_2 = 2f_1 = 2 \cdot 1 = 2$$
$$f_3 = 3f_2 = 3 \cdot 2 = 6$$
$$f_4 = 4f_3 = 4 \cdot 6 = 24$$
$$f_5 = 5f_4 = 5 \cdot 24 = 120$$ ∎

You should recognize the nth term of the sequence in Example 7 as $n!$.

■ Now work Problems 21 and 29.

Adding the First n Terms of a Sequence; Summation Notation

It is often important to be able to find the sum of the first n terms of a sequence $\{a_n\}$, namely,

$$a_1 + a_2 + a_3 + \cdots + a_n \tag{1}$$

Rather than write down all these terms, we introduce a more concise way to express the sum, called **summation notation.** Using summation notation, we would write the sum (1) as

$$a_1 + a_2 + a_3 + \cdots + a_n = \sum_{k=1}^{n} a_k$$

The symbol Σ (a stylized version of the Greek letter sigma, which is an S in our alphabet) is simply an instruction to sum, or add up, the terms. The integer k is called the **index** of the sum; it tells you where to start the sum and where to end it. Therefore, the expression

$$\sum_{k=1}^{n} a_k \tag{2}$$

is an instruction to add the terms a_k of the sequence $\{a_n\}$ from $k = 1$ through $k = n$. We read expression (2) as "the sum of a_k from $k = 1$ to $k = n$."

E X A M P L E   8   *Expanding Summation Notation*

Write out each sum.

(a) $\displaystyle\sum_{k=1}^{n} \frac{1}{k}$ (b) $\displaystyle\sum_{k=1}^{n} k!$

Solution   (a) $\displaystyle\sum_{k=1}^{n} \frac{1}{k} = \frac{1}{1} + \frac{1}{2} + \frac{1}{3} + \cdots + \frac{1}{n}$ (b) $\displaystyle\sum_{k=1}^{n} k! = 1! + 2! + \cdots + n!$ ∎

E X A M P L E   9   *Writing a Sum in Summation Notation*

Express each sum using summation notation.

(a) $1^2 + 2^2 + 3^2 + \cdots + n^2$ (b) $1 + \dfrac{1}{2} + \dfrac{1}{4} + \dfrac{1}{8} + \cdots + \dfrac{1}{2^{n-1}}$

Solution (a) The sum $1^2 + 2^2 + 3^2 + \cdots + n^2$ has n terms, each of the form k^2, and starts at $k = 1$ and ends at $k = n$. Thus,

$$1^2 + 2^2 + 3^2 + \cdots + n^2 = \sum_{k=1}^{n} k^2$$

(b) The sum

$$1 + \frac{1}{2} + \frac{1}{4} + \frac{1}{8} + \cdots + \frac{1}{2^{n-1}}$$

has n terms, each of the form $1/2^{k-1}$, and starts at $k = 1$ and ends at $k = n$. Thus,

$$1 + \frac{1}{2} + \frac{1}{4} + \frac{1}{8} + \cdots + \frac{1}{2^{n-1}} = \sum_{k=1}^{n} \frac{1}{2^{k-1}}$$ ■

The index of summation need not always begin at 1 nor end at n; for example,

$$\sum_{k=0}^{n-1} \frac{1}{2^k} = 1 + \frac{1}{2} + \frac{1}{4} + \cdots + \frac{1}{2^{n-1}}$$

Letters other than k may be used as the index. For example,

$$\sum_{j=1}^{n} j! \quad \text{and} \quad \sum_{i=1}^{n} i!$$

each represent the same sum as the one given in Example 8(b).

■ Now work Problems 37 and 47.

Next, we list some properties of sequences using summation notation.

Theorem
Properties of Sequences

If $\{a_n\}$ and $\{b_n\}$ are two sequences and c is a real number, then

1. $\displaystyle\sum_{k=1}^{n} ca_k = c \sum_{k=1}^{n} a_k$

2. $\displaystyle\sum_{k=1}^{n} (a_k + b_k) = \sum_{k=1}^{n} a_k + \sum_{k=1}^{n} b_k$

3. $\displaystyle\sum_{k=1}^{n} (a_k - b_k) = \sum_{k=1}^{n} a_k - \sum_{k=1}^{n} b_k$

4. $\displaystyle\sum_{k=1}^{n} a_k = \sum_{k=1}^{j} a_k + \sum_{k=j+1}^{n} a_k, \quad \text{when } 1 < j < n$ ■

Although we shall not prove these properties, the proofs are based on properties of real numbers.

11.1

Exercise 11.1

In Problems 1–12, write down the first five terms of each sequence.

1. $\{n\}$

2. $\{n^2 + 1\}$

3. $\left\{\dfrac{n}{n + 2}\right\}$

4. $\left\{\dfrac{2n + 1}{2n}\right\}$

5. $\{(-1)^{n+1} n^2\}$

6. $\left\{(-1)^{n-1}\left(\dfrac{n}{2n - 1}\right)\right\}$

7. $\left\{\dfrac{2^n}{3^n + 1}\right\}$

8. $\left\{\left(\dfrac{4}{3}\right)^n\right\}$

9. $\left\{\dfrac{(-1)^n}{(n + 1)(n + 2)}\right\}$

10. $\left\{\dfrac{3^n}{n}\right\}$

11. $\left\{\dfrac{n}{e^n}\right\}$

12. $\left\{\dfrac{n^2}{2^n}\right\}$

In Problems 13–20, the given pattern continues. Write down the nth term of each sequence suggested by the pattern.

13. $\dfrac{1}{2}, \dfrac{2}{3}, \dfrac{3}{4}, \dfrac{4}{5}, \ldots$

14. $\dfrac{1}{1 \cdot 2}, \dfrac{1}{2 \cdot 3}, \dfrac{1}{3 \cdot 4}, \dfrac{1}{4 \cdot 5}, \ldots$

15. $1, \dfrac{1}{2}, \dfrac{1}{4}, \dfrac{1}{8}, \ldots$

16. $\dfrac{2}{3}, \dfrac{4}{9}, \dfrac{8}{27}, \dfrac{16}{81}, \ldots$

17. $1, -1, 1, -1, 1, -1, \ldots$

18. $1, \dfrac{1}{2}, 3, \dfrac{1}{4}, 5, \dfrac{1}{6}, 7, \dfrac{1}{8}, \ldots$

19. $1, -2, 3, -4, 5, -6, \ldots$

20. $2, -4, 6, -8, 10, \ldots$

In Problems 21–34, a sequence is defined recursively. Write the first five terms.

21. $a_1 = 2; \; a_{n+1} = 3 + a_n$

22. $a_1 = 3; \; a_{n+1} = 4 - a_n$

23. $a_1 = -2; \; a_{n+1} = n + a_n$

24. $a_1 = 1; \; a_{n+1} = n - a_n$

25. $a_1 = 5; \; a_{n+1} = 2a_n$

26. $a_1 = 2; \; a_{n+1} = -a_n$

27. $a_1 = 3; \; a_{n+1} = \dfrac{a_n}{n}$

28. $a_1 = -2; \; a_{n+1} = n + 3a_n$

30. $a_1 = -1; \; a_2 = 1; \; a_{n+2} = a_{n+1} + na_n$

29. $a_1 = 1; \; a_2 = 2; \; a_{n+2} = a_n a_{n+1}$

32. $a_1 = A; \; a_{n+1} = ra_n, \; r \neq 0$

31. $a_1 = A; \; a_{n+1} = a_n + d$

34. $a_1 = \sqrt{2}; \; a_{n+1} = \sqrt{a_n/2}$

33. $a_1 = \sqrt{2}; \; a_{n+1} = \sqrt{2 + a_n}$

In Problems 35–44, write out each sum.

35. $\displaystyle\sum_{k=1}^{n} (k + 2)$

36. $\displaystyle\sum_{k=1}^{n} (2k + 1)$

37. $\displaystyle\sum_{k=1}^{n} \dfrac{k^2}{2}$

38. $\displaystyle\sum_{k=1}^{n} (k + 1)^2$

39. $\displaystyle\sum_{k=0}^{n} \dfrac{1}{3^k}$

40. $\displaystyle\sum_{k=0}^{n} \left(\dfrac{3}{2}\right)^k$

41. $\displaystyle\sum_{k=0}^{n-1} \dfrac{1}{3^{k+1}}$

42. $\displaystyle\sum_{k=0}^{n-1} (2k + 1)$

43. $\displaystyle\sum_{k=2}^{n} (-1)^k \ln k$

44. $\displaystyle\sum_{k=3}^{n} (-1)^{k+1} 2^k$

In Problems 45–54, express each sum using summation notation.

45. $1 + 2 + 3 + \cdots + n$

46. $1^3 + 2^3 + 3^3 + \cdots + n^3$

47. $\dfrac{1}{2} + \dfrac{2}{3} + \dfrac{3}{4} + \cdots + \dfrac{n}{n + 1}$

48. $1 + 3 + 5 + 7 + \cdots + (2n - 1)$

49. $1 - \dfrac{1}{3} + \dfrac{1}{9} - \dfrac{1}{27} + \cdots + (-1)^n \left(\dfrac{1}{3^n}\right)$

50. $\dfrac{2}{3} - \dfrac{4}{9} + \dfrac{8}{27} - \cdots + (-1)^{n+1} \left(\dfrac{2}{3}\right)^n$

51. $3 + \dfrac{3^2}{2} + \dfrac{3^3}{3} + \cdots + \dfrac{3^n}{n}$

52. $\dfrac{1}{e} + \dfrac{2}{e^2} + \dfrac{3}{e^3} + \cdots + \dfrac{n}{e^n}$

53. $a + (a + d) + (a + 2d) + \cdots + (a + nd)$

54. $a + ar + ar^2 + \cdots + ar^{n-1}$

55. *Growth of a Rabbit Colony* A colony of rabbits begins with one pair of mature rabbits, which will produce a pair of offspring (one male, one female) each month. Assume that all rabbits mature in 1 month and produce a pair of offspring (one male, one female) after 2 months. If no rabbits ever die, how many pairs of mature rabbits are there after 7 months? [*Hint:* A Fibonacci sequence models this colony. Do you see why?]

56. *Fibonacci Sequence* Let

$$u_n = \frac{(1 + \sqrt{5})^n - (1 - \sqrt{5})^n}{2^n \sqrt{5}}$$

define the *n*th term of a sequence.

(a) Show that $u_1 = 1$ and $u_2 = 1$.   (b) Show that $u_{n+2} = u_{n+1} + u_n$.

(c) Draw the conclusion that $\{u_n\}$ is a Fibonacci sequence.

In Problems 57 and 58, we use the fact that in some programming languages it is possible to have a function subroutine include a call to itself.

57. *Programming Exercise* Write a program that accepts integer as input and prints the number and its factorial. Use a recursively defined function; that is, use a function subroutine that calls itself.

58. *Programming Exercise* Write a program that accepts a positive integer N as input and outputs the Nth Fibonacci number. Use a recursively defined subroutine.

59. Divide the triangular array below (called Pascal's triangle) using diagonal lines as shown. Find the sum of the numbers in each of these diagonal rows. Do you recognize this sequence?

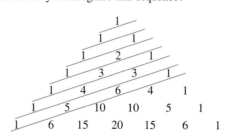

60. Investigate various applications that lead to a Fibonacci sequence. Write an essay on these applications.

11.2

Arithmetic Sequences

When the difference between successive terms of a sequence is always the same number, the sequence is called **arithmetic.** Thus, an arithmetic sequence* may be defined recursively as $a_1 = a$, $a_{n+1} - a_n = d$, or as

Arithmetic Sequence

$$a_1 = a, \qquad a_{n+1} = a_n + d \qquad\qquad (1)$$

where $a = a_1$ and d are real numbers. The number a is the first term, and the number d is called the **common difference.**

Thus, the terms of an arithmetic sequence with first term a and common difference d follow the pattern

$$a, \quad a + d, \quad a + 2d, \quad a + 3d, \quad \cdots$$

EXAMPLE 1   *Determining If a Sequence is Arithmetic*

The sequence

$$4, 7, 10, 13, \ldots$$

is arithmetic since the difference of successive terms is 3. The first term is 4, and the common difference is 3.   ■

EXAMPLE 2   *Determining If a Sequence is Arithmetic*

Show that the following sequence is arithmetic. Find the first term and the common difference.

$$\{s_n\} = \{3n + 5\}$$

Solution   The first term is $s_1 = 3 \cdot 1 + 5 = 8$. The $(n + 1)$st and nth terms of the sequence $\{s_n\}$ are

$$s_{n+1} = 3(n + 1) + 5 = 3n + 8 \quad \text{and} \quad s_n = 3n + 5$$

Their difference is

$$s_{n+1} - s_n = (3n + 8) - (3n + 5) = 8 - 5 = 3$$

Thus, the difference of two successive terms does not depend on n; the common difference is 3, and the sequence is arithmetic.   ■

EXAMPLE 3   *Determining If a Sequence is Arithmetic*

Show that the sequence is arithmetic. Find the first term and the common difference.

$$\{t_n\} = \{4 - n\}$$

Solution   The first term is $t_1 = 4 - 1 = 3$. The $(n + 1)$st and nth terms are

$$t_{n+1} = 4 - (n + 1) = 3 - n \quad \text{and} \quad t_n = 4 - n$$

*Sometimes called an **arithmetic progression.**

Their difference is

$$t_{n+1} - t_n = (3 - n) - (4 - n) = 3 - 4 = -1$$

The difference of two successive terms does not depend on n; it always equals the same number, -1. Hence, $\{t_n\}$ is an arithmetic sequence whose common difference is -1. ∎

■   Now work Problem 3.

Suppose that a is the first term of an arithmetic sequence whose common difference is d. We seek a formula for the nth term, a_n. To see the pattern, we write down the first few terms:

$$a_1 = a + 0 \cdot d$$
$$a_2 = a + d = a + 1 \cdot d$$
$$a_3 = a_2 + d = (a + d) + d = a + 2 \cdot d$$
$$a_4 = a_3 + d = (a + 2 \cdot d) + d = a + 3 \cdot d$$
$$a_5 = a_4 + d = (a + 3 \cdot d) + d = a + 4 \cdot d$$
$$\vdots$$
$$a_n = a_{n-1} + d = [a + (n - 2)d] + d = a + (n - 1)d$$

We are led to the following result:

Theorem   For an arithmetic sequence $\{a_n\}$ whose first term is a and whose common difference is d, the nth term is determined by the formula

---

nth Term of an Arithmetic Sequence

$$a_n = a + (n - 1)d \qquad (2)$$

E X A M P L E   4   *Finding a Particular Term of an Arithmetic Sequence*

Find the 13th term of the arithmetic sequence   2, 6, 10, 14, 18, ...

Solution   The first term of this arithmetic sequence is $a = 2$, and the common difference is 4. By formula (2), the nth term is

$$a_n = 2 + (n - 1)4$$

Hence, the 13th term is

$$a_{13} = 2 + 12 \cdot 4 = 50$$ ∎

Exploration: Write a program that will find the 13th term of the sequence given in Example 4. Use it to find the 20th term and the 50th term. ∎

E X A M P L E   5   *Finding a Recursive Formula for an Arithmetic Sequence*

The 8th term of an arithmetic sequence is 75, and the 20th term is 39. Find the first term and the common difference. Give a recursive formula for the sequence.

Solution   By equation (2), we know that

$$\begin{cases} a_8 = a + \ 7d = 75 \\ a_{20} = a + 19d = 39 \end{cases}$$

This is a system of two linear equations containing two variables, which we can solve by elimination. Thus, subtracting the second equation from the first equation, we get

$$-12d = 36$$
$$d = -3$$

With $d = -3$, we find $a = 75 - 7d = 75 - 7(-3) = 96$. A recursive formula for this sequence is

$$a_1 = 96, \qquad a_{n+1} = a_n - 3 \qquad \blacksquare$$

Based on formula (2), a formula for the nth term of the sequence $\{a_n\}$ in Example 5 is

$$a_n = a + (n - 1)d = 96 + (n - 1)(-3) = 99 - 3n$$

■   Now work Problems 19 and 25.

Adding the First n Terms of an Arithmetic Sequence

The next result gives a formula for finding the sum of the first n terms of an arithmetic sequence.

Theorem   Let $\{a_n\}$ be an arithmetic sequence with first term a and common difference d. The sum S_n of the first n terms of $\{a_n\}$ is

Sum of n Terms of an Arithmetic Sequence

$$S_n = \frac{n}{2}[2a + (n - 1)d] = \frac{n}{2}(a + a_n) \qquad (3)$$

Proof

$$
\begin{aligned}
S_n &= a_1 + a_2 + a_3 + \cdots + a_n \\
&= a + (a + d) + (a + 2d) + \cdots + [a + (n - 1)d] \\
&= \underbrace{(a + a + \cdots + a)}_{n \text{ terms}} + [d + 2d + \cdots + (n - 1)d] \\
&= na + d[1 + 2 + \cdots + (n - 1)] \\
&= na + d\left[\frac{(n - 1)n}{2}\right] \\
&= na + \frac{n}{2}(n - 1)d \\
&= \frac{n}{2}[2a + (n - 1)d] \qquad \text{Factor out } n/2. \\
&= \frac{n}{2}[a + a + (n - 1)d] \\
&= \frac{n}{2}(a + a_n) \qquad \qquad \text{Formula (2)} \qquad \blacksquare
\end{aligned}
$$

Formula (3) provides two ways to find the sum of the first n terms of an arithmetic sequence. Notice that one involves the first term and common difference, while the other involves the first term and the nth term. Use whichever form is easier.

E X A M P L E   6   *Finding the Sum of n Terms of an Arithmetic Sequence*

Find the sum S_n of the first n terms of the sequence $\{3n + 5\}$; that is, find

$$8 + 11 + 14 + \cdots + (3n + 5)$$

Solution   The sequence $\{3n + 5\}$ is an arithmetic sequence with first term $a = 8$ and the nth term $(3n + 5)$. To find the sum S_n we use formula (3):

$$S_n = \frac{n}{2}(a + a_n) = \frac{n}{2}[8 + (3n + 5)] = \frac{n}{2}(3n + 13)$$ ■

■   Now work Problem 33.

E X A M P L E   7   *Creating a Floor Design*

A ceramic tile floor is designed in the shape of a trapezoid 20 feet wide at the base and 10 feet wide at the top. See Figure 5. The tiles, 12 inches by 12 inches, are to be placed so that each successive row contains one less tile than the row below. How many tiles will be required?

FIGURE 5

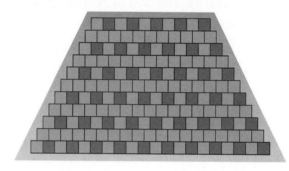

Solution   The bottom row requires 20 tiles and the top row 10 tiles. Since each successive row requires one less tile, the total number of tiles required is

$$S = 20 + 19 + 18 + \cdots + 11 + 10$$

This is the sum of an arithmetic sequence; the common difference is -1. The number of terms to be added is $n = 11$, with the first term $a = 20$ and the last term $a_{11} = 10$. The sum S is

$$S = \frac{n}{2}(a + a_{11}) = \frac{11}{2}(20 + 10) = 165$$

Thus, 165 tiles will be required.   ■

11.2

Exercise 11.2

In Problems 1–10, an arithmetic sequence is given. Find the common difference and write out the first four terms.

1. $\{n + 4\}$ **2.** $\{n - 5\}$ **3.** $\{2n - 5\}$ **4.** $\{3n + 1\}$ **5.** $\{6 - 2n\}$

6. $\{4 - 2n\}$ **7.** $\left\{\frac{1}{2} - \frac{1}{3}n\right\}$ **8.** $\left\{\frac{2}{3} + \frac{n}{4}\right\}$ **9.** $\{\ln 3^n\}$ **10.** $\{e^{\ln n}\}$

In Problems 11–18, find the nth term of the arithmetic sequence whose initial term a *and common difference* d *are given. What is the fifth term?*

11. $a = 2; d = 3$

12. $a = -2; d = 4$

13. $a = 5; d = -3$

14. $a = 6; d = -2$

15. $a = 0; d = \frac{1}{2}$

16. $a = 1; \ d = -\frac{1}{3}$

17. $a = \sqrt{2}; d = \sqrt{2}$

18. $a = 0; d = \pi$

In Problems 19–24, find the indicated term in each arithmetic sequence.

19. 12th term of 2, 4, 6, . . .

20. 8th term of $-1, 1, 3, . . .$

21. 10th term of $1, -2, -5, . . .$

22. 9th term of $5, 0, -5, . . .$

23. 8th term of $a, a + b, a + 2b, . . .$

24. 7th term of $2\sqrt{5}, 4\sqrt{5}, 6\sqrt{5}, . . .$

In Problems 25–32, find the first term and the common difference of the arithmetic sequence described. Give a recursive formula for the sequence.

25. 8th term is 8; 20th term is 44

26. 4th term is 3; 20th term is 35

27. 9th term is -5; 15th term is 31

28. 8th term is 4; 18th term is -96

29. 15th term is 0; 40th term is -50

30. 5th term is -2; 13th term is 30

31. 14th term is -1; 18th term is -9

32. 12th term is 4; 18th term is 28

In Problems 33–40, find the sum.

33. $1 + 3 + 5 + \cdots + (2n - 1)$

34. $2 + 4 + 6 + \cdots + 2n$

35. $7 + 12 + 17 + \cdots + (2 + 5n)$

36. $-1 + 3 + 7 + \cdots + (4n - 5)$

37. $2 + 4 + 6 + \cdots + 70$

38. $1 + 3 + 5 + \cdots + 59$

39. $5 + 9 + 13 + \cdots + 49$

40. $2 + 5 + 8 + \cdots + 41$

41. Find x so that $x + 3, 2x + 1$, and $5x + 2$ are terms of an arithmetic sequence.

42. Find x so that $2x, 3x + 2$, and $5x + 3$ are terms of an arithmetic sequence.

43. *Drury Lane Theater* The Drury Lane Theater has 25 seats in the first row and 30 rows in all. Each successive row contains one additional seat. How many seats are in the theater?

44. *Football Stadium* The corner section of a football stadium has 15 seats in the first row and 40 rows in all. Each successive row contains two additional seats. How many seats are in this section?

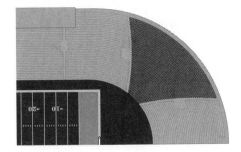

45. *Creating a Mosaic* A mosaic is designed in the shape of an equilateral triangle, 20 feet on each side. Each tile in the mosaic is in the shape of an equilateral triangle, 12 inches to a side. The tiles are to alternate in color as shown in the illustration. How many tiles of each color will be required?

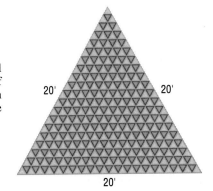

46. *Constructing a Brick Staircase* A brick staircase has a total of 30 steps. The bottom step requires 100 bricks. Each successive step requires two less bricks than the prior step.

(a) How many bricks are required for the top step?

(b) How many bricks are required to build the staircase?

 47. Make up an arithmetic sequence. Give it to a friend and ask for its 20th term.

11.3

Geometric Sequences; Geometric Series

When the ratio of successive terms of a sequence is always the same nonzero number, the sequence is called **geometric.** Thus, a geometric sequence* may be defined recursively as $a_1 = a$, $a_{n+1} / a_n = r$, or as

Geometric Sequence

$$a_1 = a, \qquad a_{n+1} = ra_n \qquad\qquad (1)$$

where $a_1 = a$ and $r \neq 0$ are real numbers. The number a is the first term, and the nonzero number r is called the **common ratio.**

Thus, the terms of a geometric sequence with first term a and common ratio r follow the pattern

$$a, \ ar, \ ar^2, \ ar^3, \ \ldots$$

EXAMPLE 1 *Determining If a Sequence is Geometric*

The sequence

$$2, 6, 18, 54, 162, \ldots$$

is geometric since the ratio of successive terms is 3. The first term is 2, and the common ratio is 3. ∎

EXAMPLE 2 *Determining If a Sequence is Geometric*

Show that the following sequence is geometric. Find the first term and the common ratio.

$$\{s_n\} = 2^{-n}$$

Solution   The first term is $s_1 = 2^{-1} = \frac{1}{2}$. The $(n + 1)$st and nth terms of the sequence $\{s_n\}$ are

$$s_{n+1} = 2^{-(n+1)} \quad \text{and} \quad s_n = 2^{-n}$$

Their ratio is

$$\frac{s_{n+1}}{s_n} = \frac{2^{-(n+1)}}{2^{-n}} = 2^{-n-1+n} = 2^{-1} = \frac{1}{2}$$

Because the ratio of successive terms is a nonzero number independent of n, the sequence $\{s_n\}$ is geometric with common ratio $\frac{1}{2}$. ∎

*Sometimes called a **geometric progression.**

E X A M P L E   3   *Determining If a Sequence is Geometric*

Show that the following sequence is geometric. Find the first term and the common ratio.

$$\{t_n\} = \{4^n\}$$

Solution   The first term is $t_1 = 4^1 = 4$. The $(n+1)$st and nth terms are

$$t_{n+1} = 4^{n+1} \quad \text{and} \quad t_n = 4^n$$

Their ratio is

$$\frac{t_{n+1}}{t_n} = \frac{4^{n+1}}{4^n} = 4$$

Thus, $\{t_n\}$ is a geometric sequence with common ratio 4. ∎

■ Now work Problem 3.

Suppose a is the first term of a geometric sequence with common ratio $r \neq 0$. We seek a formula for the nth term a_n. To see the pattern, we write down the first few terms:

$$a_1 = 1 \cdot a = ar^0$$
$$a_2 = ra = ar^1$$
$$a_3 = ra_2 = r(ar) = ar^2$$
$$a_4 = ra_3 = r(ar^2) = ar^3$$
$$a_5 = ra_4 = r(ar^3) = ar^4$$
$$\vdots$$
$$a_n = ra_{n-1} = r(ar^{n-2}) = ar^{n-1}$$

We are led to the following result:

Theorem   For a geometric sequence $\{a_n\}$ whose first term is a and whose common ratio is r, the nth term is determined by the formula

*n*th Term of a Geometric
Sequence

$$a_n = ar^{n-1}, \quad r \neq 0 \tag{2}$$

∎

E X A M P L E   4   *Finding a Particular Term of a Geometric Sequence*

Find the 9th term of the geometric sequence $2, \frac{2}{3}, \frac{2}{9}, \frac{2}{27}, \ldots$.

Solution   The first term of this geometric sequence is $a = 2$, and the common ratio is $\frac{1}{3}$. (Use $\frac{2}{3}/2 = \frac{1}{3}$, or $\frac{2}{9}/\frac{2}{3} = \frac{1}{3}$, or any two successive terms.) By formula (2), the nth term is

$$a_n = 2\left(\frac{1}{3}\right)^{n-1}$$

Hence, the 9th term is

$$a_9 = 2\left(\frac{1}{3}\right)^8 = \frac{2}{3^8} = \frac{2}{6561}$$

∎

Exploration: Write a program that will find the 9th term of the sequence given in Example 4. Use it to find the 20th term and the 50th term. ■

■ Now work Problems 25 and 33.

Adding the First n Terms of a Geometric Sequence

The next result gives us a formula for finding the sum of the first n terms of a geometric sequence.

Theorem   Let $\{a_n\}$ be a geometric sequence with first term a and common ratio r. The sum S_n of the first n terms of $\{a_n\}$ is

Sum of n Terms of a Geometric Sequence

$$S_n = a\frac{1 - r^n}{1 - r}, \qquad r \neq 0, 1 \qquad (3)$$

Proof

$$S_n = a + ar + \cdots + ar^{n-1} \qquad (4)$$

Multiply each side by r to obtain

$$rS_n = ar + ar^2 + \cdots + ar^n \qquad (5)$$

Now, subtract (5) from (4). The result is

$$S_n - rS_n = a - ar^n$$
$$(1 - r)S_n = a(1 - r^n)$$

Since $r \neq 1$, we can solve for S_n:

$$S_n = a\frac{1 - r^n}{1 - r} \qquad ■$$

E X A M P L E   5   *Finding the Sum of n Terms of a Geometric Sequence*

Find the sum S_n of the first n terms of the sequence $\left\{\left(\frac{1}{2}\right)^n\right\}$; that is, find

$$\frac{1}{2} + \frac{1}{4} + \frac{1}{8} + \cdots + \left(\frac{1}{2}\right)^n$$

Solution   The sequence $\left\{\left(\frac{1}{2}\right)^n\right\}$ is a geometric sequence with $a = \frac{1}{2}$ and $r = \frac{1}{2}$. The sum S_n that we seek is the sum of the first n terms of the sequence, so we use formula (3) to get

$$S_n = \sum_{k=1}^{n}\left(\frac{1}{2}\right)^k = \frac{1}{2} + \frac{1}{4} + \frac{1}{8} + \cdots + \left(\frac{1}{2}\right)^n$$

$$= \frac{1}{2}\left[\frac{1 - \left(\frac{1}{2}\right)^n}{1 - \frac{1}{2}}\right]$$

$$= \frac{1}{2}\left[\frac{1 - \left(\frac{1}{2}\right)^n}{\frac{1}{2}}\right]$$

$$= 1 - \left(\frac{1}{2}\right)^n \qquad ■$$

■ Now work Problem 39.

Geometric Series

Infinite Geometric Series

An infinite sum of the form

$$a + ar + ar^2 + \cdots + ar^{n-1} + \cdots$$

with first term a and common ratio r, is called an **infinite geometric series** and is denoted by

$$\sum_{k=1}^{\infty} ar^{k-1}$$

Based on formula (3), the sum S_n of the first n terms of a geometric series is

$$S_n = a\frac{1 - r^n}{1 - r} = \frac{a}{1 - r} - \frac{ar^n}{1 - r} \qquad (6)$$

If this finite sum S_n approaches a number L as $n \to \infty$, then we call L the **sum of the infinite geometric series,** and we write

$$L = \sum_{k=1}^{\infty} ar^{k-1}$$

Theorem   If $|r| < 1$, the sum of the infinite geometric series $\displaystyle\sum_{k=1}^{\infty} ar^{k-1}$ is

Sum of an Infinite Geometric Series

$$\sum_{k=1}^{\infty} ar^{k-1} = \frac{a}{1 - r} \qquad (7)$$

■

Intuitive Proof   Since $|r| < 1$, it follows that $|r^n|$ approaches 0 as $n \to \infty$. Then, based on formula (6), the sum S_n approaches $a/(1 - r)$ as $n \to \infty$.   ■

E X A M P L E   6   *Finding the Sum of a Geometric Series*

Find the sum of the geometric series $2 + \frac{4}{3} + \frac{8}{9} + \cdots$.

Solution   The first term is $a = 2$ and the common ratio is

$$r = \frac{\frac{4}{3}}{2} = \frac{4}{6} = \frac{2}{3}$$

Since $|r| < 1$, we use formula (7) to find that

$$2 + \frac{4}{3} + \frac{8}{9} + \cdots = \frac{2}{1 - \frac{2}{3}} = 6$$

■

■ Now work Problem 45.

E X A M P L E   7   *Repeating Decimals*

Show that the repeating decimal 0.999 . . . equals 1.

Solution

$$0.999\ldots = \frac{9}{10} + \frac{9}{100} + \frac{9}{1000} + \cdots$$

Thus, $0.999\ldots$ is a geometric series with first term $\frac{9}{10}$ and common ratio $\frac{1}{10}$. Hence,

$$0.999\ldots = \frac{\frac{9}{10}}{1 - \frac{1}{10}} = \frac{\frac{9}{10}}{\frac{9}{10}} = 1 \qquad \blacksquare$$

E X A M P L E   8

Pendulum Swings

Initially, a pendulum swings through an arc of 18 inches. On each successive swing, the length of the arc is 0.98 of the previous length.

(a) What is the length of arc after 10 swings?
(b) On which swing is the length of arc first less than 12 inches?
(c) After 15 swings, what total length will the pendulum have swung?
(d) When it stops, what total length will the pendulum have swung?

Solution

FIGURE 11

(a) The length of the first swing is 18 inches. The length of the second swing is $0.98(18)$ inches; the length of the third swing is $0.98(0.98)(18) = 0.98^2(18)$ inches. The length of arc of the 10th swing is

$$(0.98)^9(18) = 15.007 \text{ inches}$$

(b) The length of arc of the nth swing is $(0.98)^{n-1}(18)$. For this to be exactly 12 inches requires

$$(0.98)^{n-1}(18) = 12$$

$$(0.98)^{n-1} = \frac{12}{18} = \frac{2}{3}$$

$$n - 1 = \log_{0.98}\left(\frac{2}{3}\right)$$

$$n = 1 + \frac{\ln\left(\frac{2}{3}\right)}{\ln 0.98} = 1 + 20.07 = 21.07$$

The length of arc of the pendulum exceeds 12 inches on the 21st swing and is first less than 12 inches on the 22nd swing.

(c) After 15 swings, the pendulum will have swung the following total length L:

$$L = \underset{\text{1st}}{18} + \underset{\text{2nd}}{0.98(18)} + \underset{\text{3rd}}{(0.98)^2(18)} + \underset{\text{4th}}{(0.98)^3(18)} + \cdots + \underset{\text{15th}}{(0.98)^{14}(18)}$$

This is the sum of a geometric sequence. The common ratio is 0.98; the first term is 18. The sum has 15 terms, so

$$L = 18\,\frac{1 - 0.98^{15}}{1 - 0.98} = 18(13.07) = 235.29 \text{ inches}$$

The pendulum will have swung through 235.29 inches after 15 swings.

(d) When the pendulum stops, it will have swung the following total length T:

$$T = 18 + 0.98(18) + (0.98)^2(18) + (0.98)^3(18) + \cdots$$

This is the sum of a geometric series. The common ratio is $r = 0.98$; the first term is $a = 18$. The sum is

$$T = \frac{a}{1 - r} = \frac{18}{1 - 0.98} = 900$$

The pendulum will have swung a total of 900 inches when it finally stops.   ■

HISTORICAL FEATURE   ■ Sequences are among the oldest objects of mathematical investigation, having been studied for over 3500 years. After the initial steps, however, little progress was made until about 1600.

Arithmetic and geometric sequences appear in the Rhind papyrus, a mathematical text containing 85 problems copied around 1650 BC by the Egyptian scribe Ahmes from an earlier work (see Historical Problems 1). Fibonacci (AD 1220) wrote about problems similar to those found in the Rhind papyrus, leading one to suspect that Fibonacci may have had material available that is now lost. This material would have been in the non-Euclidean Greek tradition of Heron (about AD 75) and Diophantus (about AD 250). One problem, again modified slightly, is still with us in the familiar puzzle rhyme "As I was going to St. Ives . . . " (see Historical Problem 2).

The Rhind papyrus indicates that the Egyptians knew how to add up the terms of an arithmetic or geometric sequence, as did the Babylonians. The rule for summing up a geometric sequence is found in Euclid's *Elements* (book IX, 35, 36), where, like all of Euclid's algebra, it is presented in a geometric form.

Investigations of other kinds of sequences began in the 1500's, when algebra became sufficiently developed to handle the more complicated problems. The development of calculus in the 1600's added a powerful new tool, especially for finding the sum of infinite series, and the subject continues to flourish today.   ■

HISTORICAL PROBLEMS   ■ 1. *Arithmetic sequence problem from the Rhind papyrus (statement modified slightly for clarity)* One hundred loaves of bread are to be divided among five people so that the amounts they receive form an arithmetic sequence. The first two together receive one-seventh of what the last three receive. How many does each receive? [*Partial answer:* First person receives $1\frac{2}{3}$ loaves.]

2. The following old English children's rhyme resembles one of the Rhind papyrus problems:

> As I was going to St. Ives
> I met a man with seven wives
> Each wife had seven sacks
> Each sack had seven cats
> Each cat had seven kits [kittens]
> Kits, cats, sacks, wives
> How many were going to St. Ives?

(a) Assuming that the speaker and the cat fanciers met by traveling in opposite directions, what is the answer?

(b) How many kittens are being transported?

(c) Kits, cats, sacks, wives; how many? [*Hint:* It is easier to include the man, find the sum with the formula, and then subtract 1 for the man.]   ■

11.3

Exercise 11.3

In Problems 1–10, a geometric sequence is given. Find the common ratio and write out the first four terms.

1. $\{3^n\}$ **2.** $\{(-5)^n\}$ **3.** $\left\{-3\left(\frac{1}{2}\right)^n\right\}$ **4.** $\left\{\left(\frac{5}{2}\right)^n\right\}$ **5.** $\left\{\frac{2^{n-1}}{4}\right\}$

6. $\left\{\frac{3^n}{9}\right\}$ **7.** $\{2^{n/3}\}$ **8.** $\{3^{2n}\}$ **9.** $\left\{\frac{3^{n-1}}{2^n}\right\}$ **10.** $\left\{\frac{2^n}{3^{n-1}}\right\}$

In Problems 11–24, determine whether the given sequence is arithmetic, geometric, or neither. If the sequence is arithmetic, find the common difference; if it is geometric, find the common ratio.

11. $\{n+2\}$ **12.** $\{2n-5\}$ **13.** $\{4n^2\}$ **14.** $\{5n^2+1\}$ **15.** $\left\{3-\frac{2}{3}n\right\}$

16. $\left\{8-\frac{3}{4}n\right\}$ **17.** $1, 3, 6, 10, \ldots$ **18.** $2, 4, 6, 8, \ldots$ **19.** $\left\{\left(\frac{2}{3}\right)^n\right\}$ **20.** $\left\{\left(\frac{5}{4}\right)^n\right\}$

21. $-1, -2, -4, -8, \ldots$ **22.** $1, 1, 2, 3, 5, 8, \ldots$ **23.** $\{3^{n/2}\}$ **24.** $\{(-1)^n\}$

In Problems 25–32, find the fifth term and the nth term of the geometric sequence whose initial term a *and common ratio* r *are given.*

25. $a=2; r=3$ **26.** $a=-2; r=4$ **27.** $a=5; r=-1$

28. $a=6; r=-2$ **29.** $a=0; r=\frac{1}{2}$ **30.** $a=1; r=-\frac{1}{3}$

31. $a=\sqrt{2}; r=\sqrt{2}$ **32.** $a=0; r=1/\pi$

In Problems 33–38, find the indicated term of each geometric sequence.

33. 7th term of $1, \frac{1}{2}, \frac{1}{4}, \ldots$ **34.** 8th term of $1, 3, 9, \ldots$

35. 9th term of $1, -1, 1, \ldots$ **36.** 10th term of $-1, 2, -4, \ldots$

37. 8th term of $0.4, 0.04, 0.004, \ldots$ **38.** 7th term of $0.1, 1.0, 10.0, \ldots$

In Problems 39–44, find the sum.

39. $\dfrac{1}{4}+\dfrac{2}{4}+\dfrac{2^2}{4}+\dfrac{2^3}{4}+\cdots+\dfrac{2^{n-1}}{4}$ **40.** $\dfrac{3}{9}+\dfrac{3^2}{9}+\dfrac{3^3}{9}+\cdots+\dfrac{3^n}{9}$

41. $\displaystyle\sum_{k=1}^{n}\left(\frac{2}{3}\right)^k$ **42.** $\displaystyle\sum_{k=1}^{n}4\cdot3^{k-1}$

43. $-1-2-4-8-\cdots-(2^{n-1})$ **44.** $2+\dfrac{6}{5}+\dfrac{18}{25}+\cdots+2\left(\dfrac{3}{5}\right)^n$

In Problems 45–54, find the sum of each infinite geometric series.

45. $1+\frac{1}{3}+\frac{1}{9}+\cdots$ **46.** $2+\frac{4}{3}+\frac{8}{9}+\cdots$

47. $8+4+2+\cdots$ **48.** $6+2+\frac{2}{3}+\cdots$

49. $2-\frac{1}{2}+\frac{1}{8}-\frac{1}{32}+\cdots$ **50.** $1-\frac{3}{4}+\frac{9}{16}-\frac{27}{64}+\cdots$

51. $\displaystyle\sum_{k=1}^{\infty} 5\left(\tfrac{1}{4}\right)^{k-1}$
   52. $\displaystyle\sum_{k=1}^{\infty} 8\left(\tfrac{1}{3}\right)^{k-1}$
   53. $\displaystyle\sum_{k=1}^{\infty} 6\left(-\tfrac{2}{3}\right)^{k-1}$
   54. $\displaystyle\sum_{k=1}^{\infty} 4\left(-\tfrac{1}{2}\right)^{k-1}$

55. Find x so that x, $x + 2$, and $x + 3$ are terms of a geometric sequence.

56. Find x so that $x - 1$, x, and $x + 2$ are terms of a geometric sequence.

57. *Pendulum Swings* Initially, a pendulum swings through an arc of 2 feet. On each successive swing, the length of arc is 0.9 of the previous length.

    (a) What is the length of arc after 10 swings?
    (b) On which swing is the length of arc first less than 1 foot?
    (c) After 15 swings, what total length will the pendulum have swung?
    (d) When it stops, what total length will the pendulum have swung?

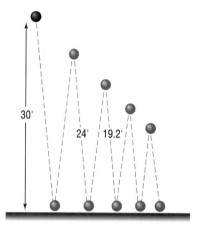

58. *Bouncing Balls* A ball is dropped from a height of 30 feet. Each time it strikes the ground, it bounces up to 0.8 of the previous height.

    (a) What height will the ball bounce up to after it strikes the ground for the third time?

    (b) What is its height after it strikes the ground for the nth time?

    (c) How many times does the ball need to strike the ground before its height is less than 6 inches?

    (d) What total distance does the ball travel before it stops bouncing?

59. *Salary Increases* Suppose you have just been hired at an annual salary of $18,000 and expect to receive annual increases of 5%. What will your salary be when you begin your fifth year?

60. *Equipment Depreciation* A new piece of equipment cost a company $15,000. Each year, for tax purposes, the company depreciates the value by 15%. What value should the company give the equipment after 5 years?

61. *Critical Thinking* You have just signed a 7 year professional football league contract with a beginning salary of $2,000,000 per year. Management gives you the following options with regard to your salary over the seven years.

    (1) A bonus of $100,000 each year
    (2) An annual increase of 4.5% per year beginning after 1 year
    (3) An annual increase of $95,000 per year beginning after 1 year

Which option provides the most money over the 7 year period? Which the least? Which would you choose? Why?

62. *A Rich Man's Promise* A rich man promises to give you $1000 on September 1, 1996. Each day thereafter he will give you $\tfrac{9}{10}$ of what he gave you the previous day. What is the first date on which the amount you receive is less than 1¢? How much have you received when this happens?

63. *Grains of Wheat on a Chess Board* In an old fable, a commoner who had just saved the king's life was told he could ask the king for any just reward. Being a shrewd man, the commoner said, "A simple wish, sire. Place one grain of wheat on the first square of a chessboard, two grains on the second square, four grains on the third square, continuing until you have filled the board. This is all I seek." Compute the total number of grains needed to do this to see why the request, seemingly simple, could not be granted. (A chessboard consists of $8 \times 8 = 64$ squares.)

64. Can a sequence be both arithmetic and geometric? Give reasons for your answer.

65. Make up a geometric sequence. Give it to a friend and ask for its 20th term.

66. Make up two infinite geometric series, one that has a sum and one that does not. Give them to a friend and ask for the sum of each series.

67. If $x < 1$, then $1 + x + x^2 + x^3 + \cdots + x^n + \cdots = 1/(1 - x)$. Make up a table of values using $x = 0.1$, $x = 0.25$, $x = 0.5$, $x = 0.75$, and $x = 0.9$ to compute $1/(1 - x)$. Now determine how many terms are needed in the expansion $1 + x + x^2 + x^3 + \cdots + x^n + \cdots$ before it approximates $1/(1 - x)$ correct to two decimal places. For example, if $x = 0.1$, then $1/(1 - x) = 10/9 = 1.111 \ldots$. The expansion requires three terms.

68. Which of the following choices, *A* or *B*, results in more money?

    A: To receive $1000 on day 1, $999 on day 2, $998 on day 3, with the process to end after 1000 days

    B: To receive $1 on day 1, $2 on day 2, $4 on day 3, for 19 days

69. You are interviewing for a job and receive two offers:

    A: $20,000 to start with guaranteed annual increases of 6% for the first 5 years

    B: $22,000 to start with guaranteed annual increases of 3% for the first 5 years

Which offer is best if your goal is to be making as much as possible after 5 years? Which is best if your goal is to make as much money as possible over the contract (5 years)?

70. Look at the figure on the right. What fraction of the square is eventually shaded if the indicated shading process continues indefinitely?

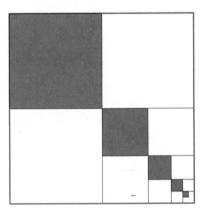

11.4

Mathematical Induction

Mathematical induction is a method for proving that statements involving natural numbers are true for all natural numbers.* For example, the statement, "$2n$ is always an even integer" can be proved true for all natural numbers by using mathematical induction. Also, the statement "the sum of the first n positive odd integers equals n^2," that is,

$$1 + 3 + 5 + \cdots + (2n - 1) = n^2 \tag{1}$$

can be proved for all natural numbers n by using mathematical induction.

Before stating the method of mathematical induction, let's try to gain a sense of the power of the method. We shall use the statement in equation (1) for this purpose by restating it for various values of $n = 1, 2, 3, \ldots$:

---

*Recall from Chapter 1 that the natural numbers are the numbers 1, 2, 3, 4, ... In other words, the terms *natural numbers* and *positive integers* are synonymous.

$n = 1$ The sum of the first positive odd integer is 1^2; $1 = 1^2$.

$n = 2$ The sum of the first 2 positive odd integers is 2^2; $1 + 3 = 4 = 2^2$.

$n = 3$ The sum of the first 3 positive odd integers is 3^2; $1 + 3 + 5 = 9 = 3^2$.

$n = 4$ The sum of the first 4 positive odd integers is 4^2; $1 + 3 + 5 + 7 = 16 = 4^2$.

Although from this pattern we might conjecture that statement (1) is true for any choice of n, can we really be sure that it does not fail for some choice of n? The method of proof by mathematical induction will, in fact, prove that the statement is true for all n.

Theorem
The Principle of Mathematical Induction

Suppose the following two conditions are satisfied with regard to a statement about natural numbers:

CONDITION I: The statement is true for the natural number 1.

CONDITION II: If the statement is true for some natural number k, it is also true for the next natural number $k + 1$.

Then the statement is true for all natural numbers.   ∎

We shall not prove this principle. However, we can provide a physical interpretation that will help us see why the principle works. Think of a collection of natural numbers obeying a statement as a collection of infinitely many dominoes (see Figure 6).

FIGURE 6

Now, suppose we are told two facts:

1. The first domino is pushed over.
2. If one of the dominoes falls over, say the kth domino, then so will the next one, the $(k + 1)$st domino.

Is it safe to conclude that *all* the dominoes fall over? The answer is yes, because, if the first one falls (Condition I), then the second one does also (by Condition II); and if the second one falls, then so does the third (by Condition II); and so on.

Now let's prove some statements about natural numbers using mathematical induction.

E X A M P L E   1   *Using Mathematical Induction*

Show that the following statement is true for all natural numbers n:
$$1 + 3 + 5 + \cdots + (2n - 1) = n^2 \tag{2}$$

Solution   We need to show first that statement (2) holds for $n = 1$. Because $1 = 1^2$, statement (2) is true for $n = 1$. Thus, Condition I holds.

Next, we need to show that Condition II holds. Suppose we know for some k that
$$1 + 3 + \cdots + (2k - 1) = k^2 \tag{3}$$

We wish to show that, based on equation (3), statement (2) holds for $k + 1$. Thus, we look at the sum of the first $k + 1$ positive odd integers to determine whether this sum equals $(k + 1)^2$:

$$1 + 3 + \cdots + (2k - 1) + (2k + 1) = \underbrace{[1 + 3 + \cdots + (2k - 1)]}_{= \, k^2 \text{ by equation (3)}} + (2k + 1)$$

$$= k^2 + (2k + 1)$$

$$= k^2 + 2k + 1 = (k + 1)^2$$

Conditions I and II are satisfied; thus, by the principle of mathematical induction, statement (2) is true for all natural numbers. ■

E X A M P L E   2   *Using Mathematical Induction*

Show that the following statement is true for all natural numbers n:

$$2^n > n$$

Solution   First, we show that the statement $2^n > n$ holds when $n = 1$. Because $2^1 = 2 > 1$, the inequality is true for $n = 1$. Thus, Condition I holds.

Next, we assume, for some natural number k, that $2^k > k$. We wish to show that the formula holds for $k + 1$; that is, we wish to show that $2^{k+1} > k + 1$. Now,

$$2^{k+1} = 2 \cdot 2^k \underset{\substack{\uparrow \\ \text{We know that} \\ 2^k > k.}}{>} 2 \cdot k = k + k \underset{\substack{\uparrow \\ k \geq 1}}{\geq} k + 1$$

Thus, if $2^k > k$, then $2^{k+1} > k + 1$, so Condition II of the principle of mathematical induction is satisfied. Hence, the statement $2^n > n$ is true for all natural numbers n. ■

E X A M P L E   3   *Using Mathematical Induction*

Show that the following formula is true for all natural numbers n:

$$1 + 2 + 3 + \cdots + n = \frac{n(n + 1)}{2} \tag{4}$$

Solution   First, we show that formula (4) is true when $n = 1$. Because

$$\frac{1(1 + 1)}{2} = \frac{1(2)}{2} = 1$$

Condition I of the principle of mathematical induction holds.

Next, we assume that formula (4) holds for some k, and we determine whether the formula then holds for $k + 1$. Thus, we assume that

$$1 + 2 + 3 + \cdots + k = \frac{k(k + 1)}{2} \quad \text{for some } k \tag{5}$$

Now, we need to show that

$$1 + 2 + 3 + \cdots + k + (k + 1) = \frac{(k + 1)(k + 1 + 1)}{2} = \frac{(k + 1)(k + 2)}{2}$$

We do this as follows:

$$1 + 2 + 3 + \cdots + k + (k + 1) = \underbrace{[1 + 2 + 3 + \cdots + k]}_{= \, \frac{k(k + 1)}{2} \text{ by equation (5)}} + (k + 1)$$

$$= \frac{k(k+1)}{2} + (k+1)$$

$$= \frac{k^2 + k + 2k + 2}{2}$$

$$= \frac{k^2 + 3k + 2}{2} = \frac{(k+1)(k+2)}{2}$$

Thus, Condition II also holds. As a result, formula (4) is true for all natural numbers. ∎

■   Now work Problem 1.

E X A M P L E   4

Using Mathematical Induction

Show that $3^n - 1$ is divisible by 2 for all natural numbers n.

Solution

First, we show that the statement is true when $n = 1$. Because $3^1 - 1 = 3 - 1 = 2$ is divisible by 2, the statement is true when $n = 1$. Thus, Condition I is satisfied.

Next, we assume that the statement holds for some k, and we determine whether the statement then holds for $k + 1$. Thus, we assume that $3^k - 1$ is divisible by 2 for some k. We need to show that $3^{k+1} - 1$ is divisible by 2. Now,

$$3^{k+1} - 1 = 3^{k+1} - 3^k + 3^k - 1$$

$$= 3^k(3 - 1) + (3^k - 1) = 3^k \cdot 2 + (3^k - 1)$$

Because $3^k \cdot 2$ is divisible by 2 and $3^k - 1$ is divisible by 2, it follows that $3^k \cdot 2 + (3^k - 1) = 3^{k+1} - 1$ is divisible by 2. Thus, Condition II is also satisfied. As a result, the statement, "$3^n - 1$ is divisible by 2" is true for all natural numbers n. ∎

Warning: The conclusion that a statement involving natural numbers is true for all natural numbers is made only after *both* Conditions I and II of the principle of mathematical induction have been satisfied. Problem 27 (below) demonstrates a statement for which only Condition I holds, but the statement is not true for all natural numbers. Problem 28 demonstrates a statement for which only Condition II holds, but the statement is *not* true for any natural number.

11.4

Exercise 11.4

In Problems 1–26, use the principle of mathematical induction to show that the given statement is true for all natural numbers.

1. $2 + 4 + 6 + \cdots + 2n = n(n+1)$

2. $1 + 5 + 9 + \cdots + (4n - 3) = n(2n - 1)$

3. $3 + 4 + 5 + \cdots + (n+2) = \frac{1}{2}n(n+5)$

4. $3 + 5 + 7 + \cdots + (2n+1) = n(n+2)$

5. $2 + 5 + 8 + \cdots + (3n - 1) = \frac{1}{2}n(3n+1)$

6. $1 + 4 + 7 + \cdots + (3n - 2) = \frac{1}{2}n(3n - 1)$

7. $1 + 2 + 2^2 + \cdots + 2^{n-1} = 2^n - 1$

8. $1 + 3 + 3^2 + \cdots + 3^{n-1} = \frac{1}{2}(3^n - 1)$

9. $1 + 4 + 4^2 + \cdots + 4^{n-1} = \frac{1}{3}(4^n - 1)$

10. $1 + 5 + 5^2 + \cdots + 5^{n-1} = \frac{1}{4}(5^n - 1)$

11. $\dfrac{1}{1 \cdot 2} + \dfrac{1}{2 \cdot 3} + \dfrac{1}{3 \cdot 4} + \cdots + \dfrac{1}{n(n+1)} = \dfrac{n}{n+1}$

12. $\dfrac{1}{1 \cdot 3} + \dfrac{1}{3 \cdot 5} + \dfrac{1}{5 \cdot 7} + \cdots + \dfrac{1}{(2n-1)(2n+1)} = \dfrac{n}{2n+1}$

13. $1^2 + 2^2 + 3^2 + \cdots + n^2 = \frac{1}{6}n(n+1)(2n+1)$

14. $1^3 + 2^3 + 3^3 + \cdots + n^3 = \frac{1}{4}n^2(n+1)^2$

15. $4 + 3 + 2 + \cdots + (5 - n) = \frac{1}{2}n(9 - n)$

16. $-2 - 3 - 4 - \cdots - (n + 1) = -\frac{1}{2}n(n + 3)$

17. $1 \cdot 2 + 2 \cdot 3 + 3 \cdot 4 + \cdots + n(n + 1) = \frac{1}{3}n(n + 1)(n + 2)$

18. $1 \cdot 2 + 3 \cdot 4 + 5 \cdot 6 + \cdots + (2n - 1)(2n) = \frac{1}{3}n(n + 1)(4n - 1)$

19. $n^2 + n$ is divisible by 2.   **20.** $n^3 + 2n$ is divisible by 3.

21. $n^2 - n + 2$ is divisible by 2.   **22.** $n(n + 1)(n + 2)$ is divisible by 6.

23. If $x > 1$, then $x^n > 1$.   **24.** If $0 < x < 1$, then $0 < x^n < 1$.

25. $a - b$ is a factor of $a^n - b^n$. [*Hint:* $a^{k+1} - b^{k+1} = a(a^k - b^k) + b^k(a - b)$]

26. $a + b$ is a factor of $a^{2n+1} + b^{2n+1}$.

27. Show that the statement "$n^2 - n + 41$ is a prime number," is true for $n = 1$, but is not true for $n = 41$.

28. Show that the formula

$$2 + 4 + 6 + \cdots + 2n = n^2 + n + 2$$

obeys Condition II of the principle of mathematical induction. That is, show that if the formula is true for some k it is also true for $k + 1$. Then show that the formula is false for $n = 1$ (or for any other choice of n).

29. Use mathematical induction to prove that if $r \neq 1$ then

$$a + ar + ar^2 + \cdots + ar^{n-1} = a\frac{1 - r^n}{1 - r}$$

30. Use mathematical induction to prove that

$$a + (a + d) + (a + 2d) + \cdots + [a + (n - 1)d] = na + d\frac{n(n - 1)}{2}$$

31. *Geometry* Use mathematical induction to show that the sum of the interior angles of a convex polygon of n sides equals $(n - 2) \cdot 180°$.

32. *The Extended Principle of Mathematical Induction* The extended principle of mathematical induction states that if conditions I and II hold, that is,

(I)   A statement is true for a natural number j.

(II)   If the statement is true for some natural number $k > j$, then it is also true for the next natural number $k + 1$.

Then the statement is true for *all* natural numbers $\geq j$.

Use the extended principle of mathematical induction to show that the number of diagonals in a convex polygon of n sides is $\frac{1}{2}n(n - 3)$. [*Hint:*   Begin by showing that the result is true when $n = 4$ (Condition I).]

33. How would you explain to a friend the principle of mathematical induction?

11.5

The Binomial Theorem

The *Binomial Theorem** is a formula for the expansion of $(x + a)^n$ for n any positive integer. If $n = 1, 2, 3$, and 4, the expansion of $(x + a)^n$ is straightforward:

$(x + a)^1 = x + a$ 2 terms, beginning with x^1 and ending with a^1

$(x + a)^2 = x^2 + 2ax + a^2$ 3 terms, beginning with x^2 and ending with a^2

$(x + a)^3 = x^3 + 3ax^2 + 3a^2x + a^3$ 4 terms, beginning with x^3 and ending with a^3

$(x + a)^4 = x^4 + 4ax^3 + 6a^2x^2 + 4a^3x + a^4$ 5 terms, beginning with x^4 and ending with a^4

*The name *binomial* derives from the fact that $x + a$ is a binomial, that is, contains two terms.

MISSION POSSIBLE

Chapter 11

GETTING THE MOST OUT OF YOUR CONTRACT

You and your band have just signed a recording contract with the nationally acclaimed recording studio, NASHBURG TENS. They've promised $200,000 a year for six years, plus a choice of one of the following four options:

    (a) a bonus of $10,000 per year
    (b) an annual increase of 4.5% per year (starting after the first year)
    (c) an annual increase of 6% per year (starting after the second year)
    (d) an annual increase of $9,500 per year (starting after the first year)

You have to tell them today which option you want to take. Your agent is out of town and out of cellular phone range, so you'll have to do the math yourselves.

1. For each option, find out what your payment would be every year for the six years. Round to the nearest dollar.
2. Identify which options are examples of arithmetic or geometric sequences.
3. For each option, find out how much the recording studio will have paid in total over the whole six years. Do you have formulas that will shorten your work?
4. Which option pays your band the most over all? Are there any considerations or circumstances that would make other options better choices even though they pay less money? Are there advantages to being paid more at the beginning of the contract? What are they?

Notice that each expansion of $(x + a)^n$ begins with x^n and ends with a^n. As you read from left to right, the powers of x are decreasing, while the powers of a are increasing. Also, the number of terms that appear equals $n + 1$. Notice, too, that the degree of each monomial in the expansion equals n. For example, in the expansion of $(x + a)^3$, each monomial $(x^3, 3ax^2, 3a^2x, a^3)$ is of degree 3. As a result, we might conjecture that the expansion of $(x + a)^n$ would look like this:

$$(x + a)^n = x^n + \underline{\quad}ax^{n-1} + \underline{\quad}a^2x^{n-2} + \cdots + \underline{\quad}a^{n-1}x + a^n$$

where the blanks are numbers to be found. This is, in fact, the case, as we shall see shortly.

First, we need to introduce a symbol.

The Symbol $\begin{pmatrix} n \\ j \end{pmatrix}$

We define the symbol $\begin{pmatrix} n \\ j \end{pmatrix}$, read "$n$ taken j at a time," as follows:

Symbol $\begin{pmatrix} n \\ j \end{pmatrix}$

If j and n are integers with $0 \le j \le n$, the symbol $\begin{pmatrix} n \\ j \end{pmatrix}$ is defined as

$$\begin{pmatrix} n \\ j \end{pmatrix} = \frac{n!}{j!(n - j)!} \tag{1}$$

Comment: On a calculator, the symbol $\begin{pmatrix} n \\ j \end{pmatrix}$ may be denoted by the key $\boxed{\text{nCr}}$ or by the key $\boxed{\text{COMB}}$.

E X A M P L E   1   *Evaluating* $\begin{pmatrix} n \\ j \end{pmatrix}$

Find:

(a) $\begin{pmatrix} 3 \\ 1 \end{pmatrix}$ (b) $\begin{pmatrix} 4 \\ 2 \end{pmatrix}$ (c) $\begin{pmatrix} 8 \\ 7 \end{pmatrix}$ (d) $\begin{pmatrix} 65 \\ 15 \end{pmatrix}$

Solution   (a) $\begin{pmatrix} 3 \\ 1 \end{pmatrix} = \dfrac{3!}{1!(3 - 1)!} = \dfrac{3!}{1!2!} = \dfrac{3 \cdot 2 \cdot 1}{1(2 \cdot 1)} = \dfrac{6}{2} = 3$

(b) $\begin{pmatrix} 4 \\ 2 \end{pmatrix} = \dfrac{4!}{2!(4 - 2)!} = \dfrac{4!}{2!2!} = \dfrac{4 \cdot 3 \cdot 2 \cdot 1}{(2 \cdot 1)(2 \cdot 1)} = \dfrac{24}{4} = 6$

(c) $\begin{pmatrix} 8 \\ 7 \end{pmatrix} = \dfrac{8!}{7!(8 - 7)!} = \dfrac{8!}{7!1!} \underset{\substack{\uparrow \\ 8! = 8 \cdot 7!}}{=} \dfrac{8 \cdot \cancel{7!}}{\cancel{7!} \cdot 1!} = \dfrac{8}{1} = 8$

(d) We use a calculator.

| Keystrokes: | 65 | | SHIFT | nCr | 15 | | = |
| Display: | | 65 | | | 15 | | 2.073747 14 |

Thus, $\binom{65}{15} = 2.073747 \times 10^{14}$. ∎

■ Now work Problem 1.

Two useful formulas involving the symbol $\binom{n}{j}$ are

$$\binom{n}{0} = 1 \quad \text{and} \quad \binom{n}{n} = 1$$

Proof

$$\binom{n}{0} = \frac{n!}{0!(n-0)!} = \frac{n!}{0!n!} = \frac{1}{1} = 1$$

You are asked to show that $\binom{n}{n} = 1$ in Problem 41 at the end of this section. ∎

Suppose we arrange the various values of the symbol $\binom{n}{j}$ in a triangular display, as shown next and in Figure 7:

$$\binom{0}{0}$$

$$\binom{1}{0} \quad \binom{1}{1}$$

$$\binom{2}{0} \quad \binom{2}{1} \quad \binom{2}{2}$$

$$\binom{3}{0} \quad \binom{3}{1} \quad \binom{3}{2} \quad \binom{3}{3}$$

$$\binom{4}{0} \quad \binom{4}{1} \quad \binom{4}{2} \quad \binom{4}{3} \quad \binom{4}{4}$$

$$\binom{5}{0} \quad \binom{5}{1} \quad \binom{5}{2} \quad \binom{5}{3} \quad \binom{5}{4} \quad \binom{5}{5}$$

This display is called the **Pascal triangle,** named after Blaise Pascal (1623–1662), a French mathematician.

FIGURE 7
Pascal triangle

```
                                        j = 0
                                 ↙
                          1          j = 1
 n = 0  →                      ↙
                                        j = 2
 n = 1  →            1     1       ↙
                                        j = 3
 n = 2  →         1     2     1    ↙
                                        j = 4
 n = 3  →      1     3     3     1 ↙
                                        j = 5
 n = 4  →   1     4     6     4     1 ↙
 n = 5  → 1     5    10    10     5     1
```

The Pascal triangle has 1's down the sides. To get any other entry, merely add the two nearest entries in the row above it. The shaded triangles in Figure 7 serve to illustrate this feature of the Pascal triangle. Based on this feature, the row corresponding to $n = 6$ is found as follows:

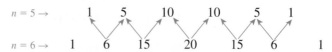

Later, we shall prove that this addition always works (see the theorem on page 718).

Although the Pascal triangle provides an interesting and organized display of the symbol $\binom{n}{j}$, in practice it is not all that helpful. For example, if you wanted to know the value of $\binom{12}{5}$, you would need to produce twelve rows of the triangle before seeing the answer. It is much faster instead to use the definition (1).

The Binomial Theorem

Now we are ready to state the **Binomial Theorem.** A proof is given at the end of this section.

Theorem   Let x and a be real numbers. For any positive integer n, we have

Binomial Theorem

$$(x + a)^n = \binom{n}{0}x^n + \binom{n}{1}ax^{n-1} + \cdots + \binom{n}{j}a^j x^{n-j} + \cdots + \binom{n}{n}a^n$$

$$= \sum_{j=0}^{n}\binom{n}{j}x^{n-j}a^j \tag{2}$$

Now you know why we needed to introduce the symbol $\binom{n}{j}$; these symbols are the numerical coefficients that appear in the expansion of $(x + a)^n$. Because of this, the symbol $\binom{n}{j}$ is called the **binomial coefficient.**

E X A M P L E   2   *Expanding a Binomial*

Use the Binomial Theorem to expand $(x + 2)^5$.

Solution   In the Binomial Theorem, let $a = 2$ and $n = 5$. Then

$$(x + 2)^5 = \binom{5}{0}x^5 + \binom{5}{1}2x^4 + \binom{5}{2}2^2 x^3 + \binom{5}{3}2^3 x^2 + \binom{5}{4}2^4 x + \binom{5}{5}2^5$$

↑
Use equation (2).

$$= 1 \cdot x^5 + 5 \cdot 2x^4 + 10 \cdot 4x^3 + 10 \cdot 8x^2 + 5 \cdot 16x + 1 \cdot 32$$

↑
Use row $n = 5$ of the Pascal triangle or formula (1) for $\binom{n}{j}$.

$$= x^5 + 10x^4 + 40x^3 + 80x^2 + 80x + 32$$

E X A M P L E   3   *Expanding a Binomial*

Expand $(2y - 3)^4$ using the Binomial Theorem.

Solution First, we rewrite the expression $(2y - 3)^4$ as $[2y + (-3)]^4$. Now we use the Binomial Theorem with $n = 4$, $x = 2y$, and $a = -3$:

$$[2y + (-3)]^4 = \binom{4}{0}(2y)^4 + \binom{4}{1}(-3)(2y)^3 + \binom{4}{2}(-3)^2(2y)^2$$

$$+ \binom{4}{3}(-3)^3(2y) + \binom{4}{4}(-3)^4$$

$$= 1 \cdot 16y^4 + 4(-3)8y^3 + 6 \cdot 9 \cdot 4y^2 + 4(-27)2y + 1 \cdot 81$$

$\uparrow$
Use row $n = 4$ of the Pascal triangle or formula (1) for $\binom{n}{j}$

$$= 16y^4 - 96y^3 + 216y^2 - 216y + 81$$

In this expansion, note that the signs alternate due to the fact that $a = -3 < 0$. ∎

■ Now work Problem 17.

E X A M P L E   4

Finding a Particular Coefficient in a Binomial Expansion
Find the coefficient of y^8 in the expansion of $(2y + 3)^{10}$.

Solution We write out the expansion using the Binomial Theorem:

$$(2y + 3)^{10} = \binom{10}{0}(2y)^{10} + \binom{10}{1}(2y)^9(3)^1 + \binom{10}{2}(2y)^8(3)^2 + \binom{10}{3}(2y)^7(3)^3$$

$$+ \binom{10}{4}(2y)^6(3)^4 + \cdots + \binom{10}{9}(2y)(3)^9 + \binom{10}{10}(3)^{10}$$

From the third term in the expression, the coefficient of y^8 is

$$\binom{10}{2}(2)^8(3)^2 = \frac{10!}{2!8!} \cdot 2^8 \cdot 9 = \frac{10 \cdot 9 \cdot 8!}{2 \cdot 8!} \cdot 2^8 \cdot 9 = 103,680$$ ∎

As this solution demonstrates, we can use the Binomial Theorem to write a particular term in an expansion without writing the entire expansion. Based on the expansion of $(x + a)^n$, the term containing x^j is

$$\binom{n}{n-j}a^{n-j}x^j \qquad (3)$$

For example, we can solve Example 4 by using formula (3) with $n = 10$, $a = 3$, $x = 2y$, and $j = 8$. Then the term containing y^8 is

$$\binom{10}{10-8}3^{10-8}(2y)^8 = \binom{10}{2} \cdot 3^2 \cdot 2^8 \cdot y^8 = \frac{10!}{2!8!} \cdot 9 \cdot 2^8y^8$$

$$= \frac{10 \cdot 9 \cdot 8!}{2!8!} \cdot 9 \cdot 2^8y^8 = 103,680y^8$$

E X A M P L E   5

Finding a Particular Term in a Binomial Expansion
Find the sixth term in the expansion of $(x + 2)^9$.

Solution A   We expand using the Binomial Theorem until the sixth term is reached:

$$(x + 2)^9 = \binom{9}{0}x^9 + \binom{9}{1}x^8 \cdot 2 + \binom{9}{2}x^7 \cdot 2^2 + \binom{9}{3}x^6 \cdot 2^3 + \binom{9}{4}x^5 \cdot 2^4$$
$$+ \binom{9}{5}x^4 \cdot 2^5 + \cdots$$

The sixth term is

$$\binom{9}{5}x^4 \cdot 2^5 = \frac{9!}{5!4!} \cdot x^4 \cdot 32 = 4032x^4$$

Solution B   The sixth term in the expansion of $(x + 2)^9$, which has ten terms total, contains x^4. (Do you see why?) Thus, by formula (3), the sixth term is

$$\binom{9}{9 - 4}2^{9-4}x^4 = \binom{9}{5}2^5x^4 = \frac{9!}{5!4!} \cdot 32x^4 = 4032x^4 \qquad \blacksquare$$

■ Now work Problems 25 and 31.

Next we show that the "triangular addition" feature of the Pascal triangle illustrated in Figure 7 always works.

Theorem   If n and j are integers with $1 \le j \le n$, then

$$\binom{n}{j - 1} + \binom{n}{j} = \binom{n + 1}{j} \qquad (4)$$

Proof

$$\binom{n}{j - 1} + \binom{n}{j} = \frac{n!}{(j - 1)![n - (j - 1)]!} + \frac{n!}{j!(n - j)!}$$

Multiply the first term by j / j and the second term by $(n - j + 1)/(n - j + 1)$.

$$= \frac{n!}{(j - 1)!(n - j + 1)!} + \frac{n!}{j!(n - j)!}$$

$$= \frac{jn!}{j(j - 1)!(n - j + 1)!} + \frac{(n - j + 1)n!}{j!(n - j + 1)(n - j)!}$$

$$= \frac{jn!}{j!(n - j + 1)!} + \frac{(n - j + 1)n!}{j!(n - j + 1)!}$$

Now the denominators are equal.

$$= \frac{jn! + (n - j + 1)n!}{j!(n - j + 1)!}$$

$$= \frac{n!(j + n - j + 1)}{j!(n - j + 1)!}$$

$$= \frac{n!(n + 1)}{j!(n - j + 1)!} = \frac{(n + 1)!}{j![(n + 1) - j]!} = \binom{n + 1}{j} \qquad \blacksquare$$

Proof of the Binomial Theorem   We use mathematical induction to prove the Binomial Theorem. First, we show that formula (2) is true for $n = 1$:

$$(x + a)^1 = x + a = \binom{1}{0}x^1 + \binom{1}{1}a^1$$

Next we suppose that formula (2) is true for some k. That is, we assume that

$$(x + a)^k = \binom{k}{0}x^k + \binom{k}{1}ax^{k-1} + \cdots + \binom{k}{j-1}a^{j-1}x^{k-j+1} + \binom{k}{j}a^jx^{k-j} + \cdots + \binom{k}{k}a^k \qquad (5)$$

Now we calculate $(x + a)^{k+1}$:

$$(x + a)^{k+1} = (x + a)(x + a)^k = x(x + a)^k + a\,(x + a)^k$$

Use equation (5).

$$= x\left[\binom{k}{0}x^k + \binom{k}{1}ax^{k-1} + \cdots + \binom{k}{j-1}a^{j-1}x^{k-j+1} + \binom{k}{j}a^jx^{k-j} + \cdots + \binom{k}{k}a^k\right]$$

$$+ a\left[\binom{k}{0}x^k + \binom{k}{1}ax^{k-1} + \cdots + \binom{k}{j-1}a^{j-1}x^{k-j+1} + \binom{k}{j}a^jx^{k-j} + \cdots + \binom{k}{k-1}a^{k-1}x + \binom{k}{k}a^k\right]$$

$$= \binom{k}{0}x^{k+1} + \binom{k}{1}ax^k + \cdots + \binom{k}{j-1}a^{j-1}x^{k-j+2} + \binom{k}{j}a^jx^{k-j+1} + \cdots + \binom{k}{k}a^kx$$

$$+ \binom{k}{0}ax^k + \binom{k}{1}a^2x^{k-1} + \cdots + \binom{k}{j-1}a^jx^{k-j+1} + \binom{k}{j}a^{j+1}x^{k-j} + \cdots + \binom{k}{k-1}a^kx + \binom{k}{k}a^{k+1}$$

$$= \binom{k}{0}x^{k+1} + \left[\binom{k}{1} + \binom{k}{0}\right]ax^k + \cdots + \left[\binom{k}{j} + \binom{k}{j-1}\right]a^jx^{k-j+1} + \cdots + \left[\binom{k}{k} + \binom{k}{k-1}\right]a^kx + \binom{k}{k}a^{k+1}$$

Because

$$\binom{k}{0} = 1 = \binom{k+1}{0}, \quad \binom{k}{1} + \binom{k}{0} \underset{(4)}{=} \binom{k+1}{1}, \quad \cdots,$$

$$\binom{k}{j} + \binom{k}{j-1} \underset{(4)}{=} \binom{k+1}{j}, \quad \cdots, \quad \binom{k}{k} = 1 = \binom{k+1}{k+1}$$

we have

$$(x + a)^{k+1} = \binom{k+1}{0}x^{k+1} + \binom{k+1}{1}ax^k + \cdots + \binom{k+1}{j}a^jx^{k-j+1} + \cdots + \binom{k+1}{k+1}a^{k+1}$$

Thus, Conditions I and II of the principle of mathematical induction are satisfied, and formula (2) is therefore true for all n. ∎

HISTORICAL FEATURE
    ■ The case $n = 2$ of the Binomial Theorem, $(a + b)^2$, was known to Euclid in 300 BC, but the general law seems to have been discovered by the Persian mathematician and astronomer Omar Khayyám (1044?–1123?), who is also well known as the author of the *Rubaiyat,* a collection of four-line poems making observations on the human condition. Omar Khayyám did not state the Binomial Theorem explicitly, but he claimed to have a method for extracting third, fourth, fifth roots, and so on. A little study shows that one must know the Binomial Theorem to create such a method.

    The heart of the Binomial Theorem is the formula for the numerical coefficients, and, as we saw, they can be written out in a symmetric triangular form. The Pascal triangle appears first in the books of Yang Hui (about 1270) and Chu Shih-chie (1303). Pascal's name is attached to the triangle because of the many applications he made of it, especially to counting and probability. In establishing these results, he was one of the earliest users of mathematical induction.

Many people worked on the proof of the Binomial Theorem, which was finally completed for all n (including complex numbers) by Niels Abel (1802–1829). ■

11.5

Exercise 11.5

In Problems 1–12, evaluate each expression.

1. $\dbinom{5}{3}$ **2.** $\dbinom{7}{3}$ **3.** $\dbinom{7}{5}$ **4.** $\dbinom{9}{7}$ **5.** $\dbinom{50}{49}$ **6.** $\dbinom{100}{98}$

7. $\dbinom{1000}{1000}$ **8.** $\dbinom{1000}{0}$ **9.** $\dbinom{55}{23}$ **10.** $\dbinom{60}{20}$ **11.** $\dbinom{47}{25}$ **12.** $\dbinom{37}{19}$

In Problems 13–24, expand each expression using the Binomial Theorem.

13. $(x + 1)^5$ **14.** $(x - 1)^5$ **15.** $(x - 2)^6$ **16.** $(x + 3)^4$

17. $(3x + 1)^4$ **18.** $(2x + 3)^5$ **19.** $(x^2 + y^2)^5$ **20.** $(x^2 - y^2)^6$

21. $(\sqrt{x} + \sqrt{2})^6$ **22.** $(\sqrt{x} - \sqrt{3})^4$ **23.** $(ax + by)^5$ **24.** $(ax - by)^4$

In Problems 25–38, use the Binomial Theorem to find the indicated coefficient or term.

25. The coefficient of x^6 in the expansion of $(x + 3)^{10}$

26. The coefficient of x^3 in the expansion of $(x - 3)^{10}$

27. The coefficient of x^7 in the expansion of $(2x - 1)^{12}$

28. The coefficient of x^3 in the expansion of $(2x + 1)^{12}$

29. The coefficient of x^7 in the expansion of $(2x + 3)^9$

30. The coefficient of x^2 in the expansion of $(2x - 3)^9$

31. The fifth term in the expansion of $(x + 3)^7$

32. The third term in the expansion of $(x - 3)^7$

33. The third term in the expansion of $(3x - 2)^9$

34. The sixth term in the expansion of $(3x + 2)^8$

35. The coefficient of x^0 in the expansion of $\left(x^2 + \dfrac{1}{x}\right)^{12}$

36. The coefficient of x^0 in the expansion of $\left(x - \dfrac{1}{x^2}\right)^9$

37. The coefficient of x^4 in the expansion of $\left(x - \dfrac{2}{\sqrt{x}}\right)^{10}$

38. The coefficient of x^2 in the expansion of $\left(\sqrt{x} + \dfrac{3}{\sqrt{x}}\right)^8$

39. Use the Binomial Theorem to find the numerical value of $(1.001)^5$ correct to five decimal places.
[*Hint:* $(1.001)^5 = (1 + 10^{-3})^5$]

40. Use the Binomial Theorem to find the numerical value of $(0.998)^6$ correct to five decimal places.

41. Show that $\dbinom{n}{n} = 1$.

42. Show that, if n and j are integers with $0 \le j \le n$, then

$$\binom{n}{j} = \binom{n}{n-j}$$

Thus, conclude that the Pascal triangle is symmetric with respect to a vertical line drawn from the topmost entry.

43. If n is a positive integer, show that

$$\binom{n}{0} + \binom{n}{1} + \cdots + \binom{n}{n} = 2^n$$

[*Hint:* $2^n = (1+1)^n$; now use the Binomial Theorem.]

44. If n is a positive integer, show that

$$\binom{n}{0} - \binom{n}{1} + \binom{n}{2} - \cdots + (-1)^n \binom{n}{n} = 0$$

45. $\binom{5}{0}\left(\frac{1}{4}\right)^5 + \binom{5}{1}\left(\frac{1}{4}\right)^4\left(\frac{3}{4}\right) + \binom{5}{2}\left(\frac{1}{4}\right)^3\left(\frac{3}{4}\right)^2 + \binom{5}{3}\left(\frac{1}{4}\right)^2\left(\frac{3}{4}\right)^3 + \binom{5}{4}\left(\frac{1}{4}\right)\left(\frac{3}{4}\right)^4 + \binom{5}{5}\left(\frac{3}{4}\right)^5 = ?$

46. *Stirling's Formula* for approximating $n!$ when n is large is given by

$$n! \approx \sqrt{2n\pi}\left(\frac{n}{e}\right)^n\left(1 + \frac{1}{12n-1}\right)$$

Calculate 12!, 20!, and 25!. Then use Stirling's formula to approximate 12!, 20!, and 25!.

11.6

Sets and Counting

Sets

A **set** is a well-defined collection of distinct objects. The objects of a set are called its **elements.** By **well-defined,** we mean that there is a rule that enables us to determine whether a given object is an element of the set. If a set has no elements, it is called the **empty set,** or **null set,** and is denoted by the symbol $\varnothing$.

Because the elements of a set are distinct, we never repeat elements. Thus, we would never write $\{1, 2, 3, 2\}$; the correct listing is $\{1, 2, 3\}$. Furthermore, because a set is a collection, the order in which the elements are listed is immaterial. Thus, $\{1, 2, 3\}$, $\{1, 3, 2\}$, $\{2, 1, 3\}$, and so on, all represent the same set.

E X A M P L E   1 *Writing the Elements of a Set*

Write the set consisting of the possible outcomes from tossing a coin twice. Use H for "heads" and T for "tails."

Solution In tossing a coin twice, we can get heads each time, HH; or heads the first time and tails the second, HT; or tails the first time and heads the second, TH; or tails each time, TT. Because no other possibilities exist, the set of outcomes is

$$\{HH, HT, TH, TT\}$$

If two sets A and B have precisely the same elements, then we say that A and B are **equal** and write $A = B$.

If each element of a set A is also an element of a set B, then we say that A is a **subset** of B and write $A \subseteq B$.

If $A \subseteq B$ and $A \ne B$, then we say that A is a **proper subset** of B and write $A \subset B$.

Thus, if $A \subseteq B$, every element in set A is also in set B, but B may or may not have additional elements. If $A \subset B$, every element in A is also in B, and B has at least one element not found in A.

Finally, we agree that the empty set is a subset of every set; that is,

$$\emptyset \subseteq A \qquad \text{for any set } A$$

E X A M P L E 2

Finding All the Subsets of a Set

Write down all the subsets of the set $\{a, b, c\}$.

Solution

To organize our work, we write down all the subsets with no elements, then those with one element, then those with two elements, and finally those with three elements. These will give us all the subsets. Do you see why?

0 ELEMENTS	1 ELEMENT	2 ELEMENTS	3 ELEMENTS
$\emptyset$	$\{a\}, \{b\}, \{c\}$	$\{a, b\}, \{b, c\}, \{a, c\}$	$\{a, b, c\}$

■ Now work Problem 21.

Intersection; Union

If A and B are sets, the **intersection** of A with B, denoted $A \cap B$, is the set consisting of elements that belong to both A and B. The **union** of A with B, denoted $A \cup B$, is the set consisting of elements that belong to either A or B, or both.

E X A M P L E 3

Finding the Intersection and Union of Sets

Let $A = \{1, 3, 5, 8\}$, $B = \{3, 5, 7\}$, and $C = \{2, 4, 6, 8\}$. Find:

(a) $A \cap B$ (b) $A \cup B$ (c) $B \cap (A \cup C)$

Solution

(a) $A \cap B = \{1, 3, 5, 8\} \cap \{3, 5, 7\} = \{3, 5\}$
(b) $A \cup B = \{1, 3, 5, 8\} \cup \{3, 5, 7\} = \{1, 3, 5, 7, 8\}$
(c) $B \cap (A \cup C) = \{3, 5, 7\} \cap [\{1, 3, 5, 8\} \cup \{2, 4, 6, 8\}]$
$\qquad\qquad\quad = \{3, 5, 7\} \cap \{1, 2, 3, 4, 5, 6, 8\} = \{3, 5\}$ ■

■ Now work Problem 5.

Usually, in working with sets, we designate a **universal set,** the set consisting of all the elements we wish to consider. Once a universal set has been designated, we can consider elements of the universal set not found in a given set.

Complement

If A is a set, the **complement** of A, denoted A', is the set consisting of all the elements not in A.

E X A M P L E 4

Finding the Complement of a Set

If the universal set is $U = \{1, 2, 3, 4, 5, 6, 7, 8, 9\}$, and if $A = \{1, 3, 5, 7, 9\}$, then $A' = \{2, 4, 6, 8\}$.   ■

Notice that: $A \cup A' = U$ and $A \cap A' = \emptyset$.

FIGURE 8

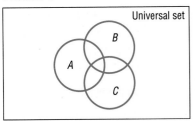

■ Now work Problem 13.

It is often helpful to draw pictures of sets. Such pictures, called **Venn diagrams,** represent sets as circles enclosed in a rectangle, which represents the universal set. Such diagrams often help us to visualize various relationships among sets. See Figure 8.

If we know that $A \subseteq B$, we might use the Venn diagram in Figure 9(a). If we know that A and B have no elements in common, that is, if $A \cap B = \varnothing$, we might use the Venn diagram in Figure 9(b).

FIGURE 9

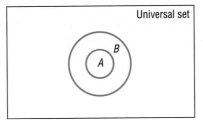
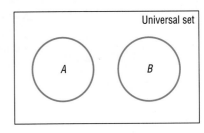

(a) $A \subseteq B$ **(b)** $A \cap B = \varnothing$

Figures 10(a), 10(b), and 10(c) use Venn diagrams to illustrate the definitions of intersection, union, and complement, respectively.

FIGURE 10

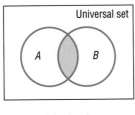
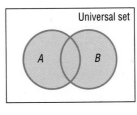
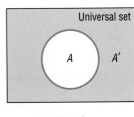

(a) $A \cap B$ **(b)** $A \cup B$ **(c)** A'

Counting

As you count the number of students in a classroom or the number of pennies in your pocket, what you are really doing is matching, on a one-to-one basis, each object to be counted with the counting numbers 1, 2, 3, . . . , n, for some number n. If a set A matched up in this fashion with the set $\{1, 2, . . . , 25\}$, you would conclude that there are 25 elements in the set A. We use the notation $n(A) = 25$ to indicate that there are 25 elements in the set A.

Because the empty set has no elements, we write

$$n(\varnothing) = 0$$

If the number of elements in a set is a nonnegative integer, we say the set is **finite.** Otherwise, it is **infinite.** We shall concern ourselves only with finite sets.

From Example 2, we can see that a set with 3 elements has $2^3 = 8$ subsets. In fact, it can be shown that a set with n elements has exactly 2^n subsets. This fact has an important application to computers, which we take up at the end of this section.

EXAMPLE 5   *Analyzing Survey Data*

In a survey of 100 college students, 35 were registered in College Algebra, 52 were registered in Introduction to Computer Science, and 18 were in both courses. How many were registered in neither course?

Solution   First, we let A = Set of students in College Algebra

          B = Set of students in Introduction to Computer Science

Then the information tells us that

$$n(A) = 35 \qquad n(B) = 52 \qquad n(A \cap B) = 18$$

FIGURE 11

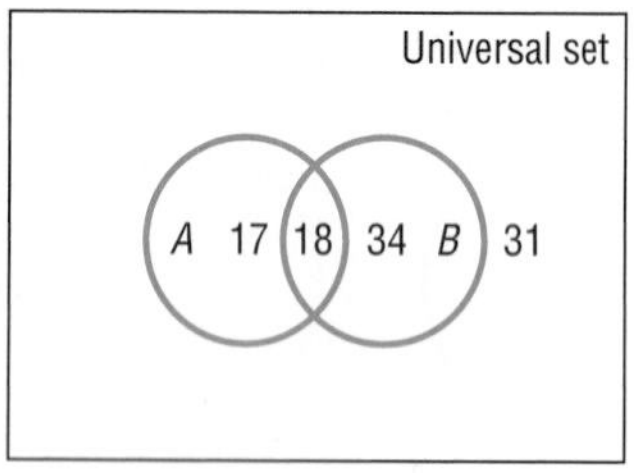

Refer to Figure 11. Do you see how the numerical entries were determined? Based on the diagram, we conclude that $17 + 18 + 34 = 69$ students were registered in at least one of the two courses. Since 100 students were surveyed, it follows that $100 - 69 = 31$ were registered in neither course. ∎

■   Now work Problem 35.

The conclusions drawn in Example 5 lead us to formulate a general counting formula. If we count the elements in each of two sets A and B, we necessarily count twice any elements that are in both A and B, that is, those elements in $A \cap B$. Thus, to count correctly the elements that are in A or B, that is, to find $n(A \cup B)$, we need to subtract those in $A \cap B$ from $n(A) + n(B)$.

Theorem   If A and B are finite sets, then

Counting Formula

$$n(A \cup B) = n(A) + n(B) - n(A \cap B) \qquad\qquad (1)$$

■

A special case of the counting formula (1) occurs if A and B have no elements in common. In this case, $A \cap B = \varnothing$ so that $n(A \cap B) = 0$.

Theorem   If two sets A and B have no elements in common, then

Addition Principle of
Counting

$$n(A \cup B) = n(A) + n(B) \qquad\qquad (2)$$

■

EXAMPLE 6   *Counting the Number of Possible Codes*

A certain code is to consist of either a letter of the alphabet or a digit, but not both. How many codes are possible?

Solution   Let the sets A and B be defined as

$$A = \text{Set of letters in the alphabet}$$
$$B = \text{Set of digits } \{0, 1, 2, \ldots, 9\}$$

Then

$$n(A) = 26 \qquad n(B) = 10$$

Because letters and digits are different, $A \cap B = \varnothing$. The number of ways either a letter or a digit can be chosen is, therefore,

$$n(A \cup B) = n(A) + n(B) = 26 + 10 = 36$$ ∎

Application to Computers

Information stored in a computer may be thought of as a series of switches, which are either on or off and are denoted by either the number 0 (off) or the number 1 (on). These numbers are the binary digits, or **bits.** A **register** holds a certain fixed number of bits. For example, the Z-80 microprocessor has 8 bit registers; the PDP-11 minicomputer has 16 bit registers, and the IBM-370 computer has 32 bit registers. Thus, a Z-80 register may hold an entry that looks like this: 01111001 (8 bits). We wish to find out how many different representations are possible in a given register.

We proceed in steps, first looking at a hypothetical 3 bit register. Look again at the solution to Example 2 and arrange all the subsets of $\{a, b, c\}$ as shown in Table 1. As the table illustrates, the number of subsets of a set with 3 elements equals the number of different representations in a 3 bit register. A set with n elements has 2^n subsets; thus, an n bit register has 2^n representations. So an 8 bit register can hold $2^8 = 256$ different symbols, a 16 bit register can hold $2^{16} = 65{,}536$ different symbols, and a 32 bit register can hold $2^{32} \approx 4.3 \times 10^9$ different symbols.

TABLE 1

a	b	c	SUBSET
0	0	0	$\varnothing$
1	0	0	$\{a\}$
0	1	0	$\{b\}$
0	0	1	$\{c\}$
1	1	0	$\{a, b\}$
0	1	1	$\{b, c\}$
1	0	1	$\{a, c\}$
1	1	1	$\{a, b, c\}$

11.6

Exercise 11.6

In Problems 1–10, use $A = \{1, 3, 5, 7, 9\}$, $B = \{1, 5, 6, 7\}$, and $C = \{1, 2, 4, 6, 8, 9\}$ to find each set.

1. $A \cup B$
2. $A \cup C$
3. $A \cap B$
4. $A \cap C$
5. $(A \cup B) \cap C$
6. $(A \cap C) \cup (B \cap C)$
7. $(A \cap B) \cup C$
8. $(A \cup B) \cup C$
9. $(A \cup C) \cap (B \cup C)$
10. $(A \cap B) \cap C$

In Problems 11–20, use U = Universal set = $\{0, 1, 2, 3, 4, 5, 6, 7, 8, 9\}$, $A = \{1, 3, 4, 5, 9\}$, $B = \{2, 4, 6, 7, 8\}$, and $C = \{1, 3, 4, 6\}$ to find each set.

11. A'
12. C'
13. $(A \cap B)'$
14. $(B \cup C)'$
15. $A' \cup B'$
16. $B' \cap C'$
17. $(A \cap C')'$
18. $(B' \cup C)'$
19. $(A \cup B \cup C)'$
20. $(A \cap B \cap C)'$
21. Write down all the subsets of $\{a, b, c, d\}$.
22. Write down all the subsets of $\{a, b, c, d, e\}$.
23. If $n(A) = 15$, $n(B) = 20$, and $n(A \cap B) = 10$, find $n(A \cup B)$.
24. If $n(A) = 20$, $n(B) = 40$, and $n(A \cup B) = 35$, find $n(A \cap B)$.
25. If $n(A \cup B) = 50$, $n(A \cap B) = 10$, and $n(B) = 20$, find $n(A)$.
26. If $n(A \cup B) = 60$, $n(A \cap B) = 40$, and $n(A) = n(B)$, find $n(A)$.

In Problems 27–34, use the information given in the figure.

27. How many are in set *A*?

28. How many are in set *B*?

29. How many are in *A* or *B*?

30. How many are in *A* and *B*?

31. How many are in *A* but not *C*?

32. How many are not in *A*?

33. How many are in *A* and *B* and *C*?

34. How many are in *A* or *B* or *C*?

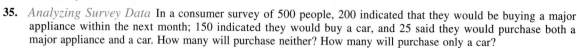

35. *Analyzing Survey Data* In a consumer survey of 500 people, 200 indicated that they would be buying a major appliance within the next month; 150 indicated they would buy a car, and 25 said they would purchase both a major appliance and a car. How many will purchase neither? How many will purchase only a car?

36. *Analyzing Survey Data* In a student survey, 200 indicated that they would attend Summer Session I and 150 indicated Summer Session II. If 75 students plan to attend both summer sessions and 275 indicated that they would attend neither session, how many students participated in the survey?

37. *Analyzing Survey Data* In a survey of 100 investors in the stock market,

> 50 owned shares in IBM
>
> 40 owned shares in AT&T
>
> 45 owned shares in GE
>
> 20 owned shares in both IBM and AT&T
>
> 20 owned shares in both IBM and GE
>
> 15 owned shares in both AT&T and GE
>
> 5 owned shares in all three

(a) How many of the investors surveyed did not have shares in any of the three companies?

(b) How many owned just IBM shares?

(c) How many owned just GE shares?

(d) How many owned neither IBM nor GE?

(e) How many owned either IBM or AT&T but no GE?

38. *Classifying Blood Types* Human blood is classified as either Rh+ or Rh−. Blood is also classified by type: A, if it contains an A antigen; B, if it contains a B antigen; AB, if it contains both A and B antigens; and O, if it contains neither antigen. Draw a Venn diagram illustrating the various blood types. Based on this classification, how many different kinds of blood are there?

39. Make up a problem different from any found in the text that requires the addition principle of counting to solve. Give it to a friend to solve and critique.

40. Investigate the notion of counting as it relates to infinite sets. Write an essay on your findings.

11.7

Permutations and Combinations

Counting plays a major role in many diverse areas, such as probability, statistics, and computer science. In this section we shall look at special types of counting problems and develop general formulas for solving them.

We begin with an example that will demonstrate a general counting principle.

E X A M P L E   1   *Counting the Number of Possible Meals*

The fixed-price dinner at a restaurant provides the following choices:

Appetizer:	soup or salad
Entree:	baked chicken, broiled beef patty, baby beef liver, or roast beef au jus
Dessert:	ice cream or cheese cake

How many different meals can be ordered?

Solution   Ordering such a meal requires three separate decisions:

CHOOSE AN APPETIZER   **CHOOSE AN ENTREE**   **CHOOSE A DESSERT**

2 choices   4 choices   2 choices

Look at the **tree diagram** in Figure 12. We see that, for each choice of appetizer, there are 4 choices of entrees. And for each of these $2 \cdot 4 = 8$ choices, there are 2 choices for dessert. Thus, there are a total of

$$2 \cdot 4 \cdot 2 = 16$$

different meals that can be ordered.

FIGURE 12

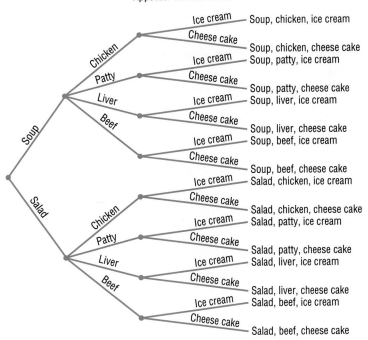

Appetizer Entree Dessert

Example 1 illustrates a general counting principle.

Theorem
Multiplication Principle
of Counting

If a task consists of a sequence of choices in which there are p selections for the first choice, q selections for the second choice, r selections for the third choice, and so on, then the task of making these selections can be done in

$$p \cdot q \cdot r \cdot \ldots$$

different ways. ■

EXAMPLE 2

Counting Airport Codes

The International Airline Transportation Association (IATA) assigns three-letter codes to represent airport locations. For example, JFK represents Kennedy International in New York. How many different airport codes are possible?

Solution

The task consists of making three selections. Each selection requires choosing a letter of the alphabet (26 choices). Thus, by the multiplication principle, there are

$$26 \cdot 26 \cdot 26 = 17{,}576$$

different airport codes. ■

In Example 2, we were allowed to repeat a letter. For example, a valid airport code is FLL (Ft. Lauderdale International Airport), in which the letter L appears twice. In the next example, such repetition is not allowed.

EXAMPLE 3

Counting without Repetition

Suppose that we wish to establish a three-letter code using any of the 26 letters of the alphabet, but we require that no letter be used more than once. How many different three-letter codes are there?

Solution

The task consists of making three selections. The first selection requires choosing from 26 letters. Because no letter can be used more than once, the second selection requires choosing from 25 letters. The third selection requires choosing from 24 letters. (Do you see why?) By the multiplication principle, there are

$$26 \cdot 25 \cdot 24 = 15{,}600$$

different three-letter codes with no letter repeated. ■

■ Now work Problems 25 and 29.

Example 3 illustrates a type of counting problem referred to as a *permutation*.

Permutation

A **permutation** is an ordered arrangement of n distinct objects without repetitions. The symbol $P(n, r)$ represents the number of permutations of n distinct objects, taken r at a time, where $r \leq n$.

For example, the question posed in Example 3 asks for the number of ways the 26 letters of the alphabet can be arranged using three nonrepeated letters. The answer is

$$P(26, 3) = 26 \cdot 25 \cdot 24 = 15{,}600$$

To arrive at a formula for $P(n, r)$, we note that the task of obtaining an ordered arrangement of n objects in which only $r \le n$ of them are used, without repeating any of them, requires making r selections. For the first selection, there are n choices; for the second selection, there are $n - 1$ choices; for the third selection, there are $n - 2$ choices; . . . ; for the rth selection, there are $n - (r - 1)$ choices. By the multiplication principle, we have

$$\begin{array}{cccc} \text{1st} & \text{2nd} & \text{3rd} & r\text{th} \\ \end{array}$$
$$P(n, r) = n \cdot (n - 1) \cdot (n - 2) \cdot \; \ldots \; \cdot [n - (r - 1)]$$
$$= n \cdot (n - 1) \cdot (n - 2) \cdot \; \ldots \; \cdot (n - r + 1)$$

This formula for $P(n, r)$ can be compactly written using factorial notation*:

$$P(n, r) = n \cdot (n - 1) \cdot (n - 2) \cdot \; \ldots \; \cdot (n - r + 1)$$

$$= n \cdot (n - 1) \cdot (n - 2) \cdot \; \ldots \; \cdot (n - r + 1) \cdot \frac{(n - r) \cdot \; \ldots \; \cdot 3 \cdot 2 \cdot 1}{(n - r) \cdot \; \ldots \; \cdot 3 \cdot 2 \cdot 1} = \frac{n!}{(n - r)!}$$

Theorem

Number of Permutations of n Distinct Objects Taken r at a Time

The number of different arrangements of n objects using $r \le n$ of them, in which

1. the n objects are distinct,
2. once an object is used it cannot be repeated, and
3. order is important,

is given by the formula

$$P(n, r) = \frac{n!}{(n - r)!} \tag{1}$$

$\blacksquare$

E X A M P L E 4

Evaluate: (a) $P(7, 3)$ (b) $P(6, 1)$

Solution

We shall work each problem in two ways.

(a) $P(7, 3) = \underbrace{7 \cdot 6 \cdot 5}_{\text{3 factors}} = 210$

$$P(7, 3) = \frac{7!}{(7 - 3)!} = \frac{7!}{4!} = \frac{7 \cdot 6 \cdot 5 \cdot 4!}{4!} = 210$$

(b) $P(6, 1) = \underbrace{6}_{\text{1 factor}} = 6$

$$P(6, 1) = \frac{6!}{(6 - 1)!} = \frac{6!}{5!} = \frac{6 \cdot 5!}{5!} = 6$$

$\blacksquare$

$\blacksquare$ Now work Problem 1.

*Recall that $0! = 1$, $1! = 1$, $2! = 2 \cdot 1, \cdots, n! = n(n - 1) \cdot \ldots \cdot 3 \cdot 2 \cdot 1$.

E X A M P L E   5   In how many ways can 5 people be lined up?

Solution   The 5 people are obviously distinct. Once a person is in line, that person will not be repeated elsewhere in the line; and, in lining up people, order is important. Thus, we have a permutation of 5 objects taken 5 at a time. We can line up the 5 people in

$$P(5, 5) = \underbrace{5 \cdot 4 \cdot 3 \cdot 2 \cdot 1}_{\text{5 factors}} = 5! = 120 \text{ ways} \qquad \blacksquare$$

■   Now work Problem 31.

Combinations

In a permutation, order is important; for example, the arrangements *ABC*, *CAB*, *BAC*, . . . are considered different arrangements of the letters *A*, *B*, and *C*. In many situations, though, order is unimportant. For example, in the card game of poker, the order in which the cards are received does not matter; it is the *combination* of the cards that matters.

Combination

> A **combination** is an arrangement, without regard to order, of *n* distinct objects without repetitions. The symbol *C(n, r)* represents the number of combinations of *n* distinct objects taken *r* at a time, where $r \le n$.

E X A M P L E   6   *Listing Combinations*

List all the combinations of the 4 objects *a, b, c, d* taken 2 at a time. What is *C*(4, 2)?

Solution   One combination of *a, b, c, d* taken 2 at a time is

$$ab$$

The object *ba* is excluded, because order is not important in a combination. The list of all such combinations (convince yourself of this) is

$$ab, \quad ac, \quad ad, \quad bc, \quad bd, \quad cd$$

Thus,

$$C(4, 2) = 6 \qquad \blacksquare$$

We can find a formula for *C(n, r)* by noting that the only difference between a permutation and a combination is that we disregard order in combinations. Thus, to determine *C(n, r)*, we need only eliminate from the formula for *P(n, r)* the number of permutations that were simply rearrangements of a given set of *r* objects. But that is easily determined from the formula for *P(n, r)* by calculating *P(r, r) = r!*. So, if we divide *P(n, r)* by *r!*, we will have the desired formula for *C(n, r)*:

$$C(n, r) = \frac{P(n, r)}{r!} = \underset{\underset{\text{Use formula (1)}}{\uparrow}}{\frac{n!/(n - r)!}{r!}} = \frac{n!}{(n - r)!r!}$$

We have proved the following result.

Theorem
Number of Combinations
of n Distinct Objects
Taken r at a Time

The number of different arrangements of n objects using $r \leq n$ of them, in which

1. the n objects are distinct,
2. once an object is used, it cannot be repeated, and
3. order is not important

is given by the formula

$$C(n, r) = \frac{n!}{(n - r)!r!} \qquad (2)$$

Based on formula (2), we discover that the symbol $C(n, r)$ and the symbol $\binom{n}{r}$ for the binomial coefficients are, in fact, the same. Thus, the Pascal triangle (see Section 9.5) can be used to find the value of $C(n, r)$. However, because it is more practical and convenient, we shall use formula (2) instead.

E X A M P L E   7

Using Formula (2)

Use formula (2) to find the value of each expression.

(a) $C(3, 1)$ (b) $C(6, 3)$ (c) $C(n, n)$ (d) $C(n, 0)$

Solution

(a) $C(3, 1) = \dfrac{3!}{(3 - 1)!1!} = \dfrac{3!}{2!1!} = \dfrac{3 \cdot 2 \cdot 1}{2 \cdot 1 \cdot 1} = 3$

(b) $C(6, 3) = \dfrac{6!}{(6 - 3)!3!} = \dfrac{6 \cdot 5 \cdot 4 \cdot 3!}{3! \cdot 3!} = \dfrac{6 \cdot 5 \cdot 4}{6} = 20$

(c) $C(n, n) = \dfrac{n!}{(n - n)!n!} = \dfrac{n!}{0!n!} = \dfrac{1}{1} = 1$

(d) $C(n, 0) = \dfrac{n!}{(n - 0)!0!} = \dfrac{n!}{n!0!} = \dfrac{1}{1} = 1$

■   Now work Problem 9.

E X A M P L E   8

Forming Committees

How many different committees of 3 people can be formed from a pool of 7 people?

Solution

The 7 people are, of course, distinct. More important, though, is the observation that the order of being selected on a committee is not significant. Thus, the problem asks for the number of combinations of 7 objects taken 3 at a time:

$$C(7, 3) = \frac{7!}{4!3!} = \frac{7 \cdot 6 \cdot 5 \cdot 4!}{4!3!} = \frac{7 \cdot 6 \cdot 5}{6} = 35$$

E X A M P L E   9

Forming Committees

In how many ways can a committee consisting of 2 faculty members and 3 students be formed if there are 6 faculty members and 10 students eligible to serve on the committee?

Solution   The problem can be separated into two parts: the number of ways the faculty members can be chosen, $C(6, 2)$, and the number of ways the student members can be chosen, $C(10, 3)$. By the multiplication principle, the committee can be formed in

$$C(6, 2) \cdot C(10, 3) = \frac{6!}{4!2!} \cdot \frac{10!}{7!3!} = \frac{6 \cdot 5 \cdot \cancel{4!}}{\cancel{4!}2!} \cdot \frac{10 \cdot 9 \cdot 8 \cdot \cancel{7!}}{\cancel{7!}3!}$$

$$= \frac{30}{2} \cdot \frac{720}{6} = 1800 \text{ ways} \qquad \blacksquare$$

■ Now work Problem 47.

Permutations with Repetition

Recall that a permutation involves counting *distinct* objects. A permutation in which some of the objects are repeated is called a **permutation with repetition.** Some books refer to this as a **nondistinguishable permutation.**

Let's look at an example.

EXAMPLE 10   *Forming Different Words*

How many different words can be formed using all the letters in the word REARRANGE?

Solution   Each word formed will have 9 letters: 3 R's, 2 A's, 2 E's, 1 N, and 1 G. To construct each word, we need to fill in 9 positions with the 9 letters:

$$\overline{1} \ \overline{2} \ \overline{3} \ \overline{4} \ \overline{5} \ \overline{6} \ \overline{7} \ \overline{8} \ \overline{9}$$

The process of forming a word consists of five tasks:

Task 1:   Choose the positions for the 3 R's.

Task 2:   Choose the positions for the 2 A's.

Task 3:   Choose the positions for the 2 E's.

Task 4:   Choose the position for the 1 N.

Task 5:   Choose the position for the 1 G.

Task 1 can be done in $C(9, 3)$ ways. There then remain 6 positions to be filled, so Task 2 can be done in $C(6, 2)$ ways. There remain 4 positions to be filled, so Task 3 can be done in $C(4, 2)$ ways. There remain 2 positions to be filled, so Task 4 can be done in $C(2, 1)$ ways. The last position can be filled in $C(1, 1)$ way. Using the Multiplication Principle, the number of possible words that can be found is

$$C(9, 3) \cdot C(6, 2) \cdot C(4, 2) \cdot C(2, 1) \cdot C(1, 1) = \frac{9!}{3! \cdot \cancel{6!}} \frac{\cancel{6!}}{2! \cdot \cancel{4!}} \frac{\cancel{4!}}{2! \cdot \cancel{2!}} \frac{\cancel{2!}}{1! \cdot \cancel{1!}} \frac{\cancel{1!}}{0! \cdot 1!} \,.$$

$$= \frac{9!}{3! \cdot 2! \cdot 2! \cdot 1! \cdot 1!} \qquad \blacksquare$$

The form of the answer to Example 10 is suggestive of a general result. Had the letters in REARRANGE each been different, there would have been $P(9, 9) = 9!$ possible words formed. This is the numerator of the answer. The presence of 3 R's, 2 A's, and 2 E's reduces the number of different words, as the entries in the denominator illustrate. We are led to the following result:

Theorem
Permutations with Repetition

The number of permutations of n objects of which n_1 are of one kind, n_2 are of a second kind, . . . , and n_k are of a kth kind is given by

$$\frac{n!}{n_1! \cdot n_2! \cdot \ldots \cdot n_k!} \qquad (3)$$

where $n = n_1 + n_2 + \cdots + n_k$. ∎

E X A M P L E   1 1

Arranging Flags

How many different vertical arrangements are there of 8 flags if 4 are white, 3 are blue, and 1 is red?

Solution

We seek the number of permutations of 8 objects, of which 4 are of one kind, 3 of a second kind, and 1 of a third kind. Using formula (3), we find that there are

$$\frac{8!}{4! \cdot 3! \cdot 1!} = \frac{8 \cdot 7 \cdot 6 \cdot 5 \cdot 4\!\!\!/!}{4\!\!\!/! \cdot 3! \cdot 1!} = 280 \text{ different arrangements} \quad \blacksquare$$

■   Now work Problem 53.

11.7

Exercise 11.7

In Problems 1–8, find the value of each permutation.

1. $P(6, 2)$ **2.** $P(7, 2)$ **3.** $P(5, 5)$ **4.** $P(4, 4)$

5. $P(8, 0)$ **6.** $P(9, 0)$ **7.** $P(8, 3)$ **8.** $P(8, 5)$

In Problems 9–16, use formula (2) to find the value of each combination.

9. $C(8, 2)$ **10.** $C(8, 6)$ **11.** $C(6, 4)$ **12.** $C(6, 2)$

13. $C(15, 15)$ **14.** $C(18, 1)$ **15.** $C(26, 13)$ **16.** $C(18, 9)$

17. List all the permutations of 5 objects a, b, c, d, and e taken 3 at a time. What is $P(5, 3)$?

18. List all the permutations of 5 objects a, b, c, d, and e taken 2 at a time. What is $P(5, 2)$?

19. List all the permutations of 4 objects 1, 2, 3, and 4 taken 3 at a time. What is $P(4, 3)$?

20. List all the permutations of 6 objects, 1, 2, 3, 4, 5, and 6 taken 3 at a time. What is $P(6, 3)$?

21. List all the combinations of the 5 objects a, b, c, d, and e taken 3 at a time. What is $C(5, 3)$?

22. List all the combinations of the 5 objects a, b, c, d, and e taken 2 at a time. What is $C(5, 2)$?

23. List all the combinations of the 4 objects 1, 2, 3, and 4 taken 3 at a time. What is $C(4, 3)$?

24. List all the combinations of the 6 objects 1, 2, 3, 4, 5, and 6 taken 3 at a time. What is $C(6, 3)$?

25. A man has 5 shirts and 3 ties. How many different shirt and tie combinations can he wear?

26. A woman has 3 blouses and 5 skirts. How many different outfits can she wear?

27. *Forming Codes* How many two-letter codes can be formed using the letters A, B, C, and D? Repeated letters are allowed.

28. *Forming Codes* How many two-letter codes can be formed using the letters A, B, C, D, and E? Repeated letters are allowed.

29. *Forming Numbers* How many three-digit numbers can be formed using the digits 0 and 1? Repeated digits are allowed.

30. *Forming Numbers* How many three-digit numbers can be formed using the digits 0, 1, 2, 3, 4, 5, 6, 7, 8, and 9? Repeated digits are allowed.

31. In how many ways can 4 people be lined up?

32. In how many ways can 5 different boxes be stacked?

33. *Forming Codes* How many different three-letter codes are there if only the letters *A, B, C, D,* and *E* can be used and no letter can be used more than once?

34. *Forming Codes* How many different four-letter codes are there if only the letters *A, B, C, D, E,* and *F* can be used and no letter can be used more than once?

35. *Arranging Letters* How many arrangements are there of the letters in the word MONEY?

36. *Arranging Digits* How many arrangements are there of the digits in the number 51,342?

37. *Establishing Committees* In how many ways can a committee of 4 students be formed from a pool of 7 students?

38. *Establishing Committees* In how many ways can a committee of 3 professors be formed from a department having 8 professors?

39. *Possible Answers on a True/False Test* How many arrangements of answers are possible for a true/false test with 10 questions?

40. *Possible Answers on a Multiple-choice Test* How many arrangements of answers are possible in a multiple-choice test with 5 questions, each of which has 4 possible answers?

41. How many four-digit numbers can be formed using the digits 0, 1, 2, 3, 4, 5, 6, 7, 8, and 9 if the first digit cannot be 0? Repeated digits are allowed.

42. How many five-digit numbers can be formed using the digits 0, 1, 2, 3, 4, 5, 6, 7, 8, and 9 if the first digit cannot be 0 or 1? Repeated digits are allowed.

43. *Arranging Books* Five different mathematics books are to be arranged on a student's desk. How many arrangements are possible?

44. *Forming License Plate Numbers* How many different license plate numbers can be made using 2 letters followed by 4 digits selected from the digits 0 through 9, if
(a) Letters and digits may be repeated?
(b) Letters may be repeated, but digits are not repeated?
(c) Neither letters nor digits may be repeated?

45. *Stock Portfolios* As a financial planner, you are asked to select one stock each from the following groups: 8 DOW stocks, 15 NASDAQ stocks, and 4 global stocks. How many different portfolios are possible?

46. *Combination Locks* A combination lock has 50 numbers on it. To open it, you turn to a number, then rotate clockwise to a second number, and then counterclockwise to the third number. How many different lock combinations are there?

47. A student dance committee is to be formed consisting of 2 boys and 3 girls. If the membership is to be chosen from 4 boys and 8 girls, how many different committees are possible?

48. *Baseball Teams* A baseball team has 15 members. Four of the players are pitchers, and the remaining 11 members can play any position. How many different teams of 9 players can be formed?

49. The student relations committee of a college consists of 2 administrators, 3 faculty members, and 5 students. There are 4 administrators, 8 faculty members, and 20 students eligible to serve. How many different committees are possible?

50. *Football Teams* A defensive football squad consists of 25 players. Of these, 10 are linemen, 10 are linebackers, and 5 are safeties. How many different teams of 5 linemen, 3 linebackers, and 3 safeties can be formed?

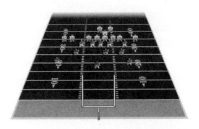

51. *Baseball* In the American Baseball League, a designated hitter may be used. How many batting orders is it possible for a manager to use? (There are 9 regular players on a team.)

52. *Baseball* In the National Baseball League, the pitcher usually bats ninth. If this is the case, how many batting orders are possible for a manager to use?

53. *Forming Words* How many different 9 letter words (real or imaginary) can be formed from the letters in the word ECONOMICS?

54. *Forming Words* How many different 11 letter words (real or imaginary) can be formed from the letters in the word MATHEMATICS?

55. *Senate Committees* The U.S. Senate has 100 members. Suppose it is desired to place each senator on exactly 1 of 7 possible committees. The first committee has 22 members, the second has 13, the third has 10, the fourth has 5, the fifth has 16, and the sixth and seventh have 17 apiece. In how many ways can these committees be formed?

56. *World Series* In the World Series the American League team (*A*) and the National League team (*N*) play until one team wins four games. If the sequence of winners is designated by letters (for example, *NAAAA* means the National League team won the first game and the American League won the next four), how many different sequences are possible?

57. *Basketball Teams* A basketball team has 6 players who play guard (2 of 5 starting positions). How many different teams are possible, assuming that the remaining 3 positions are filled and it is not possible to distinguish a left guard from a right guard?

58. *Basketball Teams* On a basketball team of 12 players, 2 only play center, 3 only play guard, and the rest play forward (5 players on a team: 2 forwards, 2 guards, and 1 center). How many different teams are possible, assuming that it is not possible to distinguish left and right guards and left and right forwards?

59. *Selecting Objects* An urn contains 7 white balls and 3 red balls. Three balls are selected. In how many ways can the 3 balls be drawn from the total of 10 balls:
(a) If 2 balls are white and 1 is red?
(b) If all 3 balls are white?
(c) If all 3 balls are red?

60. *Selecting Objects* An urn contains 15 red balls and 10 white balls. Five balls are selected. In how many ways can the 5 balls be drawn from the total of 25 balls:
(a) If all balls are red?
(b) If 3 balls are red and 2 are white?
(c) If at least 4 are red balls?

61. *Programming Exercise* When both *n* and *r* are large, finding *C(n, r)* on a computer may lead to integers too large to compute. To avoid this, we can approximate the values of *C(n, r)*. One way to do this is the following:

$$C(40, 20) = \frac{40!}{20!20!} = \frac{40 \cdot 39 \cdot 38 \cdot \ldots \cdot 21}{20 \cdot 19 \cdot 18 \cdot \ldots \cdot 1} = \frac{40}{20} \cdot \frac{39}{19} \cdot \frac{38}{18} \cdot \ldots \cdot \frac{21}{1}$$
$$\approx 2.000 \cdot 2.053 \cdot 2.111 \cdot \ldots \cdot 21.000 = 1.3784652 \times 10^{11}$$

(a) Write a program that inputs two integers *N* and *R* and computes *C(N, R)* using formula (1).
(b) Use the program to determine where overflow occurs on your computer.
(c) Write a program that inputs two integers *N* and *R* and computes *C(N, R)* by the approximation technique shown above.
(d) Compare the answers found in parts (a) and (c).

62. Make up a problem different from any found in the text that requires the multiplication principle of counting to solve. Give it to a friend to solve and critique.

63. Make up a problem different from any found in the text that requires a permutation to solve. Give it to a friend to solve and critique.

64. Make up a problem different from any found in the text that requires a combination to solve. Give it to a friend to solve and critique.

65. Explain the difference between a permutation and a combination. Give an example to illustrate your explanation.

11.8

Probability

Probability is an area of mathematics that deals with experiments that yield random results yet admit a certain regularity. Such experiments do not always produce the same result or outcome, so the result of any one observation is not predictable. However, the results of the experiment over a long period do produce regular patterns that enable us to predict with remarkable accuracy.

EXAMPLE 1   *Tossing a Fair Coin*

In tossing a fair coin, we know that the outcome is either a head or a tail. On any particular throw, we cannot predict what will happen, but, if we toss the coin many times, we observe that the number of times a head comes up is approximately equal to the number of times we get a tail. It seems reasonable, therefore, to assign a probability of $\frac{1}{2}$ that a head comes up and a probability of $\frac{1}{2}$ that a tail comes up. ∎

Probability Models

The discussion in Example 1 constitutes the construction of a **probability model** for the experiment of tossing a fair coin once. A probability model has two components: a sample space and an assignment of probabilities. A **sample space** S is a set whose elements represent all the possibilities that can occur as a result of the experiment. Each element of S is called an **outcome.** To each outcome, we assign a number, called the **probability** of that outcome, which has two properties:

1. Each probability is nonnegative.
2. The sum of all the probabilities equals 1.

Thus, if a probability model has the sample space

$$S = \{e_1, e_2, \ldots, e_n\}$$

where $e_1, e_2, \ldots, e_n$ are the possible outcomes, and if $P(e_1), P(e_2), \ldots, P(e_n)$ denote the respective probabilities of these outcomes, then

$$P(e_1) \geq 0, \quad P(e_2) \geq 0, \quad \ldots, \quad P(e_n) \geq 0 \qquad (1)$$
$$P(e_1) + P(e_2) + \cdots + P(e_n) = 1 \qquad (2)$$

Let's look at an example.

EXAMPLE 2   *Constructing a Probability Model*

An experiment consists of rolling a fair die once.* Construct a probability model for this experiment.

Solution   A sample space S consists of all the possibilities that can occur. Because rolling the die will result in one of six faces showing, the sample space S consists of

$$S = \{1, 2, 3, 4, 5, 6\}$$

Because the die is fair, one face is no more likely to occur than another. As a result, our assignment of probabilities is

$$P(1) = \tfrac{1}{6} \qquad P(2) = \tfrac{1}{6}$$

$$P(3) = \tfrac{1}{6} \qquad P(4) = \tfrac{1}{6}$$

$$P(5) = \tfrac{1}{6} \qquad P(6) = \tfrac{1}{6}$$ ■

Suppose that a die is loaded so that the probability assignments are

$$P(1) = 0, \quad P(2) = 0, \quad P(3) = \frac{1}{3}, \quad P(4) = \frac{2}{3}, \quad P(5) = 0, \quad P(6) = 0$$

This assignment would be made if the die was loaded so that only a 3 or a 4 could occur and the 4 is twice as likely as the 3 to occur. This assignment is consistent with the definition since each assignment is between 0 and 1, and the sum of all the probability assignments equals 1.   ■

■ Now work Problem 13.

EXAMPLE 3   *Constructing a Probability Model*

An experiment consists of tossing a coin. The coin is weighted so that heads (H) is three times as likely to occur as tails (T). Construct a probability model for this experiment.

Solution   The sample space S is $S = \{H, T\}$. If x denotes the probability that a tail occurs, then

$$P(T) = x \quad \text{and} \quad P(H) = 3x$$

Since the sum of the probabilities of the possible outcomes must equal 1, we have

$$P(T) + P(H) = x + 3x = 1$$
$$4x = 1$$
$$x = \frac{1}{4}$$

Thus, we assign the probabilities

$$P(T) = \frac{1}{4} \qquad P(H) = \frac{3}{4}$$ ■

■ Now work Problem 17.

*A die is a cube with each face having either 1, 2, 3, 4, 5, or 6 dots on it.

E X A M P L E   4

An experiment consists of tossing a fair die and then a fair coin. Construct a probability model for this experiment.

Solution

A tree diagram is helpful in listing all the possible outcomes. See Figure 13. The sample space consists of the outcomes

FIGURE 13

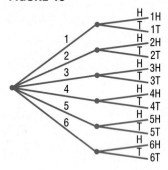

$$S = \{1H, 1T, 2H, 2T, 3H, 3T, 4H, 4T, 5H, 5T, 6H, 6T\}$$

The die and the coin are fair; thus, no one outcome is more likely to occur than another. As a result, we assign the probability $\frac{1}{12}$ to each of the 12 outcomes. ■

In working with probability models, the term **event** is used to describe a set of possible outcomes of the experiment. Thus, an event E is some subset of the sample space S. The **probability of an event** E, $E \neq 0$, denoted by $P(E)$, is defined as the sum of the probabilities of the outcomes in E. If $E = \varnothing$, then $P(E) = 0$; if $E = S$, then $P(E) = P(S) = 1$.

E X A M P L E   5

Finding the Probability of an Event

For the experiment described in Example 4, what is the probability that an even number followed by a head occurs?

Solution

The event E, an even number followed by a head, consists of

$$\overset{\text{·}}{B} = \{2H, 4H, 6H\}$$

The probability of E is

$$P(E) = P(2H) + P(4H) + P(6H) = \tfrac{1}{12} + \tfrac{1}{12} + \tfrac{1}{12} = \tfrac{1}{4}$$ ■

The next result, called the **Additive Rule,** may be used to find the probability of the union of two events.

Theorem
Additive rule

For any two events E and F,

$$P(E \cup F) = P(E) + P(F) - P(E \cap F) \qquad (3)$$

■

If E and F are disjoint, so that $E \cap F = \varnothing$, then formula (3) takes the form

Mutually Exclusive Events

$$P(E \cup F) = P(E) + P(F) \qquad (4)$$

When formula (4) applies, we say that E and F are **mutually exclusive events.**

E X A M P L E   6

Using Formulas (3) and (4)

(a) If $P(E) = 0.2$, $P(F) = 0.3$, and $P(E \cap F) = 0.1$, find $P(E \cup F)$.
(b) If $P(E) = 0.2$, $P(F) = 0.3$, and E, F are mutually exclusive, find $P(E \cup F)$.

Solution

(a) We use the Additive Rule, formula (3).

$$P(E \cup F) = P(E) + P(F) - P(E \cap F) = 0.2 + 0.3 - 0.1 = 0.4$$

(b) Since E, F are mutually exclusive, we use formula (4).

$$P(E \cup F) = P(E) + P(F) = 0.2 + 0.3 = 0.5$$ ■

A Venn diagram can sometimes be used to obtain probabilities. To construct a Venn diagram representing the information in Example 6(a), we draw two sets E and F. We begin with the fact that $P(E \cap F) = 0.1$. See Figure 14(a). Then, since $P(E) = 0.2$ and $P(F) = 0.3$, we fill in E with $0.2 - 0.1 = 0.1$ and F with $0.3 - 0.1 = 0.2$. See Figure 14(b). Since $P(S) = 1$, we complete the diagram by inserting $1 - [0.1 + 0.1 + 0.2] = 0.6$. See Figure 14(c). Now it is easy to see, for example, that the probability of F, but not E, is 0.2. Also, the probability of neither E nor F is 0.6.

FIGURE 14

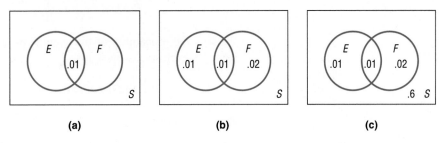

(a)   (b)   (c)

■   Now work Problem 23.

Equally Likely Outcomes

When the same probability is assigned to each outcome of the sample space, the experiment is said to have **equally likely outcomes.**

Theorem
Probability for Equally
Likely Outcomes

If an experiment has n equally likely outcomes, and if the number of ways an event E can occur is m, then the probability of E is

$$P(E) = \frac{\text{Number of ways that } E \text{ can occur}}{\text{Number of all logical possibilities}} = \frac{m}{n} \qquad (5)$$

Thus, if S is the sample space of this experiment, then

$$P(E) = \frac{n(E)}{n(S)} \qquad (6)$$

■

Based on (6), an alternative method of solution of Example 5 is

$$P(E) = \frac{n(E)}{n(S)} = \frac{3}{12} = \frac{1}{4}$$

E X A M P L E   7

Computing Probabilities for Equally Likely Outcomes

A jar contains 10 marbles; 5 are solid color, 4 are speckled, and 1 is clear.

(a) If one marble is picked at random, what is the probability it is speckled?

(b) If one marble is picked at random, what is the probability it is clear or a solid color?

Solution

The experiment is an example of one in which the outcomes are equally likely; that is, no one marble is more likely to be picked than another. If S is the sample space, then there are 10 possible outcomes in S, so $n(S) = 10$.

(a) Define the event E: speckled marble is picked. There are 4 ways E can occur. Thus,

$$P(E) = \frac{n(E)}{n(S)} = \frac{4}{10} = 0.4$$

(b) Define the events F: clear marble is picked and G: solid color marble is picked. Then there is 1 way for F to occur and 5 ways for G to occur. Thus,

$$P(F) = \frac{n(F)}{n(S)} = \frac{1}{10}, \qquad P(G) = \frac{n(G)}{n(S)} = \frac{5}{10}$$

We seek the probability of the event F or G, that is, $P(F \cup G)$. Since F, G are mutually exclusive, we use (4).

$$P(F \cup G) = P(F) + P(G) = \frac{1}{10} + \frac{5}{10} = \frac{6}{10} = 0.6$$ ■

■ Now work Problem 27.

EXAMPLE 8

The Game of Craps

In the game of "craps," two fair dice are rolled. If the total of the faces equals 7 or 11, you win. If the totals are 2, 3, or 12, you have thrown craps and you lose. In all other cases, you throw again.

(a) What is the probability that you will win?
(b) What is the probability that you will lose?
(c) What is the probability that you will need to throw again?

Solution

FIGURE 15

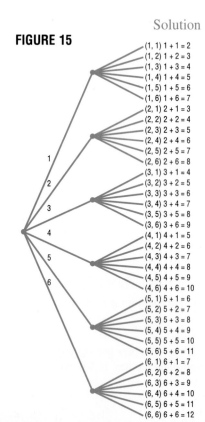

We begin by constructing a probability model for the experiment. The tree diagram in Figure 15 will help us to see all the possibilities.

Because the dice are fair, no one of the 36 possible outcomes in the sample space S is more likely to occur than any other. Thus, we have equally likely outcomes with $n(S) = 36$.

(a) The event E, "the dice total 7 or 11," consists of the outcomes

$$E = \{(1, 6), (2, 5), (3, 4), (4, 3), (5, 2), (6, 1), (5, 6), (6, 5)\}$$

Because $n(E) = 8$, we have

$$P(E) = \frac{n(E)}{n(S)} = \frac{8}{36} = \frac{2}{9} \approx 0.222$$

(b) The event F, "the dice total 2, 3, or 12," consists of the outcomes

$$F = \{(1, 1), (1, 2), (2, 1), (6, 6)\}$$

Because $n(F) = 4$,

$$P(F) = \frac{n(F)}{n(S)} = \frac{4}{36} = \frac{1}{9} \approx 0.111$$

(c) The number of possibilities that require that you throw again is $36 - n(E) - n(F) = 36 - 8 - 4 = 24$. Thus, the probability that another throw is required is

$$\frac{24}{36} = \frac{2}{3} \approx 0.667$$ ∎

∎ Now work Problem 29.

Applications Involving Permutations and Combinations

E X A M P L E   9

Computing Probabilities

Because of a mistake in packaging, 5 defective phones were packaged with 15 good ones. All phones look alike and have equal probability of being chosen. Three phones are selected.

(a) What is the probability that all 3 are defective?
(b) What is the probability that exactly 2 are defective?
(c) What is the probability that at least 2 are defective?

Solution

The sample space S consists of the number of ways 3 objects can be selected from 20 objects, that is, the number of combinations of 20 things taken 3 at a time.

$$n(S) = C(20, 3) = \frac{20!}{17! \cdot 3!} = \frac{20 \cdot 19 \cdot 18}{6} = 1140$$

Each of these outcomes is equally likely to occur.

(a) If E is the event "3 are defective," then the number of elements in E is the number of ways the 3 defective phones can be chosen from the 5 defective phones: $C(5, 3) = 10$. Thus, the probability of E is

$$P(E) = \frac{n(E)}{n(S)} = \frac{10}{1140} = 0.0088$$

(b) If F is the event "exactly 2 are defective" and 3 phones are selected, then the number of elements in F is the number of ways to select 2 defective phones from the 5 defective phones and 1 good phone from the 15 good ones. The first of these can be done in $C(5, 2)$ ways and the second in $C(15, 1)$ ways. By the Multiplication Principle, the event F can occur in

$$C(5, 2)C(15, 1) = \frac{5!}{3! \cdot 2!} \frac{15!}{14! \cdot 1!} = 10 \cdot 15 = 150 \text{ ways}$$

The probability of F is therefore

$$P(F) = \frac{n(F)}{n(S)} = \frac{150}{1140} = 0.1316$$

(c) The event G, "at least two are defective," when 3 are chosen, is equivalent to requiring that either exactly 2 defective are chosen or exactly 3 defective are chosen. That is, $G = E \cup F$. Since E and F are mutually exclusive (it is not possible to select 2 defective phones and, at the same time, select 3 defective phones), we find

$$P(G) = P(E) + P(F) = 0.0088 + 0.1316 = 0.1404$$ ■

■ Now work Problem 43.

E X A M P L E   1 0   *Tossing a Coin*

A fair coin is tossed 6 times.

(a) What is the probability of obtaining exactly 5 heads and one tail?

(b) What is the probability of obtaining between 4 and 6 heads, inclusive?

Solution   The number of elements in the sample space S is found using the Multiplication Principle. Each toss results in a head (H) or a tail (T). Since the coin is tossed 6 times, we have

$$n(S) = \underbrace{2 \cdot 2 \cdot \ldots \cdot 2}_{6 \text{ tosses}} = 2^6 = 64$$

The outcomes are equally likely since the coin is fair.

(a) Any sequence that contains 5 heads and 1 tail is determined once the position of the 5 heads (or 1 tail) is known. The number of ways we can position 5 heads in a sequence of 6 slots is $C(6, 5) = 6$. The probability of the event E: exactly 5 heads and one tail is

$$P(E) = \frac{n(E)}{n(S)} = \frac{C(6, 5)}{2^6} = \frac{6}{64} = 0.0938$$

(b) Let F be the event: between 4 and 6 heads, inclusive. To obtain between 4 and 6 heads is equivalent to the event: either 4 heads or 5 heads or 6 heads. Since each of these is mutually exclusive (it is impossible to obtain both 4 heads and 5 heads when tossing a coin 6 times), we have

$$P(F) = P(4 \text{ heads or } 5 \text{ heads or } 6 \text{ heads})$$
$$= P(4 \text{ heads}) + P(5 \text{ heads}) + P(6 \text{ heads})$$

The probabilities on the right are obtained as in part (a). Thus

$$P(F) = \frac{C(6, 4)}{2^6} + \frac{C(6, 5)}{2^6} + \frac{C(6, 6)}{2^6} = \frac{15}{64} + \frac{6}{64} + \frac{1}{64} = \frac{22}{64} = 0.3438$$ ■

HISTORICAL FEATURE   ■ Set theory, counting, and probability first took form as a systematic theory in the exchange of letters (1654) between Pierre de Fermat (1601–1665) and Blaise Pascal (1623–1662). They discussed the problem of how to divide the stakes in a game that is interrupted before completion, knowing how many points each player needs to win. Fermat solved the problem by listing all possibilities and counting the favorable ones, whereas Pascal made use of the triangle that now bears his name. As mentioned in the text, the entries in Pascal's triangle are equivalent to $C(n, r)$. This recognition of the role of $C(n, r)$ in counting is the foundation of all further developments.

The first book on probability, the work of Christian Huygens (1629–1695), appeared in 1657. In it, the notion of mathematical expectations is explored. This allows the calculation of the profit or loss a gambler may expect, knowing the probabilities involved in the game (see the Historical Problems that follow).

It is interesting to note that Girolamo Cardano (1501–1576) wrote a treatise on probability, but it was not published until 1663 in Cardano's collected works, and this was too late to have any effect on the development of the theory.

In 1713, the posthumously published *Ars Conjectandi* of Jacob Bernoulli gave the theory the form it would have until 1900. In the current century, both combinatorics (counting) and probability have undergone rapid development due to the use of computers.

A final comment about notation. The notations $C(n, r)$ and $P(n, r)$ are variants of a form of notation developed in England after 1830. The notation $\binom{n}{r}$ for $C(n, r)$ goes back to Leonhard Euler (1707–1783), but is now losing ground because it has no clearly related symbolism of the same type for permutations. The set symbols $\cup$ and $\cap$ were introduced by Giuseppe Peano (1858–1932) in 1888 in a slightly different context. The inclusion symbol $\subset$ was introduced by E. Schroeder (1841–1902) about 1890. The treatment of set theory in the text is due to George Boole (1815–1864), who wrote $A + B$ for $A \cup B$ and AB for $A \cap B$ (statisticians still use AB for $A \cap B$). ∎

HISTORICAL PROBLEMS

∎ 1. *The Problem Discussed by Fermat and Pascal* A game between two equally skilled players, A and B, is interrupted when A needs 2 points to win and B needs 3 points. In what proportion should the stakes be divided? [*Note:* If each play results in 1 point for either player, at most four more plays will decide the game.]
   (a) *Fermat's solution* List all possible outcomes that will end the game to form the sample space (for example, *ABAA*, *ABBB*, etc.). The probabilities for A to win and B to win then determine how the stakes should be divided.
   (b) *Pascal's solution* Use combinations to determine the number of ways the 2 points needed for A to win could occur in four plays. Then use combinations to determine the number of ways the 3 points needed for B to win could occur. This is trickier than it looks, since A can win with 2 points in either two plays, three plays, or four plays. Compute the probabilities and compare with the results in part (a).

2. *Huygens's Mathematical Expectation* In a game with n possible outcomes with probabilities $p_1, p_2, \ldots, p_n$, suppose that the *net* winnings are $w_1, w_2, \ldots, w_n$, respectively. Then the mathematical expectation is

$$E = p_1 w_1 + p_2 w_2 + \cdots + p_n w_n$$

The number E represents the profit or loss per game in the long run. The following problems are a modification of those of Huygens:
   (a) A fair die is tossed. A gambler wins \$3 if he throws a 6 and \$6 if he throws a 5. What is his expectation? [*Note:* $w_1 = w_2 = w_3 = w_4 = 0$]
   (b) A gambler plays the same game as in part (a), but now the gambler must pay \$1 to play. This means $w_5 = \$5$, $w_6 = \$2$, and $w_1 = w_2 = w_3 = w_4 = -\$1$. What is the expectation? ∎

11.8

Exercise 11.8

In Problems 1–6, construct a probability model for each experiment.

1. Tossing a fair coin twice

2. Tossing two fair coins once

3. Tossing two fair coins, then a fair die 5. Tossing three fair coins once

4. Tossing a fair coin, a fair die, and then a fair coin 6. Tossing one fair coin three times

In Problems 7–12, use the spinners shown below, and construct a probability model for each experiment.

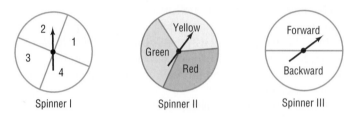

Spinner I Spinner II Spinner III

7. Spin spinner I, then spinner II. What is the probability of getting a 2 or a 4, followed by Red?

8. Spin spinner III, then spinner II. What is the probability of getting Forward, followed by Yellow or Green?

9. Spin spinner I, then II, then III. What is the probability of getting a 1, followed by Red or Green, followed by Backward?

10. Spin spinner II, then I, then III. What is the probability of getting Yellow, followed by a 2 or a 4, followed by Forward?

11. Spin spinner I twice, then spinner II. What is the probability of getting a 2, followed by a 2 or a 4, followed by Red or Green?

12. Spin spinner III, then spinner I twice. What is the probability of getting Forward, followed by a 1 or a 3, followed by a 2 or a 4?

In Problems 13–16, consider the experiment of tossing a coin twice. The table lists six possible assignments of probabilities for this experiment. Using this table, answer the following questions.

13. Which of the assignments of probabilities are consistent with the definition of the probability of an outcome?

14. Which of the assignments of probabilities should be used if the coin is known to be fair?

15. Which of the assignments of probabilities should be used if the coin is known to always come up tails?

16. Which of the assignments of probabilities should be used if tails is twice as likely as heads to occur?

17. *Assigning Probabilities* A coin is weighted so that heads is four times as likely as tails to occur. What probability should we assign to heads? to tails?

ASSIGNMENTS	SAMPLE SPACE			
	HH	**HT**	**TH**	**TT**
A	$\frac{1}{4}$	$\frac{1}{4}$	$\frac{1}{4}$	$\frac{1}{4}$
B	0	0	0	1
C	$\frac{3}{16}$	$\frac{5}{16}$	$\frac{5}{16}$	$\frac{3}{16}$
D	$\frac{1}{2}$	$\frac{1}{2}$	$-\frac{1}{2}$	$\frac{1}{2}$
E	$\frac{1}{8}$	$\frac{1}{4}$	$\frac{1}{4}$	$\frac{1}{8}$
F	$\frac{1}{9}$	$\frac{2}{9}$	$\frac{2}{9}$	$\frac{4}{9}$

18. *Assigning Probabilities* A coin is weighted so that tails is twice as likely as heads to occur. What probability should we assign to heads? to tails?

19. *Assigning Probabilities* A die is weighted so that an odd-numbered face is twice as likely as an even-numbered face. What probability should we assign to each face?

20. *Assigning Probabilities* A die is weighted so that a six cannot appear. The other faces occur with the same probability. What probability should we assign to each face?

In Problems 21–24, find the probability of the indicated event if P(A) = 0.30 and P(B) = 0.40.

21. $P(A \cup B)$ if A, B are mutually exclusive

22. $P(A \cap B)$ if A, B are mutually exclusive

23. $P(A \cup B)$ if $P(A \cap B) = 0.15$

24. $P(A \cap B)$ if $P(A \cup B) = 0.6$

In Problems 25–28, a golf ball is selected at random from a container. If the container has 9 white balls, 8 green balls, and 3 orange ones, find the probability of each event.

25. The golf ball is white.

26. The golf ball is green.

27. The golf ball is white or green.

28. The golf ball is not white.

29. What is the probability of throwing a 6 or an 8 in a game of craps? (Consult Example 8.)

30. What is the probability of throwing a 5 or a 9 in a game of craps? (Consult Example 8.)

Problems 31–34 are based on a consumer survey of annual incomes in 100 households. The following table gives the data:

Income	$0–9999	$10,000–19,999	$20,000–29,999	$30,000–39,999	$40,000 or more
Number of households	5	35	30	20	10

31. What is the probability that a household has an annual income of $30,000 or more?

32. What is the probability that a household has an annual income between $10,000 and $29,999, inclusive?

33. What is the probability that a household has an annual income less than $20,000?

34. What is the probability that a household has an annual income of $20,000 or more?

35. *Surveys* In a survey about the number of TV sets in a house, the following probability table was constructed:

Number of TV sets	0	1	2	3	4 or more
Probability	0.05	0.24	0.33	0.21	0.17

Find the probability of a house having:

(a) 1 or 2 TV sets   (b) 1 or more TV sets   (c) 3 or fewer TV sets   (d) 3 or more TV sets

(e) Less than 2 TV sets   (f) Less than 1 TV set   (g) 1, 2, or 3 TV sets   (h) 2 or more TV sets

36. *Checkout Lines* Through observation it has been determined that the probability for a given number of people waiting in line at the "5 items or less" checkout register of a supermarket is:

Number waiting in line	0	1	2	3	4 or more
Probability	0.10	0.15	0.20	0.24	0.31

Find the probability of:
(a) At most 2 people in line   (b) At least 2 people in line
(c) At least 1 person in line

37. *Winning a Lottery* In a certain lottery, there are ten balls, numbered 1, 2, 3, 4, 5, 6, 7, 8, 9, 10. Of these, five are drawn in the correct order. If you pick five numbers that match those drawn in the correct order, you win $1,000,000. What is the probability of winning such a lottery?

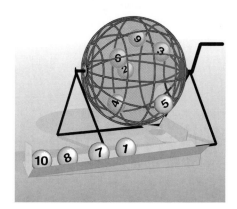

38. A committee of 6 people is to be chosen at random from a group of 14 people consisting of 2 supervisors, 5 skilled laborers, and 7 unskilled laborers. What is the probability that the committee chosen consists of 2 skilled and 4 unskilled laborers?

39. A fair coin is tossed 5 times.
    (a) Find the probability that exactly 3 heads appear.
    (b) Find the probability that no heads appear.

40. A fair coin is tossed 4 times.
    (a) Find the probability that exactly 1 tail appears.
    (b) Find the probability that no more than 1 tail appears.

41. A pair of fair dice are tossed 3 times.
    (a) Find the probability that the sum of seven appears 3 times.
    (b) Find the probability that a sum of 7 or 11 appears at least twice.

42. A pair of fair dice are tossed 5 times.
    (a) Find the probability that the sum is never 2.
    (b) Find the probability that the sum is never 7.

43. Through a mix-up on the production line, 5 defective TV's were shipped out with 25 good ones. If 5 are selected at random, what is the probability that all 5 are defective? What is the probability that at least 2 of them are defective?

44. In a shipment of 50 transformers, 10 are known to be defective. If 30 transformers are picked at random, what is the probability that all 30 are nondefective? Assume that all transformers look alike and have an equal probability of being chosen.

45. In a promotion, 50 silver dollars are placed in a bag, one of which is valued at more than $10,000. The winner of the promotion is given the opportunity to reach into the bag, while blindfolded, and pull out 5 coins. What is the probability that one of the 5 coins is the one valued at more than $10,000?

 46. Go to the library and look up the "birthday problem" in a book on probability. Write a brief essay about this problem and its solution.

Chapter Review

THINGS TO KNOW

Sequence	A function whose domain is the set of positive integers.
Factorials	$0! = 1$, $1! = 1$, $n! = n(n-1) \cdot \ldots \cdot 3 \cdot 2 \cdot 1$ if $n \geq 2$
Arithmetic sequence	$a_1 = a$, $a_{n+1} = a_n + d$, where $a =$ first term, $d =$ common difference, $a_n = a + (n-1)d$
Sum of the first n terms of an arithmetic sequence	$S_n = \dfrac{n}{2}[2a + (n-1)d] = \dfrac{n}{2}(a + a_n)$
Geometric sequence	$a_1 = a$, $a_{n+1} = ra_n$; where $a =$ first term, $r =$ common ratio, $a_n = ar^{n-1}$, $r \neq 0$
Sum of the first n terms of a geometric sequence	$S_n = a\dfrac{1 - r^n}{1 - r}$, $r \neq 0, 1$
Infinite geometric series	$a + ar + \cdots + ar^{n-1} + \cdots = \displaystyle\sum_{k=1}^{\infty} ar^{k-1}$
Sum of an infinite geometric series	$\displaystyle\sum_{k=1}^{\infty} ar^{k-1} = \dfrac{a}{1 - r}$, $|r| < 1$
Principle of mathematical induction	Condition I: The statement is true for the natural number 1.
	Condition II: If the statement is true for some natural number k, it is also true for $k + 1$
	Then the statement is true for all natural numbers.

Binomial coefficient	$\dbinom{n}{j} = \dfrac{n!}{j!(n-j)!}$	
Pascal triangle	See Figure 7.	
Binomial Theorem	$(x+a)^n = \dbinom{n}{0}x^n + \dbinom{n}{1}ax^{n-1} + \cdots + \dbinom{n}{j}a^j x^{n-j} + \cdots + \dbinom{n}{n}a^n$	

Set		Well-defined collection of distinct objects, called elements
Null set	$\varnothing$	Set that has no elements
Equality	$A = B$	A and B have the same elements
Subset	$A \subseteq B$	Each element of A is also an element of B.
Intersection	$A \cap B$	Set consisting of elements that belong to both A and B
Union	$A \cup B$	Set consisting of elements that belong to either A or B, or both
Universal set	U	Set consisting of all the elements we wish to consider
Complement	A'	Set consisting of elements of the universal set that are not in A
Finite set		The number of elements in the set is a nonnegative integer
Infinite set		A set that is not finite
Counting formula	$n(A \cup B) = n(A) + n(B) - n(A \cap B)$	
Addition principle		If $A \cap B = \varnothing$, then $n(A \cup B) = n(A) + n(B)$.
Multiplication principle		If a task consists of a sequence of choices in which there are p selections for the first choice, q selections for the second choice, and so on, then the task of making these selections can be done in $p \cdot q \cdot \ldots$ ways.
Permutation	$P(n, r) = n(n-1) \cdot \ldots \cdot [n-(r-1)]$	
	$= \dfrac{n!}{(n-r)!}$	An ordered arrangement of n distinct objects without repetition
Combination	$C(n, r) = \dfrac{P(n, r)}{r!}$	An arrangement, without regard to order, of n distinct objects without repetition
	$= \dfrac{n!}{(n-r)!r!}$	
Permutations with repetition	$\dfrac{n!}{n_1! n_2! \cdots n_k!}$	The number of permutations of n objects of which n_1 are of one kind, n_2 are of a second kind, $\ldots$, and n_k are of a kth kind, where $n = n_1 + n_2 + \cdots + n_k$
Sample space		Set whose elements represent all the logical possibilities that can occur as a result of an experiment
Probability		A number assigned to each outcome of a sample space; the sum of all the probabilities of the outcomes equals 1
Additive rule	$P(E \cup F) = P(E) + P(F) - P(E \cap F)$	
Equally likely outcomes	$P(E) = \dfrac{n(E)}{n(S)}$	The same probability is assigned to each outcome.

How To:

Write down the terms of a sequence	Find the sum of the first n terms of a geometric sequence
Use summation notation	Find the sum of an infinite geometric series
Identify an arithmetic sequence	Prove statements about natural numbers using mathematical induction
Find the sum of the first n terms of an arithmetic sequence	
Identify a geometric sequence	Apply the Binomial Theorem

Find unions, intersections, and complements of sets

Use Venn diagrams to illustrate sets

Recognize a permutation problem

Recognize a combination problem

Solve certain probability problems

Count the elements in a sample space

Draw a tree diagram

FILL-IN-THE-BLANK ITEMS

1. A(n) _____ is a function whose domain is the set of positive integers.

2. In a(n) _____ sequence, the difference between successive terms is always the same number.

3. In a(n) _____ sequence, the ratio of successive terms is always the same number.

4. The _____ _____ is a triangular display of the binomial coefficients.

5. $\binom{6}{2}$ = _____

6. The _____ of A with B consists of all elements in either A or B; the _____ of A with B consists of all elements in both A and B.

7. $P(5, 2)$ = _____; $C(5, 2)$ = _____

8. A(n) _____ is an ordered arrangement of n distinct objects.

9. A(n) _____ is an arrangement of n distinct objects without regard to order.

10. When the same probability is assigned to each outcome of a sample space, the experiment is said to have _____ _____ outcomes.

TRUE/FALSE ITEMS

T   F   **1.** A sequence is a function.

T   F   **2.** For arithmetic sequences, the difference of successive terms is always the same number.

T   F   **3.** For geometric sequences, the ratio of successive terms is always the same number.

T   F   **4.** Mathematical induction can sometimes be used to prove theorems that involve natural numbers.

T   F   **5.** $\binom{n}{j} = \dfrac{j!}{n!(n-j)!}$

T   F   **6.** The expansion of $(x + a)^n$ contains n terms.

T   F   **7.** $\displaystyle\sum_{i=1}^{n+1} i = 1 + 2 + 3 + \cdots + n$

T   F   **8.** The intersection of two sets is always a subset of their union.

T   F   **9.** $P(n, r) = \dfrac{n!}{r!}$

T   F   **10.** In a combination problem, order is not important.

T   F   **11.** In a permutation problem, once an object is used, it cannot be repeated.

T   F   **12.** The probability of an event can never equal 0.

REVIEW EXERCISES

In Problems 1–8, evaluate each expression.

1. 5!   **2.** 6!   **3.** $\binom{5}{2}$ **4.** $\binom{8}{6}$ **5.** $P(8, 3)$ **6.** $P(7, 3)$ **7.** $C(8, 3)$ **8.** $C(7, 3)$

In Problems 9–16, write down the first five terms of each sequence.

9. $\left\{(-1)^n\left(\dfrac{n+3}{n+2}\right)\right\}$

10. $\{(-1)^{n+1}(2n+3)\}$

11. $\left\{\dfrac{2^n}{n^2}\right\}$

12. $\left\{\dfrac{e^n}{n}\right\}$

13. $a_1 = 3;\quad a_{n+1} = \frac{2}{3}a_n$

14. $a_1 = 4;\quad a_{n+1} = -\frac{1}{4}a_n$

15. $a_1 = 2;\quad a_{n+1} = 2 - a_n$

16. $a_1 = -3;\quad a_{n+1} = 4 + a_n$

In Problems 17–28, determine whether the given sequence is arithmetic, geometric, or neither. If the sequence is arithmetic, find the common difference and the sum of the first n terms. If the sequence is geometric, find the common ratio and the sum of the first n terms.

17. $\{n+5\}$

18. $\{4n+3\}$

19. $\{2n^3\}$

20. $\{2n^2 - 1\}$

21. $\{2^{3n}\}$

22. $\{3^{2n}\}$

23. $0, 4, 8, 12, \ldots$

24. $1, -3, -7, -11, \ldots$

25. $3, \frac{3}{2}, \frac{3}{4}, \frac{3}{8}, \frac{3}{16}, \ldots$

26. $5, -\frac{5}{3}, \frac{5}{9}, -\frac{5}{27}, \frac{5}{81}, \ldots$

27. $\frac{2}{3}, \frac{3}{4}, \frac{4}{5}, \frac{5}{6}, \ldots$

28. $\frac{3}{2}, \frac{5}{4}, \frac{7}{6}, \frac{9}{8}, \frac{11}{10}, \ldots$

In Problems 29–34, find the indicated term in each sequence.

29. 9th term of $3, 7, 11, 15, \ldots$

30. 8th term of $1, -1, -3, -5, \ldots$

31. 11th term of $1, \frac{1}{10}, \frac{1}{100}, \ldots$

32. 11th term of $1, 2, 4, 8, \ldots$

33. 9th term of $\sqrt{2}, 2\sqrt{2}, 3\sqrt{2}, \ldots$

34. 9th term of $\sqrt{2}, 2, 2^{3/2}, \ldots$

In Problems 35–38, find a general formula for each arithmetic sequence.

35. 7th term is 31;   20th term is 96

36. 8th term is -20;   17th term is -47

37. 10th term is 0;   18th term is 8

38. 12th term is 30;   22nd term is 50

In Problems 39–44, find the sum of each infinite geometric series.

39. $3 + 1 + \frac{1}{3} + \frac{1}{9} + \cdots$

40. $2 + 1 + \frac{1}{2} + \frac{1}{4} + \cdots$

41. $2 - 1 + \frac{1}{2} - \frac{1}{4} + \cdots$

42. $6 - 4 + \frac{8}{3} - \frac{16}{9} + \cdots$

43. $\displaystyle\sum_{k=1}^{\infty} 4\left(\frac{1}{2}\right)^{k-1}$

44. $\displaystyle\sum_{k=1}^{\infty} 3\left(-\frac{3}{4}\right)^{k-1}$

In Problems 45–50, use the principle of mathematical induction to show that the given statement is true for all natural numbers.

45. $3 + 6 + 9 + \cdots + 3n = \dfrac{3n}{2}(n+1)$

46. $2 + 6 + 10 + \cdots + (4n-2) = 2n^2$

47. $2 + 6 + 18 + \cdots + 2\cdot 3^{n-1} = 3^n - 1$

48. $3 + 6 + 12 + \cdots + 3\cdot 2^{n-1} = 3(2^n - 1)$

49. $1^2 + 4^2 + 7^2 + \cdots + (3n-2)^2 = \frac{1}{2}n(6n^2 - 3n - 1)$

50. $1\cdot 3 + 2\cdot 4 + 3\cdot 5 + \cdots + n(n+2) = \dfrac{n}{6}(n+1)(2n+7)$

In Problems 51–54, expand each expression using the Binomial Theorem.

51. $(x+2)^5$

52. $(x-3)^4$

53. $(2x+3)^5$

54. $(3x-4)^4$

55. Find the coefficient of x^7 in the expansion of $(x+2)^9$.

56. Find the coefficient of x^3 in the expansion of $(x-3)^8$.

57. Find the coefficient of x^2 in the expansion of $(2x+1)^7$.

58. Find the coefficient of x^6 in the expansion of $(2x+1)^8$.

In Problems 59–66, use U = Universal set = {1, 2, 3, 4, 5, 6, 7, 8, 9}, A = {1, 3, 5, 7}, B = {3, 5, 6, 7, 8}, and C = {2, 3, 7, 8, 9} to find each set.

59. $A \cup B$ **60.** $B \cup C$ **61.** $A \cap C$ **62.** $A \cap B$

63. $A' \cup B'$ **64.** $B' \cap C'$ **65.** $(B \cap C)'$ **66.** $(A \cup B)'$

67. If $n(A) = 8$, $n(B) = 12$ and $n(A \cap B) = 3$, find $n(A \cup B)$.

68. If $n(A) = 12$, $n(A \cup B) = 30$, and $n(A \cap B) = 6$, find $n(B)$.

In Problems 69–74, use the information supplied in the figure:

69. How many are in A?

70. How many are in A or B?

71. How many are in A and C?

72. How many are not in B?

73. How many are in neither A nor C?

74. How many are in B but not in C?

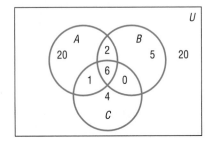

75. A clothing store sells pure wool and polyester/wool suits. Each suit comes in 3 colors and 10 sizes. How many suits are required for a complete assortment?

76. In connecting a certain electrical device, 5 wires are to be connected to 5 different terminals. How many different wirings are possible if 1 wire is connected to each terminal?

77. *Baseball* On a given day, the American Baseball League schedules 7 games. How many different outcomes are possible, assuming that each game is played to completion?

78. *Baseball* On a given day, the National Baseball League schedules 6 games. How many different outcomes are possible, assuming that each game is played to completion?

79. If 4 people enter a bus having 9 vacant seats, in how many ways can they be seated?

80. How many different arrangements are there of the letters in the word ROSE?

81. In how many ways can a squad of 4 relay runners be chosen from a track team of 8 runners?

82. A professor has 10 similar problems to put on a test with 3 problems. How many different tests can she design?

83. *Baseball* In how many different ways can the 14 baseball teams in the American League be paired without regard to which team is at home?

84. *Arranging Books on a Shelf* There are 5 different French books and 5 different Spanish books. How many ways are there to arrange them on a shelf if:
(a) Books of the same language must be grouped together, French on the left, Spanish on the right?
(b) French and Spanish books must alternate in the grouping, beginning with a French book?

85. *Telephone Numbers* Using the digits 0, 1, 2, . . . , 9, how many 7-digit numbers can be formed if the first digit cannot be 0 or 9 and if the last digit is greater than or equal to 2 and less than or equal to 3? Repeated digits are allowed.

86. *Home Choices* A contractor constructs homes with 5 different choices of exterior finish, 3 different roof arrangements, and 4 different window designs. How many different types of homes can be built?

87. *License Plate Possibilities* A license plate consists of 1 letter, excluding O and I, followed by a 4-digit number that cannot have a 0 in the lead position. How many different plates are possible?

88. Using the digits 0 and 1, how many different numbers consisting of 8 digits can be formed?

89. *Forming Different Words* How many different words can be formed using all the letters in the word MISSING?

90. *Arranging Flags* How many different vertical arrangements are there of 10 flags, if 4 are white, 3 are blue, 2 are green, and 1 is red?

91. *Forming Committees* A group of 9 people is going to be formed into committees of 4, 3, and 2 people. How many committees can be formed if:
(a) A person can serve on any number of committees?
(b) No person can serve on more than one committee?

92. *Forming Committees* A group consists of 5 men and 8 women. A committee of 4 is to be formed from this group, and policy dictates that at least 1 woman be on this committee.
(a) How many committees can be formed that contain exactly 1 man?
(b) How many committees can be formed that contain exactly 2 women?
(c) How many committees can be formed that contain at least 1 man?

93. From a box containing three 40 watt bulbs, six 60 watt bulbs, and eleven 75 watt bulbs, a bulb is drawn at random. What is the probability that the bulb is 40 watts? What is the probability that it is not a 75 watt bulb?

94. You have four $1 bills, three $5 bills, and two $10 bills in your wallet. If you pick a bill at random, what is the probability it will be a $1 bill?

95. Each of the letters in the word ROSE is written on an index card and the cards are then shuffled. What is the probability that, when the cards are dealt out, they spell the word ROSE?

96. Each of the numbers, 1, 2, . . . , 100 is written on an index card and the cards are then shuffled. If a card is selected at random, what is the probability that the number on the card is divisible by 5? What is the probability that the card selected either is a 1 or names a prime number?

97. *Computing Probabilities* Because of a mistake in packaging, a case of 12 bottles of red wine contained 5 Merlot and 7 Cabernet, each without labels. All the bottles look alike and have equal probability of being chosen. Three bottles are selected.
(a) What is the probability all 3 are Merlot?
(b) What is the probability exactly 2 are Merlot?
(c) What is the probability none is a Merlot?

98. *Tossing a Coin* A fair coin is tossed 10 times.
(a) What is the probability of obtaining exactly 5 heads?
(b) What is the probability of obtaining all heads?

99. *Constructing a Brick Staircase* A brick staircase has a total of 25 steps. The bottom step requires 80 bricks. Each successive step requires three less bricks than the prior step.
(a) How many bricks are required for the top step?
(b) How many bricks are required to build the staircase?

100. *Creating a Floor Design* A mosaic tile floor is designed in the shape of a trapezoid 30 feet wide at the base and 15 feet wide at the top. See Figure 5. The tiles, 12 inches by 12 inches, are to be placed so that each successive row contains one less tile than the row below. How many tiles will be required?

101. *Bouncing Balls* A ball is dropped from a height of 20 feet. Each time it strikes the ground, it bounces up to $\frac{3}{4}$ of the previous height.
(a) What height will the ball bounce up to after it strikes the ground for the third time?
(b) What is its height after it strikes the ground for the nth time?
(c) How many times does the ball need to strike the ground before its height is less than 6 inches?
(d) What total distance does the ball travel before it stops bouncing?

102. *Salary Increases* Your friend has just been hired at an annual salary of $20,000. If she expects to receive annual increases of 4%, what will her salary be as she begins her fifth year?

103. At the Milex tune-up and brake repair shop, the manager has found that a car will require a tune-up with a probability of 0.6, a brake job with a probability of 0.1, and both with a probability of 0.02.
(a) What is the probability that a car requires either a tune-up or a brake job?
(b) What is the probability that a car requires a tune-up but not a brake job?
(c) What is the probability that a car requires neither type of repair?

PREPARING FOR THIS CHAPTER

Before getting started on this chapter, review the following concepts:

For Section 12.1: Row operations on matrices (p. 627)

For Section 12.2: Identities (p. 441)

Factoring polynomials (Appendix A, p. 806)

Proper and improper rational functions (pp. 211, 212)

Fundamental Theorem of Algebra (p. 242; see also p. 250)

For Sections 12.3 and 12.4:   Rectangular coordinates (p. 53)

Law of Cosines (Section 8.2)

MISCELLANEOUS TOPICS

Preview   Finding the Speed and Direction of an Aircraft

A Boeing 737 aircraft maintains a constant airspeed of 500 miles per hour in the direction due south. The velocity of the jet stream is 80 miles per hour in a northeasterly direction.

(a) Find a unit vector having northeast as direction.

(b) Find a vector 80 units in magnitude having the same direction as the unit vector found in part (a).

(c) Find the actual speed of the aircraft relative to the ground.

[Example 6 in Section 12.3]

(d) Find the actual direction of the aircraft relative to the ground

[Example 4 in Section 12.4] ■

This chapter contains three topics. They are independent of each other and may be covered in any order.

The first section treats a matrix as an algebraic concept, introducing the ideas of addition and multiplication, as well as citing some of the properties of matrix algebra.

The second section, **partial fraction decomposition,** provides an application of systems of equations. This particular application is one that is used in the study of integral calculus.

Finally, the last two sections provide an introduction to the notion of a **vector** and some of its applications, an extremely important topic in engineering and physics.

Matrix Algebra

In Section 10.2, we defined a matrix as an array of real numbers and used an augmented matrix to represent a system of linear equations. There is, however, a branch of mathematics, called **linear algebra,** that deals with matrices in such a way that an algebra of matrices is permitted. In this section, we provide a survey of how this **matrix algebra** is developed.

Before getting started, we restate the definition of a matrix.

Matrix

A **matrix** is defined as a rectangular array of numbers:

$$
\begin{array}{c}
\quad\quad\quad\quad\quad j\text{th column} \\
\begin{bmatrix}
a_{11} & a_{12} & \cdots & a_{1j} & \cdots & a_{1n} \\
a_{21} & a_{22} & \cdots & a_{2j} & \cdots & a_{2n} \\
\vdots & \vdots & & \vdots & & \vdots \\
a_{i1} & a_{i2} & \cdots & a_{ij} & \cdots & a_{in} \\
\vdots & \vdots & & \vdots & & \vdots \\
a_{m1} & a_{m2} & \cdots & a_{mj} & \cdots & a_{mn}
\end{bmatrix} \quad i\text{th row}
\end{array}
$$

Each number a_{ij} of the matrix has two indices: the **row index** i and the **column index** j. The matrix shown above has m rows and n columns. The $m \cdot n$ numbers a_{ij} are usually referred to as the **entries** of the matrix.

Let's begin with an example that illustrates how matrices can be used to conveniently represent an array of information.

EXAMPLE 1

Arranging Data in a Matrix

In a survey of 1000 people, the following information was obtained:

200 males	Thought federal defense spending was too high
150 males	Thought federal defense spending was too low
45 males	Had no opinion
315 females	Thought federal defense spending was too high
125 females	Thought federal defense spending was too low
165 females	Had no opinion

We can arrange the above data in a rectangular array as follows:

	TOO HIGH	TOO LOW	NO OPINION
Males	200	150	45
Females	315	125	165

or as

$$\begin{bmatrix} 200 & 150 & 45 \\ 315 & 125 & 165 \end{bmatrix}$$

This matrix has two rows (representing males and females) and three columns (representing "too high," "too low," and "no opinion"). ∎

The matrix we developed in Example 1 has 2 rows and 3 columns. In general, a matrix with m rows and n columns is called an ***m* by *n* matrix.** Thus, the matrix we developed in Example 1 is a 2 by 3 matrix. Notice that an m by n matrix will contain $m \cdot n$ entries.

If an m by n matrix has the same number of rows as columns, that is, if $m = n$, then the matrix is referred to as a **square matrix.**

E X A M P L E 2   *Examples of Matrices*

(a) $\begin{bmatrix} 5 & 0 \\ -6 & 1 \end{bmatrix}$ A 2 by 2 square matrix   (b) $\begin{bmatrix} 1 & 0 & 3 \end{bmatrix}$ A 1 by 3 matrix

(c) $\begin{bmatrix} 6 & -2 & 4 \\ 4 & 3 & 5 \\ 8 & 0 & 1 \end{bmatrix}$ A 3 by 3 square matrix

∎

Equality and Addition of Matrices

We begin our discussion of matrix algebra by first defining what is meant by two matrices being equal and then defining the operations of addition and subtraction. It is important to note that these definitions require each matrix to have the same number of rows *and* the same number of columns as a prerequisite for equality and for addition and subtraction.

We usually represent matrices by capital letters, such as A, B, C, and so on.

Equal Matrices

Two m by n matrices A and B are said to be **equal,** written as

$$A = B$$

provided that each entry a_{ij} in A is equal to the corresponding entry b_{ij} in B.

For example,

$$\begin{bmatrix} 2 & 1 \\ 0.5 & -1 \end{bmatrix} = \begin{bmatrix} \sqrt{4} & 1 \\ \frac{1}{2} & -1 \end{bmatrix} \quad \text{and} \quad \begin{bmatrix} 3 & 2 & 1 \\ 0 & 1 & -2 \end{bmatrix} = \begin{bmatrix} \sqrt{9} & \sqrt{4} & 1 \\ 0 & 1 & \sqrt[3]{-8} \end{bmatrix}$$

$$\begin{bmatrix} 4 & 1 \\ 6 & 1 \end{bmatrix} \neq \begin{bmatrix} 4 & 0 \\ 6 & 1 \end{bmatrix} \qquad \text{Because the entries in row 1, column 2 are not equal}$$

$$\begin{bmatrix} 4 & 1 & 2 \\ 6 & 1 & 2 \end{bmatrix} \neq \begin{bmatrix} 4 & 1 & 2 & 3 \\ 6 & 1 & 2 & 4 \end{bmatrix} \qquad \text{Because the matrix on the left is 2 by 3 and the matrix on the right is 2 by 4}$$

If each of A and B is an m by n matrix (and each therefore contains $m \cdot n$ entries), the statement $A = B$ actually represents a system of $m \cdot n$ ordinary equations. We will make use of this fact a little later.

Suppose that A and B represent two m by n matrices. We define their **sum** $A + B$ to be the m by n matrix formed by adding the corresponding entries a_{ij} of A and b_{ij} of B. The **difference** $A - B$ is defined as the m by n matrix formed by subtracting the entries b_{ij} in B from the corresponding entries a_{ij} in A. Addition and subtraction of matrices are allowed only for matrices having the same number m of rows and the same number n of columns. Thus, for example, a 2 by 3 matrix and a 2 by 4 matrix cannot be added or subtracted.

E X A M P L E   3   *Adding and Subtracting Matrices*

Suppose that

$$A = \begin{bmatrix} 2 & 4 & 8 & -3 \\ 0 & 1 & 2 & 3 \end{bmatrix} \quad \text{and} \quad B = \begin{bmatrix} -3 & 4 & 0 & 1 \\ 6 & 8 & 2 & 0 \end{bmatrix}$$

Find: (a) $A + B$ (b) $A - B$

Solution   (a) $A + B = \begin{bmatrix} 2 & 4 & 8 & -3 \\ 0 & 1 & 2 & 3 \end{bmatrix} + \begin{bmatrix} -3 & 4 & 0 & 1 \\ 6 & 8 & 2 & 0 \end{bmatrix}$

$$= \begin{bmatrix} 2 + (-3) & 4 + 4 & 8 + 0 & -3 + 1 \\ 0 + 6 & 1 + 8 & 2 + 2 & 3 + 0 \end{bmatrix} \quad \text{Add corresponding entries.}$$

$$= \begin{bmatrix} -1 & 8 & 8 & -2 \\ 6 & 9 & 4 & 3 \end{bmatrix}$$

(b) $A - B = \begin{bmatrix} 2 & 4 & 8 & -3 \\ 0 & 1 & 2 & 3 \end{bmatrix} - \begin{bmatrix} -3 & 4 & 0 & 1 \\ 6 & 8 & 2 & 0 \end{bmatrix}$

$$= \begin{bmatrix} 2 - (-3) & 4 - 4 & 8 - 0 & -3 - 1 \\ 0 - 6 & 1 - 8 & 2 - 2 & 3 - 0 \end{bmatrix} \quad \text{Subtract corresponding entries.}$$

$$= \begin{bmatrix} 5 & 0 & 8 & -4 \\ -6 & -7 & 0 & 3 \end{bmatrix}$$

   Check: Enter the matrices A and B. Then find $A + B$ and $A - B$.   ∎

■   Now work Problem 1.

Many of the algebraic properties of sums of real numbers are also true for sums of matrices. Suppose that A, B, and C are m by n matrices. Then matrix addition is **commutative.** That is,

Commutative Property

$$A + B = B + A$$

Matrix addition is also **associative.** That is,

Associative Property

$$(A + B) + C = A + (B + C)$$

Although we shall not prove these results, the proofs, as the following example illustrates, are based on the commutative and associative properties for real numbers.

EXAMPLE 4   *Demonstrating the Commutative Property*

$$\begin{bmatrix} 2 & 3 & -1 \\ 4 & 0 & 7 \end{bmatrix} + \begin{bmatrix} -1 & 2 & 1 \\ 5 & -3 & 4 \end{bmatrix} = \begin{bmatrix} 2 + (-1) & 3 + 2 & -1 + 1 \\ 4 + 5 & 0 + (-3) & 7 + 4 \end{bmatrix}$$

$$= \begin{bmatrix} -1 + 2 & 2 + 3 & 1 + (-1) \\ 5 + 4 & -3 + 0 & 4 + 7 \end{bmatrix}$$

$$= \begin{bmatrix} -1 & 2 & 1 \\ 5 & -3 & 4 \end{bmatrix} + \begin{bmatrix} 2 & 3 & -1 \\ 4 & 0 & 7 \end{bmatrix} \quad ■$$

A matrix whose entries are all equal to 0 is called a **zero matrix.** Each of the following matrices is a zero matrix:

$$\begin{bmatrix} 0 & 0 \\ 0 & 0 \end{bmatrix}$$ 2 by 2 square zero matrix $$\begin{bmatrix} 0 & 0 & 0 \\ 0 & 0 & 0 \end{bmatrix}$$ 2 by 3 zero matrix $$[0 \quad 0 \quad 0]$$ 1 by 3 zero matrix

Zero matrices have properties similar to the real number 0. Thus, if A is an m by n matrix and 0 is an m by n zero matrix, then

$$A + 0 = A$$

In other words, the zero matrix is the additive identity in matrix algebra.

We also can multiply a matrix by a real number. If k is a real number and A is an m by n matrix, the matrix kA is the m by n matrix formed by multiplying each entry in A by k. The number k is sometimes referred to as a **scalar,** and the matrix kA is called a **scalar multiple** of A.

EXAMPLE 5   *Operations Using Matrices*

Suppose that

$$A = \begin{bmatrix} 3 & 1 & 5 \\ -2 & 0 & 6 \end{bmatrix} \quad B = \begin{bmatrix} 4 & 1 & 0 \\ 8 & 1 & -3 \end{bmatrix} \quad C = \begin{bmatrix} 9 & 0 \\ -3 & 6 \end{bmatrix}$$

Find: (a) $4A$ (b) $\frac{1}{3}C$ (c) $3A - 2B$

Solution   (a) $4A = 4 \begin{bmatrix} 3 & 1 & 5 \\ -2 & 0 & 6 \end{bmatrix} = \begin{bmatrix} 4 \cdot 3 & 4 \cdot 1 & 4 \cdot 5 \\ 4(-2) & 4 \cdot 0 & 4 \cdot 6 \end{bmatrix} = \begin{bmatrix} 12 & 4 & 20 \\ -8 & 0 & 24 \end{bmatrix}$

(b) $\frac{1}{3}C = \frac{1}{3} \begin{bmatrix} 9 & 0 \\ -3 & 6 \end{bmatrix} = \begin{bmatrix} \frac{1}{3} \cdot 9 & \frac{1}{3} \cdot 0 \\ \frac{1}{3}(-3) & \frac{1}{3} \cdot 6 \end{bmatrix} = \begin{bmatrix} 3 & 0 \\ -1 & 2 \end{bmatrix}$

(c) $3A - 2B = 3 \begin{bmatrix} 3 & 1 & 5 \\ -2 & 0 & 6 \end{bmatrix} - 2 \begin{bmatrix} 4 & 1 & 0 \\ 8 & 1 & -3 \end{bmatrix}$

$$= \begin{bmatrix} 3 \cdot 3 & 3 \cdot 1 & 3 \cdot 5 \\ 3(-2) & 3 \cdot 0 & 3 \cdot 6 \end{bmatrix} - \begin{bmatrix} 2 \cdot 4 & 2 \cdot 1 & 2 \cdot 0 \\ 2 \cdot 8 & 2 \cdot 1 & 2(-3) \end{bmatrix}$$

$$= \begin{bmatrix} 9 & 3 & 15 \\ -6 & 0 & 18 \end{bmatrix} - \begin{bmatrix} 8 & 2 & 0 \\ 16 & 2 & -6 \end{bmatrix}$$

$$= \begin{bmatrix} 9 - 8 & 3 - 2 & 15 - 0 \\ -6 - 16 & 0 - 2 & 18 - (-6) \end{bmatrix}$$

$$= \begin{bmatrix} 1 & 1 & 15 \\ -22 & -2 & 24 \end{bmatrix} \qquad ■$$

Comment Most graphing utilities have the capability of doing matrix algebra. In fact, most graphing calculators can handle matrices as large as 9 by 9.

Check: Enter the matrices A, B, and C. Then find $4A$, $\frac{1}{3}C$, and $3A - 2B$. ■

■   Now work Problem 5.

We list next some of the algebraic properties of scalar multiplication. Let h and k be real numbers, and let A and B be m by n matrices. Then

Properties of Scalar Multiplication	$k(hA) = (kh)A$ $(k + h)A = kA + hA$ $k(A + B) = kA + kB$

■

The proofs of these properties are based on properties of real numbers. For example, if A and B are 2 by 2 matrices, then

$$k(A + B) = k\left(\begin{bmatrix} a_{11} & a_{12} \\ a_{21} & a_{22} \end{bmatrix} + \begin{bmatrix} b_{11} & b_{12} \\ b_{21} & b_{22} \end{bmatrix} \right) = k \begin{bmatrix} a_{11} + b_{11} & a_{12} + b_{12} \\ a_{21} + b_{21} & a_{22} + b_{22} \end{bmatrix}$$

$$= \begin{bmatrix} k(a_{11} + b_{11}) & k(a_{12} + b_{12}) \\ k(a_{21} + b_{21}) & k(a_{22} + b_{22}) \end{bmatrix} = \begin{bmatrix} ka_{11} + kb_{11} & ka_{12} + kb_{12} \\ ka_{21} + kb_{21} & ka_{22} + kb_{22} \end{bmatrix}$$

$$= \begin{bmatrix} ka_{11} & ka_{12} \\ ka_{21} & ka_{22} \end{bmatrix} + \begin{bmatrix} kb_{11} & kb_{12} \\ kb_{21} & kb_{22} \end{bmatrix} = k \begin{bmatrix} a_{11} & a_{12} \\ a_{21} & a_{22} \end{bmatrix} + k \begin{bmatrix} b_{11} & b_{12} \\ b_{21} & b_{22} \end{bmatrix} = kA + kB \qquad ■$$

Multiplication of Matrices

Unlike the straightforward definition for adding two matrices, the definition for multiplying two matrices is not what we might expect. In preparation for this definition, we need the following definitions:

A **row vector** R is a 1 by n matrix:

$$R = [r_1 \quad r_2 \quad \cdots \quad r_n]$$

A **column vector** C is an n by 1 matrix:

$$C = \begin{bmatrix} c_1 \\ c_2 \\ \vdots \\ c_n \end{bmatrix}$$

The **product** RC of R times C is defined as the number

$$RC = [r_1 \quad r_2 \quad \cdots \quad r_n] \begin{bmatrix} c_1 \\ c_2 \\ \vdots \\ c_n \end{bmatrix} = r_1 c_1 + r_2 c_2 + \cdots + r_n c_n$$

Notice that a row vector and a column vector can be multiplied only if they contain the same number of entries.

E X A M P L E   6   *The Product of a Row Vector by a Column Vector*

If $R = [3 \quad -5 \quad 2]$ and $C = \begin{bmatrix} 3 \\ 4 \\ -5 \end{bmatrix}$, then

$$RC = [3 \quad -5 \quad 2] \begin{bmatrix} 3 \\ 4 \\ -5 \end{bmatrix} = 3 \cdot 3 + (-5)4 + 2(-5)$$

$$= 9 - 20 - 10 = -21 \qquad \blacksquare$$

Let's look at an application of the product of a row vector by a column vector.

E X A M P L E   7   *Using Matrices to Compute Revenue*

A clothing store sells men's shirts for \$25, silk ties for \$8, and wool suits for \$300. Last month, the store had sales consisting of 100 shirts, 200 ties, and 50 suits. What was the total revenue due to these sales?

Solution   We set up a row vector R to represent the prices of each item and a column vector C to represent the corresponding number of items sold.
Then

$$
\begin{array}{cc}
\begin{array}{c} \text{Prices} \\ \begin{array}{ccc} \text{Shirts} & \text{Ties} & \text{Suits} \end{array} \\ R = [\ 25 \quad 8 \quad 300\ ] \end{array}
&
\begin{array}{c} \text{Number} \\ \text{sold} \\ C = \begin{bmatrix} 100 \\ 200 \\ 50 \end{bmatrix} \begin{array}{l} \text{Shirts} \\ \text{Ties} \\ \text{Suits} \end{array} \end{array}
\end{array}
$$

The total revenue obtained is the product RC. That is,

$$RC = [25 \quad 8 \quad 300] \begin{bmatrix} 100 \\ 200 \\ 50 \end{bmatrix}$$

$$= \underbrace{25 \cdot 100}_{\text{Shirt revenue}} + \underbrace{8 \cdot 200}_{\text{Tie revenue}} + \underbrace{300 \cdot 50}_{\text{Suit revenue}} = \underbrace{\$19,100}_{\text{Total revenue}} \qquad \blacksquare$$

The definition for multiplying two matrices is based on the definition of a row vector times a column vector.

Product

Let A denote an m by r matrix and let B denote an r by n matrix. The **product** AB is defined as the m by n matrix whose entry in row i, column j is the product of the ith row of A and the jth column of B.

The definition of the product AB of two matrices A and B, in this order, requires that the number of columns of A equal the number of rows of B; otherwise, no product is defined:

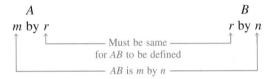

An example will help clarify the definition.

E X A M P L E   8   *Multiplying Two Matrices*

Find the product AB if

$$A = \begin{bmatrix} 2 & 4 & -1 \\ 5 & 8 & 0 \end{bmatrix} \quad \text{and} \quad B = \begin{bmatrix} 2 & 5 & 1 & 4 \\ 4 & 8 & 0 & 6 \\ -3 & 1 & -2 & -1 \end{bmatrix}$$

Solution   First, we note that A is 2 by 3 and B is 3 by 4, so the product AB is defined and will be a 2 by 4 matrix. Suppose that we want the entry in row 2, column 3 of AB. To find it, we find the product of the row vector from row 2 of A and the column vector from column 3 of B:

$$\underset{\text{Row 2 of } A}{[5 \quad 8 \quad 0]} \overset{\text{Column 3 of } B}{\begin{bmatrix} 1 \\ 0 \\ -2 \end{bmatrix}} = 5 \cdot 1 + 8 \cdot 0 + 0(-2) = 5$$

So far, we have

$$AB = \begin{bmatrix} \underline{} & \underline{} & \overset{\text{Column 3}}{\underline{}} & \underline{} \\ \underline{} & \underline{} & 5 & \underline{} \end{bmatrix} \leftarrow \text{Row 2}$$

Now, to find the entry in row 1, column 4 of AB, we find the product of row 1 of A and column 4 of B:

$$\underset{\text{Row 1 of } A}{[2 \quad 4 \quad -1]} \overset{\text{Column 4 of } B}{\begin{bmatrix} 4 \\ 6 \\ -1 \end{bmatrix}} = 2 \cdot 4 + 4 \cdot 6 + (-1)(-1) = 33$$

Continuing in this fashion, we find *AB*:

$$AB = \begin{bmatrix} 2 & 4 & -1 \\ 5 & 8 & 0 \end{bmatrix} \begin{bmatrix} 2 & 5 & 1 & 4 \\ 4 & 8 & 0 & 6 \\ -3 & 1 & -2 & -1 \end{bmatrix}$$

$$= \begin{bmatrix} \text{Row 1 of } A & \text{Row 1 of } A & \text{Row 1 of } A & \text{Row 1 of } A \\ \text{times} & \text{times} & \text{times} & \text{times} \\ \text{column 1 of } B & \text{column 2 of } B & \text{column 3 of } B & \text{column 4 of } B \\ & & & \\ \text{Row 2 of } A & \text{Row 2 of } A & \text{Row 2 of } A & \text{Row 2 of } A \\ \text{times} & \text{times} & \text{times} & \text{times} \\ \text{column 1 of } B & \text{column 2 of } B & \text{column 3 of } B & \text{column 4 of } B \end{bmatrix}$$

$$= \begin{bmatrix} 2\cdot2+4\cdot4+(-1)(-3) & 2\cdot5+4\cdot8+(-1)1 & 2\cdot1+4\cdot0+(-1)(-2) & 33\ (\text{from earlier}) \\ 5\cdot2+8\cdot4+0(-3) & 5\cdot5+8\cdot8+0\cdot1 & 5\ (\text{from earlier}) & 5\cdot4+8\cdot6+0(-1) \end{bmatrix}$$

$$= \begin{bmatrix} 23 & 41 & 4 & 33 \\ 42 & 89 & 5 & 68 \end{bmatrix}$$

Check: Enter the matrices *A* and *B*. Then find *AB*. (See what happens if you try to find *BA*.)   ■

■   Now work Problem 17.

Notice that for the matrices given in Example 8 the product *BA* is not defined, because *B* is 3 by 4 and *A* is 2 by 3. Another result that can occur when multiplying two matrices is illustrated in the next example.

E X A M P L E   9   *Multiplying Two Matrices*

If

$$A = \begin{bmatrix} 2 & 1 & 3 \\ 1 & -1 & 0 \end{bmatrix} \quad \text{and} \quad B = \begin{bmatrix} 1 & 0 \\ 2 & 1 \\ 3 & 2 \end{bmatrix}$$

find: (a) *AB*   (b) *BA*

Solution   (a) $AB = \underset{\text{2 by 3}}{\begin{bmatrix} 2 & 1 & 3 \\ 1 & -1 & 0 \end{bmatrix}} \underset{\text{3 by 2}}{\begin{bmatrix} 1 & 0 \\ 2 & 1 \\ 3 & 2 \end{bmatrix}} = \underset{\text{2 by 2}}{\begin{bmatrix} 13 & 7 \\ -1 & -1 \end{bmatrix}}$

(b) $BA = \underset{\text{3 by 2}}{\begin{bmatrix} 1 & 0 \\ 2 & 1 \\ 3 & 2 \end{bmatrix}} \underset{\text{2 by 3}}{\begin{bmatrix} 2 & 1 & 3 \\ 1 & -1 & 0 \end{bmatrix}} = \underset{\text{3 by 3}}{\begin{bmatrix} 2 & 1 & 3 \\ 5 & 1 & 6 \\ 8 & 1 & 9 \end{bmatrix}}$ ■

Notice in Example 9 that AB is 2 by 2 and BA is 3 by 3. Thus, it is possible for both AB and BA to be defined, yet be unequal. In fact, even if A and B are both n by n matrices, so that AB and BA are each defined and n by n, AB and BA will nearly always be unequal.

E X A M P L E   1 0   *Multiplying Two Square Matrices*

If

$$A = \begin{bmatrix} 2 & 1 \\ 0 & 4 \end{bmatrix} \quad \text{and} \quad B = \begin{bmatrix} -3 & 1 \\ 1 & 2 \end{bmatrix}$$

find: (a) AB (b) BA

Solution   (a) $AB = \begin{bmatrix} 2 & 1 \\ 0 & 4 \end{bmatrix}\begin{bmatrix} -3 & 1 \\ 1 & 2 \end{bmatrix}$

$$= \begin{bmatrix} 2(-3) + 1 \cdot 1 & 2 \cdot 1 + 1 \cdot 2 \\ 0(-3) + 4 \cdot 1 & 0 \cdot 1 + 4 \cdot 2 \end{bmatrix} = \begin{bmatrix} -5 & 4 \\ 4 & 8 \end{bmatrix}$$

(b) $BA = \begin{bmatrix} -3 & 1 \\ 1 & 2 \end{bmatrix}\begin{bmatrix} 2 & 1 \\ 0 & 4 \end{bmatrix}$

$$= \begin{bmatrix} (-3)2 + 1 \cdot 0 & (-3)1 + 1 \cdot 4 \\ 1 \cdot 2 + 2 \cdot 0 & 1 \cdot 1 + 2 \cdot 4 \end{bmatrix} = \begin{bmatrix} -6 & 1 \\ 2 & 9 \end{bmatrix} \qquad ■$$

The preceding examples demonstrate that an important property of real numbers, the commutative property of multiplication, is not shared by matrices. Thus, in general:

Theorem   Matrix multiplication is not commutative.   ■

■   Now work Problems 7 and 9.

Next we give two of the properties of real numbers that are shared by matrices. Assuming that each product and sum is defined, we have

Associative Property
$$A(BC) = (AB)C$$

Distributive Property
$$A(B + C) = AB + AC$$

The Identity Matrix

For an n by n square matrix, the entries located in row i, column i, $1 \le i \le n$, are called the **diagonal entries.** An n by n square matrix whose diagonal entries are 1's, while all other entries are 0's, is called the **identity matrix I_n.** For example,

$$I_2 = \begin{bmatrix} 1 & 0 \\ 0 & 1 \end{bmatrix} \qquad I_3 = \begin{bmatrix} 1 & 0 & 0 \\ 0 & 1 & 0 \\ 0 & 0 & 1 \end{bmatrix}$$

and so on.

E X A M P L E   1 1   *Multiplication with an Identity Matrix*

Let

$$A = \begin{bmatrix} -1 & 2 & 0 \\ 0 & 1 & 3 \end{bmatrix} \quad \text{and} \quad B = \begin{bmatrix} 3 & 2 \\ 4 & 6 \\ 5 & 2 \end{bmatrix}$$

Find: (a) AI_3 (b) I_2A (c) BI_2

Solution   (a) $AI_3 = \begin{bmatrix} -1 & 2 & 0 \\ 0 & 1 & 3 \end{bmatrix} \begin{bmatrix} 1 & 0 & 0 \\ 0 & 1 & 0 \\ 0 & 0 & 1 \end{bmatrix} = \begin{bmatrix} -1 & 2 & 0 \\ 0 & 1 & 3 \end{bmatrix} = A$

(b) $I_2A = \begin{bmatrix} 1 & 0 \\ 0 & 1 \end{bmatrix} \begin{bmatrix} -1 & 2 & 0 \\ 0 & 1 & 3 \end{bmatrix} = \begin{bmatrix} -1 & 2 & 0 \\ 0 & 1 & 3 \end{bmatrix} = A$

(c) $BI_2 = \begin{bmatrix} 3 & 2 \\ 4 & 6 \\ 5 & 2 \end{bmatrix} \begin{bmatrix} 1 & 0 \\ 0 & 1 \end{bmatrix} = \begin{bmatrix} 3 & 2 \\ 4 & 6 \\ 5 & 2 \end{bmatrix} = B$ ■

Example 11 demonstrates the following property: If A is an m by n matrix, then

Identity Property

$$I_mA = A \quad \text{and} \quad AI_n = A$$

If A is an n by n square matrix, then $AI_n = I_nA = A$. Thus, an identity matrix has properties analogous to those of the real number 1. In other words, the identity matrix is a multiplicative identity in matrix algebra.

The Inverse of a Matrix

Inverse Matrix   Let A be a square n by n matrix. If there exists an n by n matrix A^{-1}, read "A inverse," for which

$$AA^{-1} = A^{-1}A = I_n$$

then A^{-1} is called the **inverse** of the matrix A.

As we shall soon see, not every square matrix has an inverse. When a matrix A does have an inverse A^{-1}, then A is said to be **nonsingular.**

E X A M P L E   1 2   *Multiplying a Matrix by Its Inverse*

Show that the inverse of

$$A = \begin{bmatrix} 3 & 1 \\ 2 & 1 \end{bmatrix} \quad \text{is} \quad A^{-1} = \begin{bmatrix} 1 & -1 \\ -2 & 3 \end{bmatrix}$$

Solution   We need to show that $AA^{-1} = A^{-1}A = I_2$.

$$AA^{-1} = \begin{bmatrix} 3 & 1 \\ 2 & 1 \end{bmatrix} \begin{bmatrix} 1 & -1 \\ -2 & 3 \end{bmatrix} = \begin{bmatrix} 3 \cdot 1 + 1(-2) & 3(-1) + 1 \cdot 3 \\ 2 \cdot 1 + 1(-2) & 2(-1) + 1 \cdot 3 \end{bmatrix}$$

$$= \begin{bmatrix} 1 & 0 \\ 0 & 1 \end{bmatrix} = I_2$$

$$A^{-1}A = \begin{bmatrix} 1 & -1 \\ -2 & 3 \end{bmatrix} \begin{bmatrix} 3 & 1 \\ 2 & 1 \end{bmatrix} = \begin{bmatrix} 3 - 2 & 1 - 1 \\ -6 + 6 & -2 + 3 \end{bmatrix} = \begin{bmatrix} 1 & 0 \\ 0 & 1 \end{bmatrix} = I_2 \quad \blacksquare$$

We now show one way to find the inverse of

$$A = \begin{bmatrix} 3 & 1 \\ 2 & 1 \end{bmatrix}$$

Suppose that A^{-1} is given by

$$A^{-1} = \begin{bmatrix} x & y \\ z & w \end{bmatrix} \tag{1}$$

where x, y, z, and w are four variables. Based on the definition of an inverse, if, indeed, A has an inverse, then we have

$$AA^{-1} = I_2$$

$$\begin{bmatrix} 3 & 1 \\ 2 & 1 \end{bmatrix} \begin{bmatrix} x & y \\ z & w \end{bmatrix} = \begin{bmatrix} 1 & 0 \\ 0 & 1 \end{bmatrix}$$

$$\begin{bmatrix} 3x + z & 3y + w \\ 2x + z & 2y + w \end{bmatrix} = \begin{bmatrix} 1 & 0 \\ 0 & 1 \end{bmatrix}$$

Because corresponding entries must be equal, it follows that this matrix equation is equivalent to four ordinary equations:

$$\begin{cases} 3x + z = 1 \\ 2x + z = 0 \end{cases} \qquad \begin{cases} 3y + w = 0 \\ 2y + w = 1 \end{cases}$$

The augmented matrix of each system is

$$\begin{bmatrix} 3 & 1 & | & 1 \\ 2 & 1 & | & 0 \end{bmatrix} \quad \begin{bmatrix} 3 & 1 & | & 0 \\ 2 & 1 & | & 1 \end{bmatrix} \tag{2}$$

The usual procedure would be to transform each augmented matrix into reduced echelon form. Notice, though, that the left sides of the augmented matrices are equal, so the same row operations (see Section 10.2) can be used to reduce each one. Thus, we find it more efficient to combine the two augmented matrices (2) into a single matrix, as shown next, and then transform it into reduced echelon form:

$$\begin{bmatrix} 3 & 1 & | & 1 & 0 \\ 2 & 1 & | & 0 & 1 \end{bmatrix}$$

Now we attempt to transform the left side into an identity matrix:

$$\begin{bmatrix} 3 & 1 & | & 1 & 0 \\ 2 & 1 & | & 0 & 1 \end{bmatrix} \rightarrow \begin{bmatrix} 1 & 0 & | & 1 & -1 \\ 2 & 1 & | & 0 & 1 \end{bmatrix}$$
$$\uparrow$$
$$R_1 = -1r_2 + r_1$$

$$\rightarrow \begin{bmatrix} 1 & 0 & | & 1 & -1 \\ 0 & 1 & | & -2 & 3 \end{bmatrix} \quad (3)$$
$$\uparrow$$
$$R_2 = -2r_1 + r_2$$

Matrix (3) is in reduced echelon form. Now we reverse the earlier step of combining the two augmented matrices in (2) and write the single matrix (3) as two augmented matrices:

$$\begin{bmatrix} 1 & 0 & | & 1 \\ 0 & 1 & | & -2 \end{bmatrix} \text{ and } \begin{bmatrix} 1 & 0 & | & -1 \\ 0 & 1 & | & 3 \end{bmatrix}$$

We conclude from these matrices that $x = 1$, $z = -2$ and $y = -1$, $w = 3$. Substituting these values into matrix (1), we find that

$$A^{-1} = \begin{bmatrix} 1 & -1 \\ -2 & 3 \end{bmatrix}$$

Notice in matrix (3) that the 2 by 2 matrix to the right of the vertical bar is, in fact, the inverse of A. Also notice that the identity matrix I_2 is the matrix that appears to the left of the vertical bar. These observations and the procedures followed above will work in general.

Procedure for Finding the Inverse of a Nonsingular Matrix

To find the inverse of an n by n nonsingular matrix A, proceed as follows:

STEP 1: Form the matrix $[A|I_n]$.
STEP 2: Transform the matrix $[A|I_n]$ into reduced echelon form.
STEP 3: The reduced echelon form of $[A|I_n]$ will contain the identity matrix I_n on the left of the vertical bar; the n by n matrix on the right of the vertical bar is the inverse of A.

In other words, if A is nonsingular, we begin with the matrix $[A|I_n]$ and, after transforming it into reduced echelon form, we end up with the matrix $[I_n|A^{-1}]$.

Let's look at another example.

E X A M P L E   1 3   *Finding the Inverse of a Matrix*

The matrix

$$A = \begin{bmatrix} 1 & 1 & 0 \\ -1 & 3 & 4 \\ 0 & 4 & 3 \end{bmatrix}$$

is nonsingular. Find its inverse.

Solution   First, we form the matrix

$$[A|I_3] = \begin{bmatrix} 1 & 1 & 0 & | & 1 & 0 & 0 \\ -1 & 3 & 4 & | & 0 & 1 & 0 \\ 0 & 4 & 3 & | & 0 & 0 & 1 \end{bmatrix}$$

Next, we use row operations to transform $[A|I_3]$ into reduced echelon form:

$$\begin{bmatrix} 1 & 1 & 0 & | & 1 & 0 & 0 \\ -1 & 3 & 4 & | & 0 & 1 & 0 \\ 0 & 4 & 3 & | & 0 & 0 & 1 \end{bmatrix} \rightarrow \underset{\substack{\uparrow \\ R_2 = r_2 + r_1}}{\begin{bmatrix} 1 & 1 & 0 & | & 1 & 0 & 0 \\ 0 & 4 & 4 & | & 1 & 1 & 0 \\ 0 & 4 & 3 & | & 0 & 0 & 1 \end{bmatrix}} \rightarrow \underset{\substack{\uparrow \\ R_2 = \frac{1}{4}r_2}}{\begin{bmatrix} 1 & 1 & 0 & | & 1 & 0 & 0 \\ 0 & 1 & 1 & | & \frac{1}{4} & \frac{1}{4} & 0 \\ 0 & 4 & 3 & | & 0 & 0 & 1 \end{bmatrix}}$$

$$\rightarrow \underset{\substack{\uparrow \\ R_1 = -1r_2 + r_1 \\ R_3 = -4r_2 + r_3}}{\begin{bmatrix} 1 & 0 & -1 & | & \frac{3}{4} & -\frac{1}{4} & 0 \\ 0 & 1 & 1 & | & \frac{1}{4} & \frac{1}{4} & 0 \\ 0 & 0 & -1 & | & -1 & -1 & 1 \end{bmatrix}} \rightarrow \underset{\substack{\uparrow \\ R_3 = -1r_3}}{\begin{bmatrix} 1 & 0 & -1 & | & \frac{3}{4} & -\frac{1}{4} & 0 \\ 0 & 1 & 1 & | & \frac{1}{4} & \frac{1}{4} & 0 \\ 0 & 0 & 1 & | & 1 & 1 & -1 \end{bmatrix}}$$

$$\rightarrow \underset{\substack{\uparrow \\ R_1 = r_1 + r_3 \\ R_2 = -1r_3 + r_2}}{\begin{bmatrix} 1 & 0 & 0 & | & \frac{7}{4} & \frac{3}{4} & -1 \\ 0 & 1 & 0 & | & -\frac{3}{4} & -\frac{3}{4} & 1 \\ 0 & 0 & 1 & | & 1 & 1 & -1 \end{bmatrix}}$$

The matrix $[A|I_3]$ is now in reduced echelon form, and the identity matrix I_3 is on the left of the vertical bar. Hence, the inverse of A is

$$A^{-1} = \begin{bmatrix} \frac{7}{4} & \frac{3}{4} & -1 \\ -\frac{3}{4} & -\frac{3}{4} & 1 \\ 1 & 1 & -1 \end{bmatrix}$$ ■

You can (and should) verify that this is the correct inverse by showing that $AA^{-1} = A^{-1}A = I_3$.

Check: Enter the matrix A. Then find A^{-1}. Now find the product AA^{-1}. ■

■ Now work Problem 23.

If transforming the matrix $[A|I_n]$ into reduced echelon form does not result in the identity matrix I_n to the left of the vertical bar, then A has no inverse. The next example demonstrates such a matrix.

E X A M P L E   1 4   *Showing that a Matrix Has No Inverse*

Show that the following matrix has no inverse.

$$A = \begin{bmatrix} 4 & 6 \\ 2 & 3 \end{bmatrix}$$

Solution   Proceeding as in Example 13, we form the matrix

$$[A|I_2] = \begin{bmatrix} 4 & 6 & | & 1 & 0 \\ 2 & 3 & | & 0 & 1 \end{bmatrix}$$

Then we use row operations to transform $[A \mid I_2]$ into reduced echelon form:

$$[A|I_2] = \begin{bmatrix} 4 & 6 & | & 1 & 0 \\ 2 & 3 & | & 0 & 1 \end{bmatrix} \rightarrow \underset{\substack{\uparrow \\ R_1 = \frac{1}{4}r_1}}{\begin{bmatrix} 1 & \frac{3}{2} & | & \frac{1}{4} & 0 \\ 2 & 3 & | & 0 & 1 \end{bmatrix}} \rightarrow \underset{\substack{\uparrow \\ R_2 = -2r_1 + r_2}}{\begin{bmatrix} 1 & \frac{3}{2} & | & \frac{1}{4} & 0 \\ 0 & 0 & | & -\frac{1}{2} & 1 \end{bmatrix}}$$

The matrix $[A|I_2]$ is sufficiently reduced for us to see that the identity matrix cannot appear to the left of the vertical bar. We conclude that A has no inverse.

Check: Enter the matrix A. Try to find its inverse. What happens? ■

■ Now work Problem 51.

Solving Systems of Equations

Inverse matrices can be used to solve systems of equations in which the number of equations is the same as the number of variables.

E X A M P L E   1 5   *Using the Inverse Matrix to Solve a System of Equations*

Solve the system of equations: $\begin{cases} x + y = 3 \\ -x + 3y + 4z = -3 \\ 4y + 3z = 2 \end{cases}$

Solution   If we let

$$A = \begin{bmatrix} 1 & 1 & 0 \\ -1 & 3 & 4 \\ 0 & 4 & 3 \end{bmatrix}, \quad X = \begin{bmatrix} x \\ y \\ z \end{bmatrix}, \quad \text{and} \quad B = \begin{bmatrix} 3 \\ -3 \\ 2 \end{bmatrix}$$

then the original system of equations can be written compactly as the matrix equation

$$AX = B \qquad\qquad (4)$$

We know from Example 13 that the matrix A has the inverse A^{-1}, so we multiply each side of equation (4) by A^{-1}:

$$AX = B$$
$$A^{-1}(AX) = A^{-1}B$$
$$(A^{-1}A)X = A^{-1}B \quad \text{Associative property for multiplication}$$
$$I_3X = A^{-1}B \quad \text{Definition of inverse matrix}$$
$$X = A^{-1}B \quad \text{Property of identity matrix}$$

Now we use this to find $X = \begin{bmatrix} x \\ y \\ z \end{bmatrix}$:

$$X = \begin{bmatrix} x \\ y \\ z \end{bmatrix} = A^{-1}B = \begin{bmatrix} \frac{7}{4} & \frac{3}{4} & -1 \\ -\frac{3}{4} & -\frac{3}{4} & 1 \\ 1 & 1 & -1 \end{bmatrix} \begin{bmatrix} 3 \\ -3 \\ 2 \end{bmatrix} = \begin{bmatrix} 1 \\ 2 \\ -2 \end{bmatrix}$$

Example 13

Thus, $x = 1$, $y = 2$, $z = -2$. ■

The method used in Example 15 to solve a system of equations is particularly useful when it is necessary to solve several systems of equations in which the constants appearing to the right of the equal signs change, while the coefficients of the variables on the left side remain the same. See Problems 31–50 for some illustrations.

HISTORICAL FEATURE

■ Matrices were invented in 1857 by Arthur Cayley (1821–1895) as a way of efficiently computing the result of substituting one linear system into another (see Historical Problem 2). The resulting system had incredible richness, in the sense that a very wide variety of mathematical systems could be mimicked by the matrices. Cayley and his friend J. J. Sylvester (1814–1897) spent much of the rest of their lives elaborating the theory. The torch was then passed to G. Frobenius (1848–1917), whose deep investigations established a central place for matrices in modern mathematics. In 1924, rather to the surprise of physicists, it was found that matrices (with complex numbers in them) were exactly the right tool for describing the behavior of atomic systems. Today, matrices are used in a wide variety of applications.   ■

HISTORICAL PROBLEMS

■ 1. *Matrices and Complex Numbers* Frobenius emphasized in his research how matrices could be used to mimic other mathematical systems. Here, we mimic the behavior of complex numbers using matrices. Mathematicians call such a relationship an *isomorphism*.

$$\text{Complex number} \longleftrightarrow \text{Matrix}$$

$$a + bi \quad \longleftrightarrow \quad \begin{bmatrix} a & b \\ -b & a \end{bmatrix}$$

Note that the complex number can be read off the top line of the matrix. Thus,

$$2 + 3i \longleftrightarrow \begin{bmatrix} 2 & 3 \\ -3 & 2 \end{bmatrix} \quad \text{and} \quad \begin{bmatrix} 4 & -2 \\ 2 & 4 \end{bmatrix} \longleftrightarrow 4 - 2i$$

(a) Find the matrices corresponding to $2 - 5i$ and $1 + 3i$.

(b) Multiply the two matrices.

(c) Find the corresponding complex number for the matrix found in part (b).

(d) Multiply $2 - 5i$ by $1 + 3i$. The result should be the same as that found in part (c).

The process also works for addition and subtraction. Try it for yourself.

2. *Cayley's Definition of Matrix Multiplication* Cayley invented matrix multiplication to simplify the following problem:

$$\begin{cases} u = ar + bs \\ v = cr + ds \end{cases} \qquad \begin{cases} x = ku + lv \\ y = mu + nv \end{cases}$$

(a) Find x and y in terms of r and s by substituting u and v from the first system of equations into the second system of equations.

(b) Use the result of part (a) to find the 2 by 2 matrix A in

$$\begin{bmatrix} x \\ y \end{bmatrix} = A \begin{bmatrix} r \\ s \end{bmatrix}$$

(c) Now look at the following way to do it: Write the equations in matrix form

$$\begin{bmatrix} u \\ v \end{bmatrix} = \begin{bmatrix} a & b \\ c & d \end{bmatrix} \begin{bmatrix} r \\ s \end{bmatrix} \qquad \begin{bmatrix} x \\ y \end{bmatrix} = \begin{bmatrix} k & l \\ m & n \end{bmatrix} \begin{bmatrix} u \\ v \end{bmatrix}$$

So

$$\begin{bmatrix} x \\ y \end{bmatrix} = \begin{bmatrix} k & l \\ m & n \end{bmatrix} \begin{bmatrix} a & b \\ c & d \end{bmatrix} \begin{bmatrix} r \\ s \end{bmatrix}$$

Do you see how Cayley defined matrix multiplication?   ■

12.1

Exercise 12.1

In Problems 1–16, use the following matrices to compute the given expression.

$$A = \begin{bmatrix} 0 & 3 & -5 \\ 1 & 2 & 6 \end{bmatrix} \quad B = \begin{bmatrix} 4 & 1 & 0 \\ -2 & 3 & -2 \end{bmatrix} \quad C = \begin{bmatrix} 4 & 1 \\ 6 & 2 \\ -2 & 3 \end{bmatrix}$$

1. $A + B$ **2.** $A - B$ **3.** $4A$ **4.** $-3B$

5. $3A - 2B$ **6.** $2A + 4B$ **7.** AC **8.** BC

9. CA **10.** CB **11.** $C(A + B)$ **12.** $(A + B)C$

13. $AC - 3I_2$ **14.** $CA + 5I_3$ **15.** $CA - CB$ **16.** $AC + BC$

In Problems 17–20, compute each product.

17. $\begin{bmatrix} 2 & -2 \\ 1 & 0 \end{bmatrix}\begin{bmatrix} 2 & 1 & 4 & 6 \\ 3 & -1 & 3 & 2 \end{bmatrix}$ **18.** $\begin{bmatrix} 4 & 1 \\ 2 & 1 \end{bmatrix}\begin{bmatrix} -6 & 6 & 1 & 0 \\ 2 & 5 & 4 & -1 \end{bmatrix}$

19. $\begin{bmatrix} 1 & 0 & 1 \\ 2 & 4 & 1 \\ 3 & 6 & 1 \end{bmatrix}\begin{bmatrix} 1 & 3 \\ 6 & 2 \\ 8 & -1 \end{bmatrix}$ **20.** $\begin{bmatrix} 4 & -2 & 3 \\ 0 & 1 & 2 \\ -1 & 0 & 1 \end{bmatrix}\begin{bmatrix} 2 & 6 \\ 1 & -1 \\ 0 & 2 \end{bmatrix}$

In Problems 21–30, each matrix is nonsingular. Find the inverse of each matrix. Be sure to check your answer.

21. $\begin{bmatrix} 2 & 1 \\ 1 & 1 \end{bmatrix}$ **22.** $\begin{bmatrix} 3 & -1 \\ -2 & 1 \end{bmatrix}$ **23.** $\begin{bmatrix} 6 & 5 \\ 2 & 2 \end{bmatrix}$ **24.** $\begin{bmatrix} -4 & 1 \\ 6 & -2 \end{bmatrix}$

25. $\begin{bmatrix} 2 & 1 \\ a & a \end{bmatrix}, \ a \neq 0$ **26.** $\begin{bmatrix} b & 3 \\ b & 2 \end{bmatrix}, \ b \neq 0$ **27.** $\begin{bmatrix} 1 & -1 & 1 \\ 0 & -2 & 1 \\ -2 & -3 & 0 \end{bmatrix}$ **28.** $\begin{bmatrix} 1 & 0 & 2 \\ -1 & 2 & 3 \\ 1 & -1 & 0 \end{bmatrix}$

29. $\begin{bmatrix} 1 & 1 & 1 \\ 3 & 2 & -1 \\ 3 & 1 & 2 \end{bmatrix}$ **30.** $\begin{bmatrix} 3 & 3 & 1 \\ 1 & 2 & 1 \\ 2 & -1 & 1 \end{bmatrix}$

In Problems 31–50, use the inverses found in Problems 21–30 to solve each system of equations.

31. $\begin{cases} 2x + y = 8 \\ x + y = 5 \end{cases}$ **32.** $\begin{cases} 3x - y = 8 \\ -2x + y = 4 \end{cases}$ **33.** $\begin{cases} 2x + y = 0 \\ x + y = 5 \end{cases}$

34. $\begin{cases} 3x - y = 4 \\ -2x + y = 5 \end{cases}$ **35.** $\begin{cases} 6x + 5y = 7 \\ 2x + 2y = 2 \end{cases}$ **36.** $\begin{cases} -4x + y = 0 \\ 6x - 2y = 14 \end{cases}$

37. $\begin{cases} 6x + 5y = 13 \\ 2x + 2y = 5 \end{cases}$ **38.** $\begin{cases} -4x + y = 5 \\ 6x - 2y = -9 \end{cases}$ **39.** $\begin{cases} 2x + y = -3 \\ ax + ay = -a \end{cases} \ a \neq 0$

40. $\begin{cases} bx + 3y = 2b + 3 \\ bx + 2y = 2b + 2 \end{cases} \ b \neq 0$ **41.** $\begin{cases} 2x + y = 7/a \\ ax + ay = 5 \end{cases} \ a \neq 0$ **42.** $\begin{cases} bx + 3y = 14 \\ bx + 2y = 10 \end{cases} \ b \neq 0$

43. $\begin{cases} x - y + z = 0 \\ -2y + z = -1 \\ -2x - 3y = -5 \end{cases}$ **44.** $\begin{cases} x + 2z = 6 \\ -x + 2y + 3z = -5 \\ x - y = 6 \end{cases}$ **45.** $\begin{cases} x - y + z = 2 \\ -2y + z = 2 \\ -2x - 3y = \frac{1}{2} \end{cases}$

46. $\begin{cases} x + 2z = 2 \\ -x + 2y + 3z = -\frac{3}{2} \\ x - y = 2 \end{cases}$ **47.** $\begin{cases} x + y + z = 9 \\ 3x + 2y - z = 8 \\ 3x + y + 2z = 1 \end{cases}$ **48.** $\begin{cases} 3x + 3y + z = 8 \\ x + 2y + z = 5 \\ 2x - y + z = 4 \end{cases}$

49. $\begin{cases} x + y + z = 2 \\ 3x + 2y - z = \frac{7}{3} \\ 3x + y + 2z = \frac{10}{3} \end{cases}$ **50.** $\begin{cases} 3x + 3y + z = 1 \\ x + 2y + z = 0 \\ 2x - y + z = 4 \end{cases}$

In Problems 51–56, show that each matrix has no inverse.

51. $\begin{bmatrix} 4 & 2 \\ 2 & 1 \end{bmatrix}$ **52.** $\begin{bmatrix} -3 & \frac{1}{2} \\ 6 & 1 \end{bmatrix}$ **53.** $\begin{bmatrix} 15 & 3 \\ 10 & 2 \end{bmatrix}$

54. $\begin{bmatrix} -3 & 0 \\ 4 & 0 \end{bmatrix}$ **55.** $\begin{bmatrix} -3 & 1 & -1 \\ 1 & -4 & -7 \\ 1 & 2 & 5 \end{bmatrix}$ **56.** $\begin{bmatrix} 1 & 1 & -3 \\ 2 & -4 & 1 \\ -5 & 7 & 1 \end{bmatrix}$

In Problems 57–60, find the inverse, if it exists, of each matrix.

57. $\begin{bmatrix} 25 & 61 & -12 \\ 18 & -2 & 4 \\ 8 & 35 & 21 \end{bmatrix}$ **58.** $\begin{bmatrix} 18 & -3 & 4 \\ 6 & -20 & 14 \\ 10 & 25 & -15 \end{bmatrix}$

59. $\begin{bmatrix} 44 & 21 & 18 & 6 \\ -2 & 10 & 15 & 5 \\ 21 & 12 & -12 & 4 \\ -8 & -16 & 4 & 9 \end{bmatrix}$ **60.** $\begin{bmatrix} 16 & 22 & -3 & 5 \\ 21 & -17 & 4 & 8 \\ 2 & 8 & 27 & 20 \\ 5 & 15 & -3 & -10 \end{bmatrix}$

In Problems 61–64, use the idea behind Example 15 to solve the systems of equations.

61. $\begin{cases} 25x + 61y - 12z = 10 \\ 18x - 12y + 7z = -9 \\ 3x + 4y - z = 12 \end{cases}$ **62.** $\begin{cases} 25x + 61y - 12z = 5 \\ 18x - 12y + 7z = -3 \\ 3x + 4y - z = 12 \end{cases}$

63. $\begin{cases} 25x + 61y - 12z = 21 \\ 18x - 12y + 7z = 7 \\ 3x + 4y - z = -2 \end{cases}$ **64.** $\begin{cases} 25x + 61y - 12z = 25 \\ 18x - 12y + 7z = 10 \\ 3x + 4y - z = -4 \end{cases}$

65. *Computing the Cost of Production* The Acme Steel Company is a producer of stainless steel and aluminum containers. On a certain day, the following stainless steel containers were manufactured: 500 with 10 gallon capacity, 350 with 5 gallon capacity, and 400 with 1 gallon capacity. On the same day, the following aluminum containers were manufactured: 700 with 10 gallon capacity, 500 with 5 gallon capacity, and 850 with 1 gallon capacity.
(a) Find a 2 by 3 matrix representing the above data. Find a 3 by 2 matrix to represent the same data.
(b) If the amount of material used in the 10 gallon containers is 15 pounds, the amount used in the 5 gallon containers is 8 pounds, and the amount used in the 1 gallon containers is 3 pounds, find a 3 by 1 matrix representing the amount of material.
(c) Multiply the 2 by 3 matrix found in part (a) and the 3 by 1 matrix found in part (b) to get a 2 by 1 matrix showing the day's usage of material.
(d) If stainless steel costs Acme $0.10 per pound and aluminum costs $0.05 per pound, find a 1 by 2 matrix representing cost.
(e) Multiply the matrices found in parts (c) and (d) to determine what the total cost of the day's production was.

66. *Computing Profit* A car dealership has two locations, one in a city and the other in the suburbs. In January, the city location sold 400 subcompacts, 250 intermediate-size cars, and 50 station wagons; in February, it sold 350 subcompacts, 100 intermediates, and 30 station wagons. At the suburban location in January, 450 subcompacts, 200 intermediates, and 140 station wagons were sold. In February, the suburban location sold 350 subcompacts, 300 intermediates, and 100 station wagons.

(a) Find 2 by 3 matrices that summarize the sales data for each location for January and February (one matrix for each month).
(b) Use matrix addition to obtain total sales for the 2 month period.
(c) The profit on each kind of car is $100 per subcompact, $150 per intermediate, and $200 per station wagon. Find a 3 by 1 matrix representing this profit.
(d) Multiply the matrices found in parts (b) and (c) to get a 2 by 1 matrix showing the profit at each location.

67. Consider the 2 by 2 square matrix

$$A = \begin{bmatrix} a & b \\ c & d \end{bmatrix}$$

If $D = ad - bc \neq 0$, show that A is nonsingular and that

$$A^{-1} = \frac{1}{D} \begin{bmatrix} d & -b \\ -c & a \end{bmatrix}$$

 68. Make up a situation different from any found in the text that can be represented by a matrix.

12.2

Partial Fraction Decomposition

Consider the problem of adding the two fractions

$$\frac{3}{x + 4} \quad \text{and} \quad \frac{2}{x - 3}$$

The result is

$$\frac{3}{x + 4} + \frac{2}{x - 3} = \frac{3(x - 3) + 2(x + 4)}{(x + 4)(x - 3)} = \frac{5x - 1}{x^2 + x - 12}$$

The reverse procedure, of starting with the rational expression $(5x - 1)/(x^2 + x - 12)$ and writing it as the sum (or difference) of the two simpler fractions $3/(x + 4)$ and $2/(x - 3)$ is referred to as **partial fraction decomposition,** and the two simpler fractions are called **partial fractions.** Decomposing a rational expression into a sum of partial fractions is important in solving certain types of calculus problems. This section presents a systematic way to decompose rational expressions.

We begin by recalling that a rational expression is the ratio of two polynomials, say, P and $Q \neq 0$, that have no common factors. Recall also that a rational expression P/Q is called **proper** if the degree of the polynomial in the numerator is less than the degree of the polynomial in the denominator. Otherwise, the rational expression is termed **improper.**

Because any improper rational expression can be reduced by long division to a mixed form consisting of the sum of a polynomial and a proper rational expression, we shall restrict the discussion that follows to proper rational expressions.

The partial fraction decomposition of the rational expression P/Q depends on the factors of the denominator Q. Recall (from Section 3.5) that any polynomial whose coefficients are real numbers can be factored (over the real numbers) into products of linear and/or irreducible quadratic factors. Thus, the denominator Q of the rational expression P/Q will contain only factors of one or both of the following types:

1. *Linear factors* of the form $x - a$, where a is a real number.
2. *Irreducible quadratic factors* of the form $ax^2 + bx + c$, where a, b, and c are real numbers, $a \neq 0$, and $b^2 - 4ac < 0$ (which guarantees that $ax^2 + bx + c$ cannot be written as the product of two linear factors with real coefficients).

As it turns out, there are four cases to be examined. We begin with the case for which Q has only nonrepeated linear factors.

CASE 1: Q has only nonrepeated linear factors.

Under the assumption that Q has only nonrepeated linear factors, the polynomial Q has the form

$$Q(x) = (x - a_1)(x - a_2) \cdot \ldots \cdot (x - a_n)$$

where none of the numbers $a_1, a_2, \ldots, a_n$ are equal. In this case, the partial fraction decomposition of P/Q is of the form

$$\frac{P(x)}{Q(x)} = \frac{A_1}{x - a_1} + \frac{A_2}{x - a_2} + \cdots + \frac{A_n}{x - a_n} \qquad (1)$$

where the numbers $A_1, A_2, \ldots, A_n$ are to be determined.

We show how to find these numbers in the example that follows.

E X A M P L E   1   *Nonrepeated Linear Factors*

Write the partial fraction decomposition of $\dfrac{x}{x^2 - 5x + 6}$.

Solution   First, we factor the denominator,

$$x^2 - 5x + 6 = (x - 2)(x - 3)$$

and conclude that the denominator contains only nonrepeated linear factors. Then we decompose the rational expression according to equation (1):

$$\frac{x}{x^2 - 5x + 6} = \frac{A}{x - 2} + \frac{B}{x - 3} \qquad (2)$$

where A and B are to be determined. To find A and B, we clear the fractions by multiplying each side by $(x - 2)(x - 3) = x^2 - 5x + 6$. The result is

$$x = A(x - 3) + B(x - 2) \qquad (3)$$

or

$$x = (A + B)x + (-3A - 2B)$$

This equation is an identity in x. Thus, we may equate the coefficients of like powers of x to get

$$\begin{cases} 1 = \quad A + \ \ B & \text{Equate coefficients of } x\text{: } 1x = (A + B)x \\ 0 = -3A - 2B & \text{Equate coefficients of } x^0\text{, the constants:} \\ & \quad 0x^0 = (-3A - 2B)x^0 \end{cases}$$

This system of two equations containing two variables, A and B, can be solved using whatever method you wish. Solving it, we get

$$A = -2 \qquad B = 3$$

$\mathcal{M}$ISSION POSSIBLE

Chapter 12

KEEPING SECRETS

Your group has discovered that a rival consulting firm has been intercepting and reading the reports that you are sending one another. You need a way to encode your final decisions so that this rival firm does not take business away from you by stealing your ideas. You have decided to find a way to encode your most important messages to each other so that prying eyes are not able to read them.

Your method of coding begins with the simple one used by children, that is, assigning to each letter of the alphabet the number that represents its position in the order. For example, A = 1, B = 2, . . . , M = 13, N = 14, . . . , Z = 26. A space would be represented by 27; a period by 0. Then the message is translated into 2×1 matrices as in the example that follows.

Example: Suppose that you wish to send the message; *Choose Dealer C.* You would write this out first as letters in groups of two and then as matrices.

$$
\begin{array}{cccccccc}
Ch & oo & se & D & ea & le & r & C \\
\begin{bmatrix} 3 \\ 8 \end{bmatrix} &
\begin{bmatrix} 15 \\ 15 \end{bmatrix} &
\begin{bmatrix} 19 \\ 5 \end{bmatrix} &
\begin{bmatrix} 27 \\ 4 \end{bmatrix} &
\begin{bmatrix} 5 \\ 1 \end{bmatrix} &
\begin{bmatrix} 12 \\ 5 \end{bmatrix} &
\begin{bmatrix} 18 \\ 27 \end{bmatrix} &
\begin{bmatrix} 3 \\ 0 \end{bmatrix}
\end{array}
$$

Next you would use a coding matrix, a 2×2 square matrix with numbers chosen so that the inverse exists. We will use $\begin{bmatrix} 6 & 4 \\ -2 & 7 \end{bmatrix}$ as our coding matrix. To encode a message, you multiply the coding matrix times each matrix of the message, as follows:.

$$
\begin{bmatrix} 6 & 4 \\ -2 & 7 \end{bmatrix} \cdot \begin{bmatrix} 3 \\ 8 \end{bmatrix} = \begin{bmatrix} 50 \\ 50 \end{bmatrix}, \qquad
\begin{bmatrix} 6 & 4 \\ -2 & 7 \end{bmatrix} \cdot \begin{bmatrix} 15 \\ 15 \end{bmatrix} = \begin{bmatrix} 150 \\ 75 \end{bmatrix}, \quad \text{etc.}
$$

The resulting set of matrices would be sent along to your partner as a sequence of numbers: 50, 50, 150, 75, and so on. Your partner would use the inverse of the encoding matrix to decipher your message. Obviously, you would have to guard the encoding matrix and its inverse very carefully!

1. Find the inverse of the coding matrix above. Make sure it is correct by multiplying the two encoded matrices times your matrix to see if you can recover the original two matrices.
2. Next, create your own message (not too long!) and encrypt it using the coding matrix above. Copy the resulting code onto a separate piece of paper as a sequence of numbers.
3. Exchange coded messages with another group.
4. Use the inverse matrix to decode the message sent to you by the other group.
5. Cryptography is a science practiced by governments at war or businesses that are competitive in situations where secrecy is crucial. If your consulting firm is serious about keeping secrets, you may need to upgrade your coding procedures. For example, the message *Choose Dealer C,* given above, could be broken up into four 4×1 matrices. Then the coding matrix would have to be 4×4 with an inverse to match. A government might use a 20×20 coding matrix to decrease the probability that it could be deciphered. Create your own 4×4 coding matrix, find its inverse, and encode a message of your choosing in 4×1 matrices. Then pass the code and the key to a neighboring group for decoding.

Thus, from equation (2), the partial fraction decomposition is

$$\frac{x}{x^2 - 5x + 6} = \frac{-2}{x - 2} + \frac{3}{x - 3}$$

Check: The decomposition can be checked by adding the fractions:

$$\frac{-2}{x - 2} + \frac{3}{x - 3} = \frac{-2(x - 3) + 3(x - 2)}{(x - 2)(x - 3)} = \frac{x}{(x - 2)(x - 3)}$$

$$= \frac{x}{x^2 - 5x + 6}$$ ∎

■   Now work Problem 9.

The numbers to be found in the partial fraction decomposition can sometimes be found more readily by using suitable choices for x (which may include complex numbers) in the identity obtained after fractions have been cleared. In Example 1, the identity after clearing fractions, equation (3), is

$$x = A(x - 3) + B(x - 2)$$

If we let $x = 2$ in this expression, the term containing B drops out, leaving $2 = A(-1)$, or $A = -2$. Similarly, if we let $x = 3$, the term containing A drops out, leaving $3 = B$. Thus, as before, $A = -2$ and $B = 3$.

We use this method in the next example.

CASE 2: Q has repeated linear factors.

If the polynomial Q has a repeated factor, say, $(x - a)^n$, $n \geq 2$ an integer, then, in the partial fraction decomposition of P/Q we allow for the terms

$$\frac{A_1}{x - a} + \frac{A_2}{(x - a)^2} + \cdots + \frac{A_n}{(x - a)^n}$$

where the numbers $A_1, A_2, \ldots, A_n$ are to be determined.

E X A M P L E   2   *Repeated Linear Factors*

Write the partial fraction decomposition of $\dfrac{x + 2}{x^3 - 2x^2 + x}$.

Solution   First, we factor the denominator,

$$x^3 - 2x^2 + x = x(x^2 - 2x + 1) = x(x - 1)^2$$

and find that the denominator has the nonrepeated linear factor x and the repeated linear factor $(x - 1)^2$. By Case 1, we must allow for the term A/x in the decomposition; and, by Case 2, we must allow for the terms $B/(x - 1) + C/(x - 1)^2$ in the decomposition.

Thus, we write

$$\frac{x + 2}{x^3 - 2x^2 + x} = \frac{A}{x} + \frac{B}{x - 1} + \frac{C}{(x - 1)^2} \tag{4}$$

Again, we clear fractions by multiplying each side by $x^3 - 2x^2 + x = x(x-1)^2$. The result is the identity

$$x + 2 = A(x-1)^2 + Bx(x-1) + Cx \qquad (5)$$

If we let $x = 0$ in this expression, the terms containing B and C drop out, leaving $2 = A(-1)^2$, or $A = 2$. Similarly, if we let $x = 1$, the terms containing A and B drop out, leaving $3 = C$. Thus, equation (5) becomes

$$x + 2 = 2(x-1)^2 + Bx(x-1) + 3x$$

Now, let $x = 2$ (any choice other than 0 or 1 will work as well). The result is

$$4 = 2(1)^2 + B(2)(1) + 3(2)$$
$$2B = 4 - 2 - 6 = -4$$
$$B = -2$$

Thus, we have $A = 2$, $B = -2$, and $C = 3$.

From equation (4), the partial fraction decomposition is

$$\frac{x+2}{x^3 - 2x^2 + x} = \frac{2}{x} + \frac{-2}{x-1} + \frac{3}{(x-1)^2} \qquad \blacksquare$$

E X A M P L E   3

Repeated Linear Factors

Write the partial fraction decomposition of $\dfrac{x^3 - 8}{x^2(x-1)^3}$.

Solution   The denominator contains the repeated linear factor x^2 and the repeated linear factor $(x-1)^3$. Thus, the partial fraction decomposition takes the form

$$\frac{x^3 - 8}{x^2(x-1)^3} = \frac{A}{x} + \frac{B}{x^2} + \frac{C}{x-1} + \frac{D}{(x-1)^2} + \frac{E}{(x-1)^3} \qquad (6)$$

As before, we clear fractions and obtain the identity

$$x^3 - 8 = Ax(x-1)^3 + B(x-1)^3 + Cx^2(x-1)^2 + Dx^2(x-1) + Ex^2 \qquad (7)$$

Let $x = 0$. (Do you see why this choice was made?) Then,

$$-8 = B(-1)$$
$$B = 8$$

Now let $x = 1$ in equation (7). Then

$$-7 = E$$

Use $B = 8$ and $E = -7$ in equation (7) and collect like terms:

$$x^3 - 8 = Ax(x-1)^3 + 8(x-1)^3$$
$$+ Cx^2(x-1)^2 + Dx^2(x-1) - 7x^2$$
$$x^3 - 8 - 8(x^3 - 3x^2 + 3x - 1) + 7x^2 = Ax(x-1)^3 + Cx^2(x-1)^2 + Dx^2(x-1)$$
$$-7x^3 + 31x^2 - 24x = x(x-1)[A(x-1)^2 + Cx(x-1) + Dx]$$
$$x(x-1)(-7x + 24) = x(x-1)[A(x-1)^2 + Cx(x-1) + Dx]$$
$$-7x + 24 = A(x-1)^2 + Cx(x-1) + Dx \qquad (8)$$

We now work with equation (8). Let $x = 0$. Then

$$24 = A$$

Now, let $x = 1$ in equation (8). Then

$$17 = D$$

Use $A = 24$ and $D = 17$ in equation (8) and collect like terms:

$$-7x + 24 = 24(x - 1)^2 + Cx(x - 1) + 17x$$
$$-24x^2 + 48x - 24 - 17x - 7x + 24 = Cx(x - 1)$$
$$-24x^2 + 24x = Cx(x - 1)$$
$$-24x(x - 1) = Cx(x - 1)$$
$$-24 = C$$

We now know all the numbers A, B, C, D, and E, so, from equation (6), we have the decomposition

$$\frac{x^3 - 8}{x^2(x - 1)^3} = \frac{24}{x} + \frac{8}{x^2} + \frac{-24}{x - 1} + \frac{17}{(x - 1)^2} + \frac{-7}{(x - 1)^3} \qquad \blacksquare$$

The method employed in Example 3, although somewhat tedious, is still preferable to solving the system of five equations containing five variables that the expansion of equation (6) leads to.

■   Now work Problem 15.

The final two cases involve irreducible quadratic factors. As mentioned in Section 3.5, a quadratic factor is irreducible if it cannot be factored into linear factors with real coefficients. A quadratic expression $ax^2 + bx + c$ is irreducible whenever $b^2 - 4ac < 0$. For example, $x^2 + x + 1$ and $x^2 + 4$ are irreducible.

CASE 3: Q contains a nonrepeated irreducible quadratic factor.

If Q contains a nonrepeated irreducible quadratic factor of the form $ax^2 + bx + c$, then, in partial fraction decomposition of P/Q, allow for the term

$$\frac{Ax + B}{ax^2 + bx + c}$$

where the numbers A and B are to be determined.

EXAMPLE   4   *Nonrepeated Irreducible Quadratic Factor*

Write the partial fraction decomposition of $\dfrac{3x - 5}{x^3 - 1}$.

Solution   We factor the denominator,

$$x^3 - 1 = (x - 1)(x^2 + x + 1)$$

and find that it has a nonrepeated linear factor $x - 1$ and a nonrepeated irreducible quadratic factor $x^2 + x + 1$. Thus, we allow for the term $A/(x - 1)$ by Case 1, and we allow for the term $(Bx + C)/(x^2 + x + 1)$ by Case 3. Hence, we write

$$\frac{3x - 5}{x^3 - 1} = \frac{A}{x - 1} + \frac{Bx + C}{x^2 + x + 1} \tag{9}$$

We clear fractions by multiplying each side of equation (9) by $x^3 - 1 = (x - 1)(x^2 + x + 1)$ to obtain the identity

$$3x - 5 = A(x^2 + x + 1) + (Bx + C)(x - 1) \tag{10}$$

Now let $x = 1$. Then equation (10) gives $-2 = A(3)$, or $A = -\frac{2}{3}$. We use this value of A in equation (10) and simplify:

$$3x - 5 = -\frac{2}{3}(x^2 + x + 1) + (Bx + C)(x - 1)$$

$$3(3x - 5) = -2(x^2 + x + 1) + 3(Bx + C)(x - 1) \qquad \text{Multiply each side by 3.}$$

$$9x - 15 = -2x^2 - 2x - 2 + 3(Bx + C)(x - 1)$$

$$2x^2 + 11x - 13 = 3(Bx + C)(x - 1) \qquad\qquad \text{Collect terms.}$$

$$(2x + 13)(x - 1) = 3(Bx + C)(x - 1) \qquad\qquad \text{Factor the left side.}$$

$$2x + 13 = 3Bx + 3C$$

$$2 = 3B \qquad \text{and} \qquad 13 = 3C \qquad\qquad \text{Equate coefficients.}$$

$$B = \frac{2}{3} \qquad\qquad\qquad C = \frac{13}{3}$$

Thus, from equation (9), we see that

$$\frac{3x - 5}{x^3 - 1} = \frac{-\frac{2}{3}}{x - 1} + \frac{\frac{2}{3}x + \frac{13}{3}}{x^2 + x + 1} \qquad\blacksquare$$

■ Now work Problem 17.

CASE 4: Q contains repeated irreducible quadratic factors.

If the polynomial Q contains a repeated irreducible quadratic factor $(ax^2 + bx + c)^n$, $n \geq 2$, n an integer, then, in the partial fraction decomposition of P/Q, allow for the terms

$$\frac{A_1 x + B_1}{ax^2 + bx + c} + \frac{A_2 x + B_2}{(ax^2 + bx + c)^2} + \cdots + \frac{A_n x + B_n}{(ax^2 + bx + c)^n}$$

where the numbers $A_1, B_1, A_2, B_2, \ldots, A_n, B_n$ are to be determined.

E X A M P L E   5

Repeated Irreducible Quadratic Factor

Write the partial fraction decomposition of $\dfrac{x^3 + x^2}{(x^2 + 4)^2}$.

Solution   The denominator contains the repeated irreducible quadratic factor $(x^2 + 4)^2$, so we write

$$\frac{x^3 + x^2}{(x^2 + 4)^2} = \frac{Ax + B}{x^2 + 4} + \frac{Cx + D}{(x^2 + 4)^2} \tag{11}$$

We clear fractions to obtain

$$x^3 + x^2 = (Ax + B)(x^2 + 4) + Cx + D$$

Collecting like terms yields

$$x^3 + x^2 = Ax^3 + Bx^2 + (4A + C)x + D + 4B$$

Equating coefficients, we arrive at the system

$$\begin{cases} A = 1 \\ B = 1 \\ 4A + C = 0 \\ D + 4B = 0 \end{cases}$$

The solution is $A = 1$, $B = 1$, $C = -4$, $D = -4$. Hence, from equation (11),

$$\frac{x^3 + x^2}{(x^2 + 4)^2} = \frac{x + 1}{x^2 + 4} + \frac{-4x - 4}{(x^2 + 4)^2}$$ ∎

12.2

Exercise 12.2

In Problems 1–8, tell whether the given rational expression is proper or improper. If improper, rewrite it as the sum of a polynomial and a proper rational expression.

1. $\dfrac{x}{x^2 - 1}$ **2.** $\dfrac{5x + 2}{x^3 - 1}$ **3.** $\dfrac{x^2 + 5}{x^2 - 4}$ **4.** $\dfrac{3x^2 - 2}{x^2 - 1}$

5. $\dfrac{5x^3 + 2x - 1}{x^2 - 4}$ **6.** $\dfrac{3x^4 + x^2 - 2}{x^3 + 8}$ **7.** $\dfrac{x(x - 1)}{(x + 4)(x - 3)}$ **8.** $\dfrac{2x(x^2 + 4)}{x^2 + 1}$

In Problems 9–42, write the partial fraction decomposition of each rational expression.

9. $\dfrac{4}{x(x - 1)}$ **10.** $\dfrac{3x}{(x + 2)(x - 1)}$ **11.** $\dfrac{1}{x(x^2 + 1)}$

12. $\dfrac{1}{(x + 1)(x^2 + 4)}$ **13.** $\dfrac{x}{(x - 1)(x - 2)}$ **14.** $\dfrac{3x}{(x + 2)(x - 4)}$

15. $\dfrac{x^2}{(x - 1)^2(x + 1)}$ **16.** $\dfrac{x + 1}{x^2(x - 2)}$ **17.** $\dfrac{1}{x^3 - 8}$

18. $\dfrac{2x + 4}{x^3 - 1}$ **19.** $\dfrac{x^2}{(x - 1)^2(x + 1)^2}$ **20.** $\dfrac{x + 1}{x^2(x - 2)^2}$

21. $\dfrac{x - 3}{(x + 2)(x + 1)^2}$ **22.** $\dfrac{x^2 + x}{(x + 2)(x - 1)^2}$ **23.** $\dfrac{x + 4}{x^2(x^2 + 4)}$

24. $\dfrac{10x^2 + 2x}{(x - 1)^2(x^2 + 2)}$ **25.** $\dfrac{x^2 + 2x + 3}{(x + 1)(x^2 + 2x + 4)}$ **26.** $\dfrac{x^2 - 11x - 18}{x(x^2 + 3x + 3)}$

27. $\dfrac{x}{(3x - 2)(2x + 1)}$ **28.** $\dfrac{1}{(2x + 3)(4x - 1)}$ **29.** $\dfrac{x}{x^2 + 2x - 3}$

30. $\dfrac{x^2 - x - 8}{(x + 1)(x^2 + 5x + 6)}$ **31.** $\dfrac{x^2 + 2x + 3}{(x^2 + 4)^2}$ **32.** $\dfrac{x^3 + 1}{(x^2 + 16)^2}$

33. $\dfrac{7x + 3}{x^3 - 2x^2 - 3x}$ **34.** $\dfrac{x^5 + 1}{x^6 - x^4}$ **35.** $\dfrac{x^2}{x^3 - 4x^2 + 5x - 2}$

36. $\dfrac{x^2 + 1}{x^3 + x^2 - 5x + 3}$

37. $\dfrac{x^3}{(x^2 + 16)^3}$

38. $\dfrac{x^2}{(x^2 + 4)^3}$

39. $\dfrac{4}{2x^2 - 5x - 3}$

40. $\dfrac{4x}{2x^2 + 3x - 2}$

41. $\dfrac{2x + 3}{x^4 - 9x^2}$

42. $\dfrac{x^2 + 9}{x^4 - 2x^2 - 8}$

12.3

Vectors

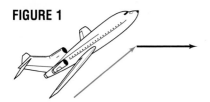

FIGURE 1

In simple terms, a **vector** (derived from the Latin *vehere,* meaning "to carry") is a quantity that has both magnitude and direction. For a vector in the plane, which is the only type we shall discuss, it is convenient to represent a vector by using an arrow. The length of the arrow represents the **magnitude** of the vector, and the arrowhead indicates the **direction** of the vector.

Many quantities in physics can be represented by vectors. For example, the velocity of an aircraft can be represented by an arrow that points in the direction of movement; the length of the arrow represents speed. Thus, if the aircraft speeds up, we lengthen the arrow; if the aircraft changes direction, we introduce an arrow in the new direction. See Figure 1. Based on this representation, it is not surprising that vectors and directed line segments are somehow related.

Directed Line Segments

If P and Q are two distinct points in the xy-plane, there is exactly one line containing both P and Q. The points on that part of the line that joins P to Q, including P and Q, form what is called the **line segment** $\overline{PQ}$. If we order the points so that they proceed from P to Q, we have a **directed line segment** from P to Q, which we denote by $\overrightarrow{PQ}$. In a directed line segment $\overrightarrow{PQ}$, we call P the **initial point** and Q the **terminal point,** as indicated in Figure 2.

FIGURE 2

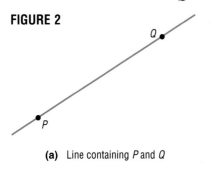

(a) Line containing P and Q

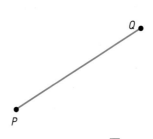

(b) Line segment $\overline{PQ}$

(c) Directed line segment $\overrightarrow{PQ}$

The magnitude of the directed line segment $\overrightarrow{PQ}$ is the distance from the point P to the point Q; that is, it is the length of the line segment. The direction of $\overrightarrow{PQ}$ is from P to Q. If a vector $\mathbf{v}$* has the same magnitude and the same direction as the directed line segment $\overrightarrow{PQ}$, then we write

$$\mathbf{v} = \overrightarrow{PQ}$$

The vector $\mathbf{v}$ whose magnitude is 0 is called the **zero vector, 0.** The zero vector is assigned no direction.

*Boldface letters will be used to denote vectors, in order to distinguish them from numbers. For handwritten work, an arrow is placed over the letter to signify a vector.

FIGURE 3

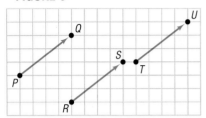

Two vectors **v** and **w** are **equal,** written

$$\mathbf{v} = \mathbf{w}$$

if they have the same magnitude and the same direction.

For example, the vectors shown in Figure 3 have the same magnitude and the same direction, so they are equal, even though they have different initial points and different terminal points. As a result, we find it useful to think of a vector simply as an arrow, keeping in mind that two arrows (vectors) are equal if they have the same direction and the same magnitude (length).

FIGURE 4

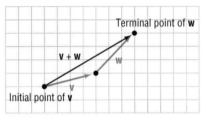

Adding Vectors

The **sum v + w** of two vectors is defined as follows: We position the vectors **v** and **w** so that the terminal point of **v** coincides with the initial point of **w**, as shown in Figure 4. The vector **v + w** is then the unique vector whose initial point coincides with the initial point of **v** and whose terminal point coincides with the terminal point of **w**.

Vector addition is **commutative.** That is, if **v** and **w** are any two vectors, then

Commutative Property

$$\mathbf{v} + \mathbf{w} = \mathbf{w} + \mathbf{v}$$

FIGURE 5

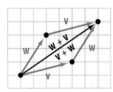

Figure 5 illustrates this fact. (Observe that the commutative property is another way of saying that opposite sides of a parallelogram are equal and parallel.)

Vector addition is also **associative.** That is, if **u**, **v**, and **w** are vectors, then

Associative Property

$$\mathbf{u} + (\mathbf{v} + \mathbf{w}) = (\mathbf{u} + \mathbf{v}) + \mathbf{w}$$

Figure 6 illustrates the associative property for vectors.

FIGURE 6

$$(\mathbf{u} + \mathbf{v}) + \mathbf{w} = \mathbf{u} + (\mathbf{v} + \mathbf{w})$$

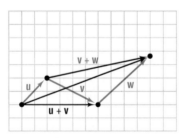

The zero vector has the property that

Identity Property

$$\mathbf{v} + \mathbf{0} = \mathbf{0} + \mathbf{v} = \mathbf{v}$$

FIGURE 7

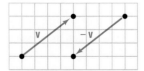

for any vector **v.**

If **v** is a vector, then $-\mathbf{v}$ is the vector having the same magnitude as **v**, but whose direction is opposite to **v**, as shown in Figure 7.

Furthermore,

Additive Inverse Property

$$\mathbf{v} + (-\mathbf{v}) = \mathbf{0}$$

If **v** and **w** are two vectors, we define the **difference v − w** as

$$\mathbf{v} - \mathbf{w} = \mathbf{v} + (-\mathbf{w})$$

Figure 8 illustrates the relationships among **v**, **w**, **v + w**, and **v − w**.

FIGURE 8

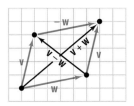

Multiplying Vectors by Numbers

When dealing with vectors, we refer to real numbers as **scalars.** Scalars are quantities that have only magnitude. Examples from physics of scalar quantities are temperature, speed, and time. We now define how to multiply a vector by a scalar.

Scalar Product

If α is a scalar and **v** is a vector, the **scalar product** $\alpha\mathbf{v}$ is defined as:

1. If $\alpha > 0$, the product $\alpha\mathbf{v}$ is the vector whose magnitude is α times the magnitude of **v** and whose direction is the same as **v**.
2. If $\alpha < 0$, the product $\alpha\mathbf{v}$ is the vector whose magnitude is $|\alpha|$ times the magnitude of **v** and whose direction is opposite that of **v**.
3. If $\alpha = 0$ or if $\mathbf{v} = \mathbf{0}$, then $\alpha\mathbf{v} = \mathbf{0}$.

FIGURE 9

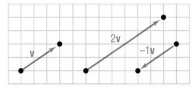

See Figure 9 for some illustrations.

For example, if **a** is the acceleration of an object of mass m due to a force **F** being exerted on it, then, by Newton's Second Law of Motion, $\mathbf{F} = m\mathbf{a}$. Here $m\mathbf{a}$ is the product of the scalar m and the vector **a**.

Scalar products have the following properties:

Properties of Scalar Products

$$0\mathbf{v} = \mathbf{0} \qquad 1\mathbf{v} = \mathbf{v} \qquad -1\mathbf{v} = -\mathbf{v}$$
$$(\alpha + \beta)\mathbf{v} = \alpha\mathbf{v} + \beta\mathbf{v} \qquad \alpha(\mathbf{v} + \mathbf{w}) = \alpha\mathbf{v} + \alpha\mathbf{w}$$
$$\alpha(\beta\mathbf{v}) = (\alpha\beta)\mathbf{v}$$

E X A M P L E   1   *Graphing Vectors*

Use the vectors illustrated in Figure 10 to graph each expression.

FIGURE 10

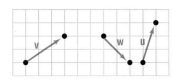

(a) **v − w**   (b) **2v + 3w**   (c) **2v − w + 3u**

Solution   Figure 11 illustrates each graph.

FIGURE 11

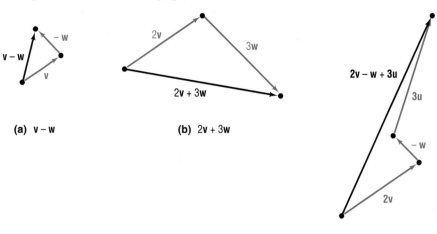

(a) **v − w**   (b) **2v + 3w**

(c) **2v − w + 3u**

■   Now work Problems 1 and 3.

Magnitudes of Vectors

If **v** is a vector, we use the symbol $\|\mathbf{v}\|$ to represent the **magnitude** of **v**. Since $\|\mathbf{v}\|$ equals the length of a directed line segment, it follows that $\|\mathbf{v}\|$ has the following properties:

Theorem   If **v** is a vector and if α is a scalar, then:

Properties of $\|\mathbf{v}\|$

(a) $\|\mathbf{v}\| \geq 0$ (b) $\|\mathbf{v}\| = 0$ if and only if $\mathbf{v} = \mathbf{0}$

(c) $\|-\mathbf{v}\| = \|\mathbf{v}\|$ (d) $\|\alpha\mathbf{v}\| = |\alpha|\|\mathbf{v}\|$ ■

Property (a) is a consequence of the fact that distance is a nonnegative number. Property (b) follows, because the length of the directed line segment $\overrightarrow{PQ}$ is positive unless P and Q are the same point, in which case the length is 0. Property (c) follows, because the length of the line segment $\overline{PQ}$ equals the length of the line segment $\overline{QP}$. Property (d) is a direct consequence of the definition of a scalar product.

Unit Vector   A vector **u** for which $\|\mathbf{u}\| = 1$ is called a **unit vector.**

To compute the magnitude and direction of a vector, we need an algebraic way of representing vectors.

Representing Vectors in the Plane

We use a rectangular coordinate system to represent vectors in the plane. Let **i** denote a unit vector whose direction is along the positive x-axis; let **j** denote a unit vector whose direction is along the positive y-axis. If **v** is a vector with initial point at the origin O and terminal point at $P = (a, b)$, then we can represent **v** in terms of the vectors **i** and **j** as

$$\mathbf{v} = a\mathbf{i} + b\mathbf{j}$$

FIGURE 12

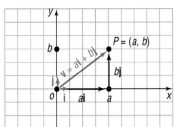

See Figure 12. The scalars a and b are called the **components** of the vector $\mathbf{v} = a\mathbf{i} + b\mathbf{j}$, with a being the component in the direction **i** and b being the component in the direction **j**.

A vector whose initial point is at the origin is called a **position vector.** The next result states that any vector whose initial point is not at the origin is equal to a unique position vector.

Theorem   Suppose that **v** is a vector with initial point $P_1 = (x_1, y_1)$, not necessarily the origin, and terminal point $P_2 = (x_2, y_2)$. If $\mathbf{v} = \overrightarrow{P_1 P_2}$, then **v** is equal to the position vector

$$\mathbf{v} = (x_2 - x_1)\mathbf{i} + (y_2 - y_1)\mathbf{j} \qquad \blacksquare$$

To see why this is true, look at Figure 13. Triangle OPA and triangle $P_1 P_2 Q$ are congruent. (Do you see why?) The line segments have the same magnitude, so $d(O, P) = d(P_1, P_2)$; and they have the same direction, so $\angle POA = \angle P_2 P_1 Q$. Since the triangles are right triangles, we have Angle–Side–Angle. Thus, it follows that corresponding sides are equal. As a result, $x_2 - x_1 = a$ and $y_2 - y_1 = b$, so **v** may be written as

$$\mathbf{v} = a\mathbf{i} + b\mathbf{j} = (x_2 - x_1)\mathbf{i} + (y_2 - y_1)\mathbf{j} \qquad (1)$$

FIGURE 13

$\mathbf{v} = a\mathbf{i} + b\mathbf{j} = (x_2 - x_1)\mathbf{i} + (y_2 - y_1)\mathbf{j}$

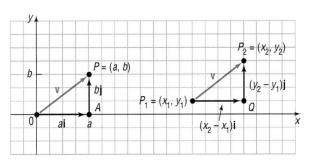

E X A M P L E   2   *Finding a Position Vector*

Find the position vector of the vector $\mathbf{v} = \overrightarrow{P_1 P_2}$ if $P_1 = (-1, 2)$ and $P_2 = (4, 6)$.

Solution   By equation (1), the position vector equal to **v** is

$$\mathbf{v} = [4 - (-1)]\mathbf{i} + (6 - 2)\mathbf{j} = 5\mathbf{i} + 4\mathbf{j}$$

FIGURE 14

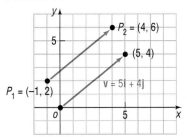

See Figure 14. $\blacksquare$

■ Now work Problem 21.

Two position vectors **v** and **w** are equal if and only if the terminal point of **v** is the same as the terminal point of **w**. This leads to the following result:

Theorem   Two vectors **v** and **w** are equal if and only if their corresponding components are equal. That is:

Equality of Vectors

> If $\mathbf{v} = a_1\mathbf{i} + b_1\mathbf{j}$ and $\mathbf{w} = a_2\mathbf{i} + b_2\mathbf{j}$,
>
> then $\mathbf{v} = \mathbf{w}$ if and only if $a_1 = a_2$ and $b_1 = b_2$.

■

Because of the above result, we can replace any vector (directed line segment) by a unique position vector, and vice versa. This flexibility is one of the main reasons for the wide use of vectors. Unless otherwise specified, from now on the term *vector* will mean the unique position vector equal to it.

Next, we define addition, subtraction, scalar product, and magnitude in terms of the components of a vector.

Let $\mathbf{v} = a_1\mathbf{i} + b_1\mathbf{j}$ and $\mathbf{w} = a_2\mathbf{i} + b_2\mathbf{j}$ be two vectors, and let α be a scalar. Then

$$\mathbf{v} + \mathbf{w} = (a_1 + a_2)\mathbf{i} + (b_1 + b_2)\mathbf{j} \tag{2}$$

$$\mathbf{v} - \mathbf{w} = (a_1 - a_2)\mathbf{i} + (b_1 - b_2)\mathbf{j} \tag{3}$$

$$\alpha\mathbf{v} = (\alpha a_1)\mathbf{i} + (\alpha b_1)\mathbf{j} \tag{4}$$

$$\|\mathbf{v}\| = \sqrt{a_1^2 + b_1^2} \tag{5}$$

These definitions are compatible with the geometric ones given earlier in this section. See Figure 15.

FIGURE 15

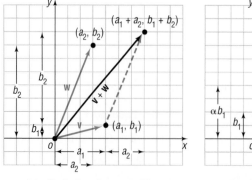

(a)   Illustration of property (2)

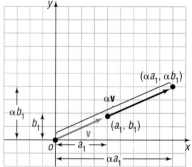

(b)   Illustration of property (4)

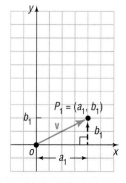

(c)   Illustration of property (5):
$\|\mathbf{v}\|$ = Distance of O to P_1
$\|\mathbf{v}\| = \sqrt{a_1^2 + b_1^2}$

Thus, to add two vectors, simply add corresponding components. To subtract two vectors, subtract corresponding components.

E X A M P L E 3

Adding and Subtracting Vectors

If $\mathbf{v} = 2\mathbf{i} + 3\mathbf{j}$ and $\mathbf{w} = 3\mathbf{i} - 4\mathbf{j}$; find:

(a) $\mathbf{v} + \mathbf{w}$ (b) $\mathbf{v} - \mathbf{w}$

Solution (a) $\mathbf{v} + \mathbf{w} = (2\mathbf{i} + 3\mathbf{j}) + (3\mathbf{i} - 4\mathbf{j}) = (2 + 3)\mathbf{i} + (3 - 4)\mathbf{j}$
$= 5\mathbf{i} - \mathbf{j}$

(b) $\mathbf{v} - \mathbf{w} = (2\mathbf{i} + 3\mathbf{j}) - (3\mathbf{i} - 4\mathbf{j}) = (2 - 3)\mathbf{i} + [3 - (-4)]\mathbf{j}$
$= -\mathbf{i} + 7\mathbf{j}$ ■

E X A M P L E 4

Forming Scalar Products

If $\mathbf{v} = 2\mathbf{i} + 3\mathbf{j}$ and $\mathbf{w} = 3\mathbf{i} - 4\mathbf{j}$, find:

(a) $3\mathbf{v}$ (b) $2\mathbf{v} - 3\mathbf{w}$ (c) $\|\mathbf{v}\|$

Solution (a) $3\mathbf{v} = 3(2\mathbf{i} + 3\mathbf{j}) = 6\mathbf{i} + 9\mathbf{j}$

(b) $2\mathbf{v} - 3\mathbf{w} = 2(2\mathbf{i} + 3\mathbf{j}) - 3(3\mathbf{i} - 4\mathbf{j}) = 4\mathbf{i} + 6\mathbf{j} - 9\mathbf{i} + 12\mathbf{j}$
$= -5\mathbf{i} + 18\mathbf{j}$

(c) $\|\mathbf{v}\| = \|2\mathbf{i} + 3\mathbf{j}\| = \sqrt{2^2 + 3^2} = \sqrt{13}$ ■

■ Now work Problems 27 and 33.

Recall that a unit vector $\mathbf{u}$ is one for which $\|\mathbf{u}\| = 1$. In many applications, it is useful to be able to find a unit vector $\mathbf{u}$ that has the same direction as a given vector $\mathbf{v}$.

Theorem For any nonzero vector $\mathbf{v}$, the vector

───────────────

Unit Vector in Direction of $\mathbf{v}$

$$\mathbf{u} = \frac{\mathbf{v}}{\|\mathbf{v}\|}$$

is a unit vector that has the same direction as $\mathbf{v}$. ■

Proof Let $\mathbf{v} = a\mathbf{i} + b\mathbf{j}$. Then $\|\mathbf{v}\| = \sqrt{a^2 + b^2}$ and

$$\mathbf{u} = \frac{\mathbf{v}}{\|\mathbf{v}\|} = \frac{a\mathbf{i} + b\mathbf{j}}{\sqrt{a^2 + b^2}} = \frac{a}{\sqrt{a^2 + b^2}}\mathbf{i} + \frac{b}{\sqrt{a^2 + b^2}}\mathbf{j}$$

The vector $\mathbf{u}$ is in the same direction as $\mathbf{v}$, since $\|\mathbf{v}\| > 0$, and

$$\|\mathbf{u}\| = \sqrt{\frac{a^2}{a^2 + b^2} + \frac{b^2}{a^2 + b^2}} = \sqrt{\frac{a^2 + b^2}{a^2 + b^2}} = 1$$

Thus, $\mathbf{u}$ is a unit vector in the direction of $\mathbf{v}$. ■

As a consequence of this theorem, if $\mathbf{u}$ is a unit vector in the same direction as a vector $\mathbf{v}$, then $\mathbf{v}$ may be expressed as

$$\mathbf{v} = \|\mathbf{v}\| \, \mathbf{u} \qquad (6)$$

This way of expressing a vector is useful in many applications.

E X A M P L E   5   *Finding a Unit Vector*

Find a unit vector in the same direction as $\mathbf{v} = 4\mathbf{i} - 3\mathbf{j}$.

Solution   We find $\|\mathbf{v}\|$ first

$$\|\mathbf{v}\| = \|4\mathbf{i} - 3\mathbf{j}\| = \sqrt{16 + 9} = 5$$

Now we multiply $\mathbf{v}$ by the scalar $1/\|\mathbf{v}\| = \frac{1}{5}$. The result is

$$\frac{\mathbf{v}}{\|\mathbf{v}\|} = \frac{4\mathbf{i} - 3\mathbf{j}}{5} = \frac{4}{5}\mathbf{i} - \frac{3}{5}\mathbf{j}$$

Check: This vector is, in fact, a unit vector because
$\left(\frac{4}{5}\right)^2 + \left(-\frac{3}{5}\right)^2 = \frac{16}{25} + \frac{9}{25} = \frac{25}{25} = 1$.

■

■ Now work Problem 43.

Applications

FIGURE 16

Resultant

$\mathbf{F}_1 + \mathbf{F}_2$ $\mathbf{F}_2$

$\mathbf{F}_1$

Forces provide an example of physical quantities that may be conveniently represented by vectors; two forces "combine" the way vectors "add." How do we know this? Laboratory experiments bear it out. Thus, if $\mathbf{F}_1$ and $\mathbf{F}_2$ are two forces simultaneously acting on an object, the vector sum $\mathbf{F}_1 + \mathbf{F}_2$ is the force that produces the same effect on the object as that obtained when the forces $\mathbf{F}_1$ and $\mathbf{F}_2$ act on the object. The force $\mathbf{F}_1 + \mathbf{F}_2$ is sometimes called the **resultant** of $\mathbf{F}_1$ and $\mathbf{F}_2$. See Figure 16.

Two important applications of the resultant of two vectors occur with aircraft flying in the presence of a wind and with boats cruising across a river with a current. For example, consider the velocity of wind acting on the velocity of an airplane (see Figure 17). Suppose that $\mathbf{w}$ is a vector describing the velocity of the wind; that is, $\mathbf{w}$ represents the direction and speed of the wind. If $\mathbf{v}$ is the velocity of the airplane in the absence of wind (called its **velocity relative to the air**), then $\mathbf{v} + \mathbf{w}$ is the vector equal to the actual velocity of the airplane (called its **velocity relative to the ground).**

FIGURE 17

(a) Velocity **w** of wind
relative to ground

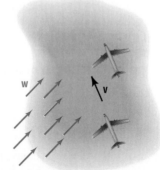

(b) Velocity **v** of airplane
relative to air

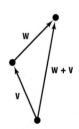

(c) Resultant **w + v** equals
velocity of airplane
relative to the ground

Our next example illustrates this use of vectors in navigation.

E X A M P L E   6   *Finding the Actual Speed of an Aircraft*

A Boeing 737 aircraft maintains a constant airspeed of 500 miles per hour in the direction due south. The velocity of the jet stream is 80 miles per hour in a northeasterly direction.

(a) Find a unit vector having northeast as direction.

(b) Find a vector 80 units in magnitude having the same direction as the unit vector found in part (a).

(c) Find the actual speed of the aircraft relative to the ground.

Solution   We set up a coordinate system in which north (N) is along the positive y-axis. See Figure 18. Let

$$\mathbf{v}_a = \text{Velocity of aircraft relative to the air} = -500\mathbf{j}$$
$$\mathbf{v}_g = \text{Velocity of aircraft relative to ground}$$
$$\mathbf{v}_w = \text{Velocity of jet stream}$$

FIGURE 18

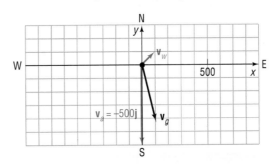

(a) A vector having northeast as direction is $\mathbf{i} + \mathbf{j}$. The unit vector in this direction is

$$\frac{\mathbf{i} + \mathbf{j}}{\|\mathbf{i} + \mathbf{j}\|} = \frac{\mathbf{i} + \mathbf{j}}{\sqrt{1 + 1}} = \frac{1}{\sqrt{2}}(\mathbf{i} + \mathbf{j})$$

(b) The velocity $\mathbf{v}_w$ of the jet stream is a vector with magnitude 80 in the direction of the unit vector $(1/\sqrt{2})(\mathbf{i} + \mathbf{j})$. Thus, from (6),

$$\mathbf{v}_w = 80\left[\frac{1}{\sqrt{2}}(\mathbf{i} + \mathbf{j})\right] = 40\sqrt{2}(\mathbf{i} + \mathbf{j})$$

(c) The velocity $\mathbf{v}_g$ of the aircraft relative to the ground is the resultant of the vectors $\mathbf{v}_a$ and $\mathbf{v}_w$. Thus,

$$\mathbf{v}_g = \mathbf{v}_a + \mathbf{v}_w = -500\mathbf{j} + 40\sqrt{2}(\mathbf{i} + \mathbf{j})$$
$$= 40\sqrt{2}\mathbf{i} + (40\sqrt{2} - 500)\mathbf{j}$$

The actual speed (speed relative to the ground) of the aircraft is

$$\|\mathbf{v}_g\| = \sqrt{(40\sqrt{2})^2 + (40\sqrt{2} - 500)^2} \approx 447 \text{ miles per hour} \qquad \blacksquare$$

We will find the actual direction of the aircraft described in Example 6 in Example 4, Section 12.4.

HISTORICAL FEATURE ■ The history of vectors is surprisingly complicated for such a natural concept. In the *xy*-plane, complex numbers do a good job of imitating vectors. About 1840, mathematicians became interested in finding a system that would do for three dimensions what the complex numbers do for two dimensions. Hermann Grassmann (1809–1877), in Germany, and William Rowan Hamilton (1805–1865), in Ireland, both attempted to find solutions.

Hamilton's system was the *quaternions,* which are best thought of as a real number plus a vector, and do for four dimensions what complex numbers do for two dimensions. In this system the order of multiplication matters; that is, **ab** ≠ **ba**. Hamilton spent the rest of his life working out quaternion theory and trying to get it accepted in applied mathematics, but he encountered fierce resistance due to the complicated nature of quaternion multiplication. In the work with quaternions, two products of vectors emerged, the scalar (or dot) and the vector (or cross) products.

Grassmann fared even worse than Hamilton; if people did not like Hamilton's work, at least they understood it. Grassmann's abstract style, although easily read today, was almost impenetrable during the previous century, and only a few of his ideas were appreciated. Among those few were the same scalar and vector products that Hamilton had found.

About 1880, the American physicist Josiah Willard Gibbs (1839–1903) worked out an algebra involving only the simplest concepts—the vectors and the two products. He then added some calculus, and the resulting system was simple, flexible, and well adapted to expressing a large number of physical laws. This system remains in use essentially unchanged. Hamilton's and Grassmann's more extensive systems each gave birth to much interesting mathematics, but little of this mathematics is seen at elementary levels. ■

12.3

Exercise 12.3

In Problems 1–8, use the vectors in the figure to the right to graph each expression.

1. **v** + **w**   2. **u** + **v**   3. 3**v**   4. 4**w**
5. **v** − **w**   6. **u** − **v**   7. 3**v** + **u** − 2**w**   8. 2**u** − 3**v** + **w**

In Problems 9–16, use the figure below to find each vector.

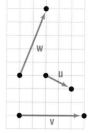

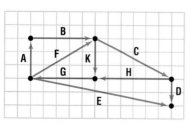

9. **x**, if **x** + **B** = **F**   10. **x**, if **x** + **D** = **E**
11. **C** in terms of **E**, **D**, and **F**   12. **G** in terms of **C**, **D**, **E**, and **K**
13. **E** in terms of **G**, **H**, and **D**   14. **E** in terms of **A**, **B**, **C**, and **D**
15. **x**, if **x** = **A** + **B** + **K** + **G**   16. **x**, if **x** = **A** + **B** + **C** + **H** + **G**
17. If $\|\mathbf{v}\| = 4$, what is $\|3\mathbf{v}\|$?   18. If $\|\mathbf{v}\| = 2$, what is $\|-4\mathbf{v}\|$?

*In Problems 19–26, the vector **v** has initial point P and terminal point Q. Write **v** in the form a**i** + b**j**; that is, find its position vector.*

19. $P = (0, 0); Q = (3, 4)$

20. $P = (0, 0); Q = (-3, -5)$

21. $P = (3, 2); Q = (5, 6)$

22. $P = (-3, 2); Q = (6, 5)$

23. $P = (-2, -1); Q = (6, -2)$

24. $P = (-1, 4); Q = (6, 2)$

25. $P = (1, 0); Q = (0, 1)$

26. $P = (1, 1); Q = (2, 2)$

In Problems 27–32, find $\|\mathbf{v}\|$.

27. $\mathbf{v} = 3\mathbf{i} - 4\mathbf{j}$

28. $\mathbf{v} = -5\mathbf{i} + 12\mathbf{j}$

29. $\mathbf{v} = \mathbf{i} - \mathbf{j}$

30. $\mathbf{v} = -\mathbf{i} - \mathbf{j}$

31. $\mathbf{v} = -2\mathbf{i} + 3\mathbf{j}$

32. $\mathbf{v} = 6\mathbf{i} + 2\mathbf{j}$

In Problems 33–38, find each quantity if $\mathbf{v} = 3\mathbf{i} - 5\mathbf{j}$ and $\mathbf{w} = -2\mathbf{i} + 3\mathbf{j}$.

33. $2\mathbf{v} + 3\mathbf{w}$

34. $3\mathbf{v} - 2\mathbf{w}$

35. $\|\mathbf{v} - \mathbf{w}\|$

36. $\|\mathbf{v} + \mathbf{w}\|$

37. $\|\mathbf{v}\| - \|\mathbf{w}\|$

38. $\|\mathbf{v}\| + \|\mathbf{w}\|$

*In Problems 39–44, find the unit vector having the same direction as **v**.*

39. $\mathbf{v} = 5\mathbf{i}$

40. $\mathbf{v} = -3\mathbf{j}$

41. $\mathbf{v} = 3\mathbf{i} - 4\mathbf{j}$

42. $\mathbf{v} = -5\mathbf{i} + 12\mathbf{j}$

43. $\mathbf{v} = \mathbf{i} - \mathbf{j}$

44. $\mathbf{v} = 2\mathbf{i} - \mathbf{j}$

45. Find a vector **v** whose magnitude is 4 and whose component in the **i** direction is twice the component in the **j** direction.

46. Find a vector **v** whose magnitude is 3 and whose component in the **i** direction is equal to the component in the **j** direction.

47. If $\mathbf{v} = 2\mathbf{i} - \mathbf{j}$ and $\mathbf{w} = x\mathbf{i} + 3\mathbf{j}$, find all numbers x for which $\|\mathbf{v} + \mathbf{w}\| = 5$.

48. If $P = (-3, 1)$ and $Q = (x, 4)$, find all numbers x such that the vector represented by $\overrightarrow{PQ}$ has length 5.

49. *Finding Ground Speed* An airplane has an airspeed of 500 kilometers per hour in an easterly direction. If the wind velocity is 60 kilometers per hour in a northwesterly direction, find the speed of the airplane relative to the ground.

50. *Finding Airspeed* After 1 hour in the air, an airplane arrives at a point 200 miles due south of its departure point. If there was a steady wind of 30 miles per hour from the northwest during the entire flight, what was the average airspeed of the airplane?

51. *Finding Speed without Wind* An airplane travels in a northwesterly direction at a constant ground speed of 250 miles per hour, due to an easterly wind of 50 miles per hour. How fast would the plane have gone if there had been no wind?

52. *Finding the True Speed of a Motorboat* A small motorboat in still water maintains a speed of 10 miles per hour. In heading directly across a river (that is, perpendicular to the current) whose current is 4 miles per hour, what will be the true speed of the motorboat?

53. Show on the graph below the force needed to prevent an object at P from moving.

54. Explain in your own words what a vector is. Give an example of a vector.

55. Write a brief paragraph comparing the algebra of complex numbers and the algebra of vectors.

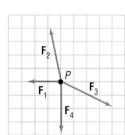

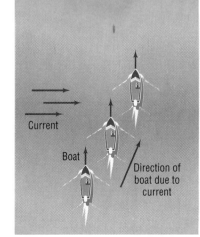

12.4

The Dot Product

The definition for a product of two vectors is somewhat unexpected. However, such a product has meaning in many geometric and physical applications.

Dot Product $\mathbf{v} \cdot \mathbf{w}$

If $\mathbf{v} = a_1\mathbf{i} + b_1\mathbf{j}$ and $\mathbf{w} = a_2\mathbf{i} + b_2\mathbf{j}$ are two vectors, the **dot product $\mathbf{v} \cdot \mathbf{w}$** is defined as

$$\mathbf{v} \cdot \mathbf{w} = a_1a_2 + b_1b_2 \qquad (1)$$

E X A M P L E   1

Finding Dot Products

If $\mathbf{v} = 2\mathbf{i} - 3\mathbf{j}$ and $\mathbf{w} = 5\mathbf{i} + 3\mathbf{j}$, find:

(a) $\mathbf{v} \cdot \mathbf{w}$ (b) $\mathbf{w} \cdot \mathbf{v}$ (c) $\mathbf{v} \cdot \mathbf{v}$

(d) $\mathbf{w} \cdot \mathbf{w}$ (e) $\|\mathbf{v}\|$ (f) $\|\mathbf{w}\|$

Solution

(a) $\mathbf{v} \cdot \mathbf{w} = 2(5) + (-3)3 = 1$ (b) $\mathbf{w} \cdot \mathbf{v} = 5(2) + 3(-3) = 1$

(c) $\mathbf{v} \cdot \mathbf{v} = 2(2) + (-3)(-3) = 13$ (d) $\mathbf{w} \cdot \mathbf{w} = 5(5) + 3(3) = 34$

(e) $\|\mathbf{v}\| = \sqrt{2^2 + (-3)^2} = \sqrt{13}$ (f) $\|\mathbf{w}\| = \sqrt{5^2 + 3^2} = \sqrt{34}$ ■

Since the dot product $\mathbf{v} \cdot \mathbf{w}$ of two vectors $\mathbf{v}$ and $\mathbf{w}$ is a real number (scalar), we sometimes refer to it as the **scalar product.**

Properties

The results obtained in Example 1 suggest some general properties.

Theorem   If $\mathbf{u}$, $\mathbf{v}$, and $\mathbf{w}$ are vectors, then

Commutative Property

$$\mathbf{u} \cdot \mathbf{v} = \mathbf{v} \cdot \mathbf{u} \qquad (2)$$

Distributive Property

$$\mathbf{u} \cdot (\mathbf{v} + \mathbf{w}) = \mathbf{u} \cdot \mathbf{v} + \mathbf{u} \cdot \mathbf{w} \qquad (3)$$

$$\mathbf{v} \cdot \mathbf{v} = \|\mathbf{v}\|^2 \qquad (4)$$

$$\mathbf{0} \cdot \mathbf{v} = 0 \qquad (5)$$

Proof   We shall prove properties (2) and (4) here and leave properties (3) and (5) as exercises (see Problems 29 and 30 at the end of this section).

To prove property (2), we let $\mathbf{u} = a_1\mathbf{i} + b_1\mathbf{j}$ and $\mathbf{v} = a_2\mathbf{i} + b_2\mathbf{j}$. Then

$$\mathbf{u} \cdot \mathbf{v} = a_1a_2 + b_1b_2 = a_2a_1 + b_2b_1 = \mathbf{v} \cdot \mathbf{u}$$

To prove property (4), we let $\mathbf{v} = a\mathbf{i} + b\mathbf{j}$. Then

$$\mathbf{v} \cdot \mathbf{v} = a^2 + b^2 = \|\mathbf{v}\|^2 \qquad ■$$

One use of the dot product is to calculate the angle between two vectors.

FIGURE 19

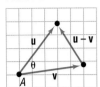

Angle between Vectors

Let **u** and **v** be two vectors with the same initial point A. Then the vectors **u**, **v**, and **u** − **v** form a triangle. The angle θ at vertex A of the triangle is the **angle between the vectors u and v.** See Figure 19. We wish to find a formula for calculating the angle θ.

The sides of the triangle are of lengths $\|\mathbf{v}\|$, $\|\mathbf{u}\|$, and $\|\mathbf{u} - \mathbf{v}\|$, and θ is the included angle between the sides of length $\|\mathbf{v}\|$ and $\|\mathbf{u}\|$. The Law of Cosines (Section 8.2) can be used to find the cosine of the included angle:

$$\|\mathbf{u} - \mathbf{v}\|^2 = \|\mathbf{u}\|^2 + \|\mathbf{v}\|^2 - 2\|\mathbf{u}\|\,\|\mathbf{v}\|\cos\theta$$

Now we use property (4) to rewrite this equation in terms of dot products:

$$(\mathbf{u} - \mathbf{v})\cdot(\mathbf{u} - \mathbf{v}) = \mathbf{u}\cdot\mathbf{u} + \mathbf{v}\cdot\mathbf{v} - 2\|\mathbf{u}\|\,\|\mathbf{v}\|\cos\theta \qquad (6)$$

and then apply the distributive property (3) twice on the left side to obtain

$$(\mathbf{u} - \mathbf{v})\cdot(\mathbf{u} - \mathbf{v}) = \mathbf{u}\cdot(\mathbf{u} - \mathbf{v}) - \mathbf{v}\cdot(\mathbf{u} - \mathbf{v})$$
$$= \mathbf{u}\cdot\mathbf{u} - \mathbf{u}\cdot\mathbf{v} - \mathbf{v}\cdot\mathbf{u} + \mathbf{v}\cdot\mathbf{v}$$
$$= \mathbf{u}\cdot\mathbf{u} + \mathbf{v}\cdot\mathbf{v} - 2\mathbf{u}\cdot\mathbf{v} \qquad (7)$$
$$\uparrow$$
$$\text{Property (2)}$$

Combining equations (6) and (7), we have

$$\mathbf{u}\cdot\mathbf{u} + \mathbf{v}\cdot\mathbf{v} - 2\mathbf{u}\cdot\mathbf{v} = \mathbf{u}\cdot\mathbf{u} + \mathbf{v}\cdot\mathbf{v} - 2\|\mathbf{u}\|\,\|\mathbf{v}\|\cos\theta$$
$$\mathbf{u}\cdot\mathbf{v} = \|\mathbf{u}\|\,\|\mathbf{v}\|\cos\theta$$

Thus, we have proved the following result:

Theorem
Angle between Vectors

If **u** and **v** are two nonzero vectors, the angle θ, $0 \le \theta \le \pi$, between **u** and **v** is determined by the formula

$$\cos\theta = \frac{\mathbf{u}\cdot\mathbf{v}}{\|\mathbf{u}\|\,\|\mathbf{v}\|} \qquad (8)$$

■

E X A M P L E   2 *Finding the Angle θ Between Two Vectors*

$$\mathbf{u} = 4\mathbf{i} - 3\mathbf{j} \quad \text{and} \quad \mathbf{v} = 2\mathbf{i} + 5\mathbf{j}$$

FIGURE 20

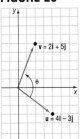

Solution We compute the quantities $\mathbf{u}\cdot\mathbf{v}$, $\|\mathbf{u}\|$, and $\|\mathbf{v}\|$:

$$\mathbf{u}\cdot\mathbf{v} = 4(2) + (-3)(5) = -7$$
$$\|\mathbf{u}\| = \sqrt{4^2 + (-3)^2} = 5$$
$$\|\mathbf{v}\| = \sqrt{2^2 + 5^2} = \sqrt{29}$$

By formula (8), if θ is the angle between **u** and **v**, then

$$\cos\theta = \frac{\mathbf{u}\cdot\mathbf{v}}{\|\mathbf{u}\|\,\|\mathbf{v}\|} = \frac{-7}{5\sqrt{29}} \approx -0.26$$

Using a calculator, we find that $\theta \approx 105°$. See Figure 20.

■

■   Now work Problem 1.

Writing a Vector in Terms of Its Magnitude and Direction

FIGURE 21

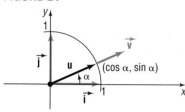

Many applications describe a vector in terms of its magnitude and direction, rather than in terms of its components. Suppose that we are given the magnitude $\|\mathbf{v}\|$ of a nonzero vector $\mathbf{v}$ and the angle α between $\mathbf{v}$ and $\mathbf{i}$. To express $\mathbf{v}$ in terms of $\|\mathbf{v}\|$ and α, we first find the unit vector $\mathbf{u}$ having the same direction as $\mathbf{v}$:

$$\mathbf{u} = \frac{\mathbf{v}}{\|\mathbf{v}\|} \quad \text{or} \quad \mathbf{v} = \|\mathbf{v}\|\mathbf{u} \tag{9}$$

Look at Figure 21. The coordinates of the terminal point of $\mathbf{u}$ are $(\cos \alpha, \sin \alpha)$. Thus, $\mathbf{u} = \cos \alpha \mathbf{i} + \sin \alpha \mathbf{j}$ and, from (9),

$$\mathbf{v} = \|\mathbf{v}\|(\cos \alpha \mathbf{i} + \sin \alpha \mathbf{j}) \tag{10}$$

EXAMPLE 3   *Writing a Vector When Its Magnitude and Direction Are Given*

A force $\mathbf{F}$ of 5 pounds is applied in a direction that makes an angle of 30° with the positive x-axis. Express $\mathbf{F}$ in terms of $\mathbf{i}$ and $\mathbf{j}$.

Solution   The magnitude of $\mathbf{F}$ is $\|\mathbf{F}\| = 5$ and the angle between the direction of $\mathbf{F}$ and $\mathbf{i}$, the positive x-axis, is $\alpha = 30°$. Thus, by (10),

$$\mathbf{F} = \|\mathbf{F}\|(\cos \alpha \mathbf{i} + \sin \alpha \mathbf{j}) = 5(\cos 30°\mathbf{i} + \sin 30°\mathbf{j})$$

$$= 5\left(\frac{\sqrt{3}}{2}\mathbf{i} + \frac{1}{2}\mathbf{j}\right) = \frac{5}{2}(\sqrt{3}\mathbf{i} + \mathbf{j}) \qquad ∎$$

EXAMPLE 4   *Finding the Actual Direction of an Aircraft*

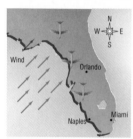

A Boeing 737 aircraft maintains a constant airspeed of 500 miles per hour in the direction due south. The velocity of the jet stream is 80 miles per hour in a north-easterly direction. Find the actual direction of the aircraft relative to the ground.

Solution   This is the same information given in Example 6 of Section 12.3. We repeat the figure from that example as Figure 22.

FIGURE 22

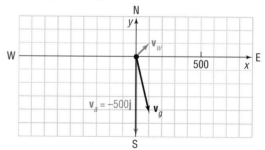

The velocity of the aircraft relative to the air is

$$\mathbf{v}_a = -500\mathbf{j}$$

The wind has magnitude 80 and direction $\alpha = 45°$. Thus, the velocity of the wind is

$$\mathbf{v}_w = 80(\cos 45°\mathbf{i} + \sin 45°\mathbf{j}) = 80\left(\frac{\sqrt{2}}{2}\mathbf{i} + \frac{\sqrt{2}}{2}\mathbf{j}\right)$$

$$= 40\sqrt{2}(\mathbf{i} + \mathbf{j})$$

The velocity of the aircraft relative to the ground is

$$\mathbf{v}_g = \mathbf{v}_a + \mathbf{v}_w = -500\mathbf{j} + 40\sqrt{2}(\mathbf{i} + \mathbf{j}) = 40\sqrt{2}\mathbf{i} + (40\sqrt{2} - 500)\mathbf{j}$$

The angle θ between $\mathbf{v}_g$ and the vector $\mathbf{v}_a = -500\mathbf{j}$ (the velocity of the aircraft relative to the air) is determined by the equation

$$\cos \theta = \frac{\mathbf{v}_g \cdot \mathbf{v}_a}{\|\mathbf{v}_g\| \, \|\mathbf{v}_a\|} = \frac{(40\sqrt{2} - 500)(-500)}{(447)(500)} \approx 0.9920$$

$$\theta \approx 7.2°$$

The direction of the aircraft relative to the ground is about 7.2° east of south.   ■

■   Now work Problem 19.

Parallel and Orthogonal Vectors

Two vectors $\mathbf{v}$ and $\mathbf{w}$ are said to be **parallel** if there is a nonzero scalar α so that $\mathbf{v} = \alpha\mathbf{w}$. In this case, the angle θ between $\mathbf{v}$ and $\mathbf{w}$ is 0 or π.

EXAMPLE 5

Determining Whether Vectors are Parallel

The vectors $\mathbf{v} = 3\mathbf{i} - \mathbf{j}$ and $\mathbf{w} = 6\mathbf{i} - 2\mathbf{j}$ are parallel, since $\mathbf{v} = \frac{1}{2}\mathbf{w}$. Furthermore, since

$$\cos \theta = \frac{\mathbf{v} \cdot \mathbf{w}}{\|\mathbf{v}\| \, \|\mathbf{w}\|} = \frac{18 + 2}{\sqrt{10}\sqrt{40}} = \frac{20}{\sqrt{400}} = 1$$

the angle θ between $\mathbf{v}$ and $\mathbf{w}$ is 0.   ■

FIGURE 23
$\mathbf{v} \cdot \mathbf{w} = 0$
$\mathbf{v}$ is orthogonal to $\mathbf{w}$

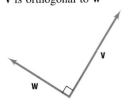

If the angle θ between two nonzero vectors $\mathbf{v}$ and $\mathbf{w}$ is $\pi/2$, the vectors $\mathbf{v}$ and $\mathbf{w}$ are called **orthogonal.***

It follows from formula (8) that if $\mathbf{v}$ and $\mathbf{w}$ are orthogonal then $\mathbf{v} \cdot \mathbf{w} = 0$, since $\cos (\pi/2) = 0$.

On the other hand, if $\mathbf{v} \cdot \mathbf{w} = 0$, then either $\mathbf{v} = \mathbf{0}$ or $\mathbf{w} = \mathbf{0}$ or $\cos \theta = 0$. In the latter case, $\theta = \pi/2$ and $\mathbf{v}$ and $\mathbf{w}$ are orthogonal. See Figure 23. If $\mathbf{v}$ or $\mathbf{w}$ is the zero vector, then, since the zero vector has no specific direction, we adopt the convention that the zero vector is orthogonal to every vector.

Theorem   Two vectors $\mathbf{v}$ and $\mathbf{w}$ are orthogonal if and only if

$$\mathbf{v} \cdot \mathbf{w} = 0$$

■

Orthogonal, perpendicular, and *normal* are all terms that mean "meet at a right angle." It is customary to refer to two vectors as being *orthogonal,* two lines as being *perpendicular,* and a line and a plane or a vector and a plane as being *normal.*

FIGURE 24

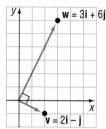

Determining Whether Two Vectors Are Orthogonal

The vectors

$$\mathbf{v} = 2\mathbf{i} - \mathbf{j} \quad \text{and} \quad \mathbf{w} = 3\mathbf{i} + 6\mathbf{j}$$

are orthogonal, since

$$\mathbf{v} \cdot \mathbf{w} = 6 - 6 = 0$$

See Figure 24.

■

■ Now work Problem 11.

Projection of a Vector onto Another Vector

FIGURE 25

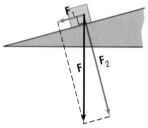

In many physical applications, it is necessary to find "how much" of a vector is applied in a given direction. Look at Figure 25. The force **F** due to gravity is pulling straight down (toward the center of Earth) on the block. To study the effect of gravity on the block, it is necessary to determine how much of **F** is actually pushing the block down the incline ($\mathbf{F}_1$) and how much is pressing the block against the incline ($\mathbf{F}_2$), at a right angle to the incline. Knowing the **decomposition** of **F** often will allow us to determine when friction is overcome and the block will slide down the incline.

Suppose that **v** and **w** are two nonzero vectors with the same initial point P. We seek to decompose **v** into two vectors: $\mathbf{v}_1$, which is parallel to **w**, and $\mathbf{v}_2$, which is orthogonal to **w**. See Figure 26(a) and 26(b). The vector $\mathbf{v}_1$ is called the **vector projection of v onto w** and is denoted by $\text{proj}_\mathbf{w}\, \mathbf{v}$.

The vector $\mathbf{v}_1$ is obtained as follows: From the terminal point of **v**, drop a perpendicular to the line containing **w**. The vector $\mathbf{v}_1$ is the vector from P to the foot of this perpendicular. The vector $\mathbf{v}_2$ is given by $\mathbf{v}_2 = \mathbf{v} - \mathbf{v}_1$. Note that $\mathbf{v} = \mathbf{v}_1 + \mathbf{v}_2$, $\mathbf{v}_1$ is parallel to **w**, and $\mathbf{v}_2$ is orthogonal to **w**. This is the decomposition of **v** that we wanted.

FIGURE 26

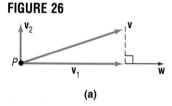

(a)

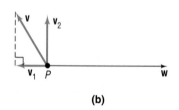

(b)

Now we seek a formula for $\mathbf{v}_1$ that is based on a knowledge of the vectors **v** and **w**. Since $\mathbf{v} = \mathbf{v}_1 + \mathbf{v}_2$, we have

$$\mathbf{v} \cdot \mathbf{w} = (\mathbf{v}_1 + \mathbf{v}_2) \cdot \mathbf{w} = \mathbf{v}_1 \cdot \mathbf{w} + \mathbf{v}_2 \cdot \mathbf{w} \tag{11}$$

Since $\mathbf{v}_2$ is orthogonal to **w**, we have $\mathbf{v}_2 \cdot \mathbf{w} = 0$. Since $\mathbf{v}_1$ is parallel to **w**, we have $\mathbf{v}_1 = \alpha\mathbf{w}$ for some scalar α. Thus, equation (11) can be written as

$$\mathbf{v} \cdot \mathbf{w} = \alpha\mathbf{w} \cdot \mathbf{w} = \alpha\|\mathbf{w}\|^2$$

$$\alpha = \frac{\mathbf{v} \cdot \mathbf{w}}{\|\mathbf{w}\|^2}$$

Thus,

$$\mathbf{v}_1 = \alpha\mathbf{w} = \frac{\mathbf{v} \cdot \mathbf{w}}{\|\mathbf{w}\|^2}\mathbf{w}$$

Theorem   If **v** and **w** are two nonzero vectors, the vector projection of **v** onto **w** is

$$\text{proj}_\mathbf{w}\, \mathbf{v} = \frac{\mathbf{v} \cdot \mathbf{w}}{\|\mathbf{w}\|^2}\mathbf{w}$$

The decomposition of **v** into $\mathbf{v}_1$ and $\mathbf{v}_2$, where $\mathbf{v}_1$ is parallel to **w** and $\mathbf{v}_2$ is perpendicular to **w**, is

$$\mathbf{v}_1 = \text{proj}_{\mathbf{w}}\, \mathbf{v} = \frac{\mathbf{v} \cdot \mathbf{w}}{\|\mathbf{w}\|^2}\mathbf{w} \qquad \mathbf{v}_2 = \mathbf{v} - \mathbf{v}_1 \qquad (12)$$

■

E X A M P L E   7   *Decomposing a Vector into Two Orthogonal Vectors*

Find the vector projection of $\mathbf{v} = \mathbf{i} + 3\mathbf{j}$ onto $\mathbf{w} = \mathbf{i} + \mathbf{j}$. Decompose **v** into two vectors $\mathbf{v}_1$ and $\mathbf{v}_2$, where $\mathbf{v}_1$ is parallel to **w** and $\mathbf{v}_2$ is orthogonal to **w**.

Solution   We use formulas (12).

$$\mathbf{v}_1 = \text{proj}_{\mathbf{w}}\, \mathbf{v} = \frac{\mathbf{v} \cdot \mathbf{w}}{\|\mathbf{w}\|^2}\mathbf{w} = \frac{1+3}{(\sqrt{2})^2}\mathbf{w} = 2\mathbf{w} = 2(\mathbf{i} + \mathbf{j})$$

$$\mathbf{v}_2 = \mathbf{v} - \mathbf{v}_1 = (\mathbf{i} + 3\mathbf{j}) - 2(\mathbf{i} + \mathbf{j}) = -\mathbf{i} + \mathbf{j}$$

FIGURE 27

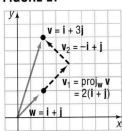

See Figure 27.

■

■   Now work Problem 13.

Work Done by a Constant Force

In elementary physics, the **work** W done by a constant force **F** in moving an object from a point A to a point B is defined as

$$W = (\text{Magnitude of force})(\text{Distance}) = \|\mathbf{F}\| \,\|\overrightarrow{AB}\|$$

FIGURE 28

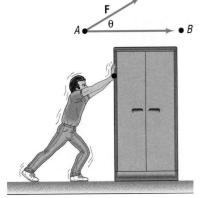

(Work is commonly measured in foot-pounds or in Newton-meters.)

In this definition, it is assumed that the force **F** is applied along the line of motion. If the constant force **F** is not along the line of motion, but, instead, is at an angle θ to the direction of motion, as illustrated in Figure 28, then the **work** W **done by F** in moving an object from A to B is defined as

$$W = \mathbf{F} \cdot \overrightarrow{AB} \qquad (11)$$

This definition is compatible with the force times distance definition given above, since

$$W = (\text{Amount of force in direction of } \overrightarrow{AB})(\text{Distance})$$

$$= \|\text{proj}_{\overrightarrow{AB}}\, \mathbf{F}\| \,\|\overrightarrow{AB}\| = \frac{\mathbf{F} \cdot \overrightarrow{AB}}{\|\overrightarrow{AB}\|^2}\|\overrightarrow{AB}\| \,\|\overrightarrow{AB}\| = \mathbf{F} \cdot \overrightarrow{AB}$$

E X A M P L E   8   *Computing Work*

Find the work done by a force of 5 pounds acting in the direction $\mathbf{i} + \mathbf{j}$ in moving an object 1 foot from $(0, 0)$ to $(1, 0)$.

Solution   First, we must express the force **F** as a vector. The force has magnitude 5 and direction $\mathbf{i} + \mathbf{j}$. The direction of $\overrightarrow{\mathbf{F}}$ therefore makes an angle of $45°$ with $\overrightarrow{\mathbf{i}}$. Thus, the force $\overrightarrow{\mathbf{F}}$ is

$$\mathbf{F} = 5(\cos 45°\mathbf{i} + \sin 45°\mathbf{j}) = 5\left(\frac{\sqrt{2}}{2}\mathbf{i} + \frac{\sqrt{2}}{2}\mathbf{j}\right) = \frac{5\sqrt{2}}{2}(\mathbf{i} + \mathbf{j})$$

The line of motion of the object is from $A = (0, 0)$ to $B = (1, 0)$, so $\overrightarrow{AB} = \mathbf{i}$. The work W is therefore

$$W = \mathbf{F} \cdot \overrightarrow{AB} = \frac{5\sqrt{2}}{2}(\mathbf{i} + \mathbf{j}) \cdot \mathbf{i} = \frac{5\sqrt{2}}{2} \text{ foot-pounds}$$ ■

■ Now work Problem 25.

EXAMPLE 9   *Computing Work*

Figure 29(a) shows a girl pulling a wagon with a force of 50 pounds. How much work is done in moving the wagon 100 feet if the handle makes an angle of 30° with the ground?

FIGURE 29

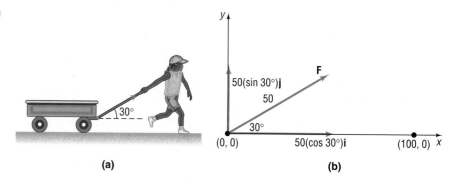

(a)   (b)

Solution   We position the vectors in a coordinate system in such a way that the wagon is moved from $(0, 0)$ to $(100, 0)$. Thus, the motion is from $A = (0, 0)$ to $B = (100, 0)$, so $\overrightarrow{AB} = 100\mathbf{i}$. The force vector $\mathbf{F}$, as shown in Figure 29(b), is

$$\mathbf{F} = 50(\cos 30°\mathbf{i} + \sin 30°\mathbf{j}) = 50\left(\frac{\sqrt{3}}{2}\mathbf{i} + \frac{1}{2}\mathbf{j}\right) = 25\sqrt{3}\mathbf{i} + 25\mathbf{j}$$

By formula (13), the work W done is

$$W = \mathbf{F} \cdot \overrightarrow{AB} = (25\sqrt{3}\mathbf{i} + 25\mathbf{j}) \cdot 100\mathbf{i} = 2500\sqrt{3} \text{ foot-pounds}$$ ■

HISTORICAL PROBLEM

■ 1. We stated in an earlier Historical Feature that complex numbers were used as vectors in the plane before the general notion of vector was clarified. Suppose that we make the correspondence

$$\text{Vector} \longleftrightarrow \text{Complex number}$$
$$a\mathbf{i} + b\mathbf{j} \longleftrightarrow a + bi$$
$$c\mathbf{i} + d\mathbf{j} \longleftrightarrow c + di$$

Show that

$$(a\mathbf{i} + b\mathbf{j}) \cdot (c\mathbf{i} + d\mathbf{j}) = \text{Real part}[\overline{(a + bi)}(c + di)]$$

This is how the dot product was found originally. The imaginary part is also interesting. It is a determinant (see Section 10.3) and represents the area of the parallelogram whose edges are the vectors. This is close to some of Hermann Grassmann's ideas and is also connected with the scalar triple product of three-dimensional vectors. ■

12.4

Exercise 12.4

In Problems 1–10, find the dot product $\mathbf{v} \cdot \mathbf{w}$ *and the cosine of the angle between* $\mathbf{v}$ *and* $\mathbf{w}$.

1. $\mathbf{v} = \mathbf{i} - \mathbf{j}, \quad \mathbf{w} = \mathbf{i} + \mathbf{j}$

2. $\mathbf{v} = \mathbf{i} + \mathbf{j}, \quad \mathbf{w} = -\mathbf{i} + \mathbf{j}$

3. $\mathbf{v} = 2\mathbf{i} + \mathbf{j}, \quad \mathbf{w} = \mathbf{i} + 2\mathbf{j}$

4. $\mathbf{v} = 2\mathbf{i} + 2\mathbf{j}, \quad \mathbf{w} = \mathbf{i} + 2\mathbf{j}$

5. $\mathbf{v} = \sqrt{3}\mathbf{i} - \mathbf{j}, \quad \mathbf{w} = \mathbf{i} + \mathbf{j}$

6. $\mathbf{v} = \mathbf{i} + \sqrt{3}\mathbf{j}, \quad \mathbf{w} = \mathbf{i} - \mathbf{j}$

7. $\mathbf{v} = 3\mathbf{i} + 4\mathbf{j}, \quad \mathbf{w} = 4\mathbf{i} + 3\mathbf{j}$

8. $\mathbf{v} = 3\mathbf{i} - 4\mathbf{j}, \quad \mathbf{w} = 4\mathbf{i} - 3\mathbf{j}$

9. $\mathbf{v} = 4\mathbf{i}, \quad \mathbf{w} = \mathbf{j}$

10. $\mathbf{v} = \mathbf{i}, \quad \mathbf{w} = -3\mathbf{j}$

11. Find a such that the angle between $\mathbf{v} = a\mathbf{i} - \mathbf{j}$ and $\mathbf{w} = 2\mathbf{i} + 3\mathbf{j}$ is $\pi/2$.

12. Find b such that the angle between $\mathbf{v} = \mathbf{i} + \mathbf{j}$ and $\mathbf{w} = \mathbf{i} + b\mathbf{j}$ is $\pi/2$.

In Problems 13–18, decompose $\mathbf{v}$ *into two vectors* $\mathbf{v}_1$ *and* $\mathbf{v}_2$, *where* $\mathbf{v}_1$ *is parallel to* $\mathbf{w}$ *and* $\mathbf{v}_2$ *is orthogonal to* $\mathbf{w}$.

13. $\mathbf{v} = 2\mathbf{i} - 3\mathbf{j}, \quad \mathbf{w} = \mathbf{i} - \mathbf{j}$

14. $\mathbf{v} = -3\mathbf{i} + 2\mathbf{j}, \quad \mathbf{w} = 2\mathbf{i} + \mathbf{j}$

15. $\mathbf{v} = \mathbf{i} - \mathbf{j}, \quad \mathbf{w} = \mathbf{i} + 2\mathbf{j}$

16. $\mathbf{v} = 2\mathbf{i} - \mathbf{j}, \quad \mathbf{w} = \mathbf{i} - 2\mathbf{j}$

17. $\mathbf{v} = 3\mathbf{i} + \mathbf{j}, \quad \mathbf{w} = -2\mathbf{i} - \mathbf{j}$

18. $\mathbf{v} = \mathbf{i} - 3\mathbf{j}, \quad \mathbf{w} = 4\mathbf{i} - \mathbf{j}$

19. *Finding the Actual Speed and Direction of an Aircraft* A DC-10 jumbo jet maintains an airspeed of 550 miles per hour in a southwesterly direction. The velocity of the jet stream is a constant 80 miles per hour from the west. Find the actual speed and direction of the aircraft.

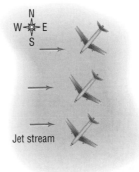

20. *Finding the Correct Compass Heading* The pilot of an aircraft wishes to head directly east, but is faced with a wind speed of 40 miles per hour from the northwest. If the pilot maintains an airspeed of 250 miles per hour, what compass heading should be maintained? What is the actual speed of the aircraft?

21. *Correct Direction for Crossing a Stream* A small stream has a constant current of 3 kilometers per hour. At what angle to a boat dock should a motorboat—capable of maintaining a constant speed of 20 kilometers per hour—be headed in order to reach a point directly opposite the dock? If the stream is $\frac{1}{2}$ kilometer wide, how long will it take to cross?

22. *Correct Direction for Crossing a Stream* Repeat Problem 21 if the current is 5 kilometers per hour.

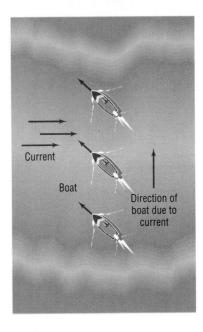

23. *Correct Direction for Crossing a Stream* A river is 500 meters wide and has a current of 1 kilometer per hour. If Sharon can swim at a rate of 2 kilometers per hour, at what angle to the shore should she swim if she wishes to cross the river to a point directly opposite? How long will it take to swim across the river?

24. An airplane travels 200 miles due west and then 150 miles 60° north of west. Determine the resultant displacement.

25. Find the work done by a force of 3 pounds acting in the direction $2\mathbf{i} + \mathbf{j}$ in moving an object 2 feet from $(0, 0)$ to $(0, 2)$.

26. *Computing Work* Find the work done by a force of 1 pound acting in the direction $2\mathbf{i} + 2\mathbf{j}$ in moving an object 5 feet from $(0, 0)$ to $(3, 4)$.

27. *Computing Work* A wagon is pulled horizontally by exerting a force of 20 pounds on the handle at an angle of 30° with the horizontal. How much work is done in moving the wagon 100 feet?

28. Find the acute angle that a constant unit force vector makes with the positive x-axis if the work done by the force in moving a particle from $(0, 0)$ to $(4, 0)$ equals 2.

29. Prove the distributive property, $\mathbf{u} \cdot (\mathbf{v} + \mathbf{w}) = \mathbf{u} \cdot \mathbf{v} + \mathbf{u} \cdot \mathbf{w}$.

30. Prove property (5), $\mathbf{0} \cdot \mathbf{v} = 0$.

31. If $\mathbf{v}$ is a unit vector and the angle between $\mathbf{v}$ and $\mathbf{i}$ is α, show that $\mathbf{v} = \cos\alpha\,\mathbf{i} + \sin\alpha\,\mathbf{j}$.

32. Suppose that $\mathbf{v}$ and $\mathbf{w}$ are unit vectors. If the angle between $\mathbf{v}$ and $\mathbf{i}$ is α and if the angle between $\mathbf{w}$ and $\mathbf{i}$ is β, use the idea of the dot product $\mathbf{v} \cdot \mathbf{w}$ to prove that

$$\cos(\alpha - \beta) = \cos\alpha\,\cos\beta + \sin\alpha\,\sin\beta$$

33. Show that the projection of $\mathbf{v}$ onto $\mathbf{i}$ is $(\mathbf{v} \cdot \mathbf{i})\mathbf{i}$. In fact, show that we can always write a vector $\mathbf{v}$ as

$$\mathbf{v} = (\mathbf{v} \cdot \mathbf{i})\mathbf{i} + (\mathbf{v} \cdot \mathbf{j})\mathbf{j}$$

34. (a) If $\mathbf{u}$ and $\mathbf{v}$ have the same magnitude, then show that $\mathbf{u} + \mathbf{v}$ and $\mathbf{u} - \mathbf{v}$ are orthogonal.
    (b) Use this to prove that an angle inscribed in a semicircle is a right angle (see the figure).

35. Let $\mathbf{v}$ and $\mathbf{w}$ denote two nonzero vectors. Show that the vector $\mathbf{v} - \alpha\mathbf{w}$ is orthogonal to $\mathbf{w}$ if $\alpha = (\mathbf{v} \cdot \mathbf{w})/\|\mathbf{w}\|^2$.

36. Let $\mathbf{v}$ and $\mathbf{w}$ denote two nonzero vectors. Show that the vectors $\|\mathbf{w}\|\mathbf{v} + \|\mathbf{v}\|\mathbf{w}$ and $\|\mathbf{w}\|\mathbf{v} - \|\mathbf{v}\|\mathbf{w}$ are orthogonal.

37. In the definition of work given in this section, what is the work done if $\mathbf{F}$ is orthogonal to $\overrightarrow{AB}$?

38. Prove the **polarization identity:** $\|\mathbf{u} + \mathbf{v}\|^2 - \|\mathbf{u} - \mathbf{v}\|^2 = 4(\mathbf{u} \cdot \mathbf{v})$

39. Make up an application different from any found in the text that requires the dot product.

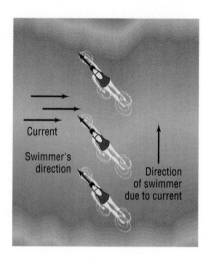

Current

Swimmer's direction

Direction of swimmer due to current

Chapter Review

THINGS TO KNOW

Matrix	Rectangular array of numbers, called entries
m by n matrix	Matrix with m rows and n columns
Identity matrix I	Square matrix whose diagonal entries are 1's, while all other entries are 0's
Inverse of a matrix	A^{-1} is the inverse of A if $AA^{-1} = A^{-1}A = I$
Nonsingular matrix	A matrix that has an inverse

Vector	Quantity having magnitude and direction; equivalent to a directed line segment $\overrightarrow{PQ}$
Position vector	Vector whose initial point is at the origin
Unit vector	Vector whose magnitude is 1
Dot product	If $\mathbf{v} = a_1\mathbf{i} + b_1\mathbf{j}$ and $\mathbf{w} = a_2\mathbf{i} + b_2\mathbf{j}$, then $\mathbf{v} \cdot \mathbf{w} = a_1a_2 + b_1b_2$.
Angle θ between two nonzero vectors $\mathbf{u}$ and $\mathbf{v}$	$\cos\theta = \dfrac{\mathbf{u} \cdot \mathbf{v}}{\|\mathbf{u}\|\,\|\mathbf{v}\|}$

How To:

Recognize equal matrices	Find the magnitude of a vector
Add and subtract matrices	Solve problems involving vectors
Multiply matrices	Find the dot product of two vectors
Find the inverse of a nonsingular matrix	Find the angle between two vectors
Solve a system of equations using the inverse of a matrix	Determine whether two vectors are parallel
Write the partial fraction decomposition of a rational expression	Determine whether two vectors are orthogonal
Add and subtract vectors	Find the vector projection of $\mathbf{v}$ onto $\mathbf{w}$
Form scalar multiples of vectors	

Fill-in-the-Blank Items

1. A matrix B for which $AB = I_n$, the identity matrix, is called the _____ of A.

2. In the algebra of matrices, the matrix that has properties similar to the number 1 is called the _____ matrix.

3. A rational function is called _____ if the degree of its numerator is less than the degree of its denominator.

4. A vector whose magnitude is 1 is called a(n) _____ vector.

5. If the angle between two vectors $\mathbf{v}$ and $\mathbf{w}$ is $\pi/2$, then the dot product $\mathbf{v} \cdot \mathbf{w}$ equals _____.

True/False Items

T   F   **1.** Every square matrix has an inverse.

T   F   **2.** Matrix multiplication is commutative.

T   F   **3.** Any pair of matrices can be multiplied.

T   F   **4.** The factors of the denominator of a rational expression are used to arrive at the partial fraction decomposition.

T   F   **5.** Vectors are quantities that have magnitude and direction.

T   F   **6.** Force is a physical example of a vector.

T   F   **7.** If $\mathbf{u}$ and $\mathbf{v}$ are orthogonal vectors, then $\mathbf{u} \cdot \mathbf{v} = 0$.

Review Exercises

In Problems 1–8, use the matrices below to compute each expression.

$$A = \begin{bmatrix} 1 & 0 \\ 2 & 4 \\ -1 & 2 \end{bmatrix} \qquad B = \begin{bmatrix} 4 & -3 & 0 \\ 1 & 1 & -2 \end{bmatrix} \qquad C = \begin{bmatrix} 3 & -4 \\ 1 & 5 \\ 5 & -2 \end{bmatrix}$$

1. $A + C$ **2.** $A - C$ **3.** $6A$ **4.** $-4B$ **5.** AB **6.** BA **7.** CB **8.** BC

In Problems 9–14, find the inverse of each matrix, if there is one. If there is not an inverse, say that the matrix is singular.

9. $\begin{bmatrix} 4 & 6 \\ 1 & 3 \end{bmatrix}$ **10.** $\begin{bmatrix} -3 & 2 \\ 1 & -2 \end{bmatrix}$ **11.** $\begin{bmatrix} 1 & 3 & 3 \\ 1 & 2 & 1 \\ 1 & -1 & 2 \end{bmatrix}$ **12.** $\begin{bmatrix} 3 & 1 & 2 \\ 3 & 2 & -1 \\ 1 & 1 & 1 \end{bmatrix}$ **13.** $\begin{bmatrix} 4 & -8 \\ -1 & 2 \end{bmatrix}$ **14.** $\begin{bmatrix} -3 & 1 \\ -6 & 2 \end{bmatrix}$

In Problems 15–24, write the partial fraction decomposition of each rational expression.

15. $\dfrac{6}{x(x-4)}$ **16.** $\dfrac{x}{(x+2)(x-3)}$ **17.** $\dfrac{x-4}{x^2(x-1)}$ **18.** $\dfrac{2x-6}{(x-2)^2(x-1)}$

19. $\dfrac{x}{(x^2+9)(x+1)}$ **20.** $\dfrac{3x}{(x-2)(x^2+1)}$ **21.** $\dfrac{x^3}{(x^2+4)^2}$ **22.** $\dfrac{x^3+1}{(x^2+16)^2}$

23. $\dfrac{x^2}{(x^2+1)(x^2-1)}$ **24.** $\dfrac{4}{(x^2+4)(x^2-1)}$

*In Problems 25–28, the vector **v** is represented by the directed line segment $\overrightarrow{PQ}$. Write **v** in the form $a\mathbf{i} + b\mathbf{j}$ and find $\|\mathbf{v}\|$.*

25. $P = (1, -2);$ $Q = (3, -6)$ **26.** $P = (-3, 1);$ $Q = (4, -2)$
27. $P = (0, -2);$ $Q = (-1, 1)$ **28.** $P = (3, -4);$ $Q = (-2, 0)$

In Problems 29–36, use the vectors $\mathbf{v} = -2\mathbf{i} + \mathbf{j}$ and $\mathbf{w} = 4\mathbf{i} - 3\mathbf{j}$.

29. Find $4\mathbf{v} - 3\mathbf{w}$.   **30.** Find $-\mathbf{v} + 2\mathbf{w}$.   **31.** Find $\|\mathbf{v}\|$.   **32.** Find $\|\mathbf{v} + \mathbf{w}\|$.
33. Find $\|\mathbf{v}\| + \|\mathbf{w}\|$.   **34.** Find $\|2\mathbf{v}\| - 3\|\mathbf{w}\|$.
35. Find a unit vector having the same direction as **v**.
36. Find a unit vector having the opposite direction of **w**.

*In Problems 37–40, find the dot product $\mathbf{v} \cdot \mathbf{w}$ and the cosine of the angle between **v** and **w**.*

37. $\mathbf{v} = -2\mathbf{i} + \mathbf{j},$ $\mathbf{w} = 4\mathbf{i} - 3\mathbf{j}$ **38.** $\mathbf{v} = 3\mathbf{i} - \mathbf{j},$ $\mathbf{w} = \mathbf{i} + \mathbf{j}$
39. $\mathbf{v} = \mathbf{i} - 3\mathbf{j},$ $\mathbf{w} = -\mathbf{i} + \mathbf{j}$ **40.** $\mathbf{v} = \mathbf{i} + 4\mathbf{j},$ $\mathbf{w} = 3\mathbf{i} - 2\mathbf{j}$

41. Find the vector projection of $\mathbf{v} = 2\mathbf{i} + 3\mathbf{j}$ onto $\mathbf{w} = 3\mathbf{i} + \mathbf{j}$.
42. Find the vector projection of $\mathbf{v} = -\mathbf{i} + 2\mathbf{j}$ onto $\mathbf{w} = 3\mathbf{i} - \mathbf{j}$.
43. Find the angle between the vectors $\mathbf{v} = 3\mathbf{i} - 4\mathbf{j}$ and $\mathbf{w} = 12\mathbf{i} - 5\mathbf{j}$.
44. Find the angle between the vectors $\mathbf{v} = \mathbf{i} - \mathbf{j}$ and $\mathbf{w} = 2\mathbf{i} + \mathbf{j}$.

45. *Actual Speed and Direction of a Swimmer* A swimmer can maintain a constant speed of 5 miles per hour. If the swimmer heads directly across a river that has a current moving at the rate of 2 miles per hour, what is the actual speed of the swimmer? (See the figure.) If the river is 1 mile wide, how far downstream will the swimmer end up from the point directly across the river?

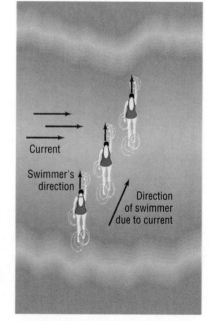

Current

Swimmer's direction

Direction of swimmer due to current

46. *Actual Speed and Direction of a Motorboat* A small motorboat is moving at a true speed of 11 miles per hour in a southerly direction. The current is known to be from the northeast at 3 miles per hour. What is the speed of the motorboat relative to the water? In what direction does the compass indicate that the boat is headed?

47. *Correct Direction for Crossing a Stream* A stream 1 kilometer wide has a constant current of 5 kilometers per hour. At what angle to the shore should a person head a boat that is capable of maintaining a constant speed of 15 kilometers per hour in order to reach a point directly opposite?

48. *Actual Speed and Direction of an Airplane* An airplane has an airspeed of 500 kilometers per hour in a northerly direction. The wind velocity is 60 kilometers per hour in a southeasterly direction. Find the actual speed and direction of the plane relative to the ground.

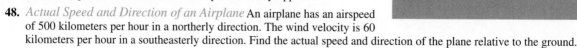

Appendix A ALGEBRA REVIEW

A.1 Polynomials and Rational Expressions
A.2 Radicals; Rational Exponents
A.3 Completing the Square; the Quadratic Formula

A.1

Polynomials and Rational Expressions

Algebra is sometimes described as a generalization of arithmetic in which letters are used to represent real numbers. We shall use the letters at the end of the alphabet, such as x, y, and z, to represent variables and the letters at the beginning of the alphabet, such as a, b, and c, to represent constants. Thus, in the expressions $3x + 5$ and $ax + b$, it is understood that x is a variable and that a and b are constants, even though the constants a and b are unspecified. As you will find out, the context usually makes the intended meaning clear.

Now we introduce some basic vocabulary.

Monomial

A **monomial** in one variable is the product of a constant times a variable raised to a nonnegative integer power. Thus, a monomial is of the form

$$ax^k$$

where a is a constant, x is a variable, and $k \geq 0$ is an integer. The constant a is called the **coefficient** of the monomial. If $a \neq 0$, then k is called the **degree** of the monomial.

Examples of monomials are as follows:

MONOMIAL	COEFFICIENT	DEGREE	
$6x^2$	6	2	
$-\sqrt{2}x^3$	$-\sqrt{2}$	3	
3	3	0	Since $3 = 3 \cdot 1 = 3x^0$
$-5x$	-5	1	Since $-5x = -5x^1$
x^4	1	4	Since $x^4 = 1 \cdot x^4$

Two monomials ax^k and bx^k with the same degree and the same variable are called **like terms.** Such monomials when added or subtracted can be combined into a single monomial by using the distributive property. For example,

$$2x^2 + 5x^2 = (2 + 5)x^2 = 7x^2 \quad \text{and} \quad 8x^3 - 5x^3 = (8 - 5)x^3 = 3x^3$$

The sum or difference of two monomials having different degrees is called a **binomial.** The sum or difference of three monomials with three different degrees is called a **trinomial.** For example,

$x^2 - 2$ is a binomial

$x^3 - 3x + 5$ is a trinomial

$2x^2 + 5x^2 + 2 = 7x^2 + 2$ is a binomial

Polynomial

A **polynomial** in one variable is an algebraic expression of the form

$$a_n x^n + a_{n-1} x^{n-1} + \cdots + a_1 x + a_0 \qquad (1)$$

where $a_n, a_{n-1}, \ldots, a_1, a_0$ are constants* called the **coefficients** of the polynomial, $n \geq 0$ is an integer, and x is a variable. If $a_n \neq 0$, it is called the **leading coefficient** and n is called the **degree** of the polynomial.

The monomials that make up a polynomial are called its **terms.** If all the coefficients are 0, the polynomial is called the **zero polynomial,** which has no degree.

Polynomials are usually written in **standard form,** beginning with the nonzero term of highest degree and continuing with terms in descending order according to degree. Examples of polynomials are the following:

POLYNOMIAL	COEFFICIENTS	DEGREE
$3x^2 - 5 = 3x^2 + 0 \cdot x + (-5)$	$3, 0, -5$	2
$8 - 2x + x^2 = 1 \cdot x^2 - 2x + 8$	$1, -2, 8$	2
$5x + \sqrt{2} = 5x^1 + \sqrt{2}$	$5, \sqrt{2}$	1
$3 = 3 \cdot 1 = 3 \cdot x^0$	3	0
0	0	No degree

Although we have been using x to represent the variable, letters such as y or z are also commonly used. Thus,

$3x^4 - x^2 + 2$ is a polynomial (in x) of degree 4.

$9y^3 - 2y^2 + y - 3$ is a polynomial (in y) of degree 3.

$z^5 + \pi$ is a polynomial (in z) of degree 5.

Algebraic expressions such as

$$\frac{1}{x} \quad \text{and} \quad \frac{x^2 + 1}{x + 5}$$

are not polynomials. The first is not a polynomial because $1/x = x^{-1}$ has an exponent that is not a nonnegative integer. Although the second expression is the quotient of two polynomials, the polynomial in the denominator has degree greater than 0, so the expression cannot be a polynomial.

---

*The notation a_n is read as "a sub n." The number n is called a **subscript** and should not be confused with an exponent. We use subscripts in order to distinguish one constant from another when a large or undetermined number of constants is required.

Adding and Subtracting Polynomials

Polynomials are added and subtracted by combining like terms.

E X A M P L E 1 *Finding Sums and Differences of Polynomials*

Find

(a) $(8x^3 - 2x^2 + 6x - 2) + (3x^4 - 2x^3 + x^2 + x)$

(b) $(3x^4 - 4x^3 + 6x^2 - 1) - (2x^4 - 8x^2 - 6x + 5)$

Solution (a) The idea here is to group the like terms and then combine them.

$$(8x^3 - 2x^2 + 6x - 2) + (3x^4 - 2x^3 + x^2 + x)$$
$$= 3x^4 + (8x^3 - 2x^3) + (-2x^2 + x^2) + (6x + x) - 2$$
$$= 3x^4 + 6x^3 - x^2 + 7x - 2$$

(b) $(3x^4 - 4x^3 + 6x^2 - 1) - (2x^4 - 8x^2 - 6x + 5)$
$$= 3x^4 - 4x^3 + 6x^2 - 1 - 2x^4 + 8x^2 + 6x - 5$$

Be sure to change the sign of each term in the second polynomial.

$$= (3x^4 - 2x^4) + (-4x^3) + (6x^2 + 8x^2) + 6x + (-1 - 5)$$

Group like terms.

$$= x^4 - 4x^3 + 14x^2 + 6x - 6$$ ∎

■ Now work Problem 1.

Multiplying Polynomials

Products of polynomials are found by repeated use of the distributive property and the laws of exponents.

E X A M P L E 2 *Finding the Product of Two Polynomials*

Find the product: $(2x + 5)(x^2 - x + 2)$

Solution *Horizontal Multiplication*

$$(2x + 5)(x^2 - x + 2) = 2x(x^2 - x + 2) + 5(x^2 - x + 2)$$

Distributive property

$$= 2x \cdot x^2 - 2x \cdot x + 2x \cdot 2 + 5 \cdot x^2 - 5 \cdot x + 5 \cdot 2$$

Distributive property

$$= 2x^3 - 2x^2 + 4x + 5x^2 - 5x + 10$$

Law of exponents

$$= 2x^3 + 3x^2 - x + 10$$

Combine like terms

Vertical Multiplication: The idea here is very much like multiplying a two-digit number by a three-digit number.

$$x^2 - x + 2$$
$$2x + 5$$

$$2x^3 - 2x^2 + 4x \quad \text{This line is } 2x(x^2 - x + 2).$$
$$(+) \quad 5x^2 - 5x + 10 \quad \text{This line is } 5(x^2 - x + 2).$$
$$2x^3 + 3x^2 - x + 10 \quad \text{The sum of the preceding two lines.} \quad \blacksquare$$

■   Now work Problem 5.

Certain products, which we call **special products,** occur frequently in algebra. In the list that follows, x, a, b, c, and d are real numbers:

Difference of Two Squares	
$$(x - a)(x + a) = x^2 - a^2$$	(2)

Squares of Binomials, or Perfect Squares	
$$(x + a)^2 = x^2 + 2ax + a^2$$	(3a)
$$(x - a)^2 = x^2 - 2ax + a^2$$	(3b)

Miscellaneous Trinomials	
$$(x + a)(x + b) = x^2 + (a + b)x + ab$$	(4a)
$$(ax + b)(cx + d) = acx^2 + (ad + bc)x + bd$$	(4b)

Cubes of Binomials, or Perfect Cubes	
$$(x + a)^3 = x^3 + 3ax^2 + 3a^2x + a^3$$	(5a)
$$(x - a)^3 = x^3 - 3ax^2 + 3a^2x - a^3$$	(5b)

Difference of Two Cubes	
$$(x - a)(x^2 + ax + a^2) = x^3 - a^3$$	(6)

Sum of Two Cubes	
$$(x + a)(x^2 - ax + a^2) = x^3 + a^3$$	(7)

The formulas in equations (2) through (7) are used often and their patterns should be committed to memory. But if you forget one or are unsure of its form, you should be able to derive it as needed.

A **polynomial in two variables** x and y is the sum of one or more monomials of the form ax^ny^m, where a is a constant called the **coefficient,** x and y are variables, and n and m are nonnegative integers. The **degree** of the monomial ax^ny^m is $n + m$. The **degree** of a polynomial in two variables x and y is the highest degree of all the monomials with nonzero coefficients that appear.

Polynomials in three variables x, y, and z and polynomials in more than three variables are defined in a similar way. Here are some examples:

$$3x^2 + 2x^3y + 5 \qquad\qquad \pi x^3 - y^2 \qquad\qquad x^4 + 4x^3y - xy^3 + y^4$$

Two variables,   Two variables,   Two variables,
degree is 4   degree is 3   degree is 4

$$x^2 + y^2 - z^2 + 4 \qquad\qquad x^3y^2z \qquad\qquad 5x^2 - 4y^2 + z^3y + 2w^2x$$

Three variables,   Three variables,   Four variables,
degree is 2   degree is 6   degree is 4

Adding and multiplying polynomials in two or more variables is handled in the same way as for polynomials in one variable.

■ Now work Problem 3.

Dividing Polynomials

The procedure for dividing two polynomials is similar to the procedure for dividing two integers. This process should be familiar to you, but we review it briefly below.

EXAMPLE 3   *Dividing Two Integers*

Divide 842 by 15.

Solution

$$
\begin{array}{r}
56 \quad \leftarrow \text{Quotient} \\
\text{Divisor} \rightarrow \quad 15\overline{)842} \quad \leftarrow \text{Dividend} \\
75 \quad \leftarrow 5 \cdot 15 \quad (\text{Subtract}) \\
\hline
92 \\
90 \quad \leftarrow 6 \cdot 15 \quad (\text{Subtract}) \\
\hline
2 \quad \leftarrow \text{Remainder}
\end{array}
$$

Thus, $\frac{842}{15} = 56 + \frac{2}{15}$.   ■

In the long division process detailed in Example 3, the number 15 is called the **divisor,** the number 842 is called the **dividend,** the number 56 is called the **quotient,** and the number 2 is called the **remainder.**

To check the answer obtained in a division problem, multiply the quotient by the divisor and add the remainder. The answer should be the dividend.

(Quotient)(Divisor) + Remainder = Dividend

For example, we can check the results obtained in Example 3 as follows:

$$(56)(15) + 2 = 840 + 2 = 842$$

To divide two polynomials, we first must write each polynomial in standard form. The process then follows a pattern similar to that of Example 3. The next example illustrates the procedure.

EXAMPLE 4   *Dividing Two Polynomials*

Find the quotient and the remainder when

$$3x^3 + 4x^2 + x + 7 \quad \text{is divided by} \quad x^2 + 1$$

Solution   Each polynomial is in standard form. The dividend is $3x^3 + 4x^2 + x + 7$, and the divisor is $x^2 + 1$.

STEP 1:   Divide the leading term of the dividend, $3x^3$, by the leading term of the divisor, x^2. Enter the result, $3x$, over the term $3x^3$, as follows:

$$
\begin{array}{r}
3x \\
x^2 + 1\overline{)3x^3 + 4x^2 + \ x + 7}
\end{array}
$$

STEP 2:   Multiply $3x$ by $x^2 + 1$ and enter the result below the dividend.

$$
\begin{array}{r}
3x \\
x^2 + 1\overline{)3x^3 + 4x^2 + \ x + 7} \\
\underline{3x^3 \qquad\quad + 3x} \qquad \leftarrow 3x \cdot (x^2 + 1) = 3x^3 + 3x
\end{array}
$$

$\uparrow$
Notice that we align the $3x$ term under the x to make the next step easier.

STEP 3:   Subtract and bring down the remaining terms.

$$
\begin{array}{r}
3x \\
x^2 + 1\overline{)3x^3 + 4x^2 + \ x + 7} \\
\underline{3x^3 \qquad\quad + 3x} \qquad \leftarrow \text{Subtract.} \\
4x^2 - 2x + 7 \qquad \leftarrow \text{Bring down the } 4x^2 \text{ and the } 7.
\end{array}
$$

STEP 4:   Repeat Steps 1 through 3 using $4x^2 - 2x + 7$ as the dividend.

$$
\begin{array}{r}
3x + 4 \\
x^2 + 1\overline{)3x^3 + 4x^2 + \ x + 7} \\
\underline{3x^3 \qquad\quad + 3x} \\
4x^2 - 2x + 7 \qquad \leftarrow \text{Divide } 4x^2 \text{ by } x^2 \text{ to get 4.} \\
\underline{4x^2 \qquad\quad + 4} \qquad \leftarrow \text{Multiply } x^2 + 1 \text{ by 4; subtract.} \\
-2x + 3
\end{array}
$$

Since x^2 does not divide $-2x$ evenly (that is, the result is not a monomial), the process ends. The quotient is $3x + 4$, and the remainder is $-2x + 3$.

Check: (Quotient)(Divisor) + Remainder
$$
\begin{aligned}
&= (3x + 4)(x^2 + 1) + (-2x + 3) \\
&= 3x^3 + 4x^2 + 3x + 4 + (-2x + 3) \\
&= 3x^3 + 4x^2 + x + 7 = \text{Dividend}
\end{aligned}
$$
Thus,

$$
\frac{3x^3 + 4x^2 + x + 7}{x^2 + 1} = 3x + 4 + \frac{-2x + 3}{x^2 + 1}
$$ ■

The process for dividing two polynomials leads to the following result:

Theorem   The remainder after dividing two polynomials is either the zero polynomial or a polynomial of degree less than the degree of the divisor. ■

■ Now work Problem 13.

Factoring

Consider the following product:

$$
(2x + 3)(x - 4) = 2x^2 - 5x - 12
$$

The two polynomials on the left are called **factors** of the polynomial on the right. Expressing a given polynomial as a product of other polynomials, that is, finding the factors of a polynomial, is called **factoring.**

We shall restrict our discussion here to factoring polynomials in one variable into products of polynomials in one variable, where all coefficients are integers. We call this **factoring over the integers.** There will be times, though, when we will want to **factor over the rational numbers** and even **factor over the real numbers.** Factoring over the rational numbers means to write a given polynomial whose coefficients are rational numbers as a product of polynomials whose coefficients are also rational numbers. Factoring over the real numbers means to write a given polynomial whose coefficients are real numbers as a product of polynomials whose coefficients are also real numbers. Unless specified otherwise, we will be factoring over the integers.

Any polynomial can be written as the product of 1 times itself or as -1 times its additive inverse. If a polynomial cannot be written as the product of two other polynomials (excluding 1 and -1), then the polynomial is said to be **prime.** When a polynomial has been written as a product consisting only of prime factors, then it is said to be **factored completely.** Examples of prime polynomials are

$$2, \quad 3, \quad 5, \quad x, \quad x + 1, \quad x - 1, \quad 3x + 4$$

The first factor to look for in a factoring problem is a common monomial factor present in each term of the polynomial. If one is present, use the distributive property to factor it out. For example,

POLYNOMIAL	COMMON MONOMIAL FACTOR	REMAINING FACTOR	FACTORED FORM
$2x + 4$	2	$x + 2$	$2x + 4 = 2(x + 2)$
$3x - 6$	3	$x - 2$	$3x - 6 = 3(x - 2)$
$2x^2 - 4x + 8$	2	$x^2 - 2x + 4$	$2x^2 - 4x + 8 = 2(x^2 - 2x + 4)$
$8x - 12$	4	$2x - 3$	$8x - 12 = 4(2x - 3)$
$x^2 + x$	x	$x + 1$	$x^2 + x = x(x + 1)$
$x^3 - 3x^2$	x^2	$x - 3$	$x^3 - 3x^2 = x^2(x - 3)$
$6x^2 + 9x$	$3x$	$2x + 3$	$6x^2 + 9x = 3x(2x + 3)$

The list of special products (2) through (7) given on page 804 provides a list of factoring formulas when the equations are read from right to left. For example, equation (2) states that if the polynomial is the difference of two squares, $x^2 - a^2$, it can be factored into $(x - a)(x + a)$. The following example illustrates several factoring techniques.

EXAMPLE 5

Factoring Polynomials

Factor completely each polynomial.

(a) $x^4 - 16$ (b) $x^3 - 1$ (c) $9x^2 - 6x + 1$ (d) $x^2 + 4x - 12$

(e) $3x^2 + 10x - 8$ (f) $x^3 - 4x^2 + 2x - 8$

Solution   (a) $x^4 - 16 = (x^2 - 4)(x^2 + 4) = (x - 2)(x + 2)(x^2 + 4)$

         ↑   ↑

   Difference of squares   Difference of squares

(b) $x^3 - 1 = (x - 1)(x^2 + x + 1)$

     ↑

 Difference of cubes

(c) $9x^2 - 6x + 1 = (3x - 1)^2$

           ↑
Perfect square

(d) $x^2 + 4x - 12 = (x + 6)(x - 2)$

           ↑
6 and -2 are factors of -12, and the sum of 6 and -2 is 4

(e) $3x^2 + 10x - 8 = (3x - 2)(x + 4)$

(f) $x^3 - 4x^2 + 2x - 8 = (x^3 - 4x^2) + (2x - 8)$

           ↑
Regroup

$$= x^2(x - 4) + 2(x - 4) = (x^2 + 2)(x - 4)$$

           ↑
Distributive property   ■

The technique used in Example 5(f) is called **factoring by grouping.**

■   Now work Problem 21.

Rational Expressions

If we form the quotient of two polynomials, the result is called a **rational expression.** Some examples of rational expressions are

(a) $\dfrac{x^3 + 1}{x}$ (b) $\dfrac{3x^2 + x - 2}{x^2 + 5}$ (c) $\dfrac{x}{x^2 - 1}$ (d) $\dfrac{xy^2}{(x - y)^2}$

Expressions (a), (b), and (c) are rational expressions in one variable, x, whereas (d) is a rational expression in two variables, x and y.

    Rational expressions are described in the same manner as rational numbers. Thus, in expression (a), the polynomial $x^3 + 1$ is called the **numerator,** and x is called the **denominator.** When the numerator and denominator of a rational expression contain no common factors (except 1 and -1), we say that the rational expression is **reduced to lowest terms,** or **simplified.**

    A rational expression is reduced to lowest terms by completely factoring the numerator and the denominator and canceling any common factors by using the cancellation property,

$$\frac{ac}{bc} = \frac{a}{b}, \qquad b \neq 0, c \neq 0$$

    We shall follow the common practice of using a slash mark to indicate cancellation. For example,

$$\frac{x^2 - 1}{x^2 - 2x - 3} = \frac{(x - 1)(x + 1)}{(x - 3)(x + 1)} = \frac{x - 1}{x - 3}$$

EXAMPLE 6   *Simplifying Rational Expressions*

Reduce each rational expression to lowest terms.

(a) $\dfrac{x^2 + 4x + 4}{x^2 + 3x + 2}$ (b) $\dfrac{x^3 - 8}{x^3 - 2x^2}$ (c) $\dfrac{8 - 2x}{x^2 - x - 12}$

Solution (a) $\dfrac{x^2 + 4x + 4}{x^2 + 3x + 2} = \dfrac{(x+2)(x+2)}{(x+2)(x+1)} = \dfrac{x+2}{x+1}, \quad x \neq -2, -1$

(b) $\dfrac{x^3 - 8}{x^3 - 2x^2} = \dfrac{(x-2)(x^2 + 2x + 4)}{x^2(x-2)} = \dfrac{x^2 + 2x + 4}{x^2}, \quad x \neq 0, 2$

(c) $\dfrac{8 - 2x}{x^2 - x - 12} = \dfrac{2(4 - x)}{(x - 4)(x + 3)} = \dfrac{2(-1)(x-4)}{(x-4)(x + 3)} = \dfrac{-2}{x + 3}, \quad x \neq -3, 4$ ∎

The rules for multiplying and dividing rational expressions are the same as the rules for multiplying and dividing rational numbers:

$$\frac{a}{b} \cdot \frac{c}{d} = \frac{ac}{bd}, \qquad \text{if } b \neq 0, d \neq 0 \tag{8}$$

$$\frac{\dfrac{a}{b}}{\dfrac{c}{d}} = \frac{a}{b} \cdot \frac{d}{c} = \frac{ad}{bc}, \qquad \text{if } b \neq 0, c \neq 0, d \neq 0 \tag{9}$$

In using equations (8) and (9) with rational expressions, be sure first to factor each polynomial completely so that common factors can be canceled. We shall follow the practice of leaving our answers in factored form.

E X A M P L E   7   *Finding Products and Quotients of Rational Expressions*

Perform the indicated operation and simplify the result. Leave your answer in factored form.

(a) $\dfrac{x^2 - 2x + 1}{x^3 + x} \cdot \dfrac{4x^2 + 4}{x^2 + x - 2}$ (b) $\dfrac{\dfrac{x + 3}{x^2 - 4}}{\dfrac{x^2 - x - 12}{x^3 - 8}}$

Solution (a) $\dfrac{x^2 - 2x + 1}{x^3 + x} \cdot \dfrac{4x^2 + 4}{x^2 + x - 2} = \dfrac{(x - 1)^2}{x(x^2 + 1)} \cdot \dfrac{4(x^2 + 1)}{(x + 2)(x - 1)}$

$= \dfrac{(x - 1)^2(4)(x^2 + 1)}{x(x^2 + 1)(x + 2)(x - 1)} = \dfrac{4(x - 1)}{x(x + 2)},$

$x \neq -2, 0, 1$

(b) $\dfrac{\dfrac{x + 3}{x^2 - 4}}{\dfrac{x^2 - x - 12}{x^3 - 8}} = \dfrac{x + 3}{x^2 - 4} \cdot \dfrac{x^3 - 8}{x^2 - x - 12}$

$= \dfrac{x + 3}{(x - 2)(x + 2)} \cdot \dfrac{(x - 2)(x^2 + 2x + 4)}{(x - 4)(x + 3)}$

$= \dfrac{(x + 3)(x - 2)(x^2 + 2x + 4)}{(x - 2)(x + 2)(x - 4)(x + 3)} = \dfrac{x^2 + 2x + 4}{(x + 2)(x - 4)},$

$x \neq -3, -2, 2, 4$ ∎

Note: Slanting the cancellation marks in different directions for different factors, as in Example 7, is a good practice to follow, since it will help in checking for errors.

■   Now work Problem 31.

If the denominators of two rational expressions to be added (or subtracted) are equal, we add (or subtract) the numerators and keep the common denominator. That is, if a/b and c/b are two rational expressions, then

$$\frac{a}{b} + \frac{c}{b} = \frac{a+c}{b} \qquad \frac{a}{b} - \frac{c}{b} = \frac{a-c}{b}, \qquad \text{if } b \neq 0 \qquad (10)$$

E X A M P L E   8   *Finding the Sum of Two Rational Expressions*

Perform the indicated operation and simplify the result. Leave your answer in factored form.

$$\frac{2x^2 - 4}{2x + 5} + \frac{x + 3}{2x + 5}, \qquad x \neq -\frac{5}{2}$$

Solution

$$\frac{2x^2 - 4}{2x + 5} + \frac{x + 3}{2x + 5} = \frac{(2x^2 - 4) + (x + 3)}{2x + 5}$$

$$= \frac{2x^2 + x - 1}{2x + 5} = \frac{(2x - 1)(x + 1)}{2x + 5} \qquad ■$$

If the denominators of two rational expressions to be added or subtracted are not equal, we can use the general formulas for adding and subtracting quotients:

$$\frac{a}{b} + \frac{c}{d} = \frac{a \cdot d}{b \cdot d} + \frac{b \cdot c}{b \cdot d} = \frac{ad + bc}{bd}, \qquad \text{if } b \neq 0, d \neq 0$$

$$\frac{a}{b} - \frac{c}{d} = \frac{a \cdot d}{b \cdot d} - \frac{b \cdot c}{b \cdot d} = \frac{ad + bc}{bd}, \qquad \text{if } b \neq 0, d \neq 0 \qquad (11)$$

E X A M P L E   9   *Finding the Difference of Two Rational Expressions*

Perform the indicated operation and simplify the result. Leave your answer in factored form.

$$\frac{x^2}{x^2 - 4} - \frac{1}{x}, \qquad x \neq -2, 0, 2$$

Solution

$$\frac{x^2}{x^2 - 4} - \frac{1}{x} = \frac{x^2(x) - (x^2 - 4)(1)}{(x^2 - 4)(x)} = \frac{x^3 - x^2 + 4}{(x - 2)(x + 2)(x)} \qquad ■$$

Least Common Multiple (LCM)

If the denominators of two rational expressions to be added (or subtracted) have common factors, we usually do not use the general rules given by equation (11), since, in doing so, we make the problem more complicated than it needs to be. Instead, just as with fractions, we apply the **least common multiple (LCM) method** by using the polynomial of least degree that contains each denominator polynomial as a factor. Then we rewrite each rational expression using the LCM as the common denominator and use equation (10) to do the addition (or subtraction).

To find the least common multiple of two or more polynomials, first factor completely each polynomial. The LCM is the product of the different prime factors of each polynomial, each factor appearing the greatest number of times it occurs in each polynomial. The next example will give you the idea.

E X A M P L E   1 0   *Finding the Least Common Multiple*

Find the least common multiple of the following pair of polynomials:

$$x(x - 1)^2(x + 1) \quad \text{and} \quad 4(x - 1)(x + 1)^3$$

Solution   The polynomials are already factored completely as

$$x(x - 1)^2(x + 1) \quad \text{and} \quad 4(x - 1)(x + 1)^3$$

Start by writing the factors of the left-hand polynomial. (Alternatively, you could start with the one on the right.)

$$x(x - 1)^2(x + 1)$$

Now look at the right-hand polynomial. Its first factor, 4, does not appear in our list, so we insert it:

$$4x(x - 1)^2(x + 1)$$

The next factor, $x - 1$, is already in our list, so no change is necessary. The final factor is $(x + 1)^3$. Since our list has $x + 1$ to the first power only, we replace $x + 1$ in the list by $(x + 1)^3$. The LCM is

$$4x(x - 1)^2(x + 1)^3$$

Notice that the LCM is, in fact, the polynomial of least degree that contains $x(x - 1)^2(x + 1)$ and $4(x - 1)(x + 1)^3$ as factors.   ∎

The next example illustrates how the LCM is used for adding and subtracting rational expressions.

E X A M P L E   1 1   *Using the LCM to Add Rational Expressions*

Perform the indicated operation and simplify the result. Leave your answer in factored form.

$$\frac{x}{x^2 + 3x + 2} + \frac{2x - 3}{x^2 - 1}, \quad x \neq -2, -1, 1$$

Solution   First, we find the LCM of the denominators:

$$x^2 + 3x + 2 = (x + 2)(x + 1)$$
$$x^2 - 1 = (x - 1)(x + 1)$$

The LCM is $(x + 2)(x + 1)(x - 1)$. Next, we rewrite each rational expression using the LCM as the denominator:

$$\frac{x}{x^2 + 3x + 2} = \frac{x}{(x + 2)(x + 1)} = \frac{x(x - 1)}{(x + 2)(x + 1)(x - 1)}$$

Multiply numerator and denominator by $x - 1$ to get the LCM in the denominator.

$$\frac{2x-3}{x^2-1} = \frac{2x-3}{(x-1)(x+1)} \underset{\uparrow}{=} \frac{(2x-3)(x+2)}{(x-1)(x+1)(x+2)}$$

Multiply numerator and denominator by $x+2$ to get the LCM in the denominator.

Now we can add by using equation (10).

$$\frac{x}{x^2+3x+2} + \frac{2x-3}{x^2-1} = \frac{x(x-1)}{(x+2)(x+1)(x-1)} + \frac{(2x-3)(x+2)}{(x+2)(x+1)(x-1)}$$

$$= \frac{(x^2-x)+(2x^2+x-6)}{(x+2)(x+1)(x-1)}$$

$$= \frac{3x^2-6}{(x+2)(x+1)(x-1)} = \frac{3(x^2-2)}{(x+2)(x+1)(x-1)}$$

∎

If we had not used the LCM technique to add the quotients in Example 11, but decided instead to use the general rule of equation (11), we would have obtained a more complicated expression, as follows:

$$\frac{x}{x^2+3x+2} + \frac{2x-3}{x^2-1} = \frac{x(x^2-1)+(x^2+3x+2)(2x-3)}{(x^2+3x+2)(x^2-1)}$$

$$= \frac{3x^3+3x^2-6x-6}{(x^2+3x+2)(x^2-1)} = \frac{3(x^3+x^2-2x-2)}{(x^2+3x+2)(x^2-1)}$$

Now we are faced with a more complicated problem of expressing this quotient in lowest terms. It is always best to first look for common factors in the denominators of expressions to be added or subtracted and to use the LCM if any common factors are found.

■ Now work Problem 35.

Mixed Quotients

When sums and/or differences of rational expressions appear as the numerator and/or denominator of a quotient, the quotient is called a **mixed quotient.** For example,

$$\frac{1+\dfrac{1}{x}}{1-\dfrac{1}{x}} \quad \text{and} \quad \frac{\dfrac{x^2}{x^2-4}-3}{\dfrac{x-3}{x+2}-1}$$

are mixed quotients. To **simplify** a mixed quotient means to write it as a rational expression reduced to lowest terms. This can be accomplished by treating the numerator and denominator of the mixed quotient separately, performing whatever operations are indicated and simplifying the results. Follow this by simplifying the resulting rational expression.

E X A M P L E   1 2   *Simplifying Mixed Quotients*

Simplify the mixed quotient: $\dfrac{1+\dfrac{1}{x}}{1-\dfrac{1}{x}}$

Solution

$$\frac{1+\dfrac{1}{x}}{1-\dfrac{1}{x}} = \frac{\dfrac{x}{x}+\dfrac{1}{x}}{\dfrac{x}{x}-\dfrac{1}{x}} = \frac{\dfrac{x+1}{x}}{\dfrac{x-1}{x}} = \frac{x+1}{x} \cdot \frac{x}{x-1}$$

$$= \frac{(x+1)x}{x(x-1)} = \frac{x+1}{x-1}$$

■

■ Now work Problem 39.

A.1

Exercise A.1

In Problems 1–10, perform the indicated operations. Express each answer as a polynomial.

1. $(10x^5 - 8x^2) + (3x^3 - 2x^2 + 6)$

2. $3(x^2 - 3x + 1) + 2(3x^2 + x - 4)$

3. $(x + a)^2 - x^2$

4. $(x - a)^2 - x^2$

5. $(x + 8)(2x + 1)$

6. $(2x - 1)(x + 2)$

7. $(x^2 + x - 1)(x^2 - x + 1)$

8. $(x^2 + 2x + 1)(x^2 - 3x + 4)$

9. $(x + 1)^3 - (x - 1)^3$

10. $(x + 1)^3 - (x + 2)^3$

In Problems 11–20, find the quotient and the remainder. Check your work by verifying that

$$(\text{Quotient})(\text{Divisor}) + \text{Remainder} = \text{Dividend}$$

11. $4x^3 - 3x^2 + x + 1$ divided by x

12. $3x^3 - x^2 + x - 2$ divided by x

13. $4x^3 - 3x^2 + x + 1$ divided by $x + 2$

14. $3x^3 - x^2 + x - 2$ divided by $x + 2$

15. $4x^3 - 3x^2 + x + 1$ divided by $x - 4$

16. $3x^3 - x^2 + x - 2$ divided by $x - 4$

17. $4x^3 - 3x^2 + x + 1$ divided by x^2

18. $3x^3 - x^2 + x - 2$ divided by x^2

19. $4x^3 - 3x^2 + x + 1$ divided by $x^2 + 2$

20. $3x^3 - x^2 + x - 2$ divided by $x^2 + 2$

In Problems 21–30, factor completely each polynomial. If the polynomial cannot be factored, say it is prime.

21. $x^2 - 2x - 15$

22. $x^2 - 6x - 14$

23. $ax^2 - 4a^2x - 45a^3$

24. $bx^2 + 14b^2x + 45b^3$

25. $x^3 - 27$

26. $x^3 + 27$

27. $3x^2 + 4x + 1$

28. $4x^2 + 3x - 1$

29. $x^7 - x^5$

30. $x^8 - x^5$

In Problems 31–34, perform the indicated operation and simplify the result. Leave your answer in factored form.

31. $\dfrac{3x - 6}{5x} \cdot \dfrac{x^2 - x - 6}{x^2 - 4}$

32. $\dfrac{9x - 25}{2x - 2} \cdot \dfrac{1 - x^2}{6x - 10}$

33. $\dfrac{4x^2 - 1}{x^2 - 16} \cdot \dfrac{x^2 - 4x}{2x + 1}$

34. $\dfrac{12}{x^2 - x} \cdot \dfrac{x^2 - 1}{4x - 2}$

In Problems 35–42, perform the indicated operations and simplify the result. Leave your answer in factored form.

35. $\dfrac{x}{x^2 - 7x + 6} - \dfrac{x}{x^2 - 2x - 24}$

36. $\dfrac{x}{x - 3} - \dfrac{x + 1}{x^2 + 5x - 24}$

37. $\dfrac{4}{x^2 - 4} - \dfrac{2}{x^2 + x - 6}$

38. $\dfrac{3}{x - 1} - \dfrac{x - 4}{x^2 - 2x + 1}$

39. $\dfrac{x - \dfrac{1}{x}}{x + \dfrac{1}{x}}$

40. $\dfrac{1 - \dfrac{x}{x + 1}}{2 - \dfrac{x - 1}{x}}$

41. $\dfrac{3 - \dfrac{x^2}{x + 1}}{1 + \dfrac{x}{x^2 - 1}}$

42. $\dfrac{3x - \dfrac{3}{x^2}}{\dfrac{1}{(x - 1)^2} - 1}$

A.2

Radicals; Rational Exponents

Square Roots

A real number is squared when it is raised to the power 2. The inverse of squaring is finding a **square root.** For example, since $6^2 = 36$ and $(-6)^2 = 36$, the numbers 6 and -6 are square roots of 36.

The symbol $\sqrt{}$, called a **radical sign,** is used to denote the **principal,** or nonnegative, square root. Thus, $\sqrt{36} = 6$.

Principal Square Root

In general, if a is a nonnegative real number, the nonnegative number b such that $b^2 = a$ is the **principal square root** of a and is denoted by $b = \sqrt{a}$.

The following comments are noteworthy:

1. Negative numbers do not have square roots (in the real number system), because the square of any real number is *nonnegative*. For example, $\sqrt{-4}$ is not a real number, because there is no real number whose square is -4.
2. The principal square root of 0 is 0, since $0^2 = 0$. That is, $\sqrt{0} = 0$.
3. The principal square root of a positive number is positive.
4. If $c \geq 0$, then $(\sqrt{c})^2 = c$. For example, $(\sqrt{2})^2 = 2$ and $(\sqrt{3})^2 = 3$.

E X A M P L E   1

Evaluating Square Roots

(a) $\sqrt{64} = 8$ (b) $\sqrt{\frac{1}{16}} = \frac{1}{4}$ (c) $(\sqrt{1.4})^2 = 1.4$ ■

Examples 1(a) and (b) are examples of **perfect square roots.** Thus, 64 is a **perfect square,** since $64 = 8^2$; and $\frac{1}{16}$ is a perfect square, since $\frac{1}{16} = (\frac{1}{4})^2$.

In general, we have

$$\sqrt{a^2} = |a| \tag{1}$$

Notice the need for the absolute value in equation (1). Since $a^2 \geq 0$, the principal square root of a^2 is defined whether $a > 0$ or $a < 0$. However, since the principal square root is nonnegative, we need the absolute value to ensure the nonnegative result.

E X A M P L E   2

Using Equation (1)

(a) $\sqrt{(2.3)^2} = |2.3| = 2.3$ (b) $\sqrt{(-2.3)^2} = |-2.3| = 2.3$ (c) $\sqrt{x^2} = |x|$

■

*n*th Roots

The **principal *n*th root of a real number *a*,** symbolized by $\sqrt[n]{a}$, is defined as follows:

Principal *n*th Root

$\sqrt[n]{a} = b$ means $a = b^n$, where $a \geq 0$ and $b \geq 0$ if n is even and a, b are any real numbers if n is odd

Notice that if a is negative and n is even then $\sqrt[n]{a}$ is not defined. When it is defined, the principal *n*th root of a number is unique.

The symbol $\sqrt[n]{a}$ for the principal *n*th root of a is sometimes called a **radical;** the integer n is called the **index,** and a is called the **radicand.** If the index of a radical is 2, we call $\sqrt[2]{a}$ the **square root** of a and omit the index 2 by simply writing $\sqrt{a}$. If the index is 3, we call $\sqrt[3]{a}$ the **cube root** of a.

E X A M P L E 3

Evaluating Principal nth Roots

$$\sqrt[3]{8} = 2 \qquad \sqrt[6]{64} = 2 \qquad \sqrt[3]{-64} = -4 \qquad \sqrt[4]{\frac{1}{16}} = \frac{1}{2}$$

because

$$8 = 2^3 \qquad 64 = 2^6 \qquad -64 = (-4)^3 \qquad \frac{1}{16} = \left(\frac{1}{2}\right)^4 \qquad \blacksquare$$

These are examples of **perfect roots.** Thus, 8 and -64 are perfect cubes, since $8 = 2^3$ and $-64 = (-4)^3$; 2 is a perfect sixth root of 64, since $64 = 2^6$; and $\frac{1}{2}$ is a perfect fourth root of $\frac{1}{16}$, since $\frac{1}{16} = \left(\frac{1}{2}\right)^4$.

In general, if $n \geq 2$ is a positive integer and a is a real number, we have

---

$$\sqrt[n]{a^n} = a, \qquad \text{if } n \text{ is odd} \qquad (1a)$$
$$\sqrt[n]{a^n} = |a|, \qquad \text{if } n \text{ is even} \qquad (1b)$$

Notice the need for the absolute value in equation (1b). If n is even, then a^n is positive whether $a > 0$ or $a < 0$. But if n is even, the principal *n*th root must be nonnegative. Hence, the reason for using the absolute value—it gives a nonnegative result.

E X A M P L E 4

Using Equations (1a) and (1b)

(a) $\sqrt[3]{4^3} = 4$ (b) $\sqrt[5]{(-3)^5} = -3$ (c) $\sqrt[4]{2^4} = 2$

(d) $\sqrt[4]{(-3)^4} = |-3| = 3$ (e) $\sqrt{x^2} = |x|$ $\blacksquare$

Properties of Radicals

Let $n \geq 2$ and $m \geq 2$ denote positive integers, and let a and b represent real numbers. Assuming that all radicals are defined, we have the following properties:

$$\sqrt[n]{ab} = \sqrt[n]{a}\sqrt[n]{b} \tag{2a}$$

$$\sqrt[n]{\frac{a}{b}} = \frac{\sqrt[n]{a}}{\sqrt[n]{b}} \tag{2b}$$

$$\sqrt[n]{a^m} = (\sqrt[n]{a})^m \tag{2c}$$

$$\sqrt[m]{\sqrt[n]{a}} = \sqrt[mn]{a} \tag{2d}$$

When used in reference to radicals, the direction to "simplify" will mean to remove from the radicals any perfect roots that occur as factors. Let's look at some examples of how the rules listed in the box are applied to simplify radicals.

EXAMPLE 5

Simplifying Radicals

Simplify each expression. Assume that all variables are positive when they appear.

(a) $\sqrt{32}$ (b) $\sqrt[3]{8x^4}$ (c) $\sqrt{\sqrt[3]{x^7}}$ (d) $\sqrt[3]{\frac{8x^5}{27y^2}}$ (e) $\frac{\sqrt{x^5y}}{\sqrt{x^3y^3}}$

Solution   (a) $\sqrt{32} = \sqrt{16 \cdot 2} = \sqrt{16}\sqrt{2} = 4\sqrt{2}$

(2a)
16 is a perfect square.

(b) $\sqrt[3]{8x^4} = \sqrt[3]{8x^3 \cdot x} = \sqrt[3]{(2x)^3 \cdot x} = \sqrt[3]{(2x)^3}\sqrt[3]{x} = 2x\sqrt[3]{x}$

Factor out   (2a)   (1a)
perfect cube.

(c) $\sqrt{\sqrt[3]{x^7}} = \sqrt[6]{x^7} = \sqrt[6]{x^6 \cdot x} = \sqrt[6]{x^6} \cdot \sqrt[6]{x} = |x|\sqrt[6]{x}$

(2d)   (1b)

(d) $\sqrt[3]{\frac{8x^5}{27y^2}} = \sqrt[3]{\frac{2^3x^3x^2}{3^3y^2}} = \sqrt[3]{\left(\frac{2x}{3}\right)^3 \cdot \frac{x^2}{y^2}} = \sqrt[3]{\left(\frac{2x}{3}\right)^3} \cdot \sqrt[3]{\frac{x^2}{y^2}} = \frac{2x}{3}\sqrt[3]{\frac{x^2}{y^2}}$

(e) $\frac{\sqrt{x^5y}}{\sqrt{x^3y^3}} = \sqrt{\frac{x^5y}{x^3y^3}} = \sqrt{\frac{x^2}{y^2}} = \sqrt{\left(\frac{x}{y}\right)^2} = \left|\frac{x}{y}\right|$

■ Now work Problems 1 and 15.

Rationalizing

When radicals occur in quotients, it has become common practice to rewrite the quotient so that the denominator contains no radicals. This process is referred to as **rationalizing the denominator.**

The idea is to find an appropriate expression so that, when it is multiplied by the radical in the denominator, the new denominator that results contains no radicals. For example,

IF RADICAL IS	MULTIPLY BY	TO GET PRODUCT FREE OF RADICALS
$\sqrt{3}$	$\sqrt{3}$	$\sqrt{9} = 3$
$\sqrt[3]{4}$	$\sqrt[3]{2}$	$\sqrt[3]{8} = 2$
$\sqrt{3} + 1$	$\sqrt{3} - 1$	$(\sqrt{3})^2 - 1^2 = 3 - 1 = 2$
$\sqrt{2} - 3$	$\sqrt{2} + 3$	$(\sqrt{2})^2 - 3^2 = 2 - 9 = -7$
$\sqrt{5} - \sqrt{3}$	$\sqrt{5} + \sqrt{3}$	$(\sqrt{5})^2 - (\sqrt{3})^2 = 5 - 3 = 2$

You are correct if you observed in this list that, after the second type of radical, the special product for differences of squares is the basis for determining by what to multiply.

E X A M P L E 6 *Rationalizing Denominators*

Rationalize the denominator of each expression.

(a) $\dfrac{4}{\sqrt{2}}$ (b) $\dfrac{\sqrt{3}}{\sqrt[3]{2}}$ (c) $\dfrac{\sqrt{x}-2}{\sqrt{x}+2}, \quad x \geq 0$

Solution (a) $\dfrac{4}{\sqrt{2}} = \dfrac{4}{\sqrt{2}} \cdot \dfrac{\sqrt{2}}{\sqrt{2}} = \dfrac{4\sqrt{2}}{(\sqrt{2})^2} = \dfrac{4\sqrt{2}}{2} = 2\sqrt{2}$

Multiply by $\dfrac{\sqrt{2}}{\sqrt{2}}$.

(b) $\dfrac{\sqrt{3}}{\sqrt[3]{2}} = \dfrac{\sqrt{3}}{\sqrt[3]{2}} \cdot \dfrac{\sqrt[3]{4}}{\sqrt[3]{4}} = \dfrac{\sqrt{3}\sqrt[3]{4}}{\sqrt[3]{8}} = \dfrac{\sqrt{3}\sqrt[3]{4}}{2}$

Multiply by $\dfrac{\sqrt[3]{4}}{\sqrt[3]{4}}$.

(c) $\dfrac{\sqrt{x}-2}{\sqrt{x}+2} = \dfrac{\sqrt{x}-2}{\sqrt{x}+2} \cdot \dfrac{\sqrt{x}-2}{\sqrt{x}-2} = \dfrac{(\sqrt{x}-2)^2}{(\sqrt{x})^2 - 2^2}$

$= \dfrac{(\sqrt{x})^2 - 4\sqrt{x} + 4}{x - 4} = \dfrac{x - 4\sqrt{x} + 4}{x - 4}$ ∎

In calculus, sometimes the numerator must be rationalized.

E X A M P L E 7 *Rationalizing Numerators*

Rationalize the numerator: $\dfrac{\sqrt{x}-2}{\sqrt{x}+1}, x \geq 0$

Solution We multiply by $\dfrac{\sqrt{x}+2}{\sqrt{x}+2}$:

$\dfrac{\sqrt{x}-2}{\sqrt{x}+1} = \dfrac{\sqrt{x}-2}{\sqrt{x}+1} \cdot \dfrac{\sqrt{x}+2}{\sqrt{x}+2} = \dfrac{(\sqrt{x})^2 - 2^2}{(\sqrt{x}+1)(\sqrt{x}+2)} = \dfrac{x-4}{x + 3\sqrt{x} + 2}$ ∎

■ Now work Problem 31.

Equations Containing Radicals

When the variable in an equation occurs in a square root, cube root, and so on, that is, when it occurs under a radical, the equation is called a **radical equation.** Sometimes a suitable operation will change a radical equation to one that is linear or quadratic. The most commonly used procedure is to isolate the most complicated radical on one side of the equation and then eliminate it by raising each side to a power equal to the index of the radical. Care must be taken, because extraneous solutions may result. Thus, when working with radical equations, we always check apparent solutions. Let's look at an example.

E X A M P L E   8

Solving Radical Equations

Solve the equation: $\sqrt[3]{2x-4} - 2 = 0$

Solution

The equation contains a radical whose index is 3. We isolate it on the left side:

$$\sqrt[3]{2x-4} - 2 = 0$$
$$\sqrt[3]{2x-4} = 2$$

Now raise each side to the third power (since the index of the radical is 3) and solve:

$$(\sqrt[3]{2x-4})^3 = 2^3$$
$$2x - 4 = 8$$
$$2x = 12$$
$$x = 6$$

Check: $\sqrt[3]{2(6)-4} - 2 = \sqrt[3]{12-4} - 2 = \sqrt[3]{8} - 2 = 2 - 2 = 0$
The solution is $x = 6$.   ■

■   Now work Problem 37.

Rational Exponents

Radicals are used to define rational exponents.

$a^{1/n}$

If a is a real number and $n \geq 2$ is an integer, then

$$a^{1/n} = \sqrt[n]{a} \qquad (3)$$

provided $\sqrt[n]{a}$ exists.

E X A M P L E   9

Using Equation (3)

(a) $4^{1/2} = \sqrt{4} = 2$ (b) $(-27)^{1/3} = \sqrt[3]{-27} = -3$
(c) $8^{1/2} = \sqrt{8} = 2\sqrt{2}$ (d) $16^{1/3} = \sqrt[3]{16} = 2\sqrt[3]{2}$ ■

$a^{m/n}$

If a is a real number and m and n are integers containing no common factors with $n \geq 2$, then

$$a^{m/n} = \sqrt[n]{a^m} = (\sqrt[n]{a})^m \qquad (4)$$

provided $\sqrt[n]{a}$ exists.

We have two comments about equation (4):

1. The exponent m/n must be in lowest terms and n must be positive.
2. In simplifying $a^{m/n}$, either $\sqrt[n]{a^m}$ or $(\sqrt[n]{a})^m$ may be used. Generally, taking the root first is preferred.

It can be shown that the laws of exponents hold for rational exponents.

E X A M P L E   1 0   *Simplifying Expressions with Rational Exponents*

Simplify each expression. Express your answer so that only positive exponents occur. Assume that the variables are positive.

(a) $\left(\dfrac{2x^{1/3}}{y^{2/3}}\right)^{-3}$ (b) $(x^{2/3}y^{-3/4})(x^{-2}y)^{1/2}$

(c) $\left(\dfrac{9x^2y^{1/3}}{x^{1/3}y}\right)^{1/2}$ (d) $\dfrac{(2x+5)^{1/3}(2x+5)^{-1/2}}{(2x+5)^{-3/4}}$

Solution   (a) $\left(\dfrac{2x^{1/3}}{y^{2/3}}\right)^{-3} = \left(\dfrac{y^{2/3}}{2x^{1/3}}\right)^3 = \dfrac{(y^{2/3})^3}{(2x^{1/3})^3} = \dfrac{y^2}{2^3(x^{1/3})^3} = \dfrac{y^2}{8x}$

(b) $(x^{2/3}y^{-3/4})(x^{-2}y)^{1/2} = (x^{2/3}y^{-3/4})[(x^{-2})^{1/2}y^{1/2}]$

$= x^{2/3}y^{-3/4}x^{-1}y^{1/2} = (x^{2/3}x^{-1})(y^{-3/4}y^{1/2})$

$= x^{-1/3}y^{-1/4} = \dfrac{1}{x^{1/3}y^{1/4}}$

(c) $\left(\dfrac{9x^2y^{1/3}}{x^{1/3}y}\right)^{1/2} = \left(\dfrac{9x^{2-(1/3)}}{y^{1-(1/3)}}\right)^{1/2} = \left(\dfrac{9x^{5/3}}{y^{2/3}}\right)^{1/2} = \dfrac{9^{1/2}(x^{5/3})^{1/2}}{(y^{2/3})^{1/2}} = \dfrac{3x^{5/6}}{y^{1/3}}$

(d) $\dfrac{(2x+5)^{1/3}(2x+5)^{-1/2}}{(2x+5)^{-3/4}} = (2x+5)^{(1/3)-(1/2)-(-3/4)}$

$= (2x+5)^{(4-6+9)/12} = (2x+5)^{7/12}$ ∎

■ Now work Problem 57.

The next two examples illustrate some algebra that you will need to know for certain calculus problems.

E X A M P L E   1 1   *Writing an Expression as a Single Quotient*

Write the expression as a single quotient in which only positive exponents appear.

$$(x^2+1)^{1/2} + x \cdot \frac{1}{2}(x^2+1)^{-1/2} \cdot 2x$$

Solution $(x^2+1)^{1/2} + x \cdot \dfrac{1}{2}(x^2+1)^{-1/2} \cdot 2x = (x^2+1)^{1/2} + \dfrac{x^2}{(x^2+1)^{1/2}}$

$= \dfrac{(x^2+1)^{1/2}(x^2+1)^{1/2} + x^2}{(x^2+1)^{1/2}}$

$= \dfrac{(x^2+1) + x^2}{(x^2+1)^{1/2}}$

$= \dfrac{2x^2+1}{(x^2+1)^{1/2}}$ ∎

E X A M P L E   1 2   *Factoring an Expression Containing Rational Exponents*

Factor: $4x^{1/3}(2x + 1) + 2x^{4/3}$

Solution   We begin by looking for factors that are common to the two terms. Notice that 2 and $x^{1/3}$ are common factors. Thus,

$$4x^{1/3}(2x + 1) + 2x^{4/3} = 2x^{1/3}[2(2x + 1) + x]$$
$$= 2x^{1/3}(5x + 2)$$ ∎

A.2

Exercise A.2

In Problems 1–20, simplify each expression. Assume that all variables are positive when they appear.

1. $\sqrt{8}$ **2.** $\sqrt[4]{32}$ **3.** $\sqrt[3]{16x^4}$ **4.** $\sqrt{27x^3}$

5. $\sqrt[3]{\sqrt{x^6}}$ **6.** $\sqrt{\sqrt{x^6}}$ **7.** $\sqrt{\dfrac{32x^3}{9x}}$ **8.** $\sqrt[3]{\dfrac{x}{8x^4}}$

9. $\sqrt[4]{x^{12}y^8}$ **10.** $\sqrt[5]{x^{10}y^5}$ **11.** $\sqrt[4]{\dfrac{x^9y^7}{xy^3}}$ **12.** $\sqrt[3]{\dfrac{3xy^2}{81x^4y^2}}$

13. $\sqrt{36x}$ **14.** $\sqrt{9x^5}$ **15.** $\sqrt{3x^2}\sqrt{12x}$ **16.** $\sqrt{5x}\sqrt{20x^3}$

17. $(\sqrt{5}\sqrt[3]{9})^2$ **18.** $(\sqrt[3]{3}\sqrt{10})^4$

19. $\sqrt{\dfrac{2x - 3}{2x^4 + 3x^3}}\sqrt{\dfrac{x}{4x^2 - 9}}$ **20.** $\sqrt[3]{\dfrac{x - 1}{x^2 + 2x + 1}}\sqrt[3]{\dfrac{(x - 1)^2}{x + 1}}$

In Problems 21–26, perform the indicated operation and simplify the result. Assume that all variables are positive when they appear.

21. $(3\sqrt{6})(2\sqrt{2})$ **22.** $(5\sqrt{8})(-3\sqrt{3})$ **23.** $(\sqrt{3} + 3)(\sqrt{3} - 1)$

24. $(\sqrt{5} - 2)(\sqrt{5} + 3)$ **25.** $(\sqrt{x} - 1)^2$ **26.** $(\sqrt{x} + \sqrt{5})^2$

In Problems 27–36, rationalize the denominator of each expression. Assume that all variables are positive when they appear.

27. $\dfrac{1}{\sqrt{2}}$ **28.** $\dfrac{6}{\sqrt[3]{4}}$ **29.** $\dfrac{-\sqrt{3}}{\sqrt{5}}$ **30.** $\dfrac{-\sqrt[3]{3}}{\sqrt{8}}$

31. $\dfrac{\sqrt{3}}{5 - \sqrt{2}}$ **32.** $\dfrac{\sqrt{2}}{\sqrt{7} + 2}$ **33.** $\dfrac{2 - \sqrt{5}}{2 + 3\sqrt{5}}$ **34.** $\dfrac{\sqrt{3} - 1}{2\sqrt{3} + 3}$

35. $\dfrac{\sqrt{x + h} - \sqrt{x}}{\sqrt{x + h} + \sqrt{x}}$ **36.** $\dfrac{\sqrt{x + h} + \sqrt{x - h}}{\sqrt{x + h} - \sqrt{x - h}}$

In Problems 37–40, solve each equation.

37. $\sqrt{2t - 1} = 1$ **38.** $\sqrt{3t + 4} = 2$ **39.** $\sqrt{15 - 2x} = x$ **40.** $\sqrt{12 - x} = x$

In Problems 41–52, simplify each expression.

41. $8^{2/3}$ **42.** $4^{3/2}$ **43.** $(-27)^{1/3}$ **44.** $16^{3/4}$ **45.** $16^{3/2}$

46. $64^{3/2}$ **47.** $9^{-3/2}$ **48.** $25^{-5/2}$ **49.** $\left(\dfrac{9}{8}\right)^{3/2}$ **50.** $\left(\dfrac{27}{8}\right)^{2/3}$

51. $\left(\dfrac{8}{9}\right)^{-3/2}$ **52.** $\left(\dfrac{8}{27}\right)^{-2/3}$

In Problems 53–60, simplify each expression. Express your answer so that only positive exponents occur. Assume that the variables are positive.

53. $x^{3/4}x^{1/3}x^{-1/2}$

54. $x^{2/3}x^{1/2}x^{-1/4}$

55. $(x^3y^6)^{1/3}$

56. $(x^4y^8)^{3/4}$

57. $(x^2y)^{1/3}(xy^2)^{2/3}$

58. $(xy)^{1/4}(x^2y^2)^{1/2}$

59. $(16x^2y^{-1/3})^{3/4}$

60. $(4x^{-1}y^{1/3})^{3/2}$

In Problems 61–66, write each expression as a single quotient in which only positive exponents and/or radicals appear.

61. $\dfrac{x}{(1+x)^{1/2}} + 2(1+x)^{1/2}$

62. $\dfrac{1+x}{2x^{1/2}} + x^{1/2}$

63. $\dfrac{\sqrt{1+x} - x \cdot \dfrac{1}{2\sqrt{1+x}}}{1+x}$

64. $\dfrac{\sqrt{x^2+1} - x \cdot \dfrac{2x}{2\sqrt{x^2+1}}}{x^2+1}$

65. $\dfrac{(x+4)^{1/2} - 2x(x+4)^{-1/2}}{x+4}$

66. $\dfrac{(9-x^2)^{1/2} + x^2(9-x^2)^{-1/2}}{9-x^2}$

In Problems 67–70, factor each expression.

67. $(x+1)^{3/2} + x \cdot \frac{3}{2}(x+1)^{1/2}$

68. $(x^2+4)^{4/3} + x \cdot \frac{4}{3}(x^2+4)^{1/3} \cdot 2x$

69. $6x^{1/2}(x^2+x) - 8x^{3/2} - 8x^{1/2}$

70. $6x^{1/2}(2x+3) + x^{3/2} \cdot 8$

A.3

Completing the Square; the Quadratic Formula

Completing the Square

We begin with a preliminary result. Suppose that we wish to solve the quadratic equation

$$x^2 = p \tag{1}$$

where $p \geq 0$ is a nonnegative number. We proceed as follows:

$$x^2 - p = 0 \qquad \text{Put in standard form.}$$
$$(x - \sqrt{p})(x + \sqrt{p}) = 0 \qquad \text{Factor (over the real numbers).}$$
$$x = \sqrt{p} \quad \text{or} \quad x = -\sqrt{p} \qquad \text{Solve.}$$

Thus, we have the following result:

$$\text{If } x^2 = p \text{ and } p \geq 0, \text{ then } x = \sqrt{p} \text{ or } x = -\sqrt{p}. \tag{2}$$

Note that if $p > 0$ the equation $x^2 = p$ has two solutions: $x = \sqrt{p}$ and $x = -\sqrt{p}$. We usually abbreviate these solutions as $x = \pm\sqrt{p}$, read as "x equals plus or minus the square root of p." For example, the two solutions of the equation

$$x^2 = 4$$

are

$$x = \pm\sqrt{4}$$

and since $\sqrt{4} = 2$, we have

$$x = \pm 2$$

The solution set is $\{-2, 2\}$.

Do not confuse the two solutions of the equation $x^2 = 4$ with the value of the principal square root of 4, which is $\sqrt{4} = 2$. The principal square root of a positive number is unique, while the equation $x^2 = p$, $p > 0$, has two solutions.

E X A M P L E   1   *Solving Equations*

Solve each equation.

(a) $x^2 = 5$ (b) $(x - 2)^2 = 16$

Solution   (a) We use the result in equation (2) to get

$$x^2 = 5$$
$$x = \pm\sqrt{5}$$
$$x = \sqrt{5} \quad \text{or} \quad x = -\sqrt{5}$$

The solution set is $\{-\sqrt{5}, \sqrt{5}\}$.

(b) We use the result in equation (2) to get

$$(x - 2)^2 = 16$$
$$x - 2 = \pm\sqrt{16}$$
$$x - 2 = \sqrt{16} \quad \text{or} \quad x - 2 = -\sqrt{16}$$
$$x - 2 = 4 \qquad\qquad x - 2 = -4$$
$$x = 6 \qquad\qquad x = -2$$

The solution set is $\{-2, 6\}$.   ■

We now introduce the method of **completing the square.** The idea behind this method is to "adjust" a quadratic expression, $ax^2 + bx + c$, so that it becomes a perfect square—the square of a first-degree polynomial. For example, $x^2 + 6x + 9$ and $x^2 - 4x + 4$ are perfect squares because

$$x^2 + 6x + 9 = (x + 3)^2 \quad \text{and} \quad x^2 - 4x + 4 = (x - 2)^2$$

How do we adjust the quadratic expression? We do it by adding the appropriate number to create a perfect square. For example, to make $x^2 + 6x$ a perfect square, we add 9.

Let's look at several examples of completing the square when the coefficient of x^2 is 1:

START	ADD	RESULT
$x^2 + 4x$	4	$x^2 + 4x + 4 = (x + 2)^2$
$x^2 + 12x$	36	$x^2 + 12x + 36 = (x + 6)^2$
$x^2 - 6x$	9	$x^2 - 6x + 9 = (x - 3)^2$
$x^2 + x$	$\frac{1}{4}$	$x^2 + x + \frac{1}{4} = \left(x + \frac{1}{2}\right)^2$

Do you see the pattern? Provided the coefficient of x^2 is 1, we complete the square by adding the square of one-half the coefficient of x:

START	ADD	RESULT
$x^2 + mx$	$\left(\dfrac{m}{2}\right)^2$	$x^2 + mx + \left(\dfrac{m}{2}\right)^2 = \left(x + \dfrac{m}{2}\right)^2$

■ Now work Problem 1.

The next example illustrates how the procedure of completing the square can be used to solve a quadratic equation.

E X A M P L E   2   *Solving a Quadratic Equation by Completing the Square*

Solve by completing the square: $x^2 + 5x + 4 = 0$

Solution   We always begin this procedure by rearranging the equation so that the constant is on the right side:

$$x^2 + 5x + 4 = 0$$
$$x^2 + 5x = -4$$

Since the coefficient of x^2 is 1, we can complete the square on the left side by adding $\left(\frac{1}{2} \cdot 5\right)^2 = \frac{25}{4}$. Of course, in an equation, whatever we add to the left side must also be added to the right side. Thus, we add $\frac{25}{4}$ to *both* sides:

$$x^2 + 5x + \tfrac{25}{4} = -4 + \tfrac{25}{4}$$
$$\left(x + \tfrac{5}{2}\right)^2 = \tfrac{9}{4}$$
$$x + \tfrac{5}{2} = \pm\sqrt{\tfrac{9}{4}}$$
$$x + \tfrac{5}{2} = \pm\tfrac{3}{2}$$
$$x = -\tfrac{5}{2} \pm \tfrac{3}{2}$$
$$x = -\tfrac{5}{2} + \tfrac{3}{2} = -1 \quad \text{or} \quad x = -\tfrac{5}{2} - \tfrac{3}{2} = -4$$

The solution set is $\{-4, -1\}$.   ■

■ Now work Problem 7.

E X A M P L E   3   *Solving a Quadratic Equation by Completing the Square*

Solve by completing the square: $2x^2 + 8x - 5 = 0$

Solution   First, we rewrite the equation:

$$2x^2 - 8x - 5 = 0$$
$$2x^2 - 8x = 5$$

Next, we divide by 2 so that the coefficient of x^2 is 1. (This enables us to complete the square at the next step.)

$$x^2 - 4x = \tfrac{5}{2}$$

Finally, we complete the square by adding 4 to each side:

$$x^2 - 4x + 4 = \tfrac{5}{2} + 4$$
$$(x - 2)^2 = \tfrac{13}{2}$$
$$x - 2 = \pm\sqrt{\tfrac{13}{2}} = \pm\tfrac{\sqrt{26}}{2}$$
$$x = 2 \pm \tfrac{\sqrt{26}}{2}$$

We choose to leave our answer in this compact form. Thus, the solution set is
$\{2 - \sqrt{26}/2,\ 2 + \sqrt{26}/2\}$. ∎

Note: If we wanted an approximation, say, to two decimal places, of these solutions, we would use a calculator to get $\{-0.55, 4.55\}$.

■   Now work Problem 11.

The Quadratic Formula

We can use the method of completing the square to obtain a general formula for solving the quadratic equation

$$ax^2 + bx + c = 0, \qquad a \neq 0$$

As in Examples 2 and 3, we rearrange the terms as

$$ax^2 + bx = -c$$

Since $a \neq 0$, we can divide both sides by a to get

$$x^2 + \frac{b}{a}x = -\frac{c}{a}$$

Now the coefficient of x^2 is 1. To complete the square on the left side, add the square of one-half the coefficient of x; that is, add

$$\left(\frac{1}{2} \cdot \frac{b}{a}\right)^2 = \frac{b^2}{4a^2}$$

to each side.
Then

$$x^2 + \frac{b}{a}x + \frac{b^2}{4a^2} = \frac{b^2}{4a^2} - \frac{c}{a}$$

$$\left(x + \frac{b}{2a}\right)^2 = \frac{b^2 - 4ac}{4a^2} \qquad (3)$$

Provided $b^2 - 4ac \geq 0$, we now can apply the result in equation (2) to get

$$x + \frac{b}{2a} = \pm\sqrt{\frac{b^2 - 4ac}{4a^2}}$$

$$x = -\frac{b}{2a} \pm \frac{\sqrt{b^2 - 4ac}}{2a} = \frac{-b \pm \sqrt{b^2 - 4ac}}{2a}$$

What if $b^2 - 4ac$ is negative? Then equation (3) states that the left expression (a real number squared) equals the right expression (a negative number). Since this occurrence is impossible for real numbers, we conclude that if $b^2 - 4ac < 0$ the quadratic equation has no *real* solution.
    We now state the *quadratic formula*.

Theorem   Consider the quadratic equation

$$ax^2 + bx + c = 0, \qquad a \neq 0$$

If $b^2 - 4ac < 0$, this equation has no real solution. If $b^2 - 4ac \geq 0$, the real solution(s) of this equation is (are) given by the **quadratic formula:**

Quadratic Formula

$$x = \frac{-b \pm \sqrt{b^2 - 4ac}}{2a} \tag{4}$$

A.3

Exercise A.3

In Problems 1–6, tell what number should be added to complete the square of each expression.

1. $x^2 - 4x$ **2.** $x^2 - 2x$ **3.** $x^2 + \frac{1}{2}x$

4. $x^2 - \frac{1}{3}x$ **5.** $x^2 - \frac{2}{3}x$ **6.** $x^2 - \frac{2}{5}x$

In Problems 7–12, solve each equation by completing the square.

7. $x^2 + 4x - 21 = 0$ **8.** $x^2 - 6x = 13$ **9.** $x^2 - \frac{1}{2}x = \frac{3}{16}$

10. $x^2 + \frac{2}{3}x = \frac{1}{3}$ **11.** $3x^2 + x - \frac{1}{2} = 0$ **12.** $2x^2 - 3x = 1$

Appendix B GRAPHING UTILITIES

B.1

The Viewing Rectangle

All graphing utilities, that is, all graphing calculators and all computer software graphing packages, graph equations by plotting points on a screen. The screen itself actually consists of small rectangles, called **pixels.** The more pixels the screen has, the better the resolution. Most graphing calculators have 48 pixels per square inch; most computer screens have 32 to 108 pixels per square inch. When a point to be plotted lies inside a pixel, the pixel is turned on (lights up). Thus, the graph of an equation is a collection of pixels. Figure 1 shows how the graph of $y = 2x$ looks on a TI85 graphing calculator.

FIGURE 1
$y = 2x$

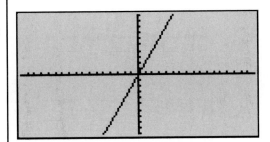

The screen of a graphing utility will display the coordinate axes of a rectangular coordinate system. However, you must set the scale on each axis. You must also include the smallest and largest values of x and y that you want included in the graph. This is called **setting the RANGE** and it gives the **viewing rectangle** or **window.**

Figure 2 illustrates a typical viewing rectangle.

FIGURE 2

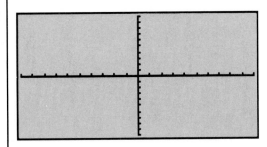

If the scale used on each axis is known, we can determine the minimum and maximum values of x and y shown on the screen by counting the tick marks. Look again at Figure 2. For a scale of 1 on each axis, the minimum and maximum values of x are -10 and 10, respectively; the minimum and maximum values of y are also -10 and 10. If the scale is 2 on each axis, then the minimum and maximum values of x are -20 and 20, respectively; the minimum and maximum value of y are -20 and 20, respectively.

Conversely, if we know the minimum and maximum values of x and y, we can determine the scales being used by counting the tick marks displayed. We shall follow the practice of showing the minimum and maximum values of x and y in our illustrations so that you will know how the RANGE was set.

To select the viewing rectangle, we must give values to the following expressions:

Xmin: the smallest value of x

Xmax: the largest value of x

Xscl: the number of units per tick mark on the x-axis

Ymin: the smallest value of y

Ymax: the largest value of y

Yscl: the number of units per tick mark on the y-axis

Figure 3 illustrates these settings for a typical screen.

FIGURE 3

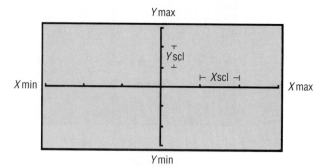

Care must be taken in selecting a viewing rectangle. If the viewing rectangle is too small, then some key parts of the graph may be excluded; if it is too large, the graph may look distorted. Figure 4 shows the graph of $y = -2x^2 + 8$ for several different viewing rectangles.

FIGURE 4

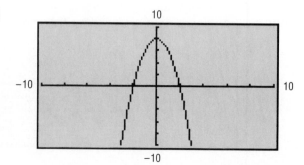

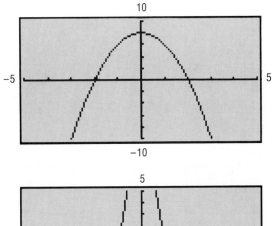

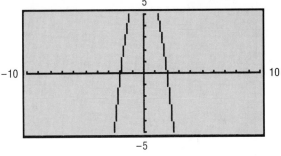

As Figure 4 illustrates, the selection of a viewing rectangle is critical for obtaining a complete graph. But how do we set it to get the best view of a graph? And, more importantly, how do we interpret what we see? This usually requires some knowledge of the properties of the equation, which we will be studying as we proceed through this book.

Some graphing utilities have a built-in function that automatically gives a complete graph. For those that do not have this feature, we discuss a way to obtain a complete graph, using the ZOOM-OUT function, in Section 2 of the Appendix.

B.1

Exercise B.1

In Problems 1–6, select a RANGE setting so that each of the given points will lie within the viewing rectangle.

1. $(-10, 5), (3, -2), (4, -1)$

2. $(5, 0), (6, 8), (-2, -3)$

3. $(40, 20), (-20, -80), (10, 40)$

4. $(-80, 60), (20, -30), (-20, -40)$

5. $(0, 0), (100, 5), (5, 150)$

6. $(0, -1), (100, 50), (-10, 30)$

In Problems 7–16, determine the RANGE settings used for each viewing rectangle.

7.

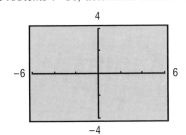

8.

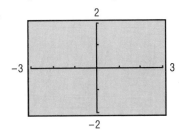

9.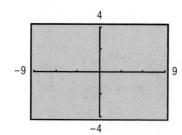

10.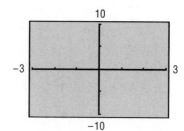

11.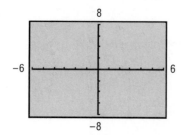

12.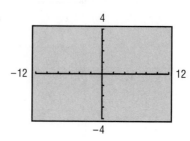

13.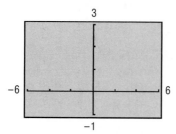

14.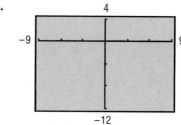

15.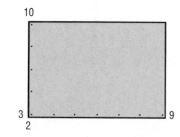

16.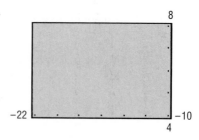

B.2

Graphing Equations Using a Graphing Utility

Most graphing utilities require the following steps in order to obtain the graph of an equation:

STEP 1: Solve the equation for y in terms of x.

STEP 2: Select the viewing rectangle.

STEP 3: Get into the graphing mode of your graphing utility. The screen will usually display $y =$ _____, prompting you to enter the expression involving x that you found in STEP 1. (Consult your manual for the correct way to enter the expression; for example, $y = x^2$ might be entered as $x\text{\^{}}2$ or as $x*x$ or as $x\ x^Y\ 2$).

STEP 4: Execute.

E X A M P L E   1   *Graphing an Equation on a Graphing Utility*

Graph the equation: $6x^2 + 3y = 24$

Solution   **STEP 1**   We solve for y in terms of x.

$$6x^2 + 3y = 24$$
$$3y = -6x^2 + 24$$
$$y = -2x^2 + 8$$

STEP 2   Select a viewing rectangle. We will use the one given next.

$$\text{Xmin} = -5$$
$$\text{Xmax} = 5$$
$$\text{Xscl}\ = 1$$
$$\text{Ymin} = -10$$
$$\text{Ymax} = 20$$
$$\text{Yscl}\ = 2$$

STEP 3   From the graphing mode, enter the expression $-2x^2 + 8$ after the prompt $y =$.

STEP 4   Execute

The screen should look like Figure 5.

FIGURE 5

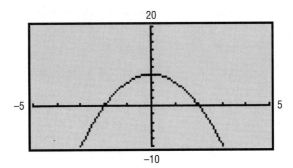

Now that we have seen the graph of $y = -2x^2 + 8$, we might want to graph it again using the settings given below:

$$\text{Xmin} = -5$$
$$\text{Xmax} = 5$$
$$\text{Xscl}\ = 1$$
$$\text{Ymin} = -10$$
$$\text{Ymax} = 10$$
$$\text{Yscl}\ = 1$$

Figure 6 shows the graph obtained. Notice the improvement over Figure 5.

FIGURE 6

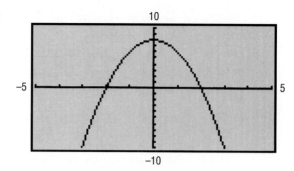

On some graphing utilities, you can scroll the graph up or down and left or right to see more of it, instead of setting a new viewing rectangle. With scrolling, the scale stays the same as it was originally; only the min/max values change.

E X A M P L E   2

Using Intercepts to Set the Viewing Rectangle

Graph: $y = x^3 - 8$

Solution   To assist in selecting the viewing rectangle, we find the intercepts:

If $x = 0$, then $y = -8$. The y-intercept is -8.

If $y = 0$, then $x^3 - 8 = 0$ or $x = 2$. The x-intercept is 2.

Since the intercepts are $(2, 0)$ and $(0, -8)$ and we want to include these points in the graph, we might use the settings

$$\text{Xmin} = -5$$
$$\text{Xmax} = 5$$
$$\text{Xscl} = 1$$
$$\text{Ymin} = -20$$
$$\text{Ymax} = 10$$
$$\text{Yscl} = 5$$

Figure 7 shows the graph.

FIGURE 7

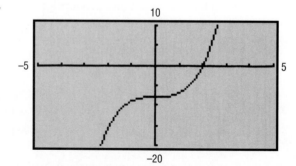

E X A M P L E   3

Using ZOOM-OUT to Obtain a Complete Graph

Graph the equation $y = x^3 - 11x^2 - 190x + 200$ using the following settings for the viewing rectangle

Xmin: -12
Xmax: 12
Xscl: 4
Ymin: -200
Ymax: 200
Yscl: 100

Solution   Figure 8 shows the graph.

FIGURE 8

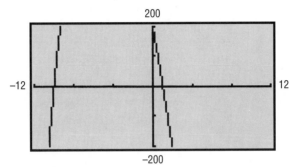

Notice how ragged the graph looks. The *y*-scale is clearly not adequate. The graph appears to have two *x*-intercepts. However, as we will discover in Chapter 5, a polynomial equation of degree 3 has either three or one *x*-intercept. Using this property of the equation, we conclude that there is a third *x*-intercept that is not shown. The graph is not complete. We can use the ZOOM-OUT function to help obtain a complete graph.

After the first ZOOM-OUT, we obtain Figure 9.

FIGURE 9

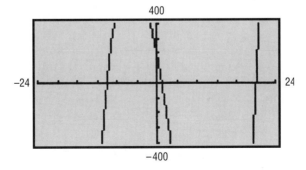

Notice that the min/max settings have been doubled, while the scales remained the same.* Also notice that we can see the three *x*-intercepts. However,

---

*On some graphing utilities, the default factor for the ZOOM-OUT function is 4, meaning that the original min/max setting will be multiplied by 4. Also, you can set the factor yourself, if you want; furthermore, the factor need not be the same for *x* and *y*.

we still cannot see the top or bottom portion of the graph. ZOOM-OUT again to obtain the graph in Figure 10.

FIGURE 10

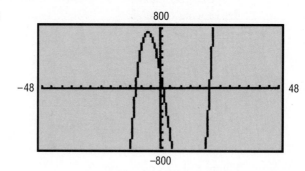

Notice that we can now see almost the entire graph with the exception of the bottom portion. Therefore ZOOMing-OUT further may not provide us with a complete graph. To get a complete graph, we instead choose to adjust the RANGE. After some experimentation, we will obtain Figure 11, a complete graph.

FIGURE 11

$y = x^3 - 11x^2 - 190x + 200$

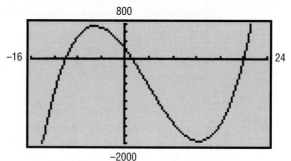

RANGE
xMin = −16
xMax = 24
xScl = 4
yMin = −2000
yMax = 800
yScl = 200

B.2

Exercise B.2

In Problems 1–20, graph each equation using the following RANGE settings:

(a) Xmin = −5   (b) Xmin = −10   (c) Xmin = −10   (d) Xmin = −5
    Xmax = 5   Xmax = 10   Xmax = 10   Xmax = 5
    Xscl = 1   Xscl = 1   Xscl = 2   Xscl = 1
    Ymin = −4   Ymin = −8   Ymin = −8   Ymin = −20
    Ymax = 4   Ymax = 8   Ymax = 8   Ymax = 20
    Yscl = 1   Yscl = 1   Yscl = 2   Yscl = 5

1. $y = x + 2$ **2.** $y = x - 2$ **3.** $y = -x + 2$ **4.** $y = -x - 2$

5. $y = 2x + 2$ **6.** $y = 2x - 2$ **7.** $y = -2x + 2$ **8.** $y = -2x - 2$

9. $y = x^2 + 2$ **10.** $y = x^2 - 2$ **11.** $y = -x^2 + 2$ **12.** $y = -x^2 - 2$

13. $y = 2x^2 + 2$ **14.** $y = 2x^2 - 2$ **15.** $y = -2x^2 + 2$ **16.** $y = -2x^2 - 2$

17. $3x + 2y = 6$ **18.** $3x - 2y = 6$ **19.** $-3x + 2y = 6$ **20.** $-3x - 2y = 6$

In Problems 21–40, graph each equation. State the viewing rectangle used and draw the graph on paper.

21. $3x + 5y = 75$ **22.** $3x - 5y = 75$ **23.** $3x + 5y = -75$ **24.** $3x - 5y = -75$

25. $y = (x - 10)^2$ **26.** $y = (x + 10)^2$ **27.** $y = x^2 - 100$ **28.** $y = x^2 + 100$

29. $x^2 + y^2 = 100$ **30.** $x^2 + y^2 = 164$ **31.** $3x^2 + y^2 = 900$ **32.** $4x^2 + y^2 = 1600$

33. $x^2 + 3y^2 = 900$ **34.** $x^2 + 4y^2 = 1600$ **35.** $y = x^2 - 10x$ **36.** $y = x^2 + 10x$

37. $y = x^2 - 18x$ **38.** $y = x^2 + 18x$ **39.** $y = x^2 - 36x$ **40.** $y = x^2 + 36x$

B.3

The TRACE, ZOOM-IN, and BOX Functions

The TRACE Function

Most graphing utilities allow you to move from point to point along the graph, displaying on the screen the coordinates of each point. This feature is called the TRACE function.

EXAMPLE 1   *Using the TRACE Function to Locate Intercepts*

Graph the equation $y = x^3 - 8$ using the following viewing rectangle. Use the TRACE function to locate various points on the graph. In particular, locate the x-intercept.

RANGE
  xMin = –5
  xMax = 5
  xScl = 1
  yMin = –20
  yMax = 10
  yScl = 5

Solution   Figure 12 shows the graph of $y = x^3 - 8$.

FIGURE 12

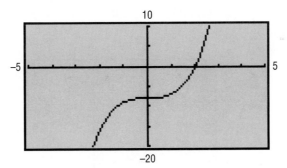

Activate the TRACE function. As you move the cursor along the graph, you will see the coordinates of each point displayed. Just before you get to the x-axis, the display will look like the one in Figure 13. (Due to differences in graphing utilities, your display may be slightly different than the one shown here).

FIGURE 13

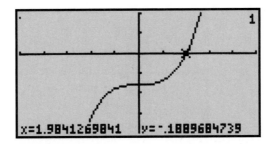

In Figure 13, the negative value of the y-coordinate indicates we are still below the x-axis. The next position of the cursor is shown in Figure 14.

FIGURE 14

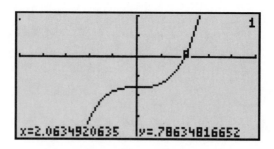

The positive value of the y-coordinate indicates we are now above the x-axis. This means that between these two points the x-axis was crossed. The x-intercept lies between 1.9841269841 and 2.0634920635.   ■

The ZOOM-IN Function

Most graphing utilities have a ZOOM-IN function that allows us to get better approximations.

EXAMPLE 2

Using ZOOM-IN

Graph the equation $y = x^3 - 8$ and use the ZOOM-IN function to improve on the approximation found in Example 1 for the x-intercept.

Solution   Using the viewing rectangle of Example 1, graph the equation and TRACE until the cursor is close to the x-intercept. See Figure 15.

FIGURE 15

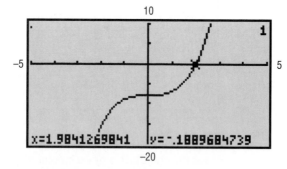

Activate the ZOOM-IN function. The result is shown in Figure 16.

FIGURE 16

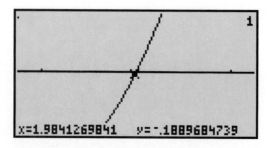

Move the cursor up. The result is shown in Figure 17.

FIGURE 17

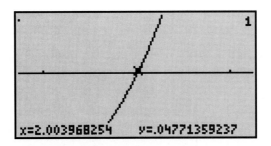

Notice the improvement. Now we know that the x-intercept lies between 1.9841269841 and 2.003968254. ZOOM-IN again. The result is shown in Figure 18(a). Figure 18(b) is the result of moving the cursor down.

FIGURE 18

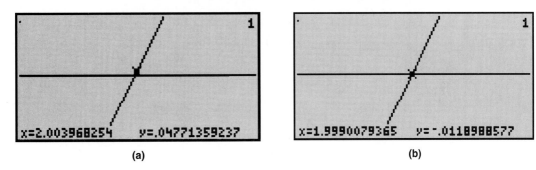

(a)   (b)

Now we know that the x-intercept lies between 1.9990079365 and 2.003968254.

The BOX Function

Most graphing utilities have a BOX function that allows you to box in a specific part of the graph of an equation.

E X A M P L E   3   *Using the BOX Function*

Graph the equation $y = x^3 - 8$ and use the BOX function to improve on the approximation found in Example 1 for the x-intercept.

Solution   Using the viewing rectangle of Example 1, graph the equation. Activate the BOX function. (With some graphing utilities, this requires positioning the cursor at one corner of the box and then tracing out the sides of the box to the diagonal corner.) See Figure 19.

FIGURE 19

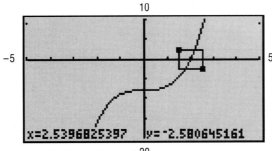

Once executed, the box becomes the viewing rectangle. The result is shown in Figure 20.

FIGURE 20

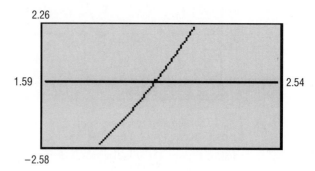

Now you can TRACE to get the approximations shown in Figure 21.

FIGURE 21

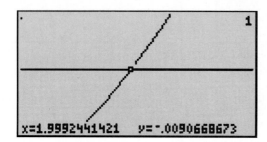 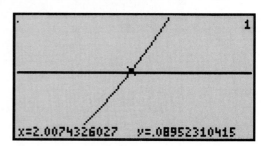

B.4

Square Screens

To get an undistorted view of slope, the same scale must be used on each axis. However, most graphing utilities have a rectangular screen. Because of this, using the same RANGE for both x and y will result in a distorted view. For example, Figure 22 shows the graph of the line $y = x$ connecting the points $(-4, -4)$ and $(4, 4)$.

FIGURE 22

Range
x Min = -4
x Max = 4
x Scl = 2
y Min = -4
y Max = 4
y Scl = 2

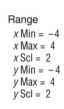 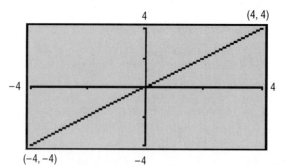

We expect the line to bisect the first and third quadrants, but it doesn't. We need to adjust the selections for Xmin, Xmax, Ymin, and Ymax so that a **square screen** results. On most graphing utilities, this is accomplished by setting the ratio of x to y at $3:2$.* In other words,

$$2(\text{Xmax} - \text{Xmin}) = 3(\text{Ymax} - \text{Ymin})$$

*Some graphing utilities have a built-in function that automatically squares the screen. For example, the TI85 has a ZSQR function that does this. Some graphing utilities require a ratio other than 3:2 to square the screen. For example, the HP48G requires the ratio of x to y to be 1:2 for a square screen. Consult your manual.

E X A M P L E   1   *Examples of Viewing Rectangles that Result in Square Screens*

(a)	Xmin = −3	(b)	Xmin = −6	(c)	Xmin = −6
	Xmax = 3		Xmax = 6		Xmax = 6
	Xscl = 1		Xscl = 2		Xscl = 2
	Ymin = −2		Ymin = −4		Ymin = −4
	Ymax = 2		Ymax = 4		Ymax = 4
	Yscl = 1		Yscl = 2		Yscl = 1

Figure 23 shows the graph of the line $y = x$ on a square screen using the viewing rectangle given in Example 1(b). Notice that the line now bisects the first and third quadrants. Compare this illustration to Figure 22.

FIGURE 23

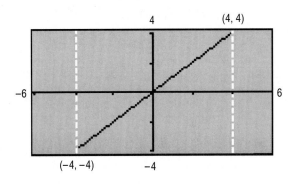

Exercise B.4

B.4

In Problems 1–8, determine which of the given RANGE settings result in a square screen.

1. Xmin = −3	**2.** Xmin = −5	**3.** Xmin = 0	**4.** Xmin = −6
Xmax = 3	Xmax = 5	Xmax = 9	Xmax = 6
Xscl = 1	Xscl = 1	Xscl = 3	Xscl = 2
Ymin = −2	Ymin = −4	Ymin = −2	Ymin = −4
Ymax = 2	Ymax = 4	Ymax = 4	Ymax = 4
Yscl = 1	Yscl = 1	Yscl = 2	Yscl = 1

5. Xmin = −6	**6.** Xmin = −6	**7.** Xmin = 0	**8.** Xmin = −6
Xmax = 6	Xmax = 6	Xmax = 9	Xmax = 6
Xscl = 1	Xscl = 1	Xscl = 1	Xscl = 2
Ymin = −2	Ymin = −4	Ymin = −2	Ymin = −4
Ymax = 2	Ymax = 4	Ymax = 4	Ymax = 4
Yscl = .5	Yscl = 1	Yscl = 1	Yscl = 2

9. If Xmin = −4, Xmax = 8, and Xscl = 1, how should Ymin, Ymax, and Yscl be selected so that the viewing rectangle contains the point (4, 8) and the screen is square?

10. If Xmin = −6, Xmax = 12, and Xscl = 2, how should Ymin, Ymax, and Yscl be selected so that the viewing rectangle contains the point (4, 8) and the screen is square?

B.5

Approximations

In practice, when we solve equations, we rarely obtain exact solutions. Usually, the best we can hope for is an approximate solution. In this section, we discuss how a graphing utility can be used to approximate solutions of equations. We shall develop an algorithm in which each successive application results in one more decimal place of accuracy.

We begin with an example that illustrates the meaning of writing a number "correct to n decimal places."

E X A M P L E   1   *Writing a Number Correct to n Decimal Places*

WRITING THE NUMBER	1/7	$\sqrt{2}$	π
Correct to 1 decimal place	0.1	1.4	3.1
Correct to 2 decimal places	0.14	1.41	3.14
Correct to 3 decimal places	0.142	1.414	3.141
Correct to 4 decimal places	0.1428	1.4142	3.1415
Correct to 5 decimal places	0.14285	1.41421	3.14159
Correct to 6 decimal places	0.142857	1.414213	3.141592 ∎

As the example illustrates, **correct to n decimal places** means the decimal that results from truncation after the nth decimal.

The following example illustrates the steps to use to approximate the solution of an equation using a graphing utility.

E X A M P L E   2   *Using a Graphing Utility to Approximate Solutions of an Equation*

Find the smaller of the two solutions of the equation $x^2 - 6x + 7 = 0$. Express the answer correct to two decimal places.

Solution   The solutions of the equation $x^2 - 6x + 7 = 0$ are the same as the x-intercepts of the function $f(x) = x^2 - 6x + 7$. We begin by graphing the function f using a scale of 1 on each axis.

Figure 24 shows the graph.

FIGURE 24

Range
  x Min = −1
  x Max = 8
  x Scl = 1
  y Min = −2
  y Max = 8
  y Scl = 1

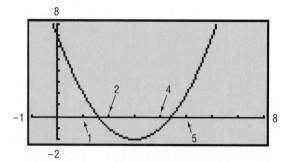

The smaller of the two x-intercepts (solutions of the equation) lies between 1 and 2. (Remember that Xscl = 1). Thus, correct to 0 decimal places, the smaller solution of the equation is $x = 1$.

Next, we BOX the graph from $x = 1$ to $x = 2$ and from $y = -1$ to $y = 1$. See Figure 25.

FIGURE 25

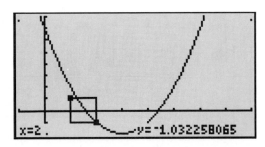

Adjust the RANGE setting so that Xscl = 0.1 and Yscl = 0.1. Graph f again. See Figure 26.

FIGURE 26

Range
  x Min = 1
  x Max = 2
  x Scl = .1
  y Min = −1.03225806452
  y Max = 1.06451612903
  y Scl = .1

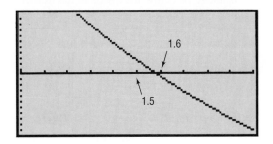

The x-intercept lies between 1.5 and 1.6. (Remember that the BOX was from $x = 1$ to $x = 2$ and Xscl = 0.1). Thus, correct to one decimal place, the smaller solution of the equation is $x = 1.5$.

Next, BOX again from $x = 1.5$ to $x = 1.6$ and from $y = -0.1$ to $y = 0.1$. See Figure 27. Adjust the RANGE so that Xscl = 0.01 and Yscl = 0.01 and graph f. See Figure 28.

FIGURE 27

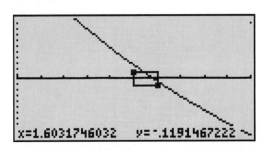

FIGURE 28

Range
  x Min = 1.5
  x Max = 1.60317460318
  x Scl = .01
  y Min = .119146722168
  y Max = .11758584807
  y Scl = .01

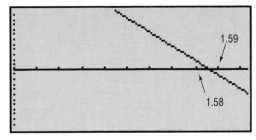

The x-intercept lies between 1.58 and 1.59. Remember that the BOX was from $x = 1.5$ to $x = 1.6$ with Xscl = 0.01. Thus, correct to two decimal places, the smaller of the two solutions is $x = 1.58$. ∎

The steps to follow for approximating solutions of equations correct to any desired number of decimals are given next.

STEP 1   To solve $f(x) = 0$, graph the function $y = f(x)$ using a viewing rectangle with Xscl = 1 and Yscl = 1.

STEP 2   If the x-intercept lies between $x = c$ and $x = c + $ Xscl, BOX the graph from $x = c$ to $x = c + $ Xscl and from $y = -$Yscl to $y = $ Yscl.

STEP 3   Adjust the RANGE so that new Xscl = old Xscl/10 and new Yscl = old Yscl/10. Graph f.

STEP 4   Repeat STEPS 2 and 3 until the desired accuracy is obtained.

If the x-intercept you are trying to approximate appears to fall on a tick mark, then you must evaluate the function at that tick mark. If the value is zero, then there is an exact solution. If the value is not zero, then you know whether the graph is above or below the x-axis at the tick mark by noting the sign of the value at the tick mark.

B.5

Exercise B.5

In Problems 1–6, write each expression as a decimal correct to three decimal places.

1. $\frac{3}{7}$ **2.** $\frac{2}{11}$ **3.** $\sqrt{5}$ **4.** $\sqrt{6}$ **5.** $\sqrt[3]{2}$ **6.** $\sqrt[3]{3}$

In Problems 7–12, use a graphing utility to approximate the smaller of the two solutions of each equation. Express the answer correct to two decimal places.

7. $x^2 + 4x + 2 = 0$ **8.** $x^2 + 4x - 3 = 0$ **9.** $2x^2 + 4x + 1 = 0$

10. $3x^2 + 5x + 1 = 0$ **11.** $2x^2 - 3x - 1 = 0$ **12.** $2x^2 - 4x - 1 = 0$

*In Problems 13–20, use a graphing utility to approximate the **positive** solutions for each equation. Express the answer correct to two decimal places.*

13. $x^3 + 3.2x^2 - 16.83x - 5.31 = 0$ **14.** $x^3 + 3.2x^2 - 7.25x - 6.3 = 0$

15. $x^4 - 1.4x^3 - 33.71x^2 + 23.94x + 292.41 = 0$ **16.** $x^4 + 1.2x^3 - 7.46x^2 - 4.692x + 15.2881 = 0$

17. $\pi x^3 - (8.88\pi + 1)x^2 - (42.066\pi - 8.88)x + 42.066 = 0$

18. $\pi x^3 - (5.63\pi + 2)x^2 - (108.392\pi - 11.26)x + 216.784 = 0$

19. $x^3 + 19.5x^2 - 1021x + 1000.5 = 0$

20. $x^3 + 14.2x^2 - 4.8x - 12.4 = 0$

ANSWERS

1. > **3.** > **5.** > **7.** = **9.** < **11.** 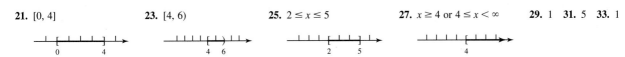 **13.** $x > 0$ **15.** $x < 2$ **17.** $x \le 1$ **19.** $2 < x < 5$

21. $[0, 4]$ **23.** $[4, 6)$ **25.** $2 \le x \le 5$ **27.** $x \ge 4$ or $4 \le x < \infty$ **29.** 1 **31.** 5 **33.** 1

35. $\frac{1}{16}$ **37.** $\frac{4}{9}$ **39.** $\frac{1}{9}$ **41.** $\frac{9}{4}$ **43.** 324 **45.** 27 **47.** 16 **49.** $4\sqrt{2}$ **51.** $-\frac{2}{3}$ **53.** y^2 **55.** $\frac{y}{x^2}$ **57.** $\frac{1}{x^3 y}$ **59.** $\frac{25y^2}{16x^2}$ **61.** $\frac{1}{x^3 y^3 z^2}$ **63.** $\frac{-8x^3}{9yz^2}$

65. $\frac{y^2}{x^2 + y^2}$ **67.** $\frac{16x^2}{9y^2}$ **69.** 13 **71.** 26 **73.** 25 **75.** 4 **77.** 24 **79.** Yes; 5 **81.** Not a right triangle **83.** Yes; 25 **85.** 72.42 **87.** 30.66

89. 111.17 **91.** 11.03 **93.** -0.49 **95.** 0.59 **97.** 9.4 in. **99.** 58.3 ft **101.** 46.7 mi **103.** 3 mi

105. $a \le b, c > 0; a - b \le 0$ **107.** $\frac{a + b}{2} - a = \frac{a + b - 2a}{2} = \frac{b - a}{2} > 0$; therefore, $a < \frac{a + b}{2}$ **109.** No; No **111.** 1
$(a - b)c \le 0(c)$
$ac - bc \le 0$ $b - \frac{a + b}{2} = \frac{2b - a - b}{2} = \frac{b - a}{2} > 0$; therefore, $b > \frac{a + b}{2}$
$ac \le bc$

113. 3.15 or 3.16

115. The light can be seen on the horizon 23.3 miles distant. Planes flying at 10,000 feet can see it 99 miles away. The ship would need to be 185.8 ft tall for the data to be correct.

Exercise 1.2

1. -1 **3.** -18 **5.** -3 **7.** -16 **9.** 0.5 **11.** 2 **13.** 2 **15.** -1 **17.** 3 **19.** No real solution **21.** $\{0, 9\}$ **23.** $\{0, 9\}$ **25.** No real solution

27. 2 **29.** 21 **31.** $\{-2, 2\}$ **33.** $-\frac{20}{39}$ **35.** 6 **37.** $\{-3, 3\}$ **39.** $\{-4, 1\}$ **41.** $\{-1, \frac{3}{2}\}$ **43.** $\{-4, 4\}$ **45.** 2 **47.** $\{-12, 12\}$ **49.** $\{-\frac{36}{5}, \frac{24}{5}\}$

51. No real solution **53.** $\{-2, 2\}$ **55.** $\{-1, 3\}$ **57.** $\{-2, -1, 0, 1\}$ **59.** $\{0, 4\}$ **61.** $\{-6, 2\}$ **63.** $\{-\frac{1}{2}, 3\}$ **65.** $\{3, 4\}$ **67.** $\frac{3}{2}$

69. $\{-\frac{2}{3}, \frac{3}{2}\}$ **71.** $\{-\frac{3}{4}, 2\}$ **73.** $\{2 - \sqrt{2}, 2 + \sqrt{2}\}$ **75.** $\left\{ \frac{5 - \sqrt{29}}{2}, \frac{5 + \sqrt{29}}{2} \right\}$ **77.** $\{1, \frac{3}{2}\}$ **79.** No real solution

81. $\left\{ \frac{-1 - \sqrt{5}}{4}, \frac{-1 + \sqrt{5}}{4} \right\}$ **83.** $\{0.59, 3.41\}$ **85.** $\{-2.80, 1.07\}$ **87.** No real solution **89.** Repeated real solution

91. Two unequal real solutions **93.** (a) 6 sec (b) 5 sec **95.** $\frac{-b + \sqrt{b^2 - 4ac}}{2a} + \frac{-b - \sqrt{b^2 - 4ac}}{2a} = \frac{-2b}{2a} = \frac{-b}{a}$ **97.** $k = -\frac{1}{2}$ or $\frac{1}{2}$

99. Solutions of $ax^2 - bx + c = 0$ are $\frac{b + \sqrt{b^2 - 4ac}}{2a}$ and $\frac{b - \sqrt{b^2 - 4ac}}{2a}$. **101.** 36

Exercise 1.3

1. $A = \pi r^2$; r = Radius, A = Area **3.** $A = s^2$; A = Area, s = Length of a side **5.** $F = ma$; F = Force, m = Mass, a = Acceleration

7. $W = Fd$; W = Work, F = Force, d = Distance **9.** $C = 150x$; C = Total cost, x = number of dishwashers

11. $11,000 will be invested in bonds and $9,000 in CD's **13.** Katy will receive $400,000, Mike $300,000, and Dan $200,000

15. The regular hourly rate is $8.50 **17.** The team got 5 touchdowns **19.** The length is 19 ft; the width is 11 ft

21. (a) The dimensions are 10 ft by 5 ft (b) The area is 50 sq ft (c) The dimensions are 7.5 ft by 7.5 ft (d) The area is 56.25 sq ft

23. Invest $31,250 in bonds and $18,750 in CD's **25.** $11,600 was loaned out at 8% **27.** The dimensions are 11 ft by 13 ft

29. The dimensions are 5 m by 8 m **31.** The dimensions of the sheet metal should be 4 ft by 4 ft

33. (a) The ball strikes the ground after 6 seconds (b) The ball passes the top of the building on its way down after 5 seconds

35. The original price was $147,058.82; purchasing the model saves $22,058.82 **37.** The bookstore paid $44.80

39. Working together, it takes 12 min **41.** Colleen needs a score of 85

43. The defensive back catches up to the tight end at the tight end's 45 yd line **45.** The border will be 2.56 ft wide

47. The dimensions of the reduced candy bar are 11.55 cm by 6.55 cm by 3 cm **49.** The border will be 2.71 ft wide

51. The current is 2.286 mi/hr **53.** Mike passes Dan 1/3 mile from the start, 2 minutes from the time Mike started to race

55. The rescue craft reaches the ship in 2 hrs   **57.** Start the auxiliary pump at 9:45 AM   **59.** The tub will fill in 1 hr
61. The most you can invest in the CD is $66,667   **63.** The average speed is 49.5 mi/hr
65. Set the original price at $40. At 50% off, there will be no profit at all.

Exercise 1.4

1. $\{x \mid x \geq 2\}$ or $[2, \infty)$ **3.** $\{x \mid x > -7\}$ or $(-7, \infty)$ **5.** $\{x \mid x \leq \frac{2}{3}\}$ or $(-\infty, \frac{2}{3}]$ **7.** $\{x \mid x < -20\}$ or $(-\infty, -20)$

   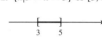

9. $\{x \mid x \geq \frac{4}{3}\}$ or $[\frac{4}{3}, \infty)$ **11.** $\{x \mid 3 \leq x \leq 5\}$ or $[3, 5]$ **13.** $\{x \mid \frac{2}{3} \leq x \leq 3\}$ or $[\frac{2}{3}, 3]$ **15.** $\{x \mid -2 < x < 5\}; (-2, 5)$

17. $\{x \mid -3 < x < 3\}; (-3, 3)$ **19.** $\{x \mid -\infty < x < -4 \text{ or } 3 < x < \infty\}; (-\infty, -4) \text{ or } (3, \infty)$ **21.** $\{x \mid -\infty < x < -1 \text{ or } 8 < x < \infty\}; (-\infty, -1) \text{ or } (8, \infty)$

23. No real solution   **25.** $\{x \mid 1 < x < \infty\}; (1, \infty)$ **27.** $\{x \mid -\infty < x < 1 \text{ or } 2 < x < 3\}); (-\infty, 1) \text{ or } (2, 3)$

29. $\{x \mid -1 < x < 0 \text{ or } 3 < x < \infty\}); (-1, 0) \text{ or } (3, -\infty)$ **31.** $\{x \mid -\infty < x < -1 \text{ or } 1 < x < \infty\}; (-\infty, -1) \text{ or } (1, \infty)$ **33.** $\{x \mid 1 < x < \infty\}); (1, \infty)$

   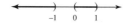

35. $\{x \mid -\infty < x < -1 \text{ or } 1 < x < \infty\}; (-\infty, -1) \text{ or } (1, \infty)$ **37.** $\{x \mid -\infty < x < -1 \text{ or } 0 < x < 1\}; (-\infty, 1) \text{ or } (0, 1)$

39. $\{x \mid -\infty < x < -1 \text{ or } 1 < x < \infty\}; (-\infty, -1) \text{ or } (1, \infty)$ **41.** $\{x \mid -\infty < x < 2\}; (-\infty, 2)$

43. $\{x \mid -\infty < x < -3 \text{ or } -1 < x < 1 \text{ or } 2 < x < \infty\}; (-\infty, -3) \text{ or } (-1, 1) \text{ or } (2, \infty)$ **45.** $\{x \mid -4 < x < 4\}; (-4, 4)$

47. $\{x \mid -\infty < x < -4 \text{ or } 4 < x < \infty\}; (-\infty, -4) \text{ or } (4, \infty)$ **49.** $\{x \mid 1 < x < 3\}; (1, 3)$ **51.** $\{t \mid -\frac{2}{3} \leq t \leq 2\}; [-\frac{2}{3}, 2]$

53. $\{x \mid -\infty < x \leq 1 \text{ or } 5 \leq x < \infty\}; (-\infty, 1] \text{ or } [5, \infty)$ **55.** $\{x \mid -1 < x < \frac{3}{2}\}; (-1, \frac{3}{2})$ **57.** $|x - 2| < \frac{1}{2}; \{x \mid \frac{3}{2} < x < \frac{5}{2}\}$

59. $|x + 3| > 2$; $\{x|-\infty < x < -5 \text{ or } -1 < x < \infty\}$ **61.** $|x - 98.6| \geq 1.5$; $x \leq 97.1°F$ or $x \geq 100.1°F$
63. From 675.48 k Whr to 2500.88 kWhr, inclusive   **65.** From \$7457.63 to \$7857.14, inclusive   **67.** Fifth test score ≥ 74
69. For t between 2 and 3 seconds, $2 < t < 3$.   **71.** Between 10 and 50 watches must be sold, $10 \leq x \leq 50$
73. $x^2 - a = (x - \sqrt{a})(x + \sqrt{a}) < 0$; $-\sqrt{a} < x < \sqrt{a}$ **75.** $\{x|-1 < x < 1\}$; $(-1, 1)$ **77.** $\{x|-\infty < x \leq -3 \text{ or } 3 \leq x < \infty\}$; $(-\infty, -3]$ or $[3, \infty)$
79. $\{x|-4 \leq x \leq 4\}$; $[-4, 4]$ **81.** $\{x|-\infty < x < -2 \text{ or } 2 < x < \infty\}$; $(-\infty, -2)$ or $(2, \infty)$ **83.** $-2 < k < 2$

Exercise 1.5

1. $8 + 5i$ **3.** $-7 + 6i$ **5.** $-6 - 11i$ **7.** $6 - 18i$ **9.** $6 + 4i$ **11.** $10 - 5i$ **13.** 37   **15.** $\frac{6}{5} + \frac{8}{5}i$ **17.** $1 - 2i$ **19.** $\frac{5}{2} - \frac{7}{2}i$ **21.** $-\frac{1}{2} + (\sqrt{3}/2)i$
23. $2i$ **25.** $-i$ **27.** i **29.** -6 **31.** $-10i$ **33.** $-2 + 2i$ **35.** 0   **37.** 0   **39.** $2i$ **41.** $5i$ **43.** $5i$ **45.** $\{-2i, 2i\}$ **47.** $\{-4, 4\}$
49. $\{3 - 2i, 3 + 2i\}$ **51.** $\{3 - i, 3 + i\}$ **53.** $\{\frac{1}{4} - \frac{1}{4}i, \frac{1}{4} + \frac{1}{4}i\}$ **55.** $\{-\frac{1}{5} - \frac{2}{5}i, -\frac{1}{5} + \frac{2}{5}i\}$ **57.** $\left\{-\dfrac{1}{2} - \dfrac{\sqrt{3}}{2}i, -\dfrac{1}{2} + \dfrac{\sqrt{3}}{2}i\right\}$
59. $\{2, -1 - \sqrt{3}i, -1 + \sqrt{3}i\}$ **61.** $\{-2, 2, -2i, 2i\}$ **63.** $\{-3i, -2i, 2i, 3i\}$ **65.** Two complex solutions   **67.** Two unequal real solutions
69. A repeated real solution   **71.** $2 - 3i$ **73.** 6   **75.** 25   **77.** $z + \bar{z} = (a + bi) + (a - bi) = 2a$; $z - \bar{z} = (a + bi) - (a - bi) = 2bi$
79. $\overline{z + w} = \overline{(a + bi) + (c + di)} = \overline{(a + c) + (b + d)i} = (a + c) - (b + d)i = (a - bi) + (c - di) = \bar{z} + \bar{w}$

Exercise 1.6

1. (a) Quadrant II
(b) Positive x-axis
(c) Quadrant III
(d) Quadrant I
(e) Negative y-axis
(f) Quadrant IV

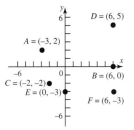

3. The points will be
on a vertical line
that is 2 units to the
right of the y-axis.

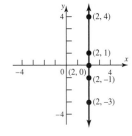

5. $\sqrt{5}$; $(1, \frac{1}{2})$ **7.** $2\sqrt{2}$; $(0, 2)$ **9.** 5; $(3, -\frac{3}{2})$ **11.** $\sqrt{85}$; $(\frac{3}{2}, 1)$ **13.** $2\sqrt{5}$; $(5, -1)$ **15.** 2.625; $(1.05, 0.7)$ **17.** $\sqrt{a^2 + b^2}$; $(\frac{a}{2}, \frac{b}{2})$

19. $d(A, B) = \sqrt{13}$
$d(B, C) = \sqrt{13}$
$d(A, C) = \sqrt{26}$
$(\sqrt{13})^2 + (\sqrt{13})^2 = (\sqrt{26})^2$
Area $= \frac{13}{2}$ square units

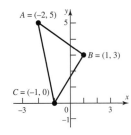

21. $d(A, B) = \sqrt{130}$
$d(B, C) = \sqrt{26}$
$d(A, C) = \sqrt{104}$
$(\sqrt{26})^2 + (\sqrt{104})^2 = (\sqrt{130})^2$
Area $= 26$ square units

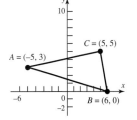

23. $(2, -4)$, $(2, 2)$ **25.** $(-2, 0)$, $(6, 0)$

27. (b) = (-3, 4)   (3, 4)

29.   (-2, 1)   (b) = (2, 1)   (a) = (-2, -1)   (c) = (2, -1)

31.   (b) = (-1, 1)   (1, 1)   (c) = (-1, -1)   (a) = (1, -1)

33. (a) = (-3, 4)   (c) = (3, 4)   (-3, -4)   (b) = (3, -4)

(c) = (-3, -4)   (a) = (3, -4)

35. (a) $(-1, 0), (1, 0)$ **(b)** x-axis, y-axis, origin   **37. (a)** $(-\frac{\pi}{2}, 0), (\frac{\pi}{2}, 0), (0, 1)$ **(b)** y-axis   **39. (a)** $(0, 0)$ **(b)** x-axis
41. (a) $(-1.5, 0), (0, -2), (1.5, 0)$ **(b)** y-axis   **43. (a)** none   **(b)** origin   **45.** $x^2 + (y - 2)^2 = 4$; $x^2 + y^2 - 4y = 0$

47. $(x - 4)^2 + (y + 3)^2 = 25$; $x^2 + y^2 - 8x + 6y = 0$ **49.** $x^2 + y^2 = 4$; $x^2 + y^2 - 4 = 0$ **51.** Center $(2, 1)$; radius 2; $(x - 2)^2 + (y - 1)^2 = 4$
53. Center $(\frac{5}{2}, 2)$; radius $\frac{3}{2}$; $(x - \frac{5}{2})^2 + (y - 2)^2 = \frac{9}{4}$ **55.** (c)   **57.** (b)
59. $r = 2$;
   $(h, k) = (0, 0)$

61. $r = 2$;
   $(h, k) = (3, 0)$

63. $r = 3$;
   $(h, k) = (-2, 2)$

65. $r = \frac{1}{2}$;
   $(h, k) = (\frac{1}{2}, -1)$

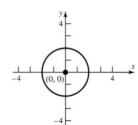

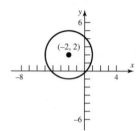

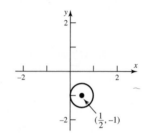

67. $r = 5$;
   $(h, k) = (3, -2)$

69.

71.

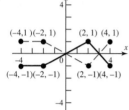

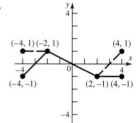

73. $(0, 0)$; symmetric with respect to the y-axis   **75.** $(0, 0)$; symmetric with respect to the origin
77. $(0, 9)$, $(3, 0)$, $(-3, 0)$; symmetric with respect to the y-axis
79. $(-3, 0)$, $(3, 0)$, $(0, -2)$, $(0, 2)$; symmetric with respect to the x-axis, y-axis, and origin   **81.** $(0, -27)$, $(3, 0)$; no symmetry
83. $(0, -4)$, $(4, 0)$, $(-1, 0)$; no symmetry   **85.** $(0, 0)$; symmetric with respect to the origin
87. (a) $(90, 0)$, $(90, 90)$, $(0, 90)$ (b) 232.4 ft   (c) 366.2 ft   **89.** $d = 50t$ **91.** $x^2 + y^2 + 2x + 4y - 4168.16 = 0$

Exercise 1.7

1. $\frac{1}{2}$ **3.** -1 **5.** Slope $= -\frac{3}{2}$ **7.** Slope $= -\frac{1}{2}$ **9.** Slope $= 0$

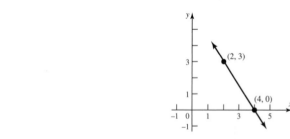

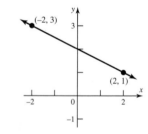

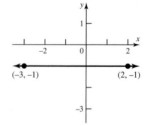

11. Slope undefined   **13.** Slope $= \dfrac{\sqrt{3} - 3}{1 - \sqrt{2}} \approx 3.06$ **15.**   **17.**

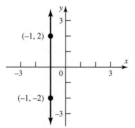

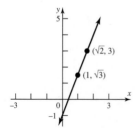

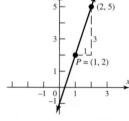

19.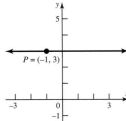

21.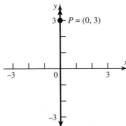

23. $x - 2y = 0$ or $y = \frac{1}{2}x$ **25.** $x + y - 2 = 0$ or $y = -x + 2$ **27.** $2x - y - 3 = 0$ or $y = 2x - 3$ **29.** $x + 2y - 5 = 0$ or $y = -\frac{1}{2}x + \frac{5}{2}$
31. $3x - y + 9 = 0$ or $y = 3x + 9$ **33.** $2x + 3y + 1 = 0$ or $y = -\frac{2}{3}x - \frac{1}{3}$ **35.** $x - 2y + 5 = 0$ or $y = \frac{1}{2}x + \frac{5}{2}$ **37.** $3x + y - 3 = 0$ or $y = -3x + 3$
39. $x - 2y - 2 = 0$ or $y = \frac{1}{2}x - 1$ **41.** $x - 2 = 0$; no slope-intercept form **43.** $2x - y + 4 = 0$ or $y = 2x + 4$ **45.** $2x - y = 0$ or $y = 2x$
47. $x - 4 = 0$ no y-intercept form **49.** $2x + y = 0$ or $y = -2x$ **51.** $x - 2y + 3 = 0$ or $y = \frac{1}{2}x + \frac{3}{2}$ **53.** $y - 4 = 0$ or $y = 4$
55. Slope $= 2$; y-intercept $= 3$ **57.** $y = 2x - 2$; Slope $= 2$; y-intercept $= -2$ **59.** Slope $= \frac{1}{2}$; y-intercept $= 2$

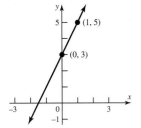
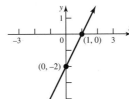
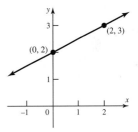

61. $y = -\frac{1}{2}x + 2$; Slope $= -\frac{1}{2}$; y-intercept $= 2$ **63.** $y = \frac{2}{3}x - 2$; Slope $= \frac{2}{3}$; y-intercept $= -2$ **65.** $y = -x + 1$; Slope $= -1$; y-intercept $= 1$

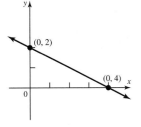
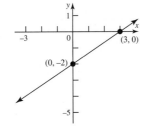
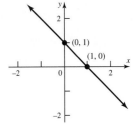

67. Slope undefined; no y-intercept **69.** Slope $= 0$; y-intercept $= 5$ **71.** $y = x$; Slope $= 1$; y-intercept $= 0$

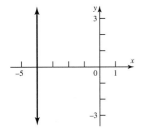
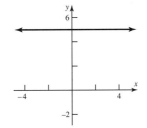
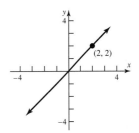

73. $y = \frac{3}{2}x$; Slope $= \frac{3}{2}$; y-intercept $= 0$

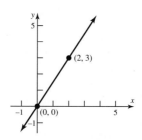

75. $y = 0$ **77.** (b) **79.** (d) **81.** $x - y + 2 = 0$ or $y = x + 2$ **83.** $x + 3y - 3 = 0$ or $y = -\frac{1}{3}x + 1$ **85.** $°C = \frac{5}{9}(°F - 32)$; approx. $21°C$
87. (a) $P = 0.5x - 100$ **(b)** \$400 **(c)** \$2400 **89.** $C = 0.10819x + 9.06$; for 300 kWhr, $C = \$41.52$; for 900 kWhr, $C = \$106.43$

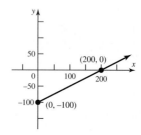

91. (a)
$$x^2 + (mx + b)^2 = r^2$$
$$(1 + m^2)x^2 + 2mbx + b^2 - r^2 = 0$$
One solution if and only if discriminant $= 0$
$$(2mb)^2 - 4(1 + m^2)(b^2 - r^2) = 0$$
$$-4b^2 + 4r^2 + 4m^2r^2 = 0$$
$$r^2(1 + m^2) = b^2$$

(b) $x = \dfrac{-2mb}{2(1 + m^2)} = \dfrac{-2mb}{2b^2/r^2} = \dfrac{-r^2m}{b}$
$$y = m\left(\dfrac{-r^2m}{b}\right) + b = \dfrac{-r^2m^2}{b} + b$$
$$= \dfrac{-r^2m^2 + b^2}{b} = \dfrac{r^2}{b}$$

(c) Slope of tangent line $= m$
Slope of line joining center to point
of tangency $= \dfrac{r^2/b}{-r^2m/b} = -\dfrac{1}{m}$

93. $\sqrt{2}x + 4y - 11\sqrt{2} + 12 = 0$ **95.** $x + 5y + 13 = 0$ **97.** All have the same slope, 2; the lines are parallel. **99.** $y = 2$

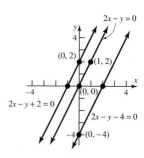

Review Exercises

1. -9 **3.** 6 **5.** $\frac{1}{5}$ **7.** 5 **9.** No real solution **11.** $\frac{11}{8}$ **13.** $\{-2, \frac{3}{2}\}$ **15.** $\left\{\dfrac{1 - \sqrt{13}}{4}, \dfrac{1 + \sqrt{13}}{4}\right\}$ **17.** $\{-3, 3\}$ **19.** No real solution

21. $\{-1, 4\}$ **23.** $\left\{\dfrac{-9b}{5a}, \dfrac{2b}{a}\right\}$ **25.** $\{x | 4 \le x < \infty\}$ **27.** $\{x | -\frac{31}{2} \le x \le \frac{33}{2}\}$ **29.** $\{x | -23 < x < -7\}$ **31.** $\{x | -4 < x < \frac{3}{2}\}$

33. $\{x | -2 < x \le 4\}$ **35.** $\{x | -\infty < x < 1 \text{ or } \frac{5}{4} < x < \infty\}$ **37.** $\{x | 1 < x < 2 \text{ or } 3 < x < \infty\}$

39. $\{x | -\infty < x < -4 \text{ or } 2 < x < 4 \text{ or } 6 < x < \infty\}$ **41.** $\{x | -\frac{3}{2} < x < -\frac{7}{6}\}$ **43.** $\{x | -\infty < x \le -1 \text{ or } 6 \le x < \infty\}$

45. $\left\{\dfrac{-1 - \sqrt{3}i}{2}, \dfrac{-1 + \sqrt{3}i}{2}\right\}$ **47.** $\left\{\dfrac{-1 - \sqrt{17}}{4}, \dfrac{-1 + \sqrt{17}}{4}\right\}$ **49.** $\left\{\dfrac{1 - \sqrt{11}i}{2}, \dfrac{1 + \sqrt{11}i}{2}\right\}$ **51.** $\left\{\dfrac{1 - \sqrt{23}i}{2}, \dfrac{1 + \sqrt{23}i}{2}\right\}$ **53.** $\dfrac{y^2}{x^2}$

55. $\dfrac{1}{x^5 y}$ **57.** $\dfrac{125}{x^2 y}$ **59.** $4 - 7i$ **61.** $3 + 2i$ **63.** $\frac{9}{10} - \frac{9}{10}i$ **65.** 1 **67.** $-46 + 9i$ **69.** $2x + y - 3 = 0$ **71.** $x + 3 = 0$

73. $x + 5y + 10 = 0$ **75.** $2x - 3y + 19 = 0$ **77.** $-x + y + 4 = 0$

79.

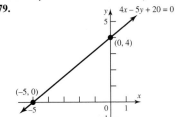

81.

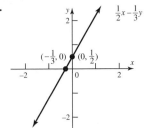

83. $\sqrt{2}x + \sqrt{3}y = \sqrt{6}$

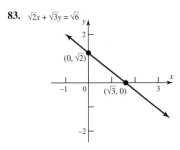

85. Center $(1, -2)$, radius $= 3$ **87.** Center $(1, -2)$, radius $= \sqrt{5}$ **89.** Intercept: $(0, 0)$; symmetric with respect to the x-axis
91. Intercepts: $(0, -1), (0, 1), (\frac{1}{2}, 0), (-\frac{1}{2}, 0)$; symmetric with respect to the x-axis, y-axis, and origin
93. Intercept: $(0, 1)$; symmetric with respect to the y-axis **95.** Intercepts: $(0, 0), (0, -2), (-1, 0)$; no symmetry
97. Yes. He could see about 229 miles. **99.** The sea plane can go as far as 616 miles.
101. (a) No **(b)** Mike wins again **(c)** Mike wins by $\frac{1}{4}$ meter **(d)** Mike should line up 5.26316 meters behind the start line **(e)** Yes

CHAPTER 2 *Exercise 2.1*

1. (a) -4 **(b)** -5 **(c)** -9 **(d)** -25 **3. (a)** 0 **(b)** $\frac{1}{2}$ **(c)** $-\frac{1}{2}$ **(d)** $\frac{3}{10}$ **5. (a)** 4 **(b)** 5 **(c)** 5 **(d)** 7
7. (a) $-\frac{1}{5}$ **(b)** $-\frac{3}{2}$ **(c)** $\frac{1}{8}$ **(d)** $\frac{7}{4}$ **9.** $f(0) = 3; f(-6) = -3$ **11.** Positive **13.** $-3, 6,$ and 10 **15.** $\{x | -6 \le x \le 11\}$
17. $-3, 6, 10$ **19.** 3 times **21. (a)** No **(b)** -3 **(c)** 14 **(d)** $\{x | x \ne 6\}$ **23. (a)** Yes **(b)** $\frac{8}{17}$ **(c)** $-1, 1$ **(d)** All real numbers
25. Not a function
27. Function **(a)** Domain: $\{x | -\pi \le x \le \pi\}$; Range: $\{y | -1 \le y \le 1\}$ **(b)** Intercepts: $(-\pi/2, 0), (\pi/2, 0), (0, 1)$ **(c)** y-axis
29. Not a function **31.** Function **(a)** Domain: $\{x | x > 0\}$; Range: all real numbers **(b)** Intercept: $(1, 0)$ **(c)** None
33. Function **(a)** Domain: all real numbers; Range: $\{y | y \le 2\}$ **(b)** Intercepts: $(-3, 0), (3, 0), (0, 2)$ **(c)** y-axis
35. Function **(a)** Domain: $\{x | x \ne 2\}$; Range: $\{y | y \ne 1\}$ **(b)** Intercept: $(0, 0)$ **(c)** None
37. All real numbers **39.** All real numbers **41.** $\{x | x \ne -1, x \ne 1\}$ **43.** $\{x | x \ne 0\}$ **45.** $\{x | x \ge 4\}$ **47.** $(-\infty, -3]$ or $[3, \infty)$
49. $(-\infty, 1)$ or $[2, \infty)$ **51.** $A = -\frac{7}{2}$ **53.** $A = -4$ **55.** $A = 8$; undefined at 3
57. (a) 15.1 m, 14.07 m, 12.94 m, 11.72 m **(b)** 2.02 sec **59.** $A(x) = \frac{1}{2}x^2$ **61.** $G(x) = 5x$
65. (a) $C(x) = 10x + 14\sqrt{x^2 - 10x + 29}, 0 < x < 5$ **(b)** $C(1) = \$72.61$ **(c)** $C(3) = \$69.60$

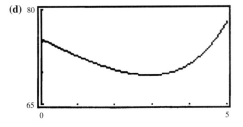

least cost: $x = 2.95$ miles

67. (a) $A(x) = (8.5 - 2x)(11 - 2x)$, **(b)** $0 \le x \le 4.25$, $0 \le A \le 93.5$ **(c)** $A(1) = 58.5$ in.^{2}, $A(1.2) = 52.46$ in.^{2}, $A(1.5) = 44$ in.^{2}
(d) $A(x) = 70$ when $x = 0.64$ in; $A(x) = 50$ when $x = 1.28$ in.

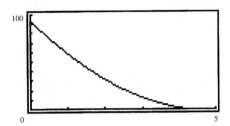

69. 1.11 sec; 0.46 sec   **71.** Function   **73.** Function   **75.** Not a function   **77.** Function   **79.** Only $h(x) = 2x$

Exercise 2.2

1. C   **3.** E   **5.** B   **7.** F
9. (a) Domain: $\{x \mid -3 \le x \le 4\}$; Range: $\{y \mid 0 \le y \le 3\}$ **(b)** Increasing on $[-3, 0]$ and on $[2, 4]$; decreasing on $[0, 2]$ **(c)** Neither
(d) $(-3, 0)$, $(0, 3)$, $(2, 0)$
11. (a) Domain: all real numbers; Range: $\{y \mid 0 < y < \infty\}$ **(b)** Inceasing on $(-\infty, \infty)$ **(c)** Neither   **(d)** $(0, 1)$
13. (a) Domain: $\{x \mid -\pi \le x \le \pi\}$; Range: $\{y \mid -1 \le y \le 1\}$
(b) Increasing on $[-\pi/2, \pi/2]$; decreasing on $[-\pi, -\pi/2]$ and on $[\pi/2, \pi]$ **(c)** Odd (symmetric with respect to the origin)
(d) $(-\pi, 0)$, $(0, 0)$, $(\pi, 0)$
15. (a) Domain: $\{x \mid x \ne 2\}$; Range: $\{y \mid y \ne 1\}$ **(b)** Decreasing on $(-\infty, 2)$ and on $(2, \infty)$ **(c)** Neither   **(d)** $(0, 0)$
17. (a) Domain: $\{x \mid x \ne 0\}$; Range: all real numbers   **(b)** Increasing on $(-\infty, 0)$ and on $(0, \infty)$ **(c)** Odd   **(d)** $(-1, 0)$, $(1, 0)$
19. (a) Domain: $\{x \mid x \ne -2, x \ne 2\}$; Range: $\{y \mid -\infty < y \le 0 \text{ and } 1 < y < \infty\}$
(b) Increasing on $(-\infty, -2)$ and on $(-2, 0]$; decreasing on $[0, 2)$ and on $(2, \infty)$ **(c)** Even   **(d)** $(0, 0)$
21. (a) Domain: $\{x \mid -4 \le x \le 4\}$; Range: $\{y \mid 0 \le y \le 2\}$ **(b)** Increasing on $[-2, 0]$ and $[2, 4]$; Decreasing on $[-4, -2]$ and $[0, 2]$ **(c)** Even
(d) $(-2, 0)$, $(0, 2)$, $(2, 0)$
23. (a) 2   **(b)** 3   **(c)** -4 **25. (a)** 4   **(b)** 2   **(c)** 5

27. (a) $-2x + 5$ **(b)** $-2x - 5$ **(c)** $4x + 5$ **(d)** $2x - 1$ **(e)** $\dfrac{2}{x} + 5 = \dfrac{5x + 2}{x}$ **(f)** $\dfrac{1}{2x + 5}$

29. (a) $2x^2 - 4$ **(b)** $-2x^2 + 4$ **(c)** $8x^2 - 4$ **(d)** $2x^2 - 12x + 14$ **(e)** $\dfrac{2 - 4x^2}{x^2}$ **(f)** $\dfrac{1}{2x^2 - 4}$

31. (a) $-x^3 + 3x$ **(b)** $-x^3 + 3x$ **(c)** $8x^3 - 6x$ **(d)** $x^3 - 9x^2 + 24x - 18$ **(e)** $\dfrac{1}{x^3} - \dfrac{3}{x}$ **(f)** $\dfrac{1}{x^3 - 3x}$

33. (a) $-\dfrac{x}{x^2 + 1}$ **(b)** $-\dfrac{x}{x^2 + 1}$ **(c)** $\dfrac{2x}{4x^2 + 1}$ **(d)** $\dfrac{x - 3}{x^2 - 6x + 10}$ **(e)** $\dfrac{x}{x^2 + 1}$ **(f)** $\dfrac{x^2 + 1}{x}$

35. (a) $|x|$ **(b)** $-|x|$ **(c)** $2|x|$ **(d)** $|x - 3|$ **(e)** $\dfrac{1}{|x|}$ **(f)** $\dfrac{1}{|x|}$

37. (a) $1 - \dfrac{1}{x}$ **(b)** $-1 - \dfrac{1}{x}$ **(c)** $1 + \dfrac{1}{2x}$ **(d)** $1 + \dfrac{1}{x - 3}$ **(e)** $1 + x$ **(f)** $\dfrac{x}{x + 1}$

39. 3   **41.** -3 **43.** $3x + 1$ **45.** $x(x + 1)$ **47.** $\dfrac{-1}{x + 1}$ **49.** $\dfrac{1}{\sqrt{x} + 1}$

51. Odd   **53.** Even   **55.** Odd   **57.** Neither   **59.** Even   **61.** Odd   **63.** At most one

65. (a) All real numbers
(b) $(0, -3)$, $(1, 0)$
(d) All real numbers

(c)

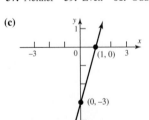

67. (a) All real numbers
(b) $(-2, 0)$, $(2, 0)$, $(0, -4)$
(d) $\{y \mid -4 \le y < \infty\}$

(c)

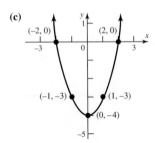

69. (a) All real numbers
(b) (0, 0)
(d) $\{y \mid -\infty < y \le 0\}$
(c)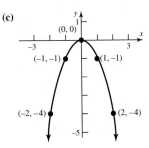

71. (a) $\{x \mid 2 \le x < \infty\}$
(b) (2, 0)
(d) $\{y \mid 0 \le y < \infty\}$
(c)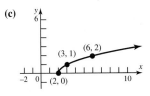

73. (a) $\{x \mid -\infty < x \le 2\}$
(b) (2, 0), (0, $\sqrt{2}$)
(d) $\{y \mid 0 \le y < \infty\}$
(c)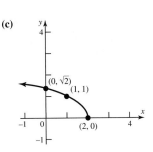

75. (a) All real numbers
(b) (0, 3)
(d) $\{y \mid 3 \le y < \infty\}$
(c)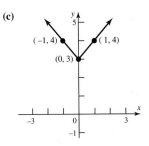

77. (a) All real numbers
(b) (0, 0)
(d) $\{y \mid -\infty < y \le 0\}$
(c)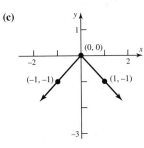

79. (a) All real numbers
(b) (0, 0)
(d) All real numbers
(c)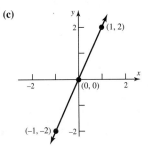

81. (a) All real numbers
(b) (0, 0), (−1, 0)
(d) All real numbers
(c)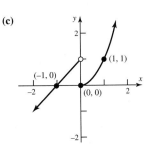

83. (a) $\{x \mid -2 \le x < \infty\}$
(b) (0, 1)
(d) $\{y \mid 0 < y < \infty\}$
(c)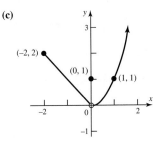

85. (a) All real numbers
(b) (0, 1)
(d) $\{-1, 1\}$
(c)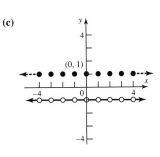

87. (a) All real numbers
(b) (x, 0) for $0 \le x < 1$
(d) Set of even integers
(c)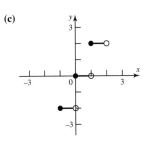

89. (a) All real numbers
(b) $(-2, 0)$, $(0, 4)$, $(2, 0)$
(d) $\{y/y \ge 0\}$
(c)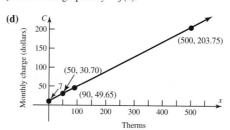

91. 2 **93.** $2x + h + 2$ **95.** $f(x) = \begin{cases} -x & \text{if } -1 \le x \le 0 \\ \frac{1}{2}x & \text{if } 0 < x \le 2 \end{cases}$ (Other answers are possible.)

97. $f(x) = \begin{cases} -x & \text{if } x \le 0 \\ -x + 2 & \text{if } 0 < x \le 2 \end{cases}$ (Other answers are possible.)

99. No; $f(-2) = 8$ and $f(2) = 6$
101. Each graph is that of $y = x^2$, but shifted vertically. If $y = x^2 + k$, $k > 0$, the shift is up k units; if $y = x^2 + k$, $k < 0$, the shift is down $|k|$ units.
103. Each graph is that of $y = |x|$, but either compressed or stretched. If $y = k|x|$ and $k > 1$, the graph is stretched; if $y = k|x|$, $0 < k < 1$, the graph is compressed.
105. The graph of $y = f(-x)$ is the reflection about the y-axis of the graph of $y = f(x)$.

107. (a) \$30.70
(b) \$203.75

(c) $C = \begin{cases} 7 + 0.47395x & \text{if } 0 \le x \le 90 \\ 15.83 + 0.37583x & \text{if } x > 90 \end{cases}$

(d)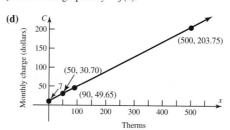

109. (a) $E(-x) = \frac{1}{2}[f(-x) + f(x)] = E(x)$ **(b)** $O(-x) = \frac{1}{2}[f(-x) - f(x)] = -\frac{1}{2}[f(x) - f(-x)] = -O(x)$
(c) $E(x) + O(x) = \frac{1}{2}[f(x) + f(-x)] + \frac{1}{2}[f(x) - f(-x)] = f(x)$ **(d)** Combine the results of parts **(a)**, **(b)**, and **(c)**.

Exercise 2.3

1. B **3.** H **5.** I **7.** L **9.** F **11.** G **13.** $y = (x - 4)^3$ **15.** $y = x^3 + 4$ **17.** $y = -x^3$ **19.** $y = 4x^3$

21.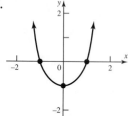

23.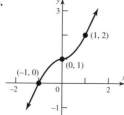

25.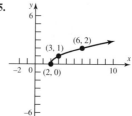

27.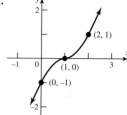

29.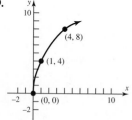

31.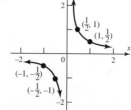

33.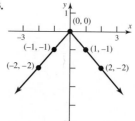

35.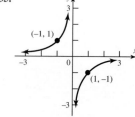

37.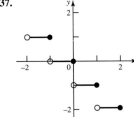

39.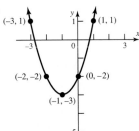

41.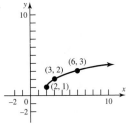

43.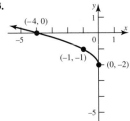

45.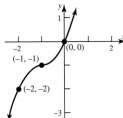

47.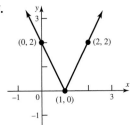

49.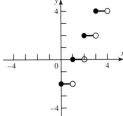

51. (a) $F(x) = f(x) + 3$

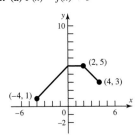

(b) $G(x) = f(x + 2)$

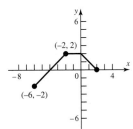

(c) $P(x) = -f(x)$

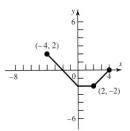

(d) $Q(x) = \frac{1}{2}f(x)$

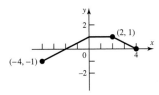

(e) $g(x) = f(-x)$

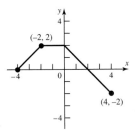

(f) $h(x) = 3f(x)$

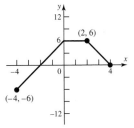

53. (a) $F(x) = f(x) + 3$

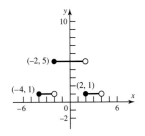

(b) $G(x) = f(x + 2)$

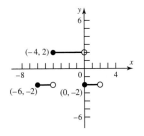

(c) $P(x) = -f(x)$

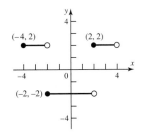

53. (d) $Q(x) = \frac{1}{2}f(x)$

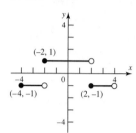

(e) $g(x) = f(-x)$

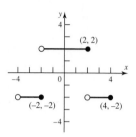

(f) $h(x) = 3f(x)$

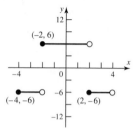

55. (a) $F(x) = f(x) + 3$

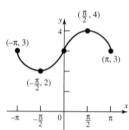

(b) $G(x) = f(x + 2)$

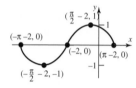

(c) $P(x) = -f(x)$

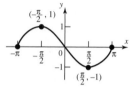

(d) $Q(x) = \frac{1}{2}f(x)$

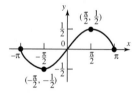

(e) $g(x) = f(-x)$

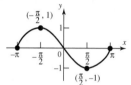

(f) $h(x) = 3f(x)$

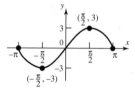

57. $f(x) = (x + 1)^2 - 1$

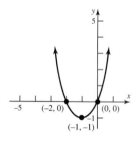

59. $f(x) = (x - 4)^2 - 15$

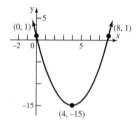

61. $f(x) = (x + \frac{1}{2})^2 + \frac{3}{4}$

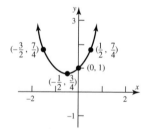

63.

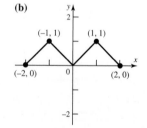

65.

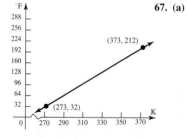

67. (a)

(b)

AN12

69. (a)

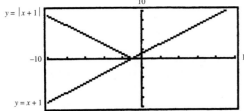

$y = |x + 1|$

10

−10

10

−10

$y = x + 1$

(b)

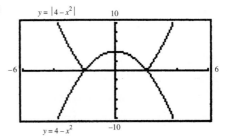

$y = |4 − x^2|$

10

−6

6

−10

$y = 4 − x^2$

(c)

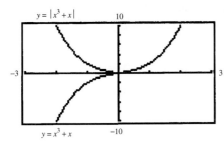

$y = |x^3 + x|$

10

−3

3

−10

$y = x^3 + x$

(d) Any part of the graph of $y = f(x)$ that lies below the x-axis is reflected about the x-axis to obtain the graph of $y = |f(x)|$

Exercise 2.4

1. (a) $(f + g)(x) = 5x + 1$; all real numbers   **(b)** $(f − g)(x) = x + 7$; all real numbers   **(c)** $(f \cdot g)(x) = 6x^2 − x − 12$; all real numbers
   (d) $\left(\dfrac{f}{g}\right)(x) = \dfrac{3x + 4}{2x − 3}$; all real numbers x except $x = \frac{3}{2}$
3. (a) $(f + g)(x) = 2x^2 + x − 1$; all real numbers   **(b)** $(f − g)(x) = −2x^2 + x − 1$; all real numbers   **(c)** $(f \cdot g)(x) = 2x^3 − 2x^2$; all real numbers
   (d) $\left(\dfrac{f}{g}\right)(x) = \dfrac{x − 1}{2x^2}$; $\{x | x \neq 0\}$
5. (a) $(f + g)(x) = \sqrt{x} + 3x − 5$; $\{x | x \geq 0\}$ **(b)** $(f − g)(x) = \sqrt{x} − 3x + 5$; $\{x | x \geq 0\}$ **(c)** $(f \cdot g)(x) = 3x\sqrt{x} − 5\sqrt{x}$; $\{x | x \geq 0\}$
   (d) $\left(\dfrac{f}{g}\right)(x) = \dfrac{\sqrt{x}}{3x − 5}$; $\{x | x \geq 0, x \neq \frac{5}{3}\}$
7. (a) $(f + g)(x) = 1 + \dfrac{2}{x}$; $\{x | x \neq 0\}$ **(b)** $(f − g)(x) = 1$; $\{x | x \neq 0\}$ **(c)** $(f \cdot g)(x) = \dfrac{1}{x} + \dfrac{1}{x^2}$; $\{x | x \neq 0\}$ **(d)** $\left(\dfrac{f}{g}\right)(x) = x + 1$; $\{x | x \neq 0\}$
9. (a) $(f + g)(x) = \dfrac{6x + 3}{3x − 2}$; $\{x | x \neq \frac{2}{3}\}$ **(b)** $(f − g)(x) = \dfrac{−2x + 3}{3x − 2}$; $\{x | x \neq \frac{2}{3}\}$ **(c)** $(f \cdot g)(x) = \dfrac{8x^2 + 12x}{(3x − 2)^2}$; $\{x | x \neq \frac{2}{3}\}$
   (d) $\left(\dfrac{f}{g}\right)(x) = \dfrac{2x + 3}{4x}$; $\{x | x \neq 0, \frac{2}{3}\}$ **11.** $g(x) = 5 − \frac{7}{2}x$

13.

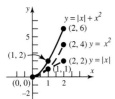

y

$y = |x| + x^2$

(2, 6)

5

(2, 4) $y = x^2$

(1, 2)

(2, 2) $y = |x|$

(1, 1)

x

(0, 0)   1   2

−2

15.

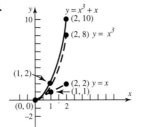

y

$y = x^3 + x$

10

(2, 10)

(2, 8) $y = x^3$

(1, 2)

(2, 2) $y = x$

(1, 1)

x

(0, 0)   1   2

−2

17. (a) 98   **(b)** 49   **(c)** 4   **(d)** 4   **19. (a)** 97   **(b)** $−\frac{163}{2}$ **(c)** 1   **(d)** $−\frac{3}{2}$ **21. (a)** $2\sqrt{2}$ **(b)** $2\sqrt{2}$ **(c)** 1   **(d)** 0
23. (a) $\frac{1}{17}$ **(b)** $\frac{1}{5}$ **(c)** 1   **(d)** $\frac{1}{2}$ **25. (a)** $\frac{3}{5}$ **(b)** $\sqrt{15}/5$ **(c)** $\frac{12}{13}$ **(d)** 0
27. (a) $(f \circ g)(x) = 6x + 3$ **(b)** $(g \circ f)(x) = 6x + 9$ **(c)** $(f \circ f)(x) = 4x + 9$ **(d)** $(g \circ g)(x) = 9x$
29. (a) $(f \circ g)(x) = 3x^2 + 1$ **(b)** $(g \circ f)(x) = 9x^2 + 6x + 1$ **(c)** $(f \circ f)(x) = 9x + 4$ **(d)** $(g \circ g)(x) = x^4$
31. (a) $(f \circ g)(x) = \sqrt{x^2 − 1}$ **(b)** $(g \circ f)(x) = x − 1$ **(c)** $(f \circ f)(x) = \sqrt[4]{x}$ **(d)** $(g \circ g)(x) = x^4 − 2x^2$

33. (a) $(f \circ g)(x) = \dfrac{1-x}{1+x}$ **(b)** $(g \circ f)(x) = \dfrac{x+1}{x-1}$ **(c)** $(f \circ f)(x) = -\dfrac{1}{x}$ **(d)** $(g \circ g)(x) = x$

35. (a) $(f \circ g)(x) = x$ **(b)** $(g \circ f)(x) = |x|$ **(c)** $(f \circ f)(x) = x^4$ **(d)** $(g \circ g)(x) = \sqrt[4]{x}$

37. (a) $(f \circ g)(x) = \dfrac{1}{4x+9}$ **(b)** $(g \circ f)(x) = \dfrac{2}{2x+3} + 3 = \dfrac{6x+11}{2x+3}$ **(c)** $(f \circ f)(x) = \dfrac{2x+3}{6x+11}$ **(d)** $(g \circ g)(x) = 4x+9$

39. (a) $(f \circ g)(x) = acx + ad + b$ **(b)** $(g \circ f)(x) = acx + bc + d$ **(c)** $(f \circ f)(x) = a^2x + ab + b$ **(d)** $(g \circ g)(x) = c^2x + cd + d$

41. $(f \circ g)(x) = f(g(x)) = f(\frac{1}{2}x) = 2(\frac{1}{2}x) = x; (g \circ f)(x) = g(f(x)) = g(2x) = \frac{1}{2}(2x) = x$

43. $(f \circ g)(x) = f(\sqrt[3]{x}) = (\sqrt[3]{x})^3 = x; (g \circ f)(x) = g(x^3) = \sqrt[3]{x^3} = x$

45. $(f \circ g)(x) = f(\frac{1}{2}(x+6)) = 2[\frac{1}{2}(x+6)] - 6 = x + 6 - 6 = x; (g \circ f)(x) = g(2x - 6) = \frac{1}{2}(2x - 6 + 6) = x$

47. $(f \circ g)(x) = f\left(\dfrac{1}{a}(x-b)\right) = a\left[\dfrac{1}{a}(x-b)\right] + b = x; (g \circ f)(x) = g(ax + b) = \dfrac{1}{a}\left(ax + b - b\right) = x$

49. $(f \circ g)(x) = 11; (g \circ f)(x) = 2$ **51.** $(f \circ (g \circ h))(x) = 5 - 3x + 4\sqrt{1 - 3x}$ **53.** $((f + g) \circ h)(x) = 9x^2 - 6x + 3 + \sqrt{1 - 3x}$ **55.** $F = f \circ g$

57. $H = h \circ f$ **59.** $q = f \circ h$ **61.** $P = f \circ f$ **63.** $f(x) = x^4; g(x) = 2x + 3$ **65.** $f(x) = \sqrt{x}; g(x) = x^2 + x + 1$ **67.** $f(x) = x^2; g(x) = 1 - \dfrac{1}{x^2}$

69. $f(x) = [[x]]; g(x) = x^2 + 1$ **71.** $-3, 3$ **73.** $S(r(t)) = \frac{16}{9}\pi t^6$ **75.** $C(N(t)) = 15000 + 800,000t - 40,000t^2$ **77.** $C = \dfrac{2\sqrt{100 - p}}{25} + 600$

79. Since f and g are odd, $f(-x) = -f(x)$ and $g(-x) = -g(x); (f \circ g)(-x) = f(g(-x)) = f(-g(x)) = -f(g(x)) = -(f \circ g)(x)$; thus, $f \circ g$ is odd.

Exercise 2.5

1. One-to-one
3. Not one-to-one
5. One-to-one

7.

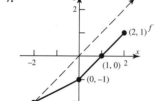

9.

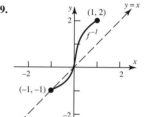

11.

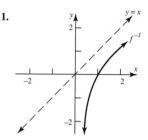

13. $f(g(x)) = f(\frac{1}{3}(x - 4)) = 3[\frac{1}{3}(x - 4)] + 4 = x; g(f(x)) = g(3x + 4) = \frac{1}{3}[(3x + 4) - 4] = x$

15. $f(g(x)) = 4\left[\dfrac{x}{4} + 2\right] - 8 = x; g(f(x)) = \dfrac{4x - 8}{4} + 2 = x$ **17.** $f(g(x)) = (\sqrt[3]{x + 8})^3 - 8 = x; g(f(x)) = \sqrt[3]{(x^3 - 8) + 8} = x$

19. $f(g(x)) = \dfrac{1}{1/x} = x; g(f(x)) = \dfrac{1}{1/x} = x$ **21.** $f(g(x)) = \dfrac{2\left(\dfrac{4x - 3}{2 - x}\right) + 3}{\dfrac{4x - 3}{2 - x} + 4} = \dfrac{8x - 3x}{-3 + 8} = x; g(f(x)) = \dfrac{4\left(\dfrac{2x + 3}{x + 4}\right) - 3}{2 - \dfrac{2x + 3}{x + 4}} = x$

23. $f^{-1}(x) = \frac{1}{3}x$
$f(f^{-1}(x)) = 3(\frac{1}{3}x) = x$
$f^{-1}(f(x)) = \frac{1}{3}(3x) = x$
Domain $f =$ Range $f^{-1} = (-\infty, \infty)$
Range $f =$ Domain $f^{-1} = (-\infty, \infty)$

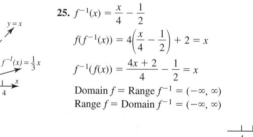

25. $f^{-1}(x) = \dfrac{x}{4} - \dfrac{1}{2}$
$f(f^{-1}(x)) = 4\left(\dfrac{x}{4} - \dfrac{1}{2}\right) + 2 = x$
$f^{-1}(f(x)) = \dfrac{4x + 2}{4} - \dfrac{1}{2} = x$
Domain $f =$ Range $f^{-1} = (-\infty, \infty)$
Range $f =$ Domain $f^{-1} = (-\infty, \infty)$

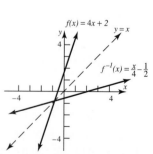

27. $f^{-1}(x) = \sqrt[3]{x+1}$
$f(f^{-1}(x)) = (\sqrt[3]{x+1})^3 - 1 = x$
$f^{-1}(f(x)) = \sqrt[3]{x^3 - 1 + 1} = x$
Domain f = Range $f^{-1} = (-\infty, \infty)$
Range f = Domain $f^{-1} = (-\infty, \infty)$

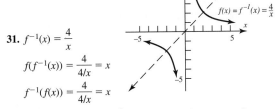

29. $f^{-1}(x) = \sqrt{x-4}$
$f(f^{-1}(x)) = (\sqrt{x-4})^2 + 4 = x$
$f^{-1}(f(x)) = \sqrt{(x^2+4) - 4} = \sqrt{x^2} = |x| = x$
Domain f = Range $f^{-1} = [0, \infty)$
Range f = Domain $f^{-1} = [4, \infty)$

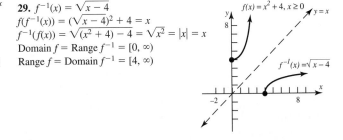

31. $f^{-1}(x) = \dfrac{4}{x}$
$f(f^{-1}(x)) = \dfrac{4}{4/x} = x$
$f^{-1}(f(x)) = \dfrac{4}{4/x} = x$
Domain f = Range f^{-1} = All real numbers except 0
Range f = Domain f^{-1} = All real numbers except 0

33. $f^{-1}(x) = \dfrac{2x+1}{x}$
$f(f^{-1}(x)) = \dfrac{1}{\dfrac{2x+1}{x} - 2} = x$
$f^{-1}(f(x)) = \dfrac{2\left(\dfrac{1}{x-2}\right) + 1}{\dfrac{1}{x-2}} = x$
Domain f = Range f^{-1} = All real numbers except 2
Range f = Domain f^{-1} = All real numbers except 0

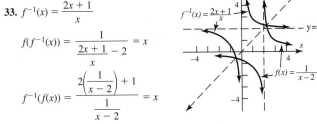

35. $f^{-1}(x) = \dfrac{2 - 3x}{x}$
$f(f^{-1}(x)) = \dfrac{2}{3 + \dfrac{2 - 3x}{x}} = x$
$f^{-1}(f(x)) = \dfrac{2 - 3\left(\dfrac{2}{3+x}\right)}{\dfrac{2}{3+x}} = x$
Domain f = Range f^{-1} = All real numbers except -3
Range f = Domain f^{-1} = All real numbers except 0

37. $f^{-1}(x) = \sqrt{x} - 2$
$f(f^{-1}(x)) = (\sqrt{x} - 2 + 2)^2 = x$
$f^{-1}(f(x)) = \sqrt{(x+2)^2} - 2 = |x+2| - 2 = x$
Domain f = Range $f^{-1} = [-2, \infty)$
Range f = Domain $f^{-1} = [0, \infty)$

39. $f^{-1}(x) = \dfrac{x}{x-2}$
$f(f^{-1}(x)) = \dfrac{2\left(\dfrac{x}{x-2}\right)}{\dfrac{x}{x-2} - 1} = x$
$f^{-1}(f(x)) = \dfrac{\dfrac{2x}{x-1}}{\dfrac{2x}{x-1} - 2} = x$
Domain f = Range f^{-1} = All real numbers except 1
Range f = Domain f^{-1} = All real numbers except 2

41. $f^{-1}(x) = \dfrac{3x+4}{2x-3}$
$f(f^{-1}(x)) = \dfrac{3\left(\dfrac{3x+4}{2x-3}\right) + 4}{2\left(\dfrac{3x+4}{2x-3}\right) - 3} = x$
$f^{-1}(f(x)) = \dfrac{3\left(\dfrac{3x+4}{2x-3}\right) + 4}{2\left(\dfrac{3x+4}{2x-3}\right) - 3} = x$
Domain f = Range f^{-1} = All real numbers except $\frac{3}{2}$
Range f = Domain f^{-1} = All real numbers except $\frac{3}{2}$

43. $f^{-1}(x) = \dfrac{-2x + 3}{x - 2}$

$$f(f^{-1}(x)) = \dfrac{2\left(\dfrac{-2x + 3}{x - 2}\right) + 3}{\dfrac{-2x + 3}{x - 2} + 2} = x$$

$$f^{-1}(f(x)) = \dfrac{-2\left(\dfrac{2x + 3}{x + 2}\right) + 3}{\dfrac{2x + 3}{x + 2} - 2} = x$$

Domain f = Range f^{-1} = All real numbers except -2
Range f = Domain f^{-1} = All real numbers except 2

45. $f^{-1}(x) = \dfrac{x^3}{8}$

$$f(f^{-1}(x)) = 2\sqrt[3]{\dfrac{x^3}{8}} = x$$

$$f^{-1}(f(x)) = \dfrac{(2\sqrt[3]{x})^3}{8} = x$$

Domain f = Range f^{-1} = $(-\infty, \infty)$
Range f = Domain f^{-1} = $(-\infty, \infty)$

47. $f^{-1}(x) = \dfrac{1}{m}(x - b)$, $m \neq 0$ **49.** No; whenever x and $-x$ are in the domain of f, two equal y values, $f(x)$ and $f(-x)$, are present.

51. Quadrant I   **53.** $f(x) = |x|$, $x \geq 0$ is one-to-one; this may be written as $f(x) = x$; $f^{-1}(x) = x$

55. $f(g(x)) = \frac{9}{5}\left[\frac{5}{9}(x - 32)\right] + 32 = x$; $g(f(x)) = \frac{5}{9}\left[\left(\frac{9}{5}x + 32\right) - 32\right] = x$ **57.** $l(T) = gT^2/4\pi^2$ **59.** $f^{-1}(x) = \dfrac{-dx + b}{cx - a}$; $f = f^{-1}$ if $a = -d$

Exercise 2.6

1. $V(r) = 2\pi r^3$ **3.** $R(x) = -\frac{1}{6}x^2 + 100x$ **5.** $R(x) = -\frac{1}{5}x^2 + 20x$ **7. (a)** $A(x) = -x^2 + 200x$ **(b)** $0 < x < 200$ **(c)** A is largest when $x = 100$ yards

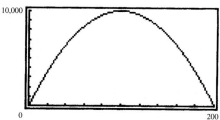

9. (a) $C(x) = x$ **(b)** $A(x) = \dfrac{x^2}{4\pi}$ **11.** $A(x) = \frac{1}{2}x^4$ **13. (a)** $d(x) = \sqrt{x^4 - 15x^2 + 64}$ **(b)** $d(0) = 8$ **(c)** $d(1) = \sqrt{50} = 7.07$

15. $d(x) = \sqrt{x^2 - x + 1}$ **17.** $d(t) = 50t$ **19. (a)** $V(x) = x(24 - 2x)^2$ **(b)** The volume is largest when x is 4

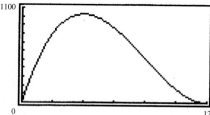

21. (a) $A(x) = 2x^2 + \dfrac{40}{x}$ **(b)** The area is smallest when x is about 2.15   **23. (a)** $A(x) = x(16 - x^2)$ **(b)** Domain: $\{x \mid 0 < x < 4\}$
(c) The area is largest for x about 2.31

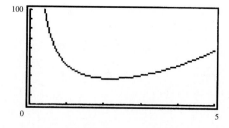

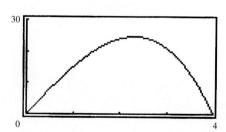

25. (a) $A(x) = 4x(4 - x^2)^{1/2}$ **(b)** $p(x) = 4x + 4(4 - x^2)^{1/2}$
   (c) The area is largest for x about 1.41

(d) The perimeter is largest for x about 1.41

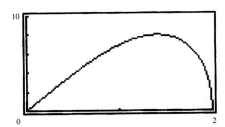

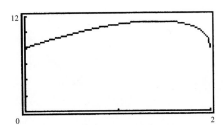

27. (a) $C(r) = 12\pi r^2 + \dfrac{4000}{r}$

29. (a) $A(x) = x^2 + \dfrac{25 - 20x + 4x^2}{\pi}$ **(b)** Domain: $\{x \mid 0 < x < 2.5\}$

   (b) The cost in least for r about 3.75 cm

(c) The area is smallest for x about 1.40 meters

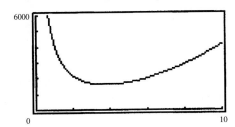

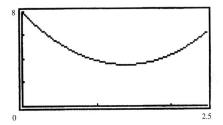

31. (a) $A(r) = 2r^2$ **(b)** $p(r) = 6r$

33. $A(x) = \left(\dfrac{\pi}{3} - \dfrac{\sqrt{3}}{4} \right) x^2$

35. $C = \begin{cases} 95 & \text{if } x = 7 \\ 119 & \text{if } 7 < x \le 8 \\ 143 & \text{if } 8 < x \le 9 \\ 167 & \text{if } 9 < x \le 10 \\ 190 & \text{if } 10 < x \le 14 \end{cases}$

37. $V(h) = \dfrac{\pi}{48}h^3$

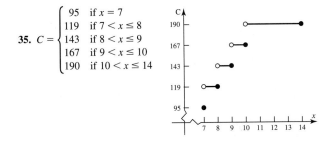

Fill-in-the-Blank Items

1. independent; dependent **2.** vertical **3.** even; odd **4.** horizontal; right **5.** $g(f(x)) = (g \circ f)(x)$ **6.** one-to-one **7.** $y = x$

True/False Items

1. T **2.** T **3.** F **4.** T **5.** F **6.** F **7.** T

Review Exercises

1. $f(x) = -2x + 3$ **3.** $A = 11$ **5. (a)** B, C, D **(b)** D **7. (a)** $f(-x) = \dfrac{-3x}{x^2 - 4}$ **(b)** $-f(x) = \dfrac{-3x}{x^2 - 4}$ **(c)** $f(x + 2) = \dfrac{3x + 6}{x^2 + 4x}$

   (d) $f(x - 2) = \dfrac{3x - 6}{x^2 - 4x}$ **9. (a)** $f(-x) = \sqrt{x^2 - 4}$ **(b)** $-f(x) = -\sqrt{x^2 - 4}$ **(c)** $f(x + 2) = \sqrt{x^2 + 4x}$ **(d)** $f(x - 2) = \sqrt{x^2 - 4x}$

11. (a) $f(x) = \dfrac{x^2 - 4}{x^2}$ **(b)** $-f(x) = -\dfrac{x^2 - 4}{x^2}$ **(c)** $f(x + 2) = \dfrac{x^2 + 4x}{x^2 + 4x + 4}$ **(d)** $f(x - 2) = \dfrac{x^2 - 4x}{x^2 - 4x + 4}$ **13.** Odd **15.** Even

17. Neither **19.** $\{x \mid x \ne -3, x \ne 3\}$ **21.** $(-\infty, 2]$ **23.** $(0, \infty)$ **25.** $\{x \mid x \ne -3, x \ne 1\}$ **27.** $[-1, \infty)$ **29.** $[0, \infty)$

31. (a) All real numbers
(b) $(0, -4)$, $(-4, 0)$, $(4, 0)$
(d) $\{y \mid -4 \le y < \infty\}$
(c)

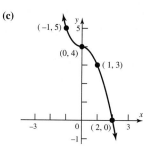

33. (a) All real numbers
(b) $(0, 0)$
(d) $\{y \mid -\infty < y \le 0\}$
(c)

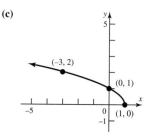

35. (a) $\{x \mid 1 \le x < \infty\}$
(b) $(1, 0)$
(d) $\{y \mid 0 \le y < \infty\}$
(c)

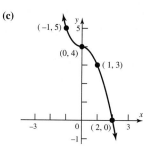

37. (a) $\{x \mid -\infty < x \le 1\}$
(b) $(1, 0)$, $(0, 1)$
(d) $\{y \mid 0 \le y < \infty\}$
(c)

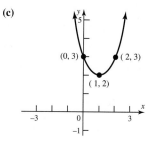

39. (a) All real numbers
(b) $(0, 4)$, $(2, 0)$
(d) All real numbers
(c)

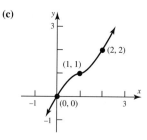

41. (a) All real numbers
(b) $(0, 3)$
(d) $\{y \mid 2 \le y < \infty\}$
(c)

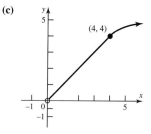

43. (a) All real numbers
(b) $(0, 0)$
(d) All real numbers
(c)

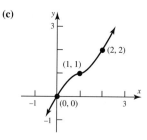

45. (a) $\{x \mid 0 < x < \infty\}$
(b) None
(d) $\{y \mid 0 < y < \infty\}$
(c)

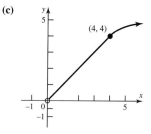

47. (a) $\{x \mid x \ne 1\}$
(b) $(0, 0)$

(d) $\{y \mid y \ne 1\}$
(c)

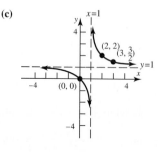

49. (a) All real numbers
(b) $-1 < x \le 0$ are x-intercepts
    0 is the y-intercept
(d) Set of integers
(c)

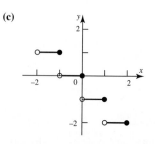

51. $f^{-1}(x) = \dfrac{2x + 3}{5x - 2}$

$$f(f^{-1}(x)) = \dfrac{2\left(\dfrac{2x + 3}{5x - 2}\right) + 3}{5\dfrac{2x + 3}{5x - 2} - 2} = x$$

$$f^{-1}(f(x)) = \dfrac{2\left(\dfrac{2x + 3}{5x - 2}\right) + 3}{5\dfrac{2x + 3}{5x - 2} - 2} = x$$

Domain f = Range f^{-1} = All real numbers except 2/5
Range f = Domain f^{-1} = All real numbers except 2/5

53. $f^{-1}(x) = \dfrac{x + 1}{x}$

$$f(f^{-1}(x)) = \dfrac{1}{\dfrac{x + 1}{x} - 1} = x$$

$$f^{-1}(f(x)) = \dfrac{\dfrac{1}{x - 1} + 1}{\dfrac{1}{x - 1}} = x$$

Domain f = Range f^{-1} = All real numbers except 1
Range f = Domain f^{-1} = All real numbers except 0

55. $f^{-1}(x) = \dfrac{27}{x^3}$

$$f(f^{-1}(x)) = \dfrac{3}{(27/x^3)^{1/3}} = x$$

$$f^{-1}(f(x)) = \dfrac{27}{(3/x^{1/3})^3} = x$$

Domain f = Range f^{-1} = All real numbers except 0
Range f = Domain f^{-1} = All real numbers except 0

57. (a) -26 (b) -241 (c) 16 (d) -1 **59.** (a) $\sqrt{11}$ (b) 1 (c) $\sqrt{\sqrt{6} + 2}$ (d) 19 **61.** (a) $\frac{1}{20}$ (b) $-\frac{13}{8}$ (c) $\frac{400}{1601}$ (d) -17

63. $(f \circ g)(x) = \dfrac{-3x + 1}{3x + 1}$; $(g \circ f)(x) = \dfrac{6 - 2x}{x}$; $(f \circ f)(x) = \dfrac{3x - 2}{2 - x}$; $(g \circ g)(x) = 9x + 4$

65. $(f \circ g)(x) = 27x^2 + 3|x| + 1$; $(g \circ f)(x) = 3|3x^2 + x + 1|$; $(f \circ f)(x) = 3(3x^2 + x + 1)^2 + 3x^2 + x + 2$; $(g \circ g)(x) = 9|x|$

67. $(f \circ g)(x) = \dfrac{1 + x}{1 - x}$; $(g \circ f)(x) = \dfrac{x - 1}{x + 1}$; $(f \circ f)(x) = x$; $(g \circ g)(x) = x$

69. (a)
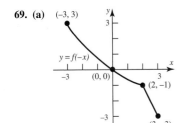
(b)
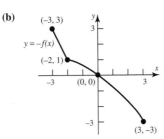
(c)
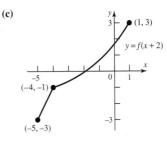

(d)
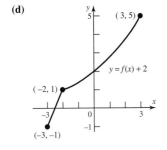
(e)
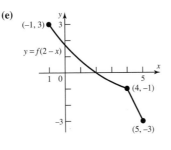
(f)
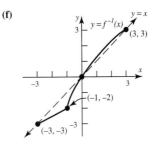

71. $T(h) = -0.0025h + 30$ **73.** $S(x) = kx(36 - x^2)^{3/2}$; Domain: $\{x|0 < x < 6\}$

CHAPTER 3 *Exercise 3.1*

1. D **3.** A **5.** B **7.** E

9.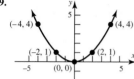

11.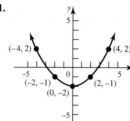

13.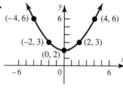

15.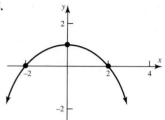

17. $f(x) = (x + 2)^2 - 2$

19. $f(x) = 2(x - 1)^2 - 1$

21. $f(x) = -(x + 1)^2 + 1$

23. $f(x) = \frac{1}{2}(x + 1)^2 - \frac{3}{2}$

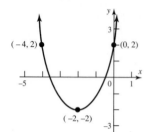

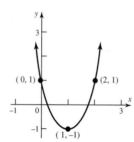

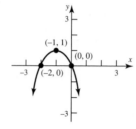

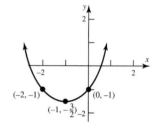

25.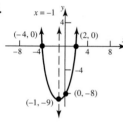

27.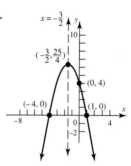

29.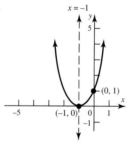

31.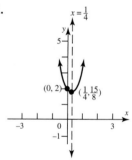

33.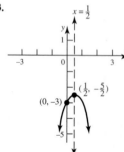

35.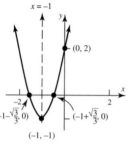

37.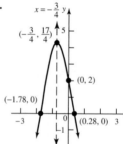

39. Minimum value; -21
41. Maximum value; 21
43. Maximum value; 13

45. Opens up;
vertex at $(-1, f(-1))$;
axis of symmetry $x = -1$

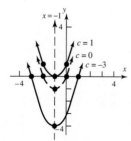

47. Each parabola opens up and passes through $(0, 1)$. Each one has the same shape.

49. Price: $500; maximum revenue: $1,000,000   **51.** 10,000 ft^2; 100 ft by 100 ft   **53.** 2,000,000 m^2 **55.** 4,166,666.7 m^2
57. (a) 219.53 ft   (b) 170.02 ft   (c) When the height is 100 ft, the projectile is 135.69 ft from the cliff   **59.** 8 PM   **61.** 18.75 m   **63.** 3 in.

65. Width $= \dfrac{40}{\pi + 4} \approx 5.6$ ft; length ≈ 2.8 ft   **67.** $x = \dfrac{16}{6 - \sqrt{3}} \approx 3.75$ ft; other side ≈ 2.38 ft   **69.** 70 members

71. $a = 6$, $b = 0$, $c = 2$;   **73.** $\dfrac{a}{2}$ **75.** $\left. \begin{array}{c} ah^2 - bh + c = y_0 \\ c = y_1 \\ ah^2 + bh + c = y_2 \end{array} \right\}$ $\left. \begin{array}{c} y_0 + y_2 = 2ah^2 + 2c \\ 4y_1 = 4c \end{array} \right\}$ Area $= \dfrac{h}{3}(2ah^2 + 6c) = \dfrac{h}{3}(y_0 + 4y_1 + y_2)$
$f(x) = 6x^2 + 2$

Exercise 3.2

1. Yes; degree 3   **3.** Yes; degree 2   **5.** No; x is raised to the -1 power.   **7.** No; x is raised to the $\frac{3}{2}$ power.   **9.** Yes; degree 4

11. 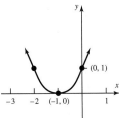   **13.** 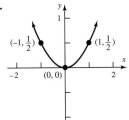   **15.** 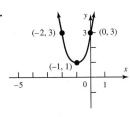   **17.**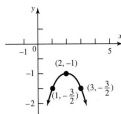

19. 7, multiplicity 1; -3, multiplicity 2; graph touches the x-axis at -3 and crosses it at 7   **21.** 2, multiplicity 3; graph crosses the x-axis at 2
23. $-\frac{1}{2}$, multiplicity 2; graph touches the x-axis at $-\frac{1}{2}$ **25.** 5, multiplicity 3; -4, multiplicity 2; graph touches the x-axis at -4 and crosses it at 5
27. No real zeros; graph neither crosses nor touches the x-axis

29. (a) x-intercept: 1; y-intercept: 1
   (b) Touches at 1
   (c) $y = x^2$
   (d) 1

   (e)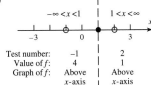

   (f)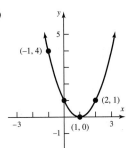

31. (a) x-intercepts: 0, 3; y-intercept: 0
   (b) Touches at 0; crosses at 3
   (c) $y = x^3$
   (d) 2

   (e)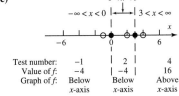

   (f)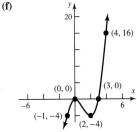

33. (a) x-intercepts: $-4, 0$; y-intercept: 0
(b) Crosses at $-4, 0$
(c) $y = 6x^4$
(d) 3

(e)

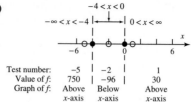

(f)

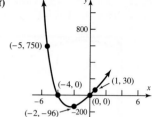

35. (a) x-intercepts: $-2, 0$; y-intercept: 0
(b) Crosses at -2; touches at 0
(c) $y = -4x^3$
(d) 2

(e)

(f)

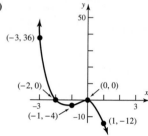

37. (a) x-intercepts: $-4, 0, 2$; y-intercept: 0
(b) Crosses at $-4, 0, 2$
(c) $y = x^3$
(d) 2

(e)

(f)

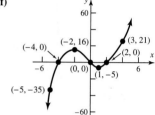

39. $f(x) = 4x - x^3 = -x(x^2 - 4) = -x(x + 2)(x - 2)$
(a) x-intercepts: $-2, 0, 2$; y-intercept: 0
(b) Crosses at $-2, 0, 2$
(c) $y = -x^3$
(d) 2

(e)

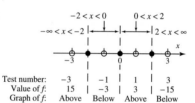

(f)

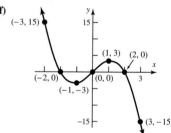

41. (a) x-intercepts: $-2, 0, 2$; y-intercept: 0
(b) Crosses at $-2, 2$; touches at 0
(c) $y = x^4$
(d) 3

(e)

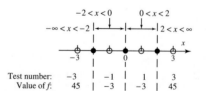

(f)

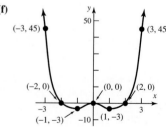

43. (a) x-intercepts: 0, 2; y-intercept: 0
(b) Touches at 0, 2
(c) $y = x^4$
(d) 3

(e)

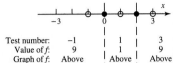

Test number:	-1	1	3
Value of f:	9	1	9
Graph of f:	Above x-axis	Above x-axis	Above x-axis

(f)

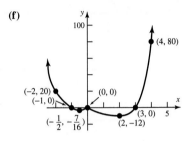

45. (a) x-intercepts: -1, 0, 3; y-intercept: 0
(b) Crosses at -1, 3; touches at 0
(c) $y = x^4$
(d) 3

(e)

Test number:	-2	$-\frac{1}{2}$	2	4
Value of f:	20	$-\frac{7}{16}$	-12	80
Graph of f:	Above x-axis	Below x-axis	Below x-axis	Above x-axis

(f)

47. (a) x-intercepts: -2, 0, 4, 6; y-intercept: 0
(b) Crosses at -2, 0, 4, 6
(c) $y = x^4$
(d) 3

(e)

Test number:	-3	-1	2	5	7
Value of f:	189	-35	64	-35	189
Graph of f:	Above x-axis	Below x-axis	Above x-axis	Below x-axis	Above x-axis

(f)

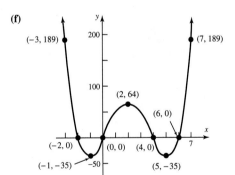

49. (a) x-intercepts: 0, 2; y-intercept: 0
(b) Touches at 0; crosses at 2
(c) $y = x^5$
(d) 4

(e)

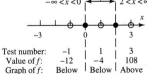

Test number:	-1	1	3
Value of f:	-12	-4	108
Graph of f:	Below x-axis	Below x-axis	Above x-axis

(f)

51. c, e, f **53.** c, e

55. *x*-intercept: 0.83
  turning points: $(-0.50, -1.53)$, $(0.20, -2.11)$

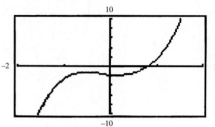

57. *x*-intercepts: -1.06, 1.61
  turning point: $(-0.41, -4.64)$

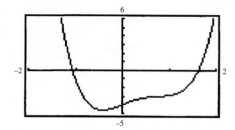

59. *x*-intercept: -0.97

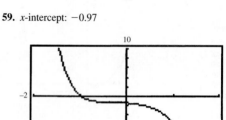

Exercise 3.3

1. All real numbers except 3   **3.** All real numbers except 2 and -4 **5.** All real numbers except $-\frac{1}{2}$ and 3   **7.** All real numbers except 2
9. All real numbers   **11.** (a) Domain: $\{x|x \neq 2\}$; Range: $\{y|y \neq 1\}$ **(b)** (0, 0)   **(c)** $y = 1$ **(d)** $x = 2$ **(e)** None
13. (a) Domain: $\{x|x \neq 0\}$; Range: all real numbers   **(b)** $(-1, 0)$, $(1, 0)$ **(c)** None   **(d)** None   **(e)** $y = 2x$
15. (a) Domain: $\{x|x \neq -2, x \neq 2\}$; Range: $\{y|-\infty < y \leq 0, 1 < y < \infty\}$ **(b)** (0, 0)   **(c)** $y = 1$ **(d)** $x = -2, x = 2$ **(e)** None

17.

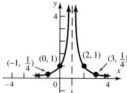

19.

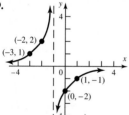

21.

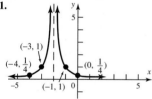

23.

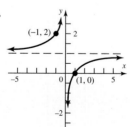

25.

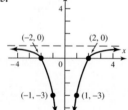

27. Horizontal asymptote: $y = 3$; vertical asymptote: $x = -4$ **29.** No asymptotes
31. Horizontal asymptote: $y = 0$; vertical asymptotes: $x = 1, x = -1$ **33.** Horizontal asymptote: $y = 0$; vertical asymptote: $x = 0$
35. Oblique asymptote: $y = 3x$; vertical asymptote: $x = 0$

37. 1. x-intercept: -1; no y-intercept
   2. No symmetry
   3. Vertical asymptotes: $x = 0$, $x = -4$
   4. Horizontal asymptote: $y = 0$, intersected at $(-1, 0)$
   5. $x < -4$: below x-axis
      $-4 < x < -1$: above x-axis
      $-1 < x < 0$: below x-axis
      $x > 0$: above x-axis

6.
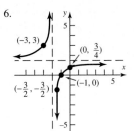

39. 1. x-intercept: -1; y-intercept: $\frac{3}{4}$
   2. No symmetry
   3. Vertical asymptote: $x = -2$
   4. Horizontal asymptote: $y = \frac{3}{2}$, not intersected
   5. $x < -2$: above x-axis
      $-2 < x < -1$: below x-axis
      $x > -1$: above x-axis

6.
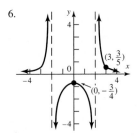

41. 1. No x-intercept; y-intercept; $-\frac{3}{4}$
   2. Symmetric with respect to y-axis
   3. Vertical asymptotes: $x = 2$, $x = -2$
   4. Horizontal asymptote: $y = 0$, not intersected
   5. $x < -2$: above x-axis
      $-2 < x < 2$: below x-axis
      $x > 2$: above x-axis

6.
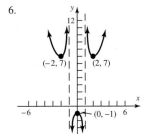

43. 1. No x-intercept; y-intercept: -1
   2. Symmetric with respect to y-axis
   3. Vertical asymptotes: $x = -1$, $x = 1$
   4. No horizontal or oblique asymptotes
   5. $x < -1$: above x-axis
      $-1 < x < 1$: below x-axis
      $x > 1$: above x-axis

6.
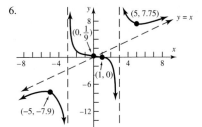

45. 1. x-intercept: 1; y-intercept: $\frac{1}{9}$
   2. No symmetry
   3. Vertical asymptotes: $x = 3$, $x = -3$
   4. Oblique asymptote: $y = x$, intersected at $\left(\frac{1}{9}, \frac{1}{9}\right)$
   5. $x < -3$: below x-axis
      $-3 < x < 1$: above x-axis
      $1 < x < 3$: below x-axis
      $x > 3$: above x-axis

6.
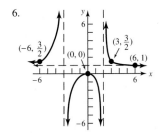

47. 1. Intercept $(0, 0)$
   2. No symmetry
   3. Vertical asymptotes: $x = 2$, $x = -3$
   4. Horizontal asymptote: $y = 1$, intersected at $(6, 1)$
   5. $x < -3$: above x-axis
      $-3 < x < 0$: below x-axis
      $0 < x < 2$: below x-axis
      $x > 2$: above x-axis

6.

49. 1. Intercept $(0, 0)$
2. Symmetry with respect to origin
3. Vertical asymptotes: $x = -2$, $x = 2$
4. Horizontal asymptote: $y = 0$, intersected at $(0, 0)$
5. $x < -2$: below x-axis
   $-2 < x < 0$: above x-axis
   $0 < x < 2$: below x-axis
   $x > 2$: above x-axis

6.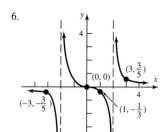

51. 1. No x-intercept; y-intercept: $\frac{3}{4}$
2. No symmetry
3. Vertical asymptotes: $x = -2$, $x = 1$, $x = 2$
4. Horizontal asymptote: $y = 0$, not intersected
5. $x < -2$: below x-axis
   $-2 < x < 1$: above x-axis
   $1 < x < 2$: below x-axis
   $x > 2$: above x-axis

6.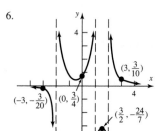

53. 1. x-intercepts: -1, 1; y-intercept: $\frac{1}{4}$
2. Symmetric with respect to y-axis
3. Vertical asymptotes: $x = -2$, $x = 2$
4. Horizontal asymptote: $y = 0$, intersected at $(-1, 0)$ and $(1, 0)$
5. $x < -2$: above x-axis
   $-2 < x < -1$: below x-axis
   $-1 < x < 1$: above x-axis
   $1 < x < 2$: below x-axis
   $x > 2$: above x-axis

6.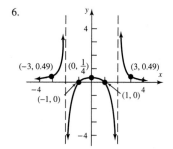

55. 1. x-intercepts: -1, 4; y-intercept: -2
2. No symmetry
3. Vertical asymptote: $x = -2$
4. Oblique asymptote: $y = x - 5$, not intersected
5. $x < -2$: below x-axis
   $-2 < x < -1$: above x-axis
   $-1 < x < 4$: below x-axis
   $x > 4$: above x-axis

6.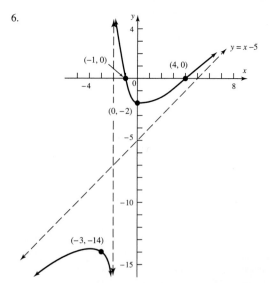

57. 1. x-intercepts: $-4, 3$; y-intercept: 3
2. No symmetry
3. Vertical asymptote: $x = 4$
4. Oblique asymptote: $y = x + 5$, not intersected
5. $x < -4$: below x-axis
$-4 < x < 3$: above x-axis
$3 < x < 4$: below x-axis
$x > 4$: above x-axis

6.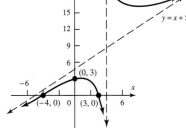

59. 1. x-intercepts: $-4, 3$; y-intercept: -6
2. No symmetry
3. Vertical asymptote: $x = -2$
4. Oblique asymptote: $y = x - 1$, not intersected
5. $x < -4$: below x-axis
$-4 < x < -2$: above x-axis
$-2 < x < 3$: below x-axis
$x > 3$: above x-axis

6.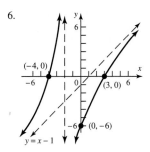

61. 1. x-intercepts: 0, 1; y-intercept: 0
2. No symmetry
3. Vertical asymptote: $x = -3$
4. Horizontal asymptote: $y = 1$, not intersected
5. $x < -3$: above x-axis
$-3 < x < 0$: below x-axis
$0 < x < 1$: above x-axis
$x > 1$: above x-axis

6.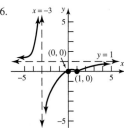

63. 4 must be a zero of the denominator; hence, $x - 4$ must be a factor.
65. No. Each of the functions is a quotient of polynomials, but not written in lowest terms. Each function is undefined for $x = 1$.
67. Minimum value: 2.00 at $x = 1.00$ **69.** Minimum value: 1.88 at $x = 0.79$

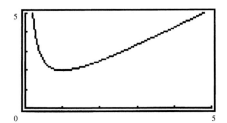   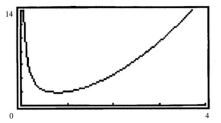

71. Minimum value: 1.75 at $x = 1.31$

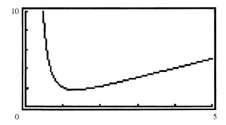

73. c, d

1. $f(2) = 12$ **3.** $f(3) = 99$ **5.** $f(-3) = -138$ **7.** $f(1) = 7$ **9.** $f(-1.1) = -0.3531$ **11.** $q(x) = x^2 + x + 4; R = 12$
13. $q(x) = 3x^2 + 11x + 32; R = 99$ **15.** $q(x) = x^4 - 3x^3 + 5x^2 - 15x + 46; R = -138$ **17.** $q(x) = 4x^5 + 4x^4 + x^3 + x^2 + 2x + 2; R = 7$
19. $q(x) = 0.1x^2 - 0.11x + 0.321; R = -0.3531$ **21.** $q(x) = x^4 + x^3 + x^2 + x + 1; R = 0$ **23.** No; $f(2) = 8$ **25.** Yes; $f(2) = 0$
27. Yes; $f(-3) = 0$ **29.** No; $f(-4) = 1$ **31.** Yes; $f(\frac{1}{2}) = 0$ **33.** 69   **35.** -4 **37.** 15   **39.** $[(3x + 2)x - 5]x + 8$
41. $[(3x - 6)x \cdot x - 5]x + 10$ **43.** $(3x \cdot x \cdot x - 82)x \cdot x \cdot x + 27$ **45.** $[(4x \cdot x - 64)x \cdot x + 1]x \cdot x - 15$ **47.** $[(2x - 1)x \cdot x + 2]x - 1$
49. 10.064   **51.** -0.1472 **53.** -105.738 **55.** -134.326 **57.** 3.8192   **59.** $k = 5$ **61.** -7 **63.** If $f(x) = x^n - c^n$, then $f(c) = c^n - c^n = 0$.
65. (a) 201,498 nanoseconds   **(b)** 101,499 nanoseconds

1. 7; 3 or 1 positive; 2 or 0 negative   **3.** 6; 2 or 0 positive; 2 or 0 negative   **5.** 3; 2 or 0 positive; 1 negative   **7.** 4; 2 or 0 positive; 2 or 0 negative
9. 5; 0 positive; 3 or 1 negative   **11.** 6; 1 positive; 1 negative   **13.** $\pm 1, \pm\frac{1}{3}$ **15.** $\pm 1, \pm 3$ **17.** $\pm 1, \pm 2, \pm\frac{1}{4}, \pm\frac{1}{2}$
19. $\pm\frac{1}{3}, \pm\frac{2}{3}, \pm 1, \pm 2$ **21.** $\pm\frac{1}{2}, \pm 1, \pm 2, \pm 4$ **23.** $\pm\frac{1}{6}, \pm\frac{1}{3}, \pm\frac{1}{2}, \pm\frac{2}{3}, \pm 1, \pm 2$ **25.** $-3, -1, 2; f(x) = (x + 3)(x + 1)(x - 2)$
27. $\frac{1}{2}; f(x) = 2(x - \frac{1}{2})(x^2 + 1)$ **29.** $-1, 1; f(x) = (x + 1)(x - 1)(x^2 + 2)$ **31.** $-\frac{1}{2}, \frac{1}{2}; f(x) = 4(x + \frac{1}{2})(x - \frac{1}{2})(x^2 + 2)$
33. $-2, -1, 1, 1; f(x) = (x + 2)(x + 1)(x - 1)^2$ **35.** $-\sqrt{2}/2, \sqrt{2}/2, 2; f(x) = 4(x + \sqrt{2}/2)(x - \sqrt{2}/2)(x - 2)(x^2 + \frac{1}{2})$ **37.** $\{-1, 2\}$
39. $\{\frac{2}{3}, -1 + \sqrt{2}, -1 - \sqrt{2}\}$ **41.** $\{\frac{1}{3}, \sqrt{5}, -\sqrt{5}\}$ **43.** $\{-3, -2\}$ **45.** $-\frac{1}{3}$
47. $(\frac{1}{2}, 0); (0, -1)$
$-\infty < x < \frac{1}{2}, f(0) = -1$, below x-axis
$\frac{1}{2} < x < \infty, f(1) = 2$, above x-axis

49. $(-1, 0); (1, 0); (0, -2)$
$-\infty < x < -1, f(-2) = 18$, above x-axis
$-1 < x < 1, f(0) = -2$, below x-axis
$1 < x < \infty, f(2) = 18$, above x-axis
(symmetric with respect to the y-axis)

51. $(-\frac{1}{2}, 0); (\frac{1}{2}, 0); (0, -2)$
$-\infty < x < -\frac{1}{2}, f(-1) = 9$, above x-axis
$-\frac{1}{2} < x < \frac{1}{2}, f(0) = -2$, below x-axis
$\frac{1}{2} < x < \infty, f(1) = 9$, above x-axis
(symmetric with respect to the y-axis)

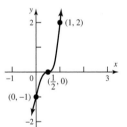

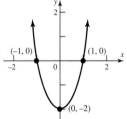

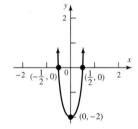

53. $(-1, 0); (-2, 0); (1, 0); (0, 2)$
$-\infty < x < -2, f(-3) = 32$, above x-axis
$-2 < x < -1, f(-\frac{3}{2}) = -\frac{25}{16}$, below x-axis
$-1 < x < 1, f(0) = 2$, above x-axis
$1 < x < \infty, f(2) = 12$, above x-axis

55. $(2, 0); (-\sqrt{2}/2, 0); (\sqrt{2}/2, 0); (0, 2)$
$-\infty < x < -\sqrt{2}/2, f(-1) = -9$, below x-axis
$-\sqrt{2}/2 < x < \sqrt{2}/2, f(0) = 2$, above x-axis
$\sqrt{2}/2 < x < 2, f(1) = -3$, below x-axis
$2 < x < \infty, f(3) = 323$, above x-axis

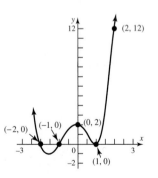

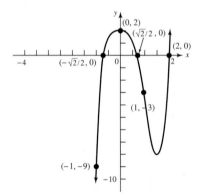

57. $\{-2, 2, -i, i\}$ **59.** $\left\{-1, 1, \dfrac{-1 - \sqrt{3}i}{2}, \dfrac{-1 + \sqrt{3}i}{2}\right\}$ **61.** $\left\{-2, 2, \dfrac{-3 - \sqrt{3}i}{2}, \dfrac{-3 + \sqrt{3}i}{2}\right\}$ **63.** $\{1, -i, i\}$ **65.** 5
67. No (use the Rational Zeros Theorem)   **69.** No (use the Rational Zeros Theorem)   **71.** 7 in.
73. All the potential rational zeros are integers. Hence, r either is an integer or is not a rational root (and is therefore irrational).

75.

$$y^3 + by^2 + cy + d = 0$$

$$\left(x - \frac{b}{3}\right)^3 + b\left(x - \frac{b}{3}\right)^2 + c\left(x - \frac{b}{3}\right) + d = 0$$

$$x^3 - \frac{3b}{3}x^2 + 3\left(\frac{b^2}{9}\right)x - \frac{b^3}{27} + b\left(x^2 - \frac{2bx}{3} + \frac{b^2}{9}\right) + cx - \frac{bc}{3} + d = 0$$

$$x^3 - \frac{b^2}{3}x + cx - \frac{b^3}{27} + \frac{b^3}{9} - \frac{bc}{3} + d = 0$$

$$x^3 + \left(c - \frac{b^2}{3}\right)x + \left(\frac{2b^3}{27} - \frac{bc}{3} + d\right) = 0$$

77. $K = \dfrac{-p}{3H}$

$$H^3 + \left(\frac{-p}{3H}\right)^3 = -q$$

$$H^6 + qH^3 - \frac{p^3}{27} = 0$$

$$H^3 = \frac{-q \pm \sqrt{q^2 + \dfrac{4p^3}{27}}}{2} \quad \text{Choose + sign}$$

$$H = \sqrt[3]{\frac{-q}{2} + \sqrt{\frac{q^2}{4} + \frac{p^3}{27}}}$$

79. $x = H + K$; now use results from Problems 77 and 78   **81.** $p = 3, q = -14; x = \sqrt[3]{7 + 5\sqrt{2}} + \sqrt[3]{7 - 5\sqrt{2}}$

83. $p = -6, q = 4; x = \sqrt[3]{-2 + \sqrt{4 - 8}} + \sqrt[3]{-2 - \sqrt{4 - 8}} = \sqrt[3]{-2 + 2i} + \sqrt[3]{-2 - 2i}$; or $x^3 - 6x + 4 = (x - 2)(x^2 + 2x - 2) = 0; x = 2$;

$$x = \frac{-2 \pm \sqrt{4 + 8}}{2} = -1 \pm \sqrt{3}$$

Exercise 3.6

1. -1 and 1 **3.** -3 and 7 **5.** -5 and 2 **7.** $f(0) = -1; f(1) = 10$ **9.** $f(-5) = -58; f(-4) = 2$ **11.** $f(1.4) = -0.17536; f(1.5) = 1.40625$
13. 1.15 **15.** 2.53 **17.** 0.21 **19.** -4.04 **21.** 1.15 **23.** 2.53 **25.** $-1.00, 0.21$ **27.** $-4.04, 0.35, 0.69$

Exercise 3.7

1. $4 + i$ **3.** $-i, 1 - i$ **5.** $-i, -2i$ **7.** $-i$ **9.** $2 - i, -3 + i$
11. Zeros that are complex numbers must occur in conjugate pairs; or a polynomial with real coefficients of odd degree must have at least one real zero.
13. If the remaining zero were a complex number, then its conjugate would also be a zero.
15. $1, -\dfrac{1}{2} + \dfrac{\sqrt{3}}{2}i, -\dfrac{1}{2} - \dfrac{\sqrt{3}}{2}i$ **17.** $-4 + i$ **19.** $-1 + 5i$ **21.** $-4 + 4i$ **23.** $-18 - 16i$ **25.** $16 - 18i$ **27.** $38 + 31i$
29. $z^3 + (-11 - 2i)z^2 + (40 + 16i)z - 48 - 32i$ **31.** $z^3 - 3z^2 + (3 - i)z - 2 + 2i$
33. $z^4 + (2i - 6)z^3 + (8 - 12i)z^2 + (6 + 18i)z - 9$

Fill-in-the-Blank Items

1. parabola; vertex   **2.** Remainder; Dividend   **3.** $f(c)$ **4.** $f(c) = 0$ **5.** zero   **6.** three; one; two; no   **7.** $\pm 1, \pm\frac{1}{2}$ **8.** $y = 1$
9. $x = -1$ **10.** $3 - 4i$

True/False Items

1. F   **2.** F   **3.** T   **4.** T   **5.** T

Review Exercises

1.

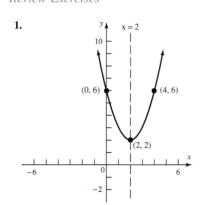

3.

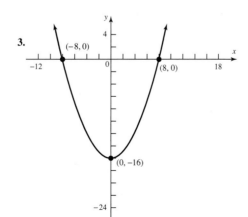

5.

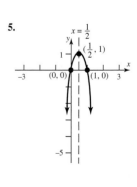

7.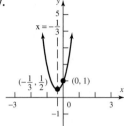

9.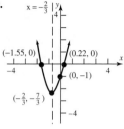

11.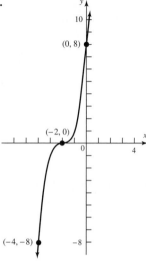

13.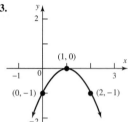

15.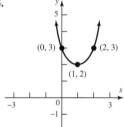

17. Minimum value; 1 **19.** Maximum value; 12 **21.** Maximum value; 16

23. (a) x-intercepts: $-4, -2, 0$; y-intercept: 0
(b) Crosses the x-axis at $-4, -2, 0$
(c) $y = x^3$
(d) 2

(e)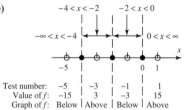

(f)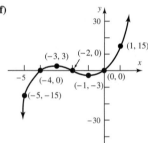

25. (a) x-intercepts: $-4, 2$; y-intercept: 16
(b) Crosses at -4; touches at 2
(c) $y = x^3$
(d) 2

(e)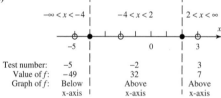

(f)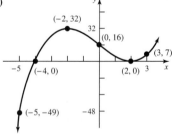

27. $f(x) = x^3 - 4x^2 = x^2(x - 4)$
(a) x-intercepts: $0, 4$; y-intercept: 0
(b) Touches at 0; crosses at 4
(c) $y = x^3$
(d) 2

(e)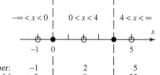

(f)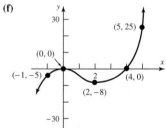

29. **(a)** *x*-intercepts: $-3, -1, 1$; *y*-intercept: 3 **(e)**
(b) Crosses at $-3, -1$; touches at 1
(c) $y = x^4$
(d) 3

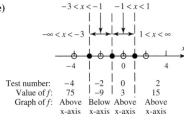

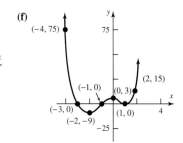

31. 1. *x*-intercept: 3; no *y*-intercept
    2. No symmetry
    3. Vertical asymptote: $x = 0$
    4. Horizontal asymptote: $y = 2$, not intersected
    5. $x < 0$: above *x*-axis
      $0 < x < 3$: below *x*-axis
      $x > 3$: above *x*-axis

6.

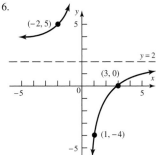

33. 1. *x*-intercept: -2; no *y*-intercept
    2. No symmetry
    3. Vertical asymptotes: $x = 0, x = 2$
    4. Horizontal asymptote: $y = 0$, intersected at $(-2, 0)$
    5. $x < -2$: below *x*-axis
      $-2 < x < 0$: above *x*-axis
      $0 < x < 2$: below *x*-axis
      $x > 2$: above *x*-axis

6.

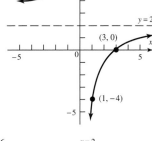

35. 1. Intercepts: $(-3, 0), (2, 0), (0, 1)$
    2. No symmetry
    3. Vertical asymptote: $x = -2, x = 3$
    4. Horizontal asymptote: $y = 1$, intersected at $(0, 1)$
    5. $x < -3$: above *x*-axis
      $-3 < x < -2$: below *x*-axis
      $-2 < x < 2$: above *x*-axis
      $2 < x < 3$: below *x*-axis
      $x > 3$: above *x*-axis

6.

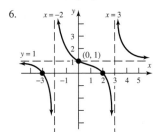

37. 1. Intercept $(0, 0)$
    2. Symmetric with respect to the origin
    3. Vertical asymptotes: $x = -2, x = 2$
    4. Oblique asymptote: $y = x$, intersected at $(0, 0)$
    5. $x < -2$: below *x*-axis
      $-2 < x < 0$: above *x*-axis
      $0 < x < 2$: below *x*-axis
      $x > 2$: above *x*-axis

6.

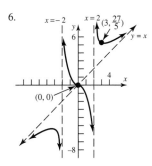

39. 1. Intercept $(0, 0)$
2. No symmetry
3. Vertical asymptote: $x = 1$
4. No oblique or horizontal asymptote
5. $x < 0$: above x-axis
   $0 < x < 1$: above x-axis
   $x > 1$: above x-axis

6.

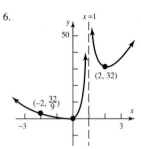

41. $q(x) = 8x^2 + 5x + 6$; $R = 10$ **43.** $q(x) = x^3 - 4x^2 + 8x - 15$; $R = 29$ **45.** $f(4) = 47,105$ **47.** 4, 2, or 0 positive; 2 or 0 negative

49. $\pm\frac{1}{12}, \pm\frac{1}{6}, \pm\frac{1}{4}, \pm\frac{1}{3}, \pm\frac{1}{2}, \pm\frac{3}{4}, \pm 1, \pm\frac{3}{2}, \pm 3$ **51.** $-2, 1, 4$; $f(x) = (x + 2)(x - 1)(x - 4)$

53. $\frac{1}{2}$, multiplicity 2; -2; $f(x) = 4\left(x - \frac{1}{2}\right)^2(x + 2)$ **55.** 2, multiplicity 2; $f(x) = (x - 2)^2(x^2 + 5)$ **57.** $\{-3, 2\}$ **59.** $\{-3, -1, -\frac{1}{2}, 1\}$
61. x-intercepts: $-2, 1, 4$
  y-intercept: 8
  Above x-axis: $-2 < x < 1, 4 < x < \infty$
  Below x-axis: $-\infty < x < -2, 1 < x < 4$

63. x-intercepts: $-2, \frac{1}{2}$
  y-intercept: 2
  Above x-axis: $-2 < x < \infty$
  Below x-axis: $-\infty < x < -2$

65. x-intercept: 2
  y-intercept: 20
  Above x-axis: all x

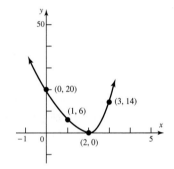

67. x-intercepts: $-3, 2$
  y-intercept: -6
  Above x-axis: $-\infty < x < -3, 2 < x < \infty$
  Below x-axis: $-3 < x < 2$

69. x-intercepts: $-3, -1, -\frac{1}{2}, 1$
  y-intercept: -3
  Above x-axis: $-\infty < x < -3, -1 < x < -\frac{1}{2}, 1 < x < \infty$
  Below x-axis: $-3 < x < -1, -\frac{1}{2} < x < 1$

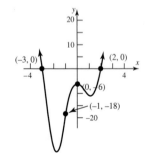

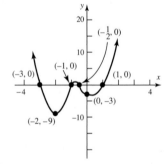

71. $f(0) = -1$; $f(1) = 1$ **73.** $f(0) = -1$; $f(1) = 1$ **75.** -2 and 2   **77.** -3 and 5   **79.** 1.52
81. 0.93   **83.** $4 - i$ **85.** $-i, 1 - i$ **87.** $f(z) = z^4 - (5 + i)z^3 + (7 + 5i)z^2 - (3 + 7i)z + 3i$
89. $f(z) = z^3 - (6 + i)z^2 + (11 + 5i)z - 6 - 6i$ **91.** $q(x) = x^2 + 5x + 6$; $R = 0$ **93.** $\{-3, 2\}$ **95.** $\{\frac{1}{3}, 1, -i, i\}$
97. $f(x) = [(8x - 3)x + 1]x - 6$; $f(1.5) = 15.75$ **99.** $f(x) = [(x - 2)x \cdot x + 1]x - 1$; $f(1.5) = -1.1875$
101. 1 is an upper bound; -2 is a lower bound   **103.** $(2, 2)$ **105.** 3.6 ft
109. **(a)** even   **(b)** positive   **(c)** even   **(d)** 0 is a zero of even multiplicity   **(e)** 8

1. **(a)** 11.212 **(b)** 11.587 **(c)** 11.664 **(d)** 11.665 **3.** **(a)** 8.815 **(b)** 8.821 **(c)** 8.824 **(d)** 8.825
5. **(a)** 21.217 **(b)** 22.217 **(c)** 22.440 **(d)** 22.459 **7.** 3.320 **9.** 0.427 **11.** B **13.** D **15.** A **17.** E
19. 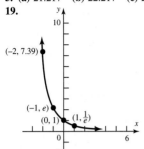 **21.** 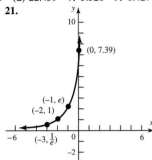 **23.** 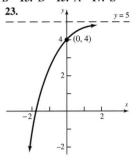 **25.**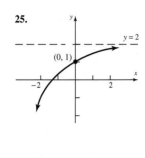

27. $\frac{1}{49}$ **29.** $\frac{1}{4}$ **31.** **(a)** 74% **(b)** 47% **33.** **(a)** 44 watts **(b)** 11.6 watts **35.** 3.35 milligrams; 0.45 milligrams
37. **(a)** 56% **(b)** 68% **(c)** 70% **(d)** $R = 40\%$ just after 6 days

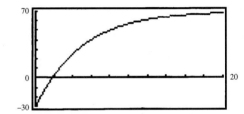

39. **(a)** 5.414 amperes, 7.5854 amperes, 10.38 amperes **(b)** 12 amperes **(c)** $I_1(t) = 12(1 - e^{-2t})$ **(d)** 3.343 amperes, 5.309 amperes,
9.443 amperes **(e)** 24 amperes **(f)** $I_2(t) = 24(1 - e^{-1/2t})$

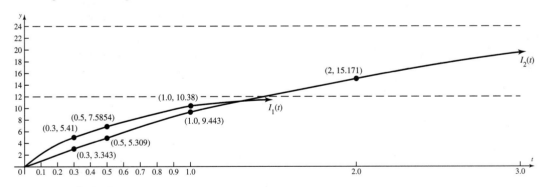

41. **(a)** 9.23×10^{-3} **(b)** 0.81 **(c)** 5 **(d)** 57.91°, 43.98°, 30.06°

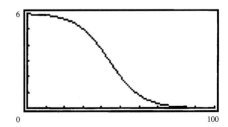

43. $n = 4$: 2.7083; $n = 6$: 2.7181; $n = 8$: 2.7182788; $n = 10$: 2.7182818
45. $\dfrac{f(x+h) - f(x)}{h} = \dfrac{a^{x+h} - a^x}{h} = \dfrac{a^x a^h - a^x}{h} = \dfrac{a^x(a^h - 1)}{h}$ **47.** $f(-x) = a^{-x} = \dfrac{1}{a^x} = \dfrac{1}{f(x)}$

49. (a) $\sinh(-x) = \frac{1}{2}(e^{-x} - e^x)$
$= -\frac{1}{2}(e^x - e^{-x})$
$= -\sinh x$

(b)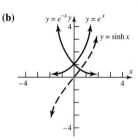

51. $f(1) = 5, f(2) = 17, f(3) = 257, f(4) = 65,537;$
$f(5) = 4,294,967,297 = 641 \times 6,700,417$

Exercise 4.2

1. $2 = \log_3 9$ **3.** $2 = \log_a 1.6$ **5.** $2 = \log_{1.1} M$ **7.** $x = \log_2 7.2$ **9.** $\sqrt{2} = \log_x \pi$ **11.** $x = \ln 8$ **13.** $2^3 = 8$ **15.** $a^6 = 3$ **17.** $3^x = 2$
19. $2^{1.3} = M$ **21.** $(\sqrt{2})^x = \pi$ **23.** $e^x = 4$ **25.** 0 **27.** 2 **29.** -4 **31.** $\frac{1}{2}$ **33.** 4 **35.** $\frac{1}{2}$ **37.** $\{x|x < 3\}$ **39.** All real numbers except 0
41. $\{x|x < -2 \text{ or } x > 3\}$ **43.** $\{x|x > 0, x \neq 1\}$ **45.** $\{x|x < -1 \text{ or } x > 0\}$ **47.** 0.511 **49.** 30.099 **51.** $\sqrt{2}$ **53.** B **55.** D **57.** A **59.** E

61. 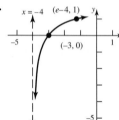 **63.** 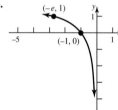 **65.** 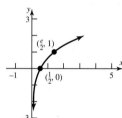 **67.**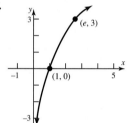

69.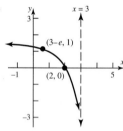

71. (a) $n = 6.93$ so 7 panes are necessary **(b)** $n = 13.86$ so 14 panes are necessary
73. (a) $d = 127.7$ so it takes about 128 days **(b)** $d = 575.6$ so it takes about 576 days
75. $h = 2.29$ so the time between injections is 2-$2\frac{1}{2}$ hours

77. 0.2695 sec; 0.8959 sec

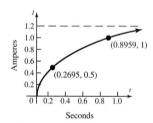

79. (a) $k = 20.07$ **(b)** 91% **(c)** 0.175 **(d)** 0.08
81. $y = 20\, e^{0.023t}$; $y = 89.2$ is predicted

Exercise 4.3

1. $a + b$ **3.** $b - a$ **5.** $a + 1$ **7.** $2a + b$ **9.** $\frac{1}{5}(a + 2b)$ **11.** $\dfrac{b}{a}$ **13.** $2\ln x + \frac{1}{2}\ln(1 - x)$ **15.** $3\log_2 x - \log_2(x - 3)$
17. $\log x + \log(x + 2) - 2\log(x + 3)$ **19.** $\frac{1}{3}\ln(x - 2) + \frac{1}{3}\ln(x + 1) - \frac{2}{3}\ln(x + 4)$ **21.** $\ln 5 + \ln x + \frac{1}{2}\ln(1 - 3x) - 3\ln(x - 4)$
23. $\log_5 u^3 v^4$ **25.** $-\frac{5}{2}\log_{1/2} x$ **27.** $-2\ln(x - 1)$ **29.** $\log_2[x(3x - 2)^4]$ **31.** $\log_a\left[\dfrac{25x^6}{(2x + 3)^{1/2}}\right]$ **33.** 2.771 **35.** -3.880

37. 5.615   **39.** 0. 874   **41.** $\log_a(x + \sqrt{x^2 - 1} + \log_a(x - \sqrt{x^2 - 1}) = \log_a[(x + \sqrt{x^2 - 1})(x - \sqrt{x^2 - 1})] = \log_a[x^2 - (x^2 - 1)] = \log_a 1 = 0$
43. $\ln (1 + e^{2x}) = \ln[e^{2x}(e^{-2x} + 1)] = \ln e^{2x} + \ln (e^{-2x} + 1) = 2x + \ln(1 + e^{-2x})$

45. $y = f(x) = \log_a x; \; a^y = x; \; \left(\dfrac{1}{a}\right)^{-y} = x; \; -y = \log_{1/a} x; \; -f(x) = \log_{1/a} x$ **47.** $f(AB) = \log_a AB = \log_a A + \log_a B = f(A) + f(B)$

49. $y = Cx$ **51.** $y = Cx(x + 1)$ **53.** $y = Ce^{3x}$ **55.** $y = Ce^{-4x} + 3$ **57.** $y = \dfrac{\sqrt[3]{C}(2x + 1)^{1/6}}{(x + 4)^{1/9}}$ **59.** 3   **61.** 1

63. If $A = \log_a M$ and $B = \log_a N$, then $a^A = M$ and $a^B = N$. Then $\log_a (M/N) = \log_a (a^A/a^B) = \log_a a^{A-B} = A - B = \log_a M - \log_a N$.

Exercise 4.4

1. $\frac{7}{2}$ **3.** $\{-2\sqrt{2}, 2\sqrt{2}\}$ **5.** 16   **7.** 8   **9.** 3   **11.** 5   **13.** 2   **15.** $\{-2, 4\}$ **17.** 21   **19.** $\frac{1}{2}$ **21.** $\{-\sqrt{2}, 0, \sqrt{2}\}$

23. $\left\{1 - \dfrac{\sqrt{6}}{3}, 1 + \dfrac{\sqrt{6}}{3}\right\}$ **25.** 0   **27.** 2   **29.** 0   **31.** $\frac{3}{2}$ **33.** 3.322   **35.** −0.088   **37.** 0.307   **39.** 1.356   **41.** 0

43. 0.534   **45.** 0.226   **47.** 2.027   **49.** $\frac{9}{2}$ **51.** 2   **53.** −1   **55.** 1   **57.** 16   **59.** −0.56   **61.** −0.70
63. 0.56   **65.** {0.39, 1.00}   **67.** 1.31   **69.** 1.30

Exercise 4.5

1. $108.29   **3.** $609.50   **5.** $697.09   **7.** $12.46   **9.** $125.23   **11.** $88.72   **13.** $860.72   **15.** $554.09   **17.** $59.71   **19.** $361.93

21. 5.35%   **23.** 26%   **25.** $6\frac{1}{4}$% compounded annually   **27.** 9% compounded monthly   **29.** 104.32 mo; 103.97 mo

31. 61.02 mo; 60.82 mo   **33.** 15.27 yrs or 15 yrs, 4 months   **35.** $104,335   **37.** $12,910.62   **39.** About $30.17 per share or $3017

41. 9.35%   **43.** Not quite. You will have $1057.60. The second bank gives a better deal, since you have $1060.62 after 1 year

45. You have $11,632.73; your friend has $10, 947.89

47. (a) Interest is $30,000   **(b)** Interest is $38,613.59   **(c)** Interest is $37,752.73. Simple interest at 12% is best

49. (a) $1364.62   **(b)** $1353.35   **51.** $4631.93

59. (a) 6.1 years   **(b)** 18.45 yrs   **(c)** $mP = P\left(1 + \dfrac{r}{n}\right)^{nt}$

$$m = \left(1 + \dfrac{r}{n}\right)^{nt}$$
$$\ln m = \ln \left(1 + \dfrac{r}{n}\right)^{nt} = nt \ln \left(1 + \dfrac{r}{n}\right)$$
$$t = \dfrac{\ln m}{n \ln \left(1 + \frac{r}{n}\right)}$$

Exercise 4.6

1. 34.7 days; 69.3 days   **3.** 28.4 yr   **5.** 94.4 yr   **7.** 5832; 3.9 days   **9.** 25,198   **11.** 9.797 g   **13.** 9727 years ago   **15.** 5:18 PM
17. 18.63°C; 25.1°C   **19.** 7.34 kg; 76.6 hr   **21.** 26.5 days.

Exercise 4.7

1. 70 decibels   **3.** 111.76 decibels   **5.** 10 W/m^2 **7.** 4.0 on the Richter scale
9. 70,794.58 mm; the San Francisco earthquake was 11.22 times as intense as the one in Mexico City.

Fill-in-the-Blank Items

1. (0, 1) and (1, a)   **2.** 1   **3.** 4   **4.** sum   **5.** 1   **6.** 7   **7.** $x > 0$ **8.** (1, 0) and (a, 1)   **9.** 1   **10.** 7

True/False Items

1. T   **2.** T   **3.** F   **4.** F   **5.** T   **6.** F   **7.** F   **8.** T

Review Exercises

1. −3   **3.** $\sqrt{2}$ **5.** 0.4   **7.** $\frac{25}{4} \log_4 x$ **9.** $\ln\left[\dfrac{1}{(x + 1)^2}\right] = -2 \ln(x + 1)$ **11.** $\log\left(\dfrac{4x^3}{[(x + 3)(x - 2)]^{1/2}}\right)$ **13.** $y = Ce^{2x^2}$

15. $y = (Ce^{3x^2})^2$ **17.** $y = \sqrt{e^{x+C}} + 9$ **19.** $y = \ln(x^2 + 4) - C$ **21.**

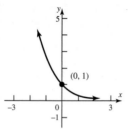

23.

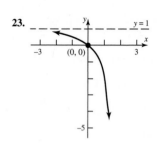

25.

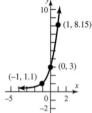

27.

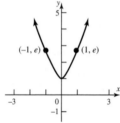

29.

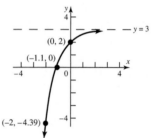

31. $\frac{1}{4}$ **33.** $\left\{\dfrac{-1 - \sqrt{3}}{2}, \dfrac{-1 + \sqrt{3}}{2}\right\}$ **35.** $\frac{1}{4}$ **37.** 4.301   **39.** $\frac{12}{5}$ **41.** 83   **43.** $\left\{-3, \frac{1}{2}\right\}$ **45.** -1 **47.** -0.609 **49.** -9.327

51. 3229.5 m   **53.** 7.6 mm of mercury   **55.** **(a)** 37.3 watts   **(b)** 6.9 decibels
57. **(a)** 71%   **(b)** 85.5%   **(c)** 90%   **(d)** About 1.6 months   **(e)** About 4.8 months
59. **(a)** 9.85 years   **(b)** 4.27 years   **61.** $41,669   **63.** 80 decibels   **65.** 24,203 years ago

CHAPTER 5   *Exercise 5.1*

1.

3.

5.

7.

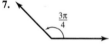

9.

11.

13. $\dfrac{\pi}{6}$ **15.** $\dfrac{4\pi}{3}$ **17.** $\dfrac{-\pi}{3}$ **19.** π **21.** $\dfrac{3\pi}{4}$ **23.** 60°   **25.** $-225°$ **27.** 90°   **29.** 15°   **31.** 120°   **33.** 5 m   **35.** 6 ft   **37.** 0.6 radian

39. $\dfrac{\pi}{3} \approx 1.047$ in.   **41.** 0.30   **43.** -0.70 **45.** 2.18   **47.** 5.93   **49.** 179.91°   **51.** 587.28°   **53.** 114.59°   **55.** 362.11°   **57.** 40.17°   **59.** 1.03°

61. 9.15°   **63.** 40°19′12″   **65.** 18°15′18″   **67.** 19°59′24″   **69.** $3\pi \approx 9.4248$ in.; $5\pi \approx 15.7080$ in.   **71.** $\omega = \frac{1}{60}$ radian/sec; $v = \frac{1}{12}$ cm/sec
73. 452.5 rpm   **75.** 37.7 in.   **77.** 2292 mph   **79.** $\frac{3}{4}$ rpm   **81.** 2.86 mph   **83.** 1.152 mi   **85.** 1037 mph

Exercise 5.2

1. $\sin \theta = \frac{4}{5}$, $\cos \theta = -\frac{3}{5}$, $\tan \theta = -\frac{4}{3}$, $\csc \theta = \frac{5}{4}$, $\sec \theta = -\frac{5}{3}$, $\cot \theta = -\frac{3}{4}$
3. $\sin \theta = -3\sqrt{13}/13$, $\cos \theta = 2\sqrt{13}/13$, $\tan \theta = -\frac{3}{2}$, $\csc \theta = -\sqrt{13}/3$, $\sec \theta = \sqrt{13}/2$, $\cot \theta = -\frac{2}{3}$
5. $\sin \theta = -\sqrt{2}/2$, $\cos \theta = -\sqrt{2}/2$, $\tan \theta = 1$, $\csc \theta = -\sqrt{2}$, $\sec \theta = -\sqrt{2}$, $\cot \theta = 1$
7. $\sin \theta = -2\sqrt{13}/13$, $\cos \theta = -3\sqrt{13}/13$, $\tan \theta = \frac{2}{3}$, $\csc \theta = -\sqrt{13}/2$, $\sec \theta = -\sqrt{13}/3$, $\cot \theta = \frac{3}{2}$
9. $\sin \theta = -\frac{3}{5}$, $\cos \theta = \frac{4}{5}$, $\tan \theta = -\frac{3}{4}$, $\csc \theta = -\frac{5}{3}$, $\sec \theta = \frac{5}{4}$, $\cot \theta = -\frac{4}{3}$

11. $\frac{1}{2}(\sqrt{2} + 1)$ **13.** 2 **15.** $\frac{1}{2}$ **17.** $\sqrt{6}$ **19.** 4 **21.** 0 **23.** 0 **25.** $2\sqrt{2} + 4\sqrt{3}/3$ **27.** -1 **29.** 1
31. $\sin(2\pi/3) = \sqrt{3}/2$, $\cos(2\pi/3) = -\frac{1}{2}$, $\tan(2\pi/3) = -\sqrt{3}$, $\csc(2\pi/3) = 2\sqrt{3}/3$, $\sec(2\pi/3) = -2$, $\cot(2\pi/3) = -\sqrt{3}/3$
33. $\sin 150° = \frac{1}{2}$, $\cos 150° = -\sqrt{3}/2$, $\tan 150° = -\sqrt{3}/3$, $\csc 150° = 2$, $\sec 150° = -2\sqrt{3}/3$, $\cot 150° = -\sqrt{3}$
35. $\sin(-\pi/6) = -\frac{1}{2}$, $\cos(-\pi/6) = \sqrt{3}/2$, $\tan(-\pi/6) = -\sqrt{3}/3$, $\csc(-\pi/6) = -2$, $\sec(-\pi/6) = 2\sqrt{3}/3$, $\cot(-\pi/6) = -\sqrt{3}$
37. $\sin 225° = -\sqrt{2}/2$, $\cos 225° = -\sqrt{2}/2$, $\tan 225° = 1$, $\csc 225° = -\sqrt{2}$, $\sec 225° = -\sqrt{2}$, $\cot 225° = 1$
39. $\sin(5\pi/2) = 1$, $\cos(5\pi/2) = 0$, $\tan(5\pi/2)$ is not defined, $\csc(5\pi/2) = 1$, $\sec(5\pi/2)$ is not defined, $\cot(5\pi/2) = 0$
41. $\sin(-180°) = 0$, $\cos(-180°) = -1$, $\tan(-180°) = 0$, $\csc(-180°)$ is not defined, $\sec(-180°) = -1$, $\cot(-180°)$ is not defined
43. $\sin(3\pi/2) = -1$, $\cos(3\pi/2) = 0$, $\tan(3\pi/2)$ is not defined, $\csc(3\pi/2) = -1$, $\sec(3\pi/2)$ is not defined, $\cot(3\pi/2) = 0$
45. $\sin 450° = 1$, $\cos 450° = 0$, $\tan 450°$ is not defined, $\csc 450° = 1$, $\sec 450°$ is not defined, $\cot 450° = 0$ **47.** 0.4695
49. 0.3839 **51.** 1.3250 **53.** 0.3640 **55.** 0.3090 **57.** 3.7321 **59.** 1.0353 **61.** 5.6713 **63.** 0.8415 **65.** 0.0175 **67.** 0.9304
69. 0.3093 **71.** $\sqrt{3}/2$ **73.** $\frac{1}{2}$ **75.** $\frac{3}{4}$ **77.** $\sqrt{3}/2$ **79.** $\sqrt{3}$ **81.** $\sqrt{3}/4$ **83.** 0 **85.** -0.1 **87.** 3 **89.** 5
91. $R \approx 310.56$ ft, $H \approx 77.64$ ft   **93.** $R \approx 19{,}542$ m, $H \approx 2278$ m   **95. (a)** 1.2 sec **(b)** 1.12 sec **(c)** 1.2 sec
97. (a) 1.9 hr **(b)** 1.69 hr **(c)** 1.63 hr **(d)** 1.67 hr **99.** 16.56 ft

Exercise 5.3

1. $\sqrt{2}/2$ **3.** 1 **5.** 1 **7.** $\sqrt{3}$ **9.** $\sqrt{2}/2$ **11.** 0 **13.** $\sqrt{2}$ **15.** $\sqrt{3}/3$ **17.** II **19.** IV **21.** IV **23.** II
25. $\tan \theta = 2$, $\cot \theta = \frac{1}{2}$, $\sec \theta = \sqrt{5}$, $\csc \theta = \sqrt{5}/2$ **27.** $\tan \theta = \sqrt{3}/3$, $\cot \theta = \sqrt{3}$, $\sec \theta = 2\sqrt{3}/3$, $\csc \theta = 2$
29. $\tan \theta = -\sqrt{2}/4$, $\cot \theta = -2\sqrt{2}$, $\sec \theta = 3\sqrt{2}/4$, $\csc \theta = -3$
31. $\tan \theta = 0.2679$, $\cot \theta = 3.7322$, $\sec \theta = 1.0353$, $\csc \theta = 3.8640$
33. $\cos \theta = -\frac{5}{13}$, $\tan \theta = -\frac{12}{5}$, $\csc \theta = \frac{13}{12}$, $\sec \theta = -\frac{13}{5}$, $\cot \theta = -\frac{5}{12}$
35. $\sin \theta = -\frac{3}{5}$, $\tan \theta = \frac{3}{4}$, $\csc \theta = -\frac{5}{3}$, $\sec \theta = -\frac{5}{4}$, $\cot \theta = \frac{4}{3}$
37. $\cos \theta = -\frac{12}{13}$, $\tan \theta = -\frac{5}{12}$, $\cot \theta = -\frac{12}{5}$, $\sec \theta = -\frac{13}{12}$, $\csc \theta = \frac{13}{5}$
39. $\sin \theta = 2\sqrt{2}/3$, $\tan \theta = -2\sqrt{2}$, $\cot \theta = -\sqrt{2}/4$, $\sec \theta = -3$, $\csc \theta = 3\sqrt{2}/4$
41. $\cos \theta = -\sqrt{5}/3$, $\tan \theta = -2\sqrt{5}/5$, $\cot \theta = -\sqrt{5}/2$, $\sec \theta = -3\sqrt{5}/5$, $\csc \theta = \frac{3}{2}$
43. $\sin \theta = -\sqrt{3}/2$, $\cos \theta = \frac{1}{2}$, $\tan \theta = -\sqrt{3}$, $\cot \theta = -\sqrt{3}/3$, $\csc \theta = -2\sqrt{3}/3$
45. $\sin \theta = -\frac{3}{5}$, $\cos \theta = -\frac{4}{5}$, $\cot \theta = \frac{4}{3}$, $\sec \theta = -\frac{5}{4}$, $\csc \theta = -\frac{5}{3}$
47. $\sin \theta = \sqrt{10}/10$, $\cos \theta = -3\sqrt{10}/10$, $\cot \theta = -3$, $\sec \theta = -\sqrt{10}/3$, $\csc \theta = \sqrt{10}$ **49.** $-\sqrt{3}/2$ **51.** $-\sqrt{3}/3$ **53.** 2
55. -1 **57.** -1 **59.** $\sqrt{2}/2$ **61.** 0 **63.** $-\sqrt{2}$ **65.** $2\sqrt{3}/3$ **67.** -1 **69.** -2 **71.** $1 - \sqrt{2}/2$ **73.** 1 **75.** 1 **77.** 0 **79.** 0.9 **81.** 9
83. 15.8 mi **85.** At odd multiples of $\pi/2$ **87.** At odd multiples of $\pi/2$ **89.** 0
91. Let $P = (x, y)$ be the point on the unit circle that corresponds to θ. Consider the equation $\tan \theta = y/x = a$. Then $y = ax$. But $x^2 + y^2 = 1$ so that
$x^2 + a^2x^2 = 1$. Thus, $x = \pm 1/\sqrt{1 + a^2}$ and $y = \pm a/\sqrt{1 + a^2}$; that is, for any real number a, there is a point $P = (x, y)$ on the unit circle for which
$\tan \theta = a$. In other words, $-\infty < \tan \theta < \infty$, and the range of the tangent function is the set of all real numbers.
93. Suppose there is a number p, $0 < p < 2\pi$, for which $\sin(\theta + p) = \sin \theta$ for all θ. If $\theta = 0$, then $\sin(0 + p) = \sin p = \sin 0 = 0$; so that $p = \pi$. If
$\theta = \pi/2$, then $\sin(\pi/2 + p) = \sin(\pi/2)$. But $p = \pi$. Thus, $\sin(3\pi/2) = -1 = \sin(\pi/2) = 1$. This is impossible. The smallest positive number p for
which $\sin(\theta + p) = \sin \theta$ for all θ is therefore $p = 2\pi$.
95. $\sec \theta = 1/(\cos \theta)$; since $\cos \theta$ has period 2π, so does $\sec \theta$
97. If $P = (a, b)$ is the point on the unit circle corresponding to θ, then $Q = (-a, -b)$ is the point on the unit circle corresponding to $\theta + \pi$. Thus,
$\tan(\theta + \pi) = (-b)/(-a) = b/a = \tan \theta$; that is, the period of the tangent function is π.
99. Let $P = (a, b)$ be the point on the unit circle corresponding to θ. Then $\csc \theta = 1/b = 1/(\sin \theta)$; $\sec \theta = 1/a = 1/(\cos \theta)$; $\cot \theta = a/b = 1/(b/a) = 1/(\tan \theta)$.
101. $(\sin \theta \cos \phi)^2 + (\sin \theta \sin \phi)^2 + \cos^2\theta = \sin^2 \theta \cos^2 \phi + \sin^2 \theta \sin^2 \phi + \cos^2 \theta = \sin^2 \theta(\cos^2 \phi + \sin^2 \phi) + \cos^2 \theta = \sin^2 \theta + \cos^2 \theta = 1$

Exercise 5.4

1. $\sin \theta = \frac{5}{13}$, $\cos \theta = \frac{12}{13}$, $\tan \theta = \frac{5}{12}$, $\csc \theta = \frac{13}{5}$, $\sec \theta = \frac{13}{12}$, $\cot \theta = \frac{12}{5}$
3. $\sin \theta = 2\sqrt{13}/13$, $\cos \theta = 3\sqrt{13}/13$, $\tan \theta = \frac{2}{3}$, $\csc \theta = \sqrt{13}/2$, $\sec \theta = \sqrt{13}/3$, $\cot \theta = \frac{3}{2}$
5. $\sin \theta = \sqrt{3}/2$, $\cos \theta = \frac{1}{2}$, $\tan \theta = \sqrt{3}$, $\csc \theta = 2\sqrt{3}/3$, $\sec \theta = 2$, $\cot \theta = \sqrt{3}/3$
7. $\sin \theta = \sqrt{6}/3$, $\cos \theta = \sqrt{3}/3$, $\tan \theta = \sqrt{2}$, $\csc \theta = \sqrt{6}/2$, $\sec \theta = \sqrt{3}$, $\cot \theta = \sqrt{2}/2$
9. $\sin \theta = \sqrt{5}/5$, $\cos \theta = 2\sqrt{5}/5$, $\tan \theta = \frac{1}{2}$, $\csc \theta = \sqrt{5}$, $\sec \theta = \sqrt{5}/2$, $\cot \theta = 2$ **11.** 30° **13.** 60° **15.** 30° **17.** $\pi/4$ **19.** $\pi/3$
21. 45° **23.** $\pi/3$ **25.** 60° **27.** $\frac{1}{2}$ **29.** $\sqrt{2}/2$ **31.** -2 **33.** $-\sqrt{3}$ **35.** $\sqrt{2}/2$ **37.** $\sqrt{3}$ **39.** $\frac{1}{2}$ **41.** $-\sqrt{3}/2$ **43.** $-\sqrt{3}$ **45.** $\sqrt{2}$
47. 0 **49.** 1 **51.** 0 **53.** 0 **55.** 1 **57. (a)** $\frac{1}{3}$ **(b)** $\frac{8}{9}$ **(c)** 3 **(d)** 3 **59. (a)** 17 **(b)** $\frac{1}{4}$ **(c)** 4 **(d)** $\frac{17}{16}$
61. (a) $\frac{1}{4}$ **(b)** 15 **(c)** 4 **(d)** $\frac{16}{15}$ **63.** 0.6 **65.** 0 **67.** 20°

69. (a) $T(\theta) = 1 + \dfrac{2}{3 \sin \theta} - \dfrac{1}{4 \tan \theta}$ **(b)** $\theta = 67.97°$ for least time; the least time is
$T = 1.62$ hrs. Sally is on the road for 0.9 hr.

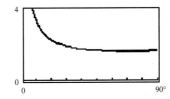

71. (a) 10 min **(b)** 20 min **(c)** $T(\theta) = 5\left(1 - \dfrac{1}{3\tan\theta} + \dfrac{1}{\sin\theta}\right)$ **(d)** 10.4 min

(e) T is least for $\theta = 70.5°$; the least time is 9.7 min; $x = 177$ ft

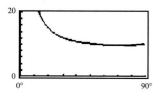

75.

θ	0.5	0.4	0.2	0.1	0.01	0.001	0.0001	0.00001
$\sin\theta$	0.4794	0.3894	0.1987	0.0998	0.0100	0.0010	0.0001	0.00001
$\dfrac{\sin\theta}{\theta}$	0.9589	0.9735	0.9933	0.9983	1.0000	1.0000	1.0000	1.0000

$\dfrac{\sin\theta}{\theta}$ approaches 1 as θ approaches 0

77. (a) $|OA| = |OC| = 1$; Angle OAC + Angle OAC + $180° - \theta = 180°$; Angle $OAC = \theta/2$

(b) $\sin\theta = \dfrac{|CD|}{|OC|} = |CD|$; $\cos\theta = \dfrac{|OD|}{|OC|} = |OD|$ **(c)** $\tan\dfrac{\theta}{2} = \dfrac{|CD|}{|AD|} = \dfrac{\sin\theta}{1 + |OD|} = \dfrac{\sin\theta}{1 + \cos\theta}$

79. $h = x\tan\theta$ and $h = (1 - x)\tan n\theta$; thus, $x\tan\theta = (1 - x)\tan n\theta$ and $x = \dfrac{\tan n\theta}{\tan\theta + \tan n\theta}$

81. (a) Area $\triangle OAC = \frac{1}{2}|OC||AC| = \frac{1}{2} \cdot \dfrac{|OC|}{1} \cdot \dfrac{|AC|}{1} = \frac{1}{2}\sin\alpha\cos\alpha$

(b) Area $\triangle OCB = \frac{1}{2}|BC||OC| - \frac{1}{2}|OB|^2\dfrac{|BC|}{|OB|} \cdot \dfrac{|OC|}{|OB|} = \frac{1}{2}|OB|^2\sin\beta\cos\beta$

(c) Area $\triangle OAB = \frac{1}{2}|BD||OA| = \frac{1}{2}|OB|\dfrac{|BD|}{|OB|} = \frac{1}{2}|OB|\sin(\alpha + \beta)$ **(d)** $\dfrac{\cos\alpha}{\cos\beta} = \dfrac{|OC|/1}{|OC|/|OB|} = |OB|$

(e) Use the hint and results from parts (a)–(d).

83. $\sin\alpha = \tan\alpha\cos\alpha = \cos\beta\cos\alpha = \cos\beta\tan\beta = \sin\beta$; $\sin^2\alpha + \cos^2\alpha = 1$, thus,

$$\sin^2\alpha + \tan^2\beta = 1$$
$$\sin^2\alpha + \dfrac{\sin^2\beta}{\cos^2\beta} = 1$$
$$\sin^2\alpha + \dfrac{\sin^2\alpha}{1 - \sin^2\alpha} = 1$$
$$\sin^2\alpha - \sin^4\alpha + \sin^2\alpha = 1 - \sin^2\alpha$$
$$\sin^4\alpha - 3\sin^2\alpha + 1 = 0$$
$$\sin^2\alpha = \dfrac{3 \pm \sqrt{5}}{2}$$
$$\sin^2\alpha = \dfrac{3 - \sqrt{5}}{2}$$
$$\sin\alpha = \sqrt{\dfrac{3 - \sqrt{5}}{2}}$$

Exercise 5.5

1. $a \approx 13.74$, $c \approx 14.62$, $\alpha = 70°$ **3.** $b \approx 5.03$, $c \approx 7.83$, $\alpha = 50°$ **5.** $a \approx 0.705$, $c \approx 4.06$, $\beta = 80°$ **7.** $b \approx 10.72$, $c \approx 11.83$, $\beta = 65°$
9. $b \approx 3.08$, $a \approx 8.46$, $\alpha = 70°$ **11.** $c \approx 5.83$, $a \approx 59.0°$, $\beta \approx 31.0°$ **13.** $b \approx 4.58$, $\alpha \approx 23.6°$, $\beta \approx 66.4°$ **15.** 1.72 in., 2.46 in.
17. 6.10 in. or 8.72 in. **19.** 23.6° and 66.4° **21.** 70 ft **23.** 985.9 ft **25.** 137 m **27.** 20.67 ft **29.** 449.36 ft **31.** 80.5° **33.** 30 ft
35. 530 ft **37.** 555 ft **39. (a)** 112 ft/sec or 76.3 mph **(b)** 82.4 ft/sec or 56.2 mph **(c)** under 18.8° **41. (a)** 130° **(b)** 103.4° **43.** 14.9°
45. (a) 3.1 mi. **(b)** 3.2 mi **(c)** 3.8 mi

47. (a) $\cos\dfrac{\theta}{2} = \dfrac{3960}{3960 + h}$ **(b)** $d = 3960\,\theta$ **(c)** $\cos\dfrac{d}{7920} = \dfrac{3960}{3960 + h}$ **(d)** 206 mi **(e)** 2990 miles

Fill-in-the Blank Items

1. angle; initial side; terminal side   **2.** radians   **3.** π **4.** complementary   **5.** cosine   **6.** standard position   **7.** 45°   **8.** 2π; π

True/False Items

1. F   **2.** T   **3.** F   **4.** T   **5.** F   **6.** F

Review Exercises

1. $3\pi/4$ **3.** $\pi/10$ **5.** 135°   **7.** −450°   **9.** $\frac{1}{2}$ **11.** $3\sqrt{2}/2 - 4\sqrt{3}/3$ **13.** $-3\sqrt{2} - 2\sqrt{3}$ **15.** 3   **17.** 0   **19.** 0   **21.** 1   **23.** 1   **25.** 1
27. −1   **29.** 1   **31.** $\cos\theta = \frac{3}{5}$, $\tan\theta = -\frac{4}{3}$, $\csc\theta = -\frac{5}{4}$, $\sec\theta = \frac{5}{3}$, $\cot\theta = -\frac{3}{4}$ **33.** $\sin\theta = -\frac{12}{13}$, $\cos\theta = -\frac{5}{13}$, $\csc\theta = -\frac{13}{12}$, $\sec\theta = -\frac{13}{5}$, $\cot\theta = \frac{5}{12}$
35. $\sin\theta = \frac{3}{5}$, $\cos\theta = -\frac{4}{5}$, $\tan\theta = -\frac{3}{4}$, $\csc\theta = \frac{5}{3}$, $\cot\theta = -\frac{4}{3}$ **37.** $\cos\theta = -\frac{5}{13}$, $\tan\theta = -\frac{12}{5}$, $\csc\theta = \frac{13}{12}$, $\sec\theta = -\frac{13}{5}$, $\cot\theta = -\frac{5}{12}$
39. $\cos\theta = \frac{12}{13}$, $\tan\theta = -\frac{5}{12}$, $\csc\theta = -\frac{13}{5}$, $\sec\theta = \frac{13}{12}$, $\cot\theta = -\frac{12}{5}$
41. $\sin\theta = -\sqrt{10}/10$, $\cos\theta = -3\sqrt{10}/10$, $\csc\theta = -\sqrt{10}$, $\sec\theta = -\sqrt{10}/3$, $\cot\theta = 3$
43. $\sin\theta = -2\sqrt{2}/3$, $\cos\theta = \frac{1}{3}$, $\tan\theta = -2\sqrt{2}$, $\csc\theta = -3\sqrt{2}/4$, $\cot\theta = -\sqrt{2}/4$
45. $\sin\theta = \sqrt{5}/5$, $\cos\theta = -2\sqrt{5}/5$, $\tan\theta = -\frac{1}{2}$, $\csc\theta = \sqrt{5}$, $\sec\theta = -\sqrt{5}/2$ **47.** $\alpha = 70°$, $b \approx 3.42$, $a \approx 9.4$
49. $a \approx 4.58$, $\alpha \approx 66.4°$, $\beta \approx 23.6°$ **51.** $\pi/3$ ft   **53.** 114.59 revolutions per hour   **55.** 839 ft   **57.** 23.32 ft   **59.** 2.15 mi

CHAPTER 6   *Exercise 6.1*

1. 0   **3.** $-\pi/2 \le x \le \pi/2$ **5.** 1   **7.** $0, \pi, 2\pi$ **9.** $\sin x = 1$ for $x = -3\pi/2, \pi/2$; $\sin x = -1$ for $x = -\pi/2, 3\pi/2$

11. 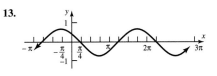   **13.** 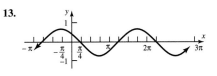   **15.**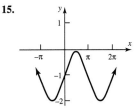

17. 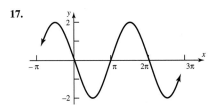   **19.** 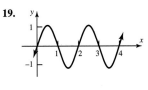   **21.**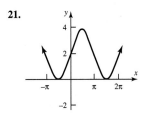

23. 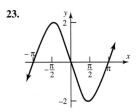   **25.**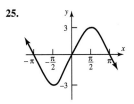

27. The graphs are the same; yes   **29.** $y = \sin\omega x$ has period $2\pi/\omega$.

Exercise 6.2

1. Amplitude = 2; Period = 2π **3.** Amplitude = 4; Period = π **5.** Amplitude = 6; Period = 2   **7.** Amplitude = $\frac{1}{2}$; Period = $4\pi/3$
9. Amplitude = $\frac{5}{3}$; Period = 3   **11.** F   **13.** A   **15.** H   **17.** C   **19.** J

AN39

21.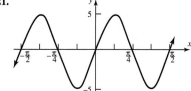

23.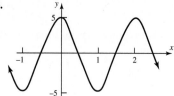

25.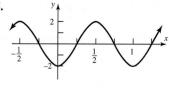

27.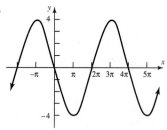

29.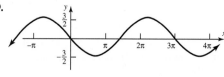

31. $y = 5 \cos \dfrac{\pi}{4} x$ **33.** $y = -3 \cos \frac{1}{2} x$ **35.** $y = \frac{3}{4} \sin 2\pi x$ **37.** $y = -\sin \frac{3}{2} x$ **39.** $y = -2 \cos \dfrac{3\pi}{2} x$ **41.** $y = 3 \sin \dfrac{\pi}{2} x$ **43.** $y = -4 \cos 3x$

45. Amplitude = 4
Period = π
Phase shift = $\pi/2$

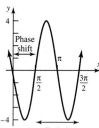

47. Amplitude = 2
Period = $2\pi/3$
Phase shift = $-\pi/6$

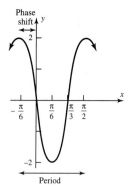

49. Amplitude = 3
Period = π
Phase shift = $-\pi/4$

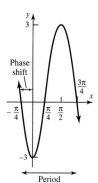

51. Amplitude = 4
Period = 2
Phase shift = $-2/\pi$

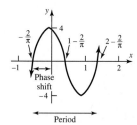

53. Amplitude = 3
Period = 2
Phase shift = $2/\pi$

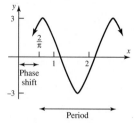

55. Amplitude = 3
Period = π
Phase shift = $\pi/4$

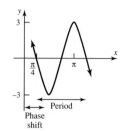

57. Period $= \frac{1}{30}$, Amplitude $= 220$

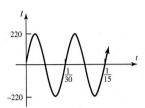

59. Period $= \frac{1}{15}$, Amplitude $= 120$, Phase shift $= \frac{1}{90}$

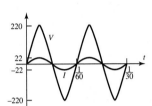

61. (a) Amplitude $= 220$, period $= \frac{1}{60}$
(b) & (e)

(c) $I = 22 \sin 120\pi t$
(d) Amplitude $= 22$, Period $= \frac{1}{60}$

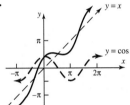

63. (a) $P = \dfrac{(V_0 \sin 2\pi ft)^2}{R} = \dfrac{V_0^2}{R} \sin^2 2\pi ft$ **(b)** $P = \dfrac{V_0^2}{R} \dfrac{1}{2}(1 - \cos 4\pi ft)$

Exercise 6.3

1.

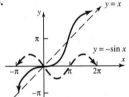

3.

5.

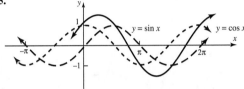

7.

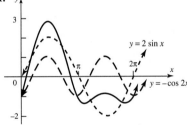

9.

11.

13.

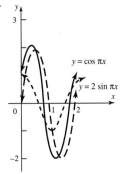

15.

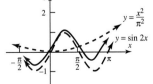

17.

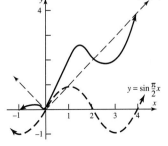

19.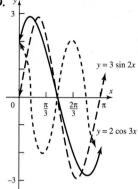

21.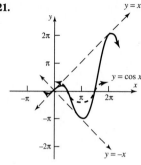

23.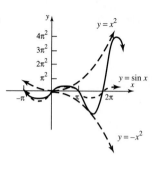

25.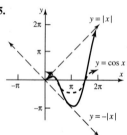

27.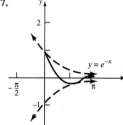

29.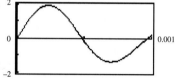

31. (a) 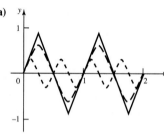 **(b)** 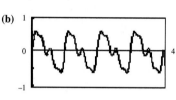 **(c)**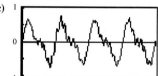

33. $d = 5 \cos \pi t$ **35.** $d = 6 \cos 2t$ **37.** $d = 5 \sin \pi t$ **39.** $d = 6 \sin 2t$

41. (a) Simple harmonic **(b)** 5 cm **(c)** $2\pi/3$ sec **(d)** $3/(2\pi)$ oscillation/sec

43. (a) Simple harmonic **(b)** 6 m **(c)** 2 sec **(d)** $\frac{1}{2}$ oscillation/sec **45. (a)** Simple harmonic **(b)** 3 m **(c)** 4π sec **(d)** $1/(4\pi)$ oscillation/sec

47. (a) Simple harmonic **(b)** 2 m **(c)** 1 sec **(d)** 1 oscillation/sec **49.** It is close to 1.

Exercise 6.4

1. 0 **3.** 1 **5.** $\sec x = 1$ for $x = -2\pi, 0, 2\pi$; $\sec x = -1$ for $x = -\pi, \pi$ **7.** $-3\pi/2, -\pi/2, \pi/2, 3\pi/2$ **9.** $-3\pi/2, -\pi/2, \pi/2, 3\pi/2$ **11.** D

13. B

15.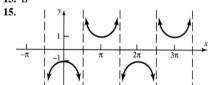

17.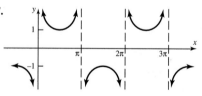

19.

21.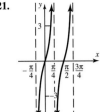

23.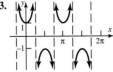

25.

27.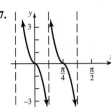

29.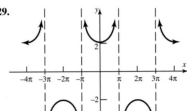

31. (a) L = 3/cos θ + 4/sin θ = 3 sec θ + 4 csc θ

(b)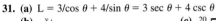

(c)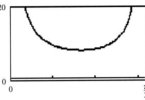

(d) L is smallest when θ = 0.83, L(0.83) = 9.86 ft.

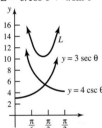

Exercise 6.5

1. 0 **3.** −π/2 **5.** 0 **7.** π/4 **9.** π/3 **11.** 5π/6 **13.** 0.10 **15.** 1.37 **17.** 0.51 **19.** −0.38 **21.** −0.12 **23.** 1.08 **25.** $\sqrt{2}/2$ **27.** $-\sqrt{3}/3$
29. 2 **31.** $\sqrt{2}$ **33.** $-\sqrt{2}/2$ **35.** $2\sqrt{3}/3$ **37.** $\sqrt{2}/4$ **39.** $\sqrt{5}/2$ **41.** $-\sqrt{14}/2$ **43.** $-3\sqrt{10}/10$ **45.** $\sqrt{5}$ **47.** 0.58 **49.** 0.10 **51.** 0.57
53. 0.43 **55.** 0.37
57. Let θ = tan⁻¹ v. Then tan θ = v, −π/2 < θ < π/2. Now, sec θ > 0 and tan² θ + 1 = sec² θ. Thus, sec θ = sec(tan⁻¹ v) = $\sqrt{1 + v^2}$.
59. Let θ = cos⁻¹ v. Then cos θ = v, 0 ≤ θ ≤ π, and tan(cos⁻¹ v) = tan θ = $\dfrac{\sin \theta}{\cos \theta} = \dfrac{\sqrt{1 - \cos^2 \theta}}{\cos \theta} = \dfrac{\sqrt{1 - v^2}}{v}$.
61. Let θ = sin⁻¹ v. Then sin θ = v, −π/2 ≤ θ ≤ π/2, and cos(sin⁻¹ v) = cos θ = $\sqrt{1 - \sin^2 \theta} = \sqrt{1 - v^2}$.
63. Let α = sin⁻¹ v and β = cos⁻¹ v. Then sin α = v = cos β, so α and β are complementary angles. Thus, α + β = π/2.
65. Let α = tan⁻¹ $\dfrac{1}{v}$. Then $\dfrac{1}{v}$ = tan α, $\dfrac{-\pi}{2} < α < \dfrac{\pi}{2}$, α ≠ 0. Let β = tan⁻¹ v. Then v = tan β, $\dfrac{-\pi}{2} < β < \dfrac{\pi}{2}$, β ≠ 0. Thus, tan α tan β = 1 so

that tan α = cot β. Thus, α + β = $\dfrac{\pi}{2}$.

67. 1.32 **69.** 0.46 **71.** −0.34 **73.** 2.72 **75.** −0.73 **77.** 2.55 **79.** 77 in. **81.** −1 ≤ x ≤ 1 **83.** −π/2 ≤ x ≤ π/2
85.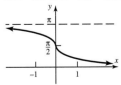

87.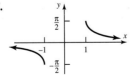

Fill-in-the-Blanks

1. y = 3 sin πx **2.** 3; π/3 **3.** y = 5x **4.** −1 ≤ x ≤ 1; −π/2 ≤ y ≤ π/2 **5.** 0 **6.** simple harmonic motion **7.** y = cos x, y = sec x
8. y = sin x, y = tan x, y = csc x, y = cot x

True/False Items

1. T **2.** T **3.** F **4.** F **5.** T

Review Exercises

1. Amplitude = 4; Period = 2π **3.** Amplitude = 8; Period = 4

5. Amplitude = 4
Period = $2\pi/3$
Phase shift = 0

7. Amplitude = 2
Period = 4
Phase shift = $-1/\pi$

9. Amplitude = $\frac{1}{2}$
Period = $4\pi/3$
Phase shift = $2\pi/3$

11. Amplitude = $\frac{2}{3}$
Period = 2
Phase shift = $6/\pi$

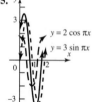

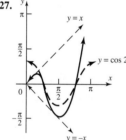

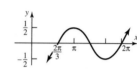

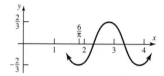

13. $y = 5 \cos \dfrac{x}{4}$ **15.** $y = -6 \cos \dfrac{\pi}{4}x$

17.

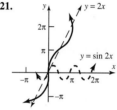

19.

21.

23.

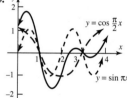

25.

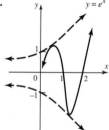

27.

29.

31.

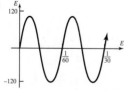

33. $\pi/2$ **35.** $\pi/4$ **37.** $5\pi/6$ **39.** $\sqrt{2}/2$ **41.** $-\sqrt{3}$ **43.** $2\sqrt{3}/3$ **45.** $\frac{3}{5}$ **47.** $-\frac{4}{3}$

49. **(a)** Simple harmonic **(b)** 6 ft **(c)** π sec **(d)** $1/\pi$ oscillation/sec

51. **(a)** Simple harmonic **(b)** 2 ft **(c)** 2 sec **(d)** $\frac{1}{2}$ oscillation/sec

53. **(a)** 120 **(b)** $\frac{1}{60}$ **(c)**

1. $\csc\theta\cdot\cos\theta=\dfrac{1}{\sin\theta}\cdot\cos\theta=\dfrac{\cos\theta}{\sin\theta}=\cot\theta$ **3.** $1+\tan^2(-\theta)=1+(-\tan\theta)^2=1+\tan^2\theta=\sec^2\theta$

5. $\cos\theta(\tan\theta+\cot\theta)=\cos\theta\left(\dfrac{\sin\theta}{\cos\theta}+\dfrac{\cos\theta}{\sin\theta}\right)=\cos\theta\left(\dfrac{\sin^2\theta+\cos^2\theta}{\cos\theta\sin\theta}\right)=\dfrac{1}{\sin\theta}=\csc\theta$

7. $\tan\theta\cot\theta-\cos^2\theta=\dfrac{\sin\theta}{\cos\theta}\cdot\dfrac{\cos\theta}{\sin\theta}-\cos^2\theta=1-\cos^2\theta=\sin^2\theta$ **9.** $(\sec\theta-1)(\sec\theta+1)=\sec^2\theta-1=\tan^2\theta$

11. $(\sec\theta+\tan\theta)(\sec\theta-\tan\theta)=\sec^2\theta-\tan^2\theta=1$ **13.** $\sin^2\theta(1+\cot^2\theta)=\sin^2\theta\csc^2\theta=\sin^2\theta\left(\dfrac{1}{\sin^2\theta}\right)=1$

15. $(\sin\theta+\cos\theta)^2+(\sin\theta-\cos\theta)^2=\sin^2\theta+2\sin\theta\cos\theta+\cos^2\theta+\sin^2\theta-2\sin\theta\cos\theta+\cos^2\theta$
$$=\sin^2\theta+\cos^2\theta+\sin^2\theta+\cos^2\theta=1+1=2$$

17. $\sec^4\theta-\sec^2\theta=\sec^2\theta(\sec^2\theta-1)=(1+\tan^2\theta)\tan^2\theta=\tan^4\theta+\tan^2\theta$

19. $\sec\theta-\tan\theta=\dfrac{1}{\cos\theta}-\dfrac{\sin\theta}{\cos\theta}=\dfrac{1-\sin\theta}{\cos\theta}\cdot\dfrac{1+\sin\theta}{1+\sin\theta}=\dfrac{1-\sin^2\theta}{\cos\theta(1+\sin\theta)}=\dfrac{\cos^2\theta}{\cos\theta(1+\sin\theta)}=\dfrac{\cos\theta}{1+\sin\theta}$

21. $3\sin^2\theta+4\cos^2\theta=3\sin^2\theta+3\cos^2\theta+\cos^2\theta=3(\sin^2\theta+\cos^2\theta)+\cos^2\theta=3+\cos^2\theta$

23. $1-\dfrac{\cos^2\theta}{1+\sin\theta}=1-\dfrac{1-\sin^2\theta}{1+\sin\theta}=1-(1-\sin\theta)=\sin\theta$ **25.** $\dfrac{1+\tan\theta}{1-\tan\theta}=\dfrac{1+\dfrac{1}{\cot\theta}}{1-\dfrac{1}{\cot\theta}}=\dfrac{\dfrac{\cot\theta+1}{\cot\theta}}{\dfrac{\cot\theta-1}{\cot\theta}}=\dfrac{\cot\theta+1}{\cot\theta-1}$

27. $\dfrac{\sec\theta}{\csc\theta}+\dfrac{\sin\theta}{\cos\theta}=\dfrac{1/\cos\theta}{1/\sin\theta}+\tan\theta=\dfrac{\sin\theta}{\cos\theta}+\tan\theta=\tan\theta+\tan\theta=2\tan\theta$ **29.** $\dfrac{1+\sin\theta}{1-\sin\theta}=\dfrac{1+\dfrac{1}{\csc\theta}}{1-\dfrac{1}{\csc\theta}}=\dfrac{\dfrac{\csc\theta+1}{\csc\theta}}{\dfrac{\csc\theta-1}{\csc\theta}}=\dfrac{\csc\theta+1}{\csc\theta-1}$

31. $\dfrac{1-\sin\theta}{\cos\theta}+\dfrac{\cos\theta}{1-\sin\theta}=\dfrac{(1-\sin\theta)^2+\cos^2\theta}{\cos\theta(1-\sin\theta)}=\dfrac{1-2\sin\theta+\sin^2\theta+\cos^2\theta}{\cos\theta(1-\sin\theta)}=\dfrac{2-2\sin\theta}{\cos\theta(1-\sin\theta)}=\dfrac{2(1-\sin\theta)}{\cos\theta(1-\sin\theta)}=\dfrac{2}{\cos\theta}=2\sec\theta$

33. $\dfrac{\sin\theta}{\sin\theta-\cos\theta}=\dfrac{1}{\dfrac{\sin\theta-\cos\theta}{\sin\theta}}=\dfrac{1}{1-\dfrac{\cos\theta}{\sin\theta}}=\dfrac{1}{1-\cot\theta}$

35. $(\sec\theta-\tan\theta)^2=\sec^2\theta-2\sec\theta\tan\theta+\tan^2\theta=\dfrac{1}{\cos^2\theta}-\dfrac{2\sin\theta}{\cos^2\theta}+\dfrac{\sin^2\theta}{\cos^2\theta}=\dfrac{1-2\sin\theta+\sin^2\theta}{\cos^2\theta}=\dfrac{(1-\sin\theta)^2}{1-\sin^2\theta}$
$$=\dfrac{(1-\sin\theta)^2}{(1-\sin\theta)(1+\sin\theta)}=\dfrac{1-\sin\theta}{1+\sin\theta}$$

37. $\dfrac{\cos\theta}{1-\tan\theta}+\dfrac{\sin\theta}{1-\cot\theta}=\dfrac{\cos\theta}{1-\dfrac{\sin\theta}{\cos\theta}}+\dfrac{\sin\theta}{1-\dfrac{\cos\theta}{\sin\theta}}=\dfrac{\cos\theta}{\dfrac{\cos\theta-\sin\theta}{\cos\theta}}+\dfrac{\sin\theta}{\dfrac{\sin\theta-\cos\theta}{\sin\theta}}=\dfrac{\cos^2\theta}{\cos\theta-\sin\theta}+\dfrac{\sin^2\theta}{\sin\theta-\cos\theta}=\dfrac{\cos^2\theta-\sin^2\theta}{\cos\theta-\sin\theta}$
$$=\dfrac{(\cos\theta-\sin\theta)(\cos\theta+\sin\theta)}{\cos\theta-\sin\theta}=\sin\theta+\cos\theta$$

39. $\tan\theta+\dfrac{\cos\theta}{1+\sin\theta}=\dfrac{\sin\theta}{\cos\theta}+\dfrac{\cos\theta}{1+\sin\theta}=\dfrac{\sin\theta(1+\sin\theta)+\cos^2\theta}{\cos\theta(1+\sin\theta)}=\dfrac{\sin\theta+\sin^2\theta+\cos^2\theta}{\cos\theta(1+\sin\theta)}=\dfrac{\sin\theta+1}{\cos\theta(1+\sin\theta)}=\dfrac{1}{\cos\theta}=\sec\theta$

41. $\dfrac{\tan\theta+\sec\theta-1}{\tan\theta-\sec\theta+1}=\dfrac{\tan\theta+(\sec\theta-1)}{\tan\theta-(\sec\theta-1)}\cdot\dfrac{\tan\theta+(\sec\theta-1)}{\tan\theta+(\sec\theta-1)}=\dfrac{\tan^2\theta+2\tan\theta(\sec\theta-1)+\sec^2\theta-2\sec\theta+1}{\tan^2\theta-(\sec^2\theta-2\sec\theta+1)}$
$$=\dfrac{\sec^2\theta-1+2\tan\theta(\sec\theta-1)+\sec^2\theta-2\sec\theta+1}{\sec^2\theta-1-\sec^2\theta+2\sec\theta-1}=\dfrac{2\sec^2\theta-2\sec\theta+2\tan\theta(\sec\theta-1)}{-2+2\sec\theta}$$
$$=\dfrac{2\sec\theta(\sec\theta-1)+2\tan\theta(\sec\theta-1)}{2(\sec\theta-1)}=\dfrac{2(\sec\theta-1)(\sec\theta+\tan\theta)}{2(\sec\theta-1)}=\sec\theta+\tan\theta$$

43. $\dfrac{\tan\theta-\cot\theta}{\tan\theta+\cot\theta}=\dfrac{\dfrac{\sin\theta}{\cos\theta}-\dfrac{\cos\theta}{\sin\theta}}{\dfrac{\sin\theta}{\cos\theta}+\dfrac{\cos\theta}{\sin\theta}}=\dfrac{\dfrac{\sin^2\theta-\cos^2\theta}{\cos\theta\sin\theta}}{\dfrac{\sin^2\theta+\cos^2\theta}{\cos\theta\sin\theta}}=\dfrac{\sin^2\theta-\cos^2\theta}{1}=\sin^2\theta-\cos^2\theta$

45. $\dfrac{\tan\theta-\cot\theta}{\tan\theta+\cot\theta}=\dfrac{\dfrac{\sin\theta}{\cos\theta}-\dfrac{\cos\theta}{\sin\theta}}{\dfrac{\sin\theta}{\cos\theta}+\dfrac{\cos\theta}{\sin\theta}}=\dfrac{\dfrac{\sin^2\theta-\cos^2\theta}{\cos\theta\sin\theta}}{\dfrac{\sin^2\theta+\cos^2\theta}{\cos\theta\sin\theta}}=\sin^2\theta-\cos^2\theta=\sin^2\theta-(1-\sin^2\theta)=2\sin^2\theta-1$

47. $\dfrac{\sec\theta+\tan\theta}{\cot\theta+\cos\theta}=\dfrac{\dfrac{1}{\cos\theta}+\dfrac{\sin\theta}{\cos\theta}}{\dfrac{\cos\theta}{\sin\theta}+\dfrac{\cos\theta\sin\theta}{\sin\theta}}=\dfrac{\dfrac{1+\sin\theta}{\cos\theta}}{\dfrac{\cos\theta+\cos\theta\sin\theta}{\sin\theta}}=\dfrac{1+\sin\theta}{\cos\theta}\cdot\dfrac{\sin\theta}{\cos\theta(1+\sin\theta)}=\dfrac{\sin\theta}{\cos\theta}\cdot\dfrac{1}{\cos\theta}=\tan\theta\sec\theta$

49. $\dfrac{1 - \tan^2 \theta}{1 + \tan^2 \theta} = \dfrac{1 - \tan^2 \theta}{\sec^2 \theta} = \dfrac{1}{\sec^2 \theta} - \dfrac{\tan^2 \theta}{\sec^2 \theta} = \cos^2 \theta - \dfrac{\sin^2 \theta/\cos^2 \theta}{1/\cos^2 \theta} = \cos^2 \theta - \sin^2 \theta = \cos^2 \theta - (1 - \cos^2 \theta) = 2\cos^2 \theta - 1$

51. $\dfrac{\sec \theta - \csc \theta}{\sec \theta \csc \theta} = \dfrac{\dfrac{1}{\cos \theta} - \dfrac{1}{\sin \theta}}{\dfrac{1}{\cos \theta} \cdot \dfrac{1}{\sin \theta}} = \dfrac{\dfrac{\sin \theta - \cos \theta}{\cos \theta \sin \theta}}{\dfrac{1}{\cos \theta \sin \theta}} = \sin \theta - \cos \theta$

53. $\sec \theta - \cos \theta = \dfrac{1}{\cos \theta} - \dfrac{\cos^2 \theta}{\cos \theta} = \dfrac{1 - \cos^2 \theta}{\cos \theta} = \dfrac{\sin^2 \theta}{\cos \theta} = \sin \theta \cdot \dfrac{\sin \theta}{\cos \theta} = \sin \theta \tan \theta$

55. $\dfrac{1}{1 - \sin \theta} + \dfrac{1}{1 + \sin \theta} = \dfrac{1 + \sin \theta + 1 - \sin \theta}{(1 + \sin \theta)(1 - \sin \theta)} = \dfrac{2}{1 - \sin^2 \theta} = \dfrac{2}{\cos^2 \theta} = 2\sec^2 \theta$

57. $\dfrac{\sec \theta}{1 - \sin \theta} = \dfrac{\sec \theta}{1 - \sin \theta} \cdot \dfrac{1 + \sin \theta}{1 + \sin \theta} = \dfrac{\sec \theta(1 + \sin \theta)}{1 - \sin^2 \theta} = \dfrac{\sec \theta(1 + \sin \theta)}{\cos^2 \theta} = \dfrac{1 + \sin \theta}{\cos^3 \theta}$

59. $\dfrac{(\sec \theta - \tan \theta)^2 + 1}{\csc \theta(\sec \theta - \tan \theta)} = \dfrac{\sec^2 \theta - 2\sec \theta \tan \theta + \tan^2 \theta + 1}{\dfrac{1}{\sin \theta}\left(\dfrac{1}{\cos \theta} - \dfrac{\sin \theta}{\cos \theta}\right)} = \dfrac{2\sec^2 \theta - 2\sec \theta \tan \theta}{\dfrac{1}{\sin \theta}\left(\dfrac{1 - \sin \theta}{\cos \theta}\right)} = \dfrac{\dfrac{2}{\cos^2 \theta} - \dfrac{2\sin \theta}{\cos^2 \theta}}{\dfrac{1 - \sin \theta}{\sin \theta \cos \theta}} = \dfrac{2 - 2\sin \theta}{\cos^2 \theta} \cdot \dfrac{\sin \theta \cos \theta}{1 - \sin \theta}$

$= \dfrac{2(1 - \sin \theta)}{\cos \theta} \cdot \dfrac{\sin \theta}{1 - \sin \theta} = \dfrac{2\sin \theta}{\cos \theta} = 2\tan \theta$

61. $\dfrac{\sin \theta + \cos \theta}{\cos \theta} - \dfrac{\sin \theta - \cos \theta}{\sin \theta} = \dfrac{\sin \theta(\sin \theta + \cos \theta) - \cos \theta(\sin \theta - \cos \theta)}{\cos \theta \sin \theta} = \dfrac{\sin^2 \theta + \sin \theta \cos \theta - \sin \theta \cos \theta + \cos^2 \theta}{\cos \theta \sin \theta}$

$= \dfrac{1}{\cos \theta \sin \theta} = \sec \theta \csc \theta$

63. $\dfrac{\sin^3 \theta + \cos^3 \theta}{\sin \theta + \cos \theta} = \dfrac{(\sin \theta + \cos \theta)(\sin^2 \theta - \sin \theta \cos \theta + \cos^2 \theta)}{\sin \theta + \cos \theta} = \sin^2 \theta + \cos^2 \theta - \sin \theta \cos \theta = 1 - \sin \theta \cos \theta$

65. $\dfrac{\cos^2 \theta - \sin^2 \theta}{1 - \tan^2 \theta} = \dfrac{\cos^2 \theta - \sin^2 \theta}{1 - \dfrac{\sin^2 \theta}{\cos^2 \theta}} = \dfrac{\cos^2 \theta - \sin^2 \theta}{\dfrac{\cos^2 \theta - \sin^2 \theta}{\cos^2 \theta}} = \cos^2 \theta$

67. $\dfrac{(2\cos^2 \theta - 1)^2}{\cos^4 \theta - \sin^4 \theta} = \dfrac{[2\cos^2 \theta - (\sin^2 \theta + \cos^2 \theta)]^2}{(\cos^2 \theta - \sin^2 \theta)(\cos^2 \theta + \sin^2 \theta)} = \cos^2 \theta - \sin^2 \theta = (1 - \sin^2 \theta) - \sin^2 \theta = 1 - 2\sin^2 \theta$

69. $\dfrac{1 + \sin \theta + \cos \theta}{1 + \sin \theta - \cos \theta} = \dfrac{(1 + \sin \theta) + \cos \theta}{(1 + \sin \theta) - \cos \theta} \cdot \dfrac{(1 + \sin \theta) + \cos \theta}{(1 + \sin \theta) + \cos \theta} = \dfrac{1 + 2\sin \theta + \sin^2 \theta + 2(1 + \sin \theta)(\cos \theta) + \cos^2 \theta}{1 + 2\sin \theta + \sin^2 \theta - \cos^2 \theta}$

$= \dfrac{1 + 2\sin \theta + \sin^2 \theta + 2(1 + \sin \theta)(\cos \theta) + (1 - \sin^2 \theta)}{1 + 2\sin \theta + \sin^2 \theta - (1 - \sin^2 \theta)} = \dfrac{2 + 2\sin \theta + 2(1 + \sin \theta)(\cos \theta)}{2\sin \theta + 2\sin^2 \theta}$

$= \dfrac{2(1 + \sin \theta) + 2(1 + \sin \theta)(\cos \theta)}{2\sin \theta(1 + \sin \theta)} = \dfrac{2(1 + \sin \theta)(1 + \cos \theta)}{2\sin \theta(1 + \sin \theta)} = \dfrac{1 + \cos \theta}{\sin \theta}$

71. $(a\sin \theta + b\cos \theta)^2 + (a\cos \theta - b\sin \theta)^2 = a^2\sin^2 \theta + 2ab\sin \theta \cos \theta + b^2\cos^2 \theta + a^2\cos^2 \theta - 2ab\sin \theta \cos \theta + b^2\sin^2 \theta$
$= a^2(\sin^2 \theta + \cos^2 \theta) + b^2(\cos^2 \theta + \sin^2 \theta) = a^2 + b^2$

73. $\dfrac{\tan \alpha + \tan \beta}{\cot \alpha + \cot \beta} = \dfrac{\tan \alpha + \tan \beta}{\dfrac{1}{\tan \alpha} + \dfrac{1}{\tan \beta}} = \dfrac{\tan \alpha + \tan \beta}{\dfrac{\tan \beta + \tan \alpha}{\tan \alpha \tan \beta}} = (\tan \alpha + \tan \beta) \cdot \dfrac{\tan \alpha \tan \beta}{\tan \alpha + \tan \beta} = \tan \alpha \tan \beta$

75. $(\sin \alpha + \cos \beta)^2 + (\cos \beta + \sin \alpha)(\cos \beta - \sin \alpha) = (\sin^2 \alpha + 2\sin \alpha \cos \beta + \cos^2 \beta) + (\cos^2 \beta - \sin^2 \alpha)$
$= 2\cos^2 \beta + 2\sin \alpha \cos \beta = 2\cos \beta(\cos \beta + \sin \alpha)$

77. $\ln|\sec \theta| = \ln|\cos \theta|^{-1} = -\ln|\cos \theta|$ **78.** $\ln|\sin \theta| - \ln|\cos \theta| = \ln\left|\dfrac{\sin \theta}{\cos \theta}\right| = \ln|\tan \theta|$

79. $\ln|1 + \cos \theta| + \ln|1 - \cos \theta| = \ln(|1 + \cos \theta|\,|1 - \cos \theta|) = \ln|1 - \cos^2 \theta| = \ln|\sin^2 \theta| = 2\ln|\sin \theta|$

Exercise 7.2

1. $\frac{1}{4}(\sqrt{6} + \sqrt{2})$ **3.** $\frac{1}{4}(\sqrt{2} - \sqrt{6})$ **5.** $-\frac{1}{4}(\sqrt{2} + \sqrt{6})$ **7.** $\dfrac{\sqrt{3} - 1}{1 + \sqrt{3}} = 2 - \sqrt{3}$ **9.** $-\frac{1}{4}(\sqrt{6} + \sqrt{2})$

11. $\dfrac{4}{\sqrt{6} + \sqrt{2}} = \sqrt{6} - \sqrt{2}$ **13.** $\frac{1}{2}$ **15.** 0 **17.** 1 **19.** -1 **21.** $-\sqrt{3}/2$ **23. (a)** $\dfrac{2\sqrt{5}}{25}$ **(b)** $\dfrac{11\sqrt{5}}{25}$ **(c)** $\dfrac{2\sqrt{5}}{5}$ **(d)** 2

25. (a) $\dfrac{4 - 3\sqrt{3}}{10}$ **(b)** $\dfrac{-3 - 4\sqrt{3}}{10}$ **(c)** $\dfrac{4 + 3\sqrt{3}}{10}$ **(d)** $\dfrac{4 + 3\sqrt{3}}{4\sqrt{3} - 3} = \dfrac{25\sqrt{3} + 48}{39}$

27. (a) $-\dfrac{1}{26}(5 + 12\sqrt{3})$ **(b)** $\dfrac{1}{26}(12 - 5\sqrt{3})$ **(c)** $-\dfrac{1}{26}(5 - 12\sqrt{3})$ **(d)** $\dfrac{-5 + 12\sqrt{3}}{12 + 5\sqrt{3}} = \dfrac{-240 + 169\sqrt{3}}{69}$

29. $\sin\left(\dfrac{\pi}{2} + \theta\right) = \sin \dfrac{\pi}{2} \cos \theta + \cos \dfrac{\pi}{2} \sin \theta = 1 \cdot \cos \theta + 0 \cdot \sin \theta = \cos \theta$

31. $\sin(\pi - \theta) = \sin \pi \cos \theta - \cos \pi \sin \theta = 0 \cdot \cos \theta - (-1)\sin \theta = \sin \theta$

33. $\sin(\pi + \theta) = \sin \pi \cos \theta + \cos \pi \sin \theta = 0 \cdot \cos \theta + (-1)\sin \theta = -\sin \theta$

35. $\tan(\pi - \theta) = \dfrac{\tan \pi - \tan \theta}{1 + \tan \pi \tan \theta} = \dfrac{0 - \tan \theta}{1 + 0} = -\tan \theta$

37. $\sin\left(\dfrac{3\pi}{2} + \theta\right) = \sin \dfrac{3\pi}{2} \cos \theta + \cos \dfrac{3\pi}{2} \sin \theta = (-1)\cos \theta + 0 \cdot \sin \theta = -\cos \theta$

39. $\sin(\alpha + \beta) + \sin(\alpha - \beta) = \sin \alpha \cos \beta + \cos \alpha \sin \beta + \sin \alpha \cos \beta - \cos \alpha \sin \beta = 2 \sin \alpha \cos \beta$

41. $\dfrac{\sin(\alpha + \beta)}{\sin \alpha \cos \beta} = \dfrac{\sin \alpha \cos \beta + \cos \alpha \sin \beta}{\sin \alpha \cos \beta} = \dfrac{\sin \alpha \cos \beta}{\sin \alpha \cos \beta} + \dfrac{\cos \alpha \sin \beta}{\sin \alpha \cos \beta} = 1 + \cot \alpha \tan \beta$

43. $\dfrac{\cos(\alpha + \beta)}{\cos \alpha \cos \beta} = \dfrac{\cos \alpha \cos \beta - \sin \alpha \sin \beta}{\cos \alpha \cos \beta} = \dfrac{\cos \alpha \cos \beta}{\cos \alpha \cos \beta} - \dfrac{\sin \alpha \sin \beta}{\cos \alpha \cos \beta} = 1 - \tan \alpha \tan \beta$

45. $\dfrac{\sin(\alpha + \beta)}{\sin(\alpha - \beta)} = \dfrac{\sin \alpha \cos \beta + \cos \alpha \sin \beta}{\sin \alpha \cos \beta - \cos \alpha \sin \beta} = \dfrac{\dfrac{\sin \alpha \cos \beta + \cos \alpha \sin \beta}{\cos \alpha \cos \beta}}{\dfrac{\sin \alpha \cos \beta - \cos \alpha \sin \beta}{\cos \alpha \cos \beta}} = \dfrac{\dfrac{\sin \alpha \cos \beta}{\cos \alpha \cos \beta} + \dfrac{\cos \alpha \sin \beta}{\cos \alpha \cos \beta}}{\dfrac{\sin \alpha \cos \beta}{\cos \alpha \cos \beta} - \dfrac{\cos \alpha \sin \beta}{\cos \alpha \cos \beta}} = \dfrac{\tan \alpha + \tan \beta}{\tan \alpha - \tan \beta}$

47. $\cot(\alpha + \beta) = \dfrac{\cos(\alpha + \beta)}{\sin(\alpha + \beta)} = \dfrac{\cos \alpha \cos \beta - \sin \alpha \sin \beta}{\sin \alpha \cos \beta + \cos \alpha \sin \beta} = \dfrac{\dfrac{\cos \alpha \cos \beta - \sin \alpha \sin \beta}{\sin \alpha \sin \beta}}{\dfrac{\sin \alpha \cos \beta + \cos \alpha \sin \beta}{\sin \alpha \sin \beta}} = \dfrac{\dfrac{\cos \alpha \cos \beta}{\sin \alpha \sin \beta} - \dfrac{\sin \alpha \sin \beta}{\sin \alpha \sin \beta}}{\dfrac{\sin \alpha \cos \beta}{\sin \alpha \sin \beta} + \dfrac{\cos \alpha \sin \beta}{\sin \alpha \sin \beta}}$

$= \dfrac{\cot \alpha \cot \beta - 1}{\cot \beta + \cot \alpha}$

49. $\sec(\alpha + \beta) = \dfrac{1}{\cos(\alpha + \beta)} = \dfrac{1}{\cos \alpha \cos \beta - \sin \alpha \sin \beta} = \dfrac{\dfrac{1}{\sin \alpha \sin \beta}}{\dfrac{\cos \alpha \cos \beta - \sin \alpha \sin \beta}{\sin \alpha \sin \beta}} = \dfrac{\dfrac{1}{\sin \alpha} \cdot \dfrac{1}{\sin \beta}}{\dfrac{\cos \alpha \cos \beta}{\sin \alpha \sin \beta} - \dfrac{\sin \alpha \sin \beta}{\sin \alpha \sin \beta}}$

$= \dfrac{\csc \alpha \csc \beta}{\cot \alpha \cot \beta - 1}$

51. $\sin(\alpha - \beta) \sin(\alpha + \beta) = (\sin \alpha \cos \beta - \cos \alpha \sin \beta)(\sin \alpha \cos \beta + \cos \alpha \sin \beta) = \sin^2 \alpha \cos^2 \beta - \cos^2 \alpha \sin^2 \beta$

$= (\sin^2 \alpha)(1 - \sin^2 \beta) - (1 - \sin^2 \alpha)(\sin^2 \beta) = \sin^2 \alpha - \sin^2 \beta$

53. $\sin(\theta + k\pi) = \sin \theta \cos k\pi + \cos \theta \sin k\pi = (\sin \theta)(-1)^k + (\cos \theta)(0) = (-1)^k \cdot \sin \theta$, k any integer

55. $\sqrt{3}/2$ **57.** $-\dfrac{24}{25}$ **59.** $-\dfrac{33}{65}$ **61.** $\dfrac{65}{63}$ **63.** $\dfrac{1}{39}(48 - 25\sqrt{3})$ **65.** $u\sqrt{1 - v^2} - v\sqrt{1 - u^2}$ **67.** $\dfrac{u\sqrt{1 - v^2} - v}{\sqrt{1 + u^2}}$

69. $\dfrac{uv - \sqrt{1 - u^2}\sqrt{1 - v^2}}{v\sqrt{1 - u^2} + u\sqrt{1 - v^2}}$

71. $\dfrac{\sin(x + h) - \sin x}{h} = \dfrac{\sin x \cos h + \cos x \sin h - \sin x}{h} = \dfrac{\cos x \sin h - (\sin x)(1 - \cos h)}{h} = \cos x \cdot \dfrac{\sin h}{h} - \sin x \cdot \dfrac{1 - \cos h}{h}$

73. $\sin(\sin^{-1} u + \cos^{-1} u) = \sin(\sin^{-1} u)\cos(\cos^{-1} u) + \cos(\sin^{-1} u)\sin(\cos^{-1} u) = (u)(u) + \sqrt{1 - u^2}\sqrt{1 - u^2} = u^2 + 1 - u^2 = 1$

75. $\tan \dfrac{\pi}{2}$ is not defined; $\tan\left(\dfrac{\pi}{2} - \theta\right) = \dfrac{\sin\left(\dfrac{\pi}{2} - \theta\right)}{\cos\left(\dfrac{\pi}{2} - \theta\right)} = \dfrac{\cos \theta}{\sin \theta} = \cot \theta$ **77.** $\tan \theta = \tan(\theta_2 - \theta_1) = \dfrac{\tan \theta_2 - \tan \theta_1}{1 + \tan \theta_1 \tan \theta_2} = \dfrac{m_2 - m_1}{1 + m_1 m_2}$

Exercise 7.3

1. (a) $\dfrac{24}{25}$ **(b)** $\dfrac{7}{25}$ **(c)** $\sqrt{10}/10$ **(d)** $3\sqrt{10}/10$ **3. (a)** $\dfrac{24}{25}$ **(b)** $-\dfrac{7}{25}$ **(c)** $2\sqrt{5}/5$ **(d)** $-\sqrt{5}/5$

5. (a) $-2\sqrt{2}/3$ **(b)** $\dfrac{1}{3}$ **(c)** $\sqrt{\dfrac{3 + \sqrt{6}}{6}}$ **(d)** $\sqrt{\dfrac{3 - \sqrt{6}}{6}}$ **7. (a)** $4\sqrt{2}/9$ **(b)** $-\dfrac{7}{9}$ **(c)** $\sqrt{3}/3$ **(d)** $\sqrt{6}/3$

9. (a) $-\dfrac{4}{5}$ **(b)** $\dfrac{3}{5}$ **(c)** $\sqrt{\dfrac{5 + 2\sqrt{5}}{10}}$ **(d)** $\sqrt{\dfrac{5 - 2\sqrt{5}}{10}}$ **11. (a)** $\dfrac{3}{5}$ **(b)** $-\dfrac{4}{5}$ **(c)** $\dfrac{1}{2}\sqrt{\dfrac{10 - \sqrt{10}}{5}}$ **(d)** $-\dfrac{1}{2}\sqrt{\dfrac{10 + \sqrt{10}}{5}}$ **13.** $\dfrac{\sqrt{2 - \sqrt{2}}}{2}$

15. $1 - \sqrt{2}$ **17.** $-\dfrac{\sqrt{2 + \sqrt{3}}}{2}$ **19.** $\dfrac{2}{\sqrt{2 + \sqrt{2}}} = (2 - \sqrt{2})\sqrt{2 + \sqrt{2}}$ **21.** $-\dfrac{\sqrt{2 - \sqrt{2}}}{2}$

23. $\sin^4 \theta = (\sin^2 \theta)^2 = \left(\dfrac{1 - \cos 2\theta}{2}\right)^2 = \dfrac{1}{4}(1 - 2\cos 2\theta + \cos^2 2\theta) = \dfrac{1}{4} - \dfrac{1}{2}\cos 2\theta + \dfrac{1}{4}\cos^2 2\theta = \dfrac{1}{4} - \dfrac{1}{2}\cos 2\theta + \dfrac{1}{4}\left(\dfrac{1 + \cos 4\theta}{2}\right)$

$= \dfrac{1}{4} - \dfrac{1}{2}\cos 2\theta + \dfrac{1}{8} + \dfrac{1}{8}\cos 4\theta = \dfrac{3}{8} - \dfrac{1}{2}\cos 2\theta + \dfrac{1}{8}\cos 4\theta$

25. $\sin 4\theta = \sin 2(2\theta) = 2 \sin 2\theta \cos 2\theta = (4 \sin \theta \cos \theta)(1 - 2 \sin^2 \theta) = 4 \sin \theta \cos \theta - 8 \sin^3 \theta \cos \theta = (\cos \theta)(4 \sin \theta - 8 \sin^3 \theta)$

27. $16 \sin^5 \theta - 20 \sin^3 \theta + 5 \sin \theta$ **29.** $\cos^4 \theta - \sin^4 \theta = (\cos^2 \theta + \sin^2 \theta)(\cos^2 \theta - \sin^2 \theta) = \cos 2\theta$

31. $\cot 2\theta = \dfrac{1}{\tan 2\theta} = \dfrac{1}{\dfrac{2 \tan \theta}{1 - \tan^2 \theta}} = \dfrac{1 - \tan^2 \theta}{2 \tan \theta} = \dfrac{1 - \dfrac{1}{\cot^2 \theta}}{2\left(\dfrac{1}{\cot \theta}\right)} = \dfrac{\dfrac{\cot^2 \theta - 1}{\cot^2 \theta}}{\dfrac{2}{\cot \theta}} = \dfrac{\cot^2 \theta - 1}{\cot^2 \theta} \cdot \dfrac{\cot \theta}{2} = \dfrac{\cot^2 \theta - 1}{2 \cot \theta}$

33. $\sec 2\theta = \dfrac{1}{\cos 2\theta} = \dfrac{1}{2 \cos^2 \theta - 1} = \dfrac{1}{\dfrac{2}{\sec^2 \theta} - 1} = \dfrac{1}{\dfrac{2 - \sec^2 \theta}{\sec^2 \theta}} = \dfrac{\sec^2 \theta}{2 - \sec^2 \theta}$ **35.** $\cos^2 2\theta - \sin^2 2\theta = \cos 2(2\theta) = \cos 4\theta$

37. $\dfrac{\cos 2\theta}{1 + \sin 2\theta} = \dfrac{\cos^2 \theta - \sin^2 \theta}{1 + 2 \sin \theta \cos \theta} = \dfrac{(\cos \theta - \sin \theta)(\cos \theta + \sin \theta)}{\sin^2 \theta + \cos^2 \theta + 2 \sin \theta \cos \theta} = \dfrac{(\cos \theta - \sin \theta)(\cos \theta + \sin \theta)}{(\sin \theta + \cos \theta)(\sin \theta + \cos \theta)} = \dfrac{\cos \theta - \sin \theta}{\cos \theta + \sin \theta}$

$= \dfrac{\dfrac{\cos \theta - \sin \theta}{\sin \theta}}{\dfrac{\cos \theta + \sin \theta}{\sin \theta}} = \dfrac{\dfrac{\cos \theta}{\sin \theta} - \dfrac{\sin \theta}{\sin \theta}}{\dfrac{\cos \theta}{\sin \theta} + \dfrac{\sin \theta}{\sin \theta}} = \dfrac{\cot \theta - 1}{\cot \theta + 1}$

39. $\sec^2 \dfrac{\theta}{2} = \dfrac{1}{\cos^2(\theta/2)} = \dfrac{1}{\dfrac{1 + \cos \theta}{2}} = \dfrac{2}{1 + \cos \theta}$

41. $\cot^2 \dfrac{\theta}{2} = \dfrac{1}{\tan^2(\theta/2)} = \dfrac{1}{\dfrac{1 - \cos \theta}{1 + \cos \theta}} = \dfrac{1 + \cos \theta}{1 - \cos \theta} = \dfrac{1 + \dfrac{1}{\sec \theta}}{1 - \dfrac{1}{\sec \theta}} = \dfrac{\dfrac{\sec \theta + 1}{\sec \theta}}{\dfrac{\sec \theta - 1}{\sec \theta}} = \dfrac{\sec \theta + 1}{\sec \theta} \cdot \dfrac{\sec \theta}{\sec \theta - 1} = \dfrac{\sec \theta + 1}{\sec \theta - 1}$

43. $\dfrac{1 - \tan^2(\theta/2)}{1 + \tan^2(\theta/2)} = \dfrac{1 - \dfrac{1 - \cos \theta}{1 + \cos \theta}}{1 + \dfrac{1 - \cos \theta}{1 + \cos \theta}} = \dfrac{\dfrac{1 + \cos \theta - (1 - \cos \theta)}{1 + \cos \theta}}{\dfrac{1 + \cos \theta + 1 - \cos \theta}{1 + \cos \theta}} = \dfrac{2 \cos \theta}{1 + \cos \theta} \cdot \dfrac{1 + \cos \theta}{2} = \cos \theta$

45. $\dfrac{\sin 3\theta}{\sin \theta} - \dfrac{\cos 3\theta}{\cos \theta} = \dfrac{\sin 3\theta \cos \theta - \cos 3\theta \sin \theta}{\sin \theta \cos \theta} = \dfrac{\sin(3\theta - \theta)}{\frac{1}{2}(2 \sin \theta \cos \theta)} = \dfrac{2 \sin 2\theta}{\sin 2\theta} = 2$

47. $\tan 3\theta = \tan(\theta + 2\theta) = \dfrac{\tan \theta + \tan 2\theta}{1 - \tan \theta \tan 2\theta} = \dfrac{\tan \theta + \dfrac{2 \tan \theta}{1 - \tan^2 \theta}}{1 - \dfrac{\tan \theta(2 \tan \theta)}{1 - \tan^2 \theta}} = \dfrac{\tan \theta - \tan^3 \theta + 2 \tan \theta}{1 - \tan^2 \theta - 2 \tan^2 \theta} = \dfrac{3 \tan \theta - \tan^3 \theta}{1 - 3 \tan^2 \theta}$

49. $\sqrt{3}/2$ **51.** $\frac{7}{25}$ **53.** $\frac{24}{7}$ **55.** $\frac{24}{25}$ **57.** $\frac{1}{5}$ **59.** $\frac{25}{7}$ **61.** 4 **63.** $-\frac{1}{4}$ **65.** A $= \frac{1}{2}h$ (base) $+ \frac{1}{2}h$ (base) $= s \cos \dfrac{\theta}{2} \cdot s \sin \dfrac{\theta}{2} = s^2 \dfrac{\sin \theta}{2} = \frac{1}{2}s^2 \sin \theta$

67. 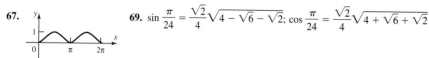 **69.** $\sin \dfrac{\pi}{24} = \dfrac{\sqrt{2}}{4}\sqrt{4 - \sqrt{6} - \sqrt{2}}$; $\cos \dfrac{\pi}{24} = \dfrac{\sqrt{2}}{4}\sqrt{4 + \sqrt{6} + \sqrt{2}}$

71. $\sin^3 \theta + \sin^3(\theta + 120°) + \sin^3(\theta + 240°) = \sin^3 \theta + (\sin \theta \cos 120° + \cos \theta \sin 120°)^3 + (\sin \theta \cos 240° + \cos \theta \sin 240°)^3$

$= \sin^3 \theta + \left(-\dfrac{1}{2}\sin \theta + \dfrac{\sqrt{3}}{2}\cos \theta\right)^3 + \left(-\dfrac{1}{2}\sin \theta - \dfrac{\sqrt{3}}{2}\cos \theta\right)^3$

$= \sin^3 \theta + \frac{1}{8}(3\sqrt{3} \cos^3 \theta - 9 \cos^2 \theta \sin \theta + 3\sqrt{3} \cos \theta \sin^2 \theta - \sin^3 \theta) - \frac{1}{8}(\sin^3 \theta + 3\sqrt{3} \sin^2 \theta \cos \theta + 9 \sin \theta \cos^2 \theta + 3\sqrt{3} \cos^3 \theta)$

$= \frac{3}{4}\sin^3 \theta - \frac{9}{4}\cos^2 \theta \sin \theta = \frac{3}{4}[\sin^3 \theta - 3 \sin \theta(1 - \sin^2 \theta)] = \frac{3}{4}(4 \sin^3 \theta - 3 \sin \theta) = -\frac{3}{4}\sin 3\theta$

73. $\dfrac{1}{2}(\ln |1 - \cos 2\theta| - \ln 2) = \ln\left(\dfrac{|1 - \cos 2\theta|}{2}\right)^{1/2} = \ln |\sin^2 \theta|^{1/2} = \ln |\sin \theta|$

75. (a) $R = \dfrac{v_0^2 \sqrt{2}}{16}(\sin \theta \cos \theta - \cos^2 \theta) = \dfrac{v_0^2 \sqrt{2}}{16}\left(\dfrac{1}{2}\sin 2\theta - \dfrac{1 + \cos 2\theta}{2}\right)$ **(b)** 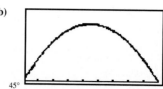 $\theta = 67.5°$ makes R largest

$= \dfrac{v_0^2 \sqrt{2}}{32}(\sin 2\theta - \cos 2\theta - 1)$

Exercise 7.4

1. $\frac{1}{2}(\cos 2\theta - \cos 6\theta)$ **3.** $\frac{1}{2}(\sin 6\theta + \sin 2\theta)$ **5.** $\frac{1}{2}(\cos 8\theta + \cos 2\theta)$ **7.** $\frac{1}{2}(\cos \theta - \cos 3\theta)$ **9.** $\frac{1}{2}(\sin 2\theta + \sin \theta)$ **11.** $2 \sin \theta \cos 3\theta$

13. $2 \cos 3\theta \cos \theta$ **15.** $2 \sin 2\theta \cos \theta$ **17.** $2 \sin \theta \sin(\theta/2)$ **19.** $\dfrac{\sin \theta + \sin 3\theta}{2 \sin 2\theta} = \dfrac{2 \sin 2\theta \cos(-\theta)}{2 \sin 2\theta} = \cos(-\theta) = \cos \theta$

21. $\dfrac{\sin 4\theta + \sin 2\theta}{\cos 4\theta + \cos 2\theta} = \dfrac{2 \sin 3\theta \cos \theta}{2 \cos 3\theta \cos \theta} = \dfrac{\sin 3\theta}{\cos 3\theta} = \tan 3\theta$ **23.** $\dfrac{\cos \theta - \cos 3\theta}{\sin \theta + \sin 3\theta} = \dfrac{-2 \sin 2\theta \sin(-\theta)}{2 \sin 2\theta \cos(-\theta)} = \dfrac{\sin \theta}{\cos \theta} = \tan \theta$

25. $\sin \theta (\sin \theta + \sin 3\theta) = \sin \theta[2 \sin 2\theta \cos(-\theta)] = 2 \sin 2\theta \sin \theta \cos \theta = \cos \theta(2 \sin 2\theta \sin \theta) = \cos \theta[2 \cdot \frac{1}{2}(\cos \theta - \cos 3\theta)]$
$\qquad = \cos \theta(\cos \theta - \cos 3\theta)$

27. $\dfrac{\sin 4\theta + \sin 8\theta}{\cos 4\theta + \cos 8\theta} = \dfrac{2 \sin 6\theta \cos(-2\theta)}{2 \cos 6\theta \cos(-2\theta)} = \dfrac{\sin 6\theta}{\cos 6\theta} = \tan 6\theta$

29. $\dfrac{\sin 4\theta + \sin 8\theta}{\sin 4\theta - \sin 8\theta} = \dfrac{2 \sin 6\theta \cos(-2\theta)}{2 \sin(-2\theta) \cos 6\theta} = \dfrac{\sin 6\theta}{\cos 6\theta} \cdot \dfrac{\cos 2\theta}{-\sin 2\theta} = \tan 6\theta(-\cot 2\theta) = -\dfrac{\tan 6\theta}{\tan 2\theta}$

31. $\dfrac{\sin \alpha + \sin \beta}{\sin \alpha - \sin \beta} = \dfrac{2 \sin \dfrac{\alpha + \beta}{2} \cos \dfrac{\alpha - \beta}{2}}{2 \sin \dfrac{\alpha - \beta}{2} \cos \dfrac{\alpha + \beta}{2}} = \dfrac{\sin \dfrac{\alpha + \beta}{2}}{\cos \dfrac{\alpha + \beta}{2}} \cdot \dfrac{\cos \dfrac{\alpha - \beta}{2}}{\sin \dfrac{\alpha - \beta}{2}} = \tan \dfrac{\alpha + \beta}{2} \cot \dfrac{\alpha - \beta}{2}$

33. $\dfrac{\sin \alpha + \sin \beta}{\cos \alpha + \cos \beta} = \dfrac{2 \sin \dfrac{\alpha + \beta}{2} \cos \dfrac{\alpha - \beta}{2}}{2 \cos \dfrac{\alpha + \beta}{2} \cos \dfrac{\alpha - \beta}{2}} = \dfrac{\sin \dfrac{\alpha + \beta}{2}}{\cos \dfrac{\alpha + \beta}{2}} = \tan \dfrac{\alpha + \beta}{2}$

35. $1 + \cos 2\theta + \cos 4\theta + \cos 6\theta = (1 + \cos 6\theta) + (\cos 2\theta + \cos 4\theta) = 2 \cos^2 3\theta + 2 \cos 3\theta \cos(-\theta) = 2 \cos 3\theta(\cos 3\theta + \cos \theta)$
$\qquad = 2 \cos 3\theta(2 \cos 2\theta \cos \theta) = 4 \cos \theta \cos 2\theta \cos 3\theta$

37. $y = 2 \sin 2061\pi t \cos 357\pi t$
39. $\sin 2\alpha + \sin 2\beta + \sin 2\gamma = 2 \sin(\alpha + \beta) \cos(\alpha - \beta) + \sin 2\gamma = 2 \sin(\alpha + \beta) \cos(\alpha - \beta) + 2 \sin \gamma \cos \gamma$
$\qquad = 2 \sin(\pi - \gamma) \cos(\alpha - \beta) + 2 \sin \gamma \cos \gamma = 2 \sin \gamma \cos(\alpha - \beta) + 2 \sin \gamma \cos \gamma = 2 \sin \gamma[\cos(\alpha - \beta) + \cos \gamma]$
$\qquad = 2 \sin \gamma\left(2 \cos \dfrac{\alpha - \beta + \gamma}{2} \cos \dfrac{\alpha - \beta - \gamma}{2}\right) = 4 \sin \gamma \cos \dfrac{\pi - 2\beta}{2} \cos \dfrac{2\alpha - \pi}{2} = 4 \sin \gamma \cos\left(\dfrac{\pi}{2} - \beta\right) \cos\left(\alpha - \dfrac{\pi}{2}\right)$
$\qquad = 4 \sin \gamma \sin \beta \sin \alpha$

41. $\qquad\qquad \sin(\alpha - \beta) = \sin \alpha \cos \beta - \cos \alpha \sin \beta$
$\qquad\qquad\qquad \sin(\alpha + \beta) = \sin \alpha \cos \beta + \cos \alpha \sin \beta$
$\qquad \sin(\alpha - \beta) + \sin(\alpha + \beta) = 2 \sin \alpha \cos \beta$
$\qquad\qquad\qquad \sin \alpha \cos \beta = \frac{1}{2}[\sin(\alpha + \beta) + \sin(\alpha - \beta)]$

43. $2 \cos \dfrac{\alpha + \beta}{2} \cos \dfrac{\alpha - \beta}{2} = 2 \cdot \dfrac{1}{2}\left[\cos\left(\dfrac{\alpha + \beta}{2} + \dfrac{\alpha - \beta}{2}\right) + \cos\left(\dfrac{\alpha + \beta}{2} - \dfrac{\alpha - \beta}{2}\right)\right] = \cos \dfrac{2\alpha}{2} + \cos \dfrac{2\beta}{2} = \cos \alpha + \cos \beta$

Exercise 7.5

1. $\pi/6, 5\pi/6$ **3.** $5\pi/6, 11\pi/6$ **5.** $\pi/2, 3\pi/2$ **7.** $\pi/2, 7\pi/6, 11\pi/6$ **9.** $3\pi/4, 7\pi/4$ **11.** $4\pi/9, 8\pi/9, 16\pi/9$ **13.** $0.4115168, \pi - 0.4115168$
15. $1.3734008, \pi + 1.3734008$ **17.** $2.6905658, 2\pi - 2.6905658$ **19.** $1.8234766, 2\pi - 1.8234766$ **21.** $\pi/2, 2\pi/3, 4\pi/3, 3\pi/2$
23. $\pi/2, 7\pi/6, 11\pi/6$ **25.** $0, \pi/4, 5\pi/4$ **27.** $\pi/4, 5\pi/4$ **29.** $0, \pi/3, \pi, 5\pi/3$ **31.** $\pi/2, 3\pi/2$ **33.** $0, 2\pi/3, 4\pi/3$
35. $0, \pi/3, 2\pi/3, \pi/2, 4\pi/3, \pi, 5\pi/3, 3\pi/2$ **37.** $0, \pi/5, 2\pi/5, 3\pi/5, 4\pi/5, \pi, 6\pi/5, 7\pi/5, 8\pi/5, 9\pi/5$ **39.** $\pi/6, 5\pi/6, 3\pi/2$ **41.** $\pi/3, 5\pi/3$
43. No real solution   **45.** No real solution   **47.** $\pi/2, 7\pi/6$ **49.** $0, \pi/3, \pi, 5\pi/3$
51.

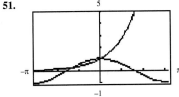

$-1.29, 0$

53.

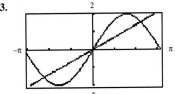

$-2.24, 0, 2.24$

55.

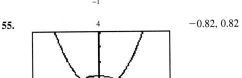

$-0.82, 0.82$

57. $-1.30, 1.97, 3.83$ **59.** 0.59 **61.** 1.25 **63.** $-1.02, 1.02$ **65.** $0, 2.14$ **67.** 1.34 **69. (a)** $60°$ **(b)** $60°$ **(c)** $A(60°) = 12\sqrt{3}$ sq in
71. $2.02, 4.91$ **73.** $28.9°$ **75.** Yes; it varies from 1.27 to 1.34 **77.** 1.47
79. If θ is the original angle of incidence and ϕ is the angle of refraction, then $(\sin\theta)(\sin\phi) = n_2$. The angle of incidence of the emerging beam is also ϕ, and the index of refraction is $1/n_2$. Thus, θ is the angle of refraction of the emerging beam.

Fill-in-the-Blank Items

1. identity; conditional   **2.** $-$ **3.** $+$ **4.** $\sin^2\theta$; $2\cos^2\theta$; $2\sin^2\theta$ **5.** $1 - \cos\alpha$

True/False Items

1. T   **2.** F   **3.** T   **4.** F   **5.** F   **6.** F

Review Exercises

1. $\tan\theta\cot\theta - \sin^2\theta = 1 - \sin^2\theta = \cos^2\theta$ **3.** $\cos^2\theta(1 + \tan^2\theta) = \cos^2\theta\sec^2\theta = 1$
5. $4\cos^2\theta + 3\sin^2\theta = \cos^2\theta + 3(\cos^2\theta + \sin^2\theta) = 3 + \cos^2\theta$
7. $\dfrac{1 - \cos\theta}{\sin\theta} + \dfrac{\sin\theta}{1 - \cos\theta} = \dfrac{(1 - \cos\theta)^2 + \sin^2\theta}{\sin\theta(1 - \cos\theta)} = \dfrac{1 - 2\cos\theta + \cos^2\theta + \sin^2\theta}{\sin\theta(1 - \cos\theta)} = \dfrac{2(1 - \cos\theta)}{\sin\theta(1 - \cos\theta)} = 2\csc\theta$

9. $\dfrac{\cos\theta}{\cos\theta - \sin\theta} = \dfrac{\cos\theta}{\dfrac{\cos\theta - \sin\theta}{\cos\theta}} = \dfrac{1}{1 - \dfrac{\sin\theta}{\cos\theta}} = \dfrac{1}{1 - \tan\theta}$

11. $\dfrac{\csc\theta}{1 + \csc\theta} = \dfrac{\dfrac{1}{\sin\theta}}{1 + \dfrac{1}{\sin\theta}} = \dfrac{1}{1 + \sin\theta} = \dfrac{1}{1 + \sin\theta} \cdot \dfrac{1 - \sin\theta}{1 - \sin\theta} = \dfrac{1 - \sin\theta}{1 - \sin^2\theta} = \dfrac{1 - \sin\theta}{\cos^2\theta}$

13. $\csc\theta - \sin\theta = \dfrac{1}{\sin\theta} - \sin\theta = \dfrac{1 - \sin^2\theta}{\sin\theta} = \dfrac{\cos^2\theta}{\sin\theta} = \cos\theta \cdot \dfrac{\cos\theta}{\sin\theta} = \cos\theta\cot\theta$

15. $\dfrac{1 - \sin\theta}{\sec\theta} = \cos\theta(1 - \sin\theta) \cdot \dfrac{1 + \sin\theta}{1 + \sin\theta} = \dfrac{\cos\theta(1 - \sin^2\theta)}{1 + \sin\theta} = \dfrac{\cos^3\theta}{1 + \sin\theta}$

17. $\cot\theta - \tan\theta = \dfrac{\cos\theta}{\sin\theta} - \dfrac{\sin\theta}{\cos\theta} = \dfrac{\cos^2\theta - \sin^2\theta}{\sin\theta\cos\theta} = \dfrac{1 - 2\sin^2\theta}{\sin\theta\cos\theta}$

19. $\dfrac{\cos(\alpha + \beta)}{\cos\alpha\sin\beta} = \dfrac{\cos\alpha\cos\beta - \sin\alpha\sin\beta}{\cos\alpha\sin\beta} = \dfrac{\cos\alpha\cos\beta}{\cos\alpha\sin\beta} - \dfrac{\sin\alpha\sin\beta}{\cos\alpha\sin\beta} = \cot\beta - \tan\alpha$

21. $\dfrac{\cos(\alpha - \beta)}{\cos\alpha\cos\beta} = \dfrac{\cos\alpha\cos\beta + \sin\alpha\sin\beta}{\cos\alpha\cos\beta} = \dfrac{\cos\alpha\cos\beta}{\cos\alpha\cos\beta} + \dfrac{\sin\alpha\sin\beta}{\cos\alpha\cos\beta} = 1 + \tan\alpha\tan\beta$

23. $(1 + \cos\theta)\left(\tan\dfrac{\theta}{2}\right) = \left(2\cos^2\dfrac{\theta}{2}\right)\dfrac{\sin(\theta/2)}{\cos(\theta/2)} = 2\sin\dfrac{\theta}{2}\cos\dfrac{\theta}{2} = \sin\theta$

25. $2\cot\theta\cot 2\theta = 2\left(\dfrac{\cos\theta}{\sin\theta}\right)\left(\dfrac{\cos 2\theta}{\sin 2\theta}\right) = \dfrac{2\cos\theta(\cos^2\theta - \sin^2\theta)}{2\sin^2\theta\cos\theta} = \dfrac{\cos^2\theta - \sin^2\theta}{\sin^2\theta} = \cot^2\theta - 1$

27. $1 - 8\sin^2\theta\cos^2\theta = 1 - 2(2\sin\theta\cos\theta)^2 = 1 - 2\sin^2 2\theta = \cos 4\theta$ **29.** $\dfrac{\sin 2\theta + \sin 4\theta}{\cos 2\theta + \cos 4\theta} = \dfrac{2\sin 3\theta\cos(-\theta)}{2\cos 3\theta\cos(-\theta)} = \tan 3\theta$

31. $\dfrac{\cos 2\theta - \cos 4\theta}{\cos 2\theta + \cos 4\theta} - \tan\theta\tan 3\theta = \dfrac{-2\sin 3\theta\sin(-\theta)}{2\cos 3\theta\cos(-\theta)} - \tan\theta\tan 3\theta = \tan 3\theta\tan\theta - \tan\theta\tan 3\theta = 0$ **33.** $\frac{1}{4}(\sqrt{6} - \sqrt{2})$ **35.** $\frac{1}{4}(\sqrt{6} - \sqrt{2})$

37. $\frac{1}{2}$ **39.** $\sqrt{\dfrac{2 - \sqrt{2}}{2 + \sqrt{2}}} = \sqrt{2} - 1$ **41. (a)** $-\frac{33}{65}$ **(b)** $-\frac{56}{65}$ **(c)** $-\frac{63}{65}$ **(d)** $\frac{33}{56}$ **(e)** $\frac{24}{25}$ **(f)** $\frac{119}{169}$ **(g)** $5\sqrt{26}/26$ **(h)** $2\sqrt{5}/5$
43. (a) $-\frac{16}{65}$ **(b)** $-\frac{63}{65}$ **(c)** $-\frac{56}{65}$ **(d)** $\frac{16}{63}$ **(e)** $\frac{24}{25}$ **(f)** $\frac{119}{169}$ **(g)** $\sqrt{26}/26$ **(h)** $-\sqrt{10}/10$
45. (a) $-\frac{63}{65}$ **(b)** $\frac{16}{65}$ **(c)** $\frac{33}{65}$ **(d)** $-\frac{63}{16}$ **(e)** $\frac{24}{25}$ **(f)** $-\frac{119}{169}$ **(g)** $2\sqrt{13}/13$ **(h)** $-\sqrt{10}/10$
47. (a) $(-\sqrt{3} - 2\sqrt{2})/6$ **(b)** $(1 - 2\sqrt{6})/6$ **(c)** $(-\sqrt{3} + 2\sqrt{2})/6$ **(d)** $(-\sqrt{3} - 2\sqrt{2})/(1 - 2\sqrt{6}) = (8\sqrt{2} + 9\sqrt{3})/23$ **(e)** $-\sqrt{3}/2$
   (f) $-\frac{7}{9}$ **(g)** $\sqrt{3}/3$ **(h)** $\sqrt{3}/2$ **49. (a)** 1 **(b)** 0 **(c)** $-\frac{1}{9}$ **(d)** Not defined   **(e)** $4\sqrt{5}/9$ **(f)** $-\frac{1}{9}$ **(g)** $\sqrt{30}/6$ **(h)** $-\sqrt{6}\sqrt{3} - \sqrt{5}/6$
51. $\pi/3, 5\pi/3$ **53.** $3\pi/4, 5\pi/4$ **55.** $3\pi/4, 7\pi/4$ **57.** $0, \pi/2, \pi, 3\pi/2$ **59.** $1.1197695, \pi - 1.1197695$ **61.** $0, \pi$ **63.** $0, 2\pi/3, \pi, 4\pi/3$
65. $0, \pi/6, 5\pi/6$ **67.** $\pi/6, \pi/2, 5\pi/6$ **69.** $\pi/2, \pi$

CHAPTER 8   *Exercise 8.1*

1. $a = 3.23$, $b = 3.55$, $\alpha = 40°$ **3.** $a = 3.25$, $c = 4.23$, $\beta = 45°$ **5.** $\gamma = 95°$, $c = 9.86$, $a = 6.36$ **7.** $\alpha = 40°$, $a = 2$, $c = 3.06$
9. $\gamma = 120°$, $b = 1.06$, $c = 2.69$ **11.** $\alpha = 100°$, $a = 5.24$, $c = 0.92$ **13.** $\beta = 40°$, $a = 5.64$, $b = 3.86$ **15.** $\gamma = 100°$, $a = 1.31$, $b = 1.31$
17. One triangle; $\beta = 30.7°$, $\gamma = 99.3°$, $c = 3.86$ **19.** One triangle: $\gamma = 36.2°$, $\alpha = 43.8°$, $a = 3.51$ **21.** No triangle
23. Two triangles; $\gamma_1 = 30.9°$, $\alpha_1 = 129.1°$, $a_1 = 9.08$ or $\gamma_2 = 149.1°$, $\alpha_2 = 10.9°$, $a_2 = 2.21$ **25.** No triangle
27. Two triangles; $\alpha_1 = 57.7°$, $\beta_1 = 97.3°$, $b_1 = 2.35$ or $\alpha_2 = 122.3°$, $\beta_2 = 32.7°$, $b_2 = 1.28$
29. (a) Station Able is 143.3 mi from the ship; Station Baker is 135.6 mi from the ship.   **(b)** Approx. 41 min
31. 1490.5 ft   **33.** 381.7 ft   **35. (a)** 169 mi   **(b)** 161.3°   **37.** 84.7°; 183.7 ft   **39.** 2.64 mi   **41.** 1.88 mi

43. $\dfrac{a+b}{c} = \dfrac{a}{c} + \dfrac{b}{c} = \dfrac{\sin\alpha}{\sin\gamma} + \dfrac{\sin\beta}{\sin\gamma} = \dfrac{\sin\alpha + \sin\beta}{\sin\gamma} = \dfrac{2\sin\dfrac{\alpha+\beta}{2}\cos\dfrac{\alpha-\beta}{2}}{2\sin\dfrac{\gamma}{2}\cos\dfrac{\gamma}{2}} = \dfrac{\sin\left(\dfrac{\pi}{2} - \dfrac{\gamma}{2}\right)\cos\dfrac{\alpha-\beta}{2}}{\sin\dfrac{\gamma}{2}\cos\dfrac{\gamma}{2}} = \dfrac{\cos\frac{1}{2}(\alpha-\beta)}{\sin\frac{1}{2}\gamma}$

45. $a = \dfrac{b\sin\alpha}{\sin\beta} = \dfrac{b\sin[180° - (\beta+\gamma)]}{\sin\beta} = \dfrac{b}{\sin\beta}(\sin\beta\cos\gamma + \cos\beta\sin\gamma) = b\cos\gamma + \dfrac{b\sin\gamma}{\sin\beta}\cos\beta = b\cos\gamma + c\cos\beta$

47. $\sin\beta = \sin(\text{Angle } AB'C) = b/(2r)$; $\dfrac{\sin\beta}{b} = \dfrac{1}{2r}$; the result follows using the Law of Sines

Exercise 8.2

1. $b = 2.95$, $\alpha = 28.7°$, $\gamma = 106.3°$ **3.** $c = 3.75$, $\alpha = 32.1°$, $\beta = 52.9°$ **5.** $\alpha = 48.5°$, $\beta = 38.6°$, $\gamma = 92.9°$
7. $\alpha = 127.2°$, $\beta = 32.1°$, $\gamma = 20.7°$ **9.** $c = 2.57$, $\alpha = 48.6°$, $\beta = 91.4°$ **11.** $a = 2.99$, $\beta = 19.2°$, $\gamma = 80.8°$
13. $b = 4.14$, $\alpha = 43.0°$, $\gamma = 27.0°$ **15.** $c = 1.69$, $\alpha = 65.0°$, $\beta = 65.0°$ **17.** $\alpha = 67.4°$, $\beta = 90°$, $\gamma = 22.6°$
19. $\alpha = 60°$, $\beta = 60°$, $\gamma = 60°$ **21.** $\alpha = 33.6°$, $\beta = 62.2°$, $\gamma = 84.3°$ **23.** $\alpha = 97.9°$, $\beta = 52.4°$, $\gamma = 29.7°$ **25.** 70.75 ft
27. (a) 12°   **(b)** 220.8 mph   **29. (a)** 63.7 ft   **(b)** 66.8 ft   **(c)** 92.5°   **31. (a)** 492.6 ft   **(b)** 269.3 ft   **33.** 342.3 ft
35. Using the Law of Cosines:

$$L^2 = x^2 + r^2 - 2xr\cos\theta$$
$$x^2 - 2rx\cos\theta + r^2 - L^2 = 0$$
$$x = 2r\cos\theta + \dfrac{\sqrt{4r^2\cos^2\theta - 4(r^2 - L^2)}}{2}$$
$$x = r\cos\theta + \sqrt{r^2\cos^2\theta + L^2 - r^2}$$

37. $\cos\dfrac{\gamma}{2} = \sqrt{\dfrac{1+\cos\gamma}{2}} = \sqrt{\dfrac{1 + \dfrac{a^2+b^2-c^2}{2ab}}{2}} = \sqrt{\dfrac{2ab + a^2 + b^2 - c^2}{4ab}} = \sqrt{\dfrac{(a+b)^2 - c^2}{4ab}} = \sqrt{\dfrac{(a+b+c)(a+b-c)}{4ab}}$

$= \sqrt{\dfrac{2s(2s - 2c)}{4ab}} = \sqrt{\dfrac{s(s-c)}{ab}}$

39. $\dfrac{\cos\alpha}{a} + \dfrac{\cos\beta}{b} + \dfrac{\cos\gamma}{c} = \dfrac{b^2+c^2-a^2}{2abc} + \dfrac{a^2+c^2-b^2}{2abc} + \dfrac{a^2+b^2-c^2}{2abc} = \dfrac{b^2+c^2-a^2 + a^2 + c^2 - b^2 + a^2 + b^2 - c^2}{2abc} = \dfrac{a^2+b^2+c^2}{2abc}$

Exercise 8.3

1. 2.83   **3.** 2.99   **5.** 14.98   **7.** 9.56   **9.** 3.86   **11.** 1.48   **13.** 2.82   **15.** 1.53   **17.** 30   **19.** 1.73   **21.** 19.90   **23.** 19.81   **25.** $5446.38
27. $A = \dfrac{1}{2}ab\sin\gamma = \dfrac{1}{2}a\sin\gamma\left(\dfrac{a\sin\beta}{\sin\alpha}\right) = \dfrac{a^2\sin\beta\sin\gamma}{2\sin\alpha}$ **29.** 0.92   **31.** 2.27   **33.** 5.44   **35.** 0.84   **37.** $A = \dfrac{1}{2}r^2(\theta - \sin\theta)$

Exercise 8.4

1.
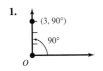

3.
(−2, 0)   O

5.
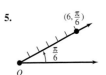

7.
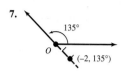

135°

O

$(-2, 135°)$

9.

$(-1, -\frac{\pi}{3})$

O

$-\frac{\pi}{3}$

11.

O $(-2, -\pi)$

$-\pi$

13.
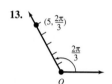

$(5, \frac{2\pi}{3})$

$\frac{2\pi}{3}$

O

(a) $(5, -4\pi/3)$
(b) $(-5, 5\pi/3)$
(c) $(5, 8\pi/3)$

15.
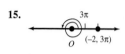

3π

$(-2, 3\pi)$

O

(a) $(2, -2\pi)$
(b) $(-2, \pi)$
(c) $(2, 2\pi)$

17.
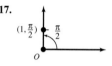

$(1, \frac{\pi}{2})$ $\frac{\pi}{2}$

O

(a) $(1, -3\pi/2)$
(b) $(-1, 3\pi/2)$
(c) $(1, 5\pi/2)$

19.
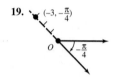

$(-3, -\frac{\pi}{4})$

O

$-\frac{\pi}{4}$

(a) $(3, -5\pi/4)$
(b) $(-3, 7\pi/4)$
(c) $(3, 11\pi/4)$

21. $(0, 3)$ **23.** $(-2, 0)$ **25.** $(-3\sqrt{3}, 3)$ **27.** $(\sqrt{2}, -\sqrt{2})$ **29.** $(-\frac{1}{2}, \sqrt{3}/2)$ **31.** $(2, 0)$ **33.** $(-2.57, 7.05)$ **35.** $(-4.98, -3.86)$ **37.** $(3, 0)$
39. $(1, \pi)$ **41.** $(\sqrt{2}, -\pi/4)$ **43.** $(2, \pi/6)$ **45.** $(2.47, -1.02)$ **47.** $(9.3, 0.47)$ **49.** $r^2 = \frac{3}{2}$ **51.** $r^2 \cos^2 \theta - 4r \sin \theta = 0$ **53.** $r^2 \sin 2\theta = 1$
55. $r \cos \theta = 4$ **57.** $x^2 + y^2 - x = 0$ or $(x - \frac{1}{2})^2 + y^2 = \frac{1}{4}$ **59.** $(x^2 + y^2)^{3/2} - x = 0$ **61.** $x^2 + y^2 = 4$ **63.** $y^2 = 8(x + 2)$
65. $d = \sqrt{(r_2 \cos \theta_2 - r_1 \cos \theta_1)^2 + (r_2 \sin \theta_2 - r_1 \sin \theta_1)^2} = \sqrt{r_1^2 + r_2^2 - 2r_1r_2 \cos(\theta_2 - \theta_1)}$

Exercise 8.5

1. Circle, radius 4, center at pole

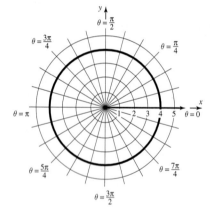

3. Line through pole, making an angle of $\pi/3$ with polar axis

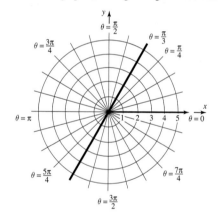

5. Horizontal line 4 units above the pole

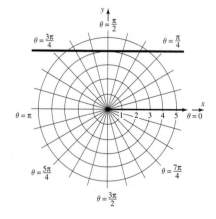

7. Vertical line 2 units to the left of the pole

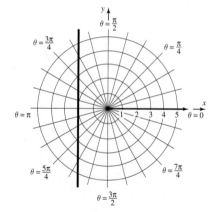

9. Circle, radius 1, center at $(1, 0)$ in rectangular coordinates

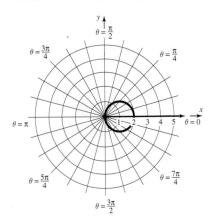

11. Circle, radius 2, center at $(0, -2)$ in rectangular coordinates

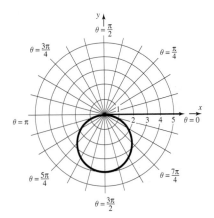

13. Circle, radius 2, center at $(2, 0)$ in rectangular coordinates

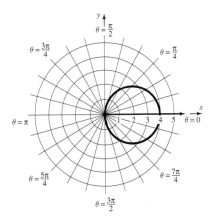

15. Circle, radius 1, center at $(0, -1)$ in rectangular coordinates

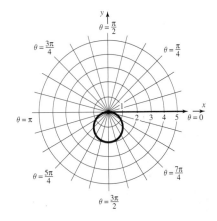

17. E **19.** F **21.** H **23.** D
25. Cardioid

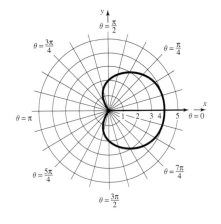

27. Cardioid

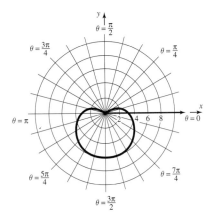

29. Limaçon without inner loop

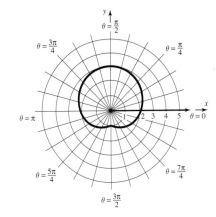

31. Limaçon without inner loop

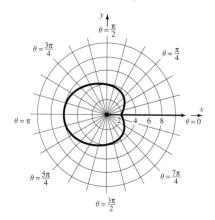

33. Limaçon with inner loop

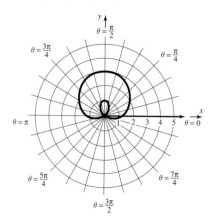

35. Limaçon with inner loop

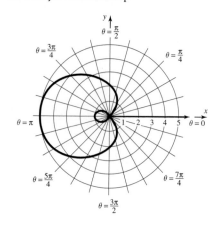

37. Rose

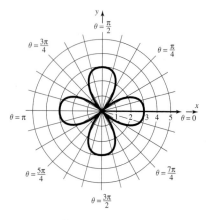

39. Rose

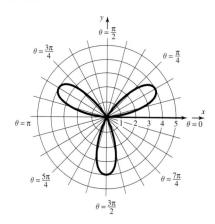

41. Lemniscate

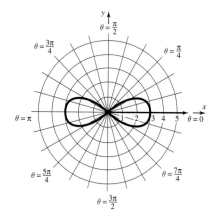

43. Spiral

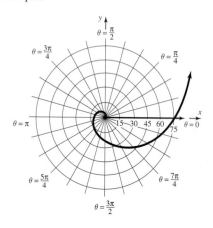

45. Cardioid

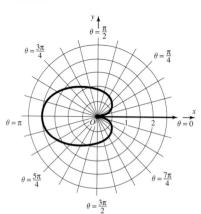

47. Limaçon with inner loop

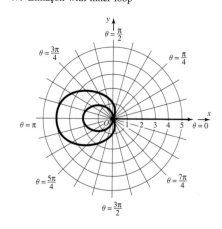

49.

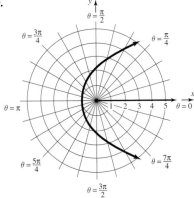

51.

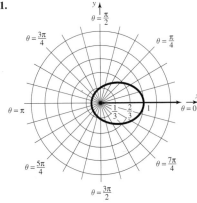

53.

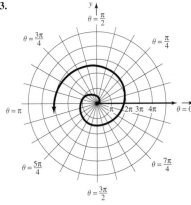

55.

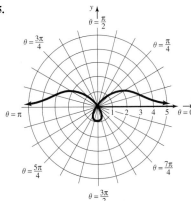

57.

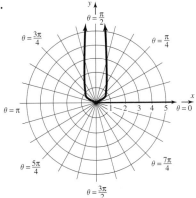

59. $r \sin \theta = a$
$y = a$

61.
$r = 2a \sin \theta$
$r^2 = 2ar \sin \theta$
$x^2 + y^2 = 2ay$
$x^2 + y^2 - 2ay = 0$
$x^2 + (y - a)^2 = a^2$
Circle, radius a, center at $(0, a)$
in rectangular coordinates

63.
$r = 2a \cos \theta$
$r^2 = 2ar \cos \theta$
$x^2 + y^2 = 2ax$
$x^2 - 2ax + y^2 = 0$
$(x - a)^2 + y^2 = a^2$
Circle, radius a, center at $(a, 0)$
in rectangular coordinates

65. (a) $r^2 = \cos \theta$: $r^2 = \cos(\pi - \theta)$ $(-r)^2 = \cos(-\theta)$
$r^2 = -\cos \theta$ $r^2 = \cos \theta$
Not equivalent; test fails   New test works

(b) $r^2 = \sin \theta$: $r^2 = \sin(\pi - \theta)$ $(-r)^2 = \sin(-\theta)$
$r^2 = \sin \theta$ $r^2 = -\sin \theta$
Test works   New test fails

Exercise 8.6

1.

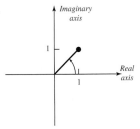

$\sqrt{2}(\cos 45° + i \sin 45°)$

3.

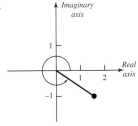

$2(\cos 330° + i \sin 330°)$

5.

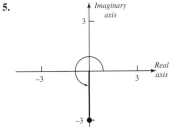

$3(\cos 270° + i \sin 270°)$

7.

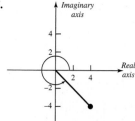

$4\sqrt{2}(\cos 315° + i \sin 315°)$

9.

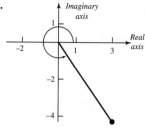

$5(\cos 306.9° + i \sin 306.9°)$

11.

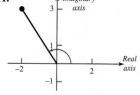

$\sqrt{13}(\cos 123.7° + i \sin 123.7°)$

13. $-1 + \sqrt{3}i$ **15.** $2\sqrt{2} - 2\sqrt{2}i$ **17.** $-3i$ **19.** $-0.035 + 0.197i$ **21.** $1.97 + 0.347i$
23. $zw = 8(\cos 60° + i \sin 60°)$; $z/w = \frac{1}{2}(\cos 20° + i \sin 20°)$ **25.** $zw = 12(\cos 40° + i \sin 40°)$; $z/w = \frac{3}{4}(\cos 220° + i \sin 220°)$
27. $zw = 4\left(\cos \dfrac{9\pi}{40} + i \sin \dfrac{9\pi}{40}\right)$; $\dfrac{z}{w} = \cos \dfrac{\pi}{40} + i \sin \dfrac{\pi}{40}$ **29.** $zw = 4\sqrt{2}(\cos 15° + i \sin 15°)$; $z/w = \sqrt{2}(\cos 75° + i \sin 75°)$
31. $32(-1 + \sqrt{3}i)$ **33.** $32i$ **35.** $\frac{27}{2}(1 + \sqrt{3}i)$ **37.** $\dfrac{25\sqrt{2}}{2}(-1 + i)$ **39.** $4(-1 + i)$ **41.** $-23 + 14.15i$
43. $\sqrt[6]{2}(\cos 15° + i \sin 15°)$, $\sqrt[6]{2}(\cos 135° + i \sin 135°)$, $\sqrt[6]{2}(\cos 255° + i \sin 255°)$
45. $\sqrt[4]{8}(\cos 75° + i \sin 75°)$, $\sqrt[4]{8}(\cos 165° + i \sin 165°)$, $\sqrt[4]{8}(\cos 255° + i \sin 255°)$, $\sqrt[4]{8}(\cos 345° + i \sin 345°)$
47. $2(\cos 67.5° + i \sin 67.5°)$, $2(\cos 157.5° + i \sin 157.5°)$, $2(\cos 247.5° + i \sin 247.5°)$, $2(\cos 337.5° + i \sin 337.5°)$
49. $\cos 18° + i \sin 18°$, $\cos 90° + i \sin 90°$, $\cos 162° + i \sin 162°$, $\cos 234° + i \sin 234°$, $\cos 306° + i \sin 306°$
51. $1, i, -1, -i$ **53.** Look at formula (8); $|z_k| = \sqrt[n]{r}$ for all k.   **55.** Look at formula (8). The z_k are spaced apart by an angle of $2\pi/n$.

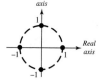

Fill-in-the-Blank Items

1. Sines   **2.** Cosines   **3.** Heron's   **4.** pole; polar axis   **5.** -2 **6.** $r = 2 \cos \theta$ **7.** the polar axis (x-axis)   **8.** magnitude or modulus; argument

True/False Items

1. F   **2.** T   **3.** F   **4.** T   **5.** F   **6.** T

Review Exercises

1. $\gamma = 100°$, $b = 0.65$, $c = 1.29$ **3.** $\beta = 56.8°$, $\gamma = 23.2°$, $b = 4.25$ **5.** No triangle   **7.** $b = 3.32$, $\alpha = 62.8°$, $\gamma = 17.2°$
9. No triangle   **11.** $c = 2.32$, $\alpha = 16.1°$, $\beta = 123.9°$ **13.** $\beta = 36.2°$, $\gamma = 63.8°$, $c = 4.56$ **15.** $\alpha = 39.6°$, $\beta = 18.5°$, $\gamma = 121.9°$
17. Two triangles: $\beta_1 = 13.4°$, $\gamma_1 = 156.6°$, $c_1 = 6.86$; $\beta_2 = 166.6°$, $\gamma_2 = 3.4°$, $c_2 = 1.02$ **19.** $a = 5.23$, $\beta = 46°$, $\gamma = 64°$ **21.** 1.93
23. 18.79   **25.** 6   **27.** 3.80   **29.** 0.32
31. $(3\sqrt{3}/2, \frac{3}{2})$

33. $(1, \sqrt{3})$

35. $(0, 3)$

37. $(3\sqrt{2}, 3\pi/4)$, $(-3\sqrt{2}, -\pi/4)$ **39.** $(2, -\pi/2)$, $(-2, \pi/2)$ **41.** $(5, 0.93)$, $(-5, 4.07)$ **43.** $3r^2 - 6r \sin \theta = 0$ **45.** $r^2(2 - 3 \sin^2 \theta) - \tan \theta = 0$
47. $r^3 \cos \theta = 4$ **49.** $x^2 + y^2 - 2y = 0$ **51.** $x^2 + y^2 = 25$ **53.** $x + 3y = 6$

55. Circle; radius 2, center at (2, 0) in rectangular coordinates

57. Cardioid

59. Limaçon with inner loop

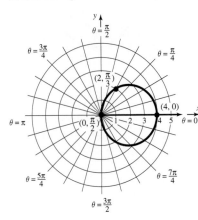
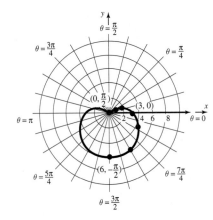
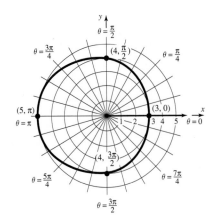

61. $\sqrt{2}(\cos 225° + i \sin 225°)$ **63.** $5(\cos 323.1° + i \sin 323.1°)$ **65.** $-\sqrt{3} + i$ **67.** $-\frac{3}{2} + (3\sqrt{3}/2)i$ **69.** $0.098 - 0.017i$

71. $zw = \cos 130° + i \sin 130°$; $z/w = \cos 30° + i \sin 30°$ **73.** $zw = 6(\cos 0 + i \sin 0)$; $\dfrac{z}{w} = \dfrac{3}{2}\left(\cos \dfrac{8\pi}{5} + i \sin \dfrac{8\pi}{5}\right)$

75. $zw = 5(\cos 5° + i \sin 5°)$; $z/w = 5(\cos 15° + i \sin 15°)$ **77.** $\frac{27}{2}(1 + \sqrt{3}i)$ **79.** $4i$ **81.** 64 **83.** $-527.1 - 335.8i$

85. 3, $3(\cos 120° + i \sin 120°)$, $3(\cos 240° + i \sin 240°)$ **87.** 204.1 mi   **89. (a)** 2.59 mi   **(b)** 2.92 mi   **(c)** 2.53 mi

91. (a) 131.8 mi   **(b)** 23.1°   **(c)** 0.2 hr   **93.** 9024 sq ft.

CHAPTER 9   *Exercise 9.2*

1. B   **3.** E   **5.** H   **7.** C

9. $y^2 = 16x$

11. $x^2 = -12y$

13. $y^2 = -8x$

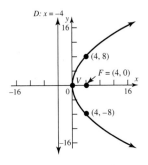
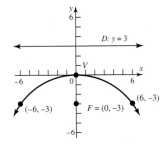
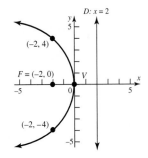

15. $x^2 = 2y$

17. $(x - 2)^2 = -8(y + 3)$

19. $x^2 = \frac{4}{3}y$

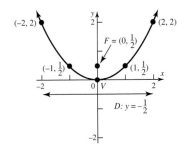
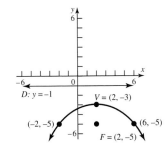
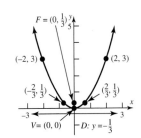

21. $(x + 3)^2 = 4(y - 3)$

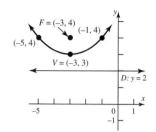

23. $(y + 2)^2 = -8(x + 1)$

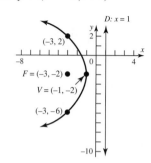

25.

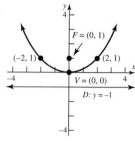

27.

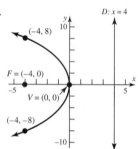

29.

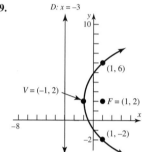

31.

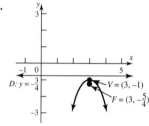

33.

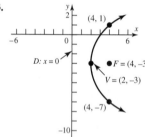

35.

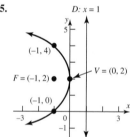

37.

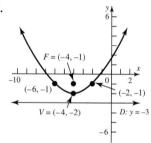

39.

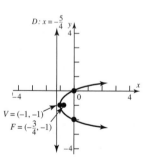

41.

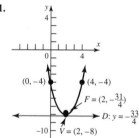

43. $(y - 1)^2 = x$ **45.** $(y - 1)^2 = -(x - 2)$ **47.** $x^2 = 4(y - 1)$ **49.** $y^2 = \frac{1}{2}(x + 2)$
51. 1.5625 ft from the base of the dish, along the axis of symmetry   **53.** 1 in from the vertex   **55.** 20 ft
57. 0.78125 ft   **59.** 4.17 ft from the base along the axis of symmetry   **61.** 24.31 ft, 18.75 ft, 7.64 ft

63. $Ax^2 + Ey = 0$

$$x^2 = -\frac{E}{A}y$$

This is the equation of a parabola with vertex at (0, 0) and axis of symmetry the y-axis. The focus is at $(0, -E/4A)$; the directrix is the line $y = E/4A$. The parabola opens up if $-E/A > 0$ and down if $-E/A < 0$.

65. $Ax^2 + Dx + Ey + F = 0, A \neq 0$

$$Ax^2 + Dx = -Ey - F$$
$$x^2 + \frac{D}{A}x = -\frac{E}{A}y - \frac{F}{A}$$
$$\left(x + \frac{D}{2A}\right)^2 = -\frac{E}{A}y - \frac{F}{A} + \frac{D^2}{4A^2}$$
$$\left(x + \frac{D}{2A}\right)^2 = -\frac{E}{A}y + \frac{D^2 - 4AF}{4A^2}$$

(a) If $E \neq 0$, then the equation may be written as
$$\left(x + \frac{D}{2A}\right)^2 = -\frac{E}{A}\left(y - \frac{D^2 - 4AF}{4AE}\right)$$
This is the equation of a parabola with vertex at
$(-D/2A, (D^2 - 4AF)/(4AE))$ and axis of symmetry parallel
to the y-axis.

(b)–(d) If $E = 0$, the graph of the equation contains no points if
$D^2 - 4AF < 0$, is a single vertical line if $D^2 - 4AF = 0$,
and is two vertical lines if $D^2 - 4AF > 0$.

Exercise 9.3

1. C **3.** B

5.
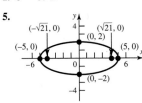

7.
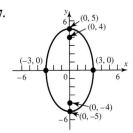

9. $\dfrac{x^2}{4} + \dfrac{y^2}{16} = 1$
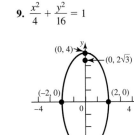

11. $\dfrac{x^2}{8} + \dfrac{y^2}{2} = 1$
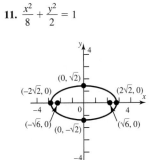

13.
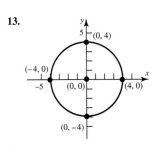

15. $\dfrac{x^2}{25} + \dfrac{y^2}{16} = 1$
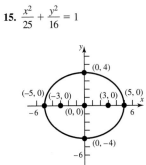

17. $\dfrac{x^2}{9} + \dfrac{y^2}{25} = 1$

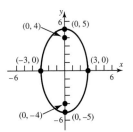

19. $\dfrac{x^2}{9} + \dfrac{y^2}{5} = 1$
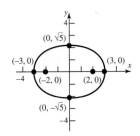

21. $\dfrac{x^2}{4} + \dfrac{y^2}{13} = 1$
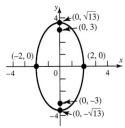

23. $x^2 + \dfrac{y^2}{16} = 1$

25. $\dfrac{(x + 1)^2}{4} + (y - 1)^2 = 1$

27. $(x - 1)^2 + \dfrac{y^2}{4} = 1$

29.
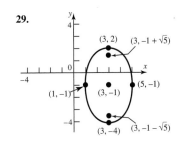

31. $\dfrac{(x+5)^2}{16} + \dfrac{(y-4)^2}{4} = 1$

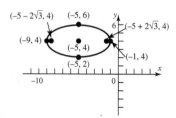

33. $\dfrac{(x+2)^2}{4} + (y-1)^2 = 1$

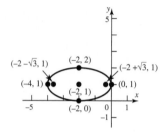

35. $\dfrac{(x-2)^2}{3} + \dfrac{(y+1)^2}{2} = 1$

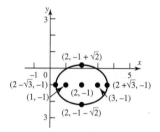

37. $\dfrac{(x-1)^2}{4} + \dfrac{(y+2)^2}{9} = 1$

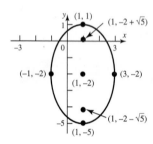

39. $x^2 + \dfrac{(y+2)^2}{4} = 1$

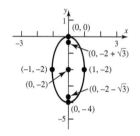

41. $\dfrac{(x-2)^2}{25} + \dfrac{(y+2)^2}{21} = 1$

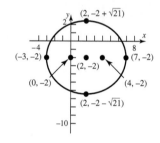

43. $\dfrac{(x-4)^2}{5} + \dfrac{(y-6)^2}{9} = 1$

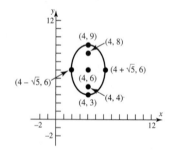

45. $\dfrac{(x-2)^2}{16} + \dfrac{(y-1)^2}{7} = 1$

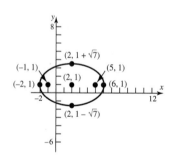

47. $\dfrac{(x-1)^2}{10} + (y-2)^2 = 1$

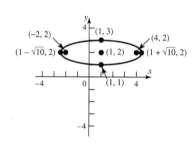

49. $\dfrac{(x-1)^2}{9} + (y-2)^2 = 1$

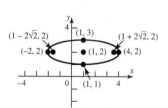

51.

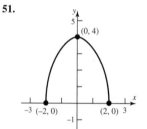

53.

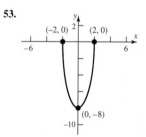

55. $\dfrac{x^2}{100} + \dfrac{y^2}{36} = 1$ **57.** 43.3 ft   **59.** 24.65 ft, 21.65 ft, 13.82 ft   **61.** 0 ft, 12.99 ft, 15 ft, 12.99 ft, 0 ft

63. 91.5 million miles; $\dfrac{x^2}{(93)^2} + \dfrac{y^2}{8646.75} = 1$ **65.** perihelion: 460.6 million miles; mean distance: 483.8 million miles; $\dfrac{x^2}{(483.8)^2} + \dfrac{y^2}{233524} = 1$

67. 30 ft

69. (a) $Ax^2 + Cy^2 + F = 0$
$Ax^2 + Cy^2 = -F$

If A and C are of the same sign and F is of opposite sign, then the equation takes the form $x^2/(-F/A) + y^2/(-F/C) = 1$, where $-F/A$ and $-F/C$ are positive. This is the equation of an ellipse with center at $(0, 0)$.

(b) If $A = C$, the equation may be written as $x^2 + y^2 = -F/A$. This is the equation of a circle with center at $(0, 0)$ and radius equal to $\sqrt{-F/A}$.

Exercise 9.4

1. B   **3.** A

5. $x^2 - \dfrac{y^2}{8} = 1$

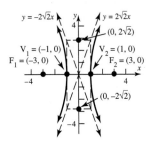

7. $\dfrac{y^2}{16} - \dfrac{x^2}{20} = 1$

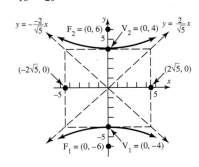

9. $\dfrac{x^2}{9} - \dfrac{y^2}{16} = 1$

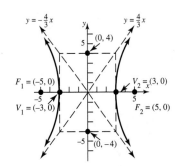

11. $\dfrac{y^2}{36} - \dfrac{x^2}{9} = 1$

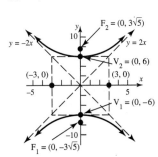

13. $\dfrac{x^2}{8} - \dfrac{y^2}{8} = 1$

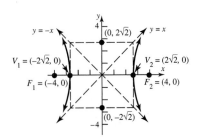

15. $\dfrac{x^2}{16} - \dfrac{y^2}{4} = 1$

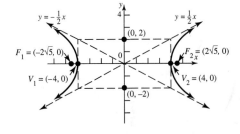

17. $\dfrac{x^2}{4} - \dfrac{y^2}{16} = 1$

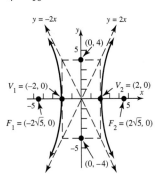

19. $\dfrac{y^2}{9} - x^2 = 1$

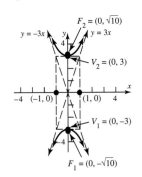

21. $\dfrac{y^2}{25} - \dfrac{x^2}{25} = 1$

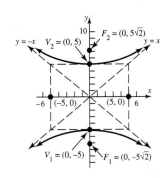

23. $x^2 - y^2 = 1$

25. $\dfrac{y^2}{36} - \dfrac{x^2}{9} = 1$

27. $\dfrac{(x-4)^2}{4} - \dfrac{(y+1)^2}{5} = 1$

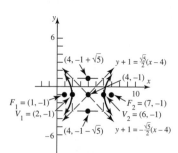

29. $\dfrac{(y+4)^2}{4} - \dfrac{(x+3)^2}{12} = 1$

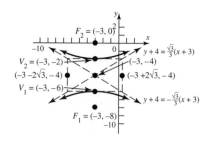

31. $(x-5)^2 - \dfrac{(y-7)^2}{3} = 1$

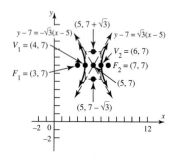

33. $\dfrac{(x-1)^2}{4} - \dfrac{(y+1)^2}{9} = 1$

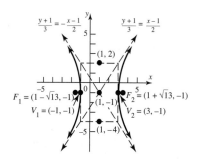

35. $\dfrac{(x-2)^2}{4} - \dfrac{(y+3)^2}{9} = 1$

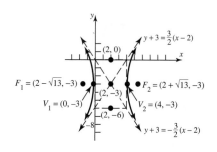

37. $\dfrac{(y-2)^2}{4} - (x+2)^2 = 1$

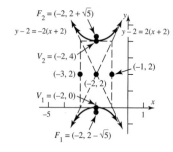

39. $\dfrac{(x+1)^2}{4} - \dfrac{(y+2)^2}{4} = 1$

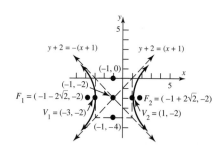

41. $(x-1)^2 - (y+1)^2 = 1$

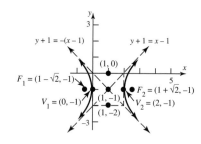

43. $\dfrac{(y-2)^2}{4} - (x+1)^2 = 1$

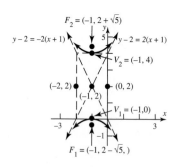

45. $\dfrac{(x-3)^2}{4} - \dfrac{(y+2)^2}{16} = 1$

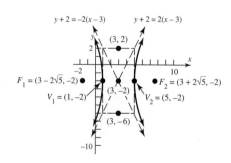

47. $\dfrac{(y-1)^2}{4} - (x+2)^2 = 1$

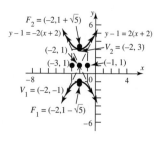

49.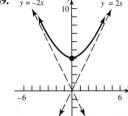

51.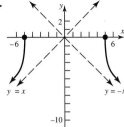

53. (a) The ship will reach shore or a point 64.66 miles from the master station.   (b) 0.00086 second   (c) (104, 50)

55. (a) 450 ft   **57.** If e is close to 1, narrow hyperbola; if e is very large, wide hyperbola

59. $\dfrac{x^2}{4} - y^2 = 1$; asymptotes $y = \pm\dfrac{1}{2}x$ $\qquad\qquad\qquad$ $y^2 - \dfrac{x^2}{4} = 1$; asymptotes $y = \pm\dfrac{1}{2}x$

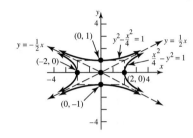

61. $Ax^2 + Cy^2 + F = 0$ $\qquad$ If A and C are of opposite sign and $F \neq 0$, this equation may be written as $x^2/(-F/A) + y^2/(-F/C) = 1$, where
$\qquad\quad Ax^2 + Cy^2 = -F$ $\qquad$ $-F/A$ and $-F/C$ are opposite in sign. This is the equation of a hyperbola with center at $(0, 0)$. The transverse axis is
$\qquad\qquad\qquad\qquad\qquad\qquad$ the x-axis if $-F/A > 0$; the transverse axis is the y-axis if $-F/A < 0$.

Exercise 9.5

1. Parabola   **3.** Ellipse   **5.** Hyperbola   **7.** Hyperbola   **9.** Circle   **11.** $x = \sqrt{2}/2(x' - y')$, $y = \sqrt{2}/2(x' + y')$
13. $x = \sqrt{2}/2(x' - y')$, $y = \sqrt{2}/2(x' + y')$ **15.** $x = \frac{1}{2}(x' - \sqrt{3}y')$, $y = \frac{1}{2}(\sqrt{3}x' + y')$ **17.** $x = \sqrt{5}/5(x' - 2y')$, $y = \sqrt{5}/5(2x' + y')$
19. $x = \sqrt{13}/13(3x' - 2y')$, $y = \sqrt{13}/13(2x' + 3y')$
21. $\theta = 45°$ (see Problem 11) $\qquad\qquad$ **23.** $\theta = 45°$ (see Problem 13) $\qquad\qquad$ **25.** $\theta = 60°$ (see Problem 15)

$x'^2 - \dfrac{y'^2}{3} = 1$ $\qquad\qquad\qquad\qquad$ $x'^2 + \dfrac{y'^2}{4} = 1$ $\qquad\qquad\qquad\qquad$ $\dfrac{x'^2}{4} + y'^2 = 1$

Hyperbola $\qquad\qquad\qquad\qquad\qquad$ Ellipse $\qquad\qquad\qquad\qquad\qquad\quad$ Ellipse
Center at origin $\qquad\qquad\qquad\qquad$ Center at $(0, 0)$ $\qquad\qquad\qquad\qquad$ Center at $(0, 0)$
Transverse axis is the x'-axis $\qquad\quad$ Major axis is the y'-axis $\qquad\qquad$ Major axis is the x'-axis
Vertices at $(\pm 1, 0)$ $\qquad\qquad\qquad\quad$ Vertices at $(0, \pm 2)$ $\qquad\qquad\qquad$ Vertices at $(\pm 2, 0)$

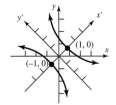 $\qquad\qquad$ 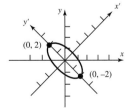 $\qquad\qquad$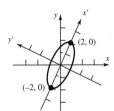

27. $\theta \approx 63°$ (see Problem 17)
$y'^2 = 8x'$
Parabola
Vertex at $(0, 0)$
Focus at $(2, 0)$

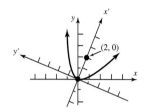

29. $\theta \approx 34°$ (see Problem 19)
$\dfrac{(x' - 2)^2}{4} + y'^2 = 1$
Ellipse
Center at $(2, 0)$
Major axis is the x'-axis
Vertices at $(4, 0)$ and $(0, 0)$

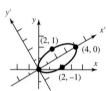

31. $\cot 2\theta = \frac{7}{24}$; $\theta \approx 37°°'$
$(x' - 1)^2 = -6(y' - \frac{1}{6})$
Parabola
Vertex at $(1, \frac{1}{6})$
Focus at $(1, -\frac{4}{3})$

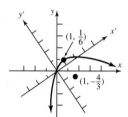

33. Hyperbola **35.** Hyperbola **37.** Parabola **39.** Ellipse **41.** Ellipse
43. Refer to equation (6): $A' = A \cos^2 \theta + B \sin \theta \cos \theta + C \sin^2 \theta$
$B' = B(\cos^2 \theta - \sin^2 \theta) + 2(C - A)(\sin \theta \cos \theta)$
$C' = A \sin^2 \theta - B \sin \theta \cos \theta + C \cos^2 \theta$
$D' = D \cos \theta + E \sin \theta$
$E' = -D \sin \theta + E \cos \theta$
$F' = F$
45. Use Problem 43 to find $B'^2 - 4A'C'$. After much cancellation, $B'^2 - 4A'C' = B^2 - 4AC$.
47. Use formulas (5) to find $d^2 = (x_2 - x_1)^2 + (y_2 - y_1)^2$. After simplifying, $(x_2 - x_1)^2 + (y_2 - y_1)^2 = (x'_2 - x'_1)^2 + (y'_2 - y'_1)^2$.

Exercise 9.6

1. Parabola; directrix is perpendicular to the polar axis 1 unit to the right of the pole.
3. Hyperbola; directrix is parallel to the polar axis $\frac{4}{3}$ units below the pole.
5. Ellipse; directrix is perpendicular to the polar axis $\frac{3}{2}$ units to the left of the pole.

7. Parabola; directrix is perpendicular to the polar axis 1 unit to the right of the pole.

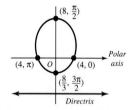

9. Ellipse; directrix is parallel to the polar axis $\frac{8}{3}$ units above the pole.

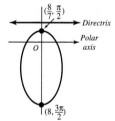

11. Hyperbola; directrix is perpendicular to the polar axis $\frac{3}{2}$ units to the left of the pole.

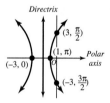

13. Ellipse; directrix is parallel to the polar axis 8 units below the pole; vertices are at $(8, \pi/2)$ and $(\frac{8}{3}, 3\pi/2)$.

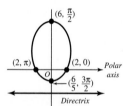

15. Ellipse; directrix is parallel to the polar axis 3 units below the pole; vertices are at $(6, \pi/2)$ and $(\frac{6}{5}, 3\pi/2)$.

17. Ellipse; directrix is perpendicular to the polar axis 6 units to the left of the pole; vertices are at $(6, 0)$ and $(2, \pi)$.

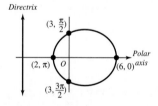

19. $y^2 + 2x - 1 = 0$ **21.** $16x^2 + 7y^2 + 48y - 64 = 0$ **23.** $3x^2 - y^2 + 12x + 9 = 0$ **25.** $4x^2 + 3y^2 - 16y - 64 = 0$
27. $9x^2 + 5y^2 - 24y - 36 = 0$ **29.** $3x^2 + 4y^2 - 12x - 36 = 0$ **31.** $r = 1/(1 + \sin \theta)$ **33.** $r = 12/(5 - 4 \cos \theta)$ **35.** $r = 12/(1 - 6 \sin \theta)$
37. Use $d(D, P) = p - r \cos \theta$ in the derivation of equation (a) in Table 5.
39. Use $d(D, P) = p + r \sin \theta$ in the derivation of equation (a) in Table 5.

Exercise 9.7

1. $x - 3y + 1 = 0$

3. $y = \sqrt{x - 2}$

5. $x = y + 8$

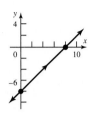

7. $x = 3(y - 1)^2$

9. $2y = 2 + x$

11. $y = x^3$

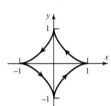

13. $\dfrac{x^2}{4} + \dfrac{y^2}{9} = 1$

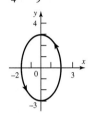

15. $\dfrac{x^2}{4} + \dfrac{y^2}{9} = 1$

17. $x^2 - y^2 = 1$

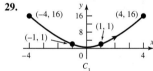

19. $x^{2/3} + y^{2/3} = 1$

21. $x = t, y = t^3;\ x = \sqrt[3]{t}, y = t$ **23.** $x = t, y = t^{2/3};\ x = t^{3/2}, y = t$ **25.** $x = 2\cos \pi t, y = -3\sin \pi t, 0 \le t \le 2$
27. $x = -2\sin 2\pi t, y = 3\cos 2\pi t, 0 \le t \le 1$
29.

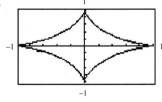

31. The orientation is from (x_1, y_1) to (x_2, y_2).

33.

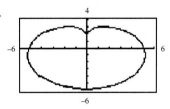

35.

Fill-in-the-Blank Items

1. parabola   **2.** ellipse   **3.** hyperbola   **4.** major; transverse   **5.** y-axis   **6.** $y/3 = x/2$; $y/3 = -x/2$ **7.** $\cot 2\theta = (A - C)/B$
8. $\frac{1}{2}$; ellipse; parallel; 4; below   **9.** ellipse

True/False Items

1. T   **2.** F   **3.** T   **4.** T   **5.** T   **6.** T   **7.** T   **8.** T   **9.** T   **10.** F

Review Exercises

1. Parabola; vertex $(0, 0)$, focus $(-4, 0)$, directrix $x = 4$
3. Hyperbola; center $(0, 0)$, vertices $(5, 0)$ and $(-5, 0)$, foci $(\sqrt{26}, 0)$ and $(-\sqrt{26}, 0)$, asymptotes $y = \frac{1}{5}x$ and $y = -\frac{1}{5}x$
5. Ellipse; center $(0, 0)$, vertices $(0, 5)$ and $(0, -5)$, foci $(0, 3)$ and $(0, -3)$
7. $x^2 = -4(y - 1)$: Parabola; vertex $(0, 1)$, focus $(0, 0)$, directrix $y = 2$
9. $\frac{x^2}{2} - \frac{y^2}{8} = 1$: Hyperbola; center $(0, 0)$, vertices $(\sqrt{2}, 0)$ and $(-\sqrt{2}, 0)$, foci $(\sqrt{10}, 0)$ and $(-\sqrt{10}, 0)$, asymptotes $y = 2x$ and $y = -2x$
11. $(x - 2)^2 = 2(y + 2)$: Parabola; vertex $(2, -2)$, focus $(2, -\frac{3}{2})$, directrix $y = -\frac{5}{2}$
13. $\frac{(y - 2)^2}{4} - (x - 1)^2 = 1$: Hyperbola; center $(1, 2)$, vertices $(1, 4)$ and $(1, 0)$, foci $(1, 2 + \sqrt{5})$ and $(1, 2 - \sqrt{5})$, asymptotes $y - 2 = \pm 2(x - 1)$
15. $\frac{(x - 2)^2}{9} + \frac{(y - 1)^2}{4} = 1$: Ellipse; center $(2, 1)$, vertices $(5, 1)$ and $(-1, 1)$, foci $(2 + \sqrt{5}, 1)$ and $(2 - \sqrt{5}, 1)$
17. $(x - 2)^2 = -4(y + 1)$: Parabola; vertex $(2, -1)$, focus $(2, -2)$, directrix $y = 0$
19. $\frac{(x - 1)^2}{4} + \frac{(y + 1)^2}{9} = 1$: Ellipse; center $(1, -1)$, vertices $(1, 2)$ and $(1, -4)$, foci $(1, -1 + \sqrt{5})$ and $(1, -1 - \sqrt{5})$

21. $y^2 = -8x$ **23.** $\frac{y^2}{4} - \frac{x^2}{12} = 1$ **25.** $\frac{x^2}{16} + \frac{y^2}{7} = 1$

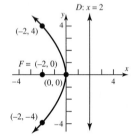   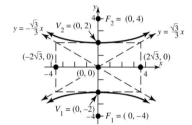   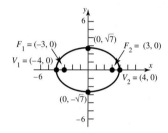

27. $(x - 2)^2 = -4(y + 3)$ **29.** $(x + 2)^2 - \frac{(y + 3)^2}{3} = 1$ **31.** $\frac{(x + 4)^2}{16} + \frac{(y - 5)^2}{25} = 1$

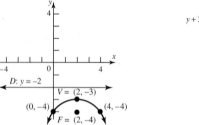   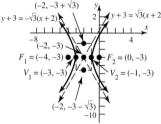   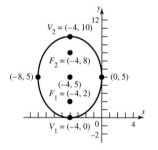

33. $\frac{(x + 1)^2}{9} - \frac{(y - 2)^2}{7} = 1$ **35.** $\frac{(x - 3)^2}{9} - \frac{(y - 1)^2}{4} = 1$

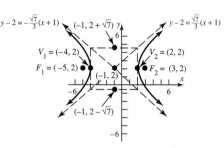   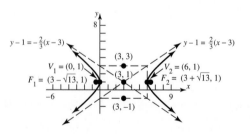

37. Parabola **39.** Ellipse **41.** Parabola **43.** Hyperbola **45.** Ellipse

47. $x'^2 - \dfrac{y'^2}{9} = 1$

Hyperbola
Center at the origin
Transverse axis the x'-axis
Vertices at $(\pm 1, 0)$

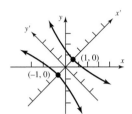

49. $\dfrac{x'^2}{2} + \dfrac{y'^2}{4} = 1$

Ellipse
Center at origin
Major axis the y'-axis
Vertices at $(0, \pm 2)$

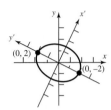

51. $y'^2 = -\dfrac{4\sqrt{13}}{13} x'$

Parabola
Vertex at the origin
Focus on the x'-axis at $(-\sqrt{13}/13, 0)$

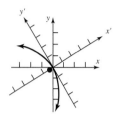

53. Parabola; directrix is perpendicular to the polar axis 4 units to the left of the pole.

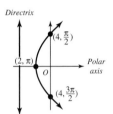

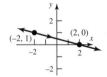

55. Ellipse; directrix is parallel to the polar axis 6 units below the pole; vertices are $(6, \pi/2)$ and $(2, 3\pi/2)$.

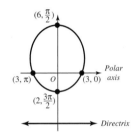

57. Hyperbola; directrix is perpendicular to the polar axis 1 unit to the right of the pole; vertices are at $(\frac{2}{3}, 0)$ and $(-2, \pi)$

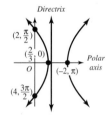

59. $y^2 - 8x - 16 = 0$ **61.** $3x^2 - y^2 - 8x + 4 = 0$

63. $x + 4y = 2$

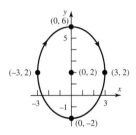

65. $\dfrac{x^2}{9} + \dfrac{(y-2)^2}{16} = 1$

67. $1 + y = x$

69. $\dfrac{x^2}{5} - \dfrac{y^2}{4} = 1$ **71.** The ellipse $\dfrac{x^2}{16} + \dfrac{y^2}{7} = 1$ **73.** $\frac{1}{4}$ ft or 3 in **75.** 19.72 ft, 18.86 ft, 14.91 ft
77. (a) 45.24 miles from the Master Station **(b)** 0.000645 second **(c)** (66, 20)

CHAPTER 10 *Exercise 10.1*

1. $2(2) - (-1) = 5$ and $5(2) + 2(-1) = 8$ **3.** $3(2) - 4(\frac{1}{2}) = 4$ and $\frac{1}{2}(2) - 3(\frac{1}{2}) = -\frac{1}{2}$ **5.** $2^2 - 1^2 = 3$ and $(2)(1) = 2$
7. $\dfrac{0}{1+0} + 3(2) = 6$ and $0 + 9(2)^2 = 36$ **9** $3(1) + 3(-1) + 2(2) = 4$, $1 - (-1) - 2 = 0$, and $2(-1) - 3(2) = -8$ **11.** $x = 6, y = 2$
13. $x = 3, y = 2$ **15.** $x = 8, y = -4$ **17.** $x = \frac{1}{3}, y = -\frac{1}{6}$ **19.** Inconsistent **21.** $x = 1, y = 2$ **23.** $x = 4 - 2y$, y is any real number
25. $x = 1, y = 1$ **27.** $x = \frac{3}{2}, y = 1$ **29.** $x = 4, y = 3$ **31.** $x = \frac{4}{3}, y = \frac{1}{5}$ **33.** $x = 8, y = 2, z = 0$ **35.** $x = 2, y = -1, z = 1$ **37.** Inconsistent
39. $x = 5z - 2$, $y = 4z - 3$ where z is any real number, or $x = \frac{5}{4}y + \frac{7}{4}$, $z = \frac{1}{4}y + \frac{3}{4}$ where y is any real number, or $y = \frac{4}{5}x - \frac{7}{5}$, $z = \frac{1}{5}x + \frac{2}{5}$ where x is any real number

41. Inconsistent   **43.** $x = 1$, $y = 3$, $z = -2$ **45.** $x = -3$, $y = \frac{1}{2}$, $z = 1$ **47.** $x = \frac{1}{5}$, $y = \frac{1}{3}$ **49.** $x = 48.15$, $y = 15.18$
51. $x = -21.47$, $y = 16.12$ **53.** $x = 0.26$, $y = 0.06$ **55.** 20 and 61   **57.** 15 ft by 30 ft   **59.** Cheeseburger \$1.55; shake \$0.85
61. 22.5 lb   **63.** Average wind speed 25 mph; average airspeed of the Piper is 175 mph   **65.** 80 \$25 sets and 120 \$45 sets   **67.** \$5.56
69. 12, 15, 21   **71.** $I_1 = \frac{10}{71}$, $I_2 = \frac{65}{71}$, $I_3 = \frac{55}{71}$ **73.** 100 orchestra, 210 main, and 190 balcony seats   **79.** $b = -\frac{1}{2}$; $c = \frac{3}{2}$ **81.** $a = \frac{4}{3}$, $b = -\frac{5}{3}$, $c = 1$
83. $x = \dfrac{b_1 - b_2}{m_2 - m_1}$, $y = \dfrac{m_2 b_1 - m_1 b_2}{m_2 - m_1}$ **81.** $y = mx + b$, x is any real number

Exercise 10.2

1. $\begin{bmatrix} 1 & -5 & | & 5 \\ 4 & 3 & | & 6 \end{bmatrix}$ **3.** $\begin{bmatrix} 2 & 3 & | & 6 \\ 4 & -6 & | & -2 \end{bmatrix}$ **5.** $\begin{bmatrix} 0.01 & -0.03 & | & 0.06 \\ 0.13 & 0.10 & | & 0.20 \end{bmatrix}$ **7.** $\begin{bmatrix} 1 & -1 & 1 & | & 10 \\ 3 & 2 & 0 & | & 5 \\ 1 & 1 & 2 & | & 2 \end{bmatrix}$ **9.** $\begin{bmatrix} 1 & 1 & -1 & | & 2 \\ 3 & -2 & 0 & | & 2 \end{bmatrix}$

11. $\begin{bmatrix} 1 & -3 & -5 & | & 2 \\ 0 & 1 & 6 & | & 1 \\ 0 & 0 & 13 & | & 36 \end{bmatrix}$ **13.** $\begin{bmatrix} 1 & -3 & 4 & | & 3 \\ 0 & 1 & -2 & | & 0 \\ 0 & 0 & 4 & | & 15 \end{bmatrix}$ **15.** $\begin{bmatrix} 1 & -3 & 2 & | & -6 \\ 0 & 1 & -1 & | & 8 \\ 0 & -5 & 108 & | & -12 \end{bmatrix}$ **17.** $\begin{bmatrix} 1 & -3 & 1 & | & -2 \\ 0 & 1 & 4 & | & 2 \\ 0 & 0 & 39 & | & 16 \end{bmatrix}$

19. $\begin{bmatrix} 1 & -3 & -2 & | & 3 \\ 0 & 1 & 6 & | & -7 \\ 0 & 0 & 64 & | & -62 \end{bmatrix}$

21. $\begin{cases} x = 5 \\ y = -1 \end{cases}$ **23.** $\begin{cases} x = 1 \\ y = 2 \\ 0 = 3 \end{cases}$ **25.** $\begin{cases} x + 2z = -1 \\ y - 4z = -2 \\ 0 = 0 \end{cases}$ **27.** $\begin{cases} x_1 = 1 \\ x_2 + x_4 = 2 \\ x_3 + 2x_4 = 3 \end{cases}$ **29.** $\begin{cases} x_1 + 4x_4 = 2 \\ x_2 + x_3 + 3x_4 = 3 \\ 0 = 0 \end{cases}$

consistent; $x = 5$, $y = -1$ 　　inconsistent 　　consistent; 　　consistent; 　　consistent;
$\qquad\qquad\qquad\qquad\qquad\qquad\qquad\qquad\qquad\quad x = -1 - 2z$, 　　$x_1 = 1$, $x_2 = 4 - x_4$, 　　$x_1 = 2 - 4x_4$,
$\qquad\qquad\qquad\qquad\qquad\qquad\qquad\qquad\qquad\quad y = -2 + 4z$, 　　$x_3 = 3 - 2x_4$, 　　$x_2 = 3 - x_3 - 3x_4$,
$\qquad\qquad\qquad\qquad\qquad\qquad\qquad\qquad\qquad\quad z$ any real number 　　x_4 any real number 　　x_3, x_4 any real numbers

31. $x = 6$, $y = 2$ **33.** $x = 2$, $y = 3$ **35.** $x = 4$, $y = -2$ **37.** Inconsistent   **39.** $x = \frac{1}{2}$, $y = \frac{3}{4}$ **41.** $x = 4 - 2y$, y is any real number
43. $x = \frac{3}{2}$, $y = 1$ **45.** $x = \frac{4}{3}$, $y = \frac{1}{5}$ **47.** $x = 8$, $y = 2$, $z = 0$ **49.** $x = 2$, $y = -1$, $z = 1$ **51.** Inconsistent
53. $x = 5z - 2$, $y = 4z - 3$, z any real number; or $x = \frac{5}{4}y + \frac{7}{4}$, $z = \frac{1}{4}y + \frac{3}{4}$, y any real number; or $y = \frac{4}{5}x - \frac{7}{5}$, $z = \frac{1}{5}x + \frac{2}{5}$, x any real number
55. Inconsistent   **57.** $x = 1$, $y = 3$, $z = -2$ **59.** $x = -3$, $y = \frac{1}{2}$, $z = 1$ **61.** $x = \frac{1}{3}$, $y = \frac{2}{3}$, $z = 1$ **63.** $x = 1$, $y = 2$, $z = 0$, $w = 1$
65. $y = 0$, $z = 1 - x$, x is any real number   **67.** $x = 2$, $y = z - 3$, z is any real number   **69.** $x = \frac{13}{9}$, $y = \frac{7}{18}$, $z = \frac{19}{18}$
71. $x = \frac{7}{5} - \frac{3}{5}z - \frac{2}{5}w$, $y = -\frac{8}{5} + \frac{7}{5}z + \frac{13}{5}w$, where z and w are any real numbers   **73.** $y = -2x^2 + x + 3$ **75.** $f(x) = 3x^3 - 4x^2 + 5$

77. $x = $ liters of 15% H_2SO_4, $y = $ liters of 25% H_2SO_4, $z = $ liters of 50% H_2SO_4: $\begin{cases} x = \frac{5}{2}z - 150 \\ y = 250 - \frac{7}{2}z \end{cases}$

15%	25%	50%	40%
0	40	60	100
10	26	64	100
20	12	68	100

79. If $x = $ Price of hamburgers, $y = $ Price of fries, $z = $ Price of colas, then $x = 2.75 - z$, $y = 0.68 + \frac{1}{3}z$, z any real number.

There is not sufficient information:

x	\$2.15	\$2.00	\$1.85
y	\$0.88	\$0.93	\$0.98
z	\$0.60	\$0.75	\$0.90

81. (a)

Amount Invested At		
7%	9%	11%
0	10,000	10,000
1,000	8,000	11,000
2,000	6,000	12,000
3,000	4,000	13,000
4,000	2,000	14,000
5,000	0	15,000

(b)

Amount Invested At		
7%	9%	11%
12,500	12,500	0
14,500	8,500	2000
16,500	4,500	4000
18,750	0	6250

(c) All the money invested at 7% provides \$2100, more than what is required.

83. $I_1 = \frac{4}{15}$, $I_2 = \frac{8}{15}$, $I_3 = \frac{4}{5}$ **85.** $I_1 = 3.5$, $I_2 = 2.5$, $I_3 = 1$
89. If $a_1 \neq 0$,

$$\begin{bmatrix} a_1 & b_1 & | & c_1 \\ a_2 & b_2 & | & c_2 \end{bmatrix} \rightarrow \begin{bmatrix} 1 & \dfrac{b_1}{a_1} & | & \dfrac{c_1}{a_1} \\ a_2 & b_2 & | & c_2 \end{bmatrix} \rightarrow \begin{bmatrix} 1 & \dfrac{b_1}{a_1} & | & \dfrac{c_1}{a_1} \\ 0 & \dfrac{-a_2 b_1}{a_1} + b_2 & | & \dfrac{-a_2 c_1}{a_1} + c_2 \end{bmatrix} \rightarrow \begin{bmatrix} 1 & \dfrac{b_1}{a_1} & | & \dfrac{c_1}{a_1} \\ 0 & \dfrac{-a_2 b_1 + b_2 a_1}{a_1} & | & \dfrac{-a_2 c_1 + c_2 a_1}{a_1} \end{bmatrix}$$

$$\rightarrow \begin{bmatrix} 1 & \dfrac{b_1}{a_1} & | & \dfrac{c_1}{a_1} \\ 0 & 1 & | & \dfrac{-a_2 c_1 + c_2 a_1}{a_1} \cdot \dfrac{a_1}{-a_2 b_1 + b_2 a_1} \end{bmatrix} \rightarrow \begin{bmatrix} 1 & \dfrac{b_1}{a_1} & | & \dfrac{c_1}{a_1} \\ 0 & 1 & | & \dfrac{-a_2 c_1 + c_2 a_1}{-a_2 b_1 + b_2 a_1} \end{bmatrix} \rightarrow \begin{bmatrix} 1 & 0 & | & \dfrac{-b_1 c_2 + b_2 c_1}{-a_2 b_1 + b_2 a_1} \\ 0 & 1 & | & \dfrac{-a_2 c_1 + c_2 a_1}{-a_2 b_1 + b_2 a_1} \end{bmatrix}$$

$$x = \frac{1}{a_1 b_2 - a_2 b_1}(c_1 b_2 - c_2 b_1) = \frac{1}{D}(c_1 b_2 - c_2 b_1), \quad y = \frac{1}{a_1 b_2 - a_2 b_1}(a_1 c_2 - a_2 c_1) = \frac{1}{D}(a_1 c_2 - a_2 c_1)$$

If $a_1 = 0$, then $a_2 \neq 0$, $b_1 \neq 0$, and

$$\begin{bmatrix} 0 & b_1 & | & c_1 \\ a_2 & b_2 & | & c_2 \end{bmatrix} \rightarrow \begin{bmatrix} a_2 & b_2 & | & c_2 \\ 0 & b_1 & | & c_1 \end{bmatrix} \rightarrow \begin{bmatrix} 1 & \dfrac{b_2}{a_2} & | & \dfrac{c_2}{a_2} \\ 0 & b_1 & | & c_1 \end{bmatrix} \rightarrow \begin{bmatrix} 1 & \dfrac{b_2}{a_2} & | & \dfrac{c_2}{a_2} \\ 0 & 1 & | & \dfrac{c_1}{b_1} \end{bmatrix} \rightarrow \begin{bmatrix} 1 & 0 & | & \dfrac{c_2}{a_2} - \dfrac{b_2 c_1}{a_2 b_1} = \dfrac{c_1 b_2 - c_2 b_1}{-a_2 b_1} \\ 0 & 1 & | & \dfrac{c_1}{b_1} = \dfrac{-a_2 c_1}{-a_2 b_1} \end{bmatrix}$$

Exercise 10.3

1. 2 **3.** 22 **5.** -2 **7.** 10 **9.** -26 **11.** $x = 6$, $y = 2$ **13.** $x = 3$, $y = 2$ **15.** $x = 8$, $y = -4$ **17.** $x = 4$, $y = -2$ **19.** Not applicable
21. $x = \frac{1}{2}$, $y = \frac{3}{4}$ **23.** $x = \frac{1}{10}$, $y = \frac{2}{5}$ **25.** $x = \frac{3}{2}$, $y = 1$ **27.** $x = \frac{4}{3}$, $y = \frac{1}{5}$ **29.** $x = 1$, $y = 3$, $z = -2$ **31.** $x = -3$, $y = \frac{1}{2}$, $z = 1$
33. Not applicable **35.** $x = 0$, $y = 0$, $z = 0$ **37.** Not applicable **39.** $x = \frac{1}{5}$, $y = \frac{1}{3}$ **41.** -5 **43.** $\frac{13}{11}$ **45.** 0 or -9 **47.** -4 **49.** 12
51. 8 **53.** 8
55. $(y_1 - y_2)x - (x_1 - x_2)y + (x_1 y_2 - x_2 y_1) = 0$
$(y_1 - y_2)x + (x_2 - x_1)y = x_2 y_1 - x_1 y_2$
$(x_2 - x_1)y - (x_2 - x_1)y_1 = (y_2 - y_1)x + x_2 y_1 - x_1 y_2 - (x_2 - x_1)y_1$
$(x_2 - x_1)(y - y_1) = (y_2 - y_1)x - (y_2 - y_1)x_1$
$y - y_1 = \dfrac{(y_2 - y_1)}{(x_2 - x_1)}(x - x_1)$

57. $\begin{vmatrix} x^2 & x & 1 \\ y^2 & y & 1 \\ z^2 & z & 1 \end{vmatrix} = x^2 \begin{vmatrix} y & 1 \\ z & 1 \end{vmatrix} - x \begin{vmatrix} y^2 & 1 \\ z^2 & 1 \end{vmatrix} + \begin{vmatrix} y^2 & y \\ z^2 & z \end{vmatrix} = x^2(y - z) - x(y^2 - z^2) + yz(y - z)$

$= (y - z)[x^2 - x(y + z) + yz] = (y - z)[(x^2 - xy) - (xz - yz)] = (y - z)[x(x - y) - z(x - y)] = (y - z)(x - y)(x - z)$

59. $\begin{vmatrix} a_{13} & a_{12} & a_{11} \\ a_{23} & a_{22} & a_{21} \\ a_{33} & a_{32} & a_{31} \end{vmatrix} = a_{13}(a_{22}a_{31} - a_{32}a_{21}) - a_{12}(a_{23}a_{31} - a_{33}a_{21}) + a_{11}(a_{23}a_{32} - a_{33}a_{22})$

$= -[a_{11}(a_{22}a_{33} - a_{32}a_{23}) - a_{12}(a_{21}a_{33} - a_{31}a_{23}) + a_{13}(a_{21}a_{32} - a_{31}a_{22})] = -\begin{vmatrix} a_{11} & a_{12} & a_{13} \\ a_{21} & a_{22} & a_{23} \\ a_{31} & a_{32} & a_{33} \end{vmatrix}$

61. $\begin{vmatrix} a_{11} & a_{12} & a_{11} \\ a_{21} & a_{22} & a_{21} \\ a_{31} & a_{32} & a_{31} \end{vmatrix} = a_{11}(a_{22}a_{31} - a_{32}a_{21}) - a_{12}(a_{21}a_{31} - a_{31}a_{21}) + a_{11}(a_{21}a_{32} - a_{31}a_{22})$

$= a_{11}a_{22}a_{31} - a_{11}a_{32}a_{21} - a_{12}(0) + a_{11}a_{21}a_{32} - a_{11}a_{31}a_{22} = 0$

Exercise 10.4

1. $x = -2$, $y = 0$

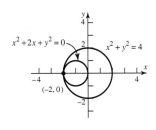

3. $x = 1$, $y = -2$; $x = 2$, $y = 1$

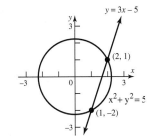

5. $x = 0$, $y = 2$; $x = 0$, $y = -2$;
$x = -1$, $y = \sqrt{3}$; $x = -1$, $y = -\sqrt{3}$

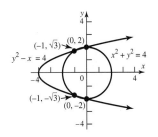

7. $x = 2, y = 2; x = -2, y = -2$

9. Inconsistent

11. $x = 3, y = 5$

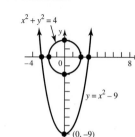

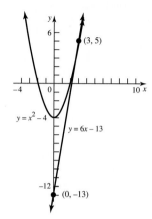

13. $x = 1, y = 4; x = -1, y = -4; x = 2\sqrt{2}, y = \sqrt{2}; x = -2\sqrt{2}, y = -\sqrt{2}$ **15.** $x = 0, y = 1; x = -\frac{2}{3}, y = -\frac{1}{3}$
17. $x = 0, y = -1; x = \frac{5}{2}, y = -\frac{7}{2}$ **19.** $x = 2, y = \frac{1}{3}; x = \frac{1}{2}, y = \frac{4}{3}$ **21.** $x = 3, y = 2; x = 3, y = -2; x = -3, y = 2; x = -3, y = -2$
23. $x = \frac{1}{2}, y = \frac{3}{2}; x = \frac{1}{2}, y = -\frac{3}{2}; x = -\frac{1}{2}, y = \frac{3}{2}; x = -\frac{1}{2}, y = -\frac{3}{2}$ **25.** $x = \sqrt{2}, y = 2\sqrt{2}; x = -\sqrt{2}, y = -2\sqrt{2}$
27. No solution; system is inconsistent **29.** $x = \frac{8}{3}, y = 2\sqrt{10}/3; x = -\frac{8}{3}, y = 2\sqrt{10}/3; x = \frac{8}{3}, y = -2\sqrt{10}/3; x = -\frac{8}{3}, y = -2\sqrt{10}/3$
31. $x = 1, y = \frac{1}{2}; x = -1, y = \frac{1}{2}; x = 1, y = -\frac{1}{2}; x = -1, y = -\frac{1}{2}$ **33.** No solution; system is inconsistent
35. $x = 2, y = 1; x = -2, y = -1; x = \sqrt{3}, y = \sqrt{3}; x = -\sqrt{3}, y = -\sqrt{3}$ **37.** $x = 3, y = 2; x = -3, y = -2; x = 2, y = \frac{1}{2}; x = -2, y = -\frac{1}{2}$
39. $x = 3, y = 1; x = -1, y = -3$ **41.** $x = 0, y = -2; x = 0, y = 1; x = 2, y = -1$ **43.** $x = 2, y = 8$ **45.** 3 and 1; -3 and -1

47. 2 and 2; -2 and -2 **49.** $\frac{1}{2}$ and $\frac{1}{3}$ **51.** 5 **53.**

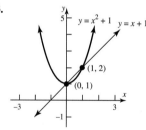

55.

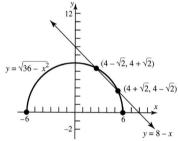

57.

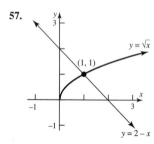

59.

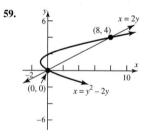

61.

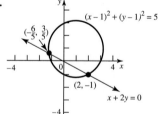

63. Solutions: $(0, -\sqrt{3} - 2), (0, \sqrt{3} - 2), (1, 0), (1, -4)$ **65.**

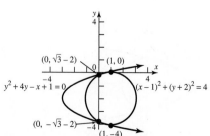

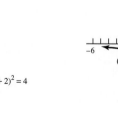

67. $x = 0.48, y = 0.61$

69. $x = -1.64$, $y = -0.89$ **71.** $x = 0.58$, $y = 1.85$; $x = 1.81$, $y = 1.05$; $x = 0.58$, $y = -1.85$; $x = 1.81$, $y = -1.05$ **73.** $x = 2.34$, $y = 0.85$
75. 5 in. by 3 in.   **77.** 2 cm and 4 cm   **79.** tortoise: 7 m/hr, hare $7\frac{1}{2}$ m/hr   **81.** 12 cm by 18 cm   **83.** $x = 60$ ft; $y = 30$ ft

85. $l = \dfrac{P + \sqrt{P^2 - 16A}}{4}$; $w = \dfrac{P - \sqrt{P^2 - 16A}}{4}$ **87.** $y = 4x - 4$ **89.** $y = 2x + 1$ **91.** $y = -\frac{1}{3}x + \frac{7}{3}$ **93.** $y = 2x - 3$

95. $r_1 = \dfrac{-b + \sqrt{b^2 - 4ac}}{2a}$

$r_2 = \dfrac{-b - \sqrt{b^2 - 4ac}}{2a}$

Exercise 10.5

1.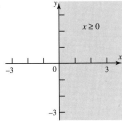

3.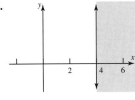

5.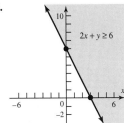

7.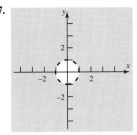

9.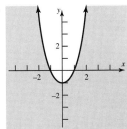

11.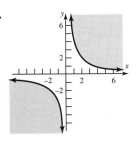

13.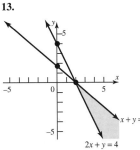

15.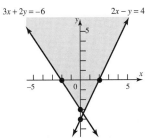

17.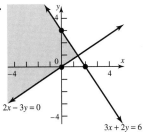

19.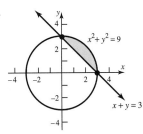

21.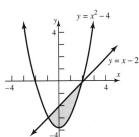

23.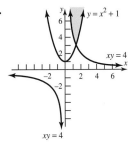

25.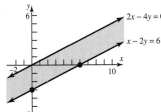

27.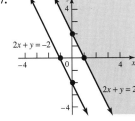

29. No solution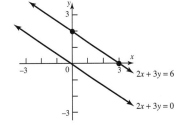

31. Bounded; corner points
(0, 0), (3, 0), (2, 2), (0, 3)

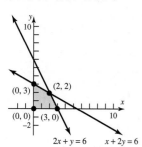

33. Unbounded; corner points (2, 0), (0, 4)

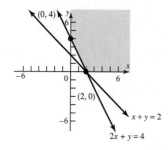

35. Bounded; corner points (2, 0), (4, 0), $(\frac{24}{7}, \frac{12}{7})$, (0, 4), (0, 2)

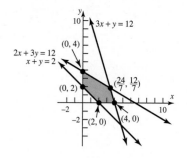

37. Bounded; corner points (2, 0), (5, 0), (2, 6), (0, 8), (0, 2)

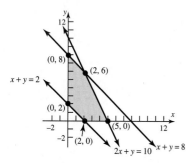

39. Bounded; corner points (10, 0), (0, 5)

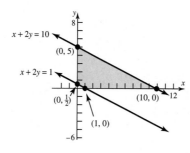

41. $\begin{cases} x \le 4 \\ x + y \le 6 \\ x \ge 0 \\ y \ge 0 \end{cases}$ **43.** $\begin{cases} x \le 20 \\ y \ge 15 \\ x + y \le 50 \\ x \le y \\ x \ge 0 \end{cases}$

45. (a) $\begin{cases} x + y \le 50{,}000 \\ x \ge 35{,}000 \\ y \le 10{,}000 \\ x \ge 0 \\ y \ge 0 \end{cases}$ (b)

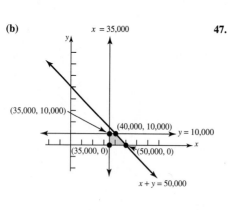

47. (a) $\begin{cases} x \ge 0 \\ y \ge 0 \\ x + 2y \le 300 \\ 3x + 2y \le 480 \end{cases}$ (b)

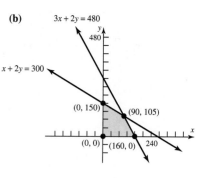

49. (a) $\begin{cases} 30x + 20y \le 1600 \\ 2x + 3y \le 150 \\ x \ge 0 \\ y \ge 0 \end{cases}$ (b)

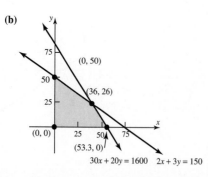

Exercise 10.6

1. Maximum value is 11; minimum value is 3   **3.** Maximum value is 65; minimum value is 4   **5.** Maximum value is 67; minimum value is 20
7. The maximum value of z is 12, and it occurs at the point $(6, 0)$ **9.** The minimum value of z is 4, and it occurs at the point $(2, 0)$
11. The maximum value of z is 20, and it occurs at the point $(0, 4)$ **13.** The minimum value of z is 8, and it occurs at the point $(0, 2)$
15. The maximum value of z is 50, and it occurs at the point $(10, 0)$ **17.** 8 downhill, 24 cross-country; $1760; $1920
19. 30 acres of soybeans and 10 acres of corn; maximum profit is $8500   **21.** $\frac{1}{2}$ hour on machine I; $5\frac{1}{4}$ hours on machine II; $182.50
23. 100 lbs. of ground beef and 50 lbs of pork; $97.50   **25.** 10 racing skates, 15 figure skates   **27.** 2 metal samples, 4 plastic samples; $34
29. (a) 10 first class, 120 coach   **(b)** 15 first class, 120 coach

Fill-in-the-Blank Items

1. inconsistent   **2.** matrix   **3.** determinants   **4.** augmented   **5.** half-plane   **6.** objective function   **7.** feasible point

True/False Items

1. F   **2.** T   **3.** F   **4.** F   **5.** T   **6.** T   **7.** T

Review Exercises

1. $x = 2, y = -1$ **3.** $x = 2, y = \frac{1}{2}$ **5.** $x = 2, y = -1$ **7.** $x = \frac{11}{5}, y = -\frac{3}{5}$ **9.** $x = -\frac{8}{5}, y = \frac{12}{5}$ **11.** $x = 6, y = -1$ **13.** $x = -4, y = 3$
15. $x = 2, y = 3$ **17.** Inconsistent   **19.** $x = -1, y = 2, z = -3$ **21.** $x = \frac{2}{5}, y = \frac{1}{10}$ **23.** $x = \frac{1}{2}, y = \frac{2}{3}, z = \frac{1}{6}$
25. $x = -\frac{1}{2}, y = -\frac{2}{3}, z = -\frac{3}{4}$ **27.** $z = -1, x = y + 1$, y any real number   **29.** $x = 1, y = 2, z = -3, t = 1$ **31.** 5   **33.** 108   **35.** -100
37. $x = 2, y = -1$ **39.** $x = 2, y = 3$ **41.** $x = -1, y = 2, z = -3$ **43.** $x = -\frac{2}{5}, y = -\frac{11}{5}; x = -2, y = 1$
45. $x = 2\sqrt{2}, y = \sqrt{2}; x = -2\sqrt{2}, y = -\sqrt{2}$ **47.** $x = \dfrac{\pm 3(-3 + \sqrt{265})}{2}, y = \dfrac{-3 + \sqrt{265}}{2}$
49. $x = \sqrt{2}, y = -\sqrt{2}; x = -\sqrt{2}, y = \sqrt{2}; x = \frac{4}{3}\sqrt{2}, y = -\frac{2}{3}\sqrt{2}; x = -\frac{4}{3}\sqrt{2}, y = \frac{2}{3}\sqrt{2}$ **51.** $x = 1, y = -1$
53. Unbounded; corner point $(0, 2)$ **55.** Bounded; corner points $(0, 0)$, $(0, 2)$, $(3, 0)$ **57.** Bounded; corner points $(0, 1)$, $(0, 8)$, $(4, 0)$, $(2, 0)$

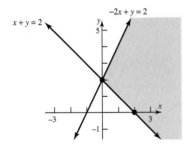

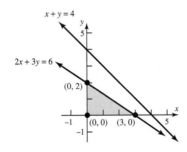

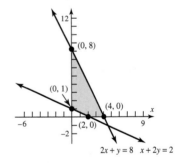

59.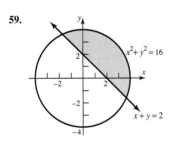

61.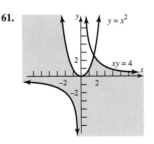

63. The maximum value is 32 when $x = 0$ and $y = 8$ **65.** The minimum value is 3 when $x = 1$ and $y = 0$
67. The maximum value is $\frac{108}{7}$ when $x = \frac{12}{7}$ and $y = \frac{12}{7}$ **69.** 10   **71.** $y = -\frac{1}{3}x^2 - \frac{2}{3}x + 1$ **73.** 70 lbs of $3 coffee and 30 lbs of $6 coffee
75. 1 small, 5 medium, 2 large   **77.** 24 ft by 10 ft   **79.** $4 + \sqrt{2}$ in and $4 - \sqrt{2}$ in   **81.** $120\sqrt{10}$ ft
83. Katy gets $10, Mike gets $20, Danny gets $5, Colleen gets $10   **85.** Katy: 4 hr; Mike: 2 hr; Danny: 8 hr
87. 35 gasoline engines, 15 diesel engines; 15 gasoline engines, 0 diesel engines

1. 1, 2, 3, 4, 5 **3.** $\frac{1}{3}, \frac{2}{4} = \frac{1}{2}, \frac{3}{5}, \frac{4}{6} = \frac{2}{3}, \frac{5}{7}$ **5.** 1, -4, 9, -16, 25 **7.** $\frac{1}{2}, \frac{2}{5}, \frac{2}{7}, \frac{8}{41}, \frac{8}{61}$ **9.** $-\frac{1}{6}, \frac{1}{12}, -\frac{1}{20}, \frac{1}{30}, -\frac{1}{42}$ **11.** $1/e$, $2/e^2$, $3/e^3$, $4/e^4$, $5/e^5$
13. $n/(n+1)$ **15.** $1/2^{n-1}$ **17.** $(-1)^{n+1}$ **19.** $(-1)^{n+1}n$ **21.** $a_1 = 2$, $a_2 = 5$, $a_3 = 8$, $a_4 = 11$, $a_5 = 14$
23. $a_1 = -2$, $a_2 = -1$, $a_3 = 1$, $a_4 = 4$, $a_5 = 8$ **25.** $a_1 = 5$, $a_2 = 10$, $a_3 = 20$, $a_4 = 40$, $a_5 = 80$ **27.** $a_1 = 3$, $a_2 = 3$, $a_3 = \frac{3}{2}$, $a_4 = \frac{1}{2}$, $a_5 = \frac{1}{8}$
29. $a_1 = 1$, $a_2 = 2$, $a_3 = 2$, $a_4 = 4$, $a_5 = 8$ **31.** $a_1 = A$, $a_2 = A + d$, $a_3 = A + 2d$, $a_4 = A + 3d$, $a_5 = A + 4d$

33. $a_1 = \sqrt{2}$, $a_2 = \sqrt{2 + \sqrt{2}}$, $a_3 = \sqrt{2 + \sqrt{2 + \sqrt{2}}}$, $a_4 = \sqrt{2 + \sqrt{2 + \sqrt{2 + \sqrt{2}}}}$, $a_5 = \sqrt{2 + \sqrt{2 + \sqrt{2 + \sqrt{2 + \sqrt{2}}}}}$

35. $3 + 4 + \cdots + (n + 2)$ **37.** $\frac{1}{2} + 2 + \frac{9}{2} + \cdots + \frac{n^2}{2}$ **39.** $1 + \frac{1}{3} + \frac{1}{9} + \cdots + \frac{1}{3^n}$ **41.** $\frac{1}{3} + \frac{1}{9} + \cdots + \frac{1}{3^n}$

43. $\ln 2 - \ln 3 + \ln 4 - \cdots + (-1)^n \ln n$ **45.** $\sum_{k=1}^{n} k$ **47.** $\sum_{k=1}^{n} \frac{k}{k+1}$ **49.** $\sum_{k=0}^{n} (-1)^k \left(\frac{1}{3^k} \right)$ **51.** $\sum_{k=1}^{n} \frac{3^k}{k}$ **53.** $\sum_{k=0}^{n} (a + kd)$ **55.** 21

59. A Fibonacci sequence

Exercise 11.2

1. $d = 1$; 5, 6, 7, 8 **3.** $d = 2$; -3, -1, 1, 3 **5.** $d = -2$; 4, 2, 0, -2 **7.** $d = -\frac{1}{3}, \frac{1}{6}, -\frac{1}{6}, -\frac{1}{2}, -\frac{5}{6}$ **9.** $d = \ln 3$; $\ln 3$, $2 \ln 3$, $3 \ln 3$, $4 \ln 3$
11. $a_5 = 14$; $a_n = 3n - 1$ **13.** $a_5 = -7$; $a_n = 8 - 3n$ **15.** $a_5 = 2$; $a_n = \frac{1}{2}(n - 1)$ **17.** $a_5 = 5\sqrt{2}$; $a_n = \sqrt{2}n$ **19.** $a_{12} = 24$ **21.** $a_{10} = -26$
23. $a_8 = a + 7b$ **25.** $a_1 = -13$; $d = 3$; $a_n = -16 + 3n$ **27.** $a_1 = -53$; $d = 6$; $a_n = -59 + 6n$ **29.** $a_1 = 28$; $d = -2$; $a_n = 30 - 2n$
31. $a_1 = 25$; $d = -2$; $a_n = 27 - 2n$ **33.** n^2 **35.** $\frac{n}{2}(9 + 5n)$ **37.** 1260 **39.** 324 **41.** $-\frac{3}{2}$ **43.** 1185 seats

45. 210 of (1st colors name) and 190 (2nd colors name)

Exercise 11.3

1. $r = 3$; 3, 9, 27, 81 **3.** $r = \frac{1}{2}$; $-\frac{3}{2}, -\frac{3}{4}, -\frac{3}{8}, -\frac{3}{16}$ **5.** $r = 2$; $\frac{1}{4}, \frac{1}{2}, 1, 2$ **7.** $r = 2^{1/3}$; $2^{1/3}, 2^{2/3}, 2, 2^{4/3}$ **9.** $r = \frac{3}{2}$; $\frac{1}{2}, \frac{3}{4}, \frac{9}{8}, \frac{27}{16}$ **11.** Arithmetic; $d = 1$
13. Neither **15.** Arithmetic; $d = -\frac{2}{3}$ **17.** Neither **19.** Geometric; $r = \frac{2}{3}$ **21.** Geometric; $r = 2$ **23.** Geometric; $r = 3^{1/2}$
25. $a_5 = 162$; $a_n = 2 \cdot 3^{n-1}$ **27.** $a_5 = 5$; $a_n = (-1)^{n-1}(5)$ **29.** $a_5 = 0$; $a_n = 0$ **31.** $a_5 = 4\sqrt{2}$; $a_n = (\sqrt{2})^n$ **33.** $a_7 = \frac{1}{64}$ **35.** $a_9 = 1$
37. $a_8 = 0.00000004$ **39.** $-\frac{1}{4}(1 - 2^n)$ **41.** $2[1 - (\frac{2}{3})^n]$ **43.** $1 - 2^n$ **45.** $\frac{3}{2}$ **47.** 16 **49.** $\frac{8}{5}$ **51.** $\frac{20}{3}$ **53.** $\frac{18}{5}$ **55.** -4
57. (a) 0.775 ft (b) 8th (c) 15.88 ft (d) 20 ft **59.** \$21,879.11
61. Option 2 results in the most: \$16,038,304; Option 1 results in the least: \$14,700,000 **63.** 1.845×10^{19} **67.** 3, 4, 8, 23, 72
69. A: \$25,250 per year in 5th year, \$112,742 total; B: \$24,761 per year in 5th year, \$116,801 total

Exercise 11.4

1. (I) $n = 1$: $2 \cdot 1 = 2$ and $1(1 + 1) = 2$
   (II) If $2 + 4 + 6 + \cdots + 2k = k(k + 1)$, then $2 + 4 + 6 + \cdots + 2k + 2(k + 1) = (2 + 4 + 6 + \cdots + 2k) + 2(k + 1)$
   $= k(k + 1) + 2(k + 1) = k^2 + 3k + 2 = (k + 1)(k + 2)$.
3. (I) $n = 1$: $1 + 2 = 3$ and $\frac{1}{2}(1)(1 + 5) = \frac{1}{2}(6) = 3$
   (II) If $3 + 4 + 5 + \cdots + (k + 2) = \frac{1}{2}k(k + 5)$, then $3 + 4 + 5 + \cdots + (k + 2) + [(k + 1) + 2] = [3 + 4 + 5 + \cdots + (k + 2)] + (k + 3) = \frac{1}{2}k(k + 5) + k + 3 = \frac{1}{2}(k^2 + 7k + 6) = \frac{1}{2}(k + 1)(k + 6)$.
5. (I) $n = 1$: $3 \cdot 1 - 1 = 2$ and $\frac{1}{2}(1)[3(1) + 1] = \frac{1}{2}(4) = 2$
   (II) If $2 + 5 + 8 + \cdots + (3k - 1) = \frac{1}{2}k(3k + 1)$, then $2 + 5 + 8 + \cdots + (3k - 1) + [3(k + 1) - 1]$
   $= [2 + 5 + 8 + \cdots + (3k - 1)] + 3k + 2 = \frac{1}{2}k(3k + 1) + (3k + 2) = \frac{1}{2}(3k^2 + 7k + 4) = \frac{1}{2}(k + 1)(3k + 4)$.
7. (I) $n = 1$: $2^{1-1} = 1$ and $2^1 - 1 = 1$
   (II) If $1 + 2 + 2^2 + \cdots + 2^{k-1} = 2^k - 1$, then $1 + 2 + 2^2 + \cdots + 2^{k-1} + 2^{(k+1)-1} = (1 + 2 + 2^2 + \cdots + 2^{k-1}) + 2^k = 2^k - 1 + 2^k = 2(2^k) - 1 = 2^{k+1} - 1$.
9. (I) $n = 1$: $4^{1-1} = 1$ and $\frac{1}{3}(4^1 - 1) = \frac{1}{3}(3) = 1$
   (II) If $1 + 4 + 4^2 + \cdots + 4^{k-1} = \frac{1}{3}(4^k - 1)$, then $1 + 4 + 4^2 + \cdots + 4^{k-1} + 4^{(k+1)-1} = (1 + 4 + 4^2 + \cdots + 4^{k-1}) + 4^k = \frac{1}{3}(4^k - 1) + 4^k = \frac{1}{3}[4^k - 1 + 3(4^k)] = \frac{1}{3}[4(4^k) - 1] = \frac{1}{3}(4^{k+1} - 1)$.
11. (I) $n = 1$: $\dfrac{1}{1 \cdot 2} = \dfrac{1}{2}$ and $\dfrac{1}{1 + 1} = \dfrac{1}{2}$
   (II) If $\dfrac{1}{1 \cdot 2} + \dfrac{1}{2 \cdot 3} + \dfrac{1}{3 \cdot 4} + \cdots + \dfrac{1}{k(k + 1)} = \dfrac{k}{k + 1}$, then $\dfrac{1}{1 \cdot 2} + \dfrac{1}{2 \cdot 3} + \dfrac{1}{3 \cdot 4} + \cdots + \dfrac{1}{k(k + 1)} + \dfrac{1}{(k + 1)[(k + 1) + 1]} =$
   $\left[\dfrac{1}{1 \cdot 2} + \dfrac{1}{2 \cdot 3} + \dfrac{1}{3 \cdot 4} + \cdots + \dfrac{1}{k(k + 1)} \right] + \dfrac{1}{(k + 1)(k + 2)} = \dfrac{k}{k + 1} + \dfrac{1}{(k + 1)(k + 2)} = \dfrac{k + 1}{k + 2}$.
13. (I) $n = 1$: $1^2 = 1$ and $\frac{1}{6} \cdot 1 \cdot 2 \cdot 3 = 1$
   (II) If $1^2 + 2^2 + 3^2 + \cdots + k^2 = \frac{1}{6}k(k + 1)(2k + 1)$, then $1^2 + 2^2 + 3^2 + \cdots + k^2 + (k + 1)^2 =$
   $(1^2 + 2^2 + 3^2 + \cdots + k^2) + (k + 1)^2 = \frac{1}{6}k(k + 1)(2k + 1) + (k + 1)^2 = \frac{1}{6}(2k^3 + 9k^2 + 13k + 6) = \frac{1}{6}(k + 1)(k + 2)(2k + 3)$.

15. (I) $n = 1$: $5 - 1 = 4$ and $\frac{1}{2}(9 - 1) = \frac{1}{2} \cdot 8 = 4$

(II) If $4 + 3 + 2 + \cdots + (5 - k) = \frac{1}{2}k(9 - k)$, then $4 + 3 + 2 + \cdots + (5 - k) + 5 - (k + 1) = [4 + 3 + 2 + \cdots + (5 - k)] + 5 - (k + 1)$
$= \frac{1}{2}k(9 - k) + 4 - k = \frac{1}{2}(-k^2 + 7k + 8) = \frac{1}{2}(8 - k)(k + 1) = \frac{1}{2}(k + 1)[9 - (k + 1)]$.

17. (I) $n = 1$: $1 \cdot (1 + 1) = 2$ and $\frac{1}{3} \cdot 1 \cdot 2 \cdot 3 = 2$

(II) If $1 \cdot 2 + 2 \cdot 3 + 3 \cdot 4 + \cdots + k(k + 1) = \frac{1}{3}k(k + 1)(k + 2)$, then
$1 \cdot 2 + 2 \cdot 3 + 3 \cdot 4 + \cdots + k(k + 1) + (k + 1)(k + 2) = [1 \cdot 2 + 2 \cdot 3 + 3 \cdot 4 + \cdots + k(k + 1)] + (k + 1)(k + 2)$
$= \frac{1}{3}k(k + 1)(k + 2) + (k + 1)(k + 2) = \frac{1}{3}(k + 1)(k + 2)(k + 3)$.

19. (I) $n = 1$: $1^2 + 1 = 2$ is divisible by 2.

(II) If $k^2 + k$ is divisible by 2, then $(k + 1)^2 + (k + 1) = k^2 + 2k + 1 + k + 1 = (k^2 + k) + 2k + 2$. Since $k^2 + k$ is divisible by 2 and $2k + 2$ is divisible by 2, therefore, $(k + 1)^2 + k + 1$ is divisible by 2.

21. (I) $n = 1$: $1^2 - 1 + 2 = 2$ is divisible by 2.

(II) If $k^2 - k + 2$ is divisible by 2, then $(k + 1)^2 - (k + 1) + 2 = k^2 + 2k + 1 - k - 1 + 2 = (k^2 - k + 2) + 2k$. Since $k^2 - k + 2$ is divisible by 2 and $2k$ is divisible by 2, therefore, $(k + 1)^2 - (k + 1) + 2$ is divisible by 2.

23. (I) $n = 1$: If $x > 1$, then $x^1 = x > 1$.

(II) Assume, for any natural number k, that if $x > 1$, then $x^k > 1$. Show that if $x > 1$, then $x^k + 1 > 1$:
$$x^{k+1} = x^k \cdot x^1 > 1 \cdot x = x > 1$$
$$\uparrow$$
$$x^k > 1$$

25. (I) $n = 1$: $a - b$ is a factor of $a^1 - b^1 = a - b$.

(II) If $a - b$ is a factor of $a^k - b^k$, show that $a - b$ is a factor of $a^{k+1} - b^{k+1}$: $a^{k+1} - b^{k+1} = a(a^k - b^k) + b^k(a - b)$. Since $a - b$ is a factor of $a^k - b^k$ and $a - b$ is a factor of $a - b$, therefore, $a - b$ is a factor of $a^{k+1} - b^{k+1}$.

27. $n = 1$: $1^2 - 1 + 41 = 41$ is a prime number.
$n = 41$: $41^2 - 41 + 41 = 1681 = 41^2$ is not prime.

29. (I) $n = 1$: $ar^{1-1} = a \cdot 1 = a$ and $a \cdot \dfrac{1 - r^1}{1 - r} = a$, because $r \neq 1$.

(II) If $a + ar + ar^2 + \cdots + ar^{k-1} = a\left(\dfrac{1 - r^k}{1 - r}\right)$, then $a + ar + ar^2 + \cdots + ar^{k-1} + ar^{(k+1)-1} = (a + ar + ar^2 + \cdots + ar^{k-1}) + ar^k =$
$a\left(\dfrac{1 - r^k}{1 - r}\right) + ar^k = \dfrac{a(1 - r^k) + ar^k(1 - r)}{1 - r} = \dfrac{a - ar^k + ar^k - ar^{k+1}}{1 - r} = a\left(\dfrac{1 - r^{k+1}}{1 - r}\right)$.

31. (I) $n = 3$: The sum of the angles of a triangle is $(3 - 2) \cdot 180° = 180°$.

(II) Assume for any k that the sum of the angles of a convex polygon of k sides is $(k - 2) \cdot 180°$. A convex polygon of $k + 1$ sides consists of a convex polygon of k sides plus a triangle (see the illustration). The sum of the angles is $(k - 2) \cdot 180° + 180° = (k - 1) \cdot 180°$. Since Conditions I and II have been met, the result follows.

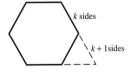

k sides

$k + 1$ sides

Exercise 11.5

1. 10 **3.** 21 **5.** 50 **7.** 1 **9.** 1.866×10^{15} **11.** 1.483×10^{13} **13.** $x^5 + 5x^4 + 10x^3 + 10x^2 + 5x + 1$
15. $x^6 - 12x^5 + 60x^4 - 160x^3 + 240x^2 - 192x + 64$ **17.** $81x^4 + 108x^3 + 54x^2 + 12x + 1$ **19.** $x^{10} + 5y^2x^8 + 10y^4x^6 + 10y^6x^4 + 5y^8x^2 + y^{10}$
21. $x^3 + 6\sqrt{2}x^{5/2} + 30x^2 + 40\sqrt{2}x^{3/2} + 60x + 24\sqrt{2}x^{1/2} + 8$ **23.** $(ax)^5 + 5by(ax)^4 + 10(by)^2(ax)^3 + 10(by)^3(ax)^2 + 5(by)^4(ax) + (by)^5$
25. 17,010 **27.** $-101,376$ **29.** 41,472 **31.** $2835x^3$ **33.** $314,928x^7$ **35.** 495 **37.** 3360 **39.** 1.00501
41. $\dbinom{n}{n} = \dfrac{n!}{n!(n - n)!} = \dfrac{n!}{n!0!} = \dfrac{n!}{n!} = 1$ **43.** $2^n = (1 + 1)^n = \dbinom{n}{0}1^n + \dbinom{n}{1}(1)(1)^{n-1} + \cdots + \dbinom{n}{n}1^n = \dbinom{n}{0} + \dbinom{n}{1} + \cdots + \dbinom{n}{n}$ **45.** 1

Exercise 11.6

1. $\{1, 3, 5, 6, 7, 9\}$ **3.** $\{1, 5, 7\}$ **5.** $\{1, 6, 9\}$ **7.** $\{1, 2, 4, 5, 6, 7, 8, 9\}$ **9.** $\{1, 2, 4, 5, 6, 7, 8, 9\}$ **11.** $\{0, 2, 6, 7, 8\}$
13. $\{0, 1, 2, 3, 5, 6, 7, 8, 9\}$ **15.** $\{0, 1, 2, 3, 5, 6, 7, 8, 9\}$ **17.** $\{0, 1, 2, 3, 4, 6, 7, 8\}$ **19.** $\{0\}$

21. $\varnothing$, $\{a\}$, $\{b\}$, $\{c\}$, $\{d\}$, $\{a, b\}$, $\{a, c\}$, $\{a, d\}$, $\{b, c\}$, $\{b, d\}$, $\{c, d\}$, $\{a, b, c\}$, $\{b, c, d\}$, $\{a, c, d\}$, $\{a, b, d\}$, $\{a, b, c, d\}$ **23.** 25 **25.** 40
27. 25 **29.** 37 **31.** 18 **33.** 5 **35.** 175; 125 **37.** (a) 15 (b) 15 (c) 15 (d) 25 (e) 40

1. 30 **3.** 120 **5.** 1 **7.** 336 **9.** 28 **11.** 15 **13.** 1 **15.** 10,400,600
17. {*abc, abd, abe, acb, acd, ace, adb, adc, ade, aeb, aec, aed*
  bac, bad, bae, bca, bcd, bce, bda, bdc, bde, bea, bec, bed
  cab, cad, cae, cba, cbd, cbe, cda, cdb, cde, cea, ceb, ced
  dab, dac, dae, dba, dbc, dbe, dca, dcb, dce, dea, deb, dec
  eab, eac, ead, eba, ebc, ebd, eca, ecb, ecd, eda, edb, edc}; 60
19. {123, 124, 132, 134, 142, 143, 213, 214, 231, 234, 241, 243, 312, 314, 321, 324, 341, 342, 412, 413, 421, 423, 431, 432}; 24
21. {*abc, abd, abe, acd, ace, ade, bcd, bce, bde, cde*}; 10 **23.** {123, 234, 124, 134}; 4 **25.** 15 **27.** 16 **29.** 8 **31.** 24 **33.** 60 **35.** 120
37. 35 **39.** 1024 **41.** 9000 **43.** $P(5,5) = 5! = 120$ **45.** $C(8,1) \cdot C(15,1) \cdot C(4,1) = 8 \cdot 15 \cdot 4 = 480$ **47.** 336 **49.** 5,209,344
51. 362,880 **53.** 90,720 **55.** 1.156×10^{76} **57.** 15 **59.** (a) 63 (b) 35 (c) 1

1. $S = \{HH, HT, TH, TT\}$; $P(HH) = \frac{1}{4}$, $P(HT) = \frac{1}{4}$, $P(TH) = \frac{1}{4}$, $P(TT) = \frac{1}{4}$
3. $S = \{HH1, HH2, HH3, HH4, HH5, HH6, HT1, HT2, HT3, HT4, HT5, HT6, TH1, TH2, TH3, TH4, TH5, TH6, TT1, TT2, TT3, TT4, TT5, TT6\}$;
each outcome has the probability of $\frac{1}{24}$.
5. $S = \{HHH, HHT, HTH, HTT, THH, THT, TTH, TTT\}$; each outcome has the probability of $\frac{1}{8}$.
7. $S = \{$1 Yellow, 1 Red, 1 Green, 2 Yellow, 2 Red, 2 Green, 3 Yellow, 3 Red, 3 Green, 4 Yellow, 4 Red, 4 Green$\}$; each outcome has the probability of $\frac{1}{12}$; thus, $P(2 \text{ Red}) + P(4 \text{ Red}) = \frac{1}{12} + \frac{1}{12} = \frac{1}{6}$.
9. $S = \{$1 Yellow Forward, 1 Yellow Backward, 1 Red Forward, 1 Red Backward, 1 Green Forward, 1 Green Backward, 2 Yellow Forward, 2 Yellow Backward, 2 Red Forward, 2 Red Backward, 2 Green Forward, 2 Green Backward, 3 Yellow Forward, 3 Yellow Backward, 3 Red Forward, 3 Red Backward, 3 Green Forward, 3 Green Backward, 4 Yellow Forward, 4 Yellow Backward, 4 Red Forward, 4 Red Backward, 4 Green Forward, 4 Green Backward$\}$; each outcome has the probability of $\frac{1}{24}$; thus, $P(1 \text{ Red Backward}) + P(1 \text{ Green Backward}) = \frac{1}{24} + \frac{1}{24} = \frac{1}{12}$.
11. $S = \{$11 Red, 11 Yellow, 11 Green, 12 Red, 12 Yellow, 12 Green, 13 Red, 13 Yellow, 13 Green, 14 Red, 14 Yellow, 14 Green, 21 Red, 21 Yellow, 21 Green, 22 Red, 22 Yellow, 22 Green, 23 Red, 23 Yellow, 23 Green, 24 Red, 24 Yellow, 24 Green, 31 Red, 31 Yellow, 31 Green, 32 Red, 32 Yellow, 32 Green, 33 Red, 33 Yellow, 33 Green, 34 Red, 34 Yellow, 34 Green, 41 Red, 41 Yellow, 41 Green, 42 Red, 42 Yellow, 42 Green, 43 Red, 43 Yellow, 43 Green, 44 Red, 44 Yellow, 44 Green$\}$; each outcome has the probability of $\frac{1}{48}$; thus, $E = \{$22 Red, 22 Green, 24 Red, 24 Green$\}$;
$P(E) = n(E)/n(S) = \frac{4}{48} = \frac{1}{12}$.
13. *A, B, C, F* **15.** *B* **17.** $\frac{4}{5}; \frac{1}{5}$ **19.** $P(1) = P(3) = P(5) = \frac{2}{9}$; $P(2) = P(4) = P(6) = \frac{1}{9}$ **21.** 0.7 **23.** 0.55 **25.** $\frac{9}{20}$ **27.** $\frac{17}{20}$ **29.** $\frac{5}{18}$ **31.** $\frac{3}{10}$
33. $\frac{2}{5}$ **35.** (a) 0.57 (b) 0.95 (c) 0.83 (d) 0.38 (e) 0.29 (f) 0.05 (g) 0.78 (h) 0.71 **37.** 0.000033068 **39.** (a) $\frac{10}{32}$ (b) $\frac{1}{32}$
41. (a) 0.00463 (b) 0.049 **43.** $\frac{1}{C(30,5)} = 7.02 \times 10^{-6}$; 0.183 **45.** 0.1

Fill-in-the-Blank Items

1. sequence **2.** arithmetic **3.** geometric **4.** Pascal triangle **5.** 15 **6.** union; intersection **7.** 20; 10 **8.** permutation **9.** combination
10. equally likely

True/False Items

1. T **2.** T **3.** T **4.** T **5.** F **6.** F **7.** F **8.** T **9.** F **10.** T **11.** T **12.** F

Review Exercises

1. 120 **3.** 10 **5.** 336 **7.** 56 **9.** $-\frac{4}{3}, \frac{5}{4}, -\frac{6}{5}, \frac{7}{6}, -\frac{8}{7}$ **11.** $2, 1, \frac{8}{9}, 1, \frac{32}{25}$ **13.** $3, 2, \frac{4}{3}, \frac{8}{9}, \frac{16}{27}$ **15.** 2, 0, 2, 0, 2 **17.** Arithmetic; $d = 6$; $\frac{n}{2}(n + 11)$
19. Neither **21.** Geometric; $r = 8$; $\frac{8}{7}(8^n - 1)$ **23.** Arithmetic; $d = 4$; $2n(n - 1)$ **25.** Geometric; $r = \frac{1}{2}$; $6[1 - (\frac{1}{2})^n]$ **27.** Neither **29.** 35
31. $\frac{1}{10^{10}}$ **33.** $9\sqrt{2}$ **35.** $5n - 4$ **37.** $n - 10$ **39.** $\frac{9}{2}$ **41.** $\frac{4}{3}$ **43.** 8
45. (I) $n = 1$: $3 \cdot 1 = 3$ and $\frac{3 \cdot 1}{2}(2) = 3$
  (II) If $3 + 6 + 9 + \cdots + 3k = \frac{3k}{2}(k + 1)$, then $3 + 6 + 9 + \cdots + 3k + 3(k + 1) = (3 + 6 + 9 + \cdots + 3k) + (3k + 3)$
    $= \frac{3k}{2}(k + 1) + (3k + 3) = \frac{3k^2}{2} + \frac{9k}{2} + \frac{6}{2} = \frac{3}{2}(k^2 + 3k + 2) = \frac{3}{2}(k + 1)(k + 2)$.
47. (I) $n = 1$: $2 \cdot 3^{1-1} = 2$ and $3^1 - 1 = 2$
  (II) If $2 + 6 + 18 + \cdots + 2 \cdot 3^{k-1} = 3^k - 1$, then $2 + 6 + 18 + \cdots + 2 \cdot 3^{k-1} + 2 \cdot 3^{(k+1)-1}$
    $= (2 + 6 + 18 + \cdots + 2 \cdot 3^{k-1}) + 2 \cdot 3^k = 3^k - 1 + 2 \cdot 3^k = 3 \cdot 3^k - 1 = 3^{k+1} - 1$.
49. (I) $n = 1$: $1^2 = 1$ and $\frac{1}{2}(6 - 3 - 1) = \frac{1}{2}(2) = 1$
  (II) If $1^2 + 4^2 + 7^2 + \cdots + (3k - 2)^2 = \frac{1}{2}k(6k^2 - 3k - 1)$, then
    $1^2 + 4^2 + 7^2 + \cdots + (3k - 2)^2 + [3(k + 1) - 2]^2 = [1^2 + 4^2 + 7^2 + \cdots + (3k - 2)^2] + (3k + 1)^2 = \frac{1}{2}k(6k^2 - 3k - 1) + (3k + 1)^2$
    $= \frac{1}{2}(6k^3 + 15k^2 + 11k + 2) = \frac{1}{2}(k + 1)(6k^2 + 9k + 2) = \frac{1}{2}(k + 1)[6(k + 1)^2 - 3(k + 1) - 1]$.
51. $x^5 + 10x^4 + 40x^3 + 80x^2 + 80x + 32$ **53.** $32x^5 + 240x^4 + 720x^3 + 1080x^2 + 810x + 243$ **55.** 144 **57.** 84 **59.** (1, 3, 5, 6, 7, 8)
61. {3, 7} **63.** {1, 2, 4, 6, 8, 9} **65.** {1, 2, 4, 5, 6, 9} **67.** 17 **69.** 29 **71.** 7 **73.** 25 **75.** 60 **77.** 128 **79.** 3024 **81.** 70 **83.** 91

85. 1,600,000 **87.** 216,000 **89.** 1260 **91. (a)** 381,024 **(b)** 1260 **93.** $\frac{3}{20}, \frac{9}{20}$ **95.** $\frac{1}{24}$ **97. (a)** 0.045 **(b)** 0.318 **(c)** 0.159
99. (a) 8 **(b)** 1100 **101. (a)** $\left(\frac{3}{4}\right)^3 \cdot 20 = \frac{135}{6}$ ft **(b)** $20\left(\frac{3}{4}\right)^n$ ft **(c)** after the 13th time **(d)** 140 ft **103. (a)** 0.68 **(b)** 0.58 **(c)** 0.32

CHAPTER 12 *Exercise 12.1*

1. $\begin{bmatrix} 4 & 4 & -5 \\ -1 & 5 & 4 \end{bmatrix}$ **3.** $\begin{bmatrix} 0 & 12 & -20 \\ 4 & 8 & 24 \end{bmatrix}$ **5.** $\begin{bmatrix} -8 & 7 & -15 \\ 7 & 0 & 22 \end{bmatrix}$ **7.** $\begin{bmatrix} 28 & -9 \\ 4 & 23 \end{bmatrix}$ **9.** $\begin{bmatrix} 1 & 14 & -14 \\ 2 & 22 & -18 \\ 3 & 0 & 28 \end{bmatrix}$ **11.** $\begin{bmatrix} 15 & 21 & -16 \\ 22 & 34 & -22 \\ -11 & 7 & 22 \end{bmatrix}$

13. $\begin{bmatrix} 25 & -9 \\ 4 & 20 \end{bmatrix}$ **15.** $\begin{bmatrix} -13 & 7 & -12 \\ -18 & 10 & -14 \\ 17 & -7 & 34 \end{bmatrix}$ **17.** $\begin{bmatrix} -2 & 4 & 2 & 8 \\ 2 & 1 & 4 & 6 \end{bmatrix}$ **19.** $\begin{bmatrix} 9 & 2 \\ 34 & 13 \\ 47 & 20 \end{bmatrix}$ **21.** $\begin{bmatrix} 1 & -1 \\ -1 & 2 \end{bmatrix}$ **23.** $\begin{bmatrix} 1 & -\frac{5}{2} \\ -1 & 3 \end{bmatrix}$

25. $\begin{bmatrix} 1 & -1/a \\ -1 & 2/a \end{bmatrix}$ **27.** $\begin{bmatrix} 3 & -3 & 1 \\ -2 & 2 & -1 \\ -4 & 5 & -2 \end{bmatrix}$ **29.** $\begin{bmatrix} -\frac{5}{7} & \frac{1}{7} & \frac{3}{7} \\ \frac{9}{7} & \frac{1}{7} & -\frac{4}{7} \\ \frac{3}{7} & -\frac{2}{7} & \frac{1}{7} \end{bmatrix}$ **31.** $x = 3, y = 2$ **33.** $x = -5, y = 10$ **35.** $x = 2, y = -1$

37. $x = \frac{1}{2}, y = 2$ **39.** $x = -2, y = 1$ **41.** $x = 2/a, y = 3/a$ **43.** $x = -2, y = 3, z = 5$ **45.** $x = \frac{1}{2}, y = -\frac{1}{2}, z = 1$ **47.** $x = -\frac{34}{7}, y = \frac{85}{7}, z = \frac{12}{7}$
49. $x = \frac{1}{3}, y = 1, z = \frac{2}{3}$ **51.** $\begin{bmatrix} 4 & 2 & | & 1 & 0 \\ 2 & 1 & | & 0 & 1 \end{bmatrix} \rightarrow \begin{bmatrix} 1 & \frac{1}{2} & | & \frac{1}{4} & 0 \\ 2 & 1 & | & 0 & 1 \end{bmatrix} \rightarrow \begin{bmatrix} 1 & \frac{1}{2} & | & \frac{1}{4} & 0 \\ 0 & 0 & | & -\frac{1}{2} & 1 \end{bmatrix}$

53. $\begin{bmatrix} 15 & 3 & | & 1 & 0 \\ 10 & 2 & | & 0 & 1 \end{bmatrix} \rightarrow \begin{bmatrix} 1 & \frac{1}{5} & | & \frac{1}{15} & 0 \\ 10 & 2 & | & 0 & 1 \end{bmatrix} \rightarrow \begin{bmatrix} 1 & \frac{1}{5} & | & \frac{1}{15} & 0 \\ 0 & 0 & | & -\frac{2}{3} & 1 \end{bmatrix}$

55. $\begin{bmatrix} -3 & 1 & -1 & | & 1 & 0 & 0 \\ 1 & -4 & -7 & | & 0 & 1 & 0 \\ 1 & 2 & 5 & | & 0 & 0 & 1 \end{bmatrix} \rightarrow \begin{bmatrix} 1 & 2 & 5 & | & 0 & 0 & 1 \\ 1 & -4 & -7 & | & 0 & 1 & 0 \\ -3 & 1 & -1 & | & 1 & 0 & 0 \end{bmatrix} \rightarrow \begin{bmatrix} 1 & 2 & 5 & | & 0 & 0 & 1 \\ 0 & -6 & -12 & | & 0 & 1 & -1 \\ 0 & 7 & 14 & | & 1 & 0 & 3 \end{bmatrix}$

$\rightarrow \begin{bmatrix} 1 & 2 & 5 & | & 0 & 0 & 1 \\ 0 & 1 & 2 & | & 0 & -\frac{1}{6} & \frac{1}{6} \\ 0 & 1 & 2 & | & \frac{1}{7} & 0 & \frac{3}{7} \end{bmatrix} \rightarrow \begin{bmatrix} 1 & 2 & 5 & | & 0 & 0 & 1 \\ 0 & 1 & 2 & | & 0 & -\frac{1}{6} & \frac{1}{6} \\ 0 & 0 & 0 & | & \frac{1}{7} & \frac{1}{6} & \frac{11}{42} \end{bmatrix}$

57. $\begin{bmatrix} 0.01 & 0.05 & -0.01 \\ 0.01 & -0.02 & 0.01 \\ -0.02 & 0.01 & 0.03 \end{bmatrix}$ **59.** $\begin{bmatrix} 0.02 & -0.04 & -0.01 & 0.01 \\ -0.02 & 0.05 & 0.03 & -0.03 \\ 0.02 & 0.01 & -0.04 & 0.00 \\ -0.02 & 0.06 & 0.07 & 0.06 \end{bmatrix}$

61. $x = 4.57, y = -6.44, z = -24.07$ **63.** $x = -1.19, y = 2.46, z = 8.27$
65. (a) $\begin{bmatrix} 500 & 350 & 400 \\ 700 & 500 & 850 \end{bmatrix}$; $\begin{bmatrix} 500 & 700 \\ 350 & 500 \\ 400 & 850 \end{bmatrix}$ **(b)** $\begin{bmatrix} 15 \\ 8 \\ 3 \end{bmatrix}$ **(c)** $\begin{bmatrix} 11,500 \\ 17,050 \end{bmatrix}$ **(d)** $[0.10 \quad 0.05]$ **(e)** \$2002.50

67. If $a \neq 0$, $\begin{bmatrix} a & b & | & 1 & 0 \\ c & d & | & 0 & 1 \end{bmatrix} \rightarrow \begin{bmatrix} 1 & \frac{b}{a} & | & \frac{1}{a} & 0 \\ c & d & | & 0 & 1 \end{bmatrix} \rightarrow \begin{bmatrix} 1 & \frac{b}{a} & | & \frac{1}{a} & 0 \\ 0 & -cb+da & | & -\frac{c}{a} & 1 \end{bmatrix}$

$\rightarrow \begin{bmatrix} 1 & \frac{b}{a} & | & \frac{1}{a} & 0 \\ 0 & 1 & | & \frac{-c}{-cb+da} & \frac{a}{-cb+da} \end{bmatrix} \rightarrow \begin{bmatrix} 1 & 0 & | & \frac{d}{ad-bc} & \frac{-b}{ad-bc} \\ 0 & 1 & | & \frac{-c}{ad-bc} & \frac{a}{ad-bc} \end{bmatrix}$. Therefore, $A^{-1} = \frac{1}{D}\begin{bmatrix} d & -b \\ -c & a \end{bmatrix}$.

If $a = 0$, $\begin{bmatrix} 0 & b & | & 1 & 0 \\ c & d & | & 0 & 1 \end{bmatrix} \rightarrow \begin{bmatrix} c & d & | & 0 & 1 \\ 0 & b & | & 1 & 0 \end{bmatrix} \rightarrow \begin{bmatrix} 1 & \frac{d}{c} & | & 0 & \frac{1}{c} \\ 0 & b & | & 1 & 0 \end{bmatrix} \rightarrow \begin{bmatrix} 1 & 0 & | & \frac{-d}{cb} & \frac{1}{c} \\ 0 & 1 & | & \frac{1}{b} & 0 \end{bmatrix}$

$\rightarrow \begin{bmatrix} 1 & 0 & | & \frac{d}{-bc} & \frac{-b}{-bc} \\ 0 & 1 & | & \frac{-c}{-bc} & 0 \end{bmatrix}$. Since $a = 0$, $D = ad - bc = -bc$, so $A^{-1} = \frac{1}{D}\begin{bmatrix} d & -b \\ -c & a \end{bmatrix}$.

Exercise 12.2

1. Proper **3.** Improper; $1 + \dfrac{9}{x^2 - 4}$ **5.** Improper; $5x + \dfrac{22x - 1}{x^2 - 4}$ **7.** Improper; $1 + \dfrac{-2(x - 6)}{(x + 4)(x - 3)}$ **9.** $\dfrac{-4}{x} + \dfrac{4}{x - 1}$ **11.** $\dfrac{1}{x} + \dfrac{-x}{x^2 + 1}$

13. $\dfrac{-1}{x - 1} + \dfrac{2}{x - 2}$ **15.** $\dfrac{\frac{1}{4}}{x + 1} + \dfrac{\frac{3}{4}}{x - 1} + \dfrac{\frac{1}{2}}{(x - 1)^2}$ **17.** $\dfrac{\frac{1}{12}}{x - 2} + \dfrac{-\frac{1}{12}(x + 4)}{x^2 + 2x + 4}$ **19.** $\dfrac{\frac{1}{4}}{(x - 1)} + \dfrac{\frac{1}{4}}{(x - 1)^2} - \dfrac{\frac{1}{4}}{x + 1} + \dfrac{\frac{1}{4}}{(x + 1)^2}$

21. $\dfrac{-5}{x + 2} + \dfrac{5}{x + 1} + \dfrac{-4}{(x + 1)^2}$ **23.** $\dfrac{\frac{1}{4}}{x} + \dfrac{1}{x^2} - \dfrac{\frac{1}{4}(x + 4)}{x^2 + 4}$ **25.** $\dfrac{\frac{2}{3}}{x + 1} + \dfrac{\frac{1}{3}(x + 1)}{x^2 + 2x + 4}$ **27.** $\dfrac{\frac{2}{7}}{3x - 2} + \dfrac{\frac{1}{7}}{2x + 1}$ **29.** $\dfrac{\frac{3}{4}}{x + 3} + \dfrac{\frac{1}{4}}{x - 1}$

31. $\dfrac{1}{x^2 + 4} + \dfrac{2x - 1}{(x^2 + 4)^2}$ **33.** $\dfrac{-1}{x} + \dfrac{2}{x - 3} + \dfrac{-1}{x + 1}$ **35.** $\dfrac{4}{x - 2} + \dfrac{-3}{x - 1} + \dfrac{-1}{(x - 1)^2}$ **37.** $\dfrac{x}{(x^2 + 16)^2} + \dfrac{-16x}{(x^2 + 16)^3}$ **39.** $\dfrac{-\frac{8}{7}}{2x + 1} + \dfrac{\frac{4}{7}}{x - 3}$

41. $\dfrac{-\frac{2}{9}}{x} - \dfrac{\frac{1}{3}}{x^2} + \dfrac{\frac{1}{6}}{x - 3} + \dfrac{\frac{1}{18}}{x + 3}$

Exercise 12.3

1.

3.

5.

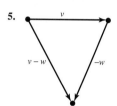

7.

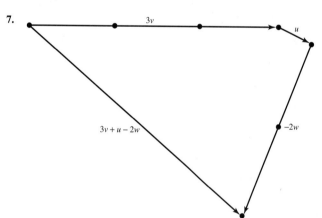

9. $\mathbf{x = A}$ **11.** $\mathbf{C = -F + E - D}$ **13.** $\mathbf{E = -G - H + D}$ **15.** $\mathbf{x = 0}$ **17.** 12 **19.** $\mathbf{v} = 3\mathbf{i} + 4\mathbf{j}$ **21.** $\mathbf{v} = 2\mathbf{i} + 4\mathbf{j}$ **23.** $\mathbf{v} = 8\mathbf{i} - \mathbf{j}$

25. $\mathbf{v} = -\mathbf{i} + \mathbf{j}$ **27.** 5 **29.** $\sqrt{2}$ **31.** $\sqrt{13}$ **33.** $-\mathbf{j}$ **35.** $\sqrt{89}$ **37.** $\sqrt{34} - \sqrt{13}$ **39.** $\mathbf{i}$ **41.** $\frac{3}{5}\mathbf{i} - \frac{4}{5}\mathbf{j}$ **43.** $\dfrac{\sqrt{2}}{2}\mathbf{i} - \dfrac{\sqrt{2}}{2}\mathbf{j}$

45. $\mathbf{v} = \dfrac{8\sqrt{5}}{5}\mathbf{i} + \dfrac{4\sqrt{5}}{5}\mathbf{j}$ or $\mathbf{v} = -\dfrac{8\sqrt{5}}{5}\mathbf{i} - \dfrac{4\sqrt{5}}{5}\mathbf{j}$ **47.** $\{-2 + \sqrt{21}, -2 - \sqrt{21}\}$ **49.** 460 kmh **51.** 218 mph

53.

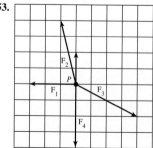

1. 0; 0   **3.** 4; $\frac{4}{5}$ **5.** $\sqrt{3} - 1$; $(\sqrt{6} - \sqrt{2})/4$ **7.** 24; $\frac{24}{25}$ **9.** 0; 0   **11.** $\frac{3}{2}$ **13.** $\mathbf{v}_1 = \text{proj}_\mathbf{w}\, \mathbf{v} = \frac{5}{2}(\mathbf{i} - \mathbf{j})$, $\mathbf{v}_2 = -\frac{1}{2}\mathbf{i} - \frac{1}{2}\mathbf{j}$
15. $\mathbf{v}_1 = \text{proj}_\mathbf{w}\, \mathbf{v} = -\frac{1}{5}(\mathbf{i} + 2\mathbf{j})$, $\mathbf{v}_2 = \frac{6}{5}\mathbf{i} - \frac{3}{5}\mathbf{j}$ **17.** $\mathbf{v}_1 = \text{proj}_\mathbf{v} = \frac{7}{5}(2\mathbf{i} + \mathbf{j})$, $\mathbf{v}_2 = \frac{1}{5}\mathbf{i} - \frac{2}{5}\mathbf{j}$ **19.** 496.7 mph; 51.5° south of west
21. 8.7° off direct heading into the current; 1.5 min   **23.** 60°; 17.32 min   **25.** 2.68 ft-lb   **27.** 1732 ft-lb
29. Let $\mathbf{u} = a_1\mathbf{i} + b_1\mathbf{j}$, $\mathbf{v} = a_2\mathbf{i} + b_2\mathbf{j}$, $\mathbf{w} = a_3\mathbf{i} + b_3\mathbf{j}$. Compute $\mathbf{u} \cdot (\mathbf{v} + \mathbf{w})$ and $\mathbf{u} \cdot \mathbf{v} + \mathbf{u} \cdot \mathbf{w}$.
31. $\cos \alpha = \dfrac{\mathbf{v} \cdot \mathbf{i}}{\|\mathbf{v}\|\,\|\mathbf{i}\|} = \mathbf{v} \cdot \mathbf{i}$; if $\mathbf{v} = x\mathbf{i} + y\mathbf{j}$, then $\mathbf{v} \cdot \mathbf{i} = x = \cos \alpha$ and $\mathbf{v} \cdot \mathbf{j} = y = \cos\left(\dfrac{\pi}{2} - \alpha\right) = \sin \alpha$.

33. $\mathbf{v} = a\mathbf{i} + b\mathbf{j}$; $\text{proj}_\mathbf{i}\, \mathbf{v} = \dfrac{\mathbf{v} \cdot \mathbf{i}}{\|\mathbf{i}\|^2}\mathbf{i} = (\mathbf{v} \cdot \mathbf{i})\mathbf{i}$; $\mathbf{v} \cdot \mathbf{i} = a$, $\mathbf{v} \cdot \mathbf{j} = b$, so $\mathbf{v} = (\mathbf{v} \cdot \mathbf{i})\mathbf{i} + (\mathbf{v} \cdot \mathbf{j})\mathbf{j}$

35. $(\mathbf{v} - \alpha\mathbf{w}) \cdot \mathbf{w} = \mathbf{v} \cdot \mathbf{w} - \alpha\mathbf{w} \cdot \mathbf{w} = \mathbf{v} \cdot \mathbf{w} - \alpha\|\mathbf{w}\|^2 = \mathbf{v} \cdot \mathbf{w} - \dfrac{\mathbf{v} \cdot \mathbf{w}}{\|\mathbf{w}\|^2}\|\mathbf{w}\|^2 = 0$ **37.** 0

Fill-in-the-Blank Items

1. inverse   **2.** identity   **3.** proper   **4.** unit   **5.** 0

True/False Items

1. F   **2.** F   **3.** F   **4.** T   **5.** T   **6.** T   **7.** T

Review Exercises

1. $\begin{bmatrix} 4 & -4 \\ 3 & 9 \\ 4 & 0 \end{bmatrix}$ **3.** $\begin{bmatrix} 6 & 0 \\ 12 & 24 \\ -6 & 12 \end{bmatrix}$ **5.** $\begin{bmatrix} 4 & -3 & 0 \\ 12 & -2 & -8 \\ -2 & 5 & -4 \end{bmatrix}$ **7.** $\begin{bmatrix} 8 & -13 & 8 \\ 9 & 2 & -10 \\ 18 & -17 & 4 \end{bmatrix}$ **9.** $\begin{bmatrix} \frac{1}{2} & -1 \\ -\frac{1}{6} & \frac{2}{3} \end{bmatrix}$ **11.** $\begin{bmatrix} -\frac{5}{7} & \frac{9}{7} & \frac{3}{7} \\ -\frac{1}{7} & \frac{1}{7} & -\frac{2}{7} \\ \frac{3}{7} & -\frac{4}{7} & \frac{1}{7} \end{bmatrix}$ **13.** Singular

15. $\dfrac{-\frac{3}{2}}{x} + \dfrac{\frac{3}{2}}{x - 4}$ **17.** $\dfrac{-3}{x - 1} + \dfrac{3}{x} + \dfrac{4}{x^2}$ **19.** $\dfrac{-\frac{1}{10}}{x + 1} + \dfrac{\frac{1}{10}x + \frac{9}{10}}{x^2 + 9}$ **21.** $\dfrac{x}{x^2 + 4} - \dfrac{4x}{(x^2 + 4)^2}$ **23.** $\dfrac{\frac{1}{2}}{x^2 + 1} + \dfrac{\frac{1}{4}}{x - 1} - \dfrac{\frac{1}{4}}{x + 1}$
25. $\mathbf{v} = 2\mathbf{i} - 4\mathbf{j}$; $\|\mathbf{v}\| = 2\sqrt{5}$ **27.** $\mathbf{v} = -\mathbf{i} + 3\mathbf{j}$; $\|\mathbf{v}\| = \sqrt{10}$ **29.** $-20\mathbf{i} + 13\mathbf{j}$ **31.** $\sqrt{5}$ **33.** $\sqrt{5} + 5 \approx 7.24$ **35.** $\dfrac{-2\sqrt{5}}{5}\mathbf{i} + \dfrac{\sqrt{5}}{5}\mathbf{j}$
37. $\mathbf{v} \cdot \mathbf{w} = -11$; $\cos \theta = -11\sqrt{5}/25$ **39.** $\mathbf{v} \cdot \mathbf{w} = -4$; $\cos \theta = -2\sqrt{5}/5$ **41.** $\text{proj}_\mathbf{v} = \frac{9}{10}(3\mathbf{i} + \mathbf{j})$ **43.** 30.5°
45. $\sqrt{29} \approx 5.39$ mph; 0.4 mi   **47.** At an angle of 70.5° to the shore

APPENDIX A   *Exercise A.1*

1. $10x^5 + 3x^3 - 10x^2 + 6$ **3.** $2ax + a^2$ **5.** $2x^2 + 17x + 8$ **7.** $x^4 - x^2 + 2x - 1$ **9.** $6x^2 + 2$ **11.** $4x^2 - 3x + 1$; remainder 1
13. $4x^2 - 11x + 23$; remainder -45 **15.** $4x^2 + 13x + 53$; remainder 213   **17.** $4x - 3$; remainder $x + 1$
19. $4x - 3$; remainder $-7x + 7$ **21.** $(x - 5)(x + 3)$ **23.** $a(x - 9a)(x + 5a)$ **25.** $(x - 3)(x^2 + 3x + 9)$ **27.** $(3x + 1)(x + 1)$
29. $x^5(x - 1)(x + 1)$ **31.** $3(x - 3)/5x$ **33.** $x(2x - 1)/(x + 4)$ **35.** $5x/[(x - 6)(x - 1)(x + 4)]$
37. $2(x + 4)/[(x - 2)(x + 2)(x + 3)]$ **39.** $(x - 1)(x + 1)/(x^2 + 1)$ **41.** $(x - 1)(-x^2 + 3x + 3)/(x^2 + x - 1)$

Exercise A.2

1. $2\sqrt{2}$ **3.** $2x\sqrt[3]{2x}$ **5.** x **7.** $\frac{4}{3}x\sqrt{2}$ **9.** x^3y^2 **11.** x^2y **13.** $6\sqrt{x}$ **15.** $6x\sqrt{x}$ **17.** $15\sqrt[3]{3}$ **19.** $\dfrac{1}{x(2x + 3)}$ **21.** $12\sqrt{3}$ **23.** $2\sqrt{3}$
25. $x - 2\sqrt{x} + 1$ **27.** $\dfrac{\sqrt{2}}{2}$ **29.** $\dfrac{-\sqrt{15}}{5}$ **31.** $\dfrac{\sqrt{3}(5 + \sqrt{2})}{23}$ **33.** $\dfrac{-19 + 8\sqrt{5}}{41}$ **35.** $\dfrac{2x + h - 2\sqrt{x(x + h)}}{h}$ **37.** $t = 1$ **39.** $x = 3$
41. 4   **43.** -3 **45.** 64   **47.** $\frac{1}{27}$ **49.** $\dfrac{27\sqrt{2}}{32}$ **51.** $\dfrac{27\sqrt{2}}{32}$ **53.** $x^{7/12}$ **55.** xy^2 **57.** $x^{4/3}y^{5/3}$ **59.** $\dfrac{8x^{3/2}}{y^{1/4}}$ **61.** $\dfrac{3x + 2}{(1 + x)^{1/2}}$
63. $\dfrac{2 + x}{2(1 + x)^{3/2}}$ **65.** $\dfrac{4 - x}{(x + 4)^{3/2}}$ **67.** $\frac{1}{2}(5x + 2)(x + 1)^{1/2}$ **69.** $2x^{1/2}(3x - 4)(x + 1)$

Exercise A.3

1. 4   **3.** $\frac{1}{16}$ **5.** $\frac{1}{9}$ **7.** $\{-7, 3\}$ **9.** $\left\{-\frac{1}{4}, \frac{3}{4}\right\}$ **11.** $\left\{\dfrac{-1 - \sqrt{7}}{6}, \dfrac{-1 + \sqrt{7}}{6}\right\}$

1. $x\text{min} = -11$
  $x\text{max} = 5$
  $x\text{scl} = 1$
  $y\text{min} = -3$
  $y\text{max} = 6$
  $y\text{scl} = 1$

3. $x\text{min} = -25$
  $x\text{max} = 45$
  $x\text{scl} = 5$
  $y\text{min} = -85$
  $y\text{max} = 45$
  $y\text{scl} = 5$

5. $x\text{min} = -5$
  $x\text{max} = 105$
  $x\text{scl} = 5$
  $y\text{min} = -10$
  $y\text{max} = 160$
  $y\text{scl} = 10$

7. $x\text{min} = -6$
  $x\text{max} = 6$
  $x\text{scl} = 2$
  $y\text{min} = -4$
  $y\text{max} = 4$
  $y\text{scl} = 2$

9. $x\text{min} = -9$
  $x\text{max} = 9$
  $x\text{scl} = 3$
  $y\text{min} = -4$
  $y\text{max} = 4$
  $y\text{scl} = 2$

11. $x\text{min} = -6$
  $x\text{max} = 6$
  $x\text{scl} = 1$
  $y\text{min} = -8$
  $y\text{max} = 8$
  $y\text{scl} = 2$

13. $x\text{min} = -6$
  $x\text{max} = 6$
  $x\text{scl} = 2$
  $y\text{min} = -1$
  $y\text{max} = 3$
  $y\text{scl} = 1$

15. $x\text{min} = 3$
  $x\text{max} = 9$
  $x\text{scl} = 1$
  $y\text{min} = 2$
  $y\text{max} = 10$
  $y\text{scl} = 2$

Exercise B.2

1. (a)   **(b)**   **(c)**   **(d)**

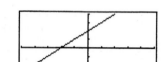 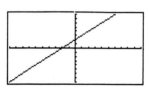 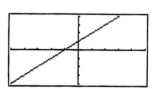 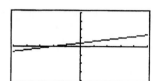

3. (a)   **(b)**   **(c)**   **(d)**

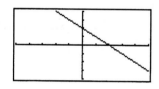 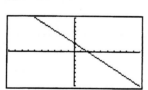 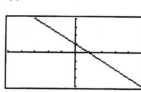 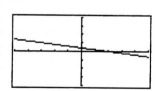

5. (a)   **(b)**   **(c)**   **(d)**

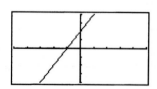 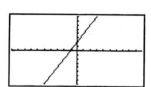 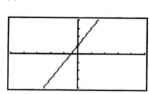 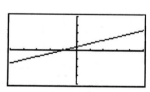

7. (a)   **(b)**   **(c)**   **(d)**

 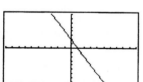 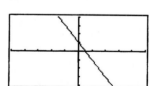 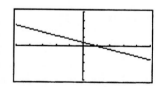

9. (a)   **(b)**   **(c)**   **(d)**

11. (a)   **(b)**   **(c)**   **(d)**

13. (a)   **(b)**   **(c)**   **(d)**

15. (a)   **(b)**   **(c)**   **(d)**

17. (a)   **(b)**   **(c)**   **(d)**

19. (a)   **(b)**   **(c)**   **(d)**

21.   **23.**

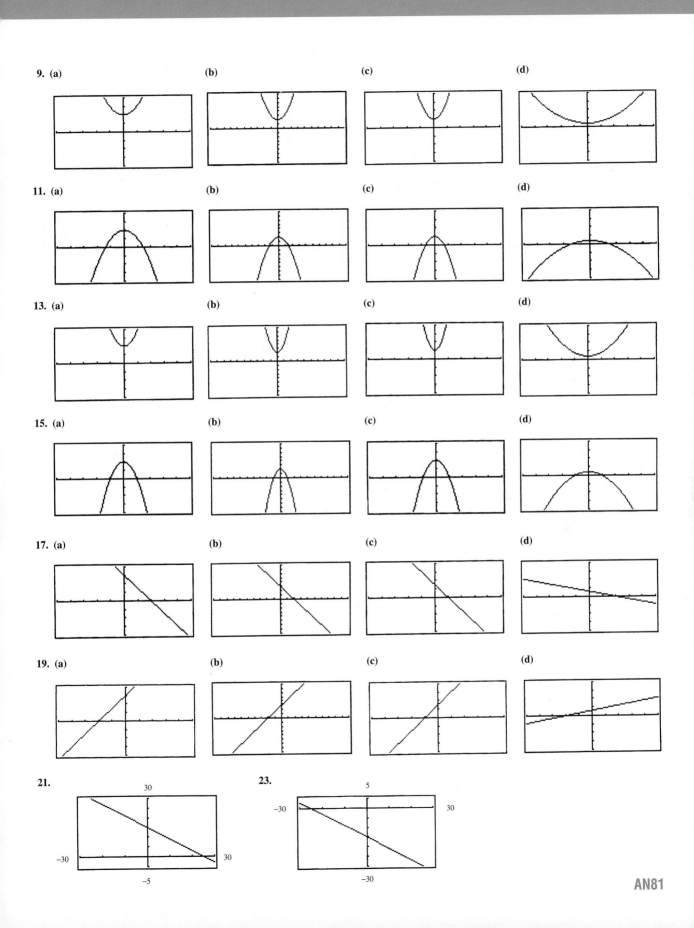

25.

27.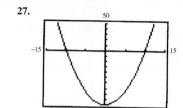

29.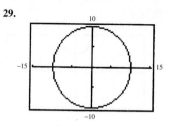

31.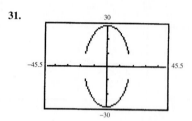

33.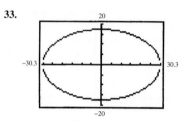

35.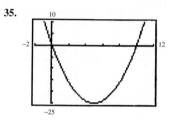

37.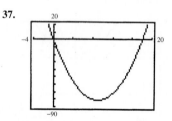

39.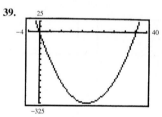

Exercise B.4

1. Yes   **3.** Yes   **5.** No   **7.** Yes
9. ymin = 4   Other answers are possible
   ymax = 12
   yscl = 1

Exercise B.5

1. 0.428   **3.** 2.236   **5.** 1.259   **7.** −3.41   **9.** −1.70   **11.** −0.28   **13.** 3.00   **15.** 4.50   **17.** 0.31, 12.30   **19.** 23.00

Index

Trigonometric Functions

Of a Real Number

$$\sin t = b \qquad \cos t = a \qquad \tan t = \frac{b}{a}, \, a \neq 0$$

$$\csc t = \frac{1}{b}, \, b \neq 0 \qquad \sec t = \frac{1}{a}, \, a \neq 0 \qquad \cot t = \frac{a}{b}, \, b \neq 0$$

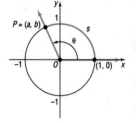

Of a General Angle

$$\sin \theta = \frac{b}{r} \qquad \cos \theta = \frac{a}{r} \qquad \tan \theta = \frac{b}{a}, \, a \neq 0$$

$$\csc \theta = \frac{r}{b}, \, b \neq 0 \qquad \sec \theta = \frac{r}{a}, \, a \neq 0 \qquad \cot \theta = \frac{a}{b}, \, b \neq 0$$

Unit circle, $x^2 + y^2 = 1$
$\theta = t$ radians; $s = t$ units

Trigonometric Identities

Fundamental Identities

$$\tan \theta = \frac{\sin \theta}{\cos \theta} \qquad \cot \theta = \frac{\cos \theta}{\sin \theta}$$

$$\csc \theta = \frac{1}{\sin \theta} \qquad \sec \theta = \frac{1}{\cos \theta} \qquad \cot \theta = \frac{1}{\tan \theta}$$

$$\sin^2 \theta + \cos^2 \theta = 1$$
$$\tan^2 \theta + 1 = \sec^2 \theta$$
$$1 + \cot^2 \theta = \csc^2 \theta$$

Even–Odd Identities

$$\sin(-\theta) = -\sin \theta \qquad \csc(-\theta) = -\csc \theta$$
$$\cos(-\theta) = \cos \theta \qquad \sec(-\theta) = \sec \theta$$
$$\tan(-\theta) = -\tan \theta \qquad \cot(-\theta) = -\cot \theta$$

Sum and Difference Formulas

$$\sin(\alpha + \beta) = \sin \alpha \cos \beta + \cos \alpha \sin \beta$$
$$\sin(\alpha - \beta) = \sin \alpha \cos \beta - \cos \alpha \sin \beta$$
$$\cos(\alpha + \beta) = \cos \alpha \cos \beta - \sin \alpha \sin \beta$$
$$\cos(\alpha - \beta) = \cos \alpha \cos \beta + \sin \alpha \sin \beta$$
$$\tan(\alpha + \beta) = \frac{\tan \alpha + \tan \beta}{1 - \tan \alpha \tan \beta}$$
$$\tan(\alpha - \beta) = \frac{\tan \alpha - \tan \beta}{1 + \tan \alpha \tan \beta}$$

Half-Angle Formulas

$$\sin \frac{\theta}{2} = \pm \sqrt{\frac{1 - \cos \theta}{2}}$$

$$\cos \frac{\theta}{2} = \pm \sqrt{\frac{1 + \cos \theta}{2}}$$

$$\tan \frac{\theta}{2} = \pm \sqrt{\frac{1 - \cos \theta}{1 + \cos \theta}}$$

Double-Angle Formulas

$$\sin 2\theta = 2 \sin \theta \cos \theta$$
$$\cos 2\theta = \cos^2 \theta - \sin^2 \theta$$

$$\cos 2\theta = 2 \cos^2 \theta - 1$$
$$\cos 2\theta = 1 - 2 \sin^2 \theta$$

$$\tan 2\theta = \frac{2 \tan \theta}{1 - \tan^2 \theta}$$

Product-to-Sum Formulas

$$\sin \alpha \sin \beta = \tfrac{1}{2}[\cos(\alpha - \beta) - \cos(\alpha + \beta)]$$
$$\cos \alpha \cos \beta = \tfrac{1}{2}[\cos(\alpha - \beta) + \cos(\alpha + \beta)]$$
$$\sin \alpha \cos \beta = \tfrac{1}{2}[\sin(\alpha + \beta) + \sin(\alpha - \beta)]$$

Sum-to-Product Formulas

$$\sin \alpha + \sin \beta = 2 \sin \frac{\alpha + \beta}{2} \cos \frac{\alpha - \beta}{2}$$

$$\sin \alpha - \sin \beta = 2 \sin \frac{\alpha - \beta}{2} \cos \frac{\alpha + \beta}{2}$$

$$\cos \alpha + \cos \beta = 2 \cos \frac{\alpha + \beta}{2} \cos \frac{\alpha - \beta}{2}$$

$$\cos \alpha - \cos \beta = -2 \sin \frac{\alpha + \beta}{2} \sin \frac{\alpha - \beta}{2}$$

Solving Triangles

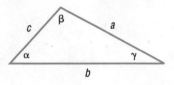

Law of Sines

$$\frac{\sin \alpha}{a} = \frac{\sin \beta}{b} = \frac{\sin \gamma}{c}$$

Law of Cosines

$$c^2 = a^2 + b^2 - 2ab \cos \gamma$$